S.CHAND'S SUCCESS GUIDES

(Questions and Answers)

INORGANIC CHEMISTRY

For B.Sc. I, II & III (Pass and Hons.) Classes

(Also covering UGC Syllabus)

S.CHAND'S SUCCESS GUIDES
(Questions and Answers)

INORGANIC CHEMISTRY

For B.Sc. I, II & III Year (Pass and Honours) Students of all Indian Universities

[Including Solved Long Answer Type, Short Answer Type Questions and Numerical Problems]

REVISED AND UPDATED EDITION

Dr. R.L. MADAN
M.Sc., Ph.D.
Former Principal, Government College Panchkula
Former Head of Chemistry Department
Government Postgraduate College
Faridabad

G.D. TULI
Former Principal
Shyam Lal College
University of Delhi, Delhi

S Chand And Company Limited
(ISO 9001 Certified Company)

S Chand And Company Limited

(ISO 9001 Certified Company)

Head Office: D-92, Sector–2, Noida – 201301, U.P. (India), Ph. 91-120-4682700

Registered Office: A-27, 2nd Floor, Mohan Co-operative Industrial Estate, New Delhi – 110 044, Phone: 011-49731800

www.**schandpublishing.com**; e-mail: **info@schandpublishing.com**

Marketing Offices:

Chennai	:	Ph: 23632120; chennai@schandpublishing.com
Guwahati	:	Ph: 2738811, 2735640; guwahati@schandpublishing.com
Hyderabad	:	Ph: 40186018; hyderabad@schandpublishing.com
Jalandhar	:	Ph: 4645630; jalandhar@schandpublishing.com
Kolkata	:	Ph: 23357458, 23353914; kolkata@schandpublishing.com
Lucknow	:	Ph: 4003633; lucknow@schandpublishing.com
Mumbai	:	Ph: 25000297; mumbai@schandpublishing.com
Patna	:	Ph: 2260011; patna@schandpublishing.com

First Edition 1999

Subsequent Editions and Reprints 2001, 2003, 2005, 2006, 2009, 2011, 2012, 2014, 2015, 2017, 2018, 2020, 2021, 2022

Reprint 2023

ISBN: 978-81-219-1857-2 **Product Code:** H6ICQ68CHEM10ENAE0XO

PRINTED IN INDIA

By Vikas Publishing House Private Limited, Plot 20/4, Site-IV, Industrial Area Sahibabad, Ghaziabad – 201 010 and Published by S Chand And Company Limited, A-27, 2nd Floor, Mohan Co-operative Industrial Estate, New Delhi – 110 044.

PREFACE TO THE REVISED EDITION

Different universities have revised their syllabi during the last couple of years, keeping in view the recommendations of University Grants Commission, New Delhi to include their Model Curriculum to the extent possible. This has necessitated the author to make substancial changes in the book. Some worth mentioning changes in the book are:

1. Matter in chapter 6 on Acids and Bases has been enhanced by including latest information on Pearson's HASB principle, correlating electronegativity with hardness – softness of acids and bases and M.O. treatment of Hard acids and hard bases.
2. There has been substantial expansion in chapter 23 on oxidation and reduction like equilibrium constant and free energy changes related to redox potential, different steps involved in redox cycle, redox stability in the solvent. Matter on Latimer and Frost diagrams has been updated. Additional matter has been Provided on pourbaix and Ellingham diagrams.
3. Topics on white cement, biofertilizers, clay and porcelain industry have been added in chapter 24 on chemical Industries.
4. Chapter 31 is enriched with new topics.
5. Topic on Least Square Method in Ch. 32 on evaluation of analytical data has been added.
6. New Chapters on volumetric analysis Ch. 33 and Gravimetric Analysis Ch. 34 have been added.

I appreciate the response to the book from the students and teaching fraternity since the first edition of the book. I hope the book in the present form will be found to be even more useful. Any constructive criticism for the improvement of the book is welcome.

Dr. R.L. MADAN
Mob. : 9971775666

PREFACE TO THE FOURTH EDITION

The author is pleased to place the fourth enlarged and updated edition of the book at the hands of students and teachers. Overwhelming response to earlier editions has encouraged me to undertake thorough revision of the text. Some of the additions that have been made in this edition are:

1. Ch. 2 Ionic Bond — Dependence of per cent ionic character of a polar covalent bond on electronegativity difference and dipole moment. Metallic Bond. Free electron, valence bond and band theories. Van der Waal's forces, Lattice defects, semi-conductors.
2. Ch. 6 — Acids and Bases. Arrhenius theory, Bronsted-Lowry Concept, Lux-Flood Theory, Basis of Hard-Hard and Soft-Soft interactions, symbiosis. Effect of polarity, dielectric constant and hydration on the strength of acids and bases.
3. Ch. 7— Metallurgies. Extraction of titanium, vanadium, chromium, molybdenum, tungsten, cobalt, platinum.
4. Ch. 8— General properties of *s*-and *p*-block elements. Alkyls and aryls of alkali metals. Complexes of alkali and alkaline metals. Role of alkali and alkaline earth metals in biological systems, fullerenes, carbides, nitrogen fixation, tetrasulphur tetranitride, fluorocarbons, extraction of Li, boron nitride, extraction of Sn, Pb and Ge, dry ice, white lead, red lead, PCl_5, P_2O_5 $KBrO_3$, KIO_3
5. Ch. 17— *d*-Block elements. Electroplating, galvanising.
6. Ch. 19—*f*-Block elements. change of oxidation state method to separate a mixture of lanthanides, La_2O_3 and LaX_3
7. Ch. 21–Coordination compounds. Calculation of magnetic moment of transition metal complexes, ligand field theory, stability of a complex ion.
8. New chapter number 30 on role of organic and inorganic reagents in inorganic analysis and chapter 31 on bioinorganic chemistry have been added.

It is earnestly hoped that the book in its present form will be found to be even more useful for the readers. Any constructive criticism for the improvement of the book is welcome.

Dr. R.L. MADAN

PREFACE TO THE THIRD EDITION

Overwhelming response from the teachers and students across the country has encouraged me to prepare the third edition of the book. New developments keep taking place in every subject and these must be included in the book to broaden the outlook of the students. Students must always keep themselves abreast of every new development that occurs in the subject.

Chapter on Evaluation of Analytical Data has been expanded to make it more comprehensive. Latimer and Frost diagrams have been added to the syllabus by some universities. This topic has been covered in Appendix 2 of the book.

Stewart has added a new dimension to the periodic classification of elements. Stewart's Table in the galaxy form has been dealt with in Appendix 3 of the book.

It is hoped that the book will be found to be more useful by the students with the addition of these topics.

Dr. R.L. MADAN

PREFACE TO THE FIRST EDITION

I feel immense pleasure in presenting Simplified Course in Physical Chemistry for B.Sc. I, II and III (Pass and Honours) students of Indian universities. I have written this book keeping in view the difficulties faced by the students, as observed by me during the last three decades. I have presented the matter in a manner that will inspire the students to read physical chemistry with interest.

The book has been written in question-answer style. This will apprise the students of the style and nature of questions that are asked in the examination and the manner of answering them appropriately. Every effort has been made to present the subject matter lucidly by using a simple and clear language. If the student makes an earnest effort, he can follow the subject-matter readily.

A large number of solved numericals have been included for a clear understanding of the subject. Numerous well-labelled structures and diagrams have been given to explain the points. Important principles, laws, concepts and results have been printed in italics or bold to help the students grasp the subject easily. Questions from university papers of various Indian Universities have been included in the text.

I thank Shri Ravindra Kumar Gupta, Director, S. Chand & Co. Ltd., for the inspiration that I received from him. I also thank the staff of this Company for their co-operation and efforts in bringing out the book in a nice form.

I sincerely hope that the book will go a long way to satisfy the long-felt need of students for a friendly book on Physical Chemistry. Any suggestions for improvement of the book are welcome and will be gratefully acknowledged.

Dr. R. L. MADAN

CONTENTS

GREEK ALPHABET TABLE

Capital	*Small*	*Pronunciation*	*English equivalent*
Α	α	alpha	a
Β	β	beta	b
L	γ	gamma	g, n
Δ	δ	delta	d
Ε	ε	epsilon	e
Ζ	ζ	zeta	z
Η	η	eta	ē
Θ	θ	theta	th
Ι	ι	iota	i
Κ	κ	kappa	k
Λ	λ	lambda	l
Μ	μ	mu	m
Ν	ν	nu	n
Ξ	ξ	xi	x
Ο	ο	omricon	o
Π	π	pi	p
Ρ	ρ	rho	r, rh
Σ	σ	sigma	s
Τ	τ	tau	t
Φ	ϕ	phi	ph
Χ	χ	chi	ch
Ψ	ψ	psi	ps
Ω	ω	omega	ō

1 ATOMIC STRUCTURE

Q. 1. What is hydrogen spectrum? How has it helped in understanding structure of atom?

Ans. Spectrum of hydrogen has helped a lot in understanding atomic structure. This spectrum can be obtained by passing the light obtained from hydrogen gas at low temperature contained in a discharge tube through the spectrograph as shown in Fig. 1.1

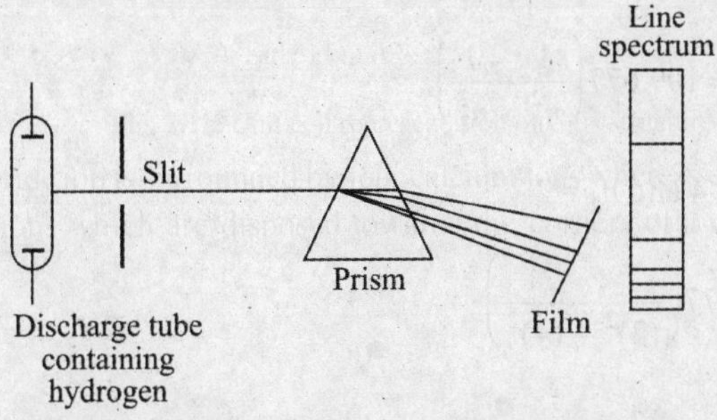

Fig. 1.1. Emission spectrum of hydrogen.

The spectrum obtained consists of a large number of sharp lines. Each line corresponds to a particular frequency of light emitted by hydrogen atoms. Lines in the emission spectrum of hydrogen are present in ultraviolet, visible and infrared regions.

The lines in the emission spectrum of hydrogen are classified into five series as follows:

(*i*) Lyman Series ..... Ultraviolet region

(*ii*) Balmer Series ..... Visible region

(*iii*) Paschen Series

(*iv*) Brackett Series } .... Infrared region

(*v*) Pfund Series

The complete spectrum of hydrogen is shown in Fig. 1.2.

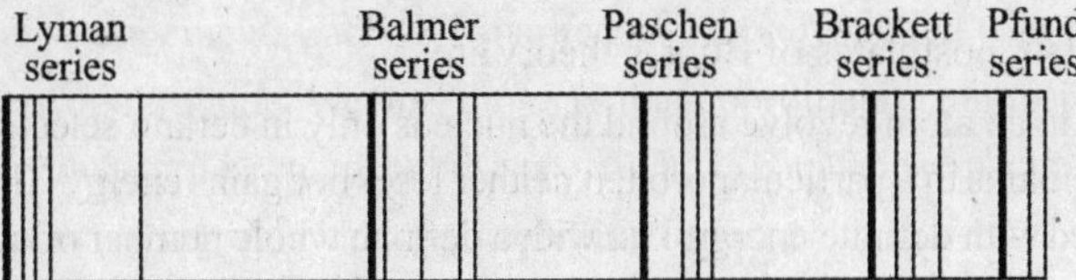

Fig. 1.2. Atomic spectrum of hydrogen.

Rydberg showed that if the spectral lines are expressed in terms of wave numbers (inverse of wavelengths), then the lines in various series obey the formula

$$\bar{\nu}(\text{cm}^{-1}) = 109.677\left(\frac{1}{n_1^2} - \frac{1}{n_2^2}\right)$$

where n_1 and n_2 are integers, such that $n_2 > n_1$. For a particular series n_1 is constant.

For Lyman Series $n_1 = 1$, $n_2 = 2, 3, 4$ ....

For Balmer Series $n_1 = 2$, $n_2 = 3, 4, 5$ ...

For Paschen Series $n_1 = 3$, $n_2 = 4, 5, 6$ ...

Q. 2. What is the wavelength of light emitted when the electron, in a hydrogen atom undergoes transition from the energy level with *n* = 4 to the energy level with *n* = 2? In which region of electromagnetic spectrum does this radiation fall? (*Vidyasagar 2007*)

Ans. According to Rydberg's formula,

$$\bar{\nu}(\text{cm}^{-1}) = 109.677\left(\frac{1}{n_1^2} - \frac{1}{n_2^2}\right)$$

In the present case $n_2 = 4$ and $n_2 = 2$

$$\bar{\nu} = 109.677\left(\frac{1}{(2)^2} - \frac{1}{(4)^2}\right)$$

$$= 109.677 \times \frac{3}{16}$$

$$= 20564 \text{ cm}^{-1}$$

$$\lambda = \frac{1}{\bar{\nu}}$$

$$= \frac{1}{20564} \text{ cm}$$

$$= 486 \text{ nm}$$

The above radiation corresponds to Balmer series and it will be in visible region..

Q. 3. Write down the postulates of Bohr model of atom.

(*Vidyasagar 2007; Calcutta 2007*)

Ans. The important postulates of Bohr's theory are:

1. The electrons in an atom revolve around the nucleus only in certain selected circular orbits. As long as the electron remains in a particular orbit, it neither loses nor gains energy. This led to the idea that each orbit is associated with definite energy, i.e., with a definite whole number of quanta of energy. The orbits, therefore, are also known as energy levels or energy shells. Bohr gave number 1, 2, 3, 4, etc., (starting from the nucleus) to these energy levels (Fig. 1.3). These are now termed as **principal quantum numbers.** The various energy levels are also designated by letters K, L, M, N, etc. The farther the energy level from the nucleus, the greater is the energy associated with it.

2. Energy is emitted or absorbed by an atom only when an electron moves from one orbit or energy level to another. The amount of energy emitted or absorbed is given by the difference of the energies associated with the two energy levels concerned. Thus, the energy of an electron cannot change continuously. It changes only abruptly as the electron jumps from one energy level to another.

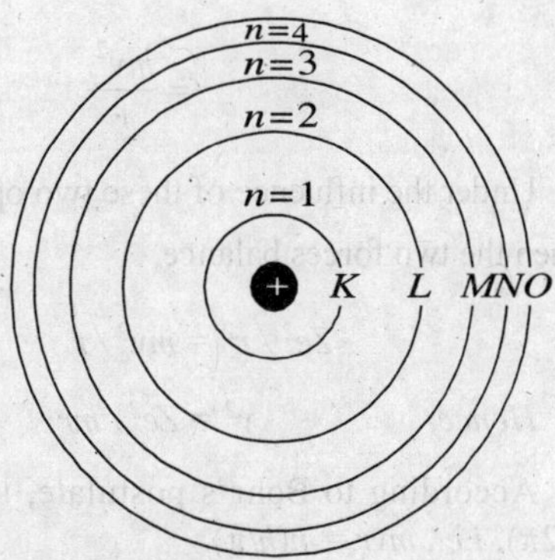

Fig. 1.3. Orbit like representation of various energy levels.

3. The angular momentum of an electron moving round the nucleus is **quantised**. An electron, like any other body moving in a circular orbit, has an angular momentum given by *mvr* where *m* is the mass and *v* is the linear velocity of the electron while *r* is the radius of the orbit. Bohr postulated that the angular momentum of an electron in an atom can have only definite or discrete values given by the expression

$$\text{Angular momentum} = mvr = n(h/2\pi)$$

where n is any interger (1, 2, 3, etc.)

Thus, the angular momentum of an electron may be h/π or simple whole number multiples of $h/2\pi$, such as $2h/2\pi$, $3h/2\pi$,... $nh/2\pi$. This principle is known as **quantisation of angular momentum.**

Q. 4. Derive Bohr's equations for the radius of the orbit and energy associated with an orbit of hydrogen atom. *(Guwahati, 2006; Calcutta 2007; Kerala, 2000)*

Ans. Bohr was able to calculate the radius of orbit and energy of the electron moving in different orbits round the nucleus in a hydrogen atom. He assumed that the electron continues to remain in its orbit because the electrostatic force of attraction exerted by the nucleus is exactly balanced by the centrifugal force arising from its circular motion.

A hydrogen atom, having one proton in the nucleus and one electron outside the nucleus, is represented in Fig. 1. 4.

Fig. 1.4. The electron revolving around the nucleus in a hydrogen atom.

Let the charge on the nucleus be + *Ze*, +*e* is the charge on one proton. For hydrogen atom, Z = 1 and for hydrogen-like atoms such as He^+ and Li^{2+}, Z = 2 and 3, respectively. The electron with charge –e revolves round the nucleus in an orbit of radius *r*. Let *v* be the tangential velocity and *m* the mass of the electron.

According to classical mechanics, the electrostatic force of attraction acting on the electron, which tends to pull the electron towards the nucleus is

$$= \frac{(Ze) \times e}{r^2} = \frac{Ze^2}{r^2} \quad ...(i)$$

However, the centrifugal force acting on the electron which tends to keep it away from its orbit is

$$= \frac{mv^2}{r} \quad \text{... (ii)}$$

Under the influence of these two opposing forces, the electron keeps on revolving in its orbit. When the two forces balance,

$$Ze^2 / r^2 = mv^2 / r \quad \text{... (iii)}$$

Hence, $$v^2 = Ze^2 / mr \quad \text{... (iv)}$$

According to Bohr's postulate, the angular momentum, *mvr*, is an integral multiple of $(h/2\pi)$, *i.e.*, $mvr = n(h/\pi)$.

Thus $$v = nh/2\pi mr \quad \text{... (v)}$$

or $$v^2 = n^2h^2/4\pi^2m^2r^2 \quad \text{... (vi)}$$

From Eqs. (*iv*) and (*vi*), we have

$$\frac{Ze^3}{mr} = \frac{n^2h^2}{4\pi^2 m^2 r^2} \quad \text{... (vii)}$$

or $$Ze^2 = \frac{n^2h^2}{4\pi^2 mr} \quad \text{... (viii)}$$

The radius of the *n*th orbit, which we will write as r_n, is given by

$$r_n = \frac{n^2h^2}{4\pi^2 Ze^2 m} \quad \text{... (ix)}$$

Since, for hydrogen atom, $Z = 1$, hence,

$$r_n = \frac{n^2h^2}{4\pi^2 me^2} \quad \text{...(x)}$$

For hydrogen atom in the ground state ($n = 1$) the radius (r_1) is designated as the Bohr's radius. Thus,

Bohr radius $$(r_1) = \frac{h^2}{4\pi^2 me^2} \quad \text{...(xi)}$$

The total energy *E* of the revolving electron is equal to the sum of the kinetic energy $\left(= \frac{1}{2}mv^2\right)$ and the potential energy ($= -Ze^2/r$).

Thus, $$E = \frac{1}{2}mv^2 - Ze^2 / r \quad \text{.... (xii)}$$

From Eq. (*iv*), we find that

$$mv^2 = \frac{Ze^2}{r} \quad \text{...(xiii)}$$

Substituting the above value of mv^2 in Eq. (*xii*), we have

$$E = \frac{Ze^2}{2r} - \frac{Ze^2}{r} = -\frac{Ze^2}{2r} \qquad \text{.... } (xiv)$$

Substituting the value of *r* from Eq. (*ix*) in Eq. (*xiv*), we get

$$E = -\frac{Ze^2}{2} \times \frac{4\pi^2 Ze^2 m}{n^2 h^2} \qquad \text{... } (xv)$$

Thus, the energy *E* of the electron in the *n*th orbit, which we shall now designate as E_n, is given by

$$E_n = \frac{2\pi^2 Z^2 me^4}{\pi^2 h^2} \qquad \text{... } (xvi)$$

This is the famous Bohr equation applicable to hydrogen and hydrogen like atoms such as He^+, Li^{2+}, Be^{3+}, etc. (For H, Z = 1; for He^+, Z = 2; for Li^{2+}, Z = 3).

For hydrogen atom Z = 1

Substituting the values for *m, e, h* and π in equation (*xvi*), we get

$$E_n = -\frac{1312}{n^2} \text{ kJ mol}^{-1} \text{ or } -\frac{2.179 \times 10^{-21}}{n^2} \text{ kJ/atom} \qquad (xvii)$$

This gives us the energy of *n*th orbit in the atom of hydrogen.

Q. 5. How does Bohr's model explain the atomic spectrum of hydrogen?

(*Kerala, 2000; Calcutta, 2005*)

Ans. According to Bohr's model electron in an atom can have only certain definite energy levels. When the electron is present in lowest possible energy level, it is said to be in **ground state**. When energy is supplied from some external source, the electron may absorb energy and jump to some higher energy level. The electron in such a state is said to be in **excited state**. The electron has tendency to come back to the ground state. When the electron jumps back to lower energy levels, it gives out energy in the form of a quantum equal to the difference of energies between the two energy levels. If E_1 and E_2 are the energies of lower and higher energy levels respectively, then frequency (ν) of the radiation emitted is given by the following relation:

$$E_2 - E_1 = h\nu$$

or

$$\nu = \frac{E_2 - E_1}{h}$$

where *h* is Planck's constant.

According to Bohr's model, in a hydrogen atom (or in any other atom) E_2 and E_1 can have only certain definite values. From this, it follows that ν can have only certain fixed values. Thus, Bohr's model explains why there are certain discrete lines in the spectrum of hydrogen.

In hydrogen atom there is one electron which is present in first orbit in ground state. When energy is supplied, such as by passing electric discharge, this electron may be excited to some higher energy level. Since in a sample of hydrogen there are large number of atoms, the electrons in different atoms absorb different amounts of energies and are excited to different higher energy levels. For example, the electron in some atoms may jump to third energy level while in others it may be raised to fourth energy level and so on. Now, from excited states the electron may return to ground state in one or more jumps downwards.

These different downwards jumps are associated with different amounts of energies and hence result in the emission of radiations of different wavelength which appear as different lines in the hydrogen spectrum. Since large number of different type of downward transitions take place simultaneously in a sample of hydrogen, therefore, large number of lines are obtained in emission spectrum of hydrogen.

The lines which arise due to the transitions from higher energy levels to first energy level are grouped as **Lyman Series**. Similarly, the lines obtained as a result of transitions of electrons from higher energy levels to second, third, fourth and fifth energy levels give rise to **Balmer, Paschen, Brackett and Pfund** series respectively as illustrated in Fig. 1.5.

Lyman series	From $n = 2, 3, 4, 5 \ldots$ to $n = 1$
Balmer series	From $n = 3, 4, 5, 6 \ldots$ to $n = 2$
Paschen series	From $n = 4, 5, 6, 7 \ldots$ to $n = 3$
Brackett series	From $n = 5, 6, 7, 8 \ldots$ to $n = 4$
Pfund series	From $n = 6, 7, 8, 9 \ldots$ to $n = 5$

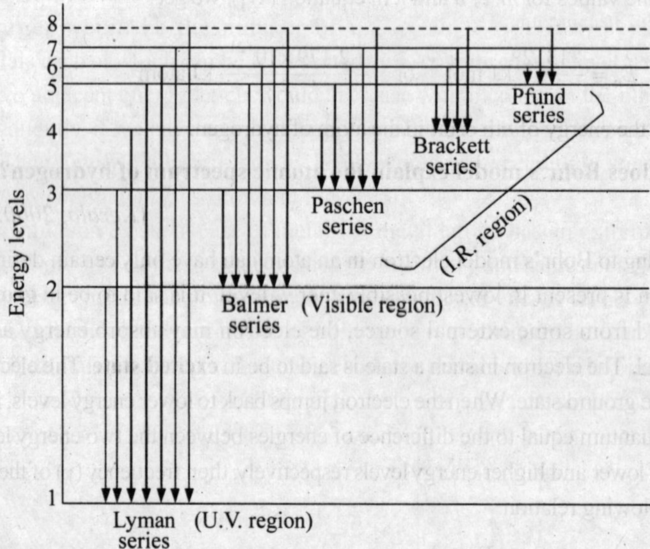

Fig. 1.5. Various spectral series in hydrogen spectrum.

Q. 6. Applying Bohr's model calculate the energy emitted when an electron in a hydrogen atom makes a transition from third energy level to first energy level. (*Calcutta, 2005*)

Planck's constant $h = 6.63 \times 10^{-34}$ Js.

Ans. For third energy level $n = 3$

and For first energy level, $n = 1$

Energy of third energy level, $E = \dfrac{-1312}{(3)^2}$ kJ mol^{-1}

Energy of first energy level, $E = \dfrac{-1312}{(1)^2}$ kJ mol^{-1}

$$\Delta E = E_2 - E_1$$

$$= -1312\left(\frac{1}{(3)^2} - \frac{1}{(1)^2}\right)$$

$$= 1166 \text{ kJmol}^{-1}$$

$$\Delta E \text{ (per atom)} = \frac{1166}{6.02 \times 10^{23}}$$

$$= 1.93 \times 10^{-21} \text{ kJ/atom}$$

$$= 1.93 \times 10^{-18} \text{ J/atom.}$$

Q. 7. Describe Sommerfeld extension of Bohr theory of structure of atom.

(*Calcutta, 2005*)

Ans. The radii and energies of orbits as calculated according to Bohr's theory were found to be in good agreement with the experimental values.

Bohr's theory could also explain successfully the positions of various series of lines in the hydrogen spectrum. The theory was, therefore, largely accepted. However, the use of spectroscopes of high resolving powers revealed the presence of fine structures in the line spectra of hydrogen and hydrogen-like atoms or ions having solitary electron. Bohr's theory could not explain the occurrence of these fine spectra. The theory was, therefore, extended by Sommerfeld by putting forward the idea of elliptical orbits.

According to Sommerfeld, *an electron revolving around a central positively charged nucleus is so influenced by the nuclear charge that it is set into motion in elliptical orbits with the nucleus situated at one of the foci* (Fig. 1. 6). Thus, while according to Bohr's theory, electrons move in circular orbits, according to Sommerfeld's modification, electrons move in **elliptical orbits.** In the latter case, there will be a major axis and a minor axis having different lengths. As the orbit broadens, the lengths of the two axes become closer and they become equal when the orbit becomes circular, as shown in Fig. 1.6. *Thus, the circular orbit is only a special case of elliptical orbits.*

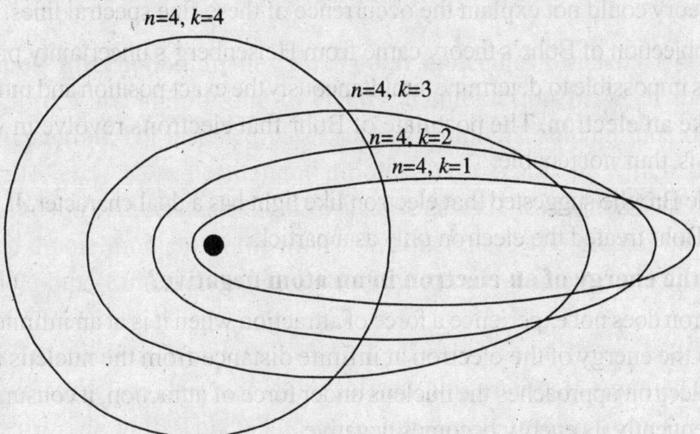

Fig. 1.6. Sommerfeld's elliptical electron orbits for $n = 4$.

The electron moving in an elliptical orbit will have its angular momentum. This momentum, according to the quantum theory, must also be quantised and thus can have only a limited number of values given by the factor $kh/2\pi$ where k is an integer known as the azimuthal quantum number. It can be shown that n, the principal quantum number, used by Bohr and k, the azimuthal quantum number, used by Sommerfeld are related to one another as

$$\frac{n}{k} = \frac{\text{length of major axis}}{\text{length of minor axis}}$$

Thus, for any given value of *n* (except 1), *k* can have more than a single value. When $k = n$, the orbit must be circular. But, as *k* becomes smaller, the orbits become elliptical with greater and greater eccentricity. It may be noted that *k* cannot be zero because in that case minor axis would be zero which would imply linear motion of the electron passing through the nucleus.

The number of possible values of *k* is equal to the principal quantum number *n*. Thus, if *n* is 4, *k* can have values varying from 4 to 1. It may have a value of 4, 3, 2 or 1. When *n is* 1, *k* can have only one value.

According to Sommerfeld's extension of Bohr theory, referred to sometimes as **Bohr-Sommerfeld theory,** the energy depends not only on *n,* the principal quantum number, but also, to some extent, on the azimuthal quantum number, *k.* Hence, the transition of an electron from one energy level n_2 to another energy level n_1 will be slightly different as it would also depend upon the various possible values of *k* in the two states. This explains the occurrence of groups of very fine lines observed in the hydrogen spectrum discovered by using highly refined spectroscopes.

Q. 8. List the shortcomings of Bohr's theory.

(*Magadh, 1994; Nagpur, 2000; Calcutta, 2005*)

Ans. The main shortcoming of Bohr's theory are:

1. Bohr's atomic model failed to account for the effect of magnetic field on the spectra of atoms or ions. It was observed that when the source of a spectrum is placed in a strong magnetic field, each spectral line is further split into a number of lines. This is called **Zeeman effect**. This observation could not be explained on the basis of Bohr's model.

2. Bohr's theory of an atom was successful in accounting for the spectra of hydrogen and hydrogen-like atoms (*e.g.*, He^+, Li^{2+}). But, it failed to explain the spectra of multi-electron atoms. The spectroscopic measurements with high precisian revealed the fine structure in the line spectra of atoms. Bohr's theory could not explain the occurrence of these fine spectral lines.

3. Another objection of Bohr's theory came from Heisenberg's uncertainty principle. According to this principle, it is impossible to determine simultaneously the exact position and momentum of a small moving particle like an electron. The postulate of Bohr that electrons revolve in well defined orbits around the nucleus is, thus not tenable.

4. In 1923, de Broglie suggested that electron like light has a dual character. It has particle as well as wave character. Bohr treated the electron only as a particle.

Q. 9. Why is the energy of an electron in an atom negative? (*Vidyasagar, 2007*)

Ans. The electron does not experience a force of attraction when it is at an infinite distance from the nucleus. Therefore the energy of the electron at infinite distance from the nucleus is arbitrarily taken as zero. When the electron approaches the nucleus under force of attraction, it consumes certain amount of energy and consequently its energy becomes negative.

Q. 10. Discuss the postulates of Wave Mechanical Model of atom (de Broglie equation).

(*Andhra 2002; Madras, 2004; Purvanchal 2007*)

Ans. Light radiations have dual behaviour and can be considered either as a wave motion or as a stream of photons. In 1924, de Broglie proposed dual bahaviour for electrons and gave an expression relating their wavelength (λ) with mass (m) and velocity (v):

$$\lambda = \frac{h}{mv} \qquad ...(1.1)$$

Davisson and Germer in 1927 showed that electron beam is diffracted by a crystal in the same way as light radiation is diffracted. This provided a proof of wave nature of electrons.

Another fundamental difficulty arose from Heisenberg's uncertainty principle which states that it is impossible to describe, simultaneously and precisely both the position and momentum of an electron (Bohr model violates the uncertainty principle, as it describes simultaneously both the position and momentum of the electron).

Wave or quantum Mechanical Model is based on the following postulates:

(*i*) Heisenberg uncertainty principle governs atomic particles. The term probability distribution is used to describe motion of an electron.

(*ii*) A stream of electrons is associated with a wave whose wavelength is given by de Broglie relationship :

$$\lambda = \frac{h}{mv}$$

(*iii*) Electron wave is compared to a stationary or standing wave formed by a vibrating string fixed between two points.

Energy values of the electron are quantized as calculated by Bohr.

Q. 11. Give the derivation of Schrodinger wave equation and define each term involved.

(Dayanand, 2002; Kanpur, 2000; Meerut, 2000; Himachal, 2000; Andhra, 2002; Guwahati, 2007; Delhi, 2007; Nagpur, 2008)

Or

Write the expression for non-time wave equation for a subatomic particle in three dimensions and explain the meaning of various terms involved in it.

(Calcutta, 2000; Kerala, 2000 : Delhi, 2003; Nagpur 2003)

Ans. Schrodinger developed a mathematical wave equation to describe the wave motion of the electron in a hydrogen atom. Here the nucleus is surrounded by a vibrating electron wave which may be mathematically compared with stationary as standing wave formed by a vibrating string fixed between two points as shown in Fig. 1.7. The electron in a hydrogen atom is under constraint imposed by the attraction of the nucleus. An electron motion can be described by an equation analogous to that used to describe a "standing or stationary" wave system whose solution in one dimension is given by the sine wave equation (1.2)

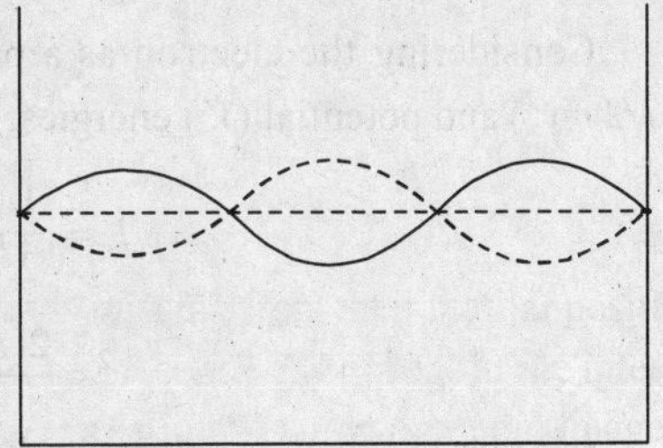

Fig. 1.7. A standing wave.

$$\Psi = A \sin \frac{2\pi x}{\lambda} \qquad ...(1.2)$$

where Ψ = amplitude of the wave of wavelength λ at a distance x from the origin A is a constant.

On double differentiation w.r.t. x, we get

$$\frac{d\Psi}{dx} = A . \frac{2\pi}{\lambda} . \cos \frac{2\pi x}{\lambda} \qquad ...(1.3)$$

$$\frac{d^2\Psi}{dx^2} = A . \frac{2\pi}{\lambda} . \frac{2\pi}{\lambda} \left[\frac{-\sin 2\pi x}{\lambda} \right]$$

$$= \frac{-4\pi^2}{\lambda^2} A \sin \frac{2\pi x}{\lambda} \qquad ...(1.4)$$

Substituting $\Psi = A \sin \frac{2\pi x}{\lambda}$ from Eq. (1.2) into Eq. (1.4), we get

$$\frac{d^2\Psi}{dx^2} = \frac{-4\pi^2}{\lambda^2}\Psi \qquad ...(1.5)$$

or
$$\frac{d^2\Psi}{dx^2} + \frac{4\pi^2}{\lambda^2}\Psi = 0 \qquad ...(1.6)$$

This is the equation for one dimensional standing wave which can be extended to describe motion of an electron in three dimensions by a second order partial differential equation (1.7)

$$\frac{\partial^2\Psi}{\partial x^2} + \frac{\partial^2\Psi}{\partial y^2} + \frac{\partial^2\Psi}{\partial z^2} + \frac{4\pi^2}{\lambda^2}\Psi = 0 \qquad ...(1.7)$$

where Ψ is a function of Cartesian coordinates *x, y, z*. This may be written more precisely as

$$\nabla^2\Psi + \frac{4\pi^2}{\lambda^2}\Psi = 0 \qquad ...(1.8)$$

where ∇^2 stands for $\frac{\partial^2\Psi}{\partial x^2} + \frac{\partial^2\Psi}{\partial y^2} + \frac{\partial^2\Psi}{\partial z^2}$

Combining Eqs. (1.1) and (1.8), we get

$$v^2\Psi + \frac{4\pi^2 m^2 v^2}{h^2}\Psi = 0 \qquad ...(1.9)$$

Considering the electron as a particle, the total energy (*E*) of the system is sum of kinetic ($1/2\ mv^2$) and potential (*V*) energies, *i.e.*,

$$E = \frac{1}{2}mv^2 + V$$

or
$$v^2 = \frac{2(E-V)}{m} \qquad ...(1.10)$$

In order to calculate values for the energy states value of v^2 (Eq. 10) is substituted into equation (1.9) to obtain a fundamental form of Schrodinger wave equation for a single particle in three dimensions.

$$\nabla^2\Psi + \frac{8\pi^2 m}{h^2}(E-V)\Psi = 0 \qquad ...(1.11)$$

or
$$\frac{\partial^2\Psi}{\partial x^2} + \frac{\partial^2\Psi}{\partial y^2} + \frac{\partial^2\Psi}{\partial z^2} + \frac{8\pi^2 m}{h^2}(E-V)\Psi = 0 \qquad ...(1.11a)$$

where, Ψ = amplitude of the wave function associated with the electron

x, y, z = Cartesian coordinates

m = mass of the electron

h = Planck's constant

E = total energy of the system

V = potential energy of the system

The potential energy term V in Schrodinger wave equation (1.11) is given by the following equation

$$V = \frac{-Ze^2}{r} \qquad ...(1.12)$$

where + Ze = charge of the nucleus

Z = atomic number

$-e$ = charge on the electron

Consequently, the equation (1.11*a*) for one electron atomic system in three dimensions may be written as

$$\frac{\partial^2\Psi}{\partial x^2}+\frac{\partial^2\Psi}{\partial y^2}+\frac{\partial^2\Psi}{\partial z^2}+\frac{8\pi^2 m}{h^2}\left[E+\frac{Ze^2}{r}\right]\Psi = 0 \qquad ...(1.13)$$

Q. 12. How will you apply Schrodinger wave equation to hydrogen atom?

(M. Dayanand 2000; Calcutta 2000; G. Nanakdev, 2000; Gujarat,2003; Nagpur, 2003; Kurukshetra, 2002)

Or

Show that Schrodinger wave equation supports Bohr's theory of hydrogen atom?

Ans. Schrodinger wave equation for one electron atomic system is

$$\frac{\partial^2\Psi}{\partial x^2}+\frac{\partial^2\Psi}{\partial y^2}+\frac{\partial^2\Psi}{\partial z^2}+\frac{8\pi^2 m}{h^2}\left[E+\frac{Ze^2}{r}\right]\Psi = 0$$

For hydrogen atom, Z= 1; therefore $V=\frac{-e^2}{r}$ and Schrodinger wave equation for H atom may be written as

$$\frac{\partial^2\Psi}{\partial x^2}+\frac{\partial^2\Psi}{\partial y^2}+\frac{\partial^2\Psi}{\partial z^2}+\frac{8\pi^2 m}{h^2}\left[E+\frac{e^2}{r}\right]\Psi = 0 \qquad ...(1.14)$$

Equation (1.14) results from the assumptions:

(*i*) that the behaviour of an electron in an atom is analogous to a system of standing wave; and

(*ii*) de Broglie relationship $\lambda = \frac{h}{mv}$ describes the wavelength of an electron.

Justification for the above equation for hydrogen atom is found in the fact that the solution of equation 1.14 gives values of energy which agree well with those obtained experimentally from atomic spectra or calculated from the equation (1.15)

$$E = \frac{-2\pi^2 me^4 Z^2}{h^2 m^2} \qquad ...(1.15)$$

given by Bohr for his model of the atom. That is, Schrodinger wave equation leads to exactly the same allowed energies as those deduced from the Bohr model. Besides that the "most probable" radius for 1*s* electron in H atom from wave mechanics comes out to be the same as determined by Bohr (0.529 Å) from equation (1.16) from Bohr model

$$r = \frac{h^2 n^2}{4\pi^2 me^2 Z} \qquad ...(1.16)$$

Q. 13. Discuss the physical significance of Ψ and Ψ^2.

(Purvanchal 2007; Delhi 2007; Nagpur 2008; Garhwal 2010)

Or

What do you understand by the expression $\Psi \Psi^* d\tau$?

Ans. In all wave equations, the square of the wave function is the property which has physical significance. For example, in case of light waves, square of the amplitude of the wave at a point is proportional to the intensity of the light wave. Similarly in electron wave motion, square of Ψ (ψ^2) should represent intensity. According to uncertainty principle, the position of electron cannot be determined with certainty. All we can say is that greater the intensity of wave function at a particular point, greater is the probability of locating the electron at that point. Thus ψ^2 may be interpreted as being proportional to electron density. When ψ^2 is high, electron density is high, that is, the probability of finding an electron is high. If ψ^2 is low, probability is low. In order to avoid imaginary values, $\psi\psi^*$ is used for ψ^2. ψ^* is a complex conjugate of ψ. We expect that if ψ extends over a finite volume in space $dx \cdot dy \cdot dz$ ($= d\tau$), then the electron would be found in that volume. So $\psi\psi^* d\tau$ gives the probability of finding the electron in the volume $d\tau$. The wave function ψ may be positive, negative or imaginary but the probability density $\psi\psi^*$ will always be positive and real. The probability of fmding the electron is finite, which means $\psi\psi^* d\tau$ must give real values. If ψ has real value, then $\psi = \psi^*$ and we can use $\psi^2 d\tau$ in place of $\psi\psi^* d\tau$.

Q. 14. What are the main conditions which Ψ must satisfy to give meaningful solution (Eigen function)?

Ans. Ψ as a wave function must obey the following conditions if it is to give meaningful results:

(*i*) Ψ must be single valued, *i.e.*, it must have only one definite value at a particular point in space.

(*ii*) Ψ must be continuous at all points in space, *i.e.*, there must not be sudden changes in the values of ψ when its variables are changed.

(*iii*) Ψ must have a finite value over the space $d\tau$ that the electron can occupy.

(*iv*) Ψ must be normalized, *i.e.*, the probability of finding the electron in the space under consideration must be unity.

(*v*) Ψ must become zero at infinity.

This means that of all the values of Ψ obtained by solving Schrodinger wave equation, only those values of Ψ are accepted which obey the above conditions. Thus only acceptable values of Ψ have significance.

The significant value of Ψ are called **Eigen functions** or **wave functions**. Eigen functions give significant values of the total energy of electron (E) called Eigen values.

Q. 15. What is the physical significance of Ψ^*? Explain mathematically.

(G. Nanakdev, 2000)

Ans. Solution of Schrodinger wave equation gives Ψ values which can be real as well as imaginary but Ψ^2 is always real and has a positive value. This is because the square of even negative values is positive. If Ψ has imaginary values, then it can be expressed as

$$\Psi = a + ib$$

where a and b are real numbers and i (known as iota) $= \sqrt{-1}$. In such a case, Ψ is made real by multiplying it by its own complex conjugate (Ψ^*) which is expressed as

$$\Psi^* = a - ib$$

The product obtained by multiplying Ψ and Ψ^* for real values of Ψ is as:

$$\Psi\,\Psi^* = (a + ib)\,(a - ib) = a^2 - (i^2 b^2)$$
$$= a^2 - (-1 \times b^2) \qquad (\because i = \sqrt{-1})$$
$$= a^2 + b^2$$

However, if Ψ is real, then $\Psi = \Psi^*$, then $\Psi\Psi^* = \Psi^2$.

Q. 16. What is the origin of quantum numbers *n,l* and *m* ?

(Nagpur, 2005; Purvanchal 2007)

Or

What do you mean by quantum number ? Name different types of quantum numbers and their significance ? *(Delhi 2002, M. Dayananad 2002; Calcutta, 2007; Garhwal, 2000)*

Ans. The three quantum numbers *n, l* and *m* and their allowed values arise as a consequence of the solution of the wave equation for hydrogen atom. These quantum numbers and their allowed values are discussed below:

(*i*) Principal or Radial Quantum Number, *n*

It arises from the solution of the radial part of Ψ and identifies the shell or level to which the electron belongs. These shells are regions in space indicating the "most probable" position of the electron from the nucleus. The quantum number *n* is restricted to have any positive integral values from 1 to infinity. The larger the value of *n,* the further is the shell from the nucleus.

$$n = 1, 2, 3, 4, 5 \ldots\ldots \infty$$

Theoretically there is no limit to the number of shells or levels but seventh shell ($n = 7$) is the highest shell occupied by the known elements.

Principal quantum number *n* alone does not give complete energy but designates the main energy level to which an electron belongs. The lowest energy state of an atom corresponds to $n = 1$ and the energy increases with increasing *n.*

(*ii*) Azimuthal or Subsidiary Quantum Number, *l*

It refers to the subshell or sublevel to which an electron belongs and describes the motion of the electron. The magnitude of the angular momentum of an electron (angular velocity × moment of inertia) is related to *l* by the expression:

$$\text{Angular momentum} = \sqrt{l(l+1)}\,\frac{h}{2\pi} \qquad \ldots(1.17)$$

where $l = 0,\ 1,\ 2,\ 3,\ \ldots\ (n-1)$

This means that the values of *l* for the subshells of a shell are governed by *n,* when $n = 1$, there is only one subshell, with the value $l = 0$, when $n = 2$, there are two subshells with the values $l = 0$, 1. When $n = 3$, there are three subshells that have the values $l = 0, 1, 2$. Similarly when $n = 4$, number of shells is 4 with values, $l = 0, 1, 2, 3$. For every value of *l*, there is a separate notation to indicate subshell. For example,

Quantum No.	$l = 0$	1	2	3	4	5 ......
Subshell Notation	$= s$	p	d	f	g	h.....

The first four notations : *s, p, d* and *f* are adjectives used earlier to identify spectral lines; sharp, principal, diffuse and fundamental but subsequent notations used for $l > 3$ proceed alphabetically *e.g., g, h ,i , j* ..... so on.

Knowing the value of *n* and finding out *l*, we designate a subshell. For example, subshell with

$n = 1, l = 0$ is called 1*s* subshell and others are given as under:

n	*l*	*Subshell notation*
2	0	2*s*
	1	2*p*
3	0	3*s*
	1	3*p*
	2	3*d*
4	0	4*s*
	1	4*p*
	2	4*d*
	3	4*f*

(*iii*) Magnetic Quantum Number, *m*

Each subshell is comprised of one or more orbitals (regions with maximum probability of finding the electron) whose number is equal to the number of ways the electrons in a subshell can orient themselves in space. The number of orientations is given by $2l + 1$. Thus, the number of orientations or orbitals is one when $l = 0$ (*s*-subshell), three for $l = 1$ (*p* -subshell), five for $l = 2$ (*d*-subshell) and seven for $l = 3$ (*f* -subshell). In other words, *s*-subshell has one *s*-orbital, *p*-subshell has three *p*-orbitals, *d*-subshell has five *d*-orbitals and so on. In the absence of electric or magnetic field, these orbitals in a subshell are degenerate, that is, they have identical energy. For example, for $l = 1$, the three orbitals P_x, P_y, and P_z are degenerate but this degeneracy is split in the presence of a magnetic field.

In this way three *p*-orbitals of different energies with the lowest energy orbital aligned with the magnetic field ($m = +1$) and five *d*-orbitals of different energies with the lowest aligned with magnetic field ($m = +2$) are obtained. Each orbital within a subshell is identified in term of the component of the angular momentum $m\frac{h}{2\pi}$ where m = magnetic quantum number. In other words, *m* is associated with different orientations of the orbital with reference to some defined directions.

For a given subshell, *m* can have values

$m = -l, -(l-1), -(l-2) \ldots.0\ldots.. \quad (l-2), (l-1), l$

Thus for $l = 0$, $m = 0$, we get an *s*-orbital.

For $l = 1$, *m* can have three values $m = -1, 0, +1$ corresponding to three *p*-orbitals; p_x, p_y, p_z.

For $l = 2$, *m* can have five values $m = -2, -1, 0, +1, +2$ corresponding to five *d*-orbitals; $d_{x^2-y^2}$, d_{xy}, d_{yz}, d_{zx}, d_{z^2}.

Some salient features of these quantum numbers are given below:

(*i*) Radial quantum number *n* and its allowed values arise from the solution of radial part of ψ, whereas *l* and *m* quantum numbers arise from solution of angular part of ψ.

(*ii*) The lowest energy state of an atom corresponds to $n = 1$ and energy increases as *n* increases.

(*iii*) The number of subshells in a shell is equal to the value of *n* where $n = 2$, there are 2 subshells ($l = 0, 1$); for $n = 3$, there are 3 subshells ($l = 0, 1, 2$) and so on.

(*iv*) The number of orbitals of a type in a given subshell is given by $2l + 1$. That is, we have one *s*-type orbital of $l = 0$; three *p*-type orbitals for $l = 1$; five *d*-type orbitals for $l = 2$ and seven *f*-type orbitals for $l = 3$.

(*v*) The total number of orbitals in a given shell $= n^2$ (n = principal quantum number). For example for $n = 2$, the total number of orbitals $= 2^2 = 4$.

Spin Quantum Number m_s

In 1896, Zeeman observed that spectral lines in the atomic spectrum get split under the influence of a strong magnetic field. This phenomenon, called the **Zeeman effect**, is attributed to the orientation of the components of the angular momentum with respect to the external magnetic field and is given by $m \frac{h}{2\pi}$, where m = magnetic quantum number.

Later on it was observed in the atomic spectra of alkali metals that the spectral lines which were earlier considered to be single lines are actually narrow doublets (two lines quite close together). An explanation to these doublet lines was offered by Uhlenbeck and Goudsmit in 1925. They proposed that the electron is also associated with rotation about its own axis (spinning). Axial spin of the electron is associated with spin angular momentum whose magnitude is given by $m_s \frac{h}{2\pi}$, where m_s is called the spin quantum number. The spin quantum number can have values $+\frac{1}{2}$ or $-\frac{1}{2}$ which arise from the direction of spin (clockwise or anticlockwise). It is also designated as ↑ or ↓. Thus, the description of an electron in an atom in terms of three quantum numbers (described earlier) is not complete and, in fact, four quantum numbers, *n, l, m* and m_s: are needed to describe the electron in an atom completely.

An electron is a spinning negative charge and a spinning charge is magnetic, therefore, a spinning electron can be considered to behave like a tiny magnet. Two electrons with opposite spins in an orbital will behave as two tiny magnets with opposite poles towards each other and thus attract each other.

Q. 17. Which sets of quantum numbers are not permissible and why?

(*i*) $n = 5, \quad l = 4, \quad m = 0, \quad m_s = \frac{1}{2}$

(*ii*) $n = 3, \quad l = 0, \quad m = -1, \quad m_s = -\frac{1}{2}$

(*iii*) $n = 3, \quad l = 1, \quad m = 2, \quad m_s = \frac{1}{2}$

(*iv*) $n = 2, \quad l = 2, \quad m = 0, \quad m_s = \frac{1}{2}$

Ans. (*i*) This is allowed.

(*ii*) It is not allowed because value of m (+ 1, ... 0 ... –1) are governed by l and when $l=0$, m cannot have values other than zero ($m = 0$).

(*iii*) It is not allowed because when $l = 1$, m has allowed values = + 1, 0, –1 only.

(*iv*) It is not allowed because values of l are governed by n. For $n = 2$, 1 can have values equal = 0, 1 only. Therefore for $n = 2$, $l = 2$ is not permissible.

Q. 18. Explain why orbitals like 1*p*, 2*d* and 3*f* are not possible.

Ans. This is because the sets of quantum numbers required for these orbitals are not allowed. 1*p* refers to $n = 1, l = 1$ which is not allowed because values of l are governed by n, that is $l = 0, 1, 2 ..... (n-1)$. Thus, for $n = 1$, l cannot have a value equal to one.

Similarly, 2*d* refers to $n = 2, l = 2$ which again is not possible. For $n = 2$; l can have values 0, 1 only.

Also 3*f* is not possible because for 3*f*, the combination $n = 3, l = 3$ is again not allowed. For $n = 3$, l can't have a value higher than 2.

Q. 19. Calculate and identify all the individual orbitals for (*i*) $n = 3$ (*ii*) $n = 4$.

(*Calcutta, 2000*)

Ans. (*i*) $n = 3$; $l = 0, 1, 2$; Total no. of orbitals = 9

These are: 3*s*; $3p_x$, $3p_y$; $3p_z$; $3d_{xy}$, $3d_{yz}$, $3d_{zx}$, $3dz^2$, $3d_{x^2-y^2}$

(*ii*) For $n = 4$; $l = 0, 1, 2, 3$. Individual orbitals are:

l	*m*	*Orbitals*
0	0	4*s*
1	± 1, 0	$4p_x$, $4p_y$, $4p_z$
2	± 2, ± 1, 0	$4d_{xy}$, $4d_{yz}$, $4d_{xz}$, $4dz^2$, $4d_{x^2-y^2}$
3	± 3, ± 2, ± 1, 0	$4f_{x^3}$, $4f_{y^3}$, $4f_{z^3}$, $4f_{xyz}$ $4f_{(z^2-y^2)}$, $4f_{(z^2-x^2)}$, $4f_{(x^2-y^2)}$

For $n = 4$; total number of orbitals = 1 + 3 + 5 + 7 = 16.

Q. 20. Draw the shapes of each of the following orbitals $2p_y, 3d_{z^2}, 3d_{x^2-y^2}, 3d_{xy}$.

Ans. Shapes of the orbitals are as shown in Fig. 1.8.

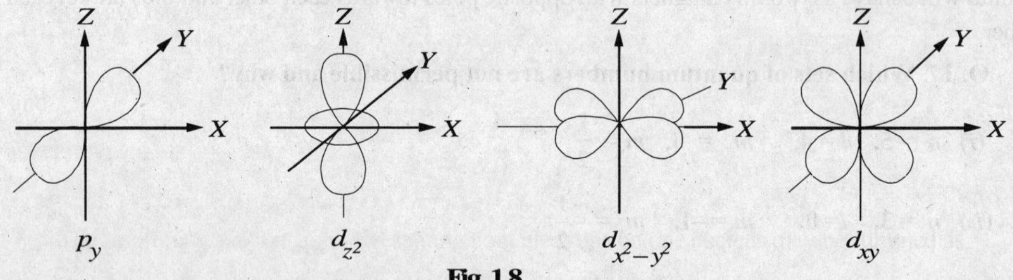

Fig. 1.8

Q. 21. Which of the following orbitals are not possible?

1*p*, 2*s*, 2*p*, 3*f*

Give reasons.

Ans. 1*p* and 3*f* orbitals are not possible. For 1*p* orbital $n = 1$ and $l = 1$. Then the value of l exceeds the permissible value of 0 ($l = n - 1$). For 3*f* orbital $n = 3$ and $l = 3$. Here l can have maximum value of 2.

Q. 22. Which of the following sets of quantum numbers are not allowable and why?

(*i*) $n = 2 \quad l = 1 \quad m = 0 \quad s = +\frac{1}{2}$

(*ii*) $n = 2$ $l = 2$ $m = -1$ $s = -\frac{1}{2}$

(*iii*) $n = 2$ $l = 1$ $m = +1$ $s = -\frac{1}{2}$

(*iv*) $n = 3$ $l = 2$ $m = +1$ $s = 0$

(*v*) $n = 2$ $l = 0$ $m = 0$ $s = +\frac{1}{2}$

Ans. Quantum numbers in (*ii*) and (*iv*) are not allowable because the values of *l* exceeds the maximum permissible value of $(n-1)$.

Q. 23. What are normalized and orthogonal wave functions?

(*Calcutta, 2000; Gujarat 2003; Mumbai, 2004*)

Or

What do you understand by the expression $\Psi\, \Psi^*\, d\tau$?

(*G.Nanakdev, 1996; Kurukshetra, 1999*)

Ans. Normalization. $\Psi\Psi^*\, d\tau$ is proportional (not equal) to the probability of finding the electron in the given volume element $d\tau$. However, if we have a value of Ψ such that the integration of $\Psi\Psi^*$ from $-\infty + \infty$, that is, over the entire space, is not proportional but equal to the total probability of finding the electron, then the probability that the electron is present within the volume element $d\tau$ becomes unity (*i.e.*, 100%). Mathematically, it is expressed as :

$$\int_{-\infty}^{+\infty} \Psi\Psi^* d\tau = 1$$

This is the condition of normalization and Ψ satisfying this expression is said to be normalised. Sometimes Ψ may not be normalized wave function but then it is possible to multiply it by a constant *N*, so that $N\Psi$ is also a solution of the wave equation. The problem is to choose the proper value of *N*, so that, $N\Psi$ is normalized. Mathematically, the expression becomes:

$$\int N\ \Psi N\ \Psi^* d\tau = 1 \qquad ...(1.18)$$

or $$N^2 \int \Psi\ \Psi^*\ d\tau = 1$$

or $$\int \Psi\ \Psi^* d\tau = N^{2-} \qquad ...(1.18(a))$$

where *N* = Normalizing constant

Orthogonality

This means that a particular wave function must be different from all other wave functions over the entire space. This is reasonable because if two particles had non-orthogonal wave function their description would not be unique as one wave function could be described in terms of the other. Therefore, the wave functions should be independent of each other, that is, orthogonal to each other.

The condition for orthogonality is that the product of the two wave functions, integrated over the entire space, must be equal to zero. This is, if Ψ_1 and Ψ_2 represent the two different acceptable wave functions of a system, then if

$$\int \Psi_1 \Psi_2^* d\tau = 0$$

the two wave functions are said to be mutually orthogonal.

If, on the other hand, they obey the equations,

$$\int \Psi_1\Psi_1^* d\tau = 1 \text{ and } \int \Psi_2\Psi_2^* d\tau = 1$$

They are said to be normalized wave functions.

Wave functions which are both normalized and orthogonal are called orthonormal functions.

Q. 24. Draw distribution curves for radial wave function of 1*s*, 2*s* and 3*s* orbitals.

(Nagpur, 2000; G. Nanakdev, 2000; Kerala, 2000)

Ans. The radial part of the wave function *R* (*r*) gives the distribution of the electron with respect to its distance *r* from the nucleus. It depends only on the value of quantum numbers *n* and *l*. It is governed principally by the exponential term:

$$e^{-Zr/na_0}$$

where r = distance from the nucleus

e = base of natural logarithm

Z = atomic numbers; $Z = 1$ for H

a_o = Bohr radius; = 0.529 Å for 1*s* in H atom

n= radial (principal) quantum number

The exact mathematical expressions (1.19 –1.20) for radial part of wave functions of 1*s*, 2*s*, and 2*p* orbitals of hydrogen atom are given in Table 1.1.

Table 1.1. Radial part of Ψ for orbitals of hydrogen atom.

Quantum *n*	*Nos.* *l*	*Orbitals*	*Radial part of the wave function*
1	0	1*s*	$R_{(r)} = \left(\frac{Z}{a_0}\right)^{3/2} e^{-Zr/a_0}$ ...(1.19)
2	0	2 *s*	$R_{(r)} = \left(\frac{Z}{2a_0}\right)^{3/2}\left(2-\frac{Zr}{a_0}\right)e^{-Zr/2a_0}$ ..(1.19(a))
2	1	2 *p*	$R_{(r)} = \frac{1}{2\sqrt{6}}\left(\frac{Z}{a_0}\right)^{3/2}\left(\frac{Zr}{a_0}\right)e^{-Zr/2a_0}$ ...(1.20)

The radial wave function may be represented as under:

1. $R_{(r)}$, which is *r* dependent part of the wave function Ψ, is plotted against distance *r* from the nucleus as shown for 1*s*, 2*s*, 2*p*, 3*s* and 3*p* orbitals for the hydrogen atom in Figs. 1.9 to 1.11.

A look at expression 1.19 (Table 1.1) indicates that $R_{(s)}$ of Ist orbital depends on the exponential term.

$$e^{-Zr/2a_0}$$

(because other terms in expression 1.19 are constant; $z = 1$ for H and $a_0 = 0.529$ Å)

This shows that $R_{(s)}$ should decrease as *r* increases. This is reflected in Fig.1.9 showing a decrease in radial wave function $R_{(s)}$ with increasing distance (*r*) from the nucleus for 1*s* orbital of H atom.

Expression 1.19*a* indicates that the radial wave function for 2*s* orbital is governed primarily by the term

$$\left[2-\frac{Zr}{a_0}\right]e^{-Zr/2a_0}$$

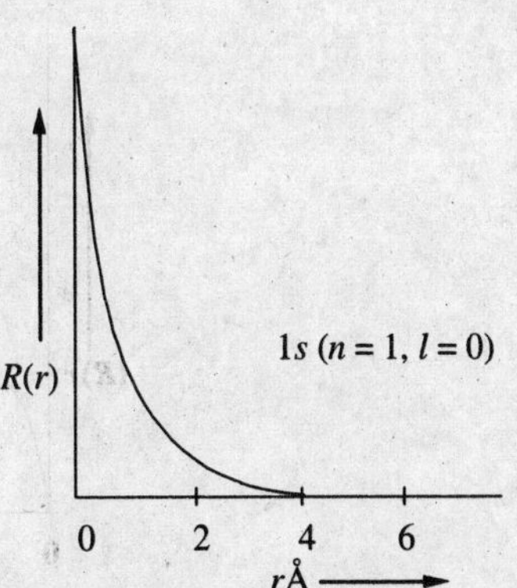

Fig. 1.9. Radial part of the wave function for 1*s* of H atom.

As r increases, the value of $R_{(s)}$ for 2s (Fig. 1.10) decreases and becomes zero at $r = \dfrac{2a_0}{Z}$ as calculated below:

$$\left[2-\frac{Z}{a_0}\cdot\frac{2a_0}{z}\right]e^{-\frac{z}{2a_0}\cdot\frac{2a_0}{z}}$$

$$R_{(s)} = (2-2)\,e^{-1} = 0$$

The point at which $R_{(s)} = 0$ is called the node. That is, the 2*s* orbital has a radial node which is a spherical surface on which $R_{(s)} = 0$. It is at a distance $= \dfrac{2a_0}{z}z$ from the nucleus.

At r greater than $2a_0$ ($Z = 1$ for H atom), the radial wave function acquires a negative value and decreases to a minimum value before increasing again to an infinitesimally small negative value as shown in Fig. 1.10 for 2*s*. The 2*s* orbital has one node.

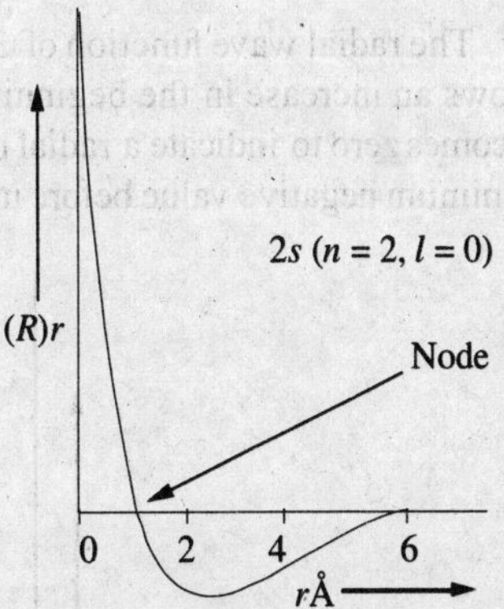

Fig. 1.10. Radial part of the wave function for 2*s* of H atom.

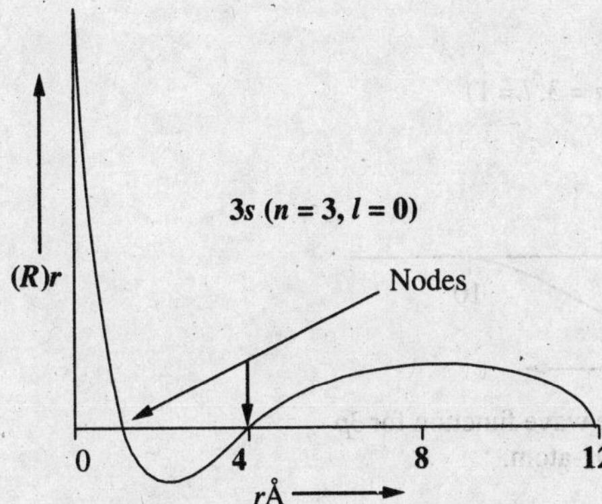

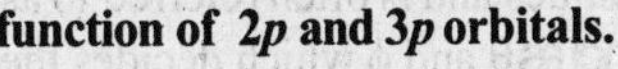

Fig. 1.11. Radial part of the wave function for 3*s* atomic orbital of H - atom.

The radial dependence of 3*s* is shown in Fig. 1.11. The radial function acquires zero value at two points and gets a positive value beyond second node. It acquires a maximum value before decreasing to an infinitesimally small positive value. Fig. 1.11 orbital has two nodes.

Orbital 1*s* has no nodes, 2*s* has one node, 3*s* has 2 nodes and so on. In general *s*-orbitals have $(n-1)$ nodes.

Q. 25. Draw distribution curves for radial wave function of 2*p* and 3*p* orbitals.

(Kerala, 2000; G. Nanakdev, 2000)

Ans. Expression 1.2 (Table 1.1) shows that the radial wave function for 2*p* orbital of H-atom ($Z = 1$) depends mainly on the term.

$$\left[\frac{r}{a_0}\right]e^{-\frac{r}{2a_0}}$$

The above expression does not become zero (except at $r = 0$) and hence the curve has no nodes as shown in Fig. 1.12. As r increases, the radial wave function $R_{(s)}$ increases to a maximum and then decreases before becoming infinitesimally small. It is positive throughout as shown in Fig. 1.12.

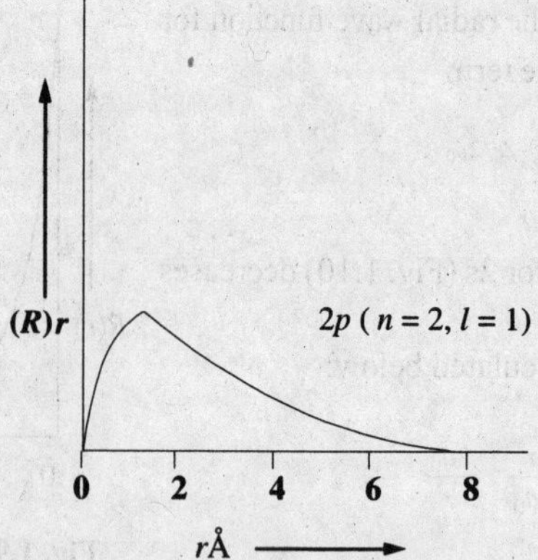

Fig. 1.12. Radial part of the wave function for 2*p* atomic orbital of H-atom.

The radial wave function of 3*p* is shown in Fig. 1.13. The value of radial wave function for 3*p* shows an increase in the beginning and after acquiring maximum value, it starts decreasing and becomes zero to indicate a radial node. Beyond that, $R_{(s)}$ decreases to become negative and acquires minimum negative value before increasing again to an infinitesimally small negative values.

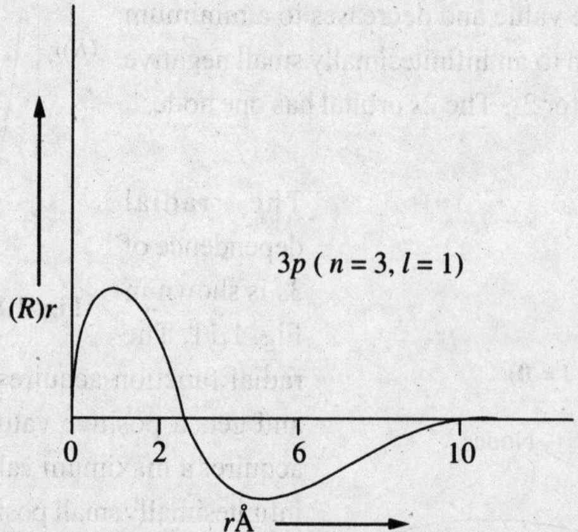

Fig. 1.13. Radial Part of the wave function for 3*p* atomic orbital of H-atom.

Q. 26. Describe the dependence of angular wave function on quantum numbers.

(Kerala, 2000)

Ans. The angular part of the wave function depends on *l* and *m* values and is often referred to as atomic orbitals. The shapes of these atom is orbitals depend upon angles θ and φ which decides the geometry of the orbitals. Mathematical expressions for angular wave functions of *s* and *p* orbitals are given in Table 1.2.

Table 1.2. Angular wave functions of hydrogen atom.

Quantum *l*	*Nos.* *m*	*Orbitals*	*Angular Wave Functions*
0	0	*s*	$\left(\frac{1}{4\pi}\right)^{1/2}$
1	0	p_z	$\left(\frac{3}{4\pi}\right)^{1/2} \cos\theta$
1	1	p_x	$\left(\frac{3}{4\pi}\right)^{1/2} \sin\theta \cos\phi$
1	–1	p_y	$\left(\frac{3}{4\pi}\right)^{1/2} \sin\theta \cos\phi$

The expression for *s*-orbital (Table 1.2) is independent of angles θ and ϕ and hence is spherically symmetrical.

(*i*) Thus the angular wave function depends on *l* and *m* values and is often referred to as atomic orbitals;

(*ii*) Shapes of atomic orbitals depends on angles θ and ϕ and determine geometry of the atom; and

(*iii*) *s*-orbitals have no angular dependence and hence spherically symmetrical.

Q. 27. State the number of nodes in $R_{(r)}$ plots of (*Kerala, 2000*)

(*i*) 3*s* **(*ii*) 2*p*** **(*iii*) 4*d*** **(*iv*) 5*f***

Ans. *s* orbitals have (*n* – 1) nodes, *p* orbitals have (*n* – 2) nodes, *d* orbitals have (*n* – 3) nodes and *f* orbitals have (*n* – 4) nodes. Based upon this generalisation

3*s* orbital has (3 – 1) *i.e.*, 2 nodes.

2*p* orbital has (2 – 2) *i.e.*, no nodes.

4*d* orbital has (4 –3) *i.e.*, 1 node.

5*f* orbital has (5 – 4) *i.e.*, 1 node.

Q. 28. Draw the plot of Ψ^2 and *r* for *s* and p orbitals.

Ans. *R*(*r*) plots of orbitals like 2*s* ,3*s* , and 3*p* show negative values of radial distribution. To avoid negative values, variation of Ψ^2 is plotted as a function of *r*. Here negative part of radial distribution becomes positive in Ψ^2 plots.

Such plots known as probability electron distribution curves are given for 1*s*, 2*s* and 2*p* orbitals of hydrogen atom in Fig. 1.14.

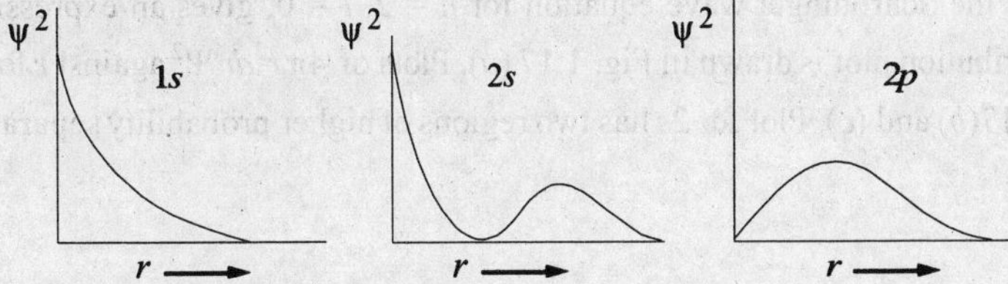

Fig. 1.14. Variation of Ψ^2 (probability of finding an electron) in ls, 2*s* and 2*p* orbitals of H-atom.

Q. 29. How do you arrive at the shape of orbitals? *(Andhra, 2002)*

Or

Draw radial probability distribution curves of 1*s*, 2*s*, 3*s* and 4*s* orbitals.

(M. Dayanand, 2000; G. Nanakdev, 2000; Kurukshetra, 2002: Nagpur; 2003; Delhi, 2007)

Or

The greatest probability of finding the electron in a small volume of 1*s* orbital, is at the nucleus. Yet the most probable distance of the electron from the nucleus is at 0.53Å. Can you agree to these two statements?

Ans. This is done by evaluating $\Psi^2\, d\tau$ in thin spherical shell, at different distance (r) from the nucleus. Considering the orbital to consist of layers of thin concentric spherical shells, one such shell of thickness $d\tau$ between distance r and $r + dr$ from origin is shown in Fig. 1.15. The volume of such a spherical shell is $4\pi r^2 dr$ and the probability of finding the electron in this volume element would be equal to $\Psi^2 \times 4\pi r^2\, dr$. A radial probability distribution curve of 1*s*, obtained by plotting relative probabilities of finding electron in this spherical shells at various distances from the nucleus, is shown in Fig. 1.16. The probability of finding the electron is highest close to the nucleus as shown in Fig. 1.14 but the volume element $4\pi r^2 dr$ is very small as r is very small and hence the product $4\,\pi\, r^2 dr \Psi^2$ is small. On the other hand, as distance increases, the volume element $4\pi r^2 dr$ increases but the probability of finding the electron Ψ^2 decreases and becomes negligible at greater distances. As shown in Fig. 1.16, the radial probability distribution of 1*s* is zero at the nucleus but increases gradually reaching a maximum at $r = r_0$ which is 52.9 pm, equal to Bohr radius for a hydrogen atom. The distance r_0 is called radius of maximum probability. Beyond r_0, the radial probability distribution of 1*s* shows a gradual decrease with value becoming infinitesimally small at larger distances from nucleus.

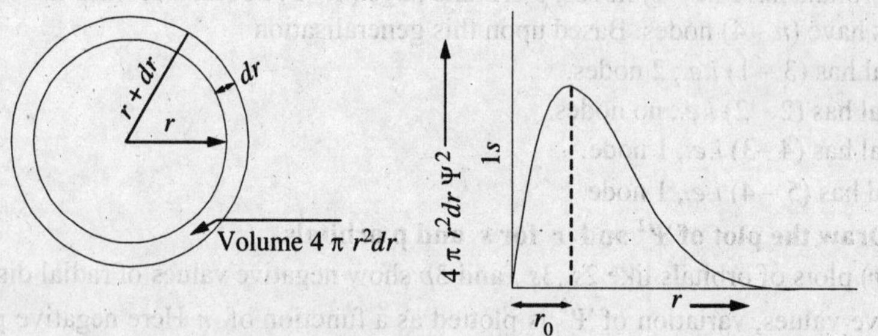

Fig. 1.15. Volume of shell of thickness *dr* around the nucleus $=4\pi r^2 dr$.

Fig.1.16. Radial probability distribution curve of 1*s*.

Solution of the Schrodinger wave equation for $n = 2$, $l = 0$, gives an expression Ψ_{2s} whose probability distribution plot is drawn in Fig. 1.17 (*a*). Plots of $4\pi\, r^2 dr\, \Psi^2$ against r for 3*s* and 4*s* are given in Fig. 1.17(*b*) and (*c*). Plot for 2*s* has two regions of higher probability separated by a node.

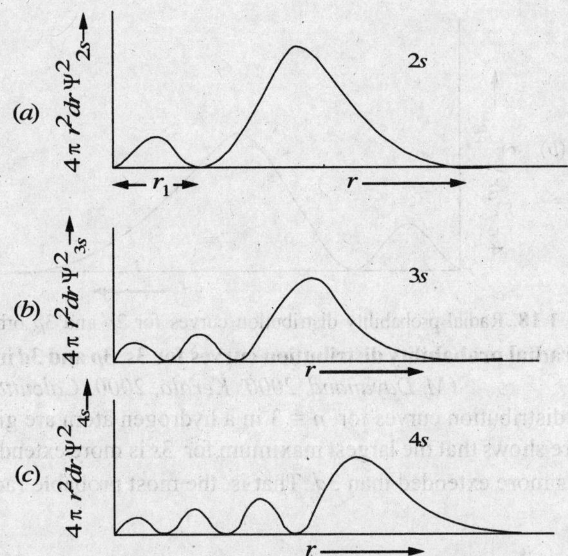

Fig. 1.17. Radial probability distribution curves for 2*s*, 3*s* and 4*s* orbitals.

For $n = 2$, the number of high probability regions =2 and number of node =1

Similarly, for $n = 3$, $l = 0$ the probability plot is as shown in Fig. 1.17 (*b*). There are three regions of high probability and two nodes. For 4*s*-orbital, there are four regions of high probability and three nodes as shown in Fig. 1.17 (*c*). It may therefore be inferred that for an *s*-orbital, the number of high probability regions is equal to *n* and number of nodes is equal to $n - 1$, where *n* is radial quantum number. It is seen from Fig. 1.17 (*a, b, c*) that as *n* increases, the distance *r* of the region of highest probability increases but the average electron density as shown by the height of the hump decreases.

Q. 30. Compare the radial probability distribution curves (plots of $4\pi r^2 dr \Psi^2$ against *r*) for 2*p* and 3*p* orbitals. (*Delhi, 2007*)

Ans. Radial or probability distribution of any 2*p*-orbital shows a rapid increase as *r* increases till a maximum is reached at r_0 which is exactly equal to the radius of 2*p*-orbital calculated by Bohr [Fig. 1.18 (*a*)]. As observed earlier, the radial distribution curve for 3*p*-orbital shows a node separating the regions of high probability [Fig. 1.18 (*b*)]. It may be stated that for any *p*-orbital, the number of regions of high probability is equal to $n - 1$ and number of nodes is equal to $n - 2$.

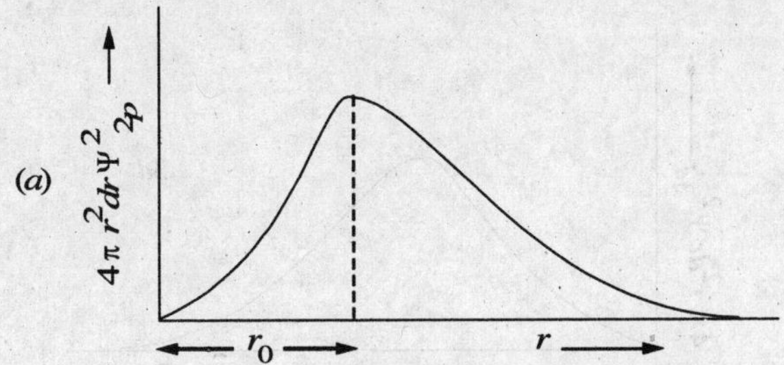

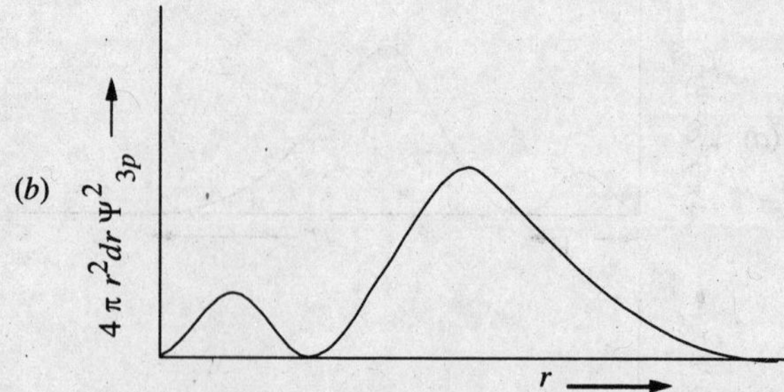

Fig. 1.18. Radial probability distribution curves for 2*p* and 3*p* orbitals.

Q. 31. Draw the radial probability distribution curves for 3*s*, 3*p* and 3*d* in hydrogen atom.
(*M. Dayanand, 2000; Kerala, 2000; Calcutta, 2000; Delhi 2007*)

Ans. Probability distribution curves for $n = 3$ in a hydrogen atom are given in Fig. 1.19 (*a*), 1.19 (*b*) and (*c*). Figure shows that the largest maximum for 3*s* is more extended in space than that of 3*p* which in turn is more extended than 3*d*. That is, the most probable radius decreases in the order 3*s* > 3*p* > 3*d*.

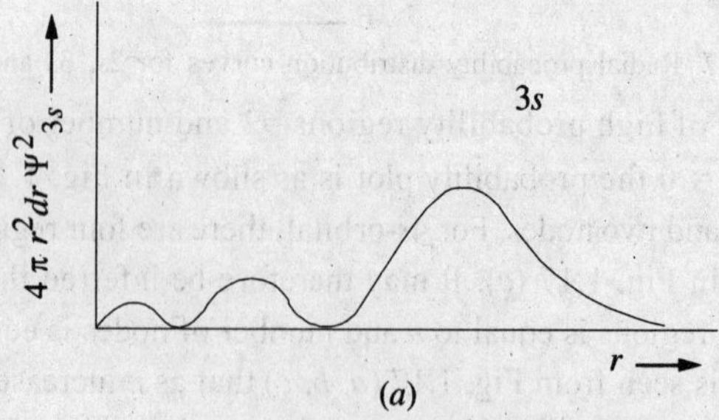

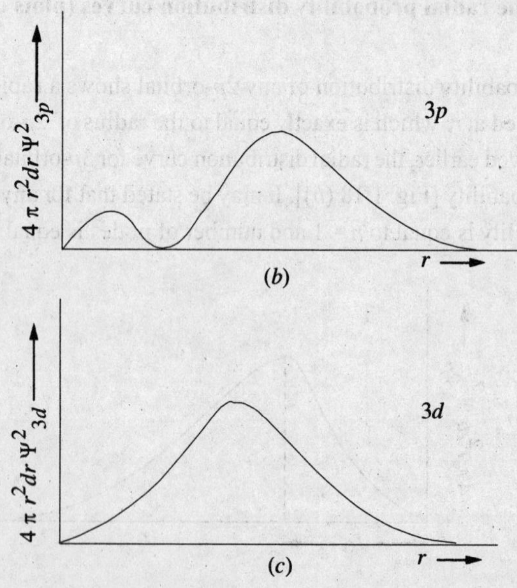

Fig. 1.19. Radial probability distribution curves for 3*s*, 3*p* and 3*d* orbitals.

Q. 32. Sketch radial distribution curves on the same scale for $n = 3, l = 0, 1, 2$.

Ans. Radial distribution curves for $3s$ ($n = 3, l = 0$) $3p$ ($n = 3, l = 1$) and $3d$ ($n = 3, l = 2$) on the same scale are given in Fig. 1.20 below.

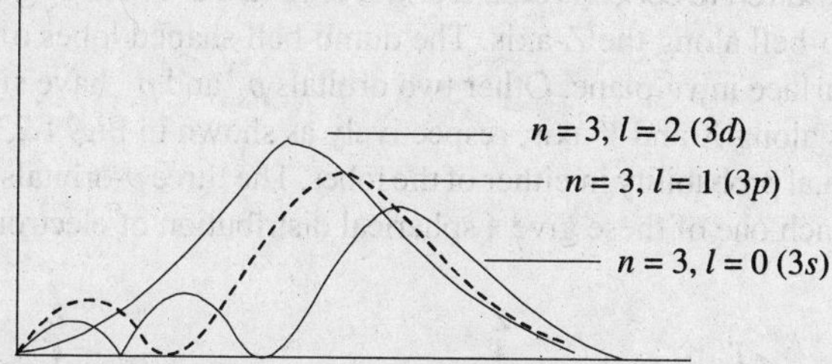

Fig. 1.20. Radial distribution curves for $3s$, $3p$ and $3d$ on the same scale.

Q. 33. Draw radial probability distribution curves for 3*s* and 3*p* orbitals and state the following in each case:

(*i*) Number of high probability region.

(*ii*) Number of nodes.

Ans. For distribution curves see Q. 32.

(*i*) Number of high probability regions for $3s = 3$

(*ii*) Number of nodes = 2

It can be seen from the curves that number of high probability regions for $3p = 2$. Number of nodes = 1.

Q. 34. Discuss the shape of *s*-orbitals. (*Kerala, 2000; Calcutta, 2007*)

Or

What is a node, nodal surface? Show it by means of a diagram for 2*s* orbital.

Ans. The expression for angular wave function for *s*-orbital is not dependent on either of the angles θ or φ. In other words, the angular part of the wave function is spherically symmetrical. That is, the probability of finding the electron is equal in all directions. Keeping in mind that atoms are three dimensional, the probability surface diagram of *s*-orbital is spherical Fig. 1.21.

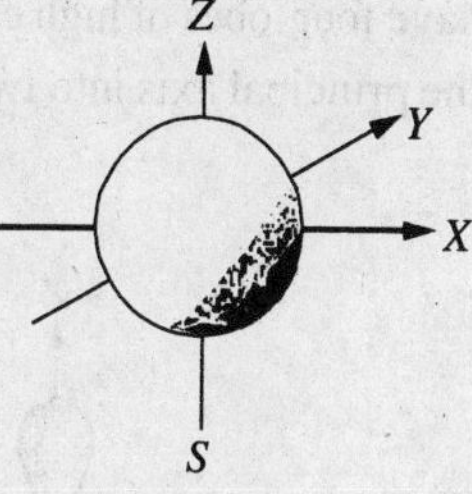

Fig. 1.21

In the radial distribution curve for 2*s*-orbital there are two regions of high probability separated by a node at $r = r_1$. Fig. 1.22 for 2*s* shows two regions of higher dot density separated by a spherical nodal surface (shown as dotted circle) at distance r_1 from the nucleus.

Such diagrams showing nodal surfaces will be avoided because by doing so we are extending the mathematical model to too much of simplification. The wave model prevents us firstly to locate the electron as a particle in one region and then explain its motion to the other region through an intervening nodal surface. For such a visualization, the electron as standing wave would be more appropriate because such a wave can exist simultaneously on both sides of nodes.

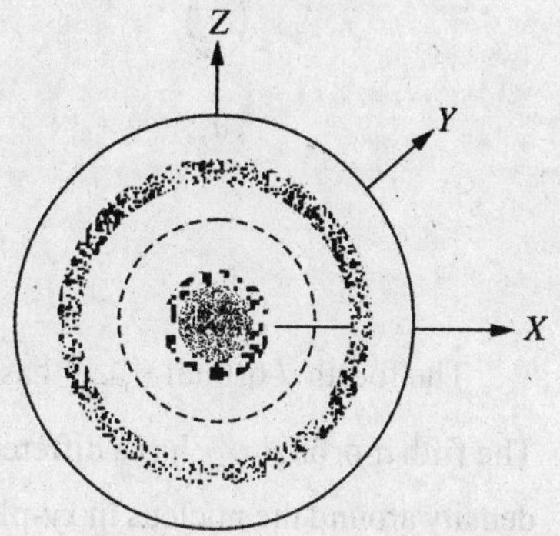

Fig. 1.22. Dot representation of electron density in 2*s* orbital.

Q. 35. Discuss the shape of *p*-orbitals.

(*Kerala, 2000; Calcutta, 2007; Vidyasagar, 2007*)

Ans. For $l = 1$, m has three values, ± 1, 0 representing the three orientations of *p*-orbitals, p_x, p_y, p_z. Angular wave function for $l = 1, m = 0$ is labelled as p_z, and the expression

$\left[\frac{3}{4}\pi\right]^{1/2}$ cosθ gives two tangent dumb-bells symmetrical about Z-axis. In other words, the angular part of the probability distribution is concentrated along Z-axis and the resulting surface of p_z electron has the form of a dumb-bell along the Z-axis. The dumb-bell shaped lobes of high probability are separated by a nodal surface in *xy*-plane. Other two orbitals p_x and p_y have similarly been found to have dumb-bell shapes along X and Y-axis, respectively as shown in Fig. 1.23. The electron in a *p*-orbital is found with equal probability in either of the lobes. The three *p*-orbitals p_x, p_y and p_z, together having an electron in each one of these give a spherical distribution of electron density.

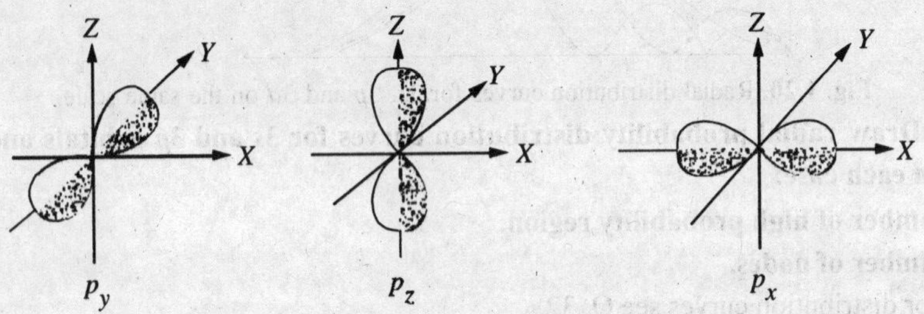

Fig. 1.23. Shape of *p*-orbitals.

Q. 36. Discuss the shape of *d*-orbitals.

(*G.Nanakdev, 2000; Kanpur, 2000; Himachal, 2000; Kurukshetra, 2002, Nagpur, 2008*)

Ans. For $l = 2$, m has five values 0, ± 1, ± 2 representing the five orientations of *d*-orbitals designated as $d_{z^2}, d_{x^2-y^2}, d_{xy}, d_{yz}, d_{zx}$, as shown in Fig. 1.24. Their angular dependence is more complicated than that of *p*-orbital. Probability surface diagrams of three orbitals d_{xy}, d_{yz} and d_{xz} have four lobes of high electron density in *xy*, *yz* and *zx* planes respectively with each lobe bisecting the principal axis into two equal halves.

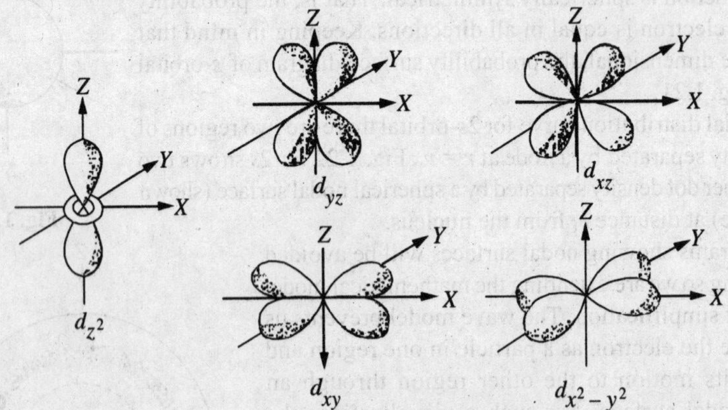

Fig. 1.24. Shapes of *d*-orbitals.

The fourth *d*-orbital $d_{x^2-y^2}$ has four lobes of electron density in *xy* plane along the *x* and *y* axes.

The fifth *d*-orbital d_{z^2} has a different shape. It has two lobes along the *z*-axis with a ring of electron density around the nucleus in *xy*-plane. The five orbitals together having an electron in each one of these give a spherical distribution of electron density. Each of the *d*-orbitals possesses two nodal planes.

Q. 37. Discuss the sign of angular wave function.

Ans. It may be pointed out that the angular part of wave function Ψ may be positive or negative depending on θ and ϕ. Angular part of the wave function for hydrogen like orbitals is given in Fig. 1.25. The *s*-orbital is positive all through because it is independent of θ and ϕ and there is no change in its shape which remains spherical. The *p*-orbital has one lobe +ve and the other –ve because the angular part of, say p_z, depends on cosθ which has +ve sign between θ – 90° and 270 – 360° and –ve sign between 90 –180 and 180 –270 along *z*-axis. The *p*-orbitals (Fig. 1.25) have spherical lobes;

θ	0	30	60	90
cosθ	1.0	0.866	0.5	0
θ		120	150	180
cosθ		– 0.5	– 0.866	–1.0

whereas the Ψ^2 plots give elongated lobes (Fig. 1.23) shapes and signs of p_y and p_x are similar to p_x as shown in Fig. 1.25. Angular dependence of *d*-orbitals gives opposite lobes with identical sign (Fig. 1.25) and the shapes of the lobes are less elogated relative to probability surface diagram (Fig. 1.24). The true significance of signs is realised when orbitals combine to form bonds in molecules.

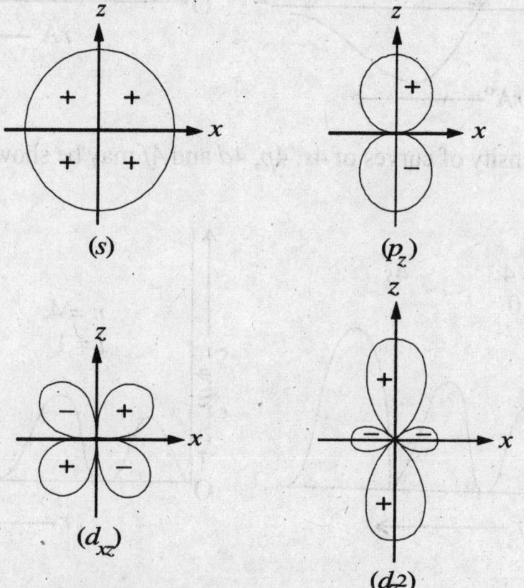

Fig. 1.25. Angular wave function of *s*, *p* and *d* orbitals.

Q. 38. (*a*) How did Schrodinger arrive at his famous wave equation for Hydrogen atom ?

(*b*) Plot radial distribution function and radial probability density curves for 4*s*, 4*p*, 4*d* and 4*f* orbitals.

(*c*) Explain as to why only '*s*' orbitals are spherical.

(*d*) Draw the shapes of 5'*d*' orbital indicating sign of wave function 4, 4, 2, 2½.

(*Delhi, 2006*)

Ans. (*a*) See Q. 11 and 12.

(*b*) Radial distribution function for 4*s*, 4p, 4*d* and 4*f* orbitals may be shown as below:

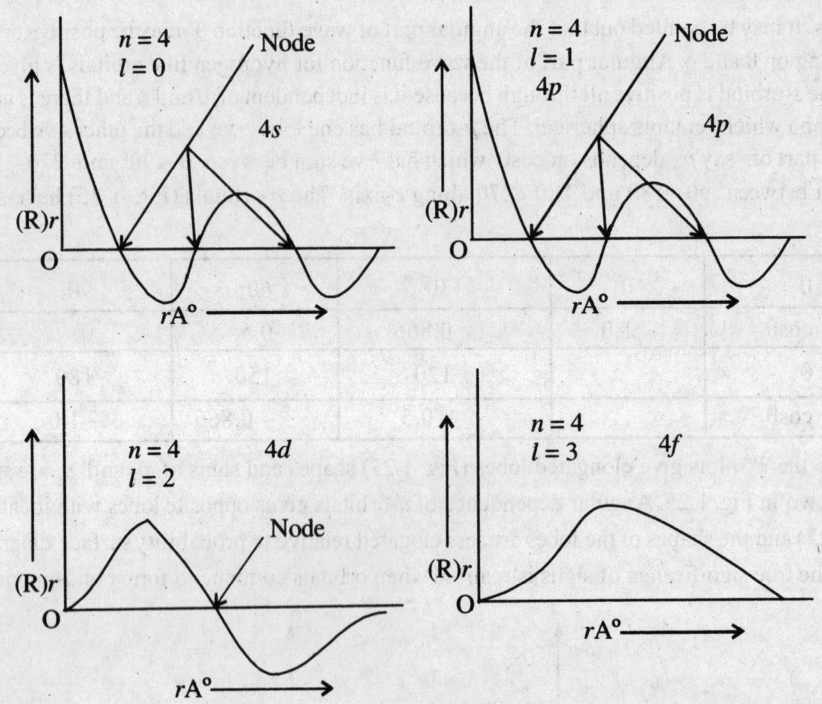

Radial probability density of curves of 4*s*, 4*p*, 4*d* and 4*f* may be shown as below:

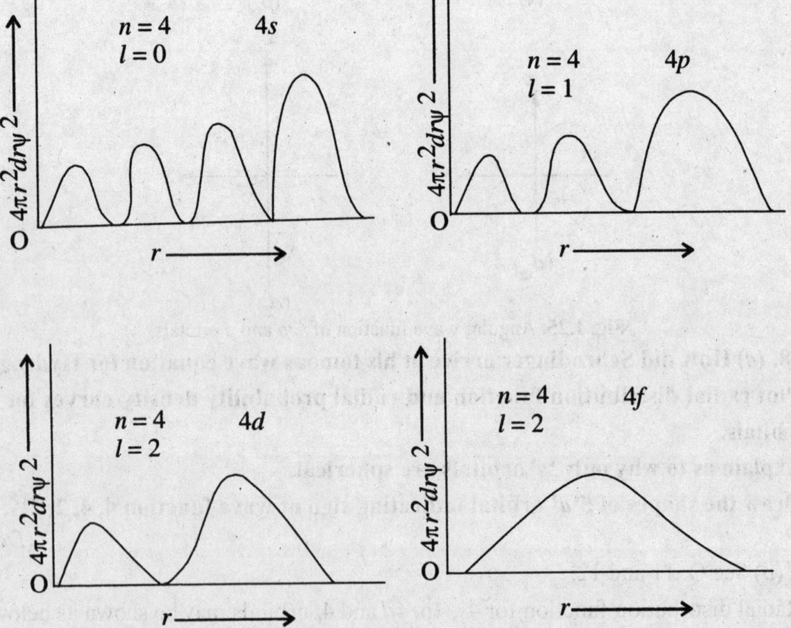

(*c*) Shape of an orbital depends on two factors *viz* radial wave function and angular wave function.

Radial wave function for s orbital is given by

$$R_{(r)} = \left(\frac{Z}{na_0}\right)^{3/2} e^{-zr/na_0}\left(2-\frac{Zr}{a_0}\right)......\left(n-\frac{Zr}{a_0}\right)$$

Angular wave function for s orbital is given by

$$R_{(\theta+\phi)} = \left(\frac{1}{4\pi}\right)^{1/2}$$

As clear from the expression of angular wave function, $R(\theta, \phi)$ is independent of (ϕ, θ) so, orbital will be symmetrical throughout and depend only on $R_{(r)}$ so, *s* orbital is spherical, as it is symmetrical.

(*d*) Shapes of *d*-orbitals indicating signs of wave function are given below :

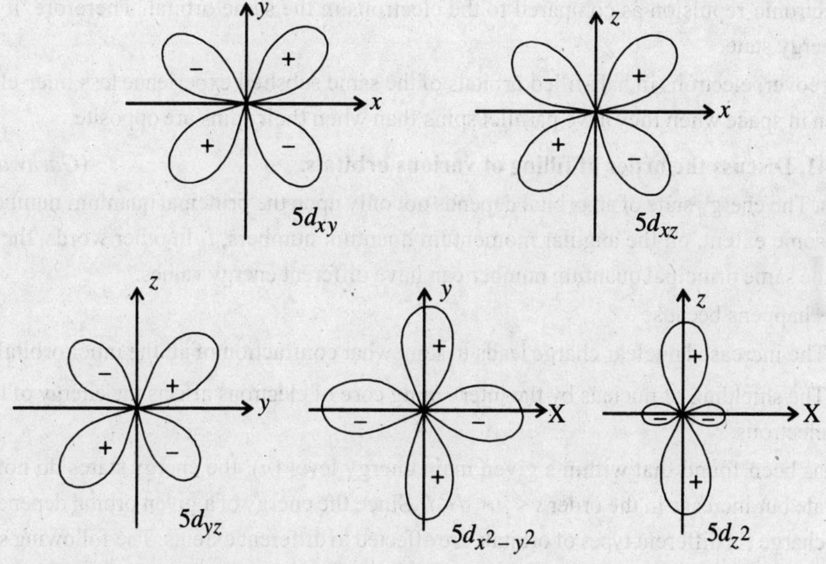

Q. 39. What do you understand by Pauli's exclusion principle?

(*Kerala, 2000; G. Nanakdev, 2000; Kumaon, 2000; Nagpur, 2000; Meerut, 2002; Rohilkhand, 2003; Vidyasagar, 2007*)

Ans. Electrons with the same spin repel each other and occupy different orbitals and if they have to occupy the same orbital, they must have opposite spins. This is the result of Pauli's exclusion principle put forward by Pauli in 1925. It states:

No two electrons in an atom can have identical set of four quantum numbers.

Another way of stating Pauli principle is as follows:

In an atom, a maximum of two electrons can occupy the same orbital.

The two electrons in the same orbital will have identical *n, l* and *m* but must have different spins (+ 1/2 and – 1/2).

Pauli exclusion principle has no mathematical explanation but in an experimental fact. Although no proof of the exclusion principle exists, there has been no contradiction to this.

Q. 40. What is Hund's rule of maximum multiplicity?

(*Nagpur, 2000; Kerala, 2000; Meerut, 2000; Calcutta,2005; Guwahati, 2006; Vidyasagar 2007; Nagpur 2008*)

Ans. It can be stated in two parts as given below:

(*a*) In an atom, no electron pairing takes place in the *p* and *d* or *f* orbitals respectively until each orbital of the given subshell contains one electron.

(*b*) The unpaired electrons present in the various orbitals of the same subshell should have parallel spins.

For example, each of three *p*-orbitals of the 2*p*-subshell gets one electron of parallel spin before any one of them receives a second electron of opposite spin.

This is because presence of the electron each in two orbitals of the same subshell involves lesser inter-electronic repulsion as compared to the electrons in the same orbital. Therefore, it leads to lower energy state.

Moreover, electrons in half-filled orbitals of the same subshell experience less inter-electronic repulsion in space when they have parallel spins than when their spins are opposite.

Q. 41. Discuss the order of filling of various orbitals. (*Garhwal, 2000*)

Ans. The energy state of an orbital depends not only upon the principal quantum number, *n,* but also, to some extent, on the angular momentum quantum numbers, *l.* In other words, the orbitals having the same principal quantum number can have different energy values.

This happens because

(*i*) The increased nuclear charge leads to somewhat contraction of all the inner orbitals.

(*ii*) The shielding of nucleus by the intervening core of electrons affects the energy of the outer electrons.

It has been found that within a given main energy level (*n*), the energy states do not remain degenerate but increase in the order $s < p < d < f$. Since the energy of a given orbital depends on the nuclear charge (*z*) different types of orbitals are affected to different extents. The following sequence is considered to be the most useful for assigning outermost or valence electrons for all the elements:

$$1s < 2s < 2p < 3s < 3p < 4s < 3d < 4p < 5s < 4d < 5p < 6s < 4s < 5d < 6p < 7s$$

An easy way to remember the order of filling of energy states is shown in Fig. 1.19. Represent a given orbital by a circle. Then draw orbitals of each principal quantum number on the same horizontal level. Draw parallel arrows and follow these as shown in Fig. 1.26, starting from 1*s*, 2*s*, 2*p* ⟶ 3*s*, 3*p* ⟶ 4*s* and so on.

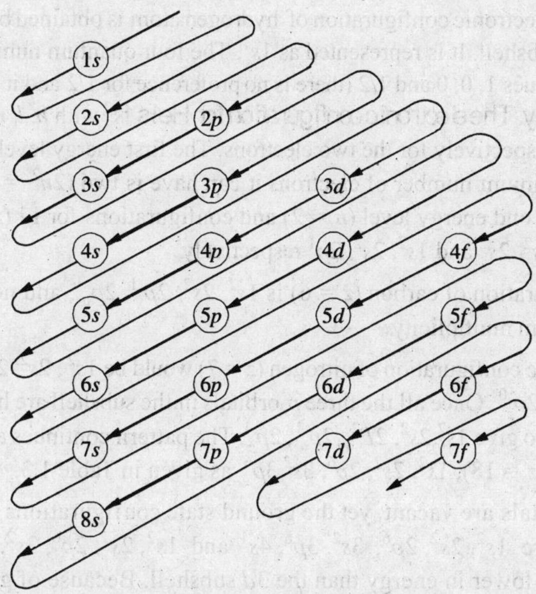

Fig. 1.26. An aid to memory for order of filling orbitals.

Q. 42. Arrange the following orbitals in order of increasing energy for a multielectron atom 1*s*, 3*s*, 5*s*, 2*p*, 4*p*, 3*d*, 4*d*, and 4*f*.

Ans. Follow Fig. 1.26. The order comes out to be: 1*s*, 2*p*, 3*s*, 3*d*, 4*p*, 5*s*, 4*d*, 4*f*

Q. 43. Discuss Aufbau principle. *(Kanpur, 2000; Meerut, 2001; Nagpur, 2005; Lucknow, 2002; M. Dayanand, 2002)*

Or

What is $(n + l)$ rule? Explain with examples. *(Indore, 2001)*

Ans. According to Aufbau principle: "In the ground state of an atom the electrons occupy the lowest energy orbitals available to them". This means that the electrons enter the orbitals in the order of increasing energies. The lowest energy available orbitals being filled up first.

Since the energy of orbitals is determined by the quantum numbers *n* and *l*, the sequence of filling the orbitals proceeds according to the following rules:

(*i*) Orbitals are filled in order of increasing value of $n + l$. For instance, 4*s* $(n + l = 4 + 0 = 4)$ is filled before 3*d* $(n + l = 3 + 2 = 5)$.

(*ii*) For the orbitals having the same value of $n + l$ the orbital having lower value of *n* is filled up first. For example, 2*p* $(n + l = 2 + 1 = 3)$ is filled before 3*s* $(n + l = 3 + 0 = 3)$ because 2*p* has lower value of *n*.

The general order of increasing energies of the orbitals is:

1*s*, 2*s*, 2*p*, 3*s*, 3*p*, 4*s*, 3*d*, 4*p*, 5*s*, 4*d*, 5*p*, 6*s*, 4*f*, 5*d*, 6*p*, 7*s*, 5*f*, 6*d*, 7*p*

In describing the electron configuration by *s*, *p*, *d* and *f* notations, one must remember that there is one orbital in an *s*-subshell, three degenerate (of equal energy) orbitals in a *p*-subshell, five degenerate orbitals in a *d*-subshell and seven degenerate orbitals in an *f*- subshell. Each orbital accommodates two electrons and a given subshell is filled before electrons are added to the next subshell. Each subshell can accommodate a maximum of $4l + 2$ electrons. As for example, an *s*-subshell $(l = 0)$ has a maximum of 2 electrons, *p*-subshell $(l = 1)$ 6 electrons *d*-subshell $(l = 2)$ 10 electrons, *p*-subshell $(l = 3)$ 14 electrons.

The ground state electronic configuration of hydrogen atom is obtained by placing the electrons in the lowest energy subshell. It is represented as $1s^1$. The four quantum numbers *n, l, m* and m_s for hydrogen atom have values 1, 0, 0 and 1/2 (there is no preference for 1/2 and it may be any orientation 1/2 or – 1/2) respectively. The electronic configuration for He is $1s^2$ with *n, l, m* and m_s values 1, 0, 0, 1/2 and 1, 0, 0, –1/2 respectively for the two electrons. The first energy level gets completely filled at He because the maximum number of electrons it can have is two ($2n^2 = 2 \times 1^2 = 2$). The next electron goes to the second energy level ($n = 2$) and configurations for Li ($z = 3$), Be ($z = 4$) and B ($z = 5$) are $1s^2$, $2s^1$ $1s^2$, $2s^2$ and $1s^2$, $2s^2$, $2p^1$ respectively.

Electronic configuration of carbon ($z = 6$) is $1s^2$, $2s^2$, $2p_x^1$, $2p_y^1$, and not $1s^2$,$2s^2$, $2p_x^2$ as per Hund's rule of maximum multiplicity.

Thus, the electronic configuration of nitrogen ($z = 7$) would be $1s^2$, $2s^2$, $2p_x^1$, $2p_y^1$, $2p_z^1$ and not $1s^1$, $2s^2$, $2p_x^2$, $2p_y^1$ and $2p_z^0$. Once all the three *p*-orbitals in the subshell are half-filled, pairing will start at oxygen ($z = 8$) to give $1s^2$,$2s^2$, $2P^2_x$,$2p_y^1$, $2p_z^1$. The pattern continues and nothing unusual is observed up to Argon ($z = 18$), $1s^2$, $2s^2$, $2p^6$, $3s^2$ $3p^6$ as given in Table 1.3.

Although 3*d*-orbitals are vacant, yet the ground state configurations of potassium ($z = 19$) and calcium (z =20) are $1s^2$, $2s^2$, $2p^6$, $3s^2$, $3p^6$, $4s^1$ and $1s^2$, $2s^2$, $2p^6$, $3s^2$, $3p^6$, $4s^2$ respectively because 4*s* subshell is lower in energy than the 3*d* subshell. Because of greater penetration, 4*s* electron faces higher nuclear charge and this effect is so pronounced that in spite of higher principal quantum number, the energy of the 4*s* subshell is lower than that of 3*d* subshell. At the next element scandium ($z = 21$), the 3*d* orbitals start filling up and this continues till 3*d* subshell is completely filled at zinc ($z = 30$).

Electronic configuration of the elements are given in Table 1.3.

Table 1.3. Ground state electronic configurations of gaseous atoms.

Z	*Symbol*	*Configuration as [Core] plus "Outermost" Orbitals*	*Z*	*Symbol*	*Configuration as [Core] plus "Outermost" Orbitals*
1.	H	$1s^1$	15.	P	[Ne] $3s^2 3p^3$
2.	He	$1s^2$, or [He]	16.	S	[Ne] $3s^2 sp^4$
3.	Li	[He] $2s^1$	17.	Cl	[Ne] $3s^2 3p^5$
4.	Be	[He] $2s^2$	18.	Ar	[Ne] $3s^2 3p^5$ or [Ar]
5.	B	[He] $2s^2 2p^1$	19.	K	[Ar] $4s^1$
6.	C	[He] $2s^2 2p^2$	20.	Ca	[Ar] $4s^2$
7.	N	[He] $2s^2 2p^3$	21.	Sc	[Ar] $3d^1 4s^2$
8.	O	[He] $2s^2 2p^4$	22.	Ti	[Ar] $3d^2 4s^2$
9.	F	[He] $2s^2 2p^5$	23.	V	[Ar] $3d^3 4s^2$
10.	Ne	[He] $2s^2 2p^6$ or [Ne]	24.	Cr	[Ar] $3d^5 4s^1$
11.	Na	[He] $3s^1$	25.	Mn	[Ar] $3d^5 4s^2$
12.	Mg	[Ne] $3s^2$	26.	Fe	[Ar] $3d^6 4s^2$
13.	Al	[Ne] $3s^2 3p^1$	27.	Co	[Ar] $3d^7 s^2$
14.	Si	[Ne] $3s^2 3s^2$	28.	Ni	[Ar] $3d^8 4s^2$

(*Contd.*)

Z	Symbol	Configuration as [Core] plus "Outermost" Orbitals	Z	Symbol	Configuration as [Core] plus "Outermost" Orbitals
29.	Cu	[Ar] $3d^{10}4s^1$	57.	La	[Xe] $5d^1 6s^2$
30.	Zn	[Ar] $3d^{10}4s^2$	58.	Ce	[Xe] $4f^2 5d^{10} 6s^2$
31.	Ga	[Ar] $3d^{10} 4s^2 4p^1$	59.	Pr	[Xe] $4f^3 6s^2$
32.	Ge	[Ar] $3d^{10}4s^2 4p^2$	60.	Nd	[Xe] $4f^4 6s^2$
33.	As	[Ar] $3d^{10} 4s^2 4p^3$	61.	Pm	[Xe] $4f^5 6s^2$
34.	Se	[Ar] $3d^{10} 4s^2 4p^4$	62.	Sm	[Xe] $4f^6 6s^2$
35.	Br	[Ar] $3d^{10} 4s^2 4p^5$	63.	Eu	[Xe] $4f^7 6s^2$
36.	Kr	[Ar] $3d^{10} 4s^2 4p^6$ or [Kr]	64.	Gd	[Xe] $4f^7 5d^1 6s^2$
37.	Rb	[Kr] $5s^1$	65.	Tb	[Xe] $4f^9 6s^2$
38.	Sr	[Kr] $5s^2$	66.	Dy	[Xe] $4f^{10} 6s^2$
39.	Y	[Kr] $4d^1 5s^2$	67.	Ho	[Xe] $4f^{11} 6s^2$
40.	Zr	[Kr] $4d^2 5s^2$	68.	Er	[Xe] $4f^{12} 6s^2$
41.	Nb	[Kr] $4d^4 5s^1$	69.	Tm	[Xe] $4f^{13} 6s^2$
42.	Mo	[Kr] $4d^5 5s^1$	70.	Yb	[Xe] $4f^{14} 6s^2$
43.	Tc	[Kr] $4d^6 5s^1$	71.	Lu	[Xe] $4f^{14} 5d^1 6s^2$
44.	Ru	[Kr] $4d^7 5s^1$	72.	Hf	[Xe] $4f^{14} 5d^2 6s^2$
45.	Rh	[Kr] $4d^8 5s^1$	73.	Ta	[Xe] $4f^{14} 5d^3 6s^2$
46.	Pd	[Kr] $4d^{10}$	74.	W	[Xe] $4f^{14} 5d^4 6s^2$
47.	Ag	[Kr] $4d^{10} 5s^1$	75.	Re	[Xe] $4f^{14} 5d^5 6s^2$
48.	Cd	[Kr] $4d^{10} 5s^2$	76.	Os	[Xe] $4f^{14} 5d^6 6s^2$
49.	In	[Kr] $4d^{10} 5s^2 5p^1$	77.	Ir	[Xe] $4f^{14} 5d^7 6s^2$
50.	Sn	[Kr] $4d^{10} 5s^2 5p^2$,	78	Pt	[Xe] $4f^{14} 5d^9 6s^1$
51.	Sb	[Kr] $4d^{10} 5s^2 5p^3$	79.	Au	[Xe] $4f^{14} 5d^{10} 6s^1$
52.	Te	[Kr] $4d^{10} 5s^2 5p^4$	80.	Hg	[Xe] $4f^{14} 5d^{10} 6s^2$
53.	I	[Kr] $4d^{10} 5s^2 5p^5$	81.	Tl	[Xe] $4f^{14} 5d^{10} 6s^2 6p^1$
54.	Xe	[Kr] $4d^{10} 5 s^2 p^6$ or [Xr]	82.	Pb	[Xe] $4f^{14} 5d^{10} 6s^2 6p^2$
55.	Cs	[Xe] $6s^1$	83.	Bi	[Xe] $4f^{14} 5d^{10} 6s^2 6p^3$
56.	Ba	[Xe] $6s^2$	84.	Po	[Xe] $4f^{14} 5d^{19} 6s^2 6p^4$

(Contd.)

Z	Symbol	Configuration as [Core] plus "Outermost" Orbitals	Z	Symbol	Configuration as [Core] plus "Outermost" Orbitals
85.	At	[Xe] $4f^{14}5d^{10}6s^2 6p^5$	96.	Cm	[Rn] $5f^7 6d^1 7s^2$
86.	Rn	[Xe] $4f^{14}5d^{10}6s^2 6p^6$ or [Rn]	97.	Bk	[Rn] $5f^8 6d^1 7s^2$
87.	Fr	[Rn] $7s^1$	98.	Cf	[Rn] $5f^{10} 7s^2$
88.	Ra	[Rn] $7s^2$	99.	Es	[Rn] $5f^{11} 7s^2$
89.	Ac	[Rn] $6d^1 7s^2$	100.	Fm	[Rn] $5f^{12} 7s^2$
90.	Th	[Rn] $6d^2 7s^2$	101.	Md	[Rn] $5f^{13} 7s^2$
91.	Pa	[Rn] $5f^2 6d^1 7s^2$	102.	No	[Rn] $5f^{14} 7s^2$
92.	U	[Rn] $5f^3 6d^1 7s^2$	103.	Lr	[Rn] $5f^{14} 6d^1 7s^2$
93.	Np	[Rn] $5f^5 6d^0 7s^2$	104.	Rf	[Rn] $5f^{14} 6d^2 7s^2$ (?)
94.	Pu	[Rn] $5f^6 7s^2$	105.	Ha	[Rn] $5f^{14} 6d^3 7s^2$ (?)
95.	Am	[Rn] $5f^7 7s^2$	106.	?	[Rn] $5f^{14} 6d^4 7s^2$ (?)

Q. 44. Why is 4*s* orbital lower in energy than 3*d* orbital?

Ans. According to Aufbau Rule, an orbital having a lower value for $(n + l)$ is lower in energy.

For 4*s* orbital $n = 4$ and $l = 0$ $\quad \therefore \quad n + l = 4 + 0 = 4$

For 3*d* orbital $n = 3$ and $l = 2$ $\quad \therefore \quad n + l = 3 + 2 = 5$

Thus 4*s* has lower energy than 3*d*.

Q. 45. Describe the following orbitals in terms of their corresponding $(n + l)$ value.

3*s*, 4*d* and 5*p* (*Purvanchal, 2007*)

Ans. For 3*s* :

$n = 3, \; l = 0 \quad \therefore n + l = 3$

For 4*d* :

$n = 4, \; l = 2 \quad \therefore n + l = 6$

For 5*p*:

$n = 5, \; l = 1 \quad \therefore n + l = 6$

Q. 46. Write electronic configuration of $_{25}Mn^{2+}$, $_{30}Zn^{2+}$, $_{64}Cd^{3+}$ and $_{13}Al^{3+}$

(*M. Dayanand, 2000*)

Ans.

Mn^{2+} [Ar] $3d^5$

Zn^{2+} [Ar]$3d^{10}$

Cd^{3+} [Xe] $5d^1$

Al^{3+} [Ne]

Q. 47. What are the limitations of Aufbau principle?

Ans. Although electron configurations based on Aufbau order of filling agree mostly with those derived experimentally, yet there are quite a few exceptions where errors in Aufbau configurations arise due to one (rarely two) misplaced electron. Some of the limitations of Aufbau order are listed here.

(*i*) The Aufbau cannot be used to predict electron configurations of atoms on ionization. In other words, it does not tell us which electrons are to be removed when an ion is formed from an atom. For example, the configuration of Fe based on Aufbau principle is $1s^2, 2s^2, 2p^6, 3s^2, 3p^6\ 4s^2, 3d^6$. It has been experimentally confirmed by spectral and magnetic studies that Fe^{2+} has configuration $1s^2, 2s^2, 2p^6, 3s^2, 3p^6, 3d^6$ and not $1s^2, 2s^2, 2p^6, 3s^2, 3p^6, 4s^2\ 3d^4$. It means that ionization results in the loss of 4*s* electrons in preference to 3*d* electrons even though the 3*d* were the last to be added in building up the configuration of Fe atom. It means that in Fe^{2+}, 3*d* has lower energy than 4*s* which is contrary to Aufbau order of filling. This can be said of any transition elements on ionization.

In general, in the process of ionization, the first electrons are lost from the subshell with the higher *n* and if *n* is the same then from subshell with the highest *l*.

(*ii*) Although $(n-1)\,d$ subshells and *ns* subshells lie quite close together, yet the former is somewhat higher in energy. Accordingly, the predicted Aufbau outer configuration of Cr (Z = 34) is ..... $3d^4, 4s^2$ but the experimentally supported configuration is ..... $.3d^5, 4s^1$.

Similarly the experimentally observed configuration of Cu, Ag, Au, are different from Aufbau configurations as shown in Table 1.4.

Table 1.4. Limitations of Aufbau order.

Element	*At. No.*	*Aufbau Confg.*	*Experimental Confg.*
Copper	29	...$3d^9 4s^2$	...$3d^{10}\ 4s^1$
Silver	47	...$4d^9\ 5s^2$	...$4d^{10}\ 5s^1$
Gold	79	...$5d^9\ 6s^2$	...$5d^{10}\ 6s^1$

The reasons for these deviations lie in the extra stability associated with half filled or completely filled subshells.

(*iii*) In the sixth period, 4 *f* and 5*d* subshells are exceedingly close in energy. At lanthanum (Z= 57), the last electron does not go to 4 *f* as predicted in Aufbau order but is added to 5*d* subshell to give it a configuration ....$4d^{10}\ 5s^2, 5p^6, 5d^1, 6s^2$. However, the next electron for cerium (Z = 58) is added to the 4 *f* subshell and the previous electron in 5*d* in lanthanum is also shifted to 4 *f* to give it a configuration ....$4d^{10}\ 5s^2\ 5p^6\ 4f^2\ 5d^0\ 6s^2$.

(*iv*) The Aufbau order configuration of zirconium (Z = 40) is ..... – $5s^2\ 4d^2$ and that the predicted order for the next element niobium Nb (Z = 41) would be ...... $5s^2\ 4d^3$. However, the observed configuration of Nb is ... $5s^1\ 4d^4$. It is understandable that the next element Molybedenum Mo (Z = 42) acquire ..... $5s^1\ 4d^5$ configuration due to higher stability associated with half-filled configuration. However, the next element technitium Tc (Z = 43) with observed configuration $5s^1\ 4d^6$ in place of predicted $5s^2\ 4d^5$ shows that the preference for half-filled or completely filled configurations is not the only controlling factor but some complex nuclear-electron and electron-electron forces also have some role to play in determining these configurations.

(*v*) In case of palladium Pd (Z = 46), the predicted Aufbau configuration is $5s^2\ 4d^8$ but the experimentally observed configuration is $5s^0\ 4d^{10}$. This is perhaps the only case where the observed configuration has two misplaced electrons.

Q. 48. Explain, giving examples, the factors governing special stability associated with half-filled and completely filled subshells.

(*G.Nanakdev, 2000; M. Dayanand 2000*)

Or

Define exchange energy and discuss its importance in influence electronic configuration.

(*Garhwal 2000*)

Ans. The Aufbau order fails to predict sometimes the correct electronic configurations in cases where energies of neighbouring subshells (*e.g.*, 4*s* ,3*d;* 5*s,* 4*d;* 4*f* ,5*d;* etc.) are quite close together and comparable. As for example, the observed configurations, $3d^5 4s^1$ and $3d^{10} 4s^1$ for Cr (Z = 24) and Cu (Z = 29) respectively, differ from respective Aufbau predicted configurations, $3d^4 4s^2$ and $3d^9 4s^2$. This is explained on the basis of "special stability associated with half-filled and completely-filled subshells".

According to Hund's rule, electrons distribute themselves in degenerate orbitals so as to keep like spins as far apart as possible. For example, the most stable carbon configuration $1s^2\ 2s^2\ 2p_x^{\ 1}\ 2p_y^{\ 1}\ 2p_z^{\ 0}$, has both the *p*-electrons in degenerate orbitals with like spins. Similarly the lowest energy configurations for nitrogen $1s^2\ 2s^2\ 2p_x^{\ 1}\ 2p_y^{\ 1}\ 2p_z^{\ 1}$ has all the three *p*-electrons with the same spin. This happens because electrons as charged particles repel one another and electron-electron repulsion is minimum when electrons with like spins in a subshell keep as far apart as possible. This situation may be visualized in two steps shown in Fig. 1.27. The first step called the charge correlation means interactions among electrons by virtue of their charges (coulomb forces) and refers to lower energy situation for two electrons having different *m* and m_s values. The next step called the spin correlation refers to further lowering of energy for like (parallel) spins. Spin correlation leading to decrease in repulsion for electrons of like spins than for electrons of different spins is called the **exchange** energy. Spin correlation and its exchange energy of stabilization are thought to be the basis for special stability of half-filled subshells. Placing two electrons in the same orbital will lead to higher repulsion than if they occupy degenerate orbitals as shown in Fig. 1.27 for configuration.

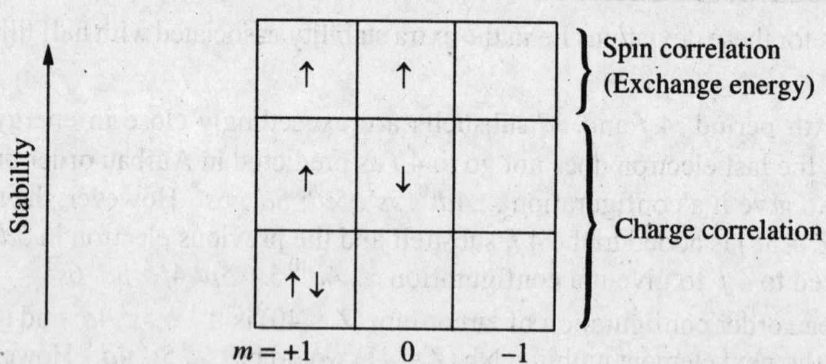

Fig 1.27. Stability for two *p*-electrons in carbon configuration.

Chromium with observed configuration ...$3d^5 4s^1$ is more stable than the Aufbau configuration ..... $3d^4\ 4s^2$ due to higher exchange energy of stabilization associated with the former arrangement having larger number of like spins.

Besides half-filled, completely filled subshells also contribute to the stability of atoms. For example, Cu, Ag, Au acquire configuration with completely-filled *d* subshells $(n-1)\ d^{10}\ ns^1$, instead of the predicted configuration $(n-1)\ d^9\ ns^2$. The observed configuration of palladium $Pd - 4d^{10}, 5s^0$ involves a transfer of two electrons from predicted configuration $4d^8\ 5s^2$ to acquire a completely-filled *d* sub-shell. The stability of noble gases is the most important example of special stability associated with completely-filled subshells.

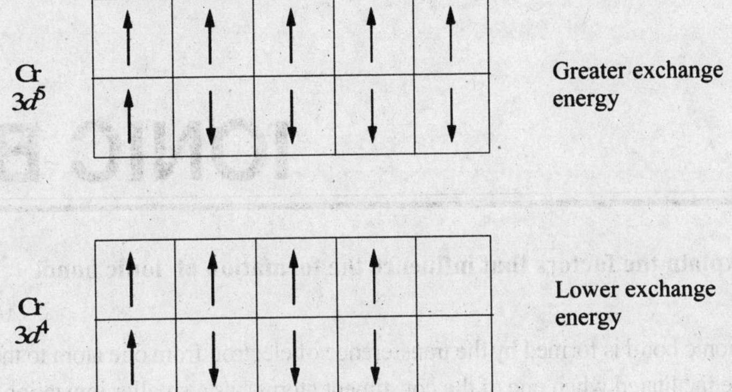

Fig. 1.28. Configuration of chromium

Q. 49. Explain anomalous configuration of Cu (Z = 29) and Cr (Z = 24). (*Nagpur 2008*)

Ans. See Q. 48.

Q. 50. How is special stability of filled and half-filled configuration related to ionisation potential and electron affinity?

Ans. Special stability associated with half-filled and completely filled subshells illustrated in the trends of ionization potential and electron affinity (eV) is given in Table 1.5. Ionization potentials and electron affinities of Be and Mg which have filled *s* subshell in the outermost shell have exceptional values that fall out of trends along a period (Table 1.5). The trend of increasing ionization potential along a period for *p*-block elements with valence configuration $ns^2\ np^x$ ($x = 1 - 6$) is followed smoothly up to nitrogen s^2p^3. The break occurs at s^2p^4, when the fourth electron with opposite spin is added to a *p*-orbital which already has an electron. This additional electron does not benefit from the exchange energy of stabilization possible for C and N. Special stability of half-filled and completely filled configuration is supported by abnormal values of electron affinity for atoms with s^2p^3 and s^2p^6 configuration.

Table 1.5. Ionization potential and electron affinity values (eV) along second and third periods.

Li 5.4	Be 9.3	B 8.3	C 11.3	N 14.5	0 13.6	F 17.4	Ne 21.6	Ionization potential trends
Na 5.1	Mg 7.6	Al 6.0	Si 8.1	P 10.5	S 10.4	Cl 13.0	Ar 15.8	
Li 0.62	Be –2.5	B 0.24	C 1.27	N 0.0	0 1.47	F 3.34	Ne –0.30	Electron affinity trends
Na 0.55	Mg –2.4	Al 0.46	Si 1.24	P 0.77	S 2.08	Cl 3.61	Ar –0.36	

Q. 51. Give two examples where observed configuration has no preference for half-filled and completely filled configuration.

Ans. In Tungsten W (Z = 74) the configuration is $5d^4\ 6s^2$.

In Technicium Tc (Z = 43) the configuration is $4d^6\ 5s^1$.

The above two elements show no preference for half-filled configuration.

2 IONIC BOND

Q. 1. Explain the factors that influence the formation of ionic bond.

(*Rohilkhand, 2003*)

Ans. An ionic bond is formed by the transference of electron from one atom to the other. Hence this process will be facilitated when one of the constituent atoms has a smaller ionization energy, so that the electron could be removed from the outermost orbit. And the second atom should have a high value of electron affinity so that it could easily accept the electron released by the other atom.

Q. 2. Describe general characteristics of ionic compounds. (*Guwahati, 2007*)

Ans. Some general characteristics of ionic compounds are decribed as under:

(*i*) Melting and Boiling Points

Melting and boiling points are dependent on the nature of binding forces in the crystalline solid. Melting involves breaking of the crystal lattice. Since electrostatic interactions between +ve and –ve ions in the crystal are generally strong, therefore their breakdown requires higher energies and hence the ionic compounds have higher melting and boiling points.

Significant variations in melting and boiling points within ionic compounds are attributed primarily to (*i*) charges on the ions, (*ii*) ionic radii.

For example, NaC1 and BaO have similar internuclear distances (NaC1 = 2.81 Å and BaO = 2.7 Å) but the former has a melting point of 800°C only while the latter melts at 1921°C. The significantly higher melting point of BaO is attributed to the stronger cation-anion interaction between the doubly charged Ba^{2+} and O^{2-} ions as compared to singly charged ions in NaC1 crystal. Melting point also depends upon the size of the ions. The larger the ions, the weaker the attractive forces and lower would be the melting points. This is illustrated if we compare melting points of NaF and NaI. The force of attraction between Na^+ and larger I^- ions will be lower than that between Na^+ and small F^- ions. As a result, NaI lattice is less stable than NaF lattice. Hence melting points of NaI (925°C) is lower than that of NaF (1270°C).

(*ii*) Hardness

The electrostatic interaction between oppositely charged ions determines hardness of an ionic compound. The electrostatic interaction between oppositely charged ions depends on:

(*i*) ionic radii (*ii*) charges on the ions

For a given ionic charge, hardness increases with decrease in interionic distance for crystals with similar structure. For example, alkaline earth oxides CaO, SrO and BaO have similar NaC1 type of structure but their interionic distance increases along the group from CaO to SrO to BaO. Consequently, their hardness decreases down the group. In short, more compact the structure, harder is the substance.

Hardness increases with increase in charge on the ions for crystals with similar interionic distances for ionic compounds having similar structure. For example, LiCl and SrO have the same NaCl type of structure and same interionic distance of 2.57 Å. Their hardness increases as we move from singly charged Li^+ and Cl^- ions to doubly charged Sr^{2+} and O_2^-. Similarly, NaCl and CaS (with similar structure and interionic distance) show an increase in hardness from NaCl to CaS because of increase in charge on ions.

***(iii)* Brittleness**

The constituent units which make up ionic solids are oppositely charged ions [Fig. 2.1 (*a*)]. It may be observed that if requisite energy is supplied to the crystal layers, the movement of layers of ions brings ions of like charges near each other [Fig. 2.1 (*b*)] and this causes strong repulsions which lead to the breakdown of the crystal. For this reason ionic crystals, in spite of their hardness, are brittle and can be powdered easily.

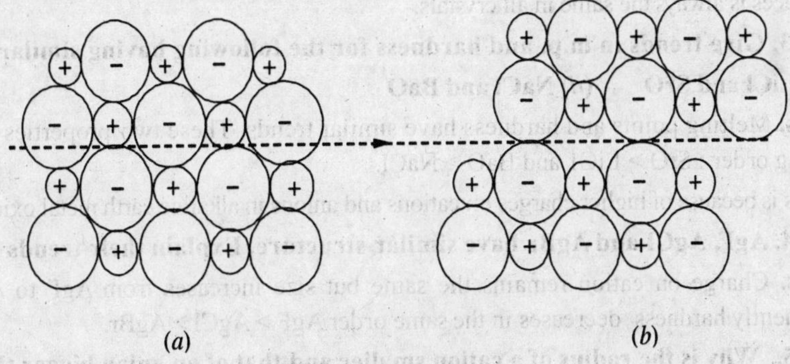

Fig. 2.1. Brittleness.

***(iv)* Cleavage**

The phenomenon of cleavage is observed in ionic crystals. That is, in the crystals the splitting of layers is easier in some directions than others. For example, in a crystal of sodium chloride, cleavage can be achieved only along planes parallel to cube faces. Any attempt to cleave such a crystal along an arbitrary plane will shatter the structure as shown in Fig. 2.2.

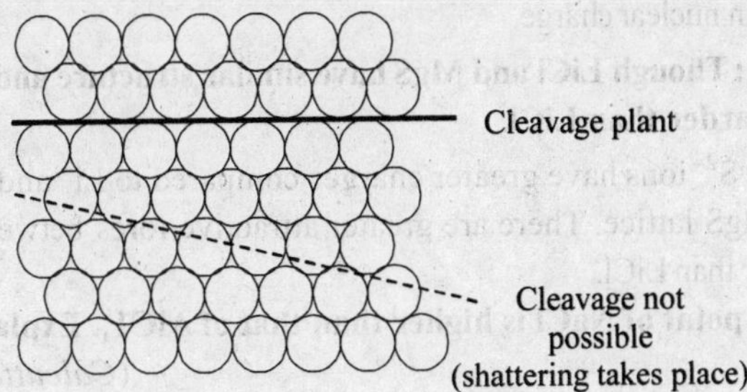

Fig. 2.2. Cleavage.

***(v)* Solubility**

Ionic compounds are soluble in highly polar solvents. Water is a good solvent for dissolving ionic compounds because

(*a*) It has a larger dipole moment which provides high hydration energy due to strong ion-solvent dipole interactions. If hydration energy is greater than the crystal energy, the lattice breaks.

(*b*) It has high dielectric constants (78.5 at 25°C) which is responsible for reducing attractive force between cations and anions in solutions. This keeps the unlike charges apart from each other.

Suitable solvents for dissolving ionic compounds should have high dipole moment and high dielectric constant.

(*vi*) Conductance

In an ionic crystal, the ions are situated in fixed positions in the lattice, that is, they are not mobile. Therefore, ionic solids do not conduct electricity in the solid state. However, in molten state or in solution the crystal lattice breaks and ions are free to move in an electric field. Thus, they conduct electricity due to increase in relative mobility of ions in fused state or in solution.

(*vii*) Sterochemistry

The constituent ions of an ionic solid possess an orderly arrangement in a definite geometric pattern which repeats itself throughout the crystal. Crystals are always bound by plane faces and these faces always meet at some fixed angles. For any particular substance, the angle between corresponding pair of faces is always the same in all crystals.

Q. 3. Give trends in m.p. and hardness for the following having similar structures:

(*i*) LiCl and SrO (*ii*) NaCl and BaO

Ans. Melting points and hardness have similar trends. These two properties decrease in the following order : SrO > LiC1 and BaO > NaCl.

This is because of higher charges on cations and anions in alkaline earth metal oxides.

Q. 4. AgF, AgCI and AgBr have similar structure. Explain their trends in hardness.

Ans. Charge on cation remains the same but size increases from AgF to AgC1 to AgBr. Consequently hardness decreases in the same order AgF > AgCl > AgBr.

Q. 5. Why is the radius of a cation smaller and that of an anion bigger than that of the atom?

Ans. When a neutral atom gains an electron to become an anion, its radius increases. This is expected due to the increase in the number of electrons which the same nuclear charge has to hold. The force of attraction on the electron decreases and so the radius increases. On the other hand, when a neutral atom loses on electron to form a cation, the radius decreases, because the nuclear charge is now acting on a fewer number of electrons. It is also expected that the size of the ions decreases in an isoelectronic series with increase in nuclear charge.

Q. 6. Explain : Though LiCl and MgS have similar structure and nearly same internuclear distance, MgS is harder than LiCl.

Ans. Mg^{2+} and S^{2-} ions have greater charges compared to Li^+ and Cl^-. Therefore packing is more compact in MgS lattice. There are greater attractive fores between the cations and anions. Hence MgS is harder than LiCl.

Q. 7. Melting point of NaCl is higher than that of $AlC1_3$. Explain.

(*Calcutta, 2007; Guwahati, 2007*)

Ans. Na^+ are smaller than $A1^{3+}$ ions. In sodium chloride, therefore, cation-anion interactions are stronger. The lattice is more compact and thus melting point of NaC1 is higher than that of $A1C1_3$. This can also be explained on the basis of covalent character in $A1Cl_3$. (See Fajan's rules ahead)

Q. 8. CsCl lattice is less stable than NaCl structure. Explain.

Ans. Distance between Cs and Cl in CsC1 is more than that between Na and Cl in NaC1. Greater distance mean smaller force of attraction between the cation and anion. Hence CsC1 will have a relatively weaker lattice or less stable structure than NaCl.

Q. 9. Give Pauling's method of calculation of ionic radius.

Ans. The radii calculated by Pauling are based on the observed internuclear distances in four crystals namely, NaF, KC1, RbBr and CsI. The set of radii calculated by this method are given in Table 2.1.

Table 2.1. Pauling set of ionic radii (pm).

Li^+	60	Be^{2+}	31	B^{3+}	20	C^{4+}	15	O^{2-}	140	F^-	136
Na^+	95	Mg^{2+}	65	Al^{3+}	50	Si^{4+}	41	S^{2-}	184	Cl^-	181
K^+	133	Ca^{2+}	99					Se^{2-}	198	Br^-	195
Rb^+	148	Sr^{2+}	113	Zn^{2+}	74			Te^{2-}	221	I^-	216
Cs^+	169	Ba^{2+}	135					–	–	–	–

Some of the factors which affect interionic distances in crystals are (*i*) electron distribution of ions, (*ii*) charges on the ions, (*iii*) crystal structure, and (*iv*) cation to anion radius ratio. The four compounds mentioned above were so chosen that all these factors remain constant. In such a situation, it was proposed by Pauling that the radius of an ion is inversely proportional to the effective nuclear charge, Z_{eff} of the ion.

$$R_{ion} \propto \frac{1}{Z_{eff}}$$

$$\propto \frac{1}{Z-S} \text{ (where } Z = \text{nuclear charge, } S = \text{screening constant)}$$

$$= \frac{C}{Z-S} \text{ (where } C \text{ is proportionality constant)}$$

Let us calculate the radii of Na^+ and F^- from the measured internuclear separation at NaF of 231 pm.

Both sodium and fluoride ions are isoelectronic and have the same value of screening constant, *i.e.*, $S = 4.5$.

Effective nuclear charge will be calculated as under:

The effective nuclear of $Na^+ = 11 - 4.5 = 6.5$

Similarly, effective nuclear of $F^- = 9 - 4.5 = 4.5$

The proportionality constant C may be calculated as

$$95 = \frac{C}{6.5} \text{ for } Na^+$$

and $$136 = \frac{C}{4.5} \text{ for } F^-$$ (See Table 2.1)

Combining the above two equations

$$\frac{C}{6.5} + \frac{C}{4.5} = 231$$

or $$C = 641$$

Using the above value of C, radius of each ion may be found out as:

$$r_{Na^+} = \frac{C}{Z_{eff}} = \frac{641}{6.5} = 95 \text{ pm}$$

and $$r_{F^-} = \frac{641}{4.5} = 136 \text{ pm}$$

Q. 10. What is radius ratio? How can it help to predict the structure of an ionic crystal?

(*Kurukshetra, 2002; Nagpur, 2003; Rohilkhand, 2003; Guwahati, 2007*)

Ans. The ratio of radius of the cation (r^+) and radius of anion (r^-) is called radius ratio.

Their radius ratio = $\frac{r^+}{r^-}$

Ionic compounds consist of +ve and –ve ions held together by electrostatic interactions in such a way that the distance between cations and anions is minimum. If each ion is considered as a hard sphere the optimum arrangement of ions for the most stable structure would be one that involves touching of the largest number of oppositely charged spheres without any crowding. Such a situation leads to maximum attraction between opposite charges and minimum repulsion between like charges. Such an arrangement of ions in a crystal is greatly influenced by the radius ratio; it determines how many ions of a given size can be held in contact with a single ion of smaller size.

The relation between radius ratio and structure is given in Table 2.2.

Table 2.2. Relation between radius ratio and structure.

Radius ratio r^+/r^-	*Coordination number*	*Structural arrangement*	*Ionic lattices*
0.155 – 0.225	3	Trigonal planar	Boron oxide
0.225 – 0.414	4	Tetrahedral	ZnS(Wurtzite, Zinc blende)
0.414–0.732	6	Octahedral	NaC1, TiO_2(Rutile)
0.732–1.00	8	Cubic	CsCl,CaF_2 (Fluorite)

A cation will fit into an octahedral arrangement of anions for radius ratio $\frac{r^+}{r^-}$ in the range 0.414 - 0.732. That is, only in such a situation a cation will be able to keep six anions in contact with each other. Smaller cations will prefer to fit into tetrahedral holes in the lattice. For radius ratio $\frac{r^+}{r^-}$ ranging between 0.225 and 0.414, tetrahedral sites will be preferred. A cubic lattice with coordination number 8 is preferred for radius ratio $\frac{r^+}{r^-}$ in the range 0.732 – 1.

Coordination number is the number of anions that surrounded a cation to give a stable lattice.

In zinc sulphide, with a radius ratio of 74/184 *i.e.*, 0.4, zinc will occupy tetrahedral holes in the close packed lattice of sulphide ions whereas in sodium chloride (radius ratio 5.2), Na^+ will be surrounded by six chloride *i.e.*, it will have octahedral arrangement.

By using radius ratio rule, it is possible to predict the cation/anion coordination number in any compound. So radius ratio is a useful measure in establishing the structure of ionic solids. In Fig. 2.3, it is shown how the arrangement of ions in a single layer is effected by the differences in the sizes of the ions.

Figure 2.3 (*a*) shows the ideal situation where M^+ fits into X^- anions.

In Fig. 2.3 (b), if the anions move closer to the size of M^+, then they would crowd each other. There would be repulsion amongst them and the energy of the system would be raised. If they hold each other apart then they are not very close to the oppositely charged M^+. Fig. 2.3 (*c*) does not allow the X^- ions to come close enough to feel any X^- – X^- repulsions. Therefore, if M^+ is too large

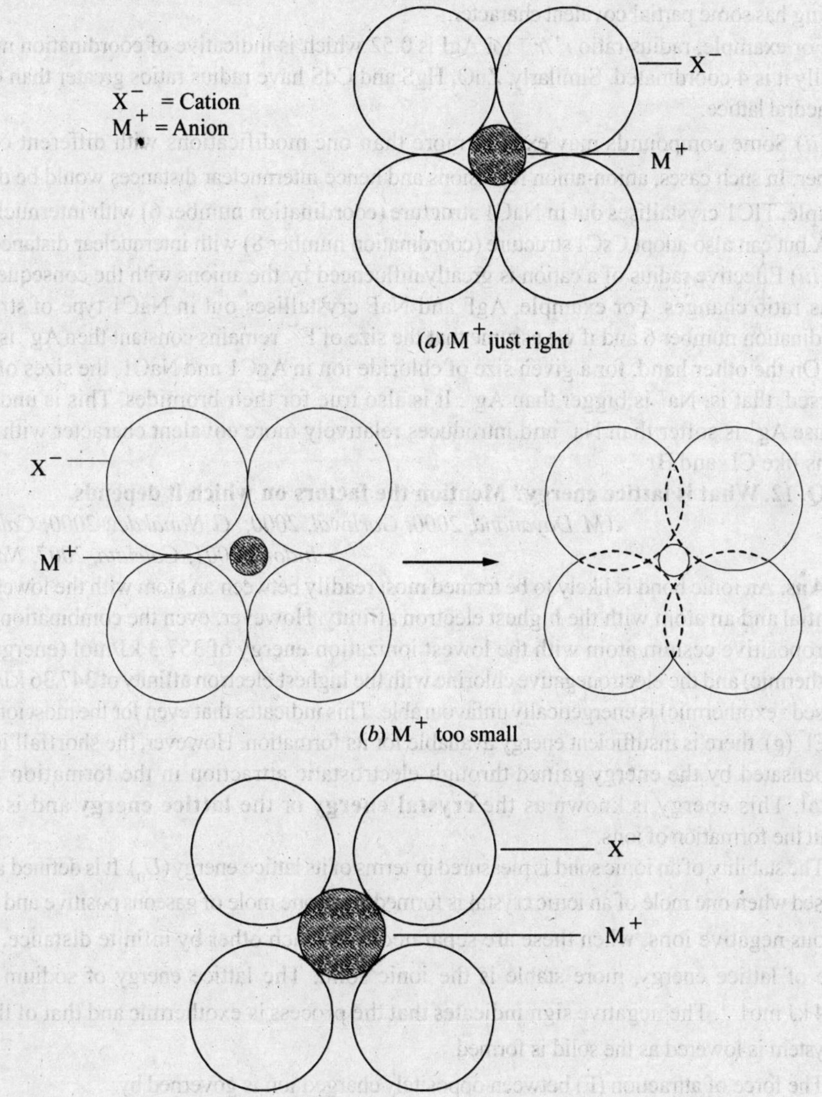

Fig. 2.3. Effect of ionic size on structure.

as in Fig. 2.3 (*c*) or too small (as in *b*), a different MX crystal packing might be obtained. If (*a*) represents the lattice of NaC1, then in situation (*c*) when M^+ is quite large, we have an arrangement in which M^+ may have more X^- as neighbours in CsCl. On the other hand, if M^+ is too small, as in (*b*), only four negative ions pack around the positive ions. ZnS provides an example of this type.

Q. 11. What are the limitations of radius-ratio rule?

Ans. Limitations of radius ratio rule are:

(*i*) An ion is considered as a sphere for determining the optimum arrangements of ions in the crystal lattice. This is far from reality and serious errors can be made if anions get polarized and the bonding has some partial covalent character.

For example, radius ratio r^+/r^- for AgI is 0.52 which is indicative of coordination number 6 but actually it is 4 coordinated. Similarly, ZnO, HgS and CdS have radius ratios greater than 0.5 but have tetrahedral lattice.

(*ii*) Some compounds may exist in more than one modifications with different coordination number. In such cases, anion-anion repulsions and hence internuclear distances would be different. For example, TIC1 crystallises out in NaC1 structure (coordination number 6) with internuclear distance 3.15Å but can also adopt CsC1 structure (coordination number 8) with internuclear distance 3.32 Å.

(*iii*) Effective radius of a cation is greatly influenced by the anions with the consequence that the radius ratio changes. For example, AgF and NaF crystallises out in NaC1 type of structure with co-ordination number 6 and if we assume that the size of F^- remains constant then Ag^+ is bigger than Na^+. On the other hand, for a given size of chloride ion in AgC1 and NaC1, the sizes of cations are reversed, that is, Na^+ is bigger than Ag^+. It is also true for their bromides. This is understandable because Ag^+ is softer than Na^+ and introduces relatively more covalent character with polarizable anions like Cl^- and Br^-.

Q. 12. What is lattice energy? Mention the factors on which it depends.

(*M. Dayanand, 2000; Garhwal, 2000; G. Nanakdev, 2000; Calcutta, 2000; Indore, 2001; Calcutta, 2007; Nagpur, 2008*)

Ans. An ionic bond is likely to be formed most readily between an atom with the lowest ionization potential and an atom with the highest electron affinity. However, even the combination of the most electropositive cesium atom with the lowest ionization energy of 357.3 kJ/mol (energy absorbed : endothermic) and the electronegative chlorine with the highest electron affinity of 347.36 kJ/mol (energy released : exothermic) is energetically unfavourable. This indicates that even for the most ionic bond *i.e.*, $Cs^+ Cl^-$ (g), there is insufficient energy available for its formation. However, the shortfall is more than compensated by the energy gained through electrostatic attraction in the formation of an ionic crystal. This energy is known as the **crystal energy** or the **lattice energy** and is sufficient to permit the formation of ions.

The stability of an ionic solid is measured in terms of its lattice energy (U_0). It is defined as the energy released when one mole of an ionic crystal is formed from one mole of gaseous positive and one mole of gaseous negative ions, when these are separated from each other by infinite distance. Greater the value of lattice energy, more stable is the ionic solid. The lattice energy of sodium chloride is – 774 kJ mol^{-1}. The negative sign indicates that the process is exothermic and that of the energy of the system is lowered as the solid is formed.

The force of attraction (F) between oppositely charged ion is governed by

(*i*) their ionic radii

(*ii*) the charges on the ions.

It is given by the equation : $F = \dfrac{q_1 q_2}{r^2}$

where q_1 and q_2 are the points charges separated by distance r is vacuum.

It may be observed that if the radii of the ions are large, the force of attraction will be less. This is better illustrated if we compare the stability of NaC1 and CsCl. The force of attraction between the large Cs^+ ions and Cl^- ions will be less than that between the smaller Na^+ ions and Cl^- ions. As a result, CsC1 is less stable than NaCl. We may say that the stability of an ionic solid increases with decreasing interionic distance for crystals with similar ionic charges.

The role played by the charges on the ions demonstrated if we compare the lattice energy of BaO and NaC1, since both have very similar internuclear distances (NaC1 = 281 and BaO = 277 pm). Since BaO consists of doubly charged Ba^{2+} and O^{2-} ions, the force of attraction is much larger than that in NaC1 which contains singly charged ions. Hence BaO is more stable than NaC1 (melting points NaC1 = 800°C, BaO = 1911°C). The stability of an ionic solid increases with increasing nuclear charge for crystals with similar interionic distances.

Q. 13. What is Madelung constant ? *(Delhi, 2007)*

Ans. The lattice energy of an ionic crystal is determined by coulombic interaction between all of its ions; attractive forces between oppositely charged ions and repulsive forces because of the interpenetration of electron clouds. The potential energy of a pair of oppositely charged ions representing force of attraction varies inversely with distance, *r*, between the two ions. (Fig. 2.4).

$$PE_{(attraction)} = \frac{Z_1 Z_2 e^2}{r}$$

where Z_1 and Z_2 are charges on the ions and *e* is electronic charge. For NaC1, $Z_1 = Z_2 = 1$ and for CaO, $Z_1 = Z_2 = 2$ but for $BaC1_2$, $Z_1 = 2$, $Z_2 = 1$.

As the ions come very close to each other, they repel one another because of interpenetration of the electron clouds. The repulsion energy is inversely proportional to the nth power of the distance *r* between the ions.

$$PE_{(repulsion)} = \frac{be^2}{r^n}$$

where *e* is the electronic charge and *n* is the constant called the Born repulsion exponent and *b* is a constant called a repulsion coefficient. The Born equation gives the net potential energy for a pair of ions as the summation of the two terms, *i.e.*, when these ions are brought together from infinite distance.

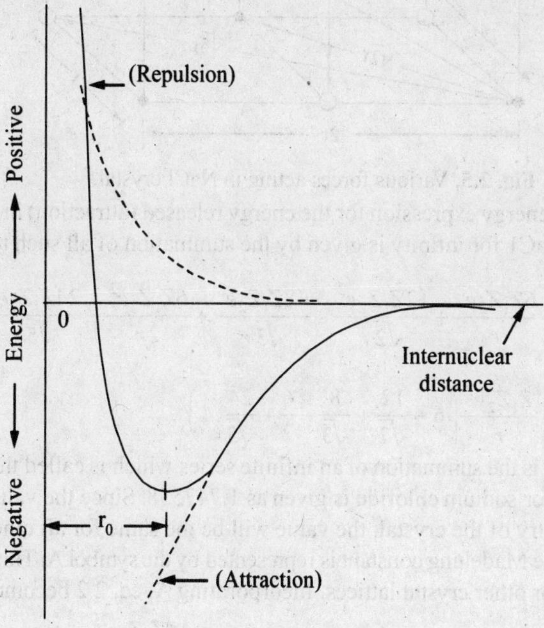

Fig. 2.4. Potential energy diagram of an ionic solid.

$$PE = \frac{Z_1 Z_2 e^2}{r} + \frac{be^2}{r^n} \qquad ...(2.1)$$

The – ve sign on attraction term indicates release of energy. It is clear from Fig. 2.4 that repulsion term increases very rapidly as the separation between the ions decreases beyond the internuclear separation. The potential energy is at a minimum at the equilibrium separation, r_0. However, in ionic crystals, no two separate ions can be considered, the entire crystal must be considered as a single entity. The force on any one ion will be determined not only by the oppositely charged ions which are directly surrounding it but also somewhat due to other ions (both positive and negative) at greater distances. To consider the forces due to all the ions, we may take a specific example, say sodium chloride. The potential energy of an ion will depend on the crystal structure of the solid.

In a crystal of sodium chloride (Fig. 2.5) the potential at a sodium ion results from:

(*a*) Six nearest Cl^- ions at a distance r;

(*b*) Twelve next nearest Na^+ ions at a distance $\sqrt{2}r$;

(*c*) Eight $(\text{next})^2$ nearest Cl^- ions at a distance $\sqrt{3}r$;

(*d*) Six $(\text{next})^3$ nearest Na^+ ions at a distance $2r$;

(*e*) Twenty-four $(\text{next})^4$ nearest Cl^- ions at a distance $\sqrt{5}r$.

These distances from a sodium ion are clearly shown in Fig. 2.5.

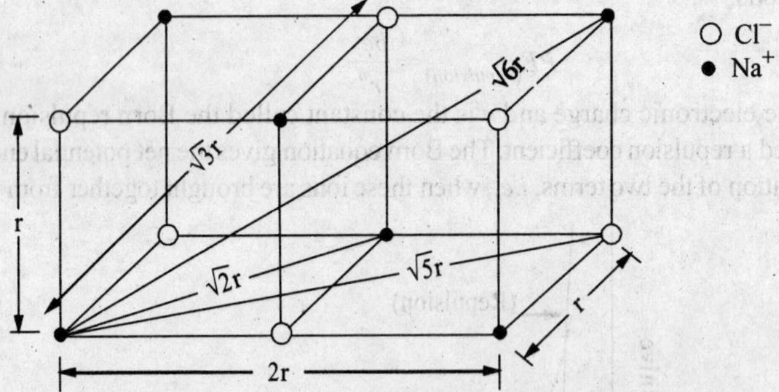

Fig. 2.5. Various forces acting in NaCl crystal.

The potential energy expression for the energy released (attraction) in bringing a sodium ion in a crystal lattice of NaCl for infinity is given by the summation of all such terms:

$$PE = -\frac{6Z_1Z_2e^2}{r} + \frac{12Z_1Z_2e^2}{\sqrt{2}\,r} - \frac{-8Z_1Z_2e^2}{\sqrt{3}\,r} + \frac{6Z_1Z_2e^2}{2r} - \frac{24Z_1Z_2e^2}{\sqrt{5}\,r} + ...$$

or

$$PE = -\frac{Z_1Z_2e^2}{r}\left(6 - \frac{12}{\sqrt{2}} + \frac{8}{\sqrt{3}} - \frac{6}{2} + \frac{24}{\sqrt{5}} ....\right) \quad ...(2.2)$$

Equation (2.2) is the summation of an infinite series which is called the Madelung constant. The summation value for sodium chloride is given as 1.747558. Since the value of this constant depends only on the geometry of the crystal, the value will be the same for all other solids which exhibit the NaCl structure. The Madelung constant is represented by the symbol A. This quantity can be calculated in the same way for other crystal lattices. Incorporating A, eq. 2.2 becomes.

$$PE = -\frac{-AZ_1Z_2e^2}{r}$$

Values of Madelung constant for some common crystal structure type are given in Table 2.3.

Table 2.3. Madelung constant (A) for some common crystal lattices.

Structure	*Formula*	*Coordination number*	*A*
Sodium chloride	NaC1	Na 6; Cl 6	1.7475
Cesium chloride	CsC1	Cs 8; Cl 8	1.7627
Zinc blende	ZnS	Zn 4; S 4	2.6081
Wurtzite	ZnS	Zn 4; S 4	1.6413
Fluorite	CaF_2	Ca 8; F 4	5.0388

Similarly, the second term in the potential energy is modified. The repulsive energy term $\frac{be^2}{r^n}$ increases very steeply as interionic distance decreases after ions come into contact. The repulsion coefficient (b) is modified in view of its approximate proportionality to the number of nearest neighbours. In case of sodium chloride, six Cl^- ions are the immediate neighbours surrounding the Na^+ ion. Therefore instead of (b), the repulsion coefficient is taken as ($6b$) for NaCl. Thus, the expression for repulsion term becomes $\frac{6be^2}{r}$. This is expressed in a more general form: $\frac{Be^2}{r^n}$, where $B = 6b$ for NaCl.

The Born repulsion exponent, n results from forces of repulsion due to the interpenetration of ionic atmosphere and generally lies between 8 and 10. It has specific values for each noble gas, or pseudo noble gas configuration, with $n = 5$ for He, 7 for Ne, 9 for Ar or Cu^+, 10 for Kr or Ag^+, 12 for Xe or Au^+.

Q.14. Derive Born-Lande equation for lattice energy. (*Calcutta, 2000*)

Ans. The net potential energy of an ion considering all types of forces due to the neighbouring ion is given by

$$PE = -\frac{Ae^2Z^2}{r} + \frac{Be^2}{r^n} \quad ...(2.3)$$

where Z replaces Z_1 and Z_2. At the equilibrium internuclear distance, r_0 the attractive and repulsive forces are exactly balanced. The potential energy is minimum at r_0. Differentiation of Eq. 2.3 for $r = r_0$ and equating it to zero we obtain the value of r_0 and B and then eliminate the latter from the potential energy expression.

Equation 2.3 may be written as

$$PE = -Ae^2 Z^2 r^{-1} + Be^2 r^{-n}$$

On differentiation, we get

$$\frac{d(PE)}{dr} = (-1)(-)A e^2 Z^2 r^{-1-1} + Be^2(-n) r^{-n-1}$$

or $$A e^2 Z^2 r^{-2} - nBe^2 r^{-n-1}$$

On applying the necessary conditions, that $\frac{d}{dr}(PE) = 0$ for $r = r_0$, we have

$$0 = A e^2 Z^2 r_0^{-2} - nBe^2 r_0^{-n-1}$$

or $$A\,e^2 Z^2 r_0^{-2} = nBe^2 r_0^{-n-1}$$

$\therefore$ $$B = \frac{AZ^2}{n} r_0^{-2+n+1}$$

or $$B = \frac{AZ^2}{n} r_0^{n-1} \quad \text{...(2.4)}$$

and similarly, we can get

$$r_0 = \left(\frac{nB}{AZ^2}\right)^{\frac{1}{n-1}} \quad \text{... (2.5)}$$

By substituting the value of B in expression 2.3, we get

$$(PE)_0 = \frac{-Ae^2Z^2}{r_0} + \frac{AZ^2e^2}{nr_0^n} r_0^{n-1}$$

$$= \frac{-Ae^2Z^2}{r_0} + \frac{AZ^2e^2}{n} r_0^{n-1-n}$$

$$= \frac{Ae^2Z^2}{r_0}\left[\frac{1}{n} - 1\right]$$

$$= \frac{-Ae^2Z^2}{r_0}\left[1 - \frac{1}{n}\right] \quad \text{...(2.6)}$$

The lattice energy U_0 is the amount of energy released when one mole of NaC1 crystal is formed from gaseous ions which are at infinite separation.

Lattice energies of ionic compounds are always negative (being energy released).

For one mole of the ion-pairs the said expression (2.6) is multiplied by Avogadro's number (N)

$$U_0 = \frac{-NAe^2Z^2}{r_0}\left[1 - \frac{1}{n}\right] \quad \text{... (2.7)}$$

This expression is known as **Born-Lande** equation for lattice energy. The charge Z is taken as unity for NaC1 and other similar uni-univalent systems and two for BaO and other divalent ionic compounds. Care should be taken in cases like say CaF_2 where the value of charge on the cation is 2 and the value of charge on the anion is 1. Consequently, Z^2 in CaF_2 would be 2.

Q. 15. Calculate the value of Madelung constant A for MgO, given

r_0 = 2.10 Å, n = 7, N = 6.02 × 10^{23}

U_0 = – 3940 kJ/mol, e = 4.8 × 10^{-10} e.s.u.

Ans. In case of MgO, the values of charges on cation or anion are 2 each. Therefore Z^2 = 4.

Substituting values and taking care of conversion of erg/mol to kJ/mole, we have

$$-3940 = \frac{-6.02\times10^{23}\times2^2\left(4.8\times10^{-10}\right)^2\times A}{2.10\times10^{-8}\times10^{10}}\left(1-\frac{1}{7}\right)$$

$$3940 = 2264.5\text{ A}$$

or $$A = 1.74$$

Q. 16. Calculate the value of Z^2 in CaF_2, given

$$A = 2.519,\ r_0 = 2.35\ \text{A},\ n = 7$$

$$N = 6.02 \times 10^{23},\ e = 4.8 \times 10^{-10}\ \text{e.s.u.}$$

and $$U_0 = -2550\ \text{kJ/mol}$$

Ans. Convert R.H.S. expression (eq. 2.7) into kJ/mol by dividing by a factor 10^{10} because 10^{10} erg = 1 kJ

$$-2550 = \frac{-6.02 \times 10^{23} \times 2.519 \times (4.8 \times 10^{-10})^2 \times Z^2}{2.35 \times 10^{-8} \times 10^{10}} \left[1 - \frac{1}{7}\right]$$

$$2550 = 1274\ Z^2$$

or $$Z^2 = 2$$

Q. 17. Calculate lattice energy of sodium chloride (in kJ/mol) from the following data:

$$A = 1.75,\ r_0 = 2.8\text{Å},\ n = 9$$

$$N = 6.02 \times 10^{23}, \qquad e = 4.8 \times 10^{10}\ \text{e.s.u.}$$

Ans. $$U = -\frac{NAe^2Z^2}{r_0}\left(1 - \frac{1}{n}\right)$$

$$= -\frac{6.02 \times 10^{23} \times 1.75 \times (4.8 \times 10^{-10})^2 \times 1^2}{2.8 \times 10^{-8}} \left[1 - \frac{1}{9}\right]$$

$$= -770.56 \times 10^{10}\ \text{ergs/mol}$$

$$U_0 = -770.56 \times 10^{10}\ \text{ergs/mol.}$$

Since $$10\ \text{ergs} = 1\ \text{J}$$

or $$10^{10}\ \text{ergs} = 1\ \text{kJ}$$

$\therefore$ $$V = -770.56\ \text{kJ/mol.}$$

Q. 18. What is Born-Haber cycle? How can we obtain lattice energy of a solid with its help? (*Kurukshetra 2002; Bangalore, 2004; Nagpur, 2005; Calcutta, 2007; Guwahati 2007; Pune 2010*)

Ans. The formation of an ionic solid *MX* from its elements may be formulated by two different paths interrelating different thermochemical quantities, as shown in Fig. 2.6.

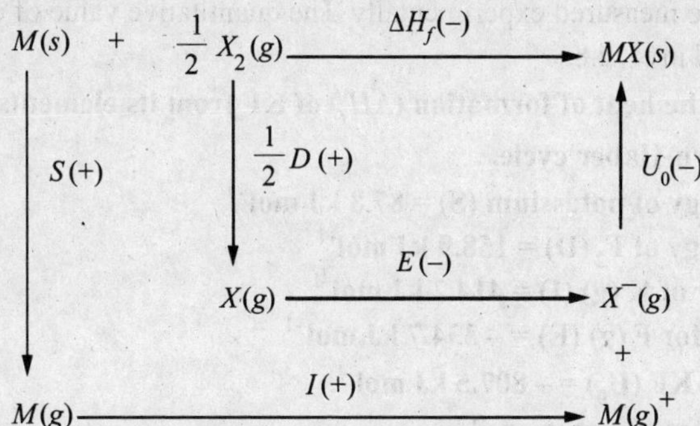

Fig. 2.6. Born-Haber cycle for the formulation of an ionic solid MX.

The various terms are

(*i*) S = sublimation energy of the metal

$$M(s) \longrightarrow M(g)$$

Since energy is required for the process (endothermic), its value is taken as a positive quantity.

(*ii*) D = Dissociation energy of X_2 gas into atoms.

$$X_2(g) \longrightarrow 2X(g)$$

This process is endothermic and its value is taken as a positive quantity.

(*iii*) I = Ionization energy, *i.e.*, energy required for the removal of an electron from an isolated gaseous atom, *M*. The process is endothermic and is taken as a positive quantity.

$$M(g) \longrightarrow M^+(g) + e^-$$

(*iv*) E = Electron affinity, *i.e.*, amount of energy released when a neutral gaseous atom gains an electron. The process is exothermic and its value is taken as a negative quantity.

$$X(g) + e \longrightarrow X^-(g)$$

(*v*) U_0 = Lattice energy; it is amount of energy released when one mole of gaseous positive and negative ion condense to form an ionic solid.

$$M^+(g) + X^-(g) \longrightarrow MX(s)$$

The process is exothermic and its value is always negative.

(*vi*) ΔH_f = Heat of formation, it is defined as the amount of energy released when one mole of a substance is formed from its element.

$$M(s) + \frac{1}{2}X_2(g) \longrightarrow MX(s)$$

Energy is released in the process and its value is taken as a negative quantity. This is a single step reaction equivalent to sum of all other steps (*i*) to (*v*).

Since the two processes are independent of each other, the two can be equated, *i.e.*, heat of formation of *MX* must be equal to the summation of all other terms.

$$\Delta H_f = S + \frac{1}{2}D + I + E + U_0$$

$$U_0 = \Delta H_f - S - \frac{1}{2}D - I - E \quad \text{... (2.8)}$$

Thus, the lattice energy of a solid can be calculated with the help of other thermochemical quantities which can be measured experimentally. The quantitative value of each term along with their sign are substituted in eq. 2.8.

Q. 19. Calculate the heat of formation (ΔH_f) of KF from its elements from the following data by the use of Born-Haber cycle.

Sublimation energy of potassium (S) = 87.8 kJ mol^{-1}

Dissociation energy of F_2 (D) = 158.9 kJ mol^{-1}

Ionization energy of K (g) (I) = 414.2 kJ mol^{-1}

Electron affinity for F (g) (E) = – 334.7 kJ mol^{-1}

Lattice energy of KF (U_0) = – 807.5 kJ mol^{-1}.

Ans. Substituting the values in eq. 2.8

$$\Delta H_f = S + \frac{1}{2}D + I + E + U_0$$

$$\Delta H_f = 87.8 + \frac{1}{2}(158.9) + 414.2 + (-334.7) + (-807.5)$$

$$= -560.8 \text{ kJ mol}^{-1}$$

This can be represented as:

$$K(s) + \frac{1}{2} F_2(g) \longrightarrow KF(s) + 560.8 \text{ kJ}$$

Q. 20. Calculate the heat of formation (ΔH_f) of MgF_2 from its elements using the Born-Haber cycle.

The thermochemical data are as follows:

Sublimation energy of magnesium, (S) = 146.4 kJ mol^{-1}

Dissociation of F_2, (D) = 158.9 kJ mol^{-1}

Ionization energy of Mg (g), I (Mg^{2+}) = 2184.0 kJ mol^{-1}

Electron affinity for F (g), E(F^-) = – 334.7 kJ mol^{-1}

Lattice energy of MgF_2 (U_0) = – 2922.5 kJ mol^{-1}

Ans. Substituting these values in the expression for MgF_2:

$$Mg(s) + F_2(g) \longrightarrow MgF_2 + 1102.6 \text{ kJ}$$

$$\Delta H_f = S + D + I + 2E + U_0$$

Here D will be used instead of $\frac{1}{2}D$ because of the following process

$$F_2(g) + D \longrightarrow 2F(g)$$

Also $2E$ is taken in place of E because of the following change

$$2F(g) + 2e \longrightarrow 2F^-(g)$$

$$\Delta H_f = 146.4 + 158.9 + 2184.0 + 2(-334.7) - 2922.5$$

$$= -1102.6 \text{ kJ mol}^{-1}$$

Q. 21. Calculate lattice energy of NaCl crystal from the following data by use of Born-Haber cycle.

Sublimation energy (S) = 108.7 kJ mol^{-1}

Dissociation energy for Cl_2 (D) = 225.9 kJ mol^{-1}

Ionization energy for Na (g) (I) = 489.5 kJ mol^{-1}

Electron affinity for Cl (g) E = – 351.4 kJ mol^{-1}

Heat of formation of NaCl (ΔH_f) = – 414.2 kJ mol^{-1} *(Delhi, 2007)*

Ans. Substituting these values in equation 2.8

$$U_0 = (-414.2) - (+108.7) - \frac{1}{2}(225.9) - (+489.5) - (-315.4)$$

$$= -414.2 - 108.7 - 112.9 - 489.5 + 351.4$$

$$= -773.9 \text{ kJ mol}^{-1}$$

The lattice energy of NaCl is –773.9 KJ mol^{-1}.

Q. 22. How can we calculate electron affinity with the help of lattice energy?

Ans. Estimation of electron affinities. Electron affinities can be measured directly only for the conversion of an isolated gaseous atom into its singly charged anion. That is,

$$O(g) + e \longrightarrow O^-(g) + E$$

However, in order to find the sum of the first and second electron affinities of the oxygen atom, that is, to know ΔH for the reaction

$$O(g) + 2e \longrightarrow O^{2-}(g)$$

we can apply Born-Haber cycle to any metal oxide whose lattice energy can be calculated. For example, the lattice energy of CaO, which has sodium chloride structure with Madelung constant A = 1.7475, interionic distance r_0= 240 pm, z = 2 and fairly accurate value of Born exponent has been calculated to be about –3389 kJ mol^{-1} using expression 2.7. The heat of formation of CaO is – 635.9 kJ mol^{-1}. The values of S, 1/2 D, and I (Ca^{2+}) are 153.9, 227.1 and 1735.1 kJ mol^{-1}; respectively. From this data we can calculate electron affinity.

Using Born-Haber cycle, $E(O^{2-})$ can be calculated as follows :

$$\Delta H_f = S + \frac{1}{2}D + I + E + U_0$$

Substituting these values, we get

$$-635.9 = 153.9 + 227.1 + 1735.1 + E\,(O^{2-}) - 3389$$

or
$$E\,(O^{2-}) = 3389 - 635.9 - 153.99 - 227.1 - 1735.1$$
$$= +\,637 \text{ kJ mol}^{-1}$$

In other words, ΔH for the reaction

$$O\,(g) + 2e \longrightarrow O^{2-}\,(g)$$

is + 637 kJ mol^{-1}.

The first electron affinity of oxygen is –118.0 kJ mol^{-1} but the addition of the second electron absorbs 774.8 kJ mol^{-1}. Their sum is + 636.8 kJ /mol. This shows that the formation of oxide ion is not favoured but the only reason why an oxide ion (O^{2-}) exists, is its high lattice energy in contact with metal ions.

Q. 23. How can we estimate heat of formation of an unknown compound from lattice energy measurement?

Ans. Heats of formation of an unknown compound can be estimated by making a reasonable guess about its lattice energy. As for example, in order to find out if the formation of argon chloride ArC1 is favourable or not, we make use of Born-Haber cycle. If argon chloride were to exist as a salt Ar^+ Cl^-, it would have NaCl structure and lattice energy close to the nearest alkali metal chloride, KC1.

or
$$U_0 = -722 \text{ kJ mol}^{-1}$$

Other thermochemical values are given in the cycle shown in Fig. 2.7.

$$\Delta H_f = \frac{1}{2}D + I + E + U_0$$
$$= 113.4 + 1526.3 - 352.8 - 693$$
$$= 593.9 \text{ kJ / mol}$$

This indicates that the formation of hypothetical solid ArC1 is highly endothermic, that is, strongly unfavourable.

It is clear that the very high ionization energy of argon (1526.3 kJ/mol) is responsible for making the reaction highly endothermic.

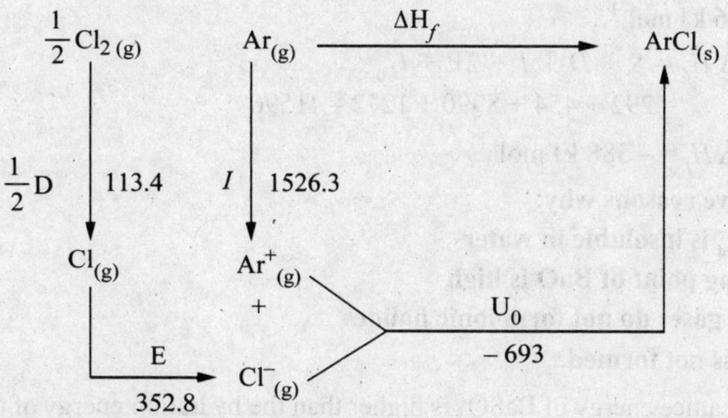

Fig. 2.7. Born-Haber cycle for the formation of ionic ArCl.

Q. 24. Describe the importance and role of lattice energy and hydration or solvation energy? (*M. Dayanand, 2000; Indore, 2001; Nagpur, 2003*)

Ans. The role of the two quantities is discussed as under:

(*i*) Greater is the lattice energy, greater the stability of an ionic solid. Because more energy will be needed to pull apart the +ve and –ve ions.

(*ii*) Salts with high lattice energies require a greater input of thermal energy to break down the crystal lattice. Consequently, salts have high melting points.

(*iii*) The magnitude of lattice energy of a solid may give us an idea about the solubility of that substance in different solvents. Ionic solids in general are insoluble in non-polar solvents such as carbon tetrachloride, whereas these are soluble in polar solvents, like water. For a solid to dissolve in water, the strong forces of attraction between its ions (lattice energy) must be overcome. This energy can be overcome by ion-solvent interactions. The solvation of ions is referred to in terms of solvation energy, which is always negative, *i.e.*, energy is released in the process. The amount of solvation energy depends upon the nature of the solvent.

The solvation energy increases if the solvent has high dipole moment/or high polarizability. As a general rule, we may say that for a solid to dissolve in water its hydration energy should be greater than the lattice energy of that solid. Therefore if

Hydration energy > Lattice energy ... (salt dissolves in water)

Hydration energy < Lattice energy ... (salt does not dissolve in water)

Q. 25. Calculate the heat of formation of ionic solid neon monochloride Ne^+Cl^- and indicate whether this hypothetical solid is favoured or not. Given:

$$U_0 = -840 \text{ kJ mol}^{-1} \quad D(Cl_2) = 226 \text{ kJ mol}^{-1}$$

$$I_0 = 2080 \text{ kJ mol}^{-1} \quad E(Cl^-) = -351 \text{ kJ mol}^{-1}$$

Ans. $\Delta H_f = \frac{1}{2}D + I + E + U_0$ Substitute the values,

$$\Delta H_f = 113 + 2080 - 351 - 840$$

$$= 1002 \text{ kJ mol}^{-1}$$

Formation of NaCl is endothermic and hence not favoured.

Q. 26. The experimental lattice energy of SnO_2 is –11596 kJ mol^{-1}. Calculate, ΔH_f of SnO_2. Given: S (Sn) = 292 kJ mol^{-1}, D (O_2) = 454 kJ mol^{-1}, I (Sn^{4+}) = 8990 kJ mol^{-1} and

$E(O^{2-}) = +636$ **kJ mol^{-1}.**

Ans. $\Delta H_f = S + D + I + 2E + U_0$

$= 292 + 454 + 8990 + 1272 - 11596$

$\Delta H_f = -588$ kJ mol^{-1}

Q. 27. Give reasons why:

(*a*) $BaSO_4$ is insoluble in water

(*b*) Melting point of BaO is high

(*c*) Noble gases do not form ionic halides

(*d*) KCl_2 is not formed.

Ans. (*a*) Lattice energy of $BaSO_4$ is higher than the hydration energy of the ions.

(*b*) Lattice energy is very high because the constituent ions have high charges.

(*c*) Ionization energy of noble gas atom is very high. Consequently the ΔH_f is not favoured.

(*d*) Second ionization energy of potassium is very high and not compensated by lattice energy of KCl_2.

Q. 28. Explain the terms : polarization, polarizing power and polarizability.

(*M. Dayanand, 2000; Indore 2001; Kurukshetra, 2002; Rohilkhand, 2003; Nagpur, 2008*)

Ans. Anions are larger in size than cations and, therefore, their electron clouds are less lightly held. A small highly charged cation shall, therefore, distort the electron cloud of the large anions in a manner that it increases the electron density between the nuclei. For example, the large iodide ion by itself is perfectly symmetrical. However when a small positively charged lithium ion comes close to the iodide ion, the electron cloud on the anion is pulled towards the positive lithium ion. The iodide ion is said to be polarized and the process is called polarization. The polarization effect produces covalence in LiI molecule because there is some electron density in between the nuclei as in the sharing of electrons in a covalent bond. The power of an ion to distort the other ion is known as its **polarizing power** and the tendency of the anion to distort is known as its **polarizability**. The polarizing power of a cation is proportional to its charge/radius ratio. This is illustrated in Fig. 2.8 when a large and soft anion comes under the influence of a small cation, the cation is able to polarize the anion (Fig. 2.8 (*b*)).

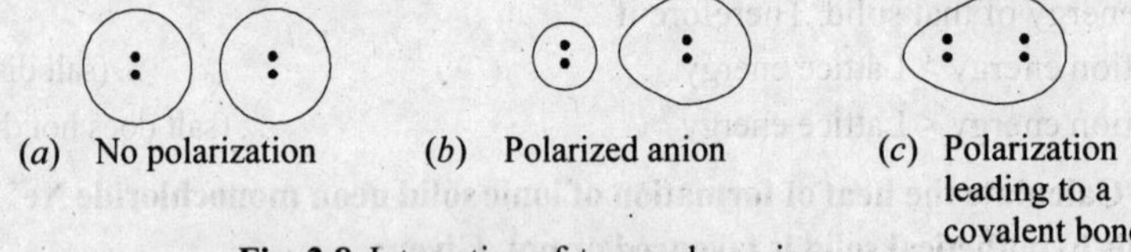

Fig. 2.8. Interaction of + ve and – ve ions.

This results in deformation of electron cloud on anion, called the polarization of anion. Due to small size of the cation, polarization of the cation is not much. Extensive polarization would be witnessed when cation penetrates the anionic electron cloud giving a covalent bond (Fig. 2.8). In compound consisting of large negative ions and small positive ions, the polarization leads to partial covalent character in ionic compounds. Examples of such ionic covalent compounds are $FeC1_3$, $A1C1_3$, LiBr etc.

Q. 29. State and explain Fajans's rules to explain covalent character in ionic compounds.

(*M. Dayanand, 2000; Himachal, 2000; Garhwal, 2000; Delhi, 2003; Vidyasagar 2007*)

Ans. Fajans explained the variation of nonpolar character in ionic compounds in terms of polarization effects. He gave qualitative ideas to offer simple approach to the problem of partial covalent character in ionic compounds. Fajans proposed the following rules to determine the extent of polarization of an anion by a cation, that is, the extent of covalent character induced in an ionic compound.

***(i)* Small, highly charged cations.** Small, highly charged cations have higher polarizing effects on anions than large, lowly charged cations. In other words, induced covalent character in ionic compounds increases with decrease in size of the cation or increase in charge on the cation. This is because very small cations or cations with higher charge have higher charge density which tends to distort or polarize the electron cloud around the anion. The greater the polarization, the higher the induced covalency. For example, Na^+ and Ca^{2+} have almost similar ionic radii but $CaCl_2$ (m.p. = 772°C) has higher covalent character than NaCl (m.p. = 800 °C) because of higher charge on Ca^{2+}. With identical charges on cations, $BeCl_2$ (m.p. = 405°C) is relatively more covalent than $CaCl_2$ (m.p. = 772°C) because of smaller size of Be^{2+} as compared to Ca^{2+}.

***(ii)* Large, highly charged anions.** Covalent character induced in ionic compound due to polarization of the anion increases with an increase in size and charge on the anion. The electron cloud on the large-sized anion is less strongly held by its nucleus and hence it is easily polarized by the approaching cation. In other words, the larger is the anion charge, the higher is its polarizability. For example, anions like Br^-, I^-, S^{2-}, Te^{2-}, As^{3-}, Sb^{3-} etc., are easily polarizable and have tendency to induce covalent character in ionic compounds. For a given cation, say Li^+, as the size of the anion, increases from F^- to I^-, the covalent character increases and consequently melting point decreases: LiF = 870 °C, LiCl = 613 °C, LiBr = 574 °C, LiI = 446 °C.

***(iii)* Cation configuration.** Cations with pseudo noble gas configuration (*i.e.*, 18 electron outer-shell) have higher polarizing power than cations with noble gas configurations. In other words, cations with $ns^2\ np^6$ configuration cause less distortion than those with $ns^2\ np^6\ nd^{10}$ configuration. This is because *d*-electrons shield the nucleus poorly with the consequence that such a cation will exert a higher nuclear charge on the interacting anion than a cation with 8 electron configuration. In other words, a cation with 18 electron configuration will lead to higher polarization of the anion. For example, Hg^{2+} (Xe) $4f^{14}\ 5d^{10}$ and Ca^{2+} (.... $3s^2\ 3p^6$) have similar ionic radii but the former is more polarizing and its compounds are more covalent than those of Ca^{2+}. For example, for a given anion (Cl^-), the melting point of $HgCl_2$ = 276°C and that of $CaCl_2$ = 772°C. Also Cu^+ (..... $3s^2\ 3p^6\ 3d^{10}$) and Na^+ (..... $2s^2\ 2p^6$) have similar ionic radii but the former is more polarizing and induces higher covalent character than the latter. This is supported by their melting points : CuCl : 430°C; NaCl= 800°C.

Q. 30. (*a*) Using Born-Haber cycle, calculate the electron affinity of chlorine from the following data :

Bond enthalpy of Cl_2 = + 240.0 kJ mole^{-1}
Enthalpy of formation of NaCl(s) = – 440.0 kJ mole^{-1}
Enthalpy of sublimation of Na(s) = + 110.0 kJ mole^{-1}
Enthalpy of ionization of Na(g) = + 480.0 kJ mole^{-1}
Enthalpy of lattice formation of NaCl(s) = – 810.0 kJ mole^{-1}

(*b*) Describe Fajan's rule. How does it explain the relative covalent/ionic nature of the following pairs of compounds :

(*i*) $FeCl_2$, $FeCl_3$;
(*ii*) LiI, CsI
(*iii*) CuCI, NaCl. *(Delhi, 2006)*

Ans. (*a*) Electron affinity of chlorine may be calculated as under :

$$\mathrm{Na}(s) + \tfrac{1}{2}\mathrm{Cl_2}(g) \xrightarrow{\Delta H_f(-)} \mathrm{NaCl}$$

$S(+)$: $\mathrm{Na}(s) \rightarrow \mathrm{Na}(g)$; $\tfrac{1}{2}D(+)$: $\tfrac{1}{2}\mathrm{Cl_2}(g) \rightarrow \mathrm{Cl}(g)$; $I(+)$: $\mathrm{Na}(g) \rightarrow \mathrm{Na}(g^+)$; $I(-)$: $\mathrm{Cl}(g) \rightarrow \mathrm{Cl}(g^-)$; $V_0(-)$: $\rightarrow \mathrm{NaCl}$

$$\Delta H_f = S + \tfrac{1}{2} D + I - E - V_0$$

Using data

$$-440 = 110 + 480 + \tfrac{1}{2} \times 240 - 810 - E$$

or $\quad E = 340$ kJ/mole.

(*b*) The Fajan's rule may be stated as

(*i*) Induced covalent character on ionic compound increases with decrease in size of the cation or increase in charge on the cation.

(*ii*) Covalent character induced in ionic compounds due to polarization of anion increases with an increase in size and charge of the anion.

(*iii*) Cations with pseudo noble gas configuration (*i.e.*, 18 e^- in outer shell) have higher polarizing power than cation with noble gas configuration.

Covalent/ionic nature of the pairs of compounds may be compared as under:

(*i*) $FeCl_2$, $FeCl_3$

As both Fe^{2+} and Fe^{3+} have almost same ionic radii but Fe^{3+} has more charge. So according to Fajan's rule $FeCl_3$ will be more covalent or less ionic than $FeCl_2$.

(*ii*) LiI, CsI

As both Li and Cs have some charge but radius of Cs is much greater than that of Li so according to Fajan' rule LiI will be more covalent.

(*iii*) CuCl, NaCl

As Cu and Na have some charge but Na is much smaller than Cu in radius so according to Fajan's rule NaCl will be more covalent.

Q. 31. How does per cent ionic character of a polar covalent bond depend upon (*i*) electronegativity difference, (*ii*) Dipole moment? (*Guwahati, 2007*)

Ans. 1. Electronegativity Difference and Per cent Ionic Character. A pure covalent bond is formed between atoms *A* and *B* when the electronegativities of the two atoms are the same. If *A* and *B* have different electronegativities, the resulting bond $\overset{-\delta}{A} \bullet \overset{+\delta}{B}$ would be a polar covalent bond because the bonding electron-pair is pulled towards the more electronegative atom causing some excess negative charge on it. As a result, some excess positive charge is developed on the less electronegative atom. If *A* is more electronegative than *B,* a polar covalent bond would be formed between *A* and *B.* Larger the difference in the electronegativities *of A* and *B,* greater would be the magnitude of δ and the higher would be the polarity of the bond.

Some empirical equations have been proposed to calculate the per cent ionic character of a covalent bond from the electronegativities of the bonding atoms. Two of these equations are the *Pauling equation* and the *Hannay-Smith* equation.

The Pauling Equation. Pauling proposed the following equation for determining the per cent ionic character of a covalent bond :

$$\text{Per cent Ionic Character} = 1 - e^{-1/4(\chi_A - \chi_B)} \qquad .....(1)$$

Where χ_A and χ_B are the electronegativities of A and B. He established the following empirical relationship between $\chi_A - \chi_B$ and the per cent ionic character :

$\chi_A - \chi_B$	Per cent ionic character
0.6	9
1.0	22
1.4	39
2.0	63
2.4	76
3.0	89

The Hannay-Smith Equation. Hannay and Smith proposed the following equation for the purpose :

$$\text{Per cent Ionic Character} = 16(\chi_A - \chi_B) + 3.5\,(\chi_A - \chi_B)^2 \qquad .....(2)$$

2. Dipole Moment and Per cent Ionic Character. Consider the bond H – X where X is a halogen atom. If the bond is purely covalent, its dipole moment would be zero. If the bond is ionic, *viz.,* H^+X^-, it implies that complete transfer of an electron has taken place from H to X and the bond would have a high value of dipole moment which is obtained by the product of the electronic charge and the internuclear distance between H and X. For a polar covalent bond which is intermediate between a pure covalent and a pure ionic bond, the dipole moment would be given by the product of the actual charge (δ) caused by polarisation and the internuclear distance. The magnitude of δ will depend upon the extent of the ionic character developed in the bond.

Thus,

$$\text{Per cent Ionic Character} = \frac{\text{Actual dipole moment of the bond}}{\text{Dipole moment of pure ionic bond}} \times 100$$

For example, in **HF,** the internuclear distance is 0.92 Å and the dipole moment for the ionic structure $\mathbf{H^+F^-}$ when an electron is completely transferred from hydrogen atom to fluorine atom is given by 4.8×10^{-10} esu × 0.92×10^{-8} cm, *i.e.,* 4.42 D. The actual dipole moment of HF is found to be 1.98 D. Per cent ionic character is calculated as under:

$$\text{Per cent Ionic Character} = \frac{1.98\,\text{D} \times 100}{4.42\,\text{D}} = 44.8$$

Q. 32. Silver halides are least soluble in water, though their lattice energy is almost the same as that of highly soluble alkali metal halides. Explain.

(G. Nanakdev, 2000; Nagpur; 2005)

OR

Ag I is covalent while NaI is covalent. Explain. *(Purvanchal 2007)*

Ans. Ag^+ can cause greater polarization of the halide because of its electronic configuration. There are *d*-electrons which cannot screen the nucleus effectively and thus pull the electrons of halide more than expected. This brings about covalency leading to the substance being insoluble in water.

Q. 33. Explain why NaCl is soluble in water but $BaSO_4$ is not.

Ans. In the case of $BaSO_4$, lattice energy is much more than hydration energy. Hence it does not dissolve.

With NaCl, hydration energy exceeds the lattice energy. So it is soluble.

Q. 34. Give the trends of thermostability of alkaline earth metal carbonates.

Ans. The thermal stability increases in the order

$$BeCO_3 < MgCO_3 < CaCO_3 < SrCO_3 < BaCO_3$$

This can be explained on the basis of decreasing polarizing power of the cation as we move from L.H.S. to R.H.S. making the substance more and more ionic or thermally stable.

Q. 35. Explain the sequence of solubility of lithium halide in ethyl alcohol.

Ans. The halides to be considered are:

LiF
LiCl
LiBr
LiI

The cation is same in all the cases. Hence the nature of the anion will decide solubility. Larger the anion, greater the polarizability hence greater the covalent character. And greater the covalent character greater the solubility in ethyl alcohol. On this basis solubility increases from LiF to LiI.

Q. 36. State, giving reasons, which cation will have greater polarizing power in the following:

(*i*) Na^+ or Mg^{2+} (*ii*) Cu^{2+} or Ca^{2+} (*iii*) Pb^{2+} or Pb^{4+}

Ans. (*i*) Mg^{2+} has greater polarizing power because of greater charge.

(*ii*) Ca^{2+} has greater polarizing power because of smaller size.

(*iii*) Pb^{4+} has greater polarizing power because of greater charge.

Q. 37. $AlCl_3$ anhydrous is covalent but $AlCl_3 . 6H_2O$ is ionic in nature. How would you account for this behaviour?

Ans. $AlCl_3$ is covalent in nature because of small and highly charged Al^{3+} ion. This polarizes the chloride ion causing covalency.

In the hydrated sample, the Al^{3+} ion get surrounding by water molecules. A lot of energy called hydration energy is released and thus Al^{3+} lose their polarizing power and remain in ionic state.

Q. 38. Give reasons why $ZnCl_2$ is soluble in organic solvents while $MgCl_2$ is insoluble.

Ans. Zn^{2+} ions have greater polarizing power, because of the presence of low shielding *d*-orbitals, as compared to Mg^{2+} ions. Thus Zn^{2+} polarizes the chloride ion and brings about covalent character and therefore $ZnCl_2$ is soluble in organic solvents.

Q. 39. Highly charged cations are rare. Explain. (*M. Dayanad, 2000*)

Ans. Highly charged cations strongly polarize the anion and bring about covalent character in the molecule. Thus the charge on the cation is neutralized and a covalent bond involving sharing of electrons or overlapping of orbitals is formed.

Q. 40. Explain the following:

(*i*) $SnCl_4$ is more covalent than $SnCl_2$

(*ii*) NaCl is more ionic than CuCl.

Ans. (*i*) $SnCl_4$ is more covalent because Sn^{4+} has greater polarizing power than Sn^{2+}.

(*ii*) NaCl is more ionic than CuCl because Cu^+ has greater polarizing power than Na^+. This is because there are *d*-electrons which poorly shield the nucleus.

The cation thus can polarize the anion more powerfully.

Q. 41. The inter-nuclear distance in HI is 1.62 Å. Assuming complete transfer of electron from H to I, calculate % ionic character and fractional charge on each ion. If the measured dipole moment is 0.38 D, also calculate the % ionic character of HI.

Given e = (electronic charge) = 4.8×10^{-10} e.s.u.

Ans. Assuming complete transfer of electron from H to I, the charge on H and I would be $+e$ and $-e$ respectively.

The dipole moment of HI in such case,

$$= q \times r$$
$$= e \times r$$
$$= (4.8 \times 10^{-10} \text{ e.s.u.})(1.62 \times 10^{-8} \text{ cm})$$
$$= 7.776 \text{ D}$$

Observed dipole moment = 0.38 D

$$\% \text{ ionic character} = \frac{0.38}{7.776} \times 100 = 4.88.$$

Q. 42. Discuss close packing of particles (atoms or ions) in crystals. (*Nagpur, 2005*)

Ans. In most of the solids, the constituent atoms are closely packed so as to utilize the maximum available space. For simplicity, we shall consider the constituents as spheres. These spheres are in contact with one another in various layers. Naturally, there will be some empty space in between these spheres. This empty space is called the *void.* Depending upon the size of the spheres and their arrangement, the voids will have different sizes. Then within the same solid, there are different types of voids like *tetrahedral* void or *octahedral* void.

Packing of particles in a crystal

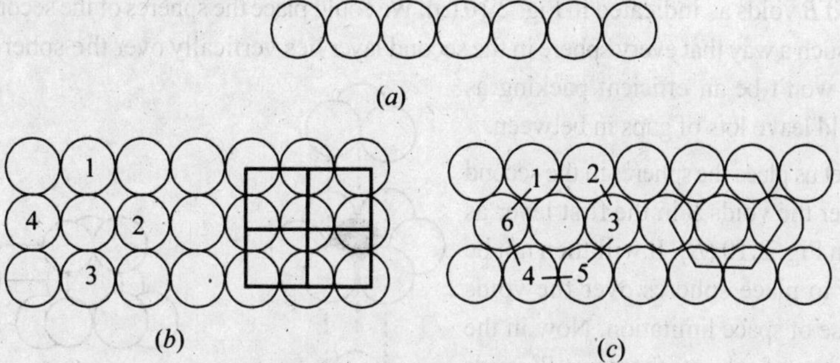

Fig. 2.9. Packing of spheres for efficient use of space. (*a*) Edge formation (*b*) Square close packing (*c*) Hexagonal close packing.

Considering the particles as spheres of equal size, we can visualize the formation of a solid as shown in Fig. 2.9.

One Dimensional Packing

We can arrange the spheres in a row [see Fig. 2.9 (*a*)], such that they touch one another. This is to utilize the space economically and to account for the high density of the solids.

Two Dimensional Packing

Then we place another row of spheres by the side of the first row. This can be done in two ways. One way is depicted in Fig. 2.9 (*b*). Second method is shown in Fig. 2.9 (*c*). The first type [Fig. 2.9 (*b*)] is called *Square Close Packing* and the second type [Fig. 2.9 (*c*)] is called *Hexagonal Close Packing.*

In square close packing, a sphere is surrounded by four spheres in the plane whereas in hexagonal close packing, a sphere is surrounded by six spheres in the plane. Thus the empty space in square close packing is more as compared to that in hexagonal close packing. Or in other words, Hexagonal Close packing is more dense than the Square Close packing.

Three Dimensional Packing

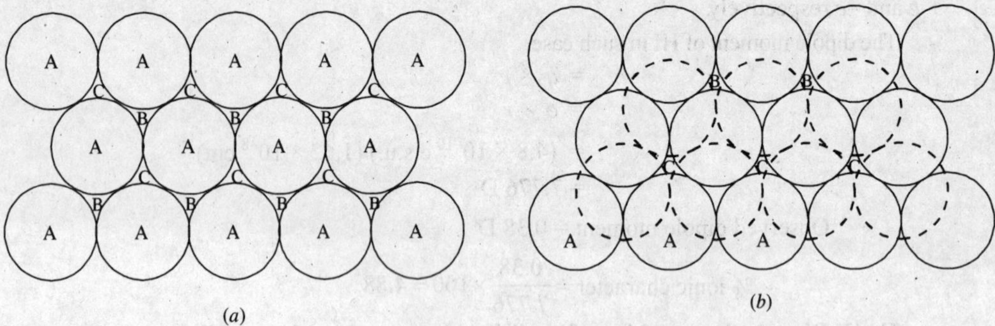

Fig. 2.10. Layers in close packing of spheres (*a*) Hexagonal close packed layer (*b*) Formation of the second layer.

Now let us form a second layer of spheres on the first layer in hexagonal close packing (see Fig. 2.10).

Spheres in the first layer are named A. There are voids in between them. These voids are *C* voids and *B* voids as indicated in Fig. 2.10 (*a*). We could place the spheres of the second layer on the first layer in such a way that every sphere in the second layer lies vertically over the sphere in the first layer. But this won't be an efficient packing as this would leave lots of gaps in between.

So let us place the spheres in the second layer over the voids *B* in the first layer as shown in Fig. 2.10 (*B*). It will then not be possible to place spheres over the voids *C* because of space limitation. Now in the second layer of spheres there are *B′* voids and there are second *C′* voids, which are in fact over the *C* voids of the first layers.

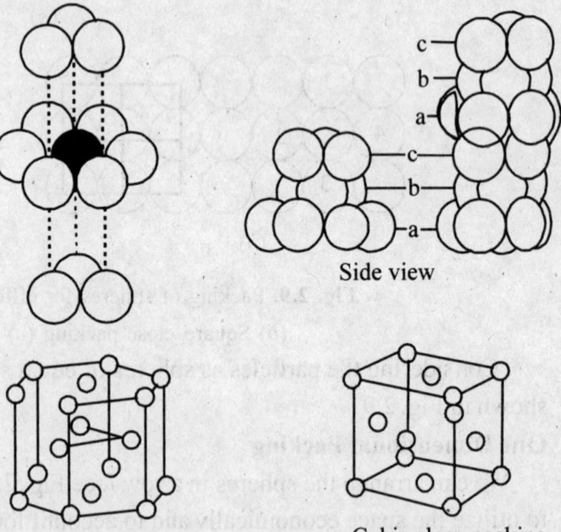

Fig 2.11. Three dimensional view of hexagonal and cubic close packed lattices.

Let us now place a third layer of spheres over the second layer. There are two possibilities. Either we place spheres on the *B′* voids or on the *C′* voids. If we place the spheres on the B^{-} voids in the third layer, these spheres in fact would be seated over the spheres of the first layer. This arrangement would be *ABABAB* ...arrangement. This type of packing is called *hexagonal close packing.*

If we place the spheres in the third layer over the C' voids then the spheres in the fourth layer will be found seated over the spheres in the first layer. This will give rise to *ABCABC*... arrangement. This type of packing is called *cubic close packing.* This is similar to face centred cubic system. Three dimensional views of hexagonal and cubic close packed lattices are shown in Fig. 2.11.

Q. 43. What is meant by body-centred cubic packing (BCC structure) ?

Ans. This is deviation from close packing. If the spheres of the first layer are loosened up, we obtain body-centred Cubic Structure. Each sphere in the second layer is in contact with four spheres of the first layer (see Fig. 2.12). The spheres in the second layer have the same packing as those in the first layer. The spheres in the third layer, having the same packing, will be lying exactly over the corresponding spheres in the first layer. Thus every sphere will be in contact with 8 spheres, four of the layer above and four of the layer below it. The coordination number is thus 8.

Fig. 2.12. Body centred cubic packing.

Alkali metals like Li, Na, K etc. crystallise in BCC structure. The structure is less closely packed than the CCP and HCP structures and the empty space is about 32%.

Q. 44. What are tetrahedral and octahedral voids ?

Ans. A void or empty space in a lattice which is surrounded by four spheres is called a tetrahedral void. For examples, the voids *B* in Fig. 2.10 are tetrahedral voids because they are surrounded by four spheres three in the first layer and the fourth in the second layer.

A void which is surrounded by six spheres is called the octahedral void. The voids C' in Fig. 2.10 are octahedral voids because they are surrounded by six spheres, three in the first layer and three in the second layer. These voids are shown in Fig. 2.13.

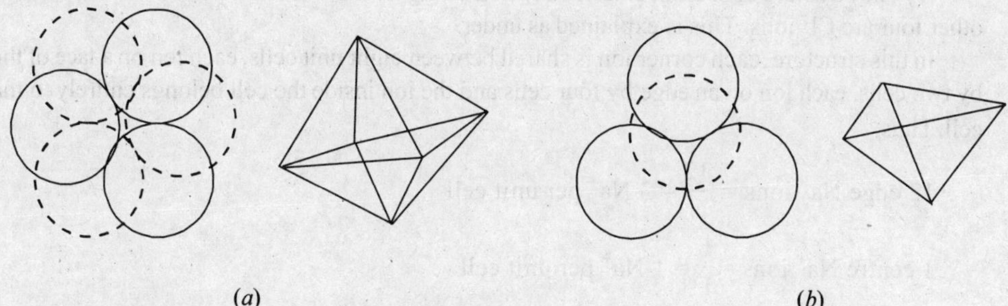

Fig. 2.13. Two types of voids (*a*) octahedral (*b*) tetrahedral.

Q. 45. What are coordination numbers in HCP and CCP ?

Ans. Every sphere in either HCP or CCP has 12 spheres directly touching it, 6 in its own layer, 3 in layer below it and 3 in the layer above it. Thus the coordination number of a sphere in HCP or CCP is 12. In other types of crystals, coordination number of 4, 6, 8 is quite common.

Q. 46. Discuss the structure of sodium chloride. *(G Nanakdev, 2000)*

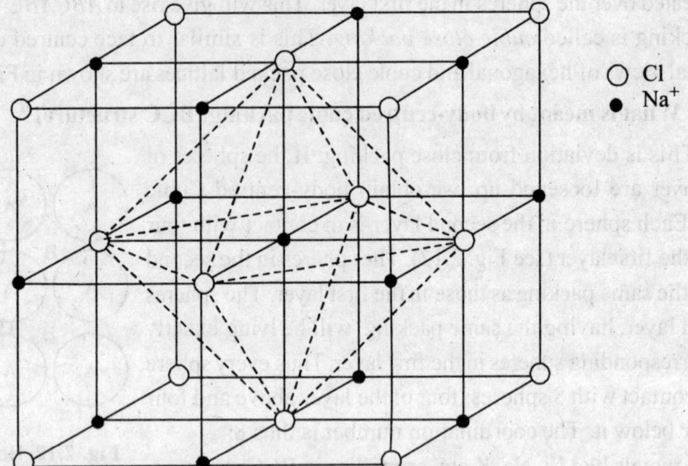

Fig. 2.14. Unit cell representation of NaCl structure.

Ans. A unit cell representation of sodium chloride is shown in Fig. 2.14. The main features of its structure are:

(*i*) Chloride ions are in ccp type of arrangement, *i.e.*, it contains chloride ions at the corners and at the centre of each face of the cube.

(*ii*) Sodium ions are so arranged that there are six chloride ions around it. Thus sodium ions occupy all the octahedral sites.

(*iii*) There is only one octahedral site for every chloride ion, thus the stoichiometry is 1:1.

(*iv*) It is obvious from the diagram that each chloride ion is surrounded by six sodium ions which are oriented towards the corners of a regular octahedron. Cations and anions are present in equivalent positions and the structure has 6 : 6 coordination.

(*v*) The structure of sodium chloride consists of eight ions in a unit cell, four are Na^+ ions and the other four are Cl^- ions. This is explained as under.

In this structure, each corner ion is shared between eight unit cells, each ion on a face of the shell by two cells, each ion on an edge by four cells and the ion inside the cell belongs entirely to that unit cell. Thus,

$$12 \text{ edge } Na^+ \text{ ions} = \frac{12}{4} = 3\ Na^+ \text{ per unit cell}$$

$$1 \text{ centre } Na^+ \text{ ions} = \frac{1}{1} = 1\ Na^+ \text{ per unit cell}$$

$$8 \text{ centre } Cl^- \text{ ions} = \frac{8}{8} = 1\ Cl^- \text{ per unit cell}$$

$$6 \text{ faces } Cl^- \text{ ions} = \frac{6}{2} = 3\ Cl^- \text{ per unit cell}$$

Q. 47. Discuss the structure of zinc blende. *(G. Nanakdev, 2000)*

Ans. A unit cell representation of zinc blende is shown in Fig. 2.15. The sulphur atoms are marked (O) and zinc atoms are marked (•). The features of its structures are described as under :

(*i*) The sulphur atoms are in ccp type of arrangement, *i.e.*, sulphur atoms are at the corners and at the centre of each face of the cube.

(*ii*) Zinc atoms are so placed that there are four sulphur atoms around it. Or it may be said that zinc atoms occupy tetrahedral sites and their coordination number is four.

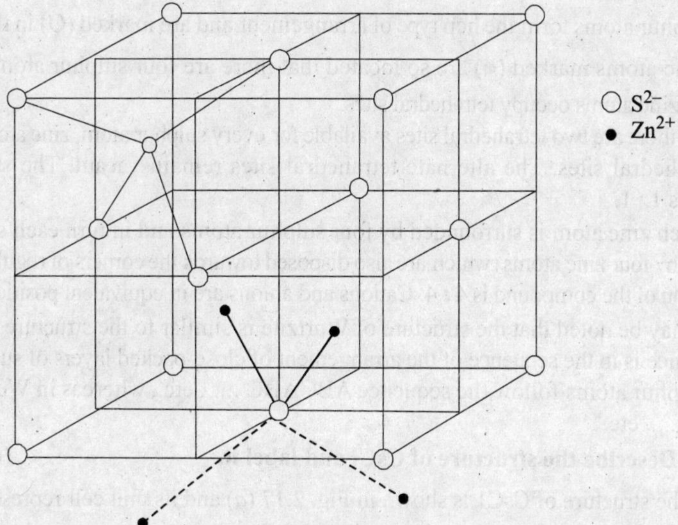

Fig. 2.15. Unit cell representation of zinc blende structure.

(*iii*) As there are two tetrahedral sites available for every sulphur atom zinc atoms occupy only half of the tetrahedral sites. So the stoichiometry of the compound is 1:1.

(*iv*) Each zinc atom is surrounded by four sulphur atoms and each sulphur atom is also surrounded by four zinc atoms which are also disposed towards the corners of a regular tetrahedron. Thus cations and anions are present in equivalent positions and the coordination of zinc blende structure is described as 4:4.

Q. 48. Discuss the Wurtzite structure of zinc sulphide. (*G. Nanakdev, 2000*)

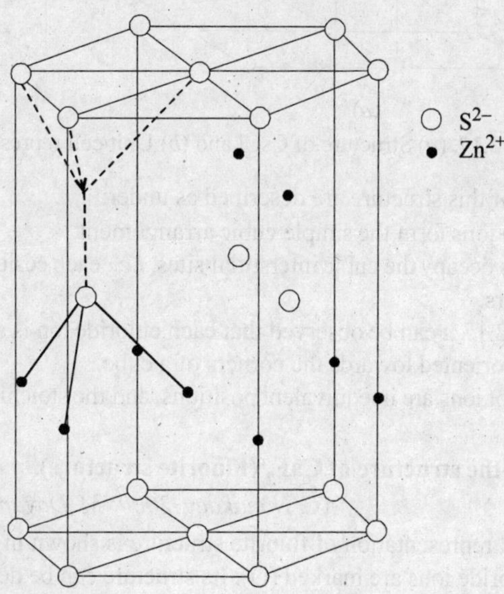

Fig. 2.16. Unit cell representation of Wurtzite structure.

Ans. It is an alternative form of ZnS. Its unit cell representation is shown in Fig. 2.16. The features of its structure are described below.

(*i*) Sulphur atoms form the hcp type of arrangement and are marked (O) in the diagram.

(*ii*) Zinc atoms marked (●) are so located that there are four sulphur atoms around each zinc atom. Thus zinc atoms occupy tetrahedral sites.

(*iii*) As there are two tetrahedral sites available for every sulphur atom, zinc atoms occupy only half of the tetrahedral sites. The alternate tetrahedral sites remain vacant. The stoichiometry of the compound is 1 : 1.

(*iv*) Each zinc atom is surrounded by four sulphur atoms and in turn each sulphur atom is also surrounded by four zinc atoms (which are also disposed towards the corners of regular tetrahedron). The co-ordination of the compound is 4 : 4. Cations and anions are in equivalent positions.

(*v*) It may be noted that the structure of Wurtzite is similar to the structure of zinc blende. The only difference is in the sequence of the arrangement of close-packed layers of sulphur atoms. In zinc blende, sulphur atoms follow the sequence ABCABC ...... etc., whereas in Wurtzite the sequence is ABABA...... etc.

Q. 49. Describe the structure of CsCl and label it. (*G. Nanakdev, 2000*)

Ans. The structure of CsC1 is shown in Fig. 2.17 (*a*) and its unit cell representation is shown in Fig. 2.17 (*b*).

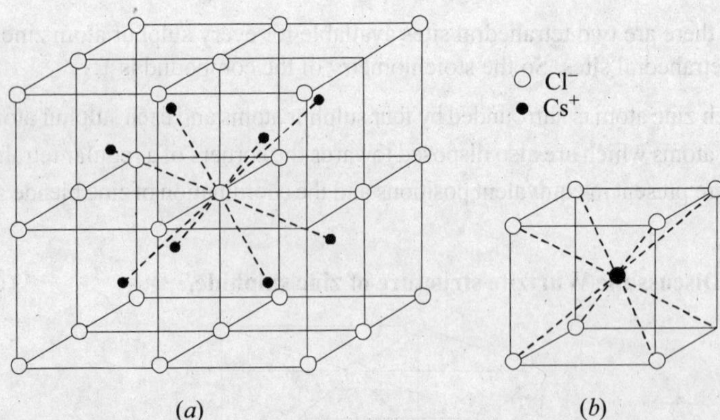

Fig. 2.17. (*a*) Structure.of CsCl and (*b*) Unit cell representation of CsCl.

Main features of this structure are described as under:

(*i*) The chloride ions form the simple cubic arrangement.

(*ii*) Cesium ions occupy the cubic interstitial sites. *i.e.*, each cesium ion has eight chloride ions as its nearest neighbours.

(*iii*) From Fig. 2.17 it can be observed that each chloride ion is also surrounded by eight cesium ions which are also oriented towards the corners of a cube.

(*iv*) Both types of ions are in equivalent positions, and the stoichiometry is 1: 1. The coordination is 8 : 8.

Q. 50. Discuss the structure of CaF_2 (Fluorite structure).

(*G. Nanakdev, 2000; M. Dayanand, 2000; Kurukshetra, 2002*)

Ans. A unit cell representation of fluorite structure is shown in Fig. 2.18. The calcium ions are marked (●) and fluoride ions are marked (O). Its structure can be described as:

(*i*) The calcium ions form the ccp arrangement, *i.e.*, these ions occupy all the corner positions and the centre of each face of the cube.

(*ii*) Fluoride ions are so located that these are four calcium ions around it. It may be said that the fluoride ions occupy tetrahedral sites and the coordination number of fluoride ion is 4.

(*iii*) As there are two tetrahedral sites available for every calcium ion, the fluoride ion occupy all the tetrahedral sites. The stoichiometry of the compound is 1: 2.

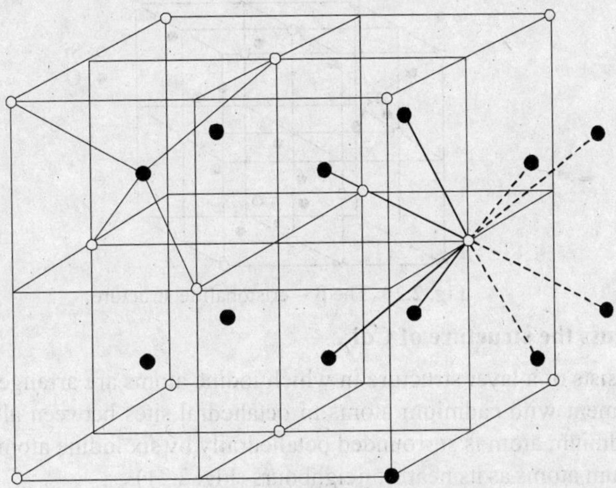

Fig. 2.18. Unit cell representation of CaF_2 structure.

(*iv*) Each fluoride ion is surrounded by four calcium ions whereas each calcium ion is surrounded by eight fluoride ions which are disposed towards the corners of a cube. The coordination of the compound is 8 : 4.

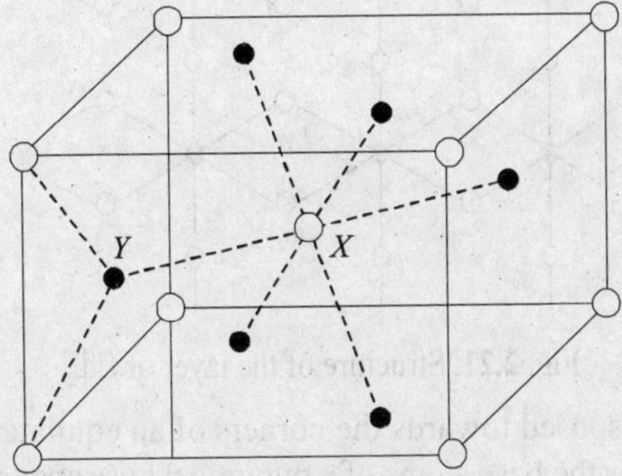

Fig. 2.19. Rutile structure.

Q. 51. Discuss the structure of TiO_2. (*Himachal,2000*)

Ans. Tetragonal unit cell of TiO_2 is shown in Fig. 2.19.

The radius ratio for r Ti^{4+} /rO^{2-} (0.64/1.33) = 0.5) lies between the values 0.414 – 0.732. It suggests the coordination number of Ti^{4+} ion as six. Each titranium ion is surrounded by six O^{2-} ions placed at the corners of an octahedron as shown for titanium ion'*X*'. The Ti^{4+} ions have a distorted body-centred cubic arrangement. Each O^-_2 marked as *Y* is surrounded trigonally by three Ti^{4+} ions placed at the corners of a triangle. We may thus say that coordination number of Ti^{4+} is 6 and that of O^{2-} is 3. We may say that rutile structure consists of an hexagonal close packed arrangement of oxide ions with Ti^{4+} ions occupying only half the octahedral holes.

Q. 52. Discuss the β-cristoballite structure of silicon dioxide.

Ans. Its structure is related to diamond. There is silicon atom covering positions occupied by carbon atoms (Fig. 2.20) in diamond, the oxygen atoms lie midway between the silicon atoms. In this arrangement a silicon atom would have four nearest neighbours (oxygen atoms) and these are arranged in a tetrahedrally arrangement. An oxygen atom has two silicon atoms arranged linearly. The coordination numbers are 4 for silicon and 2 for oxygen.

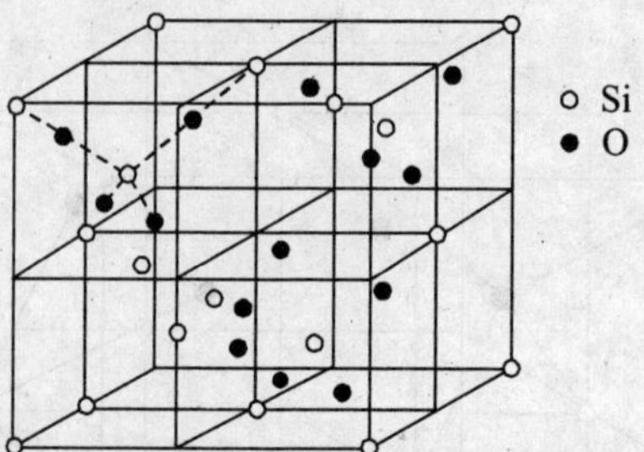

Fig. 2.20. The β – cristoballite structure.

Q. 53. Discuss the structure of CdI_2. (*M.Dayanand 2002*)

Ans. It consists of a layer structure in which iodine atoms are arranged in a hexagonal close-packed arrangement with cadmium atoms in octahedral sites between alternate layers of iodine atoms. Each cadmium atom is surrounded octahedrally by six iodine atoms and each iodine atom has three cadmium atoms as its nearest neighbours (Fig. 2.21).

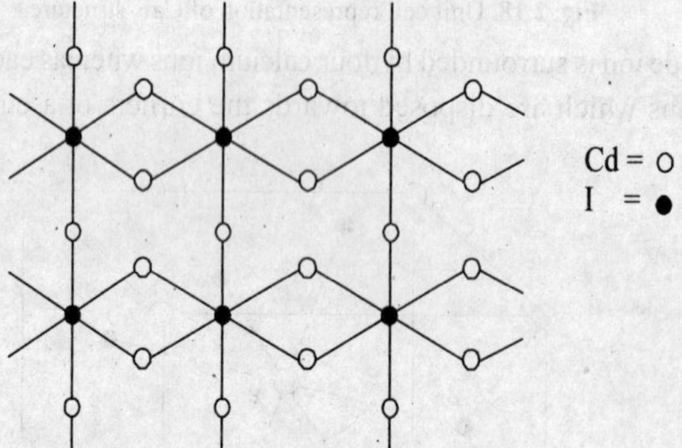

Fig. 2.21. Structure of the layer of CdI_2.

But these are not disposed towards the corners of an equilateral triangle. The three cadmium atoms may be regarded as the base atoms of a pyramid whose apex position is occupied by an iodine atom.

The arrangement of three layers I - Cd - I corresponds to the composition CdI_2 (stoichiometry is 1:1). The cadmium atoms are sandwiched between the layers of negatively charged iodine atoms, the outside layers of the sandwich structure are negative.

Q. 54. Why are CdI_2 crystals flaky?

Ans. Since one layer of negatively charged iodine atoms faces another negatively charged layer of iodine atoms without any positively charged buffer layer between them, there would be very little force of attraction and would not be held strongly. Thus CdI_2 structure is a layered structure (Fig. 2.22).

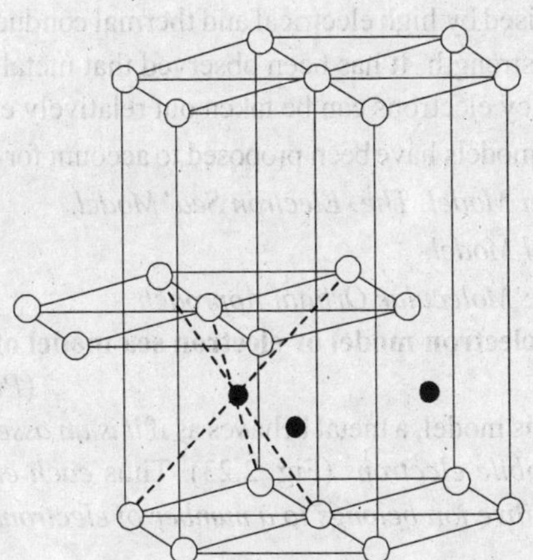

Fig. 2.22. The cadmium iodide structure.

Q. 55. How can you arrive at the structure of molecules from dipole moment data?

Ans. In case of molecules containing more than two atoms, the dipole moment not only depends upon the individual dipole moments of the bonds but also on the spatial arrangement of various bonds in the molecule. In such cases, the dipole moment of the molecule is the vector sum of the dipole moments of the various bonds. Obviously, the polar or non-polar nature of molecule will depend upon the spatial arrangement of atoms in the molecule. The bonds may be polar yet the molecule could be non-polar.

For example, carbon dioxide (CO_2) and water (H_2O) molecules are both triatomic molecules but the dipole moment of carbon dioxide is zero while that of water is 1.84 D. This can be easily explained on the basis of their structures.

In *carbon dioxide* molecule, both the carbon-oxygen bonds are polar. Since the molecule is linear, the dipole moment of C = O bond on one side of the molecule is cancelled by that on the other side of the molecule giving net zero dipole moment. On the other hand, *water* molecule is not linear, it has a *bent structure* as shown. Therefore, the dipole moment of water molecule is not zero. It is the resultant of the dipole moments of the two O — H bonds inclined at an angle of 104.5°. It has been found to be 1.84 D.

Similarly, the dipole moment of BH_3 is zero while that of NH_3 is 1.49 D, although both the molecules are tetra-atomic. This is because BH_3 has symmetrical structure in which the three B — H bonds are oriented at an angle of 120° to one another. The three bonds lie in one plane and the dipole moments of these bonds cancel one another giving net dipole moment equal to zero. On the other hand, the ammonia molecule has *pyramidal structure*. The dipole moments of three N —H bonds do not cancel one another (as they are in different planes). The molecule of ammonia shows a net value of dipole moment.

METALLIC BOND

Metals are characterised by high electrical and thermal conductivity, bright lustre, malleability, ductility and high tenstile strength. It has been observed that metals, generally, have low ionisation energies because the valency electrons can be taken out relatively easily.

The following three models have been proposed to account for the nature of bonding in metals :

1. *The Free Electron Model. The 'Electron Sea' Model.*
2. *The Valence Bond Model.*
3. *The Band Model : Molecular Orbital Approach.*

Q. 56. Discuss free electron model or electron sea model of metallic bond.

(*Purvanchal, 2007; Nagpur 2008*)

Ans. According to this model, a metal behaves as if *it is an assemblage of positive ions (kernels) immersed in a 'sea' of mobile electrons* (Fig. 2.23). Thus *each electron belongs to a number of positive ions and each positive ion belongs to a number of electrons.*

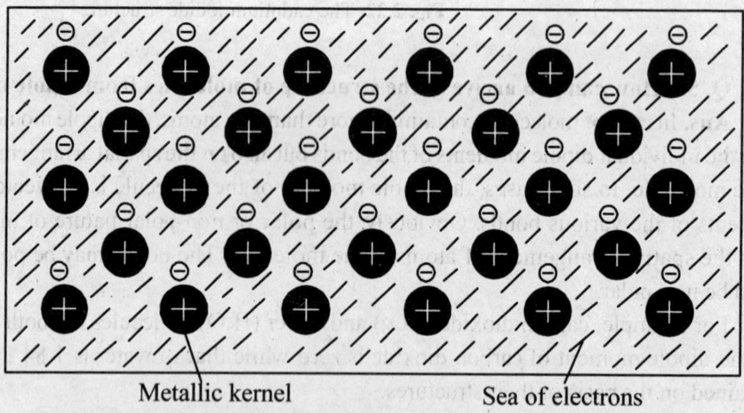

Fig. 2.23. The electron sea model for metals.

The force of attraction which binds a metal atom kernel to a number of electrons within its sphere of influence is known as a **metallic bond.**

Explanation of Metallic Properties.

Properties such as electrical and thermal conductivity, lustre, malleability, ductility, elasticity, etc., can be explained satisfactorily on the basis of free electron model.

1. The high electrical conductivity of metals, for example, is due to the presence of the mobile valency electrons. They move readily in an electric field and thus conduct electricity throughout the metal from one end to the other.

2. The high thermal conductivity is again due to the presence of these mobile electrons. If one part of a metal is heated, the electrons in that part acquire a large amount of kinetic energy. These electrons move rapidly through the crystal and pass on heat to other parts of the metal.

3. The metallic lustre can be explained as due to the presence of the highly mobile electrons. As a beam of light comprising of electromagnetic waves falls on the surface of a metal, *the electric field associated with light waves sets the electrons present on the surface of the metal into to and fro oscillations.* A moving charge always emits electromagnetic energy. Hence, oscillating electrons emit electromagnetic energy in the form of light. Thus, when light falls on a metal surface, it appears as if light is being reflected.

4. Malleability, ductility and softness. The metallic bond holding the positive ions (say, M^+) and the valency electrons is *non-directional.* In other words, the force of attraction between the M^+ ions and the valency electrons is uniform in all directions. Also the bonds holding the crystal lattice in metals are not rigid. The result is that M^+ ions can be easily moved from one lattice site to another.

The nearest neighbours can thus be changed easily and new metal bonds can be formed readily. This explains why metals are malleable, *i.e.,* they can be flattened out into thin sheets when hammered.

The ease with which the metal ions can be moved from one lattice site to another is also responsible for the fact that the metals are *ductile, i.e.,* they can be drawn into wires by very little expenditure of energy.

5. Metals have high tensile strength, *i.e., they can resist stretching without breaking.* This is due to the existence of strong electrostatic attraction between the positively charged metal ions and the 'sea' of negative electrons surrounding them.

6. Metals possess elasticity. *Elasticity is a property by virtue of which a substance can resist a deforming force and* a property by virtue of which *a substance can recover its original form soon after the removal of the deforming force.* The elasticity of metals is also due to the ease with which metal ions can move from one lattice site to another.

Q. 57. Discuss valence bond model of metallic bond. *(Andhra, 2010)*

Ans. In the crystal lattice of lithium atom, each Li atom in this lattice is surrounded by eight nearest Li atoms and six next to nearest Li atoms. With only one valence electron, each Li atom is capable of forming only one normal covalent bond with one of its nearest Li atoms. *This Li— Li covalent bond is assumed to resonate* **mainly** *between all the eight neighbours and* **to some extent** *between the six next to the nearest neighbours.* In this way, each Li atom would bind its 56 ($8 + 6 \times 8 = 56$) neighbours. The resonance of the Li – Li bonds which is similar to the resonance of the double bonds in benzene, stabilizes the metallic bond. The resonance of Li—Li bonds between a few Li atoms is shown below.

```
Li – Li  Li – Li          Li   Li – Li   Li
                           |              |
Li – Li  Li – Li    ↔     Li   Li – Li   Li
Li – Li  Li – Li          Li – Li   Li – Li
```

Greater stabilisation results from resonance of the following type :

```
                          ⊖            ⊕         ⊕            ⊖
Li – Li   Li – Li         Li   Li – Li Li        Li  Li – Li  Li
                          |            ⊖             ⊕        |
Li – Li   Li – Li    ↔    Li   Li– Li  Li   ↔   Li – Li – Li  Li
                                 ⊕     |                  ⊖
Li – Li   Li – Li         Li – Li – Li Li        Li – Li   Li Li
```

Although the valence bond model does meet some success in explaining some of the characteristics of metals, yet it is only qualitative in nature.

Q. 58. Explain band model (Molecular Orbital Approach) of metallic bonding.

Ans. When two atomic orbitals combine to form two molecular orbitals, the energy of one of the resulting molecular orbitals is higher and that of the other is lower than the energy of either of the combining atomic orbitals. The energies of the two molecular orbitals are represented in the form of two energy levels whose magnitudes vary with the internuclear distance between the two atoms, as shown in Fig. 2.24.

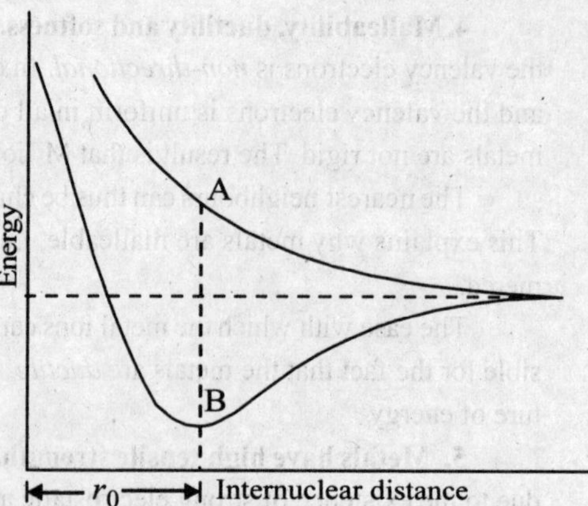

Fig. 2.24. Variation of the two molecular orbital energy levels with internuclear distance in the case of H_2 molecule. r_0 is the equilibrium internuclear distance.

The energy separation between the two molecular orbital energy levels at equilibrium internuclear distance r_0 is presented by *AB*.

If there are *n* atomic orbitals, one on each atom, in a metallic crystal, they may combine to give *n* molecular orbitals corresponding to *n* energy levels. It is essential, however, that the atomic orbitals have the same symmetry and are associated similar amount of energy. Thus, larger the number of atomic orbitals satisfying the above conditions, larger would be the number of molecular orbitals each of which can be represented by a separate energy level. The energy separation between any two adjacent energy levels would decrease with increase in the number of overlapping orbitals. Consequently, if we have an exceedingly large number of overlapping orbitals, the energy separation between any two adjacent energy levels would be exceedingly small so that we get an almost *continuous energy band,* as shown in Fig. 2.25

A similar situation exists in case of metals. A metal lattice has an extremely large number of atoms. The atomic orbitals on these atoms with same symmetry and same energy overlap resulting in an almost continuous energy band of extremely large number of energy levels of molecular orbitals. Each energy level in this band is capable of holding two electrons. The overlaping of *3s* and *3p* orbitals (valence orbitals) in the crystal lattice of **sodium metal** is shown in Fig. 2.26.

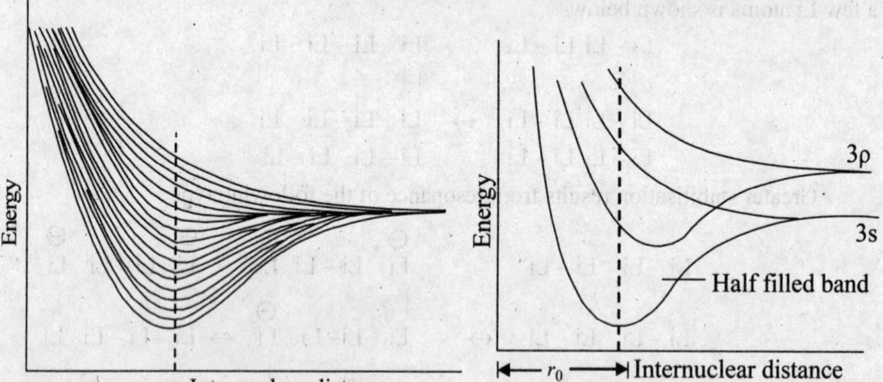

Fig. 2.25. An almost continuous energy band of energy levels of molecular orbitals obtained by overlapping of an exceedingly large number of atomic orbitals.

Fig. 2.26. The valence band of sodium metal. The dotted line indicates equilibrium internuclear distance.

Since sodium atom has only a single electron in the *3s* orbital, therefore, only half of the *3s* energy level band is occupied by electrons and even on slightest excitation, these electrons can move to the higher energy levels in this band. The mobility of these electrons makes the sodium metal a good conductor of electricity. A similar situation occurs in all the other alkali metals in which the valence energy level band is always half-filled.

The situation in the case of alkaline earth metals is, however, different. Here, the outermost valence orbital is fully occupied by electrons so that no vacant energy level of the energy band is available for the electrons to jump to. However, at the equilibrium internuclear distance, the 3*s energy band and* the 3*p energy band* overlap so that some of the vacant energy levels of 3*p band and* the filled energy levels of 3*s* band fall in the same energy range. Consequently, the 3*s* electrons can be easily excited to the 3*p energy* levels in this range. This makes the alkaline earth metals also good conductors of electricity.

In an **insulator,** there is a large energy gap between the lowest vacant energy band and the highest fully-filled energy band even at the equilibrium internuclear distance. As a result, the electrons cannot be easily excited. This makes the material non-conducting.

In **a semi-conductor,** there is a lesser energy gap between the two energy bands. Although the electrons are not very easily excited from the filled to the vacant energy bands yet these can be excited if an appropriate amount of energy is made available.

Q. 59. What are van der Waals' forces? Explain.

(*i*) Dipole-dipole interactions (*ii*) Ion-dipole interactions (*iii*) London forces

Ans. Forces of attraction exist between *neutral molecules* as well. The magnitude of these forces is maximum in case of solids and decreases on passing from solid to liquid and from liquid to gaseous states. The forces existing between the molecules of a substance are known as **intermolecular forces.** These are also designated as **van der Waals' forces.** The attractive forces which keep the molecules in position in a molecular crystal are van der Waals' forces.

The question arises, how can there be electrical attraction between neutral molecules lying close to one another ? The existence of electrical repulsion between such molecules, on the other hand, can be more readily understood as these may arise from the interaction of the electron clouds of the molecules when they are in close proximity to each other.

Thus, what could be the cause of **physical forces or molecular attraction?** The answer is as follows:

Dipole-Dipole Interaction. In the case of **polar molecules** which have permanent dipoles, the van der Waals' forces are mainly due to electrical interaction between the dipoles known as **dipole-dipole interaction.** For instance, gases such as ammonia, sulphur dioxide, hydrogen fluoride, hydrogen chloride, etc., have permanent dipoles, as a result of which there is appreciable dipole-dipole interaction between the molecules of these gases. The magnitude of this type of interaction depends upon the dipole moment of the molecule concerned. The greater the dipole moment, the greater is the dipole-dipole interaction. Because of the attractive interaction, these gases can be easily liquefied.

Ion-Dipole Interaction. If an ionic solid is to dissolve, some energy is needed to overcome the force of inter-ionic attraction. If water is the solvent, this energy is supplied by the *interaction of water molecules with the ions.* This is known as **ion-dipole interaction.** The positive pole of the water molecules reacts with the negative ion while the negative pole reacts with the positive ion. The ion-dipole interaction results in the release of a considerable amount of energy which becomes available for the separation of the oppositely charged ions. This results in the dissolution of the solute.

London Forces. But van der Waals' forces exist even in non-polar molecules such as F_2, O_2 and N_2 and even in non-polar mono-atomic molecules such as He, Ne, Ar, etc. This attraction is evident from the condensation of these gases into liquids at sufficiently high pressures and low temperatures. This is explained by London forces.

In 1930, F. London provided an explanation for the existence of forces of attraction between *non-polar* molecules, based on quantum mechanics. According to this view, *electrons of a neutral molecule keep on oscillating with respect to the nuclei of the atoms. As a result of this, at a given instant, positive charge may be concentrated in one region and negative charge in another region of the same molecule. Thus, a non-polar molecule may become momentarily self-polarised.* This polarised molecule may induce a dipole moment in a neighbouring molecule, as shown in Fig. 2.27.

The electrostatic forces of attraction between induced dipoles and the original dipoles (due to electron oscillation) are known as **London forces**. These forces are also called **dispersive forces** because the well known phenomenon of dispersion of light is also connected with these dipoles. For a pair of adjacent molecules, London forces vary inversely as the *seventh power* of the distance between them, *i.e.*, F $\propto 1/r^7$. The magnitude of these forces, evidently, would be very small.

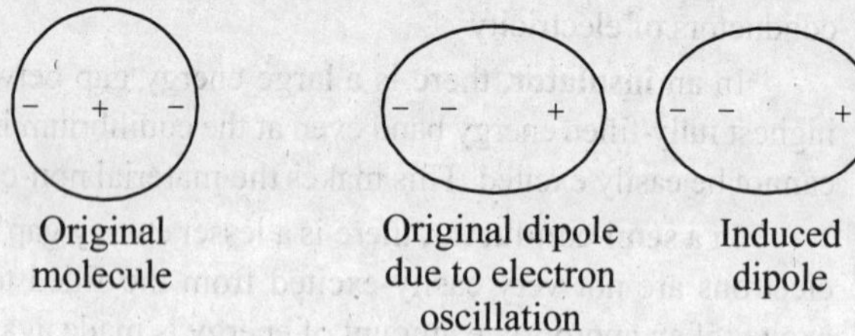

Fig. 2.27. Origin of cohesive forces in non-polar molecules.

The van der Waals' attraction in *non-polar molecules* is thus *exclusively* due to London forces. In the case of polar molecules having permanent dipoles, however, other electrostatic forces due to dipole-dipole interaction also contribute towards the van der Waals' attraction.

Factors determining van der Waals' forces. Van der Waals' forces depend largely on the following factors:

1. The Number of Electrons Present in a Molecule. Van der Waals' forces increase with increase in the number of electrons in the molecule. Thus, boiling points (T_b) of the halogens increase as we move from fluorine to chlorine, bromine and iodine. Similarly, boiling points of noble gases increase as we move from helium to radon, as shown below:

Noble gas	He	Ne	Ar	Kr	Xe	Rn
Boiling pt. (K)	4.1	27	67.2	121	165	211.2

2. Molecular Size. The larger the molecular size (or roughly the larger the molar mass), the stronger the van der Waals' forces. Thus, the boiling point of ethane, C_2H_6 (185 K) is higher than that of methane, CH_4 (112 K). In compounds having similar structures, viz. CH_4, SiH_4, GeH_4, the boiling points increase with increase in the number of electrons and size of the nolecule, as shown below:

Substance	CH_4	SiH_4	GeH_4
No. of electrons	10	18	36
Boiling point (K)	112	161	183

Q. 60. What is meant by stoichiometric crystals? Write notes on (*i*) Schottky defect (*ii*) Frenkel defect.

Ans. Stoichiometric crystals. Crystals in which the number of positive and negative ions are exactly in the ratios indicated by their chemical formulae are called stoichiometric crystals. For simplicity, we can consider the crystals of the type *AB* having equal number of A^+ and B^- ions.

1. Schottky Defect. This defect arises if *some of the lattice points are unoccupied.* The points which are unoccupied are called *lattice vacancies* or *'holes'*. The existence of two holes, one due to a missing positive ion and the other due to a missing negative ion in a crystal lattice, is illustrated in Fig. 2.28. It may be noted that the crystal, as a whole, remains neutral because the number of missing positive and negative ions is the same.

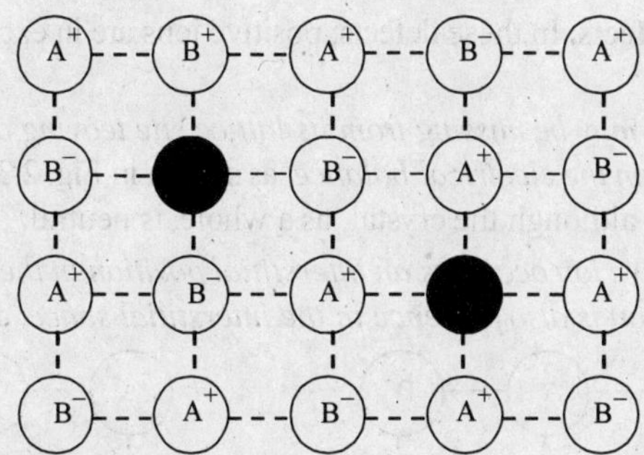

Fig. 2.28. The Schottky defect in crystals.

2. Frenkel Defect. This defect arises when *an ion occupies an interstitial position between the lattice points*. This is shown in Fig. 2.29. It is seen that one of the positive ions occupies a position in the interstitial space rather than at its own appropriate site in the lattice. A 'hole' is thus created in the lattice as shown. It may be noted again that the crystal remains neutral since the number of positive ions is the same as that of the negative ions. Positive ions being smaller are more likely to occupy interstitial positions than the negative *ions*.

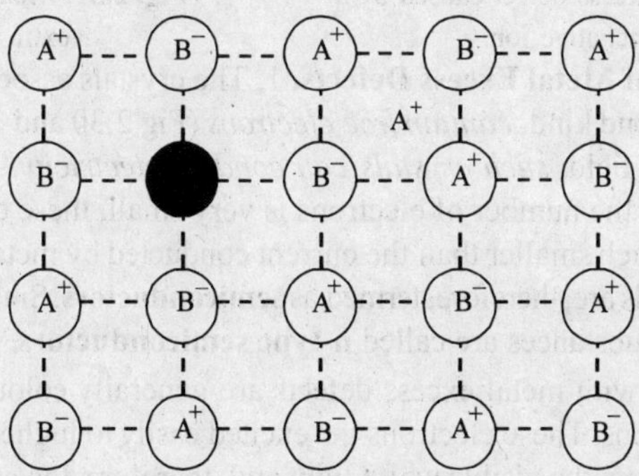

Fig. 2.29. The Frenkel defect in crystals.

In silver bromide, AgBr, some of the Ag^+ ions are generally missing from their regular positions and occupy positions in between the other ions in the lattice. The presence of Ag^+ ions in the interstitial spaces of the silver bromide crystal is responsible for the formation of a photographic image on exposure of silver bromide crystals (*i.e.* photographic plate) to light.

Lattice vacancies ('holes') occur in almost all types of ionic crystals. However, Schottky defect appears more often than Frenkel defect. The reason is that the *energy needed to form a Schottky defect is much less than that needed to form a Frenkel defect.*

Q. 61. What is meant by non-stoichiometric crystal? Write notes on (*i*) Metal excess defect (*ii*) Metal deficiency effect (*iii*) Thermal defects (*iv*) Impurity defect. (*Andhra, 2010; Pune, 2010*)

Ans. Non-stoichiometric crystals. These crystal do not obey strictly the law of constant proportions. The numbers of positive and negative ions present in such compounds are different from those expected from their ideal chemical formulae. In other words, there is excess of either positive charge or negative charge in the crystal. However, the crystal as a whole is neutral. If positive charge is in excess, it is balanced by the presence of *extra electrons*. If negative charge is in excess, it is balanced by the presence of *extra positive ions*. The crystal structure, therefore, becomes irregular and this implies the presence of defects. Defects in such crystals are called **non-stoichiometric defects.**

Metal Excess Defects. In these defects, positive ions are in excess. These defects may manifest in two ways.

1. *A negative ion may be missing from its lattice site leaving a 'hole' which is occupied by an extra electron to maintain the electrical balance,* as shown in Fig. 2.30. There is, evidently, *an excess of positive (metal) ions,* although the crystal, as a whole, is neutral.

2. *An extra positive ion occupies an interstitial position in the lattice and to maintain electrical neutrality, an electron is also presence in the interstitial space,* as shown in Fig. 2.31.

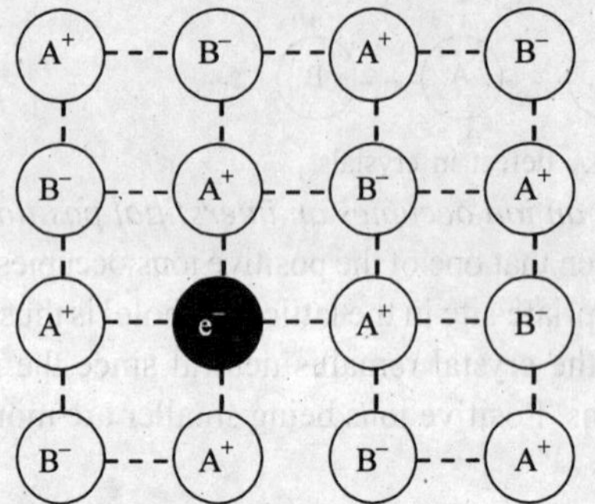

Fig. 2.30. Metal excess defect caused by a missing negative ion.

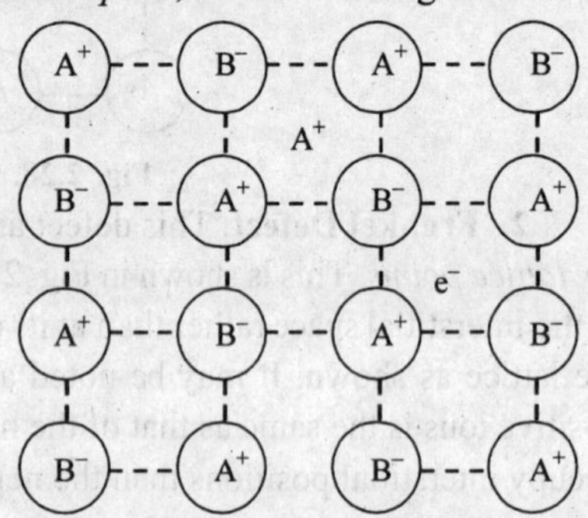

Fig. 2.31. Metal excess defect due to extra positive ion.

Consequences of Metal Excess Defects. 1. The crystals associated with metal excess defects of the first kind or second kind, *contain free electrons* (Fig 2.30 and 2.31). Since these electrons can migrate in an electric field, *such crystals can conduct electricity.* However, since the number of defects and, therefore, the number of electrons is very small, these crystals can conduct only small amounts of current, much smaller than the current conducted by metals, fused salts or salts dissolved in water. These materials are, therefore, termed as **semiconductors**. Since current, is carried by electrons in *normal way,* these substances are called **n-type semiconductors.**

2. The crystals with metal excess defects are generally coloured. This, again, is due to the presence of free electrons. These electrons get excited easily to higher energy levels by absorption of certain wavelengths from the visible white light and, therefore, the compounds appear coloured.

Metal Deficiency Defects. There are *two possible ways* in which metal deficiency defects can occur.

1. *One of the positive ions may be missing from its lattice site. The extra negative charge may be balanced by some nearby metal ion acquiring two charges instead of one.* This is represented in Fig. 2.32. *The metal,* obviously, *must be in a position to exhibit variable valency.* This type of defect, therefore, is generally found amongst the compounds of transition metals.

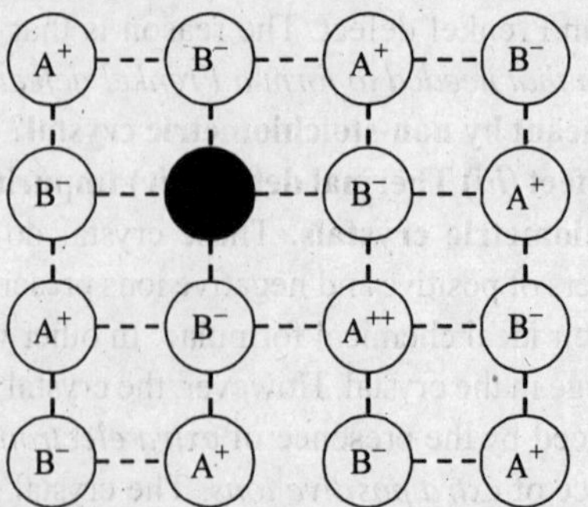

Fig. 2.32. Metal deficiency defect due to a missing positive ion.

2. An *extra negative ion may occupy an interstitial position. The extra negative charge is balanced again by means of an extra charge on one of the adjacent metal ions,* as shown in Fig. 2.33. *The metal* in this case also, *should be capable of exhibiting variable valency, i.e.,* it should be one of the transition metals. However, since the negative ions are usually very large, it is not expected that they would be able to fit into the interstitial positions. As a matter of fact, no crystal showing this type of metal deficiency defect is known.

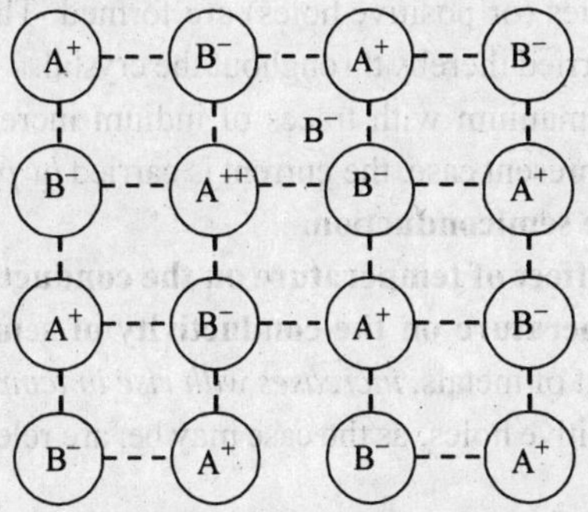

Fig. 2.33. Metal deficiency defect due to an extra negative ion.

Consequences of Metal Deficiency Defects. Crystals with metal deficiency defects can also be semiconductors. This property arises from the movement of an electron from one ion to another ion. In this way, the ion, say, A^+, changes into A^{++}. Thus, the movement of an electron from A^+ ion is *an apparent movement of A^{++} ion.* This is called *movement of positive 'hole'* and the substances showing this type of movement are known as ***p-type*** **semiconductors.**

Thermal Defects. Intrinsic Conduction. Atoms of both germanium and silicon have four electrons in the outer shell. Each atom, therefore, is covalently bonded with four neighbouring atoms through sp^3 hybrid bonds, giving rise to a highly stable tetrahedral structure. There are no free or conducting electrons. The electrical conductance, therefore, is very low. Now, suppose a sufficient amount of energy, is supplied to the crystal as a result of which *one of the covalent bonds gets broken and the electrons are set free.* The electrons, thus released, can migrate leaving behind a 'positive hole', at the site of the missing bond. *The crystal will now be able to conduct electricity* because, when electric field is applied, the electrons migrate in one direction and the 'positive holes' in the other. This type of conduction in known as **intrinsic conduction** because if happens without adding an external substance.

Impurity Defects. Certain defects in crystals happen due to the presence of chemical impurities. These are known as **impurity defects.** One addition of external metal impurities to germanium and silicon crystals, the latter act as **semiconductors.** Germanium and silicon belong to Group 14 of the periodic table. These elements, in pure state, have very low electrical conductivity. However, on adding even traces of an element belonging to Group 13 or 15, the electrical conductivity is greatly enhanced. This may be explained as follows:

Suppose, a Group 15 element, like arsenic, is added to a germanium crystal. As a germanium atom is substituted by an atom of asrenic, four of the electrons in arsenic form covalent bonds with the surrounding germanium atoms but the *fifth electron remains free.* In this way, an extra electron gets introduced into the crystal. Thus, germanium containing traces of arsenic (known as *arsenic-doped germanium)* begins to exhibit fairly high electrical conductivity. This type of conduction is known as **extrinsic conduction**. It is much greater than the intrinsic conduction discussed above. Since, in this type of conduction, current is carried by excess electrons in the *normal way,* it is called ***n*-type semiconduction.**

Now, suppose a Group III element, like indium, having only three electrons in the outer shell, is added in small traces to, say pure germanium. The atoms of indium, evidently, are not able to complete tetrahedral covalent structures because they have one electron short of the requirement. This gives rise to *electron-vacancies.*

The electron-vacant sites are known as 'positive holes' because the net charge at these sites is positive. When electric field is applied, adjacent electrons move into the positive holes and in this way other electron-vacancies (or positive holes) are formed. The migration of positive holes thus continues and current is carried thereby throughout the crystal.

Thus, doping of germanium with traces of indium increases the electrical conductivity of germanium crystal. In the present case, the current is carried *by positive holes,* therefore, this type of conduction is called ***p*-type semiconduction.**

Q. 62. What is the effect of temperature on the conductivity of semiconductors?

Ans. Effect of temperature on the conductivity of semiconductors. *The conductivity of semiconductors,* unlike that of metals, *increases with rise in temperature.* This is due to the fact that more extra electrons or positive holes, as the case may be, are released from the crystal lattice for the conduction of electricity.

Applications of semiconductors. The combination of *n* and *p-type* semiconductors (known as ***n-p* junctions**), finds interesting applications in the **fabrication of transistors**. This device can conduct electric current more easily in one particular direction than in the reverse direction and, therefore, can be used as **rectifier** for changing alternating current into direct current. The device is schematically shown in Fig. 2.34. The material on the left side of each junction is *n-type* conductor obtained by doping germanium with, say, arsenic. The minus signs represent extra electrons. The material on the right side of the junction is *p*-type conductor obtained by doping germanium with, say indium. The positive signs represent 'positive holes' arising from the deficiency of electrons at the indium impurity centres. When an external voltage is applied in such a way as to cause motion of electrons (*n*-current) from left to right and motion of positive holes (*p*-current) from right to left (Fig. 2.34), current is readily conducted. If direction of voltage is reversed so that there is separation of electrons and positive holes as shown in Fig. 2.34, there is cancellation of *n* and *p*-currents and hence the conduction stops. Thus, the *n-p* junction permits the current from an outside source to flow *in one direction only.*

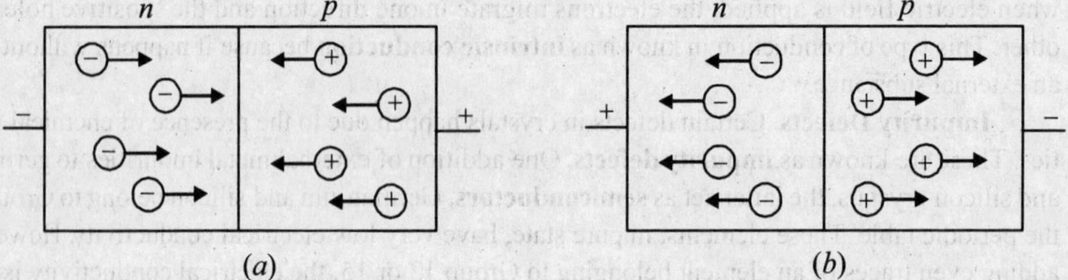

Fig. 2.34. Semiconductors acting as rectifier.

COVALENT BOND - I

Q. 1. What is valence bond theory of bonding? How does valence bond theory account for bonding in H_2 molecule? *(Lucknow, 2000)*

Ans. The main points of valence bond approach are as under:

(*i*) Atoms retain their individuality in the molecule.

(*ii*) The bond is formed due to the interaction of valence electrons as the atoms approach each other, *i.e.*, the inner atomic orbits from each atom forming the bond are undisturbed.

(*iii*) When they form a bond, only valence electrons from each bonded atom lose their identity.

(*iv*) The stability of the bond is explained by the fact that electrons forming the bond between two atoms undergo exchange between them.

This method is based on two **theorems** which are stated as:

(*i*) If Ψ_A and Ψ_B are the wave functions for two isolated independent atoms, with energies E_A and E_B, respectively, then the total wave function ψ of the system can be written as a product of the wave functions of the two atoms

$$\Psi = \Psi_A \Psi_B \quad ...(3.1)$$

and energy of the system by

$$E = E_A \pm E_B \quad ...(3.2)$$

(*ii*) If a system can be represented by a number of wave functions such as Ψ_1, Ψ_2, Ψ_3, ..., then the true wave function Ψ can be obtained by taking a linear combination of all these wave functions.

$$\Psi = N(C_1 \Psi_1 + C_2\Psi_2 + C_3\Psi_3 + ...) \quad ...(3.3)$$

where N is the normalization constant and C_1, C_2, C_3 ......, are the coefficients indicating the weight of each of the Ψ_n's. They are so adjusted as to give a state of lowest energy. The squares of the coefficients may the taken as a measure of the weights of each wave function to the total wave function.

FORMATION OF THE HYDROGEN MOLECULE

Let us now apply the concept of valence bond theory of the formation of hydrogen molecule from hydrogen atoms and consider the various forces of attraction and repulsions which may come into play.

For two hydrogen atoms which are infinitely apart when there is no interaction between them the wave function of system is:

$$\Psi = \Psi_a \Psi_b \quad ...(3.4)$$

where Ψ_a and Ψ_b represent the wave functions for two hydrogen atoms H_a and H_b respectively.

Now if we bring them close together, *i.e.*, distance close enough for interaction, the wave function for the system I, $H_a^1 . H_b^2$ can be written as :

$$\Psi_1 = \Psi_a(1) \cdot \Psi_b(2) \qquad ...(3.5)$$

where the hydrogen nuclei are represented as H_a and H_b and the electrons as 1 and 2 (Fig. 3.1).

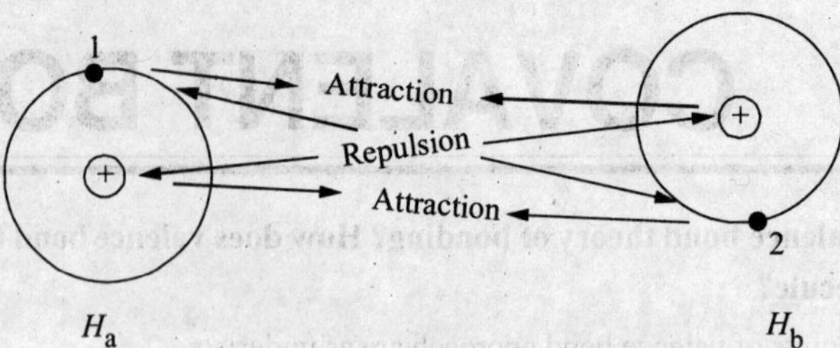

Fig. 3.1. Various interactions between two H atoms.

With the help of wave function Ψ_1, as given by eq. (3.5), the energy of the system consisting of two hydrogen atoms may be calculated as a function of internuclear distance r_{ab} by applying Schrodinger wave equation. The energy changes as a function of distance r_{ab} may be shown graphically in Fig. 3.2 (curve I).

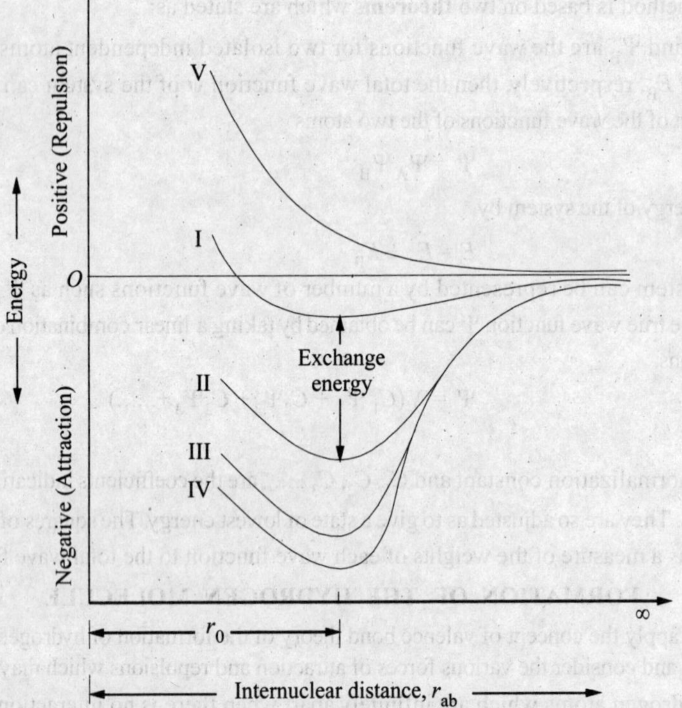

Fig. 3.2. Potential energy curves for hydrogen molecule.

It may be noted that the total energy of the two isolated hydrogen atoms is taken as zero. As the two atoms approach one another [without exchange of electrons, eq. (3.5)], the electron 1 of H_a is attracted by the nucleus of H_b and electron 2 of H_b is similarly attracted to the nucleus of H_2. As a result there is weak attraction and consequently energy of the system starts decreasing. With decrease in distance between the two atoms, forces of attraction increase rapidly. Ultimately an equilibrium internuclear distance (r_0) is reached where the system has lowest energy. At distances closer than this, stronger repulsion due to close proximity between the two nuclei sets in and consequently the energy of the system becomes positive (curve I). The minima in curve I represents that a molecule is formed but the bond energy represented by wave function Ψ_1 (about 25.2 kJ/mole) compares very poorly with the experimental value of 431.4 kJ/mole (curve IV). One may thus conclude that the wave function Ψ_1 representing structure I is not correct. Certain improvements are therefore necessary to match the theoretical and observed values.

***(a)* First Improvement : Exchange Energy.** A careful examination of the wave function Ψ_1 reveals that electron 1 is associated with atom H_a and electron 2 is associated with atom H_b. When a molecule is formed the two atoms come close together, their atomic orbitals overlap and we are no longer definite that the atoms retain their individual electrons. We cannot trace any one electron in a system according to uncertainty principle. It clearly states that we cannot exactly locate an electron but only the probability of finding the electron is given. Hence the system of two hydrogen atoms may as well be represented by structure II.

II $\quad H_a^{\,2} \cdot H_b^{\,1}$

The wave function for this may be represented as

$$\Psi_{\text{II}} = \Psi_a(2) \cdot \Psi_b(1)$$

So if both wave functions Ψ_{I} and Ψ_{II} are equally probable representations then the real wave function is likely to be a linear combination of these two wave functions.

$$\Psi = C_1\,\Psi_{\text{I}} + C_2\,\Psi_{\text{II}}$$

or

$$\Psi = C_1\,\Psi_a(1)\,{}_b(2) + C_2\Psi_a(2)\,\Psi_b(1) \qquad ...(3.6)$$

Since structures I and II are equally probable, the contribution of wave functions, Ψ_{I} and Ψ_{II} to the true wave function must be equal. As stated earlier in theorem II, the weight of each wave function is proportional to the squares of its coefficients. So we have

$$C_1^{\,2} = C_2^{\,2} \text{ or } C_1 = \pm\, C_2$$

and if for simplicity we assume $C_1 = 1$, then $C_2 = \pm 1$.

Equation (3.6) may be represented by two equations

$$\Psi_+ = \Psi_a\,(1)\,\Psi_b\,(2) + \Psi_a\,(2)\,\Psi_b(1) \qquad ...(3.7)$$

$$\Psi_- = \Psi_a\,(1)\,\Psi_b(2) - \Psi_a\,(2)\,\Psi_b\,(1) \qquad ...(3.8)$$

Equation (3.7) represented by Ψ_+ is a symmetric combination as there is no change of sign on exchange of electrons. The energy of the system calculated as a function of internuclear distance, r_{ab} by using wave function Ψ_+ results in a significant lowering of energy and is represented by curve II. The bonding energy is shown in curve II and is approximately 302.4 kJ/mole and compares well with the experimental curve IV. The additional lowering in energy *i.e.*, 302.4 – 25.2 = 277.2 kJ/mole has been obtained by virtue of exchange of electrons between the two hydrogen atoms and is known as **exchange energy.** If we plot the energy of the system using Ψ_- eq. (3.8) as function of distance r_{ab} we obtain curve V which shows that the energy curve represented by Ψ_- is an antisymmetric combination since there is a change of sign on electron interchange. In this case the two atoms approach each other with their electrons having parallel spins, Fig. 3.3 (*b*). It is clear that Ψ_+ is the wave function which represents the H_2 molecule, *i.e.*, electrons approach each other with opposite spins, Fig. 3.3 (*a*). Ψ_+ is also referred to as Ψ_{cov} as each nuclei is always in possession of one electron at all times.

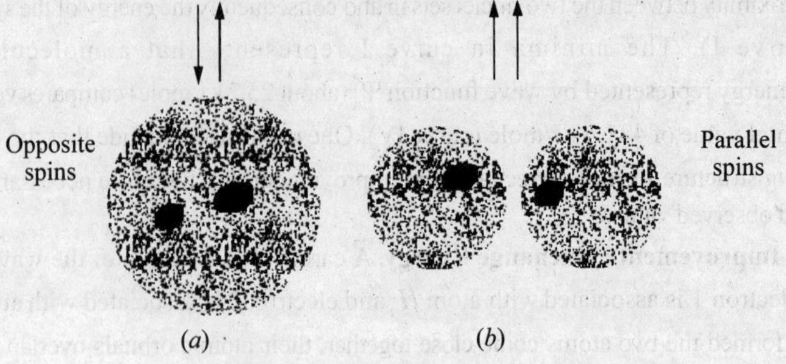

Fig. 3.3. Electrons approaching each other.

***(b)* Second Improvement : Shielding Effect.** The wave function can be improved by considering the screening effect of electrons. Since electrons shield one another by varying extents from the nuclear charge, so we must use, Z_{eff} (effective nuclear charge) rather than *Z*. When Z_{eff} is used, bonding energy of 339.4 kJ/mole is obtained.

(c) Third Improvement : Ionicity. By the use of wave function Ψ_+ (or Ψ_{cov}) in eq. (3.7), we have made a significant improvement in the choice of our wave function. But so far we have assumed that one electron is always associated with one atom. Such a situation may not always be true. There is the possibility that both electrons 1 and 2 are either on atom H_a or H_b. The wave function describing such a state would be a combination of eqs. 3.9 and 3.10.

III $\quad H_a^{1,2}$.........H_b

Its wave function may be written as

$$\Psi_{III} = \Psi_a(1)\,\Psi_a(2) \qquad ...(3.9)$$

and IV $\quad H_b^{1,2}$ ....H_a

is represented as

$$\Psi_{IV} = T_b(1)\,\Psi_b(2) \qquad ...(3.10)$$

The resulting wave function representing both electrons on one nucleus may be summed as

$$\Psi_{ion} = \Psi_a(1)\Psi_a(2) + \Psi_b(1)\,\Psi_b(2) \qquad ...(3.11)$$

and is called Ψ_{ion}. Since it represents a state in which ionic forms also contribute to the structure. If we take wave function Ψ_{cov} and Ψ_{ion} together then the real wave function may be written as

$$\Psi = \Psi_{cov} + \lambda\,\Psi_{ion}$$

or

$$\Psi = \Psi_a(1)\,\Psi_b(2) + \Psi_a(2)\,\Psi_b(1) + \lambda[\Psi_a(1)\,\Psi_a(2) + \Psi_b(1)\,\Psi_b(2)] \qquad ...(3.12)$$

where λ represents the mixing coefficient. It tells us how much Ψ_{ion} wave function is to be introduced in the real wave function Ψ. By applying the variation method, we can adjust the value of λ so as to obtain the lowest energy. This way we can measure the extent of ionic character in the covalent bond. For H_2, the value of λ is about 0.17, and the ionic contribution to the bonding energy is about 23.0 kJ/mole.

If we now calculate the energy of the system using wave function as given by eq. (3.12) we get a new energy wave which is marked III. The energy associated with curve III is 362.4 kJ/mol and is an improvement over our earlier results and is quite close to the experimental value.

This treatment which we have discussed for the formation of H_2 is essentially the pairing of two electrons, one from each bonded atom. The valence bond method is also referred to as method of electron pairing. It is obvious even from eq. (3.7) that the atoms must approach one another with the spins of their electrons in opposite collection. It follows that the formation of covalent bond is explained by the lowering of the potential energy of the electrons, resulting from the increased electron density in the internuclear space.

Q. 2. Define the terms resonance and resonance energy.

(G. Nanakdev, 2000; Garhwal, 2000; Nagpur, 2000; Calcutta, 2000; Awadh, 2000; Indore, 2001; M. Dayanand 2002; Vidyasagar, 2007)

Ans. Sometimes it is not possible to describe the properties of a substance on the basis of a single structure. We explain it by saying that the substance exists in a number of structures called **resonating** structures. The actual substance is then hybrid of these resonating structures. This phenomenon is called resonance. Each resonating structure makes a definite contribution, big or small, to the actual structure of the substance. This is explained with the help of resonating structures I, II, III of H_2.

$$\underset{\text{I}}{\underset{\text{(covalent)}}{H:H}} \leftrightarrow \underset{\text{II}}{\underset{\text{(ionic)}}{H^+ \quad :H^-}} \leftrightarrow \underset{\text{III}}{\underset{\text{(ionic)}}{H\overline{:} \quad H^+}}$$

These structures can be represented by wave functions defined by eqs. 3.7, 3.9, and 3.10. The physical significance of eq. (3.12) is that hydrogen molecule cannot be represented either by pure covalent structure eq. (3.7) or by the ionic structures eqs. (3.9 and 3.10). The double pointing arrows ($\leftrightarrow$) between these structures indicate that they are in resonance with one another. However it is not to be inferred that H_2 exhibits all of these three structures, now can it be said that hydrogen exists for a certain fraction of time in one form and for another fraction of time in some other form. The molecules have single structures which cannot be described in terms of electron dot diagram on the paper. The difficulty lies in the description and not in the molecule. The real structure is a resonance hybrid of these three structures. A resonance hybrid can be equated to a mule which exhibits the characteristics of both the donkey and the horse, but at no moment it is either of these. The resonance hybrid is more stable than any one of the contributing structures. This extra stability of the resonance hybrid as compared with the most stable of the resonating structures is known as resonance energy. It is the greatest when all the resonating structures are of the same energy.

Resonance energy = Actual bond energy — Energy of the pure covalent bond

It may be pointed out that if a molecule or ions cannot be represented by a single structure, then the real structure is expressed in terms of resonance hybrid of all possible structures.

Q. 3. What are the necessary conditions for writing resonating structures?

(Rajasthan, 1997; Nagpur, 2000)

Ans. The necessary conditions for writing resonating structures are as follows :

(*i*) The nuclei of the atoms is each structure must remain in fixed positions. Only the arrangement of electrons may change. The bond angles and bond lengths should be closer to the real structure. Only then the contribution of that structure to the hybrid will be significant. For example, the Dewar structures of benzene are less important than the Kekule structures because in Dewar structures there is one very long bond across the ring.

Kekule forms　　　　Dewar forms

The following two structures of phosphorus acid (H_3PO_3) are not resonance structures because of the displacement in the position of atoms.

$$\mathrm{H\!:\!\ddot{O}\!:\!\overset{H}{\underset{\underset{\textstyle :\ddot{O}:}{\textstyle ::}}{P}}\!:\!\ddot{O}\!:\!H} \rightleftharpoons \mathrm{H\!:\!\ddot{O}\!:\!\ddot{P}\!:\!\ddot{O}\!:\!H}\ \text{(with } :\ddot{O}:H \text{ on P)}$$

The two structures represent two different chemical species (tautomers) which are in equilibrium, but are not resonance structures.

(*ii*) Each structure must have the same number of unpaired electrons. We should not draw structures in which one form contains a single unpaired electron and another contains two unpaired electrons. For example:

$$:\dot{N}=\ddot{O}: \longleftrightarrow \ :\ddot{N}=\dot{O}: \quad \text{and not} \quad :\dot{N}=\dot{O}:$$

(*iii*) The contributing structures should all have similar energies. This rule may be illustrated by taking the example of benzene. The Dewar structures have higher energy and are, therefore, less important than the more stable Kekule forms in the combination that represents the true structure of benzene.

(*iv*) Those structures in which negative charge resides on the most electronegative atoms and positive charge on the most electropositive atom will be closest to the true structure. For example, we may write the following resonating structures of HC1.

$$\underset{\text{(covalent)}\ \mathrm{I}}{\mathrm{H:\ddot{C}l:}} \longleftrightarrow \underset{\text{(ionic)}\ \mathrm{II}}{\mathrm{H^{+}\ :\ddot{C}l:^{-}}} \longleftrightarrow \underset{\text{(ionic)}\ \mathrm{III}}{\mathrm{\bar{H}:\ \ddot{C}l^{+}}}$$

There is significant contribution from structure II (because of greater electronegative of chlorine). However, any contribution by structure III is not possible.

(*v*) The contributing structures should be so written that unlike charges reside on neighbouring atoms. Consider the resonance structures of HNO_3. The last structure does not contribute because of +ve charge on adjacent N atoms and double –ve charge on the third nitrogen atom.

$$\mathrm{H-\dot{N}=\overset{+}{N}=\bar{N}:}$$

$$\mathrm{H-\bar{N}-\overset{+}{N}\equiv N:}$$

$$\mathrm{H-\overset{+}{N}\equiv\overset{+}{N}-\ddot{N}:^{2-}}$$

(*vi*) The greater the number of covalent bonds, greater is the contribution of the resonating structures. For example, the double bonded structures of BF_3 are important because of the formation of an additional covalent bond.

$$\overset{(+)}{F} = \overset{(-)}{B}\begin{matrix} F \\ \\ F \end{matrix} \longleftrightarrow F-\overset{(-)}{B}\begin{matrix} \overset{(+)}{F} \\ \\ F \end{matrix} \longleftrightarrow F-\overset{(-)}{B}\begin{matrix} F \\ \\ \underset{(+)}{F} \end{matrix}$$

Q. 4. Write the resonating structures for CO and explain the cause of its low dipole moment.

(Rajasthan, 1997; Kurukshetra, 2002)

Ans. Following resonance structures can be written for this molecule.

$$:C::\ddot{O}: \longleftrightarrow :\overset{-}{C}:::\overset{+}{O}: \longleftrightarrow :\overset{+}{C}:\overset{-}{\underset{..}{\ddot{O}}}:$$

(covalent) I (ionic) II (ionic) III

***(i)* Bond length.** The observed bond length between carbon and oxygen in CO is 113 pm. The bond length of a normal carbon to oxygen double bond as in ketone is 122 pm and that observed for carbon to oxygen triple bond is 110 pm. It is, therefore, obvious that the bond length in carbon monoxide is shorter that that of for a carbon to oxygen double bond but slightly longer than that for a triple bond. However, bond length in carbon monoxide is closer to carbon oxygen triple bond distance. Bond length measurements suggest that contribution from resonance structure II dominates over other structures, I and III.

***(ii)* Dipole moment.** On the basis of bond length, we may say that structure II makes the maximum contribution. However, we may note that this structure suggest a partial positive charge on oxygen and partial negative change on carbon. If the electronegativities of carbon and oxygen were the same, carbon monoxide would have an appreciable dipole moment in the direction shown.

$$\underset{\longleftarrow}{:C:::O:}$$

But since oxygen is more electronegative than carbon it, therefore, draws electron density back from carbon to the oxygen atom. This force works in the opposite direction and hence carbon monoxide has very low dipole moment (0.12 D).

Q. 5. Draw resonating structures for carbon dioxide.

(G.Nanakdev, 2000., M. Dayanand, 2000; Kerala, 2000)

Ans. It may be represented as a resonance hybrid of the following three structures :

***(i)* Bond length.** On the basis of structure I the bond length in CO_2 should be equal to carbon-

$$:\ddot{O}=C=\ddot{O}: \longleftrightarrow :\overset{-}{\underset{..}{\ddot{O}}}-C\equiv\overset{+}{O}: \longleftrightarrow :\overset{+}{O}\equiv C-\overset{-}{\underset{..}{\ddot{O}}}:$$

(covalent) I (ionic) II (ionic) III

oxygen double bond distance of 122 pm. The observed bond distance is shorter than this value and is 115 pm. Therefore, there must be some contribution from structures II and III.

***(ii)* Dipole moment.** Contributions from structures II and III must be equal. If their contributions are unequal, the molecule will have dipole moment. But CO_2 has zero dipole moment.

***(iii)* Bond energy.** The energy of a carbon - oxygen double bond is 731.5 kJ/mole. Therefore, heat of formation of the molecule O = C = O (structure I) should be about 1463.0 kJ/mole. The experimental value of heat of formation is 1601 kJ/mole. The difference between these two values, *i.e.*, 138 kJ/mole suggests the extra stability acquired due to contribution by structures II and III. It may be noted here that structure I is most dominant form due to its symmetrical nature.

Q. 6. Write down the resonance structures of N_2O. *(Delhi, 2002)*

Ans. The molecule is isoelectronic with CO_2 and hence has the same linear structure. It can be

$$\overset{-}{:\ddot{N}}=\overset{+}{N}=\ddot{O}: \longleftrightarrow :N\equiv\overset{+}{N}-\overset{-}{\ddot{\underset{..}{O}}}:$$

(I) (II)

represented as resonance hybrid of two resonance structures I and II.

Each of the structures I and II separately predicts a sizeable dipole moment for the molecule N_2O. The low value (0.12 D) is explained on the cancelling effect of dipole moments in opposite directions in structures I and II.

Q. 7. Write down the resonance structure of carbonate ion.

(G.Nanakdev., 2000; Garhwal, 2000)

Ans. The Lewis dot representation of carbonate ion contains a double bond and two single bonds between central C and oxygen atoms as shown below. This structure indicates that one of the carbon oxygen bonds (C = O) should be shorter than the other two bonds. Also the angle α should be greater than the angle β. But this is contrary to observations.

X-ray studies show that all the three bonds have the same bond length and each angle is 120°.

I II III

IV

This has been explained by a combination of resonance structures I, II and III for carbon dioxide.

Thus carbonate ion may be represented by structure IV which is combination of these three structures.

Q. 8. Write down the resonance structure of

(*i*) Cl_2O

(*ii*) N_3^-

(*iii*) HNO_2

(*iv*) SCN^-

(*v*) NO_2^-

(Calcutta 2000; M.Dayanand, 2002)

Ans.

(*i*) Cl_2O

:O: (Cl—O—Cl) ⟷ $\overset{+}{O}$ (Cl—O=Cl⁻) ⟷ $\overset{+}{O}$ (⁻Cl=O—Cl)

(*ii*) N_3^-

The three structures are equally important and result in the same bond order (≅ 2) for each bond in N_3^-.

$\overset{(-)}{N}=\overset{(+)}{N}=\overset{(-)}{N}$ ⟷ $\overset{(2-)}{N}-\overset{(+)}{N}\equiv N$ ⟷ $N\equiv\overset{(+)}{N}-\overset{(2-)}{N}$

(*iii*) HNO_2

H—O—N=O ⟷ H—$\overset{+}{O}$=N—O^-

(*iv*) SCN^-

$\overset{-}{S}\equiv C-N$ ⟷ $S=C=\overset{-}{N}$

S — C bond order is between 1.5 and 2 and C—N bond order is between 2 and 3.

(*v*) NO_2^-

O=N—O^- ⟷ $^-$O—N=O

Q. 9. What are σ (sigma) and π (pi) bonds ?

(*Kerala, 2000; Nagpur, 2000; Bangalore, 2005*)

Ans. σ-covalent bond. Sigma bond is formed by end-to-end overlap of any one of these combinations of atomic orbitals. This type of overlapping results in a strong covalent bond. In these cases, bond is formed by electron cloud having maximum density on the line connecting the centres of atoms.

(*a*) *s* — *s* overlap (*b*) *s* — p_z overlap (*c*) p_z—p_z overlap

Bond formation between two hydrogen atoms to give H_2 results due to overlap of two 1*s* atomic orbitals. The covalent bond formed by the *s*-orbital electron cloud below is called a σ-bond.

1*s* 2*s* End to end overlap or H —— H σ-covalent bond

Formation of H_2 molecule.

The formation of the HF may be visualized as an example of the overlap between 1*s* and 2p_z atomic orbitals as shown below.

Formation of HF.

The formation of F_2 molecule results by the overlap of $2p_z$ atomic orbitals of fluorine atoms. This is shown in the figure below.

Formation of F_2

π bond. π bond is formed by sidewise or lateral overlap between two *p*-orbitals which have their lobes perpendicular to the molecular axis.

The sidewise overlap is not efficient and results in a weak bond. Sigma bond is always stronger than a pi-bond.

Q. 10. Write a note on three-electron (odd-electron) bond.

(*M. Dayanand, 2002; Kurukshetra, 2002; Lucknow, 2002; Nagpur, 2003; Rohilkhand, 2003*)

Ans. Pauling in 1931 explained stability of all odd molecules such as NO (nitric oxide) having odd no. of electrons 5 + 6 = 11 to existence of three-electron bond between pairs of atoms. He suggested that there are only two different arrangements (I & II) of accommodating three electrons into the two orbitals.

A: ·B ------ (I)

A· :B ------ (II)

He suggested that calculation of energy for structure I alone shows repulsion between the two atoms

(+ve potential energy) and so structure (I) does not correspond to the formation of a stable bond. Similarly, structure (II) alone also leads to an unstable arrangement. Pauling suggested that if two atoms A and B are identical or are very similar, so that the two structures I and II have very close energies, then resonance may occur between them. This resonance energy will lead to attraction between two atoms (and lowering in energy) and will lead to the formation of a stable three-electron bond between the two atoms. The resonance between structures I & II may be represented as

$$A\!:\ .B \rightleftharpoons A.\ :\!B$$

This type of a three-electron bond is represented as A$\cdot\cdot\cdot$B. It is shown by Pauling that the three-electron bond is about one-half as strong as an electron pair bond.

Condition for the formation of a three-electron bond

Pauling suggested the following conditions for the existence of a three-electron bond between atoms A and B.

(*i*) Atoms A and B must be identical or very similar.

$$A\!:\quad \cdot B \quad \text{------} \quad (I)$$
$$A\cdot\quad :\!B \quad \text{------} \quad (II)$$

(*ii*) Structures (I) and (II) between atoms A and B must have nearly the same energy, so that resonance stabilization is provided to the system.

(*iii*) Three-electron bond is formed between two different atoms A and B if the electronegativity difference between A and B does not exceed by 0.5. Thus, three-electron bonds have been found to exist between oxygen and fluorine, nitrogen and oxygen and chlorine and oxygen.

Some examples of molecules containing three-electron bond:

(*i*) Nitric oxide is the most stable of various odd molecules. The valence bond representation of NO is given below. Structure (II) is expected to be less stable than structure (I) because of distribution of unfavourable charges, *i.e.*, we are placing positive charge on electronegative oxygen atom.

$$:\!N = \dot{\ddot{O}}\!: \text{------} \quad (I)$$
$$:\!\overset{-}{\ddot{N}} = \overset{+}{\dot{O}}\!: \text{------} \quad (II)$$

However, structures I and (II) are nearly of equal stability and thus there is complete resonance between the two structures.

$$:\!\dot{N} = \ddot{O}\!: \longleftrightarrow :\!\overset{-\,\cdot\cdot}{N} = \overset{+}{\dot{O}}\!:$$

The molecule can thus be assigned the following structure consisting of a double bond and

$$:\!N \overset{\cdots}{=} O\!:$$

a three-electron bond. This structure is more stable than either of the two structures I and II. Thus molecule is described as having $2\frac{1}{2}$ bond. It is expected to show bond length between that for a double and a triple bond. The experimental value for bond length for NO is 115 pm, somewhat larger than the triple bond value of 106 pm for NO^+.

(*ii*) Nitrogen dioxide (NO_2). It may be assigned the following resonance structures.

$$\text{:}\ddot{O}{=}\dot{N}{-}\ddot{O}\text{:} \longleftrightarrow \text{:}\ddot{O}{-}\ddot{N}{=}\ddot{O}\text{:}$$

The NO bond distance was predicted to be 118 pm which agrees well with the experimentally determined value of 119 pm by infrared spectroscopy.

Q. 11. What is meant by bond energy ?

(*Meerut, 2000; Awadh, 2000; M. Dayanand, 2002; Madras,2004; Bangalore, 2005; Nagpur, 2005*)

Ans. For a diatomic molecule, say AB, the bond dissociation energy is defined as the energy needed (absorbed) to break the molecule AB into its atoms A & B. The process of bond rupture in a molecule is called **dissociation.** Therefore, the bond energy of a diatomic molecule which is identical with bond dissociation energy (D) may also be defined as the enthalpy change for dissociation of gaseous molecule at 25°C into gaseous atoms at the same temperature. As for example,

$$H_2\,(g) \longrightarrow 2H\,(g) \qquad D\,(H-H) = \Delta H = 435.9 \text{ kJ / mole}^{-1}.$$

It is observed that there is a definite trend in bond energy among successive numbers in a group. The heavier analogs generally show lower values of dissociation energy (F_2 is an exception). As far as range of dissociation energies is concerned, the diatomic alkali metals have low values and hydrogen halides have comparatively large values of dissociation energy. There is no trend along a period. It is found that for similar molecules the dissociation energy decreases as bond order decreases. In other words, for related molecules, the bond dissociation energy decreases as bond length increases. Bond dissociation energy of F_2 is exceptionally low because of high repulsion between the non-bonding electrons on the similar fluorine atoms (other halogens have longer bonds).

For polyatomic molecules, the bond dissociation energy and bond energy terms are not identical and should not be used synonymously. Bond dissociation energy (D) of a polyatomic molecule AB_n is defined as the energy required to break just one bond to form atom B and a group of atoms (radical) AB_{n-1}. Bond dissociation energies for a given bond are different for different structural and electronic arrangements in the radical.

Although the bond dissociation energy of a given bond is not independent of the environments but the variation is not very large. The bond dissociation energies may not change significantly with change in environments around a given bond. Thus, it should be possible to characterise a bond by an average bond energy (E) which is defined as the approximate energy required for breaking of the bond in any gaseous molecule. It does not include energies necessary for vaporization or sublimation.

The distinction between bond dissociation energy (D) and average bond energy (E) shall be clear from the following examples showing stepwise dissociation of gaseous water.

$$H_2O\,(g) \longrightarrow H\,(g) + OH\,(g) \quad D_1\,(H-OH) = 500 \text{ kJ}$$

$$OH\,(g) \longrightarrow H\,(g) + O\,(g) \quad D_2\,(H-O) = 424 \text{ kJ}$$

The average bond energy (E) for O – H bond is the mean of the two separate bond dissociation energies $(D_1 + D_2) / 2$, that is, (500 + 424)/2 = 462 kJ. The enthalpy change for the formation of water molecule by the reaction is –926 kJ mol^{-1}.

$$2H\,(g) + O\,(g) \longrightarrow H_2O\,(g)$$

It is negative because heat is evolved in the formation of gaseous water. As expected, it is almost twice the average bond energy (E = 462) for an O–H bond and hence in close agreement with the presence of two O–H bonds in H_2O. However, the average bond energy (E) is quite different from the bond dissociation energy D_1 or D_2.

Q. 12. Explain: The bond energy of Li_2 (113 kJ mol^{-1}) is less than that of H_2 (431 kJ mol^{-1}) though both contain a σ-bond.

Ans. The bond energy of a bond depends upon the bond length. Greater the bond length, smaller the bond energy. Li—Li bond is larger than H—H bond. Hence bond energy of Li_2 is smaller than that of H_2.

Q. 13. Write explanatory notes on:

(1) Hydrogen bond **(2) Intermolecular hydrogen bonding**

(3) Intramolecular hydrogen bonding

(*Kanpur, 2001; M.Dayanand, 2002; Kurukshetra, 2002; Patna, 2010*)

Ans. (1) Hydrogen Bond. When hydrogen is joined to a highly electronegative element, it can form an additional bond to another electronegative element. This additional bond is known as hydrogen bond. For example, in the compound KHF_2, the anion HF_2^- is believed to have structure $(FHF)^-$ in which hydrogen acts as bridge between the two fluorine atoms. The hydrogen bond is considerably less stable than the usual valence bond. In order to distinguish a hydrogen bond from other valence bond, it is written as dotted line so that HF_2^- ion becomes (F.......H— F)$^-$. In general, when a hydrogen bond links two electronegative atoms A and B as a resonance hybrid between A – H.....B and A.....H – B. The more electronegative the atom to which hydrogen is joined , and the smaller it is, the stronger is the hydrogen bond. Hydrogen bonds are formed only between highly electronegative atoms like F, O and N. The energy of the hydrógen bond varies depending on the state and the nature of the species to which hydrogen is joined. The energy of the hydrogen bond H ..... F is about 10 kcal while the energies of the O.....H and N.....H bonds are 5 and 2 kcal respectively. Evidence in support of the existence of hydrogen bonding is obtained, if we compare some of the physical properties of compounds containing hydrogen bonds.

(2) Intermolecular Hydrogen Bond. We find that the boiling points of HF and H_2O are abnormally high as compared with other members of each series (Fig. 3.4). In the series HF, HCl, HBr and HI boiling points are expected to rise due to increasing number of electrons (because of increased van der Waals forces). The unexpectedly high boiling point of H_2O in the series H_2O, H_2S, H_2Se and H_2Te is also attributed to hydrogen boiling.

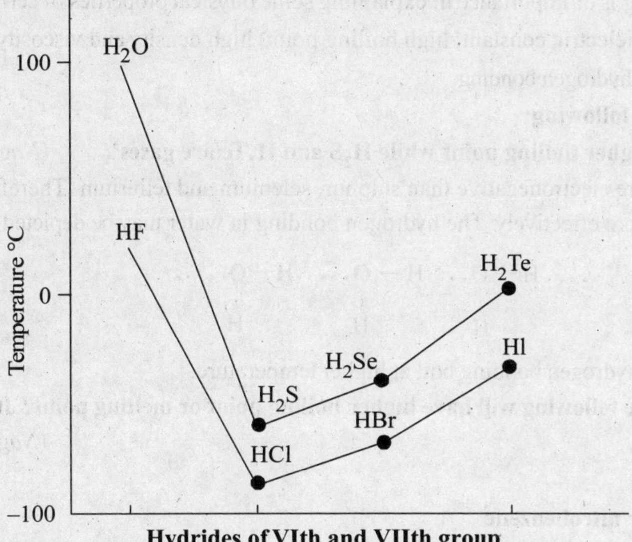

Fig. 3.4. Comparison of boiling points of hydrides of 6th and 7th group elements.

Hydrogen fluoride consists of $(HF)_n$ molecules which have the following structure.

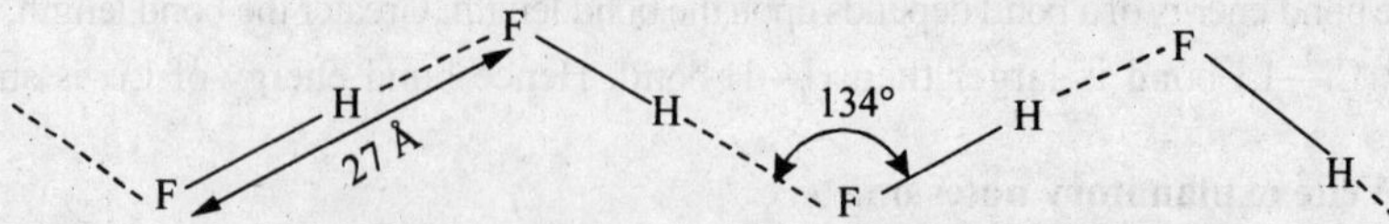

(3) **Intramolecular Hydrogen Bonding.** In many organic compounds, often the hydrogen bond joins part of a single molecule (*i.e.*, within the molecule) and in such cases intramolecular hydrogen bonding is favoured. Substances which are intramolecularly hydrogen bonded are more soluble. In nonpolar solvents (such as $CC1_4$, C_6H_6 etc.) and have lower melting and boiling points than their other isomers which are intermolecularly hydrogen bonded.

o-nitrophenol *o*-hydroxybenzaldehyde

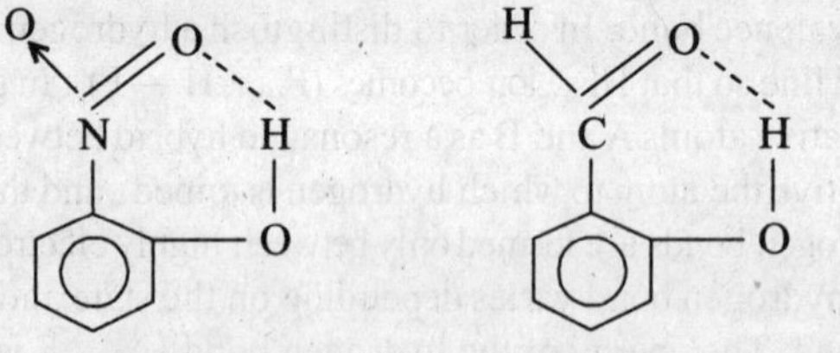

Fig. 3.5

The hydrogen bond plays important part in many processes. For example, the solubility of an uncharged organic compound in water depends mainly on the ability of a substance to form hydrogen bonds with the solvent. Thus dimethyl ether is completely miscible with water whereas dimethyl sulphide has slight solubility; sugar, an organic compound, is water soluble due to hydrogen bonding with water. Hydrogen bonding is of importance in explaining some physical properties of certain liquids. For example, the high dielectric constant, high boiling point, high density and viscosity of water is explained on the basis of hydrogen bonding.

Q. 14. Explain the following:

'H_2O has a higher boiling point while H_2S and H_2Te are gases'. (*Nagpur, 2008*)

Ans. Oxygen is more electronegative than sulphur, selenium and tellurium. Therefore oxygen forms hydrogen bond more effectively. The hydrogen bonding in water may be depicted as follows.

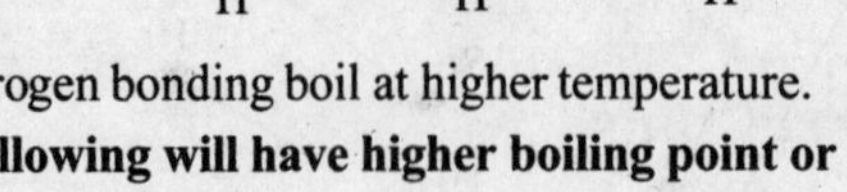

$$\ldots .\text{H}-\underset{\text{H}}{\underset{|}{\text{O}}}\ldots .\text{H}-\underset{\text{H}}{\underset{|}{\text{O}}}\ldots ..\text{H}-\underset{\text{H}}{\underset{|}{\text{O}}}\ldots ..$$

Substances having hydrogen bonding boil at higher temperature.

Q. 15. Which of the following will have higher boiling point or melting point? Justify your answer: (*Nagpur, 2005*)

(*i*) **H_2O or H_2S**

(*ii*) **Nitrophenol or nitrobenzene**

Ans. (*i*) H_2O boils at a high temperature. It has effective hydrogen bonding than H_2S because oxygen is more electronegative than sulphur.

(*ii*) *Nitrophenol* has a higher boiling point compared to nitrobenzene.

Q. 16. HF is liquid whereas HCl, HBr and HI are gases. Explain.

(Rohilkhand, 2003; Delhi, 2002)

Ans. See Q. 13

Q. 17. Methane is very slightly soluble in water, but methanol is highly soluble. Explain.

(Delhi, 2002)

Ans. Methanol forms hydrogen bond with water. No such bonding takes place between methane and water.

Q. 18. Why is boiling point of NH_3 higher than that of PH_3 ? *(Delhi, 2002)*

Ans. There is greater electronegativity difference between N and H than between P and H. Thus due to hydrogen bonding, although small, NH_3 molecule is associated and boils at a higher temperature.

Q. 19. Explain the following :

(*i*) Alcohol is a better drying agent than acetone.

(*ii*) Density of ice is less than that of water.

(*iii*) *o*-nitrophenol is less soluble in water than *p*-nitrophenol. *(Guwahati 2007)*

Ans. (*i*) Alcohol can combine with water more than with acetone due to hydrogen bonding. Therefore, alcohol is a better drying agent.

(*ii*) Ice has a cage like structure comprising hydrogen bonds. There is a lot of empty space in the structure due to which its density is much less than that of water.

(*iii*) See Q. 13 (3).

4 COVALENT BOND – II

HYBRIDIZATION

Q. 1. What is meant by hybridization of orbitals?

(*Nagpur, 2005; M. Dayanand, 2002; Rohilkhand, 2003; Garhwal, 2010*)

Ans. Ground state of configuration of carbon is

$$1s^2\, 2s^2\, 2p_x^1\, 2p_y^1$$

To account for tetravalency of carbon it is assumed that one electron from 2*s* orbital is excited to $2p_z$ orbital so that the configuration in excited state is

$$1s^2\, 2s^1\, 2p_x^1\, 2p_y^1\, 2p_z^1$$

The structure of methane consists of four equivalent bonds arranged about the carbon atom at the centre in a manner that all the angles between carbon and hydrogen are equal, its value is 109° 28′. The geometric arrangement which fulfils these requirements is a tetrahedron. In such a structure, carbon atom occupies a position at the centre of the tetrahedron and four hydrogen atoms occupy the corner positions. This arrangement is shown in Fig. 4.1

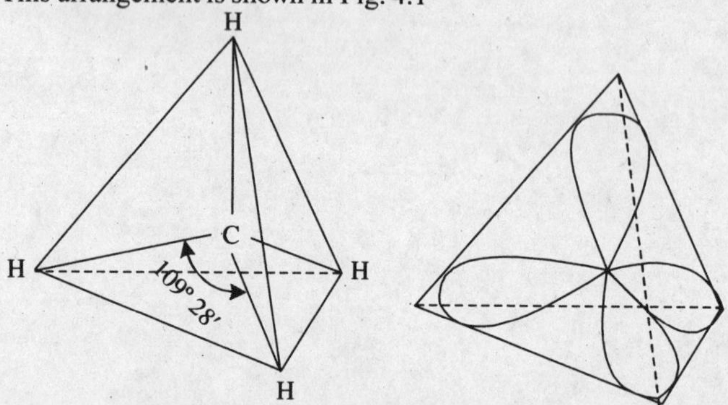

Fig. 4.1. Structure of methane.

The equivalence of four bonds can be explained by assuming that the available orbitals of carbon, the 2*s* and three 2*p* orbitals are intermixed or hybridized in a manner as to result in four equivalent orbitals which are oriented towards the four corners of regular tetrahedron. *This mixing of atomic orbitals to give equal number of symmetrically hybridized orbitals is referred to as hybridization.* The carbon in this case is said to be sp^3 hybridized and carbon hydrogen bonds are referred to as hybrid bonds.

Q. 2. What are different types of hybridization? (*Awadh, 2000; Meerut, 2000; Garhwal 2012*)

Ans. The types of hybridization are:

(*i*) sp^3 type

(*ii*) sp^2 type

(*iii*) *sp* type
(*iv*) sp^3d type
(*v*) sp^3d^2 or d^2sp^3 type
(*vi*) $sp^3\,d^3$ type

Q. 3. What is *sp* hybridization or diagonal hybridization? Discuss the shape of beryllium fluoride based on this hybridization.

(*Kumaon, 2000; Lucknow, 2000; Vidyasagar, 2007*)

Ans. Electronic configuration of beryllium in ground state is $1s^2\,2s^2$. One of the 2*s* electrons gets promoted to next available empty orbital (*i.e.*, $2p_x$) and therefore electronic configuration of beryllium in excited state is $1s^2, 2s^1, 2p_x^1$.

1s	2s	2p			
↑↓	↑↓				Beryllium atom in ground state
↑↓	↑	↑			Beryllium atom in excited state
↑↓	↑↓	↑↓			Beryllium atom in BeF_2

sp type hybridisation

Now there are two half-filled orbitals, each having single unpaired electron. But they belong to two different orbitals. One belongs to *s* and the other two *p*-orbital. Therefore, to get equivalent bonds, these two orbitals hybridise or get mixed up to form two new equivalent orbitals each of which has half *s* and half *p* character.

In the above case one *s* and one *p*-orbitals are involved to form hybridized orbitals. This type of hybridization is therefore known as *sp hybridization* or *diagonal hybridization.* The hybridized

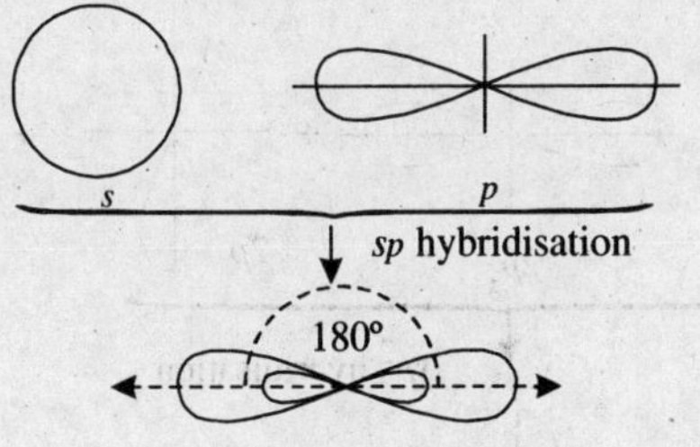

Fig. 4.2

orbitals are oriented at an angle of 180° to each other along the same axis (Fig. 4.2).

These two hybridized orbitals overlap separately with the *p*-orbitals of fluorine to form two Be-F bonds. Shape of BeF_2 molecule can be represented as shown in Fig. 4.3. There are two bonded

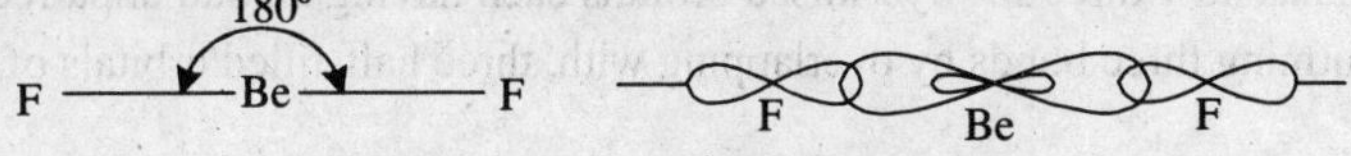

Fig. 4.3. Pictorial representation of BeF_2.

electron pairs. Therefore molecule is linear with two orbitals at an angle of 180° to each other.

Q. 4. What is sp^2 hybridization or trigonal hybridization? Describe the shape of BF_3 based upon this hybridization.

(Awadh, 2000; Kumaon, 2000; Kerala, 2000; M. Dayanand, 2002; Nagpur, 2005; Guwahati, 2006; Purvanchal 2007; Calcutta 2007)

Ans. Electronic configuration of boron in ground state is $1s^2\,2s^2\,2p_x^1$. One of the two 2*s* electrons gets promoted to next available vacant orbital ($2p_y$). The electronic configuration of boron in the excited state is $1s^2\,2s^1\,2p_x^1\,2p_y^1$.

	1s	2s	2p		
Configuration of B atom in ground state	↑↓	↑↓	↑		
Configuration of B atom in excited state	↑↓	↑	↑	↑	
	↑↓	↑↓	↑↓	↑↓	

sp^2 hybridisation

Now there are three half-filled orbitals each having odd unpaired electrons. But they belong to different orbitals. To get equivalent bonds these three orbitals hybridize to form three new equivalent hybridised orbitals. In the above case one *s* and two *p* orbitals are involved to form three hybridised orbitals. This type of hybridization is, therefore, known as *sp^2 hybridization or trigonal hybridization.*

Boron forms sp^2 hybridised orbitals. These three orbitals lie in one plane (*i.e.*, boron atom is planer) and are directed towards the three corners of an equivalent triangle, so that the angle between them is 120°.

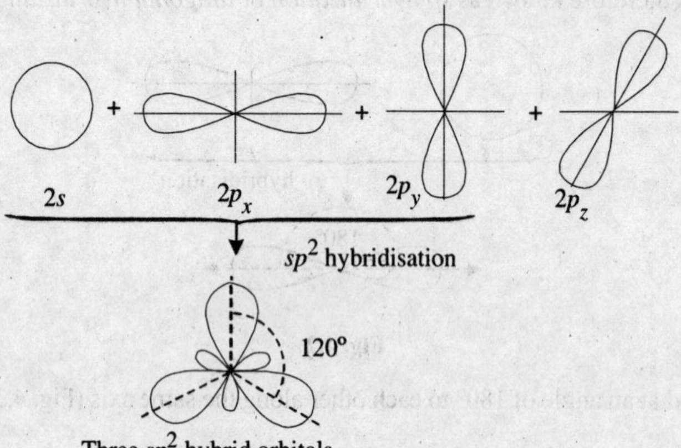

Fig. 4.4. Orientation of sp^2 hybrid orbitals.

Boron atom has now three sp^2 hybridised orbitals each having an odd unpaired electron. They are capable of forming three bonds by overlapping with, three half-filled orbitals of other atoms.

Bonding capacity of boron, therefore, is three. Boron forms compounds of type X — B(X)(X) where X = H or Cl

Shape of BF_3

Shape of BF_3 molecule can be represented as shown in Fig. 4.5. There are three bonded electron pairs. The three bonds are equivalent and 120° apart. The molecule is, therefore, triangular and planar in shape.

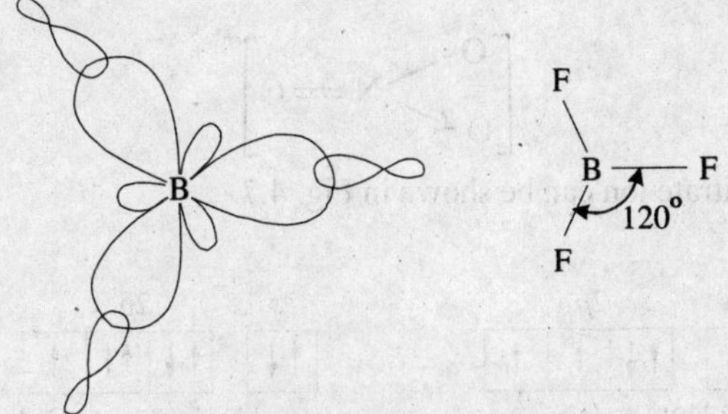

Fig. 4.5. Shape of BF_3.

Q. 5. Explain why BF_3 is triangular in shape while BeF_2 is linear. (*M. Dayanand, 2002*)

Ans. In BeF_2, Be atom is *sp* hybridized. The *sp* hybrid orbitals are linearly oriented. That is why the molecule of BeF_2 is linear.

In BeF_3, B atom is sp^2 hybridized. The sp^2 hybrid orbitals are oriented at an angle of 120° towards the corners of a triangle. That is why the shape of BF_3 is triangular.

Q. 6. Discuss the shape of stannous chloride molecule.

(*G. Nanakdev, 2000; M.Dayanand, 2002*)

Ans. Atomic no. of Sn is 50. In stannous chloride the tin (Sn) atom has electronic configuration of outermost shell as : $5s^2\ 5p_x^{\ 1}\ P_y^{\ 1}$. Two electrons in the 5*s* orbital and two electrons in *p*-orbitals participate in hybridisation. It may be noted that an orbital with paired electrons can participate in hybridisation but not in bond formation. Thus it is a case of sp^2 hybridisation.

In $SnCl_2$, two unpaired electrons in outermost shell of tin are used for bond formation with two atoms of chlorine, while one electron pair is left unused as lone pair. Thus there are three electron pairs, two bond pairs and a lone pair. According to its hybridisation it is expected to be trigonal planar (Fig. 4.6) one of the three positions is occupied by lone pair of electrons. The expected angle is 120°. But the actual angle is less than 120°. This is because of a greater repulsion between the bond pair and lone pair than between the bond pair and bond pair. (VSEPR Theory, See Q. 29).

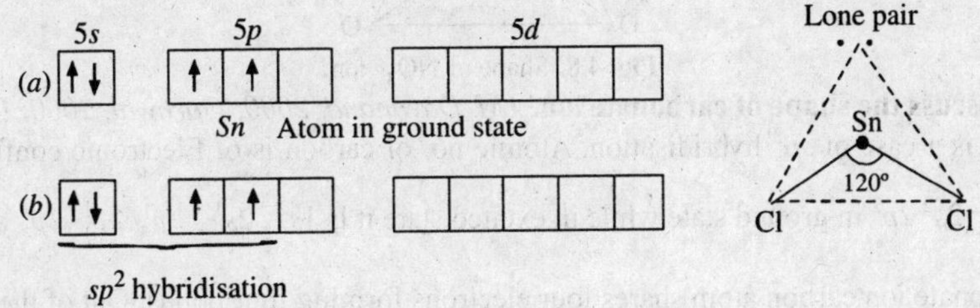

Fig. 4.6

Q. 7. Discuss the shape of nitrate ion. (*Rajasthan, 1996; G. Nanakdev, 2000; Delhi, 2007*)

Ans. It is a case of sp^2 hybridization. Atomic no. of 'N' is 7. In nitrate ion nitrogen is central atom and it has 5 electrons in outermost shell ($2s^2$, $2p^3$) while oxygen atoms have six electrons ($2s^2$, $2p^4$). The nitrate ion carries a unit negative charge. It means that one of the oxygen atoms carries an extra electron.

$$\left[\begin{array}{l} O \searrow \\ \qquad N{=}O \\ O \swarrow \end{array} \right]^-$$

The bond formation in nitrate ion can be shown in Fig. 4.7.

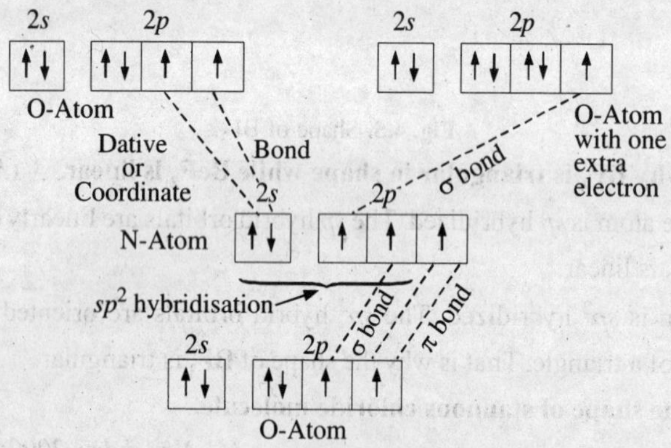

Fig. 4.7. Formation of nitrate ion.

π bond is formed by unhybridized orbitals, the $2p_x$ orbital of nitrogen is left unhybridized and it is involved in π bond formation. The rest of orbitals are sp^2 hybridised.

Total number of electron pairs used for hybridisation are three and no. of orbitals required are three *viz.*, one *s* and two *p*, *i.e.*, the hybridisation is sp^2. Thus the shape of molecule is trigonal planar with bond angle 120° as shown in Fig. 4.8.

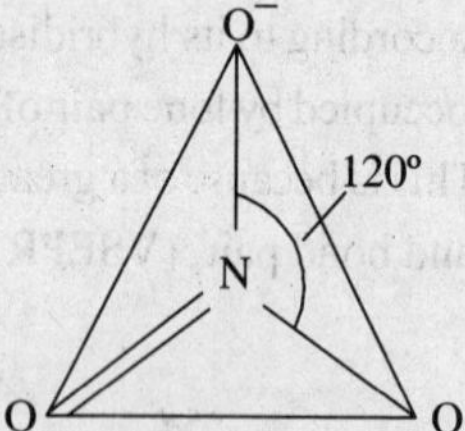

Fig. 4.8. Shape of NO_3^- ion.

Q. 8 Discuss the shape of carbonate ion. (*M. Dayanand, 2000; Garhwal, 2000; Delhi, 2006*)

Ans. It is a case of sp^2 hybridisation. Atomic no. of carbon is 6. Electronic configuration of carbon is $1s^2\ 2s^2\ 2p^2$ in ground state while in excited state it is $1s^2$, $2s^1$; $2p_x^1, 2p_y^1, 2p_z^1$ as described below.

In carbonate ion carbon atom shares four electrons forming four bonds. Out of these three are sigma (σ) and one pi (π) bond. π bonds are excluded from hybridisation and therefore $CO^{2}{}_3^-$ ion has plane triangular shape (Fig. 4.9). The bond angle is 120°. However carbon atom has no lone pair of electrons and hence the molecule has a regular geometry.

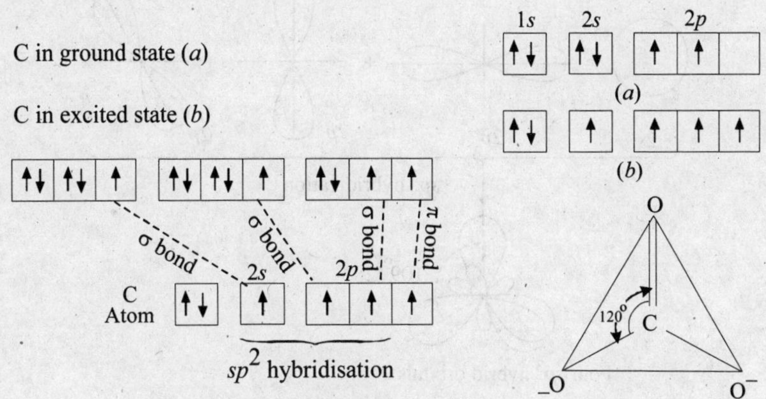

Fig. 4.9. Shape of CO_3^{2-}

As carbonate ion has two negative charges, it is implied that two oxygens have an extra electron while the third oxygen is normal. The half-filled unhybridised orbital of carbon forms a π bond with that of oxygen atom having no extra electrons and thus bonded by a double-bond to carbon atom.

Q. 9. Discuss the shape of BO_3^{3-} ion.

Ans. Atomic no. of B is 5. Electronic configuration of B in ground state is $1s^2, 2s^2, 2p^1$. In excited state one of the electrons from 2s is shifted to 2p orbital. It undergoes sp^2 hybridisation forming three sigma bonds with each of three oxygen atoms having only one half–filled p-orbital giving a plane triangular shape as shown in Fig. 4.10.

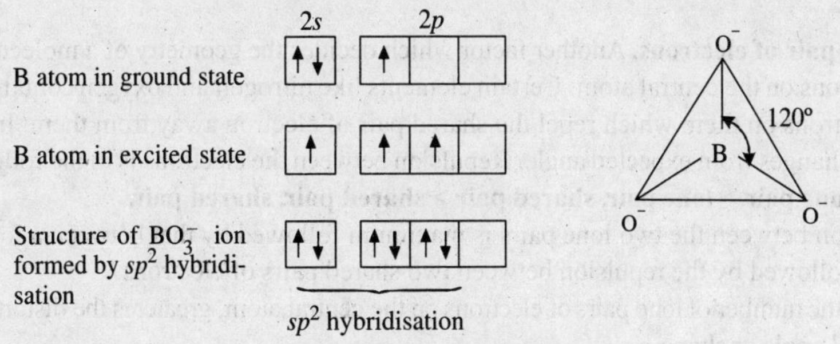

Fig. 4.10. Shape of BO_3^{3-}.

Q. 10. What is sp^3 hybridisation or tetrahedral hybridisation?

(Kumaon, 2000, Lucknow, 2000)

Ans. When one s orbital and three p orbitals combine together to produce equivalent type of orbitals, sp^3 hybridisation takes place. Four sp^3 hybridised orbitals are produced which have the same energy and shape and are oriented at an angle of 109°28' to one another. This is also known as tetrahedral hybridisation because the hybridised orbitals are oriented towards the four corners of tetrahedron with the nucleus occupying the centre of tetrahedron. In this position, hybridised orbitals experience minimum repulsive forces. This is illustrated in Fig. 4.11.

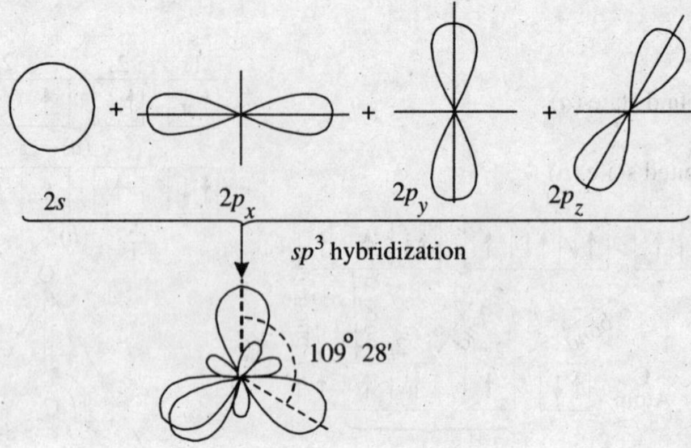

Fig. 4.11. sp^3 hybridisation.

Q. 11. What are the factors that determine the shape of molecules?

Ans. The shapes of molecules depend upon the following factors:

1. Types of hybridisation. Type of hybridisation decides the shape of molecules. If *sp* hybridisation takes place and the molecule is of the type AB_2 where the atom A is central and surrounded by two B atoms then molecule will have a linear shape B — A — B with the ∠ BAB equal to 180°.

Similarly if the molecule is of the type AB_3 with sp^2 hybridisation the atom A lies in the centre of an equilateral traingle and three B atoms occupy the vertices of the triangle. If however the molecule is of type AB_4 with sp^3 hybridisation taking place the atom A occupies the centre of tetrahedral and vertices of tetrahedron being occupied by B. This is based on the fact that the hybridised orbitals repel each other due to similar charge and occupy positions in which there are minimum repulsive forces.

2. Lone pair of electrons. Another factor which decides the geometry of a molecule is the lone pair of electrons on the central atom. Certain elements like nitrogen and oxygen contain lone pair or pairs of electrons on them which repel the shared pair of electron away from them. In such a case bond angle changes from expected angle. Repulsion between the electrons is in the following order :

Lone pair, lone pair > lone pair, shared pair > shared pair, shared pair.

Repulsion between the two lone pairs is maximum followed by that between a lone pair and a shared pair followed by the repulsion between two shared pairs of electrons.

Greater the number of lone pairs of electrons on the central atom, greater is the distortion produced in the normal angle or shape.

Q. 12. Discuss the shape of methane molecule.

Ans. At no. of carbon = 6

Electronic configuration of C in ground state

$$= 1s^2\, 2s^2\, 2p_x^1\, 2p_y^1$$

Electronic configuration of C in excited state

$$= 1s^2\, 2s^1\, 2p_x^1\, 2p_y^1\, 2p_z^1$$

There are four orbitals having unpaired electrons. These $2s$, $2p_x$, $2p_y$ and $2p_z$ orbitals will first hybridise. So it is a case of sp^3 hybridisation. Four hybridised orbitals tetrahedrally are shown in Fig. 4.12.

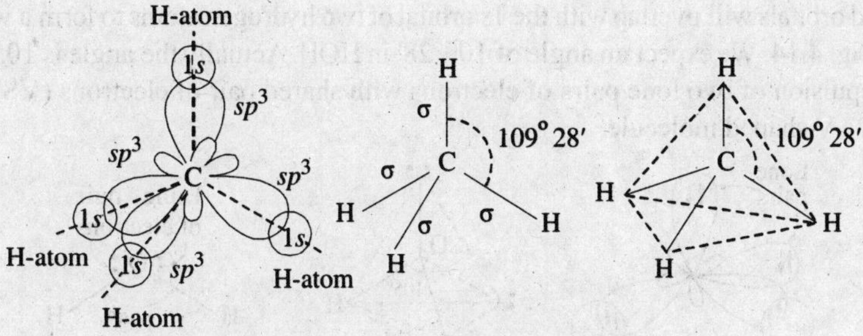

Fig. 4.12. Shape of methane molecule.

All the four hybridised orbitals overlap with the 1*s* orbital of four hydrogen atoms. All these overlappings take place along the internuclear axes. Hence four carbon-hydrogen σ bonds are formed. The angle ∠ HCH is 109° 28'.

Q. 13. Discuss the shape of ammonia molecule.

(*Awadh, 2000; Kerala, 2000; Meerut, 2000; G. Nanakdev, 2000; Kanpur, 2000; Lucknow, 2000; Bangalore, 2005; Calcutta, 2005; Guwahati, 2006; Nagpur, 2008*)

Ans. Atomic no. of nitrogen = 7

Electronic configuration of nitrogen, $1s^2\ 2s^2\ 2p_x^1\ 2p_y^1\ 2p_z^1$

Here, there is no possibility of shifting of electron from 2*s* level to 2*p* level as there is no empty 2*p* orbital. 2*s* orbital, although incapable of overlapping or forming the bond or forming the bond, participates in hybridisation. This will give rise to tetrahedral shape. Hybridised orbitals are oriented towards the corners of a tetrahedron with one corner occupied by the lone pair of electrons from 2*s* orbital. Three sp^3 hybridised orbitals overlap with the 1*s* orbitals of three hydrogen atoms (Fig. 4.13). We expect an angle of 109°28' between H — N — H in tetrahedral geometry. Actually this angle is 107°. This contraction in angle takes place because the lone pair of electrons repels the shared pair of electrons or bonded pair of electrons with a greater force (Fig. 4.13) thereby reducing the angle. (Valence Shell Electron Pair Repulsion, VSEPR Theory, See Q. 29).

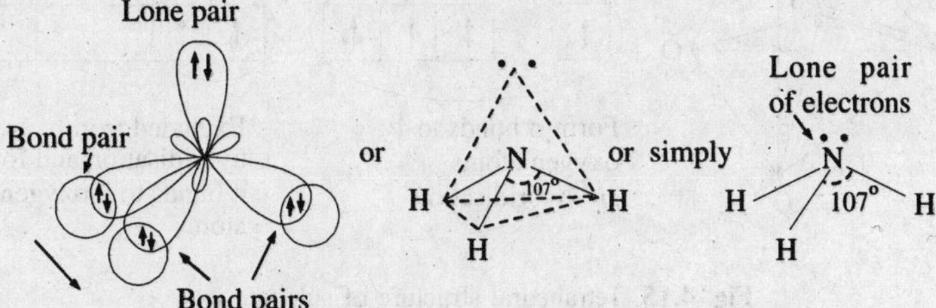

Fig. 4.13. Shape of ammonia molecule.

Thus it has pyramidal shape.

Q. 14. Discuss the shape of water molecule.

(*Awadh, 2000; Meerut, 2000; Kanpur, 2001; Bangalore, 2005*)

Ans. At. no. of oxygen = 8

Electronic configuration = $1s^2\ 2s^2\ 2p_x^2\ 2p_y^1\ 2p_z^1$

In the 2nd orbit there are four orbitals : 2*s*, $2p_x$ $2p_y$ and $2p_z$. Out of these 2*s* and $2p_x$ cannot form the bonds as they are paired. But they can participate in hybridisation. Hence it is case of sp^3 hybridisation. The hybridised orbitals are oriented along the corners of a tetrahedron. Two of the positions on the tetrahedron are occupied by lone pair of electrons from 2*s* and $2p_x$ orbitals. The other two hybridised orbitals will overlap with the 1*s* orbital of two hydrogen atoms to form a water molecule as shown in Fig. 4.14. We expect an angle of 109°28' in HOH. Actually the angle is 105°. This is due to stronger repulsion of two lone pairs of electrons with shared pair of electrons (VSEPR Theory). Thus water is a V-shaped molecule.

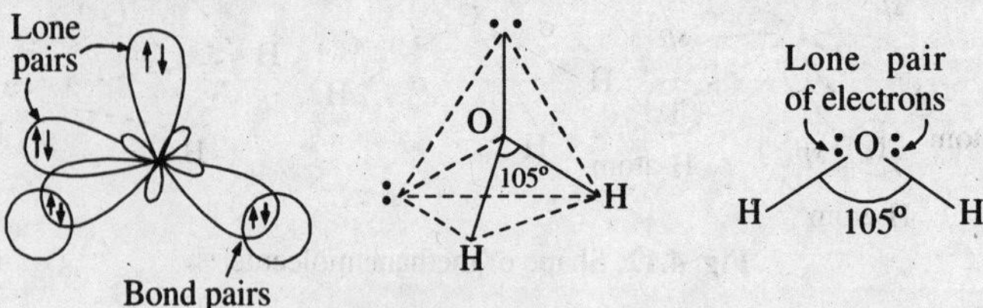

Fig. 4.14. Shape of water molecule.

Q. 15. Discuss the shape of sulphate ion. (*Garhwal, 2000; Himachal, 2000*)

Ans. Atomic no. of sulphur is 16. Its outer electronic configuration is $3s^2\ 3p^4$. The shape of molecule suggests that it is a case of sp^3 hybridisation. To have the requisite no. of half-filled orbitals one electron from *s* and one electron from *p* orbitals are excited to *d*-orbitals as shown in Fig. 4.15. It may be remembered that out of four oxygen atoms two have extra electron because the net charge on SO_4^{2-} is –2. Sulphur combines with four oxygen atoms forming four bonds, thus using four pairs of electrons. Also sulphur atom has no non-bonded electron pair. The number of hybrid orbitals used are four *viz.*, one *s* and three *p* orbitals as per sequence. Thus it is a case of sp^3 hybridisation and therefore SO_4^{2-} is tetrahedral in shape as shown in Fig. 4.15. The two half-filled *d*-orbitals form π bonds with two oxygen atoms, having no extra electron and thus bonded by double bonds to sulphur.

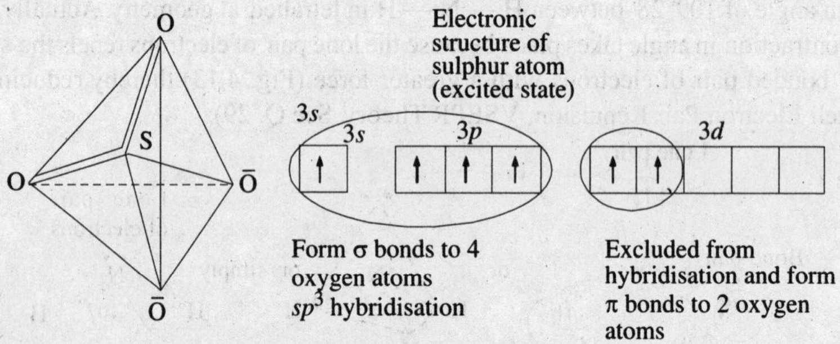

Fig. 4.15. Tetrahedral structure of sulphate ion.

Q. 16. Discuss the shape of perchlorate ClO_4^- ion.

(*G. Nanakdev, 2000; M. Dayanand, 2000; Lucknow, 2000; Garhwal, 2000*)

Ans. This is a case of sp^3 hybridisation At. no. of Cl is 17. Its outer electronic configuration is $3s^2\ 3p^5$. One electron each from $3s$, $3p_x$ and $3p_y$ are excited to $3d$ orbitals. Single electron in $3s$, $3p_x$ $3p_y$ and $3p_z$ hybridise sp^3 to give tetrahedral shape. Electrons in $3d$ orbital form π bonds. ClO_4^- ion is tetrahedral. It is because chlorine makes use of four bond orbitals and these are one *s* and three *p*-orbitals and hybridisation is sp^2 and shape is tetrahedral. Chlorine has no non-bonded electron pair in the excited state. Three half-filled *d*-orbitals form π bonds with three oxygen atoms without extra electron and thus bonded by double bonds to chlorine as shown in Fig. 4.16.

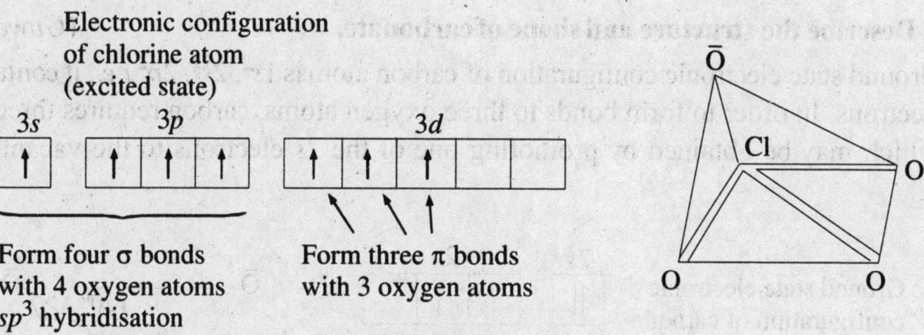

Fig. 4.16. Tetrahedral shape of perchlorate ion.

Q.17. Discuss the shape of chlorate ion (ClO_3^-).

(*M. Dayanand, 2000; G. Nanakdev, 2000; Kurukshetra, 2002*)

Ans. At. no. of Cl is 17. Its outer configuration is $3s^2\ 3p_x^2\ 3p_y^2\ 3p_z^1$. It combines with three atoms of oxygen using three of its electrons. Thus it forms three single σ bonds with oxygen using three of its electrons. Here two electrons from $3p_x$ and $3p_y$ are excited to 3*d* orbitals which don't participate in hybridisation but form π bonds with two oxygen atoms without extra electron. 3*s* and 3*p* orbitals (sp^3) hybridise. It forms three single σ bonds with oxygen using three orbitals. Its actual shape is pyramidal with one corner position occupied by a lone pair of electrons. Its expected geometry is tetrahedral as sp^3 hybridisation is involved. The two half-filled *d*-orbitals form π bonds with two oxygen atoms and thus bonded to chlorine by double bonds as shown in Fig. 4.17

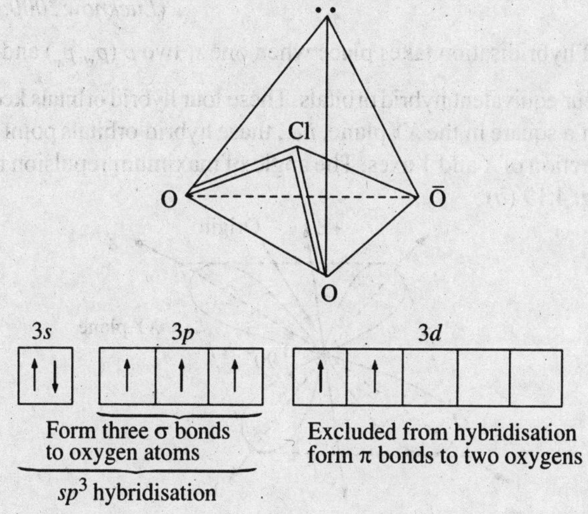

Fig. 4.17. Structure of ClO_3^- .

Q. 18. ClO_3^- and ClO_4^- ions have same number of electron pairs around central chlorine but their geometry is different. Why? (*M. Dayanand, 2000*)

Ans. sp^3 hybridisation takes place in the formation of both molecules. In ClO_3^- one tetrahedral position is occupied by a lone pair, therefore the ion gets pyramidal shape.

As expected from sp^3 hybridisation, ClO_4^- achieves tetrahedral shape.

Q. 19. Describe the structure and shape of carbonate. *(Guwahati, 2007)*

Ans. Ground state electronic configuration of carbon atom is $1s^2, 2s^2, 2p^2$ *i.e.*, it contains only two unpaired electrons. In order to form bonds to three oxygen atoms, carbon requires three unpaired electrons which may be obtained by promoting one of the 2*s* electrons to the vacant 2*p* orbitals. (Fig. 4.18).

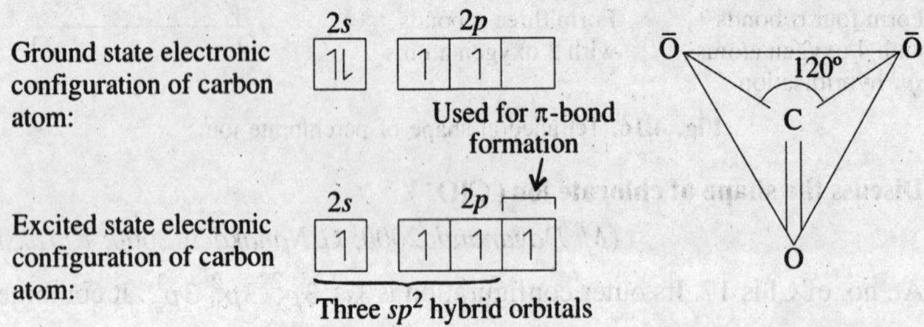

Fig. 4.18. Structure of carbonate ion.

The three sp^2 hybridised orbitals are obtained by mixing 2*s* and two 2*p* atomic orbitals. Each sp^2 hybridised orbital contains only one electron and is used to form three σ covalent bonds with three oxygen atoms. This accounts for the planner geometry of carbonate ion.

The third 2*p* orbital containing one electron overlaps with a vacant 2*p* orbitals of an oxygen atom to give a π bond between carbon and one of the three oxygen atoms.

Q. 20. What is dsp^2 square planner hybridisation? Discuss the structure of $[Ni(CN)_4]^{2-}$. *(Lucknow, 2000; M. Dayanand, 2000)*

Ans. This type of hybridisation takes place when one *s*, two *p* (p_x, p_y) and one *d* ($d_{x^2-y^2}$) orbitals mix together to give four equivalent hybrid orbitals. These four hybrid orbitals keep their lobes directed towards the corners of a square in the *XY* plane, *i.e.*, these hybrid orbitals point away from the origin in the ± *X* and ± *Y* direction of *X* and *Y* axes. The angle of maximum repulsion between their orbitals is 90° as shown in Fig. 4.19 (*a*).

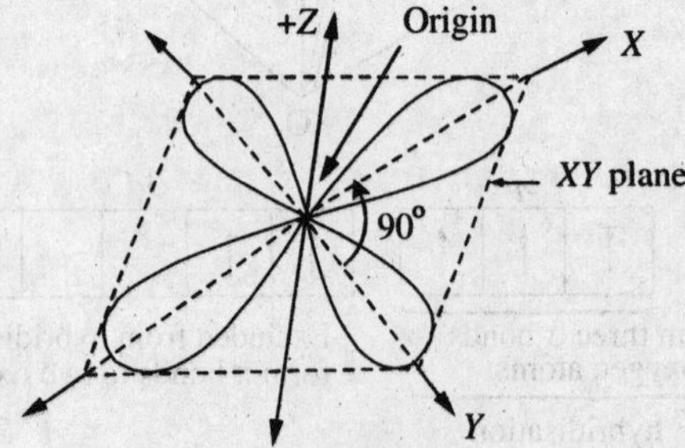

Fig. 4.19 (*a*) dsp^2 hybridisation.

Shape of $[Ni(CN)_4]^2$

Pairing of single electrons in 3*d* orbitals in Ni^{2+} takes place thereby creating one vacant 3*d* orbital.

Electron pairs from four CN^- ions are filled up in four hybrid orbitals obtained as a result of dsp^2 hybridisation as shown in Fig. 4.19 (*b*).

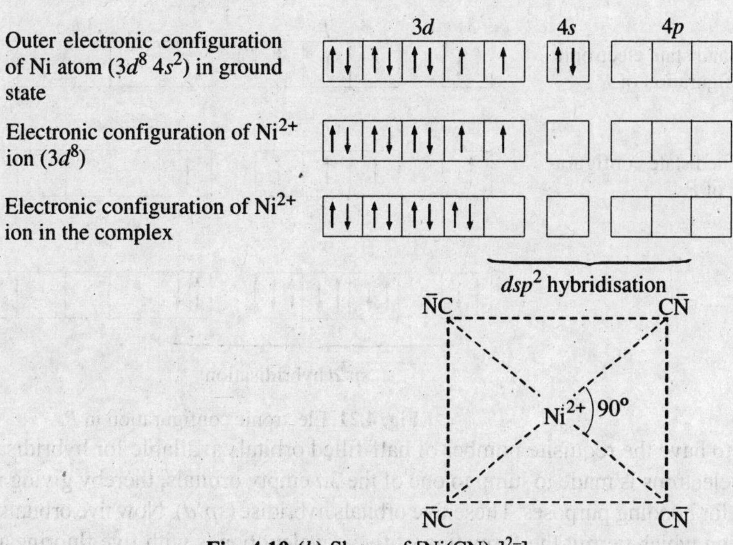

Fig. 4.19 (*b*) Shape of $[Ni(CN)_4]^{2-}]$

Q. 21. What is sp^3d hybridisation? Discuss the shape of phosphorus pentafluoride molecule.

(*Dayanand, 2000; M. Dayanand, 2002; Rohilkhand, 2003; Calcutta, 2007*)

Ans. Mixing of one *s*, three *p* and one *d* atomic orbitals gives five sp^3d hybrid orbitals and process is called sp^3d hybridisation. These are directed towards the corners of a trigonal bipyramid as shown in figure. They form three coplanner bonds with bond angles of 120°, remaining two bonds are at right angles to this plane one above and one below.

Formation of five $sp^3\ d$ hybrid orbitals is shown in Fig. 4.20.

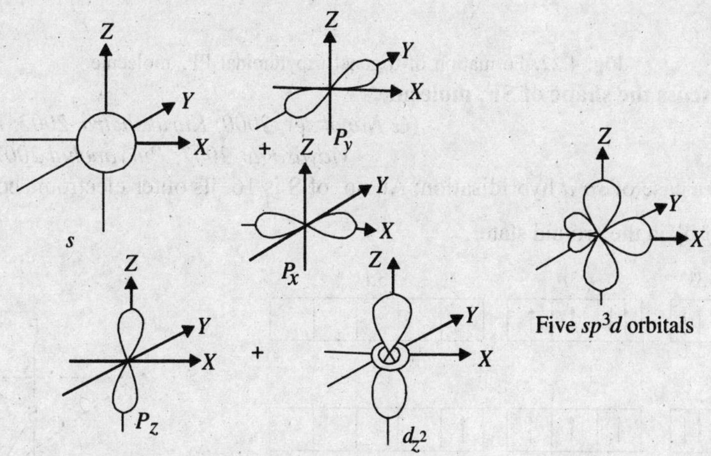

Fig. 4.20. sp^3d hybridisation.

Shape of PF_5

Here sp^3d type hybridisation is involved. At. no. of phosphorus atom is 15. Its outer electronic configuration in the ground state is $3s^2\ sp^3$ as shown in Fig. 4.21.

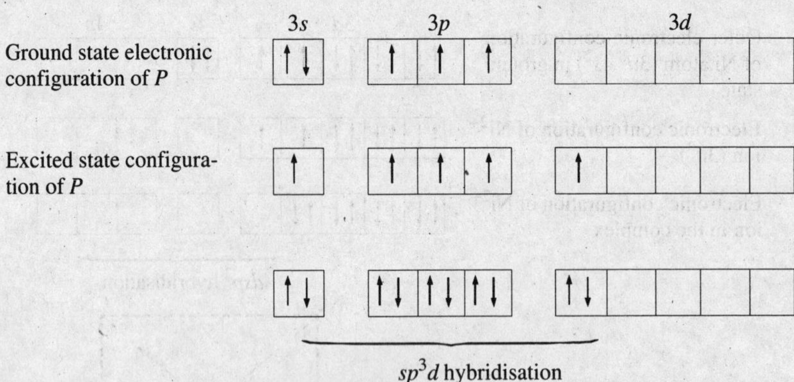

Fig. 4.21. Electronic configuration in *P*.

In order to have the requisite number of half-filled orbitals available for hybridisation purpose one of the 3*s* electrons is made to jump to one of the 3*d* empty orbitals, thereby giving rise to 5 half-filled orbitals for bonding purposes. These five orbitals hybridise (sp^3d). Now five orbitals are available for combination which permit the formation of five covalent bonds with five fluorine atoms. But all the hybridised orbitals are not equivalent. Three of these are oriented towards the three corners of an equilateral triangle making an angle of 120° between them. The bonds formed by these orbitals are known as equitorial bonds. Remaining two orbitals are oriented at right angles to the plane of first set of three orbitals. Thus PF_5 has trigonal bipyramidal shape as shown in Fig. 4.22.

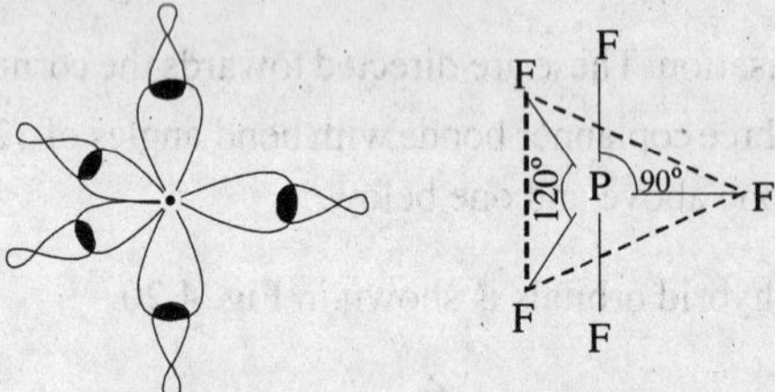

Fig. 4.22. Formation of trigonal tripyramidal PF_5 molecule.

Q. 22. Discuss the shape of SF_4 molecule.

(G. Nanakdev, 2000; Kurukshetra, 2002; Andhra, 2002; Vidyasagar 2007; Purvanchal 2007, Delhi 2007)

Ans. It is a case of sp^3d hybridisation; At. no. of S is 16, its outer electronic configuration is $3s^2\ 3p_x^2\ 3p_y^1\ 3p_z^1$ in the ground state.

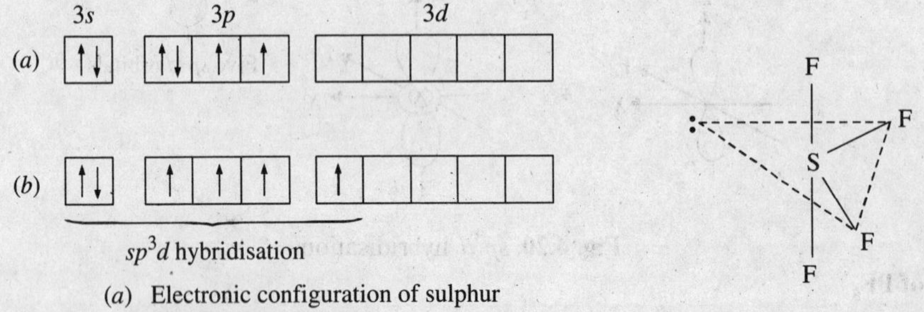

(*a*) Electronic configuration of sulphur atom in ground state.

(*b*) Sulphur atom in excited state.

Fig. 4.23. Distorted trigonal bipyramid structure of SF_4.

To have four unpaired electrons, one electron from 3*p* orbital is unpaired and promoted to 3*d* orbital. Thus five orbitals are used, four contain shared or bonded pairs of electrons and one contains a lone pair, sp^3d hybridisation occurs, giving an expected trigonal bipyramidal shape. Actually the four fluorine atoms are joined tetrahedrally to sulphur atom in an irregular manner due to presence of a lone pair (Fig. 4.23).

Q. 23. Discuss the shape of ClF_3 molecule.

(*G. Nanakdev, 2000; Nagpur, 2000; Kerala, 2000; Rohilkhand, 2003; Delhi, 2006; Purvanchal, 2007; Nagpur, 2008*)

Ans. It is a case of sp^3d hybridisation. At. no. of Cl is 17. Its outer electronic configuration in ground state is $3s^2\ 3p_x^2\ 3p_y^2\ 3p_z^1$.

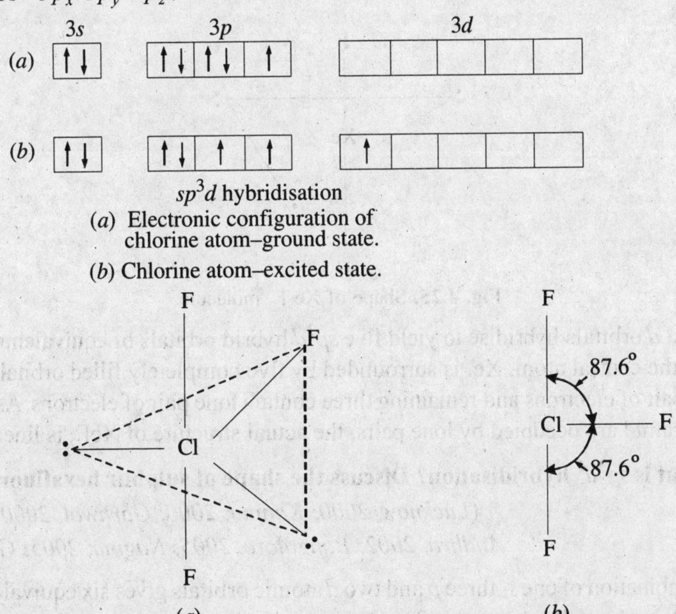

sp^3d hybridisation
(*a*) Electronic configuration of chlorine atom–ground state.
(*b*) Chlorine atom–excited state.

Fig. 4.24. Shape of ClF_3 molecule.

In the formation of ClF_3 molecule one of the paired electrons in 3*p* orbital is promoted to 3*d* orbital. The three unpaired electrons permit the formation of three covalent bonds with three fluorine atoms. If however there are only three orbitals around the central chlorine atom $C1F_3$ molecules should have been plane triangular. But this is actually not so, the molecules of chlorine trifluoride is known to have a *T*-shaped structure (Fig. 4.24). The promotion of 3*p* electrons is followed by hybridisation involving one *s*, three *p* and one *d* orbitals. This is thus a case of sp^3d hybridisation. The chlorine atom is therefore surrounded by five orbitals, three containing a bond pair of electrons while the other two contain lone pair of electrons. As two positions of the trigonal, bipyramid are occupied by lone pairs, the molecule has a *T*-shape. The angle F—Cl—F expected from this geometry is 90°. However actual angle is 87.6°. This is due to repulsion from lone pairs which result in contraction of angle.

Q. 24. In XeF_2, Xe undergoes sp^3d hybridisation but the molecule is linear. Explain.

(*Kerala, 2000; Delhi, 2006*)

Ans. It is a case of sp^3d hybridisation. Atomic number of Xe is 54. Its outer electronic configuration is $5s^2\ 5p_x^2\ 5p_y^2\ 5p_z^2$ as shown in Fig. 4.25. In order to have the requisite number of half-filled orbitals for hybridisation one of the $5p_z$ electrons is made to jump to one of the $5d$ empty orbitals. The unpaired electrons in the $5p$ and $5d$ orbitals are used in chemical bonding, the pairing electrons being provided by two fluorine atoms.

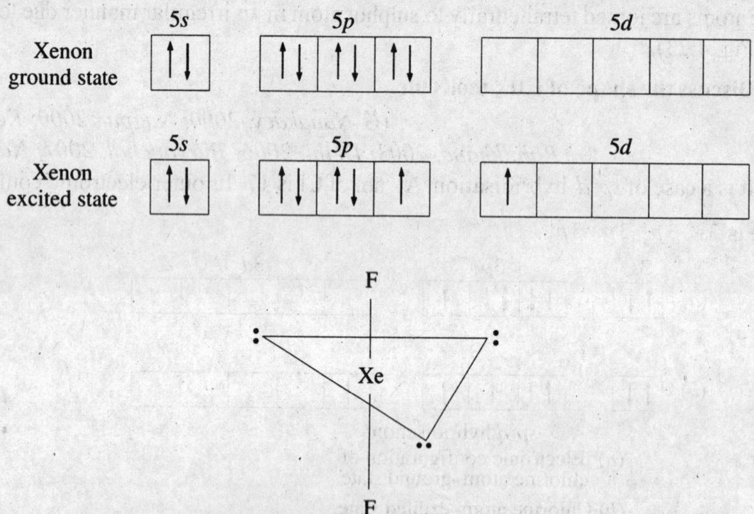

Fig. 4.25. Shape of Xe F_2 molecule.

The *s, p* and *d* orbitals hybridise to yield five sp^3d hybrid orbitals of equivalent energy. Thus in XeF_2 molecule the central atom, Xe, is surrounded by five completely filled orbitals, two of which contain shared pair of electrons and remaining three contain lone pair of electrons. As three positions in trigonal bipyramid are occupied by lone pairs, the actual structure of XeF_2 is linear (Fig. 4.25).

Q. 25. What is sp^3d^2 hybridisation? Discuss the shape of sulphur hexafluoride (SF_6).

(Lucknow, 2000; Kanpur, 2000; Garhwal, 2000; Kerala, 2000; Andhra, 2002; Bangalore, 2005; Nagpur, 2005; Guwahati, 2007)

Ans. A combination of one *s*, three *p* and two *d* atomic orbitals gives six equivalent sp^3d^2 hybrid orbitals directed octahedrally, *i.e.*, away from the origin on ± *x* and ± *y*, ± *z* directions. This type of combination is known as sp^3 d^2 hybridisation. Shapes of participating orbitals is shown in Fig. 4.26. Hybrid orbitals are oriented towards the corners of an octahedral (Fig. 4.27).

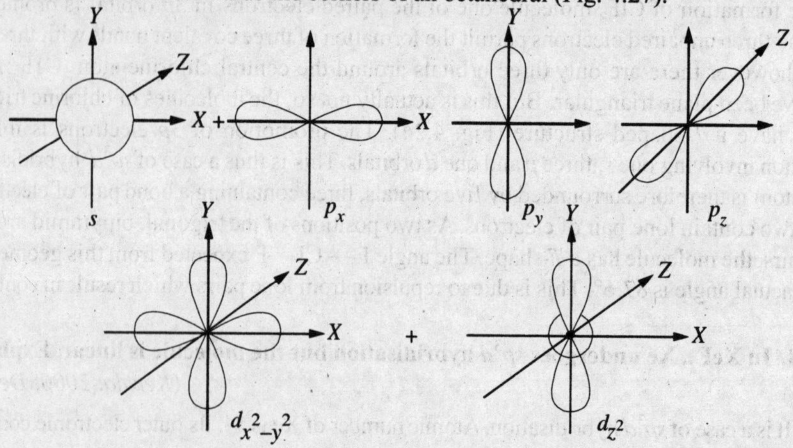

Fig. 4.26. Shapes of participating orbitals.

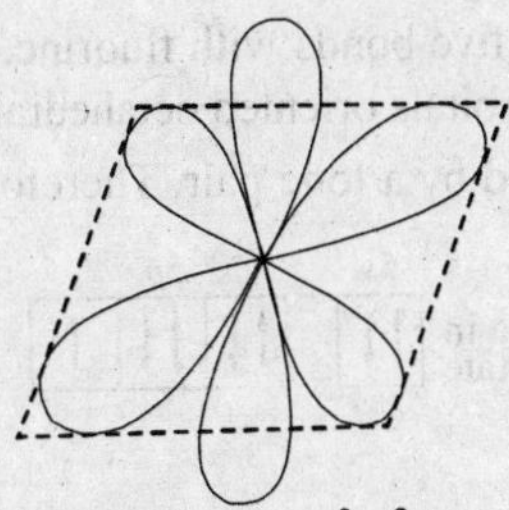

Fig. 4.27. Orientation of $sp^3 d^2$ hybrid orbitals.

Shape of SF_6

At. no. of sulphur is 16. Outer electronic configuration of central atom in the ground state is $3s^2\ 3p_x^2\ 3p_y\ 3p_z$ as shown in Fig. 4.28. Then to obtain the requisite number of half-filled orbitals for hybridisation purposes, one of the 3*s* and one of $3p_x$ electrons are made to jump to two of the 3*d* empty orbitals. Now six unpaired electrons are available in six orbitals (one *s*, three *p* and two *d*), for bond formation to give rise to SF_6 molecule.

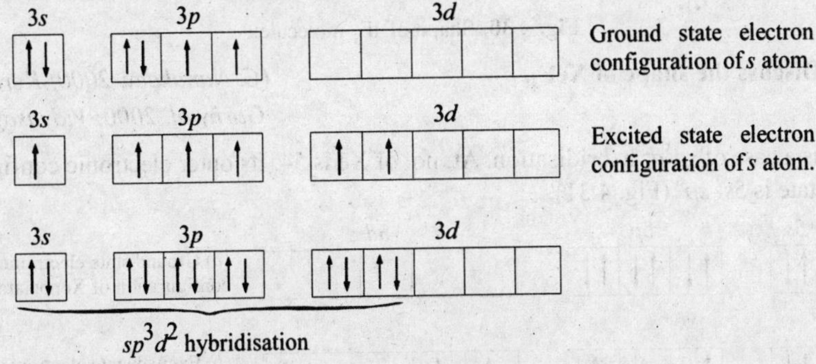

Fig. 4.28. Electronic configuration of S.

These six orbitals hybridise in sp^3d^2 type. The six hybrid orbitals are now directed towards the corners of a regular octahedron. These six hybrid orbitals overlap with the *p*-orbitals of six fluorine atoms. So SF_6 molecule has an octahedral structure as shown in Fig. 4.29.

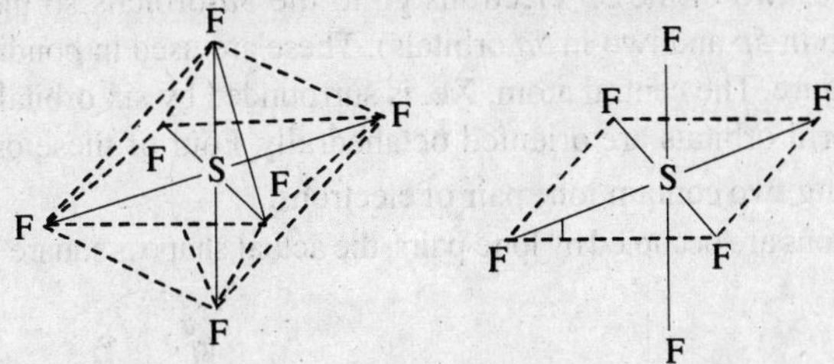

Fig. 4.29. Octahedral molecule of sulphur hexafluoride formed by $sp^3\ d^2$ hybridisation.

Q. 26. Discuss the shape of iodine pentafluoride IF_5.

(Kerala, 2000)

Ans. It is a case of sp^3d^2 hybridisation. At. no. of iodine is 53. Its outer electronic configuration is

$$5s^2\ 5p_x^2\ 5p_y^2\ 5p_z^1$$

In the excited state two electrons from 5*p* orbitals are excited to 5*d* orbitals, thereby giving five single electrons for the formation of five bonds with fluorine. $5s$, $5p_x$, $5p_y$, $5p_z$ and two *d*-orbitals (sp^3d^2) hybridise to give equivalent orbitals oriented octahedrally.

However one position is occupied by a lone pair. Therefore actual shape is square pyramidal (Fig. 4.30).

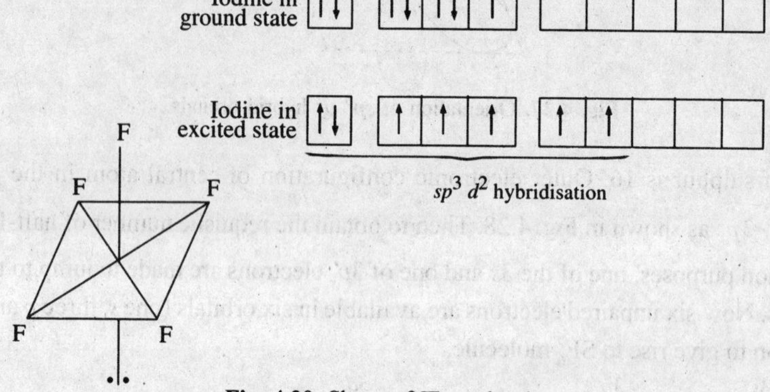

Fig. 4.30. Shape of IF_5 molecule.

Q. 27. Discuss the shape of XeF_4. (*G. Nanakdev, 2000; Kerala, 2000; Garhwal, 2000; Vidyasagar, 2007*)

Ans. It is a case of sp^3d^2 hybridisation. At. no. of Xe is 54. Its outer electronic configuration in the ground state is $5s^2\ sp^6$ (Fig. 4.31).

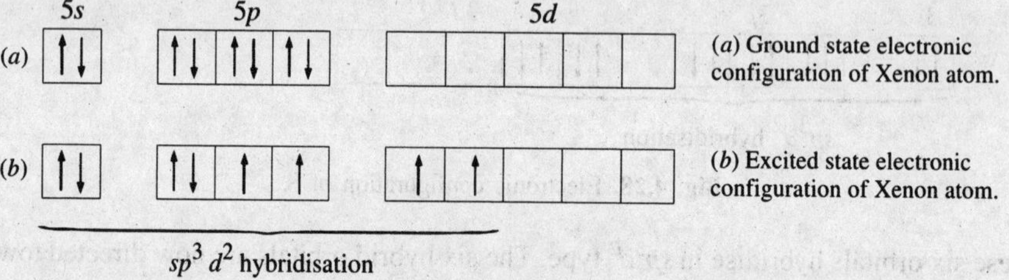

Fig. 4.31. sp^3d^2 hybridisation.

In the excited state, two of the *5p* electrons go to the *5d*-orbitals so that there are now four unpaired electrons (two in *5p* and two in *5d* orbitals). These are used in bonding with four atoms of fluorine as shown in figure. The central atom, Xe, is surrounded by six orbitals resulting from sp^3d^2 hybridisation. The hybrid orbitals are oriented octahedrally. Four of these orbitals contain shared pairs while the remaining two contain lone pair of electrons.

As the vertex positions are occupied by lone pairs, the actual shape is square planner as represented in Fig. 4.32.

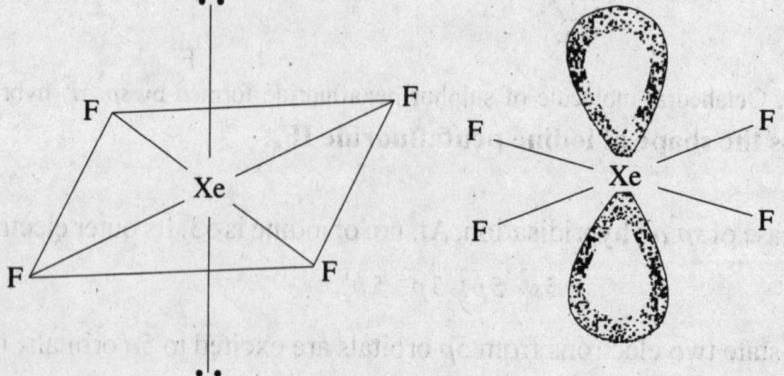

Fig. 4.32. Shape of XeF_4 molecule.

Q. 28. What is sp^3d^3 hybridisation? Describe the shape of IF_7 (Iodine heptafluoride).

(Kerala, 2000; G. Nanakdev, 2000; Guwahati, 2006)

Ans. Mixing of one *s*, three *p* and three *d* atomic orbitals gives seven sp^3d^3 hybrid orbitals which are directed towards the corners of a pentagonal bipyramid. These are not equivalent hybrid orbitals. Five of them point towards the vertices of a regular pentagon while others are perpendicular to the plane.

Shape of IF_7

It is a case of sp^3d^3 hybridisation. At. no. of iodine is 53. The outer electronic configuration of central atom in the ground state is $5s^2\ 5p_x^2\ 6p_z^2$ as shown in Fig. 4.33.

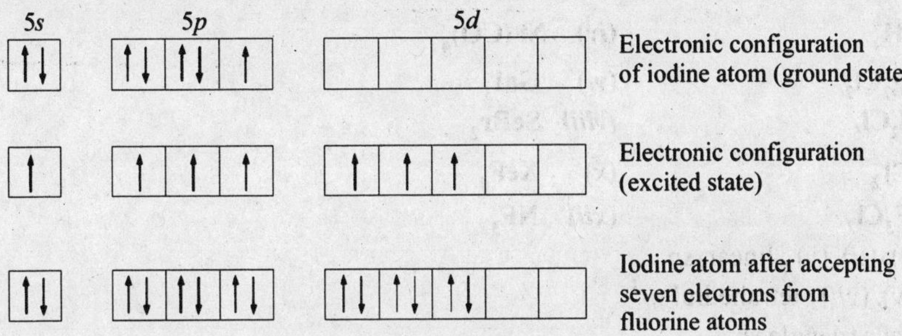

Fig. 4.33. sp^3d^3 hybridisation.

To have the requisite number of half-filled orbitals for hybridisation purposes, unpairing of 5*s* as well as of $5p_x$ and $5p_y$ electrons takes place. Thus seven unpaired electrons are available for bond formation to give rise to IF_7 molecule. The resulting molecule acquires pentagonal bipyramidal shape shown in Fig. 4.34.

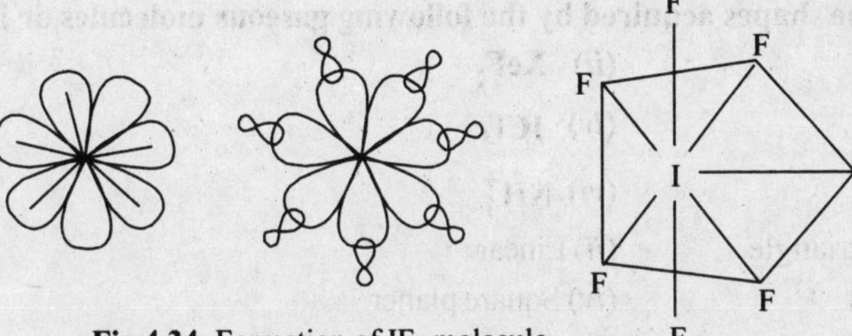

Fig 4.34. Formation of IF_7 molecule.

In this case all the seven hybrid orbitals are not equivalent. Five of them are directed towards the vertices of a regular pentagon giving an angle 72° with one another, the rest of two are oriented at right angle to the plane of the first set of five orbitals. Thus the bond angles in the molecule are of 72° and 90°.

Q. 29. Give main points of Valence Shell Electron Pair Repulsion Theory (VSEPR).

(Lucknow, 2000; Kerala, 2000; Garhwal, 2010)

Ans. The main points of VSEPR theory are as follows:

(*i*) Geometry of a molecule depends upon the total number of valence shell electron pairs (bonded or non-bonded) around the central atom in the molecule.

(*ii*) Repulsion between valence shell electron pairs determines molecular shape. In order to minimise electron-electron repulsion or to acquire a state of minimum energy or maximum stability the valence shell electron pairs keep as far apart as possible.

(*iii*) In case more than one geometrical arrangement is possible for a given number of bonded and non-bonded pairs, then the most stable arrangement is in accordance with the following:

(*a*) Lone pair-lone pair (*lp-lp*) repulsion is greater than lone pair-bond pair (*lp-bp*) repulsion which in turn is greater than bond pair-bond pair *(bp-bp)* repulsion. That is

$$lp\text{-}lp > lp\text{-}bp > bp\text{-}bp$$

(*b*) Repulsion between electron pairs at vertices greater than 115° apart can be neglected.

Q. 30. Pick up species with sp^3d^3 hybridisation.

(*i*) XeF_6 (*ii*) IF_7 (*iii*) BrF_5 (*iv*) $XeOF_4$ (*Delhi, 2007*)

Ans. XeF_6, IF_7.

Q. 31. Give shapes and hybridisations of following molecules or ions:

(*i*) $(Me)_2Be$ (*ii*) CS_2

(*iii*) BH_4^- (*iv*) $Ni(CO)_4$

(*v*) Hg_2Cl_2 (*vi*) GaI_3

(*vii*) Al_2Cl_5 (*viii*) $SeBr_2$

(*ix*) ICl_2^- (*x*) XeF_4

(*xi*) PF_3Cl_2 (*xii*) NF_3

Ans. (*i*), (*ii*), (*v*) : linear, *sp*

(*iii*), (*iv*), (*vii*) : tetrahedral sp^3

(*vi*) Plane triangle sp^2

(*viii*) Angular (V-shaped), sp^3

(*xii*) Pyramidal, sp^3

(*ix*) Linear sp^3d

(*x*) Square planner, sp^3d

(*xi*) Trigonal bipyramidal, sp^3d^2

Q. 32. Give the shapes acquired by the following gaseous molecules or ions.

(*i*) SO_3 (*ii*) XeF_2 (*Awadh, 1999*)

(*iii*) CS_2 (*iv*) ICl_4^-

(*v*) SCl_2 (*vi*) NH_4^+

Ans. (*i*) Plane triangle (*ii*) Linear

(*iii*) Linear (*iv*) Square planer

(*v*) Angular (*vi*) Tetrahedral

Q. 33. Indicate the species with sp^3 hybridisation of central atom.

(*i*) SO_4^{2-} (*ii*) $AsCl_3$

(*iii*) BF_4^- (*iv*) ClO_4^-

(*v*) ClO_2^- (*vi*) NH_4^+

(*vii*) OCl_2 (*viii*) SiX_4

Ans. All are sp^3 hybridised.

Q. 34. Molecules CH_4, NH_3 and H_2O involve the same kind of sp^3 hybridisation, yet they possess different geometry. (*Kerala, 2000*)

Ans. One of the tetrahedral positions in NH_3 is occupied by lone pair of electrons. Similarly in the case of water two positions are occupied by lone pair of electrons.

The repulsion between lone pair-bond pair is greater than that between bond pair and bond pair. As a result ∠HNH is decreased to 107°. There are still greater repulsion in the case of water. Here the ∠HOH is further decreased to 105°.

Q. 35. Explain: *lp-lp* repulsion > *lp-bp* repulsion > *bp-bp* repulsion. *(Indore, 2001)*

Ans. A bond pair is under the influence of two nuclei of bonding atoms whereas the lone pair is under the influence of only one nucleus. Hence lone pair is more active and mobile in causing repulsion with other orbitals. As a result, *lp-lp* repulsion is greater than *lp-bp* repulsion which in turn is greater than *bp-bp* repulsion.

Q. 36. Giving reason arrange the following molecules in order of increasing bond angle

H_2O, NH_3, PH_3 *(Indore, 2001; Banaglore, 2005)*

Ans. The order of increasing bond angle in

$$H_2O < PH_3 < NH_3$$

In case of water, there are two lone pair of electrons, therefore repulsion between lone pair and bond pair will be maximum and angle will be minimum.

Out of NH_3 and PH_3 repulsion between bond pair and bond pair will be smaller in case of PH_3 because of smaller electronegativity of phosphorus and larger P-H bond length and hence ∠HPH will be smaller than ∠HNH.

MOLECULAR ORBITAL THEORY

Q.37. Give the main points of molecular orbital theory of covalent bonding.

(M. Dayanand, 2002)

Ans. The main points of molecular orbital theory are:

(*i*) Electrons belonging to two bonding atoms are controlled by the entire molecule. The electrons will occupy molecular orbitals that are analogous to atomic orbitals. These molecular orbitals belong to the entire molecule.

(*ii*) In an atom there are *s, p, d* .... atomic orbitals determined by various values of quantum numbers. In a molecule we have a set of molecular orbitals σ, π, δ ..... determined by various quantum numbers.

(*iii*) Just as an atomic orbital can contain a maximum of two electrons, molecular orbitals can also accommodate only a maximum of two electrons as per Pauli's exclusion principle.

(*iv*) The various electrons occupy different orbitals according to Hund's rule of maximum multiplicity as in the case of atomic orbitals.

Q. 38. What is linear combination of atomic orbitals (LCAO)?

(Kurukshetra 2000; Nagpur, 2003)

Ans. Molecular orbitals can be visualised as being made up of appropriate atomic orbitals provided by the atoms joined together in the molecule. A very simple, reasonable and satisfactory model is obtained if it is assumed that atomic orbitals of constituents combine in a linear additive fashion to form molecular orbitals. Consider two atoms *A* and *B* each possessing a single valence electron. Let ψ_A and ψ_B represent the wave functions of atoms *A* and *B*. Let atomic orbitals A and B accommodating valence electrons overlap. The valence electrons will be in region where the orbitals overlap and therefore, they can be associated with either nucleus. Consequently, the valence electrons have a much larger volume available for their movement as compared with that available in each isolated atom. The electrons can be thought of as moving in a molecular orbital whose total volume is approximately equal to the sum of the volume of atomic orbitals of *A* and *B*. In this fashion, the wave function of molecular orbitals can be expressed as a linear combination of atomic orbitals (LCAO).

ψ_{AB} is the wave function of molecular orbital.

$$\psi_{AB} = C_A \psi_A + C_B \psi_B$$

and C_A and C_B are mixing coefficients. Value of C_A and C_B is so chosen as to get a molecular orbital of lowest energy.

In case we are considering the bonding in two similar atoms (homonuclear) then the magnitude of the coefficients C_A and C_B must be same.

In that case

$$C_A^2 = C_B^2 \text{ or } C_A = \pm C_B$$

There will be two molecular orbitals as a result of two types of combinations as given below.

$$\psi^b = \psi_A + \psi_B$$

$$\psi^* = \psi_A - \psi_B$$

Values of ψ_A and ψ_B have been taken as unity.

Thus one molecular orbital (ψ^b) can be obtained by addition of wave function of two atomic orbitals and then the other molecular orbital (ψ^*) can be obtained by the subtraction of wave function of the atomic orbitals. This is shown in Fig. 4.35.

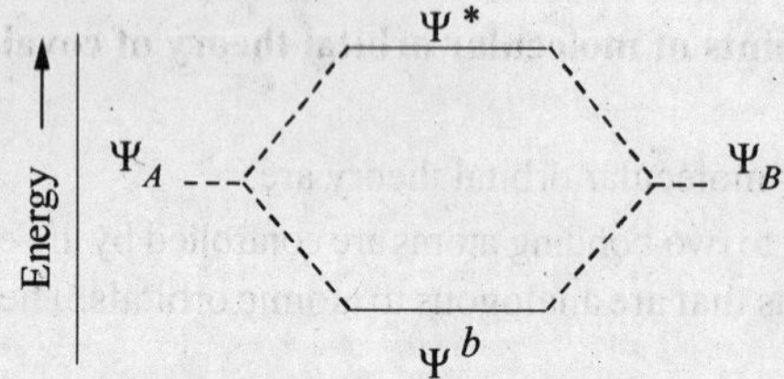

Fig. 4.35. Formation of MO.

The number of molecular orbitals is always equal to the number of atomic orbitals involved in bonding. Effective combination of atomic orbitals to give rise to stable molecular orbitals requires:

(*i*) Symmetry and relative energies of atomic orbitals ψ_A and ψ_B must be close to each other.

(*ii*) Extent of overlap between the charge clouds of ψ_A and ψ_B should be large.

(*iii*) Atomic orbitals of same sign should overlap.

Q. 39. What are bonding and antibonding molecular orbitals?

(*Nagpur, 2005*)

Ans. Bonding orbitals. The molecular orbital that is obtained by the addition overlap of two atomic orbitals is termed bonding molecular and is represented as ψ^b

$$\psi^b = \psi_A + \psi_B$$

Some features of bonding molecular orbitals are:

(*i*) It possesses lower energy than the atomic orbitals from which it is formed.

(*ii*) It has high electron density in the region between the two nuclei and this accounts for stability of bond.

(*iii*) It is formed when lobes of the combining atomic orbitals have been same sign. The magnitude of ψ^b is numerically bigger than ψ_A or ψ_B.

(*iv*) Every electron in a bonding M.O. contributes to attraction between two atoms.

Antibonding M.O. The other molecular orbital which is obtained by the subtraction overlap of two atomic orbitals is termed the antibonding molecular orbital and is represented as ψ^*.

$$\psi^* = \psi_A - \psi_B$$

Some features of antibonding molecular orbitals are:

(*i*) It possesses higher energy than the atomic orbitals from which it is formed.

(*ii*) The probability of finding the electron in between the nuclei is negligible. The electron density is concentrated only on one nucleus at any particular moment.

(*iii*) It is formed when the lobes of combining atomic orbitals have opposite signs. ψ^* is numerically smaller than ψ_A or ψ_B.

(*iv*) Every electron in an antibonding molecular orbital contributes to repulsion between two atoms.

Q. 40. Describe the combination of *s* orbitals to give bonding and antibonding molecular orbitals.

Ans. Let us consider the combination of two 1*s* atomic orbitals belonging to say two hydrogen atoms. Using LCAO approach the two atomic orbitals will overlap to give two molecular orbitals. The first combination is obtained by the addition of two atomic orbitals. The pictorial representation is shown in Fig. 4.36.

ψ^b MO $= N(\psi_{1s_a} + \psi_{1s_b})$ where N is a normalizing constant.

For simplicity its value may be taken as 1

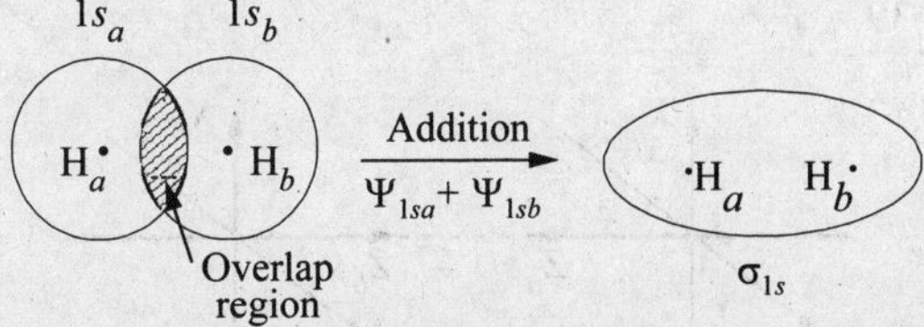

Fig. 4.36. Formation of a bonding MO obtained by the addition overlap of 1*s* atomic orbitals.

The electron in this MO will spend most of its time between the two nuclei H_a and H_b. The electron is now stabilized by two nuclear charges instead of one in an atomic orbital. The resulting MO is of lower energy than the 1*s* orbital. That is why it is called a bonding MO.

The other linear combination is obtained by subtracting the two atomic orbitals and is shown in Fig. 4.37.

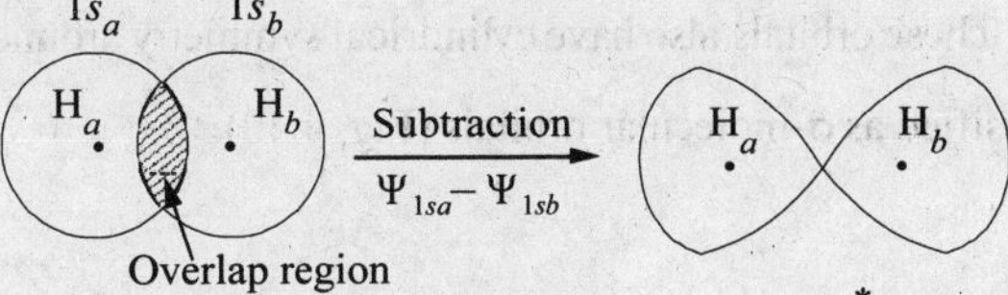

Fig. 4.37. Formation of an antibonding MO formed by the subtraction overlap of 1*s* atomic orbitals.

$\psi^*_{MO} + N^*(\psi_{1s_a} - \psi_{1s_b})$ where N^* is a normalizing constant.

The resulting molecular orbital has zero electron density (Nodal Plane) between the nuclei. The electron in this orbital is concentrated in regions away from each nuclei. The net effect is to push the nuclei apart rather than to hold them together. The resulting molecular orbital is of higher energy than the isolated 1*s* atomic orbital. The resulting MO is called an antibonding molecular orbital. The molecular orbitals discussed above have cylindrical symmetry and are called σ (sigma) molecular orbitals.

The combination of 2*s* atomic orbitals also gives rise to σ-molecular orbitals. The situation is exactly like the overlapping of 1*s* atomic orbitals. The only difference is that their sizes are different. Their overlap is illustrated in a diagram below (Fig. 4.38).

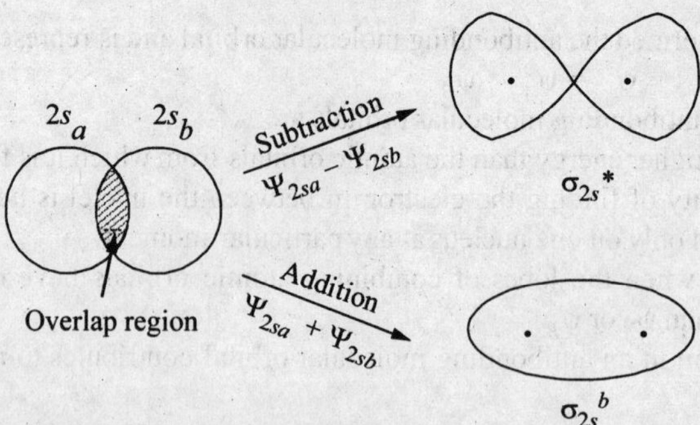

Fig. 4.38. Formation of bonding and antibonding MO's obtained by the overlap of 2*s* atomic orbitals.

Q. 41. Describe the formation of σ molecular orbitals resulting from the overlapping of *p*-orbitals along the internuclear axis. *(Delhi, 2006)*

Or

Draw the shape of MO obtained by the overlap of $p_z - p_z$ atomic orbitals.

Ans. σ-type molecular orbitals are also formed when p_z atomic orbitals of two atoms combine with each other. We are now assuming that the internuclear axis is the *z*-axis. This is illustrated in a diagram shown in Fig. 4.39.

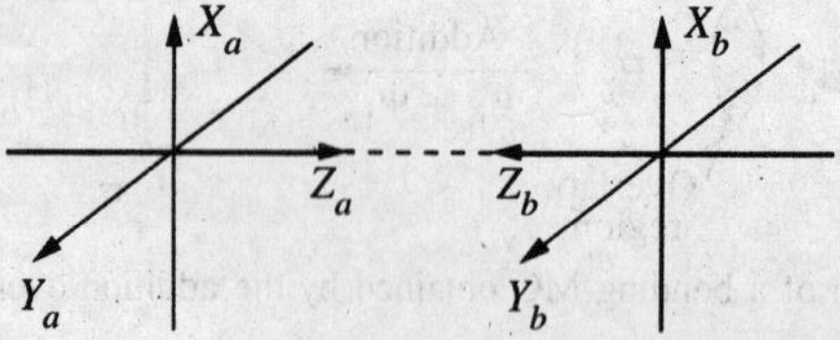

Fig. 4.39. Co-ordinate system for the overlap of 2*p* atomic orbitals.

The molecular orbital formed by the addition, overlap of $2p_z$ atomic orbital is designated as σ^b_{2pz} or simply as σ^b_z while the molecular orbital formed by the subtraction of $2p_z$ atomic orbitals is designated as $\sigma^*_{2p_z}$ or simply σ^*_z. These orbitals also have cylindrical symmetry around the molecular axis and hence they are also classified as σ-molecular orbitals (Fig. 4.40).

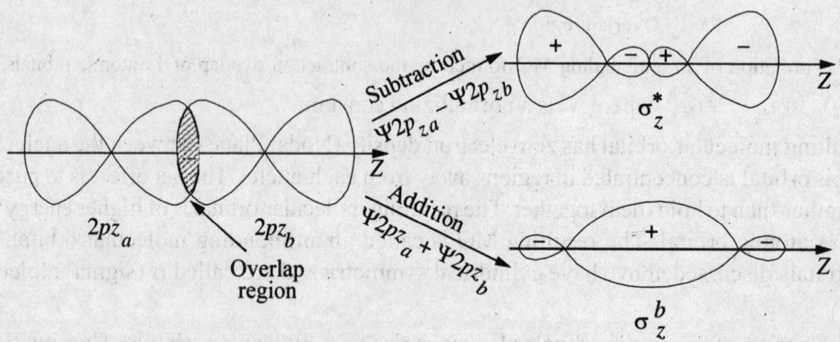

Fig. 4.40. Diagram showing the formation of bonding and antibonding MO's by the overlap of $2p_z$ atomic orbitals

Q. 42. Describe the formation of π orbitals by the lateral overlap of *p*-atomic orbitals.

(*Delhi, 2006*)

Or

Draw the shape of MO obtained by the overlap of $p_x - p_x$ (or $p_y - p_y$) atomic orbitals.

Ans. Molecular orbitals which are obtained by the lateral (or sidewise) overlap of p_x or p_y orbitals are designated as π (pi) molecular orbitals. This is illustrated by showing overlap of $2p_x$ atomic orbitals.

Overlaps containing the same algebraic signs yield bonding MO and those of opposite sign give antibonding M.O. (Fig. 4.41)

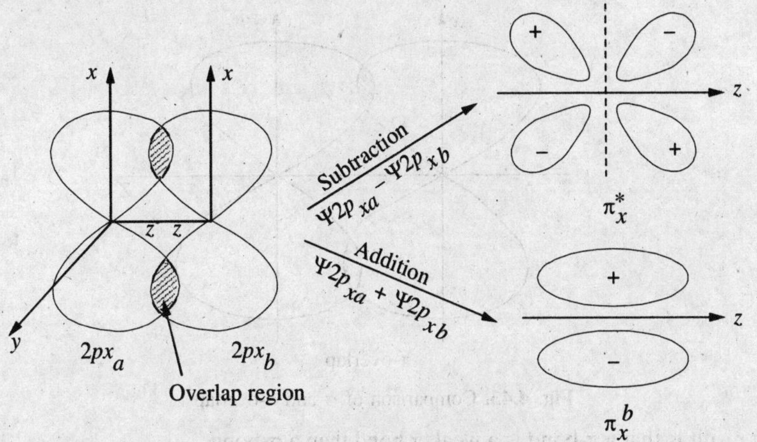

Fig. 4.41. Diagram showing the formation of bonding and antibonding MO's by the overlap of $2p_x$ atomic orbitals.

The bonding molecular orbital has (+) lobe above the *z*-axis and (–) lobe below the z-axis. The plus and minus signs refer to the sign of wave function in the various lobes and not to nuclear or electronic charges. The (+ve) sign (or upward direction) represents one phase and (–ve) sign (or downward direction) represents second phase of wave function. If we rotate this MO through 180°, it changes sign (Fig. 4.42). A molecular orbital which is antisymmetric about the molecular axis is called a π (pi) molecular orbital. The antibonding MO is also a π type M.O. They are labelled as π_x^b and π_x^* orbitals.

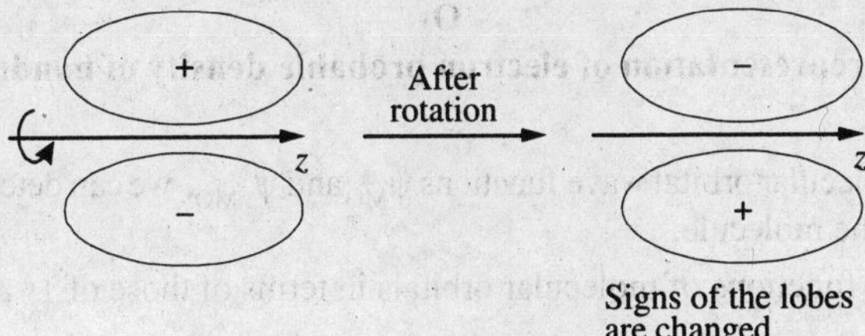

Fig. 4.42. Rotation of bonding MO.

The two $2p_y$ atomic orbitals will overlap in a similar fashion to give π_y^b and π_y^*. Both π_x and π_y molecular orbitals are equivalent and have the same energy. These orbitals are said to be degenerate, *i.e.*, orbitals with same energy.

Q. 43. Compare the electron overlap in σ and π molecular orbitals.

Ans. In a σ-overlap, the overlap region is much greater than in a π–overlap. The difference arises because in a σ-overlap, two orbitals involve head-on overlap whereas in a π–overlap, the two orbitals overlap laterally (or sidewise) as shown in diagram (Fig. 4.43). It is assumed that z-axis is the molecular axis.

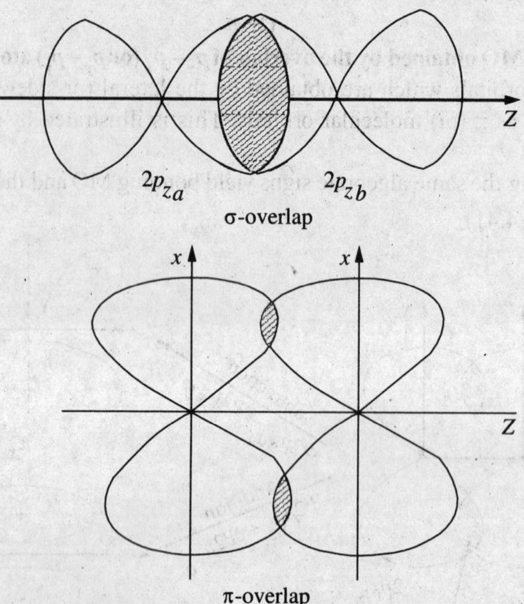

Fig. 4.43. Comparison of σ and π-overlaps.

The net result is that a π-bond is a weaker bond than a σ-bond.

Q. 44. Show by LCAO method as to which of these pairs of AO's result in positive overlap

$$s + p_x$$
$$s + p_z$$
$$p_x + p_z$$
$$p_z + p_z$$

***z*-axis may be assumed as a molecular axis.**

Ans. $s + p_z$ and $p_z + p_z$ will result in positive overlap.

Q. 45. Discuss the variation of electronic charge with internuclear distance.

Or

Show graphical representation of electron probable density of bonding and antibonding molecular orbitals.

Ans. From the molecular orbital wave functions ψ_{MO} and ψ^*_{MO}, we can determine the distribution of electron charge in the molecule.

Writing the wave functions of molecular orbitals in terms of those of 1*s* atomic orbitals of two atoms *a* and *b*.

$$\psi_{MO} = \psi_{1s_a} + \psi_{1s_b}$$
$$\psi^*_{MO} = \psi_{1s_a} - \psi_{1s_b}$$

Taking squares of both sides of the above equations

$$\psi^2_{MO} = [\psi^2_{1s_a} + \psi^2_{1s_b} + 2\psi_{1s_a} \ \psi_{1s_b}]$$
$$\psi^{*2}_{MO} = [\psi^2_{1s_a} + \psi\psi^2_{1s_b} - 2\psi_{1s_a} . \psi_{1s_b}]$$

It is observed that the bonding MO functions shows an increase in the electron charge density in the overlap region by an amount $2\psi_{1s_a}\ \psi_{1s_b}$ over that of the individual atoms as described by the functions $\psi^2_{1s_a}$ and $\psi^2_{1s_b}$.

This is represented graphically in Fig. 4.44. The dotted lines represent the charge densities of individual atoms and heavy lines represent the electron density for bonding MO.

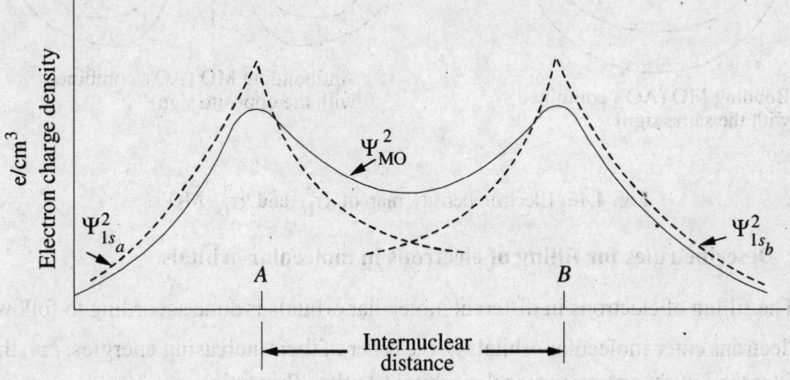

Fig. 4.44. Graph of electron density vs. distance results.

On the other hand the antibonding MO functions results in decrease in the charge density in the overlap region by an amount $\psi_{1s_a}\ \psi_{1s_b}$. This is represented in Fig. 4.45. It is also clear that bonding MO favours electron density as it is concentrated between the nuclei, whereas the antibonding MO tends to decrease the charge density between the nuclei and concentrate it in regions on the individual atoms.

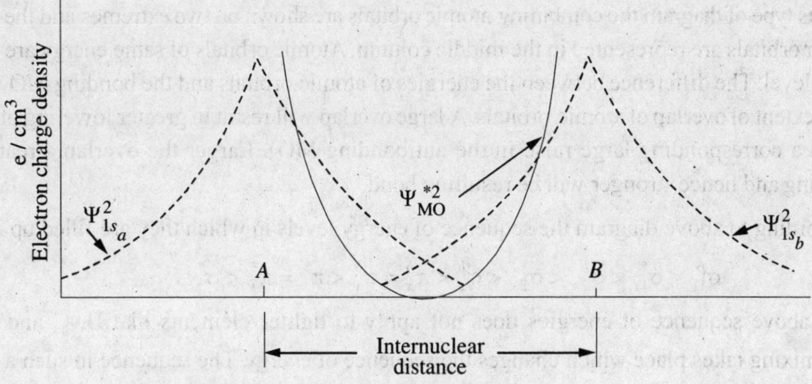

Fig. 4.45. Graph of electron density vs. internuclear density.

Electron density maps for bonding and antibonding (σ_{1s}^{b} and σ_{1s}^{*}) MO's are shown in Fig. 4.46.

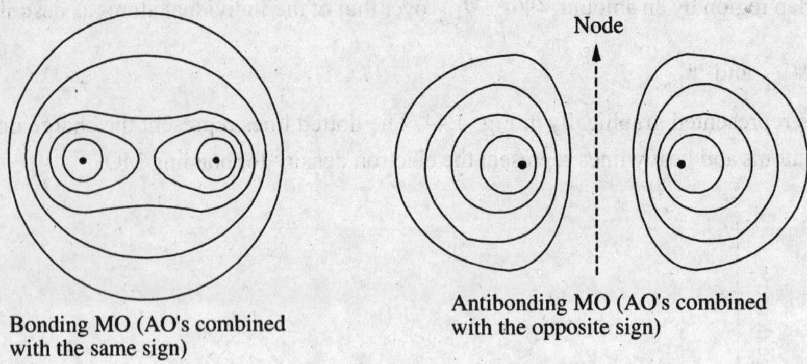

Fig. 4.46. Electron density map of σ_{1s}^{b} and σ_{1s}^{*} MO's.

Q. 46. Describe rules for filling of electrons in molecular orbitals.

Ans. The filling of electrons in different molecular orbitals is done according to following rules:

(*i*) Electrons enter molecular orbitals in the order of their increasing energies, *i.e.*, the electron will first enter the lowest energy molecular orbital (Aufbau Principle).

(*ii*) Each molecular orbital can accommodate a maximum of two electrons with their spins in opposite direction in accordance with Pauli's exclusion principle.

(*iii*) In agreement with Hund's rule whenever two equal energy molecular orbitals are available, electrons prefer to remain unpaired.

(*iv*) Electrons (equal to the sum of the electrons actually present in combining atoms) are placed one at a time into molecular orbital to obtain the electronic configuration of the molecule.

Q. 47. Give the energy level diagram of molecular orbitals. *(Pune 2010)*

Ans. Energy level diagram of MO's is shown in Fig. 4.47.

In this type of diagram the combining atomic orbitals are shown on two extremes and the resulting molecular orbitals are represented in the middle column. Atomic orbitals of same energy are shown at the same level. The difference between the energies of atomic orbitals and the bonding MO's depend upon the extent of overlap of atomic orbitals. A large overlap will result in greater lowering of bonding MO (and a corresponding large raise in the antibonding MO). Larger the overlap, greater is the overlapping and hence stronger will be resulting bond.

According to above diagram the sequence of energy levels in which they are filled up is:

$$\sigma_{1s}^{b} < \sigma_{1s}^{*} < \sigma_{2s}^{b} < \sigma_{2s}^{*} < \sigma_{z}^{b} < \pi_{x}^{b} = \pi_{y}^{b} < \pi_{x}^{*} = \pi_{y}^{*} < \sigma_{z}^{*}$$

The above sequence of energies does not apply to lighter elements like B, C and N. Here $2s - 2p_z$ mixing takes place which changes the sequence of energy. The sequence in such a case is

$$\pi_{x}^{b} = \pi_{y}^{b} < \sigma_{z}^{b} < \pi_{x}^{*} = \pi_{y}^{*} < \sigma_{z}^{*}$$

Lower molecular orbitals have the same energy sequence.

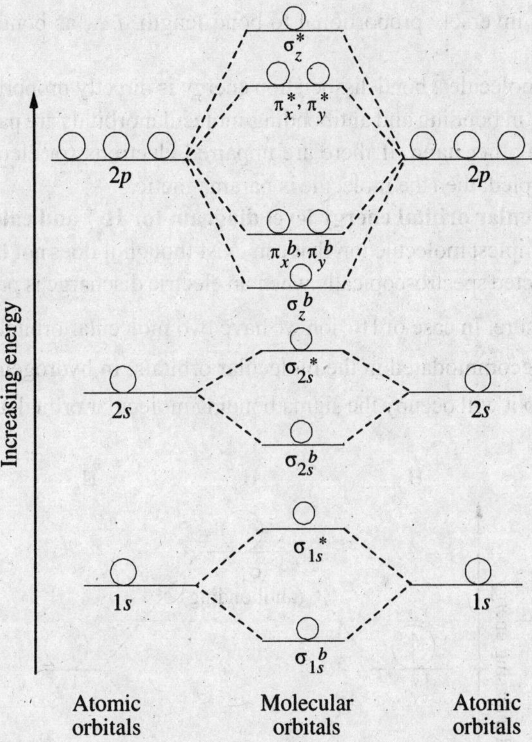

Fig. 4.47. Energy level diagram.

Q. 48. How is a bond order calculated from number of electrons in bonding and antibonding molecular orbitals? What information is conveyed by it?

Ans. The number of electrons belonging to two atoms are counted and then filled in the energy level diagram as constructed above. It is possible to predict the behaviour of molecules from their configuration. All the electrons occupying bonding molecular orbitals (N^b) and antibonding molecular orbitals (N^*) are counted separately. The following cases arise.

(*a*) $N^b > N^*$. If the number of electrons in bonding MO's are greater than the number of electrons in antibonding MO's, then the molecule would be stable.

(*b*) $N^b = N^*$. If the number of electrons in bonding MO's are equal to the number of electrons in antibonding MO's, then the molecule is unstable and it does not exist. This is because an electron in an antibonding MO cancels out the stability gained by electron in the bonding MO.

Bond order. It is defined as the number of covalent bonds present in a molecule. It is calculated as

$$\text{Bond order} = \frac{1}{2}\left\{\begin{matrix}\text{number of electrons in} \\ \text{bonding MO's}\end{matrix} - \begin{matrix}\text{Number of electrons in} \\ \text{antibonding MO's}\end{matrix}\right\}$$

Information Conveyed by Bond Order

(*i*) The number gives nature of a bond whether it is a single, double or a triple bond.

(*ii*) When the bond order comes out to be zero (*i.e.*, when the number of electrons in bonding MO's is equal to the number of electrons in antibonding MO's), then the molecule does not exist.

(*iii*) Bond order is inversely proportional to bond length, *i.e.*, as bond order increases, bond length decreases.

(*iv*) For diatomic molecules, bond dissociation energy is directly proportional to the bond order.

(*v*) If all electrons in bonding and antibonding molecular orbitals are paired, then the molecule is diamagnetic. On the other hand, if there are unpaired electrons (or electrons), *i.e.*, molecular orbitals are singly occupied, then the molecule is paramagnetic.

Q. 49. Draw molecular orbital energy level diagram for H_2^+ and calculate the bond order.

Ans. This is the simplest molecule ion that can exist though it does not have any real existence. Its presence can be detected spectroscopically when an electric discharge is passed through hydrogen gas under reduced pressure. In case of H_2^+ ion we have two molecular orbitals σ_{1s}^b and σ_{1s}^*. In total four electrons can be accommodated in the molecular orbitals. In hydrogen molecule ion we have only one electron and so it will occupy the sigma bonding molecular orbital as shown in Fig. 4.48.

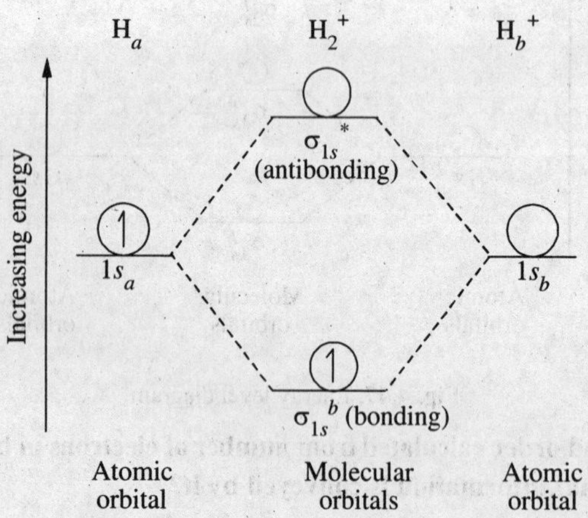

Fig. 4.48. Energy level diagram of H_2^+ ion.

Thus H_2^+ molecule ion will have the electronic configuration $[(\sigma_{1s}^b)^1]$

The bond order in $H_2^+ = 1/2\,[1 - 0] = 1/2\ \sigma$ bond

Since the bond order in H_2^+ is only one-half of a normal covalent bond, it will have low bond dissociation energy and large bond length.

Bond dissociation energy of H_2^+ = 255 kJ/mole. Bond length = 106 pm.

Q. 50. Draw molecular energy level diagram for hydrogen molecule (H_2) and calculate the bond order. (*Nagpur, 2000; G. Nanakdev, 2000; Awadh, 2000*)

Ans. The hydrogen molecule is formed by the overlap of 1*s* atomic orbitals. It shall have two molecular orbitals, one bonding (σ_{1s}^b) and the other antibonding σ_{1s}^*). The H_2 molecule has two electrons and these can be accommodated in the σ_{1s}^b molecular orbital. In accordance with Pauli exclusion principle, these electrons must have opposite spins. The energy level diagram of H_2 is shown in Fig. 4.49.

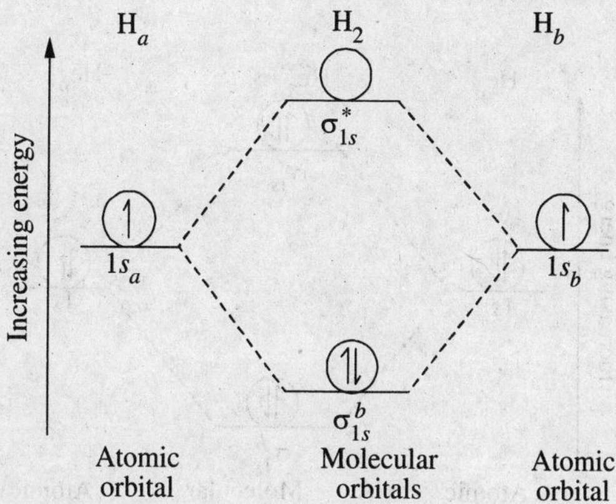

Fig. 4.49. Energy level diagram of H_2.

The ground state of H_2 is represented as $[(\sigma_{1s}^b)^2]$ and the bond order is 1/2 (2 – 0) = 1, *i.e.*, one σ-bond. Bond dissociation energy in H_2 is 431.4 kJ/mole and bond length is 77 pm.

Q. 51. Draw molecular energy level diagram in He_2^+.

Ans. Here three electrons are to be filled in the molecular orbitals (Fig. 4.50). The configuration is $\left(\sigma_{1s}^b\right)^2\left(\sigma_{1s}^*\right)$.

The bond order is: $\frac{1}{2}[2-1]=\frac{1}{2}$. This is an hypothetical ion.

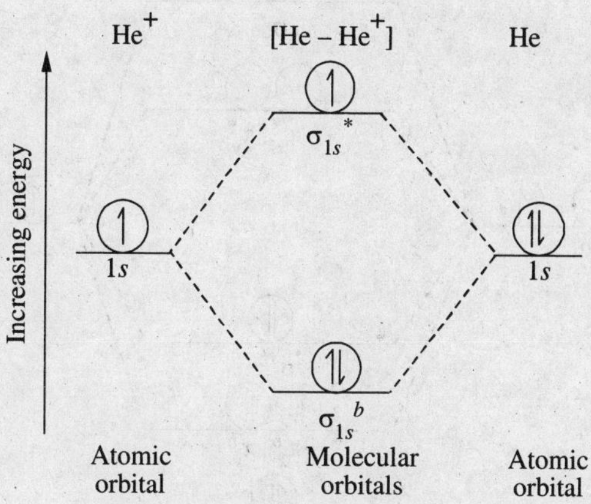

Fig. 4.50. Energy level diagram of He_2^+.

Q. 52. Draw molecular orbital energy level diagram for He_2 molecule and comment on the stability of the molecule. (*M. Dayanand, 2000*)

Ans. If a He_2 molecule is to be formed by the combination of helium ions, a total of four electrons are involved. From a combination of two 1*s* atomic orbitals we have two molecular orbitals σ_{1s}^b and σ_{1s}^*. Two electrons would go in the bonding σ_{1s}^b MO and the other two would have to go into the antibonding σ_{1s}^* MO (Fig. 4.51).

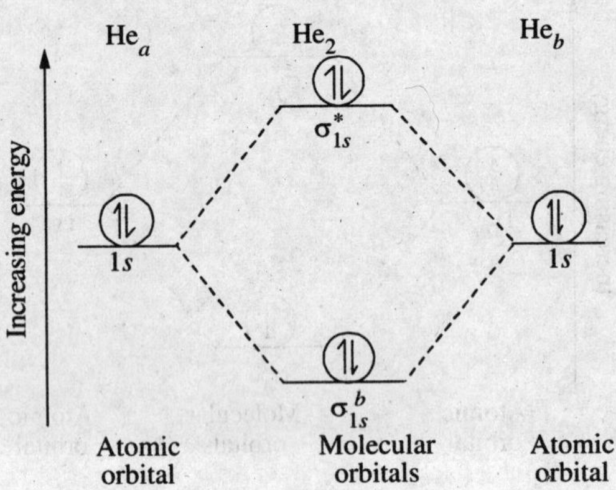

Fig. 4.51. Energy level diagram of He_2.

The electronic configuration of He_2 may be written as $[(\sigma_{1s}^{b})^2 (\sigma_{1s}^{*})^2]$.

The bond order is 1/2 [2 – 2] = 0. In effect the bonding and antibonding orbitals cancel each other.

Q. 53. Draw energy level diagram for the formation of N_2 molecule and comment on its properties.

(*Kanpur, 2001; Bangalore, 2000; Guwahati, 2007; Nagpur 2008; Andhra 2010*)

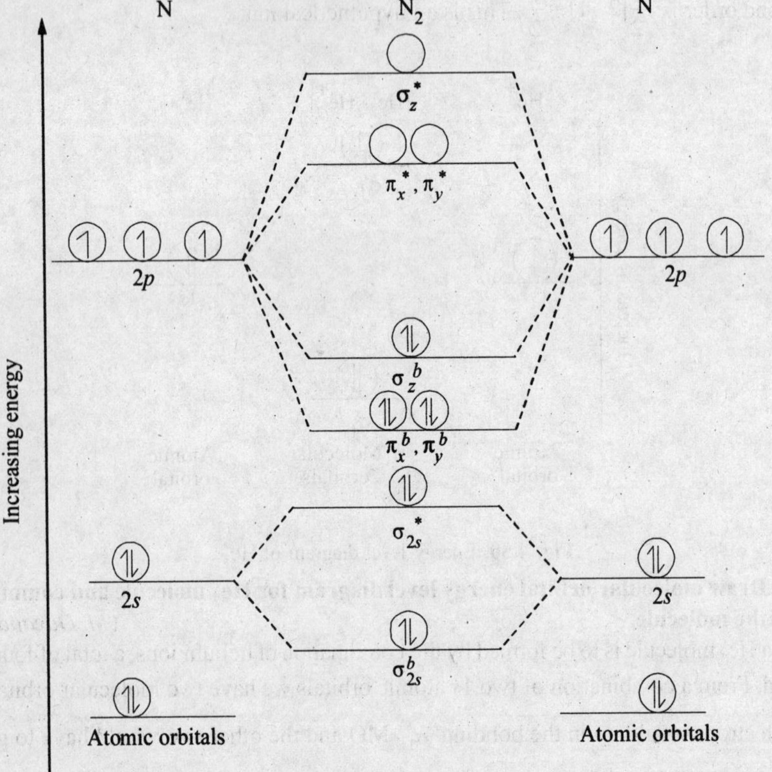

Fig. 4.52. Molecular orbital energy level diagram of hypothetical N_2 molecule.

Ans. The nitrogen atom has electronic configuration $1s^2, 2s^2, 2p^3$. The N_2 molecule is well known. Its molecular orbital configuration may be written as $[KK(\sigma_{2s}^b)^2 (\sigma_{2s}^*)^2 (\pi_x^b)^2 (\pi_y^b)^2 (\sigma_z^b)^2]$. KK signifies that the K shells of the combining atoms are full. It has eight electrons in bonding and two electrons in antibonding orbitals. Since there are six electrons in excess in bonding orbitals so the bond order in N_2 is three. The molecule has 1σ and 2π bonds. This is in conformity with its very high bond energy of 940 kJ/mole and very short bond length of 110 pm. The energy level diagram is shown in Fig. 4.52.

Q. 54. Draw molecular orbital energy level diagram for oxygen molecule (O_2). Discuss its properties. *(G.N.D.U., 2000; Kanpur, 2001; Awadh, 2000; Nagpur, 2005, Bangalore, 2003; Guwahati 2006)*

Ans. The oxygen has electronic configuration $1s^2, 2s^2, 2p^4$. The O_2 molecule is well known and shows paramagnetism (Fig. 4.53), equivalent to that of two unpaired electrons. The molecule has the following configuration:

$$[KK\,(\sigma_{2s}^b)^2\,(\sigma_{2s}^*)^2\,(\sigma_z^b)^2\,(\pi_x^b)^2(\pi_y^b)^2\,(\pi_y^*)^1\left(\pi_y^*\right)^1]$$

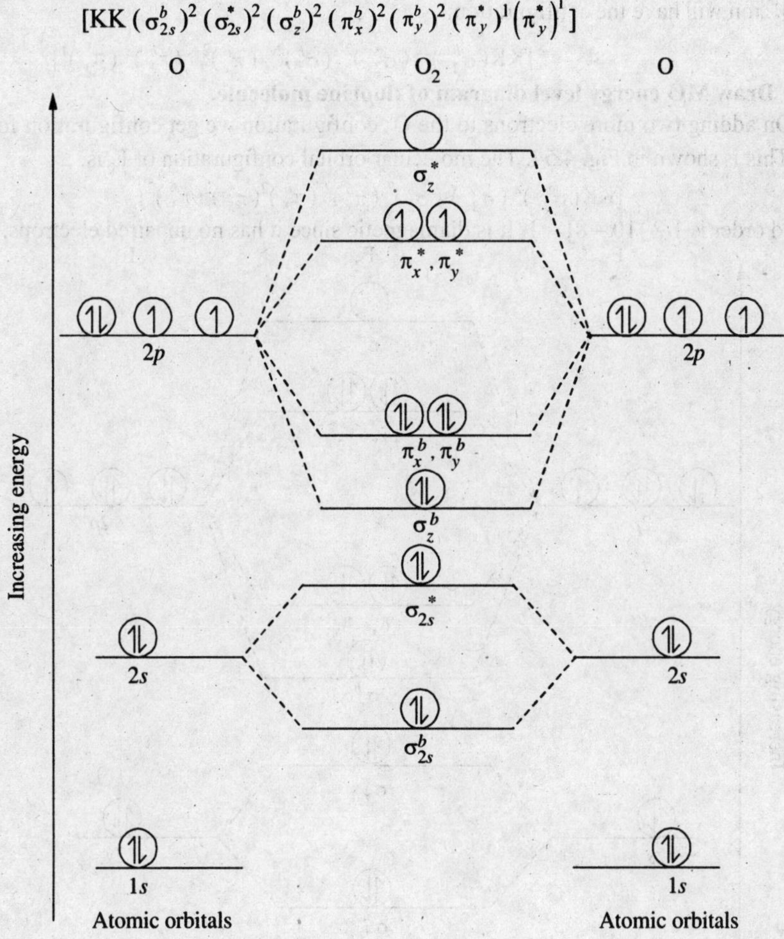

Fig. 4.53. Molecular orbital energy level diagram of O_2.

$$[\text{Bond order} = \frac{1}{2}(8-4) = \frac{1}{2}\times 4 = 2$$

The bond order is two. In accordance with Hund's rule the two electrons in the π^* orbitals have parallel spins. The molecular orbital method explains the presence of two unpaired electrons in the molecule which accounts for its paramagnetism (Fig. 4.53). The valence bond theory had failed to predict the presence of unpaired electrons in O_2 molecule.

This is in accordance with the experimental results that O_2 consists of a double bond (one σ and one π bond).

Bond dissociation energy = 493.2 kJ/mol

Bond length = 121 pm.

Q. 55. Draw MO energy level diagram for O_2^+ ion. Comment on its bond order and magnetic properties.

Ans. There are 12 – 1 = 11 electrons to be filled in the LL shells.

The bond order is = 1/2 (8 – 3) = $2\frac{1}{2}$. Thus it has a stronger bond compared to oxygen. O_2^+ is paramagnetic.

The O_2^+ ion will have the configuration:

$$[KK(\sigma_{2s}^b)^2\ (\sigma_{2s}^*)^2\ (\sigma_z^b)^2\ (\pi_x^b)^2\ (\pi_y^b)^2\ (\pi_y^*)^1]$$

Q. 56. Draw MO energy level diagram of fluorine molecule.

Ans. On adding two more electrons to the O_2 configuration we get configuration for fluorine molecule. This is shown in Fig. 4.54. The molecular orbital configuration of F_2 is:

$$[KK(\sigma_{2s}^b)^2\ (\sigma_{2s}^*)^2 (\sigma_z^b)^2\ (\pi_x^b)^2\ (\pi_y^b)^2 (\pi_x^*)^2 (\pi_y^*)^2]$$

and its bond order is 1/2 [10 – 8] = 1. It is diamagnetic since it has no unpaired electrons.

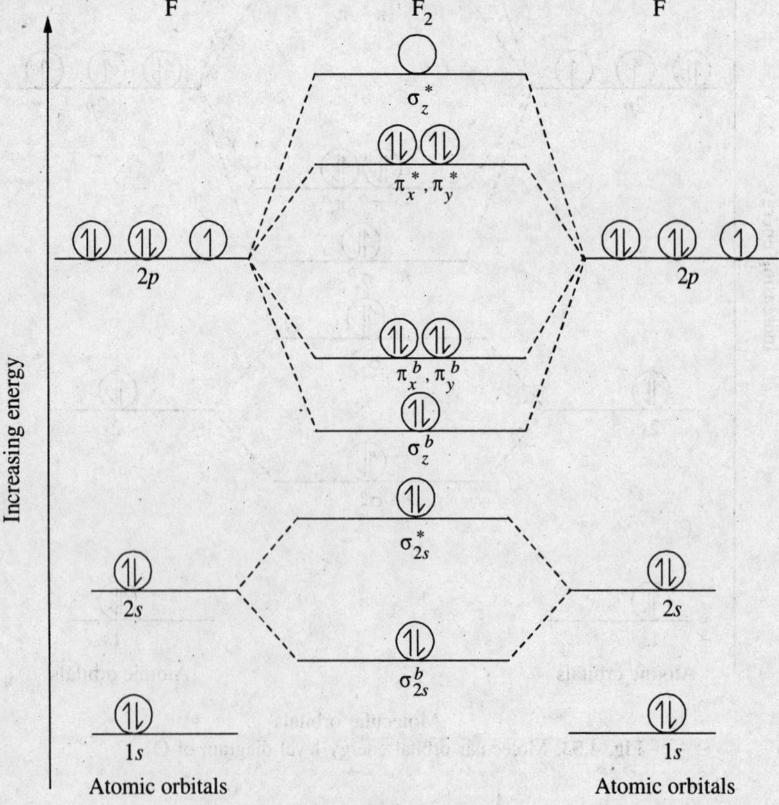

Fig. 4.54. Molecular orbital energy level diagram of F_2.

Q. 57. How does the MO theory account for the following?

(i) Bond order in O_2^+ is more than in O_2 while (*G. Nanakdev, 2000*)

(ii) Bond order in N_2^+ is less than in N_2 molecule.

Ans. (*i*) In O_2^+, there are 11 valence electrons to be filled. The outer shell molecular orbital configuration is

$$(\sigma_{2s})^2\ (\sigma_{2s}^*)^2\ (\sigma_{2pz})^2\ (\pi_{2px})^2\ (\pi_{2py})^2\ (\pi_{2px}^*)^1$$

Bond order = $\frac{1}{2}$ (8 – 3) = 2.5

In O_2, there are 12 valence electrons to be filled. The outer shells MO configuration is

$$(\sigma_{2s})^2(\sigma_{2s}^*)^2(\sigma_{2pz})^2(\pi_{2px})^2(\pi_{2py})^2(\pi_{2px}^*)^1(\pi_{2py}^*)^1$$

Bond order = $\frac{1}{2}$ (8 – 4) = 2

(*ii*) In N_2^+, there are 9 valence electrons to be filled. The configuration is

$$KK\ (\sigma_{2s})^2(\sigma_{2s}^*)^2(\pi_{2px})^2(\pi_{2py})^2(\pi_{2pz})^1$$

Bond order = $\frac{1}{2}$ (7 – 2) = 2.5

In N_2, there are 10 electrons to be filled. The configuration is

$$KK\ (\sigma_{2s})^2(\sigma_{2s}^*)^2(\pi_{2px})^2(\pi_{2py})^2(\sigma_{2pz})^2$$

Bond order = $\frac{1}{2}$ (8 – 2) = 3

Q. 58. Explain: Bond energy of N_2 molecule is higher than that of O_2.

Ans. According to MO theory *in the case of N_2 molecule*

No. of electrons in bonding MO = 8

No. of electrons in antibonding MO = 2

Bond order = $\frac{1}{2}$ (8 – 2) = 3

In case of O_2 molecule

No. of electrons in bonding MO = 8

No. of electrons in antibonding MO = 4

Bond order = $\frac{1}{2}$ (8 – 4) = 2

Greater the bond order, greater the bond energy.

Q. 59. On the basis of MO theory, explain why?

(i) N_2 molecule is diamagnetic while O_2 is paramagnetic.

(*Lucknow, 2000; Kerala, 2000*)

(ii) Bond order in N_2^+ is lower than that in N_2.

Ans. (*i*) The MO configuration of N_2 and O_2 can be written as

N_2 : KK $(\sigma_{2s})^2(\sigma_{2s}^*)^2(\pi_{2px})^2(\pi_{2py})^2(\sigma_{2pz})^2$

O_2 : KK $(\sigma_{2s})^2(\sigma_{2s}^*)^2(\sigma_{2pz})^2(\pi_{2px})^2(\sigma_{2py})^2(\pi_{2px}^*)^1(\pi_{2py}^*)^1$

N_2 does not contain unpaired electron, it is diamagnetic. Oxygen is paramagnetic as it contains two unpaired electrons.

(*ii*) Bond order in $N_2^+ = \frac{1}{2}(7-2) = 2.5$

Bond order in $N_2 = \frac{1}{2}(8-2) = 3$

Q. 60. Draw MO energy level diagram for CO molecule.

(*G. Nanakdev, 2000; Lucknow, 2000; Himachal 2000; M. Dayanand, 2002; Madras, 2004*)

Ans. In CO, the oxygen atomic orbitals are lower in energy than the carbon atomic orbitals. Due to this difference, the σ_{2s}^* orbital is higher in energy than the σ_{2px} orbital (Fig. 4.55). Carbon monoxide possesses 10 valence electrons and, therefore, has a valence configuration :

$$(\sigma_{2s}^b)^2(\sigma_{2pz}^b)^2(\pi_x^b \cdot \pi_y^b)^4(\sigma_{2s}^*)^2$$

There are three net bonding pairs (one σ and one π) describing the C ≡ O molecule.

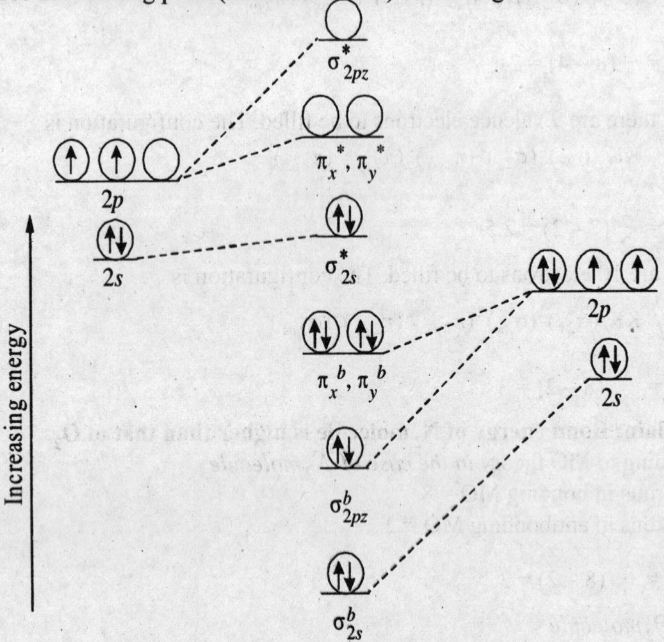

Fig. 4.55. MO diagram for carbon monoxide.

Q. 61. Write down MO electrons configuration for nitrogen oxide (NO) molecule and show that it is paramagnetic in nature.

(*Kanpur, 2001; Nagpur, 2005; Delhi 2006; Andhra 2010*)

Ans. NO (11 valence electrons) has an electron configuration :

$$(\sigma_{2s}^b)^2(\sigma_{2pz}^b)^2(\pi_x^b.\pi_y^b)^4(\sigma_{2s}^*)^2(\pi_x^*)^1$$

$$\text{Bond order in NO} = \frac{1}{2}(8-3) = 2.5$$

It is paramagnetic with a single unpaired electron. This is manifested by its easy oxidation to NO^+ (by loss of π^* electron). This is associated with an increase in bond order from $2\frac{1}{2}$ in NO to 3 in NO^+, and a decrease in bond length.

Q. 62. Draw MO energy level diagram for hydrogen fluoride molecule.

(*G. Nanakdev, 2000; Calcutta, 2005; Nagpur, 2008*)

Ans. The 2*s* atomic orbital of the fluorine atom does not interact with the 1*s* orbital of hydrogen atom because it is of low energy and is retained as atomic orbital. Only the $2p_x$ orbital of fluorine atom has proper symmetry and energy to interact with the 1*s* orbital of the hydrogen atom, giving one sigma bonding molecular orbital and one sigma antibonding molecular orbital (Fig. 4.56). The $2p_x$ and $2p_y$ orbitals of fluorine yield non-bonding molecular orbitals at the same energies because of their symmetry because they yield zero overlap with 1*s* orbital.

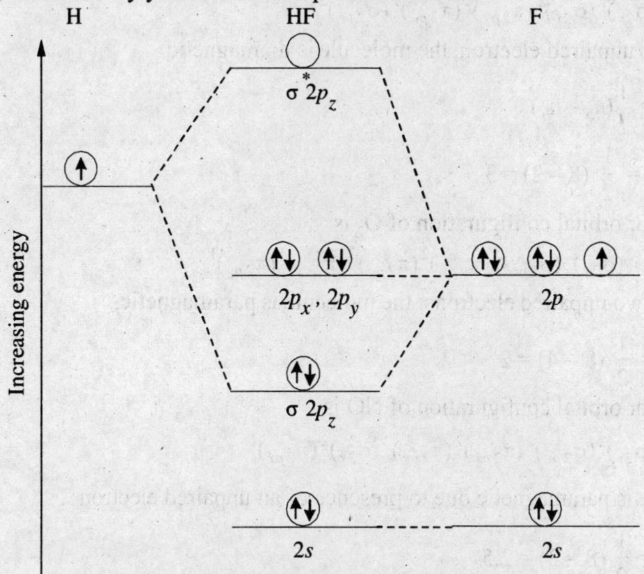

Fig. 4.56. MO diagram for hydrogen fluoride.

Q. 63. Draw MO energy level diagram of HCI molecule.

Ans. Molecular orbital energy level diagram of hydrogen chloride (Fig. 4.57) is similar to that of hydrogen fluoride. The energy of 3*s* and 3*p* atomic orbitals of the chlorine atom is lower than that of 1*s* atomic orbitals of the hydrogen atom. The 3*s* atomic orbitals of chlorine atom does not interact with the 1*s* atomic orbital of H atom because the 3*s* has a very low energy and is, therefore, retained as atomic orbital. The $2p_z$ atomic orbital of chlorine atom has proper symmetry and energy to interact with the 1*s* atomic orbital of hydrogen atom giving one sigma bonding molecular orbital and one sigma antibonding molecular orbital. The $3p_x$ and $3p_y$ orbitals of chlorine yield non-bonding molecular orbitals which are actually unchanged atomic orbitals of chlorine atom. These represent the two *p* orbitals on chlorine atom which lie perpendicular to the bond axis and hence do not overlap with 1*s* atomic orbital of the hydrogen atom.

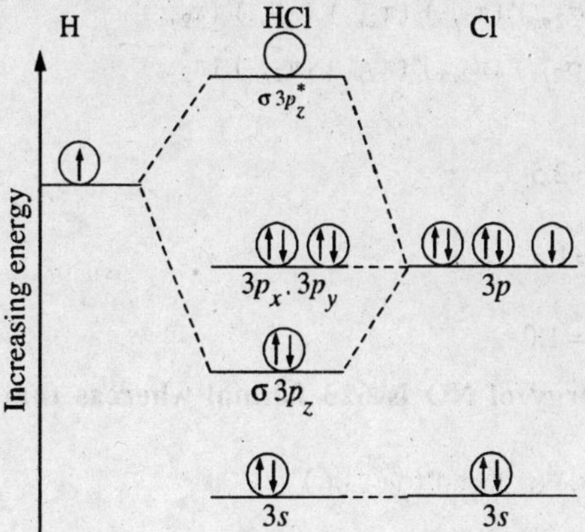

Fig. 4.57. Molecular orbital energy level diagram of HCl.

Q. 64. On the basis of MO theory, explain the magnetic character of N_2, O_2, NO and CO molecules. Also give the bond order in each case.

Ans. (*i*) Molecular orbital configuration of N_2 is

$$KK\ (\sigma_{2s})^2(\sigma^*_{2s})^2(\pi_{2px})^2(\pi^*_{2py})^2(\sigma_{2pz})^2$$

As there is no unpaired electron, the molecule is diamagnetic

$$\text{Bond order} = \frac{1}{2}(n_b - n_a)$$

$$= \frac{1}{2}(8-2) = 3$$

(*ii*) Molecular orbital configuration of O_2 is

$$KK\ (\sigma_{2s})^2(\sigma^*_{2s})^2(\sigma_{2pz})^2(\pi_{2py})^2(\pi_{2py})^2(\pi^*_{2px})^1(\pi^*_{2py})^1$$

As there are two unpaired electrons, the molecule is paramagnetic.

$$\text{Bond order} = \frac{1}{2}(8-4) = 2$$

(*iii*) Molecular orbital configuration of NO is

$$KK\ (\sigma_{2s})^2(\sigma_{2pz})^2(\pi_{2px})^2(\pi_{2py})^2(\sigma^*_{2s})^2(\pi^*_{2px})^1$$

The molecule is paramagnetic due to presence of an unpaired electron

$$\text{Bond order} = \frac{1}{2}(8-3) = 2.5$$

(*iv*) Molecular orbital configuration of CO is

$$KK\ (\sigma_{2s})^2(\sigma_{2pz})^2(\pi_{2px})^2(\pi_{2py})^2(\sigma^*_{2s})^2$$

As there are no unpaired electrons, the molecule is diamagnetic.

$$\text{Bond order} = \frac{1}{2}(8-2) = 3.$$

Q. 65. Draw MO diagram for O_2^+, O_2^- and O_2^{2-} and give the value of bond order in each case.

Ans. Molecular orbital configuration of O_2^+, O_2^- and O_2^{2-} are written as under:

O_2^+ : $KK\ (\sigma_{2s})^2(\sigma^*_{2s})^2(\sigma_{2pz})^2(\pi_{2px})^2(\pi_{2py})^2(\pi^*_{2px})^1$

O_2^- : $KK\ (\sigma_{2s})^2(\sigma^*_{2s})^2(\sigma_{2pz})^2(\pi_{2px})^2(\pi_{2py})^2(\pi^*_{2px})^2(\pi^*_{2py})^1$

O_2^{2-} : $KK\ (\sigma_{2s})^2(\sigma^*_{2s})^2(\sigma_{2pz})^2(\pi_{2px})^2(\pi_{2py})^2(\pi^*_{2px})^2(\pi^*_{2py})^2$

Bond order

(*i*) in $O_2^+ = \frac{1}{2}(8-3) = 2.5$

(*ii*) in $O_2^- = \frac{1}{2}(8-5) = 1.5$

(*iii*) in $O_2^{2-} = \frac{1}{2}(8-6) = 1.0$

Q. 66. The bond energy of NO is 623 kJ/mol whereas that of NO^+ molecule ion is 1049 kJ/mol. Justify.

Ans. NO : $KK\ (\sigma_{2s})^2(\sigma_z)^2(\pi_x)^2(\pi_y)^2(\sigma^*_{2s})^2(\pi^*_x)^1$

$$\text{Bond order} = \frac{8-3}{2} = 2\frac{1}{2}$$

NO^+: KK $(\sigma_{2s})^2(\sigma_z)^2(\pi_x)^2(\pi_y)^2(\sigma^*_{2s})^2$

Bond order = $\frac{8-2}{2} = 3$

Higher bond order of NO^+ justifies its higher bond energy than NO.

Q. 67. Using molecular orbital configurations indicate paramagnetic nature of B_2 and non-existence of He_2. *(Kerala, 2000)*

Ans. $He_2 : (\sigma_{1s})^2(\sigma^*_{1s})^2$

Bond order = $\frac{2-2}{2} = 0$

Bond order zero supports non-existence of helium molecule.

B_2 : KK $(\sigma^2_{2s})^2(\sigma^*_{2s})^2(\pi_x)^1(\pi_y)^1$

Bond order = $\frac{4-2}{2} = 1$

Q. 68. Give molecular orbital configuration of O_2, O_2^+ and O_2^-. Give order of stability with appropriate reasons. *(Pune 2010)*

Ans. O_2: KK $(\sigma_{2s})^2(\sigma^*_{2s})^2(\sigma^b_z)^2(\pi^b_x)^2(\pi^b_y)^2(\pi^*_x)^1(\pi^*_y)^1$

Bond order = $\frac{8-4}{2} = 2$

O_2^+ : KK $(\sigma^b_{2s})^2(\sigma^*_{2s})^2(\sigma^b_z)^2(\pi^b_x)^2(\pi^b_y)^2(\pi^*_x)^1$

Bond order = $\frac{8-3}{2} = 2\frac{1}{2}$

O_2^- : KK $(\sigma^b_{2s})^2(\sigma^*_{2s})^2(\sigma^b_2)^2(\pi^b_x)^2(\pi^b_y)^2(\pi^*_x)^2(\pi^*_y)^1$

Bond order = $\frac{8-5}{2} = 1.5$

Because of decreasing bond order, the stability decreases as

$$O_2^+ > O_2 > O_2^-$$

Q. 69. Give the points of similarity and difference between valence-bond and molecular orbital theory. *(G. Nanakdev, 2000; Awadh, 2000; Himachal, 2000; Kanpur, 2001)*

Ans. Points which are common to both theories:

(*i*) Both these theories are approximations.
(*ii*) Both interpret covalent bonds as orbitals embracing two atomic nuclei.
(*iii*) Both require that the combining atomic orbitals should be of nearly equal energy : they should be capable of overlapping and the combining orbitals must have the same symmetry.
(*iv*) Both predict the concentration of electron density between the nuclei.
(*v*) Both can account for directed valency.
(*vi*) Both predict the non-existence of helium molecule.

Some points of difference between the two theories:

Valence bond theory	*Molecular orbital theory*
(*i*) Electrons in molecules are localized as if they are in isolated atoms.	It treats the nuclei of the molecule as polycentric and then constructs one electron molecular orbital.
(*ii*) The total wave function in the valence bond treatment takes the contribution due to ionic terms.	Molecular orbital theory overlapping estimates the role of ionic terms in its total wave functions.

(*contd.*).

(*iii*) It could not explain the paramagnetism of O_2.	It can easily explain the paramagnetism of O_2.
(*iv*) Resonance is an integral part of valence bond theory.	Resonance has no role in molecular orbital theory.
It is simple to apply.	It is more difficult to apply.

Q. 70. Why a σ^b_{2s} orbital has a higher energy than σ^*_{1s} orbital? Why is the former MO called a bonding orbital while the latter is called an antibonding MO?

Ans. The relative energies of σ^b_{2s} and σ^*_{2s} are determined by the energies of the atomic orbitals from which they originate. The σ^b_{2s} MO is lower in energy compared to the 2*s* atomic orbital from which it is originated and is, therefore, called a bonding orbital while the σ^*_{1s}. MO is higher in energy than the 1*s* atomic orbital from which it originates and thus called an antibonding MO.

Q. 71. Draw a diagram showing the formation of a σ_{2p} bonding MO from atomic orbitals.

Ans. Formation of σ_{2p} bonding MO from atomic orbitals is shown in Fig. 4.58.

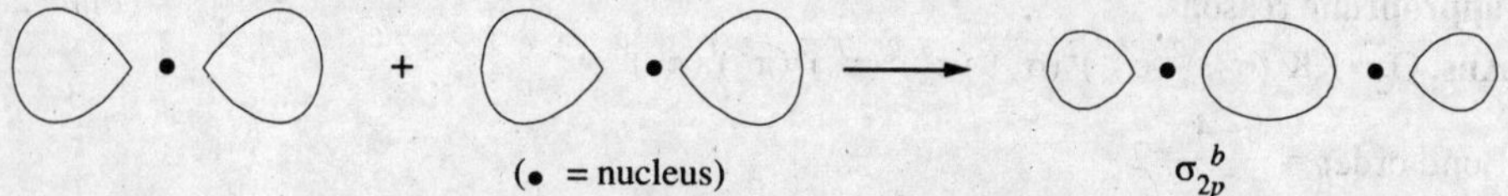

Fig. 4.58. Formation of σ_{2p}orbital.

ELECTRON - DEFICIENT COMPOUNDS

Q. 72. What are electron-deficient compounds?

Ans. An electron -deficient compound is one in which there is an insufficient number of electrons to complete the octet of central atom. These compounds contain insufficient numbers of electrons to form normal electron pairs bonds, between each pair of bonded atoms. For example, boron which has electronic configuration [He], $2s^2\ 2p^1$ forms compounds which are described as electron-deficient compounds. Boron has only three valence electrons but has four available orbitals (2*s*, $2p_x$, $2p_y$ and $2p_z$) to accommodate these electrons. Other members of boron family, namely, aluminium, gallium, indium and thallium, show tendency to lose their valence electrons to form ionic M^{3+} species in their compounds. However, boron is an exception. Due to its very small size the ionization energies are very high and the formation of ionic B^{3+} species does not take places. Boron fills its fourth available valence orbital in the following different ways:

(*i*) $p\pi - p\pi$ donation as observed in boron halides (BF_3 etc.).

(*ii*) donor-acceptor donation as in $Me_3N \rightarrow BF_3$.

(*iii*) Formation of electron-deficient compounds as in diborane (B_2H_6).

Q. 73. Describe the structure and bonding in electron-deficient diborane, B_2H_6.

(*Culcutta, 2000; Kerala, 2000; Purvanchal, 2007*)

Ans. Diborane B_2H_6 does not possess sufficient number of electrons to allow a bonding similar to ethane, C_2H_6 in which each bond consists of a shared electron pair (Fig. 4.59).

```
H        H          H        H
  \     /             \     /
H — B — B — H        H — C — C — H
  /     \             /     \
H        H          H        H
   Not possible          Possible
```

Fig. 4.59

There are twelve valency electrons in B_2H_6 (three per boron atom and one for each H atom) and for molecule to have ethane type structure a minimum of fourteen electrons would be required, since seven bonds would require seven electron pairs (*i.e.*, 2-centre, 2-electrons) bonds.

The experimentally determined structure of diborane suggests the following bonding scheme as shown in Fig. 4.60.

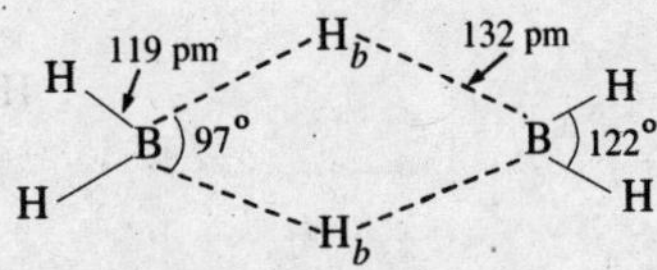

4.60. Structure of diborane.

A 3-centre, 2-electron bonds in addition to the normal electron pair (2-centre, 2-electron) bonds was proposed.

Each boron atom is assumed to use its four sp^3 hybrid orbitals and the six hydrogen atoms use their 1*s* atomic orbitals in each bonding scheme. The visualization of sp^3 hybrid orbitals by boron is suggested by the approximately tetrahedral geometry for each boron atom in the experimentally determined structure of diborane. Each of the two boron atoms then uses two of its sp^3 hybrid orbitals to overlap with the 1 *s* atomic orbitals of each of the two hydrogen atoms (terminal B —H bonds). These terminal B — H bonds are formed by the overlap of one electron from boron and one electron from hydrogen to form normal electron pair (2-centre, 2-electron, 2c -2e) bonds.

The remaining two sp^3 hybrid orbitals of each boron atoms are then arranged in a manner so that an sp^3 hybrid orbital from each boron atom and ls orbital from a bridging H atom overlap to form two multicentre orbitals as shown below in Fig. 4.61.

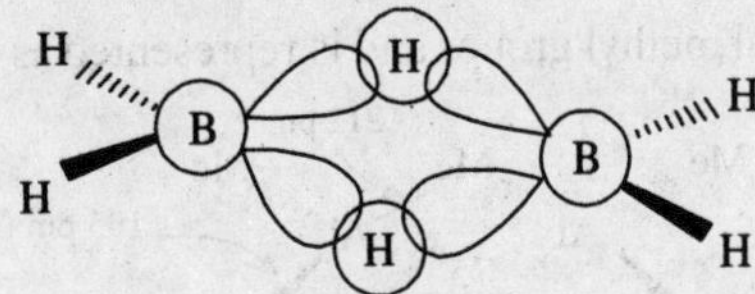

Fig. 4.61

Their overlap results in the formation of two bonding B—H—B orbitals, two non-bonding B—H—B orbitals and two antibonding B—H—B orbitals as depicted below in Fig. 4.62.

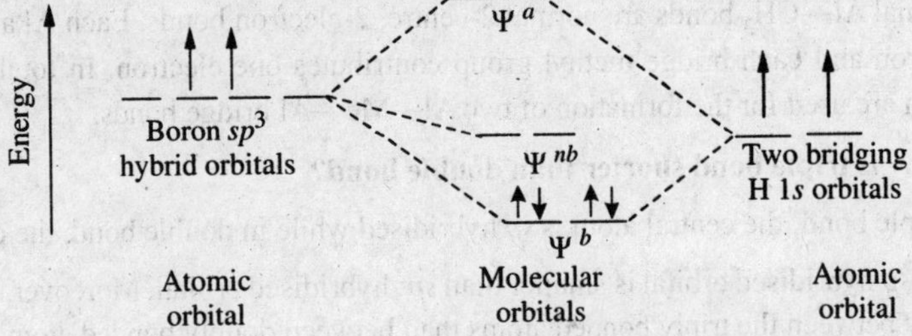

Fig. 4.62

The two lowest energy bonding B—H—B molecular orbitals are then occupied by a total of 4 electrons (one from each boron and hydrogen atom) to result in the formation of two 3-centre, 2-electron B—H—B bridge bonds. Since only two electrons are available for the formation of two B—H_b bonds among the three B—H_b—B atoms, the bond order for each B—H_b bonds comes out to be equivalent to 0.5 σ bond. Such electron-deficient 3-centre, 2-electron bonds are weak and get readily cleaved.

Q. 74. Discuss the structure of beryllium dialkyls.

Ans. X-ray analysis reveals a chain polymeric structure for dimethyl beryllium ($BeMe_2$) as depicted below in Fig. 4.63.

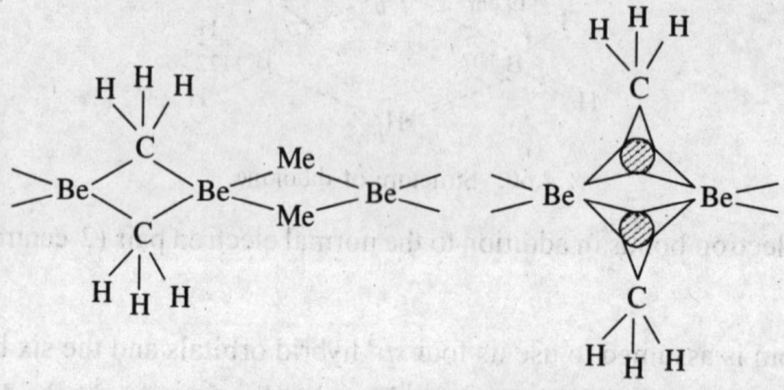

Fig. 4.63. Structure of Be $(Me)_2$.

The bonding can be explained in terms of 3-centre, 2-electron bridge bonds involving Be(CH_3) Be bonds. Each Be atom is sp^3 hybridized and has two electrons. Each bridging methyl carbon atom contributes one electron making a total of 4 - electrons for [$BeMe_2$] group. Each Be—(CH_3) — Be may then be described in terms of 3-centre, 2-electron (3c-2e) bonding as in electron-deficient diborane molecule.

Q. 75. Discuss the structure of aluminium trialkyls.

Ans. The structure of trimethyl aluminium which exists as Al_2Me_6 dimer contains two bridging methyl groups and four terminal methyl groups and is represented as under in Fig. 4.64.

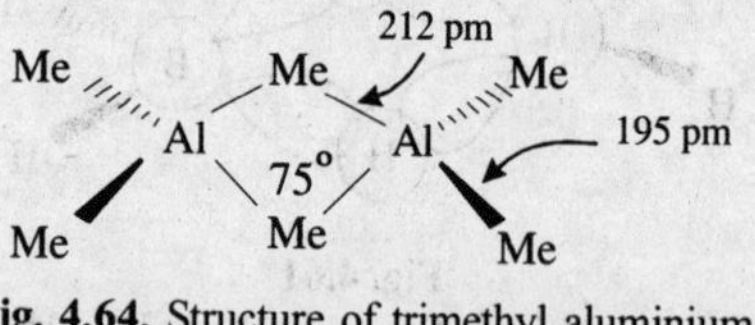

Fig. 4.64. Structure of trimethyl aluminium.

Each Al atom may be assumed as having sp^3 hybridized orbitals and has three valence electrons. The four terminal Al—CH_3 bonds are normal 2-centre, 2-electron bonds. Each Al atom is now left with one electron and each bridge methyl group contributes one electron. In total there are four electrons which are used for the formation of two Al—Me—Al bridge bonds.

Q. 76. Why is triple bond shorter than double bond? *(Delhi, 2000)*

Ans. In triple bond, the central atom is *sp* hybridised while in double bond, the central atom is sp^2 hybridised. *sp* hybridised orbital is smaller than sp^2 hybridised orbital. Moreover, there is greater electron density between the triply bonded atoms than between doubly bonded atoms.

Q. 77. Both NH_3 and H_2O molecules are formed by sp^3 hybridisation, but the bond angles are different. Explain on the basis of VSEPR theory. *(Nagpur, 2005)*

Ans. See Q. 13 and 14.

5 PERIODIC CLASSIFICATION OF ELEMENTS

Q. 1. What is meant by periodicity of element? *(Lucknow, 2000)*

Ans. If the elements are arranged in order of increasing atomic numbers, then their properties get repeated after regular intervals. The repetition of similar properties of the element after regular intervals is called **periodicity.** It is possible to explain the cause of periodicity in the light of the structure of the atom.

It has now been established that chemical properties of the elements are governed by the number of valence electrons in their atom. Therefore it is expected that atoms having similar valence shell electronic configuration will have similar properties. This is actually true. The arrangement of elements in order of increasing atomic number results in the repetition of similar electronic configuration after regular intervals.

Consider the electronic configuration of alkali metals (Table 5.1). The alkali metals have one electron in the *s* orbital of their valence shell. In other words, their electronic configuration can be written as ns^1 where n is the number of outmost energy level. It is due to the similar electronic configuration in the outermost orbit that all the alkali metals have similar properties.

Table 5.1. Electronic configuration of alkali metals.

Element	*At. No.*	*Electronic configuration*
Li	3	$1s^2; 2s^1$
Na	11	$1s^2; 2s^2 2p^6; 3s^1$
K	19	$1s^2; 2s^2 2p^6; 3s^2 3p^6; 4s^1$
Rb	37	$1,s^2; 2s^2 2p^6; 3s^2 3p^6 3d^{10}; 4s^2 4p^6; 5s^1$
Cs	55	$1s^2; 2s^2 2p^6; 3s^2 3p^6 3d^{10}; 4s^2 4p^6 4d^{10}; 5s^2 5p^6; 6s^1$
Fr	87	$1s^2; 2s^2 2p^6; 3s^2 3p^6 3d^{10}; 4s^2 4p^6 4d^{10} 4f^{14}; 5s^2 5p^6 5d^{10}; 6s^2 6p^6; 7s^1$

Hence, the cause of periodicity of the properties of elements is the repetition of similar electronic configurations after certain regular intervals.

Q. 2. Discuss the features of long form of periodic table. *(Lucknow, 2000)*

Ans. The long form periodic table is arranged in seven horizontal rows called *periods* and eighteen vertical columns called *groups* (See Table 5.2).

Periods

The seven periods are numbered from 1 to 7 starting from top and going downwards. Each period begins with an alkali metal and ends with an inert gas element. The number of elements in each period is as follows:

Table 5.2 Periodic Table of the Elements.

Alkali Metals 1	Alkaline earths 2	3	4	5	6	7	8	9	10	11 Coinage metals	12	13	14	15	Chal-cogens 16	Halo-gens 17	Noble gases 18
1008 **H** 1																	4.003 **He** 2
6.939 **Li** 3	9.012 **Be** 4	Transition metals										10.811 **B** 5	12.011 **C** 6	14.007 **N** 7	15.999 **O** 8	18.998 **F** 9	20.183 **Ne** 10
22.990 **Na** 11	24.312 **Mg** 12											26.982 **Al** 13	28.085 **Si** 14	30.974 **P** 15	32.064 **S** 16	35.453 **Cl** 17	39.94 **Ar** 18
39.102 **K** 19	40.08 **Ca** 20	44.956 **Sc** 21	47.90 **Ti** 22	50.942 **V** 23	51.996 **Cr** 24	54.938 **Mn** 25	55.847 **Fe** 26	58.933 **Co** 27	58.71 **Ni** 28	63.546 **Cu** 29	65.37 **Zn** 30	69.72 **Ga** 31	72.59 **Ge** 32	74.922 **As** 33	78.96 **Se** 34	79.904 **Br** 35	83.80 **Kr** 36
85.47 **Rb** 37	87.62 **Sr** 38	88.905 **Y** 39	91.22 **Zr** 40	92.906 **Nb** 41	95.94 **Mo** 42	(99) **Tc** 43	101.07 **Ru** 44	102.905 **Rh** 45	106.4 **Pd** 46	107.868 **Ag** 47	112.40 **Cd** 48	114.82 **In** 49	118.69 **Sn** 50	121.75 **Sb** 51	127.60 **Te** 52	126.904 **I** 53	83.80 **Xe** 54
138.905 **Cs** 55	137.34 **Ba** 56	138.91 **La** 57	178.49 **Hf** 72	180.948 **Ta** 73	183.85 **W** 74	186.2 **Re** 75	190.2 **Os** 76	192.2 **Ir** 77	195.09 **Pt** 78	196.967 **Au** 79	200.59 **Hg** 80	204.37 **Tl** 81	207.19 **Pb** 82	208.980 **Bi** 83	(210) **Po** 84	(210) **At** 85	(222) **Rn** 86
(223) **Fr** 87	(226) **Ra** 88	(227) **Ac** 89	(257) **Unq** 104	(260) **Unp** 105	(263) **Unh** 106	(265) **Uns** 107	(265) **Uno** 108	(267) **Une** 109									

Lanthanide series	140.12 **Ce** 58	140.907 **Pr** 59	144.24 **Nd** 60	(147) **Pm** 61	150.35 **Sm** 62	151.96 **Eu** 63	157.25 **Gd** 64	158.924 **Tb** 65	162.50 **Dy** 66	164.930 **Ho** 67	167.26 **Er** 68	168.934 **Tm** 69	173.04 **Yb** 70	174.97 **Lu** 71
Actinde series	232.038 **Th** 90	(231) **Pa** 91	238.03 **U** 92	(237) **Np** 93	(242) **Pu** 94	(243) **Am** 95	(247) **Cm** 96	(247) **Bk** 97	(249) **Cf** 98	(254) **Ef** 99	(253) **Fm** 100	(256) **Md** 101	(253) **No** 102	(257) **Lr** 103

Table 5.2 A.
Name of Elements, their Symbols and Atomic Numbers.

Element	Symbol	Atomic No.	Element	Symbol	Atomic No.
Actinium	Ac	89	Mercury	Hg	80
Aluminium	Al	13	Molybdenum	Mo	42
Americium	Am	95	Neodymium	Nd	60
Antimony	Sb	51	Neon	Ne	10
Argon	Ar	18	Neptunium	Np	93
Arsenic	As	33	Nickel	Ni	28
Astatine	At	85	Niobium	Nb	41
Barium	Ba	56	Nitrogen	N	7
Berkelium	Bk	97	Nobelium	No	102
Beryllium	Be	4	Osmium	Os	76
Bismuth	Bi	83	Oxygen	0	8
Boron	B	5	Palladium	Pd	46
Bromine	Br	35	Phosphorus	P	15
Cadmium	Cd	48	Platinum	Pt	78
Calcium	Ca	20	Plutonium	Pu	94
Californium	Cf	98	Polonium	Po	84
Carbon	C	6	Potassium	K	19
Cerium	Ce	58	Praseodymium	Pr	59
Cesium	Cs	55	Promethium	Pm	61
Chlorine	Cl	17	Protactinium	Pa	91
Chromium	Cr	24	Radium	Ra	88
Cobalt	Co	27	Radon	Rn	86
Copper	Cu	29	Rhenium	Re	75
Curium	Cm	96	Rhodium	Rh	45
Dysprosium	Dy	66	Rubidium	Rb	37
Einsteinium	Es	99	Ruthenium	Ru	44
Erbium	Er	68	Samarium	Sm	62
Europium	Eu	63	Scandium	Sc	21
Fermium	Fm	100	Selenium	Se	34
Fluorine	F	9	Silicon	Si	14
Francium	Fr	87	Silver	Ag	47
Gadolinium	Gd	64	Sodium	Na	11
Gallium	Ga	31	Strontium	Sr	38
Germanium	Ge	32	Sulphur	S	16
Gold	Au	79	Tantalum	Ta	73
Hafnium	Hf	72	Technetium	Tc	43
Helium	He	2	Tellurium	Te	52
Holmium	Ho	67	Terbium	Tb	65
Hydrogen	H	1	Thallium	Tl	81
Indium	In	49	Thorium	Th	90
Iodine	I	53	Thulium	Tm	69
Iridium	Ir	77	Tin	Sn	50
Iron	Fe	26	Titanium	Ti	22
Krypton	Kr	36	Tungsten	W	74
Lanthanum	La	57	Uranium	U	92
Lawrencium	Lr	103	Vanadium	V	23
Lead	Pb	82	Xenon	Xe	54
Lithium	Li	3	Ytterbium	Yb	70
Lutetium	Lu	71	Yttrium	Y	39
Magnesium	Mg	12	Zinc	Zn	30
Manganese	Mn	25	Zirconium	Zr	40
Mendelevium	Md	101			

Period	*No. of elements*
1	2
2	8
3	8
4	18
5	18
6	32
7	19

The first three periods are called *short* periods because they contain small number (2 or 8) of elements. The next four periods are called *long* periods because they contain large number (18 or 32) of elements. The seventh period is a long period but is *incomplete.*

This period consists of 26 elements, 14 of them being placed in *f*-block and 12 in *d*-block.

Elements from atomic numbers 104 to 112 have been reported recently. But these heavy elements are highly unstable. Names of these elements as suggested by American Chemical Society and IUPAC are given in the following Table. American chemical society has given the names only for elements up to 109.

Atomic number	American names	IUPAC names
104	Rutherfordium (Rf)	Unnilquadium (Unq)
105	Hahnium (Ha)	Unnilpentium (Unp)
106	Seaborgium (Sg)	Unnilhexium (Unh)
107	Bohrium (Bh)	Unnilseptium (Uns)
108	Hassium (Hs)	Unniloctium (Uno)
109	Meitnerium (Mt)	Unnilennium (Une)
110	—	Ununnilium (Uun)
111	—	Unununilium (Uuu)
112	—	Unundodium (Uud)

Thus this period is incomplete and requires 6 more elements to be discovered to complete $7p^{1-6}$.

Groups

There are 18 vertical columns in the periodic table which constitute 18 groups and are numbered as 1, 2, 3, 4, 5, 6, 7, 8, 9, 10, 11, 12, 13, 14, 15, 16, 17, and 18.

The elements belonging to groups 1, 2, 13, 14, 15, 16, 17 and 18 are known as *normal or representative elements.* Elements in between groups 2 and 13 are called *transition elements* because they represent a transition (change) from electropositive elements on the left-hand side to electronegative elements on the right-hand side of the periodic table.

Two rows of elements (each containing 14 elements) are known as *lanthanides and actinides* respectively. They are placed separately in the periodic table to avoid uncontrolled sideway expansion of the periodic table. Names of elements in alphabetical order, their symbols and at. numbers are given in Table 5.2A.

Q. 3. What are the different blocks constituting the periodic table?

(*Garhwal, 2000; Delhi, 2002*)

Ans. To simplify the study of elements the periodic table has been divided into four blocks known as *s, p, d* and *f* blocks. The division depends upon the type of atomic orbital which receives the last electron in an atom *e.g.*, magnesium ($1s^2 2s^2 2p^6 3s^2$) which receives its last electron in *s*-orbital belongs to *s*-block. The division of periodic table into four blocks is depicted in Fig. 5.1.

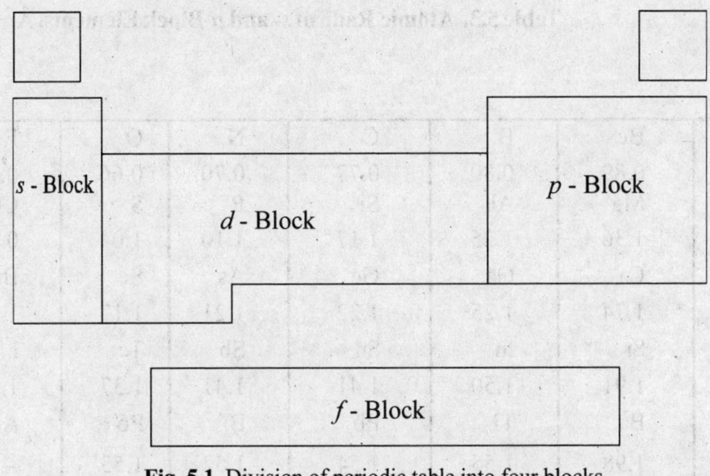

Fig. 5.1. Division of periodic table into four blocks.

s-block elements are those elements whose last electron enters the *s*-orbital of the valence shell. Therefore, elements of group 1 and group 2 constitute *s*-block elements.

p-block elements are those elements whose last electron enters the *p*-orbital of the valence shell. Therefore, elements of groups 13, 14, 15, 16, 17 and 18 constitute *p*-block.

d-block elements are those elements whose last electron enters *d*-orbitals. Therefore, elements of groups 3, 4, 5, 6, 7, 8, 9, 10, 11 and 12 constitute *d*-block. These elements are called transition elements.

f-block elements are those whose last electron enters *f*-orbital. Two series of elements at the bottom of the periodic table known as lanthanides and actinides constitute *f*-block elements. These elements are also known as inner transition elements because they constitute transition series within the transition elements.

Q. 4. How do you define atomic radius of an element? Discuss the variations in atomic radius along a period and down the group. (*Nagpur, 2003; Meerut, 2002; Patna 2010*)

Ans. The radius of an atom or ion is defined as the distance from the centre of the nucleus to the outermost shell of the atom or ion. It is impossible to isolate an individual atom and determine its radius. It is assumed that atoms behave as perfect hard spheres and in a homonuclear diatomic molecule two such spheres are just in contact with each other. The distance between the centres of two atoms in molecules can be measured by X-ray and spectral studies of molecules. The radius of an atom is calculated as one half of the distance between the centres of two atoms. If we consider a homonuclear diatomic molecules say F-F which has 1.28 Å as the internuclear distance than 1/2 the F-F distance *i.e.*, 0.64 Å will be taken as the atomic radius of fluorine.

The hybridization of atoms governs atomic radii. It has been shown that the radius of carbon atom in different states of hybridization is different. If carbon atom is sp^3 hybridized as in CH_4, then its atomic radius is 0.77 Å. sp^2 hybridized carbon atom is 0.74 Å in size and *sp* hybridized carbon is only 0.70 Å. This is attributed to the increasing *s*-character in different hybridization schemes in going from sp^3 to *sp*. This is because the 2*s* orbital has smaller radius than 2*p* orbital. The atomic radii of *s*- and *p*-block elements are given in Table 5.3.

Table 5.3. Atomic Radii of *s*- and *p*-Block Elements, Å

							He 0.93
Li 1.23	Be 0.89	B 0.80	C 0.77	N 0.70	O 0.66	F 0.64	Ne 1.12
Na 1.57	Mg 1.36	Al 1.25	Si 1.17	P 1.10	S 1.04	Cl 0.99	Ar 1.54
K 2.03	Ca 1.74	Ga 1.25	Ge 1.22	As 1.21	Se 1.17	Br 1.14	Kr 1.69
Rb 2.16	Sr 1.91	In 1.50	Sn 1.41	Sb 1.41	Te 1.37	I 1.33	Xe 1.90
Cs 2.35	Ba 1.98	Tl 1.55	Pb 1.54	Bi 1.52	Po 1.53	At —	Rn 2.2

The following observations may be made from the atomic radii values given in Table 5.3.

(*i*) The atomic radius decreases from left to right across a period in the periodic table. In a given period the alkali metal atom has the largest and the halogen atom the smallest atomic radius.

(*ii*) The atomic radius increases on descending a group in the periodic table.

It is impossible to measure the atomic radii of zero group elements. In their case only Bragg-Slater (vander Waals radii) are known and are given in Table 5.3.

The periodic trends of atomic radii can be easily accounted for in terms of the electronic configuration of these elements. Periodic trends in main group elements are shown in Fig. 5.2.

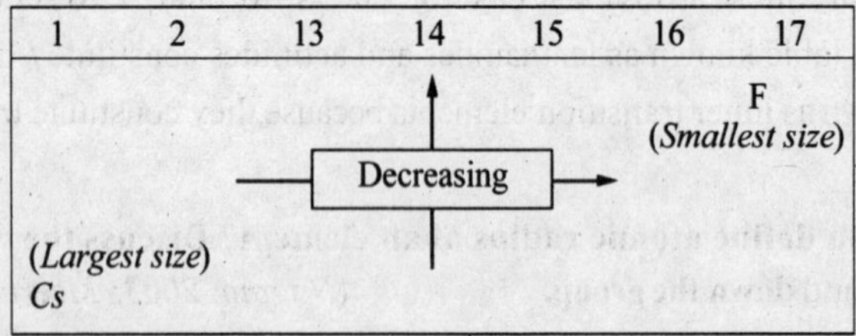

Fig. 5.2. Periodic trends of atomic radii among *s*- and *p*-block elements.

(*i*) Variation in a period. The size of atoms decreases from left to right across the periodic table. This may be explained on the basis of increasing nuclear charge. Since with an increase in the atomic number, the magnitude of nuclear charge increases while the corresponding addition of electrons takes place in the same principal quantum shell. As a result the increased nuclear charge pulls the electrons more strongly towards the nucleus thereby causing a decrease in the size of the atoms. The decrease in size is the most pronounced when a second '*s*' electron is filled, but it is less marked as we fill the *p*-electrons.

(*ii*) Variation in a group. The size of the element increases on descending in a group. This may be attributed to the addition of electrons in a new quantum shell at each step. This addition of extra shells of electron outweighs the effect of increased nuclear charge.

Q. 5. van der Waals radius is larger than the covalent radius. Explain.

(*Calcutta, 2000; Indore, 2001*)

Ans. One half the distance between nuclei of the two covalently bonded atoms of the element in a molecule is taken as the *covalent radius* of the atom of the element.

van der Waals radius is defined as the half of the distance between the nuclei of two adject atoms belonging to two neighbouring molecules of an element in the solid state.

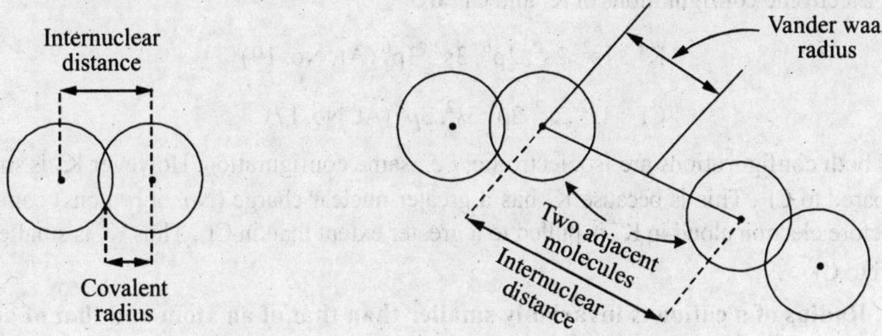

Fig. 5.3. Covalent and van der Waals radii.

Q. 6. What is ionic radius? (*Kurukshetra, 1996; Calcutta, 2005; Patna 2010*)

Ans. The distance between the nucleus and the outermost electron in an ion is called ionic radius. When an atom loses an electron, a positive ion is formed. The size of a cation is always smaller than a neutral atom. This can be attributed to the number of electrons whereas the magnitude of the nuclear charge remains the same. The nuclear charge now acts on a fewer number of electrons, attracts them more tightly and hence the electron cloud shrinks in size.

Atomic radius of Na = 1.57 Å

Ionic radius of Na^+ = 0.95 Å

When a neutral atom gains an electron, a negative ion is formed. The size of an anion is always larger than its neutral atom. In this case the same nuclear charge acts over greater number of electrons than were present in the neutral atom. The electron cloud is now less tightly held which causes an expansion in its size. The evaluation of ionic radius is dependent on several factors such as crystal structure, cation to anions radius ratio, charge on ions etc.

Periodic Trends: For an isoelectronic series (ions, which have the same number of electrons) the ionic radius decreases very significantly as the nuclear charge increases. This is to be expected because the increased nuclear charge is on the same number of electrons.

In a group the ionic radius increases as the atomic number increases. It may be remembered here that increase in size of the members of a group does not show any marked change in the radii whenever there is an intervening transition series *e.g.*, in going from Ca to Sr the first transition series intervenes. This is even more significant when we move down from In (1.50 Å) to Tl (1.55 Å) because of the intervening *f*- block elements. This is because the filling of next electron takes place in inner shells.

The ionization energy of an element is defmed as the amount of energy required to remove the most loosely bound electron from an isolated gaseous atom.

Q. 7. Explain: Increasing order of size in case of following species should be

$$Fe^{3+} < Fe^{2+} < Fe$$

Ans. When two electrons are removed from Fe atom, Fe^{2+} ion is obtained. As the positive charge on the nucleus outnumbers negative charge on the electrons, the nucleus attracts the electrons with a much greater force resulting in the shrinking of size.

On removal to a further electron to convert Fe^{2+} into Fe^{3+}, difference in the positive charge on nucleus and negative charge on the electron further increases, this increases further the pull of the nucleus on the electrons. Hence the size decreases further. Increasing order of size is therefore

$Fe^{3+} < Fe^{2+} < Fe$

Q. 8. Ionic radius of K^+ is smaller than that of Cl^-, both are isoelectronic. Explain.

Ans. Electronic configurations of K^+ and Cl^- are

$$K^+ : 1s^2, 2s^2, 2p^6, 3s^2, 3p^6 \text{ (At. No. 19)}$$

$$Cl^- : 1s^2, 2s^2, 2p^6, 3s^2, 3p^6 \text{ (At. No. 17)}$$

Thus both configurations are isoelectronic, *i.e.*, same configuration. However K^+ is smaller in size compared to Cl^-. This is because K^+ has a greater nuclear charge (No. of protons) compared to Cl^-. Therefore electron cloud in K^+ is pulled to a greater extent than in Cl^-. Thus K^+ is smaller in size compared to Cl^-.

Q. 9. Radius of a cation is invariably smaller than that of an atom and that of an anion invariably lower than that of the corresponding atom. Explain. (*Delhi, 2002; Nagpur 2005*)

Ans. After an electron is removed from an atom, the cation is obtained.

$$\underset{\text{Atom}}{M} \longrightarrow \underset{\text{Cation}}{M^+} + e$$

Positive charge in M^+ outweighs the negative charge on electron. The pull of the nucleus on the electron, therefore increases. That is why a cation is smaller than the atom. However, when an electron is added to an atom, an anion is obtained.

$$M + e^- \longrightarrow \underset{\text{Anion}}{M^-}$$

Now the electronic charge is greater than the nuclear charge. Moreover repulsion among the electrons will increase thereby increasing the size of the anions compared to that of the neutral atom.

Q. 10. What is meant by (*i*) Ionisation energy and (*ii*) Successive ionisation energy? (*Nagpur 2005*)

Ans. (*i*) *The ionisation energy of an element is defined as amount of energy required to remove the most loosely bound electron from an isolated gaseous atom.*

$$M(g) + \text{Ionisation Energy} \longrightarrow M^+(g) + e^-$$

The experimental values of I.E. of elements are determined from their spectra and are reported either in electron volts or in kcal or kJ (1 kJ = 4.184 kcal).

(*ii*) *Successive ionisation energies.* After the first electron has been removed from a neutral atom, it is possible to remove second, third or even more electrons. The energies required for the removal of subsequent electrons are known as second IE_2, **third IE_3, fourth IE_4 or in general successive ionisation energies.** This may be expressed as

$$A(g) + IE_1 \longrightarrow A^+(g) + e^- \quad \text{(first ionisation energy)}$$

$$A^+(g) + IE_2 \longrightarrow A^{2+}(g) + e^- \quad \text{(second ionisation energy)}$$

$$A^{2+}(g) + IE_3 \longrightarrow A^{3+}(g) + e^- \quad \text{(third ionisation energy)}$$

It is observed that successive ionisation energies go on increasing progressively because it becomes increasingly difficult to remove each successive electron. This is because of the following reasons. (*i*) After the removal of first electron, the atom changes into a positive ion in which the electrons are held more tightly by the nucleus. This is because the positive ion contains smaller number of electrons while the nuclear charge remains the same. Similarly, the removal of the second electron would further increase the attractive force of the nucleus on the remaining electrons and so on. This naturally leads to increases in ionisation energies. (*ii*) Another situation that may arise in many cases is that after the removal of one, two or more electrons from the atom, the next one may be present in a lower energy level. This electron would naturally require much more energy for its removal *e.g.*, after 2 electrons have been removed from *2s* orbital of Be to form Be^{2+}, the next electron lies in 1*s* orbital. Therefore, the third ionisation energy of beryllium which involves the removals of 1s electron has a much higher value than second ionisation energy.

Q. 11. What are the factors that govern ionisation energy of an element?

(*Delhi, 2003; Nagpur, 2008*)

Ans. The ionisation energy depends upon the following factors:

(*i*) *Magnitude of nuclear charge.* This attractive force of the nucleus on the electron is directly proportional to the product of the nuclear and electronic charges. Therefore as the nuclear charge increases, the electron would be held more strongly by the nucleus. In other words, more energy would have to be supplied to remove the electron from the atom. Thus the ionisation energy increases as the nuclear charge increases.

(*ii*) *Size of the atom.* As the size of the atom increases, the outer electrons are less strongly attracted by the nucleus, because the force of attraction is inversely proportional to the square of the distance between the charged particles. With the decrease in attractive force, the outer electron is less tightly held and is, therefore, easier to remove. Thus the ionisation energy decreases as the size of the atom increases.

(*iii*) *Screening effect of the inner electrons.* The inner electronic shells act as a screen between the nucleus and outer shell. This reduces the pull of the outer shell towards the nucleus. In other words, the electrons that lie between the nucleus and outermost shell have a shielding or screening effect on the electrons of outermost shell. The larger the number of electrons in the inner shells, the greater will be the screening effect and the lesser will be nuclear attraction on the outermost electrons. Thus increase in the number of inner electrons will trend to decrease I.E. It should be carefully noted that electrons in the same or similar type of orbital do not shield each other. Thus the electrons in *s*-orbital will not shield each other. Similarly, electrons in p_x, p_y and p_z orbital of the same subshell do not shield each other. It has also been found that the screening effect of *s*-electron is more than that of *p*-electron whose effect is in turn more than that of a *d*-electron and so on.

(*iv*) *Penetration effect of the electrons.* Electrons in the *s*-orbitals have the maximum probability of being nearer to the nucleus while this probability goes on decreasing in case of *p, d* and *f* electrons. In other words *s*-electrons have the greatest penetration towards the nucleus and the penetrating effect of the electron with a given principal quantum number, *n* is in the order : $s > p > d > f$. Greater the penetration of electron, greater is the ionisation energy since the electron would be bound more tightly to the nucleus in that case. Thus, for the same value of *n,* the ionisation energy would be more for *s*-electron compared to that for a *p*-electron which is turn will be greater than for *d*-electron and so on.

(*v*) *Electronic arrangement.* (*a*) Noble gases have the most stable electronic configuration having s^2p^6 arrangement of the outermost shell while all the inner shells are also complete.

(b) The arrangement in which all the occupied orbitals are completely filled as, for instance, in the case of Be ($1s^2$, $2s^2$) and Mg ($1s^2$, $2s^2$, $2p^6$, $3s^2$).

(c) The configuration in which all the orbitals of the same sublevel are exactly half filled also represents a very stable arrangement such as in the case of N ($1s^2$, $2s^2$, $2p^1_x 2p^1_y$ and $2p^1_z$) and *P* ($1s^2$, $2s^2$, $2p^6$, $3s^2$, $3p^1_x 3p^1_y$ $3p^1_z$).

The more stable the electronic arrangement the greater is the ionisation energy. That is why each noble gas has the highest ionisation energy of the element in the same period.

Q. 12. First ionisation energy of Al is lower than that of Mg. Explain.

(Indore, 2001; Delhi, 2003)

Ans. Electric configuration of Al is $1s^2, 2s^2, 2p^6, 3s^2, 3p^1$) and that of Mg is $1s^2, 2s^2, 2p^6, 3s^2$. Although we expect the I.E. to increase as we move form left to right in the periodic table, here the situation is different. Fully filled orbitals are more stable compared to half filled orbital and therefore greater energy is needed to remove the electron. Therefore Mg has a fully filled 3s orbital and it is more difficult to remove the electron from the atom. In Al, it is the electron from half filled 3*p* orbital which is to be removed, this will require comparatively less energy.

Q. 13. Discuss the variation in ionisation energy (*i*) along a period and (*ii*) down a group.

(Kumaon, 2000; Kerala, 2000)

Ans. Variation along a period. In general, the ionisation energy increases with increasing atomic number as we move from left to right along a period. This is because of the following reasons:

(*i*) The nuclear charge increases from left to right.

(*ii*) The number of main energy levels remains the same while the atomic size decreases. Due to the gradual increase in the nuclear charge and simultaneous decrease in the atomic size, the electrons get more and more firmly bound to the nucleus. Therefore it becomes increasingly difficult to remove an electron and hence the ionisation energy keeps on increasing. Some of the exceptional values observed within a period can be explained in terms of the role played by the other factors governing ionisation energy. To illustrate this let us consider the changes in the second period.

As we move from Li to Be there is an increase in the ionisation energy. This may be easily explained in terms of increased nuclear charge and smaller atomic size while other factors do not make a significant contribution.

On passing from Be to B, we observe that in spite of the increased nuclear charge and smaller atomic size, there is a slight decrease in ionisation energy.

This can be explained in terms of following factors:

(*a*) The electronic arrangement of B ($1s^2, 2s^2, p_x^1$) is less stable that than of Be ($1s^2, 2s^2$).

(*b*) The outermost electron in B is present in 2*p* orbital which is less penetrating than 2*s* orbital which contains the outermost electron of Be.

(c) The outermost electron in B is shielded by 4 electrons while that in Be is shielded only by 2 electrons.

As a result of the above factors, the outermost electron of B is not so firmly held by the nucleus and is easily removed in comparison. For similar reasons there is a slight decrease in energy in moving from Mg to Al in third period.

As we pass from B to C to N the ionisation energy keeps on increasing primarily due to increasing nuclear charge and decreasing atomic size, the combined effect of other factors not being appreciable.

Oxygen has a slightly smaller ionisation energy than N. This is primarily due to the more stable electronic arrangement of nitrogen ($1s^2, 2s^2, 2p_x^{\ 1}\ 2p_y^{\ 1}\ 2p_z^{\ 1}$) in which all the *p* orbitals are half-filled while in oxygen ($1s^2\ 2s^2, 2p_x^{\ 2}\ 2p_y^{\ 1}\ 2p_z^{\ 1}$) all the *p* - orbitals are neither half-filled not completely filled. In the third period the ionisation energy of *S* is smaller than that of P for similar reasons.

There is increase in ionisation energy in going from O to F to N as expected.

Variation along a group. The values of ionisation energies show a regular decrease in going from top to bottom in a group. This can be attributed to the following factors:

(*i*) In going from top to bottom in a group the nuclear charge increases.

(*ii*) There is a simultaneous increase in the number of energy levels and therefore, the atomic size increases.

(*iii*) There is an increase in the number of electrons that lie in between the nucleus and the outermost electron and therefore, the shielding effect in the outermost electron gradually increases.

The increase in atomic size and the shielding effect more than neutralise the effect of the increased charge. As such the electron in the higher orbital is less firmly bound to the nucleus. Consequently there is gradual decrease in the ionisation energy.

Q. 14. The Be and N in second period and Mg and P in the third period of the table have slightly higher ionisation energies than expected. Explain.

Ans. Configurations in which there are fully filled and half filled orbitals are extraordinarily more stable than others. So more energy is needed to take out electrons from the outermost shells. In Be and Mg, there are fully filled orbitals in the outermost shell. Similarly in nitrogen and phosphorus, the outermost shells have half-filled orbitals.

$_7N:\ 1s^2, 2s^2, 2p^2_x\ 2p_y^{\ 1}\ 2p_z^{\ 1},\qquad {}_{15}P:\ 1s^2, 2s^2\ 2p^6\ 3s^2\ 3p_x^{\ 1}\ 3p_y^{\ 1}\ 3p_z^{\ 1}$

Therefore these atoms have higher ionisation energy than expected.

Q. 15. Explain: Nitrogen has higher value of I.E. than oxygen. (*M. Dayanand, 2000*)

Ans. Nitrogen has more stable electronic configuration in which the orbitals are all half filled.

$_7N\ :1s^2, 2s^2\ 2p_x^{\ 1}\ 2p_y^{\ 1}\ 2p_z^{\ 1}$

$8O:1s^2, 2s^2\ 2p_x^{\ 2}\ 2p_y^{\ 1}\ 2p_z^{\ 1}$

Half-filled configuration is more stable than others. Hence greater energy is needed to extract the electron from them.

Q. 16. Calculate the amount of energy required to convert 1.5 g of potassium atoms in the gaseous state to form potassium ions. Ionisation energy of potassium is 419 kJ mole^{-1} and atomic mass of potassium is 39 a.m.u.

Ans. $\underset{(1\text{mole})}{K(g)} + 419\ \text{kJ (IE)} \longrightarrow \underset{(1\text{mole})}{K^+(g)} + \underset{(1\text{mole})}{\text{electron}}$

Thus 39 g of potassium atoms require = 419 kJ energy for ionisation

$\therefore$ 1.5 g of potassium atoms required = $\dfrac{419}{39} \times 1.5.$

$= 16.1$ **kJ**

Q. 17. Define electron affinity. What are the factors that govern the electron affinity?

(*Kumaon, 2000; Calcutta, 2000; Lucknow, 2000; Indore, 2001; Kerala, 2000*)

Ans. *The amount of energy released when a neutral atom in the gaseous state acquires an extra electron to form a negatively charged ion is known as electron affinity.* The process may be depicted as

$$A(g) + e^-(g) \longrightarrow A^-(g) + \text{Energy (Electron affinity)}$$

The electron affinity determines the tendency of an atom to add an extra electron and the firmness with which the added electron is bound to the nucleus.

Factors determining electron affinity. Electron affinity of an element depends mainly upon the following factors:

(*i*) *Nuclear charge.* More the nuclear charge of the atom, more strongly will it attract additional electron. Therefore, as electron affinity increases, the nuclear charge increases.

(*ii*) *Atomic size.* The smaller the size of atom, smaller will be the distance between the extra electron and the nucleus. Therefore, electrostatic force of attraction will be more and the electron affinity will be higher.

(*iii*) *Electronic configuration.* Atoms having stable electronic configuration (*i.e.*, those having completely filled or exactly half-filled outer orbitals) do not show much tendency to add extra electron and have either zero or very low electron affinities.

Q. 18. Discuss the variations of electron affinity in the periodic table (*i*) down the group and (*ii*) along the period. (*Kerala, 2000; Nagpur, 2003*)

Ans. *Down a group.* The atomic size as well as the nuclear charge increase on moving down the group. But the overall effect of these factors is that the effective nuclear attraction for the electron decreases. As such atom would have less tendency to attract additional electron towards itself so that its electron affinity would decrease.

Across a period. The atomic size decreases while the nuclear charge increases. Naturally, therefore, the nucleus would have increased attraction for additional electron and electron affinity would be more. But there are many exceptions to this general trend which are mostly due to the relative stabilities of electronic configurations.

Some important features of electron affinities of different elements

(*i*) *Electron affinity of fluorine is smaller than that of chlorine.* It has been pointed out above that in general electron affinities go on decreasing in moving down a group. Therefore, amongst the halogens the electron affinity of fluorine should be higher than that of chlorine. Actually the electron affinity of fluorine is lesser than that of chlorine.

This is due to the small size of fluorine atom. In fluorine the second shell is the outermost shell. Due to the small volume of this shell the addition of an extra electron results in strong electron-electron repulsions. As a result, attractive interaction between the incoming electron and the nucleus of fluorine atom is considerably depressed. This result is low electron affinity of fluorine.

(*ii*) *Electron affinities of halogens are very high.* This is because halogens have the general electronic configuration $n^2\ np^5$. As such they have a strong tendency to acquire an extra electron to change into ions having the stable s^2p^6 configuration of the next noble gas. Therefore their electron affinities are very high.

(*iii*) *Electron affinities of noble gases are zero.* This can be attributed to the fact that their electronic configurations are very stable and they have absolutely no tendency to add an extra electron in the higher energy level. Similarly, the low value of electron affinity in case of elements like Be, Mg. N and P can be explained in terms of their stable electronic arrangements marked by the presence of either only completely filled or exactly half filled orbitals of the same type. Therefore it is not easy to add an electron in their case.

Q. 19. Explain why?

(*i*) Electron affinities of halogen are very high.

(*ii*) Electron affinities of noble gases are zero.

(*iii*) Successive electron affinities have negative value.

Ans. For (*i*) and (*ii*) see question 18. (*iii*) Successive electron affinity means affinity for the addition of second, third or next electrons to the anion after one electron has been added to the neutral atom.

Actually there is a lot of resistance put up by the anions (uninegative) to the addition of more electrons because of repulsive forces. That is why successive electron affinities have negative values.

Q. 20. Explain why fluorine has lower electron affinity than chlorine?

(*Calcutta, 2000; M. Dayanand, 2000; Nagpur, 2005*)

Ans. See Q. 18.

Q. 21. How much energy in kJ is released when 1.0 g of chlorine atom are converted completely to Cl^- ions in the gaseous state, electron affinity of Cl (g) = 349 kJ and atomic mass of chlorine is 35.5 a.m.u. (*Punjab, 1987*)

Ans. $\underset{1\text{ mole}}{Cl(g)} + \underset{1\text{ mole}}{\text{electron}} \longrightarrow \underset{1\text{ mole}}{Cl^-(g)} + 349\text{ kJ}$

35.5 g of chlorine on conversion to Cl^- ions release 349 kJ of energy.

$\therefore$ 1 g of chlorine on conversion to Cl^- release $= \frac{349}{35.5} \times 1\text{kJ} = 9.83\text{k J}$

Q. 22. Discuss Pauling's scale of electronegativity.

(*Himachal, 2000; Garhwal, 2000; Lucknow, 2000; Calcutta, 2000; Kerala, 2000; Indore, 2001*)

Ans. *Electronegativity is defined as power of an atom in a molecule to attract electrons to itself.* Greater the ability of an atom to attract electron, larger is the value of electronegativity. Thus it is different from electron affmity which is quantitative property and is not the property of an isolated atom but of an atom that may be bonded to one or more atoms in a molecule or complex ion.

Pauling's Scale of Electronegativity

In 1932 Linus Pauling proposed a scale of electronegativity. He based his scale on the difference (ΔE) between the measured bond energy of a molecule A – B [$E_{(A-B)}$] and the energy expected for the purely covalent bond A – B [$E_{(A-B)\,cov.}$] that is

$$\Delta E = E_{(A-B)} - E_{(A-B)\,cov} \qquad ...(5.1)$$

According to him if the two atoms A and B have the same electronegativity value then the molecules AB is bonded by purely covalent bonds as in homonucleur molecules A_2 or B_2. Then $E_{(A-B)\,cov}$, would be the mean of $E_{(A-A)}$ and $E_{(B-B)}$, that is,

$$E_{(A-B)cov} = \frac{1}{2}[E_{(A-A)} + E_{(B-B)}] \qquad \text{(Arithmetical mean)}$$

or

$$E_{|A-B|cov} = [E_{(A-A)} \times E_{(B-B)}]^{1/2} \qquad \text{(Geometrical mean)} \qquad ...(5.2)$$

Pauling used geometric mean for his calculations. Substituting from (5.2) into (5.1),

$$\Delta E = E_{A-B} - [E_{A-A} \cdot E_{B-B}]^{1/2} \qquad ...(5.3)$$

When molecules AB is purely covalent then

$$E_{(A-B)} = E_{(A-B)cov} \quad \text{and} \quad \Delta E = 0$$

$$\Delta E = E_{(A-B)} - E_{(A-B)cov} = 0$$

$$\Delta E = E_{(A-B)} - [E_{(A-B)} \times E_{B-B}]^{1/2} = 0$$

However, if the two atoms A and B have different electronegativities the bond A – B will no longer be purely covalent and the energy $E_{(A-B)}$ would be greater than the mean of $E_{(A-A)}$ and $E_{(B-B)}$.

$$E_{(A-B)} > [E_{(A-A)} \times E_{(B-B)}]^{1/2}$$

In other words, ΔE would be greater than zero. This excess bond energy ΔE is called ionic-covalent resonance energy and arises due to the resonance stabilization of the sort

$$\underset{\text{covalent}}{A-B} \rightleftharpoons \underset{\text{ionic}}{A^+B^-} \quad \text{(B is more electronegative than A)}$$

Such a resonance hybrid of ionic and covalent characters gives rise to a strong bond than a purely covalent or purely ionic bond. The value of ΔE increases as the difference in electronegativity between the two atoms increases. This is clearly indicated on hydrohalides where the value of ΔE decreases as the electronegativity difference between the two atoms (H and $X = F$ to I) decreases.

Halide	HF	HCl	HBr	HI
ΔE (kJ/ mol^{-1})	268	42	46	5

This also indicates decrease in ionic character from HF to HI. In order to assign numerical values for the electronegativity of an element, Pauling proposed the following empirical expression to correlate ΔE with electronegativity difference.

$$0.18 \ (\Delta E)^{1/2} = \chi_A - \chi_B \qquad ...(5.4)$$

where χ_A is electronegativity of element A and χ_B is the electronegativity of element B and ΔE is expressed in kcal mole^{-1}.

However if ΔE is expressed in eV per molecule then the above expression becomes

$$0.208 \ (\Delta E)^{1/2} = \chi_A - \chi_B \qquad ...(5.5)$$

The equation valid for S.I. unit of energy is

$$0.088 \ (\Delta E)^{1/2} = \chi_A - \chi_B \qquad ...(5.6)$$

Assigning an arbitrary χ value of 2.1 for H, Pauling got a suitable range of numerical value for the element C to F in the range 2.5 to 4.0. This procedure allowed the evaluation of electronegativity of 14 elements H, C, N, O, F, Si, P, S, Cl, Ge, As, Se, Br and I. Values for others were then calculated from thermochemical data of compounds containing one of these 14 elements. Pauling values of electronegativity are given in Table 5.4. The determination of ΔE is subject to considerable errors and hence numerical values of electronegativity are subject to variation. This exposes the drawback of Pauling's method.

Table 5.4. Electronegativities of some elements.

Element	*Revised Pauling*	*Alfred-Rochow*
H	2.2	2.20
Li.	0.98	0.97
Be	1.57	1.47
B	2.04	2.01
C	2.55	2.50
N	3.04	3.07
O	3.44	3.50

(*Contd.*)

Element	*Revised Pauling*	*Alfred-Rochow*
F	3.98	4.10
Na	0.93	1.01
Mg	1.31	1.23
Al	1.61	1.47
Si	1.90	1.74
P	2.19	2.06
S	2.58	2.44
Cl	3.16	2.83
Ar	......	......
K	0.82	0.91
Ca	1.00	1.04
Zn	1.65	1.66
Ga	1.81	1.82
Ge	2.01	2.02
As	2.18	2.20
Se	2.55	2.48
Br	2.96	2.74
Kr	2.9	......
Rb	0.82	0.89
Sr	0.95	0.99
Y	1.22	1.11
Cd	1.69	1.46
In	1.78	1.49
Sn	1.96	1.72
Sb	2.05	1.82
Te	2.1	2.01
I	2.66	2.21
Xe	2.6	.....
Cs	0.79	0.86
Ba	0.89	0.97
Hg	2.00	1.44
Tl	2.04	1.44
Pb	2.33	1.55
Bi	2.02	1.67

Q. 23. Calculate electronegativity of C in C – H bond if E_{C-H}, E_{H-H} and E_{C-C} are 98.8, 104 and 83 kcal/mole respectively.

Ans. Pauling equation relating ΔE and electronegativity difference between the two elements

$$0.18 \times (\Delta E)^{1/2} = X_C - X_H$$

ΔE in the present case is obtained as under :

$$\Delta E = E_{C-H} - [E_{C-C} \cdot E_{H-H}]^{1/2}$$
$$= 98.8 - (83 \times 104)^{1/2}$$
$$= 98.8 - 92.9$$
$$= 5.9$$

Now $\quad 0.18\,(5.9)^{1/2} = X_C - 2.1$

$$X_C = 2.1 + 0.18 \times 5.9^{1/2}$$
$$= 2.1 + 0.18 \times 2.43$$
$$= 2.1 + 0.44 = 2.54$$

Q. 24. Electronegativities of hydrogen and fluorine are 2.1 and 4.0 respectively. Calculate the percentage ionic character in HF.

Ans. According to Pauling, when the electronegativity difference between two elements is 1.7 percentage of ionic character is 50.

Therefore, % of ionic character if the electronegativity difference is (4.0 – 2.1) or 1.9

$$= \frac{1.9}{1.7} \times 50$$
$$= 56$$

Q. 25. Discuss:

(*i*) Mulliken's Scale of Electronegativity

(*ii*) Alfred-Rochow Scale of Electronegativity.

Ans. Mulliken Scale of Electronegativity. According to Mulliken, electronegativity of an element is the arithmetic mean of its ionization potential and electron affinity.

$$\text{Thus electronegativity} = \frac{\text{Ionization Potential} + \text{Electron affinity}}{2}$$

It is possible to calculate the electronegativity of an element on Pauling scale if its electronegativity is known on Mulliken scale by using the relationship

$$\text{Electronegativity on Pauling scale} = \frac{\text{Electronegativity on Mulliken scale}}{3.15}$$

Alfred-Rochow Electronegativity. Alfred and Rochow proposed an empirical approach to electronegativity based on electrostatic force between the nucleus and electron. They defined electronegativity as electrostatic force (F) exerted by the effective nuclear charge (Z_{eff}) on the outer-most electron at a distance r from the nucleus. That is

$$F \propto \frac{Z_{eff}}{r^2} \qquad \text{...(5.7)}$$

where r = covalent radius of the atom in Å,

Z_{eff} is calculated according to Slater rules based on screening effect of intervening electrons. The attractive force (F) is then correlated to electronegativity value on the Pauling scale using an empirical relationship to obtain Alfred-Rochow electronegativities χ_{AR} (AR for Alfred-Rochow).

$$\chi_{AR} = 0.359 \times \frac{Z_{eff}}{r^2} + 0.744$$

The electronegativity values so obtained agree quite closely with those of Pauling's (Table 5.4).

Advantages

(*i*) Alfred-Rochow method depends on measuring covalent distance which are obtained with greater accuracy by X-ray diffraction. Therefore the X_{AR} values are quite accurate.

(*ii*) The Alfred-Rochow approach is based on the more fundamental and reasonable dependence of coulombic force on distance.

Disadvantages

(*i*) Slater rules used for calculated Z_{eff} are empirical and may not give accurate values.

(*ii*) Covalent radii are inaccurate when the bond order is not known.

Q. 26. Discuss Mulliken-Jaffee electronegativity concept.

Ans. Jaffee and his co-workers revised the Mulliken definition and suggested that electronegativity of an atom is different in different environment. Some of the factors which affect electronegativity values are as follows:

(*a*) Oxidation state. Electronegativity of an atom varies from one bonding situation to another depending upon the oxidation number of the atom in the molecule, for example, Pauling electronegativity value is 1.62 for Tl(I) whereas it is around 2.04 for T1(II). Similarly, electronegativity of S in SF_6 is higher than that in $SC1_2$ because the former has S in higher oxidation state. The higher the oxidation state, the greater the attraction for electron and higher the electronegativity.

(*b*) Hybridization. It affects electronegativity value considerably because *s*-orbital due to its higher penetration effect is more tightly held than other orbitals and hence has low energy. Consequently it will have greater tendency to attract electrons and hence show higher electronegativity values, for example, carbon in CH_4 is sp^3 hybridized which has 25% *s*-character. The electronegativity of C in CH_4 is 2.53. In ethylene $H_2C = CH_2$, carbon is sp^2 hybridized which has 33% *s*-character. The electronegativity of C in C_2H_4 is 2.75. In case of acetylene $HC \equiv CH$, carbon is *sp* hybridized which has 50% *s*-character. Thus electronegativity of C in C_2H_2 is raised and its value is referred to be 3.3.

(*c*) Charge. An atom with a positive charge will attract electrons more readily than a neutral atom while a negatively charged atom will attract electron less readily than a neutral atom.

(*d*) Bond order. As bond order of the atom whose electronegativity is under discussion increases, its electronegativity increases, for example, C in C_2H_2 is more electronegative than C in C_2H_4.

Q. 27. What do you understand by Effective Nuclear Charge and Screening Effect?

(*G. Nanakdev, 2000; Meerut, 2000; Garhwal, 2000; Nagpur, 2000*)

Ans. The relation between energy and energy shell for an electron in hydrogen atom is given by expression

$$E = -R\frac{Z^2}{n^2} \quad \text{...(5.9)}$$

For hydrogen atom ($Z = 1, n = 1$), the energy of the outermost electron calculated from eq. (5.9) agrees with the observed I.E. 1312 kJ mole^{-1}. However the calculated and observed value for atom with Z > 1 do not agree fairly. The observed ionization energy of say Li is considerably lower than that calculated from equation (5.9) because of following reasons:

(*i*) The expression does not hold good for multi-electron atom because it does not take inter-electronic repulsions into account.

(*ii*) The $2s^1$ electron does not face the nuclear charge directly. There are intervening $1s^2$ electrons which not only shield the nucleus but also repel the $2s^1$ electron resulting in removal of the electron.

Thus the inner core of electrons shield or screen the nucleus from outer electron. Consequently the outer electrons experience only a part of the nuclear charge or the effective nuclear charge faced by the outer electron is lower than that expected from total nuclear charge. In case of Li, the effective nuclear charge faced by $2s^1$ electron is not 3 but between 1 and 2 (actual value is 1.59). However if we recall the radical probability distribution curve of 2*s* orbital in hydrogen like atoms, it is apparent that 2*s* electron has definite probability of finding itself near the nucleus and hence would penetrate the inner 1*s* orbital core. This means that although 2*s* electron is shielded by the 1*s* core yet 2*s* electron penetrates it to some extent and experiences somewhat higher nuclear charge. In fact the energy of outer electron is determined by the effective nuclear charge (Z_{eff}) which is less than Z (nuclear charge) by an amount shielding constant (*S*) of the intervening core of the electrons

$$Z_{eff} = Z - S$$

Due to greater penetration, *s*-electrons tend to shield the nucleus more effectively than *p*-electrons which in turn shield the nucleus more effectively than *d*-electrons. That is, for a given *n*, the screening or shielding effect varies as

$$s > p > d > f$$

Q. 28. Describe Slater Rules for determining effective nuclear charge.

(M. Dayanand, 2000; Nagpur, 2000; Himachal, 2000; G. Nanakdev, 2000; Garhwal, 2000; Calcutta, 2005)

Ans. Slater gave a set of empirical rules for calculating the effective nuclear charge (Z_{eff}) experienced by electrons in different orbitals. As stated above, the Z_{eff} acting on a given electron is calculated by subtracting the screening constant (*S*) from the atomic number *Z* (nuclear charge).

That is: $Z_{eff} = Z - S$

The 'Slater rules' for calculating screening on shielding constant (*S*) are as follows:

(1) The electronic configuration of the element is first written in the following order and groupings: (1*s*); (2*s*, 2*p*); (3*s*, 3*p*); (3*d*); (4*s*, 4*p*); (4*d*); (4*f*); (5*s*, 5*p*); (5*d*); (5*f*) and so on.

(2) For an electron considered in a group of *s*, *p* electrons the shielding constant (*S*) is the sum of the following contributions:

(*a*) No contribution for electrons in groups beyond the one considered.

(*b*) An amount of 0.35 for each electron in the group considered (except in 1*s* group where 0.30 is used instead of 0.35).

(*c*) An amount of 0.85 for each electron in the next inner shell (*n*-1).

(*d*) An amount of 1.00 for each of all other inner shell electrons.

(3) For an electron considered in *d* or *f* group, rules 2*a* and 2*b* apply as such; however rules 2*c* and 2*d* are replaced with the rule that all other electrons lying to the left of *d* or *f* group contribute 1.00 each.

Q. 29. Calculate Z_{eff} faced by a 3*s* or 3*p* electron in phosphorus atom ($Z = 15$, $1s^2\ 2s^2\ 2p^6\ 3s^2\ 3p^3$)

Ans. Group configuration = ($1s^2$) ($2s^2$, $2p^6$) ($3s^2$, $3p^3$). Four other electrons in the group ($3s^2$, $3p^3$) contribute 4×0.35 and next inner shell contributes 8×0.85 and then the next 2×1.

For a $3s$ or $3p$ electron, $S = (2 \times 1.00) + (8 \times 0.85) + (4 \times 0.35)$

$$S = 2.00 + 6.80 + 1.40 = 10.20$$

$$Z_{eff} = 15 - 10.20 = 4.80$$

Q. 30. Calculate Z_{eff} experienced by a $2p$ electron in oxygen atom (*M. Dayanand, 2002*)

Z (8, $1s^2$, $2s^2\ 2p^4$)

Ans. Group configuration = $(1s^2)$; $(2s^2\ 2p^4)$

There are 5 other electrons in group $(2s^2, 2p^4)$ under consideration which contribute 5×0.35 towards S. Next inner shell contributes 2×0.85. Thus for a $2p$ electron

$$S = (2 \times 0.85) + 5 \times (0.35) = 3.45$$

$$Z_{eff} = (8 - 3.45) = 4.55$$

Q. 31. Calculate Z_{eff} experienced by (*i*) $5s$ electron (*ii*) $4d$ electron in Ag atom (Z = 47, $1s^2, 2s^2, 2p^6, 3s^2, 3p^6, 3d^{10}, 4s^2, 4p^6, 4d^{10}, 5s^1$).

Ans. Group configuration = $(1s^2)$, $(2s^2, 2p^6)$; $(3s^2, 3p^6)$; $(3d^{10})$ $(4s^2, 4p^6)$, $(4d^{10})$ $(5s^1)$

(*i*) **For $5s$ electron,** $S = (28 \times 1.00) + (18 \times 0.85) + (0 \times 0.35)$

$$= 43.30$$

$$Z_{eff} = 47 - 43.30 = 3.70$$

(*ii*) **For $4d$ electron,** $S = (36 \times 1.00) + (9 \times 0.35) + 0$

$$= 39.15$$

$$Z_{eff} = 47 - 39.15 = 7.85$$

Q. 32. Calculate Z_{eff} for (*i*) 4s and (*ii*) $3d$ electron in iron atom

Z = (26, $1s^2, 2s^2, 2p^6, 3s^2, 3p^6, 3d^6, 4s^2$)

(*Nagpur, 2000*)

Ans. Group configuration = $(1s^2)$;$(2s^2, 2p^6)$; $(3s^2\ 3p^6)$; $(3d^6)$; $(4s^2)$

(*i*) **For 4s electron,** $S = (10 \times 1.00) + (14 \times 0.85) + (1 \times 0.35) = 22.25$

$$Z_{eff} = 26 - 22.25 = 3.75$$

(*ii*) In case of $3d$ electron, there is no contribution from outer $4s$ group electrons inner of $3d$ contribute one each.

Q. 33. Calculate the effective nuclear charge for

(*i*) an electron in $3d$ orbital, and

(*ii*) an electron in 4s orbital of chromium (Z = 24) with electronic configuration =$1s^2, 2s^2, 2p^6, 3s^2, 3p^6, 3d^5, 4s^1$ (*M. Dayanand, 2000; G. Nanakdev, 2000; Himachal, 2000; Calcutta, 2005*)

Ans. Group configuration $(1s^2)$, $(2s^2\ 2p^6)$ $(3s^2\ 3p^6)$ $(3d^5)$, $(4s^1)$

(*i*) **For $3d$ electron.** There is no contribution from outer $4s$. Remaining 4 electrons in $3d$ contribute 4×0.35 and all others contribute one each.

$$S = (18 \times 1) + (4 \times 0.35) + (4 \times 0)$$

$$= 18 + 1.40 + 0 = 19.40$$

$$Z_{eff} = 24 - 19.40 = 4.60$$

(*ii*) **For $4s$ electron.** There are no other electron in $4s$. Consequently there is no contribution and we have $0 \times 0.35 = 0$.

Next inner shell (n = 3) has 13 electrons contributing 13×0.85, others contribute each and we have

$$S = (10 \times 1) + (13 \times 0.85) + (0 \times 0.35)$$
$$= 10 + 11.45 + 0 = 21.45$$
$$Z_{eff} = 24 - 21.45 = 2.55$$

Q. 34. Calculate Z_{eff} for valence electrons in Ca and Br.

Ans. (*a*) Electronic configuration for Ca is

$$1s^2\ 2s^2\ 2p^6\ 3s^2\ 3p^6\ 4s^2$$

This can be grouped as

$$1s^2,\ (2s^2,\ 2p^6),\ (3s^2,\ 3p^6),\ 4s^2$$

Screening constant of 4*s* electron

$$= (1 \times 0.35) + (8 \times 0.85) + (10 \times 1)$$
$$= 17.15$$

Hence $Z_{eff} = 20 - 17.15 = 2.85$

(*b*) Electronic configuration for Br is

$$1s^2\ 2s^2\ 2p^6\ 3s^2\ 3p^6\ 4s^2\ \ 3d^{10}\ 4p^5$$

This can be grouped as

$$1s^2,\ (2s^2,\ 2p^6),\ (3s^2,\ 3p^6),\ 3d^{10},\ (4s^2,\ 4p^5)$$

Screening constant of 4*p* electron

$$= (4 \times 0.35) + (10 \times 0.85) + (18 \times 1)$$
$$= 27.9$$

Hence $Z_{eff} = 35 - 27.9 = 7.1$

Q. 35. Discuss important applications of Slater rules.

Ans. 1. It justifies the filling of 4*s* orbitals in preference to $(n-1)$ *d* orbitals. For example, potassium has electronic configuration,

$$1s^2\ 2s^2\ 2p^6\ 3s^2\ 3p^6\ 4s^1$$

Here Z_{eff} for $3d = 19 - (10 \times 1.00) + (8 \times 0.85) = 2.20$

If the electronic configuration were

$$1s^2\ 2s^2\ 2p^6\ \ 3p^6\ 3d^1$$

Then Z_{eff} for $3d = 19 - [(18 \times 1.00) + 0 \times 0.35)] = 1.00$

This shows that a 4*s* electron experiences a higher Z_{eff} and hence a configuration in which 4*s* is occupied in preference to 3*d* would be more stable.

2. Although Slater rules do not allow the exact calculations of the energies of electron, yet these are useful in understanding variation in size, ionization potential and electronegativity. For example, for an outer electron in elements along the 3rd period, we notice the following trend in Z_{eff} value.

Element	P	S	C l	Ar	H
Z_{eff}	4.80	5.45	6.10	6.75	(2.2)

This shows that Z_{eff} increases along the period and attains maximum value at the noble gas which accounts for increasing ionization potential along the period with maximum at the noble gas. Ionization potential shows sudden drop after that as evidenced in Z_{eff} value of *H*.

Q. 36. 4*s* subshell is filled first and 3*d* later on while 2*p* subshell is filled first and 3*p* later on. Explain the reason.

Ans. (*i*) The problem can be discussed by taking the case of potassium (K).

It has an electronic configuration:

$1s^2\ 2s^2\ 2p^6\ 3p^6\ 4s^1$ and not $1s^2\ 2s^2\ 2p^6\ 3s^2\ 3p^6\ 3d^1$

Let us calculate effective nuclear charge on the outermost electron based on the two configurations.

Configuration I

E.N.C. for 4*s* electron = $[19 - (10 \times 1) + (8 \times 0.85)] = 2.20$

Configuration II

E.N.C. for 4*s* electron = $19 - [(18 \times 1) + (0 \times 0.35)] = 1$

Greater E.N.C means greater stability and lower energy. Hence the configuration I is observed and 4*s* subshell has lower energy and filled first.

(*ii*) 2*p* is filled first and 3*p* later on.

This can be studied by calculating the E.A.N. of last electron in atom with atomic No. 5 viz., boron. The two posibilities are:

$1s^2\ 2s^2\ 2p^1$ (Configuration I)

$1s^2\ 2s^2\ 3p^1$ (Configuration II)

E.A.N. of last *i.e.*, $2p^1$ electron (Configuration I) according to Slater Rule is

$5 - [(2 \times 0.35) + (2 \times 0.85)]$

$= 5 - [0.70 + 1.70]$

$= 5 - (2.40) = 2.6$

E.A.N. of last $3p^1$ electron (Configuration II)

$= 5 - (2 \times 0.85) + (2 \times 1)]$

$= 5 - 3.7 = 1.3$

Thus in configuration I, E.A.N of the $2p^1$ electron is higher indicating lower energy and greater stability of 2*p* orbital compared to 3*p* orbital.

Q. 37. What are the limitations of Slater Rules? (*Garhwal, 2000; Kurukshetra, 2002*)

Ans. (*i*) Both *s* and *p* orbitals are grouped together for calculating Z_{eff} whereas the radial probability distribution curves for a given *n* indicate a higher penetration or shielding effect of *s* as compared to *p*-orbital.

(*ii*) All electrons irrespective of *s, p, d,* or *f* orbitals at positions lower than the $(n-1)$ shell shield the *n* shell electron with equal contribution of 1.00. This is unjustified because different orbitals cannot have the same contribution on energy consideration.

(*iii*) Penetration by higher occupied orbitals (than the one considered) into inner orbitals is well known but even then their contribution is ignored as per Slater Rule 2*a*.

6

ACIDS AND BASES

Q. 1. Explain Arrhenius concept of acids and bases. What are the limitations of this concept? *(Ranchi, 2010)*

Ans. Svante Arrhenius, as a part of his theory of ionisation, proposed a theory of acids and bases. According to Arrhenius theory:

An acid is a substance that dissociates in aqueous solution to give hydrogen ions. Thus, HCl, HNO_3, H_2SO_4, CH_3COOH and HCN are acids because they liberate H^+ ions when dissolved in water.

$$HCl \rightleftharpoons H^+ + Cl^-$$

$$HNO_3 \rightleftharpoons H^+ + NO_3^-$$

$$H_2SO_4 \rightleftharpoons 2H^+ + SO_4^{2-}$$

$$HCN \rightleftharpoons H^+ + CN^-$$

A base is a substance that dissociates in *aqueous solution to give hydroxyl ions.* Thus, NaOH, KOH, $Ba(OH)_2$ and NH_3 are bases because they liberate (OH^-) ions when dissolved in water.

$$NaOH \rightleftharpoons Na^+ + OH^-$$

$$KOH \rightleftharpoons K^+ + OH^-$$

$$Ba(OH)_2 \rightleftharpoons Ba^{2+} + 2OH^-$$

$$NH_3 \overset{H_2O}{\rightleftharpoons} NH_4^+ + OH^-$$

Hydronium Ion (H_3O^+)

According to Arrhenius, an acid releases H^+ ions in aqueous medium. However, it has been found that H^+ ion cannot exist independently.

In fact, H^+ ion is unique among ions. It has no electron. Its radius is equal to that of nucleus (10^{-13} cm). The concentration of charge in such a small volume makes the attraction between water *and H^+ ions extremely large.* H^+ ion accepts a pair of electrons from water molecules [H – O – H] to form *hydrated proton* or **hydronium ion**.

$$H^+ + H-O-H \longrightarrow H_3O^+ \text{ or } \left\{ \begin{matrix} H \quad H \\ \diagdown \diagup \\ O \\ | \\ H \end{matrix} \right\}^{\oplus} \text{ (Hydronium ion)}$$

Dissociation of hydrocloric acid in aqueous solution may be represented as:

$$HCl + H_2O \rightleftharpoons H_3O^+ + Cl^-$$

Limitations of Arrhenius Theory to explain acid-base concept

1. Limited scope. Arrhenius concept is limited to *aqueous medium* only. It fails to explain the behaviour of acids and bases in *non-aqueous* solvents like ammonia, sulphur dioxide, alcohol, etc.

2. Proton in aqueous solutions. According to Arrhenius concept acids are those which furnish H ions in water. H^+ -ion is nothing but a proton, which is not capable of independent existence. As already discussed it may exist as hydronium ion, *i.e.* H_3O^+.

3. Acids and bases without H^+ and OH^- ions. There are a number of substances (like CO_2, SO_2, NO_2, SO_3 etc.) which do not contain H^+ion, but behave as acids. Similarly, compounds like CaO, MgO, anhydrous NH_3, Na_2CO_3 etc. which do not contain OH^- ions, behave as bases.

4. Neutralisation without H^+ and OH^- ions. Neutralization between HCl(g) and NH_3 (g) does not involve H^+ ions and OH^- ions *e.g.*

$$NH_3(g) + HCl(g) \rightarrow NH_4\,Cl(s)$$
$$CaO\,(s) + SO_2(g) \rightarrow CaSO_3(s)$$

Q. 2. Explain Bronsted-Lowry concept of acid and base. *(Pune 2010)*

Ans. J.N. Bronsted and T. M. Lowry independently and simultaneously (1923) put forward a more general concept for acids and bases. According to this concept :

(*i*) *Acid is a substance (molecule or ion) which has a tendency to donate a proton to any substance, and*

(*ii*) *A base is a substance (molecule or ion) which has a tendency to accept a proton from any other substance.*

In other words, an *acid is a proton donor and base is a proton acceptor.* In fact, acid-base reaction involves transference of proton. That is why, this concept is also called *proton transfer theory* of acids and bases. In accordance with this broader definition, an acid or base may be molecular or ionic, *e.g.,*

Molecular acids	:	HF, HC1, HNO_3, H_2SO_4, CH_3COOH, H_2S, H_3PO_4, H_2O, etc.
Ionic acids	:	H_3O^+, NH_4^+, HSO_4^-, HSO_3^-, HPO_4^{2-} etc.
Molecular bases	:	RNH_2, NH_3, H_2O, R_2NH etc.
Ionic bases	:	OH^-, S^{2-}, CO_3^{2-}, Cl^-, Br^-, NO_3^-, etc.

It is important to note that no single substance is an acid or a base, A single substance cannot donate a proton unless and until some other substance which accepts the proton is also present. This proton transfer may or may not involve any medium, *e.g.,* the reaction

$$NH_3\,(g) + HC1\,(g) \rightarrow NH_4^+\,C\bar{1}\,(s)$$

does not involve any medium. Here, HC1 has denoted a proton to NH_3. Hence, HC1 is an acid w.r.t. NH_3, which is a base.

Consider some more acid-base reactions.

$$\underset{\text{Base}}{NH_3} + \underset{\text{Acid}}{H_2O} \rightleftharpoons NH_4^- + OH^-$$

$$\underset{\text{Acid}}{CH_3COOH} + \underset{\text{Base}}{NH_3} \rightleftharpoons CH_3COO^- + NH_4^+$$

$$\underset{\text{Acid}}{HCl} + \underset{\text{Base}}{H_2O} \rightleftharpoons H_3O^+ + Cl^-$$

$$\underset{\text{Acid}}{H_2SO_4} + \underset{\text{Base}}{H_2O} \rightleftharpoons H_3O^+ + HSO_4^-$$

A careful study of the above equations reveals that water is an acid w.r.t. NH_3 whereas it is a base w.r.t. HC1 or H_2SO_4. Such substances like H_2O, HCO_3^-, HSO_4^- etc. which can act as acids (proton donor) as well as bases (proton acceptor) are called **amphoteric** or **amphiprotic** substances.

Conjugate Acid-Base pairs. When an acid loses a proton, the residual part of it will have a tendency to accept a proton. Thus, it will behave as a base. *Such pairs of substances which differ from one another by a proton are known as conjugate acid-base pairs.* Consider a general example of an acid :

$$\underset{\text{Acid}}{HA} \rightleftharpoons \underset{\text{Proton}}{H^+} + \underset{\text{Conjugate base}}{A^-}$$

Consider the reaction between HC1 and H_2O.

$$\underset{\text{Acid}}{HCl} + \underset{\text{Base}}{H_2O} \rightleftharpoons \underset{\text{Acid}}{H_3O^+} + \underset{\text{Base}}{Cl^-}$$

HC1 donates a proton to water in the forward reaction.

∴ HC1 is an acid and H_2O is a base.

In the backward reaction, H_3O^+ ion donates its proton to Cl^- ion.

∴ H_3O^+ is an Acid and, Cl^- is a base.

It is clear that HC1 loses a proton and forms Cl^- ion (which is a base). HC1 and Cl^- ions are called *conjugate acid-base pair.*

$$\underset{\text{Acid}}{HCl} \rightleftharpoons H^+ + \underset{\text{Conjugate base}}{Cl^-}$$

Similarly, H_2O (base) accepts a proton and forms H_3O^+ (acid). Here, H_2O and H_3O^+ are also conjugate base-acid pair.

$$\underset{\text{Base}}{H_2O} + H^+ \rightleftharpoons \underset{\text{Conjugate acid}}{H_3O^+}$$

In the above example, consider HC1 as acid-1 and Cl^- as base-1. Similarly, consider H_2O as base-2 and H_3O^+ as acid-2. The overall reaction between HC1 and H_2O, can, hence, be written as :

$$\underset{\text{Acid–1}}{HCl} + \underset{\text{Base–2}}{H_2O} \rightleftharpoons \underset{\text{Acid–2}}{H_3O^+} + \underset{\text{Base–1}}{Cl^-}$$

Some other examples of acid base reactions and conjugate acid- base pairs are given in the following table 6.1.

Table: 6.1. Conjugate acid-base pairs.

Acid-1		Base-2		Acid-2		Base-1
H_2SO_4	+	H_2O	$\rightleftharpoons$	H_3O^+	+	HSO_4^-
H_2O	+	NH_3	$\rightleftharpoons$	NH_4^+	+	OH^-
CH_3COOH	+	H_2O	$\rightleftharpoons$	H_3O^+	+	CH_3COO^-
H_2S	+	H_2O	$\rightleftharpoons$	H_3O^+	+	HS^-
H_3O^+	+	HCO_3^-	$\rightleftharpoons$	H_2CO_3	+	H_2O
H_3O^+	+	ClO_4^-	$\rightleftharpoons$	$HClO_4$	+	H_2O
H_3PO_4	+	H_2O	$\rightleftharpoons$	H_3O^+	+	$H_2PO_4^-$

Q. 3. Prove that all Bronsted bases may not be Arrhenius bases.

Ans. According to Arrhenius, base must furnish OH^- –ions in aqueous solution. According to *Bronsted theory,* a base must accept a proton. Consider for example carbonate ion, CO_3^{2-}.

According to Bronsted theory, CO_3^{2-} ***ion is a base as it accepts a proton to form*** HCO_3^- ***ion.***

$$CO_3^{2-} + H^+ \rightarrow HCO_3^-$$

According to Arrhenius concept, CO_3^{2-} is not a base as it does not furnish OH^- ions in aqueous solution. Hence, all Arrhenius acids are also Bronsted acids but all Bronsted bases are not Arrhenius bases.

Q. 4. Explain Lux-Flood theory of acids and bases. (*Nagpur, 2008; Pune 2010*)

Ans. This concept was proposed by Lux and later extended by Flood. It visualises the acid – base reaction in terms of the oxide ion and is mainly used for molten oxides at high temperatures. According to this concept, an acid *is a substance which accepts the oxide ion and a* **base** *is a substance which donates the oxide ion.* In other words, *an* **acid** *is an oxide ion-acceptor and a* **base** *is an oxide ion-donor.* For example, in the reaction

$$\underset{\text{Base}}{CaO} + \underset{\text{Acid}}{SiO_2} \longrightarrow CaSiO_3,$$

calcium oxide is a base and silicon dioxide is an acid. Similarly, in the reaction

$$PbO + SO_3 \longrightarrow PbSO_4,$$

PbO is a base while SO_3 is an acid.

This concept, has only a limited scope.

Q. 5. Give Lewis definition of acid and base giving suitable examples.

(*Delhi, 2002; Purvanchal 2003; Guwahati, 2002; Bangalore,2005; Nagpur, 2008; Lucknow, 2010; Ranchi, 2010*)

Ans. According to Lewis, **an acid** is defined as any molecule, ion or radical that can accept one or more electron pairs, that is, an acid is an electron pair acceptor. In other words, a substance which has vacant orbital to accommodate one or more pairs of electrons can act as a Lewis acid. A **base** is defined as a species that can donate one or more electron pairs, *i.e.,* a base is an electron pair donor. Consequently any substance with unshared pair of electrons can behave as a Lewis base. Acid base or neutralisation reactions may, therefore, be termed as donor-acceptor reactions or coordinate covalent bond formation reactions as shown below.

S. No.	*Acceptor*	*Donor (Acid)*	*Donor-acceptor compound (Base) Neutralization product*
1.	BF_3	NH_3	$H_3N \rightarrow BF_3$
2.	$SnCl_4$	$2R_2O$	$(R_2O)_2\ SnCl_4$
3.	H^+	NH_3	NH_4^+
4.	$FeCl_3$	Cl^-	$(FeCl_4)^-$
5.	SO_3	O^{2-}	SO_4^{2-}

Q. 6. What are different kinds of Lewis acids? (*Kanpur, 2000; Calcutta, 2005*)

Ans. These can be classified into the following two categories:

(*i*) **All the cations are Lewis acids** because they have tendency to accept electron pairs. Larger cations of alkali and alkaline earth metals behave as relatively weaker Lewis acid due to their lesser tendency to accept electrons. Smaller cations like H^+, Li^+, Be^{++} etc., are likely to act as stronger Lewis acids due to their higher tendency to accept electrons and form stronger bonds. Also higher charged cations have a higher tendency to accept electrons relative to the lower charged cations of a given metal. For example, Fe^{+++} behaves as a stronger Lewis acid as compared to Fe^{++}. Similarly, Sn^{4+} is a stronger Lewis acid than Sn^{2+}. Examples of some cations acting as Lewis acids are given below:

S. No.	*Lewis Acid*	*Lewis Base*	*Complex*
1	Ag^+	$2\ NH_3$	$[Ag\ (NH_3)_2]^+$
2	Cu^{2+}	$4\ Cl^-$	$[CuCl_4]^{2-}$
3	Mn^{2+}	$6\ H_2O$	$[Mn\ (H_2O)_6]^{2+}$
4	Ni^{2+}	$4\ CN^-$	$[Ni\ (CN)_4]^{2-}$

(*ii*) **Electron-deficient** compounds in which the central atom can expand its valence shell can behave as Lewis acid. BF_3, $AlCl_3$, $GaBr_3$, $BeCl_2$ are electron-deficient compounds and have tendency to accept electron pairs to complete their octet.

$$BF_3 + F^- \longrightarrow [BF_4]^-$$

Compounds like SiF_4, $SnCl_4$, $SbCl_3$, $SbCl_5$, $TeBr_4$, SO_3 etc also represent Lewis acids where central atom can expand its valence shell.

$$SnCl_4 + 2Cl^- \longrightarrow [\ SnCl_6]^{2-}$$

$$SO_3 + R_2O \longrightarrow [R_2O\ (SO_3)]$$

$$SiF_4 + 2F^- \longrightarrow [\ SiF_6]^{2-}$$

Q. 7. What are the different kinds of Lewis bases?

(Kanpur, 2000; Calcutta, 2005)

Ans. Following two types of compounds constitute Lewis bases:

(*i*) **All the anions can act as Lewis bases** because they have a tendency to donate electron pairs. For example, oxides (O^{2-}), hydroxides (OH^-), halides (X^-), amides (NH_2^-) etc., act as Lewis bases. The greater the charge density on the anion, the stronger the base.

(*ii*) Substances having **unshared or lone pairs of electrons** act as Lewis bases. For example, water (H_2O :), ammonia (NH_3), ether (R_2O :), Ketones ($R_2\ C = O$:) etc., act as electron donors and are thus Lewis bases.

Q. 8. What is extended Lewis definition of acid and base? *(Kolhapur, 2000)*

Ans. The extended Lewis acids include unsaturated species like CO_2 and CS_2, where the central atom is bonded to atoms of different electronegativities by one or more multiple bonds. Although both the molecules lack acceptor character of the central atom because carbon atom cannot extend its octet, yet these have been indirectly treated as Lewis acids. This is due to the polarization of the π-electron cloud on approach of a Lewis base towards carbon atom of CO_2 or CS_2. The polarization leads to a partial +ve charge on carbon and a partial – ve charge on the more electronegative atom like O. The +vely charged carbon then acts as an electron acceptor as shown in the reaction of CO_2 with Lewis bases.

$$O=C=O + :\ddot{O}H^- \longrightarrow {}^-O-C(=O)-O-H$$

Bicarbonate

The extended Lewis bases include substance like carbon monoxide (CO) and unsaturated hydrocarbons like alkenes, alkynes, butadienes, benzene etc. These may be treated as Lewis bases because of a lone pair on : CO or donation from above mentioned hydrocarbons to form π-complexes with metals in low oxidation state.

Q. 9. What are the limitations of Lewis concept of acid and base?

(*M. Dayanand, 2000; Kolhapur, 2000; Kanpur, 2000*)

Ans. (*i*) It does not provide us any uniform scale of acid or base strength.

(*ii*) It fails to tell us whether salt formation is related to acid - base behaviour or not.

(*iii*) Normal acid - base reactions are fast but complex formation in terms of donor - acceptor bond formation is a slow process.

(*iv*) It is very general and includes formation of $Ni(CO)_4$ and $Cr(C_6H_6)_2$ as Lewis - acid base reactions. However, bonding in these compounds is more complex and beyond the scope of Lewis concept.

Q. 10. Classify the following as:

(*i*) Lewis acids

(*ii*) Border line Lewis acids.

[CS_2, $TeCl_4$, SO_3, CO_2, Mg^{++}, $GaCl_3$] (*Nagpur, 2005*)

Ans. Lewis acids: $TeCl_4$, $GaCl_3$, Mg^{++}, SO_3

Border line Lewis acids: CS_2, CO_2.

Q. 11. Which of the following can be classified as:

(*i*) Lewis bases and

(*ii*) Border line Lewis bases?

{C_2H_4, Pyridine, CO, water, benzene, diethyl ether, acetone, 1, 3–butadiene.}

Ans. Lewis bases: Pyridine, water, diethyl ether, acetone.

Border line Lewis bases: CO, C_2H_4, benzene, 1,3-butadiene.

Q. 12. Categorise following species into:

(*i*) Lewis acid-base cases and

(*ii*) Border line cases (Extended Lewis concept):

$Cr(CO)_6$, $Fe(C_5H_5)_2$, $[Al(H_2O)_6]Cl_3$, $SbCl^-_6$, R_3NH^+, $[Co(NH_3)_6]^{3-}$, $PdCl_2(C_2H_4)$

Ans. Lewis acid-base: R_3NH^+, $SbCl^-_6$, $[Al(H_2O)_6]Cl_3$, $[Co(NH_3)_6]^{3-}$.

Border line cases $Cr(CO)_6$, $Fe(C_5H_5)_2$, $PdCl_2(C_2H_4)$.

Q. 13. Classify the following as Lewis acids or bases giving reasons:

(*i*) BF_3 (*ii*) NH_3 (*iii*) Cl^- (*iv*) Zn^{2+}

Ans. BF_3 and Zn^{2+} are Lewis acids. Electron-deficient and positively-charged species act as Lewis acids.

NH_3 and Cl are Lewis bases. Molecules with lone pair of electrons and negatively charged species act as Lewis bases.

Q. 14. Give Cady-Elsey theory (Solvent system concept) of acid and base.

Ans. An acid is defined as a solute which directly or indirectly increases the concentration of cations characteristic of the solvent and base as a solute which increases the concentration of the anions characteristic of the solvent.

$$2H_2O \rightleftharpoons \underset{\text{(acid)}}{H_3O^+} + \underset{\text{(base)}}{OH^-}$$

$$2NH_3 \rightleftharpoons \underset{\text{(acid)}}{NH^+_4} + \underset{\text{(base)}}{NH^-_2}$$

$$2SO_2 \rightleftharpoons \underset{\text{(acid)}}{SO^{2+}} + \underset{\text{(base)}}{SO_3^{2-}}$$

Thus NH_4Cl behaves as an acid in ammonia because it furnishes NH^+_4 which are cations characteristic of solvent and KNH_2 is a base in NH_3 because it gives NH_2^- which are anions characteristic of the solvent. Thus on addition of NH_4Cl in NH_3, the concentration of cations characteristic of solvent *i.e.*, NH_4^+ is increased and on addition of KNH_2 in NH_3, conc. of anions characteristic of solvent *i.e.*, NH_2^- is increased.

Consider the auto-ionization of liquid nitrogen tetroxide :

$$N_2O_4 \rightleftharpoons \underset{\text{Acid ion}}{NO^+} + \underset{\text{Base ion}}{NO_3^-}$$

According to solvent — system concept, in liquid nitrogen tetroxide, the substances which furnish nitrosyl ions (NO^+) behave as acids and those which furnish nitrate ions (NO_3^-) behave as bases. Accordingly, NOC1 behaves as an acid and $NaNO_3$ behaves as a base, in this solvent.

$$\underset{\text{Acid}}{NOCl} \xrightleftharpoons{\text{liq.}N_2O_4} \underset{\text{Solvent cation}}{NO^+} + Cl^-$$

$$NaNO_3 \xrightleftharpoons{\text{liq.}N_2O_4} NO^+ + \underset{\text{Solvant anion}}{NO_3^-}$$

Autoionization of sulphuric acid is represented as

$$2H_2SO_4 \rightleftharpoons H_3SO_4^+ + HSO_4^-$$

Thus, any species giving $H_3SO_4^+$ ion acts as an acid and any species giving HSO_4^- ion acts as a base.

Q. 15. Discuss Usanovich concept of acid and base.

(*Kurukshetra 2002; M. Dayanand, 2002; Kerala, 2000; Kolhapur, 2000*)

Ans. Usanovich defined an acid as a substance *that reacts with bases to form salts, gives up cations or takes up anions or electrons. Similarly a base is any substance or species that neutralizes an acid, gives up anions or electrons or takes up cations.*

Usanovich's definition includes all substances that can be defined as acids or bases in terms of other acid base concepts. Over and above, his definition includes all redox-reactions (electron transfer reactions) as acid-base reactions.

An example of each kind of Usanovich acid or base along with plausible explanation of the concept is given in Table 6.2.

Table 6.2. Usanovich acids and bases and their reactions.

	Acid	*Base*	*Salt*	*Justifications*
1.	HCl	NaOH	$Na^+ Cl^- + H_2O$	Salt formation
2.	HCl	NH_3	$NH^+_4Cl^-$	Acid gives up cation H^+
3.	$2NH_4Cl$	$Ba.(OH)_2$	$Ba^{2+}\ Cl_2^{2-} + 2NH_3 + H_2\ O$	Acid gives up cation NH_4^+
4.	SO_3	CaO	$Ca^{2+}\ SO^2_4{}^-$	Acid SO_3 accepts anion O^{2-}
5.	Cl_2	Cu	$Cu^{2+}\ Cl_2{}^{2-}$	Acid gains an electron
6.	$Fe\ (CN)_2$	4KCN	$K_4^{4+}[Fe\ (CN_6)]^{4-}$	Base KCN gives up anionCN^-
7.	F_2	Mg	$Mg^{2+}\ F^2_2{}^-$	Base Mg loses electrons
8.	$H_2\ SO_4$	NH_3	$(NH_4\ )^{2+}_2\ SO_4{}^{2-}$	BaseNH_3takes up cations H^+

Although Usanovich concept lays more stress on the cationic and anionic aspects of acids and bases, yet it does not ignore unsaturation in cation acid - base reaction. The acidic function is determined by the presence of coordinately unsaturated species and the basic, function by the presence of coordinately saturated species as shown below:

$$\underset{\text{(acid)}}{O=C=O} + \underset{\text{(base)}}{OH^-} \longrightarrow \underset{\text{(bicarbonate)}}{HO{-}C(=O){-}O^-}$$

$$\underset{\text{(acid)}}{S=C=S} + \underset{\text{(base)}}{R_2NH} \longrightarrow \underset{\text{(dithiocarbonic acid)}}{R_2N{-}C(=S){-}SH}$$

Q. 16. What are hard and soft acids and bases?

(Lucknow, 2002; Bhopal,2004; Patna 2010; Pune 2010)

Ans. According to Usanovich concept, cations come under the category of acids. Hard acids (cations) are those which fulfil the following conditions:

(*i*) *Cations of smaller size or lighter elements,*

(*ii*) *Cations of higher charge,*

(*iii*) *Not easily polarizable, and*

(*iv*) *Molecules or ions with lesser number of valence electrons.*

Examples of hard acids are alkali metal ions, alkaline earth metal ions, B, Al, Si ions, lighter transition metal ions such as Ti^{4+}, Cr^{3+}, Fe^{3+}, Co^{2+}, etc.

Similarly soft acids are those which satisfy the following features:

(*i*) *Cations of larger size or heavier elements*

(*ii*) *Cations of lower charge*

(*iii*) *Easily polarizable*

(*iv*) *Molecules or ions with larger number of valence electrons.*

Examples of soft acids are heavy transition metal ions (2nd and 3rd row) such as Hg^{2+}, Pd^{2+}, Cd^{2+} etc. and those in lower oxidation state such as Cu^+, Ag^+, Hg^+ etc.

According to Usanovich concept ligands come in the category of bases. Hence hard bases (ligands) are those which have the following features:

(*i*) *donor atoms of higher electronegativity;*

(*ii*) *donor atoms with low polarizability.*

Examples of hard bases are molecules or ions that have donor atoms N, O, F, Cl etc. like H_2O, NH_3, F^-, ROH.

Similarly soft bases usually have:

(*i*) *donor atoms of lower electronegativity*

(*ii*) *donor atoms of higher polarizability.*

Molecules or ions where the donor atom is P, As, S, Se etc. like R_3P, R_2S, I^- are examples of soft bases. It may be noted that hard acids have the tendency to combine with hard bases. Likewise soft acids combine with soft bases.

It may be noted that there is no sharp division between hard and soft acids or bases.

Q. 17. Discuss the basis of hard-hard and soft-soft interactions?

Ans. According to the most accepted view, the cause of *hard—hard interaction* is an electrostatic interaction. Bonding between hard acids and hard bases is predominantly ionic. The electrostatic energy between a positive and negative ion pair is inversely proportional to the internuclear distance. Therefore, smaller the ions, the lesser would be the internuclear distance and the greater would be the electrostatic attraction between the two ions. Consequently, the resulting compound would be highly stable.

Electrostatic interaction cannot explain the *soft—soft interaction* because the sizes of soft species are comparatively very large. Polarisation of the species, however, plays an important role in explaining their interactions. Most of the soft acids have 6 to 10 *d* electrons in their electronic configurations. These electrons get easily polarised favouring covalent bonding between them and the soft bases which are also easily polarisable. Thus, bonding between soft acids and soft bases is assumed to be *largely covalent.*

Considering hard —hard interactions as ionic and soft —soft interactions as covalent, Misons and coworkers (1967) proposed the following relation which can tell whether a given species is hard or soft:

$$pK = -\log K = aX + bY + c$$

where *K* is the equilibrium constant for the dissociation of acid-base; *X, Y* are parameters for the acids ; *a, b* are parameters of the bases and *c* is a constant required to adjust the *pK* values in such a way that all of them lie on the same scale.

The values of parameter *Y* for some of the cations are given below :

Hard acids	Li^+	Al^{3+}	Mg^{2+}	Na^+	Ca^{2+}	Fe^{3+}	Co^{2+}	Cs^+
	0.36	0.70	0.87	0.93	1.62	2.37	2.56	2.73

Soft acids $\begin{cases} Sn^{2+}, & Tl^{3+}, & Cu^{+}, & Pb^{2+}, & Tl^{+}, & Hg^{2+}, & Au^{+} \\ 3.17 & 3.2 & 3.45 & 3.58 & 3.78 & 4.25 & 5.95 \end{cases}$

If the value of *Y* is less than 2.8, the acid is hard and if the value is more than 3.2, the acid is soft.

The value of parameter *b* for some of the bases are given below :

OH^-	NH_3	Cl^-	Br^-	I^-	$S_2O_3^{2-}$
0.40	1.08	2.48	5.58	7.17	12.4
Hard bases			Soft bases		

The value of *b,* evidently, increases with the softness of the base. If the value is less than 3, the base is hard and if the value is more than 5, the base is soft.

In soft-soft interactions, the contribution of π bonding is also important.

Q. 18. How do we classify metal ions and ligands into acids and bases.

Ans. According to Lewis concept, an acid-base reaction involves combination of vacant orbital of an acid (A) and a filled unshared orbital on a base (B) as:

$$A + :B \longrightarrow A:B$$

It may also be written as

$$A + :B \longrightarrow A \leftarrow B$$

A strong acid and a strong base form a stable complex A : B. The co-ordination chemists studied the relative reactivities of a large number of ligands (species containing unshared pairs electrons, Lewis bases) with different ions (acting as Lewis acids), They found that certain ligand form more stable complexes with heavier ions such as Ag^+, Hg^{2+}, Pd^{2+}, Pt^{2+} with nearly full *d*-orbital electrons while others prefer to form complexes with ions such as Be^{2+}, Al^{3+}, Ti^{4+} having no *d*-electrons. Based on the preferential bonding they classified the acids and bases into the categories.

Class 1 metal ions include those of *alkali metals, alkaline earth metals and lighter transition metals in their higher oxidation states such as* Ti^{4+}, Cr^{3+}, Fe^{3+}, Co^{3+}, *etc.* as well as hydrogen ion, H^+.

Class 2 metal ions include those of heavier transition metals such as Hg^{2+}, Pd^{2+}, Pt^{2+}, Cd^{2+} and those in lower oxidation states (such as Cu^+, Ag^+, Hg^+, etc.).

Similarly ligands have also been classified as class 1 and class 2 depending upon their preference for metal ions respectively.

Class 1 ligands include those molecules or ions which have the donor atoms N, O, F, Cl, etc. These ligands combine preferably with type 1 metal ions.

Class 2 ligands include those molecules or ions which have the donor atoms P, As, S, Se, I, etc. These ligands combine preferably with type 2 metal ions. Greater the tendency of a ligand to combine with a metal ion, greater will be the stability of the complex. The following table illustrates the two types of ligands.

Those having tendency to complex with class 1 metal ions	Those having tendency to complex with class 2 metal ions
N >> P > As > Sb	N << P > As > Sb
O >> S > Se > Te	O << S < Se ~ Te
F > Cl > Br > I	F < Cl < Br < I

For example, ammonia (NH_3), amines (R_3N), water (H_2O) and fluoride ion (F^-) ligands prefer to coordinate with Ti^{4+}, Fe^{3+}, Co^{3+}, Mg^{2+} metal ions. On the other hand, phosphines (R_3P), thioethers (R_2S) and iodine ion (I^-) ligands prefer to coordinate with Pt^{2+}, Pd^{2+}, Hg^{2+}, Ag^+.

Such a classification has been proved to be very usefull in predicting and explaining stability of coordination compounds.

Pearson in 1963, introduced the terms **hard** and **soft** to describe the members of class 1 and class 2 respectively. For example, **class 1 type** metal ions are called **hard acids** while class 2 type metal ions are called **soft acids**. Similarly, **class 1 type** ligands are named **hard bases** while class 2 type ligands are named **soft bases**.

Q. 19. How do we classify metal ions and ligands into strong and weak acids and bases. Give examples.

Ans. (*a*) *The metal ions of alkali metals, alkaline earth metals, light transition metals with higher oxidation states (first row transition metal ions such as Ti^{4+}, Cr^{3+}, Fe^{3+}, etc.) and hydrogen, H^+ ion are classified as* **hard acids.**

(*b*) *The metal ions of heavier transition metals (second or third row transition metals) or the ions of lower oxidation states such as Cu^+, Hg^{2+}, Pd^{2+} are classified as* **soft acids.**

(*c*) *The ligands which prefer to form stable complexes with hard acids are classified as* **hard bases.** These ligands contain molecules having O, N or F donor atoms. For example, NH_3, H_2O, RNH_2, OH^-, F^-, etc.

(*d*) *The ligands which prefer to form stable complexes with soft acids are classified as* **soft bases.** These ligands contain molecules having P, S, As, Se, etc. as donor atoms. For example, R_3P, $S_2O_3^{2-}$, SCN^-, R_3As etc.

Thus phosphine and thioethers (soft bases) have a much greater tendency to co-ordinate with Hg^{2+}, Pd^{2+} or Pt^{2+} (soft acids) while ammonia, water or fluoride ions (hard bases) show a preference for Be^{2+}, Ti^{4+}, Co^{3+} etc.

Q. 20. Describe the characteristics of hard acids and soft acids.

Ans. Characteristics of Hard Acids:

(*i*) The hard acids are cations with small size.

(*ii*) In case of multiple oxidation states, the hard acids are cations of higher oxidation states (e.g. I^{5+}, I^{7+}, Cl^{3+}, Cl^{7+}).

(*iii*) The hard acids do not possess large number of valence electrons and, therefore, their outer electrons will not be easily distorted or removed. In other words, these ions canot be easily polarized.

To summarise, **hard acids** *are those which bind strongly to a proton. These are small in size, high in positive charge and will possess no electrons which are easily distorted or removed. In general, alkali metal ions, alkaline earth metal ions, lighter and more highly charged ions belong to this category.*

Characteristics of Soft Acids:

(*i*) The soft acids have cations of large size.

(*ii*) The cations have usually zero or low oxidation states.

(*iii*) The soft acids have large number of valence electrons and therefore, their outer electrons will be easily removed or distorted. In other words, these ions can be easily polarized.

To summarise **soft acids** *are those which bind strongly to highly polarizable or unsaturated bases. The soft acids have acceptor atom of large size, small or zero positive charge and have a number of valence electrons that are easily revoved. In general, heavier transition metal ions and low valence metal ions belong to this category.*

Having said that, there is no sharp line of demarcation between soft and hard species and a number of border line cases also exist which are termed as neither too hard nor too soft. In such cases the properties like size, oxidation states or polarizability are neither too low nor too high. These properties have intermediate values.

Q. 21. State the characteristics of hard Lewis bases and soft Lewis bases.

(Rajasthan 2010)

Ans. Characteristics of Hard Lewis Bases:

1. Hard bases contain donor atoms of high electronegativity.
2. The donor atom of hard bases is of low polarizability.
3. The hard bases are not easily oxidized.
4. Hard bases have full low energy orbitals.
5. All hard bases strongly link to the proton.

Characteristics of Soft Lewis Bases

1. Soft bases have donor atoms of low electronegativity.
2. These have donor atoms which are easily polarized.
3. The soft bases are easily oxidised.
4. They contain empty low energy orbitals.

Q. 22. State Pearson's HSAB principle and discuss the applications of this principle.

(Mumbai, 2002; Kurukshetra, 2002; Rohelkhand, 2003; Purvanchal, 2003; Calcutta, 2005; Lucknow, 2010; Garhwal 2010)

Ans. Pearson in 1963 proposed a simple rule for predicting the stability of compounds formed from acids and bases. This is known as HSAB principle. It states *hard acids prefer to bond to hard bases and soft acids prefer to bond to soft bases.*

Applications of HSAB Principle

This principle has a number of applications as discussed below.

1. Stability of complexes

We can explain the relative stability of complexes on the basis of HSAB principle. Consider the following general Lewis acid-base reaction:

$$\underset{\text{Lewis acid}}{A} + \underset{\text{Lewis base}}{:B} \longrightarrow \underset{\text{Complex}}{A:B}$$

The complex AB would be most stable if A and B are either both hard or both soft. On the other hand, the complex would be least stable if A is hard and B is soft or *vice versa*. We fund that AgI_2^- exists as a stable compound while AgF_2^- does not. The HSAB principle explains this on the basis of different nature of I^- and F^-.

Ag^+ ion is a *soft acid*. Its combines with I^- ion, a *soft base* to forms a stable complex AgI_2^- whereas its combination with F^- ion, a *hard base*, forms an unstable complex, AgF_2^-.

$$\underset{\text{Soft acid}}{Ag^+} + \underset{\text{Soft base}}{2I^-} \longrightarrow \underset{\text{Soft-soft}}{AgI_2^-} \quad \text{(stable complex)}$$

$$\underset{\text{Soft acid}}{Ag^+} + \underset{\text{Hard base}}{2F^-} \longrightarrow \underset{\text{Soft-hard}}{AgF_2^-} \quad \text{(unstable complex)}$$

The reaction between LiI and CsF to give LiF and CsI always proceeds to give LiF and CsI. It is an interesting, example of preferential combination of soft-soft and hard-hard species:

$$\underset{\text{Hard-soft}}{LiI} + \underset{\text{Soft-hard}}{CsF} \longrightarrow \underset{\text{Hard-hard}}{LiF} + \underset{\text{Soft-soft}}{CsI}$$

Similarly, the following reactions go to the right because of the combination of soft-soft and hard-hard species.

$$\underset{\text{Hard-soft}}{LiI} + \underset{\text{Soft-hard}}{CsF} \longrightarrow \underset{\text{Hard-hard}}{LiF} + \underset{\text{Soft-soft}}{CsI}$$

$$\underset{\text{Hard-soft}}{CaS} + \underset{\text{Hard}}{H_2O} \longrightarrow \underset{\text{Hard-hard}}{CaO} + \underset{\text{Soft}}{H_2S}$$

$$\underset{\text{Soft-Hard}}{HgF_2} + \underset{\text{Hard-soft}}{BeI_2} \longrightarrow \underset{\text{Hard-hard}}{BeF_2} + \underset{\text{Soft-soft}}{HgI_2}$$

2. Coordination in complexes of ambidentate ligands

The HSAB principle can predict the formation of various metal ion complexes with ambidentate ligands. An **ambidentate ligand** is a monodentate ligand which can coordinate to metal ion through more than one atoms. SCN^- ion is an ambident ligand as it can coordinates through N or S. It is known to form stable complexes of the type $[M(NCS)_4]^{2-}$ with order line acid cations of Co, Ni, Cu and Zn by coordinating through N. It forms stable complexes of the type $[M(SCN)_4]^{2-}$ with acid cations of Rh, Ir, Pd, Pt and Au by coordinating through S. Their stability can be easily explained according to HSAB priciple. A **soft acid** like Rh, Ir, Pd or Pt prefers to coordinate through softer S atom (base) to form M—SCN bond. On the other hand, a *hard acid* like Co, Ni, Cu or Zn prefers to coordinate through the *harder* N atom (base) to form the M—NCS bond. However, there are many examples of both —SCN and —NCS bonding to the same metal.

3. Classification of acids and bases as soft or hard

The acids or bases can be classfied as hard or soft depending upon their preference for hard or soft reactants. For example, we want to know whether B is a hard or soft base. We shall carry out the following reaction.

$$BH^+ + \underset{\text{Soft acid}}{CH_3Hg^+} \rightleftharpoons CH_3HgB^+ + \underset{\text{Hard acid}}{H^+}$$

If B is a soft base, then the right hand side reaction will be favoured and if it is a hard base, then left hand reaction will be favoured. For example, R_2S is a soft base, so it will combine with soft acid (CH_3Hg^+) and the forward reaction will be favoured.

$$\underset{\text{Soft base}}{R_2SH^+} + \underset{\text{Soft acid}}{CH_3Hg^+} \longrightarrow \underset{\text{Soft acid-soft base}}{CH_3Hg^+\,SR_2} + \underset{\text{Hard acid}}{H^+}$$

On the other hand, NH_3 is a hard base, it will combine with H^+ (hard acid) favouring backward reaction:

$$NH_4^+ + CH_3Hg^+ + \qquad \underset{\text{Soft acid-hard base}}{CH_3Hg^+\,NH_3} + \underset{\text{Hard acid}}{H^+}$$

or

$$CH_3Hg^+NH_3 + H^+ \longrightarrow NH_4^+ + CH_3Hg^+$$

4. Poisoning of metal catalysts

Some metals like Ni, Pt, Pd, Cr Mo, etc (soft acids) act as catalysts. These metal catalysts can be easily poisoned by substances like carbon monoxide, unsaturated hydrocarbons (olefins, arenes, dienes, alkynes), phosphorus and arsenic containing ligands (soft bases). This poisoning occurs due to the soft acid-soft base interactions between soft metal ions **and soft ligands.** These ligands are strongly adsorbed on the surface of the metal and thus block the active sites. However, such soft acid catalysts are not affected by hard bases or ligands containing N, O or F.

5. Occurrence of ores and minerals

We find that certain ores occur as sulphides while others occur as oxides, carbonates or halides. The occurrence of the ores of certain metals can be explained on the basis of HSAB principle. Hard acid metal ions such as Mg^{2+}, Ca^{2+}, Mn^{2+}, Al^{3+}, Fe^{3+}, Cr^{3+}, etc., occur mostly as *their oxides, carbonates or halides* (F^- or Cl^-). This is due to their preferred *hard acid-hard base* interactions because oxides (O^{2-}), carbonates (CO_3^{2-}) and halides (F^- or Cl^-) are *hard bases.* Similarly, soft acid metal ions such as Cu^+, Ag^+, Pb^{2+}, Hg^{2+}, Pd^{2+}, etc. occur as their sulphides because of stable combination of soft acid with soft base (S^{2-} ion). Border line acids like Ni^{2+}, Fe^{2+}, Zn^{2+}, Pb^{2+}, Co^{2+}, etc. have the possibility of occurring both as *sulphides* (soft base) as well as *oxides* or carbonates (hard bases).

Q. 23. Discuss the theories proposed by different scientists to explain HSAB principle.

(Ranchi 2010)

Ans. A number of theories have been proposed by different scientists to explain the HSAB principle. A brief description of these is given below.

1. Electrostatic interactions

A large amount of energy is liberated in the formation of an ionic bond. Hard acids and hard bases form purely ionic compounds. These *hard-hard interactions are purely ionic or electrostatic.* For example, hard acids such as Li^+, Na^+, K^+ and hard bases such as OH^-, F^-, O^{2-} form ionic compounds. The electrostatic energy between a positive ion and a negative ion is inversely proportional to the interatomic distance. Smaller the ions, smaller would be the internuclear distance and greater would be the electrostatic attraction between the ions resulting in the release of energy. Consequently, the resulted complex formed by hard acid and hard base would be highly stable.

2. Polarizing power and polarizability

In the case of soft-soft interactions, we assuming that these are covalent in nature. For example, soft acids such as Ag^+, Hg^+ usually form bond with soft base such as Cl^-. The compound AgCl is much more covalent than the corresponding compounds of alkali metals. We can also explain this in terms of the polarizing power and polarizability of *d*-electrons. Most of the soft acids are transition metals with greater number of *d*-electrons. These *d*-electrons have more polarizing power and the soft bases such as I^-, S^{2-} are easily polarizable. Therefore, bonding between soft acids and soft bases is largely covalent.

3. Electronegativity and hardness and softness

Hardness and softness of acids and bases are related to electronegativity. In general, a species with high electronegativity is hard and the one with low electronegativity is soft. For example, highly electronegative ions Li^+, Na^+, etc. are hard acids while transition metal ions, Cu^+, Ag^+, Au^+ etc., having low electronegativities are soft acids. Similarly, hard and soft bases are defined on this basis. The fact that the trifluoromethyl group (CF_3) is considerably harder than the methyl group (CH_3) and boron trifluoride (BF_3) is harder than borane (BH_3) can be explained on the concept of electronegativity. The concept of electronegativity in explaining hardness/softness of acids and bases is dealt with in detail in the following section.

4. π-bonding contibutions

In soft acid-soft base interactions, π-bonding plays an important role. The soft acids are generally metals in low oxidation states containing loosely held outer *d*-electrons. These can be easily donated to ligands. Many soft bases are π-bond aceptors and many soft-acids are π-bond donors. Soft bases include ligands of phosphorus, arsenic and unsaturated ligands such as CO. The polarizability of soft acids and soft bases further favours π-bonding (back bonding).

Q. 24. Is there any relevance of the strength of an acid or base with the hardness or softness of the species. Discuss.

Ans. **Ans.** Strength of an acid or base has nothing to do with hardness or softness of the species.

The hardness or softness simply emphasizes special stability of hard-hard and soft-soft interactions. For example, both OH^- and F^- are hard bases, yet the basicity of the hydroxide ion (OH^-) is very very large (about 10^{13} times) as compared to fluoride ion. Similarly, SO_3^{2-} and Et_3P are soft bases but the latter is much stronger base (about 10^7 times) than the former towards soft acid CH_3Hg^+. It is a limitation of HSAB principle. We cannot estimate inherent strength of an acid or a base on HSAB principle. For example, the following reaction is not expected according to HSAB principle because it does not lead to hard-hard and soft-soft interactions. The stronger softer base, SO_3^{2-} ion can displace the weak hard base, F^- ion from the hard acid, the proton, H^+ :

$$\underset{\text{Soft bases}}{SO_3^{2-}} + \underset{\text{Hard-hard}}{HF} \rightleftharpoons \underset{\text{Hard-soft}}{HSO_3^-} + \underset{\text{Hard base}}{F^-} \qquad K_{eq} = 10^4$$

However, this reaction occurs (favourable $K = 10^4$). Similarly, the *stronger hard* base, OH^- ion can displace weaker *soft base,* SO_3^{2-} from the soft acid, methyl mercury cation as

$$\underset{\text{Hard}}{OH^-} + \underset{\text{Soft-soft}}{CH_3HgSO_3^-} \rightleftharpoons \underset{\text{Soft-hard}}{CH_3HgOH} + \underset{\text{Soft base}}{SO_3^{2-}} \qquad K_{eq} = 10$$

In these cases, the strength factor plays its role. The strengths of the bases ($SO_3^{2-} > F^-$ and $OH^- > SO_3^{2-}$) are sufficient to force these reactions to the forward direction inspite of hard-soft considerations. This happens especially when the inherent strengths of bases (or acids) being compared are significantly different.

Thus, the importance of inherent strengths of acids and bases should not be ignored while considering softness-hardness factor. For example, the following reaction does not occur in the forward direction, although it results into soft-soft combination.

$$\underset{\text{Soft-hard}}{CH_3CdOH} + \underset{\text{Soft}}{SO_3^{3-}} \rightleftharpoons \underset{\text{Soft-soft}}{CH_3CdSO_3^-} + \underset{\text{Hard-base}}{OH^-}$$

This is because OH^- is much stronger base than SO_3^{2-} and therefore cannot be displaced weaker base SO_3^{2-} even though the reaction is favoured by soft-soft interactions.

Thus, *in general, the Lewis acid-Lewis base reactions are governed by two independent principles* **softness and hardness** and **strength of the acid or base.**

Q. 25. What are the limitations of HSAB principle.

Ans. HSAB principle has some limitations which are discussed as under.

1. This principle does not provide any quantitative scale of measurement.

2. According to HSAB principle, the reaction,

$$CH_3^+(g) + H_2(g) \longrightarrow CH_4(g) + H^+(g)$$

should occur because of soft-soft ($CH_3^+ — H^-$) interactions. However, this reaction does not take place. It is probably because of greater acidity of the proton (H^+) relative to CH_3^+(g) cation.

3. The hard-soft factors are independent of acidic or basic character of compounds. The factors work independently of each other. However, many examples are known which show interdependence of two concepts. For example, in some cases, hard-hard or soft-soft complexes change to more stable hard-soft system. According to HSAB principle, the following reaction is not favoured.

$$\underset{\text{(Soft base)}}{SO_3^{2-}} + \underset{\text{Hard-acid}}{HF} \longrightarrow \underset{\text{Hard-soft}}{HSO_3^-} + \underset{\text{Hard}}{F^-}$$

However, this reaction occurs in violation of HSAB principle. This is due to the fact that stronger soft base SO_3^{2-} displaces the weak hard base, F^- from the hard acid, H^+.

Q. 26. Define the term symbiosis. Give some examples to explain this phenomenon.

(Ranchi 2010)

Ans. The term symbiosis was introduced by Jorgenson in 1868.

Soft ligands have a tendency to combine with a metal ion already having soft ligands and hard ligands have a tendency to combine with a metal ion already having hard ligands. This tendency has been termed as **symbiosis.** For example, F^- ion is a hard ligand and it readily combines with BF_3 because it already has threee hard ligands.

$$BF_3 + F^- \longrightarrow BF_4^-$$

Similarly, H^- ion, a soft ligand, readily combines with BH_3 to form a stable complex BH_4^-.

$$BH_3 + H^- \longrightarrow BH_4^-$$

Due to symbiosis, the symmetrical substituted compounds are preferred than unsymmetrical substituted compounds. Therefore the compounds such as BF_3H^- and BH_3F^-, having mixed substituents (soft acid-hard base or hard acid-soft base) interact spontaneously to yield BH_4^- and BF_4^-, i.e., compounds with same substituents. Thus, the reaction always proceeds to the right.

$$\underset{\text{Hard-soft}}{BF_3H^-} + \underset{\text{Soft-hard}}{BH_3F^-} \longrightarrow \underset{\text{Hard-hard}}{BF_4^-} + \underset{\text{Soft-soft}}{BH_4^-}$$

Fluorinated methane reacts in a similar manner:

$$CH_3F + CHF_3 \longrightarrow CH_4 + CF_4$$

Q. 27. Correlate electronegativity and hardness – softness of acids and bases.

(Rajasthan 2010)

Ans. This concept provides the most authentic explanation of the phemonenon of hardness and softness of acids and base. We do not take into consideration, the electronegativity of the element but we consider the electronegativity of the ion participating in a reaction. For example, Li has a low electronegativity, but Li^+ has relatively high tendency to attract the electrons towards itself, and therefore, it has high electronegativity. This is because of small ionic size and extremely high second ionization potential. Consequently, Li^+ is a **hard acid.** On the other hand, transition metal ions in low oxidation states such as Cu^+, Hg^+, Ag^+, Cd^+ *etc.* possess large ionic size and relatively low second ionization energies and therefore possess lower values of electronegativity. Therefore, they are considered as **soft acids.** This way we can consider hard and soft bases.

This relation between hardness-softness and electronegativity helps to explain the greater hardness of trifluoro methyl group (F_3C) as compared to methyl group (H_3C) and greater hardness of BF_3 as compared to BH_3. Similarly, we can explain why F_3P is harder than H_3P which is harder than Me_3P. Ammonia (H_3N) is harder than trimethyl amine (Me_3N).

According to Mulliken-Jaffe definition, electronegativity (χ) is linearly related to the partial charge (δ) on an atom as:

$$\chi = a + b\delta$$

where parameter '*a*' is the neutral atom electronegativity of an atom. The parameter '*b*' measures *the rate of change of electronegativity with charge* and is known as **charge coefficient.**

Large soft and polarizable atoms have low values of b while small, hard and non polarizable atoms possess larger values of b.

In 1983, Parr and Pearson proposed a simple approach to quantitative measure of acid-base reactions. They have introduced the term **absolute hardness** (η) for both neutral and charged species and correlated it with Mulliken-jaffe's definition of electronegativity which they called **absolute electronegativity** (χ).

The **absolute hardness** (η) is defined as one half the difference between the ionization energy and the electron affinity (both in eV).

$$\eta = \frac{I - A}{2} \quad (1)$$

where I = ionization energy and

A = electron affinity

According to Mulliken, ionization energy, electron affinity and electronegative are related as

$$\chi = \frac{I + A}{2} \quad (2)$$

According to this concept a hard acid or base is a species which has a large difference between its ionization energy and its electron affinity.

The hardness of base B and acid A may be given as

$$\eta(B) = \eta(\text{species } B^+) = \frac{1}{2}[I(B^+) - A(B^+)] \quad (3)$$

$$\eta(A) = \eta(\text{species } A) = \frac{1}{2}[I(A) - A(A)] \quad (4)$$

In this equation, the base B is represented as B^+ in eqn. (3) and acid A as such in eqn. (4). The positive charge on the base and neutral charge on the acid can be understood as follows:

Consider the interaction of acid A and base B to form AB (electrons are donated by : B)

$$A + :B \longrightarrow \overset{\delta-}{A} : \overset{\delta+}{B}$$

In the formation of AB, electrons flow from base B to acid A and therefore, B acquires partial positive charge (δ^+) and A acquires partial negative charge (δ^-). Hardness of B refers to species going (on ionization) from δ^+ to a charge $1 + \delta^+$ while hardness of A refers to a species on ionization from δ^- to a charge of $1 + \delta^-$. If the shift of charge δ is taken as equal to 0.5, then charge on B becomes 1 + 0.5 = 1.50 and that on A becomes 1 + (–0.5) = 0.5.

Therefore, charge on B changes from 0.5 to 1.5 with an average of $\frac{1}{2}$ (0.5 + 1.5) = 1.0 while that on A changes from – 0.5 to + 0.5 with an average of $\frac{1}{2}$(0.5 – 0.5) = 0. Therefore, B is shown as B^+ and A as neutral in equations (3) and (4).

Paor and Pearson argued strongly for the use of absolute hardness parameters in treating hard-soft acid-base (HSAB) interactions.

Q.28. How do we arrive at HSAB principle based on molecular orbital theory?

Ans. There have been attempts to understand HSAB principle with the help of MO theory. According to Koopman's theorem, the ionization energy gives the energy of the highest occupied molecular orbital (HOMO) while the electron affinity gives the energy of the lowest unoccupied molecular orbital (LUMO) for a given molecule:

$$E(HOMO) = -I$$
$$E(LUMO) = -EA$$

These orbitals are involved in electronegativity and HSAB relationships. The hardness and softness of a species depends upon the average gap between the HOMO and LUMO. The separation of HOMO and LUMO is taken equal to twice the value of η. Hard species have a large HOMO-LUMO gap while soft species have small energy gap. Drago and Wayland have also proposed a quantitative approach of acid-base parameters by taking into account electrostatic and covalent factors. The presence of low lying unoccupied MOs capable of mixing with the ground state orbitals accounts for the polarizability of soft species.

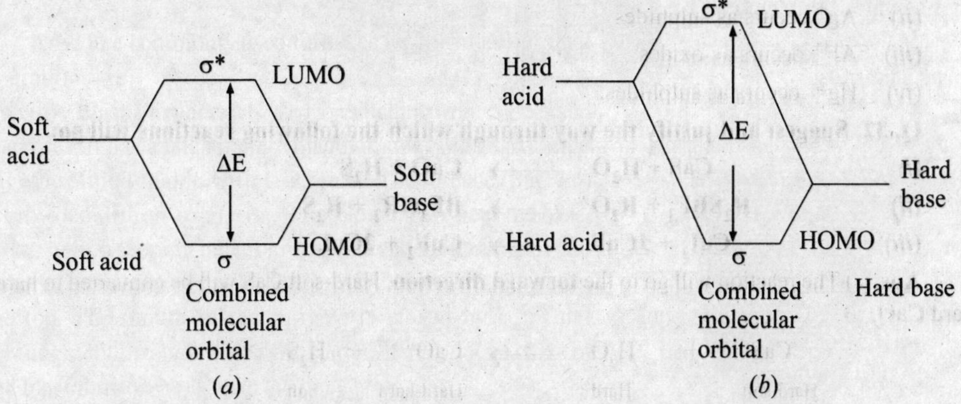

Comparison of orbital energies for (*a*) Soft-soft and (*b*) Hard-hard combinations

The combination of a soft acid with soft base is analogous to a situation in which the LUMO of the Lewis acid has almost the same energy as the HOMO of the Lewis base. The combination of orbitals of soft acid and soft base gives molecular orbital lower in energy than that of orbital of soft base. This gives a strong covalent bond

In the case of hard acid-hard base combination, there is a wide difference in the energies of bonding and antibonding molecular orbitals or LUMO and HOMO will differ greatly in energy level. This situation is not favourable for covalent bond formation. An ionic combination will result instead.

Q. 29. (*a*) Classify the following as soft or hard acids:

Cu^+, Na^+, Ti^{4+}, Ag^+, Pt^{2+}

(*b*) Classify the following as soft or hard bases:

NH_3, ROH, I^-, CO, O_2^-, C_6H_6

Ans. (*a*) Acids

Cu^+ : Soft acid Na^+ : Hard acid
Ti^{4+} : Hard acid Ag^+ : Soft acid
Pt^{2+} : Soft acid

(*b*) Bases

NH_3 : Hard base ; ROH : Hard base
I^- : Soft base ; CO : Soft base
O_2^- : Hard base ; C_6H_6 : Soft base

Q. 30. Which of the metals Ag^+, Al^{3+}, Cu^+, Ca^{2+}, Cr^{3+} and Zn^{2+} might be expected to be found in aluminosilicate minerals and which in sulphides?

Ans. Hard acids like Ca^{2+}, Al^{3+} and Cr^{3+} form aluminosilica minerals while soft acids like Cu^+, Ag^+ and Zn^{2+} form sulphides.

Q. 31. O^{2-} is a hard base and S^{2-} is a soft base. Which of the following metal ions will occur preferably as oxides or sulphides:

***(i)* Ca^{2+} *(ii)* Ag^+ *(iii)* Al^{3+} *(iv)* Hg^{2+}**

Ans. Ca^{2+} and Al^{3+} are hard acids while Ag^+ and Hg^{2+} are soft acids therefore

(*i*) Ca^{2+} occurs as oxides

(*ii*) Ag^+ occurs as sulphides

(*iii*) Al^{3+} occurs as oxides

(*iv*) Hg^{2+} occurs as sulphides.

Q. 32. Suggest and justify the way through which the following reactions will go:

(*i*) $CaS + H_2O \longrightarrow CaO + H_2S$

(*ii*) $R_2SBF_3 + R_2O \longrightarrow BF_3OR_2 + R_2S$

(*iii*) $CuI_2 + 2CuF \longrightarrow CuF_2 + 2CuI.$

Ans. (*i*) The reaction will go to the **forward direction.** Hard-soft CaS will be converted to hard-hard CaO.

CaS	+	H_2O	$\longrightarrow$	CaO	+	H_2S
Hard-soft		Hard		Hard-hard		Soft

(*ii*) The reaction will go to the **forward direction.** The hard-soft complex $BF_3.SR_2$ will be converted to a stable hard-hard complex $F_3B.OR_2$.

$BF_3.SR_2$	+	R_2O	$\longrightarrow$	$BF_3.OR_2$	+	R_2S
Hard-soft		Hard		Hard-hard		Soft

(*iii*) The reaction will go to **forward direction.** Hard-soft CuI_2 will be converted to soft-soft CuI.

CuI_2	+	$2CuF$	$\longrightarrow$	$2CuI$	+	CuF_2
Hard-soft		Soft-hard		Soft-soft		Hard-hard

Q. 33. Indicate the feasibility of the following reactions in the forward direction:

(*a*)	$C_2H_5HgSO_3^-$	+	Me_3P	$\longrightarrow$
	Soft-soft		**Soft base**	
(*b*)	CH_3CdOH	+	SO_3^{2-}	$\longrightarrow$
	Soft-hard		**Soft base**	

The base strength follows the order:

$$Me_3P >> SO_3^{2-} \text{ and } OH^- >> SO_3^{2-}.$$

Ans. (*a*) This reaction would be feasible because of much stronger Me_3P base compared to SO_3^{2-}. The inherent strength of $Me_3P >> SO_3^{2-}$ will force the reaction in forward direction inspite of formation of hard-soft pair in the product.

(*b*) It is not feasible because OH^- is much stronger base than SO_3^{2-} and hence cannot be displaced by weaker base SO_3^{2-} even though it is favoured by HSAB principle.

Q. 34. (*a*) While hard-hard interactions are generally ionic, soft-soft interactions are generally covalent? Why is it so?

(*b*) Classify the following as soft or hard bases:

(*i*) N_2H_4 (*ii*) CN^- (*iii*) CO (*iv*) F^- (*v*) O^{2-} (*vi*) C_6H_6

(*c*) Give two examples each of boarder line between hard and soft acids and bases.

Ans. (*a*) Hard acids combine with hard bases. The compounds formed by their combinations are ionic compounds. For example, hard acids ike Li^+, Na^+, H^+, combine with hard bases like OH^- , F^- or O^{2-} forming ionic compounds. This is because of big energy gap between HOMO and HUMO orbitals. The interactions between these are purely electrostatic or ionic in nature. Soft acid-soft base interactions are covalent in nature. These interactions involve bigger size of the ions.

(*b*) (*i*) N_2H_4 : Hard base (*ii*) CN^- : Soft base (*iii*) CO: Soft base

(*iv*) F^- : Hard base (*v*) O^{2-}.: Hard base (*vi*) C_6H_6 : Soft base

(*c*) Zn^{2+}, Bi^{3+} : Border line acids

NO^{2-}, Br^- : Border line bases

Q. 35. Classify the following as hard, soft or boarder line acids:

(*i*) Ag^+ (*ii*) Pt^{2+} (*iii*) Te^{4+} (*iv*) Bi^{3+} (*v*) I^{7+}

Ans.

(*i*) Ag^+ : Soft acid (*ii*) Pt^{2+} : Soft acid

(*iii*) Te^{4+} : Hard acid (*iv*) Bi^{3+}: Boarder line acid

(*v*) I^{7+} : Hard acid

Q. 36. (*a*) Predict on the basis of HSAB pribciple, whether the following reactions are feasible in the forward direction or not?

(*i*) $BF_3H^- + BH_3F^- \longrightarrow BF_4^- + BH_4^-$

(*ii*) $HgF_2 + BeI_2 \longrightarrow BeF_2 + HgI_2$

(*iii*) $CH_3Hg^+ + NH_4^+ \longrightarrow CH_3Hg^+NH_3 + H^+$

(*iv*) $CH_3HgF + HSO_3^- \longrightarrow CH_3HgSO_3^- + HF$

(*v*) $CuI_2 + 2CuF \longrightarrow CuF_2 + 2CuI$

(*vi*) $LiF + CsI \longrightarrow LiI + CsF$

(*b*) O^{2-} is a hard base and S^{2-} is a soft base. Which of the following metal ions will occur preferably as oxides or sulphides: (*i*) Ca^{2+} (*ii*) Ag^+ (*iii*) Al^{3+} (*iv*) Hg^{2+}.

Solution.

(*a*) (*i*) Feasible (*ii*) Feasible (*iii*) Not feasible

(*iv*) Feasible (*v*) Feasible (*vi*) Not feasible.

(*b*) (*i*) Ca^{2+} occurs as oxides (*ii*) Ag^+ occurs as sulphides

(*iii*) Al^{3+} occurs as oxides (*iv*) Hg^{2+} occurs as sulphides

Q. 37. Suggest and justify the way through which the following reactions will go

(*i*) $CaS + H_2S \longrightarrow CaO + H_2S$

(*ii*) $R_2SBF_3 + R_2O \longrightarrow BF_3OR_2 + R_2S$

(*iii*) $CuI_2 + 2CuF \longrightarrow CuF_2 + 2CuI$

Ans. (*i*) The reaction will go to the forward direction because hard-soft CaS will be converted to hard-hard product

$$\underset{\text{Hard-soft}}{CaS} + \underset{\text{Hard}}{H_2S} \longrightarrow \underset{\text{Hard-hard}}{CaO} + \underset{\text{Soft}}{H_2S}$$

(*ii*) The reaction will go to the forward direction because the hard-soft complex BF_3SR_2 will be converted to a stable hard-hard product.

$$\underset{\text{Hard-soft}}{BF_3.SR_2} + \underset{\text{Hard}}{R_2O} \longrightarrow \underset{\text{Hard-hard}}{BF_3OR_2} + \underset{\text{Soft}}{R_2S}$$

(*iii*) The reaction will go to forward direction because hard-soft CuI_2 will be converted to soft-soft CuI.

$$\underset{\text{Hard-soft}}{CuI_2} + \underset{\text{Soft-hard}}{2CuF} \longrightarrow \underset{\text{Soft-soft}}{2CuI} + \underset{\text{Hard-hard}}{CuF_2}$$

Q. 38. Classify the following into hard, soft and boarder line acids and bases:
Na^+, Pt^{2+}, NH_3, SCN^-, Br^-, Co^{++}

Ans. Na^+ : Hard acid — SCN^- : Soft base

Pt^{2+} : Soft acid — Br^- : Border line base

NH_3 : Hard base — Co^{++} : Border line acid

Q. 39. (*a*) Explain the action of R_2S on C_2H_5HgCl.

(*b*) AgI_2^- complex is stable but AgF_2^- is not. Explain.

Ans. (*a*) R_2S is a soft base. It will interacts with soft acid C_2H_5Hg to form a complex:

$$C_2H_5HgCl + R_2S \longrightarrow [C_2H_5Hg\ (R_2S)]^+\ Cl^-$$

(*b*) Ag^+ is a soft acid. It may combine with soft base I^- ion to give AgI_2^- (soft-soft) whereas it cannot combine with hard base F^- ion. Therefore, AgF_2^- (soft-hard) is not formed as shown in the following equations.

$$\underset{\text{Soft}}{Ag^+} + \underset{\text{Soft base}}{2I^-} \longrightarrow \underset{\text{Stable (soft-soft)}}{AgI_2^-}$$

$$\underset{\text{Soft acid}}{Ag^+} + \underset{\text{Hard base}}{2F^-} \longrightarrow \underset{\text{Unstable (soft-hard)}}{AgF_2^-}$$

Q. 40. (*a*) Explain the action of $CdCO_3$ on Na_2S solution.

(*b*) How does HSAB principle explain the validity of the following reactions:

(*i*) $BeF_2 + HgI_2 \longrightarrow HgF_2 + BeI_2$

(*ii*) $BF_3H^+ + BH_3F^- \longrightarrow BF_4^- + BH_4^+$

Ans. (*a*) $CdCO_3$ reacts with Na_2S solution to give the precipitated of CdS. This happens because S^{2-} is a soft base and prefers to combine with soft acid, Cd^{2+}.

$$CdCO_3 + Na_2S \longrightarrow Na_2CO_3 + CdS$$

(*b*) (*i*)
$$\underset{\text{Hard-hard}}{BeF_2} + \underset{\text{Soft-soft}}{HgI_2} \longrightarrow \underset{\text{Soft-hard}}{HgF_2} + \underset{\text{Hard-soft}}{BeI_2}$$

This reaction will not proceed because hard-hard and soft-soft combinations will not react to produce soft-hard combinations.

(*ii*)
$$\underset{\text{Hard-soft}}{BF_3^+H^-} + \underset{\text{Soft-hard}}{BH_3F^-} \longrightarrow \underset{\text{Hard-hard}}{BF_4^-} + \underset{\text{Soft-soft}}{BF_4^+}$$

This reaction is valid because the compounds such as BF_3H^+ and BH_3F^- having mixed substituents interact to give compounds with same substituents.

Q. 41. Calculate absolute electronegativity and absolute hardness for OH^- and Cl^- given that

OH^-	**I = 13.17**	**A = 1.83**
Cl^-	**I = 13.01**	**A = 3.62 (values are in eV)**

Solution. (*i*) OH^-

I = 13.17, A = 1.83

$$\text{Absolute electronegativity, } \chi = \frac{13.17 + 1.83}{2} = 7.50 \text{ eV}$$

$$\text{Absolute hardness} \quad \eta = \frac{13.17 - 1.83}{2} = 5.67 \text{ eV}$$

(*ii*) Cl^-

$$\text{Absolute electronegativity, } \chi = \frac{13.01 + 3.62}{2} = 8.31 \text{ eV}$$

$$\text{Absolute hardness, } \eta = \frac{13.01 - 3.62}{2} = 4.70 \text{ eV}$$

Q. 42. Which of the following does not proceed to the right?

(*i*) $HgF_2 + BeI_2 \longrightarrow BeF_2 + HgI_2$

Soft-hard, Hard-soft, Hard-hard, Soft-soft

(*ii*) $LiF + CsI \longrightarrow LiI + CsF$

Hard-hard, Soft-soft, Hard-soft, Soft-hard

Ans. The reaction (*ii*) will not proceed to the right because of unfavoured hard-soft combination.

Q. 43. Predict the feasibility of the following reactions:

(*i*) $CH_3HgOH + HSO_3^- \longrightarrow$

(*ii*) $CH_3HgOH + HF \longrightarrow$

(*iii*) $CH_3HgF + HSO_3^- \longrightarrow$

Ans. (*i*) and (*iii*) are favoured because of hard-hard and soft-soft interactions, (*ii*) not feasible because OH^- is much stronger base than F^-.

Q. 44. Classify the following bases as hard, soft or border line:

(*i*) N_2H_4 (*ii*) C_2H_4 (*iii*) NO_3^- (*iv*) C_5H_5N

(*v*) F^- (*vi*) Br^- (*vii*) I^- (*viii*) CO

(*ix*) CN^- (*x*) O^{2-}

Ans. Hard bases: N_2H_4, NO_3^-, F^-, O^{2-}

Soft bases: C_2H_4, I^-, CO, CN^-

Border line bases: C_5H_5N, Br^-

Q. 45. Which of the following metals will prefer to occur as oxides or sulphides or both types. Give reasons for your answer:

(*i*) Cr^{3+} (*ii*) Ni^{2+} (*iii*) Sn^{2+} (*iv*) Fe^{2+}

(*v*) Cu^+ (*vi*) Al^{3+} (*vii*) Ca^{2+} (*viii*) Hg^{2+}

Ans. Hard acids (metal ions) prefer to occur as oxides (O^{2-}, hard base) while soft acids (metal ions) prefer to occur as sulphides (S^{2-}, soft base). Border line acids form both oxides and sulphides.

Accordingly, Cr^{3+}, Al^{3+} and Ca^{2+} form oxides

Cu^+ and Hg^{2+} form sulphides

Sn^{2+}, Ni^{2+} and Fe^{2+} form oxides as well as sulphides

Q. 46. Which of the following reactions will proceed to right hand side?

(*i*) $CuCl_2 + 2KI \longrightarrow 2KCl + CuI + \frac{1}{2} I_2$

(*ii*) $BeI_2 + HgF_2 \longrightarrow BeF_2 + HgI_2$

(*iii*) $CdCl_2 + H_2S \longrightarrow 2HCl + CdS$

Ans. On the basis of hard-hard and soft-soft preference we can see that all the above reactions proceed towards R.H. S.

Q. 47. Discuss relative strength of acids and bases? What is meant by K_a and pK_a?

Ans. Strength of acids. The strengths of acids like HCl, HNO_3, H_2SO_4, HCN, HF can be compared by their tendencies to give protons. Acids like HCl, HNO_3, or H_2SO_4 are strong acids while CH_3COOH, HF, HCN are weak acids. Ionisation of an acid in water measures the relative strength of the acid.

For instance, ionisation of an acid HA may be represented as

$$HA \rightleftharpoons H^+ + A^-$$

The equilibrium constant for the reaction is written as

$$K_a = \frac{[H^+][A^-]}{[HA]}$$

where K_a is known as **dissociation constant** or **ionisation constant** of an acid and measures the relative strength of an acid. The stronger the acid, the larger will be its K_a.

The term pK_a, that is $-\log K_a$, is commonly used to express the dissociation of an acid.

$$pK_a = -\log K_a = \frac{1}{\log K_a}$$

Thus the higher the value of pK_a, the weaker the acid.

Strength of bases. Consider the ionisation of base MOH:

$$MOH \rightleftharpoons M^+ + OH^-$$

The equilibrium constant may be expressed as:

$$K_b = \frac{[M^+][OH^-]}{[MOH]}$$

where K_b is known as the **ionisation** or **dissociation constant** of the base. The higher the value of K_b, the stronger is the base.

K_b may be expressed as

$$pK_b = -\log K_b \frac{1}{\log K_b}$$

Larger the value of pK_b, lesser is the dissociation and weaker is the base.

Q. 48. Discuss the trends in acid strength of hydro acids.

Ans. Substances like HCl, H_2S, PH_3, and H_2Te are called hydro acids. *There is a general increase in hydro acid strength in going from left to right along a period.* For example, the following trend of acid strength is observed along the second period:

$$HF > H_2O > NH_3 > CH_4$$

This trend can be explained on the basis of increase in electronegativity as we move from left to right along a period.

$$C < N < O < F$$

With the increase in electronegativity of the central atom, the tendency to pull the electron pair from H increases, and dissociation of H becomes easier and acidic character increases.

The above trend based on the electronegativity holds true along a period. It does not hold good as we move down a group. In fact, acidity in a group increases with decrease in electronegativity as we go down the group. For example, in the hydro acids of group VII, HI is the strongest acid and HF is the weakest acid.

$$HI > HBr > HCl > HF$$

which is opposite to the trend along a period.

Similarly in group VI the acid strength increases as we move down the group with H_2O as the weakest acid.

$$H_2O < H_2S < H_2Se < H_2Te$$

and H_2Te is the strongest acid in spite of H_2O being the most polar on the basis of electronegativity.

Q. 49. Giving reasons arrange the following in order of increasing acid strength.

HF, HCl, HBr, HI *(Kurukshetra, 2002; Purvanchal, 2007)*

Ans. The order of increasing acid strength is

$$HF < HCl < HBr < HI$$

This is because bond dissociation energy of the halogen acids decreases while going from HF to HI.

Q. 50. Arrange the following in decreasing basic strength:

NH_3, PH_3 and AsH_3 *(Awadh, 2000; M. Dayanand, 2002)*

Ans. On moving down a group, the hydrides show increasing acid strength or decreasing basic strength. Hence the order of decreasing basic strength will be

$$NH_3 > PH_3 > AsH_3$$

Q. 51. Discuss the trend in acid strength of oxo-acids. *(Pune, 2010)*

Ans. Substances like $HClO_4$, H_3PO_4, HSO_4^-, containing oxygen are called oxo-acids. The strength of oxo-acids depends upon the following factors:

(*i*) *Size of the central atom*

(*ii*) *Electronegativity of the central atom in the molecule.*

(*iii*) *Inductive effects of substituents on central atom.*

Consider an oxo-acid X – O – H.

where *X* is the central atom. A small and highly electronegative atom *X* will tend to attract the shared pair of electrons more towards itself. This will create a high polarity in O – H bond thereby allowing the proton to dissociate and make the compound acidic. Thus higher the electronegativity of *X* more easily is the proton lost and greater is the acidity of oxo-acid.

As we go down a group, the electronegativity decreases as the size increases and acidic character decreases. For example, the acidic character of oxo-acids of halogens decreases in the given order as is clear from pK_a values.

	HClO	> HBrO	> HIO
pK_a	7.4	8.7	11

It has been observed that the compound XOH behaves as an acid only if X has electronegativity value greater than 1.8. For example, ClOH is an acid while LiOH is a base because electronegativity of Cl is 3.0 while that of Li is 1.0.

Presence of additional electronegative group like oxygen increases the acidity. The relative acidity increases with increase in number of oxygen atoms. For example, in the series of oxo-acids of chlorine, the acid strength increases in the order

	HClO	< $HClO_2$	< $HClO_3$	< $HClO_4$
Oxidation no. of Cl	1	3	5	7
pK_a value	7.52	1.94	–3	–10

In these compounds, the acidity increases with increasing oxidation numbers of the atom (X). All these four acids have only one protonated *oxygen i.e.*, OH group. The rest of the oxygen atoms are surrounding the central atom. With increase in the number of oxygen atoms on the central atom, the extent of electron withdrawal from the O – H bond increases. As a result, the polarity of the O – H bond or tendency for the proton formation increases.

$$Cl-OH < OCl-OH < O_2Cl-OH < O_3Cl-OH$$

The increase in acidic strength can also be explained in terms of the stability of conjugate base formed by the loss of H^+ ion from the acid.

The anions ClO^-, ClO_2^-, ClO^-_3 and ClO^-_4 formed by the loss of proton are stabilized by the charge distribution among O atoms. The charge stabilization is minimum in ClO^- and maximum in ClO^-_4. This means that ClO^- will have minimum stability and would trend to combine with H^+ thus making it the weakest acid and ClO^-_4 will have maximum stability and minimum attraction for H^+ therefore making it the strongest acid amongst the four.

$$[O \text{---} Cl]^- < [O \text{---} Cl \text{---} O]^- < [O \text{---} \overset{\overset{O}{\|}}{Cl} \text{---} O]^- < [O \text{---} \underset{\underset{O}{\|}}{\overset{\overset{O}{\|}}{Cl}} \text{---} O]^-$$

$$\xrightarrow{\text{increase in stability}}$$

Likewise, ON(OH) nitrous acid ($pK_a = 3.3$) is a weak acid but O_2N (OH) (nitric acid) is a strong ($pK_a = -1.4$) acid.

Among organic acids, the inductive effect of substituents is important. Acetic acid H_3CCOOH is a weak acid but replacement of methyl hydrogen by a more electronegative chloro group increases the acidity of the acid as shown by their pK_a values. As the number of electronegative substituents increases the acidity of the acid increases further.

Acid	pK_a
CH_3 COOH	4.8
$ClCH_2$ COOH	2.9
Cl_2 CHCOOH	1.3
Cl_3 CCOOH	0.7

Q. 52. Discuss the effect of polarity and dielectric constant on relative strengths of acids and bases.

Ans. The polarity and dielectric constant of a solvent influence considerably the acidic and basic strength of ionic species dissolved in that solvent. According to Coulomb's law, the force of attraction between two oppositely charged ions is given by

$$F = \frac{q_1 q_2}{Dr^2} \qquad .....(1)$$

where q_1 and q_2 are the charges on the ions, r is the distance between the two ions and D is the dielectric constant of the solvent. Higher the dielectric constant of the solvent, lower is the magnitude of F. In solvents with low dielectric constants, the force of attraction between the ions of the solute will be relatively large and consequently the dissociation of the solute would be small. In solvents with high dielectric constants, the force of attraction between the ions would be small and the extent of dissociation would be large. In other words, the acidic or basic strength of a species would increase with an increase in the dielectric constant of the solvent. The dielectric constant and the dipole moment are known to be directly related. Hence, the greater the polarity of a solvent, the greater would be the acidic or basic strength of the species dissolved in that solvent.

Q. 53. Discuss the effect of substituents on relative strengths of acids and bases.

Ans. The substituents present in acids and bases have a marked effect on their relative strengths. The increase or decrease in acid or base strength depends upon the nature of the substituents, their number and the type of bonding involved in them. This is illustrated below.

(*i*) Electron-releasing or Electron-withdrawing Nature of Substituents. An electron-withdrawing substituent increases and an electron-releasing substituent decreases the acid strength of a species. Thus, chloroacetic acid is stronger than acetic acid while propionic acid is weaker than acetic acid. This is because Cl is an electron-withdrawing and CH_3 is an electron-releasing group.

Similarly, an electron-releasing substituent would increase and an electron-withdrawing substituent would decrease the basic strength of a species. Thus, while NH_3 is basic, NF_3 is almost neutral. Likewise, $(CH_3)_2NH$ is more basic than CH_3NH_2 and NH_3. The CH_3 group, being electron-releasing, increases the basic character of ammonia.

The inductive effect decreases as the substituent moves away from the COOH group. This is why 1-chloro propionic acid ($CH_3CHClCOOH$) is a stronger acid than 2-chloro propionic acid ($ClCH_2CH_2COOH$).

(*ii*) Steric Effect of Substituents. Strangely enough, $(CH_3)_3N$ is less basic than $(CH_3)_2NH$ although due to the presence of an additional electron-releasing CH_3 group in trimethyl amine, it is expected to be more basic than dimethyl amine. This anomaly is due to the *steric effect*. Approach of an electron-acceptor towards the N atom to accept the electron pair is hindered because of the presence of as many as three CH_3 groups around the N atom. This restricts the donation of the electron pair of N atom in $(CH_3)_3N$ and hence reduces the basicity of $(CH_3)_3N$.

***(iii)* Type of Bonding Involved in Substituents.** The presence of a pi bond in a substituent alters the acidic or basic character of the species. For instance, although fluorine is more electron-withdrawing than chlorine or bromine yet BF_3 is *less acidic* than BCl_3 or BBr_3. This may be explained as follows : In a B – F bond in BF_3, one of the filled *p* orbitals of fluorine of proper symmetry forms a pi bond by overlapping with a vacant *p* orbital of boron, causing a shift of electronic charge from fluorine to boron. This decreases the electron-accepting power of boron in BF_3. The *p* orbitals of Cl or Br possess much higher energy and are not suited for pi bonding with boron. Therefore, the electron-accepting power of boron in BCl_3 or BBr_3 is more than that in BF_3.

The following factors also influence the relative strengths of acids and bases.

***(a)* Charge on the Species.** All other factors remaining constant, an increase in the positive charge would increase the acidic strength of the species and an increase in the negative charge would increase the basic strength of the species. This is because an increase in positive charge increases the electron-accepting capacity and an increase in negative charge increases the electron-donating capacity of the species.

***(b)* Electronegativity.** In the species HX, the higher the electronegativity (electron-pulling capacity) of X, the easier would be the release of proton (H) from HX. In other words, the acid strength (proton-releasing capacity) of HX would increase with increase in the electronegativity of X. For example, the electronegativity of non-metals of the second period are in the order C < N < O < F. The acid strength of their hydrides also varies in the same order, *i.e.*, $CH_4 < NH_3 < H_2O < HF$. Thus, within a period of the Periodic table, higher the electronegativity of a non-metal X, greater would be the acid strength of HX.

Q. 54. Discuss the effect of hydration on the strength of acid?

Ans. On the basis of electronegativities of the halogens, the strength of the halogen acids should change in the order HI < HBr < HCl < HF. But, actual order is just the reverse, *i.e.*, HF < HCl < HBr < HI. This anomaly may be explained as follows :

The ionization of HX (X = halogen) in an aqueous solution, in fact, involves a number of energy factors. We can write the following cycle similar to Born-Haber cycle, for the ionization of HX into H^+ (*aq*) and X^- (*aq*) ions.

$$\begin{array}{ccc} HX(g) + aq & \xrightarrow{\Delta H_1} & H^+(aq) + X^-(aq) \\ \downarrow \Delta_{HX} & & \uparrow \\ H(g) + X(g) & & \\ \downarrow (IE)_H \quad \downarrow (EA)_X & & \Delta H_{hydration} \\ H^+(g) \quad X^-(g) & \longrightarrow & \end{array}$$

From Hess's law,

$$\Delta H_1 = \Delta_{HX} + (IE)_H - (EA)_X - \Delta H_{hydration}$$

where ΔH_1 = Enthalphy of ionization of HX in water

Δ_{HX} = Dissociation energy of HX(g)

$(IE)_H$ = Ionization energy of H atom.

$(EA)_X$ = Electron affinity of X atom.

$\Delta H_{hydration}$ = Enthalpy of hydration of H^+ and X^- ions.

Both $(EA)_X$ and $\Delta H_{hydration}$ are exothermic processes and are written with negative signs in the Hess's law expression for ΔH_1.

Evidently, the ionization of HX, *i.e.,* the acid strength of HX would be favoured if ΔH_1 has a large negative value. To this end, while D_{HX} should be small, $(EA)_X$ and $\Delta H_{hydration}$ should be large since $(IE)_H$ is a constant factor. The enthalpies of ionization of the various halogen acids determined as above, fully support the variation of their strengths in the order HF < HC1 < HBr < HI.

Q. 55. How does the strength of acids depend upon:

(*i*) Oxidation no. of central atom.

(*ii*) Inductive effect of substituents. (*G. Nanakdev, 2000; Himachal, 2000*)

Ans. (*i*) Acid strength of an acid increases with the increase in oxidation number of central atom. For example, the acid strength increases in the order:

$$HClO < HClO_2 < HClO_3 < HClO_4$$

because the oxidation number of chlorine increases from 1 for HClO to 3 for $HClO_2$, 5 for $HClO_3$ and 7 for $HClO_4$.

(*ii*) An electron releasing group decreases the acid strength because it hinders the release of protons. For example, the acid strength for the following acids shows the order:

$$HCOOH > CH_3\,COOH > CH_3CH_2COOH$$

Thus replacements by CH_3 and CH_3CH_2 – for H – in HCOOH results in decrease of acid strength.

An electron withdrawing groups increases the acid strength because it helps in release of protons. The acid strength of the following acids exhibits the order:

$$CCl_3COOH > CHCl_2COOH > CH_2ClCOOH > CH_3COOH$$

Thus replacements of hydrogen by electron withdrawing group (Cl –) increases the acid strength. Greater the number of chlorine groups, greater the acid strength.

Q. 56. (*a*) Arrange the following in order of increasing acid strength:

(*i*) HNO_3, HPO_3, and $HAsO_3$,

(*ii*) CCl_3COOH, CBr_3COOH, CI_3COOH

(*b*) Write the conjugate base for the following:

$$HClO_4, HNO_3, H_2CO_3, H_2PO^-_4$$

Ans. (*a*) (*i*) The order of increasing acid strength is

$$HAsO_3 < HPO_3 < HNO_3$$

This is because the electronegativity of central atom increases in the order:

$$As \longrightarrow P \longrightarrow N$$

(*ii*) The order of increasing acid strength is

$$CI_3COOH < CBr_3COOH < CCl_3COOH$$

This can again be explained in terms of increase of electronegativity from I through Br to Cl.

(*b*) The conjugate base for the acids are

$$ClO_4^-, NO_3^-, HCO_3^-, HPO_4^{2-}$$

Q. 57. Which of the following bases is the weakest and why?

$$\mathbf{Cl^-, I^-, Br^-, F^-}$$

Ans. Cl^-, I^-, Br^- and F^- are conjugate bases of HC1, HI, HBr and HF respectively. Since conjugate base of strong acid is weak, I^- is expected to be weakest because it is the conjugate base of the strongest acid HI.

$$\underset{\text{Strong acid}}{HI} \longrightarrow H^+ + \underset{\substack{\text{Weak base}\\ \text{(conjugate base)}}}{I^-}$$

Q. 58. Arrange the following in order of decreasing acid strength:

HClO, $HClO_2$, $HClO_3$, and $HClO_4$ *(Delhi, 2002; M. Dayanand, 2002)*

Ans. The order is $HClO_4 > HClO_3 > HClO_2 > HClO$

This can be explained in terms of oxidation number of the central atom. Greater the oxidation number, greater the acid strength. The oxidation numbers of Cl in $HClO_4$, $HClO_3$, $HClO_2$ and HClO are 7, 5, 3 and 1 respectively. The above order can also be explained in terms of the stability of the anions (base) released by the acids. ClO^-_4 released from $HClO_4$ has the maximum stability because the negative charge can be dissipated with the greatest ease because there are 3 (maximum) other oxygens, as shown below.

$$O{=}\overset{\overset{\displaystyle O}{\|}}{\underset{\underset{\displaystyle O}{\|}}{Cl}}{-}O^- \longleftrightarrow O{=}\overset{\overset{\displaystyle O^-}{|}}{\underset{\underset{\displaystyle O}{\|}}{Cl}}{=}O \longleftrightarrow {}^-O{-}\overset{\overset{\displaystyle O}{\|}}{\underset{\underset{\displaystyle O}{\|}}{Cl}}{=}O \longleftrightarrow O{=}\overset{\overset{\displaystyle O}{\|}}{\underset{\underset{\displaystyle O^-}{|}}{Cl}}{=}O$$

There are four resonating structures of the anion released. In the case of $HClO_3$, $HClO_2$ and HClO, there will be 3, 2 and 1 resonating structures respectively of the anions released by the acids. Thus the stability of the anion will go on decreasing. A weak anion has always the tendency to combine with proton to give back the original acid in the undissociated form, thus reducing acid strength.

$$\text{Acid} \rightleftharpoons H^+ + \text{Anion}$$

Q. 59. Give reasons for the following order of acid strength:

(*i*) $HF > H_2O > NH_3$ *(M. Dayanand, 2000)*

(*ii*) $H_2Fe > H_2\,Si > H_2S > H_2O$

(*iii*) $HNO_3 > HNO_2$

Ans. (*i*) In the case of hydro acids involving elements in a period, the acid strength increases due to increase of electronegativity. HF, H_2O and NH_3 are hydro acids of F, O, and N. As electronegativity decreases in the order F > O > N, the acid strength follows the order: $HF > H_2O > NH_3$.

(*ii*) This can be explained on the basis of increasing bond dissociation energy.

(*iii*) HNO_3 > HNO_2. Nitrogen has greater oxidation number in HNO_3 than HNO_2. Therefore HNO_3 is stronger than HNO_2.

Q. 60. Give Linus Pauling's and Ricci's observations on the strength of acids.

Ans. Linus Pauling made the following observations regarding strength of acids.

(*i*) For a given polybasic acid, the successive $pK_{a_1} pK_{a_2} pK_{a_3}$ values approximate to a ratio 1 : 4 : 8 as shown below:

Acid	pK_{a1}	pK_{a2}	pK_{a3}
H_3PO_4	2.1	7.2	12.0
H_3AsO_4	2.3	7.0	13.0
$H_4P_2O_7$	0.8	2.2	6.7

(*ii*) The magnitude of pK_a is given by the formulation: $O_m E(OH)_n$.

For $m = 0$; $pK_a \geq 7$; very weak acids; *e.g.*, $B(OH)_3$

$m = 1$; $pK_a \cong 2$; weak acid; *e.g.*, $OCl(OH)$

$m = 2$; $pK_a \cong -3$; strong acid; *e.g.*, $O_2Cl(OH)$

$m = 3$; $pK_a \cong -8$; very strong acid; *e.g.*, $O_3Cl(OH)$

Observations of Ricci are given as under in the form of generalisation:

$$pK_a = 8.0 - 9.0\,m + 4.0\,n$$

where m = formal charge on the central atom E

n = the no. of non-hydroxyl oxygen atoms in the acid. Formal charge is calculated as follows:

m = no. of valence shell electrons on the central atom $-\frac{1}{2}$ (no. of electron shared with other atoms) – (no. of unshared valence shell electrons)

Thus, formal charge on chlorine in different acids can be obtained as under:

$:\ddot{\underset{\cdot\cdot}{Cl}} - OH$	$7 - \frac{1}{2}(2) - 6 = 0$
$O - \ddot{\underset{\cdot\cdot}{Cl}} - OH$	$7 - \frac{1}{2}(4) - 4 = +1$
$O_2 > Cl - OH$	$7 - \frac{1}{2}(6) - 2 = +2$
$O_3 > -Cl - OH$	$7 - \frac{1}{2}(8) - 0 = +3$

For example in $HClO_4$; $m = +3$; $n = 3$

Substituting these values in the empirical equation given above we can calculate pK_a

$$pK_a = 8.0 - 9(+3) + 4.0(3) = -7$$

Similarly for $HClO_3$

$$m = +2; \; n = 2$$

$$pK_a = 8.0 - 18 + 8 = -2$$

For $$H_3PO_4 = OP(OH)_3; \; m = 5 - 4 - 0 = 1; \; n = 1$$

$$pK_a = 8 - 9 + 4 = -3$$

The empirical values calculated according to above equation are quite close to the reported pK_a values. Thus, knowing the values of formal charge on the central atom in the oxo-acid and the number of non-hydroxyl oxygen atoms on the acid, it is possible to get an approximate of acid strength.

Q. 61. Discuss the trend in acid strength of hydrated cations.

Ans. Hydration energy is the heat change accompany the formation of 1 mole of a hydrated substance. Hydration energies of $[Ba(H_2O)_2]^{2+}$, $[Cu(H_2O)_6]^{2+}$ are approximately 30.75 and 320 kJ/mol respectively. This large variation in hydration energy is associated with acidity, that is, charge-to-size ratio of the cation. Larger value of charge-to-size ratio for the cation ($Al^{3+} > Cu^{2+} > Ba^{2+}$) is responsible for high hydration energy. There is the phenomenon of hydrolysis which takes place along with hydration as illustrated below.

$$Fe^3 + 6H_2O \xrightarrow{\text{Hydration}} [Fe(H_2O)_6]^{3+} \xrightarrow{\text{Hydrolysis}} [Fe(H_2O)_5(OH)]^{2+} + H^+$$

Small and highly charged cations with larger charge-to-size ratio can lead to the breaking of O – H bonds in the hydrated cation to yield hydrolysed species and the hydrated proton.

The larger the charge-to-size ratio (acidity) of the cation, the more polar the O – H bond in the coordinated H_2O, and the more easily a proton is released as H_3O^+. On the basis of charge-to-size ratio, the relative acid strength decreases as:

$$[Fe(H_2O)_6]^{3+} > [V(H_2O)_6]^{3+} > [Cr(H_2O)_6]^{3+} > [Sc(H_2O)_6]^{3+} > [Al(H_2O)_6]^{3-} > [La(H_2O)_6]^{3+}$$

Among divalent hydrated cations, the acid strength decreases in the order:

$$[Be(H_2O)_n]^{2+} > [Cu(H_2O)_n]^{2+} > [Mg(H_2O)_n]^{2+} > [Ca(H_2O)_n]^{2+} > [Ba(H_2O)_n]^{2+}$$

Q. 62. Explain the expected trends in the relative acid strengths of $HClO_4$, $HBrO_4$ and HIO_4.

Ans. The number of oxygen atoms on the central atom remains the same (= 3) but the electronegativity of central atom decreases as the size increases from Cl to I. Consequently, the polarity of O – H bond or the acidic character decreases in going from $HClO_4$ to HIO_4. Thus, the order of acid strength is $HClO_4 > HBrO_4 > HIO_4$.

Q. 63. Justify the relative order of acid strength

$$H_3PO_2 > H_3PO_3 > H_3PO_4$$

Ans. The structures of the acids are:

```
     O              O              O
     ||             ||             ||
     P              P              P
   / | \          / | \          / | \
  H  |  OH       H  |  OH      HO  |  OH
     H              OH             OH
```

Number of oxygen substituents on the central P atom is the same (*i. e.,* one) in all the cases but the number of OH bonds increases from one in H_3PO_2 to two in H_3PO_3 to three in H_3PO_4. Consequently, a higher polarity will be created in a single OH bond than in two OH bonds which in turn would be more than in three OH bonds. Hence the acidity changes in the order $H_3PO_2 > H_3PO_3 > H_3PO_4$. The P – H does not ionise and has no influence on the acidity of phosphorus oxy-acids.

Q. 64. Arrange the following phosphoric acids in the decreasing order of relative acid strength: H_3PO_4, $H_4P_2O_7$ and $H_5P_3O_{10}$. Give reasons for your answer.

Ans. The structures of H_3PO_4, $H_4P_2O_7$ and $H_5P_3O_{10}$ are given below:

```
     O               O        O                O        O        O
     ||              ||       ||               ||       ||       ||
     P               P        P                P        P        P
   / | \           / | \    / | \            / | \    / | \    / | \
 HO  |  OH       HO  |  O     |  OH        HO  |  O     |  O     |  OH
     OH              OH       OH               OH       OH       OH
```

Acidic strength depends upon the number of unprotonated oxygen atom on P atom (P = O) which polarise a given number of OH bonds. We have to count the number of OH groups per P = O for the three acids. In this series, the number of OH groups per P = O decreases in the order $H_3PO_4 > H_4P_2O_7 > H_5P_3O_{10}$. Therefore acidity increases in the order $H_5P_3O_{10} > H_4P_2O_7 > H_3PO_4$.

Q. 65. Explain the order of relative acidity

$O_3Mn(OH) > O_2Mn(OH)_2 > OMn(OH)_3$

Ans. The structures of acids are

```
       O                 O                 O
       ||                ||                ||
  O == Mn — O — H   O == Mn — O — H   H—O — Mn — O — H
       ||                |                 |
       O                 O                 O
                         |                 |
                         H                 H
```

The oxygen linked to Mn by double bond attracts the electrons towards itself facilitating removal of protons. Number of protons released will depend upon the number of hydroxyl groups per Mn = O group. Smaller the number of hydroxyl group per Mn = O group, greater will be proton liberation and thus greater the acid strength. The above order of acid strength is therefore justified.

Q. 66. Explain why Cl - OH is an acid whereas Na - OH is a base.

Ans. This is due to electronegativity value greater than 1.8 for Cl in HClO and less than 1.8 for Na in NaOH.

Q. 67. Explain acidity trends

(*i*) $H_3PO_4 > H_3AsO_4$ (*ii*) $H_2SO_4 > H_2SO_3$

(*iii*) $H_2S_2O_7 > H_2SO_4$ (*iv*) $[Fe(H_2O)_6]^{3+} > [Fe(H_2O)_6]^{2+}$

Ans. (*i*) Electronegativity of central As is less than that of P, therefore H_3PO_4 is a stronger acid than H_3AsO_4.

(*ii*) The number of oxygen substituents on S in H_2SO_4 is two, whereas it is one in H_2SO_3. Therefore sulphuric acid is stronger than sulphurous acid.

(*iii*) Acidity trend is explained by the structures:

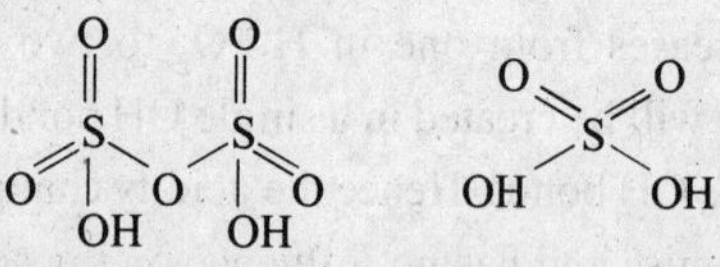

Disulphuric acid $H_2S_2O_7$ has more number of oxygen substituents on S per OH group than sulphuric acid. Therefore $H_2S_2O_7$ is stronger than H_2SO_4.

(*iv*) Larger charge-to-radius ratio of Fe^{3+} relative to Fe^{2+} determines the acid strength $[Fe(H_2O)_6]^{3+}$ is stronger than $[Fe(H_2O)_6]^{2+}$.

Q. 68. In each pair of acids, state which is stronger and why?

(*i*) H_2SO_4 and H_2SO_3

(*ii*) HF and HCl

(*iii*) CH_3COOH and HCOOH

(*iv*) C_6H_5COOH and CH_3COOH.

Ans. (*i*) H_2SO_4 is stronger than H_2SO_3 because there are more non-hydroxyl oxygens on H_2SO_4. Non-hydroxyl oxygen atoms withdraw electrons towards themselves and help in the removal of protons.

(*ii*) HCl is stronger than HF. The bond between H and fluorine becomes weaker with increase in the size of halogen. Weaker the bond, stronger the acid.

(*iii*) HCOOH is stronger than CH_3COOH. In acetic acid CH_3 – group has electron donating effect and thus would prevent release of protons.

(*iv*) C_6H_5COOH is stronger than CH_3COOH because in C_6H_5COOH, the phenyl group is electron withdrawing whereas CH_3 – group in CH_3COOH is electron donating.

Q. 69. Discuss the relative strength of Lewis acids.

Ans. Relative strength of Lewis acids depends mainly on following factors:

(*i*) Inductive effect: The electron withdrawing inductive effect makes a Lewis acid stronger and an electron releasing effect will decrease acid strength. For example, if H in BH_3 is replaced by electron withdrawing *F,* its acidic strength increases. On the other hand, if H in BH_3 is replaced by electron releasing CH_3, then its acidic strength decreases.

$$Me_3B < BH_3 < BF_3$$

(*ii*) **Resonance:** The inductive effect rule fails in explaining the relative acidity of boron halides BF_3, $BC1_3$, BBr_3. Based on electron withdrawing effect of substituents, BF_3 should have been the strongest Lewis acid and BBr_3 the weakest Lewis acid but a reverse order of acid strength is observed.

$$BBr_3 > BCl_3 > BF_3$$

This anomaly is attributed to resonance effect occurring due to the tendency of halogen atoms to contribute electron density toward boron atóm. This is known as back donation from halogen to boron as shown in Fig. 6.1.

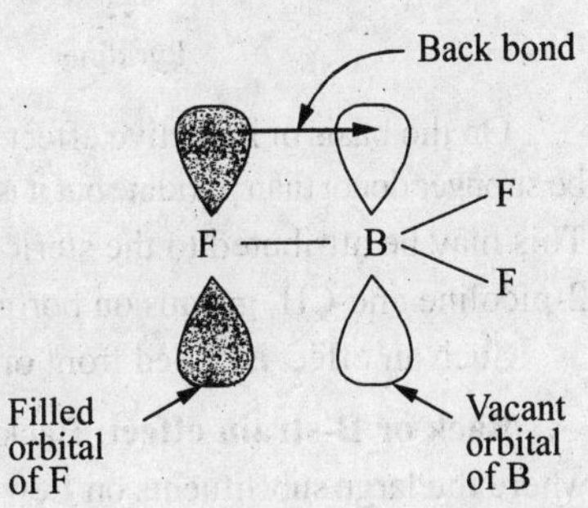

Fig. 6.1. Back bonding.

Back donation is more effective in case of BF_3 due to the easier overlap of filled 2*p* orbitals of fluorine with an empty 2*p* orbitals of boron. Since both the overlapping orbitals belong to the same energy level, therefore the overlap is more effective and leads to an additional bond known as $p\pi$–$p\pi$ back bond. This type of backbonding can oscillate among the three B – F bonds in BF_3 as shown below :

$$\overset{(+)}{F}=\overset{(-)}{B}(F)_2 \longleftrightarrow F-\overset{(-)}{B}(=\overset{(+)}{F})(F) \longleftrightarrow F-\overset{(-)}{B}(F)(=\overset{(+)}{F}) \longleftrightarrow F\cdots B(\cdots F)(\cdots F)$$

The resonance stabilized molecule BF_3 due to back donation from F to B, is no longer so electron deficient with the result that the Lewis acid character is reduced. This tendency of back donation from filled *p*-orbitals of halogen to unfilled 2*p*-orbitals of B is less pronounced in case of BCl_3 and BBr_3 due to higher energy donor 3*p*-orbitals of Cl and 4*p*-orbitals of Br relative to 2*p*-orbital of B. Consequently, resonance stabilisation of $BC1_3$ is small and that of BBr_3 is smaller. Hence the acidity order is:

$$BBr_3 > BC1_3 > BF_3$$

Similarly, B $(OR)_3$ is weaker than BR_3 due to resonance stabilization of the former.

Q. 70. Explain the order of acid strength of the following:

(*i*) $\mathbf{BF_3 < BCl_3 < BBr_3 < BI_3}$ (*G. Nanakdev, 2000*)

(*ii*) $\mathbf{Me_3B < BH_3 < BF_3}$ (*M. Dayanand, 2000*)

Ans. (*i*) Boron forms back bonding with F effectively (See Question 44).

The electrons are thus donated to B making it less electron deficient and thus weaker acid. This back bonding decreases as we move from BF_3 towards BI_3. Thus BI_3 is the strongest of these acids.

(*ii*) Methyl group (Me) is electron donating whereas F^- is electron withdrawing. Therefore BF_3 is the strongest and Me_3B is the weakest acid.

Q. 71. What are F-strain and B-strain effects? How do these affect the stability of acid-base complexes?

Ans. Front or F-strain effect: In reaction between Lewis acid and bases, such as pyridine, 2- picoline and 4-picoline with the same acid trimethylborane BMe_3, the adduct bond energies are 71, 42 and 76 kJ per mol, respectively.

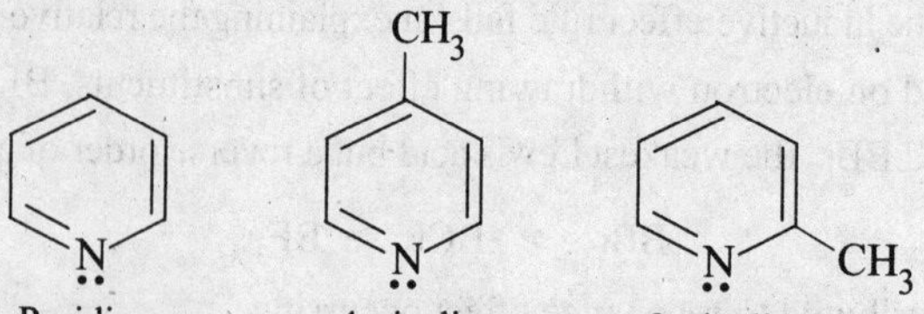

Pyridine   4-picoline   2-picoline

On the basis of inductive effect of methyl group 2-picoline should be stronger donor than pyridine but it is much weaker donor towards BMe_3. This may be attributed to the steric hindrance between CH_3 group of 2-picoline and CH_3 groups on boron in BMe_3 as shown:

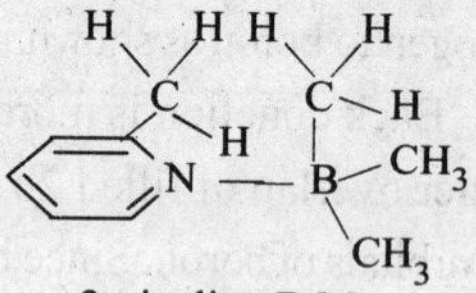

Such an effect is called front or "F-strain" effect.

2-picoline-B Me_3

Back or B-strain effect: Back or "B-strain" is observed in case where the large substituents on Lewis acids or bases are forced to fold back behind the acceptor or donor atom due to changed hybridization on compound formation. Boron is sp^2 hybridised in BR_3 (R = alkyl) but changes to sp^3 hybridization in its complexes with amines $R_3B: NR_3$. However if alkyl groups in BR_3 are larger, they may branch off to reduce its acid strength. Consequently, a large molecule like trimesityl boron is almost inert towards trialkylamines because its plane triangular (sp^2 hybridization, bond angle = 120°) geometry is forced to take tetrahedral configuration (sp^3 hybridization bond angle 109.5°) on complexation. This causes too much crowding or a considerable steric congestion due to forced fold back of mesityl groups behind boron atom in going from planar (120°) to tetrahedral (109.5°) configuration as shown in Fig. 6.2.

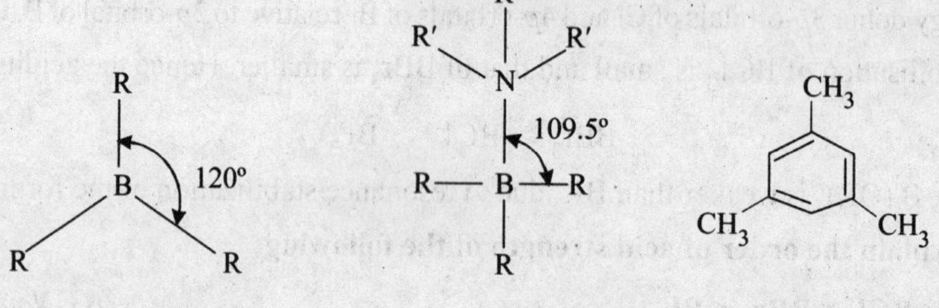

sp^2 hybridisation sp^3 hybridisation   Mesitylene

Fig. 6.2. Forced foldback of mesityl groups.

Q. 72. Why 2-Methylpyridine is weaker than pyridine?

Ans. See Q. 71.

Q. 73. Which one of these bases, pyridine, α-picoline and γ-picoline, forms strongest complex with BMe_3? Give reasons for your answer.

Ans. γ-picoline forms the strongest complex followed by pyridine followed by α -picoline. In α-picoline, complex formation is hindered because of *F*-strain.

γ-picoline has a greater basicity due to electron donating effect of methyl group. Hence it has a greater tendency for complex formation. Pyridine lies in between.

Q. 74. Discuss the order of basic strength of primary, secondary and tertiary amines.

Ans. Methyl group has electron releasing inductive effect and expected order of basic strength is:

$$R_3N > R_2NH > RNH_2$$

The observed order however is $R_2NH > RNH_2 > R_3N$

This trend could be explained on the basis of steric effect but in gaseous state the basicity of methyl amines shows a regular gradation.

$$R_3N > R_2NH > RNH$$

showing that steric effect has no role to play. This means that the reasons of lowered basicity of Me_3N lies only in the effect of solvent. When amines are dissolved in water ammonium ion is formed which undergoes hydration through hydrogen bonding (Fig. 6.3). The greater the number of hydrogen bonds formed by the ammonium ion of the base amine, the larger would be the heat of hydration and greater be the stability.

The energy of hydration of ammonium ions decreases in the order $RNH_3^+ > R_2NH_2^+ > R_3NH^+$ obtained from RNH_2, R_2NH and R_3N as shown in Fig. 6.3. This order of basicity due to solution is totally opposite of the basicity expected from inductive effect. Thus actual basicity of an amine in solution is the combined effect of inductive effect of substituents and extent of solvation of the conjugate acid of the base. As a consequence of the combined effect, dialkyl amines are the strongest bases among amines followed by primary amines and then followed by tertiary amines.

$$RNH_2 + H_2O \longrightarrow R\overset{+}{N}H_3 + O\overset{-}{H} \xrightarrow{+3H_2O} R-\overset{+}{N}(H\cdots OH_2)_2-H\cdots OH_2 + OH^-$$

$$R_2NH + H_2O \longrightarrow R_2\overset{+}{N}H_2 + O\overset{-}{H} \xrightarrow{+2H_2O} R-\overset{+}{N}(R)(H\cdots OH_2)-H\cdots OH_2 + OH^-$$

$$R_3N + H_2O \longrightarrow R_3\overset{+}{N}H + O\overset{-}{H} \xrightarrow{+H_2O} R-\overset{+}{N}(R)_2-H\cdots OH_2 + OH^-$$

Fig. 6.3. Hydrogen bonds formed by the conjugated acids: substituted ammonium ions.

Q. 75. Basicity of gaseous amines decreases in the order of $Me_3N > Me_2NH >$ MeNH but in aqueous solution of the order changes to $Me_2NH > MeNH_2 > Me_3N$. Explain this change.

Ans. See Q. 74.

Q. 76. What is meant by levelling effect and differentiating effect of *solvent?*

(*Kurukshetra, 2002*)

Ans. Solvents play an important role in altering the strength of an acid or a base. Non-aqueous solvents like ammonia, amides ($RCONH_2$), hydrazine etc., are *good proton* acceptor and hence will help acids to ionise in them. As a result of this all acids tend to become strong in a basic solvent. Such solvents are said to exert a *levelling effect* on the strength of acids. The reverse is true of acidic solvents like CH_3COOH , HCOOH, hydrogen halides etc., in which all bases tend to become strong. Such acidic solvents however tend to decrease acid strength of the acids and tend to exert differentiating effect on the strength of acids. Thus water behaves as levelling solvent for acids such as $HClO_4$, HI, HBr, HNO_3 and HCl and bases such as NaH, $NaNH_2$ and $NaOC_2H_5$. So in aqueous medium the acid strength is similar due to levelling effect of water. There will not be much difference in the acid strength due to levelling effect of the solvent. Therefore to compare their relative strengths of strong acids and bases, such a solvent should be selected that ionizes the strong acids and bases to different extent. Such a solvent is called *differentiating solvent.*

For example, acetic acid acts as differentiating solvent towards halogen acid and their ionization increases with increase in atomic number of halogen.

That is, acid strength decreases as

$$HI > HBr > HCl$$

This way, acetic acid has been used extensively to determine the relative strengths of strong acids like $HClO_4$, H_2SO_4, HNO_3 halogen acids, etc., and the following trend has been observed:

$$HClO_4 > HI > HBr > HCl > H_2SO_4 > HNO_3$$

However acetic acid exerts a levelling effect (*i.e.*, it brings basic strength on the same level) on the strength of bases. Consequently, the relative strength of ammonia, primary amines, secondary amines, tertiary amines etc., cannot be distinguished and their basic strength gets enhanced and levelled.

In general acidic solvents tend to differentiate between acids and level off bases while basic solvents tend to differentiate bases and level off acids.

Q. 77. Explain why? Acids like HCl and $HClO_4$ will have equal strength in water.

Ans. This is due to levelling effect of the solvent water.

Q. 78. Out of water and acetic acid, which one will help to decide the relative strengths of $HClO_4$ and HCl and how?

Ans. CH_3COOH. This is because water will exert a levelling effect on HCl and $HClO_4$ and both will act equally strong, whereas CH_3COOH will act as differentiating solvent for acids. The acid trend being $HClO_4 > HCl$.

Q. 79. Out of water and acetic acid, which one will help to determine relative basic strength of RNH_2, R_2NH and R_3N?

Ans. Water helps to distinguish their relative basic strength. Acetic acid cannot be used due to its levelling effect on strength of bases.

INDICATORS

Q. 80. Derive a relation between dissociation constant of the indicator and pH range.

Ans. Let the indicators which is a weak acid be represented as HIn, then the equilibrium in aqueous solution may be represented as:

$$\underset{\substack{\text{Acid colour} \\ \text{or unionised colour}}}{HIn} \rightleftharpoons \underset{\substack{\text{Base colour} \\ \text{or ionised colour}}}{H^+ + In^-}$$

HIn is a weak acid and if added into an acid solution (H^+), its ionization will be suppressed due to common ion effect and the above equilibrium will shift to left hand side with the result that the concentration of In^- ions in solution will fall. In such a case the colour of the indicator will be that of the unionised form HIn (acid colour). However if the medium is alkaline then the OH^- ions of alkali will combine with H^+ of the indicator to form undissociated H_2O. As a result, more of HIn will ionise leading to increased concentration of In^- and colour of the ionised from In^- (base colour) will dominate. Applying law of mass action to equation we can obtain indicator constant K_{In}, also known as ionization constant of the indicator, as follows:

$$K_{In} = \frac{[H^+][In^-]}{[HIn]}$$

$$H^+ = \frac{[HIn]}{[In^-]}.K_{In} \qquad ...(6.1)$$

$$H^+ = \frac{[\text{Unionised form}]}{[\text{Ionised form}]}.K_{In} \qquad ...(6.2)$$

Equation 6.2 can also be written in logarithmic form:

$$pH = pK_{In} + \log\frac{[In^-]}{[HIn]}$$

The colour of the indicator which is determined by the ratio $[In^-]/[HIn]$ is governed by the pH of the solution.

Similarly for a weak base indicator InOH ($InOH = In^+ + OH^-$ a similar expression can be written as

$$[OH^-] = \frac{[InOH]}{[In^-]}.K_{In} \qquad ...(6.3)$$

Since $[H^+][OH^-] = K_\omega$, substituting it in equation 6.3, we get

$$[H^+] = \frac{K_\omega}{K_{In}} \cdot \frac{[In^+]}{[InOH]}$$

As the pH of the solution varies, the colour of the indicator changes. The minor change cannot be noticed by the human eye. A ten-fold excess of one form over the other is required to see the colour of the predominant form.

This means that the indicator HIn will exhibit acid colour when

$$\frac{[HIn]}{[In^-]} \geq 10$$

and basic colour when,

$$\frac{[In^-]}{[HIn]} \geq 10$$

The colour of the indicator should be intermediate for ratios in between these two values and the change cannot be observed accurately by human eye. If the ratio of the two forms is taken as 10, then the expression of the full acid colour derived from equation (6.1) becomes:

$$[H^+] = 10\ K_{In}$$

$$pH = -\log 10 \,.\, K_{In}$$

$$= -1 + pK_{In}$$

and for full base colour

$$[H^+] = \frac{1}{10} K_{In}$$

$$pH = -\log \frac{1}{10} \,.\, K_{In}$$

$$pH = 1 + pK_{In}$$

This shows that the colour change interval or indicator pH range spans

from $-1 + pK_{In}$ to $1 + pK_{In}$

or indicator pH range = $pK_{In} \pm 1$.

This shows that the colour change interval of an indicator lies over approximately two pH units. In other words, within this pH range, the indicator shows a change in colour visible to the eye. If, for example, the ionization constant of HIn has a value 1×10^{-6} then such an indicator will show a colour change in the pH range 5 to 7.

Q. 81. What are universal indicators? Give an example.

(*M. Dayanand, 2002*)

Ans. Universal indicators show a colour change over a large *pH* range. A universal indicator is a mixture of several indicators with a wide range of *pH* response. It behaves like a single indicator but it is not suitable for very accurate estimation. A mixture of suitable amounts of alcoholic solution of phenolphthalein, methyl red, methyl yellow, bromothymol blue and thymol blue in sodium hydroxide sol. is yellow in colour and this sol. acts as a universal indicator showing the following colour changes at different *pH* values:

pH	2	4	6	8	10
Colour change	red	orange	yellow	green	blue

Several universal indicators with colour changes over different pH ranges are commercially available.

Q. 82. What are the main points of Ostwald theory of indicators? On the basis of this theory explain the working of

(*i*) Phenolphthalein

(*ii*) Methyl orange *(M. Dayanand, 2000)*

Ans. The main points of this theory are as follows:

(*i*) An indicator is an organic weak acid or weak base. For example, phenolphthalein is a weak acid while methyl orange, methyl red and methyl yellow are weak bases.

(*ii*) Ionisation of the indicators causes change in its colour. In other words, the unionised molecules of the indicator possess different colour than the ions.

(*iii*) Being weak electrolysis indicators are feebly ionised in sol. But, the addition of a strong acid or strong base increases their ionisation considerably which causes change in the number of coloured ions.

(*iv*) An acidic indicator such as phenolphthalein must possess a coloured anion while a basic indicator such as methyl orange must possess a coloured cation.

Working of phenolphthalein as indicator: Phenolphthalein is a weak organic acid having the formula:

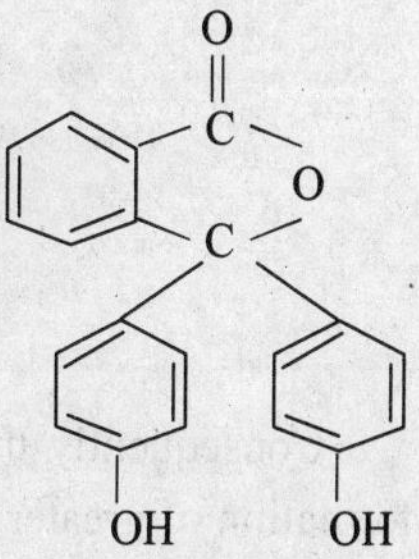

Fig. 6.4

For the sake of simplicity, its formula can be written as HPh. Being a weak acid, it exists mostly as unionised molecules. These unionised molecules are colourless while on ionisation it produces H^+ ion and Ph^- ion according to the equation:

$$\underset{\text{colourless}}{HPh} \rightleftharpoons \underset{\text{colourless}}{H^+} + \underset{\text{Pink}}{Ph^-}$$

Since the reaction is in equilibrium, the addition of an acid increases the H^+ ion concentration and shifts the equilibrium in the backward direction making the sol. colourless.Thus in acidic sol.phenolphthalien remains colourless.

However on adding a base like NaOH sol., OH^- ion are produced which react with H^+ ion to produce feebly ionised water molecules as shown below:

$$HPh \rightleftharpoons H^+ + Ph^-$$
$$NaOH \rightleftharpoons OH^- + Na^+$$
$$H^+ + OH^- \rightleftharpoons H_2O \text{ (feebly ionised)}$$

Consequently, the equilibrium of reaction is shifted in the forward direction resulting in formation of greater number of Ph^- ions which are pink coloured. Therefore in basic sol. of phenolphthalein it gives pink colour.

This theory explains satisfactorily as to why a weak base like NH_4OH cannot be titrated against a strong acid like HC1 using phenolphthalein as indicator. This is because a weak base like NH_4OH is feebly ionised and hence produces very small number of OH^- which are insufficient to shift the equilibrium of reaction towards the right. Therefore phenolphthalein fails to work as indicator for such titrations which involve weak bases.

Working of methyl orange as indicator: Methyl orange is a weak organic base having the formula

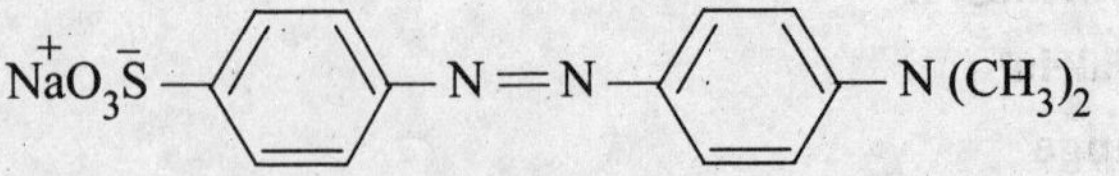

$$\overset{+}{Na}O_3\overset{-}{S}-C_6H_4-N=N-C_6H_4-N(CH_3)_2$$

For simplicity its formula can be written as MeOH. Being a weak base, it exists mostly as unionised molecules. These unionised molecules have yellow colour, while on ionisation, it produces Me^+ ions (red) and OH^- ions (colourless) according to the reaction.

$$\underset{\text{(Yellow)}}{MeOH} \rightleftharpoons \underset{\text{(Red)}}{Me^+} + \underset{\text{(Colourless)}}{OH^-}$$

Since the reaction is in equilibrium, the addition of a base like NaOH increases the concentration of OH^- ions and thus pushes the equilibrium in the backward direction making the sol. yellow in colour. Therefore in basic sol. methyl orange gives yellow colour. On the other hand, addition of an acid like HC1, H^+ ions reacts with OH^- to produce feebly ionised water molecules as shown below.

$$MeOH \rightleftharpoons Me^+ + OH^-$$
$$HCl \rightleftharpoons Cl^- + H^+$$
$$\Downarrow$$
$$H_2O \text{ (feebly ionised)}$$

Consequently, the equilibrium of the reaction is shifted in the forward direction resulting in the formation of greater number of Me^+ ions which are red coloured. Therefore in acidic sol., methyl orange gives red colour. This theory again explains satisfactorily as to why methyl orange cannot be used as indicator in neutralisation reaction involving a weak acid like acetic acid CH_3COOH. This is because a weak acid like acetic acid produces a very small number of H^+ ions which are insufficient to shift the equilibrium of reaction to the forward direction.

Q. 83. Give the main points of Quinonoid theory. On the basis of this theory, explain the working of phenolphthalein and methyl orange as indicators.

(*Himachal, 2000; Kurukshetra, 2002; Nagpur, 2003*)

Ans. The main points of this theory are as follows:

(*i*) The acid-base indicators are organic aromatic compounds which can exist in at least two tautomeric forms.

(*ii*) The tautomeric forms of the indicator exist in equilibrium with one another.

(*iii*) One of the tautomeric forms of the indicator can exist in an acid solution and the other form in the alkaline medium.

(*iv*) Depending upon the *pH* of the solution, the different tautomeric forms exhibit different colours. As the *pH* of the solution varies, the colour of the solution also varies due to conversion from one form to another.

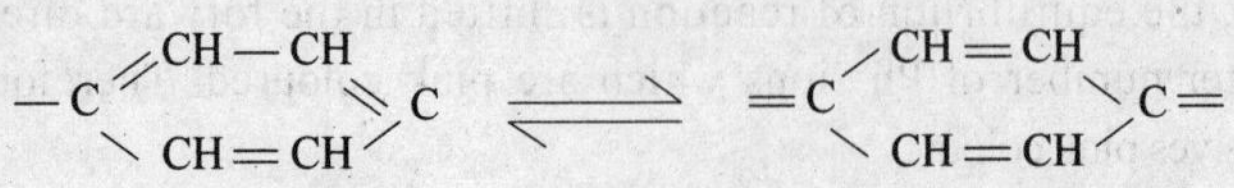

$$\underset{\text{Benzenoid form}}{-C\begin{smallmatrix}\diagup\!\!\!\!= CH-CH \diagdown\!\!\!\!= \\ \diagdown CH=CH \diagup\end{smallmatrix}C-} \rightleftharpoons \underset{\text{Quinonoid form}}{=C\begin{smallmatrix}\diagup CH=CH \diagdown \\ \diagdown CH=CH \diagup\end{smallmatrix}C=}$$

(*v*) The two tautomeric structures possessed by these compounds are benzenoid and quinonoid structures.

(*vi*) Quinonoid form usually has a darker colour than the benzenoid form.

$\underset{H^+}{\overset{OH^-}{\rightleftharpoons}}$

Benzenoid structure (colourless)
Acidic medium

Quinonoid structure (pink)
Basic medium

Working of phenolphthalein as indicator: Phenolphthalein which is a dibasic acid possesses benzenoid structure in acidic medium and is colourless. It possesses quinonoid structure in alkaline medium and gives pink colour.

Therefore, phenolphthalein gives pink colour in basic medium while it remains colourless in acidic medium.

Working of methyl orange as indicator: Methyl orange is an organic base. It has a yellow colour in benzenoid form. On the addition of an acid, it forms a cation which exists in quinonoid form having red colour.

$Na\overset{+}{O_3}\overset{-}{S}-C_6H_4-N=N-C_6H_4-N(CH_3)_2$

Benzenoid form (yellow)
Basic medium

$\underset{OH^-}{\overset{H^+}{\rightleftharpoons}}$

$Na\overset{+}{O_3}\overset{-}{S}-C_6H_4-NH-N=C_6H_4=\overset{+}{N}(CH_3)_2$

Quinonoid form (pink)
Acidic medium

Q. 84. What are advantages of Quinonoid theory over Ostwald theory?

(*M. Dayanand, 2002*)

Ans. This theory has the advantage that it could explain not only the behaviour of indicators in acid-base titrations but even in redex titrations. For example, diphenylamine is used as an indicator in dichromate titrations. In presence of reducing agent diphenylamine is colourless having benzenoid structure. On oxidation, it changes into diphenylbenzidine which has quinonoid structure and violet colour.

$2\,C_6H_5-NH-C_6H_5 \xrightarrow{\text{Oxidation}} C_6H_5-N=C_6H_4=C_6H_4=N-C_6H_5$

Diphenylamine
Benzenoid structure (colourless)

Diphenylbenzidine
Quinonoid structure (violet)

This would be difficult to explain on the basis of Ostwald theory.

Q. 85. Give a brief account of the theory of redox indicators.

Ans. Redox indicators are organic compounds which have different colours in the oxidised and the reduced forms.

Let us represent an indicator [In] in its reduced form (having colour B) and In^{n+} in its oxidised form (having colour A).

$$\underset{\substack{\text{Oxidised form} \\ \text{(Colour A)}}}{In^{n+}} + ne^- \rightleftharpoons \underset{\substack{\text{Reduced form} \\ \text{(Colour B)}}}{In}$$

Nernst equation for the above reaction can be written as

$$E = E^0 - \frac{0.059}{n} \log \frac{[In]}{[In^+]}$$

where E is the potential developed in the system.

E° is the standard potential of the indicator couple. It is essential that the reduction potential of a redox indicator should be quite close to that of the reduction potential of the redox reaction at the end point in the titration. As a result, a sharp colour change will be observed, leading to very good results.

Q. 86. Give some examples of redox indicators.

Ans. Some examples of redox indicators are:

(*i*) Diphenyl amine

(*ii*) Diphenyl benzidine

(*iii*) Methylene blue

(*iv*) Diphenyl sulphonic acid.

Q. 87. Discuss the theory of adsorption indicators for precipitation reactions.

(*Garhwal, 2000*)

Ans. In argentomeric titrations, adsorption indicator is used to locate the end point.

In Fajans method, either fluorescein or dichlorofluorescein is used as adsorption indicator.

The halide solution is taken in a titration flask and 5 to 10 drops of the fluorescein are added as indicator. The medium for this indicator should be either neutral or faintly acidic. On adding silver nitrate to the halide solution, a white ppt. is formed. At the end point, the white ppt. acquires faint reddish tinge.

It is believed that adsorption indicator, *i.e.*, sodium salt of fluorescein functions through the following steps.

1st step: Primary Adsorption. To start with, the solution in the titration flask would be rich in chloride ions. On adding silver nitrate, silver chloride is precipitated which adsorbs chloride ions and thus acquires –ve charge. This is known as primary adsorption.

2nd step: Secondary Adsorption. After primary adsorption the negatively charged silver chloride particles attract sodium ions of the indicator. This is called secondary adsorption.

Since Na^+ ions as well as Cl^- ions are colourless, their adsorption on the silver chloride precipitate does not change the colour of the precipitate.

3rd step: Adsorption Compound Formation. As soon as the medium acquired a slight excess of silver ions, silver chloride precipitate adsorbs silver ions and becomes positively charged. The positively charged silver chloride then attracts anions of the indicator and forms a pale pink adsorption compound of silver fluoresceinate on the surface of silver chloride precipitate. That is the end point.

Q. 88. What is an acid-base indicator? State with reasons what indicator would you use for the following titrations:

(*i*) NaOH vs H_2SO_4 (*ii*) NaOH vs CH_3COOH

(*iii*) Na_2CO_3 vs HCl *(Kerala, 2000; Nagpur, 2000)*

Ans. Acid-base indicators are substances which help in carrying out titrations between acids and bases. They give a particular colour in acidic medium and another colour in basic medium. Hence a sudden change in the colour is produced in the solution at the neutralization point. Phenolphthalein and methyl orange are examples of acid-base indicators.

Indicators to be used for the various acid-base combinations are given below:

(*i*) NaOH and H_2SO_4	Either of phenolphthalein and methyl orange
(*ii*) NaOH and CH_3COOH	Phenolphthalein
(*iii*) Na_2CO_3 and HCl	Methyl orange

Q. 89. Explain why methylamine is more basic than ammonia.

Ans. The structures of methylamine and ammonia are given below

$$CH_3 \rightarrow \ddot{N}H_2 \qquad H-\ddot{N}H_2$$

Q. 90. $(SiH_3)_3N$ is weaker base than $(CH_3)_3N$. Explain. *(Delhi, 2002)*

Ans. $(CH_3)_3$ group is more electron repelling than $(SiH_3)_3$ group. Therefore $(CH_3)_3N$ is a stronger base

Q. 91. Classify, giving reasons, the following either as Lewis acid or Lewis base: $AlCl_3$, S^{2-} *(Delhi, 2002)*

Ans. $AlCl_3$ is electron acceptor hence Lewis acid. S^{2-} is electron donor, hence Lewis base.

METALLURGIES

Q. 1. Discuss the occurrence of metals in the earth's crust?

Ans. Metals generally occur in combined state as under:

(*i*) *Oxides*. For example, aluminium as bauxite ($Al_2O_3 \cdot 2H_2O$), iron as haematite (Fe_2O_3), tin as tinstone (SnO_2), manganese as pyrolusite (MnO_2), copper as cuprite (Cu_2O) and zinc as zincite (ZnO).

(*ii*) *Carbonates*. For example, magnesium as magnesite ($MgCO_3$), calcium as limestone ($CaCO_3$), copper as malachite ($CuCO_3$), Cu $(OH)_2$ and zinc as calamine ($ZnCO_3$).

(*iii*) *Sulphides*. For example, copper as copper pyrites ($CuFeS_2$), iron as iron pyrites (FeS_2), lead as galena (PbS), zinc as zinc blende (ZnS) and mercury as cinnabar (HgS).

(*iv*) *Sulphates*. For example, calcium as gypsum ($CaSO_4 \cdot 2H_2O$), magnesium as epsom salt ($MgSO_4 \cdot 7H_2O$) and lead as anglesite ($PbSO_4$).

(*v*) *Halides*. For example, sodium as common salt (NaCl), silver as horn silver (AgC1) and magnesium as carnalite ($KC1 \cdot MgC1_2 \cdot 6H_2O$).

(*vi*) *Silicates*. For example, aluminium as china clay ($Al_2O_3 \cdot 2SiO_2 \cdot 2H_2O$).

(*vii*) *Phosphate*. For example, calcium as rock phosphate [Ca_3 $(PO_4)_2$].

Q. 2. Briefly describe the minerals that find existence in India? (*M. Dayanand, 2000*)

Ans. Some of the important minerals that exist in considerable quantity in the Indian soil are as under:

(*i*) *Aluminium*. Bauxite, an ore of aluminium, is found in Madhya Pradesh, Bihar, Orissa, Gujarat, Salem (Tamil Nadu) and Kashmir.

(*ii*) *Iron* is available as haematite, Fe_2O_3 in Bihar, Orissa, Madhya Pradesh and Tamil Nadu.

(*iii*) *Coal* occurs in large quantities in close proximity to iron.

(*iv*) *Copper* is available in Bihar, Orissa, Karnataka, Rajasthan and Sikkim.

(*v*) *Chromium* is found as chrome iron in Bihar, Orissa, Maharashtra and Karnataka.

(*vi*) *Gold* is found in free state in Karnataka, Andhra Pradesh and in the alluvial sands of the Ganga and the Brahmaputra.

(*vii*) *Manganese*. Next to Russia, India is the biggest producer of manganese. Its ore is available in Madhya Pradesh, Orissa, Maharashtra, Andhra Pradesh, Bihar, Rajasthan, Tamil Nadu, Karnataka and Gujarat.

(*viii*) *Limestone* is available in Satna, Katni and Rohtasgarh, marble is found in Jaipur and Madhya Pradesh.

(*ix*) *Tin* is found as tinstone in Bihar, Orissa and Rajasthan.

(*x*) *Lead and zinc* are found in Rajasthan.

(*xi*) *Magnesite* is the magnesium ore which occurs in Tamil Nadu and Sikkim.

Q. 3. What is the difference between a mineral and an ore?

Ans. Metals occur in the form of minerals in earth's crust and deep under the earth.

An ore is a mineral from which the metal can be extracted conveniently and economically.

Bauxite and clay are minerals of aluminium. But aluminium can be extracted economically from bauxite and not from clay. Thus clay is a mineral and bauxite is an ore of aluminium.

Q. 4. Comment on the statement "Every ore is a mineral but every mineral is not an ore".

(*M. Dayanand, 2000*)

Ans. See Question 3.

Q. 5. What is meant by the term "Metallurgy"? What are the broad operations carried out in a metallurgy? (*Nagpur, 2000; Bangalore, 2002*)

Ans. It is the process of extracting a metal from its ores. The ore is generally contaminated with non-metallic and rocky impurities known as gangue. The following operations are carried out for obtaining the metal in the pure form:

(1) Crushing of the ore

(2) Dressing or concentration of the ore

(3) Working up of the concentrated ore

(4) Purification or refining of the metal.

Q. 6. What is meant by "dressing" of the ore? Describe various processes carried out for the dressing or concentration of the ore. (*Nagpur, 2000; Kurukshetra, 2001; Pune, 2010*)

Ans. Dressing or concentration of the ore. Metallic ores are often found contaminated with non-metallic and rocky materials like quartz, felspar, mica and silicates. Before the ore is subjected to metallurgical processes it is necessary to remove these unwanted impurities. This operation is known as ore dressing, beneficiation or concentration.

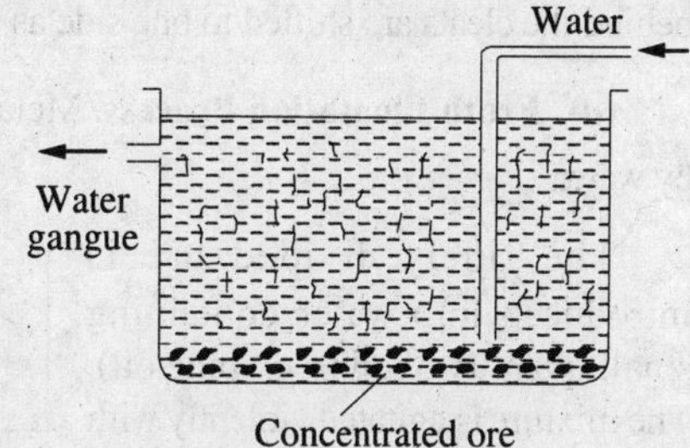

Fig. 7.1. Gravity separation.

Various methods employed for concentration are described as under:

(*a*) Gravity Separation. The ore is concentrated making advantage of the difference in the specific gravities of the metallic ore and the earthy impurifies. The ore after crushing in big jaw crushers and then grinding in a ball mill, a stamp battery or a roller mill, is agitated in a running stream of water. The heavy ore settles down rapidly while the lighter earthy material or gangue is washed away (Fig. 7.1). Oxide ores of iron and tin are concentrated by this method.

(*b*) Hydraulic Classifier. Finely-powdered ore is dropped into a conical reservoir, called Hydraulic classifier, from the top (Fig. 7.2). A powerful stream of water is introduced from the bottom of the reservoir. The gangue being lighter is carried away by the water and the heavy ore accumulates at the bottom of the cone. The conical shape of the reservoir helps in preventing the ore particles from being carried away along with the stream of water.

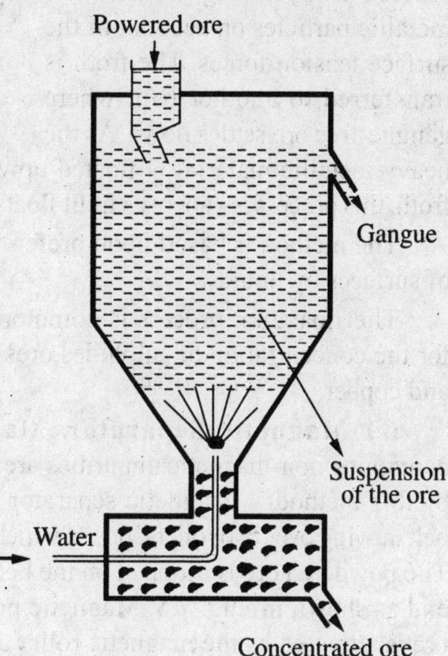

Fig. 7.2. Hydraulic classifier.

(*c*) Wilfley table method (Washing table method). Wilfley table (Fig. 7.3) is a wooden table having a

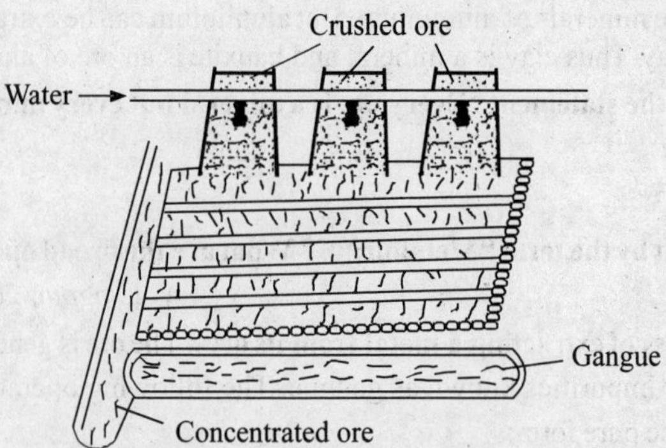

Fig. 7.3. Principle of Wilfley table.

sloping floor on which long strips called *cleats* or *riffles* are fixed.

The powdered ore is suspended in a stream of water and poured at upper end of the table which is then given a rocking motion. The heavier ore particles are obstructed by the cleats while the lighter impurities pass over and are carried away by the stream of water. The ore particles which collect behind the cleats are shifted to one side as a result of the motion of the table and are collected.

(*d*) Froth Floatation Process. Metallic sulphides are wetted by certain oil like pine oil and not by water.

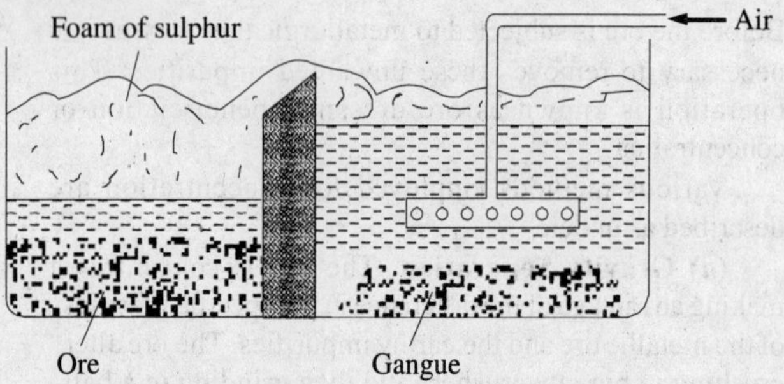

Fig. 7.4. Froth floatation process.

The finely divided ore is introduced into water containing small quantity of oil (*e.g.,* pine oil). The mixture is agitated violently with air (Fig. 7.4). A froth is formed which carries away along with it the metallic particles on account of the surface tension forces. The froth is transferred to another bath where gangue-free ore settles down. As the heavy metallic material is floated upward with the froth, this process is known as froth floatation process.

The method is based upon preferential wetting of surfaces by liquids.

The floatation process is commonly employed for the concentration of sulphides ores of lead, zinc and copper.

(*e*) Magnetic Separator. Magnetic ores containing non-magnetic impurities are concentrated by this method. A magnetic separator consists of a belt moving over two rollers, one of which is magnetic. The powdered ore is dropped on the belt at the other end as shown in Fig. 7.5. Magnetic portion of the ore is attracted by the magnetic roller and falls near the roller while the non-magnetic impurities fall farther off.

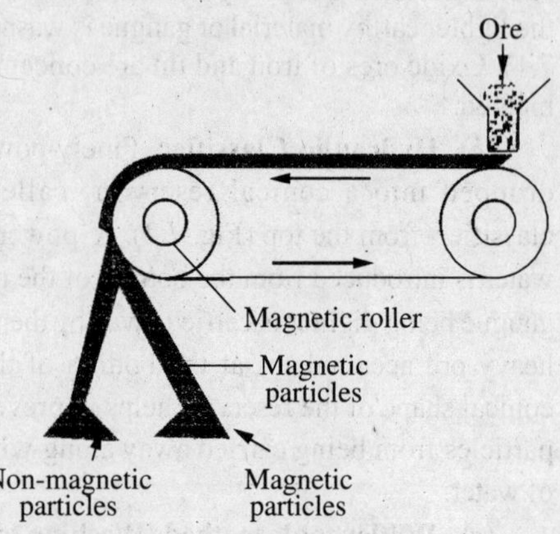

Fig. 7.5 A magnetic separator.

(f) **Leaching or Chemical Method.** The powdered ore is treated with a suitable reagent which dissolves the ore but not the impurities. The bauxite ore ($Al_2O_3 \cdot 2H_2O$) contains the main impurities of iron oxide and silicates. When it is treated with hot sodium hydroxide solution, Al_2O_3 dissolves but the impurities are left behind and filtered off

$$Al_2O_3 + 2NaOH \xrightarrow{\text{Heat}} \underset{\text{Sod. meta aluminate}}{2NaAlO_2} + H_2O$$

The filtrate is diluted and subjected to seeding with a little freshly precipitated aluminium hydroxide when most of the metal from the solution is precipitated as hydroxide.

$$NaAlO_2 + 2H_2O \xrightarrow{1475K} NaOH + Al(OH)_3$$

Q. 7. Write explanatory notes on

(i) **Calcination**

(ii) **Roasting** *(M. Dayanand, 2000; Kumaon, 2000; Delhi 2002; Lucknow, 2002)*

Ans. *(i)* Calcination. Calcination is a process in which the ore usually carbonate or hydrated oxide is subjected to the action of heat in order to expel water from hydrate of oxide and carbon dioxide from a carbonate. For example, when limestone is heated carbon dioxide is given off and it is then said to be calcined. When bauxite is calcined at high temperatures, water is removed and anhydrous alumina is left. Similarly iron ore on calcination gives anhydrous iron oxide. The following equation represents calcination:

$$ZnCO_3 \longrightarrow ZnO + CO_2$$

$$CaCO_3 \longrightarrow CaO + CO_2$$

$$Al_2O_3 \cdot 2H_2O \longrightarrow Al_2O_3 + 2H_2O$$

$$2Fe_2O_3 \cdot 3H_2O \longrightarrow 2Fe_2O_3 + 3H_2O$$

$$CuCO_3 \cdot Cu(OH)_2 \longrightarrow 2CuO + CO_2 + H_2O$$

Calcination is generally done in a reverberatory furnace.

***(ii)* Roasting.** Roasting is a process in which ores (usually sulphide ores) either alone or along with some other materials are subjected to the action of heat or heat and air at temperatures below their melting points in order to bring about chemical changes in them. Calcination is also roasting but in this case we are concerned mainly with the changes due to the expulsion of some ingredients such as water, carbon dioxide and no other chemical change occurs. But during roasting chemical changes like oxidation, chlorination etc. take place. The following equations represent roasting:

$$2PbS + 3O_2 \longrightarrow 2PbO + 2SO_2$$

$$PbS + 2O_2 \longrightarrow PbSO_4$$

$$2ZnS + 3O_2 \longrightarrow 2ZnO + 2SO_2$$

$$ZnS + 2O_2 \longrightarrow ZnSO_4$$

$$CuS + 2O_2 \longrightarrow CuSO_4$$

$$2Cu_2S + 3O_2 \longrightarrow 2Cu_2O + 2SO_2$$

Roasting is generally carried out in a reverberatory furnace or in a blast furnace.

Q. 8. What is meant by smelting? *(Kumaon, 2000)*

Ans. The extraction of a metal from its ore by a process involving melting is known as smelting. It is generally done in a reverberatory furnace or a blast furnace in a controlled supply of air. Several reducing agents such as sodium, magnesium and aluminium are used for reduction. The calcined or roasted ore is mixed with carbon (coal or coke) and heated in a reverberatory or a blast furnace. Carbon and carbon monoxide produced by incomplete combustion of carbon reduce the oxide to the free metal.

$$PbO + C \longrightarrow Pb + CO$$

$$PbO + CO \longrightarrow Pb + CO_2$$

$$SnO_2 + 2C \longrightarrow Sn + 2CO$$

Q. 9. What are flux and slag? What are various types of fluxes?

(M. Dayanand, 2000; Kurukshetra, 2000; Kerala, 2000)

Ans. The ores even after concentration contain some earthy matter called gangue which is required to be removed. There are certain substances which when mixed with furnace charge and heated combine with this earthy matter to form an easily fusible material. Such a substance is known as *flux* and the light fusible material in called *slag.* This fusible material is not soluble in molten metal. Being lighter, the slag can be easily skimmed off from the surface of the fused metal.

Fluxes are of the following two types:

(*i*) *Acidic fluxes* like silica, borax etc. These are used when the gangue is basic such as lime or other metallic oxides like MnO, FeO, etc. The chemical reaction which takes place in removing a basic gangue is as follows:

$$\underset{\text{Gangue}}{Cao} + \underset{\text{Flux}}{SiO_2} \longrightarrow CaSiO_3$$

(*ii*) *Basic fluxes* like CaO limestone ($CaCO_3$), magnesite ($MgCO_3$), haematite (Fe_2O_3) etc. These fluxes are used when the gangue is acidic like silica, P_4O_{10} etc. For example

$$\underset{\text{Gangue}}{SiO_2} + \underset{\text{Flux}}{MgCO_3} \longrightarrow MgSiO_3 + CO_2 \uparrow$$

$$P_4O_{10} + 6CaO \longrightarrow 2Ca_3(PO_4)_2$$

Q. 10. What is Bessemerisation? *(Nagpur, 2000)*

Ans. This is a process used in the metallurgy of copper. It involves the reduction of molten matte (mixture of cuprous sulphide and ferrous sulphide) obtained after smelting of copper pyrites ore in a *Bessemer converter* (Fig. 7.12), see ahead.

The molten matte is taken in Bessemer converter and heated by blowing hot blast of air and sand through tuyeres. The following reactions take place in the converter:

(*i*) Air oxidizes iron sulphide to iron oxide and SO_2

$$2FeS + 3O_2 \longrightarrow 2FeO + 2SO_2 \uparrow$$

(*ii*) Sand reacts with iron oxide to form iron silicate

$$2FeO + SiO_2 \longrightarrow FeSiO_3 \text{ (Slag)}$$

(*iii*) After the iron has been removed, hot blast of air converts a part of cuprous sulphide and cuprous oxide.

$$2Cu_2S + 3O_2 \longrightarrow 2Cu_2O + 2SO_2 \uparrow$$

(*iv*) The cuprous oxide so formed reacts immediately with remaining cuprous sulphide to form metallic copper.

$$2Cu_2O + Cu_2S \longrightarrow 6Cu + SO_2 \uparrow$$

Since the reaction is exothermic, the heat liberated maintains the crude copper in the molten state. The completion of the reaction is indicated by the appearance of green flame produced by the vaporization of copper.

Q. 11. What are different furnaces used for the metallurgical operations?

(*Nagpur, 2000; Pune, 2010*)

Ans. The commonly used furnaces for the metallurgical operations are:

(*a*) Blast Furnace. It is used for smelting iron, copper and lead ores. It has a tall structure made of steel plates rivetted together lined inside with fire bricks. The furnace (Fig. 7.6) is provided with an arrangement for blowing air near the base, a slag hole, a tapping hole for removing the molten metal and an exit for waste gases near the top. The charge is introduced at the top. The mouth of the furnace is closed by a special arrangement through which the charge is introduced. Size of the blast furnace varies with the ore.

Charging pipe
Exit for waste gases
Charge
Air blast Main
Hole for removing slag
Tapping hole

Fig. 7.6. A blast furnace.

(*b*) Reverberatory Furnace. It is used for calcination, roasting or for smelting. The charge in the powdered form along with a flux is placed on the hearth of the furnace. The fuel is placed on grate and the flames are deflected from its low sloping roof. This furnace is used in case of copper, tin, lead, and wrought iron. (Fig. 7.7).

(*c*) Muffle Furnace. A furnace in which the charge container is heated from all sides is called muffle furnace. The burnt materials including the vapours of the metal are allowed to escape through the same opening. The metal is obtained by condensing the vapours.

In Belgian process for extracting zinc, the retorts act as muffles and prolongs help in completing the condensation. The furnace is shown in Fig. 7.8.

(*d*) Vertical Retort Furnace. This furnace is used in the metallurgy of zinc and cadmium (Fig. 7.9). The charge consisting of roasted ore and powdered coke is dropped from the top into vertical retort lined with fire clay bricks to withstand a high temperature. Producer gas is used to heat the furnace.

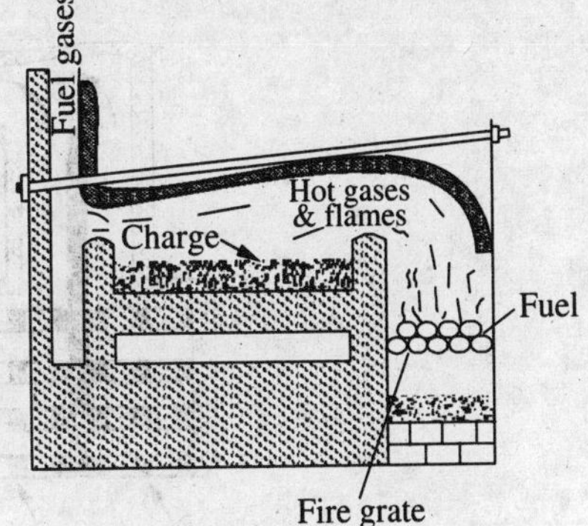

Fig. 7.7. A reverberatory furnace.

(*e*) Open-hearth Furnace. It is a large gas-fired furnace lined with silica or magnesium oxide. It is used for the manufacture of steel. Producer gas which is a mixture of carbon monoxide and nitrogen is used as a fuel. Below the hearth is a brick-checker work so arranged that the heat from the fuel gases (waste burnt gases) is used to preheat the entering gases.

Figure 7.10 illustrates working of this furnace.

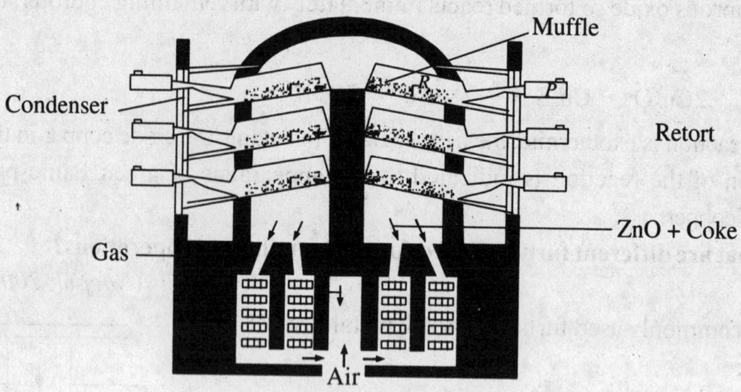

Fig. 7.8. Reduction of zinc oxide by Belgian process using muffle furnace.

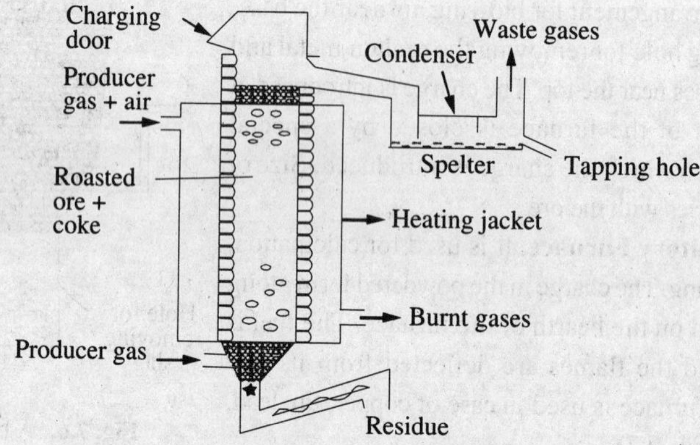

Fig. 7.9. Vertical retort furnace.

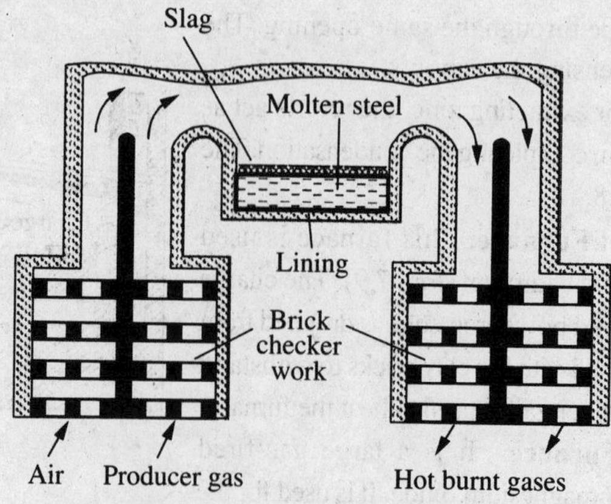

Fig. 7.10. Open hearth furnace.

***(f)* Electric Furnace.** These furnaces are mostly used for the manufacture of steel. One such furnace is shown in Fig. 7.11. This has a basic lining of dolomite ($CaCO_3 \cdot MgCO_3$).

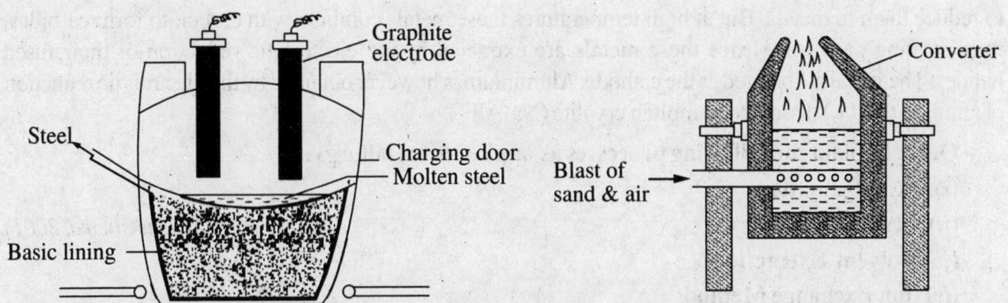

Fig. 7.11. Electric furnace for steel manufacture. **Fig. 7.12.** Bessemer converter for copper.

(g) Bessemer Converter. It is a pear-shaped furnace made of steel plates lined inside with lime or magnesium oxide. It is fixed on pivots and can be tilted in any direction. It is provided with pipes through which hot air can be blown for heating purposes (Fig. 7.12).

Q. 12. Write short notes on:

(i) Goldsmidt thermite process

(ii) Self-Reduction *(Kanpur, 2000)*

(iii) Electrolytic Reduction.

Ans. (i) Goldsmidt Thermite Process. Certain oxides are not satisfactorily reduced by carbon. In such cases aluminium is used as a reducing agent. A mixture of metallic oxide and aluminium powder is ignited in a closed crucible by means of a lighted magnesium ribbon. Metals like chromium, manganese etc. are extracted by thermite process as illustrated in Fig. 7.13.

Fig. 7.13. Alumino-thermic process for chromium.

$$Cr_2O_3 + 2Al \longrightarrow Al_2O_3 + 2Cr$$

$$3Mn_3O_4 + 8Al \longrightarrow 4Al_2O_3 + 9Mn$$

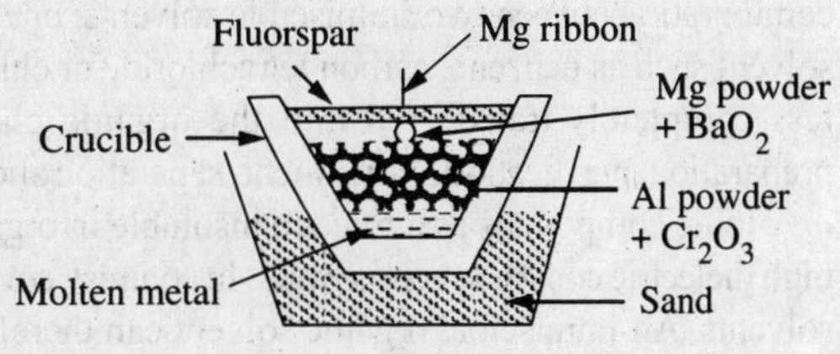

The reaction is highly exothermic and the heat produced is sufficient to melt the metal.

(ii) Self-Reduction Process. In some cases no extra reducing agent is required. A part of the sulphide ore is changed to oxide or sulphate which then reacts with the remaining part to give the metal and sulphur dioxide. Copper and lead are obtained by this method.

$$2Cu_2S + 3O_2 \longrightarrow 2Cu_2O + 2SO_2$$

$$Cu_2S + 2Cu_2O \longrightarrow 6Cu + SO_2$$

$$2PbS + 3O_2 \longrightarrow 2PbO + 2SO_2$$

$$PbS + 2O_2 \longrightarrow PbSO_4$$

$$PbSO_4 + PbS \longrightarrow 2Pb + 2SO_2$$

(ii) Electrolytic Reduction. The carbon reduction methods are not applicable for the highly electropositive and the chemically active metals such as alkali metals, alkaline earth metals, aluminium etc. The oxides of these metals are very stable and have to be heated very strongly with carbon in order to reduce them to metals. But at high temperatures these metals combine with carbon to form carbides, thus creating problem. Hence these metals are extracted by the electrolytic reduction of their fused halides. The metal is liberated at the cathode. Aluminium is however obtained by the electrolytic reduction of alumina (Al_2O_3) dissolved in molten cryolite (Na_3AlF_6).

Q. 13. Explain the following processes as applied to metallurgy:

(*i*) Amalgamation

(*ii*) Hydrometallurgy (*Kurukshetra, 2001*)

(*iii*) Solvent Extraction

(*iv*) Ion Exchange Method

Ans. (*i*) Amalgamation. This process is used for the extraction of noble metals like gold and silver from the native ores. The finely crushed ore is brought into contact with mercury which combines with the metal forming its amalgam. The metal is then recovered by distilling the amalgam.

(*ii*) Hydrometallurgy or precipitation by a more electropositive metal. Hydrometallurgy is the process of bringing the metal into solution by the action of suitable chemical reagents (*e.g.*, sodium cyanide solution or chlorine in presence of water etc.) followed by recovery of the metal by the use of a proper precipitating agent which is a more electropositive metal.

For example, poor ores of silver are suspended in a dilute solution of sodium cyanide and air is blown through it when the silver present goes into solution as the argentocyanide complex. From this solution the metal is precipitated by adding zinc turnings.

$$4Ag + 8NaCN + O_2 \text{ (air)} + 2H_2O \longrightarrow 4Na[Ag(CN)_2] + 4NaOH$$

or

$$AgCl + 2NaCN \longrightarrow Na[Ag(CN)_2] + NaCl$$

$$2Na[Ag(CN)_2] + Zn \longrightarrow Na_2[Zn(CN)_4] + 2Ag\downarrow$$

(*iii*) Solvent Extraction. Solvent extraction is the latest separation technique and has become popular because of its elegance, simplicity and speed. The method is based on preferential solubility principles.

Solvent or liquid-liquid extraction is based on the principle that a solute can distribute itself in a certain ratio between two immiscible solvents, one of which is usually water and the other an organic solvent such as benzene, carbon tetrachloride or chloroform. In certain cases, the solute can be more or less completely transferred into the organic phase. The technique can be used for purpose of preparation, purification, enrichment, separation and analysis.

Ionic compounds as a rule are insoluble in organic solvents but appreciably soluble in solvents of high dielectric constant such as water. In contrast, covalent compounds are preferentially soluble in organic solvents. An immiscible organic solvent can therefore separate a non-ionic species from an aqueous solution. Diethyl ether is one of the most important solvents in use. Very often a complexing reagent is used to convert a metal ion into a neutral molecule or a complex ion. The complexing reagents usually employed are 8-hydroxyquinoline, acetylacetone, dithiozone and dimethylglyoxime. The metal to be extracted forms a complex with one of the reagents and then is extracted by a suitable organic solvent.

(*iv*) Ion exchange method. An ion exchange material is an insoluble substance containing anionic or cationic ends. The anions or cations of the ion exchange material can be exchanged with ions in a surrounding solution without making any physical change. Certain naturally occurring silicates such as zeolite show the property of ion exchange. Removal of hardness of water by the use of zeolite involves the removal of metal ions by this method. Replacement of calcium ions in hard water by sodium ions is an application of this technique.

Several synthetic organic cation and anion exchangers have now been obtained. These are known as ion exchange resins. Those resins which contain acid groups like sulphonic, – SO_3H or carboxyl, – COOH are cation exchange resins.

$$RSO_3H + NaCl \longrightarrow RSO_3Na + HCl$$

$$2RSO_3Na + Ca^{2+} \longrightarrow (RSO_3)_2Ca + 2Na^+$$

Anion exchange resins contain basic groups such as quarternary ammonium groups and replaceable OH^- ions. An example of the anion exchange process is given below:

$$RN(CH_3)_3OH + Cl^- \longrightarrow RN(CH_3)_2Cl + OH^-$$

$$2RN(CH_3)_3Cl + SO_4^{2-} \longrightarrow [RN(CH_3)]_2SO_4 + 2Cl^-$$

The separation of rare earths by the use of ion exchange has been a great achievement of this technique.

Q. 14. What different techniques are employed for purification or refining of a metal?

(*M. Dayanand, 2000; Nagpur, 2000; Kurukshetra, 2002; Delhi, 2003*)

Ans. Different techniques employed in the refining of a metal are described as under:

(*i*) Distillation Process. The process is employed for purification of volatile metals like mercury, zinc and cadmium. The impure metal is heated in a retort and its vapours are condensed in a receiver. While the pure metal distills over and non-volatile impurities are left behind in the retort.

(*ii*) Liquation Process. The process is employed when the impurity is less fusible *i.e.*, has higher melting point than the metal itself. The impure metal is placed on the sloping hearth of a furnace and gently heated. The metal melts and moves down leaving behind the infusible material on the hearth. (Fig. 7.14)

(*iii*) Poling. The impure metal in molten state is stirred with poles of green wood. Hydrocarbon gases escaping from the poles act as reducing agents and reduce any oxide impurity in the metal. For example, blister copper contains the impurity of cuprous oxide. On poling, the oxide is reduced to metallic copper.

(*iv*) Bessemerizaton. Molten pig iron from blast furnace is taken in a Bessemer converter and the impurities are oxidised by blast of air.

(*v*) Cupellation. The impurity of lead in silver is removed by heating it in a cupel (boat-shaped shallow crucible made of bone ash) under a blast of air. Lead is easily oxidized to litharge (PbO) and is carried away by the air current while pure silver is left behind.

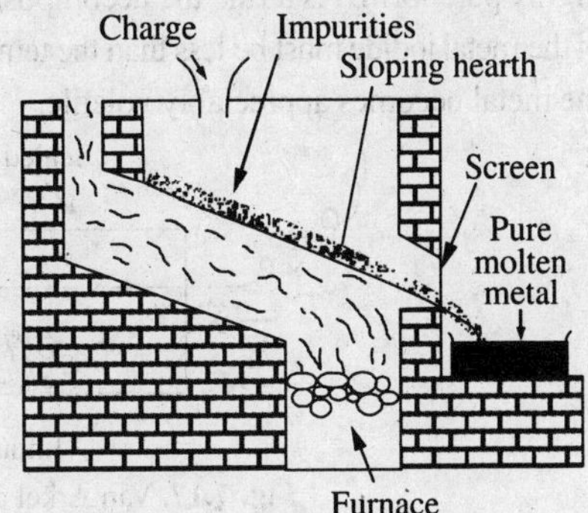

Fig. 7.14. Liquation process.

The above three techniques *viz.*, poling, cupellation and Bessemerization make use of the principle that the impurities in the metal are more oxidisable in a current of hot air.

(*vi*) Zone refining. The solidification of an impure metal crystals of the pure metal are deposited in preference to the impurities which remain in the molten form.

Based on this principle zone refming process is used to obtain metals of very high purity. The circular heater is fitted around a rod of an impure metal and is slowly moved along (Fig. 7.15). At the heated zone the rod melts. As the heater moves along further pure metal crystallises out of the metal while the impurities in the molten zone are carried away to one end of the rod. The process is repeated again and again and after several passes the impurities are completely swept to the end which is fmally discarded. Semiconductors, silicon, germanium and gallium are purified by this techqniue.

***(vii)* Electro-refining.** A large number of metals such as Cr, Sn, Cu, Ni, Zn, Au and Ag are refined by electrolytical methods. The impure metal in the form of blocks in made the anode in an electrolytic cell. The cathode is a thin plate of a pure metal. A solution of a soluble salt of the metal is used as the electrolyte. On passing the electric current the pure metal dissolves from the anode and is deposited on the cathode. The soluble impurities go into the solution while the insoluble matter called the *anode mud* settles at the bottom.

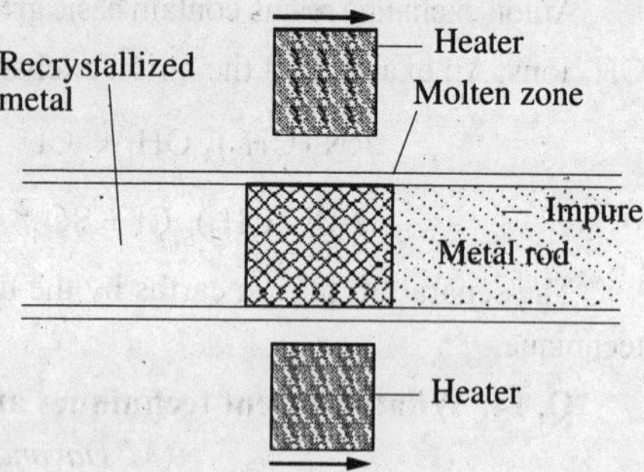

Fig. 7.15 Zone refining.

The noble metals like silver, gold and platinum are recovered from the anode mud in the electrolytic refining of copper (Fig. 7.16). The reactions can be represented as under:

$$CuSO_4 \rightleftharpoons Cu^{2+} + SO_4^{2-}$$

At cathode $Cu^{2+} + 2e^- \longrightarrow Cu\,(s)$

At anode $Cu\,(s) \longrightarrow Cu^{2+} + 2e^-$

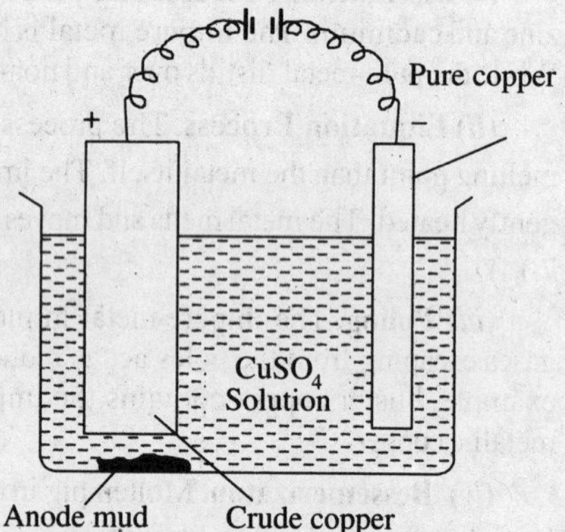

Fig. 7.16. Electro-refining of copper.

***(viii)* Van Arkel and de Boer's filament method.** This method is particularly useful for those elements which cannot be readily obtained pure by other methods, *e.g.,* titanium, zirconium and thorium. The process involves the formation of volatile iodides by heating the impure powdered metal with a little iodine in a sealed vessel (Fig. 7.17). The iodide is then decomposed by bringing it into contact with a thin tungsten wire (filament) heated electrically at a high temperature. The metal is thus deposited on the wire in a highly pure form. As a rule the decomposition temperature of the metal iodide must be less than the temperature at which the metal becomes appreciably volatile.

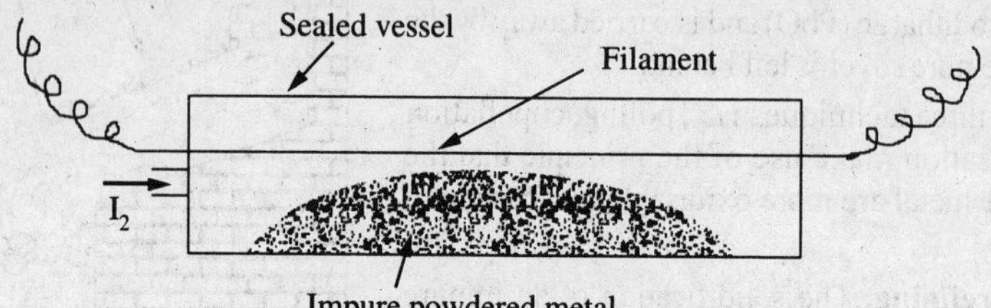

Fig. 7.17. Van Arkel and de Boer method.

***(ix)* Parting process.** This process is used to purify gold which contains copper and silver as impurities. Gold is not attacked by sulphuric acid but copper and silver dissolve in it. If the concentration of gold in the impure sample is more than 30%, the silver and copper present as impurities are also not attacked by sulphuric acid. The gold content is, therefore, lowered by adding more silver to the molten gold till the gold content is 25% (quartation). The resulting alloy is then treated with boiling concentrated sulphuric acid when copper and silver pass into solution as sulphate leaving gold undissolved. This is washed, dried and fused.

Q. 15. Differentiate between:

(*i*) Zone refining and electrolytic refining

(*Kerala, 2000; Coimbatore, 2000; Kumaon, 2000; Kurukshetra, 2001*)

(*ii*) Poling and parting.

Ans. See Q. 14.

Q. 16. Describe the following:

(*i*) Van Arkel process (Vapour phase process) (*Bangalore, 2004*)

(*ii*) Parting process

Ans. See Q. 14.

EXTRACTION OF SILVER

Q. 17. What are principal ores of silver? How is silver extracted from its ores?

(*Mumbai, 2005, Calcutta, 2005, Vidyasagar, 2007*)

Ans. Principal ores of silver are:

(*i*) Argentite or silver glance, Ag_2S

(*ii*) Horn Silver, AgCl

(*iii*) Pyargyrite or ruby silver, Ag_2S, Sb_2S

(*iv*) Silver copper glance Ag_2S Cu_2S

Silver is mainly extracted from argentite (Ag_2S) by **cyanide process.**

The process is carried out in following steps:

(*a*) Crushing and concentration. The sulphide ore is crushed is a jaw crusher and is finely powdered in ball mills. The powdered ore is concentrated by froth floatation process. It is agitated with a mixture of pine oil and water. The mixture is stirred vigorously by means of a blast of air. The particles of the ore get wetted by the oil. Frothing takes place and foam produced carry lighter ore particles to top from where they are skimmed off. Gangue particles get wetted with water and being heavier settle at the bottom of the tank.

(*b*) Treatment with sodium cyanide. The concentrated ore is treated with 0.4–0.7% sodium cyanide solution. It is agitated with a current of air for 8–10 hours. Silver sulphide goes into solution with the formation of a soluble complex of sodium argentocyanide.

$$2Ag_2S + 8NaCN + 2H_2O + O_2 \longrightarrow \underset{\text{Sod. Argentocyanide}}{4Na[Ag(CN)_2]} + 4NaOH + 2S$$

The soluble argentocyanide complex is separated from the insoluble impurities by filtration.

(*c*) Precipitation of silver. The filtrate containing complex of argentocyanide is treated with zinc dust. Zinc displaces silver from its complex and thus silver gets precipitated.

$$2Na[Ag(CN)_2] + Zn \longrightarrow \underset{\substack{\text{Sod. zinc cyanide} \\ \text{(soluble)}}}{Na_2[Zn(CN)_4]} + 2Ag\downarrow$$

(*d*) Fusion. The precipitated silver is separated by filtration, washed and dried. It is then fused with potassium nitrate or borax in a crucible. The impurities get oxidized and rise as a scum on the surface from where it is skimmed off. On cooling, compact white shining mass of silver is produced.

(*e*) Purification of silver. The silver so obtained still contains impurities like zinc, copper, gold, platinum etc. It is purified by electrolytic refining.

Electrolytic bath of the type used in the refining of copper (See Fig. 7.21) is used.

Impure silver is made as anode while pure silver strip is made as cathode. These electrodes are suspended in the electrolytic tank containing silver nitrate solution acidified with 1% nitric acid.

When an electric current is passed, silver along with impurities such as zinc, copper passes into the solution while gold which remains unoxidized gets settled at the bottom just below anode as anode mud. Silver gets deposited at the cathode. Silver thus obtained is about 99.5% pure.

EXTRACTION OF GOLD

Q. 18. What are various ores of gold? How is gold extracted and refined from its ores?

(*Coimbatore, 2000; Kurukshetra, 2002*)

Ans. Important ores of gold are:

(*i*) Sylvanite [(Au, Ag) Te_2], a telluride of gold and silver,

(*ii*) Calaverite $AuTe_2$, a telluride of gold, and

(*iii*) Bismuth auride, BeAu

The Kolar mines of Mysore produce about 99.9% of the total production of our country. Alluvial deposits are found in Bihar, Orissa and Assam.

Extraction of Gold from Alluvial Deposits

Different methods for the extraction of gold from alluvial deposits are:

(*i*) Panning. This method is applicable only to alluvial deposits containing a fairly high percentage of free gold. The alluvial sand containing fine particles of gold in big circular pans made of iron or zinc are treated with a large quantity of water. A rotatory motion is given to the pan whereby the lighter particles are washed away and heavier gold particles settle to the bottom. Gold is recovered from the heavy sand particles containing gold by chemical treatment.

Panning is not much used as a production method because the amount of gold bearing sand that can be treated per day is very small. It has been largely replaced by sluicing or placer mining.

(*ii*) Dredging. This is the most economical method for working gold bearing sands and gravels which do not need crushing and in which the amount of gold is very low. Auriferous sand from the bottom of a flowing river is taken out with the help of dredges and washed on Wilfley tables in order to remove the coarser particles. The surface of table has a number of riffles arranged lengthwise and rocking motion is given to it. The auriferous mud flows down over these riffles while the heavier gold particles are caught and made to move along these riffles and collected in a trough placed at the end (see Fig. 7.3).

(*iii*) Placer mining or sluicing. In this method the auriferous gravel and sand is introduced in an inclined channel provided with sluices and riffles (Fig. 7.18) and powerful stream of water under pressure is directed on it. The heavier gold particles are retained in the sluices due to obstruction by riffles while the lighter sand particles are washed away with water.

(*iv*) Hydraulic mining. Hydraulic mining is employed when the auriferous sand exists at the edge of a river or on a hill side. Water under high pressure is directed against the deposits when the auriferous sand flows down with water in a powerful stream. This is passed through a sluice system which consists of an inclined channel provided with a series of sluices, *i.e.,* troughs with riffles at the bottom. The water sweeps away the sand while the heavy gold particles get collected at the bottom of the sluices. It is a cheap method of mining.

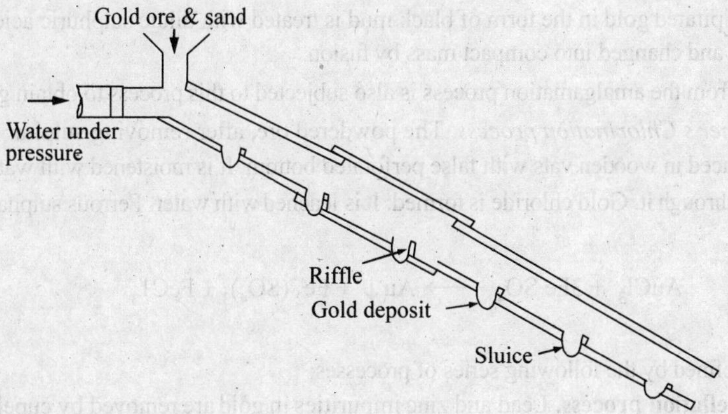

Fig. 7.18. Sluice way system used in hydraulic mining and placer mining.

Extraction of Gold from Auriferous Quartz

The extraction of gold from quartz viens involves the following steps:

***(i)* Crushing.** The big pieces of auriferous quartz are crushed and broken to pieces of 3-5 cm size in jaw or gyratory crushes.

***(ii)* Pulverization.** These pieces are then introduced in a battery of stamp mills working in series. Each stamp mill (Fig. 7.19) consists of a mortar (die) box about 125 cm long 25 cm wide and 8 cm deep. The stamp (Pestles), which is raised by a rotating cam shaft and then dropped by gravity, weighs about 1250 pounds each and falls at the rate of 96 strokes per minute through a height of about 20 cm. The powdered ore is washed out of the stamp mill by a current of water. Only particles which are fine enough can pass through the screen. The coarser particles if any are retained back and struck again.

***(iii)* Treatment of the powdered ore.** The crushed ore is subjected to one of the following three processes to obtain gold:

(*a*) *Amalgamation process.* Gold paste (slurry) is passed over amalgamated copper plates placed in a sloping position just in front of the mortar boxes. The gold combines with mercury to form an amalgam. This is periodically removed from copper plates and distilled when the mercury passes over leaving gold behind. This process can remove 70–80% of the gold. The residual ore called "tailings" is subjected to the above process again in order to recover more gold from it.

(*b*) *MacArthur and Forrest's cyanide process.* Powdered stamp mill is placed in large vats made of wood or cement provided with perforated false bottoms covered with canvas mattings. It is then treated with 0.3% solution of KCN in the presence of excess of air when the following reaction occurs.

$$4Au + 8KCN + O_2 + 2H_2O \longrightarrow 4K[Au(CN)_2] + 4KOH$$

The gold goes into solution in the form of a double cyanide of potassium and gold. From this solution gold is recovered either by the addition of zinc shaving or by electrolysis.

$$2K[Au(CN)_2] + Zn \longrightarrow K_2[Zn(CN)_4] + 2Au$$

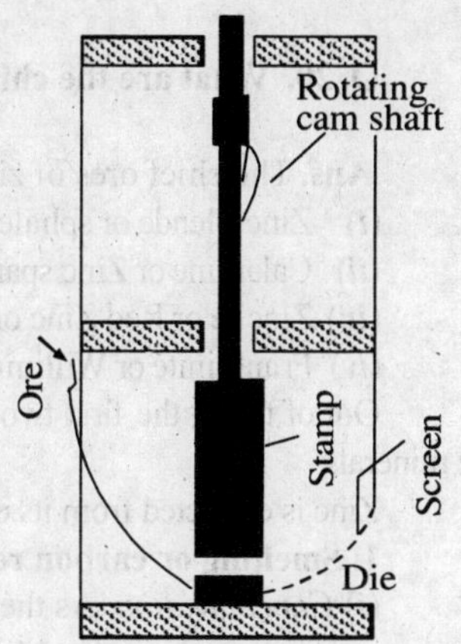

Fig. 7.19. A stamp mill.

The precipitated gold in the form of black mud is treated with dilute sulphuric acid to remove the excess of zinc and changed into compact mass by fusion.

Tailings from the amalgamation process is also subjected to this process to obtain gold efficiently.

(*c*) *Plattner's Chlorination process.* The powdered ore, after removing sulphur and arsenic by roasting, is placed in wooden vats with false perforated bottom. It is moistened with water and chlorine gas is passed through it. Gold chloride is formed. It is leached with water. Ferrous sulphate is added to it to obtain gold.

$$AuCl_3 + 3FeSO_4 \longrightarrow Au\downarrow + Fe_2(SO_4)_3 + FeCl_3$$

Refining

Gold is refined by the following series of processes:

(*a*) Cupellation process. Lead and zinc impurities in gold are removed by cupellation process. The metal along with borax and silica is fused in a cupel and air is passed over the fused mass. Impurities get oxidised. They are blown off and pure gold is left behind.

(*b*) Quartation process. Silver and copper impurities present in gold are removed by parting. Gold content is lowered by adding more silver to the impure molten gold till the gold content is about 25% of the alloy. The resulting alloy is treated with sulphuric acid. The silver and copper pass into the solution as sulphates while gold is left undissolved. This is washed, dried and fused. Silver is recovered from acidic silver sulphate solution by precipitating with copper.

$$2Ag + 2H_2SO_4 \longrightarrow Ag_2SO_4 + 2H_2O + SO_2\uparrow$$

$$Cu + 2H_2SO_4 \longrightarrow CuSO_4 + 2H_2O + SO_2\uparrow$$

(*c*) Parting with chlorine. The impure sample of gold is fused with a small amount of borax and dry chlorine gas is bubbled through it. The base metals are converted into chlorides which pass out as fumes. Gold is not affected while silver chloride forms a scum on the surface of the molten metal and can be easily skimmed off. Pure gold is left behind and is collected.

(*d*) Electrolytic refining. The impure gold is made anode in an electrolytic bath consisting of 5% gold chloride solution using some dil. hydrochloric acid. Pure gold is deposited on thin sheets of gold used as cathode while silver forms silver chloride which is removed as mud. Gold of 99.9% purity can be obtained in this way.

Q. 19. Describe the chlorination process for extraction of gold.

Ans. See Q. 18.

EXTRACTION OF ZINC

Q. 20. What are the chief ores of zinc? Give an account of metallurgy of zinc?

(*Nagpur, 2000; Kurukshetra, 2002*)

Ans. The chief ores of zinc are:

(*i*) Zinc Blende or sphalerite, ZnS

(*ii*) Calamine or Zinc spar, $ZnCO_3$

(*iii*) Zincite or Red Zinc ore, ZnO

(*iv*) Franklinite or Willimite, ZnO, Fe_2O_3

Out of these, the first two are more abundant and zinc is generally extracted from both of these minerals.

Zinc is extracted from its ores by two processes as described below.

1. Smelting or carbon reduction process. It involves the following steps:

(*i*) *Concentration.* As the ore contains iron oxide as impurity it is removed by electromagnetic separation. In case of zinc blende (or sulphide ore), the ore is first concentrated by gravity and then by froth floatation process in which the powdered ore is agitated with a mixture of pine oil and water when the ore comes to the surface along with froth and is removed while the gangue sinks to the bottom (see Fig. 7.5). The carbonate ore does not require any concentration.

(*ii*) *Roasting.* The concentrated ore is then roasted *i.e.*, heated to a high temperature in presence of excess of air in a reverberatory furnace. Zinc sulphide is oxidised to zinc oxide.

$$2ZnS + 3O_2 \longrightarrow 2ZnO + 2SO_2$$

At the same time some of it is oxidized to zinc sulphate.

$$ZnS + 2O_2 \longrightarrow ZnSO_4$$

To remove it, the roasting is done at a high temperature about 173 K when it decomposes to form zinc oxide

$$2ZnSO_4 \xrightarrow{1173\text{ K}} 2ZnO + 2SO_2 + O_2$$

If calamine is used, it decomposes to form zinc oxide in a furnace.

$$ZnCO_3 \longrightarrow ZnO + CO_2$$

(*iii*) *Reduction.* (*a*) *Belgian process.* Zinc oxide produced in the above operation is reduced with carbon (coke) by heating the mixture in fireclay retorts. In the Belgian process the retorts are fireclay tubes closed at one end (Fig. 7.20). These are so adjusted their ends slant downwards and earthenware condensers can be fitted at their mouths. The retorts are heated by gas on the regenerative system of heat economy. The temperature during the process of reduction rises a little above 1273 K. Carbon monoxide formed during the process of reduction burns at the mouth of the retorts and when the colour of the flame changes from blue to greenish white the condensers are fitted thereon.

$$ZnO + C \longrightarrow Zn + CO$$

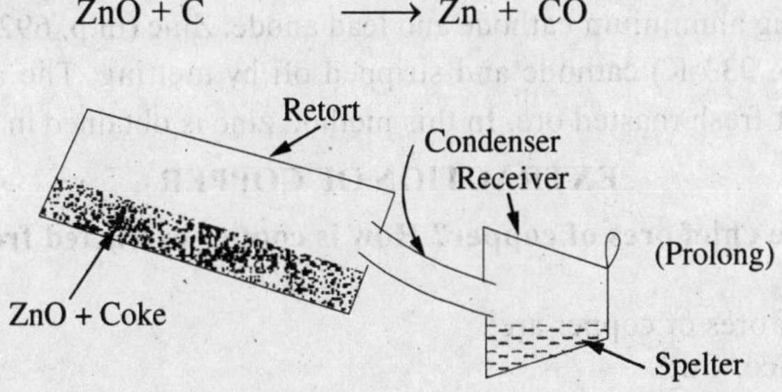

Fig. 7.20. Fireclay retort used in the Belgian process.

These are cooled by air and zinc condenses in these condensers partly as a fused metal (spelter) and partly as a powder (zinc dust).

(*b*) *Vertical retort process.* A continuous process called vertical retort process is largely used these days. In this case a vertical retort is charged with roasted ore made into small briquetts along with powdered coal or coke. It is heated by burning fuel in the surrounding jacket and also internally by producer gas at temperature of about 1673 K. Zinc obtained as a result of reduction vaporises and condenses in the condenser to give zinc spelter (See Fig. 7.9).

(*iv*) *Purification.* Zinc obtained as above contains 97 to 98% of the metal, the rest being impurities, chiefly lead, iron, cadmium arsenic and antimony.

Pure zinc is obtained from such a sample by first fusing it with nitre to remove arsenic and then submitting it to repeated fractional distillation. The boiling point of lead (1893 K) and iron (3273 K) are higher than that of zinc (1180K) whilst that of cadmium (1040K) is lower. When distillation is carried out at about 1273 K only zinc and cadmium distil over leaving iron and lead behind. Cadmium is separated from zinc by feeding it in a column maintained at above 1073 K when cadmium volatilizes off.

Zinc is also purified electrically, from a solution of $ZnSO_4$ containing a little dil. H_2SO_4 by making the impure metal the anode while the cathode consists of a sheet of pure zinc on which the metal is deposited in a chemically pure form.

2. Modern electrolytic process. The electrolytic process for the extraction of zinc involves the following steps:

(*i*) *Crushing and concentration* are done as in the first process.

(*ii*) *Roasting.* The concentrated ore is roasted at a moderate temperature (923 K) when zinc sulphide is mainly oxidized to zinc sulphate. Some zinc oxide is also obtained as given under first process.

(*iii*) *Leaching with sulphuric acid.* The roasted ore is leached with H_2SO_4 to convert zinc oxide to zinc sulphate

$$ZnO + H_2SO_4 \longrightarrow ZnSO_4 + H_2O$$

(*iv*) *Precipitation of impurities.* The solution obtained above contains Cu, Cd, As, Sb, Mn, Al and Fe as impurities. Out of these Fe, Al, As and Sb precipitated by treating the solution with milk of lime. Manganese if present is oxidized by blowing in air when it is precipitated as hydrated oxide. Copper and cadmium are precipitated by treating with zinc dust.

$$CdSO_4 + Zn \longrightarrow ZnSO_4 + Cd$$

$$CuSO_4 + Zn \longrightarrow ZnSO_4 + Cu$$

(*v*) *Electrolysis.* The zinc sulphate solution obtained above is acidified with sulphuric acid and electrolysed using aluminium cathode and lead anode. Zinc (m.p. 692.4 K) is deposited on the aluminium (m.p. 933 K) cathode and stripped off by melting. The acid solution is used again for leaching of fresh roasted ore. In this method zinc is obtained in 99.98% purity.

EXTRACTION OF COPPER

Q. 21. What are chief ores of copper? How is copper extracted from its ores?

(*Kurukshetra, 2002*)

Ans. The chief ores of copper are:

(*i*) Copper glance: Cu_2S

(*ii*) Copper pyrites: $CuFeS_2$

(*iii*) Malachite: $Cu(OH)_2 \cdot CuCO_3$

(*v*) Covellite: CuS

(*vi*) Azurite: $2CuCO_3, Cu(OH)_2$

Copper is mainly extracted from copper pyrites ($CuFeS_2$). The process for the extraction involves the following steps:

(*a*) Concentration. Copper pyrites contains large amounts of silicious impurities. The ore is crushed, sieved and concentrated by froth floatation process (See Fig. 7.4). It is agitated with a mixture of pine oil and water. The mixture is stirred vigorously with a blast of air. The particles of the ore get wetted by the oil. Frothing takes place and foam produced carry ore particles to the top from where they skimmed off while the gangue particles get wetted with water and settle at the bottom. Copper content is increased to 25–30% in this way.

(*b*) Roasting. The concentrated ore is heated strongly in a limited current of air on hearth of a reverberatory furnace (See Fig. 7.7). As a result of this treatment, the following changes occur:

(*i*) A considerable portion of sulphur present as impurity is oxidized and removed as SO_2.

$$S + O_2 \longrightarrow SO_2 \uparrow$$

(*ii*) P, As and Sb present in the ore are removed as their volatile oxides.

$$P_4 + 5O_2 \longrightarrow P_4O_{10} \uparrow$$

$$4As + 3O_2 \longrightarrow 2As_2O_3 \uparrow$$

$$4Sb + 3O_2 \longrightarrow 2Sb_2O_3 \uparrow$$

(*iii*) The pyrite is converted into mixture of cuprous and ferrous sulphide.

$$2CuFeS_2 + O_2 \longrightarrow Cu_2S + 2FeS + SO_2 \uparrow$$

(*iv*) Moisture is removed.

(c) Smelting. This process is carried out in a small blast furnace.

The roasted ore is mixed with sand (flux) and powdered coke. This mixture is heated strongly in a blast furnace (See Fig. 7.6) and hot blast air is passed through tuyeres. During smelting the following changes occur:

(*i*) Since iron has greater affinity for oxygen than sulphur, iron sulphide formed during roasting changes to ferrous oxide.

$$2FeS + 3O_2 \longrightarrow 2FeO + 2SO_2$$

(*ii*) Ferrous oxide combines with silica (flux) to form ferrous silicate (fusible slag).

$$FeO + \underset{\substack{\text{Silica} \\ \text{(flux)}}}{SiO_2} \longrightarrow \underset{\substack{\text{Ferrous silicate} \\ \text{(Fusible slag)}}}{FeSiO_3}$$

Slag being lighter forms the upper layer and is withdrawn from the upper hole.

(*iii*) Any oxide of copper formed during roasting or in blast furnace reacts with ferrous sulphide and gets changed back to cuprous sulphide.

$$2Cu_2S + 3O_2 \longrightarrow 2Cu_2O + 2SO_2 \uparrow$$

$$Cu_2O + FeS \longrightarrow \underset{\substack{\text{(Cuprous} \\ \text{sulphide)}}}{Cu_2S} + FeO$$

The cuprous sulphide along with some ferrous sulphide left after smelting process is known as matte. It contains about 45% of copper. It being heavier forms the lower layer and is withdrawn from the lower hole periodically.

(*d*) Bessemerization. The reduction of matte is carried in a pear-shaped furnace known as Bessemer converter (See Fig. 7.12) which is lined inside with magnesium oxide. The Bessemer converter is provided with tuyeres to introduce hot blast of air along with sand as flux. The converter is mounted on a horizontal axle. The converter can be tilted up and down about the axle for charging and discharging.

The molten matte is introduced in the Bessemer converter and heated by blowing hot blast of air and sand through tuyeres. The following reactions take place in the converter:

(*i*) Air oxidizes iron sulphide to iron oxide and SO_2

$$2FeS + 3O_2 \longrightarrow 2FeO + 2SO_2 \uparrow$$

(*ii*) Sand reacts with iron oxide to form iron silicate (slag)

$$FeO + SiO_2 \longrightarrow FeSiO_3 \text{ (slag)}$$

(*iii*) After the iron has been removed hot blast of air converts a part of cuprous sulphide to cuprous oxide.

$$2CuS_2 + 3O_2 \longrightarrow 2Cu_2O + 2SO_2 \uparrow$$

(*iv*) The cuprous oxide so formed reacts immediately with remaining cuprous sulphide to form metallic copper.

$$2Cu_2O + Cu_2S \longrightarrow 6Cu + SO_2 \uparrow$$

Since the reaction is exothermic, the heat liberated keeps crude copper in the molten state.

The molten mass is transferred from the converter to sand moulds and is allowed to cool.

The dissolved sulphur dioxide escapes out and blisters are formed on the surface of the solid copper known as *blister copper.* It is about 99% pure.

(*e*) Purification of Blister copper. (1) Poling process. The blister copper is melted on the hearth of a reverberatory furnace. A hot blast of air is passed through the molten mass. The following changes occur :

(*i*) S, As and Sb are removed as their volatile oxides.

(*ii*) The impurities of basic metal such as Fe, Pb, Zn, Bi, Sn either volatilize off or form a scum or slag which is skimmed off.

(*iii*) Some cuprous oxide formed is reduced to metallic copper by stirring the molten metal with green poles of wood.

A metal of about 99.5% purity can be obtained by this process.

2. Electrolytic Refining. Copper of high purity (99.95–99.99%) can be obtained by electrolytic refining of blister copper.

Electrolysis is carried out in electrolytic tank (Fig.7.21) containing copper sulphate solution (15%) acidified with sulphuric acid (5%) using an anode of blister copper and cathode of pure copper.

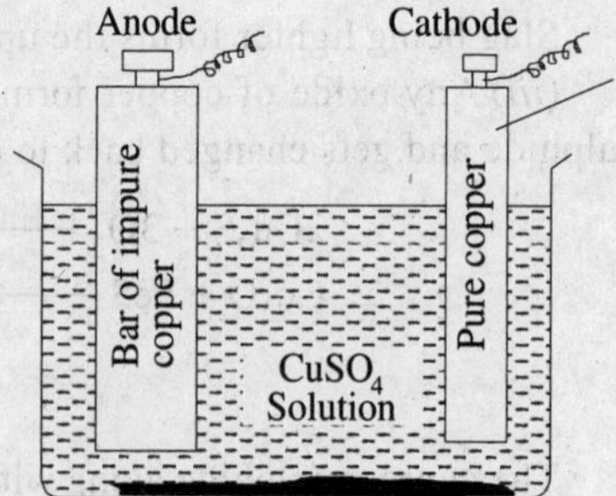

Fig. 7.21. Electrolytic refining of copper.

On passing electric current through the tank copper and other active metals such as iron, zinc, bismuth are oxidized to their corresponding cations at anode and pass into the solution. The less reactive metals such as silver and gold are not oxidized and fall to the bottom of the electrolytic cell just below the anode and are known as *anode mud.* Copper is deposited at the cathode leaving other metal ions in the solution.

$$CuSO_4 \text{ (aq)} \rightleftharpoons Cu^{2+} + SO_4^{2-}$$

At cathode:

$$Cu^{2+} + 2e^- \longrightarrow Cu$$
$$Cu - 2e^- \longrightarrow Cu^{2+}$$

Q. 22. Name two places in India where copper is found.

Ans. (1) Khetri in Rajasthan

(2) Singhbhum in Bihar.

EXTRACTION OF NICKEL

Q. 23. What are different ores of nickel? How is nickel extracted from its ores?

(*Meerut, 2000; Garhwal, 2000; Kanpur, 2000; Kurukshetra, 2001; Calcutta, 2007*)

Ans. The important ores of nickel are:

(*i*) Pentlandite (Ni, Fe) S, a copper, iron and nickel sulphide containing about 22% of nickel.

(*ii*) Garnierite Ni, $MgSiO_3 \cdot H_2O$, a double silicate of nickel and magnesium containing about 8% of nickel.

(*iii*) Kupter nickel or niccolite (False copper) NiAs

(*iv*) Nickel glance, NiAsS.

Extraction of Nickel from Sulphide Ore

Extraction of nickel from sulphide ore involves the following steps:

(*i*) Crushing and concentration. The ore is first crushed in big jaw crushers and then finely powdered in ball mills. It is then concentrated by froth floatation process using water and pine oil.

(*ii*) Roasting. To reduce sulphur content of the ore, it is then roasted in air when most of iron sulphide is oxidized to ferrous oxide and sulphur dioxide.

$$2FeS_2 + 5O_2 \longrightarrow 2FeO + 4SO_2$$

$$S + O_2 \longrightarrow SO_2$$

Copper and nickel sulphides present in the ore remain almost unaffected.

(*iii*) Smelting. The roasted ore is then introduced into a small blast furnace and smelted with coke, quartz (SiO_2) and some limestone. Most of the iron sulphide which escaped oxidation during roasting is now converted in ferrous oxide in preference of nickel and copper. The ferrous oxide formed during this and previous operations combines with silica to form ferrous silicate, which forms a fusible slag on the top of molten sulphides of Cu and Ni. The fusible slag is removed.

$$2FeS + 3O_2 \longrightarrow 2FeO + 2SO_2$$

$$FeO + SiO_2 \longrightarrow FeSiO_3 \text{ (slag)}$$

Lime removes silica as calcium silicate

$$CaO + SiO_2 \longrightarrow CaSiO_3 \text{ (slag)}$$

The heavier mass is known as **matte** and is a mixture of Ni and Cu sulphides.

(*iv*) Bessemerisation. The molten matte obtained from the above process contains about 25% of Ni and Cu along with little iron sulphide which is removed by Bessemerisation. For this purpose, the molten matte is transferred to a Bessemer converter, a pear-shaped furnace made of steel plates and lined inside with basic refractory material like that (CaO) or magnesium oxide. A blast of sand and air is blown in, when the sulphur is oxidized to sulphur dioxide and the iron to ferrous oxide the latter combining with silica to form ferrous silicate slag which is skimmed off. The Bessemerised matte obtained at this stage consists mostly of sulphides of copper and nickel, the approximate composition being about 55% nickel, 30% copper, 14% sulphur in combination with traces of iron (less than 1%).

(v) Conversion of Bessemerised matte into pure nickel. The removal of copper and subsequent conversion of the matte to pure nickel is known as refining and can be carried out by any of the following four processes:

(1) Orford Process. In this process, the separation of copper and nickel is achieved by taking advantage of the fact that copper sulphide is more soluble in sodium sulphide than nickel sulphide. The matte obtained from Bessemer converter is heated with mixture of coke and crude sodium sulphate when the latter gets reduced to sodium sulphide.

$$Na_2SO_4 + 4C \longrightarrow Na_2S + 4CO$$

The copper sulphide dissolves in fused sodium sulphide and the solution being lighter floats while the heavier nickel sulphide sinks to the bottom and is thus separated. It is then washed with water to remove any sodium sulphide present and then roasted again to convert nickel sulphide into nickel oxide. The metal is further obtained on reducing the oxide by heating with coke.

$$2NiS + SO_2 \longrightarrow 2NiO + 2SO_2$$

$$NiO + C \longrightarrow Ni + CO$$

(2) Mond's Process. The process involves the following steps:

(*i*) *Roasting.* The matte obtained from Bessemer's converter is roasted again whereby the sulphides of Ca and Ni are converted into their respective oxides.

$$2NiS + 3O_2 \longrightarrow 2NiO + 2SO_2$$

$$2CuS + 3O_2 \longrightarrow 2CuO + 2SO_2$$

(*ii*) *Leaching.* The oxide mixture is then treated with 15% sulphuric acid at 353 K which dissolves most of copper oxide leaving the nickel oxide practically undissolved. The solution is then evaporated to obtain crystals of copper sulphate which is a by-product of the process.

$$CuO + H_2SO_4 \longrightarrow CuSO_4 + H_2O$$

(*iii*) *Reduction.* The residue consisting mostly of nickel oxide NiO is heated to 573-623 K in a current of water gas ($CO + H_2$). Nickel oxide is reduced to the metallic state by water gas due to preferential reduction of nickel oxide.

$$NiO + H_2 \xrightarrow{573-623\ K} Ni + H_2O$$

$$NiO + CO \longrightarrow Ni + CO_2$$

(*iv*) *Volatilization.* The finally divided metal is kept at a temperature of 323-333 K and a current of CO is passed over it. The nickel forms a volatile nickel carbonyl (a liquid having b.p. 316.2 K).

$$Ni + 4CO \xrightarrow{323\ K} Ni + (CO)_4$$

(*v*) *Decomposition.* The gas charged with nickel carbonyl vapours is led over granulated nickel maintained at 423–453 K. The carbonyl coming in contact with heated granules undergoes decomposition depositing its nickel or granules which grow in size and are continuously removed.

$$Ni(CO)_4 \xrightarrow{453\ K} Ni + 4CO$$

The carbon monoxide released is used again for further carbonyl formation. Nickel obtained by this process is about 99.89% pure.

(3) Electrolytic Method. The Bessemerised matte is first roasted and then treated with dil. H_2SO_4 at 353 K to remove copper sulphate as in the Mond's process. The residue consisting mostly of nickel compounds is made the anode in a solution of nickel ammonium sulphate and is electrolysed with iron plate as cathode. On passing electric current nickel is deposited on the cathode and the impurities fall as anode mud.

(4) German Process. In this process the nickel-copper sulphide matte is not oxidized but heated with CO at 523 K under 200 atm pressure for about 4 days. About 85% of the nickel present is converted into carbonyl which is liquefied and then purified by fractional distillation. The purified liquid carbonyl (b.p. 316.2 K) is heated strongly in an empty tower at 493-513 K when nickel is obtained as a powder.

Extraction of nickel from silicate ore

The following procedure is employed to extract nickel from Garnierite, a double silicate of nickel and magnesium.

The ore is smelted in a blast furnace together with gypsum, limestone and coke. The gypsum gets reduced to calcium sulphide which then reacts with nickel silicate to give nickel sulphide

$$CaSO_4 + 4C \longrightarrow CaS + 4CO$$

$$NiSiO_3 + CaS \longrightarrow NiS + CaSiO_3$$

The fused product of the blast furnace separates into two layers, the upper layer consisting of slags of calcium and magnesium silicates and the lower layer of nickel sulphide matte. The slag is run off and the matte now is roasted to convert nickel sulphide into oxide. This is then smelted with coke in a crucible furnace when nickel is obtained in an almost pure condition.

$$2NiS + 3O_2 \longrightarrow 2NiO + 2SO_2$$

$$NiO + C \longrightarrow Ni + CO$$

Q.24. How is nickel obtained from Mond's Process? *(Kurukshetra, 2001)*

Ans. See Q. 23.

Q. 25. Give the composition of the following ores: (*i*) Carnallite (*ii*) Rock Phosphate (*iii*) Galena.

Ans. Carnallite is $KCl \cdot MgCl_2 \cdot 6H_2O$

Rock phosphate is $Ca_3(PO_4)_2$

Galena is PbS.

Q. 26. Write chemical formula for the following:

(*i*) Nickel glance *(Garhwal, 2000)*

(*ii*) Copper pyrites

(*iii*) Calamine *(Kurukshetra, 2001)*

Ans. Nickel glance is NiAsS

Copper pyrites is $CuFeS_2$

Calamine is $ZnCO_3$.

METALLURGY OF IRON

Q. 27. Describe various ores of iron. *(Kolhapur, 2000; Kanpur, 2000)*

Ans. The main ores of iron are :

1. Oxide Ores :

(*i*) *Magnetite, Fe_3O_4.* It is usually black in colour. It is the richest ore of iron and contains up to 70% of the metal.

(*ii*) *Haematite, Fe_2O_3.* It is red in colour.

(*iii*) *Limonite or hydrated ferric oxide, $Fe_2O_3.3H_2O$.* It has yellow, brown or red colour.

2. Carbonate Ores :

Siderite or Sepathic iron, $FeCO_3$. It is also called clay-iron stone due to the presence of excess of clay in it.

3. Sulphide Ores :

(*i*) *Iron pyrites, FeS_2* It is mainly used for the manufacture of sulphur dioxide. It is not used for extraction of iron.

(*ii*) *Chalcopyrites, $CuFeS_2$.* It is used for extraction of copper.

Huge deposits of red haematite are available in India in Mayurbhanj (Orissa), Singhbhum (Bihar), Madhya Pradesh and Mysore. India possesses about one-fourth of the total world reserves of iron ore.

Q. 28. Discuss various steps in the extraction of iron from its ore. *(Kolhapur, 2000)*

Ans. Iron is extracted from its oxide ores especially from the magnetite, haematite and limonite ores. The extraction involves the following steps:

(*i*) *Concentration of ore.* The concentration of ore is done by gravity process. The ore is crushed to small pieces and washed with water to remove silicious impurities. The washed ore is then subjected to electromagnetic separation.

(*ii*) *Calcination and roasting.* The concentrated ore is heating in excess of air. Following reactions take place:

(*a*) Moisture and carbon dioxide are removed.

$$2Fe_2O_3.3H_2O \rightarrow 2Fe_2O_3 + 3H_2O$$

$$FeCO_3 \rightarrow FeO + CO_2$$

(*b*) Sulphur, arsenic, etc., are oxidised to their oxides and are removed as volatile gases.

$$S + O_2 \rightarrow SO_2$$

$$4As + 3O_2 \rightarrow 2As_2O_3$$

(*c*) Ferrous oxide is oxidised to ferric oxide.

$$2FeO + \frac{1}{2}O_2 \rightarrow Fe_2O_3$$

(*d*) The entire mass becomes porous.

(*iii*) *Smelting.* The calcined ore is mixed with limestone ($CaCO_3$) and coke in the ratio of 8 : 1 : 4 and introduced in a blast furnace for smelting.

The charge is introduced in the furnace by lowering the cup and cone arrangement and at the same time the furnace is lit and a blast of hot air is sent upwards through the tuyeres. The temperature varies from 1870 to 520 K in the furnace from bottom to top. On the basis of variation in temperature, there are four zones where different chemical changes occur. The four zones are :

1. Combustion Zone. This is the lowest part of the furnace above hearth where the temperature is about 1770-1870 K. In this zone carbon burns in presence of hot air producing carbon dioxide and a lot of heat.

$$C + O_2 \rightarrow CO_2 + \text{Heat}$$

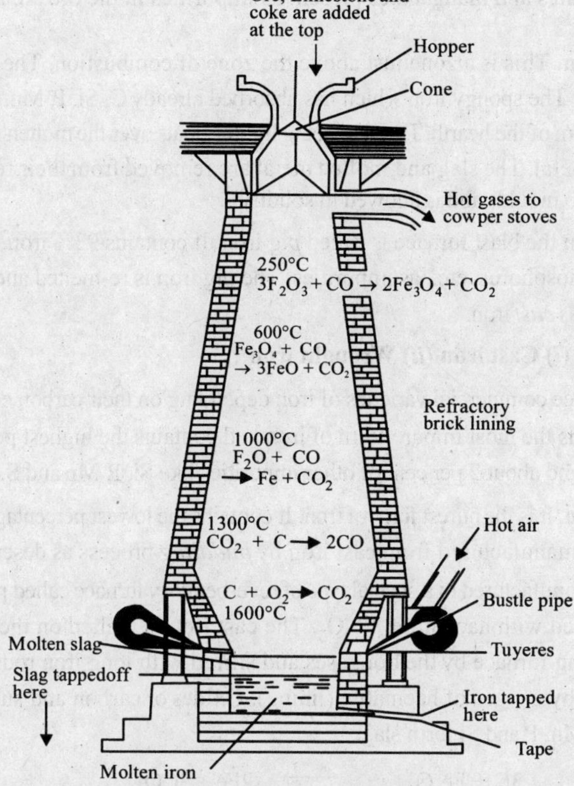

Fig. 7.22. Blast furnace.

Carbon dioxide rises upwards and meets with red hot coke. It is reduced to carbon monoxide.

$$CO_2 + C \xrightarrow{1770K} 2CO$$

2.Reduction Zone. This is the uppermost part of the furnace. The temperature varies from 520 to 970 K. The oxide ore is reduced to iron in this zone.The reduction takes place through the following stages :

$$3Fe_2O_3 + CO \xrightarrow{520\text{–}670K} 2Fe_3O_4 + CO_2$$

$$Fe_3O_4 + CO \xrightarrow{770\text{–}870K} 3FeO + CO_2$$

$$FeO + CO \xrightarrow{970K} Fe + CO_2$$

Iron formed is called *spongy* iron.

3. Slag Formation Zone. This is the central zone where the temperature varies from 1070-1270 K. The limestone present in the charge decomposes into calcium oxide and carbon dioxide.

$$CaCO_3 \xrightarrow{1270K} CaO + CO_2$$

The calcium oxide acts as a flux as it combines with silica present as an impurity (gangue) to form a fusible slag of $CaSiO_3$.

$$CaO + SiO_2 \longrightarrow CaSiO_3$$

Silicates, phosphates and manganates present as impurities in the ore are reduced to Si, P and Mn, respectively.

4. Zone of Fusion. This is a zone just above the zone of combustion. The temperature ranges between 1470-1770 K. The spongy iron which has absorbed already C, Si, P, Mn, etc. melts at 1570 K and collects at the bottom of the hearth. The slag being lighter floats over the molten iron and prevents the oxidation of molten metal. The slag and molten metal are removed from their respective holes. The molten metal is run into moulds and is allowed to solidify.

Iron obtained from the blast furnace is called *pig* iron. It contains 93% iron, 5% carbon and rest silicon, manganese, phosphorus, etc., as impurities. The pig iron is re-melted and cast or poured into mould. This is known as *cast* iron.

Q. 29. What are (*i*) Cast iron (*ii*) Wrought iron ?

Ans. There are three commercial varieties of iron depending on their carbon content:

(*i*) Cast iron. It is the most impure form of iron and contains the highest percentage of carbon, from 2.5 to 5 per cent and about 2 per cent of other impurities like Si, P, Mn and S.

(*ii*) Wrought iron. It is the purest form of iron. It contains the lowest percentage of carbon from 0.1 to 0.25 per cent. It is manufactured from cast iron by *puddling* process as described below.

Wrought iron is manufactured in a special type of reverberatory furnace called puddling furnace, the hearth of which is lined with haematite, Fe_2O_3. The cast iron is melted on the hearth of a special furnace called puddling furnace by the hot gases and stirred with long iron rods. The impurities of cast iron are oxidised by oxygen of haematite (lining). Oxides of carbon and sulphur being volatile escape while those of Mn, P and Si form slag.

$$3C + Fe_2O_3 \longrightarrow 2Fe + 3CO$$

$$3Si + 2Fe_2O_3 \longrightarrow 3SiO_2 + 4Fe$$

$$3Mn + Fe_2O_3 \longrightarrow 3MnO + 2Fe$$

$$MnO + SiO_2 \longrightarrow MnSiO_3 \ (slag)$$

$$6P + 5Fe_2O_3 \longrightarrow 3P_2O_5 + 10Fe$$

$$P_2O_5 + Fe_2O_3 \longrightarrow 2FePO_4 \ (slag)$$

With the removal of impurities, the melting point of the metal rises and it becomes a semi-solid mass. The semi-solid mass is taken out and is beaten under steam hammers to remove as much of slag as possible. This produces almost pure iron known as *wrought* iron.

Wrought iron is extremely tough, highly malleable and ductile. It softens at about 1270 K and then it can be forged and welded. On account of the presence of very small percentage of slag, it has fibrous structure and, thus, can withstand high stresses. Wrought iron is resistant towards corrosion.

It is used to make chains, nails, hooks, bolts, agricultural implements, electromagnets, bars, wires, etc.

Q. 30. What is steel ? What are its various types ? (*Vidyasagar, 2007*)

Ans. Steel is the most important commercial variety of iron. The percentage of carbon in this form of iron is intermediate between that of cast iron and wrought iron, *i.e.*, 0.25 to 2 per cent. Different varieties of steel are discussed below.

(*a*) Mild steel. They contain low percentage of carbon. Such steels show the properties of wrought iron along with elasticity and hardness.

(*b*) **Hard steels**. They contain high percentage of carbon. They are hard and brittle.

(*c*) **Special steels or alloy steels.** Steel mixed with small amount of nickel, cobalt, chromium, tungsten, molybdenum, manganese, etc., acquires special properties. Such products are called *special steels* or *alloy steels.* Some important alloy steels are described in Table 7.1.

Table 7.1. Alloy steels.

Name of the alloy steel	Metal added	Properties	Uses
1. Invar	36% Ni	Coefficient of expansion; is very small	Measuring tapes, pendulums
2 Chrome-vanadium steel	1%Cr : 0.15%V	High tensile strength	Springs, shafts, axles.
3. Manganese steel	12-15% Mn	Hard and tough	Rock crushing machinery, safes, helmets.
4. Stainless steel	11.5% Cr; 2%Ni	Resists corrosion	Common articles
5. Tungsten steel	14-20%W;3-8%Cr	Very hard	High speed tools

Q. 31. Describe Bessemer's process for the manufacture of steel.

Ans. The process is based on the fact that impurities of pig iron are completely oxidised in presence of hot air blast. This is then mixed with a known amount of spiegeleisen, an alloy of iron, manganese and carbon to obtain steel.

The process is carried out in Bessemer converter lined with silica bricks (Fig. 7.23). The molten pig iron is introduced in the converter and a blast of hot air is blown through it from the bottom and keeping the mouth of the converter vertically upwards. Silica and manganese present in pig iron are first oxidised and then combine to form slag

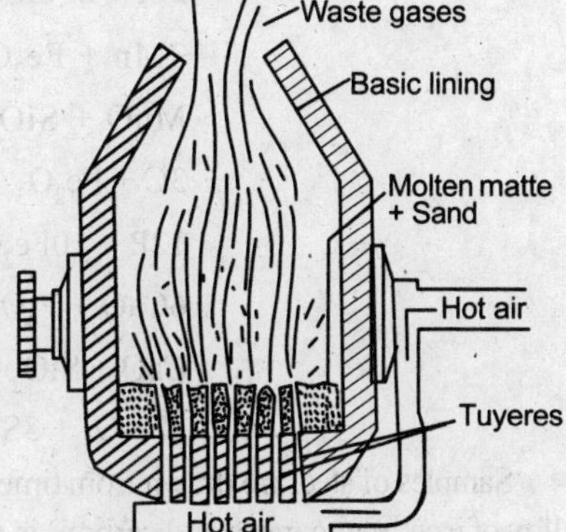

Fig.7.23. Bessemer converter.

$$Si + O_2 \rightarrow SiO_2$$

$$2Mn + O_2 \rightarrow 2MnO$$

$$2MnO + SiO_2 \rightarrow MnSiO_3 + MnSiO_3$$

In the end, carbon is oxidised to carbon monoxide which burns with blue flame at the mouth of the converter.

When whole of the carbon is oxidised, the blue flame suddenly dies out. The air supply is stopped for a while and the requisite amount of *spiegeleisen* is added. The blast is continued just for a moment to ensure complete mixing. The resulting product is the manganese steel.

When cast iron or pig iron contains phosphorus as an impurity, a basic lining of CaO or MgO is used in the Bessemer converter. Phosphorus is oxidised to P_4O_{10} which combines with CaO to form calcium phosphate as slag.

$$4P + 5O_2 \rightarrow P_4O_{10}$$

$$6CaO + P_4O_{10} \rightarrow 2Ca_3(PO_4)_2$$

This slag is used as a fertilizer and known as *Thomas slag.*

Q. 32. Describe Open Hearth Process for manufacture of steel.

Ans. In this process the furnace consists of an open hearth. The hearth is lined with silica or calcined dolomite (CaO.MgO) depending upon the nature of impurities present in pig or cast iron. Silica lining is used if the impurities are manganese, silicon, etc., and calcined dolomite lining is used if impurity of phosphorus is present. A high temperature of about 1770 K is generated by burning producer gas which works on regenerative system of heat economy. (Fig. 7.24)

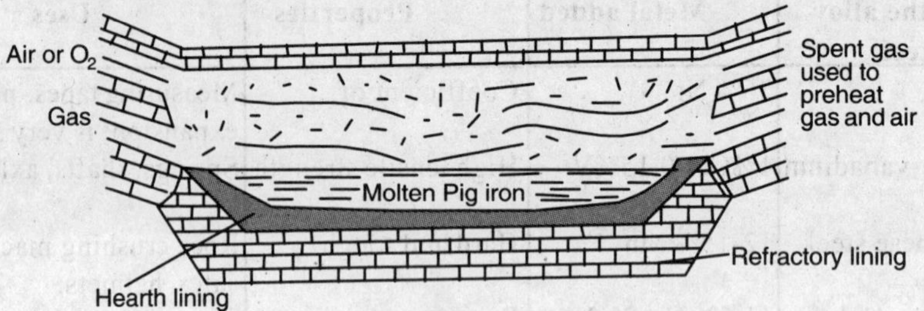

Fig.7.24. Open hearth furnace.

The charge consists of pig or cast iron, iron ore (haematite) and lime. The charge is heated on the hearth of the furnace. The impurities are oxidised by iron ore.

$$3Si + 2Fe_2O_3 \rightarrow 4Fe + 3SiO_2$$

$$3Mn + Fe_2O_3 \rightarrow 2Fe + 3MnO$$

$$MnO + SiO_2 \rightarrow MnSiO_3 \text{ (slag)}$$

$$3C + Fe_2O_3 \rightarrow 2Fe + 3CO$$

$$12P + 10Fe_2O_3 \rightarrow 3P_4O_{10} + 20Fe$$

$$6CaO + P_4O_{10} \rightarrow 2Ca_3(PO_4)_2 \text{ (slag)}$$

$$CaO + SiO_2 \rightarrow CaSiO_3 \text{ (slag)}$$

$$2Fe_2O_3 + 3S \rightarrow 3SO_2 + 4Fe$$

Samples of steel are drawn from time to time and tested for carbon content. Finally spiegeleisen (an alloy of iron, manganese and carbon) is added to the molten mass to obtain desired steel. The process takes about 8 to 10 hours for completion. The process takes longer time than Bessemer process but it has distinct advantages over the Bessemer's process.

Q. 33. Write notes on (*a*) annealing (*b*) quenching (*c*) tempering (*d*) case hardening (*e*) nitriding. (*Bangalore, 2002*)

Ans. (*a*) Annealing. It is a process of heating steel to bright redness and then cooling it very slowly. This treatment makes the steel soft and ductile. This type of steel is used in fabrication process.

(*b*) Quenching. It is a process of heating steel to bright redness and then cooling it suddenly by plunging it in water or oil. Such a steel is extremely hard and brittle. It has very low elasticity.

(*c*) Tempering. It is a process of heating the quenched steel to a temperature much below redness and then cooling it slowly. Such steel is neither so hard nor so brittle. In the process of tempering, a thin film of oxide is formed on the surface of steel. The colour of the oxide film depends on the temperature at which the quenched steel is heated.

Temperature range	Colour of the oxide film
470-500 K	Yellow
495-540 K	Brown
570 K	Blue

(d) Case hardenting. The process of producing a thin coating of hardened steel on the surface of the mild steel is called case hardening. This is done by heating the mild steel with charcoal and then plunging into oil. This produces a thin coating of hardened steel on the surface. Such a steel is resistant to wear and tear.

(e) Nitriding. The process of producing a hard coating of iron nitride on the surface of steel is called nitriding. Steel is heated in the atmosphere of dry ammonia at 770-870 K for about 3 to 4 days when a hard coating of iron nitride is produced on the surface.

METALLURGY OF MERCURY

Q. 34. Discuss the metallurgy of mercury.

Ans. Occurrence. Mercury ores are rare. It occurs in small quantities in the native state. Cinnabar, HgS, is the most important ore of mercury.

Extraction. The extraction of mercury from cinnabar ore consists of roasting and distillation. The extraction involves the following steps:

(i) Crushing and concentration. The ore is crushed and finely powdered in ball mills and concentrated by froth floatation process.

(ii) Combined roasting and distillation. There are two principal processes available which differ only in the type of furnace used. The two furnaces used are :

(a) Idrian furnace

(b) Shaft furnace

(a) Processing using Idrian furnace. The concentrated ore is taken on the perforated arches and heated by the flame as shown in Fig. 7.25.

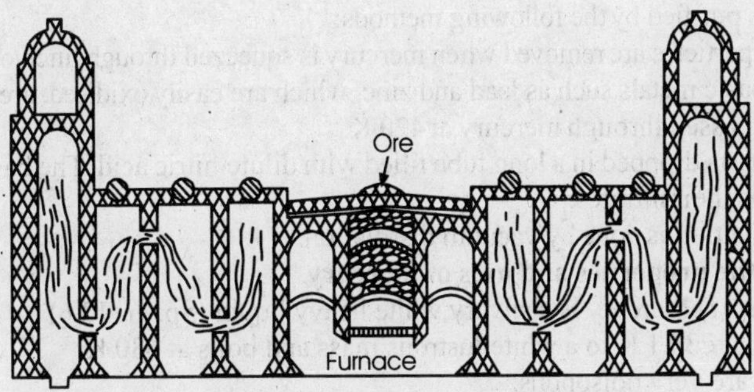

Fig. 7.25. Idrian furnace.

The mercuric sulphide undergoes oxidation with evolution of sulphur dioxide. The mercuric oxide thus formed decomposes at 570 K to give mercury vapours

$$2HgS + 3O_2 \rightarrow 2HgO + 2SO_2$$

$$2HgO \rightarrow 2Hg + O_2$$

The vapours of mercury are condensed in a series of condensers on either side of the furnace. Most of the metal collects in the first three condensers and in the last condenser, the condensation is brought about by water spray or wet straw dust.

(b) Processing using shaft furnace. The concentrated ore, mixed with charcoal, is added to the furnace by a cup cone arrangement (Fig. 7.26). There are three fireplaces communicating with the ore chamber on alternate sides of the hexagon. The furnace is heated to a red heat. Air is admitted.

Gases with mercury vapours pass through iron pipes into water cooled condensers. There is an opening in the bottom of the furnace for the removal of spent ore.

The ore is sometimes mixed with lime before roasting. This helps in the oxidation of mercuric sulphide.

$$HgS + CaO \rightarrow HgO + CaS$$
$$2HgO \rightarrow 2Hg + O_2$$

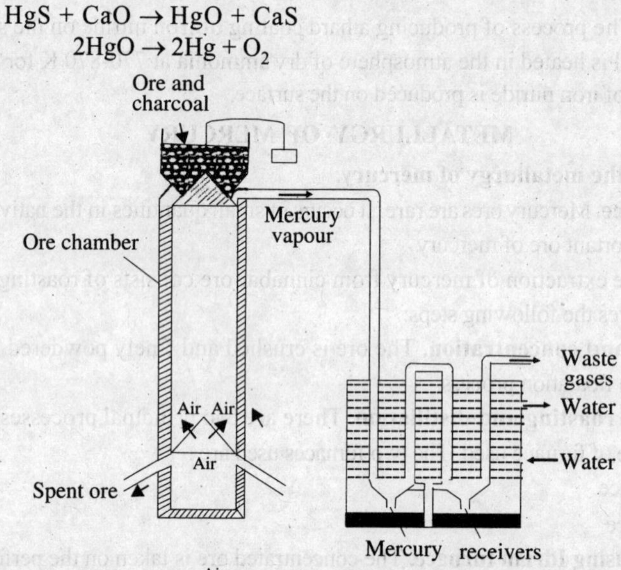

Fig. 7.26. Shaft furnace.

***(iii)* Purification.** Commercial mercury is usually contaminated with elements such as zinc, copper, lead, etc. It is purified by the following methods:

(*a*) Suspended particles are removed when mercury is squeezed through linen or chamois leather.

(*b*) The more basic metals such as lead and zinc, which are easily oxidised, are removed as *scum* when dust-free air is passed through mercury at 420 K.

(*c*) The mercury is dropped in a long tube filled with dilute nitric acid. The base metals dissolve in dilute nitric acid as their nitrates.

(*d*) Final purification is done by vacuum distillation.

Q. 35. Discuss the properties and uses of mercury.

Ans. Properties. (*i*) Mercury is a silvery white heavy liquid (Sp. gr. 13.6)

(*ii*) It solidifies at 234.1 K to a white lustrous mass and boils at 630 K.

(*iii*) Its vapours are very poisonous.

(*iv*) It has a high surface tension and forms spherical drops.

(*v*) It is not affected by air or oxygen at ordinary temperature.

(*vi*) When heated at 610 K in air, it slowly combines with oxygen forming a red scum of HgO on the surface.

(*vii*) When rubbed with sulphur, it forms mercuric sulphide, HgS.

(*viii*) It readily combines with chlorine, slowly with bromine and iodine.

(*ix*) It is not attacked by water and alkalies. Hydrochloric acid, dilute sulphuric acid and very dilute nitric acid have no action on mercury.

(*x*) Hot concentrated sulphuric acid, dilute and concentrated nitric acid react as under :

With conc. H_2SO_4 :

$$H_2SO_4 \rightarrow H_2O + SO_2 + O$$
$$Hg + H_2SO_4 + O \rightarrow HgSO_4 + H_2O$$
$$\overline{Hg + 2H_2SO_4 \rightarrow HgSO_4 + SO_2 + 2H_2O}$$

With dilute HNO_3 :

$$2HNO_3 \rightarrow H_2O + 2NO + 3\,[O]$$
$$[2Hg + O \rightarrow Hg_2O] \times 3$$
$$[Hg_2O + 2HNO_3 \rightarrow Hg_2(NO_3)_2 + H_2O] \times 3$$
$$6Hg + 8HNO_3 \rightarrow 3Hg_2(NO_3)_2 + 2NO + 4H_2O$$

With conc. HNO_3 :

$$2HNO_3 \rightarrow H_2O + 2NO_2 + O$$
$$Hg + 2HNO_3 + O \rightarrow Hg(NO_3)_2 + H_2O$$
$$Hg + 4HNO_3 \rightarrow Hg(NO_3)_2 + 2NO_2 + 2H_2O$$

(*xi*) Mercury dissolves many metals and the combinations are called amalgams. Gold, silver, tin, lead, magnesium, copper, sodium, potassium, etc., when rubbed with mercury, form amalgams.

Uses. Mercury is used

(*i*) in making thermometers, barometers, and other physical apparatus.

(*ii*) in the manufacture of caustic soda by the electrolytic process.

(*iii*) in the extraction of Ag and Au.

(*iv*) in mercury vapour lamps which give out bluish light.

(*v*) in the form of amalgams as reducing agents.

EXTRACTION OF TITANIUM

Q. 36. What are the chief ores of titanium? How is titanium extracted and purified?

Ans. Chief ores of titanium are:

1. Rutile, TiO_2
2. Ilmenite, $FeTiO_3$

Extraction from both the ores is described as under:

Extraction from rutile ore, TiO_2

Since titanium has high melting point and it reacts quite readily with air, O_2, N_2, H_2 and carbon at elevated temperatures, its extraction is a difficult task. The extraction of the metal from the rutile mineral consists of the following steps :

1. Conversion of rutile into $TiCl_4$. Rutile (TiO_2) mineral is heated with carbon in a current of Cl_2.

$$\underset{\text{Rutile}}{TiO_2} + C + 2Cl_2 \longrightarrow TiCl_4 + CO_2$$

2. Reduction of $TiCl_2$ to Ti metal. $TiCl_4$ obtained as above is reduced to Ti metal by one of the following methods:

(*a*) *Kroll process.* In this process, vapours of $TiCl_4$ are passed over molten magnesium, when $TiCl_4$ is reduced to spongy Ti metal.

$$TiCl_4 + 2Mg \xrightarrow{800°C} \underset{\substack{\text{Spongy}\\\text{Titanium}}}{Ti} + 2MgCl_2$$

The spongy metal is heated in an inert atmosphere of Ar to get a compact mass, of the metal.

(*b*) $TiCl_4$ can also be reduced to Ti metal by heating it with NaH.

$$TiCl_4 + 4NaH \xrightarrow{\Delta} Ti + 4NaCl + 2H_2$$

(*c*) *I.C.I. process.* In this process $TiCl_4$ is reduced by heating with Na in an atmosphere of Ar.

$$TiCl_4 + 4Na \xrightarrow{\text{Heat}} Ti + 4NaCl$$

Extraction from ilmenite ore, $FeTiO_3$

The extraction of titanium metal from ilmenite ore consists of the following steps:

1. Conversion of $FeTiO_3$ into TiO_2. TiO_2 can be obtained from the ilmenite ore by one the following processes :

(*a*) The powdered ilmenite ore is fused with Na_2S. The product obtained consists of FeS, TiO_2 and Na_2O.

$$\underset{\text{Ilmenitie}}{2FeTiO_3} + Na_2S \xrightarrow{\text{Fuse}} \underbrace{FeS + TiO_2}_{\text{Insoluble}} + \underset{\text{soluble}}{2Na_2O}$$

When the product consisting of FeS, TiO_2 and Na_2O is treated with H_2O, FeS and TiO_2 remain insoluble and Na_2O goes into solution. The insoluble portion containing FeS and TiO_2 is treated with H_2SO_3 in presence of atmospheric O_2. FeS is converted into soluble FeS_2O_3 but TiO_2 remains unaffected.

$$FeS + H_2SO_3 + \frac{1}{2}O_2\,(\textit{From air}) \longrightarrow \underset{\text{Soluble}}{FeS_2O_3} + H_2O$$

$$TiO_2 + H_2SO_3 + \frac{1}{2}O_2\,(\textit{From air}) \longrightarrow \text{No reaction}$$

(*b*) When ilmenite ore is heated with HCl in a current of Cl_2, TiO_2 is formed.

$$\underset{\text{Ilmenite}}{2FeTiO_3} + 4HCl + Cl_2 \xrightarrow{\Delta} 2FeCl_3 + 2TiO_2 + 2H_2O$$

(*c*) A mixture of ilmenite ore and carbon is heated in a current of Cl_2 at 360°. A mass is obtained which consists of $FeCl_3$ and $TiCl_4$. When this mass is heated at 550°C, volatile $TiCl_4$ distils over. $TiCl_4$ thus obtained is hydrolysed by H_2O to get TiO_2.

$$\underset{\text{Ilmenito}}{2FeTiO_3} + 2C + 7Cl_2 \xrightarrow{350°C} \underbrace{2FeCl_3 + 2TiCl_4}_{\downarrow\ 550°C\ \to\ TiCl_4 \text{ distils over}} + 3CO_2\uparrow$$

$$TiCl_4 + 2H_2O \xrightarrow{\text{Hydrolysis}} TiO_2 + 4HCl$$

2. Reduction of TiO_2 to Ti metal. TiO_2 obtained as above is reduced to titanium metal by either of the following methods:

(*a*) TiO_2 can be reduced to free Ti metal by heating it with carbon.

$$TiO_2 + C \longrightarrow Ti + CO_2\uparrow$$

(*b*) *By alumino thermic process.* In this method TiO_2 is mixed with Al and the mixture is ignited with a suitable fuse.

$$3TiO_2 + 4Al \longrightarrow 2Al_2O_3 + 3Ti$$

Purification of titanium metal

Titanium metal obtained from rutile or ilmenite contains a number of impurities.

Impure metal can be purified by *Van Arkel process* in which impure titanium is heated with a limited amount of I_2 in an evacuated glass apparatus. TiI_4 vapours formed are allowed to diffuse on to a tungsten filament at a very high temperature. The vapours of TiI_4 decompose and the pure Ti metal gets deposited on the W-filament.

$$\text{Impure Ti metal} + 2I_2 \xrightarrow{150\square 250°} TiI_4(g) \xrightarrow[\text{filament}]{1800° \text{ Tungsten}} \text{Pure Ti metal (gets deposited on W-filament)} + 2I_2(g)\uparrow$$

EXTRACTION OF VANADIUM

Q. 37. What are the important ores of vanadium? How is it extracted from its ores?

Ans. Important ores of vanadium are:

1. **Vanadinite, $3Pb_3(VO_4)_2.PbCl_2$ or $Pb_{10}(VO_4)_6Cl_2$ or $Pb_5(VO_4)_3Cl$**
2. **Carnotite, $K_2O.2UO_3.V_2O_5.3H_2O$ or $2K(UO_2)VO_4.3H_2O$**
3. **Patronite, V_2S_5 (Vanadium sulphide).**

Extraction

The extraction of vanadium metal from its ores consists of the following steps :

1. Preparation of ammonium metavanadate, NH_4VO_3. Ammonium metavanadate can be obtained either from carnotite or vanadinite ore as shown below :

***(a)* From carnotite ore $2K(UO_2)VO_4.3H_2O$.** Carnotite ore is boiled with Na_2CO_3 solution and concentrated. On cooling, the crystals of sodium uranyl carbonate, $Na_4(UO_2)(CO_3)_2$ are obtained and sodium orthovanadate, Na_3VO_4 is left in solution. The solution containing Na_3V0_4 is treated with NH_4Cl when ammonium metavanadate, NH_4VO_3 gets precipitated, as per the reactions given below.

$$\underset{\text{Carnotite}}{2K(UO_2)VO_4} + 7Na_2CO_3 \longrightarrow \underset{\text{crystals}}{2Na_4(UO_2)(CO_3)_3} + \underset{\text{In solution}}{2Na_3VO_4} + 2K_2CO_3$$

$$Na_3VO_4 + 3NH_4C1 \longrightarrow \underset{\text{ppt}}{NH_4VO_3} + 3NaC1 + H_2O + 2NH_3\uparrow$$

***(b)* From vanadinite ore, $Pb_5(VO_4)_3Cl$.** Powdered vanadinite ore is treated with conc. HC1 when $PbCl_2$ is obtained as precipitate and a complex salt. $[V(OH)_4]C1$ goes into solution. When the solution containing $[V(OH)_4]C1$ is boiled with NH_4C1, ammonium metavanadate, NH_4VO_3 is obtained as precipitate. The reactions are given as under:

$$\underset{\text{Vanadinite}}{Pb_5(VO_4)_3Cl} + 12HC1 \longrightarrow \underset{\text{ppt}}{5PbCl_2} + \underset{\text{In solution}}{3[V(OH)_4]C1}$$

$$[V(OH)_4]Cl + NH_4Cl \longrightarrow \underset{\text{ppt}}{NH_4VO_3} + 2HC1 + H_2O$$

***(c)* From patronite, V_2S_5.** The ore is roasted in excess of air to convert it into V_2O_5.

$$\underset{\text{Patronite}}{2V_2S_5} + 15O_2 \longrightarrow \underset{\text{Vanadium pentoxide}}{2V_2O_5} + 10SO_2$$

V_2O_5 thus obtained contains a number of impurities. To remove the impurities, V_2O_5 in fused with Na_2CO_3 and extracted with water V_2O_5 is obtained as soluble sodium orthovanadate, Na_3VO_4 while the impurities remain as residue.

$$V_2O_5 + 3Na_2CO_3 \longrightarrow 2Na_3VO_4 + 3CO_2\uparrow$$

Now to the solution containing Na_3VO_4 is added NH_4Cl, NH_4VO_3 gets precipitated.

$$Na_3VO_4 + 3NH_4Cl \longrightarrow \underset{\text{ppt}}{NH_4VO_3} + 3NaCl + H_2O + 2NH_3\uparrow$$

2. Conversion of NH_4VO_3 to V_2O_5. When NH_4VO_3 obtained as above is ignited, vanadium pentoxide (V_2O_5) is obtained.

$$2NH_4VO_3 \xrightarrow{\text{Ignite}} V_2O_5 + 2NH_3 + H_2O$$

3. Reduction of V_2O_5 to vanadium metal. Vanadium metal can be obtained by the reduction of V_2O_5 by either of the following methods

(*a*) With Al (*Aluminothermic process*)

$$3V_2O_5 + 10Al \longrightarrow 5Al_2O_3 + 6V$$

(*b*) With Ca and $CaCl_2$ (used as a flux) in a steel bomb at 900-950°C.

$$V_2O_5 + 5Ca + 5CaCl_2 \longrightarrow 2V + 5(CaO.CaCl_2)$$

Purification of vanadium metal.

Vanadium metal obtained as above is impure. The impure metal is purified by *Van Arkel method* in which the impure vanadium metal is heated with a limited amount of I_2 in an evacuated glass apparatus. VI_4 vapours formed are allowed to diffuse on a tungsten filament at a very high temperature. The vapours of VI_4 decompose and the pure vanadium metal gets deposited on the filament.

$$\text{Impure V metal} + 2I_2 \longrightarrow VI_4(g) \xrightarrow{\text{W-filament}} \underset{\text{Pure metal}}{V} + 2I_2(g)\uparrow$$

EXTRACTION OF CHROMIUM

Q. 38. What are the chief ores of chromium? How is it extracted from its ores?

Ans. Chief ores of chromium are listed below.

(*i*) Chromite or chrome ore, FeO. Cr_2O_3

(*ii*) Chrome-ochre, Cr_2O_3

(*iii*) Crocosite, Pb CrO_4

Chromium is mostly extracted from chromite

Extraction of chromium metal from chromite ore, FeO.Cr_2O_3

The extraction of Cr metal from chromite ore involves the following steps:

1. Concentration of the chromite ore. The ore is first crushed and then concentrated by *Wilfley table method* (*Gravity separation*). The crushed ore is collected in the form of a heap on a sloping floor. A stream of water is allowed to flow over the heap. The lighter impurities of earthy matter are washed down the slope while the heavier impurities remain with the ore in the heap.

2. Conversion of the chromite ore into Na_2CrO_4 (Roasting of the concentrated ore). The concentrated ore is mixed with excess of Na_2CO_3 and roasted in a reverberatory furnace at 900 – 1000°C in the presence of free supply of air. The chromite ore, FeO.Cr_2O_3 is converted into soluble sodium chromate, Na_2CrO_4.

$$\underset{\text{Chromite ore}}{4(FeO.Cr_2O_3)} + 8Na_2CO_3 + 7O_2(\textit{From air}) \xrightarrow{900-1000°C} \underset{\text{Soluble}}{8Na_2CrO_4} + \underset{\text{Insoluble}}{2Fe_2O_3} + 8CO_2\uparrow$$

The roasted mass containing Na_2CrO_4 and Fe_2O_3 is extracted with water when Na_2CrO_4 goes into solution while Fe_2O_3 is left behind as insoluble residue.

3. Conversion of Na_2CrO_4 into $Na_2Cr_2O_7$. The solution containing Na_2CrO_4 is treated with a calculated quantity of H_2SO_4 when Na_2CrO_4 is converted into sodium dichromate, $Na_2Cr_2O_7$. Less soluble Na_2SO_4 is also formed.

$$2NaCrO_4 + H_2SO_4 \longrightarrow \underset{\substack{\text{More soluble} \\ \text{(orange solution)}}}{Na_2Cr_2O_7} + \underset{\text{Less soluble}}{Na_2SO_4} + H_2O$$

The solution containing $Na_2Cr_2O_7$ (more soluble) and Na_2SO_4 (less soluble) is concentrated when the less soluble Na_2SO_4 crystallises out leaving behind more soluble $Na_2Cr_2O_7$ in solution. The solution containing $Na_2Cr_2O_7$ is further concentrated to obtain crystals of $Na_2Cr_2O_7$.

4. Conversion of $Na_2Cr_2O_7$ into Cr_2O_3. $Na_2Cr_2O_7$ obtained above can be converted into Cr_2O_3 by either of the following methods :

(*a*) $Na_2Cr_2O_7$ is heated with carbon to get sodium chromite, $NaCrO_2$ which, on treatment with H_2O, gives the precipitate of chromic oxide, Cr_2O_3.

$$Na_2Cr_2O_7 + 3C \xrightarrow{\Delta} \underset{\text{Sod. chromite}}{2NaCrO_2} + 3\,CO\uparrow$$

$$\underset{\text{Sod. chromite}}{2NaCrO_2} + H_2O \longrightarrow \underset{\text{ppt}}{Cr_2O_3} + 2NaOH$$

(*a*) $Na_2Cr_2O_7$ is heated with NH_4Cl when the precipitate of Cr_2O_3 is obtained.

$$Na_2Cr_2O_7 + 2NH_4Cl \xrightarrow{\Delta} \underset{\text{ppt}}{Cr_2O_3} + 2NaCl + 4H_2O + N_2\uparrow$$

5. Reduction of Cr_2O_3 to chromium metal. Cr_2O_3 obtained above is reduced to Cr-metal by one of the following methods.

(*a*) By passing H_2 over heated Cr_2O_3 *(Industrial preparation)*

$$Cr_2O_3 + 3H_2 \rightleftharpoons 2Cr + 3\,H_2O$$

(*b*) By heating Cr_2O_3 with Al-powder *(aluminothermite process)*

$$Cr_2O_3 + 2Al \longrightarrow Al_2O_3 + 2Cr; \quad \Delta H = -\ 468.6\ kJ.$$

(*c*) By heating Cr_2O_3 with a calculated quantity of Si in the presence of CaO which forms a slag of $CaSiO_3$ with SiO_2.

$$2Cr_2O_3 + 3Si \longrightarrow 4Cr + 3SiO_2$$

$$3SiO_2 \xrightarrow{+\,3\,CaO} \underset{\text{Slag}}{3\,CaSiO_3}$$

As the reaction is reversible, to obtain chromium in the optimal amount, water is continuously removed by absorption by Na metal.

EXTRACTION OF MOLYBDENUM

Q. 39. What are the principal ores of Molybdenum? Describe its extraction.

Ans. Principal ore of molybdenum are:

(*i*) Molybdenite, MoS_2

(*ii*) Wulfenite, Pb MoO_4

Molybdenum is mostly extracted from molybdenite.

Extraction of molybdenum from molybdenite ore, MoS_2

The extraction of molybdenum from molybdenite ore involves the following steps :

1. Concentration of the ore. Molybdenite ore is crushed in jaw-crushers and powdered in ball mills. Powdered ore is then concentrated by froth flotation process.

2. Conversion of molybdenite ore, MoS_2 into MoO_3: Roasting. Concentrated molybdenite ore is roasted in a reverberatory furnace in a free supply of air to get MoO_3.

$$2MoS_2 + 7O_2 \xrightarrow{\text{Roasting}} 2MoO_3(\text{Impure}) + 4SO_2$$

The crude MoO_3 obtained above is extracted with hot conc. NH_4OH to obtain the crystals of ammonium molybdate, $(NH_4)_2MoO_4$.

$$MoO_3\ (\textit{Impure}) + 2NH_4OH \longrightarrow (NH_4)_2\ MoO_4 + H_2O$$

Crystals of $(NH_4)_2MoO_4$, on heating, give pure MoO_3.

$$(NH_4)_2MoO_4 \xrightarrow{\Delta} M_oO_3 + 2NH_3 + H_2O$$

3. Reduction of MoO_3 into Mo metal. MoO_3 obtained above is reduced to Mo metal either by H_2, carbon or Al *(Aluminothermic process)*.

$$MoO_3 + 3H_2 \longrightarrow Mo + 3H_2O$$
$$MoO_3 + 3C \longrightarrow Mo + 3CO$$
$$MoO_3 + 2Al \longrightarrow Mo + Al_2O_3$$

Purification of Mo metal.

Mo obtained above is contaminated with impurities. To remove impurities, crude Mo metal is heated with CO at 200°C under pressure of about 200 atmospheres when $Mo(CO)_6$ is formed.

$$Mo\ (\textit{Impure}) + 6CO \longrightarrow Mo(CO)_6$$

$Mo(CO)_6$, on being heated at 360 – 410°C, gets decomposed to form pure Mo metal.

$$Mo(CO)_4 \xrightarrow{\Delta} Mo\,(\textit{pure}) + 6CO \uparrow$$

EXTRACTION OF TUNGSTEN

Q. 40. What are the important ores of tungsten? Discuss the extraction of tungsten from its ores.

Ans. Important ores of tungsten are:

(*i*) Wolframite, (Fe, Mn) WO_4

(*ii*) Scheelite, $CaWO_4$

(*iii*) Hubnerite, $MnWO_3$

Tungsten is mostly extracted from wolframite

Extraction of tungsten from wolframite ore, (Fe, Mn)WO_4

The extraction of tungsten metal from wolframite involves the following steps :

1. Concentration of the ore. The powdered wolframite ore is concentrated *by* electromagnetic separation method to remove the non-magnetic impurities.

2. Conversion of wolframite ore into WO_3. The concentrated wolframite ore is fused with Na_2CO_3 in air. This operation gives a fused mass consisting of sodium tungstate, Na_2WO_4 (soluble) and a mixture of Fe_2O_3 and MnO_3 (insoluble).

$$\underbrace{4[(Fe, Mn)\,WO_4]\,\text{or}\,(4FeWO_4 + 4MnWO_4)}_{\text{Wolframite ore}} + 8Na_2CO_3 + 3O_2\,(\text{from air})$$

$$\xrightarrow{\text{Fusion}} 8Na_2WO_4 + \underbrace{2Fe_2O_3 + 4MNO_2}_{\text{Isoluble}} + 8CO_2 \uparrow$$

Now the fused mass is leached with water when Na_2WO_4 goes into solution and the mixture of Fe_2O_3 and MnO_2 is left as residue. The solution containing Na_2WO_4, on being acidified with HCl, gives a precipitate of tungstic acid, H_2WO_4 which on ignition gives WO_3.

$$Na_2WO_4 + 2HCl \longrightarrow \underset{\text{ppt}}{H_2WO_4 \downarrow} + 2NaCl$$

$$H_2WO_4 \xrightarrow{\text{Ignition}} WO_3 + H_2O$$

3. Reduction of WO_3 into W metal. WO_3 obtained above is reduced to W metal by one of the following methods.

$$WO_3 + 3H_2 \xrightarrow{1200°C} W + 3H_2O\,; \qquad WO_3 + 3C \xrightarrow{\text{White heat}} W + 3CO$$

$$WO_3 + 3Al \longrightarrow Al_2O_3 + W$$

EXTRACTION OF COBALT

Q. 41. What are the chief ores of cobalt? Describe the extraction of cobalt from its ores.

Ans. Chief ores of cobalt are:

(*i*) Smaltite or speiss cobalt, (Ni, Co, Fe) As_2

(*ii*) Cobaltite or cobalt glance, (Co, Fe, As)S

(*iii*) Crythrite or cobalt bloom, $Co_3(AsO_4)_2.8H_2O$

(*iv*) Linnacite, $(Fe, Co, Ni)_3S_4$

Cobalt is mostly extracted from cobaltite.

Extraction from ccbaitite ore, (Co, Fe, As)S

Due to the presence of several other metals in the ores of cobalt, the extraction of cobalt is a very complex process. The extraction of cobalt from cobaltite ore, (Co, Fe, As)S involves the following steps:

1. Concentration. The ore is crushed and finely powdered and then concentrated by the Froth-Flotation process.

2. Roasting. The concentrated ore contains Fe, Co, Ni, Ag, As, Cu and S as the chief constituents. It is roasted in a free supply of air which removes S as SO_2 and As as As_2O_3 and FeS is oxidised to FeO.

$$S + O_2 \longrightarrow SO_2$$

$$4\,As + 3O_2 \longrightarrow 2As_2O_3 \uparrow$$

$$2FeS + O_2 \longrightarrow 2\,FeO + 2S$$

As_2O_3 is condensed and used as a source of As.

3. Smelting. The roasted mass is smelted in a furnace with coke and fluxing material (sand, SiO_2 and lime stone, $CaCO_3$). Here FeO produced during roasting and CaO produced from $CaCO_3$ combine with SiO_2 to form a fusible slag at the top of the molten mass.

$$CaCO_3 \longrightarrow CaO + CO_2$$

$$CaO + SiO_2 \longrightarrow \underset{\text{slag}}{CaSiO_3}\,; \qquad FeO + SiO_2 \longrightarrow \underset{\text{slag}}{FeSiO_3}$$

The molten mass obtained at the end of smelting can be separated into the following three layers.

(*i*) *Slag.* This is the top layer and consists of earthy matter present in the ore and slag of $FeSiO_3$ and $CaSiO_3$ formed during smelting. This is rejected.

(*ii*) *Bullion.* This is the bottom layer and consists of easily reducible and impure metals, mainly Ag and Cu. It is used for the recovery of Ag.

(*iii*) *Speiss.* This is the middle layer and consists of fused arsenides of Co, Ni, Fe and Cu and some As and S. This layer is subjected to further treatment.

4. Roasting the speiss with NaCl. The speiss is finely ground and roasted in air in a reverberatory furnace to convert As and S, if any, into As_2O_3 and SO_2 which are driven off. It is then mixed with NaCl and then roasted when the metals Co, Ni, Fe and Cu are converted into their chlorides.

5. Separation of Co, Ni, Fe and Cu. The roasted speiss is extracted with H_2O to obtain the aqueous solution of the chlorides. This aqueous solution is then treated with limestone ($CaCO_3$) to precipitate Fe as $FeCO_3$ and the filtrate is treated with Na_2CO_3 to precipitate Cu as $CuCO_3$. By this treatment the chlorides of Co and Ni remain unaffected and hence remain in solution. Ag (in the form of AgCl), if present, is extracted by using a dil. solution of NaCN *(Cyanide process).*

6. Separation of Ni from Co. The solution containing the chlorides of Ni and Co is carefully treated with lime, $Ca(OH)_2$ and bleaching powder, $CaOCl_2$ whereby $CoCl_2$ is converted into a precipitate of $Co(OH)_3$ and $NiCl_2$ remains in solution.

$$\underset{\text{}}{CoCl_2} + \underset{\text{Lime}}{Ca(OH)_2} \longrightarrow Co(OH)_2 + CaCl_2$$

$$2Co(OH)_2 + H_2O + \underset{\text{Bleaching powder}}{CaOCl_2} \longrightarrow \underset{\text{ppt}}{2Co(OH)_3} + CaCl_2$$

7. Formation of Co_3O_4. The precipitate of $Co(OH)_3$ obtained above is dried and ignited, when cobalto-cobaltic oxide Co_3O_4 is obtained.

$$12Co(OH)_3 \xrightarrow{\text{Ignite}} 4Co_3O_4 + 18H_2O + O_2$$

8. Reduction of Co_3O_4 to Co. Generally Co_3O_4 is used as such for the manufacture of cobalt compounds. In case cobalt metal is needed, it may be obtained by reducing Co_3O_4 by heating it with coke in an oil-fired furnace or with aluminium *(Aluminothermic process).*

$$Co_3O_4 + 4C \longrightarrow 3Co + 4CO$$

$$3Co_3O_4 + 4Al \longrightarrow 9Co + 4Al_2O_3$$

EXTRACTION OF PLATINUM

Q. 42. What are the chief ores of platinum? How is platinum extracted?

Ans. The chief ores of platinum are:

(*i*) Sperrilite, $Pt As_2$

(*ii*) Cooperite $Pt (As S)_2$

(*iii*) Braggite (Pt, Pd, Ni)S

However, platinum is extracted from the following two sources:

(*i*) Platinium-ferrous sands and gravels in the river bed. Besides Pt, these sands and gravels contain Ru, Rh, Pd, Ir. Ag, Au and Ni are present in traces as impurities.

(*ii*) Non-volatile residue obtained in the Mond's process of extraction of Ni contains Ru, Rh, Pd, Os, Ir, Pt, Cu, Ag and Au.

Extraction of platinum involves the following steps.

1. Preparation of concentrate containing Ru, Pd, Os, Ir, Pt, Ag, Au etc. metals. The concentrate can be obtained from either of the following two sources:

(*i*) Platini-ferrous sands and gravels present in the river beds contain Ru, Rh, Pd, Os, Ir, Pt, Ag, Au, Ni metals along with sand particles as impurity. Sand particles are removed by Welfley table method where a stream of water is made to flow across the table. Heavier metals are retained behind the riffles (residue) while the sand particles (gangue particles) are washed away.

The residue thus obtained contains Ru, Rh, Pd, Os, Ir, Pt (50-85%), Ag, Au, Ni etc. This residue is treated with CO gas when Ni forms volatile nickel tetracarbonyl, $Ni(CO)_4$ and is thus removed from the residue. The non-volatile residue left consists of Ru, Rh, Pd, Os, Ir, Pt, Ag,Au etc. and is called *concentrate.*

(*ii*) The non-volatile residue obtained at the end of Mond's process used in the extraction of Ni metal contains Ru, Rh, Pd, Os, Ir, Pt. Cu, Ag, Au, etc. This residue is roasted in air to convert Cu into CuO. The roasted mass obtained consists of Ru, Rh, Pd, Os, Ir, Pt, Ag, Au, and CuO. This mass is treated with H_2SO_4 when CuO gets dissolved due to the formation of soluble $CuSO_4$ while the metals remain undissolved (residue). This undissolved portion contains Ru, Rh, Pd, Os, Ir, Pt, Ag and Au.

2. Separation of Pt-metal from the concentrate. The concentrate obtained from either method is used to extract platinum metal. It is treated with hot aqua regia when Pd, Pt and Au go into solution as their soluble chloro complex compounds *viz.* $H_2[Pd^{+4}Cl_6]$, $H_2[Pt^{+4}Cl_6]$ and $H[Au^{3+}Cl_4]$ respectively while the insoluble portion consists of $RuCl_4$, Rh, Os, Ir, AgCl.

$$2HNO_3 + 8HCl + M(M = Pd \text{ or } Pt) \longrightarrow 2NOCl + 4H_2O + H_2[M^{+4}Cl_6]$$

$$Au + 3HNO_3 + 4HCl \longrightarrow H[Au^{3+}Cl_4] + 3NO_2 + 3H_2O$$

The soluble portion is treated with $FeSO_4$ solution when $H_2[Pd^{+4}Cl_6]$ and $H[Au^{3+}Cl_4]$ get reduced to $H_2[Pd^{2+}Cl_4]$ (soluble portion) and free metallic gold, Au^0 (insoluble portion) while $H_2[PtCl_6]$ (soluble portion) remains unaffected. Thus the soluble portion contains $H_2[PdCl_4]$ and $H_2[PtCl_6]$ and insoluble fraction consists of metallic gold.

$$[Pt^{+4}Cl_6]^{2-} + Fe^{2+} \longrightarrow \text{No reaction}$$

$$[Pd^{+4}Cl_6]^{2-} + Fe^{2+} + e^- \longrightarrow [Pd^{+2}Cl_4]^{2-} + Fe^{3+} + 2Cl^-$$

$$H(Au^{3+}Cl_4] + 3Fe^{2+}SO_4 \longrightarrow Au^0\downarrow + Fe^{3+}Cl_3 + Fe_2^{+3}(SO_4)_3 + HCl$$

or $$[Au^{3+}Cl_4]^- + 3Fe^{2+} \longrightarrow Au^0 + 3Fe^{3+} + 4Cl^-$$

The soluble portion containing $H_2[PdCl_4]$ and $H_2[PtCl_6]$ is treated with NH_4Cl when $H_2[PdCl_4]$ remains unaffected and hence remains in solution while $H_2[PtCl_6]$ is converted into ammonium hexachloroplatinate (IV), $(NH_4)_2[PtCl_6]$ (insoluble portion). $(NH_4)_2[PtCl_6]$ on ignition gives spongy platinum as a greyish black porous mass.

$$H_2[PtCl_6] + 2NH_4Cl \longrightarrow (NH_4)_2[PtCl_6] + 2HCl$$

$$(NH_4)_2[PtCl_6] \xrightarrow{\text{Ignite}} 2NH_4Cl + 2Cl_2 + Pt \text{ (\textit{spongy platinum})}$$

Q. 43. Discuss various forms of platinum?

Ans. Important forms of platinum are discussed as under:

1. Platinum black. *Preparation.* It is obtained by adding Zn-pieces or formic acid (HCOOH) to the solution of platinum in aqua regia ($HNO_3 + 3HC1$). First, chloroplatinic acid, $H_2[PtCl_6]$ is produced by the action of aqua regia on platinum metal.

$$[HNO_3 + 3HCl \longrightarrow NOC1 + 2H_2O + 2C1] \times 2$$

$$Pt + 4C1 \longrightarrow PtCl_4$$

$$PtCl_4 + 2HC1 \longrightarrow H_2[PtCl_6]$$

On adding : $3HNO_3 + 8HC1 + Pt \longrightarrow 2NOC1 + 4H_2O + H_2[PtCl_6]$

$H_2[PtCl_6]$ is then reduced to black platinum by nascent hydrogen produced by the action of Zn-pieces on HC1 or by HCOOH. Platinum black appears as a black precipitate.

$$Zn^+ 2HC1 \longrightarrow 2H + ZnCl_2$$

$$H_2[PtCl_6] + 4H \longrightarrow Pt\ \textit{(black platinum)} + 6HC1$$

$$H_2[Pt\ Cl_6] + 2HCOOH \longrightarrow Pt\ \textit{(black platinum)} + 6HC1 + 2CO_2$$

Uses. Platinum black is used as an absorbent of gases like H_2, O_2 etc. which can be liberated again. Platinum black is also used as an oxidising agent. It is used as a catalyst in the preparation of HCHO and in the decomposition of H_2O_2.

$$2CH_3OH + O_2\,\textit{(from air)} \xrightarrow{Pt-black} 2HCHO + 2H_2O$$

$$2H_2O_2 \xrightarrow{Pt-black} 2H_2O + O_2 \downarrow$$

Platinum black is used in automatic gas lightning. It catalyses oxidation of oil gas by air. Heat thus produced makes fine platinum wire glow which lights the gas.

2. Spongy platinum. *Preparation.* This form of platinum is obtained by igniting $(NH_4)_2[PtC1_6]$

$$(NH_4)_2[PtC1_6] \xrightarrow{Ignite} Pt\textit{(spongy platinum)} + 2C1_2 + 2NH_4C1$$

Properties. It is soft and spongy greyish-black porous mass. On fusing it in oxy-hydrogen flame in an electric furnace, it forms a compact lustrous mass. When $C1_2$ is passed through a solution of spongy platinum in conc. HCl, $H_2[PtC1_6]$ is obtained.

$$Pt\ \textit{(spongy platinum)} + 2HCl + Cl_2 \longrightarrow H_2\ [PtCl_6]$$

Uses. Spongy platinum has a great surface area and hence is used for absorbing gases. Coal gas can be ignited by spongy platinum and is, therefore, used in self-lighting Bunsen burners. Spongy platinum absorbs large quantity of O_2 and is then used as an oxidising agent.

3. Colloidal platinum. *Preparation.* It can be obtained either by *Bredig's arc method* in which an electric arc is passed between two platinum electrodes dipped in ice-cold water so that platinum particles are scattered through the solution to give a brown solution which is called colloidal platinum. It can also the obtained by converting platinum into H_2PtCl_6 by the action of aqua regia on it and then reducing H_2PtCl_6 by hydrazine ($NH_2.NH_2$) in presence of some protective colloid to get colloidal platinum.

$$H_2PtCl_6 + NH_2.NH_2 \longrightarrow Pt\ \textit{(colloidal platinum)} + N_2 + 6HCl$$

Uses. Colloidal platinum has been found to be an effective catalyst, *e.g.,* it decomposes H_2O_2 easily.

4. Platinised asbestos. *Preparation.* It is prepared as follows.

Pt metal is converted into $H_2[PtCl_6]$ by the action of aqua regia on it.

$$2HNO_3 + 8HC1 + Pt \longrightarrow 2NOCl + 4H_2O + H_2[PtCl_6]$$

Now the asbestos is soaked in 10% solution of H_2PtCl_6, dried and heated to 500°C. H_2PtCl_6 decomposes to give finely-divided platinum which is distributed throughout the asbestos fibre. The asbestos fibre having finely-divided platinum in it is called *platinised asbestos*.

Uses. It is used as a catalyst in the combination of SO_2 and O_2 to form SO_3 required in the manufacture of H_2SO_4 by contact process. Platinised asbestos absorbs large volume of O_2 and then acts as oxidising agent.

COMPOUNDS OF IRON

Q. 44. Describe preparation, properties and uses of Ferrous Sulphate (Green Vitriol), $Fe_2SO_4.7H_2O$.

Ans. Preparation. (*i*) It is obtained by dissolving scrap iron in dilute sulphuric acid.

$$Fe + H_2SO_4 \rightarrow FeSO_4 + H_2$$

The solution is crystallised by the addition of alcohol as ferrous sulphate is sparingly soluble in it.

(*ii*) It can also be prepared in the laboratory from *Kipp's waste.* The excess of sulphuric acid is neutralised by heating with a small quantity of iron fillings. The solution is then crystallised to obtain green crystals.

Manufacture. Commercially, ferrous sulphate is obtained by the slow oxidation of iron pyrites in the presence of air and moisture. The pyrites are exposed to air on large scale.

$$2FeS_2 + 2H_2O + 7O_2 \rightarrow 2FeSO_4 + 2H_2SO_4$$

The free sulphuric acid is removed by the addition of scrap iron. On crystallisation, green crystals are obtained.

Properties. (*i*) Due to atmospheric oxidation, the crystals of ferrous sulphate acquire brownish yellow colour because of formation of basic ferric sulphate.

$$4FeSO_4 + 2H_2O + O_2 \rightarrow \underset{\text{Basic ferric sulphate}}{4Fe(OH).SO_4}$$

(*ii*) The aqueous solution of ferrous sulphate is slightly acidic due to salt hydrolysis.

$$FeSO_4 + 2H_2 \rightleftharpoons \underset{\text{Weak base}}{Fe(OH)_2} + \underset{\text{Strong acid}}{H_2SO_4}$$

(*iii*) Action of heat. At 570 K, it becomes anhydrous. The anhydrous ferrous sulphate is colourless. The anhydrous salt, on strong heating, breaks up to form ferric oxide with evaluation of SO_2 and SO_3

$$\underset{\text{Green}}{2FeSO_4 7H_2O} \xrightarrow[-7H_2O]{570°C} \underset{\text{White}}{2FeSO_4} \xrightarrow[\text{temp.}]{\text{High}} Fe_2SO_3 + SO_2 + SO_3$$

(*iv*) Ferrous sulphate is a strong reducing agent.

(*a*) It turns potassium dichromate (acidified) green as dichromate is reduced to chromic salt (green).

$$K_2Cr_2O_7 + 4H_2SO_4 \rightarrow K_2SO_4 + Cr_2(SO_4)_3 + 4H_2O + 3[O]$$

$$2FeSO_4 + H_2SO_4 + O \rightarrow Fe_2(SO_4)_3 + H_2O] \times 3$$

$$6FeSO_4 + K_2Cr_2O_7 + 7H_2SO_4 \rightarrow 3Fe_2(SO_4)_3 + K_2SO_4 + Cr_2(SO_4)_3 + 7H_2O$$

(*b*) It decolourises acidified potassium permanganate.

$$2KMnO_4 + 3H_2SO_4 \rightarrow K_2SO_4 + 2MnSO_4 + 3H_20 + 5\,[O]$$

$$[2FeSO_4 + H_2SO_4 + O \rightarrow Fe_2(SO_4)_3 + H_2O] \times 3$$

$$10FeSO_4 + 2KMnO_4 + 8H_2SO_4 \rightarrow 5Fe_2(SO_4)_3 + K_2SO_4 + 2MnSO_4 + 8H_2O$$

(*c*) It reduces mercuric chloride to mercurous chloride.

$$[2HgC1_2 \rightarrow Hg_2C1_2 + 2C1] \times 3$$

$$[3FeSO_4 + 3C1 \rightarrow Fe_2(SO_4)_3 + FeC1_3] \times 2$$

$$6HgC1_2 + 6FeSO_4 \rightarrow 3Hg_2Cl_2 + 2Fe_2(SO_4)_3 + 2FeC1_3$$

(*d*) It reduces gold chloride to gold.

$$AuC1_3 + 3FeSO_4 \rightarrow Au + Fe_2(SO_4)_3 + FeC1_3$$

(*v*) A cold solution of ferrous sulphate absorbs nitric oxide forming dark brown addition compound, nitroso ferrous sulphate.

$$FeSO_4 + NO \rightarrow FeSO_4.NO$$

Nitroso ferrous sulphate
(Brown)

The NO gas is evolved when the solution is heated.

(*vi*) It combines with potassium cyanide (excess) forming potassium ferrocyanide, $K_4Fe(CN)_6$.

$$FeSO_4 + 2KCN \rightarrow Fe(CN)_2 + K_2SO_4$$

$$Fe(CN)_2 + 4KCN \rightarrow K_4Fe(CN)_6$$

$$FeSO_4 + 6KCN \rightarrow K_4Fe(CN)_6 + K_2SO_4$$

Uses. (*i*) Ferrous sulphate is used for making blue ink.

(*ii*) It is employed as a laboratory reagent.

(*iii*) It is used as a mordant in dyeing.

(*iv*) It is used as an insecticide in agriculture.

Q. 45. Give the preparation and properties of ferrous ammonium sulphate (Mohr's salt) $(NH_4)_2SO_4.FeSO_4.6H_2O$.

Ans. Preparation. The double salt is prepared by mixing saturated solutions of pure ferrous sulphate and pure ammonium sulphate in air free distilled water at 310 K and allowed to cool. A few drops of sulphuric acid and a little iron wire are added to prevent oxidation of ferrous sulphate into ferric sulphate. The salt is obtained as pale green crystals on cooling.

Properties. (*i*) It is pale green crystalline compound which does not effloresce like ferrous sulphate.

(*ii*) Chemically, it is similar to ferrous sulphate.

(*iii*) It is less readily oxidised in the solid state. It is, therefore, a better volumetric reagent in preference to ferrous sulphate.

(*iv*) All the chemical reactions observed in the case of ferrous sulphate are given by ferrous ammonium sulphate.

Q. 46. Give the preparation, properties and uses of ferric chloride.

Ans. Preparation. (*i*) Anhydrous ferric chloride is obtained by passing dry chlorine gas over heated iron filings.

The vapours are condensed in a bottle attached to the outlet of the tube.

$$2Fe + 3C1_2 \rightarrow 2FeC1_3$$

(*ii*) Hydrated ferric chloride is obtained by the action of hydrochloric acid on ferric carbonate, ferric hydroxide or ferric oxide as represented by the following equations:

$$Fe_2(CO_3)_3 + 6HCl \rightarrow 2FeCl_3 + 3H_2O + 3CO_2$$

$$Fe(OH)_3 + 3HCl \rightarrow FeCl_3 + 3H_2O$$

$$Fe_2O_3 + 6HCl \rightarrow 2FeCl_3 + 3H_2O$$

The solution on evaporation and cooling deposits yellow crystals of hydrated ferric chloride, $FeCl_3.6H_2O$.

Properties. (*i*) Anhydrous ferric chloride is a dark red deliquescent solid. It sublimes at about 570 K and its vapour density corresponds to Fe_2Cl_6. The dimer dissociates at high temprature to $FeCl_3$. The dissociation into $FeCl_3$ is complete at 1020 K. Above this temperature it breaks into ferrous chloride and chlorine.

$$Fe_2Cl_6 \xrightleftharpoons{1020\ K} 2FeCl_3 \xrightleftharpoons{Above\ 1020\ K} 2FeCl_2 + Cl_2$$

(*ii*) Anhydrous ferric chloride behaves as a covalent compound. It is soluble in non-polar solvents like ether, alcohol, etc. It is represented by chlorine bridge structure.

(*iii*) It dissolves in water. The solution is acidic in nature due to hydrolysis.

Cl, Cl, Fe, Cl, Cl, Fe, Cl, Cl

$$FeCl_3 + 3HOH \rightleftharpoons Fe(OH)_3 + 3HCl$$

(*iv*) Anhydrous ferric chloride absorbs ammonia.

$$FeCl_3 + 6NH_3 \rightarrow FeCl_3 \cdot 6NH_3$$

(*v*) Ferric chloride acts as an oxidising agent.

(*a*) It oxidises SO_2 to H_2SO_4

$$2FeCl_3 + SO_2 + 2H_2O \rightarrow 2FeCl_2 + H_2SO_4 + 2HCl$$

(*b*) It oxidises stannous chloride to stannic chloride

$$2FeCl_3 + SnCl_2 \rightarrow 2FeCl_2 + SnCl_4$$

(*c*) It liberates iodine from KI

$$2FeCl_3 + 2KI \rightarrow 2FeCl_2 + 2KCl + I_2$$

(*d*) Nascent hydrogen reduces $FeCl_3$ into $FeCl_2$

$$FeCl_3 + H \rightarrow FeCl_2 + HCl$$

(*e*) It oxidises H_2S to S.

$$2FeCl_3 + H_2S \rightarrow 2FeCl_2 + 2HCl + S$$

(*vi*) When ammonium hydroxide is added to the solution of ferric chloride, a reddish-brown precipitate of ferric hydroxide is formed.

$$FeCl_3 + 3NH_4OH \rightarrow Fe(OH)_3 + 3NH_4Cl$$

(*vii*) On heating, hydrated ferric chloride is not obtained. It is changed to Fe_2O_3 with evolution of H_2O and HCl.

$$2[FeCl_3.6H_2O] \xrightarrow{Heat} Fe_2O_3 + 6HCl + 9H_2O$$

(*viii*) When a solution of thiocyanate ions is added to ferric chloride solution, a deep red colouration is produced due to formation of a complex salt.

$$FeCl_3 + NH_4CNS \rightarrow Fe(SCN)Cl_2 + NH_4Cl$$

$$FeCl_3 + 3NH_4CNS \rightarrow Fe(SCN)_3 + 3NH_4Cl$$

(*ix*) Ferric chloride forms a complex, Prussian blue with potassium ferrocyanide.

$$4FeCl_3 + 3K_4Fe(CN)_6 \rightarrow Fe_4[Fe(CN)_6]_3 + 12KCl$$

Prussian blue
(Ferri ferrocyanide)

Uses. (*i*) It is used for making Prussian blue.

(*ii*) The alcoholic solution is used in medicine.

(*iii*) It is used as a laboratory reagent in the detection of acetates and phenols and also as an oxidising agent.

COMPOUNDS OF COPPER

Q. 47. Give the preparation, properties and uses of cupric oxide, CuO.

Ans. Preparation. (*i*) *By heating Cu_2O in air or by heating copper for a long time in air,*

$$Cu_2O + \frac{1}{2}O_2 \rightarrow 2CuO$$

$$2Cu + O_2 \rightarrow 2CuO$$

(*ii*) *By heating copper nitrate,*

$$2Cu(NO_3)_2 \rightarrow 2CuO + 4NO_2 + O_2$$

(*iii*) *By heating cupric hydroxide,*

$$Cu(OH)_2 \rightarrow CuO + H_2O$$

(*iv*) *On a commercial scale it is obtained by heating malachite which is available in nature.*

$$CuCO_3.Cu(OH)_2 \rightarrow 2CuO + CO_2 + H_2O$$

Properties. (*a*) It is black powder stable to heating up to 1370 K.

(*b*) The oxide is insoluble in water but dissolves in acids forming corresponding salts.

$$CuO + 2HNO_3 \rightarrow Cu(NO_3)_2 + H_2O$$

$$CuO + 2HCl \rightarrow CuCl_2 + H_2O$$

$$CuO + H_2SO_4 \rightarrow CuSO_4 + H_2O$$

(*c*) When heated to 1370-1470 K, it is converted into cuprous oxide with evolution of oxygen

$$4CuO \rightarrow 2Cu_2O + O_2$$

(*d*) It is reduced to metallic copper by reducing agents like hydrogen, carbon and carbon monoxide.

$$CuO + H_2 \rightarrow Cu + H_2O$$

$$CuO + C \rightarrow Cu + CO$$

$$CuO + CO \rightarrow Cu + CO_2$$

Uses. (*i*) It is used to impart green to blue colour to glazes and glass.

(*ii*) It is used in organic analysis for the estimation and detection of carbon.

Q. 48. Give the preparation, properties and uses of cupric chloride, $CuCl_2.2H_2O$.

Ans. Preparation. Anhydrous cupric chloride is obtained as a dark brown mass when copper metal is heated in excess of chlorine gas.

$$Cu + Cl_2 \rightarrow CuCl_2$$

Properties. (*i*) It is deliquescent compound and is readily soluble in water.

(*ii*) The anhydrous salt on heating forms Cu_2Cl_2 and Cl_2.

$$2CuCl_2 \rightarrow Cu_2Cl_2 + Cl_2$$

While the hydrated salt on strong heating gives CuO, Cu_2Cl_2, HCl and Cl_2.

$$3CuCl_2.2H_2O \rightarrow CuO + Cu_2Cl_2 + 2HCl + Cl_2 + 5H_2O$$

(*iii*) The aqueous solution is acidic due to hydrolysis.

$$CuCl_2 + 2H_2O \rightleftharpoons Cu\,(OH)_2 + 2HCl$$

(*iv*) A pale blue precipitate of basic cupric chloride, $CuCl_2.3Cu\,(OH)_2$ is obtained on heating NaOH.

$$CuCl_2 + 2NaOH \rightarrow Cu\,(OH)_2 + 2NaCl$$

$$CuCl_2 + 3Cu\,(OH)_2 \rightarrow CuCl_2.3Cu\,(OH)_2$$

It dissolves in ammonium hydroxide forming a deep blue solution. On evaporation of this solution deep blue crystals of tetrammine cupric chloride are obtained.

$$CuCl_2 + 4NH_4OH \rightarrow Cu\,(NH_3)_4Cl_2.H_2O + 3H_2O$$

Uses. (*i*) It is used as a crystal in Deacon's process.

(*ii*) It is also used in medicines and as an oxygen carrier in the preparation of organic dyestuffs.

Note. For cuprous chloride, see Chapter 18.

Q. 49. Describe the preparation, properties and uses of copper sulphate (Blue Vitriol), $CuSO_4.5H_2O$. (*Meerut, 2000*)

Ans. Preparation. (*i*) Copper sulphate is prepared by dissolving cupric oxide or hydroxide or carbonate in dilute sulphuric acid. The solution on evaporation and crystallisation yields blue crystals of copper sulphate.

$$CuO + H_2SO_4 \rightarrow CuSO_4 + H_2O$$

$$Cu\,(OH)_2 + H_2SO_4 \rightarrow CuSO_4 + 2H_2O$$

$$Cu\,(OH)_2\,CuCO_3 + 2H_2SO_4 \rightarrow 2CuSO_4 + 3H_2O + CO_2$$

(*ii*) On a commercial scale, it is prepared from scrap copper. The scrap is placed in a perforated lead bucket which is dipped into hot dilute sulphuric acid. Air is blown through the acid. Copper sulphate is crystallised from the solution.

$$Cu + H_2SO_4 + \frac{1}{2}\,O_2\,(air) \rightarrow CuSO_4 + H_2O$$

Properties. (*a*) It is a blue crystalline compound and is fairly soluble in water.

(*b*) **Action of heating.** $CuSO_4.5H_2O$ crystals effloresce on exposure to air and are converted into a pale blue powder, $CuSO_4.H_2O$. The monohydrate loses last molecule of water at 500 K giving the anhydrous salt, $CuSO_4$, which is white.

$$\underset{\text{Blue}}{CuSO_4 \cdot 5H_2O} \xrightleftharpoons{\text{Exposure}} \underset{\text{Pale blue}}{CuSO_4 \cdot 3H_2O} \xrightleftharpoons{370\text{ K}} \underset{\text{Bluish white}}{CuSO_4 \cdot H_2O} \xrightleftharpoons{500\text{ K}} \underset{\text{white}}{CuSO_4}$$

Anhydrous copper sulphate regains its blue colour when treated with a drop of water.

If the anhydrous salt is heated at 990 K, it decomposes into cupric oxide and sulphur trioxide

$$CuSO_4 \xrightarrow{990\ K} CuO + SO_3$$
$$\downarrow$$
$$SO_2 + \frac{1}{2}O_2$$

(*c*) **Action of alkalies.** Alkalies react with copper sulphate to form a pale blue precipitate of copper hydroxide.

$$CuSO_4 + 2NaOH \longrightarrow Cu(OH)_2 + Na_2SO_4$$

(*d*) **Action of NH_4OH:** With ammonia solution, it forms the soluble blue complex. First it forms a precipitate of $Cu(OH)_2$ which dissolves in excess of ammonia solution.

$$CuSO_4 + 2NH_4OH \longrightarrow Cu(OH)_2 + (NH_4)_2SO_4$$

$$Cu(OH)_2 + 2NH_4OH + (NH_4)_2SO_4 \longrightarrow \underset{\text{Tetrammine copper sulphate}}{Cu(NH_3)_4SO_4} + 4H_2O$$

The complex is known as Schwitzer's reagent which is used for dissolving cellulose in the manufacture of artificial silk.

(*e*) **Action of potassium cyanide.** First cupric cyanide is formed which decomposes to give cuprous cyanide and cyanogen gas. Cuprous cyanide dissolves in excess of potassium cyanide to form a complex, potassium cupro cyanide $[K_3Cu(CN)_4]$.

$$[CuSO_4 + 2KCN \rightarrow Cu(CN)_2 + K_2SO_4] \times 2$$
$$2Cu(CN)_2 \rightarrow Cu_2(CN)_2 + (CN)_2$$
$$Cu_2(CN)_2 + 6KCN \rightarrow 2K_3Cu(CN)_4$$
$$\overline{2CuSO_4 + 10KCN \rightarrow 2K_3Cu(CN)_4 + 2K_2SO_4 + (CN)_2}$$

(*f*) **Action of potassium ferrocyanide.** Reddish brown (chocolate) precipitate of cupric ferrocyanide is formed.

$$2CuSO_4 + K_4Fe(CN)_6 \rightarrow Cu_2Fe(CN)_6 + 2K_2SO_4$$

(*g*) **Action of potassium iodide.** First cupric iodide is formed which decomposes to give white cuprous iodide and iodine.

$$CuSO_4 + 2KI \rightarrow CuI_2 + K_2SO_4] \times 2$$
$$2CuI_2 \rightarrow Cu_2I_2 + I_2$$
$$\overline{2CuSO_4 + 4KI \rightarrow Cu_2I_2 + 2K_2SO_4 + I_2}$$

(*h*) **Action of H_2S :** When H_2S is passed through copper sulphate solution, a black precipitate of copper sulphide is formed.

$$CuSO_4 + H_2S \rightarrow CuS + H_2SO_4$$

(*i*) **Additions of electro-positive metals.** Electro-positive elements like zinc and iron displace copper from a solution of copper sulphate.

$$CuSO_4 + Fe \rightarrow Cu + FeSO_4$$
$$CuSO_4 + Zn \rightarrow Cu + ZnSO_4$$

(*j*) **Action of potassium sulphocyanide.** Cupric sulphocyanide is formed.

$$CuSO_4 + 2KCNS \rightarrow Cu(CNS)_2 + K_2SO_4$$

COMPOUNDS OF SILVER

Q. 50. Give the preparation, properties and uses of silver nitrate (lunar caustic), $AgNO_3$.

(Calcutta, 2005)

Ans. Preparation. It is prepared by heating silver with dilute nitric acid. The solution is concentrated and cooled when the crystals of silver nitrate are obtained.

$$3Ag + 4HNO_3 \xrightarrow{\text{Heat}} 3AgNO_3 + NO + 2H_2O$$

Properties. (*a*) It is a colourless crystalline compound, soluble in water and alcohol. It melts at 485 K.

(*b*) On heating above its melting point, it decomposes to silver nitrite and oxygen.

$$2AgNO_3 \rightarrow 2AgNO_2 + O_2$$

When heated at red heat, it further decomposes to metallic silver.

$$2AgNO_3 \rightarrow 2Ag + 2NO_2 + O_2$$

(*c*) Solutions of halides, phosphates, sulphides, chromates, thiocyanates, sulphates and thiosulphates, all give a precipitate of the corresponding silver salt with silver nitrate solution.

$$AgNO_3 + NaCl \longrightarrow \underset{\text{White}}{AgCl} + NaNO_3$$

$$AgNO_3 + NaBr \longrightarrow \underset{\text{Pale yellow}}{AgBr} + NaNO_3$$

$$AgNO_3 + NaI \longrightarrow \underset{\text{Yellow}}{AgI} + NaNO_3$$

$$3AgNO_3 + Na_3PO_4 \longrightarrow \underset{\text{Yellow}}{Ag_3PO_4} + 3NaNO_3$$

$$2AgNO_3 + Na_2S \longrightarrow \underset{\text{Black}}{Ag_2S} + 2NaNO_3$$

$$2AgNO_3 + K_2CrO_4 \longrightarrow \underset{\text{Brick red}}{Ag_2CrO_4} + 2KNO_3$$

$$AgNO_3 + NaCNS \longrightarrow \underset{\text{White}}{AgCNS} + NaNO_3$$

$$2AgNO_3 + Na_2SO_4 \longrightarrow Ag_2SO_4 + 2NaNO_3$$

$$2AgNO_3 + Na_2S_2O_3 \longrightarrow Ag_2S_2O_3 + 2NaNO_3$$

$$Ag_2S_2O_3 + H_2O \longrightarrow \underset{\text{(Black)}}{Ag_2S} + H_2SO_4$$

(*d*) In contact with organic substances it blackens due to decomposition into metallic silver. Thus, it leaves black stains when it comes in contact with skin and clothes. It produces burning sensation like caustic and leaves a black stain like the moon *Luna* on skin and is thus called *Lunar caustic.* It is decomposed by light also and is, therefore, stored in coloured bottles.

White ppt. of $Ag_2S_2O_3$ which gradually changes to black due to hydrolysis.

On account of these reactions, silver nitrate reves as an excellent laboratory reagent for the identification of various acid radicals.

(*e*) When KCN is added to silver nitrate, a white precipitate of silver cyanide appears which dissolves in excess of KCN forming a complex salt, potassium argentocyanide.

$$AgNO_3 + KCN \rightarrow AgCN + KNO_3$$

$$AgCN + KCN \rightarrow \underset{\text{Potassium argentocyanide}}{KAg(CN)_2}$$

(*f*) When treated with a solution of NaOH, it forms precipitate of silver oxide. Originally, it has brown colour but turns black when dried.

$$2AgNO_3 + 2NaOH \rightarrow Ag_2O + 2NaNO_3 + H_2O$$

(*g*) When sodium thiosulphate is added to silver nitrate, a white precipitate of silver thiosulphate appears. This precipitate, however, dissolves in excess of sodium thiosulphate forming a complex salt.

$$2AgNO_3 + Na_2S_2O_3 \rightarrow Ag_2S_2O_3 + 2NaNO_3$$

$$Ag_2S_2O_3 + 3Na_2S_2O_3 \rightarrow \underset{\text{Sodium argentothiosulphate}}{2Na_3[Ag(S_2O_3)_2]}$$

(*h*) When NH_4OH is added to silver nitrate solution, a brown precipitate of silver oxide appears which dissolves in excess of ammonia forming a complex salt.

$$2AgNO_3 + 2NH_4OH \rightarrow Ag_2O + 2NH_4NO_3 + H_2O$$

$$Ag_2O + 2NH_4NO_3 + 2NH_4OH \rightarrow 2[Ag(NH_3)_2]NO_3 + 3H_2O$$

The ammoniacal solution of $AgNO_3$ gives the following reactions :

(*i*) It reacts with acetylene to form white compound, silver acetylide.

$$2AgNO_3 + 2NH_4OH + C_2H_2 \rightarrow Ag_2C_2 + 2NH_4NO_3 + H_2O$$

(*ii*) It oxidises formaldehyde to formic acid

$$Ag_2O + HCHO \rightarrow 2Ag + HCOOH$$

(*i*) All halogen acids, except HF, precipitate silver halides from aqueous solution of $AgNO_3$

$$AgNO_3 + HX \rightarrow AgX + HNO_3$$

(*j*) Silver is readily displaced from an aqueous solution of silver nitrate by the base metals, particularly, if the solution is somewhat acidic,

$$2AgNO_3 + Cu \rightarrow 2Ag + Cu(NO_3)_2$$

$$2AgNO_3 + Zn \rightarrow 2Ag + Zn(NO_3)_2$$

Uses. (*i*) It is used as laboratory reagent for the identification of various acidic radicals especially for chloride, bromide and iodide.

(*ii*) Silver nitrate is used for making silver halides which are used in photography.

(*iii*) The ammoniacal silver nitrate solution, *i.e.*, Tollen's reagent is used in organic chemistry for testing aldehydes, reducing sugar, etc.

(*iv*) It is used extensively for the preparation of silver mirrors. The process of depositing a thin and uniform layer of silver on a clean glass surface is known as *silvering of mirrors*.

Q. 51. Give the preparation, properties and uses of silver bromide, AgBr.

Ans. Preparation. It is prepared by the reaction of silver nitrate with a soluble bromide.

$$AgNO_3 + NaBr \rightarrow AgBr + NaNO_3$$

Properties : (*a*) It is a pale yellow solid. It is insoluble in water and concentrated acids. It is partially soluble in strong solution of ammonium hydroxide due to complex formation.

$$AgBr + 2NH_4OH \rightarrow Ag(NH_3)_2Br + 2H_2O$$

(*b*) It dissolves in potassium cyanide

$$AgBr + 2KCN \rightarrow \underset{\text{Potassium argentocyanide (soluble)}}{KAg(CN)_2} + KBr$$

(*c*) It is most sensitive to light. It undergoes photoreduction.

$$2AgBr \xrightarrow{\text{Light}} 2Ag + Br_2$$

(*d*) AgBr is reduced to silver, when treated with zinc and dilute H_2SO_4.

$$Zn + H_2SO_4 \rightarrow ZnSO_4 + 2H$$

$$2AgBr + 2H \rightarrow 2Ag + 2HBr$$

Uses. It is used for making photographic films and plates because it is sensitive to light.

Q. 52. How do we measure purity of gold?

Ans. Gold is a precious and noble metal. It is very stable towards common reagents and resists the action of the atmosphere. It is not attacked by air or oxygen or water at any temperature. Since it retains its bright surface, it is used for making ornaments. Generally, small quantities of other metals, especially copper, are added to the gold to make it harder. The purity of gold is generally expressed in *carats,* pure gold being known as 24 carats. Ornaments generally available these days are made of 18 carats. This is an alloy of gold containing 18 parts gold and 6 parts other metals.

Thus, percentage of gold is 18 carats $= \frac{100}{24} \times 8 = 75$

COMPOUNDS OF ZINC

Q. 53. Give the preparation, properties and uses of zinc oxide (ZnO).

Ans. Preparation. It is obtained by the combustion of zinc or by the calcination of zinc carbonate or zinc nitrate.

$$2Zn + O_2 \rightarrow 2ZnO$$

$$ZnCO_3 \rightarrow ZnO + CO_2$$

$$2Zn (NO_3)_2 \rightarrow 2ZnO + 4NO_2 + O_2$$

Properties. (*i*) It is a white powder and turns yellow on heating and again turns white on cooling.

(*ii*) It is very light. It is insoluble in water.

(*iii*) It is an amphoteric oxide and dissolves readily in acids forming corresponding zinc salts and in alkalis forming zincates.

$$ZnO + H_2SO_4 \rightarrow ZnSO_4 + H_2O$$

$$ZnO + 2HCl \rightarrow ZnCl_2 + H_2O$$

$$ZnO + 2NaOH) \rightarrow \underset{\text{Sodium zincate}}{Na_2ZnO_2} + H_2O$$

(*iv*) When zinc oxide is heated with cobalt nitrate, a green mass is formed due to formation of cobalt zincate which is known as *Rinmann's green.*

$$2Co (NO_3)_2 \rightarrow 2CoO + 2NO_2 + O_2$$

$$ZnO + CoO \rightarrow CoZnO_2 \text{ or } CoO.ZnO$$

(*v*) When heated in hydrogen 670 K, it is reduced to metal.

$$ZnO + H_2 \rightarrow Zn + H_2O$$

It is also reduced by carbon into zinc

$$ZnO + C \rightarrow Zn + CO$$

Uses. (*i*) Zinc oxide is used as a white pigment (paint). Its covering power is less than white lead but it is superior because it is not blakened in atmosphere of hydrogen sulphide.

(*ii*) It is used in cosmetic powders and creams.

(*iii*) It is also used as zinc ointment in medicine and as a glaze in ceramics.

(*iv*) It finds use as a catalyst along with Cr_2O_3 in the manufacture of methyl alcohol from water gas.

(*v*) It is used to prepare Rinmann's green which is employed as a green pigment.

Q. 54. Give the preparation, properties and uses of zinc chloride, $ZnCl_2.2H_2O$.

Ans. Preparation. It is obtained by treating zinc oxide or zinc carbonate or zinc hydroxide with dilute hydrochloric acid. The solution on concentration and cooling yields hydrated zinc chloride crystals, $ZnCl_2.2H_2O$.

$$ZnO + 2HCl \rightarrow ZnCl_2 + H_2O$$
$$ZnCO_3 + 2HCl \rightarrow ZnCl_2 + CO_2 + H_2O$$
$$Zn(OH)_2 + 2HCl \rightarrow ZnCl_2 + 2H_2O$$

The anhydrous zinc chloride is obtained by heating zinc in the atmosphere of dry chlorine or dry HCl gas.

$$Zn + Cl_2 \rightarrow ZnCl_2$$
$$Zn + 2HCl \rightarrow ZnCl_2 + H_2$$

Properties. (*a*) Anhydrous zinc chloride is a white solid, deliquescent and soluble in water. It melts at 930 K and boils at 1000 K.

(*b*) When NaOH is added, a white precipitate of zinc hydroxide appears which dissolves in excess of sodium hydroxide forming sodium zincate.

$$ZnCl_2 + 2NaOH \rightarrow Zn(OH)_2 + 2NaCl$$
$$Zn(OH)_2 + 2NaOH \rightarrow Na_2ZnO_2 + 2H_2O$$

(*c*) When H_2S is passed through the solution, a white precipitate of zinc sulphide is formed.

$$ZnCl_2 + H_2S \rightarrow ZnS + 2HCl$$

(*d*) Anhydrous zinc chloride absorbs ammonia gas and forms an addition compound.

$$ZnCl_2 + 4NH_3 \rightarrow ZnCl_2.4NH_3$$

(*e*) On adding NH_4OH solution, a white precipitate of zinc hydroxide appears which dissolves in excess of ammonia forming a complex salt.

$$ZnCl_2 + 2NH_4OH \rightarrow Zn(OH)_2 + 2NH_4Cl$$
$$Zn(OH)_2 + 2NH_4OH + 2NH_4Cl \rightarrow \underset{\text{Tetrammine zinc chloride}}{[Zn(NH_3)_4]Cl_2} + 4H_2O$$

(*f*) Its syrupy solution when mixed with zinc oxide, ZnO, sets to a hard mass forming an oxychloride, $ZnCl_2.3ZnO$.

Uses. (*i*) The concentrated solution of zinc chloride is used for protecting timber to prevent its destruction from the action of micro-organisms.

(*ii*) It is used as dehydrating agent.

(*iii*) The mixture of syrupy zinc chloride solution and zinc oxide is used for dental filling.

(*iv*) It is also used in manufacture of dry cells.

(*v*) It is used for cleaning the surface of metals before soldering.

Q. 55. Give the preparation, properties and uses of zinc sulphate (white vitriol), $ZnSO_4.7H_2O$.

Ans. Preparation. It can be prepared by dissolving zinc metal or zinc oide or carbonate in dilute sulphuric acid. The solution on concentration and crystallisation below 312 K gives colourless crystals of zinc sulphate, $ZnSO_4.7H_2O$

$$Zn + H_2SO_4 \rightarrow ZnSO_4 + H_2$$
$$ZnO + H_2SO_4 \rightarrow ZnSO_4 + H_2O$$
$$ZnCO_3 + H_2SO_4 \rightarrow ZnSO_4 + H_2O + CO_2$$

Properties: (*a*) It is a colourless, crystallisation solid. It is freely soluble in water.

(*b*) On heating, the following changes take place:

$$ZnSO_4.7H_2O \xrightarrow[\text{Below 550K}]{312\text{ K}} ZnSO_4.6H_2O \xrightarrow{\text{Above 340 K}} ZnSO_4.H_2O$$

$$\downarrow \text{Above 550 K}$$

$$O_2 + SO_2 + ZnO \xleftarrow{1070\text{ K}} \underset{\text{(anhydrous)}}{ZnSO_4}$$

$$ZnSO_4 \xrightarrow{1070\text{ K}} ZnO + SO_3$$

$$\downarrow$$

$$SO_2 + \frac{1}{2}O_2$$

(*c*) When sodium hydroxide is added to the solution of zinc sulphate, a white precipitate of zinc hydroxide appears which dissolves in excess of NaOH forming sodium zincate.

$$Zn + SO_4 + 2NaOH \rightarrow Zn\,(OH)_2 + Na_2SO_4$$

$$Zn\,(OH)_2 + 2NaOH \rightarrow Na_2ZnO_2 + 2H_2O$$

(*d*) With alkali metal sulphates and $(NH_4)_2.SO_4$, it forms double sulphates such as $K_2SO_4.ZnSO_4.6H_2O$.

Uses. (*i*) Zinc sulphate is used as a starting material for the preparation of other zinc compounds.

(*ii*) A dilute solution of zinc sulphate is used as eye lotion.

(*iii*) The solution of zinc sulphate is used for zinc plating and in the electrolytic refining of zinc.

(*iv*) It is employed as mordant in dyeing and calico printing.

(*v*) It is used for the preparation of lithopone ($ZnS + BaSO_4$), a popular white paint.

COMPOUNDS OF MERCURY

Q. 56. Give the preparation, properties and uses of mercurous chloride, Hg_2Cl_2 (Calomel). (*Calcutta, 2007*)

Ans. Preparation. (*i*) Metallic mercury is boiled with concentrated sulphuric acid and the product, mercuric sulphate is then ground thoroughly with common salt and metallic mercury. The mixture is sublimed when Hg_2Cl_2 is separated.

$$Hg + 2H_2SO_4 \rightarrow HgSO_4 + SO_2 + 2H_2O$$

$$HgSO_4 + 2NaCl + Hg \rightarrow Hg_2Cl_2 + Na_2SO_4$$

(*ii*) By adding dilute hydrochloric acid or sodium chloride solution to a mercurous salt solution, a white precipitate of mercurous chloride is formed.

$$Hg_2\,(NO_3)_2 + 2HCl \rightarrow Hg_2Cl_2 + 2HNO_3$$

$$Hg_2\,(NO_3)_2 + 2NaCl \rightarrow Hg_2Cl_2 + 2NaNO_3$$

(*iii*) Pure mercurous chloride for medicinal purposes is prepared by heating a mixture of mercuric chloride and mercury in an iron pot. The sublimate is ground and boiled with water to remove traces of mercuric chloride.

$$HgCl_2 + Hg \rightarrow Hg_2Cl_2$$

(*iv*) It can be obtained by passing sulphur dioxide through the solution of mercuric chloride.

$$2HgCl_2 + 2H_2O + SO_2 \rightarrow Hg_2Cl_2 + H_2SO_4 + 2HCl$$

Properties. (*i*) It is an amorphous, tasteless, odourless white powder. It is sparingly soluble in water. It sublimes at 646 K.

(*ii*) It dissolves in chlorine water or aqua-regia

$$Hg_2Cl_2 + 2Cl \rightarrow 2HgCl_2$$

(*iii*) It dissolves in hot concentrated nitric acid forming mercuric chloride and mercuric nitrate.

$$3Hg_2Cl_2 + 8HNO_3 \rightarrow 3HgCl_2 + 3Hg(NO_3)_2 + 4H_2O + 2NO$$

(*iv*) It turns black when treated with ammonium hydroxide. The black substance is a mixture of mercury and mercuric amino chloride

$$Hg_2Cl_2 + 2NH_4OH \rightarrow \underset{\text{Black}}{Hg(NH_2)Cl + Hg} + NH_4Cl + 2H_2O$$

(*v*) It is reduced to mercury with stannous chloride.

$$SnCl_2 + Hg_2Cl_2 \rightarrow 2Hg + SnCl_4$$

(*vi*) On adding excess of caustic soda solution in HCl, a black precipitate of mercurous oxide is formed which is decomposed under the action of light.

$$Hg_2Cl_2 + 2NaOH \rightarrow Hg_2O + 2NaCl + H_2O$$

$$Hg_2O \rightarrow HgO + Hg$$

(*vii*) On heating, it decomposes into mercuric chloride and mercury.

$$Hg_2Cl_2 \rightarrow Hg + HgCl_2$$

Uses. (*i*) For making calomel electrodes.

(*ii*) As a purgative in medicine.

Q. 57. Describe preparation, properties and uses of mercuric chloric (corrosive sublimate), $HgCl_2$.

Ans. Preparation. (*i*) It can be prepared by dissolving mercuric oxide in hydrochloride acid

$$HgO + 2HCl \rightarrow HgCl_2 + H_2O$$

(*ii*) It can also be prepared by dissolving Hg or Hg_2Cl_2 in aqua-regia.

$$3HCl + HNO_3 \rightarrow NOCl + 2H_2O + 2Cl$$

$$Hg + 2Cl \rightarrow HgCl_2$$

$$Hg_2O_2 + 2Cl \rightarrow 2HgCl_2$$

(*iii*) The compound was earlier prepared on commercial scale by heating a mixture of mercuric sulphate with common salt. Mercuric chloride sublimes off and condenses on the cooler parts of the vessel.

$$HgSO_4 + 2NaCl \rightarrow HCl_2 + Na_2SO_4$$

(*iv*) The compound is now commercially made by heating the metal in a current of chlorine.

$$Hg + Cl_2 \rightarrow HgCl_2$$

Properties. (*i*) It is a white crystalline compound, less soluble in cold water but more soluble in hot water.

(*ii*) It behaves like a covalent compound as it is soluble in organic solvents and very slightly ionised in water.

(*iii*) It has corrosive action and is highly poisonous. It sublimes on heating. It is, therefore, known as corrosive sublimate.

(*iv*) When aqueous ammonia is added to the solution of mercuric chloride, a white precipitate of mercuric amino chloride is formed.

$$HgCl_2 + 2NH_4OH \rightarrow Hg(NH_2)Cl + NH_4Cl + H_2O$$

(*v*) It is reduced by stannous chloride first into mercurous chloride (white) and then into mercury (black).

$$2HgCl_2 + SnCl_2 \rightarrow \underset{\text{White}}{Hg_2Cl_2} + SnCl_4$$

$$2HgCl_2 + SnCl_2 \rightarrow \underset{\text{Black}}{2Hg} + SnCl_4$$

(*vi*) When H_2S is passed through its solution, it forms a black precipitate of mercuric sulphide.

$$HgCl_2 + H_2S \rightarrow HgS + 2HCl$$

(*vii*) Mercuric chloride absorbs gaseous ammonia and forms a complex salt.

$$HgCl_2 + 2NH_3 \rightarrow \underset{\text{Mercuric diammine chloride}}{Hg(NH_3)_2Cl_2}$$

(*viii*) Potassium iodide forms a scarlet precipitate of mercuric iodide when added to $HgCl_2$ solution. The precipitate of mercuric iodide dissolves in excess of potassium iodide forming a complex, K_2HgI_4.

$$HgCl_2 + 2KI \rightarrow 2KCl + HgI_2$$

$$HgI_2 + 2KI \rightarrow K_2HgI_4$$

The alkaline solution of this complex is known as *Nessler's reagent*. This reagent is used for testing ammonia. When ammonia or ammonium salts are present, the brown precipitate is obtained. This precipitate is said to be the iodide of *Millon's base*. The formation of the iodide of Millon's base can be explained as follows:

$$K_2HgI_4 \rightleftharpoons 2KI + HgI_2$$

$$HgI_2 + 2NH_3 \longrightarrow Hg\begin{matrix} \diagup NH_2 \\ \diagdown I \end{matrix} + NH_4I$$

$$Hg\begin{matrix} \diagup NH_2 \\ \diagdown I \end{matrix} + HgI_2 \longrightarrow \left[\begin{matrix} I\,Hg & & H \\ & N & \\ I\,Hg & & H \end{matrix} \right]^+ I^-$$

(Million's Base)

Uses. (*i*) For the preparation of Nessler's reagent which is used for the test of ammonia and ammonium compounds.

(*ii*) It is highly poisonous and hence used as germicide. It is used as an antiseptic.

(*iii*) It is used in dry cells.

MISCELLANEOUS QUESTIONS

Q.58. Describe the cyanide process for extraction of gold.

(*Guwahati, 2002 ; Bangalore 2002*)

Ans. See Q. 18.

Q.59. How does tempering differ from case hardening? (*Bangalore, 2002*)

Ans. See Q. 33.

Q. 60. Describe a method for concentration of sulphide ore. (*Bangalore, 2002*)

Ans. See Q. 6.

Q. 61. Write in brief the general principles involved in the extraction of metals from their ores. Write the names and molecular formula of two important ores of Ti, Ni and Cr.

(*Purvanchal, 2007*)

Ans. In general, the following steps are involved in the extraction of a metal from its ore.

1. Crushing and grinding. The ore generally occurs in big chunks. It has to be crushed and ground finely in ball-mills to proceed further.

2. Concentration of the ore. The ore is purified, washed and concentrated by one or more of the following methods depending upon impurities.

(*i*) gravity separation (*ii*) Hydraulic classifier

(*iii*) Wilfley table method (*iv*) Froth flotation method

(*v*) Magnetic separator (*vi*) Leaching

3. Calcination or roasting. If it is an oxide or carbonate, the ore is subjected to calcination (heating at high temperature). It the ore is a sulphide ore, it is subjected to roasting in the presence of air.

4. Smelting (reduction). The metal oxide obtained after calcination or roasting is subjected to smelting (heating with coal in a furnace at high temperature). The oxide is reduced to the metal.

5. Purification. The impure metal is purified by one of the following methods. The actual method varies from metal to metal and also on the extent of impurity present.

(*i*) Distillation process (*ii*) Liquation process

(*iii*) Poling (*iv*) Zone refining

(*v*) Bessemerization (*vi*) Electrorefining

(*vii*) Van Arkel method (*viii*) Parting process

Ores of Titanium

1. Rutile, TiO_2
2. Ilmenite, $FeTiO_3$

Ores of Nickel

1. Pentlandite (Ni, Fe)S containing 22% nickel
2. Garnierite Ni, Mg $SiO_3.H_2O$ Double silicate of Ni and Mg

Ores of chromium

Chromite, FeO. Cr_2O_3

Chrome ochre, Cr_2O_3

8 GENERAL CHARACTERISTICS OF *s*-AND *p*-BLOCK ELEMENTS AND THEIR COMPOUNDS

s-BLOCK ELEMENTS

Q. 1. Name the elements, along with their electronic configuration of the Ist group of periodic table. (*Nagpur, 2005*)

Ans. Names of elements along with their electronic configurations are given as under:

Element	*Electronic Configuration*
Lithium	[He] $2s^1$
Sodium	[Ne] $3s^1$
Potassium	[Ar] $4s^1$
Rubidium	[Kr] $5s^1$
Cesium	[Xe] $6s^1$
Francium	[Rn] $7s^1$

Q. 2. Name the alkaline earth elements and write their electronic configurations. (*Nagpur, 2005*)

Ans. Names of alkaline earth elements along with their electronic configurations are given below:

Element	*Electronic Configuration*
Beryllium	[He] $2s^1$
Magnesium	[Ne] $3s^1$
Calcium	[Ar] $4s^1$
Strontium	[Kr] $5s^1$
Barium	[Xe] $6s^1$
Radium	[Rn] $7s^1$

Q. 3. Why do alkali metals not form bivalent cation?

Ans. This is because second ionisation energy (*i.e.*, energy required to remove second electron) is much higher than the first ionisation energy.

Q. 4. Why do alkaline earth metals form bivalent ions though the second ionisation energy (E_2) is much higher than the first ionisation energy (E_1)?

Ans. Though E_2 is almost twice E_1 for alkaline earths yet there are no univalent positive ion [M^+] compounds of alkaline earths. This is explained on the basis of the lattice energy accompanying the formation of compounds with dispositive ions which is much greater than the lattice energy of compounds of unipositive ions. The higher lattice energy of compounds of dispositive ions more than compensates for additional energy required to remove the second electron. Moreover in aqueous solution, M^{+2} ions have more hydration energy than that of M^+ ions. The large amount of hydration energy more than counterbalances the higher value of the second ionisation energy.

Q. 5. How does the atomic radius change.

(*i*) along the period (*ii*) down the group in *s*-block elements?

Ans. The alkali metals have the largest atomic radii in their respective periods and those of alkaline earths are smaller than the preceding alkali metals. For example, in the second period, the atomic radius of Na is maximum, that of Mg comes next while those of Al, Si, P, CI are still smaller. This is explained on the basis of greater nuclear pull on the valency electrons with increasing atomic number. As we move from group 1 to 2, group 2 is one unit higher in molecular charge than that of the preceding element of group 1. With the increase in the nuclear charge, the electrons in the outer shell are pulled closer towards the nucleus. This results in a decrease in the atomic radius. Moving down the group the atomic radii increase. This is because with every succeeding element in the group, a new energy level is added up.

Q. 6. Discuss the variation in ionisation energy along a period, down the group in 1st and 2nd groups of periodic table. *(Garhwal 2010)*

Ans. *The ionization energy is defined as the amount of energy required to remove the most loosely bound electron from an isolated gaseous atom to form a gaseous positive ion.* The most loosely bound electron must lie in the outermost shell. The single electron in the valency shall of each alkali metal can be removed relatively easily. Hence the ionization energy of these elements are relatively low. On moving down the group as the atomic radius increases, the outer electron gets farther and farther away from the nucleus and therefore, the ionization energy decreases on moving down from lithium to cesium. In the case of alkaline earth metals, since there are two electrons in the *s*-orbital of outer level, two ionization energies are involved.

The first I.E. of beryllium is 895 kJ/mole,

$$\text{Be} \longrightarrow \text{Be}^{+} + e^{-}, E_1 = 895 \text{ kJ/mole}$$

and the second I.E. of beryllium is 1557 kJ/mole.

$$Be^{+} \longrightarrow Be^{2+} + e^{-}, E_2 = 1557 \text{ kJ/mole}$$

As is clear, E_2 is invariably higher than E_1. This is due to the fact that it is more difficult to remove an electron from a positively charged particle than from a neutral particle. The *s*-block elements have relatively low I.E.

It appears that the alkaline earth metals would prefer to form + 1 ions (M^+) rather than M^{2+} ions *e.g.*, Mg^{2+}, Ca^{2+} etc. However, it is the M^{2+} ions that are obtained. This is explained due to greater *lattice energy* associated with M^{2+} ion compared to that with M^+ ions. Also + 2 ions are excessively hydrated and a large amount of energy, known as *hydration energy*, is released in the process which more than counterbalances the higher value of second ionization energy.

Q. 7. Alkali metals are good reducing agents. Explain.

(Kanpur, 2000; Kurukshetra, 2001; Nagpur, 2008)

Ans. A substance that can easily impart its electrons can act as a good reducing agent. Electron donating tendency of a metal depends upon its ionisation energy. Smaller the I.E. greater is the reducing power. Since alkali metals possess smaller values of I.E., they are good reducing agents.

Q. 8. Compare the alkaline earth metals with alkali metals with respect to ionisation energy.

Ans. Alkaline earth metals have higher I.E. compared to alkali metals because of small size and greater pull of the nucleus on the outer electrons.

Q. 9. Potassium is more reactive than calcium. Explain.

Ans. Reactivity of an element depends upon the I.E. smaller the I.E. greater is the reactivity. As potassium has smaller I.E. than calcium, the former is more reactive than the latter.

Q. 10. Radius of a cation is smaller than that of its parent atom. Explain.

Ans. A cation is obtained when an electron or electrons are removed from the outermost orbit of the atom.

$$M \longrightarrow M^{n+} + ne^-$$

After the removal of the electrons, the nuclear charge becomes greater than the electronic charge. The nuclear charge pulls the outer electrons with greater force, thereby reducing the size of cation. In some cases (1 and 2 group), the outer orbit is completely removed thereby reducing the size of cation compared to parent atom.

Q. 11. First member of each group differs from other members of the same group. Explain.

(*Garhwal, 2000; Kurukshetra, 2001*)

Ans. First member of every group is smallest in size and has extraordinary large value of I.E. and small value for electron affinity. These characteristics are responsible for difference in behaviour from other members of the group. Non-availability of *d*-orbitals in the first member also accounts for difference in properties from other members.

Q. 12. Alkaline earth metals have smaller size than the corresponding alkali metals. Explain.

Ans. Alkaline earth metals have smaller size than the corresponding alkali metals because of greater nuclear charge and greater pull of the nucleus on the outer electrons.

Q. 13. Calcium forms CaF_2 and not CaF or CaF_3; why?

Ans. Calcium could form CaF because first I.E. of Ca is much smaller than second I.E. But Ca^{2+} is formed in the preference because the lattice energy of CaF_2 would be much higher than that of CaF. Thus CaF_2 forms a more stable lattice. Formation of Ca^{3+} (*i.e.*, CaF_3) does not take place because it involves a great amount of energy in removing the third electron from the inner shell.

Q. 14. I.E. of Be is larger than that of Mg. Explain.

Ans. Beryllium is a smaller atom than magnesium. The outer electrons in Be are held more strongly than in magnesium. Hence I.E. of Be is larger than that of Mg.

Q. 15. Why are alkali metals used in photoelectric cells?

Ans. A metal which can release electrons on irradiation forms a suitable substance for photoelectric cells. Since I.E. of alkali metals is small, they can easily release electrons on irradiation.

Q. 16. Discuss colours imparted to the flame by *s*-block elements. (*Kurukshetra, 2001*)

Ans. *s*-block elements impart characteristic flame colourations to the non-luminous flame. Thus all alkali metals impart a colour to the flame while all alkaline earth elements except Be and Mg also impart colrations to the flame. This is explained by the fact that in the flame, the extra energy excites the electrons in these elements to higher energy levels. When these electrons return to their original orbitals, the absorbed energy is now emitted as different frequency of visible light which determines the colour of the flame. Thus, lithium imparts a crimson red colouration while sodium a yellow, potassium, a violet and both rubidium and calcium impart a reddish violet colour to the flame. Similarly, the colour of the flame is brick-red in case of calcium, crimson in the case of strontium and apple green in the case of barium.

The colour imparted by an element depends upon the I.E. Higher the I.E., higher will be the frequency of radiation absorbed, and consequently lower will be frequency of radiation emitted *i.e.*, the colour imparted. Since Li has higher I.E., it emits radiation of lower frequency *i.e.*, red colour.

The same explanation also holds good for alkaline earths and the colours are also in accordance with the I.E. of these metals.

Beryllium and magnesium atoms are comparatively smaller and their I.E. are very high. Hence their electrons are more strongly bound and are not excited by the energy of the flame to higher energy states. These elements, therefore, do not give any colour in Bunsen flame.

Q. 17. Why only *s*-block elements give colouration in a flame and not other atoms?

Ans. Colour can be imparted to a flame if the I.E. of the element is low and the outermost electron absorbs radiation in the visible range of light. These conditions are satisfied by *s*-block elements and not others.

Q. 18. Be and Mg do not give flame colouration but Ba, Ca and Sr do. Why?

Ans. See Q. 16.

Q. 19. Discuss the variation of metallic character in the period and down the group in *s*-block elements. *(Lucknow, 2000)*

Ans. Alkali metals are the highly electropositive elements. For this reason, they form ionic compounds. Same argument applies to the alkaline earth metals. Since the I.E. and electronegativity decrease as we go down a group, the electropositive character and the metallic character increase as we go down groups (1) and (2). Thus Cs is more electropositive than Li and Ba is more electropositive than Mg. In fact Cs is the most electropositive element in the periodic table. All the *s*-block elements are soft and malleable metals. They are good conductors of heat and electricity. These are the characteristics of a metallic nature. Along the period, metallic properties decrease.

Q. 20. How do alkali metals react with water? *(Kerala, 2000)*

Ans. With water, hydrogen is liberated, *e.g.*,

$$2Na + 2H_2O \longrightarrow 2NaOH + H_2$$

The reactivity with water increases on going down from Li to Cs. This is because electropositive character increases in this order. With potassium, rubidium and caesium, the reaction is so vigorous that the evolved hydrogen catches fire. Alkaline earth metals react with water slowly. It is not certain if Be reacts with water at all.

$$Ca + 2H_2O \longrightarrow Ca(OH)_2 + H_2$$

Q. 21. Li forms oxide, Na the peroxide and potassium the superoxide. Explain. *(Himachal, 2000)*

Ans. $4Li + O_2 \longrightarrow 2Li_2O$ (Lithium oxide)

$$2Na + O_2 \longrightarrow Na_2O_2 \text{ (Sod. peroxide)}$$

$$K + O_2 \longrightarrow KO_2 \text{ (Pot. superoxide)}$$

Lithium forms only monoxide, Li_2O. In excess of oxygen sodium forms peroxide; Na_2O_2 while others form superoxides of the general formula MO_2. The superoxides contain the ion, O_2^- presumably, possessing the structure $(:\ddot{O} \cdots \ddot{O}:)^-$ which has a three electron bond. Due to the presence of one unpaired electron, the superoxides are generally coloured and paramagnetic.

The stability of superoxides increases with increasing atomic number or size. This is due to the fact that packing of larger cations and anions gives stable crystal lattices. The smaller Li^+ ion has a strong positive field around it which attracts the negative charge of O^{2-} anion so strongly that the latter ion does not bind another oxygen to form peroxide ion O_2^{2-}. On the other hand, the weak positive field around sodium and potassium ion allows the dispersal of the negative charge to form peroxide and superoxide ion respectively. This can be explained by the following reactions:

$$\underset{\text{Monoxide ion}}{O^{2-}} \xrightarrow{\frac{1}{2}O_2} \underset{\text{Peroxide ion}}{O_2^{2-}} \xrightarrow{O_2} \underset{\text{Superoxide ion}}{2(O_2^-)}$$

Q. 22. How do alkali and alkaline earth metals react with hydrogen and halogens?

Ans. With hydrogen, hydrides are formed, *e.g.*,

$$2Na + H_2 \rightarrow 2NaH$$

$$Ca + H_2 \rightarrow 2CaH_2$$

Hydrides of alkali metals are ionic or salt like of the formula $M^+ H^+$ where *M* is the alkali metal. The ease of forming hydride decreases in moving down from Li to Cs. It is due to the increase in the electropositive character due to which ionic character of the bond increases and the salt like hydride $M^+ H^-$ becomes less stable as compared to the covalent M – H hydride. In case of alkaline earth metals, beryllium does not combine with hydrogen directly to form hydride. Others form hydrides with the formula MH_2. Hydride of Mg is partly ionic in nature.

With halogens, halides are formed, *e.g.*,

$$2Na + Cl_2 \rightarrow 2NaCl$$

$$Ca + Cl_2 \rightarrow CaCl_2$$

Alkali metals form halides with halogens readily while alkaline earth metals form halides at elevated temperatures. Alkali halides are ionic having formula $M^+ X^-$ while alkaline earth halides are also ionic having the formula MX_2 except beryllium halides which are covalent in nature. Halides of Na and K have NaCl type structure in which each ion is surrounded octahedrally by six ions of the opposite charge. CsCl has body centred cubic lattice where each ion is surrounded by eight ions of the opposite charge. This is due to the larger size of metal ion which can accommodate more halide ions around it.

Q. 23. Lithium, an alkali metal, forms chloride having covalent nature. Explain.

Ans. Lithium atom being small in size holds its outer electron firmly. Thus transfer of electron to chloride to produce an ionic compound is difficult.

It prefers to combine with chlorine by sharing of electrons and forms a covalent bond.

Q. 24. Solutions of alkali metals in liquid ammonia are blue-coloured and reducing in nature. Explain. (*Himachal, 2000; Kerala, 2000*)

Or

What happens when an alkali metal is dissolved in liquid ammonia ? (*Purvanchal 2007*)

Ans. All the alkali and alkaline earth metals except Be and Mg dissolve in liquid ammonia to give deep blue solutions which are good conductors of electricity. The dissolution of the metal is accompanied by its ionisation which yields ammoniated cation and ammoniated electron. For example,

$$Na \longrightarrow Na^+ + e^-$$

$$Na^+ + xNH_3 \longrightarrow \underset{\text{Ammoniated sodium cation}}{Na(NH_3)_x^+}$$

$$e^- + yNH_3 \longrightarrow e(NH_3)_y^-$$
Ammoniated electron

As the electropositive nature of the metal ion increases, the stability of ammoniated metal cation decreases down the group. The blue colour of the solution is due to the ammoniated cation as well as the ammoniated electrons present in the solution.

Q. 25. Discuss the reducing nature of alkali metals.

Ans. The *s*-block elements show strong reducing properties. This is because of the ease with which they can lose their valency electrons because of low ionisation energy.

$$Na \longrightarrow Na^+ + e^-$$

$$Ca \longrightarrow Ca^{2+} + 2e^-$$

Some of the reducing properties of alkali metals are given below:

$$B_2O_3 + 6K \longrightarrow 2B + 3K_2O$$

$$AlCl_3 + 3Na \longrightarrow 3NaCl + Al$$

Reducing character is also related to high oxidation potential which, in fact, represents the tendency to release electrons and get oxidised. This property increases as we go down the group.

Group 1 elements have higher oxidation potential than group 2 elements. The former are therefore stronger reducing agents than the latter. The difference is more striking among the top elements of the groups. Great reducing power of lithium is due to its large heat of hydration because of small size.

Q. 26. Discuss the basic strength of hydroxide of alkali and alkaline earth metals.

Ans. The hydroxides of alkali and alkaline earth metals are basic. Alkali metal hydroxides such as NaOH are strong bases while the alkaline earth metal hydroxides such as $Ca(OH)_2$ are weak bases. This fact can be explained on the basis of I.E. of the metal, say M forming the hydroxide M–O–H. The compound M–O–H will act as an acid if the I.E. of M is high and as a base if the ionization energy of M is low.

The ionization energies of alkali metals are low, therefore, their hydroxides are basic in character. As the I.E. decreases from top to bottom in a group, the basic properties of their hydroxides increase as we move down the group. Thus, KOH is stronger than NaOH and $Sr(OH)_2$ is stronger than $Mg(OH)_2$. The ionization energies of alkaline earth metals are higher then those of alkali metals. The hydroxides of alkaline earth metals are, therefore, weaker than those of alkali metals.

If the element M is less electropositive, it would have then high I.E. Instead of releasing the electrons, it would tend to hold the electrons strongly and the bond M–O–H would not furnish OH^- and would not act as a base. If the element M is more electropositive, it would have then low I.E. It will have a strong tendency to lose electron and furnish OH^- ions.

Q. 27. How does the electronegativity of M in M–OH classify it to be an acid or base?

Ans. If M is quite electropositive, it will help in the breaking of M– O bond in the hydroxide leading to the release of hydroxide ions.

$$M - O - H \longrightarrow M^+ + OH^-$$

Consequently the hydroxide will behave as a base. If on the other hand M is electronegative, it will attract electrons on oxygen and the bond O – H will be broken giving rise to protons.

$$M - O - H \longrightarrow M - O^- + H^+$$

Consequently the hydroxide will behave as an acid.

Q. 28. What is diagonal relationship? *(Awadh, 2000; Kerala, 2000; Indore, 2001)*

Or

Why the properties of lithium resemble with those of magnesium ? *(Vidyasagar, 2007)*

Ans. The first element in any given family of the periodic table tends to resemble in certain respect the second element in the family to the right. This resemblance is known as *diagonal relationship* which is largely explained by Fajans theory. Thus lithium and magnesium have much in common and the chemistry of Be and Al is surprisingly similar in spite of the difference in preferred oxidation state. Elements behave similarly if they have similar polarising power. Polarising power of the cation is determined by the ionic charge to ionic radius ratio. This property of the cation increases as we move to the right owing to the increase in the ionic charge and decreases when we move downwards owing to increase in the ionic radius. If we move diagonally from Li to Mg; from Be to Al or from B to Si, the two changes will more or less balance one another and we will have about the same polarising power and hence similar properties.

Q. 29. Discuss diagonal relationship between lithium and magnesium. *(Coimbatore, 2000)*

Or

Why do lithium and magnesium resemble in their properties?

Ans. Lithium in many respects differs from the other alkali metals and resembles magnesium with which it is diagonally related. The resemblances are usually due to a similarity in ionic size and high polarising power of Li^+ and Mg^{2+} ions. The following points illustrate the abnormal behaviour of lithium in the alkali metal group and its similarity to magnesium.

(*i*) Both lithium and magnesium are harder metals having higher melting point than other metals in their respective groups.

(*ii*) Lithium is usually the least reactive element among the alkali metals. It is slowly attacked by water at 298 K whereas other alkali metals react vigorously. Magnesium also reacts slowly with water.

(*iii*) Lithium is the only alkali metal which combines direct with nitrogen to form the nitride Li_3N. Similarly magnesium also combines with nitrogen yielding Mg_3N_2. These are decomposed by water giving ammonia.

$$Li_3N + 3H_2O \longrightarrow 3LiOH + NH_3$$

$$Mg_3N_2 + 6H_2O \longrightarrow 3Mg(OH)_2 + 2NH_3$$

(*iv*) Lithium, when treated in air or oxygen, gives the normal oxides, Li_2O. The other alkali metals first yield the normal oxides. M_2O which combines further with oxygen to form peroxides, M_2O_2 and superoxides, MO_2. Magnesium under similar conditions forms a normal oxide MgO only.

(*v*) Lithium hydroxide, like magnesium hydroxide, is sparingly soluble in water and decomposes at red heat to form the oxide. The other alkali metal hydroxides are quite soluble and remain unaffected on heating.

$$2LiOH \xrightarrow{\text{Red heat}} Li_2O + H_2O$$

$$Mg(OH)_2 \xrightarrow{\text{Red heat}} MgO + H_2O$$

(*vi*) The solubilities of lithium salts resemble those of the corresponding magnesium salts and differ markedly from those of the other alkali metals. Thus the fluorides, carbonates and phosphates of lithium and magnesium are sparingly soluble in water but the corresponding compounds of the other alkali metals are freely soluble.

(*vii*) Because of strong polarising power of Li^+ and Mg^{2+}, these ions have great tendency to form hydrates. Thus a number of lithium and magnesium salts crystallize from water with definite number of water molecules of crystallization. For example,

$$LiCl, 2H_2O \quad \text{and} \quad LiCO_3 \cdot 3H_2O$$

$$MgC1_2 . 6H_2O \quad \text{and} \quad Mg(NO_3)_2\ 6H_2O$$

The salts of other alkali metals are not hydrated.

(*viii*) Again, because of the strong distorting effect of Li^+ and Mg^{2+} ions on the charge cloud of the anions, their oxy-compounds such as carbonates and nitrate possess low thermal stability and decompose to form the respective oxides.

$$Li_2CO_3 \xrightarrow{\text{heat}} L1_2O + CO_2$$

$$MgCO_3 \longrightarrow MgO + CO_2$$

$$4LiNO_3 \longrightarrow 2Li_2O + 4NO_2 + O_2$$

$$2Mg\,(NO_3)_2 \longrightarrow 2MgO + 4NO_2 + O_2$$

The carbonates of other alkali metals are thermally stable but the nitrates decompose at elevated temperatures yielding nitrites and oxygen.

$$2MNO_3 \longrightarrow 2MNO_2 + O_2$$

where M stands for alkali metal.

Q. 30. Discuss diagonal relationship between Be and Al. (*Calcutta, 2000*)

Or

Why do Be and Al show resemblance in properties?

(*Lucknow, 2000; Kerala, 2000*)

Ans. Beryllium and aluminium show diagonal relationship because of the following reasons:

(1) Polarizing powers of Be^{2+} and Al^{3+} are of similar magnitude.

(2) The standard oxidation potentials of beryllium and aluminium are quite close to each other.

Beryllium resembles aluminium in the following respects:

(*i*) Both Be and Al metals dissolve rapidly in 3 M H_2SO_4 but very slowly in nitric acid in which the metals are rendered passive due to formation of an inert film of oxide on the surface of metals.

(*ii*) Be and Al both react with strong alkalies liberating hydrogen and forming beryllate and aluminate ions.

$$Be + 2OH^- \longrightarrow BeO_2^{2-} + H_2$$

$$2Al + 2OH^- + 2H_2O \longrightarrow 2A1O_2^- + 3H_2$$

(*iii*) Oxides and hydroxides of both the metals are amphoteric in nature. They dissolve in acids as well as in alkalies giving their corresponding salts and beryllate and aluminate ions respectively.

$$BeO + 2\ HC1 \longrightarrow Be\ C1_2 + H_2O$$

$$BeO + 2\,NaOH \longrightarrow Na_2BeO_2 + H_2O$$

$$Al_2O_3 + 6HCl \longrightarrow 2AlCl_3 + 3H_2O$$

$$Al_2O_3 + 2NaOH \longrightarrow 2\,NaAlO_2 + H_2O$$

$$Be(OH)_2 + 2HCl \longrightarrow BeCl_2 + 2\,H_2O$$

$$Be(OH)_2 + 2NaOH \longrightarrow Na_2BeO_2 + 2H_2O$$

$$Al(OH)_3 + 3HCl \longrightarrow AlCl_3 + 3H_2O$$

$$Al(OH)_3 + NaOH \longrightarrow NaAlO_2 + 2H_2O$$

(*iv*) The anhydrous halides (except fluorides) of both Be and Al are covalent in nature and as such dissolve in organic solvents and fume in moist air.

(*v*) The carbides of Be and Al yield methane on hydrolysis.

$$Be_2C + 4H_2O \longrightarrow 2\,Be(OH)_2 + CH_4$$

$$Al_4C_3 + 12H_2O \longrightarrow 4Al(OH)_3 + 2CH_4$$

Q. 31. Why do lithium compounds show covalent character? (*Kurukshetra,2002*)

Ans. Because of the small size of the atom, the charge density on lithium is maximum. Lithium ion, therefore, possesses the greatest polarizing power out of all the alkali metal ions. It exerts a great distorting effect on a negative ion. Consequently the ion has a remarkably great tendency towards the covalent bond formation.

Q. 32. Bring out points of difference between lithium and the other alkali metals.

(*Kerala, 2000; Meerut, 2002*)

Ans. (1) Lithium possesses higher melting and boiling points.

(2) Lithium is much harder than the other alkali metals.

(3) As compared to the other alkali metals, lithium is the least reactive.

(4) It directly combines with nitrogen to form nitride while other alkali metals do not.

(5) When burnt in air or oxygen, it only forms the monoxide Li_2O while others give higher oxides also. The monoxide is less basic than the oxides of other alkali metals. Also it dissolves in water less readily.

(6) Lithium nitrate like magnesium nitrate evolves nitrogen dioxide and oxygen on heating.

$$4\,LiNO_3 \longrightarrow 2Li_2O + 4NO_2 + O_2$$

Sodium and potassium nitrates evolve only oxygen, leaving nitrites, *e.g.*,

$$2\,NaNO_3 \longrightarrow 2NaNO_2 + O_2$$

(7) Its hydroxide decomposes at red heat to Li_2O. Also it is a much weaker base and is much less soluble as compared to the hydroxides of other alkali metals. Hydroxides of others sublime unchanged.

(8) Lithium carbonate is not stable to heat and gives CO_2 while carbonates of other alkali metals are very stable towards heat.

$$Li_2CO_3 \longrightarrow Li_2O + CO_2$$

(9) *Behaviour of chlorides:* (*i*) Lithium chloride crystallizes out as $LiCl{\cdot}2H_2O$. Sodium and other alkali metal chlorides do not form hydrates at ordinary temperature.

(*ii*) Lithium chloride like $MgCl_2$ is deliquescent.

(*iii*) Lithium chloride hydrolyses in hot water to a small extent.

(10) *Solubility of salts:* (*i*) The carbonate, phosphate, oxalate and fluoride of lithium are only sparingly soluble in water. The corresponding salts of sodium and potassium are fairly soluble.

(*ii*) Lithium chloride and nitride are soluble in alcohol.

(*iii*) Lithium chloride and nitride are soluble in organic solvent.

Q. 33. Conductivity of Li^+ in aqueous solution is smaller than that of Cs^+ ion though the former is much lighter than the latter. Explain.

Ans. Li^+ having a high charge density gets excessively hydrated and thus becomes heavy, that is why its conductivity is much smaller compared to Cs^+.

Q. 34. Lithium is most extensively hydrated among alkali metals. Explain.

Ans. All the alkali metal ions have unit positive charge but the atomic volume goes on increasing from Li to Cs. Hence charge density which is the ratio of charge to volume decreases from Li to Cs. Due to maximum value of charge density on Li, it gets hydrated to the maximum extent.

Q. 35. Bring out the points of difference between beryllium and other members of the family.

Ans. (*i*) Melting point of beryllium is higher than those of rest of the members.

(*ii*) Beryllium is very hard metal while other alkaline earth metals are soft.

(*iii*) Unlike other alkaline earth metals, beryllium is not easily affected by dry air.

(*iv*) Beryllium does not liberate hydrogen from acids readily while others do so readily.

(*v*) Oxide and hydroxide of beryllium are amphoteric while others oxides are basic.

(*vi*) Beryllium does not decompose even in boiling water, while other members react with normal water.

(*vii*) Beryllium sulphate is soluble in water while sulphates of other metals are either sparingly soluble in water or insoluble in water.

(*viii*) In general, compounds of beryllium are covalent while compounds of other alkaline earth metals are ionic.

Q. 36. Describe characteristics of oxides of *s*-block elements. (*Lucknow, 2000*)

Ans. Alkali metal oxides are formed by burning the metals in air. The oxides are of three types. Monoxide Li_2O is formed by only lithium, peroxide Na_2O_2 is formed by sodium while the superoxides, MO_2, are formed by potassium and cesium. The oxides are related to each other as follows:

$$\underset{\text{Monoxide ion}}{O^{2-}} \xrightarrow{\frac{1}{2}O_2} \underset{\text{Peroxide ion}}{O_2^{2-}} \xrightarrow{O_2} \underset{\text{Superoxide ion}}{2(O_2^-)}$$

Formation of superoxides is favoured by heavier alkali members. This is due to the comparable sizes of the large cations and O_2^- anions, the packing of which gives stable crystal lattices. The superoxide contains the ion (:Ö⋯Ö:) which has a three-electron bond. Its presence in superoxides is supported by the X-ray data and paramagnetism of the compounds. The monoxides are ionic, strongly basic and highly soluble in water. Their solutions are strongly alkaline and possess M^+ and OH^- ions.

Higher oxides such as peroxides and superoxides behave as strong oxidising agents. On treatment with dilute acids they form hydrogen peroxide and oxygen respectively as shown below:

$$\underset{\text{Sodium peroxide}}{Na_2O_2} + H_2SO_4 \longrightarrow Na_2SO_4 + H_2O_2$$

$$\underset{\text{Potassium superoxide}}{4KO_2} + 2H_2SO_4 \longrightarrow 2K_2SO_4 + 2H_2O + 3O_2$$

Alkaline earth metals give two types of oxides. Normal oxide, MO, is formed either by heating the metal in oxygen or by decomposing the carbonates at high temperatures.

$$2Ca + O_2 \xrightarrow{\text{Heat}} 2\,CaO$$

$$CaCO_3 \longrightarrow CaO + CO_2$$

Normal oxides are ionic and basic in nature except beryllium oxide which is amphoteric. They dissolve in water giving hydroxides. All of them have lattice of NaCl type.

Q. 37. Compare the alkali metals with alkaline earth metals with regard to basicity of hydroxides.

Ans. Alkali metal hydroxides with the general formula M – O – H are obtained when metals react with water. These are highly soluble in water.

$$2\,M + 2H_2O \longrightarrow 2MOH + H_2$$

Since I.E. of alkali metals is very low, their hydroxides act as strong bases. The basic character of the alkali metal hydroxides increases on moving down the group as the I.E. of the metals decreases in the same order. LiOH is the weakest and CsOH is the strongest base. Alkali metal hydroxides are not decomposed on heating.

Alkaline earth hydroxides are obtained when metal monoxides (MO) react with water.

$$CaO + H_2O \longrightarrow Ca(OH)_2$$

Magnesium oxide reacts only slowly with water than those of alkali metal hydroxides. Alkaline earth hydroxides are less basic. The solubility of the alkaline earth hydroxides increases with increasing size of the metals down the group. In fact, none of them is freely soluble in water. Magnesium hydroxide is very little soluble whereas calcium hydroxide is sparingly soluble. Consequently Mg $(OH)_2$ may be prepared by the addition of an alkali to an aqueous solution of its salts.

$$\underset{\text{Solution}}{Mg\,Cl_2} + 2NaOH \longrightarrow Mg\,(OH)_2 + 2NaCl$$

Solid calcium hydroxide is known as slaked lime, a dilute solution as lime water and the milky suspension as milk of lime.

Q. 38. $Be(OH)_2$ is insoluble in water but $Ba(OH)_2$ is soluble. Explain.

Ans. Beryllium ion has a small size and a high positive charge density. According to Fajan's rule, it will exhibit covalent character and is therefore insoluble in water in contrast to $Ba(OH)_2$.

Q. 39. How are the hydrides of alkali and alkaline earth metals obtained? What are their characteristics?

Ans. Alkali and alkaline earth metals form hydrides of the type $M^+ H^-$ and $M^{2+} (H^-)_2$ respectively. These are obtained by the direct combination of metals with hydrogen at high temperatures. The stability of the hydrides decreases down the group because of increasing difference between cation and hydride sizes and consequent decreasing stability of lattices.

$$2Na + H_2 \longrightarrow 2NaH$$

$$Mg + H_2 \longrightarrow MgH_2$$

The hydrides react with water forming the corresponding hydroxides with liberation of hydrogen.

$$MH + H_2O \longrightarrow MOH + H_2$$

$$CaH_2 + 2H_2O \longrightarrow Ca(OH)_2 + 2H_2$$

All hydrides (except those of Li and Be) are ionic in nature. Hydrogen in these compounds gets liberated at the anode on hydrolysis, showing the presence of hydride ion H^- in them.

Magnesium hydride is more conveniently prepared by the action of lithium-aluminium hydride on magnesium in etheral solution.

$$2MgC1_2 + LiAlH_4 \longrightarrow 2MgH_2 + LiCl + AlCl_3$$

Q. 40. Compare the characteristics of halides of alkali and alkaline earth metals.

Ans. (1) All the alkali metals form ionic halides of the type MX where X is F, Cl, Br or I.

(2) Lithium halides, except fluoride, have some covalent character also. This is due to the polarizing effect of small lithium ion on large halogen atoms.

(3) Cesium halides crystallize in a body centred cubic lattice where the coordination number is eight while all other halides form cubic crystals and show a coordination number of six.

(4) The thermal stability of halides increases with increasing electropositive character down the group. They are all soluble in water. Lithium halides except fluoride are also soluble in organic solvents such as alcohol and ether.

(5) Alkaline earth metals form halides of the type MX_2 where X is F, Cl, Br or I. Fluorides of Ca, Sr and Ba are insoluble and precipitate from aqueous solution of the metal salt by adding a soluble fluoride. For example,

$$CaC1_2 + 2NaF \longrightarrow \underset{\text{Precipitate}}{CaF_2} + 2NaCl$$

Magnesium fluoride is obtained by dissolving magnesium in aqueous hydrofluoric acid. Hydrated chlorides $MCl_2 . 6H_2O$ where M is Mg, Ca and Sr and $BaCl_2 . 2H_2O$ crystallize from solution when metal carbonates are dissolved in dilute hydrochloride acid.

$$MCO_3 + 2HCl \longrightarrow MCl_2 + CO_2 + H_2O$$

(6) The fluorides of Mg, Ca, Sr and Ba are ionic character whereas their chlorides, bromides and iodides are appreciably covalent. Covalent character of these halides increases with decreasing size of the cation and increasing size of the anions.

Q. 41. LiCl is covalent while NaCl is ionic. Explain.

(Kurukshetra, 2002; Meerut, 2000; Madras, 2004)

Ans. Li^+ exerts polarizing effect on the large chloride ion. This brings covalent character in the compound. No such effect takes place in NaC1 which remains ionic.

Q. 42. Discuss the characteristics of carbonates and bicarbonates of alkali and alkaline earth metals.

Ans. (1) Alkali metals form carbonates of the general formula M_2CO_3. All these carbonates are white solids, soluble in water and are thermally stable. However, lithium carbonate is an exception. It is sparingly soluble and thermally unstable. On decomposition, it gives lithium oxide and carbon dioxide. Due to strong polarizing nature of small Li^+ ions, it distorts the electron cloud of the oxygen atom in $CO^{2}_{3}{}^{-}$ ion so that Li – O bond becomes stronger as compared to C – O bond. It can also be explained in terms of more stable lattice of Li_2O than $Li_2\,CO_3$.

$$Li_2\,CO_3 \xrightarrow{\text{Heat}} Li_2O + CO_2$$

(2) The carbonates of K, Rb and Cs are deliquescent. All carbonates are decomposed by acids liberating carbon dioxide. Their aqueous solutions are alkaline due to hydrolysis.

$$CO_3^{2-} + 2\,H^+ \longrightarrow CO_2 + H_2O$$

$$CO_3^{2-} + H_2O \xrightarrow{\text{Hydrolysis}} HCO_3^- + OH^-$$

(3) Solid bicarbonates are known for all the alkali metals except lithium. The stability of the bicarbonates increases with the electropositive nature of the metals down the group. All bicarbonates are white solids, soluble in water and decompose on gently heating to give carbonates and liberate carbon dioxide.

$$2MHCO_3 \xrightarrow{\text{Heat}} M_2CO_3 + H_2O + CO_2$$

The normal carbonates of calcium, strontium and barium can be precipitated by the addition of sodium carbonate solution to the solution of a salt of these metals.

$$MCl_2 + Na_2\,CO_3 \longrightarrow MCO_3 + 2\,NaCl$$

On heating, all carbonates decompose to give metal oxides and liberate carbon dioxide.

$$MCO_3 \xrightarrow{\text{Heat}} MO + CO_2$$

The thermal stability of the carbonates increases with increasing size of the metal cation down the group. Alkaline earth metal carbonates are generally insoluble but dissolve in water in presence of carbon dioxide forming bicarbonates. For example,

$$CaCO_3 + H_2O + CO_2 \longrightarrow Ca(HCO_3)_2$$

Q. 43. $BeCO_3$ is less stable than $MgCO_3$. Explain.

Ans. Be^{2+} ion is a small ion. It does not form a stable lattice with carbonate ion. That is why on heating it decomposes to BeO.

$$BeCO_3 \longrightarrow BeO + CO_2$$

Mg^{+2} forms a stable compound.

Q. 44. How are alkyls and aryls of alkali metals (organometallic compounds) obtained? What is the utility of such compounds?

Ans. Alkali metals, especially lithium, form a number of organometallic compounds which are frequently employed in the synthesis of a variety of organic compounds.

Organolithium compounds are generally synthesised as follows :

(*i*) $$Li(n\text{–}C_3H_7) + C_6H_5I \xrightarrow[\text{ether}]{\text{Solvent}} Li(C_6H_5) + n\text{–}C_3H_7I$$

(*ii*) $$4Li(C_6H_5) + Sn(CH_2-CH=CH_2)_4 \xrightarrow[\text{ether}]{\text{Solvent}} 4Li(CH_2-CH=CH_2) + Sn(C_6H_5)_4$$

(*iii*) $$2Li + C_2H_5I \xrightarrow[\text{ether}]{\text{Solvent}} LiI + Li(C_2H_5)$$

Organometallic compounds of other alkali metals such as $Na(C_6H_5)$ $Ph_2C(Na)C(Na)Ph_2$, $CsCH_2CH_2Cs$, are also known. These are generally prepared from the corresponding mercury compounds. For example,

$$2K + HgR_2 \longrightarrow Hg + 2KR$$

Applications of alkali metal alkyls and aryls

Lithium alkyls are extensively used in the synthesis of organic compounds and organometallic compounds of other elements. For example,

(*i*) $$3LiR + BCl_3 \longrightarrow 3LiCl + BR_3$$

(*ii*) $$Li(C_6H_5) + (CH_3C_6H_4)_2CO \longrightarrow (CH_3C_6H_4)_2C(C_6H_5)OLi$$

$$\downarrow H_2O$$

$$(CH_3C_6H_4)_2C(OH)C_6H_5 + LiOH$$

(Tertiary alcohols)

(*iii*) $$Li(C_6H_5) + CO_2 \xrightarrow[\text{solvent}]{\text{Nonpolar}} [C_6H_5CO_2Li]$$

$$\downarrow H_2O$$

$$C_6H_5COOH + LiOH$$

Carboxylic acid

Alkyl and aryl compounds of lithium are polymeric having considerable covalent character in their Li – C bonds whereas such organometallic compounds of other alkali metals have predominantly ionic character so that these can be represented as M^+R^-. The alkyl and aryl sodium, potassium, etc., are much more reactive than the corresponding alkyls and aryls of lithium.

Q. 45. Discuss complexes of alkali metals.

Ans. Alkali metal ions with the exception of Li^+ have very little polarising power due their large size and only a unit positive charge. As a result, they have only a weak tendency to form complexes with unidentate ligands. There is concrete evidence for the existence of $[M(H_2O)_4]^+$ (where M is Li^+, Na^+, and K^+) and $[M'(H_2O)_6]^+$ (where M′ is Rb^+ and Cs^+) in aqueous solutions. Also, lithium iodide reacts with liquid ammonia to produce $[Li(NH_3)_4]^+$.

However, alkali metal ions form reasonably stable complexes with some of the bidentate and polydentate ligands such as salicylaldehyde, β-diketones, nitrophenols, α-nitroso β-naphthol, etc. Thus, when NaOH is added to a solution of salicylaldehyde, a neutral complex is formed which is soluble in organic solvents.

$$C_6H_4(OH)(CHO) + NaOH \xrightarrow{-H_2O} C_6H_4(\bar{O}\overset{+}{Na})(CHO) \xrightarrow{\text{excess ligand}} \text{complex}$$

OH, CHO + NaOH; − H2O; ŌNa⁺, CHO; excess ligand; O, H–C=O, Na, C=O, H, O–H

Similarly, the anhydrous sodium derivative of acetylacetone takes up water to produce a hydrated complex which is also soluble in organic solvents.

H_3C, C—O, HC ⊖, C—O, H_3C + Na^+ + $2H_2O$ ⟶ H_3C, C—O, HC, Na, OH_2, OH_2, C—O, H_3C

The 'Wrap Around' Complexes. Polydentate ligands such as *crown ethers* and *cryptates* are the most important ligands which form *highly stable complexes with alkali metal ions*. These ligands are also *highly selective* in complex formation. These ligands wrap the metal ions completely and protect them from coming into contact with the solvent.

Crown ethers Cryptates

The cause of selectivity lies in the size of the 'hole' available in a polydentate ligand of the type mentioned above. Only that alkali metal ion which *exactly fits into* the 'hole', forms stable complex. The cations which are either too small or too big to fit in the hole, would not form stable complexes. For example, the 'hole' in cryptate-222 is just of the right size for the K^+ ion to fit in whereas Li^+ and Na^+ ions are *too small* to make contact with the oxygen atoms of the ligand. On the other hand, Rb^+ and Cs^+ ions are *too big* in size to fit in the hole without causing steric strain. The K^+ complex of cryptate-222 is thus most stable. Similarly, Na^+ complex of crystal -221 is most stable out of all the alkali metal ion complexes formed with this ligand.

Q. 46. Discuss complex formation by alkaline earth metals.

Ans. Be, becuase of its small size, forms stable complexes with halogens, as, for example, $[BeF_3]^-$ and $[BeF_4]^{2-}$.

The complex ion, $[BeF_3]^-$, involves sp^2 hybridisation and therefore, has a **planar trigonal** structure, as shown in Fig. 8.1

The complex ion $[BeF_4]^{2-}$ involves sp^3 hybridisation and hence has a **tetrahedral** structure (Fig. 8.2)

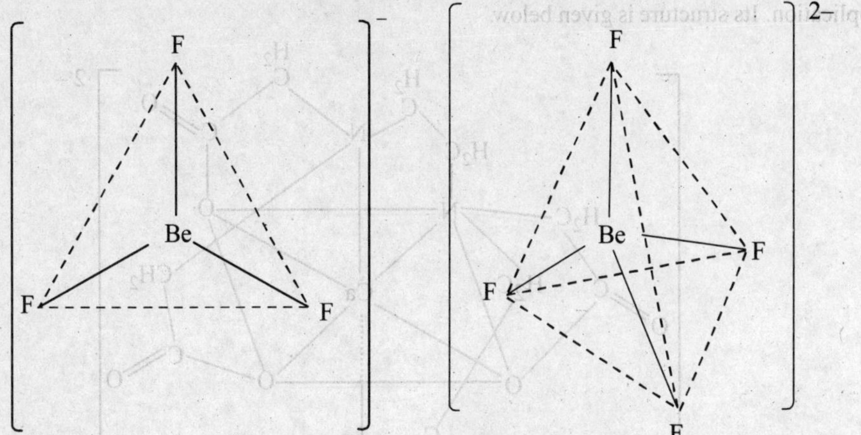

Fig. 8.1. Planar trigonal structure of $[BeF_3]^-$ complex ion.

Fig. 8.2. Tetrahedral structure of $[BeF_4]^{2-}$ complex ion.

Complex Compound of Beryllium with Oxalic Acid. Beryllium also forms a complex compound with oxalic acid. The complex anion $[Be(OX_2]^{2-}$ is known to have a **tetrahedral** structure, as shown in Fig. 8.3.

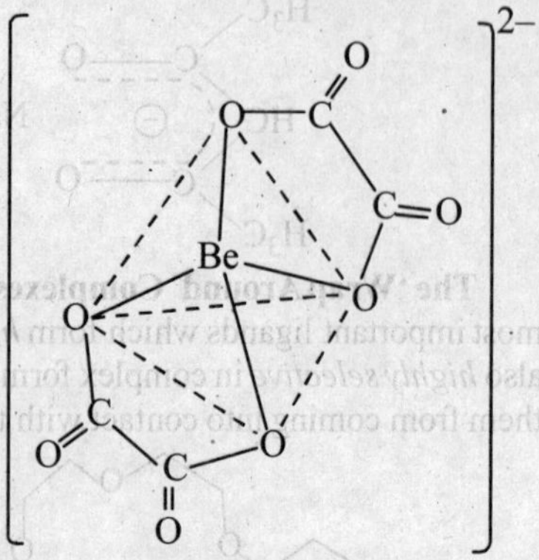

Fig. 8.3. Structure of beryllium oxalate, $[Be(Ox)_2]^{2-}$ ion

Complex of Magnesium. Magnesium forms tetrahedral halide complexes of the type $[(C_2H_5)_4N]_2[MgC1_4]$. The chelates of magnesium with EDTA and pyrophosphates are also known. However, the most important complex of magnesium is the naturally occurring **chlorophyll,** the green pigment present in plants, which photosensitises the synthesis of carbohydrates in nature from CO_2 and H_2O as, for example,

$$6CO_2 + 6H_2O \xrightarrow[\text{chlorophyll}]{hv} C_6H_{12}O_6$$

The macrocyclic ligand involved is derived from the parent tetrapyrrole molecule porphylin. The structure of chlorophyll is shown below.

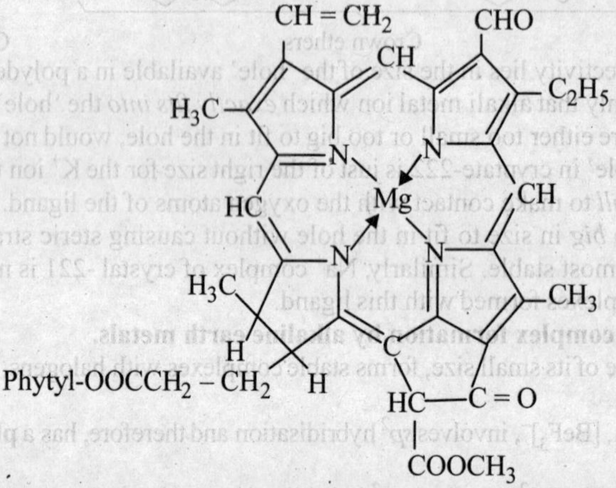

Complexes of Calcium, Strontium and Barium. Amongst the complexes of Ca^{2+}, Sr^{2+} and Ba^{2+}, the water soluble complex of Ca^{2+} with EDTA is the most important because of its analytical application. Its structure is given below.

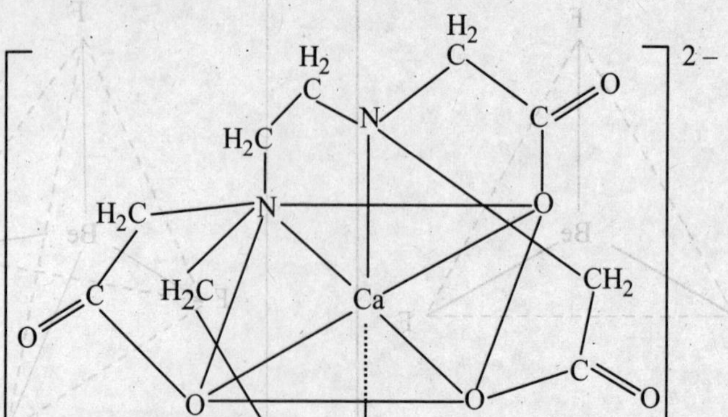

Q. 47. Discuss the role of Mg^{2+} and Ca^{2+} in biological systems.

Ans. Magnesium is an important constituent of chlorophyll which initiates the process of photosynthesis in green plants.

Magnesium ions are concentrated more in intracellular than in extracellular fluids in animal bodies. The presence of Mg^{2+} ions is a necessary requirement for the activation of phosphate transfer enzymes which participate in the energy releasing biochemical process occurring in animal body. Magnesium ions are also required for carbohydrate metabolism.

Mg^{2+} and Ca^{2+} ions are vital for the transmission of electrical impulses along the nerve fibre and for the contraction of muscles.

Calcium ions are essential for the formation of bones and teeth and for maintaining rythmn of heart and for clotting of blood. The enamel on teeth is also a double salt of calcium, $3Ca_3(PO_4)_2.CaF_2$.

p-BLOCK ELEMENTS

Q. 48. What are *p*-block elements?

Ans. All the elements belonging to groups 13, 14, 15, 16, 17 and 18 in which *p*-orbitals are being progressively filled are collectively called *p*-block elements since their properties depend upon the electrons present in *p*-orbitals.

Variations of physical and chemical properties. The physical properties of the elements in a group show gradual variations while there prevails similarity in their chemical properties. Again, the properties of elements in a row show no similarity, but show gradual variations of the various properties, with increasing atomic number.

Q. 49. How does the valency of *p*-block elements vary in the period? (*Calcutta, 2007*)

Ans. As we pass from left to right in a period, the number of electrons in the outermost orbital increases from 3 to 8. It can be shown that the valency of an element in a group is either the number of electrons in the outermost shell or eight minus the valence electrons. Thus the elements of the 13 group which have three electrons in the outermost shell have valency equal to three. Elements of 15 group have valency equal to (8–5) *i.e.*, 3. Phosphorus also exhibits a valency of 5 in some cases when it shares all its 5 valency electrons with other atoms. Oxygen is in group 16 and its valency is 8–6. *i.e.*, 2. Fluorine is in group 17. It needs to gain one electron more either by electrovalant or covalent bond formation. Thus its valency is 8 – 7 =1.

Q. 50. Define atomic radius.

Ans. The atoms in diatomic molecules are taken as perfect solid spheres just in contact with each other. Distance between the centres of any two adjacent atoms in a moleucle is taken as the sum of their radii and half of this is taken as the atomic radius.

Q. 51. Write notes on

(*a*) van der Waals radius (*b*) Covalent radius (*c*) Ionic radius

Ans. (*a*) *van der Waals radius* is regarded as one-half of the distance between the nuclei of two adjacent atoms belonging to two neighbouring molecules of an element in the solid state (Fig. 8.4). In fact, it is the distance between two non-bonded atoms of two adjacent molecules of the same elements.

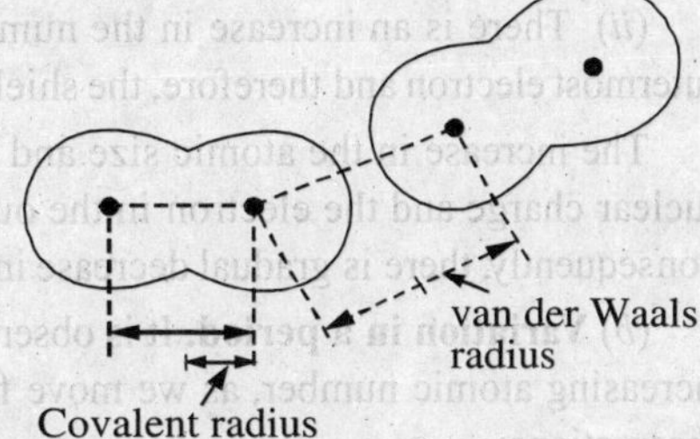

Fig. 8.4. Covalent and van der Waals radii.

(*b*) *Covalent radius* is regarded as the one-half of the distance between the nuclei of two covalently bonded atoms of the same element in a molecule. This is used for non-metals. A covalent radius is smaller than a van der Waals radius because of the fact that in the formation of a covalent bond, the atoms have to come closer to each other. This is the reason why covalent bonds are much stronger than van der Waals forces.

(*c*) *Ionic radius* is regarded as the distance from the nucleus of an ion up to which it extends its influence on electron cloud. This is used for radii of ions in ionic crystals. If the ions in a crystal lattice are regarded as spheres, the internuclear distance betw4en them is determined from X-ray measurements. Thus by knowing the radius of one ion that of the other ion can be calculated.

Q. 52. Discuss the variation of atomic radius in a period and in a group in *p*-block elements?

(*Kumaon, 2000*)

Ans. *Variation in a period.* Atomic radii decrease from left to right across a period in the periodic table. This is explained on the basis of increasing nuclear charge which pulls the electrons inside and thus decreases the atomic size and radius.

Variation in a group. On descending a group in a periodic table, such as moving from B to Al to Ga to In and finally to Tl the size of the atoms increases due to the effect of extra shells of electrons being added. This effect is so large that it outweighs the effect of increased nuclear charge. Hence the radius of the atom increases in moving from top to bottom in a group.

Q. 53. Radius of the anion is always smaller and that of the anion is always bigger than the parent atom.

Ans. (*i*) *The radius of a + ve ion is smaller than that of the corresponding atom.* This is because a +ve ion is formed by the loss of electrons. If the whole of the outer shell of electrons is removed the positively charged particle left behind is smaller in size. Even when one or more electrons are removed from the outermost shell, the ratio of the +ve charges on the nucleons to the number of orbital electrons is increased. Thus the effective nuclear charge increases and the electrons are pulled inwards resulting in the decrease of the size of +ve ion.

(*ii*) *The radius of a –ve ion is bigger than that of the corresponding atom.* This is because in the formation of a –ve ion, electrons are added to the atom whereas the nuclear charge remains the same. As a result, the effective nuclear charge is reduced and the electron cloud tends to move away from the nucleus. The –ve ions, therefore, are bigger in size as compared with the corresponding atoms.

Q. 54. Discuss the variation of ionization energy in a group and along a period in *p*-block elements.

Or

Oxygen has lower I.E. than nitrogen.

(*Awadh, 2000; Guwahati 2002; Nagpur, 2008*)

Ans. (*a*) **Variation in a group.** The values of I.E. reveal that there is a regular decrease in going from top to bottom in a group.

(*i*) There is simultaneous increase in the number of energy levels and therefore, the atomic size increases.

(*ii*) There is an increase in the number of electrons that lie in between the nucleus and the outermost electron and therefore, the shielding effect on the outermost electron gradually increases.

The increase in the atomic size and the shielding effect dominate the effect of the increased nuclear charge and the electron in the outer orbital is rendered less firmly bound to the nucleus. Consequently, there is gradual decrease in ionization energy.

(*b*) **Variation in a period.** It is observed that in general, the ionization energy increases with increasing atomic number, as we move from left to right along a period. We have the following observations:

(*i*) The nuclear charge increases from left to right.

(*ii*) The number of main energy levels remains the same while the atomic size decreases. Both these factors tend to increase the ionization energy.

As we move from B to C and to N, ionization energies keep on increasing due to increasing nuclear charge and decreasing atomic size, the combined effect of other factors not being appreciable. Oxygen has a slightly smaller ionization energy than N. This is primarily due to the more stable electronic arrangement of nitrogen in which all the orbitals are half-filled while the next *p*-electron in oxygens are somewhat less tightly held. Half-filled and fully filled electronic configurations are more stable than others.

The expected increase in ionization energy is observed in going from O to F and to Ne.

Q. 55. What is electron affinity? Discuss the variation of electron affinity in a group and in a period in *p*-block elements. *(Nagpur, 2008)*

Ans. Energy released in the process in which an electron is added to an isolated neutral gaseous is called electron affinity.

$$\underset{\text{Neutral atoms}}{X + \quad e^-} \longrightarrow \underset{\text{Anion}}{X^-} + \text{Energy}$$

So electron affinity is a measure of tightness with which an additional electron is bound to atom. The greater the energy released in the process of taking up the extra electron, the greater is the electron affinity.

It is observed that:

(*i*) Electron affinity generally decreases on moving down the group. This is expected on account of increase in the size of the atom on moving down the group. Due to increase in the size of atoms, the effective nuclear attraction for electrons decreases and hence there is less tendency to attract additional electrons with increase in atomic number.

(*ii*) Electron affinity generally increases as we move from left to right in a period. This is due to the increase in the nuclear charge which results in greater attraction for electrons. In the second period, fluorine has a maximum value of electron affinity. As the electron affinity increases the element becomes less metallic and more non-metallic, *i.e.*, the electron affinities of metals are low while those of non-metals are high.

(*iii*) The halogens have high electron affinity. This is due to the fact that the addition of an electron leads to a stable electronic configuration of $s^2 p^6$ of the valence shell.

(*iv*) Chlorine possesses high value of E.A. than fluorine which is contrary to expectation. The unexpected low value of fluorine as compared with chlorine may be associated with strong inter-electronic repulsive force between the electron charge cloud of fluorine atom and the added electron in the relatively small and compact 2*p* sub-level. Thus due to the small size and non-availability of *d*-orbitals in fluorine atom, there is a high electron density in the *p* sub-level which exerts a strong electron-electron repulsion on the incoming electron. Hence, the electron affinity of fluorine is lower than that of chlorine.

(*v*) The electron affinities of noble gases have zero values since their outermost shells have completed ns^2 and np^6 orbitals and the electron added has to go to a new orbital with a higher value of *n* (principal quantum number).

Similarly the E.A. of beryllium and magnesium also have zero values since the 2*s* and 3*s* orbitals are completely filled respectively and the atoms do not accept an additional electron. Hence also the added electron will have to go to the sub-shell *p* of considerably high energy. Sometimes their E.A. are given a –ve sign.

(*vi*) Nitrogen and phosphorus have lower value of electron affinities. This is due to the fact that their *p*-subshells are half-filled and so they are extra stable. This decreases the possibility of adding an electron.

Q. 56. What is electronegativity? Discuss the variation of electronegativity in a group and in a period. (*G. Nanakdev, 2000; Awadh, 2000; Kumaon, 2000; Meerut, 2002; Nagpur, 2005*)

Ans. *The tendency of an atom to attract the shared electrons in a molecule towards itself is called the electronegativity of the atom.* It is different from electron affinity which is a measure of the tendency of an isolated atom to attract electrons. Electron affinity is thus property of an isolated atom, whereas electronegativity is a property of a bonded atom. It therefore varies with the change in the oxidation state of the element.

For example Pb (II) has electronegativity 1.87 and Pb (IV) has electronegativity value 2.33.

This is clear that the électronegativity increases with the increase in oxidation number or formal charge on an atom. An increase in the oxidation number increases the attraction for electrons and hence, there is an increase in electronegativity also.

Electronegativity increases on moving from left to right in a period. This is due to the progressive decrease in the atomic size and increase in the effective nuclear charge as a result of which the binding electrons are attracted more strongly by the nucleus.

Electronegativity decreases on moving down the group. This is due to the increase in size of the atom so that the binding electron is removed further away from the nucleus and therefore is held less tightly. With the help of the electronegativity values, it is possible to predict the type of bond that may result between any two atoms. If two atoms have similar electronegativities, *i.e.*, there is a similar tendency to attract electrons, the bond between them will be covalent. Similarly, a large difference in electronegativity between two elements leads to an ionic bond.

When the electronegativity difference is about 1.9 the bond is 50% ionic. For a large difference, the bond is more ionic than covalent. If the difference is less than 1.9, bond is more covalent than ionic.

Q. 57. Chlorine possesses high value of electron affinity than that of fluorine.

Ans. See Q. 55.

Q. 58. Electron affinity of Cl is greater than that F but reverse is true for their electronegativity. Explain. (*Lucknow, 2000; Kurukshetra, 2002; Madras, 2004*)

Ans. Electronegativity of an element is different from E.A. It is the tendency of an atom to attract the shared pair of electrons towards itself. Its value decreases down the group. In the group of halogens, since *F* is at the top, it has the maximum electronegativity. In a normal case, even the E.A. would also be maximum for the topmost element in a group. There is a special reason why F has a lower value of electron affinity. This is because in small fluorine atom, strong repulsive forces come into play between the electrons of *F* and the extra electron that would come to it.

Q. 59. Explain why second electron affinities are negative.

Ans. Second E.A. means energy released when a mononegative ion gains another electron to produce dinegative ion.

$$X^- + e^- \longrightarrow X^{2-} + \text{Energy}$$

The second electron faces strong repulsion from the electrons of mononegative ion. Therefore instead of release of energy, absorption of energy takes place. Thus second E.A. is negative.

Q. 60. Al forms an ion $[Al F_6]^{3-}$ but B does not form $[BF_6]^{3-}$ ion. (*Himachal, 2000*)

Ans. $$Al^3 + 6F^- \longrightarrow [Al F_6]^{3-}$$

Al forms the ion $[Al F_6]^{3-}$ because there are empty *d*-orbitals in Al to accommodate the F^- ligands.

B does not form $[BF_6]^{3-}$ because there are no empty *d*-orbitals in B to accommodate the F^- ligands.

Q. 61. The electron affinity for conversion of sulphur atom to sulphide anion is negative. Explain.

Ans. See Q. 59.

Q. 62. What is meant by polarizing power and polarizability? What are Fajan's rules?

(*G. Nanakdev, 2000; Nagpur 2008*)

Ans. If two atoms A^+ and B^- are brought together to the equilibrium distance, the type of bond between them depends on interaction between ions. The +ve ion attracts the electrons on the -ve ion and repels the nucleus, thus distorting or polarizing the -ve ion. If the polarization is quite small, then the ionic bond is formed. If it is more, electrons are drawn from the –ve ion to the +ve ion and appreciable covalent bond formation takes place.

The extent of ion distortion depends on both the power of an ion to distort the other, *i.e., polarizing power*, and susceptibility of an ion to distortion, *i.e., polarizability*. It is the cation which causes polarization and it is the anion which gets polarized. Generally speaking, the polarizing power increases as ions become smaller and more highly charged. The polarizability of a –ve ion is greater than that of a +ve ion since the additional electrons are less firmly bound to the nucleus because of the difference in the effective nuclear charge. This is illustrated in Fig. 8.5.

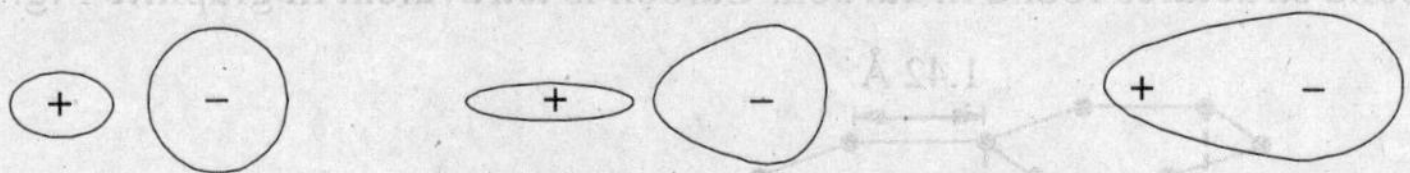

Fig. 8.5. Polarization of ions.

The above observations form the basis of *Fajan's rules*. **Fajan's rules state that covalent bonding is favoured by (*i*) small +ve ion, (*ii*) large –ve ion and (*iii*) large charge on either ion**. In moving along a row of elements, the size of the cations decreases and due to greater effective charge, the polarizing power increases.

For example, degree of covalence increases in the order NaCl, $MgCl_2$ and $AlCl_3$.

On moving down a group the size of the cation increases and in accordance with the Fajan's Rule, the polarizing power decreases. For example, degree of covalence decreases in the order $BeCl_2$, $MgCl_2$ and $CaCl_2$. However, with the increase in the size of the anion on moving down a group, its polarizability and hence the covalent character in the compounds increases. Thus, among the halides of a given cation, bromides and iodides are generally more covalent than the corresponding chlorides and fluorides.

Q. 63. Discuss the variation in metallic character of *p*-block elements (*i*) in a period and (*ii*) down the group. (*Himachal, 2000*)

Ans. Metals have a tendency to lose electrons, if supplied with energy.

$$M + \text{Energy} \longrightarrow M^+ + e^-$$

The stronger is this tendency, the more electropositive and more metallic an element is. Since ionisation energy increases from left to right across the periodic table, the possibility of delocalization of valence electrons becomes less and less. Accordingly, the metallic character of elements from left to right across the periodic table changes to non-metallic character. On moving down a group, the I.E. and the electronegativity of elements generally decrease which accounts for the gradual change from non-metallic to metallic character down a group. Fig. 8.6 illustrates the gradual transition from metallic to non-metallic character. The elements B, Si, Ge, As and Te are metalloids which have metals on their left and non-metals to their right. The noble gases have completed outermost shells which explains their chemical inactivity. They have very high ionisation energy and low electron affinity. It is for the above reasons that these elements exist in atomic state. The atoms of these noble gases are held together by van der Waals forces. These are monoatonic elements.

Li	Be	B	C	N	O	F	Ne
Na	Mg	At	Si	P	S	Cl	Ar
K	Ca	Ga	Ge	As	Se	Br	Kr
Rb	Sr	In	Sn	Sb	Te	I	Xe
Cs	Ba	Ti	Pb	Si	Po	At	Rn

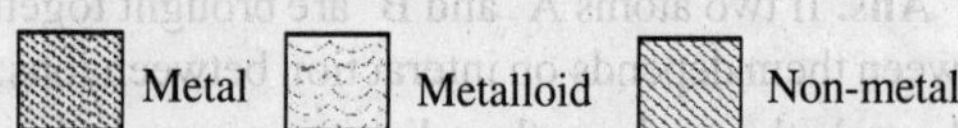

Fig. 8.6. Transition from metallic to non-metallic character.

Q. 64. What are (*i*) network solids and (*ii*) molecular solids?

Ans. (*i*) If an element has bonding capacity to form more than two covalent bonds, as is the case with atoms like C, silicon etc., they are likely to form network solids. Often the atoms of these elements may get bonded in different ways leading to allotropic modifications. Fig. 8.7 illustrates the network solid structures found in carbon. Carbon is tetravalent in graphite Fig. 8.7 (*a*). Three

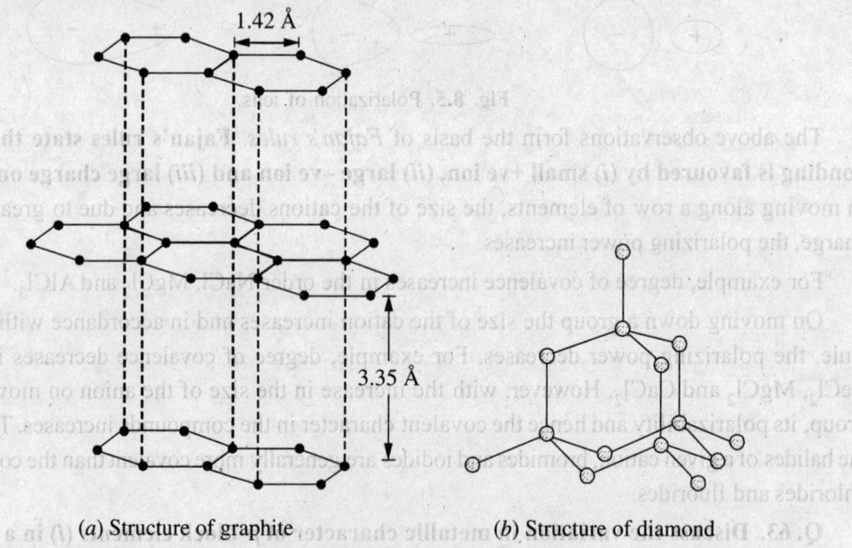

(*a*) Structure of graphite (*b*) Structure of diamond

Fig. 8.7. Network solids.

valencies of each carbon are satisfied with three carbons in the same plane while the fourth valency is completed by bonding to carbons above and below the plane. Because of this structure different layers can slip over one another. In diamond Fig. 8.7 (*b*) every carbon is tetrahedrally linked to four other carbons in a continuous manner. Because of this strict directional bonding diamond is very hard in nature.

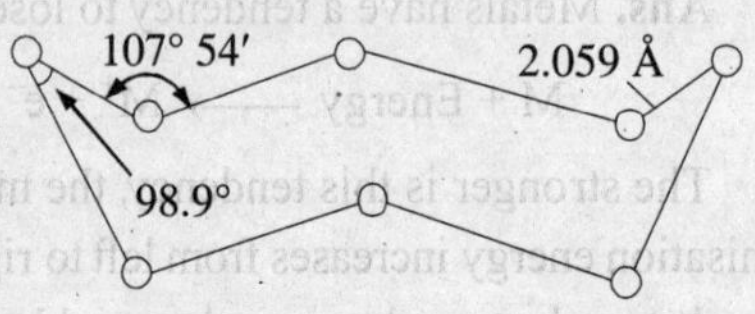

Fig. 8.8. Molecular solids, S_8 molecule.

The bonding capacity of elements on the right of those which form network solids (B, C, Si) decreases and hence they form molecular solids, *e.g.*, S_8 molecule. The molecular solids so formed are held together by van der Waals forces in the solid state of these elements (Fig. 8.8).

Q. 65. Graphite is a conductor of electricity but diamond is not. Explain.

(*Awadh, 2000*)

Ans. This can be explained in terms of structure of graphite and diamond (Fig. 8.7).

Graphite conducts electricity because of loose electrons binding different layers. In diamond, there are no such loose electrons. There are four covalent bonds formed by each carbon atom in a network structure. Therefore diamond does not conduct electricity.

Q. 66. SiO_2 is a solid but CO_2 is a gas. Explain.

(*Awadh, 2000; Himachal, 2000; Meerut, 2000; Kurukshetra, 2002*)

Ans. Silicon is tetravalent. It forms a network solid with oxygen having structure :

```
      |        |        |
      O        O        O
      |        |        |
.... Si — O — Si — O — Si ....
      |        |        |
      O        O        O
      |        |        |
```

Various atoms are bound to each other by covalent bonds which require a large amount of energy to break. Carbon dioxide molecules are held to one another by weak van der Waals forces. It is a gas at room temperature.

Q. 67. Discuss the variation in oxidising and reducing properties of *p*-block elements.

(*Kumaon, 2000; Kurukshetra, 2001*)

Ans. *Oxidation is a process in which electrons are lost and reduction is process in which electrons are gained Thus an oxidizing agent is a substance which can take up electrons while a reducing agent is a substance which supplies electrons.*

It can be understood that elements with low ionisation energy will have greater tendencies to lose electrons than those with higher I.E.

Since the I.E. increases and oxidation potential of *p*-block elements decreases with rising atomic number in a period, the reducing properties decrease from group 13 to group 17 and their oxidizing properties naturally increase in the same order.

In the third period, for instance, while the first element sodium of group 1 is a strong reducing agent, the last element, chlorine of group 17, is a strong oxidizing agent, the intermediate elements, silicon, phosphorus, sulphur, have only an intermediate character.

In a group, the I.E. decreases in moving from the top to bottom. Hence the reducing character increases and oxidizing character decreases as we move down a column. Thus, amongst the elements of group 14, while the first element, carbon, is a weak reducing agent, the last element, lead, is a strong reducing agent.

As a general rule, it may be stated that an element which is highly metallic in nature is also a strong reducing agent and an element which is highly non-metallic in character is a strong oxidizing agent.

As most of chemical reactions are carried out in aqueous solution, the oxidizing and reducing properties of the elements in solution are better understood in terms of standard electrode potential, E° (volts). The more negative the value of the reduction potential, the greater is the tendency of the metal to ionize, *i.e.*, the greater is power to act as reducing agent. Similarly, the powerful oxidizing agent has a large positive value of the reduction potential.

In the following changes:

$$Al \rightleftharpoons Al^{3+} + 3e^-;\ E° = 1.70\ V \text{ and}$$

$$2F^- \rightleftharpoons F_2 + 2e^-;\ E° = -2.80\ V$$

aluminium acts as a powerful reducing agent whereas fluorine acts as powerful oxidizing agent.

Q. 68. Al is a good reducing agent. Explain.

Ans. A substance that can easily donate electrons can acts as a reducing agent. Al has a lower I.E. and a +ve oxidation potential, therefore it is a good reducing agent.

Q. 69. What is inert pair effect?

(*Awadh, 2000; Himachal, 2000; Kurukshetra, 2000; Guwahati, 2002*)

Ans. Where several +ve oxidation states occur among a family of representative elements, they normally differ by two units and the lower oxidation states tend to become more stable as one goes down a group in the periodic table. For example, in group 13, thallium, the heaviest member in the family, is quite unstable in the +3 state and is readily reduced to the more stable +1 state. Likewise in group 14 the Pb^{2+} ion is more stable than Pb^{4+} ion in group 14. The greater stability of lower oxidation state among the heavier *p*-block elements is attributed to tendency of ns^2 electrons to remain paired. That is, these electrons do not take part in bond formation. This is known as **Inert Pair Effect.**

Q. 70. Discuss the variation in basic and acidic character of hydroxides of *p*-block elements.

Ans. Let us consider an element M and its compound M–O–H. The following observation can be made:

(*i*) If the ionisation energy of the element M is low, it tends to lose electron/electrons to the oxygen atom, resulting in the weakening of the M–O bond as

$$M-O-H \longrightarrow M^+ + OH^-$$

So the element 'M' in this case has basic nature.

(*ii*) If the I.E. of element M is high, it draws electrons from the OH group and holds the oxygen atom firmly to itself. Consequently the OH bond becomes weak and breaks up as:

$$M-O-H \longrightarrow M-O^- + H^+$$

Element 'M' in this case has acidic character. It may be noted that

(*a*) There is a gradual decrease in basic character and increase in acidic character of the hydroxides of second and third row of elements.

(*b*) The intermediate elements form acidic or basic solutions of intermediate strength.

(*c*) Halogens (except F) form strongly acidic solutions of oxy-acids.

Q. 71. Why is $(CH_3)_3$ SiOH a stronger acid than $(CH_3)_3$ COH?

Ans. $(CH_3)_3\ SiO^-$ is stabilized due to the presence of vacant 3*d*-orbitals on silicon. Negative charge is easily dispersed into the vacant 3*d*-orbitals. This stabilization is not possible in $(CH_3)_3$ COH because of the absence of *d*-orbitals in carbon.

Q. 72. Discuss the diagonal relationship between B and Si. (*Nagpur, 2005*)

Or

Boron and silicon show similar properties. Explain.

Ans. Boron, the first member of group 13, differs from other members in certain respects and resembles silicon, the second member of group 14, to which it is diagonally related. The various points of similarities are:

(*i*) Both boron and silicon are semiconductors of electricity in their crystalline form and possess high melting points.

(*ii*) Both boron and silicon form oxides such as B_2O_3 and SiO_2 respectively which are acidic in nature. They dissolve in alkalies to form borates and silicates.

$$B_2O_3 + 2\,NaOH \longrightarrow \underset{\text{Sodium metaborate}}{2NaBO_2} + H_2O$$

$$SiO_2 + 2NaOH \longrightarrow \underset{\text{Sodium metasilicate}}{Na_2SiO_3} + H_2O$$

(*iii*) Boron and silicon form hydroxides which are weak acids with no basic properties.

(*iv*) Boron and silicon halides (except BF_3) readily hydrolyse to give their corresponding acids.

$$BCl_3 + 3H_2O \xrightarrow{\text{Hydrolysis}} \underset{\text{Boric acid}}{H_3BO_3} + 3HCl$$

$$SiCl_4 + 4H_2O \xrightarrow{\text{Hydrolysis}} \underset{\text{Orthosilicic acid}}{H_4SiO_4} + 4HCl$$

(*v*) Hydrides of boron and silicon known as boranes and silanes, respectively, are covalent in nature and have similar methods of preparation.

$$\underset{\text{Magnesium bromide}}{Mg_3B_2} + 6HCl \longrightarrow 3MgCl_2 + \underset{\text{Borane}}{B_2H_6}$$

$$\underset{\text{Magnesium silicide}}{Mg_2Si} + 4HCl \longrightarrow 2MgCl_2 + \underset{\text{Silane}}{SiH_4}$$

(*vi*) The nitrides of both boron and silicon decompose steam and liberate ammonia.

$$\underset{\text{Boron nitride}}{BN} + \underset{\text{Steam}}{3H_2O} \longrightarrow H_3BO_3 + NH_3$$

$$\underset{\text{Silicon nitride}}{Si_3N_4} + \underset{\text{Steam}}{12H_2O} \longrightarrow 3H_4SiO_4 + 4NH_3$$

(*vii*) The alkoxides $B(OR)_3$ and $Si(OR)_4$ may be prepared similarly by refluxing benzene solution of the anhydrous chlorides with alcohol as:

$$BCl_3 + \underset{\text{Anhydrous ethanol}}{3C_2H_5OH} \xrightarrow{\text{Reflux}} B(OC_2H_5)_3 + 3HCl$$

$$SiCl_4 + 4C_2H_5OH \xrightarrow{\text{Reflux}} Si(OC_2H_5)_4 + 4HCl$$

Q. 73. Discuss the variation in melting point and boiling point along the period and down the group in *p*-block element.

Ans. The melting and boiling points depend on the nature of chemical bonding between atoms of these elements. It may be noted that the elements on the extreme left of the row are metals and involve metallic bonds. The atoms for their separation require small energy to separate the metallic bonds which are not as strong as covalent bonds and hence have lower melting and boiling points.

The network solids having two-dimensional or three-dimensional giant molecules, *e.g.*, C, Si, etc., on the other hand involve covalent bonds. Each atom may have two or more than two bonds to break away, which explains the high melting and boiling points of network solids. To the right of these elements are N, P, S and Cl. These elements form molecular solids. In these molecules, the atoms are strongly covalent bonded but the molecules are attracted towards each other by weak van der Waals forces. Since it is the molecule which breaks away from the solid, it requires little thermal energy to overcome the van der Waals forces. Accordingly, the melting points and boiling points of these elements on the extreme right of the periodic table are lowest.

In a group the metallic character increases and therefore the m.p. and b.p. decrease on moving down the group.

Q. 74. Discuss the stability of In^+ and Tl^+ ions.

Ans. Indium and thallium belong to group 13 containing B, Al, Ga, In and Tl. Both indium and Tl form trivalent ions, In^{3+} and Tl^{3+}, but they also form monovalent ions, In^+ and Tl^+, respectively. Loss of electrons from an atom of an element can take place in two ways. Look at the electronic configuration of In given below:

	In	2,	8,	18,	18,	3	$(5\,s^2\,p^1)$
I	In^{3+}	2,	8,	18,	18		
II	In^+	2,	8,	18,	18,	2	

Either all its *s* and *p* electrons $(5s^2\,p^1)$ may be lost as in I forming an atom in which its oxidation number equals three or the outermost *s*-electrons are not lost, only *p*-electron is lost and the oxidation number equals 1 (as in II). In this case, the pair of *s*-electrons is said to have become inert.

Similarly for thallium, the formation of ions can be shown as follows:

$$Tl = 2, 8, 18, 32, 18, 3\ (6s^2\,p^1)$$

(At. no. = 81)

$$Tl^{3+} = 2, 8, 18, 32, 18$$

$$Tl^+ = 2, 8, 18, 32, 18, 2$$

It has been observed that compounds of Tl^+ are more stable than those of Tl^{3+}. In fact Tl^{3+} salts are oxidizing agents. On the other hand, In^+ ions are immediately converted into In^{3+} ions and the metal (In) in presence of water. Gallium forms no monovalent ions. This explains the tendency of the elements in a group to show more inert pair effect as we move down the group.

As a matter of fact Tl^+ ion has the same electronic configuration as that of the Pb^{2+} ion and this explains the similarities between certain Tl^+ and Pb^{2+} compounds. Tl^+ chloride, bromide, iodide, sulphate and sulphide are only slightly soluble like the corresponding Pb^{2+} compounds.

Q. 75. Discuss the oxidation states of Sn and Pb.

Ans. Sn and Pb both form the expected tetravalent ions but they also form divalent ions as shown below:

$$Sn = 2, 8, 18, 18, 4\ (5\,s^2\,p^2)$$

(At. no. = 50)

$$Sn^{4+} = 2, 8, 18, 18$$

$$Sn^{2+} = 2, 8, 18, 18, 2\ (5s^2)$$

Pb $= 2, 8, 18, 32, 18, 4\ (6s^2 p^2)$

(At. no. = 82)

$Pb^{4+} = 2, 8, 18, 32, 18$

$Pb^{2+} = 2, 8, 18, 32, 18, 2\ (6s^2)$

As the inert pair effect is more marked in the heavier element, *i.e.*, Pb, the Pb^{2+} ion is more stable than the Pb^{4+}. On the other hand, Sn^{4+} ion is more stable than the Sn^{2+} ion. This explain why SnO and $SnCl_2$ are reducing agents whereas PbO_2 and $PbCl_4$ are oxidizing agents. Moreover, $PbBr_4$ and PbI_4 do not exist but corresponding compounds of Sn can be obtained. The non-existence of $PbBr_4$ and PbI_4 is due to the oxidation of Br^- and I^- ions by Pb (IV) ion to Br_2 and I_2 and in the process Pb (IV) is reduced to Pb (II) ion.

$$PbBr_4 \longrightarrow PbBr_2 + Br_2$$

$$PbI_4 \longrightarrow PbI_2 + I_2$$

The lower oxidation states are more ionic since the radius of M^{2+} is greater than M^{4+} and according to Fajan's rules, the smaller the ion, the greater is the tendency to covalency. The reduction potential values indicate that Pb (IV) is a stronger oxidant than Sn (IV), Pb^{4+} / Pb^{2+}, + 1.46 volts. Sn^{4+} / Sn^{2+}, + 0.188 volt) and conversely, Sn (II) is superior to Pb (II) as a reducing agent.

Q. 76. Explain why $PbBr_4$ and PbI_4 do not exist.

Ans. See. Q. 75.

Q. 77. Discuss the oxidation states of Sb and Bi.

Ans. Both Sb and Bi form trivalent as well as pentavalent ions.

$Sb = 2, 8, 18, 18, 5\ (5s^2p^3)$

(At. no. = 51)

$Sb^{5+} = 2, 8, 18, 18$

$Sb^{3+} = 2, 8, 18, 18, 2\ (5s^2)$

$Bi = 2,8, 18, 32, 18, 5\ (6s^2 p^3)$

(At. no. = 83)

$Bi^{5+} = 2, 8, 18, 32, 18$

$Bi^{3+} = 2, 8, 18, 32, 18, 2\ (6s^2)$

The inert pair effect, however, is more marked in the case of Bi which is the heaviest element of this group. Thus, Bi forms only the trichloride, $BiCl_3$, but not the pentachloride, $BiCl_5$, whereas P, As and Sb are trivalent as well as pentavalent.

Q. 78. Bring out points of similarity and dissimilarity between boron and aluminium?

Ans. Points of similarity. (*i*) *Covalent character.* Both B and Al form covalent compounds, *e.g.*, BCl_3 and $AlCl_3$. Their chlorides are volatile and hygroscopic. They are hydrolysed by water.

$$BCl_3 + 3H_2O \longrightarrow B(OH)_3 + 3HCl$$

$$AlCl_3 + 3H_2O \longrightarrow Al(OH)_3 + 3HCl$$

(*ii*) *Nitrides.* Both B and Al form nitrides which react with water to give ammonia.

$$BN + 3H_2O \longrightarrow H_3BO_3 + NH_3$$

$$AlN + 3H_2O \longrightarrow Al(OH)_3 + NH_3$$

(*iii*) *Oxides.* Both B and Al form oxides when they are heated in air.

$$6B + 3O_3 \longrightarrow 3B_2O_3$$

$$4Al + 3O_2 \longrightarrow 2Al_2O_3$$

(*iv*) *Action of alkalis.* Both B and Al react with alkalies to evolve hydrogen.

$$2B + 6NaOH \longrightarrow \underset{\text{Sodium borate}}{2Na_3BO_3} + 3H_2$$

$$2Al + 2H_2O + 2NaOH \longrightarrow \underset{\text{Sod. meta-aluminate}}{2NaAlO_2} + 3H_2$$

(*v*) *Halides as Lewis acids.* Halides of both elements such as BCl_3 and $AlCl_3$ behave as Lewis acids, because boron and aluminium atoms are electron deficient.

Points of Dissimilarity

Due to small size, high electronegativity and high I.E., boron differs from aluminium and other members of the group in the following respects:

(*i*) *Melting point.* Boron is quite hard with high m.p. while other members of the group are soft with low m.p.

(*ii*) *Electrical conductivity.* Boron is a bad conductor while other members are good conductors of electricity.

(*iii*) *Metallic character.* Because of smaller size and higher I.E. boron is non-metal while other members of the group are metals.

(*iv*) *Nature of oxides and hydroxides.* Oxides and hydroxides of boron are acidic in nature.

$$B_2O_3 + 3H_2O \longrightarrow \underset{\text{Boric acid}}{2H_3BO_3}$$

Hydroxides and oxides of Al are amphoteric in nature.

$$Al_2O_3 + 6HCl \longrightarrow 2AlCl_3 + 3H_2O$$

$$Al_2O_3 + 2NaOH \longrightarrow \underset{\text{Sod. meta-aluminate}}{2NaAlO_2} + H_2O$$

On the other hand, other members form basic oxides and hydroxides.

(*v*) *Action of acids.* Boron does not displace hydrogen from acids but other members do. For example,

$$2Al + 6HCl \longrightarrow 2AlCl_3 + 3H_2$$

With conc. HNO_3 boron is oxidized to boric acid while Al becomes passive.

(*vi*) *Action of steam.* Water or steam has no action on boron but other members decompose steam to give out hydrogen.

$$2Al + 6H_2O \longrightarrow 2[Al(OH)_3] + 3H_2$$

Q. 79. Discuss the trends in acidic nature and basic nature of the oxides of *p*-block elements in a group and in a period.

Ans. As we move in a period, in *p*-block, from left to right, the basic character of the oxides decreases (as the metallic character of the elements decreases) while acid character increases. On the other hand, in a group the basic character of oxides increases as we descend a group. Metallic character of elements increases.

The behaviour of oxides of third row and group 13 elements is given below.

III Row	Oxides :	Na_2O	MgO	Al_2O_3	SiO_2	P_2O_5	SO_2	Cl_2O_7
	Nature :	Strongly basic	Basic	Amphoteric	Weakly acidic	Acidic	More acidic	Most acidic
13 Group	Oxides :	B_2O_3	Al_2O_3	Ga_2O_3	In_2O_3	Tl_2O		
	Nature :	Acidic	Amphoteric	Weakly basic	Basic	Strongly basic		

We observe that acid character of 3rd row elements increases from Na to Cl and basic character of oxides of group 13 increases from B to Tl.

Q. 80. Categorise the following oxide in terms of acidic, basic, and amphoteric character:

BeO, B_2O_3, MgO, SiO_2 and N_2O_5

Ans. Basic oxide—MgO

Acidic oxides – B_2O_3, SiO_2, N_2O_5

Amphoteric oxide – BeO

Q. 81. Discuss hydride formation of *p*-block elements. Compare the acid strength of these hydrides.

Ans. All the elements of the *p*-block react with hydrogen to form covalent hydrides, which are volatile. There is however some ionic character in them depending upon electronegativity of the element. As electronegativity increases along the period, the covalent hydrides become more and more acidic. Thus the acidity of hydrides of second period elements CH_4, NH_3, H_2O and HF increases with the electronegativity of the non-metal present from C to F. The acidity can also be related to increasing stability of their anions as shown below:

$$\xrightarrow[\text{Increasing stability}]{CH_3^- < NH_2^- < OH^- < F^-}$$

In group 13, elements form characteristic covalent hydrides.

The groups 16 and 17 hydrides (H_2O, H_2S, H_2Se and HF, HCl, HBr, HI) show an increase in acidity on going down the group which is against normal expectation based on electronegativity concept. Here the major factor is the heat of dissociation which shows a fall down a group in the hydrides of these groups, thus causing an increase in the acid strength of the hydrides. Ionic hydrides are useful reducing agents and usually their reducing power decreases across a period and increases down a group.

Q. 82. Compare the properties of halides of *p*-block elements.

Ans. All the *p*-block elements form halides. Some of the halides show strong ionic character. There is a gradation from purely ionic halides through those of intermediate character to those which are clearly covalent. The halides of the alkali metal (group 1) are ionic in character. However, as we move along a period with decrease in ionic radius, *i.e.*, with increase in the charge radius ratio, the covalent character of the halides increases. Thus amongst chlorides of 3rd period elements, *viz.*,

NaCl, $MgCl_2$, $AlCl_3$ and $SiCl_4$, the first one is completely ionic while the last one is essentially covalent. On moving down a group, the electropositive nature of the element increases and thus the ionic nature of halides increases.

Some of the metals form halides in several oxidation states. In such cases, the halides formed in the lower oxidation states are ionic while those in the higher oxidation states tend to be covalent. Thus while $PbCl_2$ is largely ionic, $PbCl_4$ is largely covalent.

The size of the halide ion is another important factor in determining the character of the halides of elements. Thus, while AlF_3 is essentially ionic, $AlCl_3$ has an intermediate character and $AlBr_3$ and AlI_3 exist as covalent dimers. This behaviour can be explained by Fajans's rules.

Most of the ionic halides of *p*-block elements are soluble in water. There are, however, a few exceptions as for example $PbCl_2$. The covalent halides do not show any trend in their solubility. If a given metal forms all the four halides, MF_n, MCl_n, MBr_n, MI_n, solubility decreases in the order

$$I^- > Br^- > Cl^- > F^-$$

This is also because ionic character decreases from iodide to fluoride.

Q. 83. Which of the following is ionic? Justify your answer:

(*i*) $SnCl_2$ (*ii*) $SnCl_4$.

Ans. $SnCl_2$ is ionic and $SnCl_4$ is covalent because Sn^{4+} has large polarizing power. It polarizes chloride ions and induces covalent character according to Fajan's rules.

Q. 84. CCl_4 is not hydrolysed with water while $SiCl_4$ is readily hydrolysed. Explain.

Ans. $SiCl_4$ gets hydrolysed with the help of empty 3*d*-orbitals which accommodate electron pairs of H_2O to form intermediate compound which further decomposes into SiO_2.

There is no empty *d*-orbital in carbon. Hence there is no possibility of such a reaction.

$$SiCl_4 + H_2O \longrightarrow \left[Cl_3(Cl)Si \leftarrow :OH_2 \right] \longrightarrow SiO_2 + 4HCl$$

Intermediate compound

Q. 85. A metallic element M forms two stable chlorides with formulae MCl_3 and MCl_5 respectively. Where the metal M is to be placed in the periodic table?

Ans. Metal M belongs to group 15. The configuration in the outermost orbit is $ns^2\ np^3$.

MCl_5 will be obtained when all the five electrons are removed whereas MCl_3 will be obtained when *s* electrons remain inert and only the electrons belonging to *p*-orbital are removed.

Q. 86. A metallic element M forms two stable chlorides with formulae MCl and MCl_3 respectively. Where the element M is to be placed in the periodic table?

Ans. Metal M is to be placed in group 13. These elements have the configuration $ns^2\ np^1$ in the outermost orbit. MCl and MCl_3 are obtained when one electron in the *p*-orbital and all the three electrons respectively are removed.

Q. 87. PbI_4 is unstable while SnI_4 is stable. Explain.

Ans. In Sn, + 4 oxidation state is more stable. That is why SnI_4 is stable.

In Pb, + 2 oxidation state is more stable because of inert pair effect in lead. It decomposes to more stable compound PbI_2 as follows:

$$PbI_4 \longrightarrow PbI_2 + I_2$$

Q. 88. Fluorine cannot be liberated from fluoride by a chemical reagent. Explain.

Ans. Oxidation potential for the reaction

$$2F^- \longrightarrow F_2 \text{ is } -2.80 \text{ volts}$$

This is the maximum oxidation potential in the periodic table. There is no reagent which has oxidation potential more than this. Hence liberation of fluorine from fluoride is not possible.

Q. 89. HF is the weakest acid among hydrogen halides. Explain.

Ans. Bond dissociation energy in H – X, where X stands for halogen, decreases from F to I. Thus HF has the strongest bond and most difficult to break. Therefore HF is the weakest acid.

Q. 90. What is meant by $d\pi - p\pi$ bond?

Ans. A bond that is formed between lateral overlap of a *p*-and *d*-orbital is called $d\pi - p\pi$ bond. For example, 3*d* orbitals are involved in $d\pi - p\pi$ multiple bonding between sulphur and oxygen in SO_2 and SO_3 as shown in Fig. 8.9.

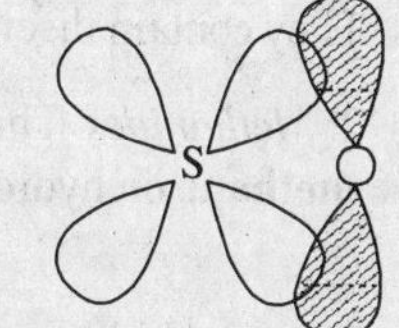

Fig. 8.9. $d\pi$ -p π bonding.

Q. 91. Write a detailed note on fullerenes.

Ans. Prof. Kroto of Susex University (U.K.) and Prof. Smalley of Rice University (USA), while conducting experiments on carbon, vaporised graphite with the help of lasers. To their surprise, some atoms which were vaporised by lasers, regrouped together to form hollow shell-like structures. In such structures, sixty or even more carbon atoms joined together to produce cage-like structures. This form of carbon (allotrope) was given the name **Buckminster fullerene** after the name of a famous architect **Buckminster fuller,** who shaped famous Geodesic dome.

This discovery has opened new vistas in the chemistry of carbon compounds. It has been found that *Buckminster fullerene* agglomerate which is of the shape of a base-ball is big enough to accommodate atoms of potassium to form $C_{60}K$, which is grey in colour. It will give new dimensions to the chemisry of carbides (alkali metal and others). Such carbides show super-conductive properties (no resistance to the flow of current) which are so vital in this age of power-crisis. Lot of further research is going on. Such carbides ($C_{60}K$) are highly coloured and thus paint a rosy picture about their success as *materials*.

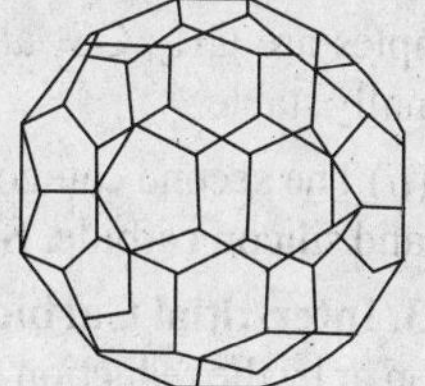

Fig. 8.10. Fullerene

Graphites have sheet structures, in which the carbon atoms occupy hexagonal positions, different sheets being held together by weak van der Waals' forces. On evaporation by lasers, the carbon atoms arrange themselves in mixed pentagons and hexagons as shown in Fig. 8.10 maintaining the basic sp^2 hybridisation scheme and producing ball-like hollow shells. More recently even bigger fullerenes containing 70, 76 and 78 carbons have been reported.

Q. 92. Give a detailed account of carbides.

Ans. Metal carbides can be discussed under the following heads:

1. Salt-like Carbides. Carbides of this type are simple compounds of carbon with more electropositive elements, *i.e.,* with metals of Groups 1, 2, 3. The metal, its oxide or hydride is heated strongly in an electric arc with carbon to get the required carbide. For example,

Be_2C is obtained by heating BeO and C at 1900-2000°C.

$$2BeO + 3C \xrightarrow{1900-2000°C} Be_2C + 2CO$$

Calcium carbide is obtained by heating lime and coke at 2200 — 2250°C.

$$CaO + 3C \longrightarrow CaC_2 + CO$$

Aluminium carbide is prepared by heating Al and C in an electric furnace.

$$4Al + 3C \longrightarrow Al_4C_3$$

In general, these compounds exist as transparent crystals and in the solid state they are non-conductors of electricity.

They are readily hydrolysed in water yielding different hydrocarbons. These carbides are further classified according to the hydrocarbon which they produce on hydrolysis.

1. *Acetylides.* The carbides which yield **acetylene** on hydrolysis are known as **acetylides.** MgC_2, BeC_2 and CaC_2 fall in this category.

$$CaC_2 + 2H_2O \longrightarrow Ca(OH)_2 + C_2H_2$$

They contain discrete C_2^{2-}, *i.e.,* $(C \equiv C)^{2-}$ units.

2. *Methanides.* The carbides such as aluminium and beryllium carbides, Al_4C_3 and Be_2C, which give **methane** on hydrolysis, are known as **methanides.**

$$Al_4C_3 + 12H_2O \longrightarrow 4Al(OH)_3 + 3CH_4$$

These carbides contain C^{4-} ions.

3. *Allylides.* Magnesium carbide, Mg_2C_3, is the only carbide which yields **allylene** on hydrolysis.

$$Mg_2C_3 + 4H_2O \longrightarrow 2Mg(OH)_2 + \underset{\text{Allylene}}{CH_3C \equiv CH}$$

2. Covalent Carbides. These carbides contain carbon covalently bonded with elements of comparable electronegativity such as hydrogen, sulphur and chlorine. The covalent carbides can be further divided into two categories.

(*i*) The first category consists of covalent carbides which contain *small discrete molecules.* These are formed by combination of carbon with hydrogen or the elements of Groups VI and VII. The examples are CH_4, CS_2 and CCl_4. These carbides are either gases or volatile liquids and are not thermally stable.

(*ii*) The second category consists of giant molecules. The common examples are boron carbide, B_4C and silicon carbide, SiC. These are thermally stable, hard and chemically inert solids.

3. Interstitial Carbides. These carbides are obtained, by the direct combination of metal and carbon or by the reduction of metal oxide with carbon at high temperature close to 2000°C. Transition metals like Ti, Zr, Hf, V, Mb, Ta and Mo form these carbides. The metals retain their close-packed lattice strcutures and *incorporate carbon atom in their octahedral interstices* to give MC and M_2C type carbides. The common examples are WC, TaC, V_2C, Nb_2C, Ta_2C, W_2C. These carbides melt at very high temperatures. They are extremely hard refractory materials and have melting points in the range of 3000–4000°C. They are very hard and brittle. They are good conductors of electricity and are chemically inert. Due to their extensive hardness and chemical inertness WC and TaC are used for the manufacture of high speed cutting tools.

Uses of Carbides in Industry.

1. Silicon carbide is used as *abrasive* under the name *carborundum* and also as a *refractory material*. It is also very resistant to abrasion and corrosion. It also finds applications in atomic reactors as it provides a good material for making containers for fuel elements at high temperatures.

2. Tungsten carbide, WC, is used in making tools and dies.

3. Boron carbide, B_4C, which is harder than silicon carbide, is being used now increasingly as an abrasive. It is also used as a shield against radioactive radiations.

Q. 93. What is meant by fixation of nitrogen? What are the different ways to achieve it?

Ans. The conversion of atmospheric nitrogen into useful nitrogenous compounds by natural or artificial methods is called fixation of nitrogen. Nitrogen present in these nitrogenous compounds is called *fixed* or *combined* nitrogen.

Natural Methods of the Fixation of Nitrogen

1. *By lightning discharges.* The nitrogen and oxygen present in air combine together to form nitric oxide under the influence of lightning discharges. Nitric oxide gets oxidised by excesss of oxygen present in the atmosphere to form nitrogen peroxide which dissolves in water to form nitric acid. This acid is washed down by rain into the soil, where it reacts with limestone and alkalis of the soil to form nitrates and is stored there as plant food.

$$N_2 + O_2 \xrightarrow{\text{Lightning}} 2NO \quad (\textit{Nitric oxide})$$

$$4NO + 3O_2 \longrightarrow 2N_2O_5 \quad (\textit{Nitrogen pentoxide})$$

$$N_2O_5 + H_2O \longrightarrow 2HNO_3$$

$$2HNO_3 + CaCO_3 \longrightarrow Ca(NO_3)_2 + CO_2 + H_2O$$

2. *By symbiotic bacteria.* The atmospheric nitrogen is being constantly transferred to the soil through certain bacteria called *symbiotic bacteria.* They grow in small nodules in the roots of plants belonging to the family of *Leguminacea* (pea, gram etc.) and directly assimilate atmospheric nitrogen and convert it into products useful for plant growth.

Artificial Methods Used for the Fixation of Nitrogen

1. *Fixation of nitrogen as HNO_3.* Under the influence of high tension electric arc where the temperature is high, nitrogen of the air combines with oxygen to form nitric oxide. It combines with more of oxygen to form nitrogen peroxide. This may be absorbed in water in presence of excess of air to give nitric acid which may be used for the manufacture of nitrogenous fertilisers.

$$N_2 + O_2 \rightleftharpoons 2NO \quad (\textit{Nitric oxide})$$

$$2NO + O_2 \longrightarrow 2NO_2 \quad (\textit{Nitrogen peroxide})$$

$$\underset{\text{Nitrogen Peroxide}}{4NO_2} + 2H_2O + \underset{\text{From air}}{O_2} \longrightarrow \underset{\text{Nitric acid}}{4HNO_3}$$

2. *Fixation of nitrogen as NH_3 and ammonium salts.* A mixture of nitrogen (manufactured by liquefaction of air) and hydrogen in the ratio 1 : 3 is compressed to a pressure of 200 — 500 atmospheres and is passed over a catalyst (finely divided iron + molybdenum) heated to about 550°C. This forms the *Haber's process* for the manufacture of ammonia which then can be converted into ammomium salts by treatment, with suitable acids.

$$N_2 + 3H_2 \underset{\text{high press}}{\overset{\text{catalyst}}{\rightleftharpoons}} 2NH_3$$

3. *Fixation of nitrogen as calcium cyanamide, $CaCN_2$.* Nitrogen gas obtained by the evaporation of liquid air is passed over calcium carbide heated to 800 — 1000°C. A mixture of calcium cyanamide and carbon is obtained which is extensively used as a fertilizer under the name of *nitrolim.*

$$\underset{\text{Calcium carbide}}{CaC_2} + N_2 \longrightarrow [\underset{\text{Calcium cynamide}}{CaCN_2} + C] \; (\textit{Nitrolim})$$

4. *Fixation of nitrogen as nitrides.* Nitrogen combines with magnesium and aluminium at high temperature to give nitrides which are employed as a source of ammonia. These nitrides are decomposed

by H_2O and NH_3 is evolved.

$$3Mg + N_2 \longrightarrow \underset{\substack{\text{Magnesium}\\ \text{nitride}}}{Mg_3N_2}$$

$$2Al + N_2 \longrightarrow \underset{\substack{\text{Aluminium}\\ \text{nitride}}}{2AlN}$$

$$AlN + 3H_2O \longrightarrow Al(OH)_3 + NH_3$$

Q. 94. How is tetrasulphur tetranitride, S_4N_4 prepared? Give its structure.

Ans. Preparation. It can be prepared by one of the following reactions:

1. $10S + 4NH_3 \longrightarrow S_4N_4 + 6H_2S$
2. $4NH_4Cl + 6S_2Cl_2 \xrightarrow{160°C} S_4N_4 + 8S + 16HCl$
3. $16NH_3 + 6S_2Cl_2 \xrightarrow[CCl_4]{50°C} S_4N_4 + 8S + 12NH_4Cl$

Properties

1. S_4N_4 is an orange compound with a melting point of 178°C.
2. It explodes when struck or heated.
3. It forms addition compounds $S_4N_4 . BF_3$ and $S_4N_4 . SbCl_5$ with BF_3 and $SbCl_5$, respectively.
4. S_4N_4 gets hydrolysed by dilute and concentrated alkali solutions.

$$2S_4N_4 + 6OH^- \text{ (dilute)} + 9H_2O \longrightarrow S_2O_3^{2-} + 2S_3O_6^{2-} + 8NH_3$$

$$S_4N_4 + 6OH^- \text{ (concentrated)} + 3H_2O \longrightarrow S_2O_6^{2-} + 2SO_3^{2-} + 4NH_3$$

Structure of S_4N_4. Tetrasulphur tetranitride has a cradle-like structure as shown in Fig. 8.11

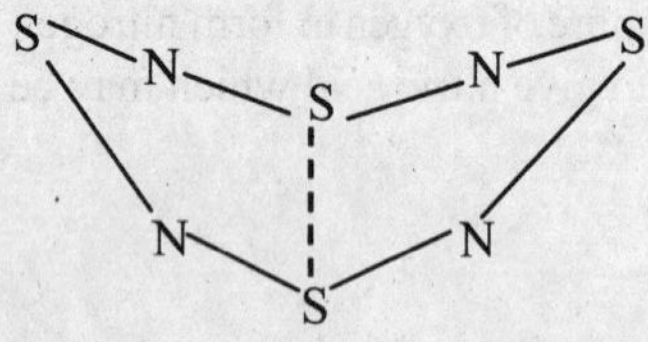

Fig. 8.11. Structure of S_4N_4

All the S – N bond distances are equal (1.60Å) which are shorter than the calculated S – N single bond distance of 1.75 Å. This indicates some double bond character in S—N bonds. The trans annular S – S distances are shorter than twice the van der Waals' radius for S indicating that there may be a weak association between the trans annular S atoms. The actual structure of S_4N_4 is a resonance hybrid of a number of the following types of structures :

$$\begin{array}{ccccccccccccccc}
N & = & S & = & N & & N & - & S & - & N & & N & = & \overset{+}{S} - \overset{-}{N} & & N - S = N \\
| & & & & | & & \| & & & & \| & & | & & | & & \| \qquad\;\; | \\
S & & & & S & \leftrightarrow & S & & & & S & \leftrightarrow & S & & S & \leftrightarrow & S \text{———} S \;\; \text{etc.} \\
| & & & & | & & \| & & & & \| & & | & & | & & | \qquad\;\; \| \\
N & = & S & = & N & & N & - & S & - & N & & N^- & - & \overset{+}{S} = N & & N = S - N
\end{array}$$

Q. 95. What are fluorocarbons? Give methods of preparation, properties, and uses.

Ans. Fluorocarbons are hydrocarbons or other organic molecules in which most of the hydrogen atoms are replaced by fluorine atoms. These compounds have wide industrial applications.

Preparation. Fluorocarbons can be prepared by the following methods.

1. By heating organic halides with anhydrous HF in the presence of SbF_5 at 50°-100°C temperature and 50-500 atm pressure.

$$2CCl_4 + 3HF \xrightarrow[50-100°C \; 50-500\,atm]{SbF_5} CCl_2F_2 + CCl_3F + 3HCl$$

2. By fluorinating oxygen-containing organic compounds with SF_4.

$$R(R')C{=}O + SF_4 \longrightarrow R(R')CF_2 + SOF_2$$

3. By heating metal fluorides (MF_X) such as LiF, NaF, CsF, HgF_2, ZnF_2, SbF_3, etc., with organic halides.

$$RCl + MF \xrightarrow{\Delta} \underset{\text{Fluorocarbon}}{RF} + MCl$$

The difference in the lattice energies of MF and RCl pushes the reaction forward. Within a group, the above difference increases with increase in the size of the metal M. Hence, within a group, the fluorides with bigger metal ions make the reaction more facile. Accordingly, CsF is a better fluorinating agent than NaF.

4. By electrolysing organic compounds in liquid HF in steel cells with Ni anodes and steel cathodes. For example,

$$C_8H_{18} \xrightarrow[\text{in liquid HF}]{\text{Electrolysis}} C_8F_{18}$$

$$(C_4H_9)_3N \xrightarrow[\text{in liquid HF}]{\text{Electrolysis}} (C_4F_9)_3N$$

5. By reacting organic compounds with fluorine in the presence of a catalyst such as copper gauze. The reaction is carried out in the presence of an inert gas such as helium or nitrogen.

$$C_6H_6 + 9F_2 \xrightarrow[\text{Cu at 256°C}]{N_2 \text{ or He}} C_6F_{12} + 6HF$$

Properties. 1. Since bond energy of C – F bond (486 kJ mol^{-1}) is very high fluorocarbons are highly stable compounds.

2. Fluorine atom has a small size so that the replacement of H atoms by F atoms does not produce any appreciable strain in the molecule. Replacement of H atoms by other halogens causes appreciable strain in the resulting molecules because of larger differences in the sizes of H and halogen atoms. As a result, the fluoro compounds are more stable than the other halogen compounds.

3. Fluorocarbons are stable towards hydrolysis. This is because carbon does not have the *d* orbitals which could permit the formation of the 5-coordinated intermediate by necleophilic attack of H_2O molecule.

4. Fluorocarbons are not attacked by acids, alkalies and most other chemicals.

5. Completely fluorinated compounds have very low boiling points. The intermolecular forces in these compounds are also very low as a result of which these compounds have very low coefficients of friction.

6. Fluorine being more electronegative than carbon, pulls the electronic charge of the bond pair of electrons in C—F bond towards itself. Carbon atom in C—F bond thus develops a partial positive charge and can be considered as *oxidised.* The fluorocarbons are not oxidised at all because the partial positive charge on the carbon atom of fluorocarbons makes further loss of electrons from it extremely difficult.

7. The fluorine atoms in fluorocarbons protect the carbon atoms effectively making them less vulnerable to attack by other reagents. The fluoro compounds are thus not very reactive.

Uses. 1. $CF_3CHBrCl$ and some similar other compounds are used as anaesthetics.

2. $CHClF_2$ is used for preparing tetrafluoro ethylene which is further converted into teflon by polymerisation.

$$2CHClF_2 \xrightarrow[\text{Polymerisation}]{500-1000°C} CF_2=CF_2+2HCl$$

$$nCF_2=CF_2 \longrightarrow \underset{\text{Teflon}}{-(CF_2-CF_2)_n-}$$

3. Fluorocarbons are used as lubricants because of their very low coefficients of friction.
4. Chlorofluorocarbons such as CCl_3F, $CClF_3$ and CF_2Cl_2 are used as refrigerants.

EXTRACTION OF LITHIUM

Q. 96. How will you extract lithium from its ores? *(Calcutta, 2007)*

Ans. The chief ores of lithium are:

1. *Lepidolite or lithia mica* $(Li, Na, K)_2Al_2(SiO_3)_3(F, OH)_3$
2. *Petalite*, $LiAl(Si_2O_5)_4$,
3. *Spodumene*, $LiAl(SiO_3)_2$
4. *Triphylite*, $(Li,Na)_3PO_4.(Fe,Mn)_3(PO_4)_2$

The procedure for the extraction of lithium from its ores involves the following steps.

1. Preparation of lithium chloride. The minerals are first of all converted into lithium chloride. The mineral is finely powdered and boiled with sulphuric acid. The insoluble silica (SiO_2) thus formed is filtered off. The solution is treated with the requisite amount of sodium carbonate to precipitate aluminium and iron. Excess sodium carbonate is then added to the filtrate to precipitate lithium as carbonate which is filtered and dissolved in hydrochloric acid to obtain lithium chloride which is purified by extraction with alcohol.

2. Electrolysis of lithium chloride. A mixture of dry lithium chloride and potassium chloride is fused and electrolysed in an electrolytic cell shown in Fig. 8.12. The addition of potassium chloride increases the conductivity of lithium chloride and lowers its melting point. The cell is operated at a temperature close to 400°C and a voltage of 8-9 volts is applied.

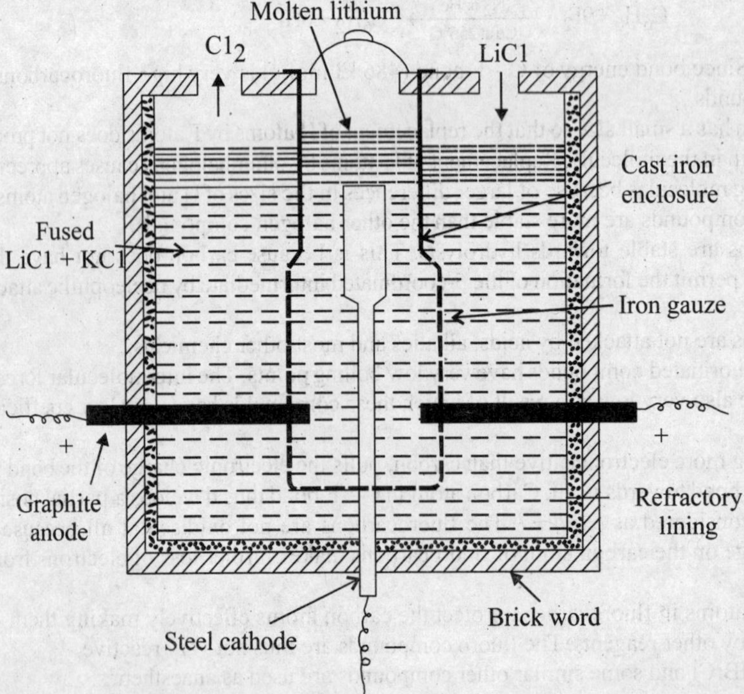

Fig. 8.12 Electrolytic preparation of lithium.

The reactions taking place at the electrodes are:

$$LiCl \rightleftharpoons Li^+ + Cl^-$$
$$Li^+ + e^- \longrightarrow Li \text{ (at cathode)}$$
$$2Cl^- - 2e^- \longrightarrow Cl_2 \text{ (at anode)}$$

Chlorine gas liberated at the anode which leaves the cell through an exit and the molten lithium metal rises to the surface of the fused electrolyte and collects in the cast iron enclosure surrounding the cathode.

Q. 97. Describe preparation and properties of lithium carbonate(Li_2CO_3). (*Calcutta, 2007*)

Ans. It is prepared by adding ammonium carbonate to a solution of a soluble salt of lithium in the presence of ammonia.

$$2LiCl + (NH_4)_2CO_3 \xrightarrow{NH_3} Li_2CO_3 + 2HN_4Cl$$

Lithium carbonate, being sparingly soluble, gets precipitated. It is filtered and dried.

Properties

1. It is white stable powder which decomposes to oxide on heating to above 600°C ($Li_2CO_3 \longrightarrow Li_2O + CO_2$). Li_2CO_3 dissociates into Li_2O more easily than other alkali metal carbonates. This is due to the fact that Li_2O is more stable than other alkali metal oxides. Li^+ ion because of small size has greater polarising power than the other alkali metal cations. It, therefore, polarises the electron charge cloud of O^{2-} ion towards itself to a considerable extent. This allows partial overlap between the orbitals of Li^+ and O^{2-} ions introducing some covalent character in Li—O bond apart from the ionic character in Li_2O. This gives extra stability to Li_2O. The other alkali metal oxides, M_2O, are almost entirely ionic in character.

2. Lithium carbonate is only sparingly soluble in water like the carbonates of calcium and magnesium. On passing carbon dioxide through a suspension of lithium carbonate, the latter dissolves due to the formation of soluble lithium bicarbonate.

Uses

1. The solution of lithium bicarbonate is used in medicine as a remedy for gout and rheumatism under the name *lithia water.*

2. Lithium carbonate is used in glass industry.

Q. 98. Discuss briefly extracion of magnesium from its ores.

Ans. Important ores of magnesium are : *magnesite,* $MgCO_3$; *dolomite,* $MgCO_3.CaCO_3$; *camallite,* $KCl.MgCl_2.6H_2O$; *asbestos,* $CaMg_3(SiO_3)_4$, etc. It is an essential constituent of chlorophyll, the green colouring matter of plants. Halides of magnesium are found in sea water.

Extraction

The metal is usually extracted by the electrolysis of *fused* magnesium chloride or magnesium oxide. Carnallite ore is used as the source of $MgCl_2$ and magnesite ore is calcined to get MgO. $MgCl_2$ can also be obtained from sea water in which magnesium is present as $MgSO_4$ and $MgCl_2$. These salts are first precipitated as $Mg(OH)_2$ by treatment with slaked lime and $Mg(OH)_2$ is then converted to $MgCl_2$ by treatment with HCl.

From MgO, the metal can also be obtained by its reduction by carbon, silicon or aluminium at about 2000°C in vacuum. The metal is distilled away and condensed.

$$3MgO + 2Al \longrightarrow 3Mg + Al_2O_3$$

The reactions may the represented as under:

With fused $MgCl_2$:

$$MgCl_2 \longrightarrow Mg^{2+} + 2Cl^-$$
$$Mg^{2+} + 2e \longrightarrow Mg \text{ (At cathode)}$$
$$2Cl^- \longrightarrow Cl_2 + 2e^- \text{(At anode)}$$

With fused MgO

$$MgO \longrightarrow Mg^{2+} + O^{2-}$$
$$Mg^{2+} + 2e^- \longrightarrow Mg \text{ (At cathode)}$$
$$O^{2-} \longrightarrow O + 2e \text{ (At anode)}$$
$$O + O \longrightarrow O_2$$

Q. 99. Discuss method of preparation, properties uses and structure of calcium carbide, CaC_2.

Ans. Preparation. It is manufactured by heating lime with powdered coke at about 3000°C in an electric furnace.

$$CaO + 3C \xrightarrow{3000^\circ C} CaC_2 + CO$$

The furnace is made of fire-clay bricks and has a thick layer of graphite at the bottom. This serves as one electrode. A block of graphite suspended from the middle serves as the second electrode. Calcium carbide formed in the molten state is taken out from a tapping hole provided near the bottom of the furnace.

Properties

1. Calcium carbide is a hard crystalline solid. The pure product is white but the commercial product is generally grey, the depth of the colour increases with the extent of impurities.
2. It reacts with water yielding acetylene.

$$CaC_2 + 2H_2O \longrightarrow Ca(OH)_2 + C_2H_2$$

3. It reacts with nitrogen at red heat to yield calcium cyanamide.

$$CaC_2 + N_2 \longrightarrow CaNCN + C$$

The mixture of calcium cyanamide and carbon is used as such as a fertilizer under the name **nitrolim.** It gets hydrolysed supplying ammonia to the soil :

$$CaNCN + 3H_2O \longrightarrow CaCO_3 + 2NH_3$$

Uses.

Calcium carbide is used for the preparation of acetylene gas and for the manufacture of calcium cyanamide which is used as a fertilizer.

Structure. Calcium carbide is an ionic solid consisting of Ca^{2+} and $(C \equiv C)^{2-}$ ions. Its structure is similar to that of sodium chloride because of similar radius ratio (0.52). Thus each Ca^{2+} is surrounded by six carbide ions which lie at the corners of a regular octahedron.

Q. 100. Describe the preparation, uses and structure of silicon carbide or carbon silicide, Si C (also called carborundum) (*Vidyasagar 2007, Calcutta, 2007*)

Ans. Preparation. 1. It is prepared by heating a mixture of coke and SiO_2 in an electric furnace.

$$SiO_2 + 3C \longrightarrow SiC + 2CO$$

2. It may also be prepared by passing acetylene over heated silicon.

$$2Si + C_2H_2 \longrightarrow 2SiC + H_2$$

Manufacture (Acheson's process)

A mixture of sand (54%), coke (34%), sawdust (10%) and salt (2%) is heated in a furnace made of fire-bricks at 1500–2200°C. The furnace is provided with carbon electrodes consisting of sixty carbon rods at either end. A heavy current is passed for 36 hours. The following reaction takes place.

$$SiO_2 + 3C \xrightarrow{1500\text{-}2000^\circ C} SiC + 2CO \uparrow$$

At the end of operation, black crystals of SiC are removed, crushed and washed successively with H_2SO_4 and NaOH solutions to remove impurities.

The salt acts as flux and saw dust increases the porosity of the charge.

Uses. 1. Silicon carbide is extremely hard and hence used as an abrasive for cutting and grinding glasses.

2. SiC has a low coefficient of thermal expansion and high thermal stability. It is used for making crucibles for melting different metals.

3. Because of high melting point, it is used in furnace lining.

4. SiC is used as a high temperature semi-conductor.

5. It is used in the fabrication of grinding wheels for sharpening metallic tools.

Structure : There is an infinite three dimensional array of silicon and carbon in which each carbon is tetrahedrally surrounded by four silicon atoms and each silicon is tetrahreadely surrounded by four carbon atoms.

— C — Si — C — Si —
— Si — C — Si — C —
— C — Si — C — Si —
— Si — C — Si — C —

Structure of SiC

Q. 101. Give the preparation and properties of boron trifluoride, BF_3.

Ans. Preparation. It is obtained by heating boron trioxide or borax with calcium fluoride and concentrated sulphuric acid.

$$B_2O_3 + 3CaF_2 + 3H_2SO_4 \longrightarrow 3CaSO_4 + 3H_2O + 2BF_3$$
$$Na_2B_4O_7 + 6CaF_2 + 8H_2SO_4 \longrightarrow 2NaHSO_4 + 6CaSO_4 + 7H_2O + 4BF_3$$

2. Pure BF_3 can be obtained by thermal decomposition of diazonium tetrafluoroborate.

$$C_6H_5N_2BF_4 \xrightarrow{\Delta} C_6H_5F + N_2 + BF_3$$

Properties

1. It is a colourless pungent gas (m.p. –127.1°C and b.p. – 99.9°C).
2. It is exceedingly soluble in water and is, therefore, collected over mercury.
3. It combines with water forming two hydrates, $BF_3.H_2O$ and $BF_3.2H_2O$.
4. It gets hydrolysed in aqueous solution yielding boric acid and hydrofluoric acid.

$$BF_3 + 3H_2O \longrightarrow H_3BO_3 + 3HF$$

5. If boron trifluoride is passed in excess, it dissolves in the hydroflouric acid formed, yielding **hydrofluoboric acid, $H[BF_4]$.**

$$BF_3 + HF \longrightarrow H[BF_4]$$

6. Boron trifluoride, being an electron-deficient compound, can accept a lone pair of electrons from ammonia, hydrogen sulphide and also a fluoride ion and thus behaves as a *Lewis acid.* Tendency to accept electrons makes BF_3 useful as a catalyst in a number of reactions such as alkylation and acylation, in organic chemistry.

Q. 102. Discuss relative strength of BF_3, BCl_3 and $B\,Br_3$ as Lewis acids.

Ans. The relative strengths of trihalides of boron to accept a pair of electrons and thus to act as Lewis acids have been determined by measuring their heats of formation as well as by studying their infra-red spectra. The power to accept an electron pair has been found to decrease in the order:

$$BF_3 < BC1_3 < BBr_3$$

This order is just the reverse of what is expected from the electronegativities of hologens. We can explain it on the basis of **back bonding.**

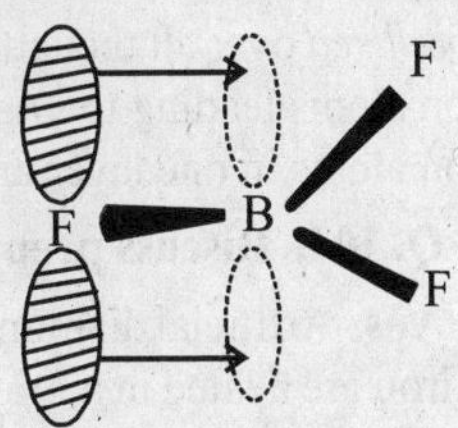

Fig. 8.13. Back bonding

Halogen have a tendency to back-donate their electron to the boron atom.

In BF_3, 2 *p* orbitals of all fluorine atom are fully-filled while one of the 2ρ orbitals of B is vacant. The 2ρ orbitals of F involved in the formation of B—F bond overlap laterally thus transferring electrons of fluorine into the vacant 2ρ orbital of *B*. Thus an additonal $p\pi - p\pi$ bond is established between *B* and *F* atoms, as shown in Fig. 8.13.

As a result of back donation of electrons from F to B, electron deficiency of boron atom gets compensated and Lewis character (tendency to attract electrons) of BF_3 decreases.

As we move from BF_3 to BCl_3 and $B\,Br_3$, tendency to form $\rho\pi - \rho\pi$ bond decreases. The reason for this is the energy difference between the 2p orbital of *B* and 3p orbital of *Cl* and between the 2*p* orbital of *B* and 4*p* orbital of *Br*. In the case of BF_3, it is the 2*p* orbitals of *B* and *F* of equivalent energies.

Therefore tendency to attract electrons or Lewis acid character shows the gradation.

$$BF_3 < BCl_3 < BBr_3$$

Q. 103. Write a short note on boron nitride, BN.

Ans. Boron nitride is obtained by the interaction of ammonia and diborane. Boron nitride is a white crystalline solid and resembles graphite in its structure, as shown in Fig. 8.14.

A boron atom bonded together with a nitrogen atom has 8 valence electrons, 3 from boron and 5 from nitrogen. This is the same number of electrons as are present in two adjacent carbon atoms in graphite. The structure of boron nitride, therefore, is very similar to that of graphite shown in Fig. 8.14(*b*). The B – N bonds in boron nitride are formed by the overlapping of sp^2 hybrid orbitals of boron and nitrogen atoms. The remaining electrons form π bonds.

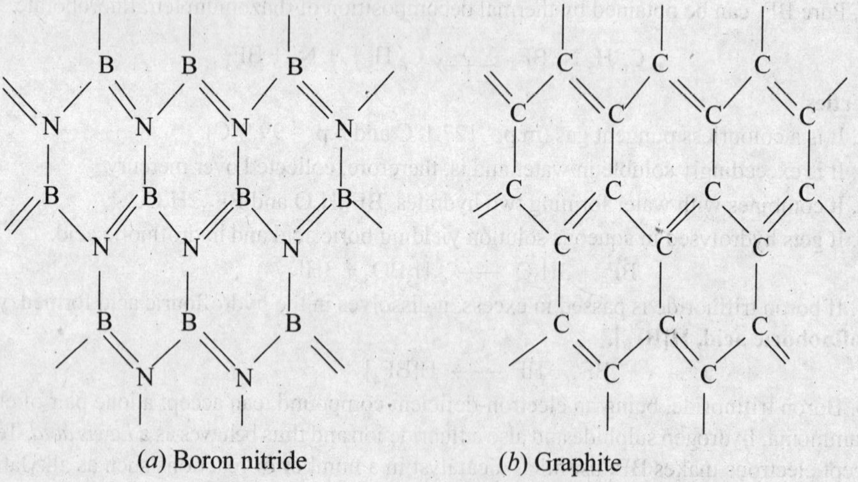

(*a*) Boron nitride (*b*) Graphite

Fig. 8.14. Similarity in the structure of boron nitride and graphite.

However there is one difference between the two. π electrons in graphite are *completely delocalised* over all the lattice. While π electrons in boron nitride *are only partially delocalised* over boron atoms leading to *unequal* B—N *bond lengths.* The layers are arranged in such a manner that boron atoms in one layer are immediately above the nitrogen atoms in the adjacent layer.

Q. 104. Discuss preparation of artificial diamonds and artificial graphite.

Ans. Artificial diamonds are prepared by Moissan's method. In this method, pure sugar charcoal and iron are heated in a graphite crucible to a temperature of about 3000°C in an electric arc furnace. Iron dissolves an appreciable amount of carbon. The graphite crucible containing the molten mass is *cooled rapidly.* As a result of the solidification of the external surface of the iron, an enormous pressure is exerted on the interior mass. Under these conditions, the carbon is thrown out in the form of *minute diamonds*. The mass is cooled and treated with hydrochloric acid to dissolve the iron. The residual carbon is found to contain a number of small diamonds.

Artificial graphite. Large amounts of graphite are prepared artificially by Acheson Process. A mixture of sand and powdered anthracite (or coke) is heated strongly in an electric furnace having two carbon electrodes connected by a thin carbon rod. An alternating current is passed and heating is continued for 24–30 hours. Silicon carbide first formed decomposes into graphite and silicon, the latter volatilises off leaving behind graphite.

$$SiO_2 + \underset{\text{Coke}}{3C} \longrightarrow 2CO + SiO$$

$$SiC \longrightarrow Si + \underset{\text{Graphite}}{C}$$

Q. 105. Describe extraction of silicon.

Ans. Occurrence: Silicon is found in sand, quartz of various forms, flint, opal, etc. It is also found as silicates, *e.g., feldspar* ($KAlSi_2O_8$), *mica* [$KH_2Al(SiO)_3$], in rocks and clays.

Preparation:

1. In laboratory, silicon may be prepared by heating dry powdered silica with magnesium or aluminium.

$$SiO_2 + 2Mg \longrightarrow Si + 2MgO$$

Magnesium oxide and any unreacted silica are removed by treatment with hydrofluoric acid in polythene vessel. Silicon left behind is a *brown amorphous powder.*

2. On a commercial scale, it is prepared by heating a mixture of finely powdered sand or quartz and carbon in an electric furnace.

$$SiO_2 + 2C \longrightarrow Si + 2CO$$

Silicon formed in this way is of *crystalline grey* variety.

Q. 106. Write down the name and formula of the ore of tin. Discuss the principle of extraction of tin from its ore. *(Calcutta 2007)*

Ans. Important ore of tin is cassiterite, SnO_2. It is also called tin stone. This ore contains 0.5–10% of the metal as SnO_2. In India, small quantities of this ore occur in Hazrat Bagh (Bihar) and Orissa.

Extraction. Extraction of tin from tin stone involves the following steps:

1. Crushing and concentration of the ore. The ore is crushed and powdered in ball mills.

Impurities of walframite (Fe, Mn) WO_4 which are magnetic in nature are removed by electromagnetic separation. Impurities of silicon and earthy matter present in tin stone are removed by wilflay table method and hydraulic classifier method (For details of these methods, refer to chapter on metallurgy)

2. Roasting. The concentrated ore is heated strongly in an inclined revolving furnace with free supply of air, when the impurities of S, As and Pb are converted into their oxides. These oxides being volatile leave the furnace. Impurities of iron and copper (as sulphides) are oxidised to their sulphates

$$CuS + 2O_2 \xrightarrow{\Delta} CuSO_4$$

$$FeS + 2O_2 \xrightarrow{\Delta} FeSO_4$$

The roasted mass is extracted with water. Copper sulphate and ferrous sulphate dissolve leaving behind SnO_2 which is separated by filtration.

3. Smelting (Reduction of SnO_2 to Sn). This oxide is mixed with coal and heated in a reverberatory furnace at 1200–1300°C. A little lime (CaO) or fluorspar is added as a flux.

SnO_2 is reduced to metallic tin

$$SnO_2 + 2C \rightarrow Sn + 2CO \uparrow$$

Some stannous oxide (SnO) is also formed as a result of reduction with coal.

$$SnO_2 + C \longrightarrow \underset{\text{stannous oxide}}{SnO} + CO$$

SnO reacts with silicon to form a slag of $SnSiO_3$

$$SnO + SiO_2 \longrightarrow \underset{\text{(slag)}}{SnSiO_4}$$

Lime reacts with silica to form a slag of $CaSiO_3$

$$\underset{\text{(Basic flux)}}{CaO} + \underset{\text{(Acidic impurity)}}{SiO_2} \longrightarrow CaSiO_3\,\text{(slag)}$$

With the removal of slags, the impurities are removed and tin metal is obtained.

4. Refining of impure tin. Tin obtained as above is not pure. It contains impurities of Cu, Fe, W etc and also some unreduced SnO_2. These impurities are removed by the following methods:

(*i*) Liquation process

(*ii*) Poling process

(*iii*) Electrolytic process

For details of these processes, refer to chapter on metallurgy.

Q. 107. Name an ore of lead. Describe the process of extraction of lead from its ore.

Ans. Important ore of lead is galena, PbS.

Extraction of lead from galena

Air Reduction Process (For ores rich in lead content) It involves the following steps:

1. Concentration. After crushing and powdering in the ball-mill the ore is concentrated by froth-flotation process.

2. Roasting. The concenrated ore is taken to the hearth of a reverberatory process, where it is heated to a moderate temperature (to avoid volatilisation of lead into lead oxide). Some part of galena is oxidised to lead oxide and the rest to lead sulphate.

$$2PbS + 3O_2 \longrightarrow PbO + 2SO_2$$

$$PbS + 2O_2 \longrightarrow PbSO_4$$

Now the flow of air is reduced and excess of galena is added to the reaction mixture and temperature is raised. PbO and $PbSO_4$ formed earlier are reduced to lead metal

$$2PbO + PbS \longrightarrow 3Pb + SO_2$$

$$PbSO_4 + PbS \longrightarrow 2Pb + 2SO_2$$

Carbon reduction process (For ores poor in lead content)

The following steps are involved

1. Concentration. The crushed ore is concentrated by Froth-Flotation Process

2. Roasting. The concentrated ore is roasted in the presence of air in a sinterer made of sheet metal and lined inside with fire bricks. Galena is oxidised to PbO

$$2PbS + 3O_2 \longrightarrow 2PbO + 2SO_2$$

Lead oxide (PbO) contains impurities of silica (SiO_2). It is mixed with coke and lime (as basic flux) and heated in a furnace. The following reactions take place.

$$2PbO + C \longrightarrow 2Pb + CO_2$$
$$CaO + SiO_2 \longrightarrow CaSiO_3 \text{ (slag)}$$

Slag floats on molten lead and can be skimmed off. A noteworthy point is formation of $PbSiO_3$, by the combination of PbO and SiO_2 is prevented because CaO is more basic than PbO and only the former (CaO) combines with SiO_2 to form the slag. Thus wastage of Pb (as Pb SiO_3) is prevented.

Q. 108. Write a short note on the extraction of germanium.

Extraction of Germanium. The only relatively rich ore of germanium is *germanite,* GeO_2, which contains nearly 6 per cent of the metal. The ore is strongly heated with hydrochloric acid when the chloride $GeCl_4$ distils off. This is hydrolysed to give pure dioxide, GeO_2, which is dried and then reduced by hydrogen or carbon to give the element.

Reactions:

$$GeO_2 + 4HCl \longrightarrow GeCl_4 + 2H_2O$$
$$GeCl_4 + 4H_2O \longrightarrow Ge(OH)_4 + 4HCl$$
$$Ge(OH)_4 \xrightarrow[-2H_2O]{\Delta} GeO_2$$

Q. 109. Give the preparation and properties of carbon monoxide, CO.

Ans. Preparaion: 1.Carbon monoxide is obtained by burning carbon in a limited supply of oxygen.

2. Pure CO can be obtained by heating formic acid with sulphuric acid, the latter acting as a dehydrating agent.

$$HCOOH \xrightarrow{H_2SO_4} CO + H_2O$$

Properties.

1. It is a colourless odourless toxic gas.
2. Its toxic nature is due to its ability to form a complex with haemoglobin in the blood.
3. Carbon monoxide is an essential constituent of gaseous fuels like water gas, producer gas, etc.

The bonding in carbon monoxide may be represented as

$$:C:::O: \text{ or } C \equiv O:$$

Two pairs of electrons are considered to be shared between carbon and oxygen resulting in a double bond. The oxygen may then be considered to be donating its lone pair of electrons to carbon giving a coordinate bond. The properties of CO can be explained satisfactorily if it is maintained that carbon monoxide is a resonance hybrid of three structures, as shown :

$$:C::\ddot{O}: \leftrightarrow :\overset{-}{C}:::\overset{+}{O}: \leftrightarrow :\overset{+}{C}:\ddot{\overset{-}{O}}:$$

Q. 110. Write a note on dry ice (solid carbon dioxide).

Ans. Commonly known as **dry ice,** is a soft white, snow-like substance with specific gravity 1.53 and melting point – 56.6°C at 5.2 atm pressure.

Preparation

It is manufactured by evaporating liquid carbon dioxide at 20°C and 70 atm pressure to a pressure of less than 5.2 atm. Solid carbon dioxide is used as a refrigerant under the commercial name *drikold.* It is highly suitable for this purpose as it leaves no residue. It is used as a substitute of ice in cold drinks and in the preparation of ice-cream. Mixed with ether or acetone it provides freezing mixtures with exceedingly low temperatures. It is extensively used in the transport of perishable food-stuffs. It provides cold as well as an inert atmosphere.

It is also used in hospitals for surgical operations of sores.

Artificial rain-making has also been accomplished by spraying small pellets of dry ice from aeroplanes over clouds to produce condensation.

The bonding in carbon dioxide is represented as

$$:\ddot{O}::\ddot{C}::\ddot{O}: \quad \text{or} \quad :\ddot{O}=C=\ddot{O}:$$

However, carbon-oxygen bond length in carbon dioxide is 1.15 Å while the value to be expected for a double bond, C = O, is about 1.20 Å and for a triple bond, C ≡ O, is 1.10 Å. This shows that the simple double bond formula is inadequate.

It is now maintained that resonance occurs between the three possible structures given below :

$$:\overset{-}{\ddot{\underset{..}{O}}}-C\equiv\overset{+}{O}: \quad \leftrightarrow: \quad \ddot{O}=C=\ddot{O}: \quad \leftrightarrow: \quad \overset{+}{O}\equiv C-\underset{..}{\ddot{O}}:$$

Q. 111. Write notes on the following:

(*i*) White lead, $Pb(OH)_2$. 2Pb CO_3 (*ii*) Red lead, Pb_3O_4

Ans. White Lead, $Pb(OH)_2.2PbCO_3$. It can be prepared in laboratory by mixing alkali carbonate with a solution of lead salt.

$$4Na_2CO_3 + 3Pb(NO_3)_2 + 2H_2O \longrightarrow \underset{\text{Ppt. of white lead}}{Pb(OH)_2.2PbCO_3} + 2NaHCO_3 + 6NaNO_3$$

There are three commercial processes for its manufacture.

1. *Dutch Process.* This is a very old method and is based on the action of acetic acid with metallic lead in the presence of carbon dioxide and moist air.

Small perforated buckets of lead are placed in earthenware pots having perforated shelves in which vinegar or dilute (about 2.5 per cent) acetic acid is placed. The pots are partly covered with wooden planks arranged in rows. The space in between the rows is filled by spent tan bark and horse dung.

Basic lead acetate formed first is decomposed by the carbon dioxide produced during the fermentation of tan bark and horse dung. The following reactions are probably involved.

$$2Pb + O_2(\text{air}) + 2H_2O \longrightarrow 2Pb(OH)_2$$

$$2Pb(OH)_2 + 2CH_3COOH \longrightarrow \underset{\text{Basic lead acetate}}{Pb(OH)_2.(CH_3COO)_2Pb} + 2H_2O$$

$$Pb(OH)_2 + Pb(OH)_2.(CH_3COO)_2Pb + 2CO_2 \longrightarrow \underset{\text{White lead}}{Pb(OH)_2.2PbCO_3} + 2CH_3COOH$$

2. *Carter's Process.* In this process, molten lead is atomised in wooden cylinders by blowing in a jet of air or superheated steam. The wooden cylinders keep rotating about their horizontal axes. A spray of dilute acetic acid and a current of air mixed with carbon dioxide are admitted into the cylinders. Here they meet the atomised lead and chemical reactions like those which occur in Dutch process take place between lead, acetic acid and carbon dioxide.

3. *Chamber Process.* This process has been developed in recent years. Here sheets of lead are suspended horizontally in large chambers into which carbon dioxide and acetic acid vapours are introduced. The reactions take place as in the above processes .

Uses.

White lead mixed with linseed oil is used as white paint. It has a large covering power. But it has one serious drawback that it gets blackened if there is hydrogen sulphide in the atmosphere.

Sublimed white lead. It is also used as a white pigment. Its approximate composition is $PbSO_4$ 75%, PbO 20% and ZnO 5%. It is prepared by roasting galena (PbS) mixed with a small amount of zinc sulphide, in a strong blast of air. The fumes given out during the process are collected and condensed in a separate chamber. The sublimate is the *sublimed white lead.* It has a good covering power and is not so readily darkened by hydrogen sulphide.

2. Red Lead, Pb_3O_4. It is prepared by heating lead monoxide to 473 – 480°C in air for a long time.

$$6PbO + O_2 \longrightarrow 2Pb_3O_4$$

Red lead is a brilliant red powder. On heating it becomes violet and finally black. However, on cooling, the original red colour is restored.

Q. 112. Give the preparation and uses of silicon gel.

Ans. Preparation. Silica gel may be prepared by hydrolysing silicon tetrachloride or tetrafluoride. It may also be prepared by adding a mineral acid such as hydrochloric acid to a fairly concentrated solution of sodium silicate or by adding, gradually, a solution of ammonium sulphate to that of sodium silicate.

$$SiF_4 + 4H_2O \xrightarrow[-4HF]{} Si(OH)_4 \xrightarrow{\Delta} SiO_2 + 2H_2O$$

$$Na_2SiO_3 + 2HCl \longrightarrow 2NaCl + H_2O + SiO_2$$

$$(NH_4)_2SO_4 + NaSiO_3 \longrightarrow Na_2SO_4 + 2NH_3 + SiO_2 + H_2O$$

In each case, silica separates out as a gel. It is, a colloidal system in which water is dispersed in a solid. The gel is filtered and washed to remove the soluble impurities. It is then dried by heating at 110°C in an air-oven.

Uses. 1. When exposed to moist atmosphere it adsorbs moisture. Hence, dry silica gel is used as an excellent drying agent. Every time, after drying, it can be used over again.

2. It can also adsorb sulphur compounds and is used in removing them from petroleum.

3. When silica gel is fused and cooled, it changes into a transparent solid known as *silica glass.* It is superior to ordinary glass since it softens at a lower temperature and can be more easily shaped into vessels such as crucibles, dishes, etc. It has low coefficient of expansion and, therefore, an article made of silica can withstand sudden changes in temperature.

Q. 113. Give the preparation and properties of phoshorus pentachloride, PCl_5.

Ans. Preparation. It can be prepared by the action of chlorine or sulphur chloride on phosphorus trichloride.

$$PCl_3 + Cl_2 \longrightarrow PCl_5$$

$$3PCl_3 + S_2Cl_2 \longrightarrow PCl_5 + 2PSCl_3$$

Phosphorus pentachloride is a yellowish white crystalline compound having a sharp odour. It sublimes on heating at 160°C and melts at 45°C under pressure.

Chemical Properties. 1. Phosphorus pentachloride dissociates on heating into phosphorus trichloride and chlorine.

$$PCl_5 \rightleftharpoons PCl_3 + Cl_2$$

2. It is violently hydrolysed by water giving phosphorus oxochloride or phosphoric acid depending upon the quantity of water.

$$PCl_5 + H_2O \xrightarrow{\text{Insufficient water}} POCl_3 + 2HCl$$

$$PCl_5 + 4H_2O \xrightarrow{\text{Excess of water}} H_3PO_4 + 5HCl$$

3. It reacts with P_4O_{10} to form phosphorus oxytrichloride

$$6PCl_5 + P_4O_{10} \longrightarrow 10POCl_3$$

4. Hydrogen reduces phosphorus pentachloride to trichloride.

$$6PCl_5 + H_2 \longrightarrow PCl_3 + 2HCl$$

5. It reacts with acetic acid and ethyl alcohol as under:

$$CH_3COOH + PCl_5 \longrightarrow \underset{\text{Acetyl chloride}}{CH_3COCl} + POCl_3 + HCl$$

$$C_2H_5OH + PCl_5 \longrightarrow \underset{\text{Ethyl chloride}}{C_2H_5\,Cl} + POCl_3 + HCl$$

6. It combines with KF as follows:

$$PCl_5 + 6KF \longrightarrow KPF_6 + 5KCl$$

7. Phosphorus pentachloride combines with Cl^- acceptors to form tetrahedral $[PCl_4]^+$ ion.

$$PCl_5 + 6KF \longrightarrow [PCl_4]^+[BCl_4]^-$$

Q. 114. Give the method of preparation and properties of phosphorus pentoxide.

Ans. Preparation. It is obtained by burning phosphorus in an excess of dry air or O_2.

$$4P + 5O_2 \xrightarrow{\text{Burn}} 2P_2O_5$$

Properties

1. Pure P_2O_5 is odourless while impure form has a garlic odour due to the presence of P_2O_3.

2. *Action of water:* Orthophosphoric acid is formed.

$$P_2O_5 + 3H_2O \longrightarrow 2H_3PO_4$$

3. *Dehydrating property* : P_2O_5 can combine with one molecule of water (H_2O) and form HPO_3, this oxide removes H_2O molecule from inorganic and organic compounds and renders them anhydrous. Thus, P_2O_5 can remove H_2O molecule from H_2SO_4, HNO_3, CH_3CONH_2 (acetamide), CH_3COOH, C_2H_5OH etc., and thus acts as a dehydrating agent. Examples are given below:

$$H_2SO_4 \xrightarrow{-H_2O} SO_3 + H_2O$$

$$P_2O_5 \longrightarrow 2HPO_3$$

$$H_2SO_4 + P_2O_5 \longrightarrow SO_3 + 2HPO_3$$

Similarly,

$$2HNO_3 + P_2O_5 \longrightarrow N_2O_5 + 2HPO_3$$

$$CH_3CONH_2 + P_2O_5 \longrightarrow CH_3CN + HPO_3$$

$$2CH_3COOH + P_2O_5 \longrightarrow (CH_3CO)_2O + 2HPO_3$$

$$C_2H_5OH + P_2O_5 \longrightarrow H_2C{=}CH_2 + 2HPO_3$$

4. *Action of NH_4OH. Ammonium metaphosphate, NH_4PO_3* is formed.

$$P_2O_5 + 2NH_4OH \longrightarrow 2NH_4PO_3 + H_2O$$

5. *Action of basis oxides.* When P_2O_5 reacts with basic oxides like CaO, phosphate is obtained.

$$3\,CaO + P_2O_5 \longrightarrow Ca_3(PO_4)_2; \quad 3Na_2O + P_2O_5 \longrightarrow 2Na_3PO_4$$

6. *Action of H_2O_2.* When P_2O_5 is treated with 30% H_2O_2 at low temperature and then the reaction mixture is diluted with H_2O and cooled, a solution of peroxy-monophosphoric acid, H_3PO_5 is obtained.

$$P_2O_5 + 2H_2O_2 + H_2O \xrightarrow[\text{Temperature}]{\text{Low}} 2H_3PO_5\,(solution)$$

7. *Reduction by carbon.* P_2O_5 is reduced to red P when heated with carbon.

$$P_2O_5 + 5C \longrightarrow 2P + 5CO$$

Uses. P_2O_5 is used (*i*) a drying and dehydrating agent (*ii*) for the preparation of orthophosphoric and (H_3PO_4) and peroxy-monophosphoric acid (H_3PO_5).

Q. 115. Give the method of preparation and properties of SeO_2.

Ans. Preparation. This is prepared as follows :

$$Se + 4HNO_3 \longrightarrow SeO_2 + 2H_2O + 4NO_2$$

Properties. 1 *Physical state.* SeO_2 is a white volatile solid which sublimes on heating (sublimation temperature = 315°C).

2. *Stability.* SeO_2 is less stable than SO_2.

3. *Acidic/amphoteric character.* SeO_2 is soluble in water and gives acids.

$$H_2O + SeO_2 \longrightarrow H_2SeO_3 \text{ (Selenious acid)}$$

H_2SeO_3 can be isolated in the crystalline state. Due to the production of H_2SeO_3, SeO_2 shows acidic character and is known as anhydride of H_2SeO_3.

Q. 116. Give the preparation, properties and uses of sodium ammonium hydrogen phosphate or microcosmic salt, $Na(NH_4)HPO_4.4H_2O$.

Ans. Preparation : This phosphate is also called microcosmic salt since it is present in the urine of the *"microcosm" i.e.,* man. It is prepared by dissolving disodium hydrogen phosphate (Na_2HPO_4) and NH_4Cl in a little hot water, filtering off the precipitate of NaCl and crystallising the solution.

$$Na_2HPO_4 + NH_4Cl \longrightarrow Na(NH_4)HPO_4 \text{ (solution)} + NaCl \text{ (ppt.)}$$

Properties 1. It is colourless and crystalline substance having needle-like crystals.

2. *Action of heat.* When heated strongly, it loses NH_3 and H_2O and forms a transparent glass bead of sodium metaphosphate, $NaPO_3$ in the molten state.

$$Na(NH_4)HPO_4 \xrightarrow{\Delta} NH_3 + H_2O + NaPO_3 \text{ (in molten state)}$$

3. When molten $NaPO_3$ is cooled rapidly, it gets polymerised and forms sodium hexametaphosphate. $(NaPO_3)_6$ or $Na_6P_6O_{18}$. *(calgon).*

$$3NaPO_3 \xrightarrow[\text{Polymerisation}]{\text{Cool}} (NaPO_3)_6 \text{ or } Na_6P_6O_{18}$$

Uses.

The bead of $NaPO_3$ obtained as above can combine with metallic oxides like CuO etc., and form coloured metallic sodium orthophosphate like $CuNaPO_4$.

$$CuO + NaPO_3 \text{ (bead)} \longrightarrow CuNaPO_4 \text{ (blue)}$$

The formation of coloured metallic sodium orthophosphate can be used to detect the coloured basic radicals like Cu^{2+}, Co^{2+} etc., present in a given inorganic mixture. Another use of $Na(NH_4)HPO_4$ is in testing SiO_2 with which the salt gives a cloudy bead in which the particles of SiO_2 are floating.

Q. 117. Give the methods of preparation and properties of potassium bromate, $KBrO_3$.

Ans. Preparation. It is prepared:

1. by adding a large quantity of Br_2 to a hot concentrated solution of an alkali

$$3Br_2 + 6KOH \longrightarrow 5KBr + KBrO_3 + 3H_2O$$

In the above reaction Br_2 undergoes disproportionation into Br^- and Br O_3^- ions. $KBrO_3$ is much less soluble than KBr and hence can be separated by cooling.

2. by passing Cl_2 into an alkaline solution of KBr.

$$KBr + 6\,KOH + 3Cl_2 \longrightarrow KBrO_3 + 6KCl + 3H_2O$$

Properties: 1. It forms hexagonal crystals which are sparingly soluble in H_2O.

2. *Action of heat.* $KBrO_3$ gets decomposed on heating and gives KBr and O_2.

$$2KBrO_3 \longrightarrow 2KBr + 3O_2$$

3. *Action of KBr on $KBrO_3$ in presence of HCl.* When a mixture of $KBrO_3$ and a bromide (*e.g.*, KBr) is acidified with HCl, Br_2 is evolved.

$$KBrO_3 + 5KBr + 6HCl \longrightarrow 3Br_2 + 6KCl + 3H_2O$$

Uses.

The above reaction makes the basis for preparing standard solution of bromine. Bromine is a volatile substance and hence its standard solution cannot be prepared merely by dissolving bromine in water. It is, therefore, prepared by dissolving $KBrO_3$ in H_2O and then treating the solution with excess of KBr in presence of HCl.

Q. 118. Give the methods of preparation and properties of potassium iodate, KIO_3.

Ans. Preparation. It is prepared:

1. by the action of I_2 on hot and conc. KOH solution.

$$3I_2 + 6KOH \longrightarrow 5KI + KIO_3 + 3H_2O$$

In the above reaction I_2 undergoes disproportionation into I^- and IO_3^- ions.

2. by heating a conc. solution of $KClO_3$ with I_2 in presence of little HNO_3.

$$2KClO_3 + I_2 \longrightarrow 2KIO_3 + Cl_2$$

Properties. 1. It forms monoclinic (or cubic) crystals, which are sparingly soluble in cold water, but soluble in hot water.

2. *Action of KI on KIO_3 in presence of HCl.* When a mixture of KIO_3 and an iodide (*e.g.*, KI) is acidified with HCl, I_2 is liberated.

$$KIO_3 + 5KI + 6HCl \longrightarrow 3I_2 + 6KCl + 3H_2O$$

The above reaction makes the basis for standardising sodium thiosulphate solution.

3. *Action of heat.* On heating it melts at 560° C and then decomposes at a somewhat higher temperature into KI and O_2.

$$2KIO_3 \longrightarrow KI + 3O_2$$

Due to this property, KIO_3 is used as an oxidising agent also.

On thermal decompostion, KIO_3 also gives potassium metaperiodate (KIO_4) and potassium iodide (KI).

$$4KIO_3 \xrightarrow{\Delta} 3KIO_4 + KI$$

Q. 119. Which allotrope of carbon is a non-conductor of electricity and why ?

(*Vidyasagar,* 2007)

Ans. Diamond is non-conductor of electricity. Its structure is an infinite network of carbon atoms in which each carbon is tetrahedrally surrounded by four carbon atoms. No free electrons are available. Thus, there is no scope for conductance of electricity.

Q. 120. Nitrogen cannot form nitrogen penta chloride while phosphorus can. Explain.

(*Calcutta, 2007*)

Or

Nitrogen forms NCl_3 whereas phosphorus forms PCl_3 and PCl_5 both. Explain.

(*Purvanchal, 2007*)

Ans. Electronic configurations of N and P are given below :

N $1s^2 2s^2 2px^1 2py^1 2pz^1$

P $1s^2 2s^2 2p^6 3s^2 3px^1 3py^1 3pz^1$

In the case of N, only three half filled orbitals are available for bonding. Of course, 2s orbital which is paired also participates in hybridisation (sp^3). Thus N forms three bonds to give NCl_3.

In the case of P, *d* orbitals are also available. One of the electrons in 3*s* orbital is excited to 3*d* orbitals. Now there are five orbitals which are half filled. They hybridise (sp^3d) to give five equivalent orbitals. Thus, P can form PCl_5.

9

NUCLEAR CHEMISTRY

Q. 1. What are isotopes? How are they represented? *(Kerala, 2000; Nagpur, 2000)*

Ans. Isotopes are different atoms of the same element having the same atomic number and hence the same chemical properties and the same position in periodic table but differing in atomic weight (mass number).

The phenomenon of isotopes was first discovered by Soddy in 1916.

Representation of isotopes. Each isotope is represented by its appropriate symbol along with its mass number at the left upper end (superscript) and atomic number at its lower left end. Thus three isotopes of oxygen are written as ${}^{16}_{8}O$, ${}^{17}_{8}O$, ${}^{18}_{8}O$. Similarly, two isotopes of carbon of mass number 12 and 13 written as ${}^{12}_{6}C$ and ${}^{13}_{6}C$ respectively.

Soddy recognised the presence of such particles which are identical in every respect, except mass and radioactive properties. They have different atomic weight but same atomic number and, therefore, occupy the same place in the periodic table.

Lead was the first non-radioactive element which was shown to exhibit isotropy. The end product of each radioactive series was found to be an isotope of lead with mass number 206 (uranium series), 207 (actinium series) and 208 (thorium series) as shown in Table 9.1. Thus the ordinary lead (At. wt = 207.2) is merely a mixture of the three isotopes.

Table 9.1. Radioactive series.

Sr. No.	*Series*	*Parent*	*Stable end product*
1.	Uranium series	${}^{238}_{92}U$	${}^{206}_{82}Pb$
2.	Actinium series	${}^{235}_{92}U$	${}^{207}_{82}Pb$
3.	Thorium series	${}^{232}_{90}Th$	${}^{208}_{82}Pb$

Chlorine has two isotopes with mass numbers 35 and 37. The former contains 17 protons and 18 neutrons in the nucleus while the latter contains 17 protons and 20 neutrons. The atomic number in each case is 17 and so their electronic configuration is also the same. They differ only in the number of neutrons contained in their respective nuclei. Two isotopes of chlorine are shown in Fig. 9.1.

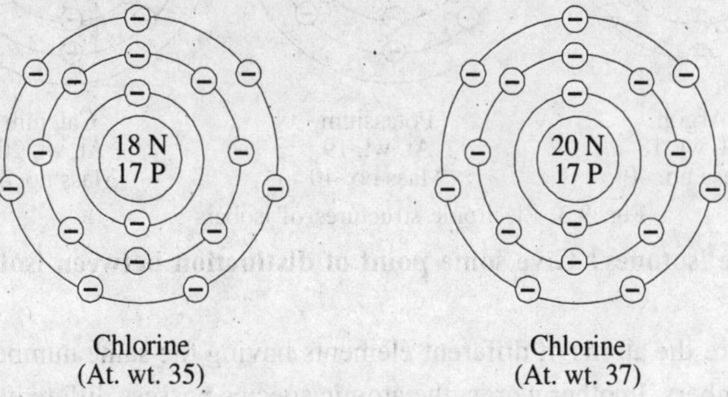

Fig. 9.1. Isotopes of chlorine.

Similarly, hydrogen has three isotopes of mass numbers 1, 2 and 3. Isotopes of hydrogen are represented in Fig. 9.2.

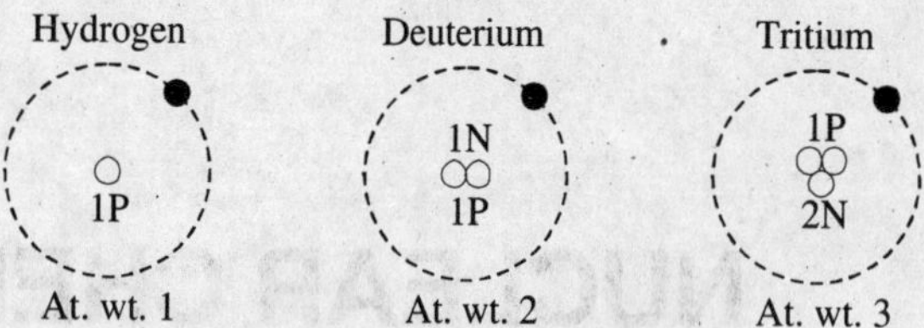

Fig. 9.2. Isotopes of hydrogen.

Q. 2. What are isobars? Explain giving electronic structures.

(*Kerala, 2000*)

Ans. Isobars are atoms of different elements having the same mass number but different atomic numbers. In other words, nuclei of two isobars contain the same total numbers of nucleons (protons + neutrons) but have unequal number of protons. For example, argon, potassium and calcium having atoms of the same mass number 40: ${}^{40}_{18}Ar$, ${}^{40}_{19}K$, ${}^{40}_{20}Ca$, where 18, 19 and 20 shown as subscripts are the atomic numbers of the three elements respectively are isobars. Since the atomic numbers are different, their chemical properties are also different. A few more examples of isobars are given below:

(*i*) ${}^{76}_{32}Ge$ and ${}^{76}_{34}Se$

(*ii*) ${}^{210}_{82}Pb$, ${}^{210}_{83}Bi$ and ${}^{210}_{84}Po$

(*iii*) ${}^{228}_{88}Ra$, ${}^{228}_{89}Ac$ and ${}^{228}_{90}Th$

A radioactive element on emission of a β-particle produces an isobar.

The structures of isobars of argon ${}^{40}_{18}Ar$, potassium ${}^{40}_{19}K$ and calcium ${}^{40}_{20}Ca$ are given below in Fig. 9.3.

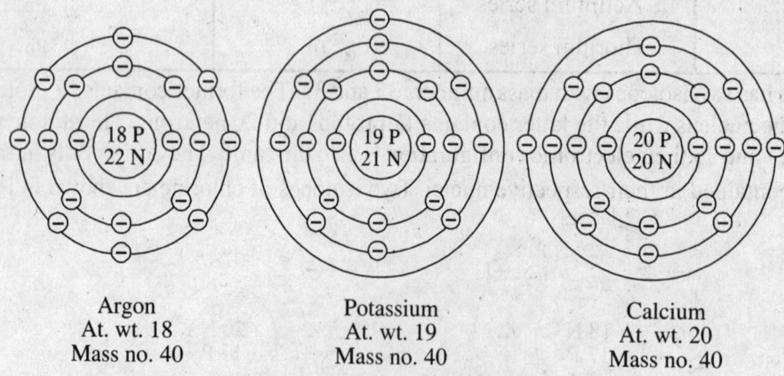

Fig. 9.3. Electronic structures of isobars

Q. 3. What are isotones? Give some point of distinction between isotopes, isobars and isotones.

(*Kerala, 2000*)

Ans. Isotones are the atoms of different elements having the same numbers of neutrons but different atomic numbers. In other words, the atomic species possess different number of protons and also different mass number.

$^{30}_{14}Si$, $^{31}_{15}P$, $^{32}_{16}S$ are isotones since they all contain 16 neutrons. The following Table 9.2 shows distinction between isotopes, isobars and isotones.

Table 9.2. Distinction between isotopes, isobars and isotones.

Atomic Spaces	*Atomic numbers* Z	*Mass numbers* A	*Neutrons* N	*Examples*
Isotope	Same	Different	Different	$^{35}_{17}Cl$, $^{37}_{17}Cl$
Isobar	Different	Same	Different	$^{3}_{1}H$, $^{3}_{2}H$
Isotone	Different	Different	Same	$^{2}_{1}H$, $^{3}_{2}He$

Q. 4. Describe different methods used for separation of isotopes.

(Kerala, 2000; Kurukshetra, 2001)

Ans. Different methods used for isotopes separation are described as under:

1. The gaseous diffusion. This method is applicable to gases. Two gaseous molecules of molecular weight M_1 and M_2 can be separated by gaseous diffusion (Fig. 9.4). According to Graham's law of gaseous diffusion, the rates of diffusion of the two gaseous species are given by the relation

$$\frac{r_1}{r_2} = \sqrt{\frac{M_2}{M_1}}$$

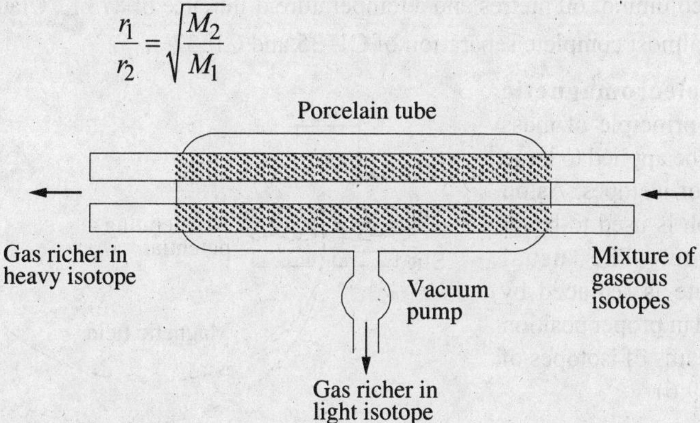

Fig. 9.4. Apparatus for the separation of gaseous isotopes.

Applied to isotopes this would be given by an ideal separation factor of

$$S = \sqrt{\frac{M_2}{M_1}}$$

where M_1 and M_2 are the molecular weights of the two isotopes or of the compounds in which they occur. Aston used this principle first for the partial separation of isotopes of neon. Harkins applied the same principle of gaseous diffusion for separating the isotopes of chlorine from hydrogen chloride gas. After a laborious series of experiments a slight enrichment of the heavier chlorine isotopes was achieved. Thus chlorine of atomic weight 35.498 was obtained as compared to the normal value 35.467. Harkin' s method of diffusion at atmospheric pressure is less effective in the separation of isotopes than the method of diffusion at low pressure first applied successfully by Hertz. Hertz also introduced the cascade system (many successive diffusion stages) with a view to further improve the rate of separation.

2. The thermal diffusion. The method is based essentially on the principle that if two gases of different molecular weights are enclosed between two surfaces at different temperatures, the heavier gas will tend to diffuse to the cooler end and the lighter gas towards the hotter end. This effect occurs side by side with ordinary diffusion which tends to restore a uniform concentration in the gaseous mixture. The apparatus used in this method consists of a vertical tube about 3 metres in length and 1 cm in diameter. It is cooled by water on the outside and heated electrically on the inside by means of platinum wire stretched axially along the tube. When a temperature gradient is set up between the wire and the cooled walls of the tube, the lighter isotopic species diffuse preferentially to the hotter centre of the tube. The convection currents in the tube carry the lighter species upwards and the heavier cold species downwards (Fig. 9.5). The convection currents thus further enhance the isotopic separation. With column of 60 metres and a temperature difference of 873 K, Clausius and Dickel accomplished an almost complete separation of C1–35 and C1–37.

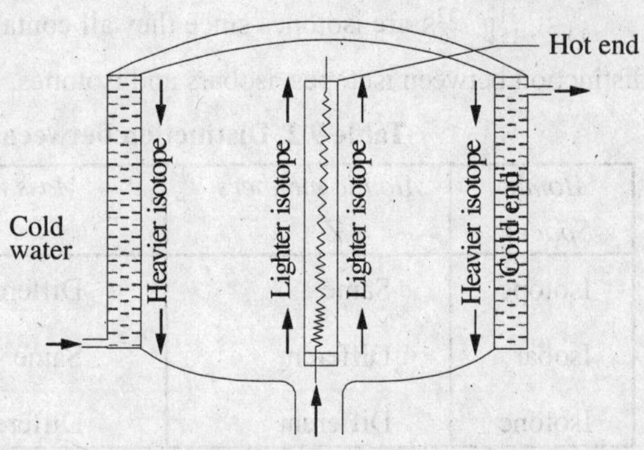

Fig. 9.5. Thermal diffusion of gases.

3. The elecromagnetic separation. The principle of mass spectrograph can be applied to large scale separation of isotopes. Aston mass spectrograph is used to bring about separation. The usual photographic plate is replaced by receptacles placed in proper position to receive the streams of isotopes of the element (Fig. 9.6).

The electromagnetic method is the only method which achieves 100% separation of isotopes.

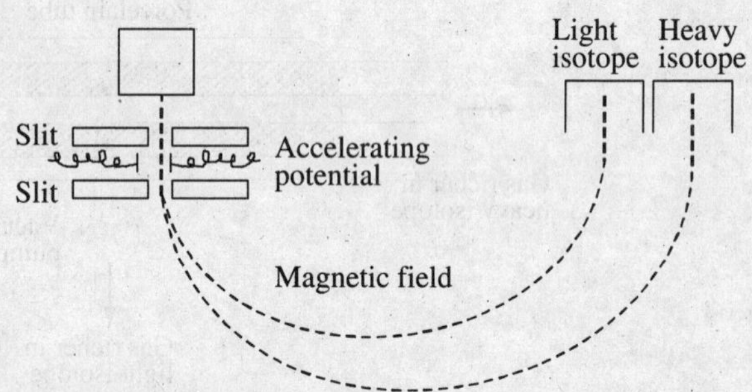

Fig. 9.6. Mass spectrographic method of separation of isotopes.

Appreciable amounts of U^{235} required for experiment purposes were separated electromagnetically by Nier in 1940. Now-a-days we utilize this method for the separation of isotopes of a large number of elements. In all the methods of separation other than the electromagnetic, a single process produces two fractions of the material with different concentrations. The extent to which a mixture of two *isotopes may be separated in one stage of a particular process is called the separation factor.* It is the ratio of the relative concentration of the desired isotopes after processing to its relative concentration before processing. If n_1 and n_2 are the numbers of light and heavy species in a mixture before processing and N_1 and N_2 the corresponding numbers after processing in a one stage process, the separation factor S may be represented mathematically as

$$S = \frac{N_1 / N_2}{n_1 / n_2}$$

The higher the value of S, the greater is the efficiency of separation. In most of the processes available for the separation of isotopes, S is only slightly greater than unity and consequently, the given process has to be carried out in several single-separation stages to achieve the desired extent of separation.

4. The fractional distillation. This method of separating isotopes is based upon the difference in the vapour pressures or boiling points of liquids. The rate at which an isotope evaporates is inversely proportional to the square roots of its mass so that on fractional distillation the *condensate* tends to become richer in the lighter isotopes and the residual liquid in the heavier isotopes. Bronsted and Hevesy (1921) used this principle to partially separate the isotopes of mercury by distillation in a high vacuum. The apparatus used by them is shown in Fig. 9.7. Mercury is evaporated in partial vacuum at 313 to 333 K between the walls of a Dewar flask. The inner vessel is cooled with liquid air placed about 1 to 2 cm above mercury. The lighter isotopes evaporate preferentially and condense on the cold surface. After about one-fourth of mercury has evaporated the condensate and residue are separated again. The condensate is melted and then partially evaporated. On repeating the process several times a slight enrichment of the lighter isotopes of mercury occurs in the condensate and of the heavier isotopes in the residual liquid.

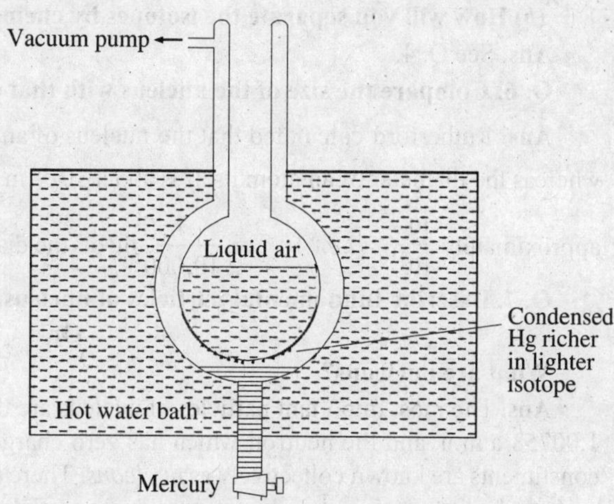

Fig. 9.7. Apparatus for the separation of mercury isotopes.

5. The electrolytic method. Urey and Washburn (1932) observed that the residual water becomes enriched in the heavier isotope on continuous electrolysis of water. Lewis and co-workers succeeded in preparing about 0.1 cc of water containing only heavier isotopes of hydrogen of mass 2 by continuous electrolysis of a very large amount of water using an old commercial electrolytic cell. The electrolytic productions of O^{18} and D_2O are now carried out commercially where electric power is cheap. The electrolytic method has not been very successful in separating the isotopes of the elements other than hydrogen.

6. The chemical exchange method. Although the chemical properties of the isotopes of an element are similar, the rates at which the reactions occur can differ appreciably. The principle of chemical exchange has been employed for the separation of isotopes. Thus when ammonia gas containing N^{15} isotope is bubbled upwards through a column in which ammonium nitrate solution is flowing downwards, the following exchange reaction takes place:

$$N^{15}H_3 \text{ (gas)} + N^{14}H_4NO_3 \text{ (soln.)} \longrightarrow N^{14}H_3 \text{ (gas)} + N^{15}H_4NO_3 \text{ (soln.)}$$

An enrichment of about 70% of N^{15} has been achieved by Thode and Urey (1939) using the above exchange reaction.

7. The centrifugal method. If a mixture of two isotopes of the same element in gaseous or liquid state is subjected to a gravitational field, the heavier isotope will tend to concentrate in the direction of the field and consequently, separation of the isotopes will occur provided there is no mixing to counteract this movement. This is the underlying principle of this method. The extent of separation in this method depends on the mass difference between two isotopes whereas in the diffusion method it is dependent upon the square of the ratio of the masses. The separation factor is, therefore, better in the centrifugal method than in the diffusion method.

Q. 5. (*a*) Describe the principle involved, apparatus and other experimental details of thermal diffusion method for separation of isotopes.

(*b*) How will you separate the isotopes by chemical exchange method?

Ans. See Q. 4.

Q. 6. Compare the size of the nucleus with that of the atom.

Ans. Rutherford calculated that the nucleus of an atom has an effective diameter of 10^{-15} m whereas the diameter of an atom itself is about 10^{-10} m . It follows that the diameter of a nucleus is approximately $\frac{10^{-15}}{10^{-10}}$, *i.e.*, 10^{-5} or $\frac{1}{10,000}$th of the diameter of an atom.

Q. 7. Describe fundamental particles of nucleus. *(Kurukshetra, 2002)*

Or

What are nucleons?

Ans. The most important particles of nucleus are the proton which has +1 charge and mass of 1.00758 a.m.u. and the neutron which has zero charge and mass of 1.00893 a.m.u. The nuclear constituents are known collectively as *nucleons*. There are a number of other fundamental particles such as electron, anti-proton, positron (all mass particles) and neutrino, photons and the gravitation (energy particles) which are stable. Their characteristics are given in Table 9.3. Among unstable fundamental particles are mesons and V-particles besides neutrons. There are listed in Table 9.4.

Table 9.3. Characteristics of certain stable fundamental particles.

Particle	*Symbol*	*Charge*	*Mass a.m.u.*
Proton	P_1, H^1	+	1.00758
Electron	e^-, β^-	–	0.0005486
Positron	e^+, β^+	+	0.0005486
Anti-proton	$\bar{p}$	–	1.00758
Neutrino	ν	0	< 0.00002
Photon	γ	0	0
Gravitation	G	0	0

Table 9.4. Characteristics of certain unstable fundamental particles.

Particle	*Symbol*	*Charge*	*Mass a.m.u.*
Neutron	n, ${}_0n^1$	0	1.00893
Negative mu meson	μ^-	–	0.1152
Positive mu meson	μ^+	+	0.1152
Negative pi meson	π^-	–	0.1514
Positive pi meson	π^+	+	0.1514
Neutral pi meson	π^0	0	0.1454

(Contd.)

Particle	*Symbol*	*Charge*	*Mass a.m.u.*
Tau meson	τ	$\pm$	0.5349
Kappa meson	κ	$\pm$	0.6035
Zeta meson	ζ	$\pm$	0.3018
Negative chi meson	χ^-	–	0.7680
Positive chi meson	χ^+	+	0.7680
Negative V-particle	V^-	–	1.2069
Positive V-particle	V^+	+	1.2069
Neutral V-particle V_1^0		0	1.2014
Neutral V-particle V_2^0		0	0.4643

Q. 8. Define the terms: atomic number and mass number. What are the notations used for them?

Ans. *Atomic number.* It is the number of protons in the nucleus of an atom.

Mass number. It is the sum of the number of protons and neutrons in the nucleus of an atom

Z = Number of protons in the nucleus
= Atomic number

A = Total number of nucleons, *i.e.*, protons and neutrons
= Mass number

N = Number of neutrons in the nucleus = A – Z

Mass number A is an integer with a value close to the atomic weight.

Thus a nuclide is a nucleus containing a particular number of neutrons and protons. The total number of protons and neutrons together represent the mass of nuclide and is known as mass number.

Mass Number = No. of protons + No. of neutrons.

Q. 9. The average mass of chlorine is 35.5. The mass numbers of two isotopes are 35 and 37. Give the proportion in which two isotopes are present in ordinary chlorine.

Ans. Let the proportion of Cl–35 and Cl–37 be 1: x

Then the average atomic mass would be

$$\frac{(35\times 1)+(37\times x)}{1+x}$$

But this is given to be 35.5

Hence $$\frac{(35\times 1)+(37\times x)}{1+x} = 35.5$$

or $$\frac{35+37x}{1+x} = 35.5$$

$$35 + 37x = 35.5(1+x)$$

$$35 + 37x = 35.5 + 35.5\,x$$

$$1.5\,x = 0.5$$

$$x = 1/3$$

Therefore the ratio is 1:1/3 or 3:1.

Q. 10. Describe the Meson theory of stability of nucleus.

Ans. Protons have a positive charge. In any nucleus containing two or more protons there will be electrostatic repulsion between the like charges. In a stable nucleus, the attractive forces are greater than the repulsive forces. In an unstable nucleus, the repulsive forces exceed the attractive forces and spontaneous fission occurs.

The first successful attempt for explaining the nuclear forces was made by Yukawa of Japan in 1935. He put forward his *meson theory* and postulated the existence of a nuclear particle π meson (called Pion) which acts as a glue binding the protons and neutrons together in the nucleus. The π meson may be positive (π^+), negative (π^-) or neutral (π^0). The charged π meson has a mass 273 times that of an electron and a charge –1 or +1. These mesons hold the protons and neutrons together by continuously exchanging position between neutrons and protons in the nucleus. Exchange of π^- and π^+ mesons accounts for binding energy between neutrons and protons. Thus a proton would release a π^+ meson and would become a neutron, whereas a neutron would release a –ve π meson and become a proton or vice versa. This exchange process is represented by

$$\underset{\text{proton}}{p} \rightleftharpoons \underset{\text{neutron}}{n} + x^{+1}$$

and

$$n \rightleftharpoons p + \pi^-$$

The exchange of π meson between neutrons and protons occurs at the rate of 10^{24} transfers/sec.

This concept of exchange force is similar to the exchange forces resulting in the formation of a covalent bond.

The neutral π-meson is uncharged particle with a mass 264 times that of an electron. The binding force between like particles *i.e.*, exchange of neutron-neutron ($n - n$) and proton–proton ($p - p$) involves neutral π mesons. The exchange reactions involving neutral π mesons are shown below:

$$p \rightleftharpoons p + \pi^0$$

$$n \rightleftharpoons n + \pi^0$$

Thus the two protons or two neutrons are held together by the exchange of π^0.

The attractive force between $p - n$, $n - n$ and $p - p$ are probably all similar in strength. The continual transfer of mesons means that the particles representing neutrons and protons are constantly changing. The transformation of neutrons into protons and vice versa are first order reactions. The rates of the reactions depend on the relative numbers of neutrons and protons present. In a stable nucleus, these two changes are in equilibrium.

Q. 11. Discuss *n/p* ratio for stability of nucleus. What is zone of stability?

(*Lucknow, 2000*)

Ans. The stability of nucleus depends on the number of protons and neutrons present. Nuclei with even number of protons and even number of neutrons seem to be particularly stable.

Light elements up to atomic number 20 form most stable nuclei when the number of protons p in the nucleus is equal to the number of neutrons n, *i.e.*, $n/p = 1$. However as the nuclear charge exceeds 20, the ratio n/p for stability exceeds unity. In such case the number of neutrons is greater than that of protons. With increase in the number of protons the force of repulsion between them increases and this tends to lower the stability of nucleus. The force of repulsion can be diminished to some extent if the number of neutrons is more. This would increase the size of nucleus and hence the proton would lie farther apart and suffer from smaller repulsion.

As the value of n/p increases beyond 1.6, *i.e.*, beyond atomic number 92, the nuclei become so large that they undergo spontaneous fission.

The variation in the number of neutrons with number of protons in stable non-radioactive nuclei is shown in Fig. 9.8. The stable nuclei lie within the shaded area which is called the **zone of stability.** Nucleus that lies outside the zone of stability will therefore be unstable. Nuclei that fall above the stability zone have more neutrons while those lying below have more protons. Both of these situations cause instability. Such nuclei would attain stability by undergoing change that would produce a nucleus with n to p ratio within the stability zone. Two cases arise depending on the n/p ratio of the nuclei:

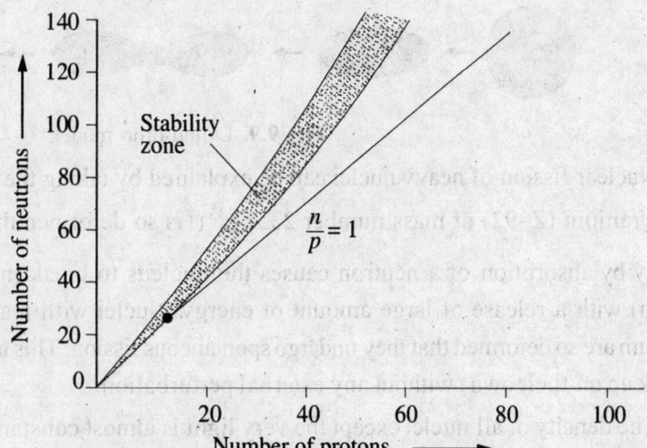

Fig. 9.8. The relation between number of neutrons and protons in stable nuclei.

(*a*) *A nucleus that possesses a lower n/p ratio would achieve stability by an increase of neutrons, a decrease of protons or both simultaneously.*

(*b*) *A nucleus that possesses a higher n/p ratio would involve a decrease of neutrons, an increase of proton or a simultaneous alteration of both particles to attain stability.*

The nuclei of radioactive element lie above and below the zone of stability (also known as stability belt). Hence they undergo spontaneous disintegration to emit α- and β–particles. Loss of each α-particle is equivalent to loss of two protons and two neutrons and loss of each β-particle is equivalent to increase of one proton and decrease of one neutron.

Q. 12. Discuss liquid drop model (L.D.M.) to explain stability of nucleus. (*Guwahati, 2002*)

Ans. This model introduced by Niels Bohr is based on the following assumptions:

(*i*) *Nucleus is like a droplet of incompressible and homogeneous liquid and all nuclei have same density.*

(*ii*) *Interaction between nucleus is strong.*

(*iii*) *Nuclear force is independent of spin and charge, i.e., there is no difference in the magnitude of the force between n – n, p – p and n – p.*

(*iv*) *Nuclear force is having a short range character and is effective between nearest neighbours. The nucleus is described in terms of a 'liquid drop'. A small liquid drop is almost spherical. It is held together by short range forces, i.e., the attraction between neighbours. Molecules at the edge of drop feel the attractive force on one side only which is known as the surface tension effect. A large liquid drop becomes elongated and with a little disturbance breaks into pieces.*

In a similar way, a nucleus is held together by short range forces (the exchange of π mesons). The surface tension effect ensures that small nuclei are spherical. As the mass of nucleus increases, the repulsion between protons increases more rapidly than the attractive forces. To minimize the repulsive force the shape of the nucleus is deformed just like the elongation of a large liquid drop. The two ends of the drop will have positive charge leading to a dumb-bell shaped structure, finally breaking into two parts as shown in Fig. 9.9.

Fig. 9.9. Liquid-drop model.

Nuclear fission of heavy nuclei can be explained by taking the example of uranium.

Uranium (Z–92) of mass number 235, $^{235}_{92}U$ is so deformed that the addition of a little extra energy by absorption of a neutron causes the nucleus to break into two smaller nuclei (nuclear fusion) with a release of large amount of energy. Nuclei with mass numbers larger than that of uranium are so deformed that they undergo spontaneous fission. This means that the nuclei disintegrate (break up on their own) without any external perturbation.

The density of all nuclei except the very light is almost constant, thus the volume of nucleus is directly proportional to the number of nucleons present.

Q. 13. Discuss shell model to explain stability of nucleus.

Or

What do you mean by magic number? What is their importance in the stability of nuclei?

(*Calcutta, 2000*)

Ans. Nuclear shell model is based on the assumption that each nucleon moves independently in an average potential generated by the remaining nucleons. This means that the interaction among the nucleons is weak so that mean free path of nucleons is larger than the nuclear dimension. In other words, nucleons move inside the nucleus without any collision. This goes against the assumption of L.D.M. that interactions among nucleons are strong so that mean free path of the nucleons is smaller than the nuclear dimension.

The electrons are arranged in different shells and orbitals which may be described by four quantum numbers. The nucleons are similarly thought to be arranged in shells, corresponding to different energy levels. When the nucleons occupy the lowest energy levels, this corresponds to the ground state. Under different conditions the nucleons may occupy different (higher or excited) energy levels. Usually the population of these higher nuclear energy *levels* is so shortlived that it cannot be observed. Thus the properties of the nucleus depend only on the number of neutrons and protons and not on the energy levels they occupy.

In a few cases the excited nuclear states have a measurable life and when this occurs nuclear isomers are said to exist. Nuclear isomers are simple nuclei with the same number of neutrons and protons but whose energies differ. It follows that the masses of the isomers differ by a very small amount corresponding to the difference in energy.

Many nuclei are unstable even when they correspond to the ground state. Unstable nuclei decompose by emitting various particles and electromagnetic radiation and this is called radioactive decay. Over 1,500 unstable nuclei are known. If no radioactive decomposition can be detected, the nucleus is said to be stable.

The shell model is supported by a periodicity in nuclear properties. Certain combinations of neutrons and protons are particularly stable.

(*a*) *Elements of even atomic number are more stable and more abundant than neighbouring elements of odd atomic number. This is known as Harkin's rule. The rule applies almost universally but H is an exception.*

(*b*) *Elements of even atomic number are richer in isotopes and never have less than three stable isotopes. Elements with odd atomic numbers often have only one stable isotope and never have more than two.*

(*c*) *There is a tendency for the number of neutrons and the number of protons in the nucleus to be even.*

This suggests that nucleons may be paired in the nucleus analogous to the pairing of electrons in atomic and molecular orbitals. If two protons spin in opposite directions, the magnetic fields they produce will mutually cancel each other and develop attractive forces. The small amount of binding energy thus generated is sufficient to stabilize the nucleus. However, this is not the most important source of energy in the nucleus.

This model gains acceptability because we find that certain nuclei are extra stable, and this is attributable to a filled shell. Nuclei with 2, 8, 20, 28, 50, 82 or 126 neutrons or protons are particularly stable and have a large number of isotopes. These numbers are termed *'magic numbers'*. The numbers 114, 164 and 184 should also be included in the series of magic numbers. When both the number of protons and the number of neutrons are magic numbers, the nucleus is very stable, for example, Pb is very stable and has 82 protons, and 208 – 82 = 126 neutrons. The emission of γ-rays by the nucleus is readily explained by the shell model. If nucleons in an excited state fall to a lower nuclear energy level, they will emit energy as γ-rays.

Q. 14. Explain the terms (*i*) Binding energy and (*ii*) Mass defect.

(*M. Dayanand, 2000; Kerala, 2000; Kurukshetra, 2001; Calcutta, 2007*)

Ans. Binding energy. When a nucleus is formed from its constituent protons and neutrons, a large amount of energy is released. This energy is equal in magnitude to binding energy of the nucleus, since the binding energy is defined as the "energy required to break up a nucleus into its individual nucleons, *i.e.*, its protons and neutrons.

Mass defect. The mass of atomic nucleus is always less than the combined masses of the protons and the nuetrons from which it is formed. The difference is called the mass defect. The mass defect is related to the binding energy holding the neutrons and protons together in the nucleous. The energy equivalent of the loss in mass is known as the binding energy of the nucleus.

Helium nucleus consists of two protons and two neutrons. It is found that the mass of helium is smaller than the combined mass of two protons and two neutrons by an amount ΔM where ΔM is the mass defect.

ΔM can be calculated by substituting the masses of the helium nuclei, protons and neutrons. Binding energy can be obtained by using the relationship

$E = (\Delta M) \times C^2$ where C is the velocity of light.

Q. 15. Give Einstein's mass energy relation.

Ans. Einstein in 1905 proposed that mass can be converted into energy and the mass-energy equivalence is given by the relation

$$E = mc^2$$

where E is energy in ergs, m is the decrease in mass of a body in grams and c is the velocity of light 2.9979×10^{10} cm/sec. Thus loss of mass results in the liberation of energy. Since the velocity of light has a very high value, a small amount of mass can be converted into a large amount of energy.

The energy equivalent of 1 g of matter.

$$E = 1 \times (2.9979 \times 10^{10})^2 = 8.9874 \times 10^{20} \text{ ergs}$$

$$= 8.9874 \times 10^{20} \times 10^{-7} \text{ Joules}$$

$= 8.9874 \times 10^{13}$ Joules

Nuclear energies are generally expressed in million electron volts (mega electron volts) (MeV)

1 MeV = 1.6022×10^{-6} ergs, hence for

$$1 \text{ g of matter } E = \frac{8.9874 \times 10^{20}}{1.6022 \times 10^{-6}} = 5.6094 \times 10^{26} \text{ MeV}$$

Atomic mass unit (a.m.u.) is another unit of mass and energy which is frequently used in nuclear science. 1 a.m.u. is equal to the twelfth of the mass of C^{12} atom. On this scale, 1 a.m.u. = 1.6606×10^{-24} g

$$E = (1.6606 \times 10^{-24})(2.9979 \times 10^{10})^2 \text{ ergs}$$

$$= 1.4924 \times 10^{-3} \text{ ergs}$$

$$= \frac{1.4924 \times 10^{-3}}{1.6022 \times 10^{-6}} = 931.14692 \text{ MeV} = 931.5 \text{ MeV}$$

POINTS TO REMEMBER

1. 1 standard a.m.u. = 1.6606×10^{-24} g = 1.6606×10^{-27} kg or 1 a.m.u. = 931.5 MeV and 1 eV = 1.6022×10^{-12} ergs.
2. 10^7 ergs = 1 Joule
3. 1 MeV = 1.6022×10^{-13} J = 1.6022×10^{-6} ergs
4. E (in MeV) = 931.5 × Δ M (mass in a.m.u.)
5. E (in ergs) = m (gram) × C^2 (cm/sec)
6. 1 eV= 1.6022×10^{-19} Joules
7. C = 2.9979×10^{10} cm/sec = 2.9979×10^8 m/sec.

Q. 16. Calculate the binding energy per nucleon of $^{16}_{8}O$

Given mass of $^{16}_{8}O$ = 15.9949 a.m.u.

Mass of $^{1}_{1}H$ = 1.007825 a.m.u.

Mass of $^{1}_{0}n$ = 1.008665 a.m.u.

Ans. Mass of 1 proton + 1 electron = mass of 1 H atom

= 1.007825 a.m.u.

Mass of 8 protons + 8 electrons = 8 × 1.007825

= 8.0626 a.m.u.

Mass of 8 neutrons = 8 × 1.008665

= 8.069320 a.m.u.

Total mass of $8p + 8e + 8n$ = 8.06260 + 8.06932

= 16.13192 a.m.u.

Actual mass of O_8^{16} = 15.9949 a.m.u.

ΔM = 16.13192 – 15.9949 = 0.13702

Binding energy E = Δm × 931.5 MeV.

= 0.13702 × 931.5 = 127.63 MeV.

Binding energy per nucleon = $\frac{127.63}{16}$ = 7.977

Q. 17. Mass of $^{7}_{3}$Li, $^{6}_{3}$Li and $^{1}_{0}$n are 7.016, 6.0151 and 1.0087 a.m.u. respectively. Calculate the binding energy of neutron in $^{7}_{3}$Li in MeV.

Ans. $Li_3^6 + n_0^1 \rightarrow Li_3^7$

Mass of Li_3^6 = 6.0151 a.m.u.

Mass of n_0^1 = 1.0087 a.m.u.

Total mass = 6.0151 + 1.0087 = 7.0238 a.m.u.

But mass of Li_3^7 = 7.016 a.m.u.

$$\Delta m = 7.0238 - 7.0160 = 0.0078 \text{ a.m.u.}$$

$$E = \Delta m \times 931.5 = 0.0078 \times 931.5 = 7.2657 \text{ MeV}$$

Q. 18. Calculate the binding energy of an α-particle. Express the result in MeV.

Mass of proton = 1.0078 a.m.u.

Mass of neutron = 1.0089 a.m.u.

Mass of α-particle = 4.0084 a.m.u.

Ans. Mass of 2 protons and 2 neutrons

= 2 × 1.0078 + 2 × 1.0089 = 4.0334 a.m.u.

Mass of α-particle = 4.0084 a.m.u.

Loss of mass = 4.0334 − 4.0084

= 0.0250 a.m.u.

∴ Binding energy = 0.0250 × 931.5

= 23.288

Q. 19. Calculate the mass defect and the binding energy of the helium nucleus having a mass of 4.0039 a.m.u. given that masses of proton and neutron are 1.00758 and 1.00893 a.m.u. respectively.

Ans. Since helium nucleus is formed from 2 neutrons and 2 protons

∴ ΔM or mass defect = [2(1.00758) + 2(1.00893) − 4.0039]

= 0.02912 a.m.u.

or = $(0.02912 \times 1.6606 \times 10^{-24})$ g

Binding energy $E = \Delta mc^2$

$= (0.02912 \times 1.6606 \times 10^{-24}) \times (2.9979 \times 10^{10})^2$

where 2.9979×10^{10} cm/sec is velocity of light

$= 4.3460 \times 10^{-5}$ ergs

1 MeV = 1.6022×10^{-6} ergs

$$\therefore \quad E = \frac{4.3460 \times 10^{-5}}{1.6022 \times 10^{-6}} = 27.13 \text{ MeV}$$

Alternatively binding energy can be calculated by multiplying the mass defect in a.m.u. by 931.5 to obtain E in MeV.

$$\text{Binding energy per nucleon} = \frac{27.13}{4} = 6.78 \text{ MeV}$$

Q. 20. What is the importance of binding energy curve?

Ans. More important than the total binding energy is the energy per nucleon E/A which is plotted in Fig. 9.10 against mass number A. Binding energy per nucleon is known as mean finding energy.

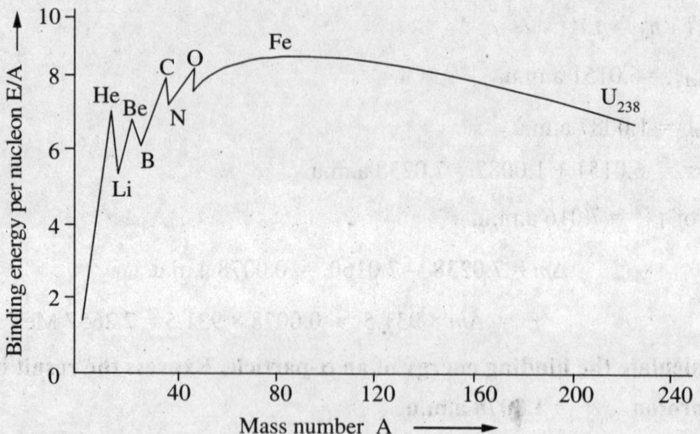

Fig. 9.10. Variation of binding energy per nucleon with mass number.

Mean binding energy is helpful to compare the stability of nuclei of different elements. Except for fluctuations at the lower mass numbers, the binding energy increases with A, goes through a maximum at A = 60 and then decreases. Thus nuclei of maximum stability have mass numbers of about 60. The elements with low or high mass number become more stable by acquiring mass number of about 60. Because of the maximum in the binding energy per particle that occurs near mass number 60 the fission of a very heavy nucleus to a pair of nuclei of mass number approximately 60 is a process that releases energy. Similarly the fusion of two of the lighter nuclei is also accompanied by release of energy.

Q. 21. What is meant by packing fraction? (*M. Dayanand, 2000; Kurukshetra, 2001*)

Ans. With the use of mass spectrograph very precise determination of atomic masses becomes possible. Aston found that in most cases there was a slight deviation from the whole number for the mass of an isotope. The nearest integral number called the mass number is indeed the sum total of the numbers of neutrons and protons. The deviation from the whole number varies from nucleus to nucleus and Aston expressed it in terms of packing fraction which is defined as

$$\text{Packing fraction} = \frac{\text{Isotopic mass} - \text{mass number}}{\text{Mass number}} \times 10^4$$

The packing fraction thus indicates average mass gain or loss per nucleon for the isotope in question as compared with the condition existing in the O^{16} nucleus. Packing fractions are numerically of the order of magnitude of 10^{-4} and are customarily multiplied by 10^4.

A negative value of the packing fractions means the isotopic mass is less than the nearest integer, *i.e.*, nearest whole number. It means that some of the mass has been converted into energy in the formation of that particular nucleus. Such nuclei therefore would be stable. For example, in case of isotopes of chlorine the isotopic mass is 34.980 whereas the mass number is 35. Hence

$$\text{Packing fraction} = \frac{34.980 - 35}{35} \times 10^4 = -5.7$$

But a positive value of the packing fraction shows that the resulting nucleus has a tendency towards instability. However it is not strictly correct particularly for those elements having small atomic weights. This is due to the fact that the masses of the protons and neutrons of which the nuclei are composed are somewhat greater than unity. In general the lower the packing fraction of an element, the greater is the binding energy and the greater is the stability of the nucleus.

When the packing fractions of different elements are plotted against their mass numbers, then the packing fractions nearly fall on or very close to the curve as shown in Fig. 9.11.

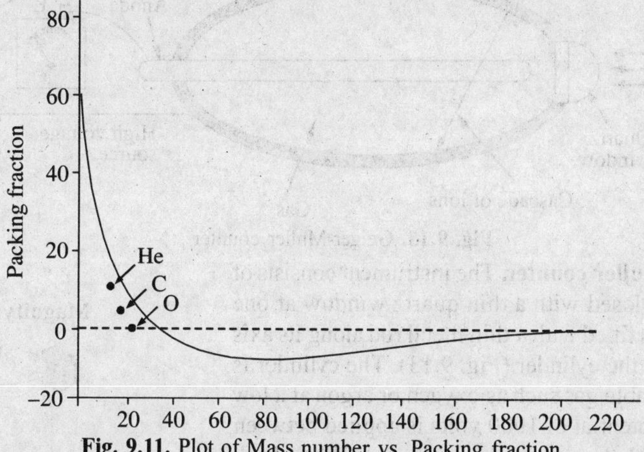

Fig. 9.11. Plot of Mass number vs. Packing fraction.

Only He (4), carbon (12) and oxygen (16) show some departure from the general behaviour. It is seen that the transition elements between chromium and zinc have the lowest values of packing fractions thereby indicating maximum stabilities of their nuclei. With elements of mass number more than 200, the packing fraction increases considerably and their nuclei therefore become more and more unstable. This agrees with the occurrence of spontaneous radioactive disintegration among such elements.

Q. 22. Write a note on natural radioactivity.

Ans. Henri Becquerel discovered that photographic plates wrapped in black paper that had been placed near uranium salts were fogged or blackened. He explained that the fogging was caused by radiations emanating from the uranium salts.

Subsequently, it was shown that radiations were similar to X-rays, can ionise air and is emitted from the elements as well as its salts and is not affected by temperature or source of uranium. This spontaneous emission of radiations by an element naturally without any external aid is called *Natural Radioactivity.* The elements which exhibit this behaviour are known as *radioactive elements.*

Q. 23. Describe different methods for detection and measurement of radioactivity.

(*Guwahati, 2002*)

Ans. The radioactivity of a substance can be detected and measured by the ionisation of gases caused by alpha and beta particles and also indirectly by gamma rays. Some methods commonly used are described below:

(1) Wilson cloud chamber method. The apparatus consists of a chamber with a glass window and is closed by movable piston (Fig. 9.12). It is filled with dust-free air and water vapour at saturation pressure. The piston is

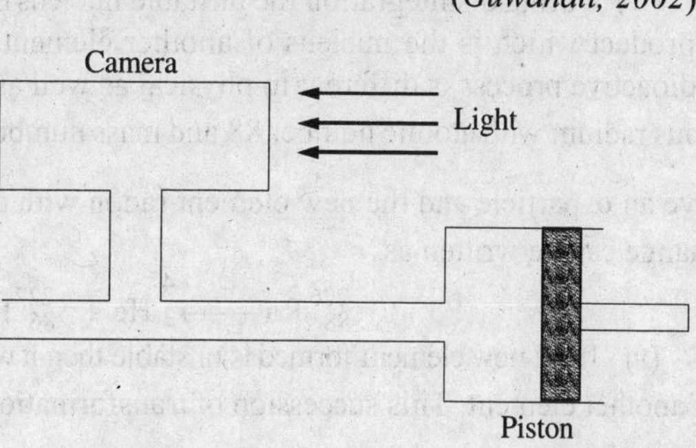

Fig. 9.12. Wilson cloud chamber

moved up to bring about cooling by sudden expansion. This causes condensation of water vapour in the form of extremely minute droplets. If alpha or beta particles from the radioactive source enter the cylinder at this stage through the window, the water droplets condense on them and the path can be photographed. This leads to the detection of alpha and beta particles.

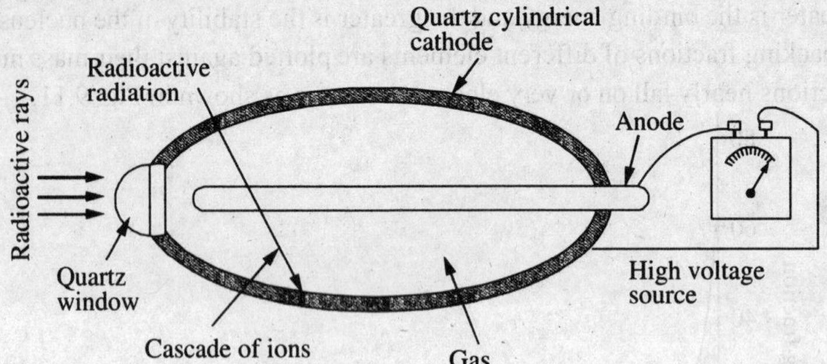

Fig. 9.13. Geiger-Muller counter.

(2) Geiger-Muller counter. The instrument consists of a copper cylinder closed with a thin quartz window at one end. The cylinder is fitted with a thin metal rod along its axis and insulated from the cylinder (Fig. 9.13). The cylinder is fitted with an ionizable gas such as oxygen or argon at a low pressure. A high potential of 1000 volts is applied between the rod (anode) and the outer tube (cathode). When high energy α and β-particles enter the metallic cylindrical tube through the quartz window they produce enough ions to cause a pulse of a current to flow between the rod and the tube. The pulse can be amplified and recorded on a current meter.

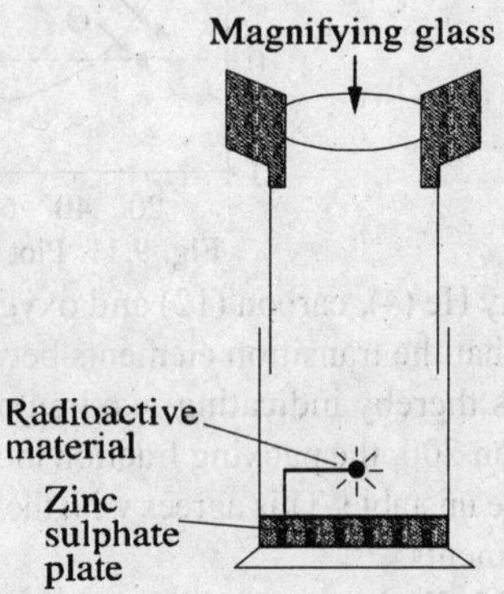

Fig. 9.14. Scintillation counter.

(3) Scintillation counter. The high energy particles are allowed to fall on a screen coated with zinc sulphide containing a trace of thallium. Each optical flash activates an electrically powdered photoelectric cell and this arrangement can record radiations up to a million per second (Fig. 9.14).

Q. 24. Discuss theory of radioactive disintegration with suitable examples.

(*Calcutta, 2007*)

Ans. Various points of theory of disintegration are given as under:

(*i*) Radioactivity is the property of the nucleus of an atom.

(*ii*) Radioactive elements possess unstable nucleus.

(*iii*) The unstable nuclei of radioactive elements are undergoing a process of disintegration.

(*iv*) During disintegration the unstable nucleus of a radioactive atom gives α or β-particles and a product which is the nucleus of another element. The new element formed as a result of the radioactive process is different in physical as well as chemical properties from its parent element. Thus radium with atomic number 88 and mass number 226 which can be written as $^{226}_{88}\text{Ra}$ breaks to give an α-particle and the new element radon with atomic number 86 and mass number 222. The change can be written as

$$^{226}_{88}\text{Ra} \longrightarrow {}^{4}_{2}\text{He} + {}^{222}_{86}\text{Rn}$$

(*v*) If the new element formed is unstable then it will also emit α or β-particles with the production of another element. This succession of transformation will go on till a stable product is formed.

(*vi*) γ-rays are secondary effects of radioactive disintegration.

To understand the nature of products that will be formed, let us consider the following cases.

Case I. If the *n/p* ratio for a nucleus lies above the stability belt, it means that it contains too many neutrons as compared to protons (high *n/p* ratio). In order to attain stability, this ratio must be lowered. It can be achieved by the emission of β-particle in which one neutron will be changed to proton and the *n/p* ratio will come closer to the stability belt. For example,

$${}^{1}_{0}n \longrightarrow {}^{1}_{1}p + {}^{0}_{-1}e + V \quad \text{(neutrino)}$$

$${}^{14}_{6}C \longrightarrow {}^{14}_{7}N + {}^{0}_{-1}e + V \quad \text{(neutrino)}$$

In the above process a β-particle is emitted by C^{14} and consequently *n/p* ratio falls.

Case II. If the *n/p* ratio for a nucleus lies below the stability belt, it means that it contains too many protons in comparison to neutrons (low *n/p* ratio). In order to come nearer to the stability belt, this ratio must be increased. It can be achieved by the emission of a positron which converts a proton into a neutron.

$${}^{1}_{1}p \longrightarrow {}^{1}_{0}n + {}^{0}_{1}e + V^{-} \quad \text{(anti-neutrino)}$$

$${}^{11}_{6}C \longrightarrow {}^{11}_{5}Be + {}^{0}_{1}e + V^{-} \quad \text{(anti-neutrino)}$$

In this process a positron is emitted by C^{11} and hence the ratio of *n/p* rises.

In such cases where *n/p* ratio is low, the nuclear stability may be achieved in yet another way, *viz.*, by the entry of an orbital electron into the nucleus. In this process which is usually called an *orbital electron capture or k-electron capture,* a proton is converted into a neutron.

$${}^{1}_{1}p + {}^{0}_{-1}e \longrightarrow {}^{1}_{0}n$$

Here the net change is the same as in a positron decay. This electron capture is not common but occurs in nuclei where the *n/p* ratio is low and the nucleus has insufficient energy for positron emission. Some typical examples are:

$${}^{7}_{4}Be + {}^{0}_{-1}e \longrightarrow {}^{7}_{3}Li$$

$${}^{40}_{19}K + {}^{0}_{-1}e \longrightarrow {}^{40}_{18}Ar$$

Q. 25. What is meant by rate of radioactive disintegration? Derive an equation for the rate of radioactive disintegration. (*Calcutta, 2007*)

OR

Show that radioactive disintegration follows first order kinetics.

Ans. Let the No. of atoms of radioactive substance at the beginning $= N_o$

No of atom left after the time $t = N$

Rate of reaction (disintegration) $\propto N$

or $\quad \frac{dN}{dt} \propto N \quad$ or $\quad \frac{dN}{dt} = \lambda N \quad$ ...(1)

where λ = Disintegration constant

Equation (1) may be written as

$$\frac{dN}{N} = \lambda dt \quad \text{...(2)}$$

Integrating eq. (2), we get

$$-\ln N = \lambda t + C \quad ...(3)$$

where C is constant of integration.

when $t = 0$, $N = N_o$, substituting in eq. (3)

$$-\ln N_o = C$$

Substituting the value of C in eq (3)

$$-\ln N = \lambda t - \ln N_o \quad \text{or} \quad \lambda t = \ln N_o - \ln N$$

$$\text{or} \quad \lambda t = \ln \frac{N_o}{N} \quad \text{or} \quad \lambda = \frac{1}{t} \ln \frac{N_o}{N}$$

Converting natural log (ln) into log to the base 10, we get

$$\lambda = \frac{2.303}{t} \ln \frac{N_o}{N}$$

This equation is called radioactive disintegration rate equation

Q. 26. What is half-life of a radioactive element? How is the disintegration constant related to half-life. *(Calcutta 2007, Vidyasagar, 2007)*

Ans. Half life of a radioactive element, represented by $t_{1/2}$ or $t_{0.5}$ is the time required for the disintegration of one half of the original amount.

At $t = t_{0.5}$, the disintegration equation (Q 25) can be written as

$$\lambda = \frac{2.303}{t_{0.5}} \log \frac{N}{N/2} \quad \text{or} \quad \lambda = \frac{2.303}{t_{0.5}} \log 2$$

$$\text{or} \quad \lambda = \frac{2.303}{t_{0.5}} \times 0.3010 \quad \text{or} \quad \lambda = \frac{0.693}{t_{0.5}} \quad \text{or} \quad t_{0.5} = \frac{0.693}{\lambda}$$

Q. 27. What is average life of a radioactive element? How is it related to disintegration constant ? *(Vidyasagar, 2007)*

Ans. It will be found from the disintegration rate equation

$$\lambda = \frac{2.303}{t} \log \frac{N_o}{N}$$

that the whole of the radioactive material will never disintegrate theoretically. Thus, it is meaningless to talk about total decay period. Therefore, a term *average life* has been introduced. It is the reciprocal of disintegration constant.

$$\lambda = \frac{0.693}{t_{0.5}} \quad \text{or} \quad \frac{1}{\lambda} = \frac{t_{0.5}}{0.693}$$

$$\text{or} \quad \text{Average life} \left(= \frac{1}{\lambda}\right) = \frac{t_{0.5}}{0.693} = 1.44\, t_{0.5}$$

Q. 28. Explain the terms: *(i)* K-electron capture, *(ii)* Neutrino and *(iii)* Anti-neutrino.

Ans. *(i)* K-electron capture. When an electron from the nearest orbital, *i.e.*, K-shell orbital is absorbed by the nucleus to convert a proton into a neutron without emitting any particle, the process is known as an orbital electron capture. Since it is most usually a K-electron which is captured by the nucleus, the process is also known as K-electron capture. Usually an electron from higher energy level L, M, N etc., drop back to fill the vacancy in K-shell.

***(ii)* Neutrino.** This elementary particle explains the conservation of angular momentum in β-decay and is postulated to balance the spin. It has zero charge and almost zero mass. It is represented as V.

***(iii)* Anti-neutrino.** The existence of an anti-neutrino (V) was also postulated to balance the spin and to distinguish between $\beta^-\left({}^{0}_{-1}e\right)$ and $\beta^+\left({}^{0}_{+1}e\right)$ decay. It was further suggested that neutrino (V) and anti-neutrino ($\bar{V}$) have relationship of the type existing between an electron and a positron.

Q. 29. Write a note on radioactive equilibrium. *(Vidyasagar, 2007)*

Ans. Consider a radioactive substance A (parent) which breaks down to give a second element B (daughter) which in turn disintegrates to give a third one C.

$$A \longrightarrow B \longrightarrow C$$

B is said to be in radioactive equilibrium with A when its decay is exactly compensated by its production from A. When the equilibrium is reached, the amount of B remains constant because it is being formed and disintegrated at the same rate. It is also called steady state. If K_1 and K_2 represent the disintegration constants of A and B; N_1 and N_2 the number of atoms of radioactive elements (parent and daughter) present, then at equilibrium the rate of formation of the daughter from its parent is equal to its own rate of disintegration.

$$\left[\frac{dn}{dt}\right]_{parent} = \left[\frac{dn}{dt}\right]_{daughter}$$

$$K_1 N_1 = K_2 N_2$$

$$\frac{N_1}{N_2} = \frac{K_2}{K_1} = \frac{\lambda_1}{\lambda_2}$$

where λ_1 and λ_2 denote average life periods of parent and daughter element species respectively.

Hence at equilibrium A and B are present in the ratio of their average life periods. They will also be present in the ratio of their half-life periods. This also implies that when radioactive elements are in equilibrium, the one with a high average life is present in greater amount than the other with shorter average life. Hence at equilibrium, the following relations hold good:

(*i*) The number of atoms of each radioactive element present is inversely proportional to its decay constant.

(*ii*) Directly proportional to its half-life.

(*iii*) If the equilibrium mixture contains an element of an extremely long life and another of shorter measurable half-life, it is possible to calculate the former from the relative amount of both species present. For example, all uranium ores contain radium and uranium in the atomic ratio.

$$\frac{N_{Ra}}{N_U} = 3.48 \times 10^{-7}$$

From the above relation, it follows that

$$\frac{N_U}{t_{1/2}U} = \frac{N_{Ra}}{t_{1/2}Ra} \quad \text{or} \quad t_{1/2}\ U = \frac{N_U \times t_{1/2}Ra}{N_{Ra}}$$

since $t_{1/2}$ Ra is 90 years

$$t_{1/2}\ U = \frac{1}{3.48} \times 10^{-7} \times 1590 \text{ years} = 4.5 \times 10^9 \text{ years.}$$

Q. 30. Write in detail about Artificial Disintegration or Transmutation of Elements. *(Punjab, 2000)*

Ans. Transmutation of element means conservation of one element into another by artificial means. Nitrogen on bombardment with helium nuclei (α particles) gives oxygen as shown by the following equation:

$$^{14}_{7}N + ^{4}_{2}He \longrightarrow ^{17}_{8}O + ^{1}_{1}H$$

In the process protons are produced.

In 1932 Chadwick identified a new particle obtained from the bombardment of beryllium by α–particles. It had a unit mass and no charge. It was named neutron $^{1}_{0}n$.

$$^{9}_{4}Be + \underset{(\alpha\text{–particle})}{^{4}_{2}He} \longrightarrow ^{12}_{6}C + \underset{(\text{neutrons})}{^{1}_{0}n}$$

This is how the neutron was discovered. Being a neutral particle, it can penetrate into the nuclei of even the heavier elements and has been successfully used for artificial transmutations. Accelerated protons ($^{1}_{1}H$), deuterons ($^{2}_{1}H$), *i.e.,* heavy hydrogen, have also been used as projectiles for the disruption of the stable nuclei. Since α-particles carry double positive charge, they are not good projectiles for bombarding of atoms as they are repelled by nucleus. An α-particle, proton or a deuteron, before it can be captured by a nucleus, must possess sufficient kinetic energy to overcome the repulsive force that develops as the +ve particle approaches the +ve nucleus. For this purpose, several instruments for accelerating these particles have been developed. One of the most effective apparatus invented by E.O. Lawrence is known as cyclotron which can impart energy up to 400 MeV (Million electron volt). Synchotron is a more powerful modern version.

A large number of transmutations have been carried out by means of alpha particle, protons, neutrons and deuterons. A few of these are given below in the form of equations. At the end of reaction two particles are given in the brackets. The first particle refers to the particle used for bombardment and the second particle refers to the new particle produced.

(*i*) Transmutation by alpha particles ($^{4}{}_{2}He$), α

(*a*) $^{12}_{6}C + ^{4}_{2}He \longrightarrow ^{15}_{8}O + ^{1}_{0}n$ (α, n – type)

(*b*) $^{27}_{13}Al + ^{4}_{2}He \longrightarrow ^{30}_{14}Si + ^{1}_{1}H$ (α, p – type)

(*ii*) Transmutation by protons ($^{1}_{1}H$), *p*

(*a*) $^{27}_{13}Al + ^{1}_{1}H \longrightarrow ^{24}_{12}Mg + ^{4}_{2}He$ (p, α – type)

(*b*) $^{7}_{3}Li + ^{1}_{1}H \longrightarrow ^{4}_{2}He + ^{4}_{2}He$ (p, α – type)

(*c*) $^{11}_{5}B + ^{1}_{1}H \longrightarrow ^{11}_{6}C + ^{1}_{0}n$ (p, n – type)

(*d*) $^{23}_{11}Na + ^{1}_{1}H \longrightarrow ^{23}_{12}Mg + ^{1}_{0}n$ (p, n – type)

(*iii*) Transmutation by deuterons ($^{2}{}_{1}H$), d

(*a*) $^{31}_{17}Cl + ^{2}_{1}H \longrightarrow ^{32}_{18}Ar + ^{1}_{0}n$ (d, n – type)

(*b*) $^{31}_{15}P + ^{2}_{1}H \longrightarrow ^{32}_{15}P + ^{1}_{1}H$ (d, p – type)

(*c*) $^{12}_{6}C + ^{2}_{1}H \longrightarrow ^{10}_{5}B + ^{4}_{2}He$ (d, α – type)

***(iv)* Transmutation by neutrons ($_0^1n$), n.** It is brought about by high-speed neutrons. It may result in the emission of a proton, an α-particule or another neutron.

(*a*) $^{16}_{8}O + ^{1}_{0}n \longrightarrow ^{13}_{6}C + ^{4}_{2}He$ (n, α – type)

(*b*) $^{14}_{7}N + ^{1}_{0}n \longrightarrow ^{11}_{5}B + ^{4}_{2}He$ (n, α – type)

(*c*) $^{27}_{13}Al + ^{1}_{0}n \longrightarrow ^{27}_{12}Mg + ^{1}_{1}H$ (n, p – type)

(*d*) $^{14}_{7}N + ^{1}_{0}n \longrightarrow ^{14}_{6}C + ^{1}_{1}H$ (n, p – type)

(*e*) $^{39}_{19}K + ^{1}_{0}n \longrightarrow ^{38}_{19}K + 2^{1}_{0}n$ (n, 2n – type)

In some cases only γ-radiation is emitted.

$$^{23}_{11}Na + ^{1}_{0}n \longrightarrow ^{24}_{11}Na + \gamma - \text{radiation (Energy)}$$

Q. 31. Complete the following:

(i) $^{30}_{15}P \longrightarrow ^{30}_{14}Si + \ldots\ldots$

(ii) $^{37}_{17}Cl + \ldots\ldots \longrightarrow ^{38}_{18}Ar + ^{1}_{0}n$

(iii) $^{235}_{92}U + ^{1}_{0}n \longrightarrow ^{141}_{56}Ba + ^{92}_{36}Kr + \ldots\ldots$

(iv) $^{27}_{13}Al + \ldots\ldots \longrightarrow ^{30}_{15}P + ^{1}_{0}n$ (*Kerala, 2000*)

Ans. (*i*) $^{30}_{15}P \longrightarrow ^{30}_{14}Si + ^{0}_{+1}e$

(*ii*) $^{37}_{17}Cl + \underset{\text{Deuteron}}{^{2}_{1}D} \longrightarrow ^{38}_{18}Ar + ^{1}_{0}n$

(*iii*) $^{235}_{92}U + ^{1}_{0}n \longrightarrow ^{141}_{56}Ba + ^{92}_{36}Kr + 3^{1}_{0}n$

(*iv*) $^{27}_{13}Al + \underset{\alpha-\text{particle}}{^{4}_{2}He} \longrightarrow ^{30}_{15}P + ^{1}_{0}n$

Q. 32. Give a detailed description of artificial or induced radioactivity. (*Kerala, 2000*)

Ans. Artificial or induced radioactivity is a phenomenon in which a new radioactive isotope of an element is formed by the artificial transmutation of a stable element.

The artificial radioactive element behaves in the same fashion as the naturally occurring radioactive series. The radioactive substance produced in the first step disintegrates further to give a stable element. Many radioactive isotopes can be produced by the bombardment of stable isotopes by protons, α-particles and neutrons. Some examples are given below.

(1) Induced radioactivity by bombardment with α-particles

$$^{24}_{12}Mg + ^{4}_{2}He \longrightarrow ^{27}_{14}Si + ^{1}_{0}n$$

$$ {}^{27}_{14}Si \longrightarrow {}^{27}_{13}Al + {}^{0}_{1}e $$
Stable Positron

(2) Induced radioactivity by bombardment with neutrons (${}^{1}_{0}n$). Both high-speed and slow-speed neutrons are used for this purpose. Commonly, slow-speed neutrons are more effective in producing nuclear reactions than high-speed neutrons. Three types of non-fission reactions have been produced by neutrons:

***(a)* Radioactive capture.** The *(n, r)* is the most common nuclear reaction and has been carried out for isotopes of all elements when slow-speed neutrons are bombarded on the target. For example,

$$ {}^{40}_{18}Ar + \underset{\text{slow}}{{}^{1}_{0}n} \longrightarrow \underset{\text{Radioactive}}{{}^{41}_{18}Ar} + \gamma $$

$$ {}^{41}_{18}Ar \longrightarrow {}^{41}_{19}K + {}^{0}_{-1}e $$

***(b)* Emission of neutrons.**

$$ {}^{39}_{19}K + {}^{1}_{0}n \longrightarrow {}^{38}_{19}K + 2{}^{1}_{0}n $$

$$ {}^{38}_{19}K \longrightarrow {}^{38}_{18}Ar + {}^{0}_{+1}e $$

***(c)* Emission of charged particle.** These are favoured with large kinetic-energy neutrons or high -speed or fast-moving neutrons.

$$ {}^{35}_{17}Cl + \underset{\text{fast}}{{}^{1}_{0}n} \longrightarrow \underset{\text{Radioactive}}{{}^{35}_{16}S} + {}^{1}_{1}H $$

$$ {}^{35}_{16}S + \longrightarrow {}^{35}_{17}Cl + {}^{0}_{-1}e $$

(3) Induced radioactivity by bombardment with deuterons (${}^{2}_{1}H$) or d.

(i) $$ {}^{10}_{5}B + {}^{2}_{1}H \longrightarrow \underset{\text{Radioactive}}{{}^{11}_{6}C} + {}^{1}_{0}n;\quad {}^{11}_{6}C \longrightarrow {}^{11}_{5}B + {}^{0}_{1}e $$

(ii) $$ {}^{12}_{6}C + {}^{2}_{1}H \longrightarrow {}^{13}_{7}N + {}^{1}_{0}n $$

$$ {}^{13}_{7}N \longrightarrow {}^{13}_{6}C + {}^{0}_{1}e $$

(4) Induced radioactivity by bombardment with protons (${}^{1}_{1}H$)

$$ {}^{7}_{3}Li + {}^{1}_{1}H \longrightarrow \underset{\text{Radioactive}}{{}^{7}_{4}Be} + {}^{1}_{0}n $$

$$ {}^{7}_{4}Be \longrightarrow \underset{\text{Stable}}{{}^{7}_{3}Li} + \underset{\text{Positron}}{{}^{0}_{1}e} $$

Curie and Joliot explained that the nucleus formed in the artificial transmutation reaction is unstable, which exhibits radioactivity and then undergoes disintegration. This explanation was tested chemically by separating the target element from the product and it was found that the latter exhibited radioactivity.

Q. 33. What are different types of nuclear reactions depending upon transformation of nucleus? (*Punjab, 2000; Kanpur, 2001; Kurukshetra, 2002; Guwahati, 2002*)

Ans. On the basis of overall transformation, the various types of nuclear reactions may be classified as:

***(i)* Capture reactions.** These reactions occur when the bombarding particle is absorbed by the target nucleus but no massive particle (α, n, p etc.) other than the product nucleus is liberated. However γ-radiation may be emitted. Slow neutron capture is a very common reaction and is known for many nuclei.

$$^{238}_{92}U + ^{1}_{0}n \longrightarrow ^{239}_{92}U + \gamma$$

A shorthand notation known as Bethe's reaction notation is used for the representation of the reaction as $^{238}_{92}U\ (n, \gamma)\ ^{238}_{92}U$.

***(ii)* Particle-particle reactions.** In these reactions along with the product nucleus, a massive particle (p, n, α etc.) is also produced. Such reactions are most common. Examples are:

$$^{12}_{6}C(\alpha, n)^{15}_{8}O,\ ^{27}_{13}Al\ (n, p)\ ^{27}_{12}Mg$$

***(iii)* Fission reactions.** These reactions take place when the target nucleus is of heavy mass. The product nucleus formed after bombarding breaks into two nearly equal parts and a large amount of energy along with several neutrons is liberated. Neutron-induced fission of $^{235}_{92}U$ and $^{239}_{94}Pu$ is quite important.

$$^{235}_{92}U + ^{1}_{0}n \longrightarrow ^{141}_{56}Ba + ^{92}_{36}K + 3^{1}_{0}n$$

***(iv)* Spallation reactions.** Spallation (meaning crumbling) reactions are very much similar to fission reactions but such reactions leave a single residual nucleus as the end product and such reactions are not self-sustaining. As a result of the very high energy projectiles drastic nuclear reactions occur resulting in the formation of large numbers of light fragments. The light fragments correspond to decrease in mass number up to 30 units and in charge up to 14 units along with the main product. Nuclear particles such as protons, neutrons etc., are also emitted in large number.

An example of spallation reactions is bombardment of $^{238}_{92}U$ with α particle of energy 400 MeV to produce Ra^{223} with the release of 6 protons and 13 neutrons.

$$^{238}_{92}U\ (\alpha, 6p, 13n)\ ^{223}_{88}Ra$$

***(v)* Fusion reactions.** In these reactions certain light nuclei may be fused together to produce heavier nuclei, under special circumstances. Tremendous amount of energy is liberated in this process. An example is fusion of hydrogen nuclei into helium.

$$^{1}_{1}H + ^{1}_{1}H + 2^{1}_{0}n \longrightarrow ^{4}_{2}He + \text{Energy}$$

Q. 34. Categorise various types of nuclear reactions on the basis of nature of bombarding particles. *(Kurukshetra, 2001)*

Ans. On the basis of nature of the bombarding particle, the various types of nuclear reactions may be classified as:

***(i)* Proton-induced reactions.** In these reactions, the coulombic repulsions are less but the smaller mass of the protons renders it difficult to penetrate the nuclear energy barrier. The *(p, α)*, *(p, n)*, *(p, d)* and *(p, γ)* types of reactions are:

$$^{14}_{7}N\ (p, \alpha),\ ^{11}_{16}C;\ ^{11}_{5}B\ (p, n),\ ^{11}_{6}C;\ ^{9}_{4}Be\ (p, d),\ ^{8}_{4}Be;\ ^{14}_{7}N\ (p, \gamma)\ ^{15}_{8}O$$

***(ii)* Deuteron-induced reactions.** Deuterons are the most effective of the positively charged particles. On account of a comparatively large mass defect the deuteron is inherently energetic.

Further increase in energy may be done by cyclotron acceleration. A (d, p) reaction is similar to a simple (n, γ) reaction. The deuteron is first broken up outside the target nucleus to proton and neutron. Then only the neutron brings about the reactions. The following three (*d*, α), (*d*, *n*) and (*d*, *p*) types of reactions are:

$$^{16}_{8}O(d, \alpha)\ ^{14}_{7}N;\ ^{9}_{4}Be\ (d, n),\ ^{10}_{5}B;\ ^{23}_{11}Na\ (d, p)\ ^{21}_{11}Na$$

***(iii)* Alpha-induced reactions.** In these reactions there are large coulombic repulsions between the positively charged α-particles. α-particles are first accelerated sufficiently in cyclotron or some other accelerating device. The two (α, p) and (α, n) common types of reactions are:

$$^{14}_{7}N(\alpha, p)\ ^{17}_{8}O;\ ^{10}_{5}B\ (\alpha, n)\ ^{13}_{7}N$$

***(iv)* Neutron-induced reactions.** Neutron being a neutral particle does not face any coulombic repulsion and therefore, the neutron reactions depend only upon the energy of the bombarding particle. Generally, all the light as well as heavy elements interact with neutrons. On bombarding, lighter elements generally eject some particles such as α-particles or protons while heavier elements generally capture them and form their own isotopes. The five (*n*, γ), (*n*, α), (*n*, *p*), (*n*, 2*n*) and (*n*, fission) types of reactions are:

$$^{238}_{92}U(n, \gamma)\ ^{239}_{92}U,\ ^{7}_{3}Li\ (n, \alpha)\ ^{3}_{1}H,\ ^{14}_{7}N\ (n, p)\ ^{14}_{6}C,$$

$$^{238}_{92}U(n, 2n)\ ^{237}_{92}U,\ ^{235}_{92}U\ (n, \text{fission})\ \text{fission products.}$$

Q. 35. What is meant by Q-value of nuclear reactions? How is it calculated?

(*Kurukshetra, 2002; M. Dayanand, 2000*)

Ans. In all nuclear reactions, there is either an increase in the mass or decrease in the mass. This decrease or increase in mass ΔM, called mass defect during the reactions, is converted into energy Q. This is measured in MeV (Million electron volts). It means that the nuclear reactions, like chemical reactions, are always accompanied by a release or absorption of energy. This energy is expressed by adding the term Q to the right hand side of the equation.

For example, in the ordinary nuclear reactions studied by Rutherford in 1919

$$^{4}_{2}He + \ ^{14}_{7}N \longrightarrow \ ^{17}_{8}O + \ ^{1}_{1}H + Q\ \text{(energy)}$$

The quantity Q is called the energy of the reaction or more frequently just "the Q of the reaction". The Q can be positive or negative. Positive Q shows the energy is released and the reaction is said to be *exoergic* and negative Q shows energy absorption and the reaction is *endoergic*.

There is an important difference between a chemical reaction and a nuclear reaction. In case of chemical reactions, we always consider macroscopic amount of material undergoing reactions and consequently heats of reactions are usually given per mole or occasionally per gram of one of the reactants. In case of nuclear reactions, we usually consider single processes and the Q values are, therefore, given per nucleus transformed.

In the case of the above reactions, Q can be calculated from the mass of the particles as follows:

Mass on the left hand side: $^{4}_{2}He = 4.0026$

$$^{14}_{7}N = 14.0031$$

$$= 18.0057 \text{ a.m.u.}$$

Mass on the right hand side: $^{1}_{1}H = 1.0078$

$^{17}_{8}O = 16.991$

$= 18.0069$ a.m.u.

R.H.S. gives an increase of mass of 0.0012 a.m.u. Thus an amount of energy equivalent to 0.0012 mass unit has to be supplied to make the reactions energetically possible.

Hence $Q = -0.0012 \times 931.5 = -1.1178$ MeV

$[\because 1 \text{ a.m.u.} = 931.5 \text{ MeV}]$

This reactions is, therefore, endoergic as Q is –ve.

Q. 36. Calculate the Q-value for the reaction:

$$^{235}_{92}U + ^{1}_{0}n \longrightarrow ^{98}_{42}Mo + ^{136}_{54}Xe + 2^{1}_{0}n$$

The exact masses of reactants and products are

$^{235}_{92}U = 235.044$ a.m.u; $^{1}_{0}n = 1.009$ a.m.u.

$^{98}_{42}Mo = 97.905$ a.m.u.; $^{136}_{54}Xe = 135.917$ a.m.u.

Ans. ΔM = Mass of products – Mass of reactants

$= [97.905 + 135.917 + 2.018] - (235.044 + 1.009]$ a.m.u.

$= -0.213$ a.m.u.

Thus there is a decrease in mass equal to 0.213 a.m.u. Hence energy will be released during the reaction. Hence Q is positive.

Energy Released $= 0.213 \times 931.5$ MeV

$= 198.41$ MeV.

Q. 37. What is meant by nuclear fission? (*Kanpur, 2001; Vidyasagar, 2007; Patna, 2010*)

Ans. *The process of splitting up of a heavy nucleus into two, more or less, equal halves, is called Nuclear Fission.* This is accompanied by liberation of some neutrons and enormous energy.

Otto Hahn and Strassmann (1939) observed that when uranium was irradiated with neutrons, isotopes of intermediate mass number were obtained. They proved by analysis that one of the products of this nuclear reaction was $^{144}_{56}Ba$. Thus,

$$^{235}_{92}U + ^{1}_{0}n \longrightarrow ^{144}_{56}Ba + ^{92}_{36}Kr$$

These isotopes of Ba and Kr were radioactive and gave more stable isotopes by the release of more neutrons.

$$^{235}_{92}U + ^{1}_{0}n \longrightarrow ^{141}_{56}Ba + ^{92}_{36}Kr + 3\,^{1}_{0}n$$

When fission takes place, the nucleus breaks into two fragments of comparable sizes.

The difference in the actual masses of product substances and the parent substances is converted into energy. Thus in the following case,

$$\underset{235.0439}{{}^{235}_{92}U} + \underset{1.00886}{{}^{1}_{0}n} \longrightarrow \underset{140.9136}{{}^{141}_{56}Ba} + \underset{91.8976}{{}^{92}_{36}Kr} + \underset{3 \times 1.00866}{3\,{}^{1}_{0}n} + Q$$

The total mass on the L.H.S. is 236.05276 a.m.u. and on the R.H.S. is 235.8371 a.m.u. Hence there is a decrease of mass by 0.21566 a.m.u. Since 1 a.m.u is equivalent to 931.5 MeV energy, an amount of 0.21566 × 93.15 or 200.88729 MeV of energy is produced. Thus in the fission of one nucleus of ^{235}U about 200 MeV energy is released. This is the Q value for this fission reaction. It may be understood that nuclear fission involves a net loss of mass. This excess mass is transformed into energy according to Einstein equation $E = mc^2$. The magnitude of energy is enormous. It has been calculated that the energy released by the fission of one kg of natural uranium is equal to that available from 20,000 kg of coal.

Chain reactions. It has been observed that the fission of ^{235}U is very rapid and releases large amount of energy. In the reaction

$${}^{235}_{92}U + {}^{1}_{0}n \longrightarrow {}^{141}_{56}Ba + {}^{92}_{36}Kr + 3\,{}^{1}_{0}n$$

we see that three neutrons are produced as a result of fission of one ^{235}U nucleus. It may, however, be added that U-235 is known to split at least in 30 different ways. These three neutrons will cause further fission of three more ^{235}U nuclei producing nine neutrons as shown in Fig. 9.15. These nine neutrons will cause further fission of nine ^{235}U nuclei producing twenty seven neutrons and so on.

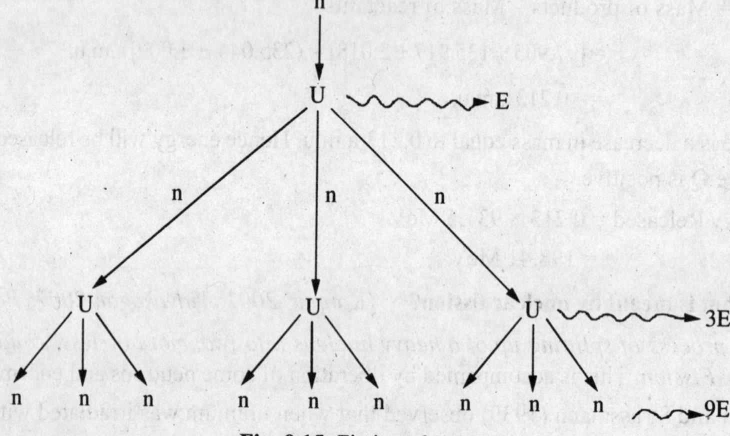

Fig. 9.15. Fission of U 235.

This process is known as *chain reaction or auto catalytic reaction.* Absorption of neutron by non-fissionable material reduces the efficiency of chain reaction.

In order to ensure an efficient chain reaction moderators are used. Moderators are substances which slow down the speeds of the fast neutrons obtained after fission very rapidly so that the chances of their capture (of neutron) by non-fissionable material becomes very small. Graphite and heavy water are used as moderators.

If E be the energy released per fission, the total energy liberated from the chain reactions goes on multiplying in each step and ultimately becomes enormous. This large amount of energy could be exploited for useful purposes.

Q. 38. Write a note on plutonium bomb.

Ans. Bombarding of ^{235}U with neutrons liberates a tremendous amount of energy. This raises the prospects of the utilisation of atomic energy as a source of power. But we are faced with a difficulty. Ordinary uranium consists of 99.28% of ^{235}U, 0.71% of ^{235}U and about of 0.01% of ^{234}U. It is very difficult to isolate ^{235}U (fissionable material) from the above isotopic mixture. It was found that plutonium, which does not occur in nature but which can be prepared from the most abundant isotopes ^{238}U, undergoes fission readily like ^{235}U. Plutonium can be prepared from ^{238}U by neutron bombardment as follows:

$$^{238}_{92}U + ^{1}_{0}n \longrightarrow ^{239}_{92}U \quad \text{(Half-life of 23 minutes)}$$

$$^{239}_{92}U \longrightarrow ^{239}_{93}Np + ^{0}_{-1}e \quad \text{(Half-life of 2-3 days)}$$

$$^{239}_{93}Np \longrightarrow ^{239}_{94}Pu + ^{0}_{-1}e \quad \text{(Half-life of } 2.4 \times 10^4 \text{ years)}$$

The $^{238}_{92}U$, on capture of a neutron, is converted into $^{239}_{92}U$ which is radioactive and changes over with emission of a β-particle into the transuranic element $^{239}_{93}Np$. $^{239}_{93}Np$ is also radioactive and emits a β-particle to give plutonium of atomic number 94, *i.e.*, $^{239}_{94}Pu$.

Plutonium, which is chemically different from uranium, can be separated easily. It has an additional advantage because its half-life period is very long of the order of about 24,000 years. The problem of release of nuclear energy through nuclear fission was solved with the discovery of Pu. It is estimated that energy released by a gram of plutonium is equal to that of 30 tons of T.N.T. This forms the basis of the manufacture of atom bombs.

Q. 39. What is nuclear fusion? *(Kanpur, 2001; Vidyasagar, 2007)*

Ans. It is the process in which two or more light nuclei are made to combine to give a heavier nuclei. During this process also some mass is lost and is converted into energy.

Principal source of stellar energy, *i.e.*, energy of the sun, is the nuclear energy released, during the fusion of hydrogen nuclei to form helium nuclei (4H → He) (see Fig. 9.16).

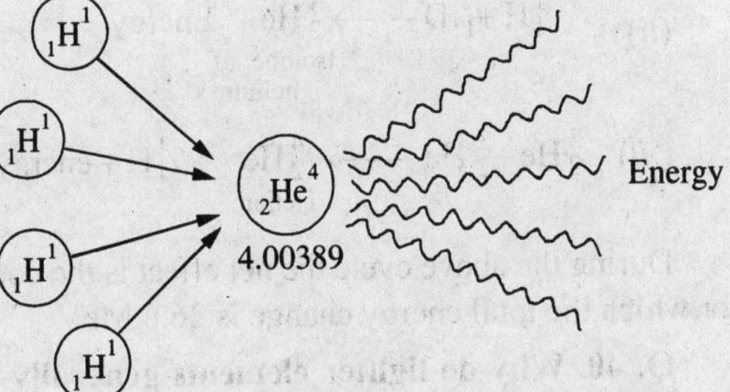

Fig. 9.16. Principle of solar energy.

$$4^{1}_{1}H \longrightarrow ^{4}_{2}He + 2^{0}_{1}e + \gamma$$

$$4^{1}_{1}H = 4 \times 1.00813 = 4.03252 \text{ a.m.u.};$$

$$^{4}_{2}He = 4.00389 \text{ a.m.u.}$$

From four hydrogen atoms with atomic weight 4 × 1.00813 = 4.00389 a.m.u., an atom of helium is obtained with atomic weight 4.00389 a.m.u. The balance of mass 0.02863 a.m.u. is converted into energy according to the relation.

Energy equivalent to 0.02863 mass unit = 0.02863 × 931.5 MeV = 26.6688 MeV.

Weizacker and Bethe in 1939 showed that a closed carbon and nitrogen cycle is responsible for the origin of energy in stars including our sun, *i.e.*, the stellar energy, as given below:

$$^{12}_{6}C + ^{1}_{1}H \longrightarrow ^{13}_{7}N + \gamma$$

$$^{13}_{7}N \longrightarrow ^{13}_{6}C + ^{0}_{+1}e + \gamma$$

$$^{13}_{6}C + ^{1}_{1}H \longrightarrow ^{14}_{7}N + \gamma$$

$$^{14}_{7}N + ^{1}_{1}H \longrightarrow ^{15}_{8}O + \gamma$$

$$^{15}_{8}O \longrightarrow ^{15}_{7}N + ^{0}_{+1}e + \gamma$$

$$^{15}_{7}N + ^{1}_{1}H \longrightarrow ^{12}_{6}C + ^{4}_{2}He + \gamma$$

By cancellation:

$$4^{1}_{1}H \longrightarrow ^{4}_{2}He + 2\,^{0}_{+1}e + \text{energy}$$

The net reaction in the process is a closed chain comprising the fusion of four protons to give a helium nucleus, two positrons and release of vast amount of energy (26.6 MeV/cycle).

The process of nuclear fusion has a very high activation energy and takes place, therefore, at very high temperature, of the order of million of degrees centigrade, ($> 10^6$ K) which only exists in the interior of the sun. It is for this reason that such processes are, therefore, called **Thermonuclear Reactions**. Once a fusion reaction is initiated the process goes on since there are both H and He in the sun and since the temperature in the interior is high enough to produce transmutation. This suggests that the process of producing the sun's energy may be due to following cycles:

(*i*) $^{1}_{1}H + ^{1}_{1}H \longrightarrow ^{2}_{1}D + ^{0}_{+1}e + \text{energy}$ (Positron)

(*ii*) $^{1}_{1}H + ^{2}_{1}D \longrightarrow ^{3}_{2}He + \text{Energy}$ (Isotope of helium)

(*iii*) $^{3}_{2}He + ^{3}_{2}He \longrightarrow ^{4}_{2}He + 2^{1}_{1}H + \text{energy}$ (Helium)

During the above cycle the net effect is the conversion of four protons into the *helium* nucleus, for which the total energy change is 26.6 MeV.

Q. 40. Why do lighter elements generally undergo fusion while heavier elements show nuclear fission?

Ans. Lighter elements have smaller values of *n/p*. They have a tendency to increase this *n/p* value. This can be achieved through fusion. For example,

$$4^{1}_{1}H \longrightarrow ^{4}_{2}He + 2\,^{0}_{+1}e + \gamma$$

In helium *n/p* ratio is 1 whereas in hydrogen there are no neutrons and *n/p* value is zero. Heavier elements have a higher value of *n/p* and they have a tendency to reduce this value. They achieve it through fission in which certain number of neutrons are ejected thus lowering the *n/p* value of the products.

Q. 41. Write notes on (*i*) Hydrogen bomb and (*ii*) Cobalt bomb. *(Pune 2010)*

Ans. (*i*) Hydrogen bomb. The manufacture of hydrogen bomb is based upon the principle of nuclear fusion. The energy released in the process is about 1000 times that of an atomic bomb. However, this reaction takes place at very high temperature, of the order of that prevailing in the interior of the sun. It is believed that in a hydrogen bomb, a mixture of heavy isotopes of hydrogen, *i.e.*, deuterium or tritium, is enclosed in a space surrounding an ordinary atomic bomb. The transformation and possibilities are believed to be of the following four types.

These are:

(*i*) $^{2}_{1}H + ^{2}_{1}H \longrightarrow ^{3}_{2}He + ^{1}_{0}n + 3.25 \text{ MeV}$

(*ii*) $^{2}_{1}H + ^{2}_{1}H \longrightarrow ^{3}_{1}H + ^{1}_{1}H + 4.0 \text{ MeV}$

(*iii*) $^{3}_{1}H + ^{2}_{1}H \longrightarrow ^{4}_{2}He + ^{1}_{0}n + 17.6 \text{ MeV}$

(*iv*) $^{2}_{1}H + ^{3}_{2}He \longrightarrow ^{4}_{2}He + ^{1}_{1}H + 18 \text{ MeV}$

The reactions (*iii*) and (*iv*) are actually the reaction products of (*i*) and (*ii*) so that the overall conversion of six deuterons is as follows:

$$6^{2}_{1}H \longrightarrow 2^{4}_{2}He + 2^{1}_{0}n + 43 \text{ MeV}$$

For the fusion of 6 deuterons, as we see above, the total energy produced is 43 MeV. Thus the energy produced per deuteron is 7.16 MeV. Therefore, for equal mass of material much more energy is released in the fusion than in fission.

In hydrogen bomb, there is no limitation of the critical mass of fusible material, hence any amount of fusible material can be used. Thus there is no limit for the explosive power of the hydrogen bomb.

In hydrogen bomb, the fusion between deuterium nuclei and tritium nuclei is made to take place. The essential conditions for carrying out this fusion reactions are high temperature, of the order of millions of degrees and high pressure of the order of millions of atmospheres. These conditions can be attained by exploiting an atom bomb. Hence, for the preparation of hydrogen bomb, a suitable quantity of a mixture of deuterium and tritium is combined with an atom bomb. The atom bomb first explodes and provides the essential conditions of temperature and pressure for the fusion reaction. The fusion reaction then starts and releases enormous amount of energy. Thus, this hydrogen bomb is a fission fusion bomb.

(*ii*) Cobalt bomb. When hydrogen bomb is encased in a sheath of cobalt, it is called cobalt bomb. During explosion, the cobalt sheath is converted into gigantic radioactive cloud producing lethal radiations. It destroys everything which is covered by it. Moreover, its effect extends to thousands of miles away because the wind takes the cloud to distant places. The effect is prolonged to five year span due to its long life period. Thus cobalt bomb is more dangerous than the hydrogen bomb.

Q. 42. Describe important applications of radioactivity.

Ans. Important applications of radioactivity are described as under:

(1) Atomic structure revealed. The famous experiment of scattering of α-particles by Rutherford was done by radioactive material. This led to the concept of nucleus. The artificial disintegration of atoms increased our knowledge about the nature of nucleus, *i. e.*, presence of protons or neutrons in it. Moreover, many more fundamental particles like positron, mesons etc. have been discovered as a result of work on radioactivity.

(2) New source of energy. Nuclear fusion has opened a new source of atomic energy at our disposal. The controlled fission chain reactions can be used to run various industries. Research is going on to achieve controlled fusion reactions which will give even more amount of energy.

(3) Discovery of new elements and new isotopes. The discovery of transuranic elements has been achieved by artificial radioactivity. Moreover, radioactive isotopes of other stable elements have been discovered by the artificial transmutation of element.

(4) Preparation of luminiscent points. If a radioactive substance like radio thorium is mixed with fluorescent material like ZnS, it exhibits luminiscence, so that it becomes visible in the dark. Such mixtures constitute luminiscent paints. These are used to make the pointers and dials of watches visible during night.

(5) Sterilization. Antibiotics are sterilized by radiations from radioactive substances. Partial sterilization of food is also done for its preservation.

(6) Radiotherapy. Radiations from Co^{60} sources have been used in the treatment of cancer. Sometimes γ-radiations from Au^{198} are also used.

(7) Determination of Avogadro number. The γ-rays from radioactive material have been used to determine Avogadro number. The value comes out to be 6.02×10^{23} from this method and agrees well with those found by other methods.

(8) Radioactive dating. This is the process of determining the age of material. The age of earth can be estimated by this method. We can distinguish ordinary lead and the lead obtained from radioactive material with the help of mass spectrograph. Thus we can fmd out the ratio of Pb^{206} to that of uranium in the oldest mineral. Now if we determine the overall rate of disintegration in the mineral, then the age of the mineral can be obtained. It is assumed that the mineral originated a little later than the earth. The age of the earth calculated by this method comes out to be 2-3 billion years which is in close proximity with that obtained by other methods.

(9) Tracer applications. In these applications, the radioactive isotopes are used as a label to study the behaviour of some material. These are the most important applications of radioactive isotopes in various branches of applied science.

(10) Measurement of thickness of material. A source emitting β– or γ–rays is placed on one side of the material whose thickness is to be measured and the detector on the other side. The force experienced by the radiation is a function of the thickness though which it is passed. Hence from the measurement of force the thickness of the absorbing material can be obtained.

(11) Measurement of the level of liquids. A rodioactive source is placed on the surface of liquid in a closed container. The detector outside the vessel will detect the level of the radioactive source, which will be the level of liquid inside the container. This technique is used to know the level of oil in refineries.

(12) Mutation induced by radiations. If the plants and seeds are irradiated by radioactive source, the rate of mutation is greatly increased and development of stains is achieved in a fraction of time taken otherwise.

(13) Preparation of nuclear batteries. Radioactive energy is converted into electrical energy in these batteries. Two electrodes are used in an evacuated chamber. One of the electrodes A is cathode with β-emitter. The flow of β-particle inside the chamber results in the passage of current in the external circuit connected with these electrodes.

Q. 43. What is meant by tracer and tracer technique?

(M. Dayanand, 2000; Indore, 2001)

Ans. The majority of elements occur as mixture of isotopes the proportion of which ramains constant throughout the course of physical, chemical and biological changes. Some of these isotopes are radioactive. The radioactive isotopes of a given element behave chemically in the same way as non-radioactive isotopes. Hence if a radioactive isotope is added to its non-radioactive variety the mixture will behave in the same way chemically. Since radioactive isotopes can always be detected easily, they may be used as indicators or tracers in various fields. The process in which a radioactive isotope is mixed with its non-radioactive variety and the activity of radioactive isotope is noted after physical, chemical or biological changes, is called **radioactive tracing or tracer technique**. The success of radioactive tracing lies in the selection of a suitable nuclide.

Q. 44. Discuss various applications of tracer (radio isotopes) technique.

(M. Dayanand, 2000; Indore, 2001; Kurushetra, 2002; Calcutta, 2007)

Ans. Various applications of tracer technique are discussed as under:

(1) Diffusion. The process of diffusion has been studied by tracer method. The rate of diffusion of lead has been determined by mixing radio lead with ordinary lead and allowing the mixture to diffuse into inactive lead. It is found that at 473 K diffusion of lead is about ten times slower than that of tin in lead.

(2) In quantitative analysis. Isotopic dilution analysis is applied for the determination of the amount of substance which is present in a concentration that cannot be measured by physical or chemical methods. The substance under investigation is mixed in a known ratio with its active form, and the mixture is introduced into the system. Any change in the ratio will be due to the dilution caused by the substance originally present in the system. Knowing the change in the ratio and by measuring the activity, the amount of the substance can be determined.

(3) Solubility. Solubility of sparingly soluble salts can be determined accurately by this method. For example, solubility of lead chromate can be determined as follows: A known weight of radioactive lead is mixed with a weighed amount of inactive lead and the mixture is converted into lead chromate and a saturated solution of the salt in water is prepared. A known volume of the solution is evaporated to dryness and by determining the radioactivity of the residue the concentration of lead chromate in water is calculated.

(4) Exchange phenomena. The study of several exchange reactions using tracers has given valuable information regarding reactivity, chemical bond types, effect of solvent on properties and structure of ions and compounds.

It was thought that when a salt is in contact with its saturated solution, there is continuous exchange of molecules between the liquid and the salt along boundary surface. This has been verified by tracer technique. Crystals of activated lead chloride were placed in a saturated solution of inactive lead chloride. The saturated solution after some time was found to be radioactive.

It has been shown by tracer technique that the two sulphur atoms of thiosulphate are not equivalent. Inactive sulphite is converted into thiosulphate by treating with radio-sulphur. On adding dilute acid to the product the sulphurous acid that is regenerated is found to be inactive.

$$Na_2SO_3 + S^* \longrightarrow Na_2SS^*O_3$$

$$Na_2SS^*O_3 + 2HCl \longrightarrow 2NaCl + H_2SO_3 + S^*$$

(5) Discovery of new compounds. The possibility of preparing hydrides of bismuth and lead was established by tracer method. When magnesium filing coated with radio bismuth are treated with dilute acid, the gas evolved is found to be radioactive. This gas cannot be pure hydrogen, since the acid, which is source of hydrogen, is inactive. Therefore it is inferred that the gas is bismuth hydride which contains radio bismuth

$$3Mg + 6HCl \longrightarrow 3MgCl_2 + 6H$$

$$2Bi + 6H \longrightarrow 2Bi^*H_3 \text{ (* = Radioactive)}$$

Detection of this hydride would have been almost impossible otherwise.

(6) Radiometric titrations. Radioactive tracers may also be used to determine the end-point of a titrations where the reaction product is either insoluble or extractable. For example, if Ag is to be estimated with standard HCl solution Ag^{110} tracer is added. The specific activity of the solution is determined after each addition of the known volume of HCl solution. As the titration proceeds, Ag is removed from the solution as AgCl precipitate so that the specific activity of the solution is lowered. A plot of activity versus volume of HCl gives a line, the extrapolation to zero activity gives the end-point of titration.

(7) The completeness of precipitation. Lead salt containing a known amount of radioactive lead is precipitated from the solution. In order to test for complete precipitation, the amount of activity remaining in solution is determined after evaporating it to dryness. This measurement will give the amount of lead which has not been precipitated.

(8) Reaction mechanism. Radioactive isotopes find wide applications in the study of the mechanism of reactions. The oxidation of chlorite to perchlorate by hyprochlorous acid labelled with ^{33}Cl is found to proceed thus:

$$2HCl^{*}O + NaC1O_2 \longrightarrow NaClO_4 + 2HCl^{*}$$

The study of the reduction of chlorate to chloride using Cl^{33} as tracer shows that the reaction proceeds in the following steps:

$$ClO_3^- \longrightarrow ClO_2^- \longrightarrow ClO^- \longrightarrow Cl^-$$

Radiocarbon, ^{34}C, is extensively used in the study of the mechanism of organic reactions. There are a number of reactions in which carbon atoms or groups of carbon atoms migrate from one part of the molecule to another. By labelling the migrating groups with radiocarbon at suitable position it is possible to understand the course of the reactions.

(9) Applications in agriculture. In recent years tremendous amount of work has been done to study rates and mechanisms of absorption of essential elements by the plants from various compounds. These have resulted in the improvement of crops and efficient use of fertilizers. ^{32}P has extensively been used in such studies. Tritium ($^{3}_{1}H$) has been used as a tracer to measure the rate of

movement and distribution of water absorbed by roots of the plants. ^{14}C labelled CO_2 has been used in the investigation of respiration by plants.

(10) Applications in medicine. In medical science the radioactive tracing is mainly used to locate disorder in the human body and to cure diseases. ^{131}I is used to study the functioning of thyroid glands, and to locate brain tumours. ^{58}Co has been used in the study of function of vitamin B_{12} in the

body. ^{24}Na is used to fmd out whether the blood of a patient is reaching the wound or not. Similarly, ^{32}P is used to fmd out whether phosphorus is being absorbed by the bones of a patient with a bone fracture. Apart from this, radium and radio cobalt are used to cure cancer.

(11) Radiocarbon dating. It is based on the radioactivity of ^{14}C isotope and hence known as radiocarbon dating. The isotope $^{14}_{6}C$ is radioactive with half-life period 5568 years. The C^{14} isotope is continually being produced in the upper atmosphere by the bombardment of nitrogen atoms by cosmic rays where it combines with oxygen to form $C^{14}O_2$. Gradually, $C^{14}O_2$ mixes with the non-radioactive CO_2 in the atmosphere and is absorbed by plants during their photosynthesis. Animals by eating plants and secreting waste products continually have an intake and output of ^{14}C. But as soon as a plant or animals dies, its ^{14}C intake terminates. The ^{14}C present in the dead species then undergoes radioactive decay. It follows that any material which is constructed from plants or animals will have ^{14}C incorporated in them. The amount of ^{14}C present leads to the age of the materials.

(12) Applications in biology. A number of biological processes both in plants and animals have been studies using the tracer technique. Mechanism of photosynthesis in plants was established using tracer technique. A small amount of C^*O_2 was introduced with ordinary CO_2 in the plant. It was found that the oxygen which is evolved comes from the water molecules. Accordingly the reaction for photosynthesis can be written as

$$6\,CO_2^* + 12H_2O \longrightarrow C_6H_{12}O^*_6 + 6\,H_2O^* + 6O_2$$

(13) Applications in industry. Using tracer technique the movement of mud and sand in river and harbours has been investigated. ^{131}I has been used to trace the direction and velocity of underground water. ^{24}Na can be used to locate leaks in underground water pipes. Cracks are inspected by putting them in radioactive salt-baths and the fmding out with Gieger Muller counter whether any penetration has taken place in the metal.

An important application of the tracer technique is that one pipeline can be used for the transportation of several kinds of petroleum products with little mixing at the interface of liquids. A radioactive material is added between the products at the source end of the pipeline. A Gieger counter at the receiving end placed outside the pipe promptly indicates the arrival of the new product.

Q. 45. What is a nuclear reactor? Give the design of nuclear reactor.

Ans. Nuclear reactor is a device to produce the nuclear energy in a controlled way to be used for peaceful purposes. It works on the same principle as the atomic bomb but under controlled conditions. All the neutrons produced are not allowed to carry out chain reaction.

The essential parts of a reactor (Fig. 9.17) are:

1. Core. It consists of the rods of fissionable material such as ^{235}U or ^{239}Pu.

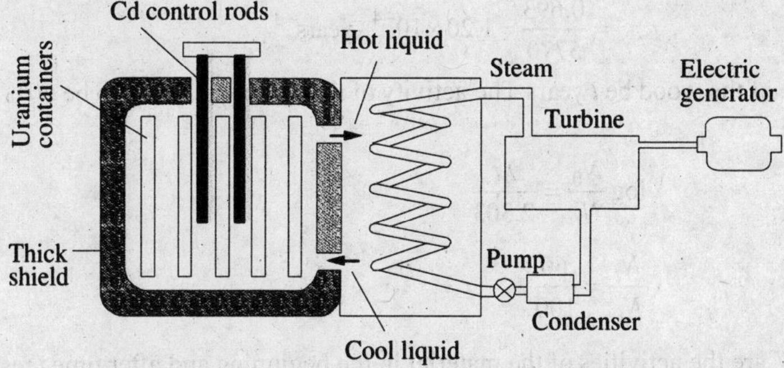

Fig. 9.17. A diagrammatic view of a nuclear reactor.

Conversion on of ^{235}U into plutonium is effected in a nuclear reactor. Uranium is purified from other elements but not separated into isotopes. The uranium rods are "canned" in aluminium cylinders to protect them against corrosion by the water that may be used as a cooling agent.

2. Moderator. A reactor using graphite bricks as moderator consists of cubic blocks of graphite provided with slots into which are inserted metallic rods of natural uranium 'canned' in aluminium cylinders. Some ^{235}U atoms undergo fission with productions of fast neutrons. Since slow neutrons are captured by ^{238}U, neutrons are slowed down by collisions with graphite which acts as a moderator. ^{238}U absorbs slow neutrons forming plutonium. The rods are removed from time to time and plutonium is separated by chemical processes.

In some reactions heavy hydrogen in the form of heavy water is used as moderator instead of graphite, for example, in the atomic pile at Bombay. If graphite is used as a moderator, the reactor is usually called an atomic pile and if heavy water is used as moderator, the reactor is called a swimming pool type reactor. In the swimming pool type reactor, the moderator is heavy water in a tank in which the uranium rods are suspended.

(3) Control rods. The rate of the reaction may be controlled by varying the amount of uranium inserted in the pile or the reactor and also by the insertion of rods of cadmium or of boron steel which absorb neutrons. If the temperature of the reactor appears to be going higher than regarded safe for the reactor, control rods of cadmium or of boron steel are pushed into the pile. Cadmium is a very good absorbent of neutrons. Absorption of neutrons breaks up the chain processes and the nuclear fissions can thus be kept under control and can even be stopped.

(4) Coolant. The heat generated in the reactor is removed constantly from it by circulating in it a liquid which is known as coolant. Usually a low melting alloy of sodium and potassium metals is employed. The alloy takes heat to the heat exchanger for raising steam. Steam generates electricity.

(5) Protective screen. Since neutrons have a very damaging influence on human beings and other living organisms, protective walls are needed around the reactor. They may be of lead glass or of special concrete about 3 metres thick.

Q. 46. A piece of wood from an archaeological source shows a C–14 activity which is 60% of the activity found today.

Calculate the age of the sample. ($t_{1/2}$ for $^{14}_{6}C$ = 5770 years)

Ans.
$$\lambda = \frac{0.693}{t_{1/2}}$$

$$= \frac{0.693}{5770} = 1.20 \times 10^{-4} \text{ years}^{-1}$$

Let the age of the wood be t years. The activity of the sample is found to be 60% of the activity of fresh sample.

$$\log \frac{N_0}{N_t} = \frac{\lambda t}{2.303} \qquad \text{... }(i)$$

Also,
$$\frac{N_t}{N_0} = \frac{60}{100}$$

where N_0 and N_t are the activities of the material in the beginning and after time t respectively.

or
$$\frac{N_0}{N_t} = \frac{100}{60} = 1.6666$$

Substituting the values in eq. (i)

$$\log 1.666 = \frac{1.2 \times 10^{-4} \times t}{2.303}$$

or $$0.2218 = \frac{1.2\times10^{-4}\times t}{2.303}$$

or $$t = \frac{2.303\times0.2218}{1.2\times10^{-4}} = 4256.7 \text{ years}$$

Q. 47. The final product of ^{238}U is ^{206}Pb. A sample of Pitch blende contains 0.0453 g of ^{206}Pb for every gram of ^{238}U present in it. Supposing that the mineral pitch blende formed at the time of formation of the earth did not contain any ^{206}Pb, calculate the age of earth (Half-life period of $^{238}U = 4.5\times10^9$ years)

Ans. Amount of ^{238}U present now, *i.e.*, $(a-x) = 1$ g

Amount of ^{238}U originally present (a) = Amount of ^{238}U now present + Amount of ^{238}U changed into 0.0453 g of ^{206}Pb

$$= 1+\frac{238}{206}\times0.0453 = 1.05234 \text{ g}$$

Suppose t is the time in which 1.05234 g of ^{238}U changed into 1 g of ^{238}U.

Using the equation, $$t = \frac{2.303}{k}\log\frac{a}{a-x}$$

But $$k = \frac{0.693}{t_{1/2}} = \frac{0.693}{4.5\times10^9} = 1.54\times10^{10} \text{ years}^{-1}$$

Substituting the values, $$t = \frac{2.303}{1.54\times10^{-10}} = \log\frac{1.05234}{1}$$

$$= 3.29\times10^8 \text{ years}$$

Q. 48. (*a*) Complete the following reactions :

(*i*) ${}_{13}Al^{27} + {}_{2}He^{4} \longrightarrow \ldots\ldots\ldots + {}_{0}n^{1}$

(*ii*) ${}_{7}N^{14} + \ldots\ldots\ldots \longrightarrow {}_{6}C^{14} + {}_{1}H^{1}$

(*b*) F^{18} is found to undergo 90% radioactive decay in 356 minutes. Calculate its half life.

(*Vidyasagar, 2007*)

Ans. (*a*) (*i*) ${}_{13}Al^{27} + {}_{2}He^{4} \longrightarrow {}_{15}p^{30} + {}_{0}n^{1}$

(*ii*) ${}_{7}N^{14} + {}_{0}n^{1} \longrightarrow {}_{6}C^{14} + {}_{1}n^{1}$

(*b*) Apply the relation

$$\lambda = \frac{2.303}{t}\log\frac{N_o}{N}$$

After the time 356 minutes, 90% matter has decayed or 10% matter has been left. Substituting the values for t and N

$$\lambda = \frac{2.303}{356}\log\frac{N_o}{0.1N_o} \quad \text{or} \quad \lambda = \frac{2.303}{356}\log 10$$

or $\lambda = \frac{2.303}{356}$

Half-life is related to disintegration constant as

$$t_{0.5} = \frac{0.693}{\lambda} \text{ or } \frac{0.693}{2.303} \times 356$$

or $t_{0.5} = 107$ minutes

Q. 49. Half-life of ^{232}Th is 1.4×10^{10} years and that of its daughter element ^{228}Ra is 7 years. What is the weight of ^{228}Ra in equilibrium with 1 g of ^{232}Th. (*Calcutta 2007*)

Ans. For two radioactive species in equilibrium

$$\lambda_A N_A = \lambda_B N_B$$

or $$\lambda_{Th} N_{th} = \lambda_{Ra} N_{Ra}$$

λ_A and N_A stand for disintegration constant and concentration of A. λB and N_B are corresponding values for B.

or $$\frac{0.693}{t_{0.5}\,(\text{Th})}\, 1 \text{ g} = \frac{0.693}{t_{0.5}\,(\text{Ra})} \times N_B$$

or $$\frac{0.693}{1.4 \times 10^{10}} = \frac{0.693}{7} \times N_B$$

or $$N_B = \frac{7}{1.4 \times 10^{10}} = 5 \times 10^{-10}\text{ g}$$

Q. 50. What are the differences between fission and fusion ? (*Calcutta, 2007*)

Ans. See Q. 37 and 39.

Q. 51. (*a*) State the laws of radioactive disintegration.

(*b*) Derive a relative between disintegration constant and half-life of a radioactive element.

(*c*) Calculate the disintegration constant and average life of a radioactive element if its half life is 30 days.

(*d*) Complete the following artificial nuclear reactions :

(*i*) $_4Be^9 + {}_1H^1 \longrightarrow$ .... + $_2He^4$

(*ii*) $_{13}Al^{27} + {}_0n^1 \longrightarrow$ .... + $_1H^1$

(*iii*) $_5B^{10}(\alpha, p) \longrightarrow$ ....+............

(*Vidyasagar, 2007*)

Ans. (*a*) See Q. 25

(*b*) See Q. 26

(*c*) Relation between average life and half-life is:

Average life = $1.44 \times t_{0.5}$

or Average life = 1.44 × 30 = 43.2 days

Relation between disintegration constant and half life is

$$\lambda = \frac{0.693}{t_{0.5}} \quad \text{or} \quad \lambda = \frac{0.693}{30} = 0.2231 \text{ day}^{-1}$$

(*d*) (*i*) $_4Be^9 + {_1H^1} \longrightarrow {_3Li^6} + {_2He^4}$

(*ii*) $_{13}Al^{27} + {_0n^1} \longrightarrow {_{12}Mg^{27}} + {_1H^1}$

(*iii*) $_5B^{10} + {_2He^4} \longrightarrow {_6C^{13}} + {_1H^1}$

Q. 52. State and explain Soddy's group displacement law. (*Calcutta, 2007*)

Ans. The atomic number of an element is decreased by two units and atomic mass is decreased by four units when it emits an α-particle

$$_a^bM \xrightarrow{-\alpha \text{ particle}} {_{a-2}^{b-4}M'}$$

where the metal M has the atomic number *a* and mass number *b*

The element is shifted two positions to the left in the periodic table as a result of α-emission

Also

The atomic number of an element is increased by one unit while the mass number remains unchanged when it emits a β-particle

$$_a^bM \xrightarrow{-\beta \text{ particle}} {_{a+1}^{b}M'}$$

The position of the element is shifted one position to the right as a result of β-emission.

10 COMPOUNDS OF BORON AND SILICON

Q. 1. Describe the preparation, properties and structure of metaboric acid HBO_2.

Ans. Preparation. Metaboric acid is prepared by heating orthoboric acid to 373 K.

$$H_3BO_3 \xrightarrow[-H_2O]{373\,K} \underset{\text{Metaboric acid}}{HBO_2}$$

Dehydration of orthoboric acid takes place resulting in metaboric acid.

Properties. Metaboric acid is a crystalline solid. On heating to 413 K, it dehydrates first to tetraboric acid, $H_2B_4O_7$ and on further heating at red heat it finally gives boric anhydride B_2O_3, also called boron trioxide or boric oxide.

$$4HBO_2 \xrightarrow[-H_2O]{413K} \underset{\text{Tetraboric acid}}{H_2B_4O_7} \xrightarrow[-H_2O]{\text{Red heat}} \underset{\text{Boron trioxide}}{2B_2O_3}$$

Fig. 10.1. Structure of metaboric acid.

Structure. Cyclic structure of metaboric acid is shown in Fig. 10.1.

Q. 2. Describe the preparation, properties and structure of orthoboric acid, H_3BO_3 or $B(OH)_3$. (*Himachal, 2000; Kurukshetra, 2001; Delhi, 2002*)

Ans. Preparation. (*i*) Orthoboric acid, also called 'boric acid', is obtained on hydrating B_2O_3.

$$B_2O_3 + 3H_2O \longrightarrow 2H_3BO_3$$

(*ii*) Boric acid can also be processed from natural sources. Steam jets of the volcanic vapour containing traces of boric acid on condensation and concentration yield boric acid.

(*iii*) Borax on treating with concentrated HCl or sulphuric acid gives flaky crystals on cooling.

$$Na_2B_4O_7 + 2HCl + 5H_2O \longrightarrow 2NaCl + 4H_3BO_3$$

(*iv*) Colemanite is also another rich source of boric acid. Sulphur dioxide is passed through powdered colemanite suspended in boiling water when $CaHSO_3$ and H_3BO_3 are formed.

$$Ca_2B_6O_{11} + 2SO_2 + 9H_2O \longrightarrow 2CaSO_3 + 6H_3BO_3$$

$$2CaSO_3 + 2H_2O + 2SO_2 \longrightarrow 2Ca\,(HSO_3)_2$$

Boric acid crystallizes out on cooling and calcium bisulphite remains in the solution.

(*v*) It is also formed by the hydrolysis of various compounds of boron such as halides and hydrides.

$$BCl_3 + 3H_2O \longrightarrow H_3BO_3 + HCl$$

$$B_2H_6 + 6H_2O \longrightarrow 2H_3BO_3 + 6H_2$$

Physical properties. Boric acid forms white needle-like crystals which are moderately soluble in water with the absorption of heat. Solubility of boric acid in water increases markedly with temperature.

Chemical properties. (*i*) **Action of heat.** Boric acid on heating at 373 K loses water to give metaboric acid which on further heating at 413 K loses more water to form tetraboric acid. At still higher temperature (Red heat) it swells up giving a glassy mass of boron trioxide.

$$\underset{}{4H_3BO_3} \xrightarrow[-4H_2O]{373\,K} \underset{\text{Metaboric acid}}{4HBO_2} \xrightarrow[-H_2O]{413\,K} \underset{\text{Teraboric acid}}{H_2B_4O_7} \xrightarrow[-H_2O]{\text{Red heat}} \underset{\text{Boron trioxide}}{2B_2O_3}$$

(*ii*) Alcohols react with $B(OH)_3$ in the presence of concentrated H_2SO_4 giving corresponding borates.

$$B(OH)_3 + 3C_2H_5OH \xrightarrow{\text{Con.}H_2SO_4} \underset{\text{Ethyl borate}}{B(OC_2H_5)_3} + 3H_2O$$

(*iii*) It behaves as a weak monobasic acid:

$$B(OH)_3 \rightleftharpoons H_3BO_3 \rightleftharpoons H^+ + H_2BO_3^- \rightleftharpoons H_2O + BO_2^-$$

Thus, on titration with NaOH, it gives sodium metaborate salt.

$$H_3BO_3 + NaOH \rightleftharpoons NaBO_2 + 2H_2O$$

The end of point in this titration is not sharp due to excessive hydrolysis of $NaBO_2$. On complexing with polyhydroxy compounds such as glycerol or sugars, the acid strength is increased. This results in a large increase in the H_3O^+ ion concentration in aqueous H_3BO_3 solution to the following reaction:

$$\underset{\substack{\text{Basic}\\\text{acid}}}{B(OH)_3} + \underset{\text{Glycerol}}{\begin{matrix} | \\ HO-C- \\ | \\ HO-C- \\ | \\ -C- \\ | \end{matrix}} \longrightarrow \underset{\text{Complex compound}}{\left[\begin{matrix} OH & & O-C- \\ & B & | \\ OH & & O-C- \end{matrix}\right]} + H_3O^+$$

The acid strength of the acid is enhanced as a result of complexation of BO_2^- group by polyhydroxy compound. This reaction is used for estimating boric acid tetrimetrically using NaOH in the presence of phenolphthalein as indicator.

(*iv*) Boric acid on fusion with metal oxides gives metal borates.

$$B(OH)_3 + MO \xrightarrow{\text{Fusion}} M - \text{borates}$$

where M stands for a bivalent metal.

(*v*) When fused with NH_4HF_2, boric acid gives ammonium borofluoride which on heating with boric anhydride gives boron trifluoride.

$$B(OH)_3 \xrightarrow[\text{Fusion}]{NH_4HF_2} NH_4BF_4 \xrightarrow[\text{Heat}]{B_2O_3} BF_3$$

Structure. Boric acid has a layer structure involving triangular BO_3 groups joined by hydrogen bonds. In fact its structure contains $B(OH)_3$ units linked together by unsymmetrical hydrogen bonds to form layers of nearly hexagonal symmetry as shown in Fig. 10.2.

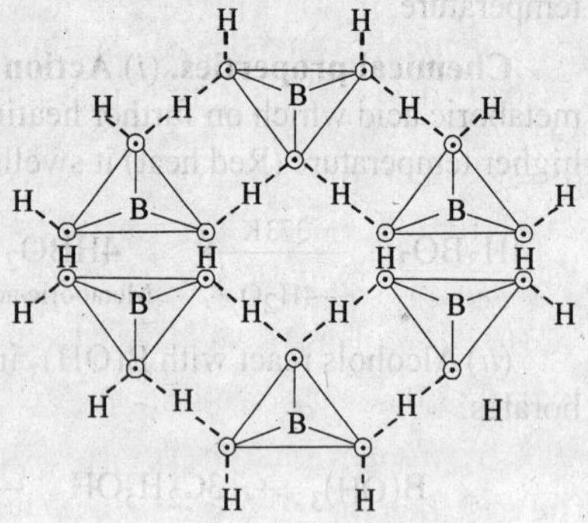

Fig. 10.2. Structure of boric acid

Q. 3. Under what conditions boric acid can be titrated against sodium hydroxide using phenolphthalein as indicator. *(Delhi, 2002)*

Ans. The end point in this titration is not sharp due to excessive hydrolysis of $NaBO_2$ obtained as under:

$$\underset{\text{Boric acid}}{H_3BO_3} + NaOH \rightleftharpoons NaBO_2 + 2H_2O$$

Boric acid is complexed with a polyhydroxy compound such as glycerol or glycol. This helps in the release of H^+ as H_3O^+ and a sharp end point is possible. Also see Q. 2 under chemical properties of boric acid.

Q. 4. $B(OH)_3$ is an acid but $Al(OH)_3$ is amphoteric. Explain. *(Kurukshetra, 2001)*

Ans. $B(OH)_3$ is a weak monobasic acid because it accepts – OH ion rather than donating H^+ ions. $Al(OH)_3$ is basic due to the following ionisation:

$$Al(OH)_3 \rightleftharpoons Al^{3+} + 3OH^-$$

It is also weakly acidic due to an acceptance of OH^- ion.

$$Al(OH)_3 + OH^- \rightleftharpoons [Al(OH)_4]^-$$

Q. 5. Discuss the action of heat on boric acid. *(Kurukshetra, 2001)*

Ans. See. Q. 2.

Q. 6. Orthoboric acid is a very weak acid but in the presence of glycerol or ethylene glycol, it acts as a strong acid. Explain. *(Himachal, 2000)*

Ans. See Q. 3.

Q. 7. What are different kinds of borates? Explain giving their structure.

Ans. Borates can be classified into the following types:

(i) **Metaborates.** B_3O_3 a six-membered ring which is the common feature of these borates. Generally two oxygen atoms of each BO_3 unit are shared in the various metaborates. Both the ring [Fig. 10.3 (*a*)] and chain [Fig. 10.3 (*b*)] structures are possible in various metaborates, *e.g.*, $K_3B_3O_6 \cdot CaB_2O_4$.

(ii) **Orthoborates.** Discrete anions which contain a B:O ratio of 1:3 are called orthoborates. In such borates only triangular BO_3 units are present. Discrete planar BO_3^{3-} ions are found in the following borates. In BO_3, $Ca_3(BO_3)_2$ and $Mg_3(BO_3)_2$ as shown in Fig. 10.4.

(iii) **Pyroborates.** In the pyroborate ion, which occurs in $Mg_2B_2O_5$ and $Co_2B_2O_5$, two triangular BO_3 units share an oxygen atom. But the ions in general are not perfectly planar (Fig. 10.5). Chains of BO_3 units, similarly linked, are also found in calcium metaborate (Fig. 10.6).

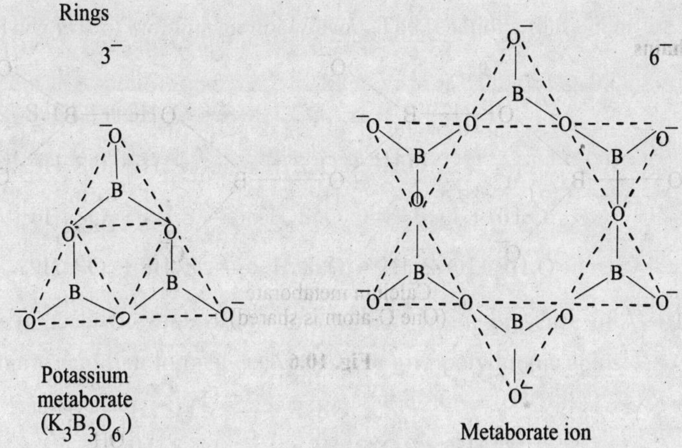

Fig. 10.3 (*a*)

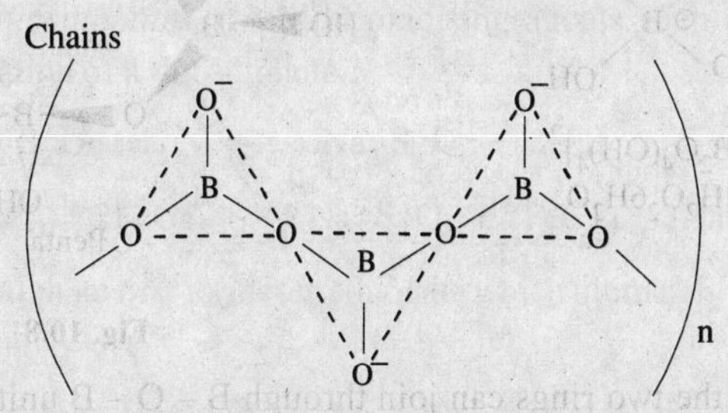

(Two O – atoms in BO_3 group are shared to give chain)
Fig. 10.3 (*b*)

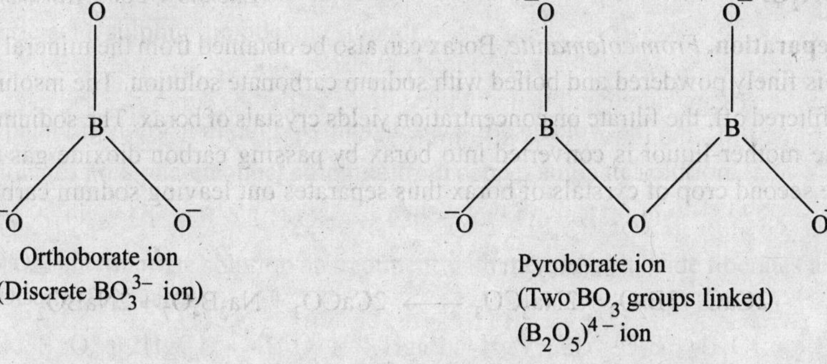

Fig. 10.4

Fig. 10.5

***(iv)* Perborate.** Compounds containing a structural unit in which two tetrahedral B atoms are bridged by two peroxo groups are called perborates, *e.g.*, sodium perborate ($NaBO_3 \cdot 4H_2O$ or $NaBO_2H_2O_2 \cdot 3H_2O$) as shown in Fig. 10.7.

***(v)* Complex borates.** The ortho, meta- and pyroborate anions, all contain only three co-ordinate boron atoms. Oxy-anions containing only four co-ordinate B atoms are not known. However, a large number of compounds having three and four co-ordinate boron atoms have been discovered.

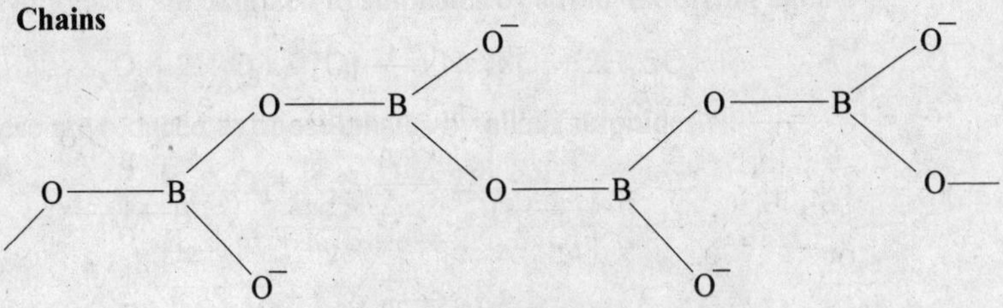

Calcium metaborate
(One O-atom is shared)

Fig. 10.6

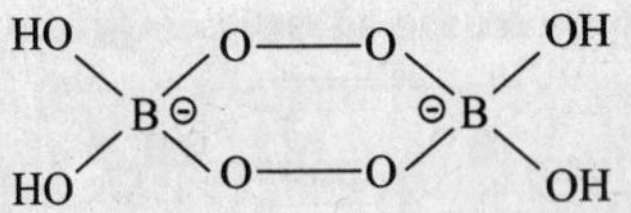

Peroxoborate ion $[B_2O_4(OH)_4]^{2-}$
as in $Na_2B_2O_4.2H_2O_26H_2O$

Fig. 10.7

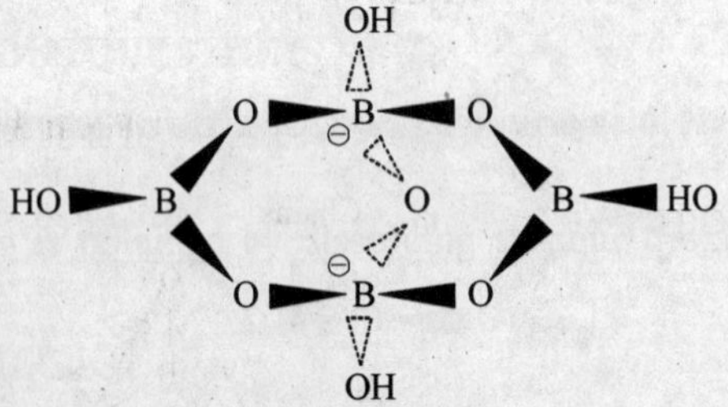

Penta

Fig. 10.8

In such compounds the two rings can join through B – O – B unit (borax : $Na_2B_4O_7 \cdot 10H_2O \equiv Na_2 [B_4O_5 (OH)_4] \, 8H_2 O$). Both trigonal BO_3 and tetrahedral BO_4 groups are present in the complex borates (Fig. 10.8).

Q. 8. Describe method of preparation, properties and uses of borax or sodium tetraborate, $Na_2B_4O_7$. $10H_2O$. (*Lucknow, 2000;Kurukshetra, 2002*)

Ans. Preparation. *From colamanite.* Borax can also be obtained from the mineral colemanite. The mineral is finely powdered and boiled with sodium carbonate solution. The insoluble calcium carbonate is filtered off, the filtrate on concentration yields crystals of borax. The sodium metaborate present in the mother-liquor is converted into borax by passing carbon dioxide gas through the solution. The second crop of cyrstals of borax thus separates out leaving sodium carbonate in the solution.

$$2CaO \cdot 3B_2O_3 + 2Na_2CO_3 \longrightarrow 2CaCO_3 + Na_2B_4O_7 + 2NaBO_2$$

$$4NaBO_2 + CO_2 \longrightarrow Na_2B_4O_7 + Na_2CO_3$$

Physical properties. Two forms of borax are crystallized from the solution depending upon the temperature. Ordinary borax, $Na_2B_4O_7 \cdot 10H_2O$, is monoclinic and crystallizes below 335 K; above this temperature octahedral borax, $Na_2B_4O_75H_2O$, separates out. It is obtained in the form of large colourless crystals moderately soluble in cold water but more soluble in hot water.

Chemical properties. (*i*) Aqueous solution of borax is alkaline due to hydrolytic decomposition which produces sodium hydroxide and boric acid.

$$Na_2B_4O_7 + 3H_2O \rightleftharpoons 2NaBO_2 + 2H_3BO_3 \quad \text{(Concentrated solution)}$$

$$NaBO_2 + 2H_2O \rightleftharpoons NaOH + H_3BO_3 \quad \text{(dilute solution)}$$

***(ii)* Borax bead test.** It swells into a white mass on heating with loss of water. On further heating it melts to form a transparent glassy solid called 'borax bead'. The borax bead further can dissolve many metal oxides to form metaborates which impart characteristic colours to the solid mass. This forms the basis of borax bead test for the qualitative analysis of various cations. In the reducing flame, the carbon reduces higher metaborates into lower ones or the free metal and hence the change in colour.

$$Na_2B_4O_7 \cdot 10H_2O \xrightarrow[-10H_2O]{\text{Heat}} Na_2B_4O_7 \xrightarrow{\text{Heat}} \underset{\text{(Glassy bead)}}{2NaBO_3 + B_2O_3}$$

$$CuO + B_2O_3 \longrightarrow \underset{\text{Blue}}{Cu + (BO_2)_2} \xrightarrow[\text{flame,C}]{\text{Heat, reducing}} \underset{\text{Colourless}}{CuBO_2} \xrightarrow[\text{flame,C}]{\text{Heat, Red.}} \underset{\text{Red colour opaque}}{Cu}$$

$$CoO + B_2O_3 \longrightarrow \underset{\text{Blue}}{Co(BO_2)_2}$$

Uses. (1) It is an important ingredient of a variety of detergents, soaps, water-softening compounds, adhesives, cosmetics and glazed paper.

(2) Borax is used in the manufacture of optical and hard-glasses, enamels and glazes for pottery.

(3) It is used as an antiseptic in making medicinal soaps and in the preservation of food.

(4) As a flux in soldering.

Q. 9. Give the chemistry of borax bead test. Why is it applicable to coloured salts?

(*G. Nanakdev, 2000; Lucknow, 2000; Kurukshetra, 2002*)

Ans. See Q. 8 under borax bead test.

Q. 10. Blue borax bead of copper first changes to colourless and then to red on heating in a reducing flame of the burner. Explain.

Ans. See Q. 8 under borax bead test.

Q. 11. An aqueous solution of borax acts alkaline. Explain.

Ans. Aqueous solution of borax is alkaline because of the following reactions:

$$Na_2B_4O_7 + 3H_2O \longrightarrow NaBO_2 + 3H_3BO_3$$

$$NaBO_2 + 2H_2O \rightleftharpoons \underset{\text{Strong alkali}}{NaOH} + H_3BO_3$$

Q. 12. Write note on *(i)* Sodium metaborate and *(ii)* Sodium peroxoborate.

Ans. *(i)* Sodium metaborate $NaBO_2 \cdot 4H_2O$. It can be obtained by the action of caustic soda solution on boric acid or borax.

$$H_3BO_3 + NaOH \longrightarrow NaBO_2 + 2H_2O$$

$$Na_2B_4O_7 + 2NaOH \longrightarrow 4NaBO_2 + H_2O$$

Sodium metaborate crystallizes in colourless needles. Its structure consists of plane triangular BO_3^{3-} units which are joined together giving a variety of polymeric chain and ring structures.

***(ii)* Sodium peroxoborate, $NaBO_3 4H_2O$.** Sodium peroxoborate is prepared either by the action of boric acid on sodium peroxide or by adding ice-cold solution of hydrogen peroxide to an alkaline solution of borax. It can also be prepared by electrolysis of sodium carbonate and borax with platinum guaze anode. This compound is widely used in the preparation of tooth powders, washing powders and for rapid oxidation of vat dyes. It is also used as bleaching agent and a disinfectant.

Q. 13. Describe the method of preparation and properties of hydrides of boron.

(Kanpur, 2000; Awadh, 2000; Purvanchal, 2007)

Ans. Preparation. Some common methods of preparation of hydride are given as under.

***(i)* By pyrolysis of diborane.** Diborane is a source for obtaining the higher hydrides mainly B_4H_{10}, B_5H_{11}, B_5H_9 and $B_{10}H_{14}$. Normally diborane required for the purpose is made by treating sodium borohydride, $NaBH_4$, with concentrated H_2SO_4.

$$2NaBH_4 + H_2SO_4 \longrightarrow Na_2SO_4 + B_2H_6$$

$$B_2H_6 \xrightarrow{373K} B_4H_{10}$$

$$B_2H_6 \xrightarrow{393K} B_5H_{11}$$

$$B_2H_6 \xrightarrow{453K} B_5H_9 + B_{10}H_{14}$$

***(ii)* By reduction of boron halide.**

$$LiA1H_4 + BF_3 \xrightarrow{Ether} [Li\,BH_4 + AlF_3] \times 3$$

$$3\,LiBH_4 + BF_2 \longrightarrow 2B_2H_6 + 3LiF$$

$$\overline{4BF_3 + 3LiA1H_4 \longrightarrow 2B_2H_6 + 3A1F_3 + 3LiF}$$

***(iii)* By hydrolysis of magnesium boride.** Magnesium boride on dissolving in aqueous acids yields mainly B_4H_{10}, B_5H_9 and B_6H_{10}.

$$\underset{\text{Magnesium boride}}{2Mg_3B_2} + 12HC1 \longrightarrow 6MgC1_2 + B_4H_{10} + H_2$$

***(iv)* Diborane can also be obtained by action of hydrogen on boric oxide.**

$$B_2O_3 + 6H_2 \xrightarrow[Al_2Cl_6]{\text{Al or}} B_2H_6 + 3H_2O$$

***(v)* By passing electric discharge either on lower boranes or on a mixture of boron halide and hydrogen.**

$$2B_5H_9 \xrightarrow{\text{Discharge}} 2B_5H_8 + H_2$$

$$2BC1_3 + 6H_2 \longrightarrow B_2H_6 + 6HC1$$

Properties of diborane. *(i)* Combustion. It bums in air or oxygen producing large amount of heat.

$$B_2H_6 + 3O_2 \longrightarrow B_2O_3 + 3H_2O;\ \Delta H = -201.7\text{ kJ mol}^{-1}$$

***(ii)* Action of alkalies.** Diborane undergoes hydrolysis in the presence of water or alkalies.

$$B_2H_6 + 2KOH + 2H_2O \longrightarrow 2KBO_2 + 6H_2$$

(*iii*) $B_2H_6 \xrightarrow[\text{High temp.}]{\text{Excess } NH_3} BN$

(*iv*) $B_2H_6 \xrightarrow{HCl} B_2H_5Cl$

(*v*) $B_2H_6 \xrightarrow{BCl_3} B_2H_5Cl$

(*vi*) $B_2H_6 \xrightarrow[298\,K]{Cl_2} BCl_3$

(*vii*) $B_2H_6 \xrightarrow{H_2O} B(OH)_3 + H_2$

(*viii*) $B_2H_6 \xrightarrow{CO} COBH_3$

(*ix*) $B_2H_6 \xrightarrow[373-523\,K]{H_2}$ Higher hydrides

(*x*) $B_2H_6 \xrightarrow[473\,K]{NH_3}$ Borozole

Q. 14. Complete the following reaction and write the structure of the product formed:

$$B_2H_6 \xrightarrow[\text{High temp.}]{\text{Excess } NH_3}$$

Ans.

$$B_2H_6 + NH_3 \longrightarrow$$ (ring: H–N, H–B, B–H, H–N, N–H, B–H)

Borazine

Q. 15. Discuss the structure of diborane.

(*Bangalore, 2005; Rohilkhand, 2003; Madras, 2004; Calcutta, 2005; Nagpur, 2008; Garhwal, 2010*)

Ans. Structure of diborane presents an interesting case since there are not enough valence electrons to form the expected number of covalent bonds. This is an *electron deficient compound.*

Since boron atom has three electrons in the outermost orbit in the excited state, therefore, it can link itself with three hydrogen atoms. Thus while each of the boron atoms in diborane can link itself to three hydrogen atoms, there are no electrons left to form a bond between the two boron atoms.

```
    H       H
    ..      ..
H : B       B : H
    ..      ..
    H       H
```

To account for this anomaly, various structures for diborane have been proposed. These can be discussed under two main types:

1. Ethane-like structures. It is assumed that all the six hydrogen atoms in B_2H_6 are equivalent as in case of ethane, C_2H_6. Such structures can be visualised by taking into account the phenomenon of resonance. The following three different possibilities of resonance are:

***(i)* Covalent resonance.** This gives two possible structures.

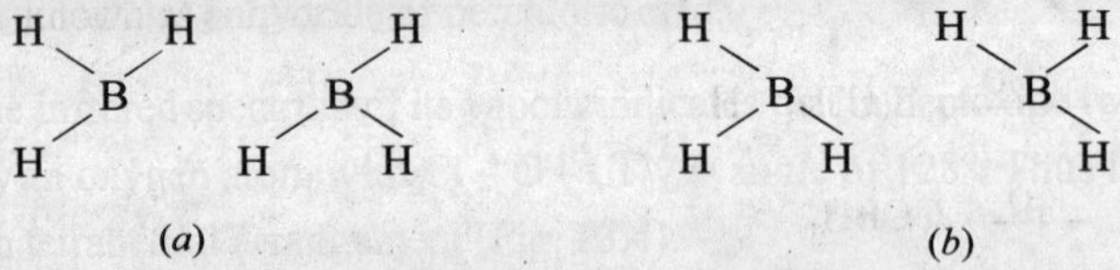

***(ii)* Covalent-ionic resonance.** It gives the following structure.

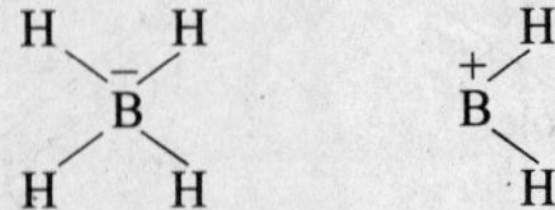

***(iii)* One-electron bond and two-electron bond resonance.** Here two possible structures obtained are:

```
  H   H               H   H
  |   |               .   |
H·B — B·H         H — B — B — H
  |   |               |   .
  H   H               H   H
   (a)                 (b)
```

Experimental evidence has shown that in B_2H_6 molecule, two of the six hydrogen atoms behave differently from the others and also there is a hindered rotation around *B-B* linkage in the molecule. These facts rule out the possibility of an ethane-like structure and indicates the existence of an ethylene-like structure.

2. *Hydrogen bridge structure* (Ethylene-type structure). Diborane, $B_2 H_6$, has been found to prossess a bridge structure in which each B atom is bonded to two *H* atoms (called terminal H atoms) by regular electron-pair bonds. The resulting BH_2 fragments are co-planar and are bridged by two H atoms. The bridging H atoms are in a plane perpendicular to the plane of BH_2 fragments, *i.e.*, one lies above and one below the plane to the rest of the molecule and prevent rotation between the two boron atoms. The various bond distances and angles are shown in Fig. 10.9.

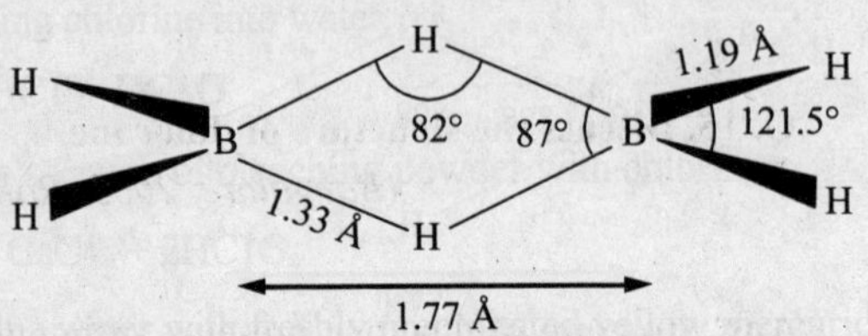

Fig. 10.9. Bridge structure of diborane.

In view of the bridge structure, diborane has 8 bonds but there are 12 electrons available for bonding (3 per B and 1 per H), hence, all the bonds in the molecule cannot be electron-pair bonds which would require 16 electrons for the structure of Fig. 10.9. Four of the terminal H-B bonds are normal single bonds as evidenced by bond angles and infrared stretching frequencies. The other two hydrogen atoms and four electrons are in a different environment.

The bond lengths between boron and hydrogen at the bridges is 1.33 Å while that of terminal positions is 1.19 Å. The terminal B-H distances are the same as the bond lengths measured in non-electron deficient compounds. So the electron deficiency is associated with the bridge groups.

A number of bonding theories have been proposed to account for the nature of the bridge in diborane structure. These are:

(*a*) **Normal hydrogen bond concept.** This concept was proposed by Buraway that the bridge between the boron atoms involves normal hydrogen bonds as shown below:

H H H
B B
H H H

However, this concept is not acceptable due to a very small difference in the electronegativity values of B (1.00) and H (2.1) atoms.

(*b*) **Protonated double bond concept.** R. S. Pitzer (1945) suggested the concept of a double bond between the two boron atoms. Each boron atom undergoes sp^2 hybridization as follows:

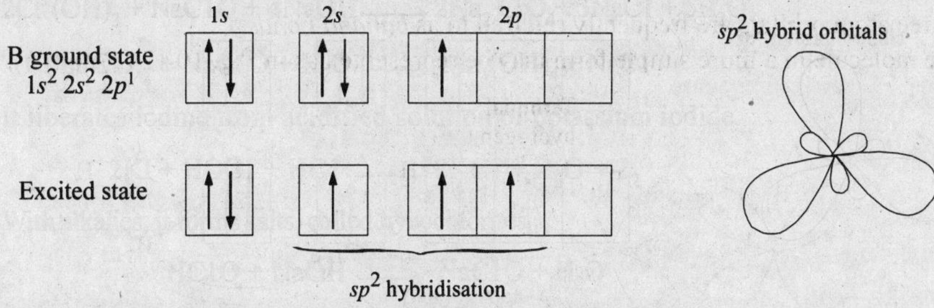

One of the sp^2 hybrid orbitals of each boron atoms overlaps to form one B-B σ bond. There is one unhybridized empty *p* orbital associated with each boron atom which lies in a plane perpendicular to the sp^2 hybrid orbitals. Sideways overlapping of these *p* orbitals gives a π bond between the two boron atoms. The two, so called different hydrogen atoms, provide electrons to this π bond and are, consequently, converted into protons which lie in the electron charge cloud of the π bond as shown in Fig.10.10. In this way the four hydrogen atoms will be coplanar with the two boron atoms while the other two will lie above and below this plane. This structure is consistent with the results of infrared and electron diffraction studies and also explains hindered rotation around B-B axis. This structure envisages acidic properties which are, however, not given by the compound.

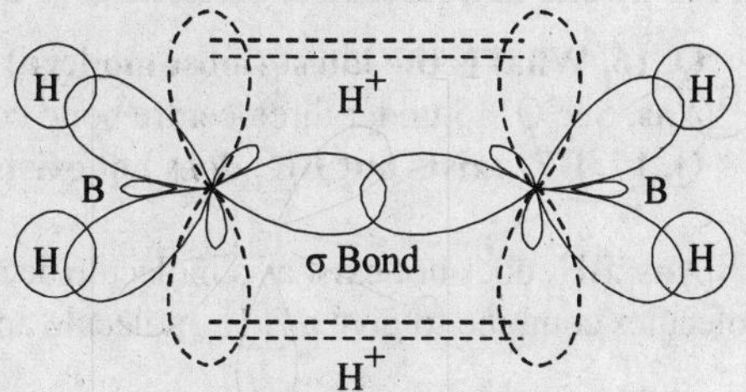

Fig. 10.10. Structure of diborane.

(c) Three-centre bond concept. This concept, also known as the molecular orbital concept, was proposed by Longuet-Higgins and Professor William N. Lipscomb. They modified Pitzer's concept in such a way that the double bond no longer involved protons, rather it involved two similar 'three centre bonds'.

It is supposed that boron atoms undergo sp^3 hybridisation. Each boron then utilizes two such orbitals containing one electron each for bonding with terminal hydrogens. There are now two sp^3 hybrid orbitals and one electron left with each boron atom. Also there are two 1*s* orbitals containing one electron each belonging to two hydrogen atoms not yet bonded. Thus, in all, there are six atomic orbitals and four electrons available which can form two three centre bonds known as the hydrogen bridges between the boron atoms as shown in Fig. 10.11.

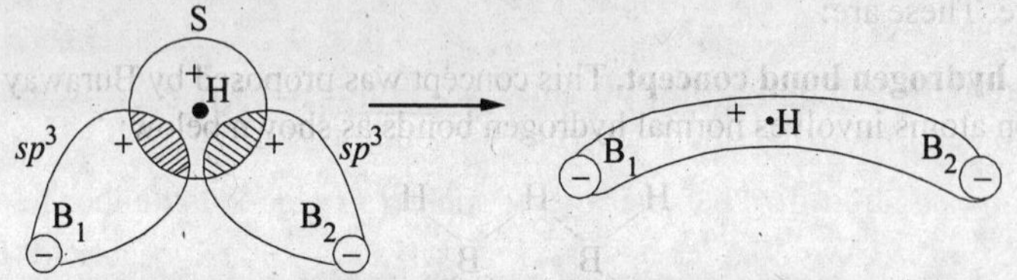

Fig. 10.11. Three-centred bonds in diborane.

This is known as three-centre electron pair bond B-H-B. The molecule of diborane contains two such bonds. Due to repulsion between the two hydrogen nuclei, the delocalised orbitals or bridges are bent away from each other in the middle. This is why the hydrogen bridge orbitals is shifted upwards in the middle. The other bridge orbital will be shifted downwards. This is the reason that these three atom orbitals are frequently referred to as *banana bonds*.

The molecule in a more simple form may be represented as in Fig. 10.12 (*a*) and (*b*).

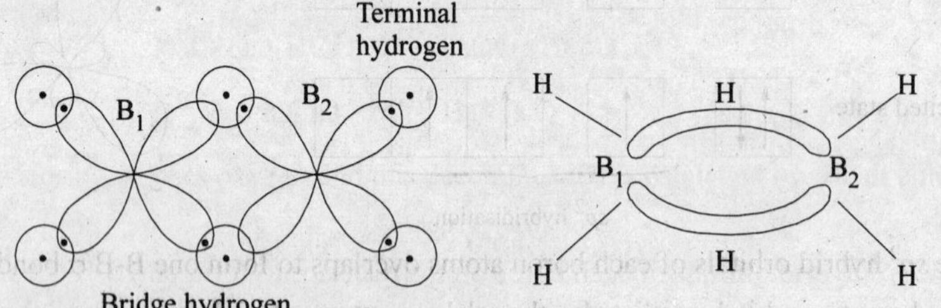

(*a*) Overlapping of atomic orbitals in B_2H_6 (*b*) Structure of B_2H_6 with three-centre bond or banana bond

Fig. 10.12.

Q. 16. What is the latest (most modern) concept of structure of diborane?

Ans. See Q. 15 under three-centre bond concept.

Q. 17. BF_3 exists but BH_3 does not exist. Assign reason.

(*M. Dayanand, 2000*)

Ans. BH_3 does not exist as a molecule because boron in BH_3 has a vacant 2*p* orbital. Two BH_3 molecules combine to give a B_2H_6 molecule and boron thus removes its electron deficiency.

In BF_3 the vacant $2p$ orbital of boron forms π bond by overlapping with filled $2p$ orbital of fluorine. There is a possibility of overlap with all the three fluorine atoms as depicted below.

F — B (F)(F) or $F - B(=F^+)(F)$

$F - \bar{B}(=F^+)(F) \longleftrightarrow \overset{+}{F}=\bar{B}(F)(F) \longleftrightarrow F - \bar{B}(F)(=F^+)$

Such π bond formation is not feasible with BH_3 as there are no filled orbitals in hydrogen.

Q. 18. Describe methods of preparation of borazine (borazole) and borazine derivatives.

(*Nagpur, 2003; Purvanchal, 2007*)

Ans. Methods of preparation of borazine and their derivatives are discussed as under:

(*i*) Borazine was first synthesized by pyrolysis of the adduct of diborane and ammonia at 453-473 K.

$$3B_2H_6 + 6NH_3 \longrightarrow \underset{\text{Adduct}}{3(B_2H_6)\ (NH_3)_2} \xrightarrow[453-473K]{\text{Heat}} \underset{\text{Borazine}}{2B_3N_3H_6} + 12H_2$$

or $\quad 3\,[(NH_4)_2^+ (B_2H_4)^{2-}]$

(*ii*) Borazole can also be prepared by the direct reaction of diborane with ammonia at 523-573 K under one atmospheric pressure. The yield is only 50%. The reaction involves the formation of the borine complex $H_3B \leftarrow NH_3$ which loses hydrogen and polymerises as shown below:

$$3B_2H_6 + 6NH_3 \xrightarrow[\text{1atm-pressure}]{523-573\text{K}} \underset{\text{borine complex}}{6H_3B \longleftarrow NH_3}$$

$$\downarrow -6H_2$$

$$\underset{\substack{\text{Borazole} \\ (50\%\text{ yield})}}{2B_3N_3H_6} \xleftarrow{\text{Polymerization}} 6HB = NH \xleftarrow{-6H_2} 6H_2B - NH_2$$

(*iii*) It can be obtained by heating together ammonium chloride and boron trichloride at 413-423 K in presence of chlorobenzene. The initial product is B-chloroborazine which is reduced with sodium borohydride in polyether to give borazole or borazine.

$$3NH_4Cl + 3BCl_3 \xrightarrow[C_6H_5Cl]{413-423\text{K}} 9HCl + \underset{\text{Trichloroborazine}}{Cl_3B_3N_3H_3}$$

$$\downarrow \text{Reduction} \quad NaBH_4 \text{ in polymer}$$

$$\underset{\text{Borazole}}{B_3N_3N_6} + \frac{3}{2} B_2H_6 + 3NaCl$$

Preparation of derivatives. B or N-substituted borazines may be prepared by appropriate substitution on the starting materials.

(*a*) Preparation of tri-B-methyl borazoles. (*i*) Tri-B-methyl borazole can be prepared by the reaction of boron trimethyl with ammonia followed by heating the addition compound at 603 K and 30 atmospheric pressure.

$$2B(CH_3)_3 + 3NH_3 \longrightarrow 3[(CH_3)_3B \leftarrow NH_3] \xrightarrow[\text{20 atom pressure}]{\text{Heat 603 K}} \underset{\text{B·B' B''– trimethy borazole}}{(CH_3)_3B_3N_3H_3}$$

(*ii*) These are also formed when borazole is treated with boron trimethyl.

$$B_3N_3H_6 + B(CH_3)_3 \longrightarrow \underset{\text{B · B' B''– trimethyl borazole}}{(CH_3)_3B_3N_3H_3}$$

(*b*) Preparation of tri-N-methyl borazoles. (*i*) Tri-*N*-methyl borazole can be prepared by reaction of monomethyl ammonium chloride with $LiBH_4$.

$$3CH_3NH_3Cl + 3LiBH_4 \longrightarrow \underset{\text{N, N',N''– trimethyl borazole}}{(CH_3)_3N_3H_3B_3} + 3LiCl + 9H_2$$

(*ii*) Tri-N-alkyl borazines can be prepared by the following method in good yield.

$$3RNH_3Cl + 3BCl_3 \longrightarrow Cl_3B_3NR_3 + 9HCl$$

$$\downarrow \text{Reduction NaBH}_4$$

$$(RNBH)_3 + 3B_2H_6 + 3NaCl$$

Q. 19. Write down the properties of borazine. (*Nagpur, 2008*)

Ans. (*i*) At ordinary temperature, both air and oxygen have no action on borazine.

(*ii*) Reaction with water. It hydrolyses in excess of water giving H_2, H_3BO_3 and NH_3. Hydrolysis becomes fast above 423 K.

$$B_3N_3H_6 + 3H_2O \longrightarrow 3B(OH)_3 + NH_3 + 3H_2$$

(*iii*) Addition reactions. (*a*) Borazine reacts with two moles of bromine at 273 K, the resulting compound decomposes at 333 K to give dibromoborazine.

$$\underset{\text{Borazine}}{B_3N_3H_6} \xrightarrow[273\text{ K}]{+2Br_2} B_3N_3H_6Br_4 \xrightarrow[333\text{ K}]{-2HBr} \underset{\text{B-dibromo-borazine}}{B_3N_3H_4Br_2}$$

(*b*) Borazine, in contrast to benzene, readily undergoes addition reactions with hydrogen halides, methyl iodide, methanol and water. In the last two cases very mild conditions must be used or disruption of the ring occurs. One mole of borazine adds three moles of the above molecules. The more negative group in the adding molecule normally goes to boron atom.

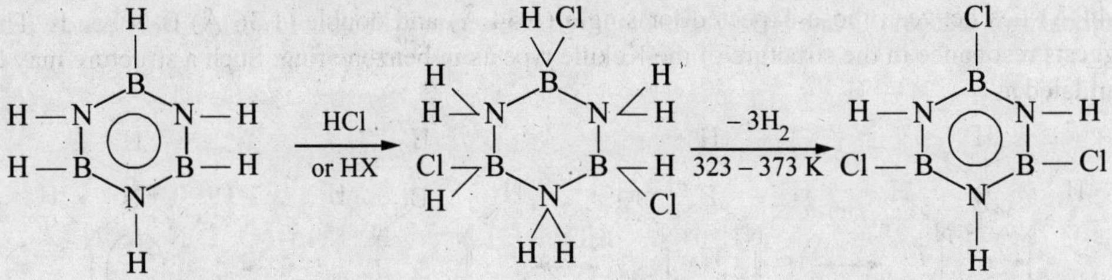

HCl or HX ; $-3H_2$, 323 – 373 K

Borazine → B-trichloroborazole

(*iv*) **Substitution reactions.** Borazine reacts with boron trimethyl, $B(CH_3)_3$ B give to trimethyl borazine (about 60% yield) together with small quantities (about 5%) of the mono and dimethyl derivatives and a non-volatile product of composition BNH

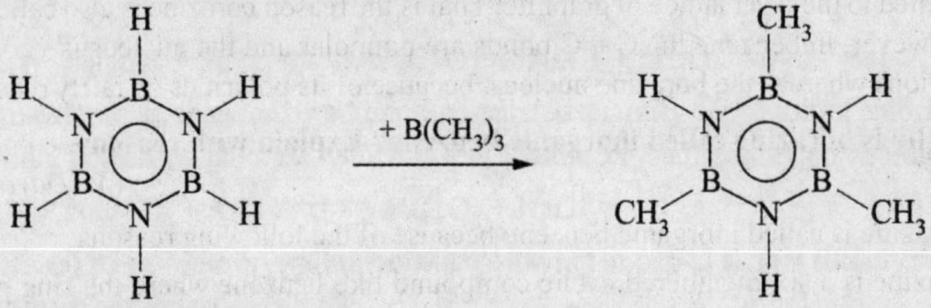

$+ B(CH_3)_3$

Borazine → B-trimethyl borazine

(*v*) **Pyrolysis of borazine** by passing a silent electric discharge through it yields compounds similar to naphthalene.

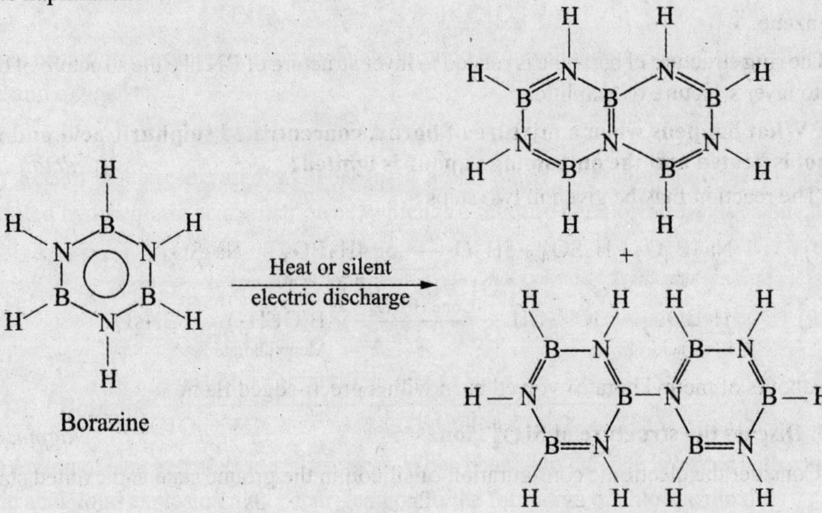

Heat or silent electric discharge

Borazine

Q. 20. Discuss the structure of borazine.

(*Kerala, 2000; Rohilkhand 2003; Nagpur, 2008*)

Ans. The properties of borazine are in conformity with a six-membered ring structure of alternating boron and nitrogen atoms.

(*i*) Structural studies of borazine including electron diffraction show the planar ring structure exactly comparable to that found in benzenes with equal B-N bond lengths (144 Å) and equal angles (120°).

(*ii*) Infrared and Raman spectra studies also confirm the planar structure. The bond distance (1.44 Å) lies between those expected for single (1.54 Å) and double (1.36 Å) B-N bonds. This suggests resonance in the structure of the Kekule type as in benzene ring. Such a structure may be fomulated as

(*iii*) This ring structure is related to the layer lattice of boron nitride as the ring structure of benzene is related to the layer lattice of graphite. That is the reason borazine is also called 'inorganic benzene'. However, in benzene the C = C bonds are nonpolar and the nucleus is very resistant to addition reactions whereas the borazine nucleus, because of its polarities, is fairly reactive.

Q. 21. Why is borazine called inorganic benzene? Explain with reasons.

(*M. Dayanand, 2000*)

Ans. Borazine is called inorganic benzene because of the following reasons:

(*i*) Borazine is a six-membered cyclic compound like benzene where the ring positions are occupied by boron and nitrogen in alternate positions. Like benzene, there is a system of conjugated single and double bonds.

(*ii*) Borazine undergoes addition and substitution reactions with HC1 and Br_2 etc., like in the case of benzene.

(*iii*) The ring structure of borazine is related to layer structure of BN like the structure of benzene is related to layer structure of graphite.

Q.22. What happens when a mixture of borax, concentrated sulphuric acid and methyl alcohol is heated and the outcoming vapour is ignited? (*Calcutta, 2007*)

Ans. The reaction may be given in two steps :

(*i*) $\underset{\text{Borax}}{Na_2B_4O_7} + H_2SO_4 + 5H_2O \longrightarrow \underset{\text{Boric acid}}{4H_3BO_3} + Na_2SO_4$

(*ii*) $\underset{\text{Boric acid}}{H_3BO_3} + 3CH_3OH \xrightarrow{\text{Conc } H_2SO_4} \underset{\text{Methyl borate}}{B(OCH_3)_3} + 3H_2O$

Vapours of methyl borate evolved burn with a green-edged flame.

Q. 23. Discuss the structure of SiO_4^{4-} ion. (*Nagpur, 2003*)

Ans. Consider the electronic configuration of silicon in the ground state and excited state.

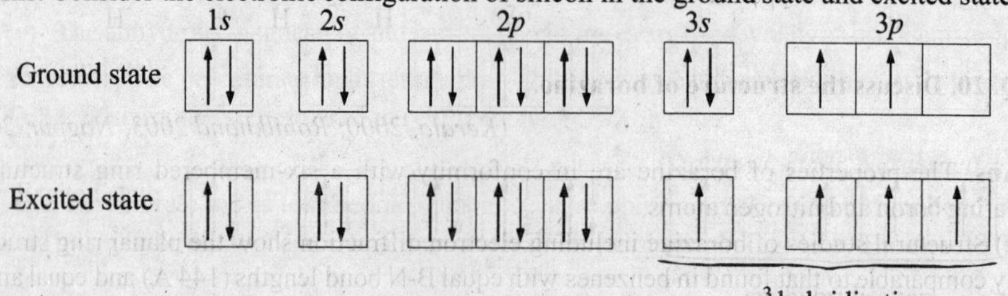

The four unpaired electrons in the four sp^3 hybrid orbitals form covalent bonds with oxygen ions to give SiO_4^{4-} tetrahedron. The negative charge on SiO_4^{4-} is due to the fact that any oxygen (which is not a bridge atom) has to pick up an electron from some metal to complete its octet. The negative charge is being balanced by the metal cations such as, Li^+, K^+, Ca^{2+}, Al^{3+}, etc. These serve to hold the solid together by ionic attraction.

The tetrahedral structure of the SiO_4^{4-} ion can also be represented by projection as shown on the extreme right of Fig. 10.13. This is obtained when the tetrahedron is viewed from above. The oxygen is shown as a small open circle and the silicon a small dot. The silicon is shown inside and the oxygen at the apex. Thus in silicates the basic unit of structure is SiO_4 tetrahedron. These tetrahedra occur singly or by sharing oxygen atoms in small groups, in small cyclic groups, infinite chain or infinite sheets.

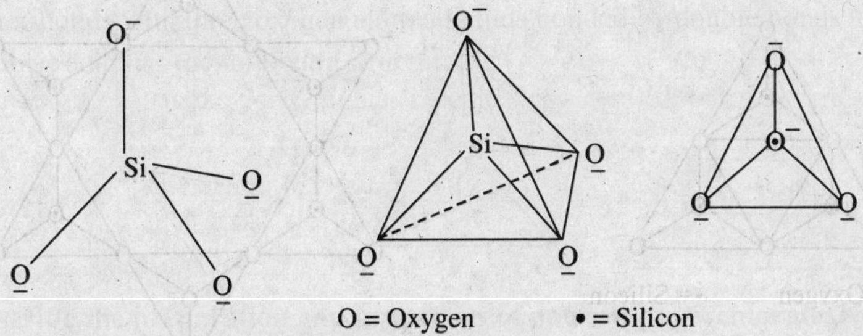

Fig. 10.13. Tetrahedral structure of SiO_4^{4-} ion.

Q. 24. Classify silicates into different types. Give the composition and structure of each type of silicate. (*Kurukshetra, 2002; Guwahati, 2002; Andhra, 2002; Bangalore, 2005; Nagpur, 2008*)

Ans. The silicates are classified into the following different types depending upon the sharing of oxygen atoms between the adjacent SiO_4^{4-} tetrahedra and the substitution of Al^{3+} for Si^{4+} ion in the structure:

(*i*) Orthosilicates
(*ii*) Pyrosilicates
(*iii*) Cyclic silicates
(*iv*) Chain silicates
(*v*) Sheet silicates
(*vi*) Three-dimensional silicates.

Each type is discussed separately as under:

(*i*) Orthosilicates. In the ortho silicates, discrete individual SiO_4^{4-} tetrahedra (Fig. 10.13) are present and there is no sharing of oxygen atoms between adjacent tetrahedra but the oxygens present are coordinated to the metal cations. The important examples are willemite, Zn_2SiO_4 and phenacite, Be_2SiO_4, the four negative charges on the tetrahedron being balanced by Zn^{2+} and Be^{2+} cations.

(*ii*) Pyro silicates. When two tetrahedral units share one oxygen atom between them $(Si_2O_7)^{6-}$ ion is obtained (Fig. 10.14). The silicates obtained are known as pyro silicates.

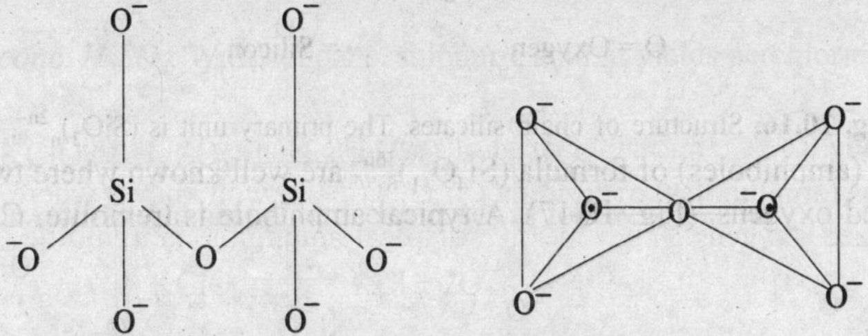

Fig. 10.14. Structure of pyrosilicates.

This is the simplest of the condensed silicate ions. Thortveitite, $Sc_2(Si_2O_7)$ is an example of this type of silicates.

***(iii)* Cyclic silicates.** If two oxygens per tetrahedron are shared ring structures (Fig. 10.15) may be formed with the general formula $(SiO_3)_n^{2n-}$. The cyclic ion $Si\,O_9^{6-}$ where $n = 3$ occurs in wollastonite, $Ca_3(Si_3O_9)$ and benitoite $BaTi(Si_3O_9)$. Another example is Beryl $Be_3Al_2(Si_6O_{18})$. It contains $Si_6O_{18}^{12-}$ cyclic ion where $n = 6$. The twelve negative charges are being balanced by Be^{2+} and Al^{3+} cations. The Si_6O_{18} units are superimposed one above the other leaving channels, hence, the mineral is permeable to gases which have small molecules. In benitoite, Ti^{4+} and Ba^{2+} ions are 6-coordinate but in beryl the Be^{2+} ions are 4-coordinate and Al^{3+} ions are 6-coordinate.

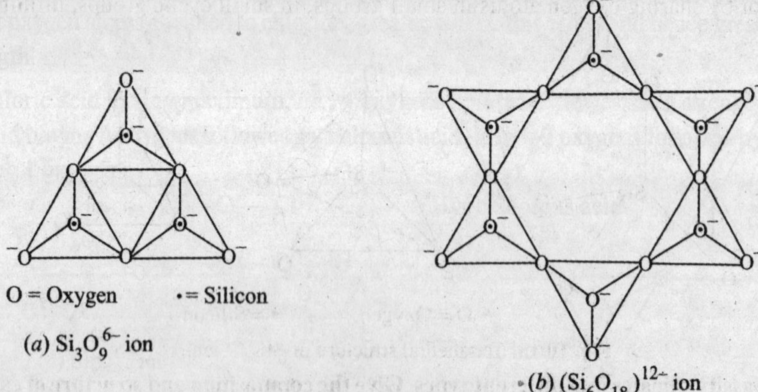

Fig. 10.15. Structures of cyclic silicates.

***(iv)* Chain silicates.** The sharing of two oxygen atoms per tetrahedron can also result in the formation of linear simple chains of indefinite length with two types of compositions, *i.e.*,

(*a*) Metasilicate having the general formula $(SiO_3)_n^{2n-}$ where the chains are single and occur in pyroxenes, and

(*b*) Metatetrasilicate having the general formula $(Si_4O_{11})_n^{6n-}$ where the chains are double and occur in amphiboles. The single chains can be represented by Fig. 10.16. Important examples of pyroxenes are spodumene $LiAl(SiO_3)_2$ enstatite, $MgSiO_3$, diopside. $CaMg(SiO_3)_2$ and jadeite, $NaAl(SiO_3)_2$.

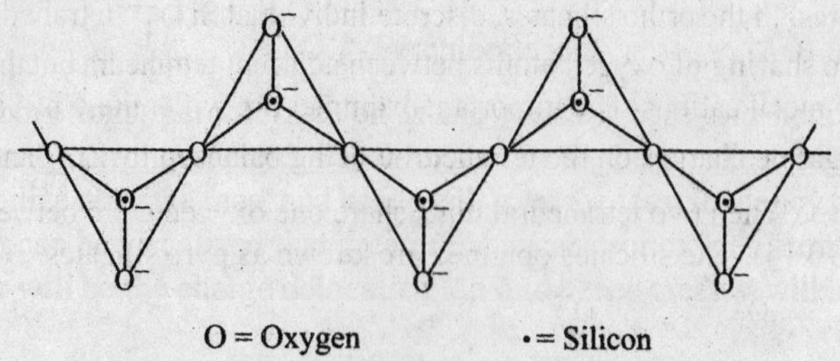

Fig. 10.16. Structure of chain silicates. The primary unit is $(SiO_3)_n^{2n-}$.

Double chains (amphiboles) of formula $(Si_4O_{11})_6^{6n-}$ are well known where two simple chains are joined by shared oxygens. (Fig. 10.17). A typical amphibole is tremolite, $Ca_2Mg_5(Si_4O_{11})\cdot(OH)_2$.

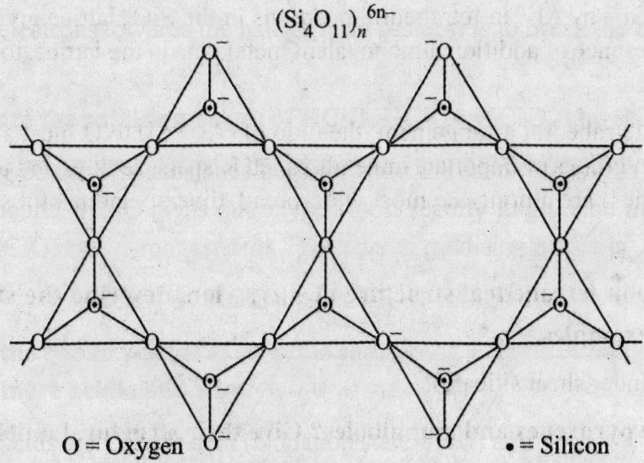

Fig. 10.17. Double chain silicates (Amphiboles).

(*v*) Sheet silicates. The sharing of three oxygen atoms by each tetrahedron gives rise to an infinite two-dimensional sheet having the empirical formula, $(Si_2O_5)_n^{2n-}$. The silicates having this type of structure are called as sheet silicates (Fig. 10.18). The sheets are bound together by the

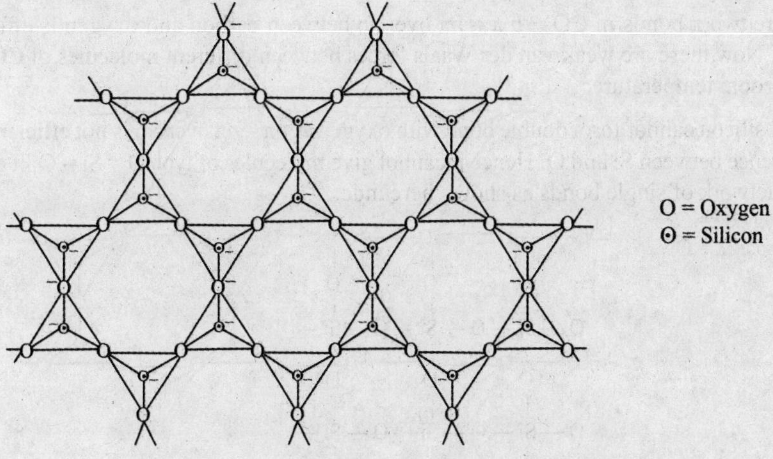

Fig. 10.18.

electrostatic forces of the cations which lie between them. These forces are not as strong as the bonds within each Si — O sheet. Such substances might thus be expected to cleave readily into thin sheets.

Examples of sheet-like-silicates are mica and different types of clay minerals such as talc. $Mg_3(OH)_2(Si_4O_{10})$ Kaolin, $Al_2(OH)_4(Si_2O_5)$ and pyrophillite, $Al_2(OH)_2(Si_4O_{10})$.

(*vi*) Three-dimensional framework silicates. If all the four oxygen atoms of SiO_4^{4-} tetrahedron are shared with other tetrahedra, a three-dimensional lattice is formed with a general formula (SiO_2) as shown in Fig. 10.19 in the planer form for simplicity.

```
   |         |         |
— Si — O — Si — O — Si —
   |         |         |
   O         O         O
   |         |         |
— Si — O — Si — O — Si —
   |         |         |
   O         O         O
   |         |         |
— Si — O — Si — O — Si —
   |         |         |
```

Fig. 10.19. Crystal structure of SiO_2

Replacement of Si^{4+} by Al^{3+} in tetrahedral positions in the SiO_2 lattice gives $(Si, Al)_nO_{2n}$ unit which requires the presence of additional monovalent metal ions in the lattice, to preserve electrical neutrality.

Also to preserve tetrahedral arrangement, the ratio of (Al + Si) to O must always be 1:2. Such isomorphism results in groups of important minerals called felspars, zeolites and ultramarines, which (excepting ultramarines) are among the most widespread, diverse and useful silicate minerals in nature.

Q. 25. Based upon tetrahedral structure of SiO_4^{4-} ion, describe the structure of sheet-silicates giving two examples. (*Delhi, 2002*)

Ans. See Q. 24 under sheet silicates.

Q. 26. What are pyroxenes and amphiboles? Give their structural units with examples.

Ans. See Q. 24 under chain silicates.

Q. 27 Assign reason: CO_2 is a gas while SiO_2 is a solid at room temperature.

Ans. Carbon dioxide has a structure

$$O=C=O$$

There are two π bonds in $CO_2 \cdot p\pi - p\pi$ overlap between carbon and oxygen is efficient and quite strong. Now there are weak van der Waals forces between different molecules of CO_2. Hence it is a gas at room temperature.

In SiO_2, silicon cannot form double bond with oxygen as $p\pi - p\pi$ overlap is not efficient because of size difference between Si and O. Hence it cannot give molecules of type O = Si = O. It completes its octet by network of single bonds as shown hereunder:

```
      |        |        |
      O        O        O
      |        |        |
O — Si — O — Si — O — Si —
      |        |        |
      O        O        O
      |        |        |
O — Si — O — Si — O — Si —
      |        |        |
      O        O        O
      |        |        |
```

As a result, it is a high-boiling solid.

Q. 28. Write notes on

(*i*) Felspar

(*ii*) Zeolites

(*iii*) Ultramarine (*Guwahati, 2002*)

Ans. (*i*) Felspar. Felspars are silicates in which Si^{4+} is partly replaced by Al^{3+} in the tetrahedron position giving rise to units $(Si, Al)_n O_{2n}$.

Felspars constitute some two-thirds of the igneous rocks and are, in fact, the most abundant of all the rock forming minerals. They are divided into two groups according to the symmetry of their structure.

(*a*) *Orthoclase felspars, e.g.*, orthoclase, $KAlSi_3O_8$ and

(*b*) *Plagioclase felspars, e.g.*, albite, $NaA1Si_3O_8$.

In orthoclase, K^+ ion has the appropriate size to fit into the lattice while in plagioclase, Na^+ is smaller and causes distortion. Therefore, orthoclase are more symmetrical.

(*ii*) Zeolites are the most important three-dimensional silicates. They have a much open structure than the felspar which permits them to take up or lose water quite readily from the channels in the honeycomb-like structure. A variety of other molecules such as CO_2, NH_3, C_2H_2OH may also be absorbed instead of water. Zeolites act as molecular sieves by absorbing molecules small enough to enter cavities but not those which are too big to enter. Zeolites also act as natural ion exchangers. For example, permutit, water softeners, are sodium zeolites. They take Ca^{2+} ions from hard water and replace them by Na^+ thereby softening the water. The permutit softener is regenerated by treatment with brine (NaCl-solution), when the reverse process takes place.

$$\underset{\text{Permutit}}{Na_2Z} + \underset{\text{from hard water}}{CaSO_4} \longrightarrow \underset{\text{Softening of water}}{CaZ + Na_2SO_4} \qquad \text{where } Z = \text{Zeolite ion}$$

$$CaZ + \underset{\text{(Brine)}}{2NaCl} \longrightarrow \underset{\text{Regeneration of permutit}}{Na_2Z + CaCl_2}$$

(*iii*) Ultramarines are produced synthetically. Apart from $(Si, Al)_n O_{2n}$, frameworks and balancing cations, there are additional negative ions such as Cl^-, SO_4^{2-} or S_2^{2-} (persulphide ion) also present in these materials. Most of the ultramarines are coloured and are used as pigments. Examples are, ultramarine $[Na_8 (A1_6 Si_6 O_{24}) S_2]$, sodalite, $[Na_8 (A1_6 Si_6 O_{24}) C1_2]$ and noselite $[Na_8 (A1_6 Si_6 O_{24}) SO_4]$. Ultramarines have basket-like structures and all baskets are joined to each other. The various cations and anions occupy the cavities in the lattice and can exchange their positions.

Q. 29. Give the methods of preparation, properties and uses of silicones.

(*Andhra, 2002; Mumbai, 2002; Delhi, 2002; Nagpur, 2003; Garhwal 2010*)

Ans. Methods of preparation of silicones are given as under:

(*i*) Hydrolysis of trialkyl monochlorosilane. R_3SiC1 gives hexaalkylsiloxane, *i.e.*, silicone,

$$R_3SiCl \xrightarrow{H_2O} R_3SiOH$$

$$R-\overset{\displaystyle R}{\underset{\displaystyle R}{Si}}-\boxed{OH + H}O-\overset{\displaystyle R}{\underset{\displaystyle R}{Si}}-R \longrightarrow \underset{\text{Silicone}}{R-\overset{\displaystyle R}{\underset{\displaystyle R}{Si}}-O-\overset{\displaystyle R}{\underset{\displaystyle R}{Si}}-R}$$

(*ii*) Hydrolysis of dialkyl dichlorosilane. R_2SiC1_2 gives rise to straight chain linear polymers and since an active OH group is left at each end of the chain, polymerisation continues and the chain increases in length.

$$R_2SiCl_2 \xrightarrow{H_2O} R_2Si(OH)_2$$

$$\begin{array}{c} R \\ | \\ HO - Si - O\boxed{H + OH} - \\ | \\ R \end{array}\begin{array}{c} R \\ | \\ Si - O\boxed{H + HO} - \\ | \\ R \end{array}\begin{array}{c} R \\ | \\ Si - O\boxed{H + OH} - \\ | \\ R \end{array}\begin{array}{c} R \\ | \\ Si - OH \\ | \\ R \end{array}$$

$$\longrightarrow \begin{array}{c} R \quad\quad R \quad\quad R \quad\quad R \\ HO - Si - O - Si - O - Si - O - Si - OH \\ R \quad\quad R \quad\quad R \quad\quad R \end{array}$$

(*iii*) The hydrolysis of the alkyl trichlorosilane. $RSiCl_3$ gives a complex cross-linked polymer.

$$R - SiCl_3 \xrightarrow{H_2O} R - \begin{array}{c} OH \\ | \\ Si \\ | \\ OH \end{array} - OH$$

$$\begin{array}{c} OH \quad\quad\quad\quad\quad R \\ R - Si - OH + HO - Si - OH \\ OH \quad\quad\quad\quad\quad OH \\ + \quad\quad\quad\quad\quad + \\ OH \quad\quad\quad\quad\quad OH \\ R - Si - OH + HO - Si - R \\ OH \quad\quad\quad\quad\quad OH \end{array} \longrightarrow \begin{array}{c} | \\ O \quad\quad\quad R \\ R - Si - O - Si - O - \\ O \quad\quad\quad O \\ R - Si - O - Si - R \\ O \quad\quad\quad O \\ | \quad\quad\quad\quad | \end{array}$$

Silicone

Alkyl chlorosilanes required for the process are prepared by the following reaction:

(*i*) $SiCl_4 + CH_3MgCl \longrightarrow CH_3SiCl_3 + MgCl_2$

$CH_3SiCl_2 + CH_3MgCl \longrightarrow (CH_3)_2SiCl_2 + MgCl_2$

(*ii*) $Si + 2CH_3Cl \xrightarrow[\text{Catalyst}]{\text{Cu}} (CH_3)_2SiCl_2$

Properties. The amount of cross-linking as well as the nature of the alkyl group determines the properties of the polymer. Materials ranging from oils which remain fluid at very low temperatures to rubber-like solids are known. Properties generally characteristic to the silicones are:

(*i*) Water-repellent in character,

(*ii*) High dielectric strength,

(*iii*) Resistance to oxidation and thermal decompositon, and

(*iv*) Resistance to attack by many organic reagents. The nature of alkyl group and extent of cross-linking determine the properties of the polymer.

Uses of silicones. These polymers are used in waterproofing textiles, in glassware, as lubricants and anti-foaming agents. Silicone rubbers are useful since they retain their elasticity from 180-525 K and because they are good electrical insulators.

11 OXIDES AND OXY-ACIDS OF NITROGEN AND PHOSPHORUS

OXIDES OF NITROGEN

Q. 1. Describe the preparation, properties and structure of nitrous oxide.

(*Kurukshetra, 2001*)

Ans. Preparation. N_2O is obtained by

(*i*) Heating ammonium nitrate, NH_4NO_3 or preferably heating a mixture of $NH_4Cl + NaNO_3$ as NH_4NO_3 is liable to explode on direct heating.

$$NH_4NO_3 \xrightarrow[525-535\text{ K}]{\Delta} N_2O + 2H_2O$$

(*ii*) Warming sulphamic acid with conc. HNO_3.

$$\underset{}{HSO_3NH_2} + \underset{\text{Conc.}}{HNO_3} \longrightarrow N_2O + H_2SO_4 + H_2O$$

(*iii*) Adding a concentrated solution of sodium nitrite to a warm solution of hydroxylamine sulphate.

$$(NH_3OH)_2^+ (SO_4)^{2-} + 2NaNO_2 \longrightarrow 2N_2O\ 4H_2O\ Na_2SO_4$$

Properties. (*i*) N_2O has a faint sweet smell.

(*ii*) It is a colourless gas.

(*iii*) Its aqueous solution is neutral and its conductivity is identical with that of pure water.

(*iv*) At high temperature (~ 573 K) it acts as a powerful oxidizing agent towards organic compounds because of the following reaction:

$$2N_2O \longrightarrow 2N_2 + O_2$$

(*v*) Nitrous oxide is relatively unreactive, being inert to halogens, alkali metals and ozone at room temperature.

(*vi*) It is used as a light general anaesthetic as it produces laughter when inhaled. Hence the name "laughing gas".

Structure. Spectroscopic studies show that the molecule is linear and unsymmetrical and has a low dipole moment. The molecule can be represented by the structure

$$N \xrightarrow{1.13\text{ Å}} N \xrightarrow{1.19\text{ Å}} O$$

Thus the observed molecular length is 2.32 Å.

On the V.B. theory N_2O is resonance hybrid of the two canonical forms.

$$\overset{-}{:\ddot{N}}=\overset{+}{N}=\ddot{O}: \longleftrightarrow :N\equiv\overset{+}{N}-\overset{..}{\underset{..}{:O:}}^{-}$$

Q. 2. What happens when nitrous oxide is passed over hot sodamide at 200°C? Give the chemical equation. (*Vidyasagar 2007*)

Ans. Nitrous oxide reacts with molten alkali metal amides to give azides.

$$NaNO_2 + N_2O \xrightarrow{200^\circ C} \underset{\text{Sodium azide}}{NaN_3} + H_2O$$

Q. 3. Describe the preparation, properties and structure of nitric oxide (NO).

Ans. Preparation. (*i*) Nitric oxide is obtained by the reduction of dil. HNO_3 with copper.

$$3Cu + 8HNO_3 \xrightarrow{\Delta} 3Cu(NO_3)_2 + 2NO + 4H_2O$$

(*ii*) Commercially, it is obtained by catalytic oxidation of ammonia with air (Ostwald's process).

$$4NH_3 + 5O_2 \xrightarrow[1075-1175\text{ K}]{Pt} 4NO + 6H_2O$$

(*iii*) Reasonably pure NO is obtained by the aqueous reaction as follows:

$$2NaNO_2 + 2FeSO_4 + 3H_2SO_4 \longrightarrow Fe_2(SO_4)_3 + 2NaHSO_4 + 2H_2O + 2NO\uparrow$$

Properties. (*i*) It is a very stable colourless gas but dissociates at 1275 K. The liquid or compressed gas slowly changes into N_2O_3 and $N_2O_3 + N_2O$.

$$4NO \longrightarrow N_2O_3 + N_2O$$

(*ii*) It reacts instantaneously with oxygen forming brown NO_2.

$$2NO + O_2 \longrightarrow 2NO_2$$

(*iii*) It reacts with F_2, Cl_2 and Br_2 (not with I_2) to give the nitrosyl halide XNO.

$$2NO + Cl_2 \longrightarrow 2NOCl$$

(*iv*) It is oxidized to HNO_3 by strong oxidizing agents, *e.g.*, $KMnO_4$. It is reduced to N_2O by SO_2 and to NH_2OH by chromous ion in acid solution.

$$2NO + SO_2 + H_2O \longrightarrow N_2O + H_2SO_4$$

(*v*) NO is freely soluble in $FeSO_4$ solution to give the brown nitrosyl cationic complex $[Fe(H_2O)_5NO]^{2+}$. This complex is utilized in the qualitative tests of nitric oxide, nitrites and nitrates.

Structure. Nitric oxide is an 'odd molecule' since it has an unpaired electron (total electrons = 11) and shows paramagnetism. However its behaviour does not point to an odd molecule since it is not coloured (although in the solid and liquid phases it has a blue colour) and the molecule does not show the marked tendency to associate through electron pair. Some association occurs in the liquid and in the solid state which has been found to consist of loose dimers having structures shown in Fig. 11.1.

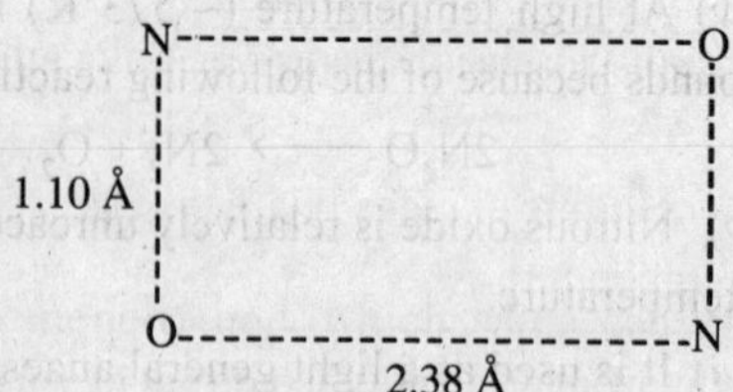

Fig. 11.1. Structure of nitric oxide dimer.

In comparison with other odd molecules, NO is chemically rather inactive. These characteristics suggest an unusual structure.

The bond distance of 1.10Å lies between values of 1.18 Å and 1.06 Å for double and triple bonds suggesting resonance. Two electronic structures are probable.

$$\underset{\text{I}}{:\dot{N}::\ddot{O}:} \longleftrightarrow \underset{\text{II}}{:\ddot{N}::\dot{O}:}$$

The structure of I is more probable in view of the unfavourable charge distribution in structure II. However, small stability difference permits resonance between the two structures, imparting an effective three-electron bond structure to the material.

:N⁝⁝O: or :N≡O:⁺

The arrangement stabilizes the molecules sufficiently with respect to structure I and that is why the unusual properties mentioned above result. The small dipole moment (0.16 D) and its dielectic constant (0.2×10^{-18} e.s.u.) are also in arrangement with this structure.

The structure of NO molecule is explained by the MO theory as follows: The MO configuration for NO is $\sigma 1s^2\ \sigma^*1s^2\ \sigma 2s^2\ \sigma^*2s^2\ \sigma 2p_z^2\ \pi 2p_x^2\ \pi 2p_y^2\ \pi^*2p_x^1$ (Fig. 11.2).

Bond order of NO is: $\frac{1}{2}$ (Total bonding electrons–total antibonding electrons)

$$= \frac{10-5}{2} = 2\frac{1}{2}$$

structure of NO is, therefore,

:N ⁝⁝⁝ O:

which contains one σ bond and two π bonds with an unshared electron in the antibonding MO $\pi^* 2p_x$ orbital.

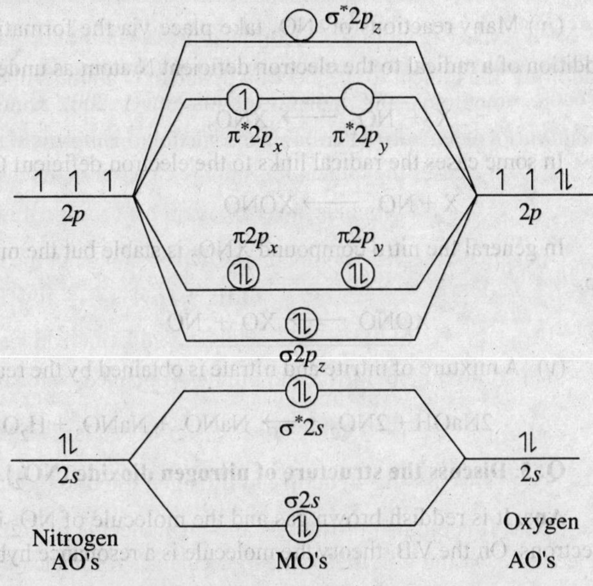

Fig. 11.2. MO diagram of NO molecule.

Q. 3. Give the methods of preparation and properties nitrogen dioxide (NO_2).

Ans. Methods of preparation.

(*i*) *By oxidation of nitric oxide in air or by reduction of nitric acid by metal:*

$$2NO + O_2 \longrightarrow 2NO_2$$

$$Cu + 4HNO_3 \longrightarrow Cu(NO_3)_2 + 2NO_2 + 2H_2O$$

(*ii*) *By heating lead or copper nitrate and passing the vapours through a cooled tube where NO_2 liquefies and O_2 passes on.*

$$2Cu(NO_3)_2 \longrightarrow 2CuO + \underset{\text{(Red-brown gas)}}{4NO_2} + O_2$$

$$2Pb(NO_3)_2 \longrightarrow 2PbO + 4NO_2 + O_2$$

The gas condenses to a brown liquid which turns pale on cooling and eventually becomes a colourless solid due to dimerisation to N_2O_4.

Physical properties. (*i*) NO_2 is a brown gas.

(*ii*) It is an odd electron molecule and is, therefore, paramagnetic.

(*iii*) At ordinary temperature, it exists in equilibrium with its colourless diamagnetic dimer, N_2O_4.

Chemical properties. (*i*) It reacts with water as follows:

$$2NO_2 + H_2O \longrightarrow HNO_3 + HNO_2$$

Thus it is the mixed anhydride of these two acids. The HNO_2 decomposes particularly on warming.

$$3HNO_2 \longrightarrow HNO_3 + 2NO + 2H_2O$$

(*ii*) It decomposes on heating as follows:

$$2NO_2 \xrightarrow{\Delta} 2NO + O_2$$

(*iii*) At high temperatures the system acts as a powerful oxidizing agent.

(*iv*) Many reactions of NO_2 take place via the formation of intermediate compound by the addition of a radical to the electron deficient N atom as under:

$$X + NO_2 \longrightarrow XNO_2$$

In some cases the radical links to the electron deficient O atom as under:

$$X + NO_2 \longrightarrow XONO$$

In general the nitro compound XNO_2 is stable but the nitrite compound XONO tends to break up.

$$XONO \longrightarrow XO + NO$$

(*v*) A mixture of nitrite and nitrate is obtained by the reaction with alkalies.

$$2NaOH + 2NO_2 \longrightarrow NaNO_2 + NaNO_3 + H_2O$$

Q. 5. Discuss the structure of nitrogen dioxide (NO_2).

Ans. It is reddish brown gas and the molecule of NO_2 is the only molecule that contains 17 electrons. On the V.B. theory the molecule is a resonance hybrid of the following canonical forms:

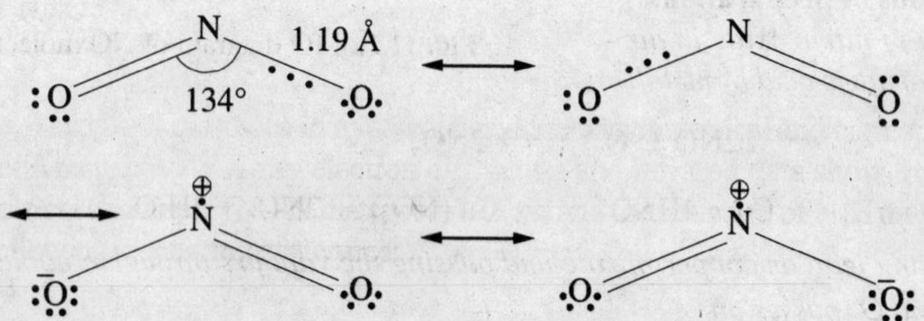

The monomer NO_2 is angular with O—N—O angle of 132° and N—O bond distance 1.20 Å.

NO_2 is an 'odd molecule', *i.e.*, it contains odd number of electrons and the odd electron appears to be located mainly on the *N*-atom. Its properties like red-brown colour and ready dimerisation to colourless and diamagnetic nature of N_2O_4 support the view that it is an odd molecule. It is further suggested that the structure contains a double bond, a single bond, a three-electron bond. By dimerisation it is able to convert itself from an odd molecule to even molecule.

Q. 6. NO_2 readily dimerizes while NO does not. Explain.

Ans. In NO, the electron cloud spreads over the whole molecules. Therefore there is no possibility of dimerisation.

In NO_2, the unpaired electron is localised on nitrogen which results into dimerisation of NO_2 into N_2O_4.

Q. 7. Describe the preparation, properties and structure of nitrogen tetroxide (N_2O_4).

Ans. Methods of preparation. N_2O_4 is prepared by the action of nitrogen dioxide on itself.

$$\underset{\text{Brown paramagnetic}}{NO_2 + NO_2} \longrightarrow \underset{\text{(Colourless diamagnetic)}}{N_2O_4}$$

Properties. (*i*) N_2O_4 is a strong oxidizing agent in aqueous solution.

(*ii*) Liquid N_2O_4 mixed with an organic solvent is very reactive. It dissolves relatively noble metals to form nitrates often solvated with N_2O_4. Thus copper reacts vigorously with N_2O_4 in ethyl acetate to give crystalline Cu $(NO_3)_2 . N_2O_4$ from which anhydrous volatile cupric nitrate (at 425–475 K) is obtained.

(*iii*) Dinitrogen tetroxide is an important non-aqueous solvent. The electrical conductivity of the liquid is quite low. It forms molecular addition compounds with a great variety of nitrogen, oxygen and aromatic donor compounds.

Structure. The dimer of NO_2 is known to exist in three isomeric forms. The most stable is the planar $O_2N — NO_2$ molecule as shown below.

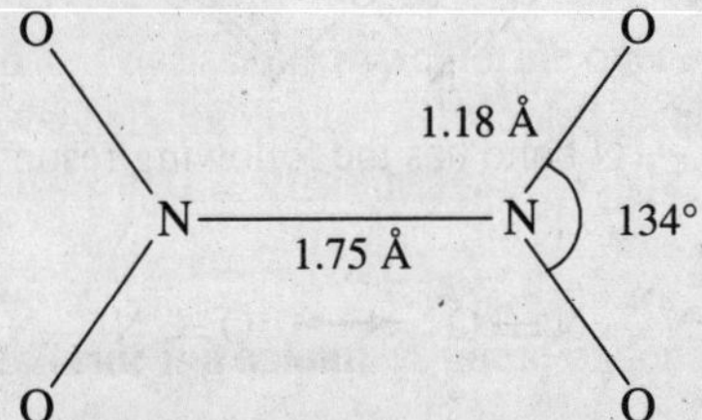

The planar N_2O_4 molecule has several unusual properties like

(*i*) the greater stability of the planar molecule over a twisted one, since the former maximizes O •••• O repulsions.

(*ii*) the unusually long N — N bond 1.75 Å compared to 1.47 Å in $H_2N.NH_2$.

None of the explanations offered so far have been adequate.

Q. 8. Describe the preparation, properties and structure of dinitrogen trioxide or nitrogen sesquioxide N_2O_3.

Ans. Preparation. N_2O_3 is prepared by condensing NO and NO_2 together in an equimolar ratio at 253 K as given by the following equation:

$$NO + NO_2 \xrightarrow{253\ K} N_2O_3$$

A mixture of nitric oxide and nitrogen dioxide in the correct proportion to form N_2O_3 is obtained by warming nitric acid with arsenious oxide.

$$2H_2O + 2HNO_3 + As_2O_3 \xrightarrow{\Delta} 2H_3AsO_4 + N_2O_3$$

N_2O_3 exists in pure form only in the solid state at low temperatures. In the liquid and vapour states it is largely dissociated.

$$N_2O_3\ (g) \longrightarrow NO\ (g) + NO_2\ (g)$$

Properties. (*i*) The gas reacts with alkali forming nitrite.

$$N_2O_3 + 2NaOH \longrightarrow 2NaNO_2 + H_2O$$

(*ii*) It is an intensely blue liquid and a very pale blue solid having m.p. 171 K. It is an acidic oxide and is the anhydride of nitrous acid HNO_2.

$$N_2O_3 \xrightarrow{H_2O} 2HNO_2$$

Structure. The solid N_2O_3 is believed to have two forms. One form is unstable with the structure ON—O—NO and the other is relatively more stable with a long N—N bond. Either of the structures ON—O—NO or ON—NO_2 would explain the ready thermal decomposition of N_2O_3. If ordinary ^{14}NO is mixed with $^{15}NO_2$, there is a complete exchange of N atoms as shown below:

$$^{14}NO + {}^{15}NO_2 = O^{14}N \quad O \longrightarrow {}^{15}NO = O^{14}NO + {}^{15}NO$$

The above rapid exchange reaction shows that the two *N* atoms are equivalent which favours the first structure ON — O — NO having an angular shape as shown below.

O=N—O—N=O (angular)

The second form with a long N — N bond has the following resonance structure:

$$O \leftarrow N(=O) - \ddot{N} = O \longleftrightarrow O = \ddot{N} - N(=O) \rightarrow O$$

According to spectroscopic studies supported with molecular concept on bonding, the other form of N_2O_3 is assigned the following structure with a π only N—N bond.

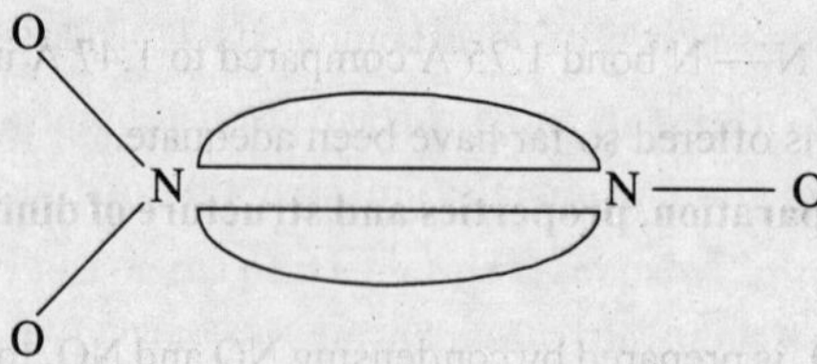

Q. 9. Describe the preparation, properties and structure of dinitrogen pentoxide.

Ans. Preparation. N_2O_5 is prepared.

(*i*) *By action of phosphoryl chloride on warm* $AgNO_3$.

$$3AgNO_3 + POCl_3 \longrightarrow Ag_3PO_4 + 3NO_2Cl$$

$$AgNO_3 + NO_2Cl \longrightarrow AgCl + NO^+_2\, NO^-_3$$

(*ii*) *By distilling* HNO_3 *with* P_4O_{10}, *i.e., by dehydration of nitric acid with* P_4O_{10}.

$$4HNO_3 + P_4O_{10} \xrightarrow{\Delta} 4HPO_3 + 2N_2O_5$$

Properties. (*i*) It is an anhydride of HNO_3. It dissolves in water giving:

$$N_2O_5 + H_2O \longrightarrow 2HNO_3$$

(*ii*) It is deliquescent, readily producing nitric acid by the above reaction.

(*iii*) With alkalies, it gives nitrate as

$$N_2O_5 + 2KOH \longrightarrow 2KNO_3 + H_2O$$

Structure. The gaseous N_2O_5 appears to have a structure of the type:

$$\begin{array}{ccccc} O & & & & O \\ & \diagdown\!\!\diagdown & & \diagup\!\!\diagup & \\ & N & -O- & N & \\ & \swarrow & & \searrow & \\ O & & & & O \end{array}$$

Optical and X-ray evidence shows that the solid N_2O_5 in its stable form is ionic as nitronium nitrate $NO_2^+NO_3^-$ but when the gas is condensed on a surface at 90 K, the molecular form persists for several hours. On warming to 200 K, however, it rapidly rearranges to $NO_2^+NO_3^-$.

OXIDES OF PHOSPHORUS

Q. 10. Describe the preparation, properties and structure of phosphorus pentoxide P_4O_{10}.

(*Kurukshetra, 2002; Nagpur, 2008*)

Ans. Preparation. The correct formula of phosphorus pentoxide is P_4O_{10}. It is usually the main product of burning of phosphorus. However under proper conditions with excess of oxygen, P_4O_{10} is the only product

$$P_4 + 5O_2 \longrightarrow P_4O_{10}$$

Properties. (*i*) It is a white crystalline material which sublimes at 630 K, and 1 atm pressure.

(*ii*) The most important chemical property of P_4O_{10} is its affinity for water. It is one of the most effective drying agents known at temperature below 372 K. It reacts with water to form a mixture of phosphoric acid whose composition depends upon the quantity of water.

$$\underset{}{P_4O_{10}} \xrightarrow{H_2O} \underset{\text{Metaphosphoric acid}}{(HPO_3)_x} \xrightarrow{H_2O} \underset{\text{Triphosphoric acid}}{H_5P_3O_{10}} \xrightarrow{H_2O} \underset{\text{Pyrophosphoric acid}}{H_4P_2O_7} \xrightarrow{H_2O} \underset{\text{Orthophosphoric acid}}{H_3PO_4}$$

In can extract the element of water from many other substances, *e.g.*, it converts pure HNO_3 to N_2O_3 and H_2SO_4 to SO_3. It also dehydrates many organic compounds *e.g.*, converting amides to nitriles.

(*iii*) The fusion of P_4O_{10} with basic oxides gives solid phosphates of various types.

Structure. Phosphorus pentoxide exists in three crystalline modifications. [Hexagonal (H), Tetragonal (T) and Orthorhombic (O)]. It also exists in amorphous and glassy forms. The Hexagonal (H) form is obtained on sublimation. It consists of P_4O_{10} molecule with the structure shown in Fig. 11.3. In this structure, the P atoms are at the corners of a tetrahedron with six oxygen atoms along the edges and the remaining four lying along extended three-fold axes of the tetrahedron.

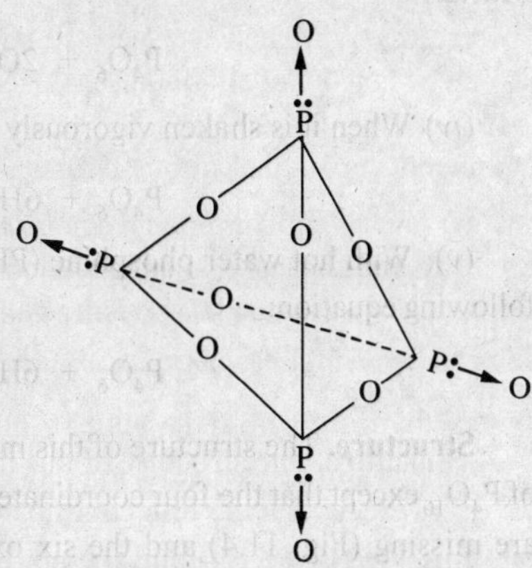

Fig. 11.3. Structure of P_4O_{10}.

Each P atom forms three covalent bonds to oxygen atoms. There are five outer electrons around a P atom so that every phosphorus atom has a lone pair. In P_4O_{10}, each P forms an additional coordinate bond by donating this lone pair to an oxygen atom. The twelve P – O distance between P atoms and shared O atoms are 1.62Å each of which is nearly P – O single bond distance but the other four P – O distances are only 1.39 Å. (Coordinate bonds are shorter than a single bond.) This is explained by $p\pi - d\pi$ back bonding because of the lateral overlap of full p orbitals of oxygen with empty d-orbitals of phosphorus. Gaseous state has the same structure.

On heating, the hexagonal form at 675 -775 K is converted into orthorhombic modification (O) which appears to be an infinite sheet type of polymer made up of interlocking rings but based upon the PO_4 tetrahedra. Further heating to still higher temperatures gives a tetragonal modification (T) in which conjugated sheets of P_4O_{10} molecules are found. Rapid heating of the hexagonal form gives a liquid which polymerises rapidly to a glass.

Q. 11. Write notes on (*i*) phosphorus dioxide $(PO_2)_n$ and (*ii*) phosphorus trioxide (P_4O_6).

(*Kurukshetra, 2002; Nagpur, 2008*)

Ans. (*i*) Phosphorus dioxide $(PO_2)_n$. Preparation. It is a solid subliming at 453 K in vacuum. It is obtained by heating phosphorus oxide in a sealed tube to 713 K.

$$nP_4O_6 \longrightarrow 3\,(PO_2)_n + nP$$

Properties. (*i*) It dissolves in H_2O to give a mixture of H_3PO_4 and H_3PO_3 in equal quantities. It is not well characterised.

(*ii*) It can be sublimed at high temperature to give transparent crystals which are deliquescent and very soluble in water. The compound is commonly called the tetroxide and probably has a molecular formula P_2O_4.

Phosphorus trioxide (P_4O_6). Preparation. It is a colourless, volatile compound (m.p. 297 K, b.p. 448 K) which is formed in about 50% yield when white P is burnt in a limited supply of oxygen and purified by distillation.

Properties. (*i*) On heating above 483 K, P_4O_6 decomposes into red P and another oxide P_2O_4.

(*ii*) It reacts vigorously with chlorine and bromine to give the oxy-halides and with iodine in a sealed tube to give P_2I_4.

(*iii*) It is stable to oxygen at room temperature but slow oxidation to P_4O_{10} takes place on contact with air.

$$P_4O_6 + 2O_2 \longrightarrow P_4O_{10}$$

(*iv*) When it is shaken vigorously with cold water, phosphorus acid, H_3PO_3, is formed.

$$P_4O_6 + 6H_2O \longrightarrow 4H_3PO_3$$

(*v*) With hot water phosphine (PH_3) and phosphoric acid (H_3PO_4) are formed as given by the following equation:

$$P_4O_6 + 6H_2O \longrightarrow PH_3 + 3H_3PO_4$$

Structure. The structure of this molecule is very similar to that of P_4O_{10} except that the four coordinated oxygen atoms at the apexes are missing (Fig. 11.4) and the six oxygen atoms are located just outside the centre of the six edges. This fact is well supported by electron diffraction data.

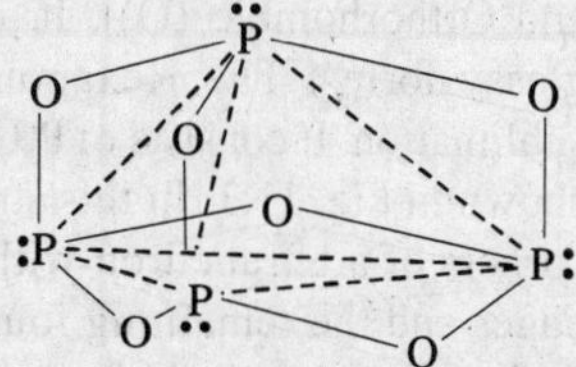

Fig. 11.4. Structure of P_4O_6

OXY-ACIDS OF NITROGEN

Q. 12. Describe preparation, properties and structure of hyponitrous acid ($H_2N_2O_2$).

Ans. Preparation. This material is known mainly in the forms of its salts (hyponitrites), although the free acid has been obtained as a white crystalline compound by reacting the silver salt with HCl in ether and evaporating the resulting ethereal solution.

$$Ag_2N_2O_2 + 2HCl \longrightarrow 2AgCl\downarrow + H_2N_2O_2$$

Properties. (*i*) The free acid is a white crystalline solid which decomposes readily on heating. Its aqueous solutions are more stable but lose nitrous oxide slowly.

$$H_2N_2O_2 \longrightarrow H_2O + N_2O$$

(*ii*) The acid is also soluble in alcohols, ether, chloroform and benzene.

(*iii*) Its aqueous solution is a weak acid and its alkali metal salts are extensively hydrolysed.

Structure. Exact structure of hyponitrous acid is not known but it appears to have an arrangement given below.

$$H:\ddot{\underset{\cdot\cdot}{O}}:\ddot{N}::\ddot{N}:\ddot{\underset{\cdot\cdot}{O}}:H \quad \text{or} \quad HO-N=N-OH$$

Because of the double bond, cis-transisomerism is possible. Hantzsch believed the acid to have the *trans* form as given below.

$$\begin{array}{ll} & :N-OH \\ & \;\;\| \\ HO-N: & \end{array}$$

This structure is supported by infrared spectroscopy of silver salt and dipole moments of its esters.

Hyponitrites. Hyponitrites are obtained by treating hydroxylamine NH_2OH with amyl nitrite in alcohol containing sodium ethoxide. Reduction of nitrites with sodium amalgam also gives hyponitrites.

$$2NH_2OH + \underset{\text{Amyl nitrite}}{C_5N_{11}NO_2} + 2C_2H_5ONa \longrightarrow Na_2N_2O_2 + C_5H_{11}OH + NH_3 + 2C_2H_5-OH$$

$$2NaNO_2 + 4H \xrightarrow[\text{amalgam}]{\text{Sodium}} \underset{\text{Sodium hyponitrite}}{Na_2N_2O_2} + 2H_2O$$

Hyponitrites possess both reducing and oxidising properties in acidic and alkaline solutions depending on conditions. Strong oxidising agents will convert hyponitrites to nitrates.

Q. 13. How do you obtain nitroxylic acid? Give its structure.

Ans. Preparation. This acid has not been obtained in the free state. But its sodium salt of yellow colour ($Na_4N_2O_4$) has been prepared by reducing $NaNO_2$ with sodium in liquid ammonia.

$$2NaNO_2 + 2Na \xrightarrow[\text{aammonia}]{\text{In liquid}} Na_4N_2O_4$$

Structure. The structure acid is as given below:

$$\begin{array}{ccc} HO & & OH \\ & N-N & \\ HO & & OH \end{array}$$

Q. 14. Give the methods of preparation of nitrous acid (HNO_2). Discuss its properties and structure.

Ans. Preparation. Free nitrous acid has not been isolated although there is indication of its existence in the vapour state. Aqueous solutions of nitrous acid are easily made by acidifying solution of nitrites. For example,

$$Ba(NO_2)_2 + H_2SO_4 \longrightarrow BaSO_4 + 2HNO_2$$

$$KNO_2 + HCl \longrightarrow KCl + HNO_2$$

Properties. **(*i*)** Nitrous acid is a weak monobasic acid. The solution has a bluish colour and decomposes rapidly when heated according to the equation

$$3HNO_2 \longrightarrow HNO_3 + H_2O + 2NO$$

(*ii*) Decomposition. Due to its unstable nature, it decomposes rapidly and all higher temperature gives nitric oxide and nitrogen dioxide.

$$2HNO_2 \longrightarrow NO + NO_2 + H_2O$$

(*iii*) Oxidising agent. Since nitrous acid can easily lose oxygen according to the equation

$$2HNO_2 \longrightarrow 2NO + H_2O + [O]$$

it behaves as an oxidising agent. It oxidizes:

(*a*) *Moist sulphur dioxide to sulphuric acid*

$$2HNO_2 \longrightarrow 2NO + H_2O + [O]$$

$$SO_2 + H_2O + [O] \longrightarrow H_2SO_4$$

$$2HNO_2 + SO_2 \longrightarrow 2NO + H_2SO_4$$

(*b*) *Hydrogen sulphide to sulphur*

$$2HNO_2 \longrightarrow 2NO + H_2O + O$$

$$H_2S + O \longrightarrow H_2O + S$$

$$2HNO_2 + H_2S \longrightarrow 2NO + 2H_2O + S\downarrow$$

(*c*) *Potassium iodide to iodine*

$$2HNO_2 \longrightarrow H_2O + 2NO + [O]$$

$$2KI + H_2O + O \longrightarrow 2KOH + I_2$$

$$2HNO_2 + 2KI \longrightarrow 2NO + 2KOH + I_2$$

(*d*) *Stannous chloride to stannic chloride*

$$2HNO_2 \longrightarrow 2NO + H_2O + [O]$$

$$SnCl_2 + 2HCl + [O] \longrightarrow SnCl_4 + H_2O$$

$$2HNO_2 + SnCl_2 + 2HCl \longrightarrow 2NO + H_2O + SnCl_4$$

(*iv*) Reducing properties. Nitrous acid acts as a reducing agent towards strong oxidising agents as it readily takes up oxygen to form nitric acid.

$$HNO_2 + [O] \longrightarrow HNO_3$$

(*a*) *Acidified potassium permanganate is decolourised due to reduction*

$$2KMnO_4 + 3H_2SO_4 \longrightarrow K_2SO_4 + 2MnSO_4 + 3H_2O + 5[O]$$

$$HNO_2 + [O] \longrightarrow HNO_3] \times 5$$

$$2KMnO_4 + 3H_2SO_4 + 5HNO_2 \longrightarrow K_2SO_4 + 2MnSO_4 + 3H_2O + 5HNO_3$$

(*b*) *Acidified potassium dichromate is turned green due to reduction*

$$K_2Cr_2O_7 + 4H_2SO_4 \longrightarrow K_2SO_4 + Cr_2(SO_4)_3 + 4H_2O + 3[O]$$

$$HNO_2 + [O] \longrightarrow HNO_3] \times 3$$

$$K_2Cr_2O_7 + 4H_2SO_4 + 3HNO_2 \longrightarrow K_2SO_4 + Cr_2(SO_4)_3 + 4H_2O + 3HNO_3$$

(*v*) Action with amino compounds. With organic compounds containing amino ($-NH_2$) group, it reacts to replace the amino group by hydroxyl group ($-OH$) and evolving nitrogen. For example,

$$\begin{matrix} HO:N:O \\ C_2H_5:N:H_2 \\ \text{Ethyl amine} \end{matrix} \xrightarrow{(HNO_2)} \underset{\text{Ethyl alcohol}}{C_2H_5OH} + N_2\uparrow + H_2O$$

$$\begin{matrix} HO:N:O \\ CH_3CO:N:H_2 \end{matrix} \xrightarrow{(HNO_2)} \underset{\text{Acetic acid}}{CH_3COOH} + N_2 + H_2O$$

with aromatic primary amines it forms diazonium compound at 273 K.

$$C_6H_5NH_2 + HNO_2 + HCl \xrightarrow{273\text{ K}} \underset{\text{Benzene diazonium chloride}}{C_6H_5N_2Cl} + 2H_2O$$

Q. 15. Nitrous acid can behave as oxidising agent. Explain. (*Vidyasagar 2007*)

Ans. See Q. 14 (*iii*).

Q. 16. Write down preparation, properties and structure of nitrites.

Ans. Preparation and properties. Nitrites of alkali metals are best prepared by reducing the nitrates with carbon, lead, iron etc.

$$NaNO_3 + Pb \longrightarrow NaNO_2 + PbO$$

Nitrous acid and nitrites act as oxidizing agents with the formation of N_2O or NO, although powerful oxidizing agents such as $KMnO_4$ and Cl_2 oxidise NO_2^- to NO_3^-. The nitrite ion is a good co-ordinating agent (ligand) and because lone pairs of electrons are present in both N and O atoms, either can form a coordinate bond, so that isomerism occurs between M $\leftarrow NO_2$ and M $\leftarrow$ ONO.

Structure. X-ray and Raman data indicate that nitrite ion has a bent structure N– O bond distance being 1.13 Å and O – N – O bond angle 120 to 130°. A resonance seems likely as given below:

$$:N(=O)(-O^-) \longleftrightarrow :N(-O^-)(=O)$$

In organic nitro compounds, N – O bond distance is 1.21 Å and O – N – O bond angle 127°. According to molecular orbital theory, N forms sp^2 hybrid orbitals, one orbital accommodates an unshared pair and the other two, being used in σ bonding to the oxygens. The p_z orbitals of N and O are then used for π bond formation. (Fig. 11.5).

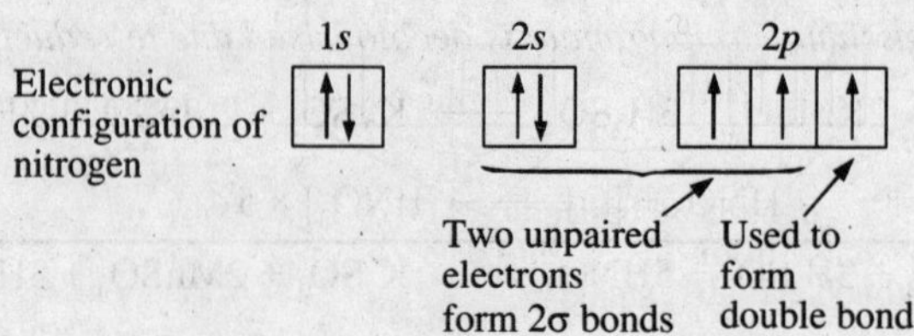

Fig. 11.5. Structure of nitrite ion.

Q. 17. Describe preparation, properties and structure of nitric acid. *(Garhwal 2010)*

Ans. Preparation. Nitric acid is prepared in the laboratory by heating a mixture of sodium nitrate and conc. H_2SO_4.

$$NaNO_3 + H_2SO_4 \longrightarrow NaHSO_4 + HNO_3$$

Manufacture of nitric acid by Ostwald's process. Nitric acid is obtained on a large scale by the oxidation of ammonia by this process. The process was developed in Germany during World War I. It involves the following steps:

(*i*) Ammonia, manufactured by Haber's process, is mixed with about 8 times its volume of air or oxygen and the mixture is passed over platinum gauze heated electrically to about 1075 K when it is oxidized to nitric oxide.

$$4NH_3 + 5O_2 \xrightarrow[Pt]{1075\ K} 4NO + 6H_2O$$

(*ii*) Nitric oxide obtained after being cooled to 343-355 K is made to combine with excess of oxygen present to form nitrogen dioxide.

$$2NO + O_2 \longrightarrow 2NO_2$$

(*iii*) The nitrogen dioxide is then finally absorbed in water in the presence of oxygen to give nitric acid.

$$4NO_2 + 2H_2O + O_2 \longrightarrow 4HNO_3$$

Details of the process. (*a*) **Catalyst chamber.** Mixture of ammonia (1 part) and pure dry air (8 parts) is introduced at the top of steel chamber (Fig. 11.6) provided with platinum wire gauze cases closed at the base by a silica plate. It is about 22 cm in diameter and 33 cm in length. The platinized wire cage is heated electrically to 975-1075 K.

The gaseous mixture of ammonia and air is then forced out through the walls of the platinum gauze when ammonia gets oxidised to nitric oxide.

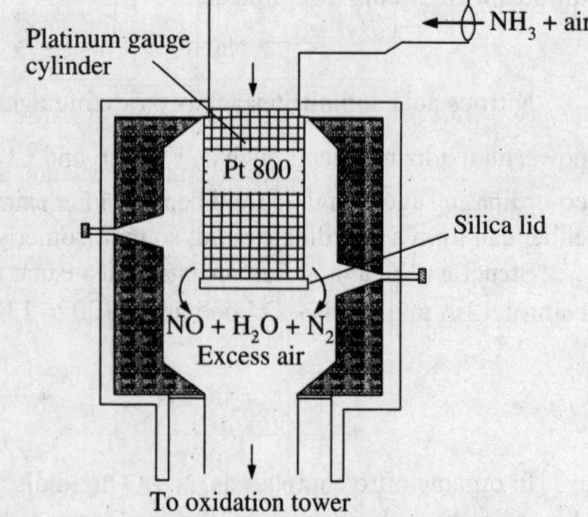

Fig. 11.6. Catalyst chamber for the oxidation of ammonia.

$$4NH_3 + 5O_2 \longrightarrow 4NO + 6H_2O + 920\,kJ$$

The yield is about 90-95%.

Since the reaction is exothermic, the heat of the reaction is sufficient to maintain the temperature of the catalyst at 1075 K once the reaction has started.

(b) Cooling arrangement. The gases coming out of the catalyst chamber are passed through cooling pipes to cool them below 350 K.

This low temperature of gases (NO + excess of air) is essential for the oxidation of nitric oxide to nitrogen dioxide.

(c) Oxidation chamber. It is a cylindrical vessel made of sheet steel lined inside with acid-proof bricks. The cooled gaseous mixture of nitric oxide and air is introduced into this chamber. Nitric oxide is oxidised into nitrogen dioxide by the oxygen of the air.

$$2NO + O_2 \longrightarrow 2NO_2$$

This oxidation is favoured below 350 K since NO_2 formed is decomposed above 350 K.

(d) Absorption tower. The gases from the oxidation chamber are introduced into the absorption tower which is made of stainless steel packed with broken acid-proof quartz. Water is sprayed from the top of the tower.

Nitrogen dioxide is absorbed in water in the presence of oxygen to give nitric acid which is collected from the base of the tower.

$$4NO_2 + 2H_2O + O_2 \longrightarrow 4HNO_3$$

The complete diagram for Ostwald process is shown in Fig. 11.7.

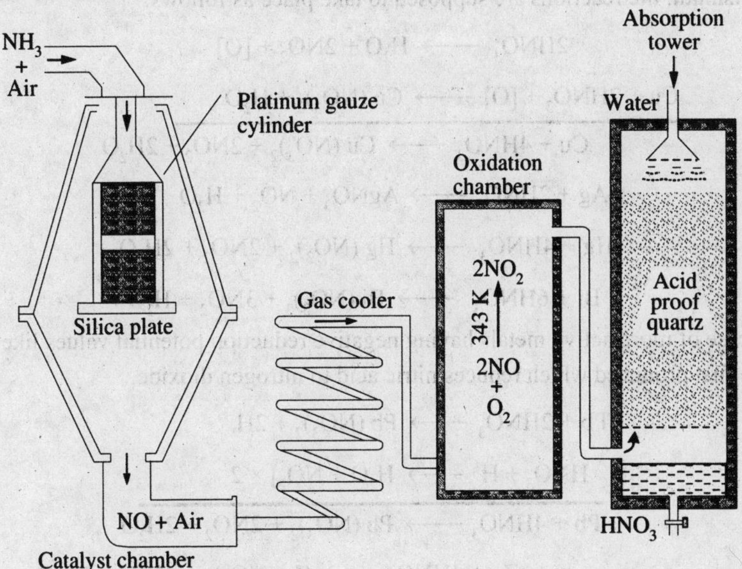

Fig. 11.7. Ostwald's process for manufacture of nitric acid.

Properties. (*i*) Anhydrous nitric acid is a colourless liquid boiling at 357 K and freezing to a crystalline solid at 231.4 K.

(*ii*) The normal concentrated aqueous acid (~ 70% by weight) is colourless but often becomes yellow as a result of photochemical decomposition to give NO_2

$$2HNO_3 \xrightarrow{h\nu} 2NO_2 + H_2O + \frac{1}{2}O_2$$

(*iii*) **Action on metals.** Fuming nitric acid contains dissolved NO_2. HNO_3 is a powerful oxidizing agent. Metals like Au, Pt, Rh and Ir are not attacked while Al, Fe and Cr are rendered passive. Passivity is due to the formation of a very thin film of oxide on the metal surface. Magnesium alone can liberate H_2 and that also in the beginning. Non-metals are usually oxidized by conc. HNO_3 to oxides or oxy acids. The oxidizing power of HNO_3 depends upon its concentration and below a certain concentration the acid has no oxidizing power.

The products of reaction include reaction products of nitric acid like nitrogen dioxide (NO_2), Nitric (NO), nitrous oxide (N_2O) and ammonium nitrate (NH_4NO_3).

$$2HNO_3 \xrightarrow[-H_2O]{-O} 2NO_2 \xrightarrow{-2O} 2NO \xrightarrow{-O} N_2O \xrightarrow[-H_2O]{+8H} 2NH_3 \xrightarrow{+2HNO_3} 2NH_4NO_3$$

The stage to which the reduction proceeds depends upon temperature, concentration and nature of the metal.

According to their reactivity, metals may be grouped under three categories:

(*a*) *Less active metals* having positive reduction potential values, *e.g.*, copper, silver, mercury and bismuth.

(*b*) *More active metals* having negative reduction potential values, *e.g.*, lead, zinc, tin and iron.

(*c*) *Noble metals* like gold and platinum.

[A] Action of concentrated nitric acid. It attacks most of the metals forming their nitrates, nitrogen dioxide and water, notable exceptions are iron, chromium, nickel, aluminium, gold and platinum.

In the case of less active metals having positive reduction potential values like copper, mercury, silver and bismuth, the reactions are supposed to take place as follows:

$$2HNO_3 \longrightarrow H_2O + 2NO_2 + [O]$$

$$Cu + 2HNO_3 + [O] \longrightarrow Cu(NO_3)_2 + H_2O$$

$$Cu + 4HNO_3 \longrightarrow Cu(NO_3)_2 + 2NO_2 + 2H_2O$$

Similarly,

$$Ag + 2HNO_3 \longrightarrow AgNO_3 + NO_2 + H_2O$$

$$Hg + 4HNO_3 \longrightarrow Hg(NO_3)_2 + 2NO_2 + 2H_2O$$

$$Bi + 6HNO_3 \longrightarrow Bi(NO_3)_3 + 3NO_2 + H_2O$$

In the case of more active metals having negative reduction potential values like lead or zinc, hydrogen is first produced which reduces nitric acid to nitrogen dioxide.

$$Pb + 2HNO_3 \longrightarrow Pb(NO_3)_2 + 2H$$

$$HNO_3 + H \longrightarrow H_2O + NO_2] \times 2$$

$$Pb + 4HNO_3 \longrightarrow Pb(NO_3)_2 + 2NO_2 + 2H_2O$$

Similarly,

$$Zn + 4HNO_3 \longrightarrow Zn(NO_3)_2 + 2NO_2 + 2H_2O$$

The reaction becomes more brisk as the reactivity of the metal increases. Tin, instead of forming a nitrate, is oxidized to its oxy-acid.

$$2HNO_3 \longrightarrow H_2O + 2NO_2 + O] \times 2$$

$$Sn + 2[O] + H_2O \longrightarrow H_2SnO_3$$

$$Sn + 4HNO_3 \longrightarrow \underset{\text{Metastannic acid}}{H_2SnO_3} + 4NO_2 + H_2O$$

Iron, chromium, nickel and aluminium become passive, *i.e.*, they lose their normal reactivity after being dipped in concentrated nitric acid. This is due to the formation of a thin protective layer of the oxide on the surface of the metal which prevents further action of the acids. Gold and platinum are too noble to react and are not attacked at all by the acid.

[B] Action of dilute nitric acid. The action of dilute nitric acid on less active metals like copper, mercury and silver gives nitric oxide. The mechanism of the reaction is represented by the following equations:

$$2HNO_3 \longrightarrow H_2O + 2NO + 3\,[O]$$

$$Cu + 2HNO_3 + [O] \longrightarrow Cu\,(NO_3)_2 + H_2O\,] \times 3$$

$$3Cu + 8HNO_3 \longrightarrow 3Cu\,(NO_3)_2 + 2NO + 4H_2O$$

Similarly, $6Hg + 8HNO_3 \longrightarrow 3Hg_2\,(NO_3)_2 + 2NO + 4\,H_2O$

$$3Ag + 4HNO_3 \longrightarrow 3Ag\,NO_3 + 2H_2O + NO$$

Dilute nitric acid has no action on antimony and bismuth. In case of more active metals, nascent hydrogen is first produced and it reduces the acid to nitric oxide, nitrous oxide or ammonia depending upon the metal, concentration of the acid and temperature. Ammonia produced further combines with nitric acid to produce ammonium nitrate.

Thus with lead, nitric oxide is produced.

$$Pb + 2HNO_3 \longrightarrow Pb\,(NO_3)_2 + 2H] \times 3$$

$$3\,[H] + HNO_3 \longrightarrow 2H_2O + NO] \times 2$$

$$3\,Pb + 8HNO_3 \longrightarrow 3Pb\,(NO_3)_2 + 4H_2O + 2NO$$

In case of zinc, dilute acid gives nitrous oxide, while very dilute acid gives ammonium nitrate.

$$Zn + 2HNO_3 \longrightarrow Zn\,(NO_3)_2 + 2H] \times 4$$

$$8[H] + 2HNO_3 \longrightarrow N_2O + 5\,H_2O$$

$$\underset{\text{(Dilute)}}{Zn + 10\,HNO_3} \longrightarrow 4Zn\,(NO_3)_2 + N_2O + 5H_2O$$

$$Zn + 2HNO_3 \longrightarrow Zn\,(NO_3)_2 + 2\,[H] \ldots\times\ 4$$

$$8[H] + HNO_3 \longrightarrow NH_3 + 3H_2O$$

$$NH_3 + HNO_3 \longrightarrow NH_4\,NO_3$$

$$\underset{\text{V. dil. acid}}{4Zn + 10HNO_3} \longrightarrow 4Zn\,(NO_3)_2 + NH_4\,NO_3 + 3\,H_2O$$

Tin and iron also react similarly producing ammonium nitrate in each case.

$$4Sn + 10HNO_3 \longrightarrow 4Sn\,(NO_3)_2 + NH_4\,NO_3 + 3H_2O$$

$$4Fe + 10HNO_3 \longrightarrow 4Fe\,(NO_3)_2 + NH_4\,NO_3 + 3H_2O$$

Magnesium and manganese are the only metals which produce hydrogen with very dilute cold acid. (1-2%)

$$Mg + 2HNO_3 \longrightarrow Mg(NO_3)_2 + H_2\uparrow$$

$$Mn + 2HNO_3 \longrightarrow Mn\,(NO_3)_2 + H_2\uparrow$$

[C] Formation of aqua regia. It is a mixture of one part of concentrated nitric acid and three parts of concentrated hydrochloric acid and is used for dissolving noble metals like gold and platinum as their chlorides.

$$HNO_3 + 3HCl \longrightarrow 2H_2O + \underset{\text{Nitrosyl chloride}}{NOCl} + \underset{\text{Nascent chlorine}}{2[Cl]}$$

$$Pt + 4[Cl] \longrightarrow \underset{\text{Platinum chloride}}{PtCl_4}$$

$$Au + 3[Cl] \longrightarrow \underset{\text{Gold chloride}}{AuCl_3}$$

***(iv)* Action on non-metals and metalloids.** Carbon, sulphur, phosphorus, iodine (non-metals), arsenic and antimony (metalloids) are oxidized by conc. nitric acid to form carbonic acid (H_2CO_3), sulphuric acid (H_2SO_4), phosphoric acid (H_3PO_4), iodic acid (HIO_3), arsenic acid (H_3AsO_4) and antimonic acid (H_3SbO_4) respectively as depicted below.

$$2HNO_3 \longrightarrow 2NO_2 + H_2O + O$$

Thus

$$2HNO_3 \longrightarrow 2NO_2 + H_2O + [O] \times 2$$

$$C + 2O + H_2O \longrightarrow H_2CO_3$$

$$C + 4HNO_3 \longrightarrow H_2CO_3 + 4NO_2 + H_2O$$

$$S + 6HNO_3 \longrightarrow H_2SO_4 + 2H_2O + 6NO_2$$

$$P_4 + 20\,HNO_3 \longrightarrow 4H_3PO_4 + 4H_2O + 20\,NO_2$$

Similarly,

$$I_2 + 10HNO_3 \longrightarrow 10NO_2 + 4H_2O + 2HIO_3$$

$$As + 5HNO_3 \longrightarrow H_3AsO_4 + H_2O + 5NO_2$$

$$Sb + 5HNO_3 \longrightarrow H_3SbO_4 + H_2O + 5NO_2$$

***(v)* Oxidising properties.** Conc. HNO_3 and even dil. HNO_3 oxidizes some compounds, such as H_2S to S, SO_2 to H_2SO_4, ferrous salts to ferric salts.

Dilute acid loses its oxygen as:

$$2HNO_3 \longrightarrow 2NO + H_2O + 3O$$

Thus *(a)*

$$2HNO_3 \longrightarrow H_2O + 2NO + 3O$$

$$SO_2 + O + H_2O \longrightarrow H_2SO_4] \times 3$$

$$3SO_2 + 2HNO_3 + 2H_2O \longrightarrow 3H_2SO_4 + 2NO$$

(b)

$$2HNO_3 \longrightarrow 2NO + H_2O + 3O$$

$$H_2S + O \longrightarrow H_2O + S] \times 3$$

$$3H_2S + 2HNO_3 \longrightarrow 3S + 4H_2O + 2NO$$

(c)

$$2HNO_3 \longrightarrow 2NO + H_2O + 3O$$

$$2FeSO_4 + H_2SO_4 + O \longrightarrow Fe_2(SO_4)_3 + H_2O] \times 3$$

$$6\,FeSO_4 + 3\,H_2SO_4 + 2HNO_3 \longrightarrow 2Fe_2(SO_4)_3 + 2NO + 4H_2O$$

(*vi*) It oxidizes saw dust, turpentine oil and sugar. Saw dust catches fire when hot nitric acid is poured over it. Turpentine oil also burns when poured in fuming nitric acid. Sugar is oxidized to oxalic acid.

$$\underset{\text{Sugar}}{C_{12}H_{22}O_{11}} + \underset{\substack{\text{(From}\\\text{nitric acid)}}}{18[O]} \longrightarrow \underset{\substack{\text{Oxalic}\\\text{acid}}}{6(COOH)_2} + 5H_2O$$

(*vii*) It reacts with glycerine giving glycerine trinitrate and with toluene giving trinitrotoluene (T.N.T.) which are used as explosives.

(*v*) It replaces one or more hydrogen atoms of an aromatic compound, *e.g.*, benzene, phenol etc., by a nitro group ($-NO_2$). The process is called nitration. The reaction takes place in the presence of H_2SO_4.

$$\underset{\text{Benzene}}{C_6H_6} + HONO_2 \xrightarrow{H_2SO_4} \underset{\text{Nitrobenzene}}{C_6H_5NO_2} + H_2O$$

HNO_3 has the ability to nitrate many organic compounds especially in the presence of conc. H_2SO_4. This property results from the dissociation of the acid to produce the nitronium ion, NO_2^+.

$$2HNO_3 \longrightarrow NO_2^+ + NO_3^- + H_2O$$

$$\underset{\text{Benzene}}{C_6H_6} + NO_2^+ \longrightarrow \left[C_6H_6NO_2\right]^+ \longrightarrow \underset{\text{Nitrobenzene}}{C_6H_5NO_2} + H^+$$

Structure. Being a monobasic acid it may be structurally represented as below:

$$\underset{\text{I}}{H-O-N{\overset{\displaystyle O}{\nearrow\!\!\!\!/\!/}}\!\!\searrow O} \longleftrightarrow \underset{\text{II}}{H-O-N{\overset{\displaystyle O}{\nearrow}}\!\!\searrow\!\!\!\!\backslash\!\backslash O}$$

The gaseous HNO_3 molecule has a planar triangular structure (Fig. 11.8) as suggested by electron diffraction data.

H–O 0.96 Å; O–N 1.405 Å; N–O 1.206 Å; ∠HON 102°; ∠ONO 116°, 130°, 114°

Fig. 11.8. Structure of HNO_3

Q. 18. Discuss the structure of nitrate (NO_3^-) ion.

Ans. Nitrates of practically all metallic elements are water-soluble and are frequently hydrated. Some nitrates can be obtained anhydrous and some of them sublime [*e.g.*, Cu $(NO_3)_2$] without decomposition. Alkali nitrates on decomposition gives nitrites while NH_4NO_3 gives N_2O and H_2O. Al and Zn in alkaline solution reduce them to NH_3. The nitrate ion is planar and symmetrical as shown in Fig. 11.9.

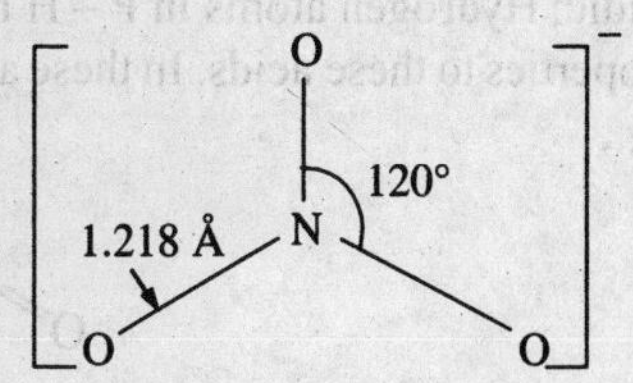

Fig. 11.9. Structure of NO_3^- ion.

The structure may be explained by V.B. theory as a resonance hybrid.

$$O \leftarrow \overset{O}{\underset{}{N}} - O^- \;\longleftrightarrow\; O = \overset{O^-}{N} \rightarrow O \;\longleftrightarrow\; O \leftarrow \overset{O}{\overset{\|}{N}} - O^-$$

Q. 19. Illustrate with two examples each the reducing power of nitrous acid and oxidising power of nitric acid.

Ans. See Q. 13 under reducing properties and Q. 17 under oxidising properties.

Q. 20. Give the names and structures of various oxy-acids of nitrogen.

Ans. Names of various oxy-acids of nitrogen are:

(1) $H_2 N_2 O_2$ Hyponitrous acid
(2) $H_2 NO_2$ Hydronitrous acid
(3) HNO_2 Nitrous acid
(4) HNO_3 Nitric acid
(5) HNO_4 Pernitric acid

For structures, see Questions 12-17.

Q. 21. Give the method of preparation and properties of pernitric acid.

Ans. Preparation. It is obtained by dissolving N_2O_5 in anhydrous hydrogen peroxide.

$$N_2O_5 + H_2 O_2 \longrightarrow HNO_3 + HNO_4$$

Properties. (*i*) It is unstable and explosive and resembles bleaching powder in smell.

(*ii*) It oxidises aniline to nitrobenzene and liberates bromine from a solution of potassium bromide.

Structure. It has been assigned the following structure:

$$HO - O - \underset{\underset{O}{\downarrow}}{N} = O$$

OXY-ACIDS OF PHOSPHORUS

Q. 22. Write a note on oxy-acids of phosphorus. (*Lucknow, 2000; Nagpur, 2003*)

Ans. The following are the main oxy-acids of phosphorus:

(1) Hypophosphorus acid H_3PO_2
(2) Phosphorus acid $H_3 PO_3$
(3) Hypophosphoric acid $H_4 P_2 O_6$
(4) Orthophosphoric acid $H_3 PO_4$
(5) Pyrophosphoric acid $H_4 P_2 O_3$
(6) Metaphosphoric acid HPO_3 $(HPO_3)_x$

All oxy-acids of P have P– OH groups where the hydrogen atom is ionisable and is therefore acidic; Hydrogen atoms in P – H bonds are not ionised. The presence of this group gives reducing properties to these acids. In these acids, the P is bonded in a tetrahedral manner with four sp^3 bonds, *e.g.*,

H, P, O, O, OH — P (OH)$_3$ or H_3PO_3

O, P, OH, OH, OH — O = P (OH)$_3$ or H_3PO_4

The acid strength depends upon the number of unprotonated oxygen atoms on P atom *i.e.*, P = O groups which polarise a given number of OH bonds. The number of OH groups per P = O, if decreases, the acidity would increase in that order. Thus the relative order of acid strength in the acids is $H_3PO_2 > H_3PO_3 > H_3PO_4$. The number of OH groups per P = O increases from one in H_3PO_2 to two in H_3PO_3 and three in H_3PO_4. Smaller the number of – OH groups per P = O group, greater the acid strength.

Q. 23. Describe the preparation, properties and structure of hypophosphorus acid (H_3PO_2).

Ans. Preparation. It is prepared by heating phosphorus with barium hydroxide solution till there is no evolution of phosphine.

$$8P + 3Ba(OH)_2 + 6H_2O \longrightarrow 3Ba(H_2PO_2)_2 + 2PH_3 \uparrow$$

To this barium salt, if sulphuric acid is added in calculated amounts and the solution filtered, free hypophosphorus acid comes down in the filtrate.

$$Ba(H_2PO_2)_2 + H_2SO_4 \longrightarrow BaSO_4 \downarrow + 2H_3PO_2$$

Properties. (*i*) It is a colourless crystalline monobasic acid, soluble in water. It forms only one series of salts.

$$H_3PO_2 \longrightarrow H^+ + H_2PO_2^-, \; Ka = 1 \times 10^{-2}$$

(*ii*) On heating strongly, it decomposes with the evolution of phosphine.

(*a*) At 415 K

$$3H_3PO_2 \xrightarrow{415\ K} \underset{\text{Phosphorus acid}}{2H_3PO_3} + PH_3 \uparrow$$

(*b*) At 435 K

$$2H_3PO_3 \xrightarrow{435\ K} \underset{\text{Orthophosphoric acid}}{H_3PO_3} + PH_3 \uparrow$$

(*iii*) It is a reducing agent and precipitates metals from solutions of their salts, *e.g.*, silver is precipitated from silver nitrate solution.

$$H_3PO_2 + 2H_2O + 4AgNO_3 \longrightarrow 4Ag \downarrow + 4HNO_3 + H_3PO_4$$

Similarly, mercuric chloride is reduced to mercurous chloride and subsequently to mercury as follows:

$$H_3PO_2 + 4HgCl_2 + 2H_2O \longrightarrow 2Hg_2Cl_2 + 4HCl + H_3PO_4$$

$$H_3PO_2 + 2Hg_2Cl_2 + 2H_2O \longrightarrow 4Hg + 4HCl + H_3PO_4$$

(*iv*) When treated with zinc and dilute HCl, it is reduced to phosphine

$$H_3PO_2 + 4[H] \xrightarrow{Zn/HCl} 2H_2O + PH_3 \uparrow$$

Structure. Strong reducing character indicates greater number of P – H bonds. Since it forms only one series of salts, it is monobasic acid. Hence the acid is given the following structure:

$$H-\overset{\overset{O}{\uparrow}}{\underset{\underset{H}{|}}{P}}-OH \quad \text{or} \quad H-\overset{\overset{O}{\|}}{\underset{\underset{H}{|}}{P}}-OH$$

Presence of PH_2 groups in hypophosphorus acid is confirmed by NMR studies.

Q. 24. Describe the methods of preparation, properties and structure of phosphorus acid (H_3PO_3). (*Nagpur, 2005*)

Ans. Preparation. (*i*) This acid is prepared by the action of cold water on phosphorus trioxide or trichloride.

$$P_2O_3 + 3H_2O \longrightarrow 2H_3PO_3$$

$$PCl_3 + 3H_2O \longrightarrow H_3PO_3 + 3HCl$$

The solution obtained is heated to 450 K. It is then allowed to cool, when the crystals of the acid separate out.

(*ii*) It can also be prepared by passing chlorine gas through water containing molten white phosphorus. The chloride formed gets hydrolysed to phosphorus acid.

$$\underset{\text{Molten}}{2P} + 3Cl_2 \longrightarrow 2PCl_3$$

$$PCl_3 + 3H_2O \longrightarrow H_3PO_3 + 3HCl$$

Properties. (*i*) Phosphorus acid is a colourless crystalline deliquescent solid, with garlic odour. It is freely soluble in water. It is a dibasic (diprotic) acid and forms two series of salts, *e.g.*, NaH_2PO_3 and Na_2HPO_3 but not of type Na_3PO_3.

The dissociation constants are as given below:

$$H_3PO_3 \rightleftharpoons H^+ + H_2PO_3^- \;\ldots\ldots\ldots\; K_{a_1} = 1.6\times10^{-2}$$

$$H_2PO_3^- \rightleftharpoons H^+ + HPO_3^{2-} \;\ldots\ldots\ldots\; K_{a_2} = 2\times10^{-7}$$

(*ii*) It decomposes at 475 K into phosphoric acid and phosphine.

$$4H_3PO_3 \longrightarrow 3H_3PO_4 + PH_3\uparrow$$

(*iii*) It is easily oxidized by air in the presence of iodine into phosphoric acid

$$3H_3PO_3 + O_2 \xrightarrow{\text{Iodine}} 2H_3PO_4$$

(*iv*) Phosphorus acid is a strong reducing agent. It precipitates copper, silver and gold from solutions of their salts and reduces mercuric salts to mercurous salts.

$$CuSO_4 + H_3PO_3 + H_2O \longrightarrow Cu\downarrow + H_3PO_4 + H_2SO_4$$

$$2AgNO_3 + H_3PO_3 + H_2O \longrightarrow 2Ag\downarrow + 2HNO_3 + H_3PO_4$$

$$2AuCl_3 + 3H_3PO_3 + 3H_2O \longrightarrow 2Au + 6HCl + 3H_3PO_4$$

$$2HgCl_2 + H_3PO_3 + H_2O \longrightarrow Hg_2Cl_2 + 2HCl + H_3PO_4$$

(*v*) Chlorine is reduced to HCl

$$Cl_2 + H_3PO_3 + H_2O \longrightarrow 2HCl + H_3PO_4$$

Structure. It is a dibasic acid and contains two –OH groups directly united to the central phosphorus atom

$$\begin{array}{c} O \\ \uparrow \\ H-P-OH \\ | \\ OH \end{array} \quad \text{or} \quad \begin{array}{c} O \\ \| \\ H-P-OH \\ | \\ OH \end{array}$$

This is in keeping with the failure of the molecule of phosphorus acid to release third H^+. The non-ionizable hydrogen is directly linked with phosphorus.

Q. 25. Describe the preparation, properties and structure of hypophosphoric acid ($H_4P_2O_6$).

Ans. Preparation. This acid is formed along with phosphoric and phosphoric acids, on treating phosphorus with a limited supply of moist air. The mixture of acids is then treated with lead acetate solution. Lead salt of hypophosphoric acid formed, being insoluble, gets precipitated which is then decomposed by passing H_2S to give free hypophosphoric acid.

$$Pb_2P_2O_6 + 2H_2S \longrightarrow 2PbS + H_4P_2O_6$$

Properties. (*i*) It is a colourless crystalline tetrabasic solid. It is hygroscopic in nature.

(*ii*) Action of heat on heating it above is m.p., it decomposes to give phosphorus acid and metaphosphoric acid.

$$H_2P_2O_6 \longrightarrow H_3PO_3 + HPO_3$$

(*iii*) It is readily hydrolyzed giving phosphorus acid and phosphoric acid.

$$H_4P_2O_6 + H_2O \longrightarrow H_3PO_3 + H_3PO_4$$

Structure. Since it is a tetrabasic acid, it indicates the presence of four – OH groups linked to the two phosphorus atoms. The presence of P – P bond is indicated by the X-ray diffraction.

$$\begin{array}{c} O \quad O \\ \uparrow \quad \uparrow \\ HO-P-P-OH \\ | \quad | \\ OH \ OH \end{array} \quad \text{or} \quad \begin{array}{c} O \quad O \\ \| \quad \| \\ HO-P-P-OH \\ | \quad | \\ OH \ OH \end{array}$$

Q. 26. Describe the preparation, properties and structure of orthophosphoric acid (H_3PO_4). (*Nagpur, 2005*)

Ans. Preparation. It is commonly known as phosphoric acid and is prepared by the following methods:

(*i*) By the hydrolysis of phosphorus pentachloride with excess of water.

$$PCl_5 + 4H_2O \longrightarrow H_3PO_4 + 5HCl$$

(*ii*) By dissolving phosphorus pentaoxide in hot water.

$$P_2O_5 + 3H_2O \longrightarrow 2H_3PO_4$$

(*iii*) *On commercial scale,* it is prepared by the action of dilute H_2SO_4 on phosphoric rocks or bone ash.

$$Ca_3(PO_4)_2 + 3H_2SO_4 \longrightarrow 3CaSO_4 + 2H_3PO_4$$

(*iv*) *Laboratory method.* It is prepared by heating phosphorus with conc. HNO_3

$$P + 5HNO_3 \longrightarrow H_3PO_4 + H_2O + 5NO_2$$

Properties. (*i*) Orthophosphoric acid is a colourless, syrupy liquid.

(*ii*) Orthophosphoric acid is tribasic acid and forms three series of salts, *e.g.*, sodium dihydrogen phosphate, NaH_2PO_4 (primary phosphate); disodium hydrogen phosphate, Na_2HPO_4 (secondary phosphate) and normal sodium phosphate Na_3PO_4 (tertiary phosphate).

It is a weak acid and undergoes ionization in three steps:

$$H_3PH_4 \rightleftharpoons H^- + H_2PO_4^- \text{.......} K_{a_1} = 7.5 \times 10^{-3}$$

$$H_2PO_4^- \rightleftharpoons H^+ + HPO_4^{2-} \text{.........} K_{a_2} = 6.2 \times 10^{-8}$$

$$HPO_4^{2-} \rightleftharpoons H^+ + PO_4^{3-} \text{.........} K_{a_3} = 1 \times 10^{-13}$$

A solution of Na_2HPO_4 is practically neutral to litmus, that of NaH_2PO_4 is acidic while that of Na_3PO_4 alkaline to litmus. Phosphoric acid can be titrated to the first stage of ionization against one equivalent of alkali in presence of methyl orange and against two equivalent (2nd state) in presence of phenolphthalein. The titration to the third stage is not feasible since the ionization is extremely small ($K_{a_3} = 1 \times 10^{-13}$)

(*iii*) It undergoes dehydration on heating

(*a*) **At 523 K** Pyrophosphoric acid is obtained.

$$2H_3PO_4 \xrightarrow{525\text{ K}} H_4P_2O_7 + H_2O$$

(*b*) **At 875K** It forms metaphosphoric acid.

$$H_3PO_4 \xrightarrow{875\text{ K}} HPO_3 + H_2O$$

Uses. (*i*) Phosphoric acid is used in place of sulphuric acid in certain reactions where the oxidizing property of latter causes difficulty.

(*ii*) Phosphoric acid is employed in the preparation of phosphates which are employed as fertilizers.

(*iii*) It is used in medicines.

Structure. Since the acid is tribasic, the structure of phosphoric acid is understood to be as follows:

```
        O                        O
        ↑                        ||
  HO — P — OH     or     HO — P — OH
        |                        |
        OH                       OH
```

The pure acid and its crystalline hydrated salts have tetrahedral PO_4 groups (shown in Figs. 11.10, 11.11) connected by hydrogen bonds.

The orthophosphate anions have the following structures:

The hydrogen bonding is responsible for the syrupy nature of phosphoric acid.

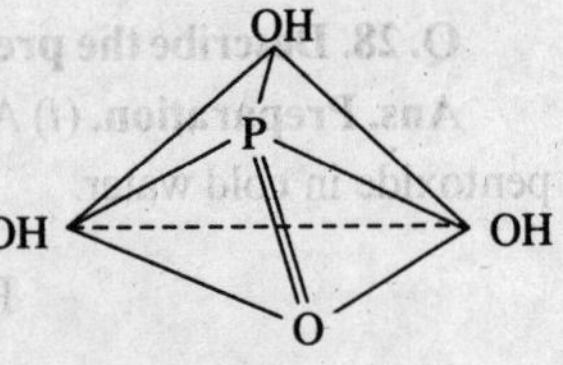

Fig. 11.10. Structure orthophosphoric acid.

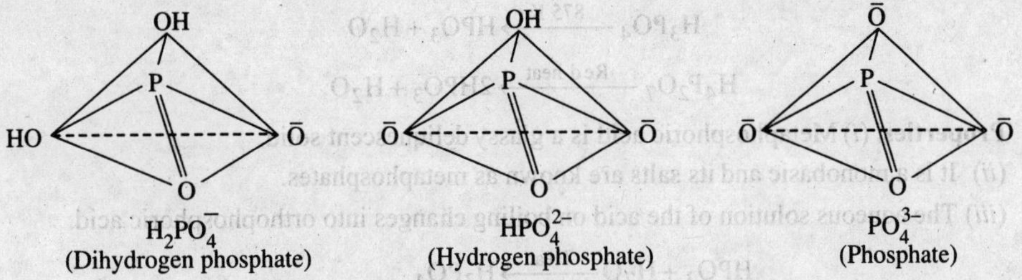

Fig. 11.11. Structures of $H_2PO_4^-$, HPO_4^{2-} and PO_4^{3-}.

Q. 27. Describe the preparation, properties and structure of pyrophosphoric acid.

(Kerala, 2000)

Ans. Preparation. This is obtained by heating orthophosphoric acid to 525 K.

$$2H_3PO_4 \xrightarrow{525\ K} H_4P_2O_7 + H_2O$$

Properties. (*i*) Acidic nature. It is a tetrabasic acid, but only two series of salts are commonly known, the normal salts *e.g.* $Na_4P_2O_7$ and diacid salts, *e.g.*, $Na_2H_2P_2O_7$.

(*ii*) Action with water. When boiled with water, it changes into orthophosphoric acid.

$$H_4P_2O_7 + H_2O \xrightarrow{Boil} 2H_3PO_4$$

(*iii*) Action of heat. On heating it to red-heat, it yields metaphosphoric acid.

$$H_4P_2O_7 \xrightarrow{Red\ heat} 2HPO_3 + H_2O$$

Structure. Pyrophosphoric acid, being tetrabasic, is given the following structure.

$$HO-\underset{OH}{\underset{|}{\overset{O}{\overset{\uparrow}{P}}}}-O-\underset{OH}{\underset{|}{\overset{O}{\overset{\uparrow}{P}}}}-OH \quad \text{or} \quad HO-\underset{OH}{\underset{|}{\overset{O}{\overset{\|}{P}}}}-O-\underset{OH}{\underset{|}{\overset{O}{\overset{\|}{P}}}}-OH$$

X-ray evidence shows that pyrophosphoric acid and its anion have a structure based on tetrahedral PO_4 units linked together through an oxygen atom common to both (Fig. 11.12).

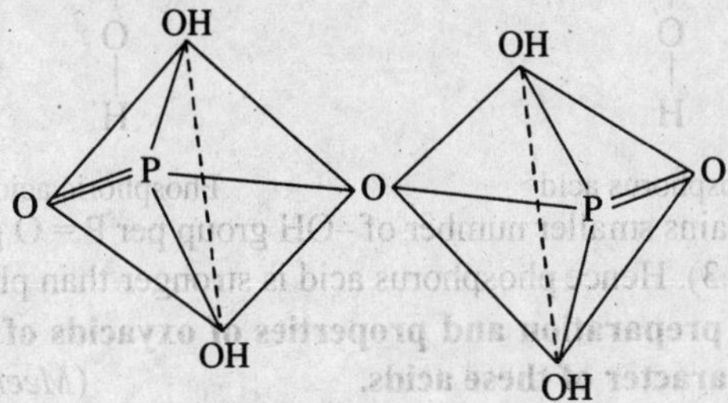

Fig. 11.12. Structure of pyrophosphoric acid $H_4P_2O_7$.

Q. 28. Describe the preparation, properties and structure of metaphosphoric acid.

Ans. Preparation. (*i*) An aqueous solution of the acid can be prepared by dissolving phosphorus pentoxide in cold water.

$$P_4O_{10} + 2H_2O \longrightarrow 4HPO_3$$

(*ii*) It can be prepared by heating orthophosphoric acid or pyrophosphoric acid.

$$H_3PO_4 \xrightarrow{875\ K} HPO_3 + H_2O$$

$$H_4P_2O_7 \xrightarrow{\text{Red heat}} 2HPO_3 + H_2O$$

Properties. (*i*) Metaphosphoric acid is a glassy deliquescent solid.

(*ii*) It is a monobasic and its salts are known as metaphosphates.

(*iii*) The aqueous solution of the acid on boiling changes into orthophosphoric acid.

$$HPO_3 + H_2O \xrightarrow{\text{Heat}} H_3PO_4$$

Structure. Since it is monobasic acid, it contains one – OH group linked to the central phosphorus atom. The acid is represented by the following structure.

$$O = \overset{\overset{O}{\uparrow}}{P} - OH$$

or

$$O = \overset{\overset{O}{\|}}{P} - OH$$

Metaphosphate ion (PO_3^-) does not exist as such but in polymerised forms as linear chain or as cyclic trimetaphosphate.

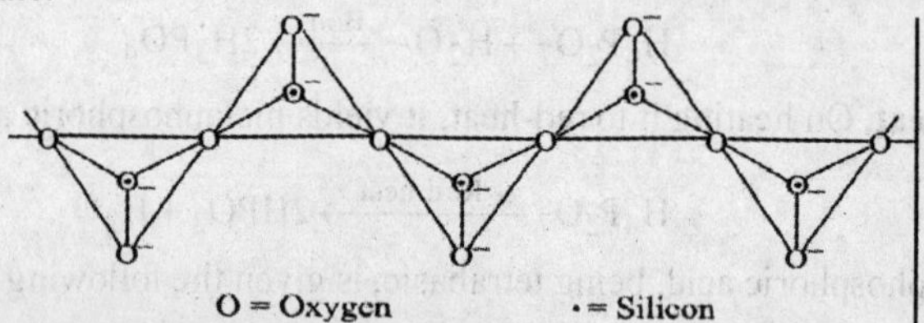

Q. 29. Which is stronger H_3PO_3 or H_3PO_4?

Ans. Strength of a phosphorus oxyacid depends upon the number of – OH groups per P = O group. It is the P = O group which induces polarisation and helps in the release of protons from – OH groups.

$$H - \overset{\overset{O}{\|}}{\underset{\underset{\underset{H}{|}}{O}}{\underset{|}{P}}} - O - H \qquad\qquad H - O - \overset{\overset{O}{\|}}{\underset{\underset{\underset{H}{|}}{O}}{\underset{|}{P}}} - O - H$$

Phosphorus acid Phosphoric acid

Phosphorus acid contains smaller number of –OH group per P = O group (*i.e.*, 2) compared to phosphoric acid (here it is 3). Hence phosphorus acid is stronger than phosphoric acid.

Q. 30. Describe the preparation and properties of oxyacids of phosphorus. Discuss the basicity and reducing character of these acids. (*Meerut, 2000; Lucknow, 2000*)

Ans. See Qs. 23, 24, 25, 26 and 27.

12 OXIDES AND OXY-ACIDS OF SULPHUR

OXIDES OF SULPHUR

Q. 1. Describe the preparation and properties of sulphur dioxide (SO_2).

Ans. Sulphur dioxide, an acidic oxide, can be prepared by the following methods:

(*a*) *Action of an acid on a metal sulphite, metabisulphite or thiosulphate.*

$$\underset{\text{Sulphite}}{SO_3^{2-}} + 2H^+ \longrightarrow H_2O + SO_2$$

$$\underset{\text{Metabisulphite}}{S_2O_5^{2-}} + 2H^+ \longrightarrow H_2O + 2SO_2$$

$$\underset{\text{Thiosulphate+}}{S_2O_3^{2-}} + 2H^+ \longrightarrow H_2O + SO_2 + S$$

(*b*) *Burning of sulphur and metal sulphide in air or oxygen.*

$$S + O_2 \longrightarrow SO_2$$

$$4FeS_2 + 11O_2 \longrightarrow 2Fe_2O_3 + 8SO_2$$

$$2ZnS + 3O_2 \longrightarrow 2ZnO + 2SO_2$$

(*c*) *Action of concentrated H_2SO_4 on copper, silver or mercury.*

$$Cu + 2H_2SO_4 \longrightarrow CuSO_4 + 2H_2O + SO_2$$

(*d*) *On an industrial scale.* It is produced as a by-product during the roasting of sulphide.

Properties. (*i*) Sulphur dioxide is a colourless gas with a sharp, choking smell of burnt sulphur.

(*ii*) It liquefies readily under reduced pressure or by cooling in a freezing mixture. The liquid is used as non-aqueous solvent for various reactions.

$$2SO_2 \rightleftharpoons SO^{2+} + SO_3^{2-}$$

$$SOCl_2 + Na_2SO_3 \xrightarrow[SO_2]{\text{Liquid}} 2NaCl + 2SO_2$$

(*iii*) SO_2 is soluble in water and is present in aqueous solution largely in hydrated form $SO_2 \cdot xH_2O$. The acidity of SO_2 solutions is due to the equilibrium:

$$SO_2 \cdot xH_2O \rightleftharpoons HSO_3^- (aq) + H_3O^+ + (x-2) H_2O$$

(*iv*) The aqueous solution of SO_2 is acidic and acts as a reducing agent. The acid so produced acts *in situ* and has never been isolated. The acid H_2SO_3 is dibasic in nature. The ionization constants of H_2SO_3 acid at 298 K are:

$$SO_2 + 2H_2O \rightleftharpoons HSO_3^- + H_3O^+ \quad (K_{a_1} = 1.7 \times 10^{-2})$$

$$HSO_3^- + H_2O \rightleftharpoons SO_3^{2-} + H_3O^+ \quad (K_{a_2} = 6.2 \times 10^{-8})$$

$$H_2SO_3 + H_2O \rightleftharpoons SO_4^{2-} + 4H^+ + 2e^- \quad (E° = +0.20 \text{ V})$$

Thus an equation solution of SO_2 liberates iodine from an iodate solution, decolourizes iodine and an acid solution of SO_2, reduces $KMnO_4$ to Mn^{2+} and ferric salts are reduced to ferrous state.

$$2KIO_3 + 5SO_2 + 4H_2O \longrightarrow I_2 + 3H_2SO_4 + 2KHSO_4$$

$$I_2 + SO_2 + 2H_2O \longrightarrow 2HI + H_2SO_4$$

$$2KMnO_4 + 5SO_2 + 2H_2O \longrightarrow K_2SO_4 + 2MnSO_4 + 2H_2SO_4$$

$$2FeCl_3 + SO_2 + 2H_2O \longrightarrow 2FeCl_2 + 2HCl + H_2SO_4$$

(*v*) In certain reactions SO_2 can act as an oxidising agent. For example, metals such as iron, tin and magnesium burn in SO_2 to give mixed sulphides and oxides. In very strong acidic solution SO_2 can oxidise Fe^{2+}, Sn^{2+}, Cu^+ or Hg_2^{2+}:

$$3Mg + SO_2 \longrightarrow MgS + 2MgO$$

$$4FeCl_2 + SO_2 + 4HCl \longrightarrow 4FeCl_3 + 2H_2O + S$$

$$6SnCl_2 + 2SO_2 + 8HCl \longrightarrow 5SnCl_4 + 4H_2O + SnS_2$$

SO_2 is reduced to S by H_2S also.

$$2H_2S + SO_2 \longrightarrow 2H_2O + 3S$$

(*vi*) It combines with heated sodium peroxide and lead peroxide to form sulphates.

$$Na_2O_2 + SO_2 \longrightarrow Na_2SO_4$$

$$PbO_2 + SO_2 \longrightarrow PbSO_4$$

(*vii*) The gas combines with chlorine in presence of sunlight giving sulphuryl chloride.

$$SO_2 + Cl_2 \longrightarrow SO_2Cl_2$$

(*viii*) Sulphur dioxide gives rise to sulphites and bisulphites, on passing through alkalies.

$$2KOH + SO_2 + H_2O \longrightarrow K_2SO_3 + 2H_2O$$

$$KOH + SO_2 + H_2O \longrightarrow KHSO_3 + H_2O$$

(*ix*) SO_2 possesses weak bleaching properties in the presence of moisture. The bleaching is due to the reduction of the colour of substances by nascent hydrogen. It can bleach straw, wool, paper, flowers etc.

Q. 2. Discuss the structure of sulphur dioxide molecule.

Ans. Sulphur dioxide is an angular molecule with an O – S – O angle of 119.5° and a high dipole moment of 1.6 D. The S – O bond length of 1.46 Å indicates some double bond character in the molecule. The structure of SO_2 involves the following resonance forms (Fig. 12.1):

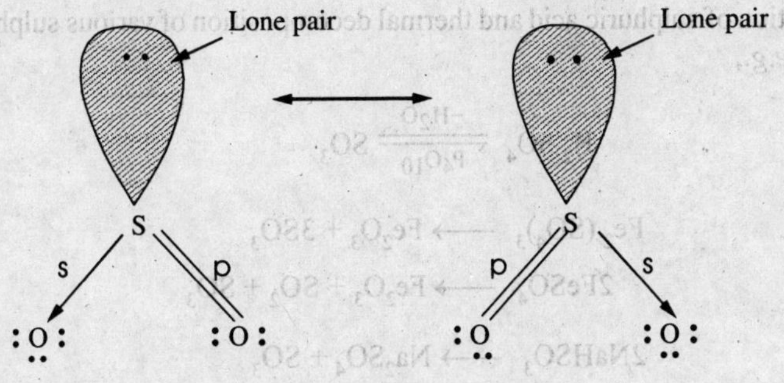

Fig. 12.1. Structure of sulphur dioxide.

Sulphur dioxide has a triangular structure. Configuration of sulphur in excited state is: $3s^2$, $3p_x^{\ 1}$, $3p_y^{\ 1}$, $3p_z^{\ 1}$, $3d^1$ (Fig. 12.2). Sulphur gives sp^2 hybridisation arising out of one lone pair of electrons of 3*s* orbital and two lone electrons of *p*-orbitals $3p_x^{\ 1}$ & $3p_y^{\ 1}$. These two half-filled sp^2 hybrid orbitals of sulphur each overlap with one of the *p*-orbitals of two oxygen atoms ($2s^2$, $2p_x^{\ 1}$, $2p_y^{\ 1}$, $2p_z^{\ 2}$), *i.e.,* $p - sp^2$ overlapping forming two σ (sigma) bonds. The bond angle of 119.5° is quite close to angle of 120° predicted by sp^2 hybridization. This is due to repulsion between the lone pair and bond pair which is more than that between two bond pairs.

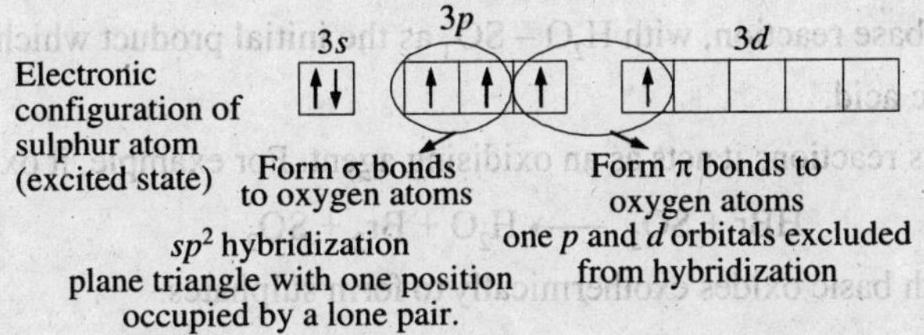

Fig. 12.2. Hybridisation scheme of sulphur in sulphur dioxids.

The sulphur atom has still two unhybridized half-filled orbitals, one *p*-orbital and one *d*-orbital. Similarly each oxygen atom has one half-filled *p*-orbital. So one *p*-orbital of the one oxygen atom overlaps with one *p*-orbital of sulphur atom forming $p\pi - p\pi$ bond by sidewise overlap. Similarly, one *p*-orbital of the second oxygen atom overlaps with the one half-filled *d*-orbital of sulphur atom forming $p\pi - d\pi$ bond. In this way *p* and *d*-orbitals of S and O get filled. It may be noted above that one oxygen atom is bonded by $sp^2 - p$ and $p\pi - p\pi$ bond and the other oxygen atom by $sp^2 - p$ and $p\pi - d\pi$ bond to the sulphur atom. This should give two different bonds but in fact both S – O bonds are identical. This is explained in terms of resonating structures of sulphur dioxide.

Q. 3. Describe the preparation and properties of sulphur trioxide.

Ans. Preparation. (*i*) Sulphur trioxide is usually prepared by the direct oxidation of sulphur or SO_2. Generally, the mixture of SO_2 and O_2 is bubbled through concentrated H_2SO_4 to dry the mixed gases and is passed over a heated platinised asbestos.

$$2SO_2 + O_2 \underset{773\text{ K}}{\overset{\text{Pt}}{\rightleftharpoons}} 2SO_3 \qquad (\Delta H = -193.2 \text{ kJ mol}^{-1})$$

Other catalysts such as V_2O_5 or NO can also be used.

(*iii*) Dehydration of sulphuric acid and thermal decomposition of various sulphates also produce sulphur trioxide, *e.g.*,

$$H_2SO_4 \xrightleftharpoons[P_4O_{10}]{-H_2O} SO_3$$

$$Fe_2(SO_4)_3 \longrightarrow Fe_2O_3 + 3SO_3$$

$$2FeSO_4 \longrightarrow Fe_2O_3 + SO_2 + SO_3$$

$$2NaHSO_3 \longrightarrow Na_2SO_4 + SO_3$$

Properties. (*i*) Sulphur trioxide dissolves in water with evolution of heat forming H_2SO_4.

$$SO_3 + H_2O \longrightarrow H_2SO_4 \qquad (\Delta \mathbf{H} = -156.7 \text{ kJ mol}^{-1})$$

(*ii*) It dissolves in concentrated H_2SO_4 producing oleum or pyrosulphuric acid which deposits as colourless crystals on cooling.

$$SO_3 + H_2SO_4 \longrightarrow H_2S_2O_7$$

(*iii*) It forms adducts with tertiary organic amines, $R_3N \rightarrow SO_3$, pyridine and with many other donor atoms and molecules.

(*iv*) In many of its reactions SO_3 behaves as a strong Lewis acid. The water reaction may also be taken as Lewis acid-base reaction, with $H_2O - SO_3$ as the initial product which later rearranges to $(HO)_2SO_2$, sulphuric acid.

(*v*) In some of its reactions it acts as an oxidising agent. For example, it oxidizes HBr to Br_2.

$$HBr + SO_3 \longrightarrow H_2O + Br_2 + SO_2$$

(*vi*) It reacts with basic oxides exothermically to form sulphates.

$$CaO + SO_3 \longrightarrow CaSO_4$$

$$BaO + SO_3 \longrightarrow BaSO_4$$

(*vii*) With hydrochloric acid it gives chlorosulphonic acid

$$SO_3 + HCl \longrightarrow ClSO_3H$$

Q. 4. Discuss the structure of sulphur trioxide. (*Kurukshetra, 2001*)

Ans. The structure of sulphur trioxide molecule is believed to be angular and it is a resonance hybrid of the following three structures as shown below in Fig. 12.3.

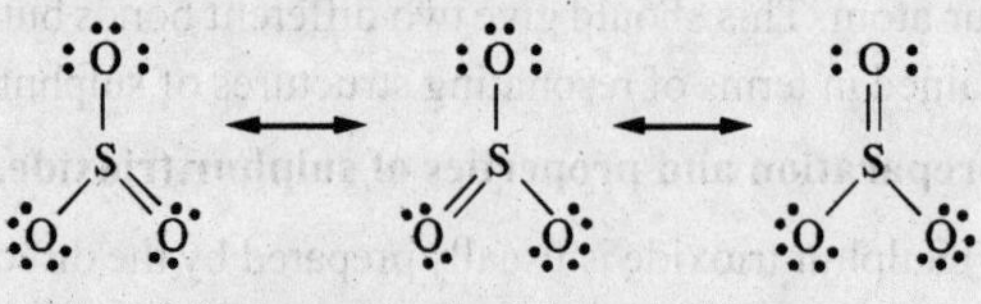

Fig. 12.3: Structure of sulphur trioxide.

Sulphur trioxide gas has a plane triangular structure. As shown in Fig. 12.4, sulphur atom in excited state has two half-filled orbitals. Each of its sp^2 hybrid orbitals overlaps with *p*-orbital of three oxygen atoms forming three σ (sigma) bonds. The angle is 120° as predicted. The S – O double bonds arise from one $p\pi - p\pi$ bond and two $p\pi - d\pi$ bonds between three oxygen atoms and sulphur. The length of S – O bond is 1.43 Å. The double bond character accounts for bond shortening.

Since all the three bonds involve similar bond pair of electrons, the bond angles are exactly identical, *i.e.*, 120°.

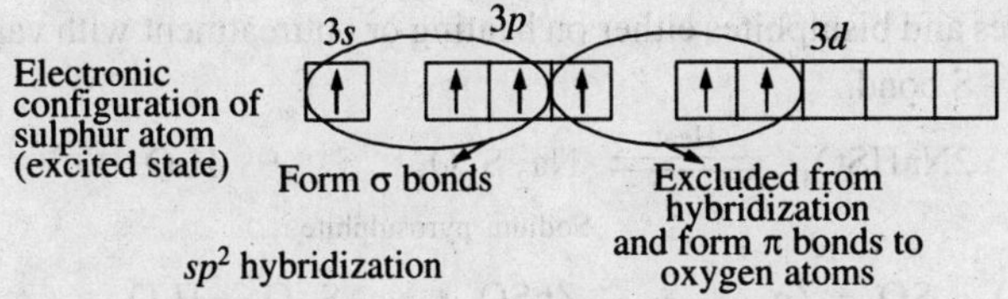

Fig. 12.4. Hybridisation scheme of sulphur in sulphur trioxide.

In the solid state, SO_3 polymerises to puckered rings or more stable layer structure as shown in Fig. 12.5.

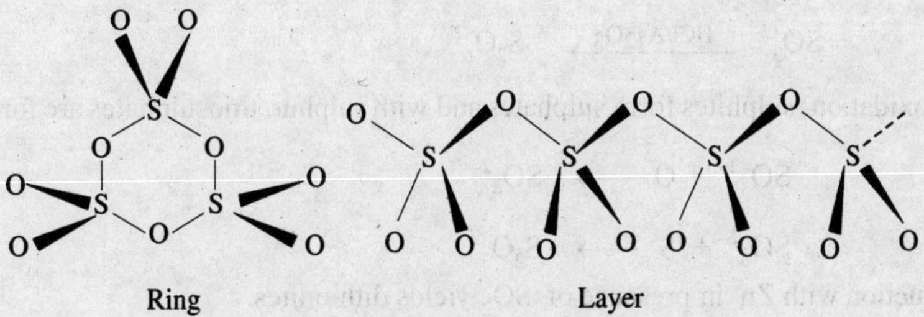

Fig. 12.5. Ring and layer structures of sulphur trioxide.

OXY-ACIDS OF SULPHUR

Q. 5. Give the method of preparation and properties of sulphurous acid (H_2SO_3).

(Delhi, 2003)

Ans. Preparation. Sulphur dioxide dissolves readily in water giving a solution containing sulphurous acid.

$$SO_2 + H_2O \longrightarrow H_2SO_3$$

The acid is dibasic and ionizes as

$$H_2SO_3 \rightleftharpoons HSO_3^- + H^+ \quad (Ka_1 = 1.3 \times 10^{-2})$$

$$HSO_3^- \rightleftharpoons SO_3^{2-} + H^+ \quad (Ka_2 = 10^{-7})$$

It forms two types of salts, bisulphites and sulphites. The bisulphite, HSO_3^- ion is present in greater concentration in the solution. Any attempt to isolate sulphurous acid in the free state makes two HSO_3^- ions condense with the elimination of water and a pyrosulphite ion is formed.

$$2NaHSO_3 \rightleftharpoons Na_2S_2O_5 + H_2O$$

or

$$2HSO_3^- \rightleftharpoons S_2O_5^{2-} + H_2O$$

Properties. Properties of sulphurous acid are explained by those of sulphites and bisulphites.

(*i*) Sulphites and bisulphites on treatment with acids give SO_2.

$$Na_2SO_3 + HCl \longrightarrow 2NaCl + H_2O + SO_2$$

(*ii*) Both sulphites and bisulphites either on heating or on treatment with various reagents form anions containing S – S bond.

$$2NaHSO_3 \xrightleftharpoons{\text{Heat}} \underset{\text{Sodium pyrosulphite}}{Na_2S_2O_5} + H_2O$$

$$2NaHSO_3 + SO_2 + Zn \longrightarrow ZnSO_3 + Na_2S_2O_4 + H_2O$$

$$Na_2SO_3 + S \longrightarrow Na_2S_2O_3$$

$$Na_2SO_3 + SO_2 \longrightarrow Na_2S_2O_5$$

$$SO_3^{2-} \xrightarrow{MnO_2} S_2O_6^{2-}$$

$$SO_3^{2-} \xrightarrow{HCl/As_2O_3} S_2O_6^{2-}$$

(*iii*) On oxidation, sulphites form sulphates and with sulphur, thiosulphates are formed.

$$SO_3^{2-} + O \longrightarrow SO_4^{2-}$$

$$SO_3^{2-} + S \longrightarrow S_2O_3^{2-}$$

(*iv*) Reduction with Zn in presence of SO_2 yields dithionites.

$$SO_3^{2-} + SO_2 \xrightarrow{Zn} S_2O_4^{2-}$$

(*v*) Moist SO_2 (or H_2SO_3) and sulphites behave as reducing agents.

$$SO_3^{2-} + H_2O \longrightarrow SO_4^{2-} + 2H^+ + 2e^-$$

$$Cl_2 + SO_3^{2-} + H_2O \longrightarrow 2Cl^- + SO_4^{2-} + 2H^+$$

$$2Fe^{3+} + SO_3^{2-} + H_2O \longrightarrow 2Fe^{2+} + SO_4^{2-} + 2H^+$$

$$Cr_2O_7^{2-} + 3SO_3^{2-} + 8H^+ \longrightarrow 2Cr^{3+} + 3SO_4^{2-} + 4H_2O$$

(*vi*) Sulphurous acid itself gets reduced in the presence of strong reducing agents, *e.g.*, iodides, hydrogen sulphide.

$$2I^- \longrightarrow I_2 + 2e^-$$

$$H_2SO_3 + 4H^+ + 4e^- \longrightarrow S + 3H_2O$$

Q. 6. Discuss the structure of sulphite ion (SO_3^{2-}).

Ans. The sulphite ion (SO_3^{2-}) is the hybrid of three resonating structures as shown below.

$$\left[O{=}S\begin{matrix}\nearrow O\\ \searrow O\end{matrix}\right]^{2-} \longleftrightarrow \left[O{\leftarrow}S\begin{matrix}\nearrow O\\ \searrow\!\!= O\end{matrix}\right]^{2-} \longleftrightarrow \left[O{\leftarrow}S\begin{matrix}= O\\ \searrow O\end{matrix}\right]^{2-}$$

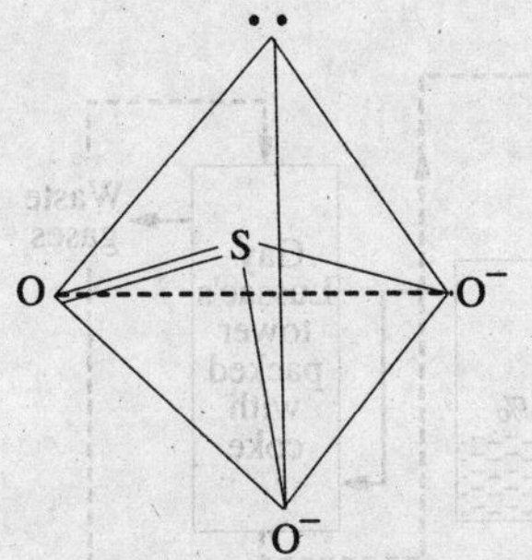

Fig. 12.6. Pyramidal structure of SO_3^{2-}.

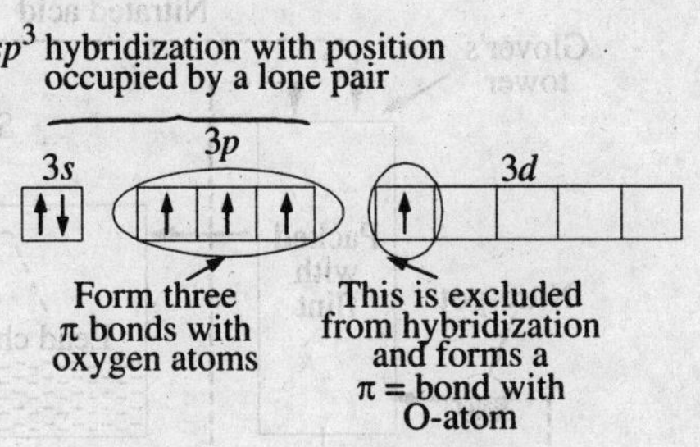

Fig. 12.7. Hybridisation scheme.

It has pyramidal structure (Fig. 12.6) involving sp^3 hybridisation in which one hybrid orbital is occupied by a lone pair ($3s^2$) and three sp^3 hybrid orbitals are half-filled. The unhybridized half-filled *d*-orbital forms a $p\pi - d\pi$ bond with oxygen atom. All the S – O bonds are identical (1.39 Å) which has been confirmed by X-ray analysis. Fig. 12.7 gives the hybridisation scheme of sulphur in SO_3^{2-}. Overlapping of various orbitals of S and O is explained in Fig. 12.8. Out of three oxygens in SO_3^{2-}, two may be visualised as having one extra electron, *i.e.*, one negative charge to account for two negative charges on sulphite ion. The lone pair distorts tetrahedral structure into a pyramidal one.

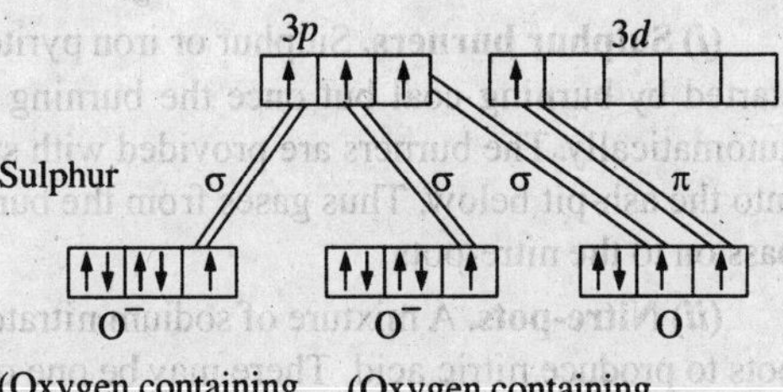

Fig. 12.8. Overlapping of orbitals in SO_3^{2-}.

Q. 7. Describe the manufacture of sulphuric acid by lead chamber process.

(Kerala, 2000)

Ans. Theory. This method involves oxidation of sulphur dioxide to sulphur trioxide in the presence of oxides of nitrogen as catalysts. Sulphur trioxide is then dissolved in water to form sulphuric acid. This process involves the following reactions:

(*i*) Sulphur dioxide is obtained by burning sulphur or iron pyrites.

$$S + O_2 \longrightarrow SO_2\uparrow$$

$$4FeS_2 + 11O_2 \longrightarrow 2Fe_2O_3 + 8SO_2\uparrow$$

(*ii*) Sulphur dioxide is next made to react with nitrogen dioxide to give sulphur trioxide and nitric oxide. Nitric oxide then takes up oxygen from the air and forms nitrogen dioxide which is again used for the oxidation of more of sulphur dioxide.

$$SO_2 + NO_2 \longrightarrow SO_3 + NO$$

$$2NO + O_2 \longrightarrow 2NO_2$$

The process is repeated. The oxides of nitrogen simply act as a carrier of oxygen from air to sulphur dioxide.

(*iii*) Sulphur trioxide then dissolves in water to produce sulphuric acid.

$$SO_3 + H_2O \longrightarrow H_2SO_4$$

Plant. The plant for the manufacture of sulphuric acid by chamber process is shown in Fig. 12.9. The various parts of the plant and their functions are described as under.

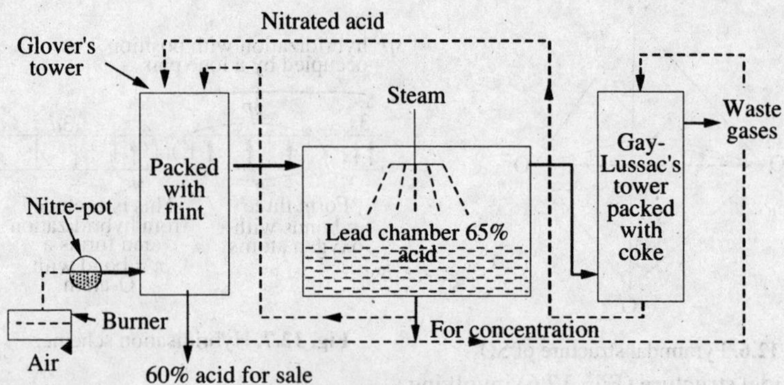

Fig. 12.9. Lead-chamber process.

***(i)* Sulphur burners.** Sulphur or iron pyrites are burnt here to produce sulphur dioxide. Fire is started by burning coal but once the burning of sulphur or the pyrites has started, it continues automatically. The burners are provided with stirrers operated mechanically, to push the burnt ash into the ash-pit below. Thus gases from the burners, containing sulphur dioxide and excess of air, pass on to the nitre-pots.

***(ii)* Nitre-pots.** A mixture of sodium nitrate and concentrated sulphuric acid is heated in these pots to produce nitric acid. There may be one or more nitre-pots. The gases while passing through them take up vapours of nitric acid which are, subsequently, reduced to the oxides of nitrogen.

$$NaNO_2 + H_2SO_4 \longrightarrow NaHSO_4 + HNO_3$$

$$4HNO_2 + SO_2 \longrightarrow H_2SO_4 + 6NO_2$$

Nitric oxide can also be produced by the oxidation of ammonia in the presence of platinum gauze at 1073 K,

$$4NH_3 + 5O_2 \longrightarrow 4NO\uparrow + 6H_2O$$

Nitric oxide (NO) is oxidized to nitrogen dioxide (NO_2) by oxygen contained in air.

$$2NO + O_2 \longrightarrow 2NO_2$$

***(iii)* Glover's tower.** The mixture of gases from nitre-pots enters the Glover's tower from the bottom. The tower is made of lead and lined inside with acid-resisting bricks. It is packed with flintstone. Dilute acid from chambers and nitrated sulphuric acid from the Gay-Lussac's tower flow through Glover's tower. The functions of Glover's tower are:

(*a*) To cool down the hot gases from the burners and the nitre-pots from about 673 K to about 343 K.

(*b*) To concentrate the chamber acid from 60-70% to about 80% by using the heat of the upgoing gases.

(*c*) To recover the oxides of nitrogen from the nitrated sulphuric acid.

$$2NO \cdot HSO_4 + H_2O \longrightarrow 2H_2SO_4 + NO_2\uparrow + NO\uparrow$$

(*d*) To provide additional space for the conversion of sulphur dioxide into sulphuric acid.

It has been noticed that the reaction here is much more vigorous than even in the lead chambers. The acid collecting at the base of the tower is thus concentrated. A part of it is used in the Gay-Lussac's tower for the recovery of oxides of nitrogen from the waste gases.

***(iv)* Lead chambers.** The gases from the Glover's tower enter the lead chambers, which are generally three in number (only one chamber is shown in Fig. 12.9). The gases pass through them slowly and get sufficent time to react. The chambers are made of lead because of its resistance to the action of sulphuric acid of 60% –70% concentration at the temperature prevailing in the chambers. The chambers may be rectangular or cylindrical and are supported on wooden frames. Each chamber is provided at the top with an inlet for introducing steam or for spraying water. It is here that oxidation of sulphur dioxide to sulphur trioxide in the presence of oxides of nitrogen largely takes place and sulphur trioxide dissolves in water to form sulphuric acid as described in the theory above. The acid formed collects at the base of chambers from where it is drawn out from time to time and is 60-70% strong. A part of the acid is sent to the Glover's tower while the rest goes for concentration.

(*v*) Gay-Lussac's tower. The outgoing gases from the chambers mostly contain air and oxides of nitrogen. The latter are recovered for use again before the waste gases are let off through the chimney. This is done by dissolving them in concentrated sulphuric acid in the Gay-Lussac's tower. The tower is lined with lead and packed with coke. From the top, concentrated sulphuric acid is sprayed slowly. It dissolves the oxides of nitrogen forming nitrated sulphuric acid.

$$NO + NO_2 + 2H_2SO_4 \longrightarrow \underset{\text{Nitrated sulphuric acid}}{2NO \cdot HSO_4} + H_2O$$

The acid coming out of the tower at the base is carried to the top of the Glover's tower for denitration.

Purification of the chamber acid. The chamber acid contains (*i*) arsenious oxide (from pyrites), (*ii*) lead sulphate (from lead chamber by the action of sulphuric acid) and (*iii*) oxides of nitrogen as impurities. These can be removed as follows:

(*i*) **Removal of lead sulphate.** On diluting the acid with water, lead sulphate separates out as precipitate and can be filtered off.

(*ii*) **Removal of arsenious oxide.** Hydrogen sulphide is passed through the acid when arsenious oxide is precipitated as arsenious sulphide. It is removed by filtration.

$$As_2O_3 + 3H_2S \longrightarrow As_2S_3 \downarrow + 3H_2O$$

(*iii*) **Removal of oxides of nitrogen.** The filtrate is distilled with ammonium sulphate when oxides of nitrogen are eliminated as nitrogen.

$$(NH_4)_2\,SO_4 + NO + NO_2 \longrightarrow 2N_2\uparrow + H_2SO_4 + 3H_2O$$

Concentration of the chamber acid. Chamber acid which contains about 65% of sulphuric acid is concentrated by the following two methods:

(*i*) **Cascade process.** In this case, a number of silica vessels are arranged one above the other on the steps of acid-proof bricks. Heating is done by gases from a furnace below. The dilute acid flows continuously into the vessel at the top and then it overflows into the next vessel below and so on. The acid becomes more and more concentrated as it flows down and the temperature continuously increases. Concentrated acid containing 95% of sulphuric acid is obtained from the lowest vessel.

(*ii*) **Gaillard tower.** Here hot gases are passed near the bottom of a tower made of acid-proof bricks. Dilute acid is sprayed from the top of the tower and it gets concentrated to about 95 per cent by the hot gases.

Q. 8. Describe Contact process for the manufacture of sulphuric acid.

Ans. Theory. Here sulphur dioxide produced by burning sulphur or iron pyrites is oxidized to sulphur trioxide at a temperature of 673-723 K in the presence of a catalyst.

Sulphur trioxide is next dissolved in concentrated sulphuric acid to give oleum which is then diluted with required quantity of water to produce sulphuric acid.

The reactions involved in the process are:

$$(i) \quad S + O_2 \longrightarrow SO_2 \uparrow$$

$$4FeS_2 + 11\,O_2 \longrightarrow 2Fe_2O + 8\,SO_2 \uparrow$$

$$(ii) \quad 2SO_2 + O_2 \rightleftharpoons 2SO_3 + 193.2 \text{ kJ}$$

$$(iii) \quad SO_3 + H_2SO_4 \longrightarrow H_2S_2O_7 \text{ (oleum)}$$

$$H_2S_2O_7 + H_2O \longrightarrow 2H_2SO_4$$

The reaction (*ii*), *i.e.*, the oxidation of sulphuric dioxide to trioxide, is reversible and exothermic and is accompanied by decrease in volume.

The following conditions have been found to favour the forward reaction:

(*i*) An excess of oxygen should be used.

(*ii*) *High pressure.* It has been found that a pressure of about two atmospheres is quite sufficient.

(*iii*) *Low temperature.* The optimum temperature has been found to be between 673 K and 723 K. At this temperature, the yield of sulphur trioxide is maximum.

(*iv*) Originally finely-divided platinum in the form of platinized asbestos was used, but now it has been replaced by the much cheaper vanadium pentoxide (V_2O_5). Vanadium pentoxide is not only cheaper than platinum but is also not poisoned by the impurities present in the reacting gases.

(*v*) The gases should be completely pure and dry. Sulphur dioxide produced by burning iron pyrites contains arsenic as an impurity which, even if present in traces, poisons the platinum catalyst.

Plant. **The plant used for the manufacture of sulphuric acid by the Contact process is shown in Fig. 12.10. The various parts of the plant and their functions are:**

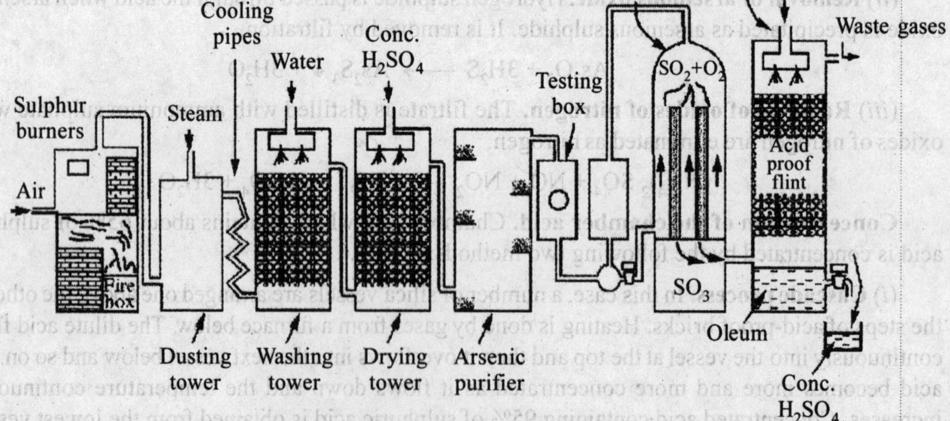

Fig. 12.10. Plant for the manufacture of sulphuric acid by Contact process.

(*i*) Burners. Here sulphur or iron pyrites are burnt in a free excess of air to produce sulphur dioxide.

$$S + O_2 \longrightarrow SO_2$$

$$4FeS_2 + 11O_2 \longrightarrow 2Fe_2O_3 + 8SO_2$$

(*ii*) Purification unit. The mixture of gases from the pyrites burners contains dust particles and arsenic oxide as impurities. These have to be removed, otherwise they will poison the catalyst. They are removed in the purification unit which consists of the following parts:

(*a*) *Dust chamber or Dust remover.* Here steam is injected from the top of the chamber which helps the dust particles to settle down.

(*b*) *Cooling pipes.* The gases are cooled to about 373 K by passing them through pipes cooled by air.

(*c*) *Washing tower.* Here the gases rise up a chamber and are made free from acid mist and any soluble impurities by a descending stream of water.

(*d*) *Drying tower.* The tower is packed with flint or coke and concentrated sulphuric acid is sprayed from the top. It comes in contact with the ascending gases and removes water vapours present in them. The gases leaving the tower at the top are free from moisture.

(*e*) *Arsenic purifier.* This is a small chamber fitted with shelves containing freshly-precipitated ferric hydroxide. Here arsenic oxide present in the gases is absorbed by ferric hydroxide.

(*iii*) Testing box. Pure and dry gases are then passed through the testing box to make sure that the purification of the gases is complete. A strong beam of light is sent into the box and is viewed at right angles. Fine particles of impurities, if present, will scatter light and make its path visible (Tyndall effect). If the gases are found to be pure, they are passed on to the contact chamber.

(*iv*) Contact chamber or converter. Contact chambers or converters of various types are used depending upon the nature of the catalyst employed. Badische converter consists of a cylindrical iron chamber fitted with vertical iron pipes (Fig. 12.11). Each pipe is packed with platinized asbestos or vanadium pentoxide (V_2O_5).

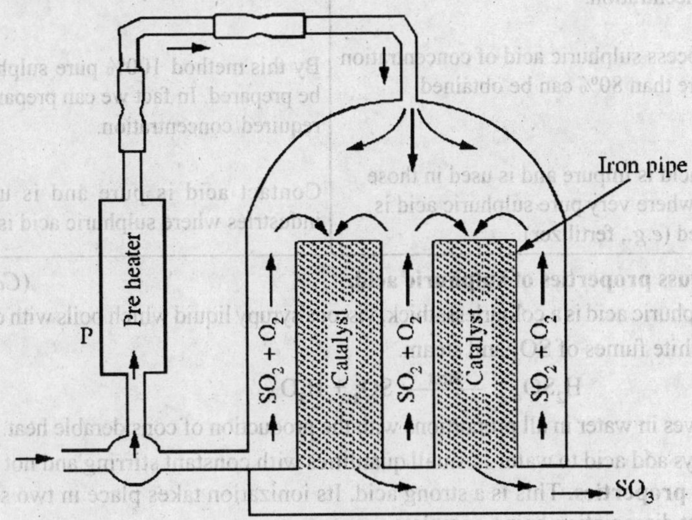

Fig. 12.11. Contact chamber.

The purified gases pass through a preheater, P, and get heated to about 723 K and then enter the converter. The gases first pass upwards around the vertical pipes (as shown by arrows) and then downward through the catalyst. As the gases pass through, the catalyst, sulphur dioxide, is oxidized to sulphur trioxide by the oxygen of the air.

$$2SO_2 + O_2 \longrightarrow 2SO_3 + 193.2 \text{ kJ}$$

Since the reaction is exothermic, no heating is required in the preheater once the oxidation reaction has commenced. The gases are then passed directly into the converter.

(*v*) Absorption tower. The absorption tower consists of a cylindrical tall tower made of steel. It is packed with acid-proof flint. Sulphur trioxide from the contact chamber is passed up in the absorption tower where a stream of concentrated sulphuric acid is sprayed from the top. Sulphur trioxide is absorbed by sulphuric acid when oleum is obtained. This is then diluted with water to obtain acid of a desired concentration, say 100%.

$$SO_3 + H_2SO_4 \longrightarrow H_2S_2O_7 \text{ (oleum)}$$

$$H_2S_2O_7 + H_2O \longrightarrow 2H_2SO_4$$

Water is not used for absorption directly, since it forms a dense fog of sulphuric acid which is difficult to condense and causes great inconvenience to workers.

Q. 9. Draw a comparison between the Chamber and the Contact processes for the manufacture of sulphuric acid. *(Calcutta, 2000)*

Ans. Various points of comparison between the chamber and contact processes are given below:

S.No.	*Chamber Process*	*Contact Process*
1.	Gaseous oxides of nitrogen act as catalyst.	Solid vanadium pentoxide or platinized asbestos is used as catalyst.
2.	$SO_2 - O_2$ mixture needs no purification.	$SO_2 - O_2$ mixture needs thorough purification.
3.	Chamber plant needs a large space.	Contact plant needs a small space.
4.	The acid obtained is dilute (65%) and needs further concentration.	Acid of any desired concentration is obtained.
5.	By this process sulphuric acid of concentration of not more than 80% can be obtained	By this method 100% pure sulphuric acid can be prepared. In fact we can prepare acid of any required concentration.
6.	Chamber acid is impure and is used in those industries where very pure sulphuric acid is not required (*e.g.*, fertilizer)	Contact acid is pure and is useful for all industries where sulphuric acid is required.

Q. 10. Discuss properties of sulphuric acid. *(Calcutta, 2000)*

Ans. (*i*) Sulphuric acid is a colourless, thick, viscous syrupy liquid which boils with decomposition to form dense white fumes of SO_3 and steam.

$$H_2SO_4 \xrightarrow{\text{Heat}} SO_3 + H_2O$$

(*ii*) It dissolves in water in all proportions with the production of considerable heat. For preparing dilute acid always add acid to water in small quantities with constant stirring and not water to acid.

(*iii*) **Acidic properties.** This is a strong acid. Its ionization takes place in two stays, *i.e.*, it is disbasic: the first dissociation being complete.

$$H_2SO_4 + H_2O \longrightarrow H_3O^+ + HSO_4^- \quad K_{a_1} = 10^3$$

The second hydrogen ion is less easily removed and dissociation takes place to the extent of about 10%.

$$HSO_4^- + H_2O \longrightarrow H_3O^+ + SO_4^{2-} \quad K_{a_2} = 10^{-8}$$

(*iv*) **Oxidizing properties.** Concentrated H_2SO_4 functions as an oxidizing agent.

$$SO_4^{2-} + 4H^+ + 2I^- \longrightarrow 2H_2O + SO_2 + I_2$$

$$C + 2H_2SO_4 \longrightarrow 2H_2O + CO_2 + 2SO_2$$

$$S + 2H_2SO_4 \longrightarrow 2H_2O + 3SO_2$$

$$3H_2S + H_2SO_4 \longrightarrow 4H_2O + 4S$$

$$4P + 8H_2SO_4 \longrightarrow 4H_3PO_4 + 2H_2O + 7SO_2 + S$$

(*v*) **Action on metals.** With metals, it is possible to obtain SO_2, H_2S and S in addition to the metallic sulphate; further H_2S on reacting with metal sulphate can give metal sulphide.

(*a*) Most metals are attacked by concentrated H_2SO_4. With metals low down in the electrochemical series, *e.g.*, copper, SO_2, is the reduction product while H_2S is formed with more reactive metals like zinc.

$$Pb + 2H_2SO_4 \longrightarrow PbSO_4 + 2H_2O + SO_2$$

$$Cu + 2H_2SO_4 \longrightarrow CuSO_4 + 2H_2O + SO_2$$

$$\underset{98\%\ \text{acid}}{Zn + 2H_2SO_4} \longrightarrow ZnSO_4 + 2H_2O + SO_2$$

$$\underset{90\%\ \text{acid}}{Zn + 5H_2SO_4} \longrightarrow ZnSO_4 + 4H_2O + H_2S$$

Gold and platinum are not attacked. Lead reacts only superficially on account of the insolubility of lead sulphate formed by the reaction on its surface.

(*b*) Copper, silver, lead, gold, platinum, bismuth and mercury with $E°$ value from + 0.34 V to + 0.80 V are not attacked by dilute H_2SO_4.

(*c*) Metals with $E°$ values above hydrogen liberate hydrogen from the dilute acid, *e.g.*, Mg, Zn, iron and Al.

$$Mg + 2H^+ \longrightarrow Mg^{2+} + H_2$$

(*vi*) **Displacement reactions.** Being relatively less volatile, concentrated H_2SO_4 will displace other acids, such as nitric and hydrochloric acids, from their salts.

$$NO_3^- + H_2SO_4 \rightleftharpoons HSO_4^- + HNO_3$$

$$Cl^- + H_2SO_4 \rightleftharpoons HCl + HSO_4^-$$

$$NaNO_3 + H_2SO_4 \rightleftharpoons NaHSO_4 + HNO_3$$

$$NaCl + H_2SO_4 \rightleftharpoons NaHSO_4 + HCl$$

(*vii*) **Dehydration.** It is used for drying all gases except ammonia and hydrogen sulphide with which it reacts. For the same reason, it is also used as a strong dehydrating agent. Thus:

(*a*) Ethyl alcohol gives ethylene.

$$C_2H_5OH \xrightarrow[H_2SO_4]{170°C} \underset{\text{Ethylene}}{C_2H_4} + H_2O$$

(*b*) Sugar (sucrose) is charred to give sugar charcoal.

$$C_{12}H_{22}O_{11} \xrightarrow{H_2SO_4} 12C + 11H_2O$$

(*c*) D.D.T. is obtained by dehydrating a mixture of trichloroacetaldehyde (chloral) and chlorobenzene.

$$\underset{\text{Chlorobenzene}}{2C_6H_5Cl} + \underset{\text{Chloral}}{CCl_3CHO} \xrightarrow{H_2SO_4} \underset{\text{D.D.T.}}{Cl-C_6H_4-\overset{\overset{CCl_3}{|}}{\underset{\underset{H}{|}}{C}}-C_6H_4-Cl} + H_2O$$

(*d*) Oxalic and formic acids lose water forming different products.

$$\underset{\text{Oxalic acid}}{\begin{matrix}COOH\\ |\\ COOH\end{matrix}} \xrightarrow{H_2SO_4} CO\uparrow + CO_2\uparrow + H_2O$$

$$\underset{\text{Formic acid}}{HCOOH} \xrightarrow{H_2SO_5} CO\uparrow + H_2O$$

(*viii*) **Addition reaction with organic compounds.** Alkenes undergo addition reactions. Addition products on dilution and heating give alcohols.

$$\underset{\text{Butene}}{CH_3-CH_2-CH{=}CH_2} + H_2SO_4 \longrightarrow CH_3-CH_2-\underset{\underset{OSO_3H}{|}}{CH}-CH_3$$

$$\xrightarrow{H_2O} \underset{\text{Butanol}}{CH_3-CH_2-\underset{\underset{OH}{|}}{CH}-CH_2} + H_2SO_4$$

(*ix*) **Action with phosphorus pentachloride.** The two – OH groups of sulphuric acid may be replaced by phosphorus pentachloride.

$$SO_2\begin{matrix}\diagup OH\\ \diagdown OH\end{matrix} + PCl_5 \longrightarrow SO_2\begin{matrix}\diagup OH\\ \diagdown Cl\end{matrix} + POCl_3 + HCl$$

$$SO_2\begin{matrix}\diagup OH\\ \diagdown Cl\end{matrix} + PCl_5 \longrightarrow SO_2\begin{matrix}\diagup Cl\\ \diagdown Cl\end{matrix} + POCl_3 + HCl$$

(*x*) **Sulphonation.** Sulphuric acid is employed as a sulphonating agent for organic compounds:

$$\underset{\text{Benzene}}{C_6H_6} + SO_2(OH)_2 \longrightarrow \underset{\text{Benzene Sulphonic acid}}{C_6H_6-SO_2OH} + H_2O$$

(*xi*) **Action with potassium permanganate and potassium dichromate.** Conc. sulphuric acid on heating with these compounds gives oxygen.

$$2KMnO_4 + 6H_2SO_4 \longrightarrow 2K_2SO_4 + 4MnSO_4 + 6H_2O + 5O_2\uparrow$$

$$2K_2Cr_2O_7 + 8H_2SO_4 \longrightarrow 2K_2SO_4 + 2Cr_2(SO_4)_3 + 8H_2O + 3O_2\uparrow$$

(*xii*) **Action with potassium chlorate.** On careful warming with potassium chlorate, conc. H_2SO_4 gives chlorine dioxide, a yellow-coloured explosive gas.

$$2KClO_3 + 3H_2SO_4 \longrightarrow 2KHSO_4 + \underset{\text{Chlorine dioxide}}{2ClO_2} + \underset{\text{Perchloric acid}}{HClO_4} + H_2O$$

(*xiii*) **Action with potassium ferrocyanide.** On heating with concentrated sulphuric acid, potassium ferrocyanide gives carbon monoxide.

$$K_4[Fe(CN)_6] + 3H_2SO_4 \longrightarrow 2K_2SO_4 + FeSO_4 + 6HCN$$

$$HCN + 2H_2O \longrightarrow HCOOH + NH_3] \times 6$$

$$HCOOH \longrightarrow H_2O + CO] \times 6$$

$$2NH_3 + H_2SO_4 \longrightarrow (NH_4)_2SO_4] \times 3$$

$$K_4[Fe(CN)_6] + 6H_2SO_4 + 6H_2O \longrightarrow 2K_2SO_4 + FeSO_4 + 3(NH_4)_2SO_4 + 6CO\uparrow$$

(*xiv*) **Action with dioxides and peroxides.** Dioxides on heating with conc. H_2SO_4 give oxygen and are reduced while peroxides give hydrogen peroxides.

$$\underset{\text{Manganese dioxide}}{2MnO_2} + 2H_2SO_4 \longrightarrow 2MnSO_4 + 2H_2O + O_2\uparrow$$

$$\underset{\text{Barium peroxide}}{BaO_2} + H_2SO_4 \longrightarrow BaSO_4 + H_2O_2\uparrow$$

Q. 11. List important uses of sulphuric acid.

Ans. Some of the most important uses of the sulphuric acid are:

(*i*) As a laboratory reagent and as dehydrating agent for drying purpose.

(*ii*) In the manufacture of fertilizers like ammonium sulphate and calcium superphosphate.

(*iii*) In the manufacture of dyes and drugs.

(*iv*) In petroleum refining or removing objectionable sulphur and tarry matter.

(*v*) In storage batteries.

(*vi*) In the manufacture of paints, explosives and textiles.

(*vii*) In the manufacture of cellulose products of cotton, wool, linen, sugar, photographic film, rubber etc.

(*viii*) In the manufacture of chemicals such as hydrochloric acid, nitric acid, alums, phosphoric and sulphates etc.

(*ix*) In metallurgical industries.

(*x*) *As a solvent.* Its solvent properties are due to its high dielectric constant, the polarity of its molecules and its ability to form strong hydrogen bonds.

Anhydrous H_2SO_4, like water or liquid SO_2, shows autoprotolysis:

$$2H_2SO_4 \rightleftharpoons H_3SO_4^+ + HSO_4^-$$

Water, nitric acid, orthophospheric acid, alcohols, carboxylic acids, ketones, amines, etc., are protonated in sulphuric acid and act as bases.

$$H_2O + H_2SO_4 \rightleftharpoons H_3O^+ + HSO_4^-$$

$$H_3PO_4 + H_2SO_4 \rightleftharpoons P(OH)_4^+ + HSO_4^-$$

$$(CH_3)_2\,CO + H_2SO_4 \rightleftharpoons (CH_3)_2\,C^+\,OH + HSO_4^-$$

$$HNO_3 + 2H_2\,SO_4 \rightleftharpoons NO_2^+ + H_3O^+ + 2HSO_4^-$$

Q. 13. Discuss the structure of sulphuric acid (H_2SO_4).

Ans. Sulphuric acid can be represented by the following structures.

```
        O                      O
        ↑                      ‖
HO — S — OH      or     HO — S — OH
        ↓                      ‖
        O                      O
```

X-ray analysis has proved that the sulphate, SO_4^{2-} ion is tetrahedral which is as expected from the orbitals used. Four oxygen atoms are located at the four corners of a regular tetrahedron, with sulphur at the centre (Fig. 12.12).

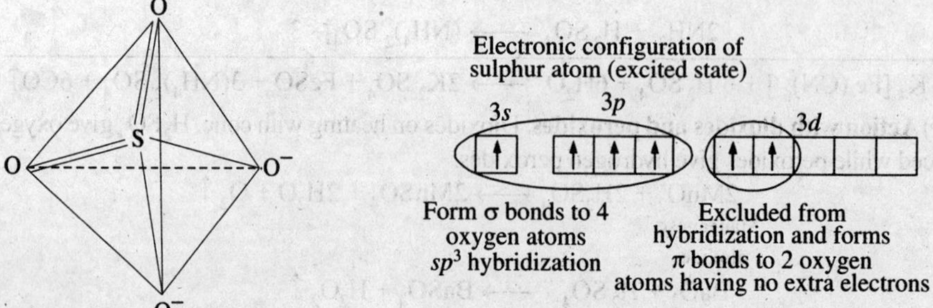

Fig. 12.12. Structure of H_2SO_4.

Fig. 12.13. Hybridisation scheme of sulphur in SO_4^{2-}.

The observed S – O bond distance is 1.44Å whereas the radius sum (between single and double bond distances) is 1.78 Å (calculated). On the other hand, S = O would require a bond distance of 1.57 Å. This indicates that S – O bond exists in resonance with S = O. According to molecular orbital theory, it involves sp^3 hybridization of four half-filled orbitals, one 3*s*- and three 3*p*-orbitals of S atom (Fig. 12.13).

They give four $p - sp^3$ sigma (σ) bonds with each of the four oxygen atoms. The two oxygen atoms are bonded by double bonds to sulphur. These are the π bonds formed by two half-filled *d*-orbitals of *S* atom with two oxygen atoms having no extra electron.

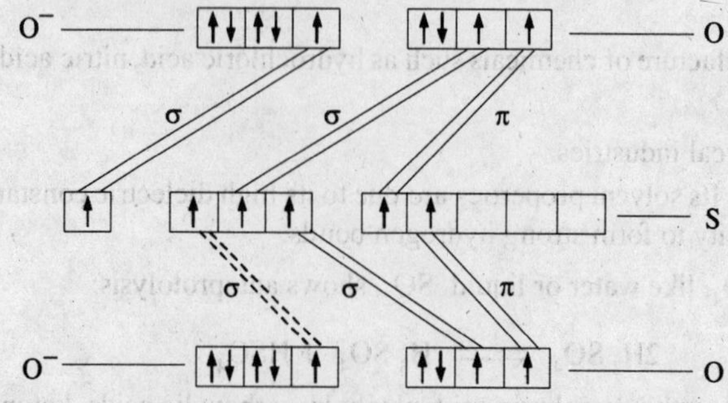

Fig. 12.14. Overlapping of orbitals of sulphur and oxygen in SO_4^{2-}.

It may be noted that out of four oxygens in SO_4^{2-}, two oxygen carry one extra electron each to account for the double negative charge on SO_4^{2-}.

Q. 13. Describe the preparation, properties and structure of peroxomonosulphuric acid, H_2SO_5 (Caro's acid).

(*M. Dayanand, 2000; Awadh, 2000; Calcutta, 2000; Guwahati, 2002; Kurukshetra, 2002; Rohilkhand, 2003*)

Ans. Preparation. (*i*) Caro's acid is also obtained by the action of conc. H_2O_2 on sulphuric acid or chloro-sulphuric acid (1:1).

$$H_2O_2 + H_2SO_4 \longrightarrow HOOSO_2OH + H_2O$$
$$H_2O_2 + HSO_3Cl \longrightarrow \underset{\text{Caro's acid}}{HOOSO_2OH} + HCl$$

(*ii*) Peroxomonosulphuric acid is obtained by hydrolysis of peroxodisulphuric acid (Marshall's acid).

$$HO-\overset{O\uparrow}{\underset{O\downarrow}{S}}-O-O-\overset{O\uparrow}{\underset{O\downarrow}{S}}-OH + H_2O \longrightarrow H_2SO_4 + \underset{\text{Caro's acid}}{HO-O-\overset{O\uparrow}{\underset{O\downarrow}{S}}-OH}$$

Properties. (*i*) Caro's acid is a crystalline white solid having m.p. 318 K.

(*ii*) It is a monobasic acid and gives salts like $KHSO_5$.

(*iii*) It behaves as a powerful oxidizing agent and possesses specific oxidizing properties. It quickly liberates iodine from iodides, oxidises aniline to nitroso benzene, sulphur dioxide to sulphur trioxide, ferrous salts to ferric salts, hydrogen chloride to chlorine etc. Reaction with aniline can be represented as:

$$C_6H_5NH_2 \xrightarrow{H_2SO_5} C_6H_5NO$$

Caro's acid precipitates peroxides from salts of Cu, Ag, Mn etc.

Structure. The structure of H_2SO_5 is based on its synthesis from H_2O_2 (1 mol) and HSO_3Cl (1 mol) as

$$HOO-\overset{O\uparrow}{\underset{O\downarrow}{S}}-OH \quad \text{or} \quad HOO-\overset{O\,\|}{\underset{\|\,O}{S}}-OH$$

Peroxomonosulphuric acid

Q. 14. Describe the preparation, properties and structure of peroxodisulphuric acid $H_2S_2O_8$. (Marshall's acid) (*Kanpur, 2000; Delhi 2002, Guwahati, 2002; Calcutta, 2005; Nagpur 2008*)

Ans. Preparation. (*i*) Pure anhydrous peroxodisulphuric acid is obtained by the action of anhydrous H_2O_2 (1 mol) upon chlorosulphuric acid (2 mols).

$$\underset{\text{Chlorosulphuric acid}}{HO-SO_2-}\boxed{Cl+H}-O-O-\boxed{H+Cl}-SO_2-OH \longrightarrow \underset{\text{Peroxodisulphuric acid}}{HO-SO_2-O-O-SO_2-OH} + 2HCl$$

From the solution, *HCl is* removed under vacuum whereby colourless crystals of $H_2S_2O_8$ are obtained.

(*ii*) $H_2S_2O_8$ is obtained by the electrolysis of 50% H_2SO_4 at low temperature and high current density. Anodic oxidation of sulphuric acid is done with a small smooth platinum wire anode surrounded by a glass tube (Fig. 12.15). A current of about one ampere is passed during the electrolysis.

Hydrogen is liberated at the cathode and $H_2S_2O_8$ is obtained at the anode.

At the cathode:

$$H^+ + e^- \longrightarrow H,$$

$$H + H \longrightarrow H_2$$

At the anode:

$$2HSO_4^- - 2e^- \longrightarrow H_2S_2O_8$$

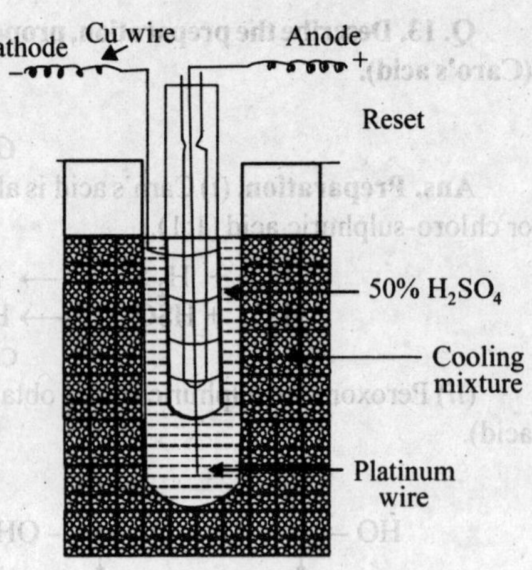

Fig. 12.15. Electrolytic preparation of peroxodisulphuric acid.

Properties. (*i*) Peroxosulphuric acid is a colourless crystalline solid having m.p. 338 K

(*ii*) It undergoes hydrolysis giving Caro's acid.

$$H_2S_2O_8 + H_2O \longrightarrow H_2SO_5 + H_2SO_4$$

(*iii*) Many metals such as Zn etc. dissolve forming their sulphates without evolution of any gas.

$$K_2S_2O_8 + Zn \longrightarrow K_2SO_4 + ZnSO_4$$

(*iv*) Its salt potassium peroxosulphate, $K_2S_2O_8$ on heating evolves sulphur trioxide and oxygen whereas the solution of the salt gives only oxygen on heating.

$$2K_2S_2O_8 \longrightarrow 2K_2SO_4 + 2SO_3 + O_2$$

$$2K_2S_2O_8 + 2H_2O \longrightarrow 2K_2SO_4 + 2H_2SO_4 + O_2$$

(*v*) Amphoteric metals, Cr, As, etc., form their oxy-acids.

$$3K_2S_2O_6 + Cr + 4H_2O \longrightarrow 3K_2SO_4 + H_2CrO_4 + 3H_2SO_4$$

(*vi*) Peroxodisulphate ion is one of the most powerful and useful oxidizing agents.

$$S_2O_8^{2-} + 2e^- \rightleftharpoons 2SO_4^{2-} \quad (E° = +2.01 \text{ volts})$$

It liberates iodine from potassium iodide, oxidizes Cr^{3+} to $Cr_2O_7^{2-}$, Fe^{2+} to Fe^{3+}, Mn^{2+} to MnO_4^- etc.

$$K_2S_2O_8 + 2KI \longrightarrow 2K_2SO_4 + I_2$$

$$K_2S_2O_8 + 2FeSO_4 \longrightarrow Fe_2(SO_4)_3 + K_2SO_4$$

$$5K_2S_2O_8 + 2MnSO_4 + 8H_2O \longrightarrow 2KMnO_4 + 4K_2SO_4 + 8H_2SO_4$$

Structure. X-ray studies show that peroxodisulphate ion, $S_2O_8^{2-}$, the structure $^-O_3S-O-O-SO_3^-$ with approximately tetrahedral angles about each sulphur atom. Also as the acid is prepared from H_2O_2 and $ClSO_3$ reacting in the molecular ratio 1:2, the structure is represented as below:

$$HO-\overset{\overset{O}{\uparrow}}{\underset{\underset{O}{\downarrow}}{S}}-O-O-\overset{\overset{O}{\uparrow}}{\underset{\underset{O}{\downarrow}}{S}}-OH \quad \text{or} \quad HO-\overset{\overset{O}{\|}}{\underset{\underset{O}{\|}}{S}}-O-O-\overset{\overset{O}{\|}}{\underset{\underset{O}{\|}}{S}}-OH$$

Q. 15. Describe the preparation, properties and uses of thiosulphuric acid, $H_2 S_2 O_3$.

Or

Describe the methods of preparation, properties and uses of sodium thiosulphate.

(*Lucknow, 2000; Awadh, 2000; Meerut, 2000; Calcutta, 2005; Vidyasagar, 2007*)

Ans. Preparation. (*i*) The free acid is quite unstable at ordinary temperature but it has been isolated at 195 K from the reaction between sulphur trioxide and hydrogen sulphide in the presence of ether.

$$SO_3 + H_2S \longrightarrow H_2S_2O_3$$

(*ii*) It can be prepared in the absence of solvent by the reaction

$$\underset{\text{Chlorosulphuric acid}}{ClSO_3H} + H_2S \longrightarrow H_2S_2O_3 + HCl$$

(*iii*) The alkali thiosulphates are manufactured as

$$Na_2SO_3 + S \xrightarrow{\text{Biol}} Na_2S_2O_3$$

Preperties. (*i*) Free thiosulphuric acid is unstable and decomposes into sulphur as soon as it is liberated from its salt solution by adding an acid.

$$H_2S_2O_3 \longrightarrow SO_2 + S + H_2O$$

Thus its reactions are mainly due to $S_2O_3^{2-}$, thiosulphate ion.

(*ii*) The thiosulphate ion, $S_2O_3^{2-}$, behaves as a reducing agent of moderate strength.

$$2S_2O_3^{2-} \longrightarrow S_4O_6^{2-} + 2e^-$$

(*iii*) Dilute acids decompose thiosulphate solutions yielding free sulphur and sulphur dioxide.

$$Na_2S_2O_3 + 2HCl \longrightarrow 2NaCl + SO_2 + S + H_2O$$

(*iv*) Iodine solution is decolourised by thiosulphate solution giving tetrathionate.

$$2Na_2S_2O_3 + I_2 \longrightarrow Na_2S_4O_6 + 2NaI$$

This reaction forms the basis of iodimetry in volumetric analysis.

(*v*) An aqueous solution of chlorine oxidizes thiosulphate to sulphate.

$$Na_2S_2O_3 + H_2O + Cl_2 \longrightarrow Na_2SO_4 + S + 2HCl$$

This reaction is responsible for the use of thiosulphate as an antichlor agent.

(*vi*) It gives a purple colour with ferric chloride solution which rapidly vanishes due to the reduction of ferric iron to ferrous state.

$$Fe^{3+} + 2S_2O_3^{2-} \longrightarrow \underset{\text{Purple}}{[Fe(S_2O_3)_2]^-}$$

$$[Fe(S_2O_3)_2]^- + Fe^{3+} \longrightarrow 2Fe^{2+} + S_4O_6^{2-}$$

(*vii*) Silver nitrate solution reacts with thiosulphate to give a white precipitate of silver thiosulphate which rapidly changes to yellow, orange, brown and finally to black silver sulphide,

$$Na_2S_2O_3 + 2AgNO_3 \longrightarrow \underset{\text{White}}{Ag_2S_2O_3} + 2NaNO_3$$

$$\underset{\text{White ppt.}}{Ag_2S_2O_3} \xrightarrow{H_2O} \underset{\text{Black}}{Ag_2S} + H_2SO_4$$

(*viii*) In presence of excess of sodium thiosulphate, a soluble sodium argentothiousulphate complex is formed.

$$2Na_2S_2O_3 + AgNO_3 \longrightarrow Na_3[Ag(S_2O_3)_2] + NaNO_3$$

Uses. (*i*) Alkali thiosulphates are mainly used in photography for fixing, *i.e.*, to dissolve out unreacted AgBr from emulsion.

$$AgBr + 2Na_2S_2O_3 \longrightarrow \underset{\text{Soluble complex}}{Na_3[Ag(S_2O_3)_2]} + NaBr$$

(*ii*) It is used as an antichlor.

(*iii*) In laboratory, it is used in volumetric estimation of iodine.

Q. 16. Discuss the structure of thiosulphuric acid $H_2S_2O_3$ and thiosulphate ion.

Ans. Thiosulphuric acid may be regarded as derived from sulphuric acid by the replacement of one of the – OH groups by a – SH group. Thus the structure of the acid can be represented as:

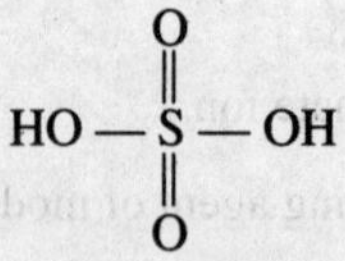

$$\underset{\text{Sulphuric acid}}{HO-\overset{\overset{O}{\|}}{\underset{\underset{O}{\|}}{S}}-OH} \qquad \underset{\text{Thiosulphuric acid}}{HS-\overset{\overset{O}{\|}}{\underset{\underset{O}{\|}}{S}}-OH}$$

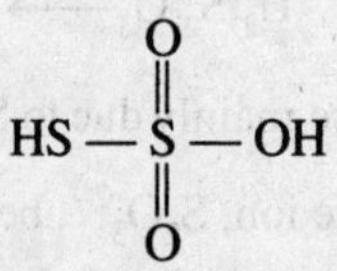

Similarly, the $S_2O_3^{2-}$ ion is a structural analogue of the SO_4^{2-} ion which has the structure as shown in Fig. 12.16.

In such a structure of $S_2O_3^{2-}$ both sulphur atoms are not equivalent, this has been established by tracer technique. When thiosulphate is prepared from a SO_3^{2-} ion and radioactive S and then decomposed by acid to *S* and SO_3^{2-} ion, the final radioactivity appears in the free sulphur. If both sulphur atoms were equivalent in the thiosulphate, the radioactivity should have been distributed between the two products.

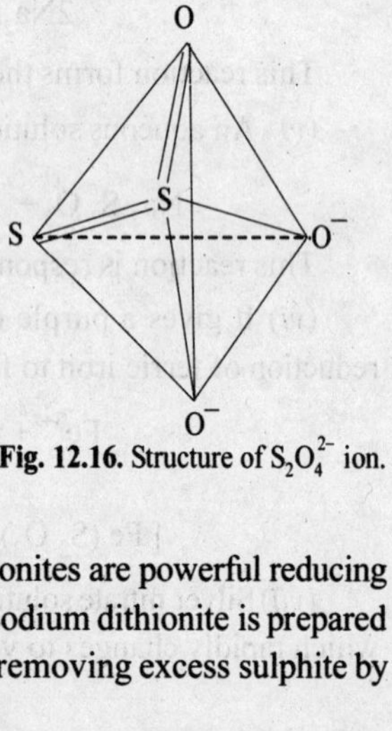

Fig. 12.16. Structure of $S_2O_4^{2-}$ ion.

Q. 17. Discuss the preparation, properties and structure of dithionous acid, $H_2S_2O_4$.

Ans. Preparation. Dithionous or hyposulphurous or hydrosulphurous acid may be prepared in aqueous solution by reducing sulphurous acid solution containing sulphites with amalgamated Zn or zinc dust but such solutions decompose rapidly according to the equation:

$$2S_2O_4^{2-} + H_2O \longrightarrow S_2O_3^{2-} + 2HSO_3^-$$

Properties. Solutions of dithionous acid and of its salts such as dithionites are powerful reducing agents. For reducing purpose Sodium dithionite is ordinarily employed. Sodium dithionite is prepared by reducing sodium bisulphite solution with zinc dust and separated by removing excess sulphite by precipitation with $Ca(OH)_2$.

Structure. Cryoscopic measurements show dithionite ion to be $S_2O_4^{2-}$. Raman data indicate the presence of sulphur-sulphur single bond in the dithionite ion. The structure of the dithionite ion (Fig. 12.17) has several remarkable features. The oxygen atoms, which must bear considerable negative charge, are closely placed together by the eclipsed configuration and by small value of the angle α which would be 35° for sp^3 tetrahedral hybridization at the sulphur atom. Secondly, the S – S distance is much longer than the S – S bonds in bisulphides, polysulphides, etc., (2.0 – 2.15 Å). The long bond is believed to be due to weakening by repulsion of lone pairs on sulphur atoms resulting from *dp*-hybridized bonding. The weak bonding is consistent with the strong reducing properties.

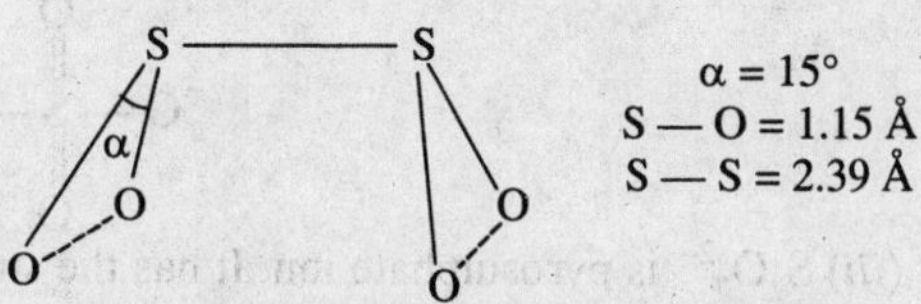

Fig. 12.17. Structure of Dithionite ion.

Q. 18. Describe the preparation, properties and structure of dithionic acid ($H_2S_2O_6$).

Ans. Preparation. (*i*) Dithionic acid and its salts are best prepared by oxidation of SO_2 or sulphite solutions. This can be done electrically or by chemical oxidizing agent such as Mn (IV) oxide, permanganate ion, iron (III) or cobalt (III) hydroxide.

(*ii*) *Laboratory preparation* involves treatment of aqueous suspension of Mn (IV) dioxide with SO_2 below 283 K to form Mn (II) sulphate and dithionate in solution. Removal of manganese and sulphate ions is done by adding Ba $(OH)_2$; $BaS_2O_6 \cdot 2H_2O$ is then crystallized.

$$2MnO_2 + 3SO_2 \xrightarrow{<283k} MnS_2O_6 + MnSO_4$$

or
$$MnO_2 + 2SO_3^{2-} + 4H^+ \longrightarrow Mn^{2+} + S_2O_6^{2-} + 2H_2O$$

$$MnSO_4 + Ba(OH)_2 \longrightarrow \underset{\text{ppt.}}{BaSO_4} + \underset{\text{ppt.}}{Mn(OH)_2}$$

$$MnS_2O_6 + Ba(OH)_2 \longrightarrow \underset{\text{Soluble}}{BaS_2O_6} + Mn(OH)_2$$

On treating the aqueous solutions of this salt with H_2SO_4, free acid solution is obtained.

Properties. Dithionic acid is moderately a strong and stable dibasic acid which decomposes slowly in concentrated solutions and on warming. Dithionates are stable to oxidizing agents and mild reducing agents, but on heating in acid solution they give both H_2SO_4 and H_2SO_3.

$$H_2S_2O_6 + H_2O \longrightarrow H_2SO_4 + H_2SO_3$$

Structure. The dithionate ion has the structure $^-O_3S-SO_3^-$ with approximately tetrahedral angles about each sulphur and shortening of the S – O bonds (1.43 Å) again suggests considerable double bond character. Dithionic acid and dithionates have the structures shown in Fig. 12.18.

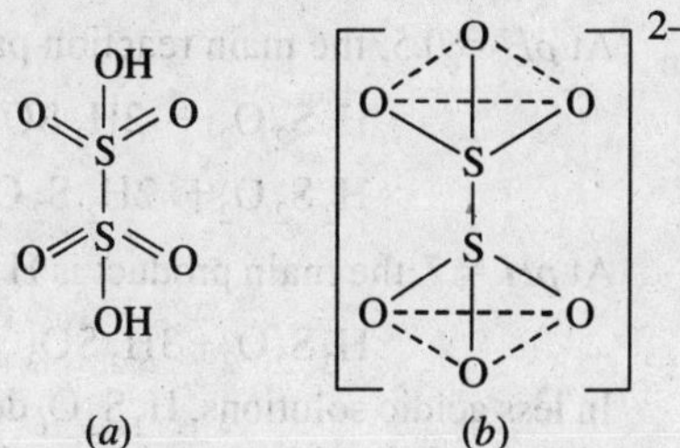

Fig. 12.18. Structure of (*a*) dithionic acid and (*b*) dithionates

Q. 19. Name and draw the structures of the following anions: $S_2O_3^{2-}$, $S_2O_6^{2-}$ and $S_2O_7^{2-}$.

Ans. (*i*) $S_2O_3^{2-}$ is thiosulphate ion. It has the structure

$$\begin{array}{c} O^- \\ \| \\ {}^-O-S-S \\ \| \\ O \end{array}$$

(*ii*) $S_2O_6^{2-}$ is dithionate ion. It has the structure

$$\begin{array}{c} O \quad O \\ \| \quad \| \\ {}^-O-S-S-O^- \\ \| \quad \| \\ O \quad O \end{array}$$

(*iii*) $S_2O_7^{2-}$ is pyrosulphate ion. It has the structure

$$\begin{array}{c} O \qquad O \\ \| \qquad \| \\ {}^-O-S-O-S-O^- \\ \| \qquad \| \\ O \qquad O \end{array}$$

Q. 20. Discuss the structure of polythionic acids ($H_2S_xO_6$) ($x > 2$) and polythionates.

Ans. Compounds like $H_2S_3O_6$ (trithionic acid), $H_2S_4O_6$ (tetrathionic acid) and $H_2S_5O_6$ (pentathionic acid) are called polythionic acid and their salts are called polythionates. Structures of trithionic acid and the trithionate ion are given in Fig.12.19.

Structures of higher polythionic acid and polythionate ions can be obtained by increasing the sulphur chain with terminal groups (SO_3) remaining the same on both sides.

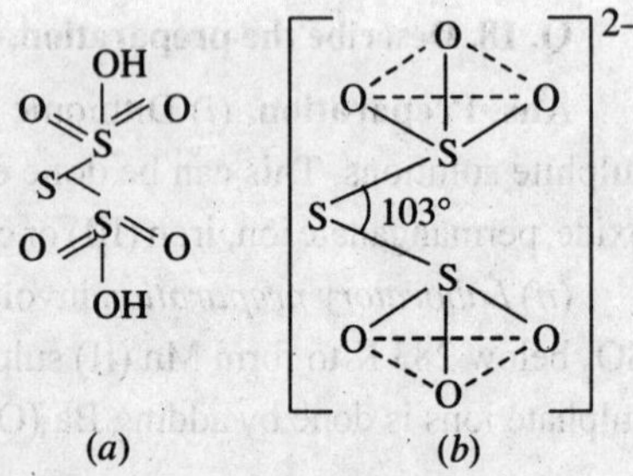

Fig. 12.19. (*a*) Structure of trithionic acid (*b*) Structure of $S_3O_6^{2}$

Q. 21. Describe methods of preparation and properties of polythionic acid. (*Garhwal, 2000*)

Ans. Preparation. Polythionates are prepared in a number of ways:

(*i*) *By passing hydrogen sulphide through saturated aqueous solution of sulphur dioxide* till its smell disappears and milky solution, called Wackenroder's liquid or solution, is produced. This solution contains a mixture of polythionic acids and free sulphur along with $H_2S_2O_3$, H_2SO_4 etc. The various reactions are:

$$H_2S + H_2SO_3 \longrightarrow H_2S_2O_2 + H_2O$$

$$H_2SO_3 + 2H_2S \longrightarrow 3H_2O + 3S$$

At $pH = 0.5$, the main reaction products are $H_2S_4O_6$ and $H_2S_6O_6$.

$$H_2S_2O_2 + 2H_2SO_3 \longrightarrow H_2S_4O_6 + 2H_2O$$

$$H_2S_2O_2 + 2H_2S_2O_3 \longrightarrow H_2S_6O_6 + 2H_2O$$

At $pH = 7$, the main product is $H_2S_3O_6$.

$$H_2S_2O_2 + 3H_2SO_3 \longrightarrow H_2S_2O_3 + H_2S_3O_6 + 2H_2O$$

In less acidic solutions, $H_2S_2O_6$ decomposes giving a mixture of lower polythionic acids. From the mixture of polythionic acids, each acid is isolated on the basis of difference in solubilities of their potassium salts. The order of solubility is

$$K_2S_3O_6 > K_2S_6O_6 > K_2S_4O_6 = K_2S_5O_6$$

(*ii*) *By the hydrolysis of sulphur monochloride.* The reactions that occur are represented as follows:

$$S_2Cl_2 + 3H_2O \longrightarrow H_2SO_3 + H_2S + 2HCl$$

$$3H_2SO_3 + H_2S \longrightarrow H_2S_4O_6 + 3H_2O$$

$$6H_2SO_3 + 2H_2S \longrightarrow H_2S_3O_6 + H_2S_5O_6 + 6H_2O$$

$$9H_2SO_3 + 3H_2S \longrightarrow H_2S_2O_6 + 2H_2S_5O_6 + 9H_2O$$

(*iii*) *By the acid decomposition of this thiosulphate* in the presence of As (III) ion such as AsO_3^{3-}. Thiosulphuric acid first formed decomposes to give polythionic acids.

$$H_2S_2O_3 \longrightarrow H_2SO_3 + S$$

$$5H_2S_2O_3 \longrightarrow 2H_2S_5O_6 + 3H_2O$$

$$3H_2SO_3 + 2H_2S_5O_6 \longrightarrow 3H_2S_2O_3 + H_2S_3O_6 + H_2S_4O_6$$

(*iv*) *By oxidation of thiosulphates.* (*a*) Mild oxidising agents such as iodine, Fe^{3+} produce tetrathionate from the oxidation of a thiosulphate.

$$2Na_2S_2O_3 + I_2 \longrightarrow Na_2S_4O_6 + 2NaI$$

$$2Na_2S_2O_3 + 2FeCl_3 \longrightarrow Na_2S_4O_6 + 2FeCl_2 + 2NaCl$$

(*b*) Using H_2O_2, a thiosulphate first forms tetrathionate which ultimately gives trithionate, formed *in situ* by the action of alkali.

$$2Na_2S_2O_3 + H_2O_2 \longrightarrow Na_2S_4O_6 + 2NaOH$$

$$Na_2S_4O_6 + 2NaOH + 3H_2O_2 \longrightarrow Na_2S_3O_6 + Na_2SO_4 + 4H_2O$$

Properties. The free polythionic acids are known only in solution. The solutions of tetra and pentathionic acids are quite stable. On standing, the solutions of these acids gradually decompose giving sulphuric acid, sulphur dioxide and sulphur.

$$H_2S_xO_6 \longrightarrow H_2SO_4 + SO_2 + (x-2)\,S \qquad (x = 3-6)$$

A few typical reactions of alkali polythionates are given below:

(*i*) Trithionates precipitate copper sulphide from copper sulphate solution.

$$K_2S_3O_6 + CuSO_4 + 2H_2O \longrightarrow CuS + K_2SO_4 + 2H_2SO_4$$

(*ii*) An alkali polythionate solution on treatment with mercuric chloride liberates an equivalent amount of free sulphuric acid.

$$2K_2S_xO_6 + 2HgCl_2 + 4H_2O \longrightarrow HgCl_2 \cdot HgS + 2(x-3)S + 4KCl + 4H_2SO_4$$

(*iii*) Potassium cyanide reacts with trithionate yielding a mixture of thiosulphate, sulphate and free hydrocyanic acid

$$K_2S_3O_6 + 2KCN + H_2O \longrightarrow K_2S_2O_3 + K_2SO_4 + 2HCN$$

Higher thionates give thiocyanates as the additional products:

$$K_2S_4O_6 + 3KCN + H_2O \longrightarrow K_2S_2O_3 + K_2SO_4 + 2HCN + KCNS$$

$$K_2S_5O_6 + 4KCN + H_2O \longrightarrow K_2S_2O_3 + K_2SO_4 + 2HCN + 2KCNS$$

(*iv*) Polythionates are oxidized to sulphates by strong oxidizing agents

$$Na_2 S_3 O_6 + 2H_2O + 4\,[O] \longrightarrow Na_2 SO_4 + 2H_2 SO_4$$

(*v*) These are reduced to thiosulphates by alkali sulphides

$$K_2 S_5 O_6 + K_2 S \longrightarrow 2K_2 S_2 O_3 + 2S$$

$$K_2 S_4 O_6 + K_2 S \longrightarrow 2K_2 S_2 O_3 + S$$

$$K_2 S_3 O_6 + K_2 S \longrightarrow 2K_2 S_2 O_3$$

(*vi*) Alkali sulphites reduce tetra, penta and hexathionates to trithionate together with thiosulphates.

$$Na_2 S_4 O_6 + Na_2 SO_3 \longrightarrow Na_2 S_2 O_3 + Na_2 S_3 O_6$$

MISCELLANEOUS QUESTIONS

Q. 22. Describe the preparation, properties and structure of pyrosulphuric acid, $H_2 S_2 O_7$, oleum (Fuming sulphuric acid).

Ans. Preparation. (*i*) Pyrosulphuric acid is obtained by dissolving sulphur trioxide in concentrated sulphuric acid.

$$H_2 SO_4 + SO_3 \longrightarrow H_2 S_2 O_7$$

(*ii*) Alkali metal pyrosulphate can be obtained by heating alkali bisulphite.

$$2NaHSO_4 \xrightarrow{\Delta} Na_2 S_2 O_7 + H_2O$$

Properties. (*i*) It is a viscous brownish liquid, fumes in air and gradually loses sulphur trioxide. Thus it changes into ordinary sulphuric acid on warming.

$$H_2 S_2 O_7 \xrightarrow{\text{Heat}} H_2 SO_4 + SO_3$$

(*ii*) Pyrosulphuric acid gives sulphuric acid when water is added into it.

$$H_2 S_2 O_7 + H_2O \longrightarrow 2H_2 SO_4$$

(*iii*) It is an excellent sulphonating agent used in organic chemistry.

Structure. The structure of the acid is given below:

```
     O         O                        O         O
     ↑         ↑                        ||        ||
H — O — S — O — S — OH      or     HO — S — O — S — OH
     ↓         ↓                        ||        ||
     O         O                        O         O
```

Pyrosulphuric acid

Q. 23. What are peroxyacids of sulphur? Describe the preparation of Marshall's acid by anodic oxidation method. How does Caro's acid react with (*i*) KI (*a*) $FeSO_4$? (*Nagpur, 2000*)

Ans. See Q. 6, 14 and 15

Q. 24. Give an account of the oxy-acids and peracids of sulphur with special reference to (*a*) their basicity (*b*) salt formation (*c*) structures (*Andhra, 2000*)

Ans. See Qs. 6 – 15

13 OXIDES AND OXY-ACIDS OF CHLORINE

OXIDES OF CHLORINE

Q. 1. Give the preparation, properties and structure chlorine monoxide Cl_2O.

(*Himachal, 2000*)

Ans. Preparation. (*i*) It is prepared by passing chlorine are freshly cooled precipitated yellow mercuric oxide. (Previously heated to 575-675 K).

$$2HgO + Cl_2 \longrightarrow HgCl_2 \cdot HgO + Cl_2O$$

(*ii*) It may also be prepared by distilling a fairly concentrated solution of hypochlorous acid at low pressure.

$$2HClO \xrightarrow[(H_2O)]{Distil} Cl_2O$$

Properties. (*i*) It is a brownish-yellow gas with a characteristic odour.

(*ii*) It undergoes decomposition readily on heating, sparking or on coming into contact with organic matter.

$$2Cl_2O \longrightarrow 2Cl_2 + O_2$$

(*iii*) As it dissolves in water yielding a golden-yellow solution of hypochlorous acid, it is regarded as the anhydride of hypochlorous acid.

$$Cl_2O + H_2O \rightleftharpoons 2HClO$$

(*iv*) It converts most metals into a mixture of oxides and chlorides.

$$2Zn + Cl_2O \longrightarrow ZnCl_2 + ZnO$$

(*v*) With NaOH, it gives sodium hypochlorite.

$$Cl_2O + 2NaOH \rightleftharpoons 2NaClO + H_2O$$

(*vi*) It is a strong oxidizing agent, it oxidizes hydrochloric acid to chlorine.

$$Cl_2O + 2HCl \longrightarrow H_2O + 2Cl_2$$

Fig. 13.1. Structure of chlorine monoxide.

Structure. The molecule is angular and symmetrical with O – Cl distance 1.71 Å and Cl – O – Cl angle of 111° (Fig. 13.1). In Cl_2O the bonding electrons are nearer to oxygen since this is more electronegative. Repulsion between bond pairs now exceeds that between lone pairs. Hence the angle is greater than the tetrahedral angle.

Q. 2. Give the preparation, properties, and structure of chlorine dioxide ClO_2.(*Kerala, 2000*)

Ans. Preparation. (*i*) It is prepared by heating a mixture of potassium chlorate and oxalic acid crystals to 335K.

$$2KClO_3 + H_2C_2O_4 \xrightarrow{\Delta} K_2CO_3 + H_2O + 2ClO_2 + CO_2$$

(*ii*) It may be obtained by bubbling sulphur dioxide through an acidified solution of sodium chlorite.

$$2NaClO_3 + SO_2 + H_2SO_4 \longrightarrow 2NaHSO_4 + 2ClO_2$$

(*iii*) It may be obtained by warming a mixture of conc. sulphuric acid and potassium chlorate.

$$KClO_3 + H_2SO_4 \longrightarrow KHSO_4 + HClO_3] \times 3$$

$$3HClO_3 \longrightarrow HClO_4 + 2ClO_2 + H_2O$$

$$3KClO_3 + 3H_2SO_4 \longrightarrow 3KHSO_4 + 2ClO_2 + HClO_4 + H_2O$$

(*iv*) It is prepared by the action of chlorine on silver chlorate.

$$2AgClO_3 + Cl_2 \xrightarrow{365\ K} 2AgCl + 2ClO_2 + O_2$$

Properties. (*i*) It is a dark-yellow, unstable and pungent smelling gas.

(*ii*) It is a strong chlorinating agent and its bleaching power is very high as compared to chlorine.

(*iii*) It is a powerful oxidizing agent. Phosphorus and carbonaceous matter ignite in it.

$$ClO_2 + 4H_3O^+ + 5e^- \rightleftharpoons Cl^- + 6H_2O\ ;\ E^\circ = 1.50\ \text{volt}$$

(*iv*) It explodes violently on the smallest provocation especially in the presence of organic matter.

$$2ClO_2 \longrightarrow Cl_2 + 2O_2$$

(*v*) It dissolves in potassium hydroxide solution according to the equation.

$$2ClO_2 + 2KOH \longrightarrow \underset{\text{Pot. chlorite}}{KClO_2} + \underset{\text{Pot. chlorate}}{KClO_3} + H_2O$$

(*vi*) It dissolves in water giving a mixture of chlorus acid and chloric acid.

$$2ClO_2 + H_2O \longrightarrow HClO_2 + HClO_3$$

Structure. It has a symmetrical bent structure with Cl – O bond distance of 1.49 Å and O – Cl – O angle of 117° (Fig. 13.2).

It is an odd molecule (total valency electrons = 19). However, it has no tendency to dimerise because the electron is less localized on the central atom chlorine due to strong electronegativity of oxygen.

Cl
1.49 Å
117°
O O

Fig. 13.2. Structure of chlorinedioxide.

The bond lengths are notably shorter than in a single bond and resonance is thought to occur between two structures.

This structure involves two coordinate bonds and a three-electron bond which may occupy either position as shown in Fig. 13.3. The three-electron bond accounts for the shorter bond lengths and is about half as strong as normal covalent or coordinate bond. Its paramagnetic nature is accounted for by the three-electron bonds.

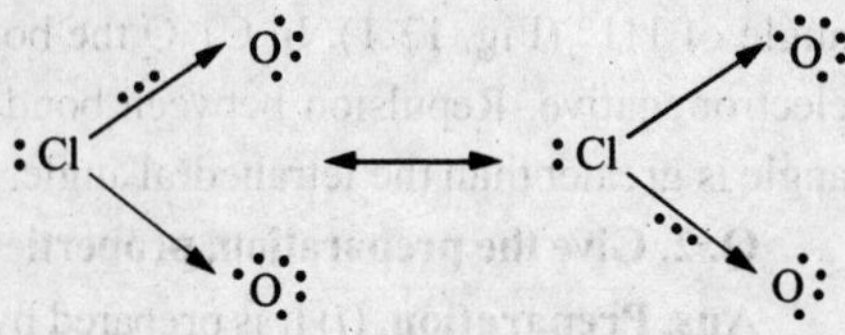

Fig. 13.3. Three-electron bonds in ClO_2.

Q. 3. Give the preparation, properties and structure of dichlorine hexoxide, Cl_2O_6.

(*Andhra, 2002*)

Ans. Dichlorine hexoxide is prepared by action of chlorine dioxide on ozone. When the mixed gases are cooled with ice, brown solution of chlorine dioxide in dichlorine hexoxide is obtained. The chlorine dioxide can be removed by distillation.

$$2ClO_2 + 2O_3 \longrightarrow Cl_2O_6 + 2O_2$$

Properties. (*i*) It is a red liquid which freezes at 276.5 K.

(*ii*) It decomposes even at 273 K to a mixture of other oxides of chlorine and ultimately to chlorine and oxygen.

$$Cl_2O_6 \longrightarrow 2ClO_2 + O_2$$

$$2ClO_2 \longrightarrow Cl_2 + 2O_2$$

(*iii*) When it is added to liquid water, an explosion occurs. However, if it is mixed with water vapour it yields a mixture of chloric acid and perchloric acid showing that it is a mixed anhydride of these acids.

$$Cl_2O_6 + H_2O \longrightarrow HClO_3 + HClO_4$$

(*iv*) With hydrogen fluoride, it reacts as follows:

$$Cl_2O_6 + HF \longrightarrow HClO_4 + \underset{\text{Chloryl fluoride}}{ClO_2F}$$

(*v*) It forms chlorate and perchlorate with alkalies.

$$Cl_2O_6 + 2NaOH \longrightarrow NaClO_3 + NaClO_4 + H_2O$$

or

$$Cl_2O_6 + 2OH^- \longrightarrow ClO^-_3 + ClO^-_4 + H_2O$$

Structure. Liquid dichlorine hexoxide is diamagnetic whereas gaseous dichlorine hexoxide exists as the paramagnetic molecule ClO_3. Thus chloride hexoxide may be regarded as a mixture of Cl_2O_6 and ClO_3.

$$Cl_2O_6 \rightleftharpoons 2ClO_3$$

O←, O←, O← (three O atoms) = Cl — Cl = →O, →O, →O (three O atoms)

It has been suggested that the structure of Cl_2O_6 contains two trigonal pyramides joined at the apices.

Paramagnetic behaviour of gaseous compound is due to structural species ClO_3 whereas Cl_2O_6 exhibits diamagnetic properties.

Q. 4. Give the preparation, properties and structure of dichloro heptoxide, Cl_2O_7.

Ans. Preparation. (*i*) It may be obtained by the dehydration of anhydrous or concentrated perchloric acid with phosphorus pentoxide at 263 K followed by vacuum distillation.

$$4HClO_4 + P_4O_{10} \longrightarrow 2Cl_2O_7 + 4HPO_3$$

(*ii*) It can also be obtained by heating potassium perchlorate with chlorosulphonic acid, $HOClSO_2$ under reduced pressure.

Properties. (*i*) It is a colourless oily liquid (b.p. 355 K). It explodes when struck or in contact with a flame.

(*ii*) It is most stable oxide of chlorine and also is the weakest oxidizing agent of all the chlorine oxides.

(*iii*) It reacts with alkalies to perform perchlorates.

$$Cl_2O_7 + 2NaOH \longrightarrow 2NaClO_4 + H_2O$$

(*iv*) With water it gives perchloric acid.

$$Cl_2O_7 + H_2O \longrightarrow 2HClO_4$$

It is, therefore, known as anhydride of perchloric acid.

Structure. The infrared spectrum of its vapour indicates that in heptoxide two ClO_3 groups are linked together by an oxygen atom with Cl – O – Cl777• angle of 128°. Thus it has the structure $O_3ClO\,ClO_3$ with tetrahedral arrangement (Fig. 13.4).

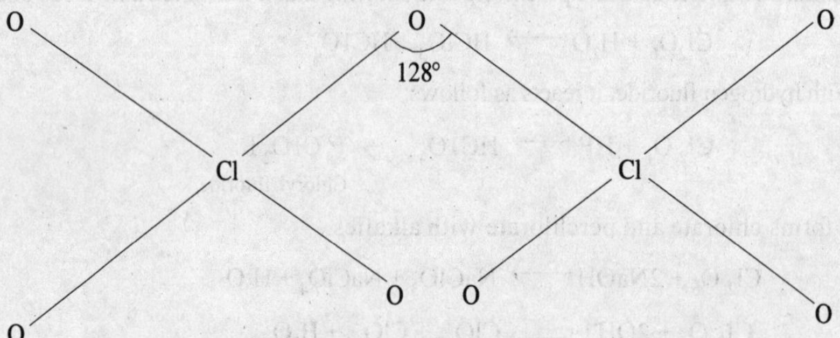

Fig. 13.4. Structure of dichlorine heptoxide (Cl_2O_7).

Q. 5.Why is the angle of OF_2 smaller than in Cl_2O ?

Ans. Electronegativity of fluorine is greater while that of chlorine is smaller than that of oxygen. In Cl_2O, the bonded electron pair is drawn towards oxygen. Repulsion between bond pair exceeds that between lone pairs. Therefore the angle crosses the tetrahedral angle of 109°. But the situation is reverse in the case of OF_2. Hence the angle is less in this case.

OXY-ACIDS OF CHLORINE

Q. 6. Give the methods of preparation and properties of hydrochlorous acid (HClO).

Ans. Preparation. (*i*) It is prepared by passing chlorine into water.

$$Cl_2 + 2H_2O \longrightarrow HCl + HClO$$

(*ii*) It may also be obtained by saturating the solution of bleaching powder with chlorine.

$$CaOCl_2 + Cl_2 + H_2O \longrightarrow CaCl_2 + 2HClO$$

(*iii*) A convenient method is to shake chlorine water with freshly precipitated yellow mercuric oxide.

$$2Cl_2 + 2HgO + H_2O \longrightarrow Hg_2Cl_2 + 2HClO$$

(*iv*) When chlorine gas is passed through a suspension of calcium carbonate in water, hypochlorous acid is obtained.

$$CaCO_3 + H_2O + 2Cl_2 \longrightarrow CaCl_2 + CO_2 + 2HClO$$

(*v*) On a large scale, it is prepared by passing carbon dioxide into a suspension of bleaching powder and filtering off calcium carbonate.

$$CaOCl_2 + CO_2 + H_2O \longrightarrow CaCO_3 + 2HClO$$

Properties. (*i*) The dilute solution of hypochlorous acid is colourless whereas its concentrated solution is yellow in colour. It undergoes decomposition into hydrochloric acid and oxygen.

$$2HClO \longrightarrow 2HCl + O_2$$

(*ii*) With magnesium, it liberates hydrogen.

$$2HClO + Mg \longrightarrow Mg(OCl)_2 + H_2$$

(*iii*) On distillation, it undergoes decomposition to form water and Cl_2O.

$$2HClO \longrightarrow Cl_2O + H_2O$$

(*iv*) As it liberates nascent oxygen, it acts as an oxidizing agent.

$$HClO \longrightarrow HCl + O$$

Thus it oxidizes an alkaline solution of $Cr(OH)_3$ into chromate and manganous hydroxide into hydrated manganic hydroxide.

$$2Cr(OH)_3 + NaClO + 4NaOH \longrightarrow 2Na_2CrO_4 + 3NaCl + 5H_2O$$

$$Mn(OH)_2 + NaClO + H_2O \longrightarrow Mn(OH)_4 + NaCl$$

(*v*) It liberates iodine from acidified solution of potassium iodide.

$$2KI + HOCl + HCl \longrightarrow 2KCl + H_2O + I_2$$

(*vi*) With alkalies, it forms salts, called hypochlorites.

$$HClO + NaOH \longrightarrow NaClO + H_2O$$

Q. 7. Discuss the structure of hypochlorous acid and hypochlorite ion (ClO^-).

Ans. The structure of hypochlorous acid is as follows:

H — O — Cl or H • ×Ö× •C̈l: (with × × above and below O and dots around Cl)

The structure of hypochlorite ion is represented by a distorted tetrahedron as shown in Fig. 13.5.

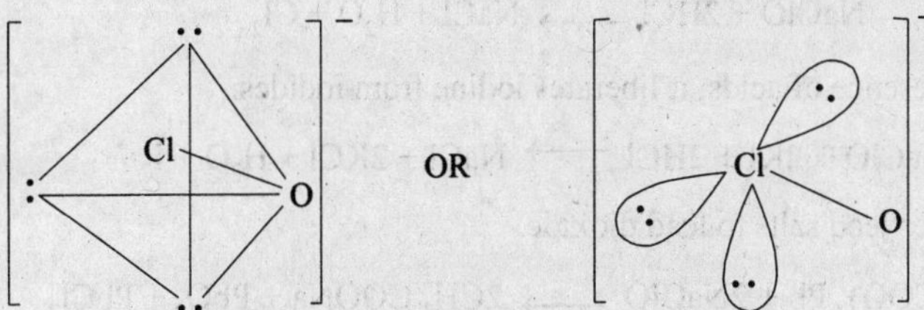

Fig. 13.5. Structure of hypochlorite ion (ClO^-).

The structure is based on sp^3 hybridisation of chlorine atom. These sp^3 hybrid orbitals form fairly weak σ bonds since the *s* and *p* levels have appreciable energy gap. The ion is stable due to strong $p\pi - d\pi$ bonding between a full 2p orbitals on O and vacant *d*-orbitals on the chlorine atom. (Fig. 13.6).

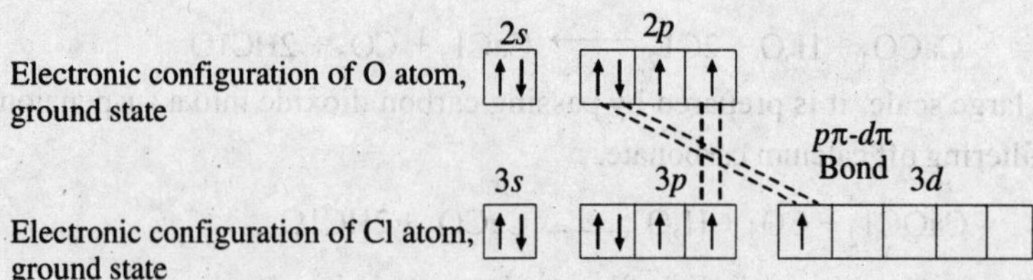

Fig. 13.6. Overlapping of orbitals in ClO^-.

Q. 8. Give the method of preparation and properties of sodium hypochlorite (NaC1O).

Ans. Preparation. (*i*) It is prepared by passing chlorine gas into a cold dilute solution of sodium hydroxide or carbonate.

$$3NaOH + Cl_2 \longrightarrow NaCl + NaClO + H_2O$$

The solution of sodium hypochlorite cannot be concentrated by heating because it decomposes into chlorate and chloride.

$$3NaClO \longrightarrow NaClO_3 + 2NaCl$$

(*ii*) On large scale, it is prepared by the electrolysis of brine solution. The chlorine liberated at anode reacts with sodium hydroxide at the cathode to form sodium hypochlorite.

Properties. (*i*) On heating, sodium hypochlorite decomposes.

$$2NaOC1 \longrightarrow NaClO_3 + 2NaCl$$

(*ii*) With dilute acid, sodium hypochlorite liberates hypochlorous acid and, thus, acts as a strong oxidizing and bleaching agent.

$$NaC1O + HCl \longrightarrow NaC1 + HClO$$

$$HClO \longrightarrow HC1 + O$$

(*iii*) On storing, it loses oxygen and this decomposition is catalyzed by cobalt chloride.

$$NaClO \xrightarrow{CoCl_2} NaCl + O$$

(*iv*) It oxidizes arsenites to arsenates.

$$Na_3AsO_3 + NaClO \longrightarrow Na_3AsO_4 + NaCl$$

(*v*) It oxidizes hydrochloric acid to chlorine

$$NaClO + 2HCl \longrightarrow NaCl + H_2O + Cl_2$$

(*vi*) In the presence of acids, it liberates iodine from iodides.

$$NaClO + 2KI + 2HCl \longrightarrow NaCl + 2KCl + H_2O + I_2$$

(*vii*) It oxidizes lead salts to lead dioxide.

$$2(CH_3COO)_2Pb + 2NaClO \longrightarrow 2CH_3COONa + PbO_2 + PbCl_2$$

Uses. It is used for bleaching cotton fabrics.

Q. 9. Give the preparation, properties and structure of chlorous acid ($HClO_2$).

Ans. Preparation. Aqueous solution of chlorous acid can be prepared by treating a suspension of barium chlorite with sulphuric acid and filtering the solution, chlorous acid is stronger acid than

HOCl with $K_{a1} = 10^{-2}$. Chlorites are made from ClO_2 by reaction with peroxides.

$$Ba(ClO_2)_2 + H_2SO_4 \longrightarrow 2HClO_2 + BaSO_4$$

$$HO^-_2 + OH^- + 2ClO_2 \longrightarrow 2ClO_2^- + H_2O + O_2$$

A chlorite can be prepared by treating chlorine dioxide with an alkali.

$$2KOH + 2ClO_2 \longrightarrow KClO_2 + KClO_3 + H_2O$$

Properties. Chlorous acid undergoes auto-oxidation at room temperature to form a mixture of HClO and $HClO_3$.

$$2HClO_2 \longrightarrow HClO + HClO_3$$

The alkali chlorites possess a caustic taste and bleach vegetable colours.

Structure. The structure of $HClO_2$ is H – O – Cl → O. The structure of chlorite ion is shown in Fig. 13.7.

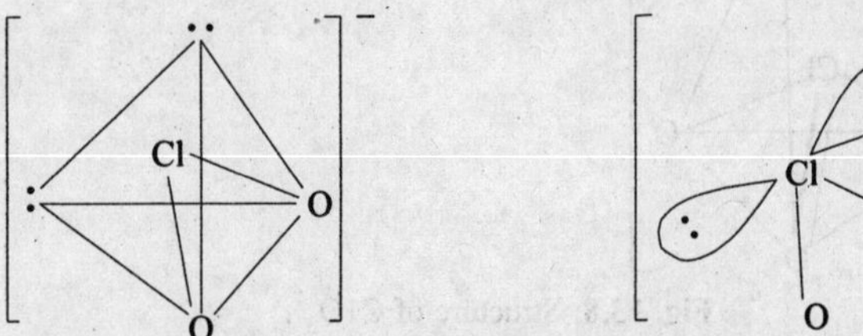

Fig. 13.7. Structure of ClO_2^-

Q. 10. Give the preparation and properties of chloric acid ($HClO_3$).

Ans. Preparation. (*i*) An aqueous solution of chloric acid is prepared by the action of dilute sulphuric acid on barium chlorate.

$$Ba(ClO_3)_2 + H_2SO_4 \longrightarrow BaSO_4 + 2HClO_3$$

The insoluble barium sulphate can be removed by fitration and the excess of sulphuric acid is precipitated by baryta water. The filtrate is concentrated in a vacuum desiccator over concentrated sulphuric acid until its composition reaches $HClO_3 . 7H_2O$ without decomposing the chloric acid.

(*ii*) It may also be prepared by the action of hydrofluorosilicic acid on potassium chlorate.

$$2KClO_3 + H_2SiF_6 \longrightarrow K_2SiF_6 + 2HClO_3$$

Insoluble potassium flurosilicate is removed by filtration.

Properties. (*i*) Concentrated chloric acid is a colourless and pungent smelling liquid.

(*ii*) If its dilute solution is allowed to stand, chloric acid decomposes according to the following equation:

$$3HClO_3 \longrightarrow HClO_4 + 2O_2 + Cl_2 + H_2O$$

(*iii*) Its concentrated solution is a strong oxidizing and bleaching agent. It is stronger than chlorous acid and it dissolves certain metals such as Zn, Fe, Al to form the metal chlorates and liberates hydrogen.

Q. 11. Discuss the structure of chloric acid and chlorate ion (ClO^-_3).

(*Kurukshetra, 2001*)

Structure. The chloric acid has been assigned the following structure.

$$H-O-Cl \begin{matrix} \nearrow O \\ \searrow O \end{matrix}$$

The ClO^-_3 ion is pyramidal in shape, *i.e.*, the structure is tetrahedral with one position occupied by a lone pair of electrons (Fig. 13.8).

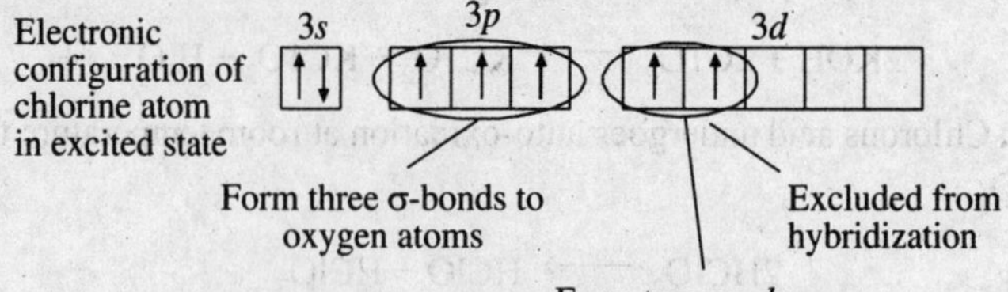

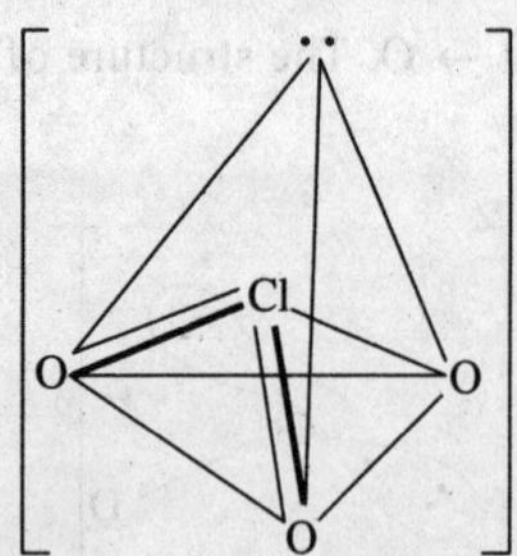

Fig. 13.8. Structure of ClO_3^-.

Q. 12. Give the methods of preparation, properties and uses of potassium chlorate $KClO_3$.

Ans. Preparation. On a small scale it is prepared by passing chlorine through a boiling solution of caustic potash.

$$2KOH + Cl_2 \longrightarrow KCl + KClO + H_2O\,] \times 3$$

$$2KClO \longrightarrow KClO_3 + 2KCl$$

$$6\,KOH + 3Cl_2 \longrightarrow KClO_3 + 5KCl + 3H_2O$$

The potassium chlorate can be removed by fractional crystallization.

Manufacture. Potassium chlorate is manufactured by the following methods:

***(i)* Old process.** In this method, a current of chlorine gas is passed through hot milk of lime which is placed in a cast iron tank. In the reaction, calcium chlorate is formed. Free lime is allowed to settle and decanted. Liquor containing calcium chlorate is treated with a calculated quantity of potassium chloride. Thus potassium chlorate is formed which can be separated by fractional crystallization.

$$2Ca\,(OH)_2 + 2Cl_2 \longrightarrow CaCl_2 + Ca\,(ClO)_2 + 2H_2O] \times 3$$

$$3Ca\,(ClO)_2 \longrightarrow Ca\,(ClO_3)_2 + 2CaCl_2$$

$$6Ca\,(OH)_2 + 6Cl_2 \longrightarrow 5CaCl_2 + Ca\,(ClO_3)_2 + 6H_2O$$

$$Ca\,(ClO_3)_2 + 2KCl \longrightarrow 2KClO_3 + CaCl_2$$

***(ii)* Modern method.** Electrolysis of a hot concentrated solution of common salt is carried out in an undivided cell. On passing the electric current, hydroxyl ions are formed at cathode by the reduction of water and react with Na^+ to form caustic soda. On the other hand, chlroine is liberated at the anode. As the electrolytic cell is undivided, it means that NaOH and Cl_2 react to form sodium chlorate.

$$NaCl \rightleftharpoons Na^+ + Cl^-$$

At cathode:

$$3H_2O + 2e^- \longrightarrow H_2 + 2OH^-$$

$$Na^+ + OH^- \rightleftharpoons NaOH$$

At anode:

$$Cl^- \longrightarrow Cl + e^-$$

$$Cl + Cl \longrightarrow Cl_2$$

$$6NaOH + 3Cl_2 \longrightarrow NaClO_3 + 5NaCl + 3H_2O$$

Potassium chlorate is obtained by adding a calculated quantity of potassium chloride to sodium chlorate solution. Potassium chlorate, which is less soluble, crystallizes out on cooling.

$$NaClO_3 + KCl \longrightarrow KClO_3 + NaCl$$

Properties. (*i*) It is a white crystalline substance having m.p. 643 K. It is slightly soluble in cold water but readily in hot water.

(*ii*) Action of heat. Potassium chlorate, when heated just above its m.p. 643 K, is converted to mixture of potassium chloride and potassium perchlorate.

$$4KClO_3 \longrightarrow KCl + 3KClO_4$$

When heated in the presence of manganese dioxide, it decomposes below its m.p. into potassium chloride and oxygen.

$$2KClO_3 \xrightarrow{MnO} 2KCl + 3O_2$$

(*iii*) Action of concentrated hydrochloric acid. When potassium chlorate is heated with concentrated hydrochloric acid, euchlorine, which is a mixture of chlorine dioxide and chlorine, is produced.

$$KClO_3 + HCl \longrightarrow KCl + HClO_3] \times 2$$

$$2HClO_3 \longrightarrow 2ClO_2 + H_2O + O$$

$$2HCl + O \longrightarrow Cl_2 + H_2O$$

$$2KClO_3 + 4HCl \longrightarrow 2KCl + Cl_2 + 2ClO_2 + 2H_2O$$

(*iv*) Action of concentrated sulphuric acid. When potassium chlorate is heated with concentrated sulphuric acid, loud explosion may occur, leading to the formation of chlorine dioxide.

$$KClO_3 + H_2SO_4 \longrightarrow KHSO_4 + HClO_3] \times 3$$

$$2HClO_3 \longrightarrow 2ClO_2 + H_2O + O$$

$$HClO_3 + O \longrightarrow HClO_4$$

$$2KClO_3 + 3H_2SO_4 \longrightarrow 3KHSO_4 + HClO_4 + \underset{\text{Chlorine dioxide}}{2ClO_2} + H_2O$$

(v) Action of iodine. When a solution of $KClO_3$ is heated with iodine in the presence of a little nitric acid, potassium iodate is formed.

$$2KClO_3 + I_2 \longrightarrow 2KIO_3 + Cl_2$$

(vi) Reduction. When potassium chlorate is warmed with sulphurous acid, it is quantitatively reduced to potassium chloride.

$$KClO_3 + 3H_2SO_3 \longrightarrow KCl + 3H_2SO_4$$

Uses. (*i*) For the preparation of oxygen in the laboratory.

(*ii*) As an oxidizing agent in fireworks, matches, explosives and qualitative analysis.

(*iii*) In calico-printing.

(*iv*) In medicine for sore throat.

Q. 13. Give the methods of preparation, properties and uses of perchloric acid $HClO_4$.

(*Kerala, 2000*)

Ans. Preparation. Aqueous solution of the acid can be made from NH_4ClO_4 and dilute HNO_3 or from $NaClO_4$ and conc. HCl.

$$NaClO_4 + HCl \longrightarrow NaCl + HClO_4$$

Commercial concentrated acid is 72% by weight.

Properties. (*i*) It is a very strong acid in aqueous solution with $K_{a_1} > 10^8$.

(*ii*) Perchloric acid can react explosively with organic matter.

$$4HClO_4 \longrightarrow 2Cl_2 + 7O_2 + 2H_2O$$

(*iii*) Anhydrous perchloric acid is a colourless, mobile liquid boiling at 363 K. It is very hygroscopic, fumes in moist air and dissolves in water with evoluation of heat.

(*iv*) Metals like Zn and Fe dissolve in dilute solutions of the acid forming their perchlorates.

$$Zn + 2HClO_4 \longrightarrow Zn(ClO_4)_2 + H_2$$

Uses. (*i*) It finds important applications as an oxidant in analytical chemistry.

(*ii*) It is used for quantitative destruction of organic matter in a wide variety of materials such a protein, fats, oils, waxes, rubber, leather etc.

(*iii*) Perchlorates of electropositive metals are soluble in water. Silver perchlorate is unique in that it is soluble in water and also in hydrocarbon solvents, such as benzene.

(*iv*) The insolubility of $KClO_4$ is used to detect potassium in qualitative analysis.

(*v*) The anhydrous magnesium and barium perchlorates are used widely as excellent desiccants.

Structure. The perchlorate ion is tetrahedral. This is because chlorine exhibits *sp*^{3} hybridisation.

Q. 14. Discuss the structure of perchlorate ion (ClO^-_4).

(*Garhwal, 2000*; *Kurukshetra, 2001*)

Ans. Perchlorate ion is tetrahedral. This is because chlorine makes use of four bond orbitals obtained by *sp*^{3} hybridisation and shape is tetrahedral. This is shown in Fig. 13.9. The three half-filled

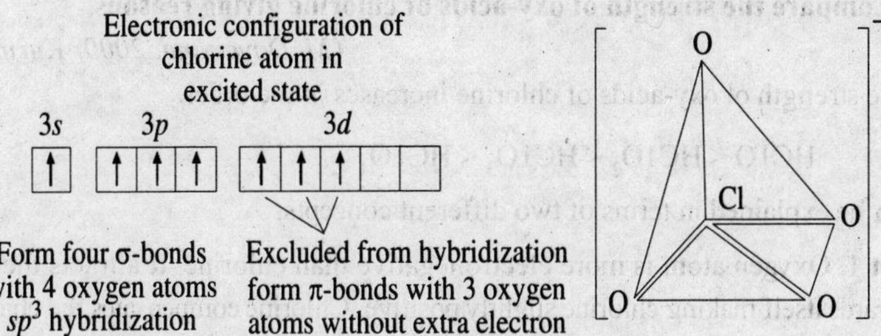

Fig. 13.9. Shape of perchlorate ion (ClO_4^-)

d orbitals form π-bonds with three oxygen atom and thus bonded by double bonds to chlorine.

Perchloric acid has the following structure:

$$\begin{array}{c} O \\ \uparrow \\ H-O-Cl\rightarrow O \\ \downarrow \\ O \end{array}$$

Q. 15. Describe the preparation and properties of potassium perchlorate, $KClO_4$.

Ans. Preparation. (*i*) Potassium chlorate, when heated just above its m.p. 643 K, is converted to a mixture of potassium chloride and potassium perchlorate.

$$4KClO_3 \longrightarrow KCl + 3KClO_4$$

Potassium perchlorate which is sparingly soluble in water crystallizes out completely leaving potassium chloride in solution.

(*ii*) On a large scale, it is manufactured by the electrolytic (anodic) oxidation of a concentrated aqueous solution of sodium chlorate at room temperature. Chlorate ion are discharged at the anode which react with water to form perchloric acid. Sodium hydroxide is formed at the cathode. The two react to form perchloric acid. Sodium hydroxide is formed at the cathode. The two react to form sodium perchlorate. This is treated with a calculated quantity of potassium chloride when sparingly soluble potassium perchlorate is precipitated.

$$ClO^-_3 + H_2O \longrightarrow HClO_4 + H^+$$

$$HClO_4 + NaOH \longrightarrow NaClO_4 + H_2O$$

$$NaClO_4 + KCl \longrightarrow KClO_4 + NaCl$$

Properties. (*i*) It is a colourless crystalline solid. It is more stable than chlorate and is sparingly soluble in water.

(*ii*) *Action of conc. H_2SO_4.* With hot conc. sulphuric acid, it yields perchloric acid.

$$KClO_4 + H_2SO_4 \longrightarrow KHSO_4 + HClO_4$$

(*iii*) *Action of heat.* On heating it decomposes as follows:

$$KClO_4 \xrightarrow{\Delta} KCl + 2O_2\uparrow$$

Q. 16. Compare the strength of oxy-acids of chlorine giving reasons.

(*M. Dayanand, 2000; Kurukshetra, 2001*)

Ans. The strength of oxy-acids of chlorine increases in the order.

$$HClO < HClO_2 < HClO_3 < HClO_4$$

This can be explained in terms of two different concepts.

Concept 1. Oxygen atom is more electronegative than chlorine. It attracts the shared pair of electrons towards itself making chlorine slightly positive. Chlorine compensates the charge by drawing the electron pair between hydrogen and chlorine. This helps in the release of protons. Greater the number of oxygen atoms attached to chlorine, greater will be this effect and hence greater will be the acid strength.

Perchloric acid having maximum, *i.e.*, 4 oxygens will have greatest acid strength followed by chloric acid having 3 oxygens followed by chlorous acid having 2 oxygens followed by perchlorous acid having 1 oxygen.

H — O — Cl Hypochlorous acid

H ➤— O — Cl ➤— O Chlorous acid

O
↑
H ➤➤— O — Cl ➤— O Chloric acid

O
↑
H ➤➤➤— O — Cl ➤— O Hypochloric acid
↓
O

Concept 2. An acid is a strong acid if it has a weak conjugate base. Conjugate bases of the above four acids are

ClO^- Hypochlorite ion

ClO^-_2 Chlorite ion

ClO^-_3 Chorate ion

ClO^-_4 Perchlorate ion

Out of the above ions, perchlorate ion is the weakest base and hence perchloric acid is the strongest acid. Perchlorate ion is the weakest base because it can stabilize itself the most and hence will have the least tendency to accept a proton. It can be visualized that greater the number of oxygen atoms in the ion, greater will be the charge delocalisation and hence greatest will be the stability.

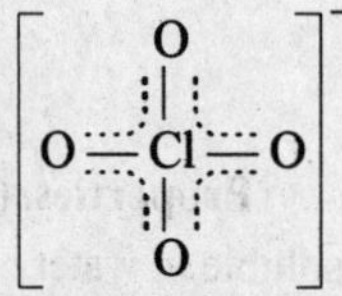

Fig. 13.10.

Q. 17. Compare the acid strength of the following: hypofluorous acid, hypochlorous acid, hypobromous acid, hypoiodous acid.

Ans. The acid strength follows the order:

$$HIO < HBrO < HClO < HFO$$

This can be explained in terms of electronegativity of the halogen. Greater the electronegativity, greater the pull of electrons towards the halogen and easier it is to break the bond between oxygen and hydrogen.

Q. 18. Compare the oxidising power of HClO, $HClO_2$, $HClO_3$ and $HClO_4$.

Ans. Oxidising power of a substance depends upon the ease with which it can lose oxygen to give a stable compound. HClO gives out oxygen most readily to give the most stable compound (HCl) and hence HClO is the strongest acid. The order of oxidising power is

$$HClO_4 < HClO_3 < HClO_2 < HClO$$

Q. 19. Name the two oxy-acids of chlorine containing + 5 and + 7 oxidation states. Which of the two will be more acidic and why?

Ans. The two acids are Chloric acid (oxidation state = + 5) and

Hypochloric acid (oxidation state = + 7)

Greater the oxidation state, greater the power with which it can attract electrons and hence greater the acid strength. Accordingly hypochloric acid ($HClO_4$) is more acidic than chloric acid ($HClO_3$).

14 SOME MORE COMPOUNDS OF NITROGEN, PHOSPHORUS SULPHUR AND HALOGENS

Q. 1. Describe the preparation, properties, structure and use of hydrazine NH_2NH_2 (diamide). (*Himachal, 2000; Garhwal, 2000; Kurukshetra, 2001; Andhra, 2002; Nagpur, 2003*)

Ans. Preparation. (*i*) *Raschig's synthesis.* Hydrazine is obtained by boiling a concentrated solution of ammonia with sodium hypochlorite solution.

$$NH_3 + NaOCl \longrightarrow \underset{\text{Chloramine}}{NH_2Cl} + NaOH \qquad (a)$$

$$NH_2Cl + NH_3 + NaOH \longrightarrow NH_2NH_2 + NaCl + H_2O \qquad (b)$$

$$2ClNH_2 + N_2H_4 \longrightarrow 2NH_4Cl + N_2 \qquad (c)$$

The reaction (*c*) is catalyzed by traces of heavy metal ions such as Fe^{3-} and Cu^{2+} ions. This is suppressed by addition of glue or gelatin which complexes with the metal ions.

The reaction product is acidified with sulphuric acid to crystallize out hydrazine sulphate $NH_2NH_2 \cdot H_2SO_4$. It is then treated with liquid ammonia when ammonium sulphate gets precipitated.

$$N_2H_4H_2SO_4 + 2NH_3 \longrightarrow N_2H_4 + (NH_4)_2SO_4 \downarrow$$

(*ii*) In laboratory hydrazine can also be prepared by the application of a Hofmann rearrangement of urea.

$$H_2N-\underset{\underset{O}{\|}}{C}-NH_2 \xrightarrow{OCl^-} H_2N-\underset{\underset{O}{\|}}{C}-NHCl \overset{OH^-}{\rightleftharpoons}$$

$$H_2N-\underset{\underset{O}{\|}}{C}-NCl \longrightarrow [O{=}C{=}N{-}NH_2] \xrightarrow{H_2O} H_2N-NH_2 + H_2O$$

Properties. (*i*) N_2H_4 (anhydrous) is colourless, fuming liquid having m.p. 274.8 K and b.p. 386.5 K. It is completely soluble in water and is also soluble in alcohol. It is strongly hygroscopic, its aqueous solution is alkaline.

(*ii*) Hydrazine is diacid base and can form salts with one or two equivalents of an acid. Thus, with hydrochloric acid, it gives hydrazine monochloride, $N_2H_5^+\,Cl^-$ and hydrazine dichloride $N_2H^{2+}_6 \cdot 2Cl^-$.

(*iii*) Like NH_3, it can also form co-ordinate complexes with both Lewis acids and metal ions such as Ni^{2+} and Co^{2+}; it is because nitrogen atoms have a lone pair of electrons. Hydrazine is potentially a bidentate ligand but generally functions as a monodentate donor because of the electrostatic and steric considerations.

(*iv*) At 473-773 K it decomposes to NH_3 and N_2 and if in contact with SiO_2, hydrogen is also evolved. In contact with Pt or W, only N_2 and H_2 are evolved.

$$3N_2H_4 \longrightarrow N_2\uparrow + 4NH_3\uparrow$$

(*v*) It bums in air to give nitrogen, the reaction is highly exothermic.

$$N_2H_4 + O_2 \longrightarrow N_2 + 2H_2O + 622 \text{ kJ}$$

(*vi*) When an electric spark is passed it decomposes to NH_3 and N_2.

(*vii*) It dissolves Li and Na to give blue solution but in a short time hydrogen and $NH_2NH \cdot Na$ are obtained. NH_2NHNa explodes in contact with moisture and oxygen.

(*viii*) When mixed with hydrogen peroxide, it is used as an excellent rocket fuel. This reaction has increased the importance of hydrazine on account of its complete combustion to nitrogen and water yielding a mixture of compounds of low average molecular weights.

$$NH_2NH_2 + 2H_2O_2 \longrightarrow N_2 + 4H_2O$$

(*ix*) It acts as a powerful reducing agent. In basic medium it behaves as follows:

$$N_2H_4 + 4OH^- \longrightarrow N_2 + 4H_2O + 4e^-$$

It reduces ferrocyanide into ferrocyanide, cupric salts to cuprous and iodine into HI.

$$N_2H_4 + 4[Fe(CN)_6]^{3-} \longrightarrow 4[Fe(CN)_6]^{4-} + 4H_2O + N_2\uparrow$$

$$N_2H_4 + 4CuO \longrightarrow 2Cu_2O + 2H_2O + N_2\uparrow$$

$$N_2H_4 + 2I_2 \longrightarrow 4HI + N_2\uparrow$$

(*x*) It also reduces permanganates into manganese salt, ferric salt to ferrous and iodates to iodides.

$$5N_2H_2 + 4KMnO_4 + 6H_2SO_4 \longrightarrow 2K_2SO_4 + 4MnSO_4 + 16H_2O + 5N_2\uparrow$$

$$N_2H_4 + 4FeCl_3 \longrightarrow 4FeCl_2 + 4HCl + N_2\uparrow$$

$$3N_2H_4 + 2IO_3^- \longrightarrow 2I^- + 6H_2O + 3N_2\uparrow$$

The powerful reducing properties of N_2H_4 are due to great stability of N_2 molecule.

(*xi*) In the presence of powerful reducing agents in solution, it is forced to act as an oxidizing agent.

$$N_2H_4 + Zn + 2HCl \longrightarrow 2NH_3 + ZnCl_2$$

(*xii*) It oxidises glyoxalic acid into oxalic acid

$$NH_2NH_2 + H_2O + \begin{array}{c}CHO\\|\\COOH\end{array} \longrightarrow 2NH_3 + \begin{array}{c}COOH\\|\\COOH\end{array}$$

(*xiii*) N_2H_4 precipitates Au, Ag and Pt. from their salts. It also precipitates metallic *Hg* from a weak acidic solution of $HgCl_2$.

$$4AuCl_3 + 3N_2H_4 \longrightarrow 4Au + 3N_2 + 12HCl$$

$$4AgNO_3 + N_2H_4 \longrightarrow 4Ag + N_2 + 4HNO_3$$

$$PtCl_4 + N_2H_4 \longrightarrow Pt + N_2 + 4HCl$$
$$2HgCl_2 + N_2H_4 \longrightarrow 2Hg + N_2 + 4\,HCl$$

(*xiv*) It produces hydrazoic acid on reaction with nitrous acid.

$$N_2H_4 + HNO_2 \longrightarrow 2H_2O + N_3H$$

(*xv*) With aldehydes and ketones, it forms condensation products called hydrazones.

$$RCH{=}\boxed{O + H_2}N.NH_2 \longrightarrow RCH{=}N.NH_2 + H_2O$$

$$R_2C{=}\boxed{O + H_2}NH_2 \longrightarrow R_2C{=}N.NH_2 + H_2O$$

Structure of hydrazine. Raman data and high dipole moment (1.85 D) of hydrazine indicate the following unsymmetrical nonplanar structure (Fig. 14.1).

The lack of rotation of NH_2 groups is due to the tendency of lone pair of electrons to be as far apart as possible.

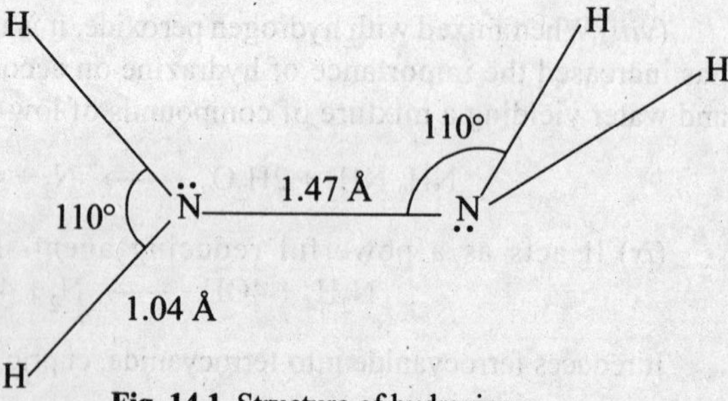

Fig. 14.1. Structure of hydrazine

Uses. (1) It has recently attracted considerable attention owing to its unusual characteristics as rocket fuel.

(2) Hydrazine and its phenyl derivative find important applications in the characterisation of carboxyl group in organic compounds.

(3) It is a good solvent for non-metals like S, P etc.

(4) It is used as an effective deoxidant for boiler water.

Q. 2. Discuss the structure of hydrazine. *(Delhi, 2002)*

Ans. See Q.1.

Q. 3. Describe the preparation, properties, uses and structure of hydroxylamine (NH_2OH). *(Calcutta, 2000; Andhra, 2002; Delhi, 2002; Kurukhetra, 2002)*

Ans. Preparation. (*i*) By electrolytic reduction of nitric acid. Hydroxylamine may be prepared by the electrolytic reduction of nitric acid in 50% sulphuric acid at an amalgamated lead cathode. The nascent hydrogen at the cathode reduces nitric acid to nitrous acid and finally to hydroxylamine:

At the cathode

$$\left[\begin{array}{l} HNO_3 + 2H \longrightarrow HNO_2 + H_2O \\ HNO_2 + 4H \longrightarrow NH_2OH + H_2O \end{array}\right]$$

$$Pb^+ \left| \begin{array}{c} 50\% \\ H_2SO_4 \\ \, \end{array} \right| \begin{array}{c} 50\% \\ H_2SO_4 \\ +HNO_3 \end{array} \Big|^- \quad \text{Pb beaker (amalgam)}$$

Lead anode is separated by a porous pot from an amalgamated lead beaker which acts as cathode. The whole assembly is cooled with ice. 50% H_2SO_4 is kept in each compartment and 50% HNO_3 is added drop by drop at the cathode.

(ii) By interaction of acid sulphites and nitrites in solution. The usual method of its preparation consists in reducing sodium nitrite with sulphur dioxide in presence of sodium carbonate at a low temperature (278 K). A sodium sulphonate of hydroxylamine is first obtained which is then hydrolysed to hydroxylamine by warming with a little suphuric acid.

The reaction may also be supposed to occur in three stages as follows:

$$(a)\quad NaNO_2 + 3NaHSO_3 \longrightarrow \underset{\substack{\text{Hydroxylamine} \\ \text{sodium sulphonate}}}{HO \cdot N(SO_3Na)_2} + Na_2SO_3 + H_2O$$

$$(b)\quad HON(SO_3Na)_2 + H_2O \longrightarrow HONH(SO_3Na) + NaHSO_4$$

$$(c)\quad HO \cdot NE(SO_3Na) + H_2O \longrightarrow HONH_2 + NaHSO_4$$

(iii) It is also obtained by the reduction of ethylnitrate by nascent hydrogen.

$$C_2H_5NO_3 + 6H \xrightarrow{\text{Sn/HCl}} NH_2OH + H_2O + C_2H_5OH$$

(iv) **By reduction with nascent hydrogen.** Hydroxylamine is obtained by reducing nitric oxide, NO, with nascent hydrogen. A current of NO is passed through a mixture of granulated tin and strong HC1.

$$NO + 3H \longrightarrow NH_2OH$$

Anhydrous hydroxylamine. Preparation. It is obtained by treating a solution of hydroxylamine hydrochloride in methyl alcohol with a solution of sodium methoxide in methyl alcohol.

$$CH_3ONa + NH_2OH \cdot HCl \longrightarrow CH_3OH + NaC1 + NH_2OH$$

The sodium chloride is filtered off and the methyl alcohol separated by distillation.

Properties. (i) NH_2OH, like hydrazine, is a weaker base than NH_3 as lone pair is not available to the same extent as in NH_3.

(ii) It is unstable and decomposes even at room temperature.

$$3NH_2OH \longrightarrow N_2 + NH_3 + 3H_2O$$
$$4NH_2OH \longrightarrow 2NH_3 + N_2O + 3H_2O$$

(iii) NH_2OH in the pure from acts as a weak monoacid base. In water, it acts as a weak base. It forms a series of coordinate compounds through a lone pair present on nitrogen.

(iv) It gives the following reducing properties:

(1) It reduces Fehling's solution

$$4CuO + 2NH_2OH \xrightarrow[\text{medium}]{\text{Alkaline}} 2Cu_2O + 3H_2O + N_2O$$

(2) It reduces gold chloride to gold.

$$4AuC1_3 + 6NH_2OH \xrightarrow[\text{medium}]{\text{Acidic}} 4Au + 3N_2O + 12HC1 + 3H_2O$$

(3) It reduces $FeCl_3$ to $FeCl_2$.

$$4FeCl_3 + 2NH_2OH \xrightarrow[\text{medium}]{\text{Acidic}} 4FeCl_2 + N_2O + H_2O + 4HCl$$

(4) Similarly, it reduces silver nitrate to silver.

$$AgNO_3 \xrightarrow[\text{acidic medium}]{NH_4OH} Ag$$

(5) $$2HgCl_2 + 2NH_2OH \xrightarrow[\text{medium}]{\text{Alkali}} 2Hg + N_2O + H_2O + 4HCl$$

(*v*) Hydroxylamine salt on heating with nitric acid evolve nitric oxide.

$$NH_2OH + HNO_3 \longrightarrow 2NO + 2H_2O$$

(*vi*) It gives the following oxidising properties:

(1) $$NH_2OH + 2Fe(OH)_2 + H_2O \xrightarrow[\text{medium}]{\text{Alkali}} 2Fe(OH)_3 + NH_3$$

(2) $$NH_2OH + 2HI \xrightarrow[\text{medium}]{\text{Acidic}} NH_3 + I_2 + H_2O$$

(3) $$Na_2AsO_3 + NH_2OH \xrightarrow[\text{medium}]{\text{Alkali}} Na_3SO_4 + NH_3$$

(4) $$2NH_2OH + 2Cl_2 \longrightarrow N_2O + H_2O + 4HCl$$

(5) $$NH_2OH + H_2O_2 \longrightarrow NH_3 + O_2 + H_2O$$

(*vii*) With aldehydes and ketones, hydroxylamine gives oximes and therefore, it is an important reagent in organic chemistry to detect carbonyl group.

$$(CH_3)_2C{=}\boxed{O + H_2}NOH \longrightarrow (CH_3)_2C{=}NOH + H_2O$$

Acetone → Acetoxime

(*viii*) When hydroxylamine reacts with acids, salts such as hydrochloride or sulphate are obtained.

$$NH_2OH \cdot HCl, (NH_2OH)_2 H_2SO_4$$

Uses. Hydroxylamine is used to make cyclohexanone oxime, which when converted to caprolactum and polymerized gives nylon 6. Thus it finds use in the preparation of synthetic fibres.

Structure. Hydroxylamine may be considered to exist in two tautomeric forms:

$$\underset{\text{Feeble acid}}{NH_2-OH} \rightleftharpoons \underset{\text{Basic}}{NH_3O}$$

$$H-\ddot{N}(OH)-H \rightleftharpoons H-N^{++}(O^{-})(H)-H$$

Q. 4. How is hydroxylamine prepared in laboratory? Compare its properties with those of ammonia. *(Bangalore, 2004)*

Ans. See Q. 3.

Q. 5. (*a*) Give an example to show that hydrazine behaves as a reducing agent.

(*b*) State with equation what happens when hydroxylamine is allowed to react with $FeCl_3$, solution. *(Vidyasagar, 2007)*

Ans. See Q. 1 (*ix*), (*x*) Q. 2 (*iv*).

Q. 6. Describe methods of preparation, properties and structure of hydrazoic acid (N_3H).

(M. Dayanand, 2000; Kurukshetra, 2001; Guwahati, 2002; Andhra, 2002; Delhi, 2002; Calcutta, 2005; Bangalore, 2005)

Ans. Preparation. *From hydrazine.* Many other oxidizing agents attract hydrazine to form small amounts of HN_3 or azides.

(*a*) Hydrazine salts when treated with nitrous acid give hydrazoic acid.

$$NH_2\,NH_2\cdot HCl + NaNO_2 \longrightarrow NH_2NH_2HNO_2 + NaCl$$

$$NH_2\,NH_2\,HNO_2 \longrightarrow N_3H + 2H_2O$$

(*b*) By shaking nitrogen trichloride with hydrazine salts in benzene solution.

$$NH_2\,NH_2 + NCl_3 \longrightarrow N_3H + 3HCl$$

(*c*) By the careful oxidation of hydrazine with nitric acid or hydrogen peroxide.

$$3NH_2\,NH_2 + 5O \longrightarrow 2N_3H\cdot 5H_2O$$

From sodamide: The sodium salt is prepared by the following reactions:

$$3NaNH_2 + NaNO_3 \xrightarrow{448\,K} NaN_3 + 3\,NaOH + NH_3$$

$$2NaNH_2 + N_2O \xrightarrow{463\,K} NaN_3 + NaOH + NH_3$$

The product mixture is dissolved in dilute sulphuric acid and distilled when a solution of hydrazoic acid distils over. The acid is then separated from water by distilling in vacuum.

Free acid can also be obtained in solution by the reaction:

$$N_2H_5^+ + HNO_2 \xrightarrow{Aq.sol} HN_3 + H^+ + 2H_2O$$

Properties. (*i*) Free N_3H, obtained by distillation from aqueous solution, is a colourless liquid (b.p. 310 K), dangerously toxic and explosive:

$$3N_3H \longrightarrow 4N_2 + NH_3$$

(*ii*) It is readily soluble in alcohol and in water (a weak acid in water K_a at 298 K = 1.8×10^{-5}).

(*iii*) HN_3 is stronger than $CH_3\,COOH$ but is not stronger than mineral acids.

(*iv*) It is stable in aqueous solution but unstable in pure form. Light can initiate the decomposition of the acid into nitrogen and ammonia.

(*v*) Salts of heavy metals known as azides are generally explosive; those of Pb, Hg, Ag and Ba explode on being struck sharply and are used in detonation caps.

(*vi*) Ag N_3 is obtained as white curdy precipitate by passing the vapours of the acid into a solution of Ag NO_3. It is sparingly soluble in water but soluble in ammonia like silver chloride.

$$AgNO_3 + HN_3 \longrightarrow AgN_3 + HNO_3$$

$$AgN_3 + 2NH_4OH \longrightarrow [Ag(NH_3)_2N_3] + 2H_2O$$

(*vii*) In its reaction with Zn, Al and Fe it forms azides and ammonia is liberated. With magnesium H_2 is evolved.

$$Zn + 3HN_3 \longrightarrow Zn(N_3)_2 + N_2 + NH_3$$

$$3N_3H + Mg \longrightarrow (N_3)_2\,Mg + H_2$$

(*viii*) Azides of electropositive metals are not explosive and in fact decompose smoothly and quantitatively when heated to 574 K or higher temperatures giving nitrogen, *e.g.*,

$$2NaN_3\,(s) \rightleftharpoons 2Na\,(l) + 3N_2\,(g)$$

(*ix*) Azide ion, like thiocyanate ion gives a red colour with Fe^{3+} as

$$Fe^{3+} + 3N_3^- \longrightarrow \underset{\text{Red colour}}{Fe(N_3)_3}$$

(*x*) *Oxidising action.* When N_3H reacts with substances like arsenic or sodium amalgam it is reduced to N_2H_4 or NH_3. With finely-divided platinum it also gives ammonia.

$$N_3H \xrightarrow{H_2/Pt} NH_3$$

(*xi*) *Reducing action.* N_3H can be oxidized by stronger oxidizing agents. It can also be oxidized by HNO_2 and hypochlorous acid.

(1) With $KMnO_4$

$$2N_3H \xrightarrow[KMnO_4]{(O)} 3N_2 + H_2O$$

(2) With HNO_2

$$N_3H + OHNO \longrightarrow N_2 + N_2O + H_2O$$

Structure. The N – N distances in hydrazoic acid are unsymmetrical and correspond to a double and a triple bond respectively X-ray electron diffraction and infrared data show that the azide ion itself is symmetrical and linear (N – N distance = 1.15 Å) and the structure of N_3H may be represented as hybrid of following resonating structures:

$$\left[H:\ddot{N}=\overset{+}{N}=\overset{-}{\ddot{N}}: \longleftrightarrow :N\equiv\overset{+}{N}:\overset{-}{\underset{\cdot\cdot}{\ddot{N}}}:H \longleftrightarrow H:\overset{-}{\underset{\cdot\cdot}{\ddot{N}}}:N\equiv N:\right]$$

In covalent azides the symmetry is lost

HN₃: H–N 1.10 Å, angle 110°; N–N 1.24 Å; N–N 1.13 Å

CH₃N₃: CH₃–N 1.47 Å, angle 135°; N–N 1.26 Å; N–N 1.10 Å

HN_3 $\qquad$ CH_3N_3

Q.7. What is the action of $KMnO_4$ on hydrazoic acid?

Ans. $$2N_3H \xrightarrow[\text{[O]}]{KMnO_4} 3N_2 + H_2O$$

HALIDES OF PHOSPHORUS

Q. 8. Give the preparation and properties of phosphorus trifluoride, PF_3.

Ans. Preparation. PF_3 is generally prepared by the flourination of phosphorus compounds.

$$PCl_3 + AsF_3 \longrightarrow PF_3 + AsCl_3$$

$$2Cu_3P + 3PbF_2 \longrightarrow 2PF_3 + 6Cu + 3Pb$$

Properties. (*a*) PF_3, a colourless gas, is the least reactive of the phosphorus trihalides.

(*b*) PF_3 forms complexes with transition metals.

(*c*) Unlike other trihalides, it is hydrolysed only slowly by water but is attacked rapidly by alkalies.

(*d*) Like CO, it is highly poisonous because of the formation of haemoglobin complex.

Q. 9. Give the preparation and properties of phosphorus trichloride, PCl_3.

Ans. Preparation. It is prepared by passing dry chlorine over red phosphorus gently heated in a retort. The trichloride distills over and is collected in a cooled receiver. It is purified by redistilling over white phosphorus to remove the excess of chlorine.

$$P_4 + 6Cl_2 \longrightarrow 4PCl_3$$

Properties. Phosphorus trichloride is a colourless liquid of density 1.6 constantly foming in the moist air.

(1) It boils at 349 K and freezes at 161.5 K.

(2) It has strong pungent smell.

(3) *Action of water.* It is decomposed by water forming hydrochloric and phosphorus acids.

$$PCl_3 + 3H_2O \longrightarrow H_3PO_3 + 3HCl$$

(4) *Reducing nature.* (*i*) *Action of air.* It slowly combines with atmospheric oxygen to form the oxy-chloride.

$$2PCl_3 + O_2 \rightleftharpoons 2POCl_3$$

(*ii*) *Action of chlorine.* It combines with chlorine to form phosphorus pentachloride PCl_5.

$$PCl_3 + Cl_2 \rightleftharpoons PCl_5$$

(*iii*) *Action of SO_3.* It combines with SO_3 producting $POCl_3$ and SO_2.

$$PCl_3 + SO_3 \longrightarrow POCl_3 + SO_2$$

Uses. It is extensively used in organic chemistry for replacing the hydroxyl group of alochols, acids etc. by chlorine.

$$PCl_3 + 3C_2H_5OH \longrightarrow 3C_2H_5Cl + H_3PO_3$$
$$\text{Ethyalcohol} \qquad\qquad \text{Ethyl chloride}$$

$$PCl_3 + 3CH_3COOH \longrightarrow 3CH_3COCl + H_3PO_3$$
$$\text{Acetic acid} \qquad\qquad \text{Acetyl chloride}$$

Structure. Structural data indicates that the phosphorus trihalides are pyramidal molecules with the bond angles decreasing steadily from near tetrahedral in PF_3 to 98° in PI_3.

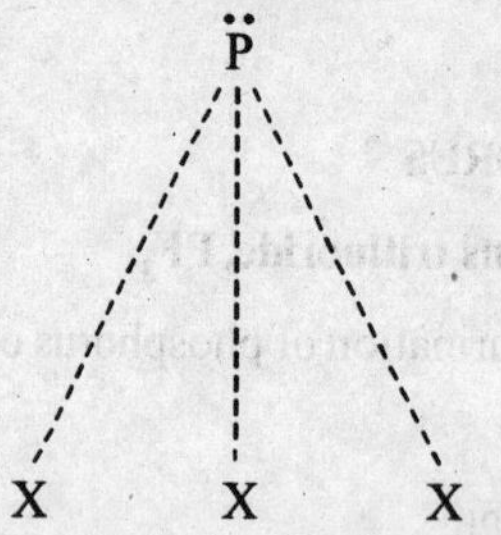

X	$d(PX)$, Å	$\angle XPX$
F	1.52	104
Cl	2.00	102
Br	2.23	100
I	2.47	98

These data suggest that σ bond formed between phosphorus and the halogen atom incorporates sp^3 hybridized phosphorus orbitals. In the case of PF_3, the hybridization is nearly sp^3 but the degree of *s* character in the hybrid orbital decreases down the series.

Fluorine < chlorine < bromine < iodine

P – X bond distance increases as we have from F to I. This is expected in terms of increasing atomic radius from F to I but the bond distances are shorter than expected. This is explained by overlapping between filled *p* orbitals of the halogen and the empty *d* orbitals of phosphorus.

Also the decreasing ∠XPX can be explained in terms of smaller repulsions between the bond pairs in PX_3 as the bond-length P – X increases.

Q. 10. Describe the preparation and properties of phosphorus pentafluoride.

Ans. Preparation. PF_5 can be prepared by treating phosphorus pentoxide with calcium or asemc fluoride.

$$6P_4O_{10} + 10\,CaF_2 \xrightarrow{573-673K} 4PF_5 + 10Ca_3(PO_3)_2$$

$$3PC1_5 + 5AsF_3 \longrightarrow 3PF_3 + 5AsC1_3$$

It is a strong Lewis acid and forms complexes with amines, ethers and other bases as well as F^- in which P becomes 6 coordinated. It is a good catalyst for ionic polymerization.

Q. 11. Describe preparation, properties, uses and structure of phosphorus pentachloride ($PC1_5$). *(Kurukshetra, 2002)*

Ans. Preparation. 1. Phosphorus pentachloride is prepared by passing an excess of dry chlorine into liquid trichloride. Chlorine reacts with phosphorus trichloride and solid phosphorus pentachloride is formed (See figure on next page).

$$PC1_3 + C1_2 \longrightarrow PC1_5$$

2. It may also be prepared by passing chlorine into a well-cooled solution of phosphorus in carbon disulphide and crystallizing out the product formed.

Properties. (1) Phosphorus pentachloride is a pale yellow crystalline solid having a pungent smell.

(2) It strongly fumes in moist air as it has great affinity for water.

(3) It sublimes at 373 K and the vapours when heated dissociate into phosphorus trichloride and chlorine. The reaction is complete at 373 K.

$$PCl_5 \longrightarrow PCl_3 + Cl_2$$

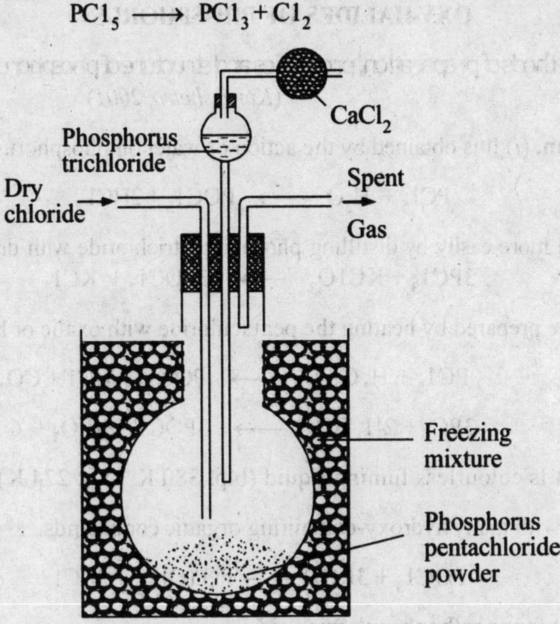

Fig. 14.2

(4) *Action of water.* It is decomposed by water in two stages forming phosphorus oxy-chloride $POCl_3$ and phosphoric and H_3PO_4 respectively.

$$PCl_5 + H_2O \longrightarrow POCl_3 + 2HCl$$

$$POCl_3 + 3H_2O \longrightarrow H_3PO_4 + 3HCl$$

$$\overline{PCl_5 + 4H_2O \longrightarrow H_3PO_4 + 5HCl}$$

(5) *Action with hydroxy compounds.* It reacts with organic compounds containing the hydroxy group, *e.g.*, acids, alcohols etc.

$$\underset{\text{Acetic acid}}{CH_3COOH} + PCl_5 \longrightarrow \underset{\text{Acetyl chloride}}{CH_3COCl} + POCl_3 + HCl$$

$$\underset{\text{Ethyl alcohol}}{C_2H_5OH} + PCl_5 \longrightarrow \underset{\text{Ethyl chloride}}{C_2H_5Cl} + POCl_3 + HCl$$

Uses. It is used for detecting the *OH* group in organic substances.

Structure. In the vapour state both PF_5 and PCl_5 exist as trigonal bipyramidal molecules which is consistent with electron pair repulsion arguments. Experiments on the condensed state of PCl_5 reveal different species. X-ray analysis of the solid compound shows that crystal consists of tetrahedral PCl_4^+ and octahedral PCl_6^- ions. The σ bonds in the tetrahedral PCl_4^+ species involve sp^3 hybrid phosphorus orbitals, whereas those in the octahedral PCl_6^- ions require d^2sp^3 hybrids. In the solid state PBr_5 is best formulated as $(PBr_4^+)Br^-$. Probably bromide ions are too large to form stable PBr^- ions.

Q. 12. Write the structure of PCl_5 in solid and vapour states. Compare the structure with PBr_5 in the solid state. (*Lucknow, 2000*)

Ans. See Q.11.

OXY-HALIDES OF PHOSPHORUS

Q. 14. Give the methods of preparation, properties and structure of phosphorus oxychloride ($POCl_3$).

(*Kurukshetra, 2001*)

Ans. Preparation. (*i*) It is obtained by the action of water on phosphorus pentachloride.

$$PCl_5 + H_2O \longrightarrow POCl_3 + 2HCl$$

(*ii*) It is obtained more easily by distilling phosphorus trichloride with dry potassium chlorate.

$$3PCl_3 + KClO_3 \longrightarrow 3POCl_3 + KCl$$

(*iii*) It may also be prepared by heating the pentachloride with oxalic or boric acid.

$$PCl_5 + H_2C_2O_4 \longrightarrow POCl_3 + HCl + CO_2 + CO$$

$$3PC_5 + 2H_3BO_3 \longrightarrow 3POCl_3 + B_2O_3 + 6HCl$$

Properties. (*i*) It is colourless fuming liquid (b.p. 380 K, m.p. 274 K).

(*ii*) It also reacts with many hydroxy-containing organic compounds.

$$POCl_3 + 3ROH \longrightarrow PO(OH)_3 + 3RCl$$

(*iii*) With water, it forms orthophosphoric acid.

$$POCl_3 + 3H_2O \longrightarrow H_3PO_4 + 3HCl$$

Structure. The following structure has been assigned to phosphorus (V) oxychloride.

Cl
Cl > P → O
Cl

Q.15. Write a note on interhalogen compounds.

(*Lucknow, 2000; Kanpur, 2000; Kurukshetra, 2001; Mumbai, 2002; Guwahati, 2002; Rohilk hand, 2003; Nagpur, 2005*)

Ans. The binary compounds formed by halogens among themselves are known as interhalogen compounds. The halogens form various compounds which are binary combination of halogens among themselves. With the exception of BrCI, ICI, ICl_3 and IBr the compounds are all halogen fluorides. It will be noted that all the interhalogen compounds are of the type AX_n. where *n* is an odd number and X is always the lighter and smaller halogen than A. The value of *n* is always odd. This is odd number of unpaired electrons available in the ground state, first, second and third excited states of the halogen atoms forming the interhalogen compounds.

These compounds can be divided into the following four types.

AX type	*AX_3 type*	*AX_5 type*	*AX_7 type*
ClF	ClF_3	BrF_5	IF_7
BrF	BrF_3	IF_5	
BrCl	ICl_3		
ICl.			
IBr			
IF (unstable)			

[Here *A* represents the larger halogen atom and X the smaller one]

The bonds in these compounds are essentially covalent because of the small electronegativity difference. The compounds formed in the AX and AX_3 group are those where the electronegativity difference is not too great; the higher types AX_5 and AX_7 are shown by large atoms such as Br and I combining with small atoms such as F, because it is possible to pack more small atoms round a larger one.

Q. 16. Why are interhalogens more reactive than halogens?

Ans. Interhalogen compounds, say of the type ClF, BrF and ClF_3, have polarity in the bonds whereas halogens, say F_2 or Cl_2, have non-polar bonds. A polar bond is easier to break compared to a nonpolar bond, that is the reason why interhalogen compounds are more reactive. The bonds in such compounds can be easily broken.

INTERHALOGENS OF AX TYPE

Q. 17. Describe the methods of preparation, properties and uses of iodine monochloride ICl. *(Guwahati, 2002)*

Ans. Preparation. (*i*) It can be prepared by the direct combination of liquid chlorine with solid iodine under controlled conditions of temperature and pressure.

$$\underset{\text{Solid}}{I_2} + \underset{\text{Liquid}}{Cl_2} \longrightarrow 2ICl$$

(*ii*) It may also be obtained by heating iodine with potassium chlorate.

$$KClO_3 + I_2 \longrightarrow KIO_3 + ICl$$

Properties. (*i*) It is a reddish oil and is readily hydrolyzed by water.

$$ICl + H_2O \rightleftharpoons HCl + HOI$$

(*ii*) It is also decomposed by sodium hydroxide.

$$ICl + NaOH \longrightarrow NaCl + HOI$$

(*iii*) When electricity is passed through iodine monochloride, iodine is liberated at the cathode and chlorine at the anode, indicating that the compound is a chloride of monovalent positively charged iodine.

$$2ICl \longrightarrow I^+ + [ICl_2]^-$$

(*iv*) With alkali metal halides it gives polyhalides.

$$KCl + ICl \longrightarrow KICl_2$$

Uses. (*i*) It is used to iodinate organic compounds, though chlorination may also occur depending on the condition of reaction.

$$\text{Salicylic acid} \xrightarrow{\text{ICl vapour}} \text{Chlorination}$$

$$\text{Salicylic acid} \xrightarrow[\text{nitrobenzene}]{\text{IClin}} \text{Iodination}$$

(*ii*) It is used as Wij's reagent in the estimation of iodine number which is a measure of unsaturation of fats and oils.

$$-CH=CH- + ICl \longrightarrow -\underset{\displaystyle I}{\underset{|}{CH}}-\underset{\displaystyle Cl}{\underset{|}{CH}}-$$

Q. 18. Describe the preparation and properties of iodine monobromide, IBr.

Ans. Preparation. It is prepared by direct combination of bromine with iodine.

$$I_2 + Br_2 \longrightarrow 2IBr$$

Properties. (*i*) It crystallizes at 315°C to a black solid.

(*ii*) Iodine monobromide undergoes decomposition when heated.

$$2I\,Br \rightleftharpoons I_2 + Br_2$$

Q. 19. Describe the preparation and properties of chlorine monofluoride, ClF.

(*Nagpur, 2003*)

Ans. Preparation. (*i*) It is best prepared by mixing Cl and ClF_3 in equimolar ratios.

$$Cl_2 + ClF_3 \longrightarrow 3ClF$$

(*ii*) It is also obtained by heating fluorine and chlorine together in a copper vessel at 425 K.

(*iii*) It is obtained when chlorine reacts with hydrogen flouride at liquid air temperature.

$$Cl_2 + 2HF \longrightarrow 2ClF + H_2$$

Properties. (*i*) Chlorine monofluoride dissociates on strong heating

$$2ClF \longrightarrow Cl_2 + F_2$$

(*ii*) It reacts with a number of metals forming their fluorides even more vigorously than fluorine.

$$2Na + 2ClF \longrightarrow NaF + Cl_2$$

Q. 20. Give the preparation and properties of (*a*) Bromine-monofluoride, (*b*) Bromine monofluoride.

Ans. (*a*) Bromine monofluoride. Preparation. It is obtained when bromine trifluoride and bromine vapours are mixed together.

$$BrF_3 + Br_2 \longrightarrow 3BrF$$

Properties. (*i*) It is a reddish brown liquid b.p. 296 K.

(*ii*) It is unstable and decomposes readily into bromine trifluoride and bromine.

$$3BrF \longrightarrow BrF_3 + Br_2$$

(*b*) Bromine monochloride, BrCl. Preparation. BrCl is obtained in traces when chlorine and bromine vapours are mixed together. However when chlorine gas is passed into liquid bromine covered with water, the formation of a solid hydrate, $BrCl4H_2O$, has been reported.

Properties. With CsI it forms polyhalide as

$$CsI + BrCl \longrightarrow CsIBrCl$$

INTERHALOGENS OF AX_3 TYPE

Q. 21. Give the method of preparation, properties and structure of iodine trichloride, ICl_3.

(*Nagpur, 2003; Mumbai, 2004*)

Ans. Preparation. It is obtained by action of excess of chlorine on iodine or by treating the monochloride with chlorine.

$$I_2 + 3Cl_2 \longrightarrow 2ICl_3$$

$$ICl + Cl_2 \rightleftharpoons ICl_3$$

Properties. (*i*) It forms fluffy orange powder which is deliquescent in air and has a penetrating pungent smell.

(*ii*) With alkali metal halides it gives polyhalides.

$$KCl + ICl_3 \longrightarrow KICl_4$$

(*iii*) It dissociates at 341 K into ICl and Cl_2.

$$ICl_3 \rightleftharpoons ICl + Cl_2$$

Structure. In the solid state, two molecules combine to form the dimer $(ICl_3)_2$ which is a flat molecule.

Cl Cl Cl
I I
Cl Cl Cl

Q. 22. Give the preparation and properties of bromine trifluoride, BrF_3.

(*Nagpur, 2003*)

Ans. Preparation. It is obtained by mixing bromine and fluorine vapours or by the action of ClF_3 on bromine at 238 K.

$$Br_2 + 3F_2 \longrightarrow 2BrF_3$$
$$2ClF_3 + Br_2 \longrightarrow 2BrF_3 + Cl_2$$

Properties. (*i*) Bromine trifluoride is a greenish-yellow fuming liquid having b.p. 400 K.

(*ii*) With alkali fluorides it forms polyhalides.

$$KF + BrF_3 \longrightarrow KBrF_4$$

(*iii*) It converts many metal oxides, halides and even metals to their respective fluorides.

$$6MO + 4BrF_3 \longrightarrow 6MF_2 + 2Br_2 + 3O_2$$

where M is a divalent metal.

(*iv*) It ionizes considerably as

$$2BrF_3 \rightleftharpoons BrF_2^+ + BrF_4^-$$

Q. 23. What are various interhalogen compounds formed by bromine? Discuss the structures of BrF_3 and BrF_5^-. (*Bangalore, 2005*)

Ans. Various interhalogen compounds of bromine are:

(1) Iodine bromide IBr.

(2) Bromine chloride BrCl

(3) Bromine fluoride BrF

(4) Bromine trifluoride BrF_3

(5) Bromine pentafluoride BrF_5

Structure of BrF_3. This molecule involves sp^3d hybridisation. The T shape structure can be illustrated as

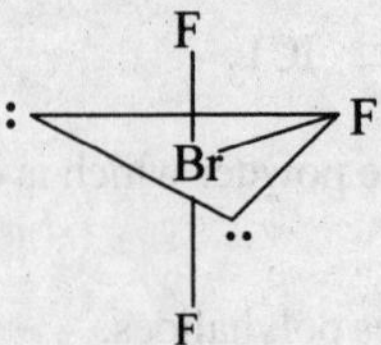

Structure of BrF_5. This molecule involves $sp^3 d^2$ hybridisation. Square pyramidal structure can be represented as

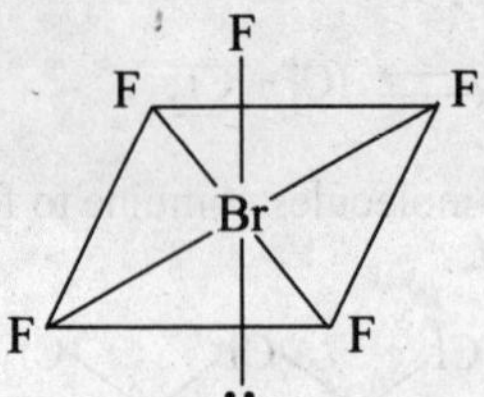

Q. 24. Give the preparation and properties of chlorine trifluoride ClF_3.

(*Mumbai, 2002; Bangalore, 2004*)

Ans. Preparation. It is prepared by passing a mixture of chlorine monofluoride and fluorine (excess) through a nickel tube maintained at 523 K.

$$ClF + F_2 \longrightarrow ClF_3$$

Properties. It is colourless gas which condenses to a green liquid and freezes to a white solid.

Q. 25. Discuss the structure of chlorine trifluoride (ClF_3).

Ans. The structure of ClF_3 is explained as follows. Structural studies show that the molecule is *T*-shaped with bond angle about 90°. Precisely it is 87°– 40'. The distortion is due to the repulsion between lone pairs of electrons. Two bond lengths are equal but different from the third, which is to be expected because a trigonal bipyramid is not a regular shape (Fig. 14.3).

	$3s$	$3p$	$3d$
Electronic configuration of chlorine atom in ground state	↑↓	↑↓ ↓↑ ↑	
Electronic configuration in excited state	↑↓	↑↓ ↑ ↑	↑

Three unpaired electrons form bonds with 3 fluorine atoms

sp^3d hybridization trigonal bipyramid with two positions occupied by lone pairs

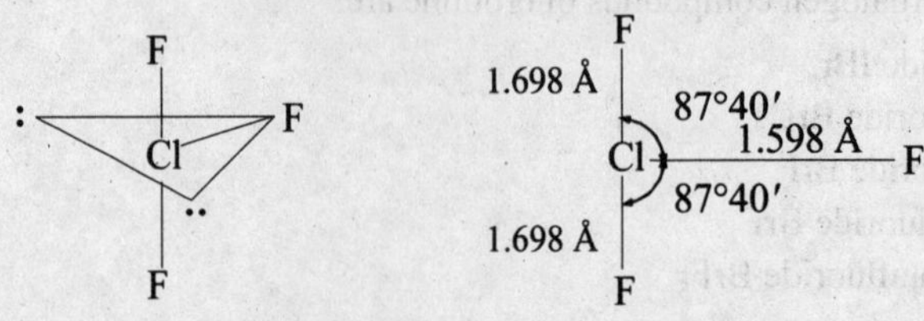

Fig. 14.3. Structure of ClF_3.

Q. 26. Give the preparation and properties of bromine pentafluoride, BrF_5.

Ans. Preparation. It is obtained by the action of bromine on excess of fluorine (1:5 molar ratio)

$$Br_2 + 5F_2 \longrightarrow 2BrF_5$$

Properties. (*i*) It is colourless liquid at ordinary temperature and is highly reactive.

(*ii*) Many metals and non-metals catch fire when they are dropped into the liquid.

(*iii*) Hydrolysis of bromopentafluoride gives HF and $HBrO_3$.

$$BrF_5 + 3H_2O \longrightarrow 5HF + HBrO_3$$

Q. 27. Give the preparation and properties of iodine pentafluoride IF_5.

(*Mumbai, 2002; Guwahati, 2002*)

Ans. Preparation. (*i*) It is obtained by heating silver fluoride with iodine.

$$10AgF + 6I_2 \longrightarrow 2IF_5 + 10AgI$$

(*ii*) It is also obtained by the direct combination of the elements under controlled condition or by the action of fluorine pentoxide.

$$I_2 + 5F_2 \longrightarrow 2IF_5$$

$$2I_2O_5 + 10F_2 \longrightarrow 4IF_5 + 5O_2$$

Properties. (*i*) Liquid FI_5 conducts electricity and self-ionizes.

$$2IF_5 \rightleftharpoons IF_4^+ + IF_6^-$$

(*ii*) It undergoes hydrolysis as

$$IF_5 + 3H_2O \longrightarrow 5HF + HIO_3$$

(*iii*) Complex salts are formed by heating IF_5 with alkali halides.

$$KI + IF_5 \longrightarrow K[IF_6]$$

Q. 28. Discuss the structure of iodine pentafluoride. (*Mumbai, 2004*)

Ans. IF_5 is formed by sp^3d^2 hybridisation and hence is octahedral in structure as shown below (Fig. 14.4).

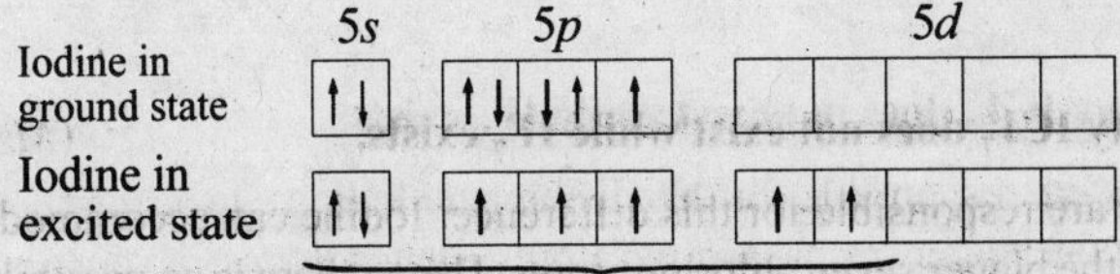

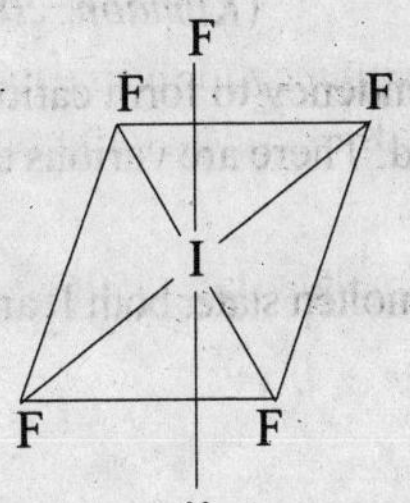

sp^3d hybridization octahedral with one position occupied by lone pair

Square pyramid

Fig. 14.4. Structure of IF_5.

The presence of one lone pair in the molecule converts its shape from regular octahedron to square pyramid.

INTERHALOGENS OF AX_7 TYPE

Q. 29. Describe the preparation, properties and structure of iodine heptafluoride, IF_7.
(*Mumbai, 2002; Guwahati, 2002; Madras, 2004; Bangalore, 2005*)

Ans. Preparation. It is formed by combination of the pentafluoride with fluorine at 525-550 K.

$$IF_5 + F_2 \longrightarrow IF_7$$

Properties. (*i*) It undergoes hydrolysis giving HF and H_5IO_6.

$$IF_7 + 6H_2O \rightarrow 7HF + H_5IO_6$$

(*ii*) It is a powerful fluorinating agent.

Structure of IF_7. This compound has a pentagonal bipyramidal structure as it is formed by sp^3d^3 hybridisation as shown in Fig. 14.5.

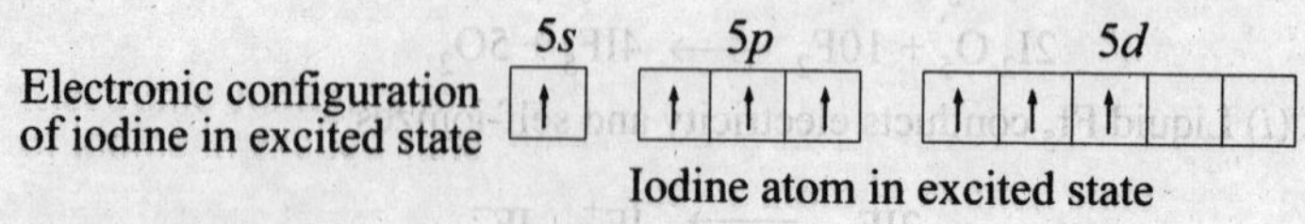

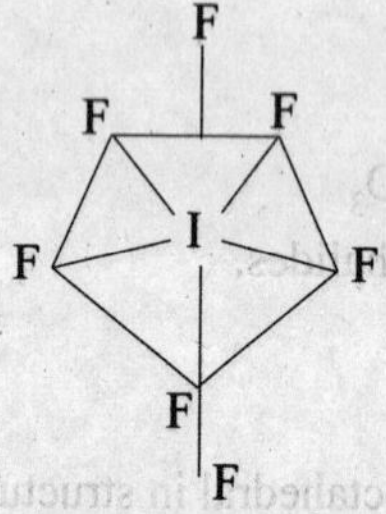

Fig. 14.5. Structure of IF_7.

Q. 30. Explain why ICl_7 does not exist while IF_7 exists. (*M. Dayanand, 2000*)

Ans. Steric factors are responsible for this difference. Iodine can accommodate seven fluorine atoms around it but not the bigger seven chlorine atoms. Hence there is no possibility of formation of ICl_7.

Q. 31. Give evidence to show that iodine behaves as a cation.

(*Kumaon, 2000; Nagpur, 2008*)

Ans. Iodine shows a strong basic character. It has the tendency to form cations. Positive iodine compounds are many and their structures are well established. There are various evidences for existence of I^+ and I^{3+}.

(*a*) **Evidence for I^+.** (*i*) ICl conducts electricity when in molten state, both I_2 and Cl_2 are liberated at the anode but only I_2 at the cathode.

The ionization is probably of following type:

$$2ICl \rightleftharpoons I^+ + ICl_2^-$$

At cathode

$$2I^+ + 2e^- \longrightarrow I_2$$

At anode

$$2ICl_2^- - 2e^- \longrightarrow I_2 + 2Cl_2$$

(*ii*) If the electrolysis of molten ICN or a solution of ICN in pyridine is carried out, iodine is liberated at the cathode to confirm that iodine forms positive ions. Thus, this suggests the more simple type of ionization of ICN as

$$ICN \rightleftharpoons I^+ + CN^-$$

(*b*) **Evidence for I^{3+}.** (*i*) In the case of ICl_3, I_2 and Cl_2 are liberated at both electrodes. This is explained by ionisation of ICl_3 into ICl_2^+ and ICl_4^-.

(*ii*) If I_2 is treated with fuming HNO_3 and acetic anhydride, iodine triacetate, $I(CH_3COO)_3$ is formed. If electricity is passed through the solution of $I(CH_3COO)_3$ using silver electrodes, one equivalent of AgI is deposited at the cathode for the passage of every three Faradays of electricity. This proves the ionization :

$$I\,(acetate)_3 \rightleftharpoons I^{3+} + 3\,(acetate^-)$$

(*iii*) If I_2 is oxidized with conc. HNO_3 in the presence of acetic anhydride and phosphoric acid, iodide phosphate IPO_4 is formed.

$$I_2 + Ac_2O + H_3PO_4 \xrightarrow{\text{Conc. } HNO_3} IPO_4 + \text{Acetic acid}$$

If iodine phosphate ionizes, I^{3+} could be expected,

POLYHALIDES

Q. 32. What are polyhalides? Discuss the structure of ICl_2^- and ICl_4^-.

(*Rohilkhand, 2003; Nagpur, 2008*)

Ans. Halides often associate with molecules of halogens or interhalogens to form polyhalide ions, *e.g.*, the solubility of iodine in water is greatly increased by the addition of iodide ions, due to the formation of the tri-iodide ion, I_3^-

$$I_2 + I^- \longrightarrow I_3^-$$

More complex ions such as I_5^-, I_7^-, I_9^- combined with large metal ion are known.

A number of polyhalides are known which contain 2 or 3 different halogens, *e.g.*, $K[ICl_2]$, $K[ICl_4]$, $Cs[IBrCl]$. These are formed from metal halides and interhalogens.

$$KCl + ICl \longrightarrow K[ICl_2]$$

$$KCl + ICl_3 \longrightarrow K[ICl_4]$$

The structures of polyhalides are known. The trihalides, *e.g.*, $K[ICl_2]$ contain a linear trihalide ion ICl_2^-. This is shown in Fig. 14.6. Structure of pentahalide ion $[ICl_4]^-$ is illustrated in Fig. 14.7. Although sp^3d^2 hybridisation of iodine orbitals takes place giving rise to octahedral geometry, the actual shape is square planar because two positions are occupied by lone pairs.

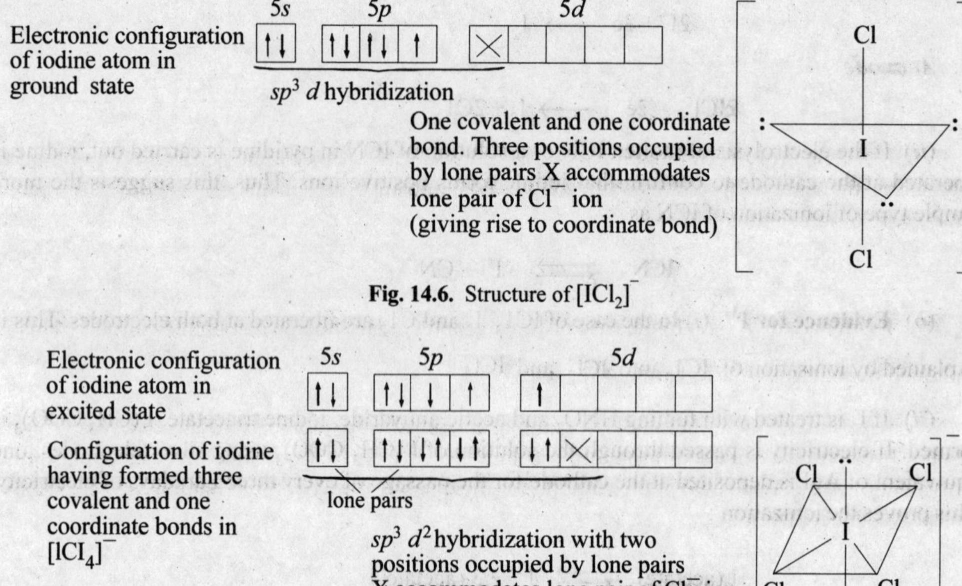

Fig. 14.6. Structure of $[ICl_2]^-$

Fig. 14.7. Structure of $[I\,Cl_4^-]$ ion

Q. 33. ***I_2* is usually insoluble in water. It dissolves readily in a solution of KI. Discuss the structure of product in this solution.**

Ans. Iodine dissolves in a solution of potassium iodide due to the following reaction:

$$I_2 + KI \rightleftharpoons KI_3$$

The structure of I_3^- is as follows:

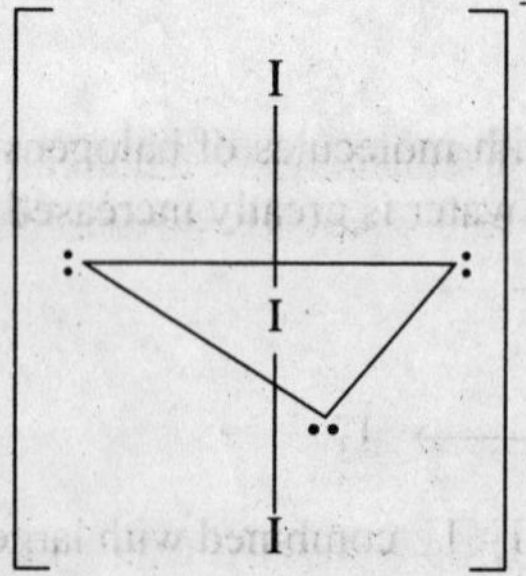

Q. 34. What are pseudohalogens and pseudohalides? Draw a resemblance between pseudohalogens and halogens. (*M. Dayanand, 2000; Delhi, 2002; Kurukshetra, 2002*)

Ans. A molecule consisting of two or more than two electronegative atoms which resemble the halogens is called pseudohalogen or halogenoid. Important pseudohalogens are cyanogen $(CN)_2$ thiocyanogen $(SCN)_2$, selenocyanogen $(SeCN)_2$ and azido carbon disulphide $(SCSN_3)_2$. These pseudo-halogens give rise to anions which resemble the halide ions in their behaviour and are called pseudohalides or halogenoid ions. The important pseudohalides are cyanide CN^-, thiocyanate SCN^-, cyanate OCN^-, selenocyanate $SeCN^-$, azido dithiocarbonate $SCSN_3^-$, iso-cyanate ONC^- and azide N_3^-.

Pseudohalogens or pseudohalide ions resemble halogens or halide ion respectively as revealed by the following reactions:

(1) Pseudohalogens like halogens combine with hydrogen to form monobasic acids (*e.g.*, HCN, HC1).

(2) The Ag^+, Pb^{2+} and $Hg^{2}{}_{2}{}^{+}$ salts of pseudohalogen and halogen acids for example (AgCN, AgC1) are insoluble in water.

(3) They react with alkalies to form simple and oxypseudohalides.

$$\underset{\text{Azido carbon disulphide}}{(SCSN_3)_2} + 2KOH \longrightarrow KSCSN_3 + KOSCSN_3 + H_2O$$

(4) They form covalent compounds which resemble covalent halides.

Covalent halogenoids	$CO(N_3)_2$	ICN	$Si(NCS)_4$
Covalent halides	$CoC1_2$	IC1	$SiCl_4$

(5) They add to carbon-carbon double bonds like halogens

$$H_2C = CH_2 + (SCN)_2 \longrightarrow \underset{SCN}{H_2C} - \underset{SCN}{CH_2}$$

$$H_2C = CH_2 + Br_2 \longrightarrow \underset{Br}{H_2C} - \underset{Br}{CH_2}$$

(6) Like halogens, pseudohalogens are also oxidizing agents.

(7) Pseudohalogens may be obtained either by chemical oxidation of the pseudohalogen acid or by thermal decomposition of higher valent pseudohalides. Halogens are also prepared similarly.

$$4HSCN + MnO_2 \longrightarrow Mn(SCN)_2 + 2H_2O + (SCN)_2$$

$$4HC1 + MnO_2 \longrightarrow MnC1_2 + C1_2 + 2H_2O$$

$$Pb(SCN)_4 \xrightarrow{\text{Heating}} Pb(SCN)_2 + (SCN)_2$$

$$PbCl_4 \xrightarrow{\text{Heating}} PbC1_2 + C1_2$$

(8) They form a large number of complexes similar to halide complexes, *e.g.*, $[Cu(CN)_4]^{2-}$ and $[Cu\,C1_4]^{2-}$; $[Co(SCN)_4]^{2-}$ and $[COC1_4]^{2-}$ etc.

(9) Pseudohalogens like halogens can form interhalogenoid compounds such as CNSCN, CNN_3 and mixed halogen pseudohalogen compounds such as CNCl, CNBr, CNI, $C1N_3$, BrN_3, IN_3 etc.

(10) They form polyhalogenoids which are similar to polyhalides, for example, $K(SeCN)_3$, $NH_4(SeCN)_3$ which are similar to KI_3.

Q. 35. Give the methods of preparation and properties of cyanogen $(CN)_2$.

(Mumbai, 2002)

Ans. Preparation. (*i*) It is obtained when potassium cyanide reacts with copper sulphate solution.

$$2CuSO_4 + 4KCN \longrightarrow Cu_2(CN)_2 + (CN)_2 + 2K_2SO_4$$

(*ii*) It may be obtained when mercuric cyanide is heated with mercuric chloride.

$$Hg(CN)_2 + HgC1_2 \longrightarrow Hg_2C1_2 + (CN)_2$$

Properties. (*i*) Cyanogen is a colourless gas, soluble in water.

(*ii*) On cooling, it readily condenses to a colourless liquid (b.p. 252 K) which freezes to a white solid (m.p. 245 K).

(*iii*) Cyanogen undergoes polymerisation to give insoluble paracyanogen $(CN)_x$ when heated to 773 K.

(*iv*) It reacts with hydrogen sulphide gas to form dithioxamide.

$$(CN)_2 + 2H_2S \longrightarrow H_2N\overset{\overset{S}{\|}}{C} - \overset{\overset{S}{\|}}{C}NH_2$$

Dithioxamide

(*v*) Cyanogen gets hydrolysed by water slowly producing ammonium formate, urea and ammonium oxalate as the final products.

(*vi*) In alkaline medium, cyanogen reacts readily to give cyanide and cyanate.

$$(CN)_2 + 2OH^- \longrightarrow CN^- + OCN^- + H_2O$$

Structure. Electron diffraction studies establish the following structure for cyanogen.

$$:N \equiv C - C \equiv N:$$

$$H_2N - \overset{\overset{O}{\|}}{C} - \overset{\overset{O}{\|}}{C} - NH_2 \xleftarrow{H_2O} (CN)_2 \xrightarrow{H_2O} HCN + HOCN$$

Oxamide ← Cyanogen

Oxamide $\xrightarrow{H_2O}$ $(NH_4)_2C_2O_4$ (Ammonium oxalate)

HCN → $NH_4O\overset{\overset{O}{\|}}{C} - H$ (Ammonium formate)

HOCN $\xrightarrow{H_2O}$ NH_4OCN (Ammonium cyanate) $\xrightarrow{H_2O}$ $Co(NH_2)_2$ (Urea)

Q. 36. Write notes on

(*i*) Ionic cyanides

(*ii*) Covalent cyanides

(*iii*) Complex cyanides

Ans. (*i*) *Ionic cyanides* are formed by alkali metals (except Li) thallium and alkaline earth metals. At ordinary temperature Na, K and Rb compounds have NaCl lattice arrangement whereas Cs and Tl compounds have CsCl structure. In the cyanide ion, the C – N bond distance is 1.05 Å and the bond is about 57% ionic in character. It involves the resonance forms.

$$[:C:::N:]^- \longleftarrow \left[\overset{+}{:C}::\ddot{N}:^-\right]^-$$

Cyanide ion may be linked to metal ion through carbon, nitrogen or both.

(*ii*) *Covalent cyanides* are formed by Ag^+, Pt^{2+}, Ni^{2+}, Cu^{2+} etc. Ag CN is insoluble in water but soluble in excess of cyanide ion due to complex formation as $[Ag(CN)_2]^-$. X-ray diffraction data on AgCN indicates the presence of infinite chain of alternating Ag and CN groups, the Ag atom being joined by cyanide group both through C and N as under:

$$\cdots\cdots -Ag - C \equiv N - Ag - \cdots\cdots$$

(*iii*) *Complex cyanides* are formed by transition metals of 2, 4, 6 and 8 coordination number. Metal derivatives of complex cyanides are also formed by transition metal ion especially of 3*d*-series.

Q. 37. Give the methods of preparation, properties and structure of thiocyanogen, $(SCN)_2$.

Ans. Preparation. (*i*) It may be prepared by the reaction of bromine on silver thiocyanate dissolved in ether.

$$2AgSCN + Br_2 \longrightarrow (SCN)_2 + 2AgBr$$

(*ii*) It is best prepared by the oxidation of lead thiocyanate with bromine.

$$Pb\,(SCN)_2 + Br_2 \longrightarrow (SCN)_2 + Pb\,Br_2$$

Properties. (*i*) Water hydrolyses it to thiocyanic, hydrocyanic and sulphuric acids.

$$3(SCN)_2 + 4H_2O \longrightarrow SHSCN + HCN + H_2\,SO_4$$

(*ii*) It shows oxidizing properties similar to halogens.

$$(SCN)_2 + 2I^- \longrightarrow 2SCN^- + I_2$$

$$(SCN)_2 + 2S_2\,O_3^{2-} \longrightarrow 2SCN^- + S_4O_3^{2-}$$

$$(SCN)_2 + H_2S \longrightarrow 2H^+ + 2SCN^- + S$$

$$(SCN)_2 + 2\,CuSCN \longrightarrow 2Cu^{2+} + 4SCN^-$$

Structure. The dimer $(SCN)_2$ has a symmetrical structure

$$N \equiv C - S - S - C \equiv N$$

Q.38. Give the preparation, properties and structure of thiocyanate, SCN^-.

Ans. Preparation. Thiocyanate ion, SCN^-, is prepared by fusing cyanide ion with sulphur.

$$CN^- + S \xrightarrow{\text{Fusion}} SCN^-$$

Properties. (*i*) It is a colourless ion and resembles the chloride ion to a considerable extent.

(*ii*) Most of the metal thiocyanates except Ag^+, Hg^{2+}_2, Cu^+ and Au^+ thiocyanates are soluble in water.

(*iii*) Thiocyanate ion is decomposed by heating with (1:1) sulphuric acid giving COS (carbonyl-sulphide)

$$SCN^- + 2H_2\,SO_4 + H_2O \longrightarrow COS + (NH_4)\,HSO_4 + HSO^-_4$$

(*iv*) With silver nitrate, a precipitate of silver thiocyanate is formed which is insoluble in water and dilute acids.

$$Ag\,NO_3 + SCN^- \longrightarrow AgSCN + NO_3^-$$

(*v*) Thiocyanate ion gives a characteristic and very sensitive reaction with Fe^{3+} ion forming red coloured ferric thiocyanate.

$$Fe^{3+} + 3SCN^- \longrightarrow \underset{\text{Blood red colour}}{Fe\,(SCN)_3}$$

(*vi*) With iodine or bromine, lead tetraacetate or manganese dioxide, the thiocyanate ion is oxidized to thiocyanogen $(SCN)_2$ as follows:

$$2SCN^- + Br_2 \longrightarrow (SCN)_2 + 2Br^-$$

Structure. Thiocyanate ion is linear and has the following structure;

Raman spectra of SCN^- ion strong the presence of $C \equiv N$ indicating that structure which I is the only structure which can be most favourable one. However some other properties of thiocyanate ion can be explained in terms of structure II.

Q. 39. Describe the methods of preparation, properties and structure of selenocyanogen $(SeCN)_2$. (*Mumbai, 2002*)

$$\underset{\text{I}}{:\ddot{S}\ :C:::N:} \quad \text{or} \quad \underset{\text{II}}{:\ddot{S}::C::\ddot{N}^-}$$

Ans. Preparation. (*i*) It may be prepared by the reaction of iodine with AgSeCN dissolved in ether.

$$2Ag\,SeCN + I_2 \longrightarrow 2AgI + (SeCN)_2$$

(*ii*) It may also be prepared by decomposition of lead (IV) selenocyanate. In this method lead (IV) acetate is heated with KSeCN. Lead (IV) selenocyanate is formed which being unstable decomposes to yield Pb $(SeCN)_2$ and Se $(CN)_2$.

$$Pb\,(CH_3COO)_4 + 4KSeCN \longrightarrow Pb\,(SeCN)_4 + 4CH_3COOK$$

$$Pb\,(SeCN)_4 \longrightarrow Pb\,(SeCN)_2 + (SeCN)_2$$

Properties. (*i*) Its molecular weight in benzene agrees with dimeric formula $(SeCN)_2$.

(*ii*) It is quite stable when kept dry and under pressure.

(*iii*) It is a weak oxidising agent.

(*iv*) It reacts with water as follows:

$$2\,(Se\,CN)_2 + 3H_2O \longrightarrow H_2\,SeO_3 + 3HSeCN + HCN$$

Q. 40. Give the preparation, properties and structure of azido carbon disulphide $(SCSN_3)_2$.

Ans. Preparation. It may be prepared by oxidation of potassium azido dithiocarbonate $KSCN_3$ with hydrogen peroxide. The potassium salt required for the purpose is obtained by the interaction of a solution of potassium azide with carbon disulphide as follows:

$$KN_3 + CS_2 \longrightarrow KSCSN_3$$

$$2KSCSN_3 + H_2O_2 \longrightarrow (SCSN_3)_2 + 2KOH$$

Properties. (*i*) It is white crystalline solid, sparingly soluble in water.

(*ii*) The reaction of $(SCSN_3)_2$ with alkalies takes place as under:

$$(SCSN_3)_2 + 2OH^- \longrightarrow SCSN^-_3 + OSCSN^-_3 + H_2O$$

$$3OSCS\,N_3^- \longrightarrow 2SCS\,N_3^- + SCSN_3\,O_3^-$$

(*iii*) It is unstable and may decompose violently on shock or impact.

$$(SCSN_3)_2 \longrightarrow 2N_2 + 2S + (SCN)_2$$

Structure. Probable structure of azido carbon disulphide is given below.

$$N \equiv N = N - \underset{\underset{\displaystyle S}{\|}}{C} - S - S - \overset{\overset{\displaystyle S}{\|}}{C} - N = N \equiv N$$

Q. 41. Give the preparation, properties and structure of cyanate, OCN^- ion.

Ans. Preparation. The cyanate ion is obtained by mild oxidation of aqueous cyanide ion with metal oxides.

$$CN^- (aq) + PbO (s) \longrightarrow OCN^- (aq) + Pb (s)$$

Properties. (*i*) It gives a bright blue colour with cobalt chloride solution acidified with acetic acid which is due to the formation of a complex ion $[Co(OCN)_4]^{2-}$.

$$CoCl_2 + 4OCN^- \longrightarrow \underset{\text{Bright blue colour}}{[Co(OCN)_4]^{2-}} + 2Cl^-$$

Structure. The cyanate ion is linear and can be represented by the following structure:

$$[:\ddot{O}::C::\dot{N}:]^- \quad \text{or} \quad [O=C=N]^-$$

Q. 42. Give the preparation and properties of oxy-cyanogen $(OCN)_2$.

Ans. Preparation. It is prepared by treating silver cyanate with iodine in carbon disulphide medium at a very low temperature.

$$2AgOCN + I_2 \longrightarrow 2AgI + (OCN)_2$$

Properties. (*i*) It is obtained as a white crystalline solid m.p. 261 K by the above method.

(*ii*) Oxycyanogen reacts with non-metals giving the compound as $Si(OCN)_4$, $P(OCN)_3$ and $Sb(OCN)_3$.

(*iii*) In the presence of dilute acids, it liberates iodine from iodides.

$$(OCN)_2 + 2I^- + 4H^+ + 2H_2O \longrightarrow 2NH_4^+ + I_2 + 2CO_2$$

Q. 43. Write a note on azide ion, N_3^-.

Ans. Azide ion, N_3^-, is derived from its hydrogen compound, hydrazoic acid HN_3. The Ag^+, Pb^{2+}, Hg_2^{2+}, Cu^+, Tl^+ azides are insoluble in water whereas those of alkali metals are soluble. Azides of a number of heavy metals explode on shock and lead azide is used as a detonator. The azide ion N_3^- forms a variety of complexes with metal ion, *e.g.*, $[Cu(NH_3)_4](N_3)_2$, $[(Cu(en)_2](N_3)_2$, $[Cu(NH_3)_2(N_3)_2]$, $K_2[Cu(N_3)_4]$, where en = ethylene diamine.

Structure. The structure of azide ion is symmetrical and linear with N – N distance = 1.15 Å. It exists as hybrid of a number of reasonating structures.

$$:\overset{-}{\underset{}{\dot{N}}}::\overset{+}{N}::\overset{-}{\dot{N}}: \longleftrightarrow :N:::\overset{+}{N}:\overset{2-}{\underset{..}{\ddot{N}}}: \longleftrightarrow :\overset{2-}{\underset{..}{\ddot{N}}}:\overset{+}{N}:::N:$$

Q. 44. Describe the interhalogens of AX, AX_3, AX_7 types. (*Kurukshetra, 2001*)

Ans. See Qs. 17, 21 and 29.

ORGANOMETALLIC COMPOUNDS*

Q. 1. Describe methods of preparation, properties and structures of alkyl and aryl lithium.

(*Awadh, 2000; Indore, 2001; Lucknow, 2010; Rajasthan 2010*)

Ans. Preparation. Organolithium compounds can be prepared as follows:

(*i*) *By direct reaction of Li with alkyl or arylchloride in benzene or petroleum.*

$$2Li + RCl \xrightarrow{\text{Solvent}} LiR + LiCl \qquad (R = \text{alkyl or aryl})$$

$$2Li + C_2H_5Cl \xrightarrow{\text{Solvent}} C_2H_5Li + LiCl$$

(*ii*) *By metal-halogen exchange in ether.* Lithium aryls are best prepared by this method using *n*-butyl lithium and an aryl iodide.

$$n - C_4H_9Li + ArI \xrightarrow{\text{Ether}} LiAr + n\text{-}C_4H_9I \ (Ar = \text{aryl})$$

$$n - C_4H_9Li + \text{(3-bromopyridine)} \xrightarrow{\text{Ether}} \text{(3-lithiopyridine)} + n - C_4H_9Br$$

(*iii*) Vinyl-allyl and other unsaturated derivatives can be obtained by transmetallation.

$$4Li\,Ph + Sn\,(CH = CH_2)_4 \xrightarrow{\text{Ether}} 4LiCH = CH_2 + SnPh_4$$

(*iv*) Methyllithium is also prepared by exchange through the interaction of $n–C_4H_9Li$ and CH_3I in hexane at low temperatures where it precipitates as insoluble white crystals.

$$n - C_4H_9Li + CH_3I \xrightarrow{\text{hexane}} CH_3Li + n - C_4H_9I$$

(*v*) *By metal-hydrogen exchange (metallation) in ether.*

$$LiPh + \underset{\text{Furan}}{C_4H_4O} \xrightarrow{Et_2O} \underset{\text{Furan lithium}}{LiC_4H_3O} + C_6H_6 \quad (Ph = \text{phenyl})$$

$$n - C_6H_9Li + \text{Fe}(C_5H_5)_2 \xrightarrow{\text{Ether}} \text{Fe}(C_5H_5)(C_5H_4Li) + n - C_4H_{10}$$

(*vi*) *By deprotonation of alkynes in liquid ammonia solution.* Lithium acetylides and dicarbides are obtained by this method using Li metal and acetylene. For example,

---

*Also see chapter 16.

$$Li + HC \equiv CH \xrightarrow{\text{Liq. NH}_3} LiC \equiv CH + 1/2\, H_2$$

$$2Li + HC \equiv CH \xrightarrow{\text{Liq. NH}_3} Li_2C_2 + H_2$$

(*vii*) *By metal-metal exchange in petroleum or benzene.* Reaction between an excess of *Li* and an organomercury compound is a useful alternative when isolation of the product is required, rather than its direct use in further synthetic work. For example,

$$2Li + Hg\,R_2 \xrightarrow[\text{benzene}]{\text{Petrol or}} 2LiR + Hg$$

Properties. (*i*) Organolithium compounds tend to be thermally unstable and most of them decompose to LiH and an alkene on standing at room temperature or above.

(*ii*) They are generally liquids or low melting solids and are found in molecular association.

(*iii*) Organolithium compounds are typical covalent substances, soluble in hydrocarbons or other non-polar liquids.

(*iv*) They react rapidly with oxygen and are usually spontaneously flammable in air, with liquid water and with water vapours.

(*v*) LiMe and LiBu are usually synthetic reagents and have been increasingly used in organic syntheses as polymerization catalysts, alkylating agents and precursors to metalated organic reagents.

Some of the reactions with different compounds are given below:

(*a*) Reactions with alkyl iodides give compounds having C – C bonds. For example,

$$LiR + R'I \longrightarrow LiI + RR'$$

(*b*) Reactions with proton donor give the corresponding hydrocarbons. For example,

$$LiR + H^+ \longrightarrow Li^+ + RH$$

(*c*) Reactions with metal carbonyls give aldehydes and ketones. For example,

$$LiR + [Ni(CO)_4] \longrightarrow Li^+\left[R-\overset{\overset{O}{\|}}{C}-Ni(CO)_3\right]^-$$

The intermediate unstable acyl nickel carbonyl complex can be attacked by electrophiles such as H^+ or R′ Br to give aldehydes or ketones by solvent induced reductive elimination.

$$Li^+[R-\overset{\overset{O}{\|}}{C}-Ni(CO)_3]^- \xrightarrow[\text{Solvent}]{H^+} Li^+ + \underset{\text{Aldehyde}}{R-\overset{\overset{O}{\|}}{C}-H} + [(\text{Solvent})\,Ni(CO)_3]$$

$$Li^+[R-\overset{\overset{O}{\|}}{C}-Ni(CO)_3]^- \xrightarrow[\text{Solvent}]{R'Br} LiBr + \underset{\text{Ketone}}{R-\overset{\overset{O}{\|}}{C}-R'} + [(\text{Solvent})\,Ni(CO)_3]$$

(*d*) Reaction with halogens gives parent alkyl (or aryl) halide. For example,

$$LiR + X_2 \longrightarrow LiX + RX$$

(*e*) Reaction with CO gives symmetrical ketones. For example.

$$2LiR + 3CO \longrightarrow 2LiCO + R_2CO$$

(*f*) Reaction with ethers gives alkenes; the organometallic reacts here as a very strong base (proton acceptor). For example,

$$LiR + \underset{H}{\overset{\diagdown}{\underset{\diagup}{C}}}-\underset{OR'}{\overset{|}{C}}- \longrightarrow LiOR' + RH + \diagdown C = C \diagup$$

(*g*) Metal-halogen exchange reactions give other organometallic compounds. For example,

$$3LiR + BCl_3 \longrightarrow 3LiCl + BR_3$$

$$4LiR + SnCl_4 \longrightarrow 4LiCl + SnR_4$$

$$3LiR + P(OEt)_3 \longrightarrow 3LiOEt + PAr_3$$

(*h*) Reaction of lithium aryls (behaving as typical carbanion) in non-polar solvents with CO_2 gives carboxylic acid and with aryl ketones gives tertiary carbinols. For example,

$$LiAr + CO_2 \longrightarrow ArCO_2Li \xrightarrow{H_2O} LiOH + ArCO_2H$$

$$LiAr + Ar'_2LO \longrightarrow [Ar'_2C(Ar)OLi] \xrightarrow{H_2O} LiOH + Ar'_2C(Ar)OH$$

(*i*) Reactions with N, N'- disubstituted amides give aldehydes and ketones. For example,

$$LiR + HCONMe'_2 \longrightarrow LiNMe_2 + RCHO$$

$$LiR + R'CONMe_2 \longrightarrow LiNMe_2 + R'COR$$

Structure. Lithium alkyls are polymeric and generally exist as tetrameric units such as $(LiCH_3)_4$. In lithium methyl there is a tetrahedral set of four lithium atoms with a methyl group placed symmetrically above each Li_3 face (Fig. 15.1). The structure clearly involves a methyl group bridging three lithium atoms. The C-atoms have coordination no. 7 being bonded directly to three H and four Li atoms. The Li – C distance is 231 pm. The Li-Li distance in the structure is 268 pm which is identical with the value of 267.3 pm for a gaseous Li_2 molecule and smaller than the value of 304 pm in Li metal.

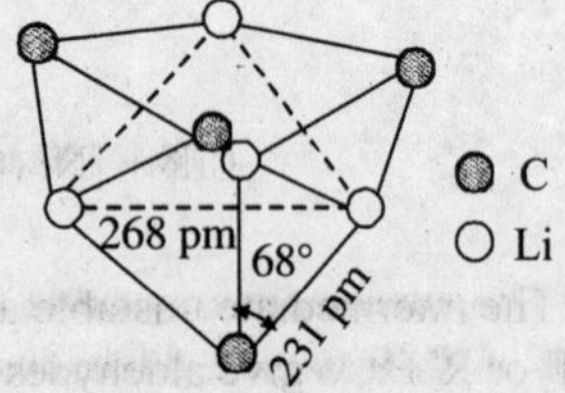

Fig. 15.1. Structure of tetrameric lithium methyl $(LiCH_3)_4$.

Q. 2. Give the preparation, properties and structure of organometallic compounds of beryllium.

Ans. Preparation. (*i*) Beryllium diaryls are made by reaction of a lithium aryl in a hydrocarbon with $BeCl_2$ in diethyl ether in which the LiCl formed is insoluble.

$$2LiC_6H_5 + BeCl_2 \xrightarrow{Ether} 2LiCl + Be(C_6H_5)_2$$

(*ii*) Beryllium dialkyls, BeR_2 (where R = Me, Et, Pr, Bu) can be made by reacting lithium alkyls or Grignard reagent with $BeCl_2$ in ethereal solution.

$$2LiMe + BeCl_2 \xrightarrow{Ether} BeMe_2 + 2LiCl$$

$$2MeMgCl + BeCl_2 \xrightarrow{Ether} BeMe_2 + 2MgCl_2$$

(*iii*) They are best mode in a pure state by heating Be metal with mercury dialkyl in ether solution.

$$Be + HgMe_2 \xrightarrow{Ether} BeMe_2 + Hg$$

The alkyl can be collected by sublimation or distillation in a vacuum.

Properties. (*i*) Beryllium dialkyls are colourless solid or viscous liquids of high reactivity which catch fire in air and are explosively hydrolysed by water.

(*ii*) Dimethylberyllium is a polymeric chain, the higher alkyls are progressively less highly polymerized. Diethyl and diisopropylberyllium are dimeric in benzene but the *t*-butyl compound is monomeric.

(*iii*) Dimethylberyllium reacts with strong donors such as Et_2O, Me_3N or Me_2S which breakdown the polymeric structure of the organometallic compound.

$$BeMe_2 + Me_3N \longrightarrow Me_3N \cdot BeMe_2$$

(*iv*) They readily undergo exchange reaction in solution and the equilibrium lies to the right.

$$BeR_2 + BeX_2 \rightleftharpoons 2RBeX$$

(*v*) Beryllium alkyls give coloured complexes with 2, 2'-bipyridine.

$$Be(C_2H_5)_2 + 2,2'\text{–bipyridine (bipy)} \longrightarrow \text{bipy } Be(C_2H_5)_2$$

(*vi*) Beryllium alkyls yield alkoxides with an excess of alcohol.

$$Be(C_2H_5)_2 + 2ROH \longrightarrow Be(OR)_2 + 2C_2H_6$$

Structure. In the solid state beryllium dimethyl has chain polymeric structure containing chains of *Be* atoms bridged by methyl groups giving tetrahedral arrangement around *Be*. The bridging Be – C – Be angle is 66°, C – Be – C angle 114° and C – Be distance 193 pm. Each carbon atom has a co-ordination number of five being bonded to three H atoms and two Be atoms as shown in Fig. 15.2. In the vapour phase Be Me_2 appears to be principally the dimer, Be_2Me together with smaller amounts of monomer and trimer. The monomer has the expected linear Me – Be – Me structure whereas the dimer and trimer species have the bridge structure shown in Figs. 15.3 and 15.4 respectively.

HHH C 193 pm 66° 114° Be Be Me Me Be C HHH

Fig. 15.2. Chain polymeric structure of Be $(CH_3)_3$ in solid state.

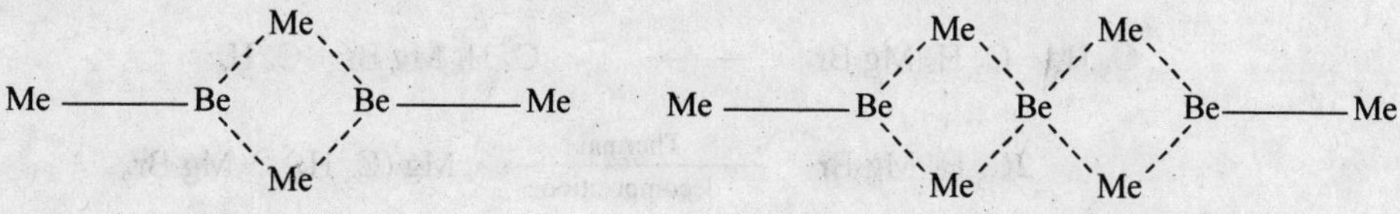

Fig. 15.3. Structure of Be_2Me_4 dimer in vapour phase.

Fig. 15.4. Structure of Be_3Me_6 trimer in vapour phase.

Q. 3. Give the preparation and properties of organometallic compounds of magnesium (Grignard reagent) dialkyl, diaryl and cyclopentadienide magnesium

(*Punjab, 2000; Indore, 2001*)

Ans. Preparation. (*i*) Grignard reagents are made by the slow addition of an alkyl or arylhalide (the order of reactivity of RX is I > Br > CI and alkyl > aryl) to a continuously stirred mixture of magnesium turnings in an absolutely dry organic solvent such as diethyl ether.

$$Mg + RBr \xrightarrow{\text{Dry ether}} \underset{\text{Grignard reagent}}{R\,Mg\,Br}$$

(R = alkyl or aryl)

Very reactive magnesium metal can be prepared by reducing magnesium halides with potassium in presence of KI. For example,

$$MgX + K \xrightarrow{KI} KX + Mg$$

and the use of this magnesium makes it easier to prepare Grignard reagents.

R Mg F (R = Me, Et, Bu, Ph) can be prepared by reacting MgR_2 with mild fluorinating agents such as BF_3, Bu_3 SnF or SiF_4.

(*ii*) Magnesium dialkyls and diaryls can be conveniently made by the reaction of LiR (or LiAr) on Grignard reagents.

$$LiR + RMg\,X \xrightarrow{\text{Dry ether}} MgR_2 + LiX$$

(*iii*) Dialkyls or diaryls are best made by the dry reaction of HgR_2 or $HgAr_2$ on Mg metal in presence of ether. For example,

$$HgR_2 + Mg\ (\text{excess}) \xrightarrow{\text{Dry ether}} MgR_2 + Hg$$

The dialkyl or diaryl is then extracted with an organic solvent and separated from Hg.

(*iv*) Diethyl magnesium can be prepared on an industrial scale by reacting alkenes at 373 K under pressure with MgH_2 or with Mg in the presence of H_2. For example,

$$2C_2H_4 + H_2 + Mg \xrightarrow{\text{Pressure}} Mg\,Et_2$$

(*v*) Magnesium cyclopentadienide can be prepared by direct action of cyclopentadiene (C_5H_6) vapour on hot Mg or by thermal decomposition of C_2H_5MgBr which in turn is made by the action of cyclopentadiene on C_2H_5MgBr in solution. For example,

$$\underset{\text{Hot Cyclopentadiene vapour}}{2C_5H_6} + Mg \xrightarrow{773\text{–}873\text{ K}} \underset{\text{Magnesium cyclopentadienide}}{Mg(C_5H_5)_2} + H_2$$

$$C_5H_6 + C_2H_5MgBr \longrightarrow C_5H_5MgBr + C_2H_6$$

$$2C_5H_5MgBr \xrightarrow[\text{decomposition}]{\text{Thermal}} Mg(C_5H_5)_2 + MgBr_2$$

(*vi*) A related class of compounds such as alkyl-magnesium alkoxides can be prepared by the reaction of MgR_2 with an alcohol or ketone or by reaction of Mg metal with the appropriate alcohol and alkyl chloride in methyl cyclohexane solvent. For example,

$$4Mg\,Et_2 + 4Bu^t\,OH \longrightarrow (Et\,MgO\,But^t)_4 + 4C_2H_6 \quad (t = \text{tertiary})$$

$$2Mg\,Me_2 + 2Ph_2\,CO \xrightarrow{Et_2O} (MeMgOCMePh_2 \cdot Et_2\,O)_2$$

$$6\,Mg + 6\,Bu^n\,Cl + Pr^i\,OH \longrightarrow (Bu^n\,MgOPr^i)_3 + 3MgCl_2 + 3C_4H_{10} \begin{bmatrix} n = \text{normal} \\ i = \text{iso} \end{bmatrix}$$

Properties of dialkyl and diaryl magnesium. (*i*) $MgPh_2$ is insoluble in benzene, dissolves in ether to give the monomeric complex $MgPh_2$. $2Et_2O$ and pyrolyses to give Ph_2 and Mg metal.

$$MgPh_2 + 2Et_2O \longrightarrow Mg\,Ph_2\,2Et_2\,O$$

$$Mg\,Ph_2\,2Et_2O \xrightarrow{555K} Ph_2 + Mg$$

(*ii*) $MgMe_2$ is a white polymeric solid which is insoluble in hydrocarbon and only slightly soluble in ether.

(*iii*) Mg Et_2 and higher homologues are very similar to Mg Me_2 except that they decompose at a lower temperature to give the corresponding alkene and Mg H_2.

(*iv*) Mg Me_2 dissolves in $Al_2\,Me_6$ to give species with CH_3 bridges. For examples,

$$Me_2Mg + (Me_3Al)_2 \rightleftharpoons Me_2Al \langle {}^{Me}_{Me} \rangle Mg \langle {}^{Me}_{Me} \rangle AlMe_2$$

(*v*) Dicyclopentadienyl magnesium $[Mg\,(\eta^5 - C_5H_5)_2]$ is very reactive towards air, moisture, CO_2 and CS_2 and reacts with transition metal halides to give transition element cyclopentadienyls.

Properties of Grignard reagents. (*i*) Grignard reagents are all very reactive, iodides (R – MgI) are the most reactive and chlorides the least reactive. Alkyl Grignard compounds are usually more reactive than aryl Grignard compounds.

(*ii*) Reaction with formaldehyde gives alcohols.

$$RMg\,Br + HCHO \xrightarrow[H_2O]{+\,acid} R\,CH_2\,OH + Mg\,(OH)\,Br$$

(*iii*) Reaction with oxygan gives primary alcohol.

$$RMgBr + O_2 \xrightarrow{+\,acid} ROH + Mg\,(OH)\,Br$$

(*iv*) Reaction with aldehydes gives secondary alcohol.

$$RMgBr + R - CHO \xrightarrow{+\,acid} R_2\,CHOH + Mg\,(OH)\,Br$$

(*v*) Reaction with ketone gives tertiary alcohols

$$RMgBr + R_2C = O \xrightarrow{+H_2O} R_3\,C - OH + Mg\,(OH)\,Br$$

(*vi*) Reaction with carbon dioxide gives carboxylic acids.

$$RMgBr + CO_2 \xrightarrow[H_2O]{+\,acid} R - COOH + Mg\,(OH)\,Br$$

(*vii*) Reaction with sulphur gives thiols (R – SH) and thioethers (R_2S).

$$RMgBr + S \xrightarrow{H_2O} RSH + Mg(OH)Br$$

(*viii*) All Grignard reagents are rapidly hydrolysed by water to give the parent hydrocarbon. For example,

$$2RMg\,Br + 2H_2O \longrightarrow 2RH + Mg\,(OH)_2 + Mg\,Br_2$$

(*ix*) Reaction with iodine gives alkyl iodides.

$$RMg\,Br + I_2 \longrightarrow RI + Mg\,(Br)\,I$$

(*x*) Reaction with lithium alkyls gives magnesium alkyls.

$$RMg\,Br + LiR \longrightarrow MgR_2 + Li\,Br$$

(*xi*) Reaction with $BeCl_2$ gives beryllium alkyls.

$$2RMg\,Br + BeCl_2 \longrightarrow BeR_2 + 2Mg\,(Br)\,Cl$$

(*xii*) Reaction with BCl_3 gives boron alkyls.

$$3RMg\,Br + BCl_3 \longrightarrow BR_3 + 3Mg\,(Br)\,Cl$$

(*xiii*) Reaction with $SiCl_4$ gives alkyl and aryl chlorosilanes.

$$RMg\,Br + SiCl_4 \longrightarrow RSiCl_3 + R_2SiCl_2,\ R_3\,SiCl,\ R_4Si$$

Such compounds are commercially important in the manufacture of silicones.

(*xiv*) Aromatic Grignard reagents undergo free radial self-coupling reactions when treated with MCl_2 (M = Cr, Mn, Fe, Co, Ni)

$$2\,PhMgCl + Cr\,Cl_2 \longrightarrow 2\,MgBrCl + (C_6H_5)–Cr–(C_6H_5)$$

$$\longrightarrow (C_6H_5–C_6H_4)Cr \xrightarrow{\Delta} Ph_2 + Cr$$

(*xv*) Reaction with proton donor gives alkanes

$$RMg\,Br + H_2O \longrightarrow RH + Mg\,(OH) + Mg\,(OH)\,Br$$

Q. 4. Give the structures of Grignard reagents magnesium alkyl and magnesium cyclopentadienide.

Ans. Structure of Grignard reagents. X-ray diffraction studies of certain crystalline Grignard reagents show that the basic Grignard structure is RMgX.*n* (solvent) (Fig. 15.5). In the structure of

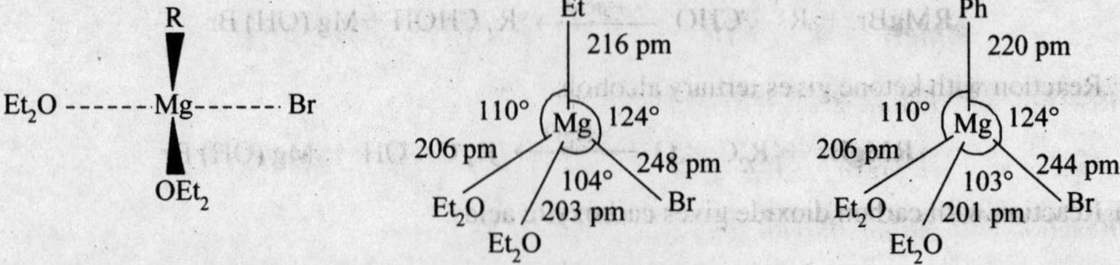

Fig. 15.5. Tetrahedral structure of RMgBr 2 (Et_2O).

Fig. 15.6. Structure of $C_2H_5MgBr.2(Et_2O)$.

Fig. 15.7. Structure of $C_6H_5MgBr.\ 2(Et_2O)$.

$C_2H_5MgBr \cdot 2(Et_2O)$ and $C_6H_5 \cdot MgBr \cdot 2(Et_2O)$ the Mg atom is tetrahedrally surrounded by C, Br and two oxygen atoms of the ether as shown in Figs. 15.6 and 15.7 respectively.

Structure of magnesium alkyls. For dimethylmagnesium, $Mg(CH_3)_2$, the structure is that of a chain polymer, similar to that of $Be(CH_3)_2$. [Fig. 15.2] with bridging methyl groups but again tetrahedral Mg. The corresponding dimensions for $MgMe_2$ being Mg – C 234 pm, angles Mg – C – Mg = 75° and C – Mg – C = 105°.

Structure of magnesium cyclopentadienide $Mg(C_5H_5)_2$. $Mg(C_5H_5)_2$ has a 'sandwich' structure similar to that of ferrocene but with $C_5H_5^-$ and Mg^{2+} ions and distances Mg – C as 230 pm, C– C as 139 pm and C– H as 103 pm (Fig. 15.8).

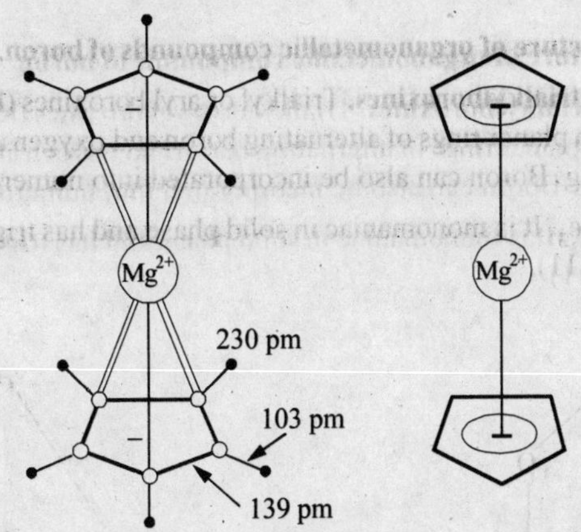

Fig. 15.8. Sandwich structure of $Mg(C_5H_5)_2$.

Q. 5. Describe the methods of preparation and properties of organo compounds of boron.

Ans. Preparation. (*i*) Alkyl borons can be easily made by hydroboration.

$$6C_2H_4 + B_2H_6 \longrightarrow 2B(C_2H_5)_3$$

(*ii*) The alkyl and aryl borons can be made from boron halides by lithium or Grignard reagents.

$$RMgBr + BCl_3 \longrightarrow BR_3$$

(*iii*) Boronic acids $RB(OH)_2$ can be made in various ways as explained below.

$$BF_3 \cdot O(C_2H_5)_2 + C_6H_5MgBr \longrightarrow C_6H_5BF_2 \xrightarrow{H_2O} C_6H_5B(OH)_2$$

(*iv*) Sodium tetraphenyl borate, $Na[B(C_6H_5)_4]$ is obtained when boron halides are treated with four equivalents of alkylation agents.

$$BCl_3 + 4C_6H_5MgBr \longrightarrow B(C_6H_5)_4^-$$

Properties. (*i*) Lower boron alkyls are reactive substance but boron aryls are stable.

(*ii*) The compound $Li(BMe_4)$, m.p. 463 K is readily soluble in benzene and can be sublimed. In solution only an ion pair, $Li^+ BMe_4^-$, is present. It is polymeric in the solid state.

(*iii*) Boronic acids $RB(OH)_2$ are quite stable and water soluble. Their acidities depend on the nature of the alkyl or aryl group. Dehydration of boronic acid by heat yields a boronic anhydride or trialkyl or aryl boroxine. For example,

$$3RB(OH)_2 \xrightarrow{\text{Heat}} \underset{\text{Trialkyl or triaryl boroxines}}{(RBO)_3} + 3H_2O$$

(*iv*) Sodium tetraphenylborate is soluble in water and is stable in weakly acid solution. It gives insoluble precipitates with longer cation such as K^+, Rb^+ Me_4N^+ that are suitable for gravimeric analysis. The ion $[B(C_6H_5)_4]$ can also act as a ligand wherein one phenyl ring is bound in an arene complex as in Fig. 15.9.

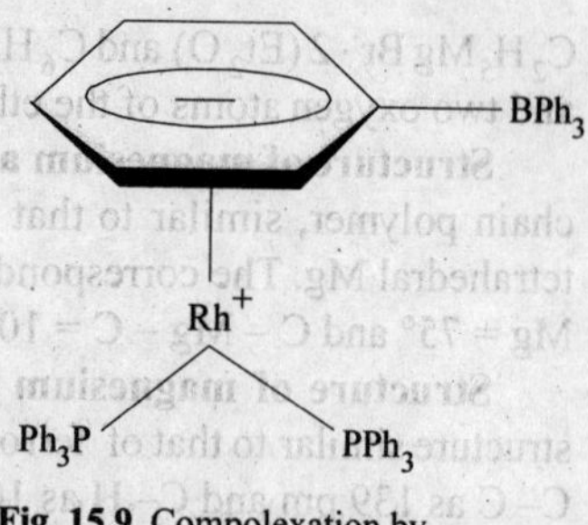

Fig. 15.9. Compolexation by $[B(C_6H_5)_4]^-$

Q. 6. Discuss the structure of organometallic compounds of boron.

Ans. (1) Structure of trialkylboroxines. Trialkyl or aryl boroxines $(RBO)_3$ have trimeric, cyclic structures (Fig. 15.10) with planar rings of alternating boron and oxygen atoms. The alkyl groups are also in the plane of the ring. Boron can also be incorporated into numerous carbon ring systems.

(2) Structure of BMe_3. It is monomaniac in solid phase and has trigonal planar shape with *B–C* distance 158 pm (Fig. 15.11).

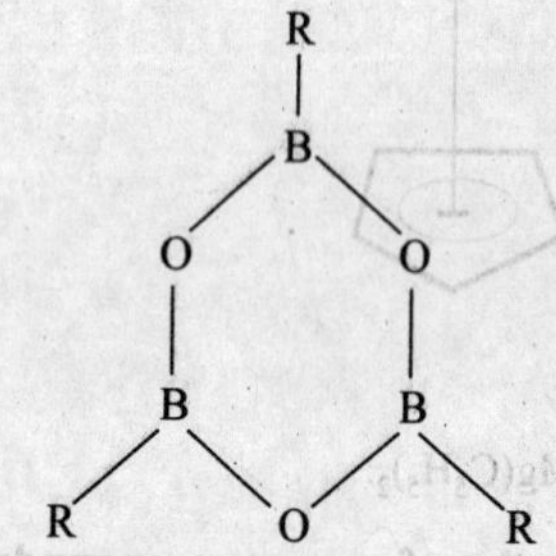

Fig. 15.10. Trimeric cyclic struture of trialkyl (or aryl) boroxines.

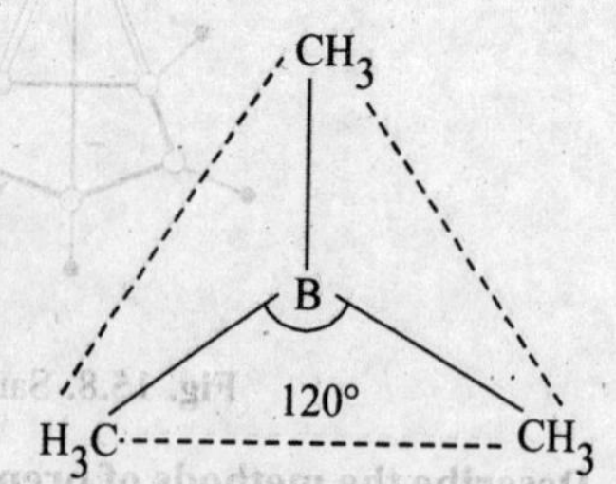

Fig. 15.11. Trigonal planar structure of BMe_3.

(3) Structure of $Li(BMe_4)$. In $Li(BMe_4)$, an unusual type of methyl bridge occurs in addition to the conventional type found in $(BeMe_2)_n$. There are linear Li – C– B chains with Li – C = 212 pm and C – B = 151 pm.

Q. 7. Describe the methods of preparation and properties of organometallic compounds of aluminium. (*Gujarat, 2003; Lucknow, 2010; Andhra; 2010*)

Ans. Preparation. Organoaluminium compounds may be prepared by the following methods:

(*i*) *By the reaction of Grignard reagents with $AlCl_3$.*

$$RMgCl + AlCl_3 \longrightarrow RAlCl_2, R_2AlCl, R_3Al$$

one, two or three alkyl groups could combine with one Al.

(*ii*) *By the reaction of aluminium with an organomercury compound.*

$$2Al + 3HgMe_2 \xrightarrow{363\text{ K}} Al_2Me_6 + 3Hg$$

Al_2Ph_6 can be prepared similarly using $HgPh_2$ in boiling toluene or by the reaction of LiPh on Al_2Cl_6.

$$6LiPh + Al_2Cl_6 \longrightarrow Al_2Ph_6 + 6LiCl$$

(*iii*) *On the industrial scale Al* is alkylated by means of *RX* giving sesquichloride $R_3Al_2Cl_3$

$$2Al + 3RCl \xrightarrow[\text{AlCl}_3 \text{ or AlR}_3]{\text{Trace of I}_2} R_3Al_2Cl_3$$

(*iv*) *Higher trialkyls* are readily prepared on an industrial scale by the alkene route in which H_2 adds to *Al* in the presence of performed AlR_3 acting as catalyst (Ziegler catalyst) to give a dialkyl-aluminium hydride which then readily adds to the alkene.

$$2Al + 3H_2 + 2Al_2Et_6 \xrightarrow{425\ K} (6\ Et_2AlH) \xrightarrow[345\ K]{6\ CH_2CH_2} 3Al_2Et_6$$

Properties. (*i*) Aluminium trialkyls and triaryls are highly reactive volatile liquids or low melting solids, which burn in air and react violently with water.

(*ii*) They react with protonic reagents to liberate alkanes.

$$Al_2R_6 + 6HX \longrightarrow 6RH + 2AlX_3$$

(*iii*) Organoaluminium compounds give alkene insertion reactions known as growth reactions to synthesize unbranched long chain primary alcohols and alkenes. Here alkenes insert into the Al – C bond of monomeric AlR_3 (acting as catalyst) at ~425 K and 100 atom to give long chain derivatives whose composition can be closely controlled by the temperature, pressure and contact time.

$$Al(C_2H_5)_3 \xrightarrow{C_2H_4} Al(CH_2CH_2C_2H_5)(C_2H_5)_2 \xrightarrow{nC_2H_4} Al[(CH_2CH_2)_nC_2H_5]_3$$

Long chains of length of 14–20 C atoms can be synthesised industrially in this way and then converted to unbranched aliphatic alcohols after oxidizing with air and then hydrolysing with water for use in the synthesis of biodegradable detergents.

$$Al(CH_2CH_2R)_3 \xrightarrow[(ii)\ H_3O^+]{(i)\ O_2} 3RCH_2CH_2OH$$

(*iv*) Al_2R_6 (or AlR_3) react readily with ligands to form adducts and behave as Lewis acids. For example,

$$AlR_3 + L\ (\text{ligand}) \longrightarrow Li\ Al\ R_3\ (\text{adduct})$$

(*L* = amines, phosphines, ethers and thioethers)

(*v*) Aluminium alkyls also combine with lithium alkyls to give mixed alkyls.

$$(C_2H_5)_3Al + LiC_2H_5 \xrightarrow{\text{In benzene}} Li\ Al(C_2H_5)_4$$

(*vi*) They react with halides or alkoxides of elements like B, Ca, Si, Sn etc which are less electropositive than Al to form other organometallics.

$$MX_n \longrightarrow MR_n + \frac{n}{3}AlX_3$$

(*vii*) Organoaluminium compounds bring about low pressure polymerization of ethene and propene in presence of organometallic mixed catalyst. A Ziegler-Natta catalyst is a mixture of $TiCl_4$ and $Al_2\ Et_6$ in heptane giving a brown suspension which rapidly absorbs and polymerises ethene even at room temperature and atmospheric pressure. Polyethene produced in this way has superior qualities.

Q. 8. Discuss the structure of organoaluminium compounds. (*Lucknow, 2000*)

Ans. In the solid and gas phase $AlMe_3$ is dimeric and has the methyl bridged structure. $Al_2\ Ph_6$ exhibits the similar dimeric structure. In both cases the bridging Al-C distance is 10% longer than the terminal Al – C distance. Solutions of $Al_2\ Me_6$ show only one proton *nmr* signal at room temperature due to the rapid interchange of bridging and terminal Me groups. In $Al_2\ Me_6$ (Fig. 15.12) the terminal Al – C distances are 195 pm and the bridging distances 212 pm, the bridging Al – C – Al angle is 76°. In general AlR_3 molecules are bridged dimers in the solid state except when R is very bulky such as tertiary butyl. The crystal structure of $Al_2\ Ph_6$ (Fig. 15.3) shows phenyl as bridging groups and the terminal and bridging. Al– C distances being 196 pm and 218 pm respectively. The bridging Al – C – Al angle is measured to be 77°.

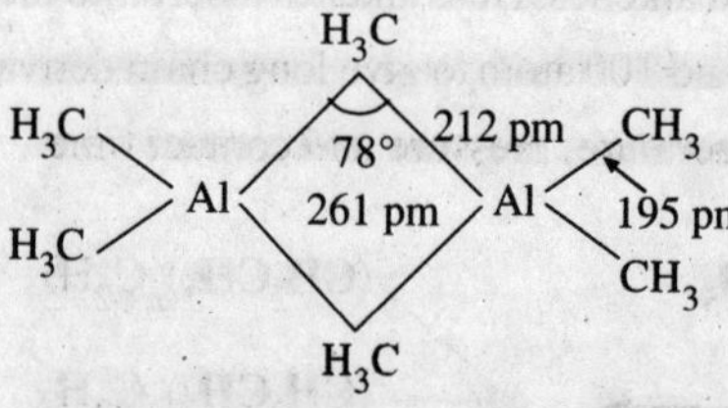

H_3C, H_3C, Al, 78°, 212 pm, 261 pm, Al, CH_3, 195 pm, CH_3, H_3C, H_3C

Fig. 15.12. Structure of $Al_2(CH_3)_6$.

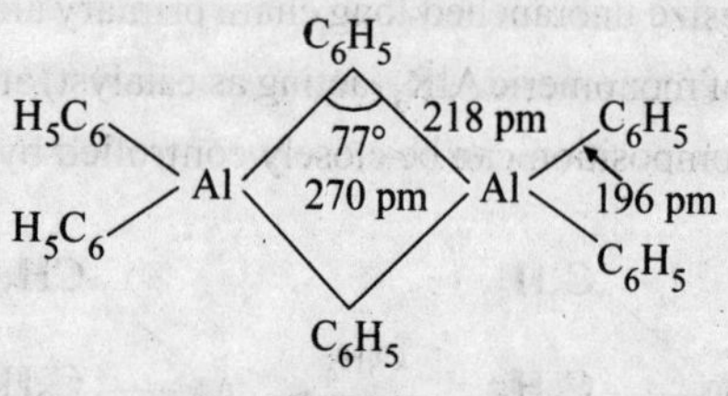

C_6H_5, H_5C_6, H_5C_6, Al, 77°, 218 pm, 270 pm, Al, C_6H_5, 196 pm, C_6H_5, C_6H_5

Fig. 15.13. Structure of $Al_2(C_6H_5)_6$

Q. 9. What are organometallic compounds ? How will you synthesize the following ?

(i) CH_3 Li (ii) $(C_2H_5)_2$ Be (iii) B $(CH_3)_3$ (*Kurukshetra, 2002*)

Ans. See Qs. 1, 2 and 5.

16 SOME MORE ORGANOMETALLIC COMPOUNDS AND π-ACID COMPLEXES

Q. 1. What are different types of organometallic compounds ?

(Punjab, 2000; Ranchi, 2010; Andhra; 2010)

Ans. The organometallic compounds may be classified into the following types depending upon the nature of metal-carbon bond.

1. Ionic compounds. Such compounds are formed when the negative charge on the hydrocarbon anion is delocalised over several carbon atoms in the aromatic or unsaturated ring. $K^+ C_5 H_5^-$ is a common example of this type where delocalisation of negative charge over the five carbon atoms of cyclopentadienyl ring gives rise to a stable complex.

These compounds exhibit properties expected from ionic compounds. They are insoluble in non-polar solvents.

2. Covalent compounds. Such compounds have the metal bonded to an organic part by normal two-electron sigma covalent bonds. Such compounds give properties expected from covalent compounds. They are soluble in organic solvents and insoluble in water. Electronegativity difference between the metal atom and carbon determines the polarity of the carbon-metal bond.

Common examples of this type of compounds are:

$$Pb(C_2H_5)_2, (CH_3)_2Cd, (C_6H_5)_2Zn, (CH_3)_3Al$$

3. Electron deficient compounds. Formation of this type of compounds cannot be explained on the basis of formation of usual two electrons–two centre bonds with carbon and metal atom. Compounds of Li, Be, Mg and Al with bridging alkyl groups appear in this category of compounds. Dimeric trialkyl aluminium (Al_2R_6), polymeric dimethyl beryllium $(Be Me_2)_n$ and diethyl magnesium $(Mg Et_2)_n$ are some examples of this type of compounds. These compounds have high charge to mass ratio and thus have strongly polarizing cation. As a result, polar covalent bond results.

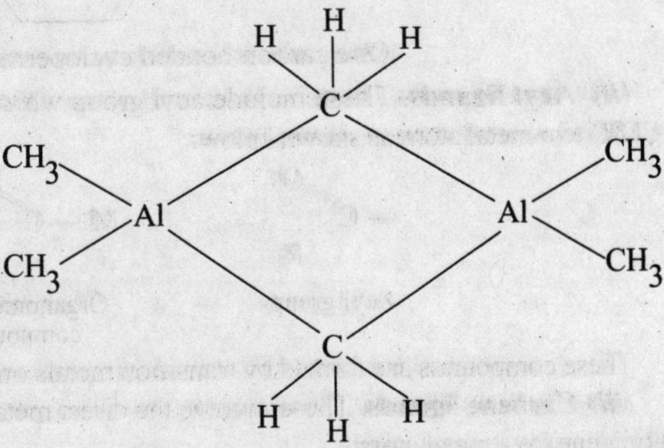

Fig. 16.1. Structure of dimeric Al_2Me_6.

Such polar molecules have tendency to associate and give polymeric structures. Trialkyl aluminium exists as dimeric in which the alkyl group is present as bridge (Fig. 16.1).

4. Transition metal organometallic compounds. These organometallic compounds include transition metals bonded to unsaturated organic compounds in which the transition metal forms bonds to more than one carbon atoms of the same organic compound. The interaction occurs between the π-orbitals of the organic ligands with the suitable *d* or *p* orbitals of the metal atom.

The common examples of ligands forming such type of organometallic compounds are (*i*) alkene (2-electron donors), (*ii*) butadiene (4-electron donors), (*iii*) cyclopentadiene (5-electron donors) and (*iv*) benzene ring (6-electron donors). The common structures of dibenzene chromium $(C_6H_6)_2Cr$ and ferrocene $(C_5H_5)_2$ Fe are shown in Fig. 16.2. Their structures are known as sandwich structures in which metal atom lies in between two rings.

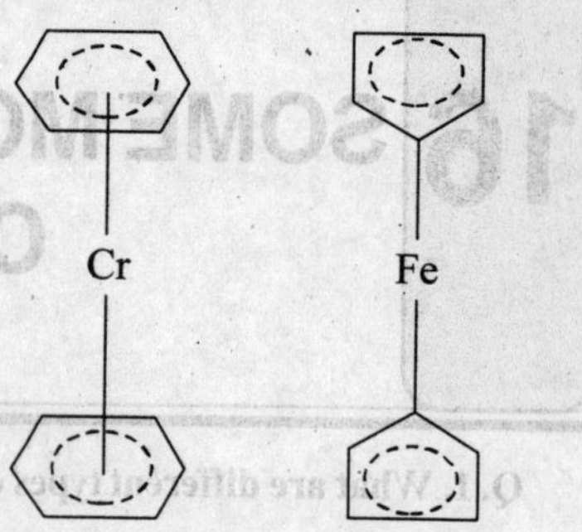

Fig. 16.2. Structures of (*a*) $(C_6H_6)_2$ Cr and (*b*) $(C_5H_5)_2$ Fe.

Q. 2. How are different ligands in an organometallic compounds classified?

Ans. An atom, ion or molecule which is capable of donating a pair of electrons to the metal atom is called a *ligand.* The number of atoms of the ligand which are within bonding distance of the metal atom is called its *hapticity.* It is written by using the symbol η (eta).

The ligands may be classified on the basis of number of carbon atoms involved in bonding to the metal atom, *i.e.*, hapticity. The hapticity ranges from 1 to 8.

1. One carbon bonded ligands. These are the molecules in which the carbon atom of the ligand is bonded directly to the metal atom. Such ligands are also called **monohapto ligands** (one electron donors), η^1. For example, $-CH_3$ group attaches by a single M – C bond. These ligands are of the following different types:

(*i*) Hydrocarbon ligands. These include alkyl $(-CH_3)$, aryl $(-C_6H_5)$, $(-CR=CR_2)$, σ-cyclopentadienyl $(-C_5H_5)$, alkynyl $(-C\equiv CR)$ group.

For example, cyclopentadiene, C_5H_5, combines with alkali metals to form $(\sigma - C_5H_5)$ M compound as shown below:

H M

(One carbon bonded cyclopentadienyl compound).

(*ii*) Acyl ligands. These include acyl group which involves direct bonding of the acyl group (–COR) to a metal atom as shown below:

$-C(=O)R$	$M-C(=O)R$
Acyl group	Organometallic compound

These compounds are formed by transition metals only.

(*iii*) Carbene ligands. These involve the direct metal carbon-bond of the carbon atom of carbene (alkylidene) to a metal atom.

$\leftarrow :CR_2$	$M \leftarrow :CR_2$
Carbene ligand	Organometallic compound

2. Two carbon bonded ligands. These are the molecules in which two carbon atoms of the ligands are bonded to the metal atom. These are called **dihapto (η^2) ligands** (or two-electron donors).

Examples are alkenes or alkynes in which both the carbon atoms at each end of the multiple bond are involved in forming the metal-carbon bonds. The bonding in these ligands is indicated either by two dotted lines to the participating carbon atom or a single solid line from the centre of the participating carbon atoms of the ligand to the metal atom.

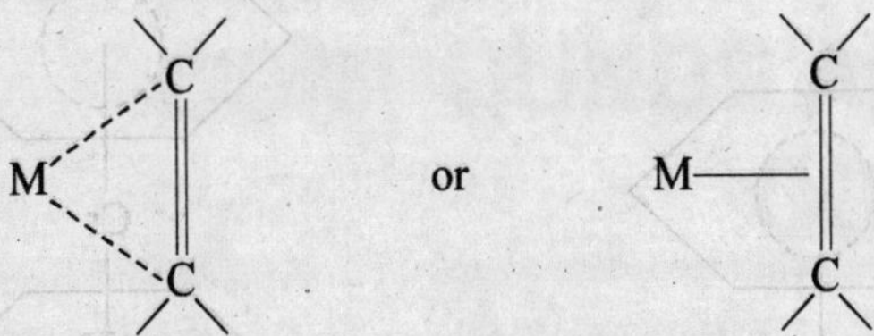

3. Three carbon bonded ligands. *These are the molecules in which three carbon atoms of the ligands are bonded to the metal atom.* These are called **trihapto ligands** (η^3). Examples are allyl group (C_3H_5) known as π -allyl ligands as shown below. These may be distinguished from σ-allyl complexes in which the allyl group is bonded through one carbon atom only (monohapto).

π - Allyl complex

σ- Allyl complex

4. Four carbon bonded ligands. These ligands have four carbon atoms which may be bonded to the metal atom. Therefore, these are called **tetrahapto** (4-electron donors) or **η^4 ligands**. These include acyclic ligands such as 1,3-butadiene and cyclic ligands such as cyclobutadiene.

5. Five carbon bonded ligands. These ligands are molecules in which five carbon atoms of the ligands are bonded to the metal atom. These are called **pentahapto (η^5) ligands**. The common examples are acyclic dienyls and cyclic dienyls (*e.g.*, cyclopentadienyl).

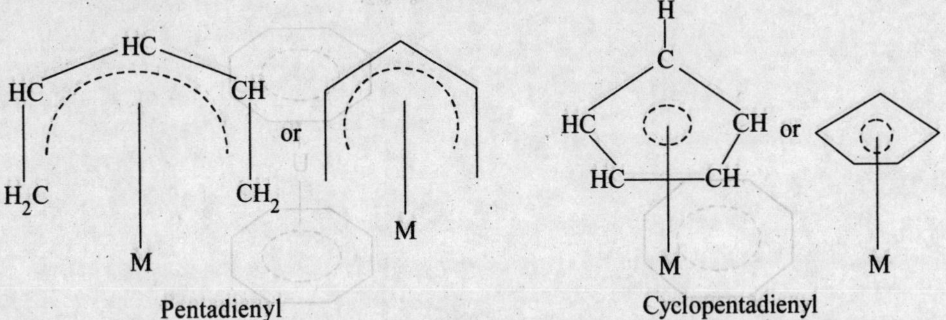

Pentadienyl

Cyclopentadienyl

6. Six carbon bonded ligands. The molecules in which six carbon atoms of the ligand are bonded to the metal atom. The common example is benzene. The common example of an organometallic compound is dibenzene chromium $(C_6H_6)_2Cr$. This is also a *sandwich compound* in which Cr atom is present in between two benzene rings.

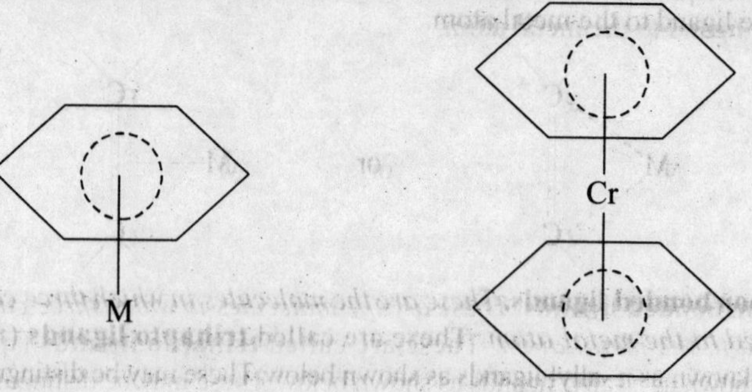

Benzene as hexahapto ligand Dibenzene chromium $(C_6H_6)_2Cr$

7. Seven carbon bonded ligands. These are molecules in which seven carbon atoms act as electron donors by participating all the seven carbon atoms in bond formation. The common example is cycloheptadienyl $(-C_7H_7)$ as shown below:

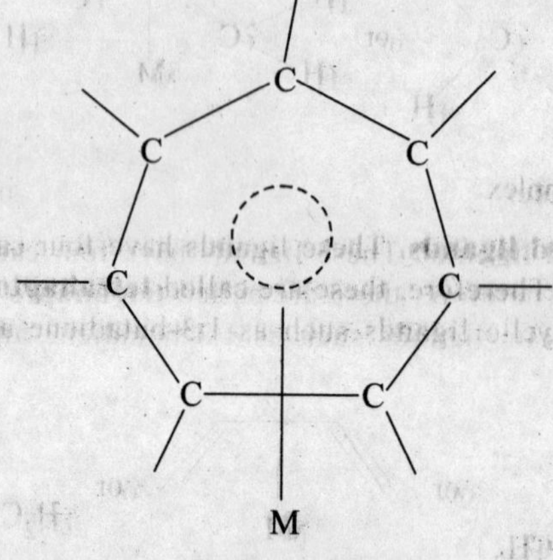

Cycloheptadienyl

8. Eight carbon bonded ligands. These are molecules in which eight carbon atoms are involved in bond formation. These are also called **octahapto ligands**. The common example is cyclo-octatetraene (C_8H_8). The common example of the complex is $U(C_8H_8)_2$.

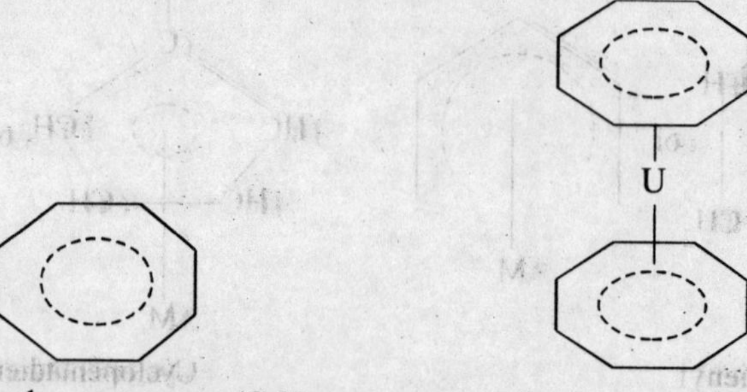

Cyclo-octatetraene (C_8H_8) Dicyclo-octatetraeneuranium

Q. 3. State and explain EAN rule as applied to organometallic compounds.

(Kerala, 2000; Lucknow, 2010; Andhra, 2010)

Ans. Sidgwick suggested that metal ion accepts electron pairs from the ligands until it achieves the next noble gas configuration. This is called **effective atomic number rule.** *The total number of electrons possessed by the central metal ion and the electrons gained by it from ligands* is called the **effective atomic number** (EAN). Thus, according to effective atomic number rule, the effective atomic number in a complex should be equal to 36 (electrons in Kr), 54 (electrons in Xe) and 86 (electrons in radon).

The rule may be illustrated by taking the following examples:

(*i*) Tetracarbonyl nickel (0), $Ni(CO)_4$

Atomic number of Ni = 28

Electrons in Ni(0) in $Ni(CO)_4$	= 28
Four CO groups give two electrons each (4 × 2)	= 8
EAN of $[Ni(CO)_4]$	= 36

Thus, the complex $[Ni(CO)_4]$ obeys EAN rule.

(*ii*) Pentacarbonyl iron (0), $Fe(CO)_5$

Atomic number of Fe = 26

Electrons in Fe(0)	= 26
Five CO groups give two electrons each (5 × 2)	= 10
EAN	= 36

EAN rule can explain the stability of metal carbonyls but it cannot be the *sole criteria to form stable carbonyls.* Metals with odd number of electrons cannot obey EAN rule because total number of electrons will be odd, no matter how many carbonyls are added. For example, hexacarbonyl vanadium $V(CO)_6$ is a stable complex though it does not obey EAN rule as:

Atomic No. of V = 23

Electrons in V(0)	= 23
6 CO groups give two electrons each (6 ×2)	= 12
EAN	= 35

Therefore, the complex $V(CO)_6$ does not obey EAN rule.

In order to satisfy EAN rule, there are some **options** for the metal ions. These are:

(*i*) They may form carbonylate anion such as $[M(CO)_n]^-$ by gaining an electron from a reducing agent or they may form cationic species, $[M(CO_n)]^+$ by losing an electron. For example, the anions $[V(CO)_6]^-$, $[Mn(CO)_5]^-$, $[Co(CO)_4]^-$ obey EAN rule. For example,

$[V(CO)_6]^-$

Atomic number of V = 23

Electrons in V (–1)	= 24
Six CO groups give two electrons each (6 ×2)	= 12
EAN	= 36

Thus, $V(CO)_6^-$ obeys EAN rule.

(*ii*) They may form single covalent bond with an atom or group having a single unpaired electron, *e.g.*, hydrogen (H*) or chlorine (Cl*) [H M$(CO)_n$] or [M $(CO)_n$ Cl]. For example,

Mn $(CO)_5$ Cl

Electrons in Mn(0)	= 25
5 CO groups give two electrons each (5× 2)	= 10
Cl*	= 1
EAN	= 36

∴ EAN for Mn $(CO)_5$ Cl = 36

Similarly, HCo $(CO)_4$

Electrons in Co(0)	= 27
4 CO groups give two electrons each (4 × 2)	= 8
H	= 1
EAN	= 36

(*iii*) If no other species is available with which the metal carbonyl containing odd number of electrons can interact, it can dimerize resulting in pairing of electrons. This will lead to the formation of metal-metal bond. While counting the EAN, the electron pair shared between two metal atoms forming metal-metal bond is counted on both the metals.

For example, manganese forms a stable dinuclear carbonyl, Mn_2 $(CO)_{10}$ having metal-metal bond, $(CO)_5$ Mn – Mn $(CO)_5$. Let us calculate EAN for this complex.

Mn_2 $(CO)_{10}$

Electrons in 2 Mn (2 × 25)	= 50
10 CO groups give two electrons each (10 × 2)	= 20
One Mn–Mn bond	= 2
EAN for Mn_2 $(CO)_{10}$	= 72
∴ EAN per Mn atom	= 36 (Kr)

EAN rule for other Organometallic Compounds

EAN rule can also be applied to other organometallic compounds. For this purpose, number of electrons donated by various ligands are taken into consideration.

Let us illustrate the calculation of EAN for some complexes.

(*a*) Alkyl groups as 1 electron donors

CH_3 Re$(CO)_5$

Electrons in Re (Rhenium)	= 75
5(CO) give two electrons each (5 × 2)	= 10
CH_3	= 1
EAN	= 86

(*b*) Olefins and alkynes acting as 2-electron donors

[Mn$(CO)_5$ C_2H_4]+

Electrons in Mn^+	= 24
5 (CO) give two electrons each	= 10
Contribution from C_2H_4	= 2
EAN	= 36

(c) π-allyls acting as 3. electron donors

$Mn(\pi-C_3H_5)(CO)_4$

Electrons in Mn = 25

Contribution of $\pi-C_3H_5 = 3$

4 (CO) contribute = 8

EAN = 36

$Co(\pi-C_3H_5)(CO)_3$

Electrons in Co = 27

Contribution of $\pi-C_3H_5 = 3$

3(CO) donate = 6

EAN = 36

(d) π -cyclo pentadienes as 5-electron donors

$Co(\pi-C_5H_5)(CO)_2$

Electrons in Co = 27

Contribution of 2 $(\pi-C_5H_5) = 5$

2 (CO) donate = 4

EAN = 36

$Fe(\pi C_5H_5)_2$

Electrons in Fe = 26

Contribution of 2 $(\pi-C_5H_5) = 10$

EAN = 36

Thus, it has been observed that most of the organometallic compounds obey EAN. There are some exceptions also in which EAN rule is not obeyed. For example,

$Co(\pi-C_5H_5)_2$

Electrons in Co = 27

Contribution of 2 $(\pi-C_5H_5) = 10$

EAN = 37

$Cr(C_6H_6)(CO)_4$

Electrons in Cr = 24

Contribution of $C_5H_5 = 6$

Contribution of 4 (CO) = 8

EAN = 38

Q. 4. Show that EAN rule is obeyed in $Fe(CO)_5$ while in $V(CO)_6$ it is not obeyed.

Ans. See Q. 3.

Q. 5. Discuss the nomenclature of organometallic compounds.

Ans.1. Nomenclature of simple compounds. The simple alkyl or aryl organometallic compounds of metals are named by writing the name of the metal after the name of organic group. For example,

CH_3 Li	Methyl lithium
$(C_2H_5)_2$ Zn	Diethyl zinc
(C_2H_5) MgBr	Ethyl magnesium bromide
C_6H_5 MgC1	Phenyl magnesium chloride

2. Nomenclature of carbonyls. (*i*) The compounds containing CO as ligands are called *metal* carbonyls. In case the metal has zero oxidation state, it may not be mentioned.

For example,

$Ni(CO)_4$	Tetracarbonylnickel
$Mn_2(CO)_{10}$	Decacarbonyldimanganese
$Fe_2(CO)_9$	Nonacarbonyldiiron
$Co_2(CO)_8$	Octacarbonyldicobalt
$[V(CO)_6]^-$	Hexacarbonylvanadate ion

(*ii*) If the ligands act as bridges between two metal atoms, the Greek letter mu (μ) is written before their names. The prefix μ is repeated before the name of each kind of bridging ligand. For example,

$[(CO)_3Co(CO)_2Co(CO)_3]$	Di-μ-carbonyl bis (tricarbonyl cobalt)
$[(CO)_3Fe(CO)_3Fe(CO)_3]$	Tri-μ–carbonyl bis (tricarbonyl iron)

(*iii*) When the metal carbonyls contain metal-metal bonds, these may be classified as *symmetrical* or *unsymmetrical.* The symmetrical metal carbonyls are named by the use of multiple prefixes (bis, tris, etc.). In case of unsymmetrical metal carbonyls, one central metal atom and its ligands are treated as a ligand on the other central metal atom. For example,

$[(CO)_4Co-Co(CO)_4]$ (symmetrical)	Bis (tetracarbonylcobalt)
$[(CO)_4Co-Re(CO)_5]$ (unsymmetrical)	Pentacarbonyl (tetracarbonyl cobalto) rhenium

3. Nomenclature of σ and π bonded ligands. To distinguish between one carbon bonded ligand and multiple carbon bonded ligands, the notations σ and π are used. For example, cyclopentadiene (C_5H_5) is referred as $\sigma-C_5H_5$ (*e.g.*, LiC_5H_5) when it behaves as one carbon bonded ligand. However, when it behaves as five carbon bonded ligand, it is referred as π-C_5H_5 (*e.g.* $\pi(C_5H_5)_2Fe$). Similarly, alkyl group is referred as σ-allyl or π–allyl depending upon whether it behaves as one electron donor or three-electron donor.

In the case of unsaturated molecules or groups, the prefix η is used. For example, a one carbon bonded ligand is specified by monohapto (or η^1), two carbon bonded ligand as dihapto (or η^2), three carbon bonded ligand as trihapto (or η^3), a four carbon bonded ligand as tetrahapto (or η^4) and so on. According to latest IUPAC convention, the η notation is recommended.

For example,

$Fe(C_5H_5)_2$	Bis (η^5-cyclopentadienyl) iron
$Cr(C_6H_6)_2$	Bis (η^6-benzene) chromium
$Co(CO)_3(\pi-C_3H_5)$	(η^3-allyl) tricarbonyl cobalt
$(C_6H_6)Cr(CO)_3$	(η^6-benzene) tricarbonyl chromium
$Fe_2(CO)_4(C_5H_5)_2$	Bis (η^5-cyclopentadienyl) tetracarbonyldiiron
$Fe(CO)_2(\sigma-C_5H_5)(\pi-C_5H_5)$	Dicarbonyl (η^1-cyclopentadienyl) (η^5-cyclopentadienyl) iron
$Fe(CO)_3(C_4H_6)$	(η^4-butadiene) tricarbonyl iron
$Mn(CO)_5(-CH_2-CH=CH_2)$	(η^3-allyl) tricarbonyl iron

Q. 6. Discuss (*i*) ionic bonding and (*ii*) σ-covalent bonding in organometallic compounds.

Ans. 1. Ionic bonding. This type of bonding is observed in organometallic compounds of electropositive metals like sodium, potassium, rubidium, cesium, magnesium etc. In such compounds, the hydrocarbon contains carbon carrying negative charge which is strongly attracted by the positively charged metal ion by electrostatic forces of attraction. In some cases, the negative charge is delocalised over the ring containing different number of carbon atoms. For example, in $Mg^{2+}(C_5H_5^-)_2$, the negative charge of cyclopentadienide anion, $C_5H_5^-$, is delocalised over the ring of five carbon atoms.

2. σ-covalent bonding. This type of bonding is generally observed in organometallic compounds of representative elements. It is generally observed in metal-alkyl bonds. For example, $(CH_3)_4Sn$, C_4H_9Li, $(CH_3)_6Al_2$, etc. The polarity of covalent bond between metal and alkyl group depends upon the difference in electronegativity between metal atom and carbon atom.

Transition metals can also form metal-carbon σ bonds with organic groups but they are relatively unstable. They can easily decompose even at room temperature. For example, the alkyl derivatives of transition metals such as $Ti(CH_3)_4$, $Nb\ CH_3)_5$, $W(CH_3)_2$ are well known but they are unstable and explode even below room temperature. The stability of the η-bonded covalent compounds can be enhanced if some additional π-bonding ligands are also present. For example, the presence of ligands such as PR_3, CO, amines, etc. increases the stability of these compounds. For example, tetramethyl titanium undergoes spontaneous decomposition violently at –23° C but the addition of dipyridine or diamine increases its stability so that it does not decompose up to room temperature.

Q. 7. Discuss the factors responsible for kinetic instability of transition metal sigma bonded organometallic compounds.

Ans. The following factors are responsible for kinetic instability of organometallic compounds containing metal-alkyl bond:

1. **Formation of reactive species due to cleavage.** The cleavage of metal-carbon bond in transition metal alkyls give reactive alkyl radicals or ions. The reactive species formed react with each other or solvent to give chain reactions. This behaviour is in sharp contrast to other inorganic metal-ligand bonds such as $M-OH_2$, $M-NR_2$, $M-X$ (X = halogen), etc. where a cleavage leads to relatively less reactive species like H_2O, R_2N^- or X^- respectively.

2. **Cleavage products form strong bonds.** When the cleavage of metal-carbon bond occurs, they combine among themselves to form strong bonds such as C—C, C—O, C—N, C—X (X = halogen) etc. depending upon the nature of organometallic compound. During the formation of new bonds, the cleavage reaction is highly exothermic and consequently, it is thermodynamically favoured. For example, metal-alkyl bond undergoes cleavage as

$$M-R \longrightarrow \underbrace{\dot{M} \quad + \quad \dot{R}}_{\text{Reactive Intermediates}}$$

The formation of alkyl radical leads to subsequent reactions.

3. **Hydrogen transfer from alkyl to metal.** The higher alkyl organometallic compounds are generally unstable particularly those which carry β-hydrogen. β-hydride elimination reaction provides an opportunity for the decomposition of metal alkyls into a metal hydride and olefin. (β-elimination reaction many be represented as

$$M-\overset{\overset{\displaystyle H}{|\alpha}}{\underset{\underset{\displaystyle H}{|}}{C}}-\overset{\overset{\displaystyle H}{|\beta}}{\underset{\underset{\displaystyle H}{|}}{C}}-R \rightleftharpoons M-H+CH_2=CHR$$

Q. 8. What do you understand by β-elimination in transition metal alkyls? How can it be avoided? Explain giving example.

Ans. In higher alkyl organometallic compounds, the hydrogen present on carbon atom at β-position to the metal is labile (removable). It brings about decomposition of the compound into metal hydride and an alkene. This is called β-elimination.

The β-elimination is responsible for the instability of alkyl organometallic compounds. Therefore the blockage of β-elimination reaction can bring about remarkable stability in metal alkyls. Therefore, the alkyls which do not contain replaceable hydrogen atom on β-carbon are highly stable. For example,

$M-CH_2-CH_3$, $M-CH_2-Ar$, $M-CH_2-NR_2$ etc.

Q.9. Discuss metal-carbon multiple bonding.

Ans. If multiple bonding occurs between metal and carbon, the bond strength is increased. For example, carbenes, $=CR_2$ and carbynes, $\equiv CR$ form double and triple bonds with the metal atom. It has been observed that in carbene compounds, the metal-carbon bond is somewhat shorter than normal metal-alkyl σ bond but larger than metal-carbon double bond. This suggests that the bond between metal-carbon in carbenes is in between single and double bond.

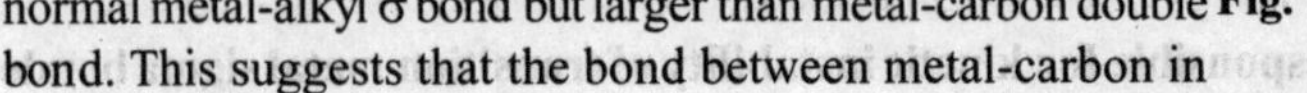

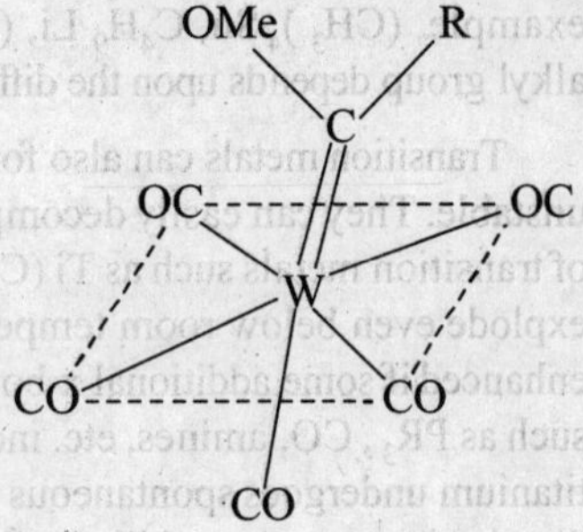

Fig. 16.3. Structure of $(CO)_5WCR$ (OMe).

The first carbene complex $(CO)_5$ W[: C(R) OMe)] was synthesised by E.O. Fisher in 1964. It has been found to have octahedral arrangement with one of the positions occupied by the carbene, = CR (OMe) (Fig. 16.3).

Metal carbynes have triple bonds. For example, $(CO)_4$ M (≡ CR)X where M = Cr, Mo or W, R = Me, Et or Ph, X = Cl, Br, I). The crystal structure of these types of complexes shows that ≡ CR group occupies position *trans* to halo ligand in an octahedral geometry (Fig. 16.4).

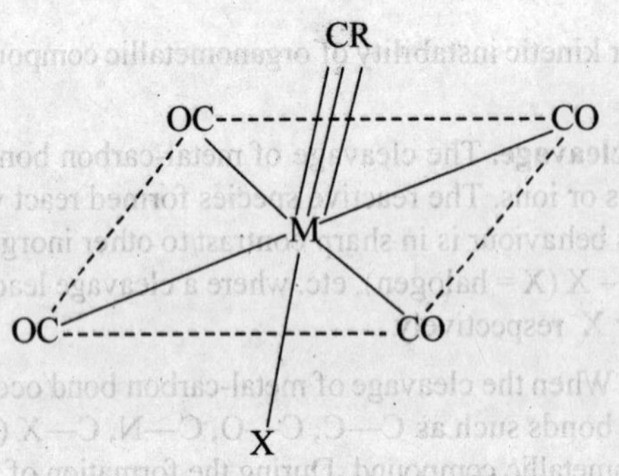

Fig. 16.4. Structure of $(CO)_4$ M (=CR)X

Q.10. Describe the methods of preparation of ferrocene.

Ans. Ferrocene is prepared by the following methods:

1. By reaction of iron halide with sodium cyclopentadienide. In this method cyclopentadiene is first converted into sodium cyclopentadienide by treating it with finely dispersed sodium in tetrahydrofuran (THF).

$$C_5H_6 + Na \xrightarrow{THF} C_5H_5\,Na + \frac{1}{2}H_2$$

Sodium cyclopentadienide is then treated with ferrous chloride in THF.

$$2C_5H_5Na + FeCl_2 \longrightarrow Fe(C_5H_5)_2 + 2NaCl$$

2. By reaction of iron halide with cyclopentadiene in the presence of strong base. Ferrocene is prepared by treating iron (II) chloride with cyclopentadiene in the presence of strong base like diethylamine. The strong base helps to eliminate HCl produced during the reaction:

$$2C_5H_6 + FeCl_3 + 2(C_2H_5)_2NH \longrightarrow Fe\,(C_5H_5)_2 + 2(C_2H_5)_2\,NH_2Cl$$

3. By treating iron (II) chloride with Grignard compound. Ferrocene can be prepared by treating iron (II) chloride with Grignard compound of cyclopentadiene.

$$FeCl_2 + 2\,(C_5H_5)\,MgBr \longrightarrow Fe\,(C_5H_5)_2 + MgCl_2 + MgBr_2$$

Laboratory preparation. In the laboratory, ferrocene is prepared by reacting iron (II) chloride with Grignard compound of cyclopentadiene.

$$C_5H_6 + KOH \longrightarrow KC_5H_5 + H_2O$$

$$2KC_5H_5 + FeC1_2 \longrightarrow (C_5H_5)_2Fe + 2KC1$$

Experimental details. Cyclopentadiene needed for the preparation of ferrocene does not exist as monomer at room temperature. It exists as dimer because of easy dimerisation of the monomer at room temperature. In order to get monomer (C_5H_6), the dimer is *cracked* at 180° C in an apparatus shown in Fig. 5.5. The apparatus consists of a flask fitted with a vertical air-condenser, thermometer, a water condenser and a receiver flask.

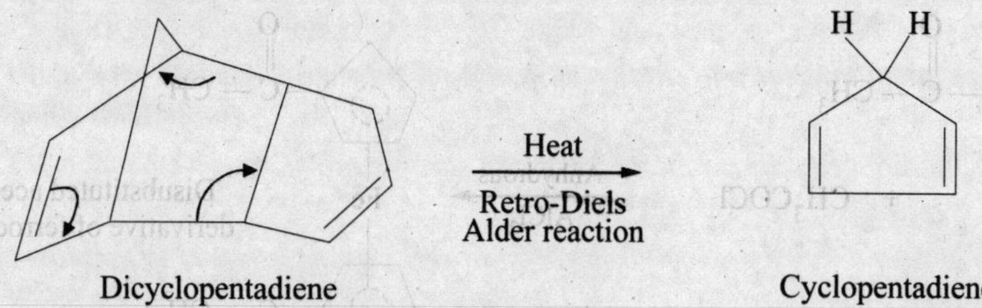

Dicyclopentadiene is taken in the flask and heated slowly to 180°C in an oil bath. After initial frothing, the monomer starts distilling off at 42°—44°C and is collected in the receiver flask. The receiver flask is placed in an ice bath to prevent dimerisation.

Cyclopentadiene is then converted to potassium cyclopentadienide (KC_5H_5), which on treatment with fresh FeCl gives ferrocene.

$$C_5H_6 + KOH \longrightarrow C_5H_5K + H_2O$$

$$2KC_5H_5 + FeC1_2 \longrightarrow \underset{\text{Ferrocene}}{(C_5H_5)_2Fe} + 2KC1$$

The water formed in the reaction is removed by using an excess of KOH. Since potassium cyclopentadienide is sensitive to air, therefore, the reaction is carried out in an inert atmosphere of nitrogen.

Q.11. Discuss properties of ferrocene. *(Calcutta, 2005)*

Ans. Physical properties. 1. It is an orange crystalline solid.

2. It is soluble in organic solvents like benzene but insoluble in water.

3. Its melting point is 174°C.

Chemical Properties

1. Acetylation. Ferrocene undergoes acetylation reactions with acetyl chloride in the presence of anhydrous $A1C1_3$. Acetic anhydride in the presence of phosphoric acid can also be used.

This reaction occurs more readily with ferrocene than with benzene. Since the reaction involves the electrophilic attack by the CH_3CO^+ group, therefore, it indicates that there is greater availability of negative charge on the π-cyclopentadienyl rings than on benzene.

$$(C_5H_5)_2Fe + CH_3COCl \xrightarrow[AlCl_3]{Anhydrous} (C_5H_5)Fe(C_5H_4COCH_3)$$

Monosubstituted derivative

$$(C_5H_5)Fe(C_5H_4COCH_3) + CH_3COCl \xrightarrow[AlCl_3]{Anhydrous} Fe(C_5H_4COCH_3)_2$$

Disubstituted acetyl derivative of ferrocene

2. Friedel Craft's alkylation. Ferrocene reacts with ethylene in the presence of anhydrous $AlCl_3$ undergoing alkylation of the ring.

$$(C_5H_5)_2Fe + C_2H_4 \xrightarrow[AlCl_3]{Anhydrous} (C_5H_5)Fe(C_5H_4CH_2CH_3)$$

3. Mannich condensation. Ferrocene reacts with formaldehyde in the presence of dimethylamine to give mono-substituted ferrocene. This reaction is called **Mannich condensation reaction.**

$$(C_5H_5)_2Fe + H-\overset{H}{\overset{|}{C}}=O + (CH_3)_2NH \longrightarrow (C_5H_5)Fe(C_5H_4-CH_2-N(CH_3)_2)$$

4. Sulphonation. Ferrocene reacts with chlorosulphuric acid or sulphuric acid in presence of acetic anhydride to form mono-substituted and hetero-substituted derivatives.

$$(C_5H_5)_2Fe + HOSO_3H \xrightarrow{(CH_3CO)_2O} C_5H_5 - Fe - C_5H_4.SO_3H + H_2O$$

5. Reaction with mercuric acetate. Ferrocene reacts with mercuric acetate (HgOAc) to give mono and disubstituted products.

The substituted derivatives of mercuric acetate react with iodide to give iodine derivatives. *This provides an indirect method for carrying out iodination of ferrocene.*

$$Fe(C_5H_5)_2 + Hg(OAc)_2 \xrightarrow{CH_3OH} C_5H_5-Fe-C_5H_4HgOAc + Fe(C_5H_4HgOAc)_2$$

$$C_5H_5-Fe-C_5H_4HgOAc \xrightarrow{I_2} C_5H_5-Fe-C_5H_4I \qquad Fe(C_5H_4HgOAc)_2 \xrightarrow{I_2} Fe(C_5H_4I)_2$$

6. Ferrocene does not undergo direct nitration and halogenation reactions because it gets oxidised to ferricinium ion, Fe $(C_2H_5)_2^+$.

7. Reaction with alkyl lithium. Ferrocene reacts with n-buty lithium to give mono and dilithium substituted derivatives.

$$Fe(C_5H_5)_2 \xrightarrow{n\text{-}BuLi} \underset{\text{(mono substituted)}}{C_5H_5-Fe-C_5H_4Li} \xrightarrow{n\text{-}BuLi} \underset{\text{(Disubstituted)}}{Fe(C_5H_4Li)_2}$$

The mono and dilithium derivatives are very useful intermediates in the synthesis of many ferrocenyl derivatives. Some of these are given below:

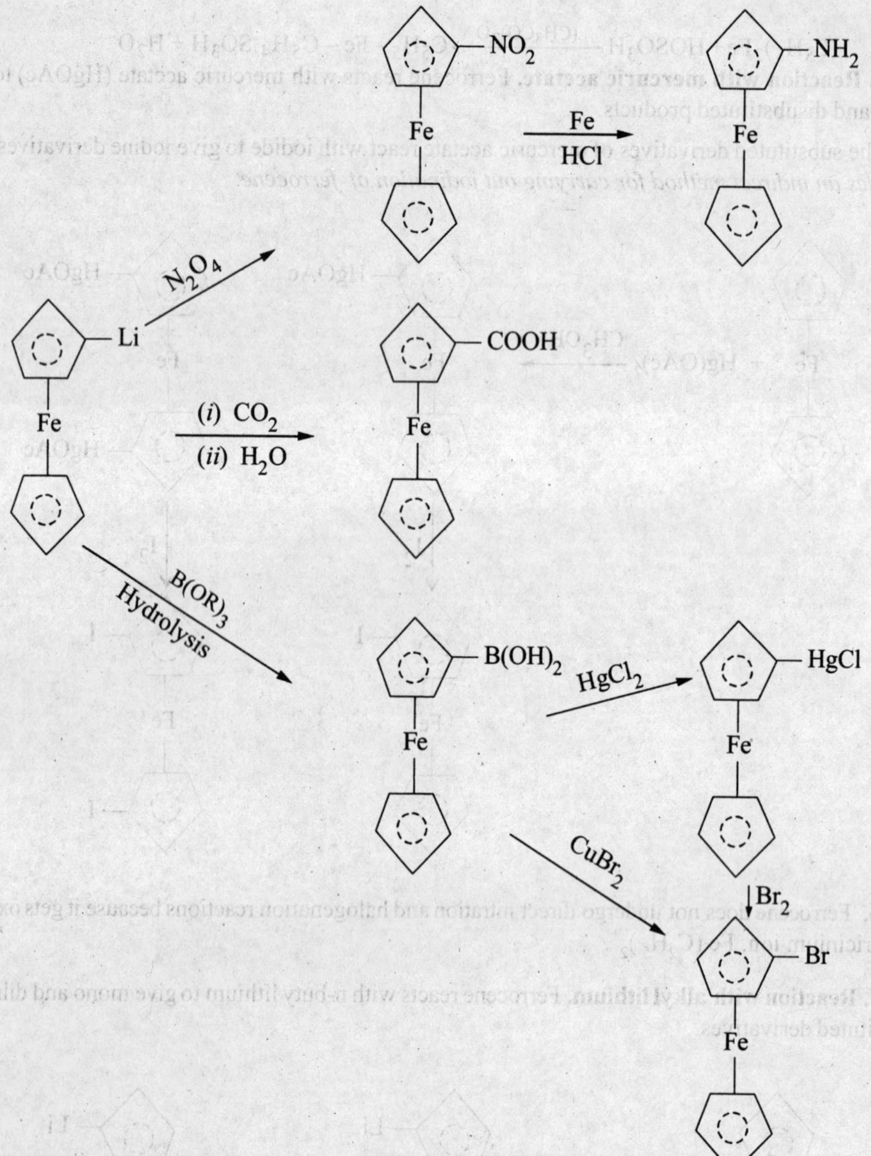

Q. 12. Discuss the structure of ferrocene. Why is it diamagnetic?

Ans. Ferrocene has a structure in which the iron atom is sandwiched between two C_5H_5 organic rings. The planes of the rings are parallel so that all the carbon atoms are at the same distance from the iron atom.

X-ray diffraction studies have shown that ferrocene has two types of arrangements of the cyclopentadiene rings known as **staggered** or **eclipsed**. It has been observed that in gas phase the structure of ferrocene is eclipsed while at low temperatures the structure of ferrocene is staggered. *The staggered configuration in the solid phase exists because of the crystal packing forces so that carbon-carbon and hydrogen-hydrogen repulsions between the two rings are minimum.* The energy barrier between staggered and eclipsed ferrocene configurations is only 4 kJ mol^{-1}.

Fe Fe

Staggered Eclipsed

The bonding in ferrocene can be explained on the basis of valence bond theory and molecular orbital theory as discussed below:

Valence bond theory. Though valence bond theory does not give satisfactory description of ferrocene molecule, yet it explains the diamagnetic character of a stable 18 electron outer configuration of the molecule. The outer electronic configuration of Fe(II) is $3d^6$ having six electrons in the $3d$ orbitals [Fig. 16.5 (*a*)]. To make available two *d*-orbitals for accommodating electron pairs from the ligands, the six electrons get paired up in three orbitals [Fig. 16.5 (*b*)]. The six vacant orbitals (two $3d$, one $4s$ and three $4p$) get hybridised forming six d^2sp^3 hybrid orbitals which accept 12 electrons; six electrons from each $C_5H^-_5$ ring from two C_5H_5 rings. Thus, the molecule is diamagnetic as observed.

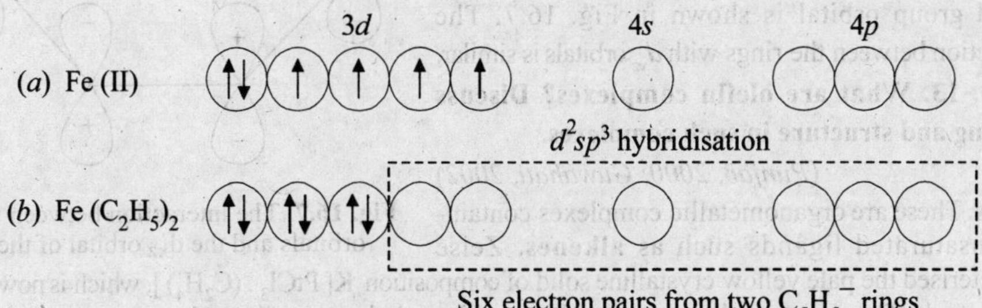

Fig. 16.5. Valence bond description for ferrocene.

Molecular orbitals theory. The exact bonding in ferrocene can be explained on the basis of molecular orbital theory. A brief idea can be obtained about the nature of bonding as described below:

The basic bond in ferrocene is a single bond between the π-orbitals of C_5H_5 ring and the iron atom. The bond is formed by the overlap of d_{xz} and d_{yz} orbitals on Fe (the *z*-axis is taken as the molecular axis) with the delocalized $p\pi$ aromatic orbital from each pentadienyl ring.

In pentadienyl ring, the carbon atom involves sp^2 hybridisation and forms three σ bonds (one with H atom and two with neighbouring C atoms(Fig. 16.6). As a result, one $p\,\pi$ orbital is left on each carbon atom containing one electron each. These five atomic orbitals combine with one another to give five molecular orbitals. These five MOs of σ ring can combine with similar MOs of the second ring to form 10 molecular orbitals called *ligand group molecular orbitals.* These ligand group molecular orbitals combine with appropriate orbitals having same symmetry of iron (five 3*d*, one 4*s* and three 4*p*) to give molecular orbitals. During their combinations 19 molecular orbitals are formed, out of which 9 are bonding and 10 are antibonding MOs. The ferrocene molecule has 18 electrons as:

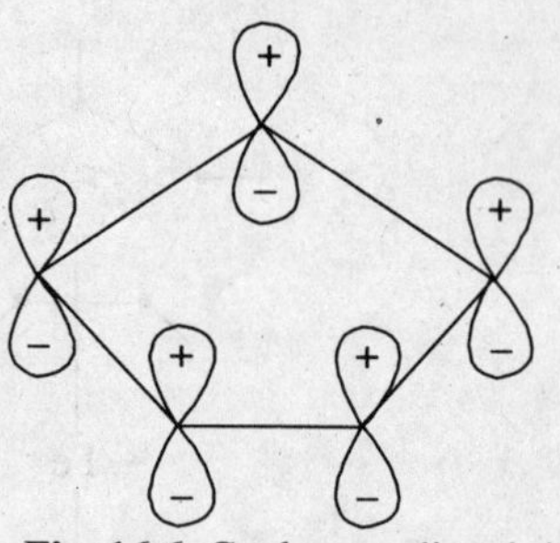

Fig. 16.6. Cyclopentadienyl ring.

8 from iron ($3d^6\ 4s^2$) + 2 × 5 from cyclopentadiene rings = 18

These 18 electrons are arranged in the nine bonding and non-bonding molecular orbitals. As a result the 10 anti-bonding MOs remain vacant.

In the ferrocene molecule, the most significant bonds are the bonds of π symmetry formed between Fe atom and the two C_5H_5 rings. These are obtained by the overlap of d_{xz} and d_{yz} atomic orbitals of Fe with appropriate ligand group molecular orbitals of cyclopentadiene rings. For example, the formation of a π-bond between d_{xz} orbital of Fe and the appropriate ligand group orbital is shown in Fig. 16.7. The interaction between the rings with d_{yz} orbitals is similar.

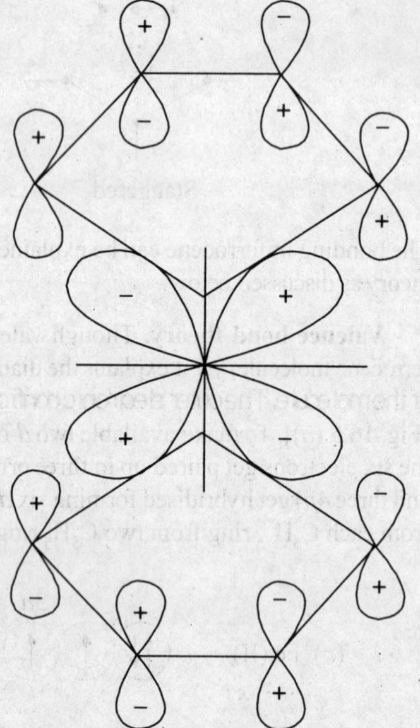

Fig. 16.7. The interaction between the ring orbitals and the d_{xz} orbital of the ion.

Q. 13. What are olefin complexes? Discuss bonding and structure in such complexes.

(*Punjab, 2000; Guwahati, 2002*)

Ans. These are organometallic complexes containing unsaturated ligands such as **alkenes.** Zeise characterised the pale yellow crystalline solid of composition K[$PtCl_3$. (C_2H_4)], which is now known as Zeise salt. This compound was prepared by passing ethylene gas through an aqueous solution of potassium tetrachloroplatinate (II).

$$PtCl_4^{2-} + C_2H_4 \longrightarrow [PtCl_3(C_2H_4)]^- + Cl^-$$

This was among the earliest olefin complexes prepared.

It was later on found that certain metal halides or ions formed complexes when treated with a variety of olefins.

Structure and Bonding in Metal Olefin Complexes

The structure of the anion of the Zeise salt was well established in 1975. The significant features of the structure are:

(*i*) The three Cl atoms and the middle point of the ethylene double bond form a square plane (Fig. 16.8).

(*ii*) The platinum atom is present in the centre of the square and C = C double bond of the ethylene molecule is perpendicular to the plane containing Pt and three Cl atoms.

(*iii*) The C_2H_4 group is significantly distorted from planarity.

The metal atom has filled (shown shaded) and empty (shown unshaded) orbitals. The bonding occurs in two ways:

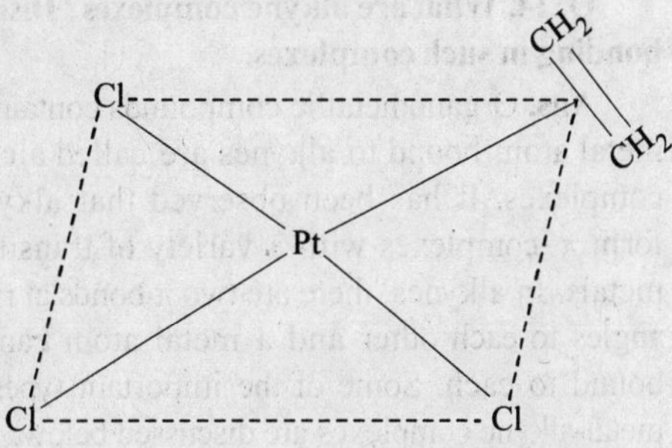

Fig. 16.8 Structure of $[PtCl_3(C_2H_4)]^-$ complex..

(*i*) *The filled π–molecular orbitals of the ethylene molecule are donated to the suitably direct vacant orbitals of the metal to form alkene* ⟶ metal sigma bond. This is shown in Fig. 16.9 (*a*).

(*ii*) *A back bond is formed by the overlap of the filled metal hybrid orbitals with the vacant π MO of the olefin.* This leads to a π-bond. The back bonding will strengthen the Pt-olefin bond but it should weaken the C — C bond in alkene.

This type of bonding is shown in Fig. 16.9 (*b*).

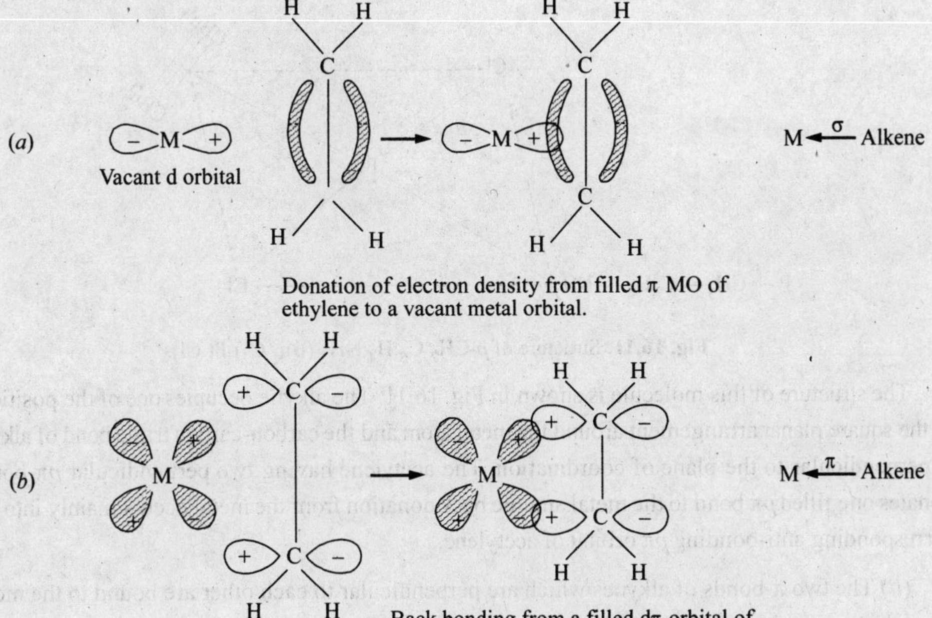

Fig. 16.9. Stepwise bonding in metal-olefin complex.

The complete bonding picture is shown in Fig. 16.10.

It may be noted that the metal olefin bond is essentially electroneutral because donation and back bonding balance the electron density. Initially the alkene ligand donates two electrons from its π-bonding MO into the vacant orbital of the metal M. Depending upon the type of alkene-complex formed, the excess electron density accumulating on M is reduced.

Q. 14. What are alkyne complexes? Discuss bonding in such complexes.

Ans. Organometallic compounds containing metal atom bound to alkynes are called alkyne complexes. It has been observed that alkynes form π–complexes with a variety of transition metals. In alkynes, there are two π bonds at right angles to each other and a metal atom can be bound to each. Some of the important types of metal-alkyne complexes are discussed below:

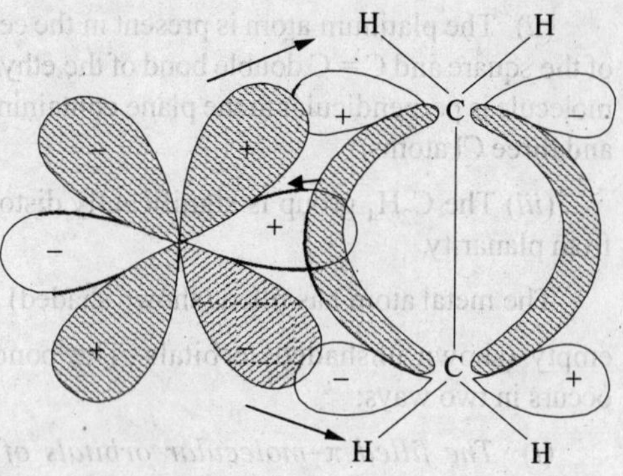

Fig. 16.10. Bonding in metal olefin complex.

(*i*) The alkyne may coordinate to only one metal atom and function like alkenes. In these molecules, alkyne molecule occupies one of the positions like that of ethylene in Zeise salt. The carbon-carbon triple bond of the alkyne is perpendicular to the plane of coordination. These compounds may be prepared by the following methods:

$$PtCl_2 + {}^tBuC \equiv C\,Bu^t + p - CH_3C_6H_4NH_2 \longrightarrow p\text{–}\,CH_2\,C_6H_4NH_2\,(Bu_2\,C_2)\,PtCl_2$$

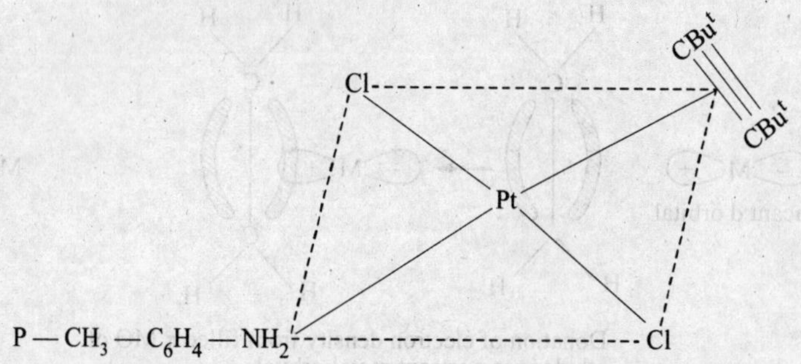

Fig. 16.11. Structure of p-$CH_2\,C_6\,H_4\,NH_2\,(Bu_2\,C_2)\,Pt\,Cl_2$.

The structure of this molecule is shown in Fig. 16.11. The alkyne occupies one of the positions of the square planar arrangement around the metal atom and the carbon-carbon triple bond of alkyne is perpendicular to the plane of coordination. The acetylene having two perpendicular $p\pi$-bonds donates one filled $p\pi$ bond to the metal and the back donation from the metal occurs mainly into the corresponding anti-bonding $p\pi$ orbital of acetylene.

(*ii*) The two π-bonds of alkynes which are perpendicular to each other are bound to the metal atom through each of their π-bonds. In these cases, the alkynes act as 4-electron donors by using both the π-bonds and act as bridging ligands. These types of complexes are prepared by the displacement of the bridging carbonyl groups with alkynes at higher temperatures. For example,

$$Co_2(CO)_8 + PhC \equiv CPh \xrightarrow{\text{Benzene}} PhC \equiv CPh[Co(CO)_3]_2$$

$$(\pi C_5H_5)_2\,Ni_2(CO)_2 + PhC \equiv CPh \xrightarrow{\text{Reflux}} (\eta^5 - C_5H_5)_2\,Ni_2(PhC \equiv CPh)$$

The structures of these complexes are given in Fig. 16.12. In these structures, the two acetylenic carbons are kept between the two cobalt atoms so that two π-bonds of acetylene, which are perpendicular to each other, interact with a cobalt atom.

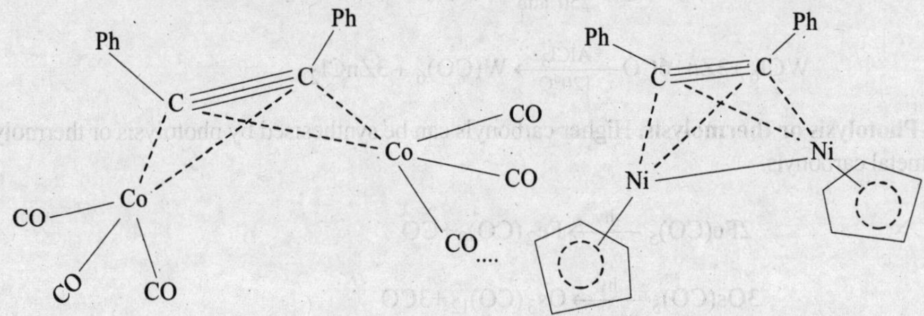

Fig. 16.12. Metal-alkyne complexes.

Q. 15. Define π-acidity. Give one example.

Ans. *The ability of a ligand to accept electron density into vacant π-orbitals is called π-acidity.* Complexes of metals with carbon monoxide are called metal carbonyls. Oxidation state of metal in these complexes is either zero or a low positive value.

There are no attractive interactions between the metal and the ligands as is possible with the positively charged metal ion. It is the main characteristic of CO ligand that it can stabilize low oxidation states. This is due to the fact that it possesses vacant π-orbitals in addition to lone pairs. The formation of a sigma bond by the donation of a lone pair of electrons into the suitable vacant metal orbitals leads to excessive negative charge on the metal (in zero or negative oxidation state). To counter the accumulation of negative charge on the metal, π-bond is formed by the back donation of electrons from the filled metal orbitals into the vacant π-type orbitals on the ligand. This also supplements the σ-bond. This ability of the ligand (CO) to accept electron density into vacant π orbitals is called π-acidity. Therefore, CO is also called a π-acceptor ligand and the metal carbonyls are referred as complexes of π-acceptor (or π acid) ligands.

Q. 16. Give the methods of preparation of metal carbonyls.

(*Rohilkhand, 2003; Madras, 2004; Andhra, 2010*)

Ans. Some of the methods of preparation of metal carbonyls are described as under:

1. Direct synthesis. The metal carbonyls can be prepared by the direct reaction of carbon monoxide with the finely divided metal. For example, nickel carbonyl is prepared by the direct reaction of finely divided nickel metal with CO at 25°C under one atmospheric pressure.

$$Ni + 4CO \xrightarrow{25°C} Ni(CO)_4$$

Some other metal carbonyls are also prepared by the direct reaction at high temperatures and pressures.

$$Fe + 5CO \xrightarrow[100\ atm]{200°C} Fe(CO)_5$$

$$Mo + 6CO \xrightarrow[250\ atm]{200°C} Mo(CO)_6$$

2. Reduction. This is the most commonly used method for preparing metal carbonyls. In this method, the metal salts are reduced in the presence of CO. For example,

$$CrCl_3 + Al + 6CO \xrightarrow[C_6H_6]{AlCl_3} Cr(CO)_6 + AlCl_3$$

$$2CoCO_3 + 8CO + 2H_2 \xrightarrow[250\ atm]{150°C} Co_2(CO)_8 + 2CO_2 + 2H_2O$$

$$WCl_6 + 3Zn + 6CO \xrightarrow[120°C]{AlCl_3} W(CO)_6 + 3ZnCl_2$$

3. Photolysis or thermolysis. Higher carbonyls can be synthesised by photolysis or thermolysis of lower metal carbonyls.

$$2Fe(CO)_5 \xrightarrow{hv} Fe_2(CO)_9 + CO$$

$$3Os(CO)_5 \xrightarrow{hv} Os_3(CO)_{12} + 3CO$$

Q. 17. Describe chemical properties of metal carbonyls.

Ans. 1. Displacement reactions. Metal carbonyls undergo displacement reactions in which one or more CO ligands may be displaced in contact with electron donors (Lewis bases) such as pyridine, phosphines (R_3P), isocyanides (RNC), etc. One molecule of CO is replaced by one electron pair donor. Therefore, a six electron donor (like benzene) could displace three CO molecules. Some of the common displacement reactions are:

$$Fe\,(CO)_5 + Ph_3\,P \longrightarrow Fe(CO)_4\,PPh_3 + CO$$

$$Fe\,(CO)_5 + 2Ph_3\,P \longrightarrow Fe\,(CO)_3\,(PPh_3\,)_2 + 2CO$$

$$Cr(CO)_6 + C_6H_6 \xrightarrow{-3CO} (C_6H_6)\,Cr\,(CO)_3$$

2. Formation of carbonyl anionic complexes. The metal carbonyls can be converted into anions known as carbonylate ions by the following methods:

(*i*) Metal carbonyls react with bases to give carbonylate ion

$$Fe\,(CO)_5 + OH^- \longrightarrow [Fe\,(CO)_4\,]^{2-} + H^+ + CO_2$$

$$Fe_2\,(CO)_9 \xrightarrow[Et_4\ NI]{KOH,\ MeOH} (Et_4N)_2\,[Fe_2\,(CO)_8]$$

(*ii*) Reduction of metal carbonyls with reducing agents produce carbonylate anions. The reduction is carried out by alkali metal amalgams, hydride reagents and Na/K alloy in basic solvents including liquid ammonia.

$$Co_2(CO)_8 \xrightarrow[THF]{Na/Hg} Na[Co(CO)_4]$$

$$Cr(CO)_6 \xrightarrow[THF]{Na/Hg} Na_2[Cr(CO)_5]$$

$$Mn_2(CO)_8 \xrightarrow[THF]{KH} 2K[Mn(CO)_5]$$

(*iii*) Many substituted carbonyl anions can be obtained by displacement of carbon monoxide from a metal carbonyl.

$$Me_4NI + Mo(CO)_6 \longrightarrow Me_4N[Mo(CO)_5I] + CO$$

$$Mo(CO)_6 + NaB_3H_8 \longrightarrow Na[Mo(CO)_5B_3H_8] + CO$$

(*iv*) Metal carbonylate anions can be prepared by the reaction with CH_3 or C_6H_5 radicals.

$$(CO)_5Cr(CO) + CH_3 \longrightarrow \left[(CO)_5Cr-C\genfrac{}{}{0pt}{}{\nearrow O}{\searrow CH_3}\right]^- \xrightarrow[-N_2]{H^+, CH_2N_2} (CO)_5Cr{=}C\genfrac{}{}{0pt}{}{\diagup OCH_3}{\diagdown CH_3}$$

3. Formation of cationic carbonyl complexes. These may be obtained in the formation of complex halides by the reaction of carbon monoxide and a Lewis acid such as $AlCl_3$ or BF_3. For example,

$$Mn(CO)_5Cl + CO + AlCl_3 \longrightarrow [Mn(CO)_6^+][AlCl_4]^-$$

$$Re(CO)_5Cl + AlCl_3 + CO \longrightarrow [Re(CO)_6][AlCl_4]^-$$

Q. 18. Discuss bonding in linear metal carbonyls (M – C – O). *(Lucknow, 2010; Rajasthan, 2010)*

Ans. Carbon monoxide has a triple bond with lone pair of electrons on both C and oxygen atom as

$$:C \equiv O:$$

(*i*) There is a dative overlap of the filled orbital of carbon (of CO) and suitable empty orbital of the metal forming a dative sigma bond (M ← CO). This is shown in Fig. 16.13 (*a*).

(*ii*) There is a π-overlap involving donation of electrons from filled metal *d*-orbitals into vacant antibonding π molecular orbitals. This results into the formation of M → CO bond. This is also called *back donation or back bonding* [Fig. 16.13 (*b*)].

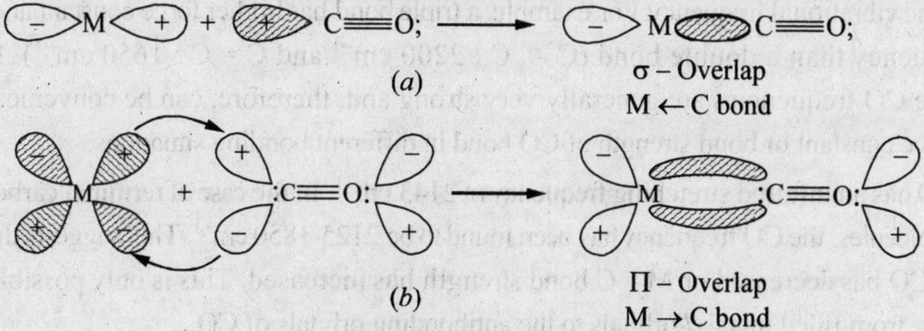

Fig. 16.13. Bonding in metal carbonyls.

The bonding in metal carbonyls is shown in Fig. 16.13. In these figures, the shaded orbitals represent filled orbitals.

The formation of σ dative bond tends to increase the electron density on the metal atom. At the same time, the formation of π bond from metal to carbon tends to decrease the electron density on metal. Then the formation of π bond increases the strength of M—CO bond. This accounts for the fact that CO is a very weak Lewis base towards non-transition metal halides like BX_3, AlX_3, etc., but forms very strong complexes with transition metals.

Evidence in Support of Bonding

1. The formation of back bonding from metal to CO molecule results in decrease in electron density on metal. This is supported by the dipole moment studies. It has been observed that the dipole moment of M—C bond is only very low, about 0.5 D.

2. The back bonding from metal to CO is expected to increase M—C bond strength with corresponding weakening of C ≡ O. This is due to the fact that electrons from back bonding fill the antibonding MOs of CO. As a result, the bonding ability of CO will decrease. Therefore, as the M—C bond becomes stronger, the C ≡ O bond becomes weaker. Therefore, the multiple bonding should be evidenced by the shorter M—C bond as compared to M—C single bonds and longer C—O bonds as compared to normal C ≡ O triple bonds. It has been observed that the bond length in CO molecule is 1.128 Å and in metal carbonyls, it is of the order of 1.15 Å.

Q. 19. How does vibrational spectra help to understand bonding in carbonyls.

Ans. Infrared spectroscopy has given valuable support for bonding in metal carbonyls. These studies provide information regarding bond orders of M—C and C ≡ O bonds. The decrease in C—O bond order or force constant is estimated by studying the CO stretching frequency in infrared spectroscopy.

The infrared spectra is characterised by frequency of vibration, which is related to force constant (k) as

$$\nu = \frac{1}{2\pi}\sqrt{\frac{k}{\mu}}$$

where μ is reduced mass of bonded atoms with mass m_1 and m_2 and is given as

$$\mu = \frac{m_1 m_2}{m_1 + m_2}$$

The force constant is a measure of bond strength. The larger the force constant, the stronger the bond and higher the vibrational frequency. For example, a triple bond has higher force constant and therefore higher frequency than a double bond (C ≡ C : 2200 cm^{-1} and C = C : 1650 cm^{-1}). In infrared spectrum the CO frequencies are generally very strong and, therefore, can be conveniently used to study the force constant or bond strength of CO bond in different bonding situations.

Free CO has an infrared stretching frequency of 2143 cm^{-1}. In the case of terminal carbonyl groups in neutral molecules, the CO frequency has been found to be 2125-1850 cm^{-1}. This suggests that the bond strength of CO has decreased or M—C bond strength has increased. This is only possible if there is back bonding from filled metal *d*-orbitals to the antibonding orbitals of CO.

Q. 20. Discuss the structure of mononuclear, binuclear and polynuclear metal carbonyls.

(*Gujarat, 2003; Lucknow, 2010; Pune, 2010*)

Ans. Mononuclear metal carbonyls. The mononuclear metal carbonyls have all terminal CO groups. The shapes are those expected from their formulae. Hexacarbonyls like $Cr(CO)_6$ have regular octahedral structures. Tetracarbonyls like $Ni(CO)_4$ have tetrahedral structure. The pentacarbonyls like $Fe(CO)_5$, $Os(CO)_5$ have trigonal bipyramidal shape. The shapes of some mononuclear carbonyls are shown in Fig. 16.14.

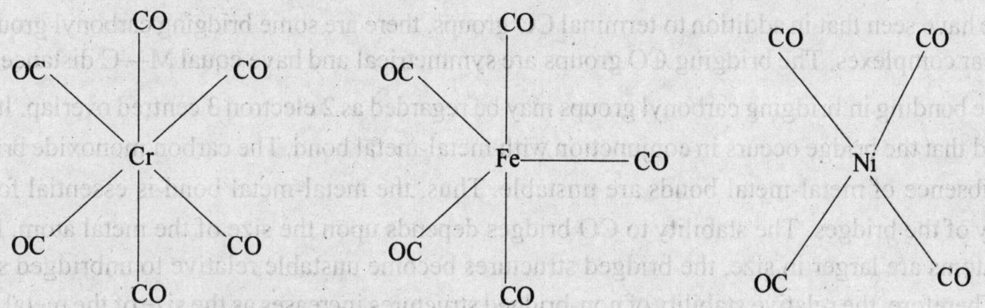

Fig. 16.14. Structures of some mononuclear carbonyls.

Binuclear Metal Carbonyls

The binuclear metal carbonyls have different types of structures. They show metal-metal bonding and bridging CO groups in their molecules. Their structures are described below:

(*i*) The binuclear metal carbonyls, $M_2(CO)_{10}$ where M = Mn, Ti or Re have metal-metal bonding in their structures. Each metal has octahedral arrangement of CO groups. The structure of $Mn_2(CO)_{10}$ is shown in Fig. 16.15.

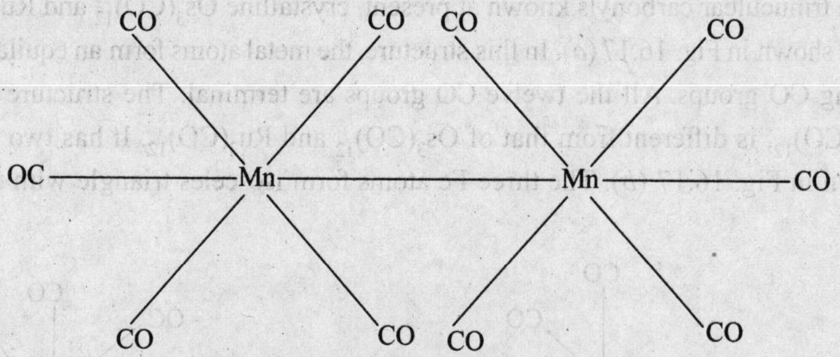

Fig. 16.15. Structure of $Mn_2(CO)_{10}$.

(*ii*) The binuclear carbonyls such as $Fe_2(CO)_9$, $Os_2(CO)_9$ and $Co_2(CO)_8$ have bridging carbonyl groups in addition to metal-metal bonds. $Fe_2(CO)_9$ has three bridging CO groups placed symmetrically around the Fe—Fe bond [Fig. 16.16 (*a*)]. Each Fe atom is also bonded to three terminal CO groups.

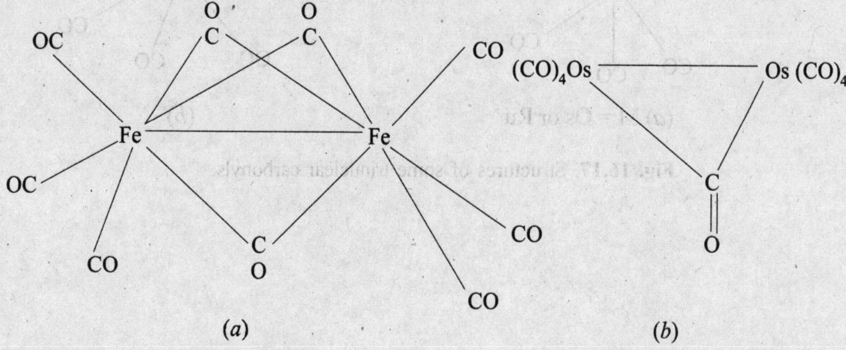

Fig. 16.16. Structure of bridged binuclear carbonyls.

$Os_2(CO)_9$ has only one bridging carbonyl group and each Os atom has four terminal CO groups [Fig. 16.16 (*b*)].

We have seen that in addition to terminal CO groups, there are some bridging carbonyl groups in dinuclear complexes. The bridging CO groups are symmetrical and have equal M—C distances.

The bonding in bridging carbonyl groups may be regarded as 2 electron 3 centred overlap. It may be noted that the bridge occurs in conjunction with metal-metal bond. The carbon monoxide bridges in the absence of metal-metal bonds are unstable. Thus, the metal-metal bond is essential for the stability of the bridges. The stability to CO bridges depends upon the size of the metal atom. If the metal atoms are larger in size, the bridged structures become unstable relative to unbridged structures. Therefore, the relative stability of non-bridged structures increases as the size of the metal atom increases. For example, the smaller Fe atom in $Fe_2(CO)_9$ has 3 bridging CO groups whereas bigger osmium in $Os_2(CO)_9$ has one bridging CO group.

Polynuclear Metal Carbonyls

The structures of polynuclear metal carbonyls are more complex. They contain metal-metal bonds and terminal and bridging carbonyl groups. The structures of some polynuclear carbonyl are discussed below:

Among the trinuculear carbonyls known at present, crystalline $Os_3(CO)_{12}$ and $Ru_3(CO)_{12}$ have the structures as shown in Fig. 16.17 (*a*). In this structure, the metal atoms form an equilateral triangle with no bridging CO groups. All the twelve CO groups are terminal. The structure of trinuclear carbonyls, $Fe_3(CO)_{12}$, is different from that of $Os_3(CO)_{12}$ and $Ru_3(CO)_{12}$. It has two bridging CO groups as shown in Fig. 16.17 (*b*). The three Fe atoms form isosceles triangle with three Fe—Fe bonds.

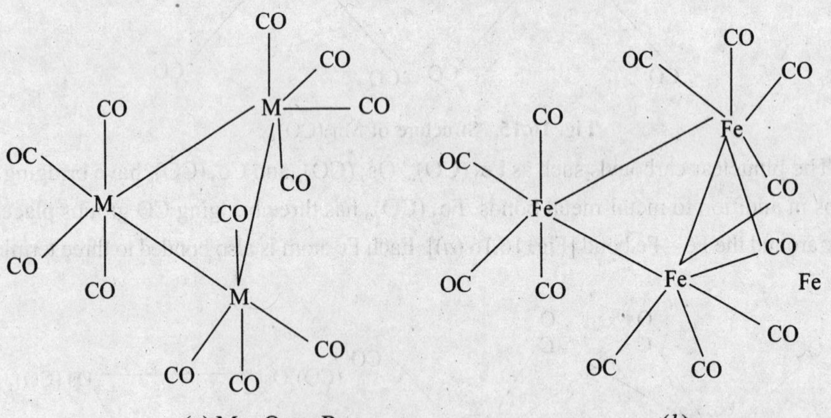

(*a*) M = Os or Ru (*b*)

Fig. 16.17. Structures of some trinuclear carbonyls.

17 THE *d*-BLOCK ELEMENTS (TRANSITION METALS)

Q. 1. What are *d*-block or transition elements?

(Purvanchal, 2007; Nagpur 2008; Pune, 2010)

Ans. The elements which involve the filling of *d*-orbitals in the penultimate shell are called *d*-block elements. In the periodic table, they are placed in between the s- and the *p*-block elements and their properties represent a change from the electropositive *s*-block elements to the electronegative *p*-block elements. That is why they are commonly called transition elements.

Transition elements of *d*-block Elements.

III B	IV B	V B	VI B	VII B	← VIII B →			IB	II B	Old numbering
3	4	5	6	7	8	9	10	11	12	New numbering
21 Sc	22 Ti	23 V	24 Cr	25 Mn	26 Fe	27 Co	28 Ni	29 Cu	30 Zn	← 3d series
39 Y	40 Zr	41 Nb	42 Mo	43 Tc	44 Ru	45 Rh	46 Pd	47 Ag	48 Cd	← 4d series
57 La	72 Hf	73 Ta	74 W	75 Re	76 Os	77 Ir	78 Pt	79 Au	80 Hg	← 5d series
89 Ac	104 Ku	105 Ha								← 6d series

The *d*-block of the periodic table (shown above) consists of four series, called as first, second, third and the fourth (*3d, 4d, 5d,* and *6d*) transition series. The first three series are complete containing 10 elements each and the fourth one is an incomplete series. All these elements are placed in eight sub-groups, viz., III B, IV B, V B, VI B, VII B, VIII B, I B and II B as shown in the periodic table along with the atomic numbers. Lately, these groups are also called 3rd, 4th, 5th.......12th groups. Elements of sub-group II B, *viz.* Zn, Cd and Hg, are not included in transition elements as they do not have a partly filled *d*- sub-shell in their elementary state. They do not show the properties of transition elements to any appreciable extent except for their ability to form complexes with ligands such as ammonia, amines, and halide ions, etc. However, they are generally studied with the *d*-block elements in order to maintain a simple classification of elements.

Q. 2. Justify the following statement:

Transition metals are less reactive than alkali and alkaline earth metals. *(Kumaon, 2000)*

Ans. Transition metals are less reactive than the alkali and alkaline earth metals because:

(*i*) transition metals are smaller in size compared to alkali and alkaline earth metals; and

(*ii*) transition metals have higher ionisation energies.

Q. 3. Cu^{2+} ($3d^9$) is more stable than Cu^+($3d^{10}$).

Ans. An ion carrying a higher charge and smaller size is associated with greater lattice and hydration energy. That is why Cu^{2+} is more stable than Cu^+. It is expected on the basis of electronic configuration that Cu^+ showed be more stable because it has completely filled orbitals. But the lattice and hydration energy factors dominate and as a result Cu^{2+} is more stable.

Q. 4. What justification is there for classifying Zn, Cd and Hg as transition elements?

Ans. Although there are no incomplete *d*-orbitals in Zn, Cd and Hg, these elements have been retained as transition elements for the sake of continuity.

Q. 5. Discuss the electronic configuration of *d*-block elements.

(*Bhopal, 2004; Purvanchal, 2007; Nagpur 2008*)

Ans. Electronic configurations of all the *d*-block elements are given below for different series.

(*i*) First transition series or 3*d* series. The electronic configurations for 3*d* series (1st transition series) elements are given below in Table 17.1. Progressive filling of electrons in 3*d* orbitals takes place.

Table 17.1

Element	*At. No.*	*Electronic Configurations*
Sc	21	$1s^2, 2s^2\,2p^6, 3s^2\,3p^6\,3d^1, 4s^2$ Or [Ar] $3d^1, 4s^2$
Ti	22	$1s^2, 2s^2\,2p^6, 3s^2 3p^6\,3d^2, 4s^2$
V	23	$1s^2, 2s^2\,2p^6, 3s^2\,3p^6\,3d^3, 4s^2$
Cr	24	$1s^2, 2s^2\,2p^6, 3s^2\,3p^6\,3d^5, 4s^1$
Mn	25	$1s^2, 2s^2\,2p^6, 3s^2 3p^6\,3d^5, 4s^2$
Fe	26	$1s^2, 2s^2\,2p^6, 3s^2 3p^6\,3d^6, 4s^2$
Co	27	$1s^2, 2s^2\,2p^6, 3s^2 3p^6\,3d^7, 4s^2$
Ni	28	$1s^2, 2s^2\,2p^6, 3s^2 3p^6\,3d^8, 4s^2$
Cu	29	$1s^2, 2s^2 2p^6, 3s^2\,3p^6\,3d^{10}, 4s^1$
Zn	30	$1s^2, 2s^2\,2p^6, 3s^2 3p^6 3d^{10}, 4s^2$

(*ii*) Second transition series or 4*d* series. In these elements filling of 4*d* orbitals takes place in the penultimate shell. The electronic configurations are given in Table 17.2.

Table 17.2

Element	*At. No.*	*Electronic Configurations*
Y	39	$1s^2, 2s^2\,2p^6, 3s^2\,3p^6\,3d^{10}, 4s^2\,4p^6\,4d^1, 5s^2$
Zr	40	$1s^2, 2s^2\,2p^6, 3s^2\,3p^6\,3d^{10}, 4s^2\,4p^6\,4d^2, 5s^2$
Nb	41	$1s^2, 2s^2\,2p^6, 3s^2\,3p^6\,3d^{10}, 4s^2\,4p^6\,4d^4, 5s^1$
Mo	42	$1s^2, 2s^2\,2p^6, 3s^2\,3p^6\,3d^{10}, 4s^2\,4p^6\,4d^5, 5s^1$
Tc	43	$1s^2, 2s^2\,2p^6, 3s^2\,3p^6\,3d^{10}, 4s^2\,4p^6\,4d^5, 5s^2$
Ru	44	$1s^2, 2s^2\,2p^6, 3s^2\,3p^6\,3d^{10}, 4s^2\,4p^6\,4d^7, 5s^1$
Rh	45	$1s^2, 2s^2\,2p^6, 3s^2\,3p^6\,3d^{10}, 4s^2\,4p^6\,4d^8, 5s^1$
Pd	46	$1s^2, 2s^2\,2p^6, 3s^2\,3p^6\,3d^{10}, 4s^2\,4p^6\,4d^{10}, 5s^1$
Ag	47	$1s^2, 2s^2\,2p^6, 3s^2\,3p^6\,3d^{10}, 4s^2\,4p^6\,4d^{10}, 5s^1$
Cd	48	$1s^2, 2s^2\,2p^6, 3s^2\,3p^6\,3d^{10}, 4s^2\,4p^6\,4d^{10}, 5s^2$

***(iii)* Third transition series or 5*d* series.** In these elements filling of 5*d* orbitals takes place in the penultimate shell. The electronic configurations are given in Table 17.3.

Table 17.3

Element	*At. No.*	*Electronic Configurations*
La	57	$1s^2, 2s^2 2p^6, 3s^2 3p^6 3d^{10}, 4s^2 4p^6 4d^{10}, 5s^2\ 5p^6\ 5d^1, 6s^2$
Hf	72	$1s^2, 2s^2 2p^6, 3s^2 3p^6 3d^{10}, 4s^2 4p^6 4d^{10}\ 4f^{14}, 5s^2\ 5p^6\ 5d^2, 6s^2$
Ta	73	$1s^2, 2s^2 2p^6, 3s^2 3p^6 3d^{10}, 4s^2 4p^6 4d^{10}\ 4f^{14}, 5s^2\ 5p^6\ 5d^3, 6s^2$
W	74	$1s^2, 2s^2 2p^6, 3s^2 3p^6 3d^{10}, 4s^2 4p^6 4d^{10}\ 4f^{14}, 5s^2\ 5p^6\ 5d^4, 6s^2$
Re	75	$1s^2, 2s^2 2p^6, 3s^2 3p^6 3d^{10}, 4s^2 4p^6 4d^{10}\ 4f^{14}, 5s^2\ 5p^6\ 5d^5, 6s^2$
Os	76	$1s^2, 2s^2 2p^6, 3s^2 3p^6 3d^{10}, 4s^2 4p^6 4d^{10}\ 4f^{14}, 5s^2\ 5p^6\ 5d^6, 6s^2$
Ir	77	$1s^2, 2s^2 2p^6, 3s^2 3p^6 3d^{10}, 4s^2 4p^6 4d^{10}\ 4f^{14}, 5s^2\ 5p^6\ 5d^7, 6s^2$
Pt	78	$1s^2, 2s^2 2p^6, 3s^2 3p^6 3d^{10}, 4s^2 4p^6 4d^{10}\ 4f^{14}, 5s^2\ 5p^6\ 5d^8, 6s^2$
Au	79	$1s^2, 2s^2 2p^6, 3s^2 3p^6 3d^{10}, 4s^2 4p^6 4d^{10}\ 4f^{14}, 5s^2\ 5p^6\ 5d^{10}, 6s^1$
Hg	80	$1s^2, 2s^2 2p^6, 3s^2 3p^6 3d^{10}, 4s^2 4p^6 4d^{10}\ 4f^{14}, 5s^2\ 5p^6\ 5d^{10}, 6s^2$

***(iv)* Fourth transition series or 6*d* series.** In these elements filling of 6*d* orbitals takes place. It is an incomplete series and includes elements of the incomplete 7th period. The electronic configurations are given in Table 17.4.

Table 17.4

Element	*At. No.*	*Electronic Configurations*
Ac	89	$1s^2, 2s^2\ 2p^6, 3s^2\ 3p^6\ 3d^{10}, 4s^2\ 4p^6\ 4d^{10}\ 4f^{14}, 5s^2\ 5p^6\ 5d^{10}, 6s^2\ 6p^6\ 6d^1, 7s^2$
Ku	104	$1s^2, 2s^2\ 2p^6, 3s^2\ 3p^6\ 3d^{10}, 4s^2\ 4p^6\ 4d^{10}\ 4f^{14}, 5s^2\ 5p^6\ 5d^{10}\ 5f^{14}, 6s^2\ 6p^6\ 6d^2, 7s^2$
Ha	105	$1s^2, 2s^2\ 2p^6, 3s^2\ 3p^6\ 3d^{10}, 4s^2\ 4p^6\ 4d^{10}\ 4f^{14}, 5s^2\ 5p^6\ 5d^{10}\ 5f^{14}, 6s^2\ 6p^6\ 6d^3, 7s^2$

Q. 6. Elements of transition series show common physical and chemical properties. Explain.

Ans. In these elements, the number of electrons in the outermost shell remains the same. Only the penultimate shell is being expanded from 8 to 18 by the addition of *d*-electrons. Their electronic configurations are not very regular. This is due to the fact that energy difference of the electrons in the $(n-1)$ *d* and *ns* levels are not significant and electrons between these levels may change positions mutually. Since chemical properties depend, to a greater extent, on the number of electrons in the outermost shell, which remains the same, the variation in chemical properties amongst these elements of a given period are very small. As is expected, they have physical and chemical properties in common.

Q. 7. Give electronic configurations of the following: (*i*) Mo,(*ii*) Pd,(*iii*) W, (*iv*) Hg.

Ans. See Tables 17.2 and 17.3.

Q. 8. Name the elements of the first transition series and give their electronic configurations.

Ans. See Table 17.1.

Q. 9. "4*s* subshell is filled prior to 3*d* but on ionisation, 4*s* electrons are removed first". Justify.

Ans. 4*s* orbital is spherical and hence has a greater penetrating power as compared to 3*d* orbital which is having a double dumb-bell shape. Therefore 4*s* orbital is associated with lower energy compared to 3*d* and hence the filling of electron takes place first in 4*s* and then in 3*d*.

However, when it comes to removing the electrons, removal first takes place in 4*s*. This is because once the 3*d* orbital has been filled, its energy falls because of greater effective nuclear charge, making it more stable than 4*s* orbital. Hence electrons are first removed from 4*s* orbital which acquires greater energy.

Q. 10. First ionisation energy of copper is higher than those of alkali metals while second and third energies are lower than those of alkali metals. Explain.

Ans. Copper has a smaller size compared to the alkali metals in their period. Due to smaller size and due to poor shielding effect of 3*d* orbital, Cu has a higher first ionisation energy. Second and third I.E. are lower compared to those of alkali metals due to loss of electrons from diffused *d*-orbitals, *d*-electrons have smaller screening effect and give rise to higher effective nuclear charge compared to *s* and *p* electrons.

Q. 11. Why the second ionisation energy for Cu and Cr are anomalously high?

(Nagpur, 2000)

Ans. As a general rule, second ionisation energy is normally higher than the first ionisation energy because after the removal of one electron, the size of the ion shrinks. The outer electrons are pulled with a greater force towards the nucleus and therefore a greater energy is needed to take out the second electron. In the case of Cu and Cr, this effect is much more pronounced. Consider the configuration of Cu and Cr in the outermost orbital.

Cr : .... $3s^2\ 3p^6\ 3d^5\ 4s^1$

Cu : .... $3s^2\ 3p^6\ 3d^{10}\ 4s^1$

whereas it is the 4*s* electron which is involved in first ionisation energy, it is 3*d* electron which is involved in second ionisation energy. It is much more difficult to remove a *d*-electron because of weaker screening effect of remaining *d*-electrons. Moreover, the stable half-filled and fully-filled *d*-orbital configuration of Cr and Cu respectively is distributed if the second electron is to be removed. This requires much more energy and hence the anomaly.

Q.12. Describe the electronic configuration of Cr and Cu.

(Meerut, 2000; Guwahati, 2006; Nagpur,2008)

Ans. The unexpected electronic configuration of Cr (Ar) $3d^5\ 4{,}s^1$ (in place of (Ar) $3d^4\ 4s^2$) and Cu (Ar) $3d^{10}\ 4s^1$ (in place of (Ar) $3d^9\ 4s^2$) can be explained on the basis of half-filled and completely filled sub-shells which are found to be more stable arrangements. The extra stability of half-filled sub-shells results from the occupation of each orbital by one electron resulting in spreading and equal distribution of charge around an atom. In case of copper, it appears that Cu (Ar) $3d^{10}\ 4s^1$ with filled 3*d* sub-shell and half-filled 4*s* sub-shell is more stable than (Ar) $3d^9\ 4s^2$. Another important factor which explains the stability of these configurations is their large exchange energy.

The electrons in various orbitals of the same sub-shell or in different sub-shell (having similar energy) can exchange their respective positions. During this exchange process a small amount of energy is released. The released energy is called Exchange energy. It has been found that exchange energy is maximum when all the orbitals of the same sub-shell are either completely filled or exactly half-filled. So the configurations of Cr (Ar) $3d^5 4s^1$ and Cu (Ar) $3d^{10}\ 4s^1$ have extra stability.

Q. 13. Discuss the electronic configurations of ion of transition metals.

Ans. Ions of the transition elements are formed by the loss of electrons from the atoms. The configuration obtained after the loss of electrons is the electronic configuration of the ion. Whether the electrons will be lost from the $(n-1)$ *d*-orbital or *ns* -orbital depends upon the stability of the electronic configuration of the ion. For example, consider the first transition series. As 4*s* orbitals are occupied first and 3*d*-orbitals later, we may expect that 4*s* orbitals are more stable than 3*d* and hence the elements of first transition series would be expected to form ions by the loss of 3*d* electrons. In fact ions are formed by the loss of 4*s* electrons first. This is because that prior to occupation of electrons, the 4*s* orbital is energetically more stable than the 3*d* orbitals. But, once the 3*d* orbitals are occupied by electrons, these repel the 4*s* electrons to the higher energy level. Thus, when transition metal atoms form ions, they lose electrons from the 4*s* orbitals before the 3*d* orbitals.

Q. 14. In moving from Ar to K, the electron enters 4*s* orbital instead of 3*d*. Why?

Ans. In argon (Ar), the filling of 3*p* orbitals is complete. As 4*s* orbital is associated with smaller energy, compared to 3*d* orbital, the next electron goes to 4*s* orbital.

Q. 15. Which of the two among Nb, V and Ta are chemically closer and why?

Ans. Vanadium (V) and Tantalum (Ta) are closer to each other because of similar configuration in the outermost orbit.

V .... $3d^3\,4s^2$

Ta .... $5d^3\,6s^2$

Q. 16. Explain: K_{19} and Ca_{20} involve filling of 4*s* while the elements Sc_{21} to Zn_{30} involve the filling of 3*d* orbitals.

Ans. From Ca_{20} onwards, the electrons don't go to 4*p* orbitals but to 3*d* orbitals. This is because the next in order of energy are 3*d* orbitals and not 4*p* orbitals. This continues up to Zn_{30} after which the filling takes place in 4*p* orbitals.

Q. 17. Describe general characteristics of *d*-block elements.

(Kanpur, 2000; Meerut 2002; Bangalore, 2004)

Ans. Some important properties of *d*-block elements are given below:

(*i*) They are practically all hard, strong, high melting, high boiling metals and conduct heat and electricity well.

(*ii*) They form alloys with one another and with other metallic elements.

(*iii*) They are all ductile metals.

(*iv*) They form interstitial compounds with non-metals like C, B, N, H, etc.

(*v*) Because of unpaired electrons they form paramagnetic compounds.

(*vi*) Most of the elements have great tendency to form stable complex ions.

(*vii*) Many of them are electropositive enough to dissolve in mineral acids.

(*viii*) They exhibit catalytic properties and are wonderful catalysts for hydrogenation, oxidation, dehydration etc.

(*ix*) The transition elements from compounds with Lewis acid ligands like CO, NO etc. in which the elements exist in low, zero or even formally negative oxidation states.

(*x*) They exhibit variable oxidation state and their ions and compounds are coloured.

Q.18. Discuss the variation in atomic radii in a period and down the group in case of transition elements. *(Rohilkhand, 2003; Nagpur, 2005; Calcutta, 2005; Pune 2009)*

Ans. Along a period. The atomic radius of an element decreases while moving from left to right in a period. This can be understood in terms of increases in nuclear charge with consequent contraction in size.

However, the decrease is very small within some range in the period. For example, in the first transition series, the decrease in atomic radius is very small from chromium to nickel. The same thing happens in other transition series. Another remarkable observation is that the atomic radius starts increasing after that. For example, the value increases from nickel through copper to zinc.

As a matter of fact, two opposing forces are simultaneously taking place. In moving from left to right, nuclear charge increases leading to decrease in atomic radius. At the same time, repulsion takes place between the electron population resulting in increase in the atomic radius. Net value of atomic radius would be determined by the combined effect of these opposing phenomena. It is found that the radius decreases but very slightly from Cr to Ni. After that repulsive forces outweigh and an increase in radius is observed.

Along a group. While there is some increase in atomic radius in moving from first to second member, the value remains unchanged while moving from second to third member. This can be explained in terms of Lanthanide contraction which is described in detail in the chapter on *f*-block elements.

Q. 19. Discuss the variation in atomic volume and density of transition elements in a period and a group.

Ans. Atomic volume shows the same trend in the period and the group as the atomic radius. Thus atomic radius volume decreases in the period up to a certain point and then increases thereafter up to the end of the period. Reasons are the same as given in Q.18.

In a group the volume increases first and then becomes constant.

Density is the ratio of mass and volume. We observe that the density increases (Density = Mass/volume) in the period as the atomic mass increases and volume shrinks. Towards the end of the period, the trend is reversed. Because of interelectronic repulsion, the atomic volume increases, thereby reducing density.

Down the group, the density increases because there is not much variation in atomic volume whereas atomic mass increases.

Q. 20. Discuss the variation in melting and boiling points of transition elements.

Ans. Transition metals have high melting and boiling points. This is due to strong overlapping between $(n-1)d$ orbitals and formation of covalent bonds by the *d*-electrons. Greater the number of unpaired *d*- electrons, greater the possibility of covalent bond formation, and hence higher the melting and boiling point.

Zn, Cd and Hg (melting points 693, 593, and 234.6 K respectively) are notable exceptions because in these elements the $(n-1)d$ orbitals are complete and covalent bond formation in *d*-orbitals is not expected.

The melting and boiling points of these elements increase down the group due to increasing atomic mass. And in a period melting and boiling points reach a maximum near group VB, and VIB. In moving across a transition series from left to right, the number of unpaired *d*-electrons rises to maximum of five in group VI beyond which electron pairing begins. The number of unpaired *d*-electrons in each atom determines the magnitude of binding forces by forming covalent bonds with unpaired electrons of neighbouring atoms.

Q. 21. Discuss the trend in ionisation energy of transition elements. (*Nagpur, 2005*)

Ans. Ionization energies. The ease with which an electron may be removed from a transition metal atom is called its ionization energy (I.E.). In the transition series the ionization energies increase along the period but the increase is not in a regular manner. However, the increase in I.E. is not as pronounced as in the *s*- and *p*-block elements in the same period. Table 17.5 shows the ionization energies of the 3*d* series. No regular increase in I.E. is observed.

Table 17.5. Ionization energies of the elements of 3*d* series

Elements	Ionization Energies ($kJ\ mol^{-1}$)		
(3d series)	*First*	*Second*	*Third*
Sc	631	1245	2450
Ti	656	1320	2721
V	650	1376	2873
Cr	652	1635	2994
Mn	717	1513	3258
Fe	762	1564	2963
Co	758	1647	3237
Ni	736	1756	3400
Cu	745	1961	3560
Zn	906	1736	3838

This may be explained in terms of screening effect of the *d*-electron. The increase in the I.E. expected on account of increase in nuclear charge is almost offset by the extra screening of the nucleus provided by the inner *d*-electrons. The ionization energies of the transition elements are intermediate between those of *s*- and *p*-block elements. This suggests that the transition metals are less electropositive than *s*-block elements.

The *d*-block elements form ionic as well as covalent compounds. Whenever a transition metal gives two series of salts, the lower one is an ionic compound and the higher one is a covalent compound.

From Table 17.5 we find that second I.E. increases almost regularly from Sc to Zn. Values for Cr and Cu are, however, much higher than those of their neighbours on the left as well as on the right. This is due to the stable configuration of Cr^+ ($3d^5$-exactly half filled) and Cu^+ ($3d^{10}$-completely filled). Removal of one electron means they change from a more stable to less stable configuration.

Q. 22. Why zinc, cadmium and mercury have low melting points unlike other *d*-block elements?

Ans. See Q. 20.

Q. 23. Explain the trends in variation of:

(*i*) melting and boiling points, (*ii*) atomic radius in the first transition series.

Ans. See Questions 18 and 20.

Q. 24. Mercury is a liquid metal. Explain.

Ans. As there are no vacant *d*-orbitals in Hg, there is no possibility of mutual overlapping of unpaired *d*-orbitals. Hence the melting point is considerably low and it is a liquid at room temperature.

Q. 25. Transition metals are less electropositive than *s*-block elements. Explain.

Ans. This is so because transition metals have smaller size and higher ionisation energy compared to *s*-block elements. High ionisation energy implies less electropositive nature.

Q. 26. Cu, Ag and Au are typical transition elements while Zn, Cd and Hg are not. Comment.

(*Kumaon, 2000*)

Ans. The electronic configurations of the elements and their ions are as under:

Cu ....$3d^{10}4s^1$ cu^{2+}....................$3d^9$

Ag ....$4d^{10}5s^1$ Ag$^+$ ....$4d^{10}$

Au....................$5d^{10}6s^1$ Au^{3+} .... $5d^8$

Zn....................$3d^{10}4s^2$ Zn^{2+} ....$3d^{10}$

Cd....................$4d^{10}5s^2$ Cd^{2+} ....$4d^{10}$

Hg....................$5d^{10}6s^2$ Hg^{2+} ....$5d^{10}$

As zinc, cadmium, mercury salts (ions) have completely filled *d*-orbitals, they do not reflect, typical transition behaviour.

In the case of Cu^{2+} and Au^{3+}, there are incomplete *d*-orbitals, hence typical transition characteristics.

Q. 27. The first ionisation energy of Hg is higher than that of Cd. Explain.

Ans. The presence of 4*f* orbitals brings about poor shielding effect which increases the effective nuclear charge on the outer valence electrons. Therefore it requires much higher energy to pull out the electrons and thus first ionisation energy of Hg is higher than that of Cd.

Q. 28. Melting points of *d*-block elements in a series reach a maximum upon the middle, thereafter show a decrease. Explain.

Ans. Melting point of a substance depends upon the atomic size and lattice energy. Smaller the size of particles, closer the packing and greater the lattice energy and higher the m.p. In a period, the atomic size decreases to some point and it increases thereafter. Therefore the lattice energy and m.p. increase up to this point and decline after that.

Q. 29. Discuss magnetic properties of transition elements.

(*Guwahati, 2002; Purvanchal, 2003; Nagpur, 2005; Lucknow, 2010; Ranchi 2010*)

Ans. Transition elements and their compounds have been classified into the following three categories on the basis of their behaviour in the magnetic field.

(*i*) *Paramagnetic substances* are those which are weakly attracted by the magnetic field and this property is associated with the unpaired electrons in an atom or a compound.

(*ii*) *Diamagnetic substances* are those which are slightly repelled by the magnetic field and this property is associated with the paired electrons in an atom or a compound.

(*iii*) *Ferromagnetic substances* are those which are strongly attracted in the magnetic field, *i.e.*, they are more paramagnetic in character and this property is associated with the substances which retain their magnetism even after the removal of magnetic fields. (Due to this property ferromagnetic substances differ from paramagnetic substances which lose their paramagnetism after removing the magnetic field.) The most familiar example of ferromagnetic substances is iron. Cobalt and nickel are also ferromagnetic.

The paramagnetic or diamagnetic nature of a substance can be measured by weighing the substances (on a Gouy magnetic balance) in absence of magnetic field and in presence of magnetic field as shown in Fig. 17.1.

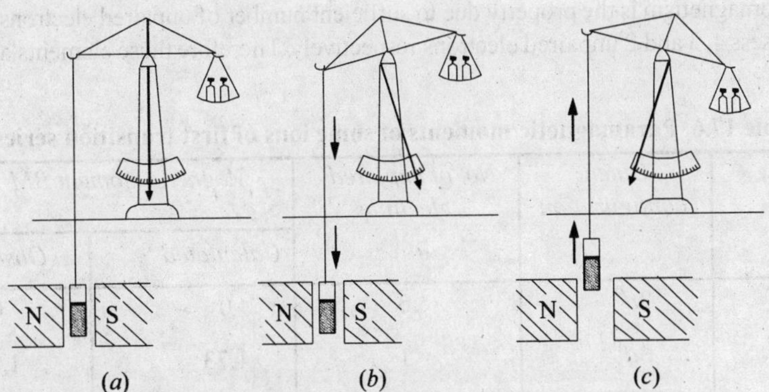

Fig. 17.1. (*a*) Gouy Balance weighing in the absence of magnetic field.
(*b*) Substance attracted by the magnetic field — paramagnetic.
(*c*) Substance repelled by the magnetic field — diamagnetic.

First the substance is weighed in air and then after applying magnetic field. If the weight of the substance increases in a magnetic field, it means that it is attracted by the magnetic field and hence it is paramagnetic. If the weight decreases, it means that it is repelled by the magnetic field and hence it is diamagnetic.

The magnetic moment, which is associated with the number of unpaired electrons and arises due to spin and orbital motion of the electron, can be calculated from the difference in weight of the substance in the Gouy balance. The magnetic moment gives important information about the presence of unpaired electrons in atoms, ions or molecules. In some cases even the structures of transition metal complexes are indicated by it.

Most of the transition elements and their compounds are paramagnetic, *i.e.*, they have a number of unpaired electrons. The relationship between the number of unpaired electrons. *n,* in an atom, ion or molecule and its magnetic moment, μ, in Bohr magnetons is given by the *spin only* formula.

$$\mu s \text{ or } \mu = \sqrt{n(n+2)} \text{ BM}$$

where n = number of unpaired electrons.

$$1\text{BM} = \frac{eh}{4\mu\, mc}$$

where e is the electronic charge, h is the Planck's constant, m is the mass of the electron and c is the velocity of light.

The magnetic field produced by unpaired electrons is partly due to their spin and partly to their orbital motion. The magnetic moment μ is given by

$$\mu_s + \mu_L = \sqrt{4s(s+1) + L(L+1)} \quad \text{...(17.1)}$$

(where s = sum of the spin quantum numbers, and L = sum of the orbital angular momentum quantum numbers)

But it has been found that in a number of compounds including first row of the transition elements, the orbital contribution (μ_L) is quenched or suppressed by the electric fields of the surrounding atoms (*i.e.*, $L = 0$) so that the magnetic moment of the species is practically equal to the value due to spin alone, *i.e.*, $\mu_s = \sqrt{4s(s+1)}$, or in terms of unpaired electrons per atom *n;*

$$\mu = \sqrt{n(n+2)} \quad \text{... (17.2)}$$

The magnetic moments of some ions of the first transition series are given in Table 16.6. The values have been calculated using the spin only formula.

Q. 30. How do you explain the ferromagnetic properties of iron, cobalt and nickel?

(*Garhwal, 2000*)

Ans. Ferromagnetism is the property due to sufficient number of unpaired electrons. Iron, cobalt and nickel possess 4, 3 and 2 unpaired electrons respectively. Therefore these elements are ferromagnetic.

Table 17.6. Paramagnetic moments of some ions of first transition series.

Ions	*Outer configuration*	*No. of unpaired electrons*	*Magnetic moment BM*	
			Calculated	*Observed*
Sc^{3+}, Ti^{4+}	$3d^0$	0	0	0
Ti^{3+}, V^{4+}	$3d^1$	1	1.73	1.75
Ti^{2+}, V^{3+}	$3d^2$	2	2.84	2.76
V^{2+}, Cr^{3+} Mn^{4+}	$3d^3$	3	3.87	3.86
Cr^{2+}, Mn^{3+}	$3d^4$	4	4.90	4.80
Mn^{2+}, Fe^{3+}	$3d^5$	5	5.92	5.96
Fe^{2+}, Co^{3+}	$3d^6$	4	4.90	5.0–5.5
Co^{2+}	$3d^7$	3	3.87	4.4–5.2
Ni^{2+}	$3d^8$	2	2.84	2.9 – 3.4
Cu^{2+}	$3d^9$	1	1.73	1.4 – 2.2
Cu^+, Zn^{2+}	$3d^{10}$	0	0	0

From Table 17.6 we find that the observed magnetic moments agree fairly well with the calculated values. In the later half of the series, the observed values slightly exceed the spin values only. These differences can be explained in terms of some contribution from the orbital magnetic moments also. For the second and third row elements, the orbital motion is not prevented or quenched and therefore a more complex formula (eq. 17.1) is required.

Cause of Magnetic Behaviour

An electron revolves round the nucleus and it also spins around its own axis to produce magnetic field like the magnetic field produced by electric current flowing through a loop of wire. Each electron acts as a tiny magnet having a definite value of magnetic moment. Diamagnetic substances have no unpaired electrons so they have zero magnetic moment. The external magnetic field, however, induces in them a small magnetic moment which is opposed to the external magnetic field. Thus, diamagnetic substances repel lines of force and show a slight decrease in weight. Paramagnetic substances, however, attract lines of force in the applied magnetic field.

Q. 31. Calculate the number of unpaired electrons in the following:

$$[Fe(H_2O)_6]^{3+}, [Co(NH_3)_6]^{3+} [Fe(CN)_6]^{3-}$$

Which of the above ions do you expect to be coloured?

Ans. Considering the charge on H_2O, NH_3 and CN^- as 0, 0 and –1 the charge on the metals can be calculated as +3, +3 and +3. Thus the ions in question are

Fe^{3+} : [Ar] $3d^5$

Co^{3+} : [Ar] $3d^6$

Fe^{3+} : [Ar] $3d^5$

Thus Fe^{3+} and Co^{3+} contain 5 and 4 unpaired electrons; all of them are coloured.

Q. 32. Why Ti^{3+} (aq) ions are paramagnetic and reddish violet coloured?

Ans. The electronic configurations of Ti and Ti^{3+} are as follows:

Ti $3d^2 4s^2$

Ti^{3+} $3d^1$

Ti^{3+} is paramagnetic due to the presence of one unpaired electron in the presence of ligand water, *d*-orbitals are split into two sets having slightly different energies. The energy difference corresponds to radiation of lower energy. Thus the single electron can absorb radiation of lower energy (red colour) to get excited to higher *d*-level. The transmitted light will be of higher energy having violet colour.

Q. 33. Transition metals are less reactive than alkali and alkaline earth metals. Explain.

Ans. Reactivity of metals. The ionization energies of transition elements are fairly high on account of their small size of the atoms. They have high heat of sublimation due to the presence of covalent bonding. In other words, they require a large amount of energy for a change from the solid to vapour state. Further the metal ions do not get hydrolysed easily, *i.e.*, they have low heat of solvation. All these factors retard the tendency to react. That is why, the *d*-block elements have rather a tendency to remain unreactive or 'noble'. This tendency is quite pronounced in gold and platinum in the third transition series.

On the other hand, alkali and alkaline earth metals possess small ionisation energies. Chemical reactivity is dependent upon the ionisation energy. Hence alkali and alkaline earth metals are more reactive.

Q. 34. Discuss oxidation states of transition elements.

(Rohilkhand 2003; Nagpur, 2008; Ranchi, 2010)

Ans. Oxidation states. One of the most striking features of the *d*-block elements is that they exhibit a variety of oxidation states in their compounds. Some common oxidation states of *d*-block elements are given in Table 17.7.

Table 17.7. Oxidation states of *d*-block elements 3*d* or first transition series.

Element	*Sc*	*Ti*	*V*	*Cr*	*Mn*	*Fe*	*Co*	*Ni*	*Cu*	*Zn*
Oxidation States				+1					+ 1	
	(+2)	+2	+ 2	+ 2	+ 2	+ 2	+ 2	+ 2	+ 2	+ 2
	+3	+3	+ 3	+ 3	(+ 3)	+3	+ 3	(+ 3)	(+ 3)	
		+ 4	+ 4	(+ 4)	+ 4	(+ 4)	(+ 4)			
			+ 5							
				+ 6	(+ 6)	(+ 6)				
					+7					

4*d* or Second transition series.

Element	*Y*	*Zr*	*Nb*	*Mo*	*Tc*	*Ru*	*Rh*	*Pd*	*Ag*	*Cd*
Oxidation States									+1	
						+2		+2	(+2)	+2
	+3		+3	+3		+3	+3	(+3)	(+3)	
		+4		+4	+4	+4	+4	+4		
			+5	+5		+5				
				+6	(+6)	+6	+6			
					+7	(+7)				
						(+8)				

5*d* or Third transition series.

Element	*Ea*	*Hf*	*Ta*	*W*	*Re*	*Os*	*Ir*	*Pt*	*Au*	*Hg*
Oxidation States									+1	+1
				(+2)		(+2)	+2	+2		+2
	+3			(+3)	(+3)	(+3)	(+3)	(+3)	+3	
		+4	(+4)	+4	+4	+4	+4	+4		
			+5	+5	+5	+5				
				+6	+6	+6	+6			
					+7					
						+8				

The values reported in the brackets are for less common oxidation states. These states are exhibited only in a few complexes and are unstable. From Table 17.7, it can be seen that many elements exhibit oxidation states which differ from each other by one unit only. In contrast to this, oxidation state of an *s*-block element is equal to its group number or two less than the group number. For example, oxidation states of Sn and Pb are +2 and + 4 while those of As, Sb and Bi are +3 and +5.

There is very small difference between the energies of electron in *ns* and $(n-1)d$ orbitals. Thus both *ns* and $(n-1)$ *d* orbitals participate in compound formation showing variable oxidation states. Along with the *s* electrons, 1, 2, or more *d* electrons may be removed or shared. The variable oxidation states shown by the first transition elements may be related to their actual electronic configuration.

The outer electronic configuration of Sc is $3d^14s^2$. It shows an oxidation state equal to +2 if both 4*s* electrons are used for bonding and +3 when two *s* and one *d*-electrons are involved.

Similarly, Ti ($3d^24s^2$) exhibits oxidation state +2 when it uses only its two *s*-electrons and +3 when one *d*–electron is used in addition to two *s*–electrons and +4 when it uses 2 *s*-electrons and 2*d*-electrons for bonding. On the same ground we can account for the different oxidation states of V, Cr and Mn. The lower oxidation states are generally obtained by loss of *s* electrons, for example, in case of Mn($3d^54s^2$), it readily loses the two 4*s* electrons to form Mn^{2+}. But the higher oxidation states are formed by sharing 4*s* and all or some of the 3*d* electrons. Manganese can exhibit a maximum oxidation state of +7 or almost any positive state between +2 and +7. Thus the correlation between electronic structure and minimum and maximum oxidation state for first five transition elements can be studied.

In the remaining five elements, *i.e.*, Fe ($3d^6 4s^2$), Co ($3d^7 4s^2$), Ni ($3d^8 4s^2$), Cu ($3d^{10} 4s^1$) and Zn ($3d^{10} 4s^2$), the number of $3d$ electrons is more than 5. The minimum oxidation state in each case equals to the number of $4s$- electrons, but the maximum oxidation state is not at all related to the total of *ns* and $(n-1)d$ electrons. Their maximum oxidation state is largely confined to +2 or +3 as in the case of iron. The + 4, + 5, and + 6 oxidation states are only rare. The +8 oxidation state is important only with Ru and Os. The different pattern in oxidation state of Fe and the corresponding two elements in $4d$ or $5d$ series (*i.e.*, Ru and Os) is attributed mainly to increased size.

Low oxidation state. In order to save the low oxidation states from oxidation by atmospheric oxygen, syntheses are often carried out in inert nitrogen atmosphere and special solvents (completely deaerated, *i.e.*, air is completely removed) like tetrahydrofuran, dichloromethane and anhydrous ethanol are used. Organic ligands such as CO, C_2H_4, $\dot{N}O$, PR_3 etc., are used to form complexes. It is the characteristic feature of these ligands that they form complexes with *d*-block elements only when they are in zero, –1 or +1 oxidation states.

This is due to the fact that the donor atom of these ligands possesses vacant orbital in addition to the lone pair, *e.g.*, CO (*i.e.*,: C : : $\ddot{O}$:) has only six electrons on carbon hence it has one vacant orbital in addition to the lone pair. The ligands first form a normal σ covalent bond by donating its lone pair to the metal atom. The vacant orbitals of the ligand then accept electrons from the filled metal orbital to form a π bond. This is called metal to ligand π bonding in addition to σ bonding. The ability of these ligands to accept back metal electrons is responsible for the stability of these complexes in their lower oxidation states.

High oxidation states. When a metal ion occurs in an oxidation state higher than the one of common occurrence, then it is called the higher oxidation state.

As the oxidation state increases, the metal ions become more acidic, *i.e.*, electron greedy. Hence these elements with high oxidation states can form compounds with the ligands which are highly electronegative and basic, *e.g.*, F^- , Cl^-, NH_3 etc. This is because the small size, highly charged ligands can form strong σ bonds by electron donation which are stable enough for independent existence.

A suitable organic ligand can also stabilise the higher oxidation states by forming stable complexes. For example, Ag (II) has been obtained as cationic complex with pyridine $[Ag(Py)_4]^{2+}$ and with dipyridyl $[Ag(dipy)_2]^{2+}$, etc.

Q. 35. Discuss the oxidation states of Cr. Which of the oxidation states are most stable and why? (*Lucknow, 2010*)

Ans. Chromium exhibits oxidation states of +1, +2, +3 and +6. It also exhibits less common oxidation state of +4. Electronic configurations of chromium in various oxidation state are:

Cr (0)	$3d^5 4s^1$
Cr (I)	$3d^5$
Cr (II)	$3d^4$
Cr (III)	$3d^3$
Cr (IV)	$3d^2$
Cr (VI)	$3p^6$

Q. 36. Sc^{3+} is more stable than Sc^{2+}. Explain.

Ans. Sc^{3+} possesses greater hydration energy and lattice energy because of greater nuclear charge and smaller size compared to Sc^{2+}.

Q. 37. Zn forms only Zn^{2+} and not Zn^{3+}. Why?

Ans. In the formation of Zn^{3+}, an electron from the *d*-orbitals has to be removed which requires abnormally higher amount of energy because the completely filled ($3d^{10}$) stable configuration is disturbed.

Q. 38. Discuss standard electrode potentials of transition elements.

Ans. Standard reduction potentials of some transition metals in acid solution at 298 K are given in the table below.

Element	*E° (volts)*	Element	*E° (volts)*
$Sc^{3+} + 3e^- \longrightarrow Sc$	– 2.08	$Fe^{2+} + 2e^- \longrightarrow Fe$	– 0.44
$Mn^{2+} + 2e^- \longrightarrow Mn$	– 1.18	$Co^{2+} + 2e^- \longrightarrow Co$	– 0.28
$Zn^{+2} + 2e^- \longrightarrow Zn$	– 0.76	$Ni^{2+} + 2e^- \longrightarrow Ni$	– 0.5
$Cr^{3+} + 3e^- \longrightarrow Cr$	– 0.74	$Cu^{2+} + 2e^- \longrightarrow Cu$	+ 0.34

It is seen that standard reduction potential of transition element is generally negative, *i.e.*, lower than that of hydrogen (taken as zero, $2H^+ + 2e^- \rightarrow H_2$). Copper is an exception in above cases. All these elements with negative reduction potential liberate hydrogen from dilute acids.

$$M + 2H^+ \longrightarrow M^{2+} + H_2\,(g)$$

The rates of evolution of hydrogen are, however, very slow in some cases. Some of these metals (*e.g.*, Cr) are unreactive due to formation of the protective fihn of oxide on their surface.

Q. 39. Though transition metals possess high electrode potentials, they are not good reducing agents. Explain. (*M.Dayanand, 2000; Nagpur, 2000; Kumaon, 2000*)

Ans. The reducing capacity of a metal depends on its tendency to be oxidised easily to change into ions. $M \longrightarrow M^{2+} + 2e^-$

The transition elements possess this property due to their sufficiently negative standard reduction potentials and should be good reducing agents. But they are not good reducing agents as compared to the elements of groups I, II and III. This is due to their less tendency to form ions because of (*i*) their high ionization energies, (*ii*) low heat of hydration of their ions, and (*iii*) their high heats of sublimation.

Q. 40. Discuss catalytic properties of transition elements.

(*M. Dayanand, 2000; Rohilkhand, 2003; Bhopal, 2004; Nagpur, 2008*)

Ans. Many transition elements and their compounds are used as a catalyst in the laboratory and in industry. Some of the important catalysts are given below:

(*i*) Finely divided iron and molybdenum are used in the manufacture of ammonia (Haber's process).

(*ii*) Finely divided nickel and palladium are used in the manufacture of vanaspati as well as in hydrogenation reaction in the laboratory.

(*iii*) Zinc chromite is used in the synthesis of methyl alcohol.

(*iv*) Platinum and vanadium pentaoxide are used in the manufacture of ammonia (Haber's process).

(v) Fenten's reagent ($FeSO_4 + H_2O_2$) is used for the oxidation of alcohols to aldehydes.

(vi) Ziegler-Natta (catalyst Ti Cl_4) is used for the manufacture of polythene.

The different factors contributing to catalytic action are given below:

(i) The catalytic properties of the transition elements are probably due to the variable oxidation states, due to whch they form unstable intermediate products with various reactants in some cases. These intermediate products decompose to give the final product, regenerating the catalyst (intermediate compound formation theory of catalysis).

$$\underset{\text{Reactant}}{A + B + \text{Catalyst}} \longrightarrow \underset{\substack{\text{Unstable inter-}\\\text{mediate compound}}}{\text{A.B. Catalyst}} \longrightarrow \underset{\text{Product}}{AB + \text{Catalyst}}$$

(ii) In some cases the finely divided metals or their compounds provide a large surface area for adsorption of the reactants. The adsorbed reactants react faster due to their closer contact (adsorption theory), on the surface of the catalyst giving rise to products.

Q. 41. Discuss complex formation in the case of transition elements.

(Kanpur, 2000; Rohilkhand, 2003; Purvanchal, 2007)

Ans. The transition metal ions have a great tendency to form complex ions with certain molecules (*e.g.*, NH_3 CO, H_2O and 'en' etc.) or ions (*e.g.*, F^-, Cl^-, N^-, etc.). The molecules or ions attaching themselves with the cation are called ligands. The ligands contain lone pairs of electrons which are easily donated to the transition metal ions. This is in contrast to the *s*- and *p*-block elements due to the following reasons:

(i) Transition metal cations due to their small size possess high charge density. These can, therefore, accept lone pairs of electrons from the ligands.

(ii) They have vacant *d*-orbitals of suitable energy to accept lone pairs of electrons to form complex compounds.

s and *p*-block elements do not meet the above requirements. The number of ligands which are attached to a metal ion by coordinate bonds to form a complex ion is called coordination number of the metal ion. The complexes are also called coordination compounds. The complex ions have in general linear, square planar, tetrahedral or octahedral structure depending upon the nature of hybridization of the metal ion orbitals.

All transition metal cations, however, do not possess the same capacity for complex formation. There are wide variations between different series but within a series certain trends towards increased stability of complex can be given.

(i) In each transition series, the stability of the complexes increases with increasing atomic number of the element and decreasing size of its atoms.

(ii) When a transition metal atom exhibits more than one oxidation state, the complex with higher oxidation state has greater stability.

These points can be explained in terms of greater attraction of metal ion for ligand electrons. For example, Co^{2+} does not form a stable complex with ammonia, whereas the corresponding Co^{3+} complex, $[Co(NH_3)_6]^{3+}$, is highly stable.

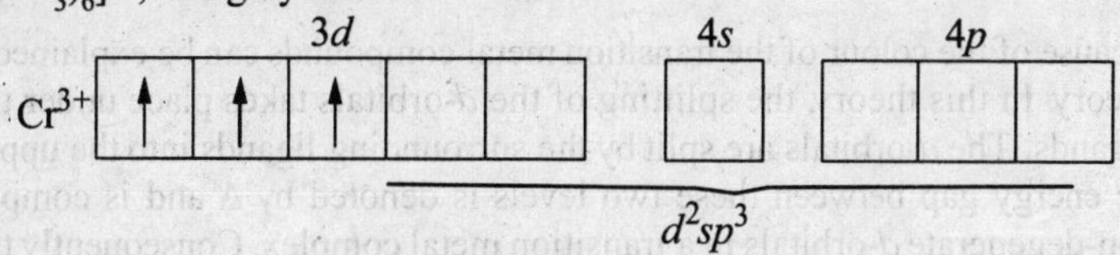

The formation of a complex, *e.g.*, $[Cr(NH_3)_6]^{3+}$, can be represented as —

$$Cr^{3+} + 6NH_3 \longrightarrow [Cr(NH_3)_6]^{3+}$$

The electronic configuration of Cr is (Ar) $3d^5 4s^1$ and that of Cr^{3+} is (Ar) $3d^3 4s^0$.

The six empty orbitals of Cr^{3+} (*i.e.*, two 3*d*, one 4*s* and three 4*p*) exhibit d^2sp^3 hybridization. The 6 hybridized orbitals have same energy to accept six pairs of electrons from six molecules of ammonia to form $[Cr(NH_3)_6]^{3+}$. Formation of complexes such as $[Ni(CO)_1]$, $[Ni(CN)_4)^{2-}$, $[Fe(CN_6)]^{4-}$ and $[Fe(CN)_6]^{3+}$ etc., has been explained in detail in the chapter on coordination compounds.

Iron in hemoglobin (a red colouring matter in blood) and cobalt in vitamin B_{12} are the examples of typical complexes of biological importance.

Q. 42. Transition metal compounds are generally coloured. Explain.

(*Nagpur, 2000; Guwahati, 2002; Lucknow, 2010; Pune, 2010*)

Ans. Compounds of *s*- and *p*-block elements are almost white whereas ionic and covalent compounds of transition elements are usually coloured. Colour, in general, is associated with the incomplete electron sub-shell and the ability of the compound to promote an electron from one energy level to another by absorption of light of a particular wavelength.

Colour in transition metals is associated with incomplete $(n-1)d$ sub-shell, *i.e.*, it is a property of unpaired *d*-electrons. In transition elements, *d*-electrons are promoted to a higher energy level within the *d*- sub-shell by absorption of a small amount of energy. This energy is obtained by absorption of light in the visible region. The rest of the light is no longer white, *i.e.*, it, appears coloured. For example, if red light is absorbed then the transmitted light contains an excess of the light of blue colour. The compound appears blue. For example, Cu^{2+} ion ($3d^9$) absorbs red light from the visible range, for promotion of 3*d* electrons, the ions reflect blue light and appear deep blue.

Anhydrous cobalt compounds also absorb red light for excitation of 3*d*-electrons. When red light is eliminated from white light, the colour appears blue (*i.e.*, blue is the combination of all the remaining colours except red). If all the incident light falling on a substance is absorbed, the substance appears black. If all the incident light is reflected, then the substance appears white. If only a particular wavelength is absorbed from the incident light, then the substance has a characteristic colour.

Q. 43. Why are salts of zinc, cadmium and mercury white?

Ans. In simple compounds of zinc, cadmium and mercury (having $Zn^{2+} - 3d^{10}$; $Cd^{2+} - 4d^{10}$ and $Hg^{2+} - 5d^{10}$) there is no possibility of promotion of electrons within *d* level due to their completely filled sub-shells and hence their compounds are white.

Q. 44. Why are compounds of *s* and *p* block elements not coloured?

Ans. In *s*- and *p*-block elements the *d*- orbitals are either missing or completely full. There cannot be any *d-d* transitions, and the energy to promote an *s*- or *p*- electron to a higher energy level is much greater and corresponds to ultraviolet regions. These compounds absorb no radiations from the visible region and hence the compounds of *s*- and *p*-block elements do not appear coloured in the visible region.

Q. 45. Explain the colour of transition metal complexes on the basis of crystals field theory.

(*Nagpur, 2008*)

Ans. The cause of the colour of the transition metal compounds can be explained on the basis of crystal field theory. In this theory, the splitting of the *d*-orbitals takes place under the influence of surrounding ligands. The *d*-orbitals are split by the surrounding ligands into the upper e_g and lower t_{2g} levels. The energy gap between these two levels is denoted by Δ and is comparatively small between the non-degenerate *d*-orbitals in a transition metal complex. Consequently the excitation of an electron from a lower to a higher level can be achieved by absorption of visible light. This causes the complex to appear coloured.

Fig. 17.2 shows the *d*-orbitals in Ti^{3+} are split into e_g and t_{2g} orbitals of differing energies when the ligands H_2O molecules approach the cation. The single electron in t_{2g} orbital absorbs green yellow light and is excited to e_g orbital and thus appears purple. Colours exhibited by hydrated ions of the first transition series are given in Table 17.8.

Table 17.8. Colours of some hydrated ions.

Ions	*Outer electronic configuration*	*Colour*	*No. of unpaired electrons*
Sc^{3+}	$3d^0$	colourless	0
Ti^{+}	$3d^1$	purple	1
V^{3+}	$3d^2$	green	2
Cr^{3+}	$3d^3$	violet	3
Mn^{3+}	$3d^4$	violet	4
Fe^{3+}	$3d^5$	yellow	5
Mn^{2+}	$3d^5$	pink	5
Fe^{2+}	$3d^6$	green	4
Co^{2}	$3d^7$	pink	3
Ni^{2+}	$3d^8$	green	2
Cu^{2+}	$3d^9$	blue	1
Cu^{+}	$3d^{10}$	colourless	0
Zn^{2+}	$3d^{10}$	colourless	0

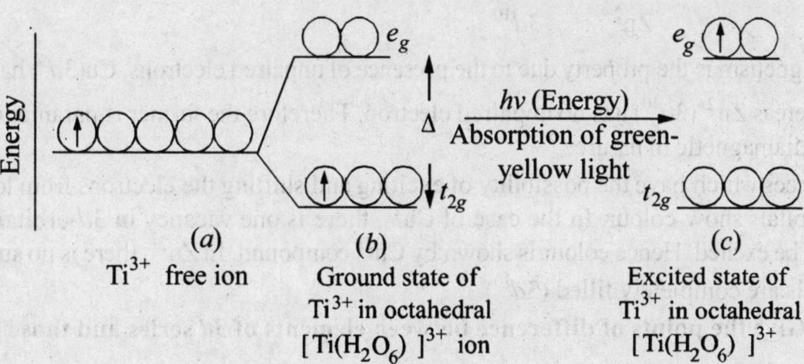

Fig. 17.2. Relative energy levels for the five 3*d*-orbitals of the free and hydrated Ti^{3+} ion.

Q. 46. Explain the formation of

(*a*) Interstitial compounds

(*b*) Non-stoichiometric compounds

(*c*) Alloys. (*Kanpur, 2000*)

Ans. (*a*) Interstitial compound formation. The transition metals appear to be unique in their ability to react with small atoms of non-metals to give interstitial compounds. The important compounds are hydrides, nitrides, carbides and borides of these elements. Interstitial compounds are those in which the non-metal atoms like nitrogen, carbon etc., occupy holes (interstices) in metal structure.

(*b*) Formation of non-stoichiometric compounds. The feature of the transition or *d*-block elements is the existence of non-stoichiometric compounds, *i.e.*, compounds of indefinite composition. For example, in FeO, the ratio is not 1:1 but the composition varies between $Fe_{0.94}O$ and $Fe_{0.84}O$. The compounds of transition elements with oxygen, sulphur, selenium and tellurium are generally non-stoichiometric.

(*c*) Alloy formation. Transition metals mix freely with each other in the molten state and on cooling a solution of different metals result in the alloy, *e.g.*, chromium dissolves in nickel to form Cr–Ni alloy. Manganese dissolves in iron to form manganese steels. The alloy formation is explained on the basis of similar sizes of the atoms of these metals which allow the atoms of one metal to take up the positions in the crystal lattice of the other.

Q. 47. Why do transition elements :

(*i*) Show variable oxidation states. (*M. Dayanand, 2000; Kurukshetra, 2000*)

(*ii*) Form a large number of coordination or complex compounds. (*M. Dayanand, 2000*)

(*iii*) Give coloured and paramagnetic ions. (*S; M. Dayanand, 2000*)

(*iv*) Exhibit good catalytic properties. (*M. Dayanand, 2000*)

Ans. See Questions 29, 34, 40, 41 and 42.

Q.48. Cu^{2+} ions are coloured and paramagnetic while Zn^{2+} ions are colourless and diamagnetic. Explain. (*Nagpur, 2005*)

Ans. Consider the electronic configurations of Cu^{2+} and Zn^{2+} in the outermost orbit.

Cu^{2+} : $3d^9$

Zn^{2+} : $3d^{10}$

Paramagnetism is the property due to the presence of unpaired electrons. Cu($3d^9$) has an unpaired electron whereas Zn^{2+}($3d^{10}$) has no unpaired electron. Therefore the former is paramagnetic whereas the latter is diamagnetic in nature.

Substances which have the possibility of exciting and shifting the electrons from lower levels to vacant *d*-orbitals show colour. In the case of Cu^{2+}, there is one vacancy in 3*d*-orbital to which an electron can be excited. Hence colour is shown by Cu^{2+} compound. In Zn^{2+}, there is no such possibility as 3*d*-orbitals are completely filled ($3d^{10}$).

Q. 49. Give the points of difference between elements of *3d* series and those of 4*d* and 5*d* series.

Ans. (*i*) Atomic/Ionic size. The atomic as well as ionic radii of the 3*d* series elements are always smaller than the 4*d* and 5*d* series elements but due to lanthanide contraction, the atomic as well as ionic radii of the 4*d* and 5*d* series elements almost have identical size.

(*ii*) Coordination number. The elements of the first series rarely show coordination number greater than six. Coordination numbers 1, 8 or higher are quite common with elements of second and third series. This may be due to their bigger size, so that compounds with higher coordination number may be formed.

(*iii*) Reactivity of elements. The reactivity of the transition elements generally decreases in going from first transition series to the third transition series. The elements of the first transition series are much more reactive than those of the second and third series. The less reactivity may be due to their higher ionization energies, lower hydration energies, higher sublimation energies, etc.

***(iv)* Lower oxidation state.** The elements of the first transition series have at least one stable lower oxidation state which furnishes stable hydrated cations and oxy salts. On the other hand, the second and third series elements form compounds in their higher oxidation states. The hydrated cations and oxy salts for 4*d* and 5*d* elements generally do not exist.

***(v)* Maximum oxidation state.** The maximum oxidation state in each series is exhibited by iron-group elements. Iron shows a maximum of oxidation state of +6 in FeO_4^{2-} and Ru and Os exhibit maximum oxidation state of +8 in their tetraoxide, MO_4.

***(vi)* Nature of complexes.** The four coordinate complexes of the 3*d* series elements, with few exceptions, are tetrahedral; for example, Ni^{2+}, Co^{2+}, Cu^{2+} form tetrahedral as well as square planar complexes. The four coordinate complexes of 4*d* and 5*d* series elements are square planar.

The elements of 3*d* series form both low and high spin complexes but due to high crystals field splitting energies elements of 4*d* and 5*d* series form only low-spin complexes.

***(vii)* Stability of complexes.** Elements of the 3*d* series form more stable complexes with N, O and F donors, whereas the elements of 4*d* and 5*d* series form more stable complexes with P, S and heavier halogen donors.

***(viii)* Application of spin-only formula.** The magnetic data for the 3*d* series elements can be interpreted by spin, only formula. On the other hand, the magnetic data for 4*d* and 5*d* series elements cannot be interpreted by this formula only.

Q. 50. Cupric salts are blue in colour while cuprous salts are colourless. Explain.

(Purvanchal, 2003)

Ans. Electronic configurations of Cu^{2+} and Cu^{+} in the outermost orbit are:

$$Cu^{2+} : \quad 3d^9$$
$$Cu^{+} : \quad 3d^{10}$$

There is a possibility of excitation of electron from one *d*-level to another *d*-level in Cu^{2+} with the absorption of light from visible range. The transmitted light is blue in colour. Hence Cu^{2+} salts are blue in colour. In Cu^{+} salts, there is no possibility of excitation of the electron, hence no colour is shown by such salts.

Q. 51. Cu (I) is diamagnetic whereas Cu (II) is paramagnetic.

Ans. Magnetic properties of substance are determined by the number of unpaired electrons in the ion. In Cu (II) ion the outermost orbital config. is $3d^9$. Thus there is one unpaired electron. Hence it is paramagnetic. In Cu (I), the outermost orbital config. is $3d^{10}$. There is no unpaired electron and therefore Cu (I) is diamagnetic.

Q. 52. In moving down a transition series, the atomic sizes increase from 3*d* to 4*d* but there is hardly any increase from 4*d* to 5*d*. Explain.

Ans. Atomic size decreases while moving from 3*d* to 4*d* due to increased nuclear charge. In moving from 4*d* to 5*d*, increase in nuclear charge is balanced by screening effect of inner orbitals.

Q. 53. What is electoplating? What is the purpose of electoplating? What are the requisites of a good electrolyte? *(Calcutta, 2007)*

Ans. A process in which a superior metal is deposited over an inferior metal (called base metal) by the electrolysis of the aqueous solution of a suitable electrolyte of the metal which has to be deposited on the base metal, is called electroplating.

Purpose of electroplating

Electroplating is done for the following purposes:

1. *For protecting metals from corrosion.* Metals like Fe and Cu get easily corroded by the atmospheric air, mositure and CO_2 and hence are liable to be destroyed. Their life is considerably increased when they are coated with metals like Ni and Cr, which protect them from corrosion and give them a good look.

2. *For decorating.* Au and Ag are deposited on base metal articles in order to improve the appearance of the latter.

3. Sometimes suitable metal is deposited on the worn out parts of machinery to repair them.

Requisites of good electrolyte

The electrolyte used in the electrolytic bath depends on the nature of the metal to be deposited. For example in Cu-plating, when a course erystalline deposit is required, copper sulphate acidified with H_2SO_4 may be used. But, for finer deposits, an alkaline solution of potassium tetracyanocopper (I), $K_3[Cu(CN)_4]$ is used. The electrolyte used should be cheap, highly soluble, a good conductor, stable towards hydrolysis, oxidation and reduction, and should possess a good covering power. Generally a mixture of two or more substance is used as electrolyte.

Q. 54. Discuss nickel plating, copper plating, silver plating and gold plating?

Ans. Nickle plating. In nickel plating of a steel article, an aqueous solution of $NiSO_4$ is used as an electrolyte, which is taken in the electrolytic tank. The steel article is made the cathode and the pure nickel metal plate is made the anode.

Reactions. When electric current of a suitable voltage is passed through the electrolytic solution, the following reactions take place:

(*a*) *Ionization of $NiSO_4$* : $NiSO_4 \rightleftharpoons Ni^{2+} + SO_4^{2+}$

(*b*) *Reactions taking place at cathode* (*steel article*). Ni^{2+} ions from the electrolyte move towards cathode (*i.e.,* towards the steel article), where they pick up electrons and are converted into neutral Ni atoms, which get deposited on the article (cathode).

$$Ni^{2+} + 2e^- \longrightarrow Ni$$

(*c*) *Reactions taking place at anode* (*pure nickel plate*). SO_4^{2-} ions their negative charges by losing electrons at the anode and become neutral SO_4 radicals, which combine with pure nickle metal of the anode, to from $NiSO_4$, which goes into the electrolytic solution.

$$SO_4^{2-} - 2e^- \longrightarrow SO_4$$
$$SO_4 + Ni \longrightarrow NiSO_4$$

Thus we see that Ni, from the electrolyte, gets deposited on the article (cathode) and an equivalent amount of Ni from the anode goes into solution of the electrolyte of $NiSO_4$ and hence the strength of $NiSO_4$ solution (electrolyte) remains unchanged.

2. Copper plating. When a coarse crystalline deposit of copper is required, $CuSO_4$ acidified with H_2SO_4 is used as the electrolyte, and if finer deposit of copper is needed, an alkaline solution of $K_3[Cu(CN)_4]$ is used as the electrolyte.

The reactions that take place are similar to those in the case of nickel plating.

3. Silver plating. The article to be silver plated is made the cathode, while the anode consists of pure silver. $K[Ag(CN)_2]$ solution is used as the electrolyte. This complex compound dissociates in water as:

$$K[Ag(CN)_2] \longrightarrow K^+ + [Ag(CN_2]^-$$

Since $[Ag(CN)_2]^-$ ion formed as above is highly stable, it dissociates to a very small extent and hence gives Ag^+ ions of very low concentration. A steady film of Ag gets deposited on the cathode (article).

$$[Ag(CN)_2]^- \rightleftharpoons Ag^+ + 2CN^-$$

$$Ag^+ + e^+ \longrightarrow Ag$$

This silver gets deposited more slowly and uniformly than when a solution of $AgNO_3$ is used as the electrolyte. The concentration of Ag^+ ions obtained from $[Ag(CN)_2]^-$ ion (which is very low) remains almost constant as the electrolysis proceeds.

4. Gold plating. Here the electrolyte used is the solution of $K[Au(CN)_2]$ or $K[Au(CN)_4]$. Like $[Ag(CN)_2]^-$ ions, $[Au(CN_2)]^-$ ions are also very stable and hence give Au^+ ions at very low concentration.

$$K[Au(CN)_2] \longrightarrow K^+ + [Au(CN)_2]^-$$

$$[Au(CN)_2]^- \rightleftharpoons Au^+ + 2CN^-$$

The concentration of Au^+ ions remains constant as the electrolysis proceeds.

Q. 55. What is galvanising? What are the different processes to perform galvanising?

Ans. Galvanising is process in which iron articles are coated with a thin layer of zinc metal. On account of its resistance to rusting and corrosion, the layer of zinc prevents the iron articles from rusting. Zinc coating protects iron mechanically, as well as, chemically. The surface of iron does not come directly in touch with air and moisture, because of the layer of zinc. It prevents even that part of an article from rusting, from which the zinc has been slightly detached due to wear and tear. This is due to the fact that zinc and iron will form a galvanic cell and zinc being more electropositive would react first and dissolve alone.Thus, galvanised iron remains free from rust, so long as even a small amount of zinc is present.

Different processes to carry out galvanising are:

1. Galvanising. The iron articles on which Zn coating is required, is first cleaned with hydrochloric acid. The process is known as "pickling". This is then thoroughly washed with water. Then this article is dipped in molten zinc, covered with sal ammoniac. NH_4Cl, as flux and taken out immediately. The coating of zinc is made uniform by pressing through rollers.

2. Sherardising. In this process, iron article, after cleaning with dilute hydrochloric acid, is heated for several hours, in closed drums and then zinc dust is sprayed on them, so that the upper surface of the article gets a coating of an alloy of zinc and iron. This coating is rust proof just like galvanisation.

3. Metallizing. It is not possible to impart a coating of zinc to large structure, like ships, tanks, bridges, by dipping them in molten zinc. Such objects are coated with zinc by spraying zinc over them.

18 SOME IMPORTANT COMPOUNDS OF TRANSITION ELEMENTS

Q. 1. Give the preparation, properties and structure of titanium nitrate $Ti(NO_3)_4$.

(Punjab, 2000)

Ans. Preparation. Titanium (IV) nitrate can be prepared by the action of N_2O_5 on titanium (IV) chloride.

$$TiCl_4 + 2N_2O_5 + 2H_2O \longrightarrow Ti(NO_3)_4 + 4HCl$$

Properties. 1. It is a white volatile compound.

2. It melts at 58°C.

3. It is highly reactive towards organic substances. While reacting with organic substances, it causes inflammation or explosion. The highly reactive nature of titanium nitrate is probably due to the presence of very reactive nitrate (NO_3) radical in its molecule.

Structure. The structure of $Ti(NO_3)_4$ is shown in Fig. 18.1. In its structure two oxygen atoms from each nitrate group are bonded to titanium atom exhibiting its bidentate nature. The bidentate nitrate groups form a tetrahedron around titanium bonded through two oxygen atoms giving a dodecahedral structure. In this compound titanium attains a coordination number of 8.

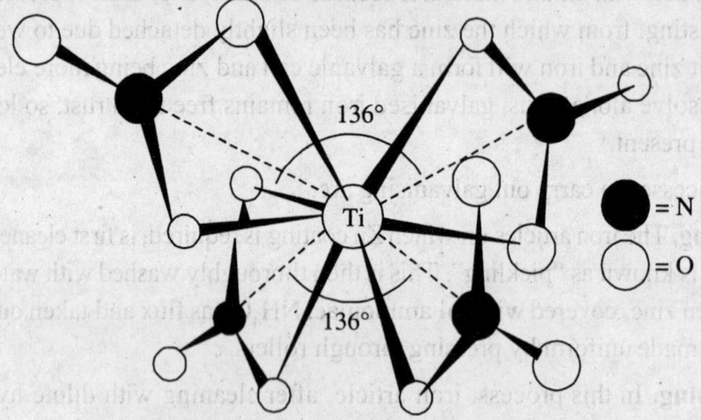

Fig. 18.1. Structure of titanium (IV) nitrate.

Q. 2. Give the preparation, properties, structure and uses of titanium dioxide, TiO_2.

(Lucknow 2010, Meerut, 2000)

Ans. Preparation. It is prepared by one of the following processes:

1. From ilmenite ($FeTiO_3$). The powdered ilmenite is digested with excess of conc. H_2SO_4 when $FeSO_4$, $Fe_2(SO_4)_3$ and $TiOSO_4$ (titanyl sulphate) are formed. The resulting product is extracted with water and the solution is then treated with scrap iron to reduce ferric sulphate to ferrous sulphate. The solution is filtered and transferred to vacuum evaporator. During this process crystals of $FeSO_4 . 7H_2O$ are formed which are separated and the solution containing $TiOSO_4$ is boiled with excess water to get hydrated titanium dioxide. The precipitate is filtered, washed and calcined in a rotary kiln to get white pigment grade titanium dioxide. The overall reaction is given below.

$$TiOSO_4 + H_2O \longrightarrow TiO_2 + H_2SO_4$$

This process is also called **sulphate process**.

2. Chloride process. In this process, rutile is heated with chlorine and coke at 900°C forming $TiCl_4$. It is volatile and can be separated from impurities. $TiCl_4$ is heated with oxygen forming pure TiO_2 and Cl_2.

$$\underset{\text{Rutile}}{2TiO_2} + 3C + 4Cl_2 \longrightarrow 2TiCl_4 + 2CO + CO_2$$

$$TiCl_4 + 2O_2 \xrightarrow{\text{Heat}} TiO_2 + 2Cl_2$$

The chlorine is reused. Alternatively, $TiCl_4$ can be hydrolysed completely to form TiO_2.

$$TiCl_4 + 4H_2O \longrightarrow TiO_2 + 2H_2O$$

Properties. 1. It is a white solid which turns yellow on heating.

2. It is insoluble in water.
3. It is reduced with hydrogen to form Ti_2O_3.

$$2TiO_2 + H_2 \longrightarrow Ti_2O_3 + H_2O$$

However, on reduction with Ca, Mg or Al, it forms titanium metal.

$$TiO_2 + 2Ca \longrightarrow Ti + 2CaO$$

$$TiO_2 + 2Mg \longrightarrow Ti + 2MgO$$

4. Amphoteric character. It is an amphoteric oxide and dissolves both in acids and bases forming titanates and titanyl compounds. For example, it dissolves in conc. H_2SO_4 forming titanyl sulphate.

$$TiO_2 + H_2SO_4 \longrightarrow \underset{\text{Titanyl sulphate}}{TiOSO_4} + H_2O$$

However, it is insoluble in HCl, HNO_3 or aqua regia. It is soluble in HF forming haxafluoro complex, H_2TiF_6.

$$TiO_2 + 6HF \longrightarrow H_2TiF_6 + 2H_2O$$

It reacts with sodium hydroxide to form sodium titanate.

$$TiO_2 + 2NaOH \longrightarrow \underset{\text{Sodium titanate}}{Na_2TiO_3} + H_2O$$

5. Action with halogens. On passing chlorine or bromine through a mixture of TiO_2 and coke at 950°C, it forms titanium (IV) tetrabromide.

$$2TiO_2 + 3C + 4Cl_2 \xrightarrow{950°C} 2TiCl_4 + CO_2 + 2CO$$

$$2TiO_2 + 3C + 8Br_2 \xrightarrow{950°C} 2TiBr_4 + CO_2 + 2CO$$

6. Action with titanium. On fusion with titanium metal, it produces TiO.

$$TiO_2 + Ti \longrightarrow 2TiO$$

7. Action with carbon. On heating with coke at 1100°C, it forms titanium carbide.

$$TiO_2 + 3C \xrightarrow{1100°C} \underset{\text{Titanium carbide}}{TiC} + 2CO$$

Structure. Its structure is known as rutile structure. It has coordination number 6 :3 in which each titanium is surrounded octahedrally by six oxygen atoms while each oxygen atom is surrounded by three titanium atoms disposed towards the corners of an equilateral triangle as shown in Fig. 18.2.

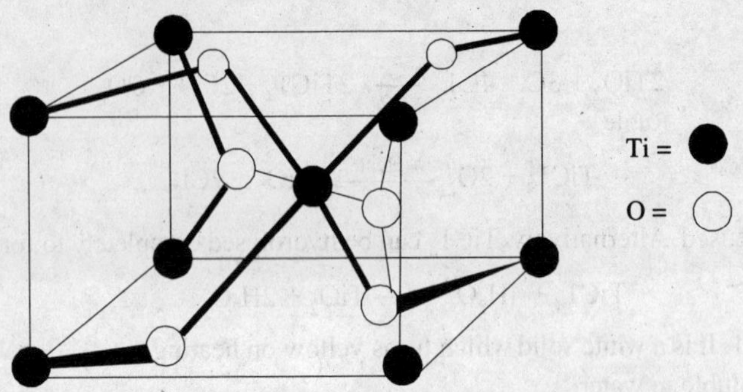

Fig. 18.2. Rutile (TiO_2) structure.

Uses. 1. It is used in large quantities as a white pigment in paints and cosmetic industry. Its properties like high opacity, inertness, good covering power, non-poisonous nature and resistance to temperature make it a very useful compound.

TiO_2 has replaced the earlier pigments such as white lead $2PbCO_3.Pb(OH_2)$ and $BaSO_4$ for this purpose.

2. It is used as an additive in the manufacture of fibre glass.

3. It is used as a filler in plastics and rubber.

4. It has a high refractive index and, therefore, its crystals have brilliance, even more than diamond. However, because of its softness it is not well-suited for use in jewellery.

5. It is used in enamelling of porcelain goods.

Q. 3. How are TiO_2 and $Ti(NO_3)_4$ structurally represented? Give any two reactions to show the Lewis acid character of $TiCl_4$.

Ans. See Qs. 1 and 2 for structures. Reactions of $TiCl_4$ with water and oxygen indicate Lewis acid character of $TiCl_4$.

Q. 4. Give the preparation, properties and uses of vanadyl chloride, $VOCl_2$.

Ans. Preparation. 1. It can be prepared by the reduction of $VOCl_3$ with hydrogen at 300°C.

$$2VOCl_3 + H_2 \xrightarrow{300°C} 2VOCl_2 + 2HCl$$

2. Pure compound is prepared by heating $VOCl_3$ with metallic zinc in a sealed tube.

$$2VOCl_3 + Zn \longrightarrow 2VOCl_2 + ZnCl_2$$

3. It can be prepared by the action of HCl on V_2O_5.

$$V_2O_5 + 6HCl \longrightarrow 2VOCl_2 + 3H_2O + Cl_2$$

4. It can be prepared by the hydrolysis of VCl_4.

$$VCl_4 + H_2O \longrightarrow VOCl_2 + 2HCl$$

Properties. 1. The pure compound exists as grassy green crystals.

2. Hydrolysis. It is hydrolysed to yellow brown $VO(OH)_2$ with dilute sodium carbonate solution:

$$VOCl_2 + 2H_2O\,(aq) \xrightarrow{Na_2CO_3(aq)} VO\,(OH)_2\,(aq) + 2HCl$$

3. Action with alkalies. It reacts with strong alkalies to form vanadates.

$$VOCl_2 + 4NaOH \longrightarrow \underset{\text{Sodium vanadate}}{Na_2VO_3} + 2NaCl + 2H_2O$$

Uses. Vanadyl chloride is used commercially in dyeing under the name *Divanadyl tetrachloride.* It is a strong reducing agent.

Structure. $VOCl_2$ has been found to contain double (V = O) bond formed by the overlap of π electrons of oxygen into the vacant *d*-orbitals of vanadium. It has trigonal planar structure:

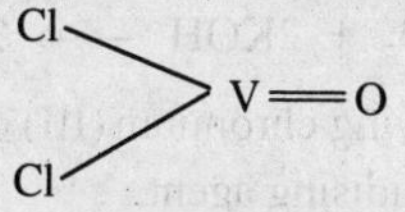

Cl
V═O
Cl

Q. 5. How is $VOCl_2$ obtained from $VOCl_3$? Given equation and experimental conditions only.

Ans. See Q. 4.

Q. 6. Describe the preparation, properties, uses and structure of chromous acetate, $Cr(CH_3COO)_2.2H_2O$.

Ans. Preparation. It is prepared as sparingly soluble red crystalline precipitate when a solution of chromous chloride is added to a fairly concentrated solution of sodium acetate.

$$2Cr\,Cl_2 + 2CH_3\,COONa \longrightarrow Cr\,(CH_3\,COO)_2 + 2NaCl$$

Properties. It is sparingly soluble in water but readily dissolves in acids and therefore affords a method to prepare various soluble chromous salts.

Uses. A solution of this salt in hydrochloric acid is employed to determine the proportion of oxygen in a gas mixture.

Structure. Chromous acetate exists as a dimeric molecule and has the formula

$$[(CH_3COO)_2\,Cr]_2 \cdot 2H_2O$$

H₃C, CH₃, C, O, Cr, H₂O, OH₂, 1.97 Å, 2,.46 Å, 2.20 Å

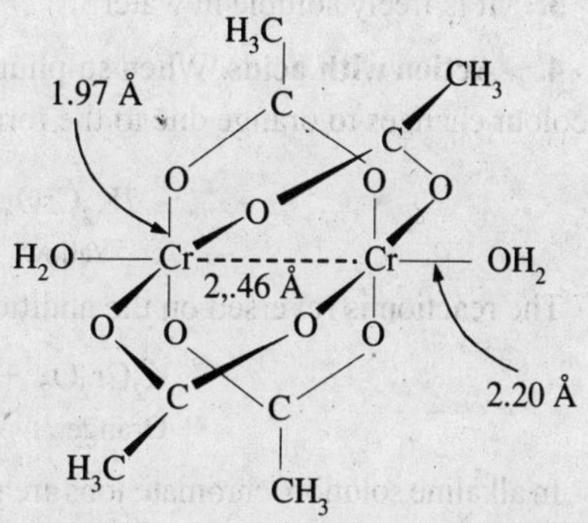

Fig. 18.3. Dimeric structure of chromous acetate.

Its structure is similar to the acetate of Cu (II) involving a very short [2.46 Å] metal-metal bond and has the H_2O molecule as end groups [Fig. 18.3].

Q. 7. Chromium (II) acetate monohydrate is diamagnetic. Explain.

Ans. This compound exists as a dimeric molecule and has the formula

$$[(CH_3\,COO)_2\,Cr]_2 \cdot 2H_2O$$

Interaction between two metal ions is so strong that four unpaired electrons of each Cr^{2+} become fully paired and therefore the compound becomes diamagnetic.

Q. 8. Describe methods of preparation, properties, uses and structure of potassium chromate, K_2CrO_4. *(Garhwal, 2000; Purvanchal, 2003)*

Ans. Preparation. A mixture of powdered chromite ore, excess of potassium carbonate and a little lime is heated in the presence of air.

$$\underset{\text{Chromite ore}}{8FeCr_2O_4} + 8K_2CO_3 + 7O_2 \xrightarrow{\text{Lime}} \underset{\text{Potassium chromate}}{8K_2CrO_4} + 2Fe_2O_3 + 8CO_2$$

The fused mass is extracted with water to get potassium chromate while ferric oxide, Fe_2O_3, is left behind.

2. It can be prepared by treating aqueous solution of potassium dichromate with potassium carbonate or potassium hydroxide.

$$K_2Cr_2O_7 + K_2CO_3 \longrightarrow 2K_2CrO_4 + CO_2$$

$$K_2Cr_2O_7 + 2KOH \longrightarrow 2K_2CrO_4 + H_2O$$

3. It can also be prepared by dissolving chromium (III) oxide, Cr_2O_3, or hydroxide, $Cr(OH)_3$, in excess of alkali in the presence of an oxidising agent.

$$2Cr(OH)_3 + 4KOH + 3O \longrightarrow 2K_2CrO_4 + 5H_2O$$

$$Cr_2O_3 + 4KOH + 3O \longrightarrow 2K_2CrO_4 + 2H_2O$$

Properties. 1. It occurs as lemon yellow crystalline solid. Its crystals do not contain water of crystallisation.

2. Its crystals are isomorphous with potassium sulphate, K_2SO_4.

3. It is freely soluble in water.

4. Action with acids. When sulphuric acid is added to yellow solution of potassium chromate, its colour changes to orange due to the formation of potassium dichromate.

$$\underset{\text{Yellow}}{2K_2CrO_4} + H_2SO_4 \longrightarrow \underset{\text{Orange}}{K_2Cr_2O_7} + K_2SO_4 + H_2O$$

The reaction is reversed on the addition of an alkali.

$$\underset{\text{Orange}}{K_2Cr_2O_7} + 2KOH \longrightarrow \underset{\text{Yellow}}{2K_2CrO_4} + H_2O$$

In alkaline solution, chromate ions are present while in acidic solution dichromate ions are present.

5. Action with barium, lead and silver salts. (*i*) With barium chloride solution, it gives yellow precipitate of barium chromate.

$$BaCl_2 + K_2CrO_4 \longrightarrow \underset{\text{Yellow ppt}}{BaCrO_4} + 2KCl$$

(*ii*) With lead nitrate solution, it gives yellow precipitate of lead chromate.

$$Pb(NO_3)_2 + K_2CrO_4 \longrightarrow \underset{\text{Yellow ppt}}{PbCrO_4} + 2KNO_3$$

(*iii*) With silver nitrate solution, a brick red precipitate of Ag_2CrO_4 is obtained.

$$2AgNO_3 + K_2CrO_4 \longrightarrow \underset{\text{Silver chromate}}{Ag_2CrO_4} + 2KNO_3$$

6. Oxidising properties. The acidic solution of potassium chromate acts as an oxidising agent. It oxidises (*i*) *Hydrogen sulphide to sulphur*

$$2K_2CrO_4 + 5H_2SO_4 \longrightarrow 2K_2SO_4 + Cr_2(SO_4)_3 + 5H_2O + 3O$$
$$H_2S + O \longrightarrow H_2O + S] \times 3$$
$$2K_2CrO_4 + 5H_2SO_4 + 3H_2S \longrightarrow 2K_2SO_4 + Cr_2(SO_4)_3 + 8H_2O + 3S$$

(*ii*) *Sulphur dioxide to sulphuric acid*

$$2K_2CrO_4 + 5H_2SO_4 \longrightarrow 2K_2SO_4 + Cr_2(SO_4)_3 + 5H_2O + 3O$$
$$SO_2 + H_2O + O \longrightarrow H_2SO_4] \times 3$$
$$2K_2CrO_4 + 3SO_2 + 2H_2SO_4 \longrightarrow 2K_2SO_4 + Cr_2(SO_4)_3 + 2H_2O$$

(*iii*) *Hydrochloric acid to chlorine*

$$2K_2CrO_4 + 5H_2SO_4 \longrightarrow 2K_2SO_4 + Cr_2(SO_4)_3 + 5H_2O + 3O$$
$$2HCl + O \longrightarrow H_2O + Cl_2] \times 3$$
$$2K_2CrO_4 + 5H_2SO_4 + 6HCl \longrightarrow 2K_2SO_4 + Cr_2(SO_4)_3 + 8H_2O + 3Cl_2$$

(*iv*) *Liberates iodine from potassium iodide*

$$2K_2CrO_4 + 5H_2SO_4 \longrightarrow 2K_2SO_4 + Cr_2(SO_4)_3 + 5H_2O + 3O$$
$$2KI + H_2SO_4 + O \longrightarrow K_2SO_4 + H_2O + I_2] \times 3$$
$$2K_2SO_4 + 8H_2SO_4 + 6KI \longrightarrow 5K_2SO_4 + Cr_2(SO_4)_3 + 8H_2O + 3I_2$$

Uses. (*i*) As an indicator in silver nitrate titrations.

(*ii*) In the manufacture of pigments, basic lead chromate $PbCrO_4 \cdot PbO$ and basic zinc chromate $ZnCrO_4 . ZnO$.

(*iii*) As a reagent for testing silver, lead and barium ions.

Structure of Chromate Ion CrO_4^{2-}

Four oxygens around chromium are arranged in a tetrahedral manner as shown in Fig. 18.4.

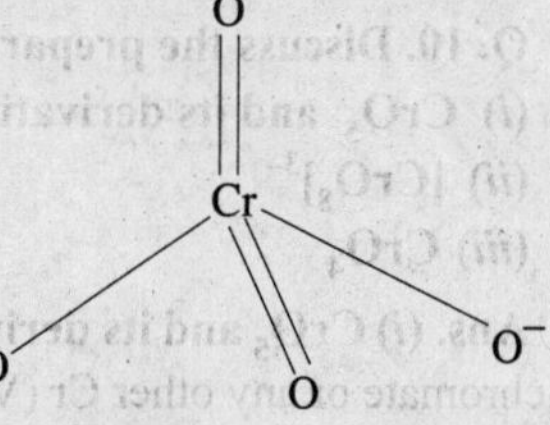

Fig. 18.4. Structure of chromate ion.

Q. 9. Give the preparation, properties, uses and structure of chromyl chloride, CrO_2Cl_2.

(*Garhwal; 2000*)

Ans. Preparation. 1. It is prepared by heating a mixture of sulphuric acid, potassium dichromate and potassium chloride.

$$K_2Cr_2O_2 + 4KCl + 3H_2SO_4 \xrightarrow{\Delta} 2CrO_2Cl_2 + 3K_2SO_4 + 3H_2O$$

2. It can also be obtained by treating chromium (VI) oxide with gaseous hydrogen chloride.

$$CrO_3 + 2HCl \longrightarrow CrO_2Cl_2 + H_2O$$

Properties. 1. It is a deep red liquid boiling at 116°C under atmospheric pressure.

2. It is soluble in carbon disulphide and nitrobenzene.

3. **Action of heat.** It decomposes on passing through red hot tube.

$$4CrO_2Cl_2 \longrightarrow 2Cr_2O_3 + 4Cl_2 + O_2$$

4. **Action with sodium chloride.** When vapours of chromyl chloride are passed into concentrated solution of a sodium chloride, a chlorochromate is formed.

$$CrO_2Cl_2 + NaCl + H_2O \longrightarrow \underset{\text{Sodium chloro-chromate}}{NaCrO_3Cl} + 2HCl$$

5. **Action with alkalies.** With alkalies, it forms chromates.

$$CrO_2Cl_2 + 4NaOH \longrightarrow Na_2CrO_4 + 2NaCl + 2H_2O$$

6. **Action with water.** It is hydrolysed by water giving hydrochloric acid and chromic acid.

$$CrO_2Cl_2 + 2H_2O \longrightarrow 2HCl + H_2CrO_4$$

7. **As an oxidising agent.** Chromyl chloride is a powerful oxidising agent. It oxidises toluene to benzaldehyde.

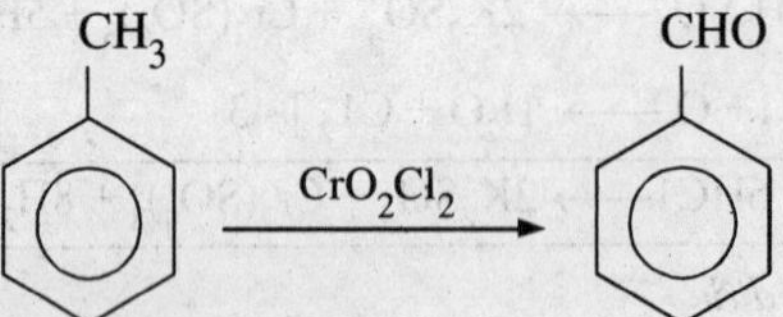

This reaction is known as Etard's reaction.

Uses. (*i*) it is used in quantitative analysis for detection and confirmation of chloride radical.

(*ii*) It is used in Etard's reaction.

Structure. It has a tetrahedral structure as shown in Fig. 18.5.

O, Cr, Cl, O, Cl

Fig. 18.5 Structure of chromyl chloride.

Q. 10. Discuss the preparation, properties and structure of

(*i*) CrO_5 and its derivatives

(*ii*) $[CrO_8]^{3-}$

(*iii*) CrO_4 (*Punjab, 2000*)

Ans. (*i*) CrO_5 and its derivatives. When hydrogen peroxide is added to an acidified solution of a dichromate or any other Cr (VI) compound, a deep blue colour is obtained due to the formation of $CrO(O_2)_2$. This colour fades away and the compound decomposes rapidly in aqueous solution to Cr (III) ions and oxygen. However, the peroxo compound can be extracted into ether, where it remains stable. On the addition of pyridine to the ether solution, it forms the adduct Py.CrO_5. Blue colour in the above reaction is due to the formation of chromium peroxide, CrO_5 and the following reactions occur:

$$\underset{\text{Orange}}{Cr_2O_7^{2-}} + 2H^+ + 4H_2O_2 \longrightarrow \underset{\text{Blue}}{2CrO_5} + 5H_2O$$

$$\underset{\text{Yellow}}{CrO_4^{2-}} + 2H^+ + 2H_2O_2 \longrightarrow \underset{\text{Blue}}{2CrO_5} + 3H_2O$$

The CrO_5 is essentially $CrO(O_2)$ and has the structure shown in Fig. 18.6 (*a*). This structure is pentagonal in which chromium is at the centre of pentagon. In ether, it forms hexacoordinated complex with ether and gets stabilized [Fig. 18.6 (*b*)]. On adding pyridine to the etheral solution, it forms solid pyridine adduct, Py.CrO_5.

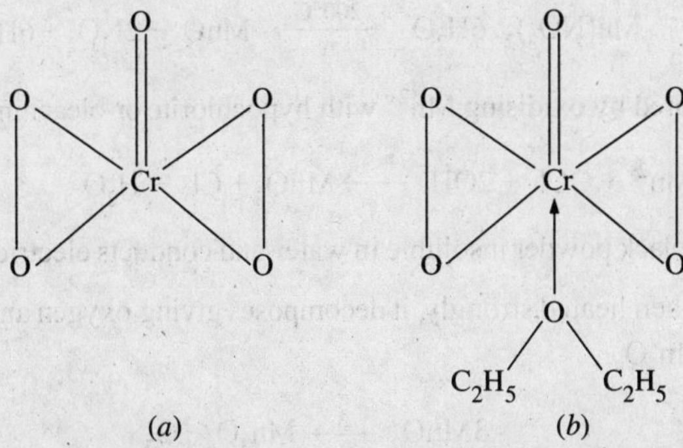

Fig. 18.6. Structures of (*a*) CrO_5 (*b*) CrO_5 –ether complex.

***(ii)* Compounds containing $[CrO_8]^{3-}$ ion.** On adding 30% H_2O_2 solution to alkaline ammonium dichromate or potassium dichromate, a red brown compound $(NH_4)_3CrO_8$ ammonium peroxochromate or K_3CrO_8, is formed. This compound contains the tetraperoxo species $[Cr(O_2)_4]^{3-}$. This contains chromium in (V) oxidation state and is paramagnetic (magnetic moment = 1.8 BM) indicating the presence of one unpaired electron.

The structure of this ion is *dodecahedral* in which the eight oxygen atoms form a regular dodecahedron around chromium (V) ion as shown in Fig. 18.7.

Fig. 18.7. Structure of $[Cr_3O_8]^{3-}$ ion.

***(iii)* Compounds containing CrO_4.** When ammonium peroxochromate $(NH_4)_3CrO_5$ is heated to 50°C, a reddish brown complex $CrO_4.3NH_3$ is formed. This contains chromium in (IV) oxidation state and is paramagnetic (magnetic moment = 2.8 BM) indicating the presence of two unpaired electrons.

The structure of $CrO_4.3NH_3$ is pentagonal bipyramidal as shown in Fig. 18.8 (a). On adding KCN to the complex $CrO_4.3NH_3$ another complex $CrO_4(CN)_3^{2-}$ is obtained. Its structure is shown in Fig. 18.8 (*b*).

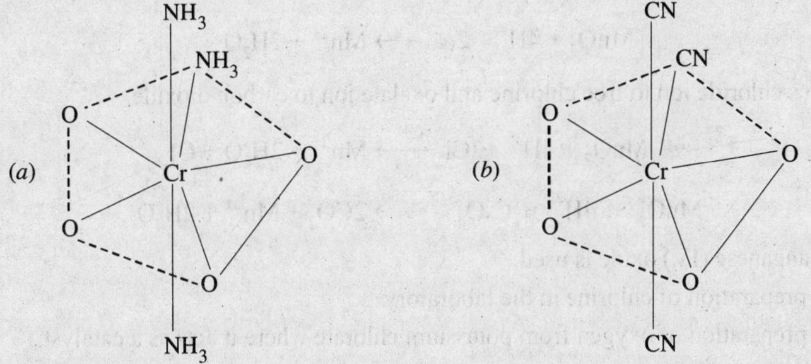

Fig. 18.8. (*a*) Structure of $CrO_4.3NH_3$ (*b*) Structure of $CrO_4(CN)_3^{3-}$.

Q. 11. Describe preparation, properties and uses of manganese dioxide, MnO_2.

Ans. Preparation. 1. It occurs as pyrolusite in nature.

2. It may be prepared by heating manganese nitrate at 200°C.

$$Mn(NO_3)_2 . 6H_2O \xrightarrow[\Delta]{200°C} MnO_2 + 2NO_2 + 6H_2O$$

3. It can also be prepared by oxidising Mn^{2+} with hypochlorite or bleaching powder.

$$Mn^{2+} + OCl^- + 2OH^- \longrightarrow MnO_2 + Cl^- + H_2O$$

Properties. 1. It is a black powder insoluble in water and conducts electricity.

2. **Action of heat.** When heated strongly, it decomposes giving oxygen and leaving a residue of trimanganese tetroxide, Mn_3O_4.

$$3MnO_2 \xrightarrow{\Delta} Mn_3O_4 + O_2$$

3. **Action with acids.** When heated with cold concentrated hydrochloric acid, it gives Mn (II) chloride and chlorine.

$$MnO_2 + 4HCl \longrightarrow MnCl_2 + 2H_2O + Cl_2$$

When heated with concentrated sulphuric acid, manganese (II) sulphate and oxygen are produced.

$$2MnO_2 + 2H_2SO_4 \longrightarrow 2MnSO_4 + 2H_2O + O_2$$

Nitric acid does not react with manganese dioxide.

4. **Action with alkalies.** On fusion with alkalies in the presence of air or oxidising agents such as $NaNO_3$ or $NaClO_3$, green mass of K_2MnO_4 is formed.

$$2MnO_2 + 4KOH + O_2 \longrightarrow \underset{\substack{\text{Potassium manganate} \\ \text{Green mass}}}{2K_2MnO_4} + 2H_2O$$

5. **Reduction.** It is reduced to manganese by aluminium.

$$2MnO_2 + 4Al \longrightarrow 3Mn + 2Al_2O_3$$

6. **As an oxidising agent.** Manganese dioxide in acid solution acts as a strong *oxidising agent* when it is reduced to Mn (II) ion.

$$MnO_2 + 4H^+ + 2e^- \longrightarrow Mn^{2+} + 2H_2O$$

It oxidises chloride ion to free chlorine and oxalate ion to carbon dioxide.

$$MnO_2 + 4H^+ + 2Cl^- \longrightarrow Mn^{2+} + 2H_2O + Cl_2$$

$$MnO_2 + 4H^+ + C_2O_4^{2-} \longrightarrow 2CO_2 + Mn^{2+} + 2H_2O$$

Uses. Manganese (IV) oxide is used

1. in the preparation of chlorine in the laboratory.
2. in the preparation of oxygen from potassium chlorate where it acts as a catalyst.
3. for colouring glass.
4. in making dry cells (batteries).
5. in the production of ferromanganese.

Q. 12. Describe the preparation, properties, uses and structure of potassium permanganate, ($KMnO_4$). (*Lucknow, 2000; Meerut, 2000; Kanpur, 2001; Kerala, 2000; Purvanchal, 2003; Calcutta, 2007; Vidyasagar, 2007*)

Ans. Preparation. 1. Potassium permanganate is prepared by fusing MnO_2 or pyrolusite with KOH or K_2CO_3 in presence of atmospheric oxygen or any other oxidizing agent, *e.g.*, KNO_3. The fused mass is green due to the formation of potassium manganate.

$$2MnO_2 + 4KOH + O_2 \xrightarrow{\text{heat}} 2K_2MnO_4 + 2H_2O$$

$$2MnO_2 + 2K_2CO_3 + O_2 \longrightarrow \underset{\textbf{Potassium manganate}}{2K_2MnO_4} + 2CO_2$$

The fused mass is extracted with water and treated with a current of ozone, chlorine or CO_2 when K_2MnO_4 is oxidized to potassium permanganate. The purple solution thus obtained is concentrated when it deposits dark purple, needle-like crystals of $KMnO_4$.

$$2K_2MnO_4 + O_3 + H_2O \longrightarrow 2KMnO_4 + 2KOH + O_2$$

$$2K_2MnO_4 + Cl_2 \longrightarrow 2KMnO_4 + 2KCl$$

CO_2 neutralizes the alkali formed and facilitates the reaction.

$$3K_2MnO_4 + 2CO_2 \longrightarrow 2KMnO_4 + 2K_2CO_3 + MnO_2$$

2. It is manufactured by the electrolytic oxidation of a basic solution of potassium manganate. The manganate solution as placed in a cell provided with a rotating nickel wire-netting anode and an inlet pipe made of iron for the liquor which acts as cathode (Fig. 18.9). The oxygen liberated at the anode converts manganate into permanganate.

$$MnO_4^{2-} - e^- \longrightarrow MnO_4^-$$

Hydrogen is evolved in the cathodic compartment.

$$\underset{\text{Fresh water ionisation}}{2H^+} + 2e^- \longrightarrow 2H \rightarrow H_2$$

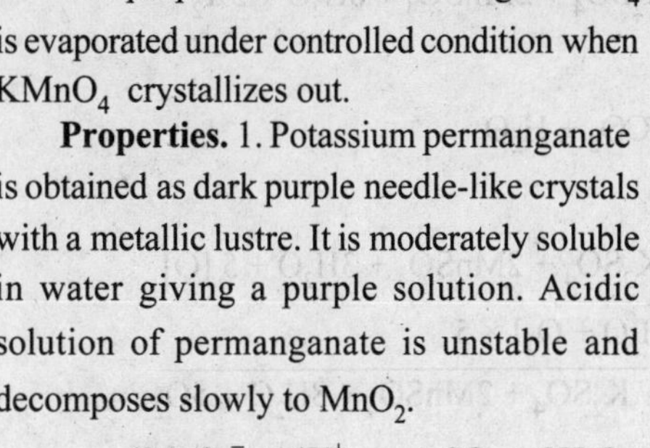

The purple solution containing $KMnO_4$ is evaporated under controlled condition when $KMnO_4$ crystallizes out.

Properties. 1. Potassium permanganate is obtained as dark purple needle-like crystals with a metallic lustre. It is moderately soluble in water giving a purple solution. Acidic solution of permanganate is unstable and decomposes slowly to MnO_2.

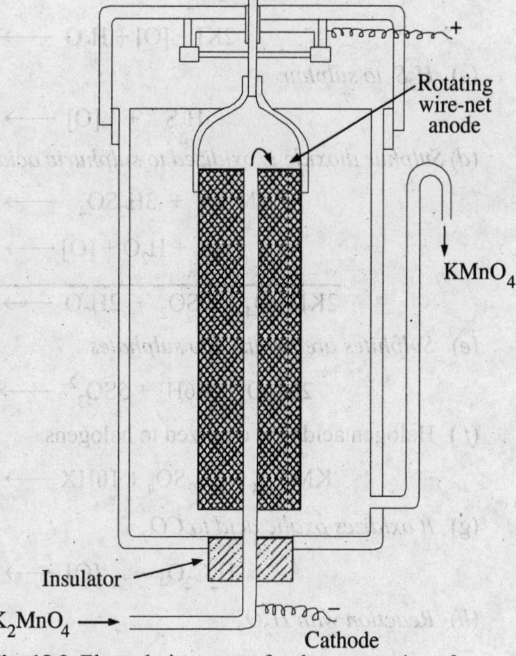

Fig. 18.9. Electrolytic process for the preparation of $KMnO_4$.

$$4MnO_4^- + 4H^+ \longrightarrow 3O_2 + 2H_2O + 4MnO_2\downarrow$$

2. Action of Heat.

$$2KMnO_4 \xrightarrow{\text{Heat}} \underset{\text{Pot. manganate}}{K_2MnO_4} + \underset{\text{Manganese dioxide}}{MnO_2} + O_2$$

3. With conc. H_2SO_4, it gives [Mn (VII)] oxide Mn_2O_7, a dark brown highly explosive liquid which further decomposes into MnO_2 and oxygen.

$$2KMnO_4 + H_2SO_4 \longrightarrow Mn_2O_7 + K_2SO_4 + H_2O$$

$$2Mn_2O_7 \longrightarrow 4MnO_2 + 3O_2$$

4. Potassium permanganate acts as a powerful oxidising agent in neutral, alkaline or acidic solution.

In acidic medium or solution. The main reaction is

$$2KMnO_4 + 3H_2SO_4 \longrightarrow K_2SO_4 + 2MnO_4 + 3H_2O + 5\,[O]$$

or $$2MnO_4 + 6H^+ \longrightarrow 2Mn^{2+} + 3H_2O + 5\,(O)$$

The oxidation reactions of acidified $KMnO_4$ are catalysed by the Mn (II) ion.

(*a*) *It oxidizes ferrous salts to ferric salts.*

$$2FeSO_4 + H_2SO_4 + [O] \longrightarrow Fe_2(SO_4)_3 + H_2O$$

(*b*) *It liberates halogens from halides.*

$$2KI + [O] + H_2O \longrightarrow 2KOH + I_2$$

(*c*) *H_2S to sulphur.*

$$H_2S + [O] \longrightarrow H_2O + S$$

(*d*) *Sulphur dioxide is oxidized to sulphuric acid.*

$$2KMnO_4 + 3H_2SO_4 \longrightarrow K_2SO_4 + 2MnSO_4 + 3H_2O + 5\,[O]$$

$$[SO_2 + H_2O + [O] \longrightarrow H_2SO_4] \times 5$$

$$\overline{2KMnO_4 + 5SO_2 + 2H_2O \longrightarrow K_2SO_4 + 2MnSO_4 + 2H_2SO_4}$$

(*e*) *Sulphites are oxidized to sulphates.*

$$2MnO_4^- + 6H^+ + 5SO_3^{2-} \longrightarrow 2Mn^{2+} + 5SO_4^{2-} + 3H_2O$$

(*f*) Halogen acids are oxidized to halogens.

$$KMnO_4 + 3H_2SO_4 + 10HX \longrightarrow K_2SO_4 + 2MnSO_4 + 8H_2O + 5X_2$$

(g) *It oxidizes oxalic acid to CO_2.*

$$H_2C_2O_4 + [O] \longrightarrow 2CO_2 + H_2O$$

(*h*) *Reaction with H_2O_2*

$$2KMnO_4 + 3H_2SO_4 \longrightarrow K_2SO_4 + 2MnSO_4 + 3H_2O + 5\,[O]$$

$$H_2O_2 + [O] \longrightarrow H_2O + O_2] \times 5$$

$$\overline{2KMnO_4 + 3H_2SO_4 + 5H_2O_2 \longrightarrow K_2SO_4 + 2MnSO_4 + 8H_2O + 5O_2}$$

(*i*) *Ethyl alcohol is oxidized to acetaldehyde*

$$CH_3CH_2OH + [O] \xrightarrow{KMnO_4} CH_3CHO + H_2O$$

In alkaline solution. Main reaction is

$$2KMnO_4 + 2KOH \longrightarrow 2K_2MnO_4 + H_2O + [O]$$

(*a*) *It oxidizes ethylene to ethylene glycol.*

$$C_2H_4 + [O] + H_2O \longrightarrow HOCH_2 \cdot CH_2OH$$

(*b*) *It oxidizes toluene to benzoic acid*

$$C_6H_5CH_3 + 3[O] \longrightarrow C_6H_5COOH + H_2O$$

(*c*) *It oxidizes KI to* KIO_3.

$$KI + 3[O] \longrightarrow KIO_3$$

In neutral solution. Main reaction is

$$2KMnO_4 + H_2O \longrightarrow 2KOH + 2MnO_2 + 3[O]$$

(*a*) *It oxidizes* H_2S *to sulphur.*

$$H_2S + [O] \longrightarrow H_2O + S$$

(*b*) *Sodium thiosulphate is oxidized to sodium sulphate.*

$$8KMnO_4 + 3Na_2S_2O_3 + H_2O \longrightarrow 8MnO_2 + 3Na_2SO_4 + 3K_2SO_4 + 2KOH$$

(*c*) *It oxidizes* $MnSO_4$ *to* MnO_2,

$$MnSO_4 + H_2O + [O] \longrightarrow MnO_2 + H_2SO_4$$

Uses. 1. Potassium permanganate is used as oxidizing agent.

2. It is also used in volumetric estimation of ferrous salts, oxalic acid or oxalates, hydrogen peroxide etc.

3. It finds application as a disinfectant.

4. It is used in qualitative analysis for detecting halides, sulphites, oxalates etc.

5. It is employed in the manufacture of saccharin, benzoic acid, and acetaldehyde.

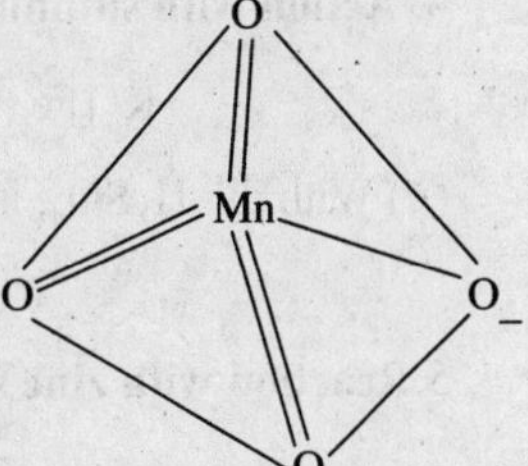

Fig. 18.10. Structure of MnO_4^- ion.

Structure. The structure of permanganate ion, MnO_4^-, is roughly tetrahedral as shown in Fig. 18.10. The Mn – O bond length is 1.63 Å and O–Mn–O bond angle is close to the tetrahedral angle of 109°.

Q. 13. How is $KMnO_4$ prepared? Comment upon its oxidising ability in different media.

Ans. See Q. 12.

Q. 14. Describe preparation, properties, uses and structure of potassium ferrocyanide, $K_4[Fe(CN)_6]$. (*Garhwal, 2000; Kanpur, 2001; Calcutta, 2007, Vidyasagar, 2007*)

Ans. Preparation. 1. It is obtained during the purification of coal gas. Coal gas contains some cyanogon $(CN)_2$ as an impurity. The gas is passed over hydrated iron oxide.

$$(CN)_2 + H_2O \longrightarrow 2HCN + \frac{1}{2}O_2$$

$$Fe(OH)_2 + 2HCN \longrightarrow Fe(CN)_2 + 2H_2O$$

$$Fe(OH)_3 + 3HCN \longrightarrow Fe(CN)_3 + 3H_2O$$

$$3Fe(CN)_2 + 4Fe(CN)_3 \longrightarrow \underset{\substack{\text{Ferric ferrocyanide} \\ \text{(Prussian blue)}}}{Fe_4[Fe(CN)_6]_3}$$

Ferric ferrocyanide formed is treated with $Ca(OH)_2$ when calcium ferrocyanide is formed.

$$Fe_4[Fe(CN)_6]_3 + 6Ca(OH)_2 \longrightarrow \underset{\text{Soluble}}{3Ca_2[Fe(CN)_6]} + \underset{\text{ppt.}}{4Fe(OH)_3}$$

It is extracted with water and then treated with K_2CO_3 when potassium ferrocyanide is obtained on crystallisation.

$$Ca_2[Fe(CN)_6] + 2K_2CO_3 \longrightarrow K_4[Fe(CN)_6] + 2CaCO_3$$

2. It is manufactured by heating fused KCNS with iron in the absence of air. Potassium ferrocyanide is obtained by concentrating the solution.

$$6KCNS + 6Fe \longrightarrow 6KCN + 6FeS$$

$$6KCN + FeS \longrightarrow K_2S + K_4[Fe(CN)_6]$$

Properties. 1. Potassium ferrocyanide is a yellowish crystalline substance and it is soluble in water.

2. **Action of heat.** On heating above 373 K, it decomposes as

$$K_4[Fe(CN)_6] \longrightarrow 4KCN + FeC_2 + N_2$$

3. **Reducing action.** It is a reducing agent and is oxidized by Cl_2, H_2O_2 or acidified $KMnO_4$.

$$2K_4[Fe(CN)_6] + Cl_2 \longrightarrow 3K_3[Fe(CN)_6] + 2KCl$$

4. **Action with sulphuric acid.** (*i*) With conc. H_2SO_4, it produces carbon monoxide

$$K_4[Fe(CN)_6] + 6H_2SO_4 + 6H_2O \longrightarrow 2K_2SO_4 + 3(NH_4)_2SO_4 + FeSO_4 + 6CO$$

(*ii*) With dil. H_2SO_4, it gives hydrogen cyanide.

$$K_4[Fe(CN)_6] + 3H_2SO_4 \longrightarrow 2K_2SO_4 + FeSO_4 + 6HCN$$

5. **Reaction with zinc salts.**

$$K_4[Fe(CN)_6] + 2ZnSO_4 \longrightarrow \underset{\substack{\text{Zinc ferrocyanide} \\ \text{white ppt.}}}{Zn_2[Fe(CN)_6]} + 2K_2SO_4$$

6. **Reaction with ferric salts.** With ferric salts, it produces Prussian blue colour.

$$Fe^{3+} + K_4[Fe(CN)_6] \longrightarrow \underset{\substack{\text{III} \quad \text{II} \\ \text{Soluble Prussian blue}}}{K\{Fe[Fe(CN)_6]\}} + 3K^+$$

7. **Reaction with copper salts.**

$$K_4[Fe(CN)_6] + 2CuSO_4 \longrightarrow \underset{\substack{\text{Cupric ferrocyanide} \\ \text{[Reddish brown ppt.]}}}{Cu_2[Fe(CN)_6]} + 2K_2SO_4$$

Uses. 1. It is used in making Prussian blue pigment.

2. It is used for making blue prints.

3. It is used as a reagent in the laboratory for testing Fe^{3+}, Cu^{2+} and Zn^{2+} ions.

Structure. Structure of ferrocyanide ion $[Fe(CN)_6]^{4-}$ is octahedral, shown in Fig. 18.11. This structure is explained below by valence bond theory of complexes. The compound is diamagnetic as there is no unpaired electron present.

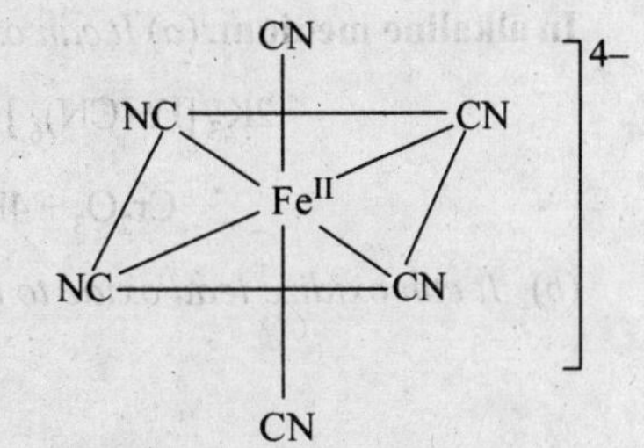

Fig. 18.11. Octahedral structure of $[Fe(CN)_6]^{4-}$ ion.

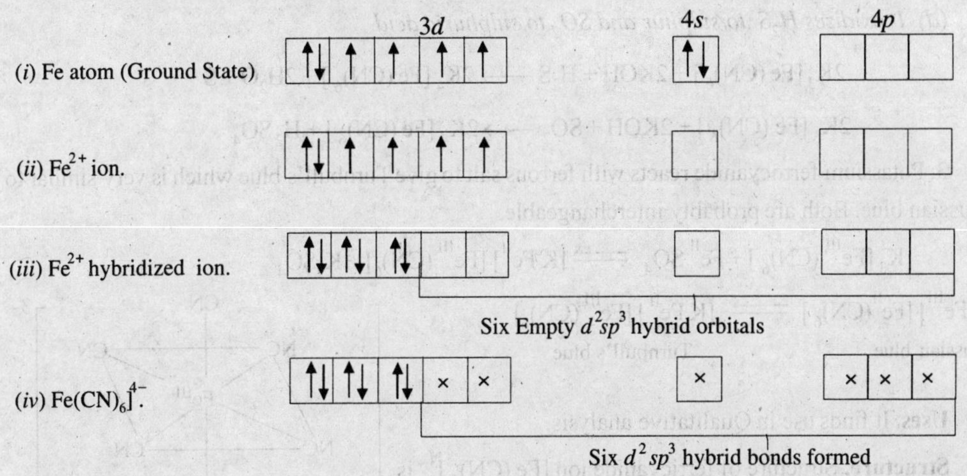

Q. 15. Describe preparation, properties and structure of potassium ferricyanide.

Ans. Preparation. 1.On passing chlorine gas through a solution of potassium ferrocyanide, potassium ferricyanide is obtained.

$$2K_4[Fe(CN)_6] + Cl_2 \longrightarrow 2K_3[Fe(CN)_6] + 2KCl$$

The ferricyanide is extracted from the mixture by fractional crystallisation.

Properties. 1. It forms dark-red anhydrous crystals, giving brown yellow solution with water.

2. **Oxidizing properties.** It is a powerful oxidizing agent both in acid and alkaline media.

Oxidation in Acidic Medium

(*a*) It *oxidizes H_2S to sulphur*

$$3H_2S + 6K_3[Fe(CN)_6] \longrightarrow 4K_4[Fe^{II}(CN)_6] + [K_2Fe^{II}][Fe^{II}(CN)_6] + 6HCN + 3S$$

(*b*) *It evolves chlorine from hydrochloric acid.*

$$K_3[Fe(CN)_6] + 3HCl \longrightarrow 3KCl + H_3[Fe(CN)_6]$$

$$2H_3[Fe(CN)_6] + 2HCl \longrightarrow 2H_4[Fe(CN)_6] + Cl_2\uparrow$$

In alkaline medium. (*a*) *It can oxidize chromic oxide to potassium chromate.*

$$2K_3[Fe(CN)_6] + 2KOH \longrightarrow 2K_4[Fe^{II}(CN)_6] + H_2O + O$$

$$Cr_2O_3 + 4KOH + 3O \longrightarrow 2K_2CrO_4 + 2H_2O$$

(*b*) *It can oxidize lead oxide to dioxide and manganous oxide to dioxide.*

$$PbO + O \longrightarrow PbO_2$$

$$MnO + O \longrightarrow MnO_2$$

(*c*) *Hydrogen peroxide is reduced to water and oxygen.*

$$2K_3[Fe(CN)_6] + H_2O_2 + 2KOH \longrightarrow 2K_4[Fe(CN)_6] + H_2O + O_2$$

(*d*) *It oxidizes H_2S to sulphur and SO_2 to sulphuric acid.*

$$2K_3[Fe(CN)_6] + 2KOH + H_2S \longrightarrow 2K_4[Fe(CN)_6] + 2H_2O + S$$

$$2K_3[Fe(CN)_6] + 2KOH + SO_2 \longrightarrow 2K_4[Fe(CN)_6] + H_2SO_4$$

3. Potassium ferrocyanide reacts with ferrous salt to give Turnbull's blue which is very similar to Prussian blue. Both are probably interchangeable.

$$K_3[Fe^{III}(CN)_6] + Fe^{II}SO_4 \rightleftharpoons [KFe^{II}][Fe^{III}(CN)_6] + K_2SO_4$$

$$\underset{\text{Prussian blue}}{[KFe^{III}][Fe^{II}(CN)_6]} \rightleftharpoons \underset{\text{Turnbull's blue}}{[KFe^{II}][Fe^{III}(CN)_6]}$$

Uses. It finds use in Qualitative analysis.

Structure. Structure of ferricyanide ion $[Fe(CN)_6]^{3-}$ is octahedral (Fig. 18.12). This structure is explained below. The compound is paramagnetic because of the presence of one unpaired electron.

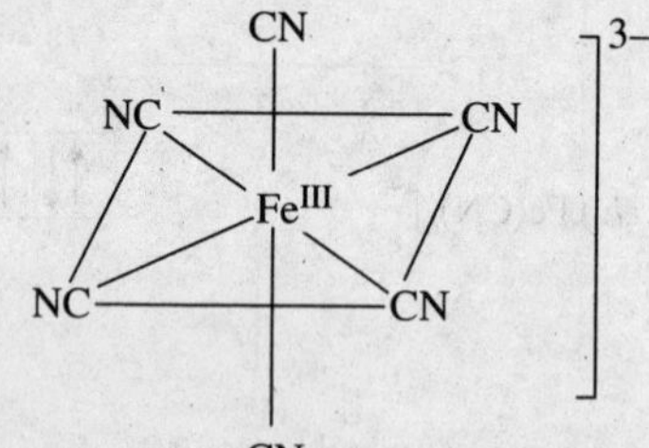

Fig. 18.12. Octahedral structure of $[Fe(CN)_6]^{3-}$ ion.

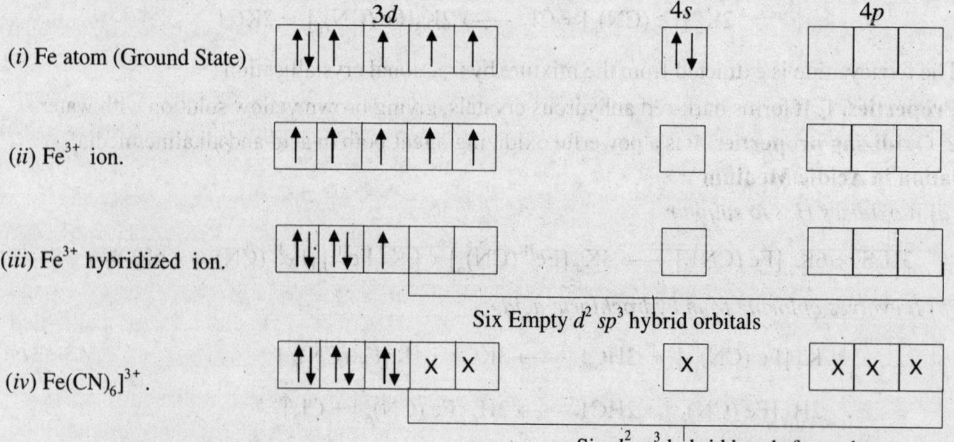

Q. 16. What happens when (give equations also)

(*i*) H_2S is passed through acidified solution of $K_3Fe(CN)_6$?

(*ii*) Alkali is added to a solution of $KMnO_4$?

Ans. See Qs. 15 and 12.

Q. 17. Assign reasons for the following:

(*i*) Chromium (II) acetate monohydrate is diamagnetic

(*ii*) Potassium ferrocyanide is used for testing copper salts.

(*iii*) Sodium cobaltinitrite is used in the detection of K^+ ions.

Ans. (*i*) Chromium (II) acetate monohydrate is diamagnetic because of coupling of spins of unpaired electrons on two chromium atoms in the dimeric structure of chromium acetate.

(*ii*) $K_4 Fe (CN)_6 + 2 Cu SO_4 \longrightarrow \underset{\substack{\text{Cupric ferrocyanide} \\ \text{(Reddish brown ppt.)}}}{Cu_2 [Fe (CN)_6]} + 2K_2 SO_4$

(*iii*) $Na_3 [Co (NO_2)_6] + K^+ \longrightarrow \underset{\text{Yellow ppt.}}{K [Co (NO_2)_6]} + Na^+$

Q. 18. Give one method of preparation of $K_4Fe(CN)_6$ and $K_3Fe(CN)_6$. What happens when $K_4Fe(CN)_6$] reacts with (*i*)H_2S and (*ii*) SO_2 in alkaline medium?

Ans. See. Q. 14.

Q. 19. Describe preparation, properties, uses and structure of Prussian blue, $KFe^{III}[Fe^{II}(CN)_6]$, Turnbull's blue and Berlin green.

Ans. ***Preparation Prussian blue.*** It is prepared by the action of Fe^{3+} ion ($FeCl_3$) with potassium ferrocyanide solution.

$$Fe Cl_3 + K_4 [Fe (CN)_6 \longrightarrow \underset{\text{Prussian blue}}{KFe^{III} [Fe^{II} (CN)_6]} + 3KCl$$

The reaction may also be written as

$$4Fe Cl_3 + 3K_4 [Fe (CN)_6] \longrightarrow Fe_4^{III} [Fe^{II} (CN)_6] + 12 KCl$$

Properties. 1. It has deep blue colour and is soluble in water. The deep blue colour is due to absorption of visible light causing electron transfer between Fe (II) and Fe (III) states.

2. It undergoes oxidation with nitric acid or chlorine and forms green pigment known as Berlin green having the formula $Fe^{III} [Fe(CN)_6]$.

Uses. It is used

1. as a blue dye for fabrics.
2. as a dye for blue print paper and for ink.

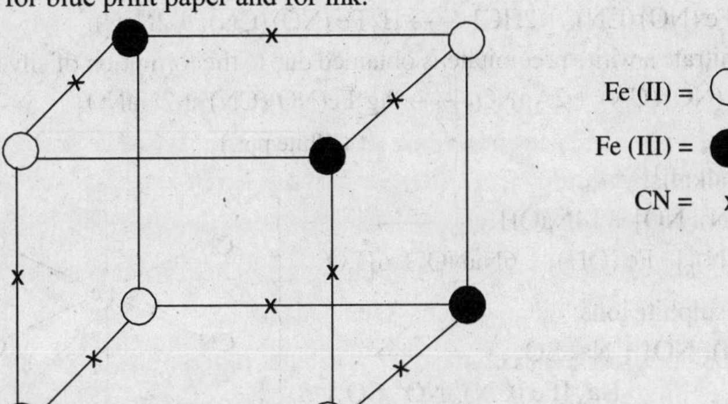

Fig. 18.13. Structure of $KFe[Fe(CN)_6]$, Prussian blue.

Berlin's green

when $KFe^{III}[Fe^{II}(CN)_6]$ is oxidised by means of Cl_2 or HNO_3, it gives a green pigment called Berlin's green, $Fe^{III}[Fe^{III}(CN)_6]$ ferriferricyanide.

$$2KFe^{III}[Fe^{II}(CN)_6] + Cl_2 \longrightarrow 2KCl + \underset{\text{Berlin's green}}{Fe^{III}[Fe^{III}(CN)_6]}$$

Structure. It has simple cubic structure in which iron atoms occupy each corner of the cube and CN^- ions occupy the middle of each side of the cube (Fig. 18.13). In this arrangement, half of the iron atoms have six carbon atoms as the nearest neighbours while the other half have six nitrogen atoms as the nearest neighbours. Potassium atoms occupy only half the cubic centres.

Turnbull's blue

Another compound chemically identical to Prussian blue is Turnbull's blue. It is obtained by the action of potassium ferricyanide with aqueous ferrous sulphate.

$$4K_3[Fe^{III}(CN)_6] + 3Fe^{II}SO_4 \longrightarrow \underset{\text{Turnbull's blue}}{Fe_4^{III}[Fe^{II}(CN)_6]_3} + 6KCN + 3K_2SO_4$$

Recent studies accept Prussian blue and Turnbull's blue as same compound.

Q. 20. Describe the preparation and properties of sodium nitroprusside, $Na_2[Fe(NO)(CN)_5]\cdot 2H_2O$. *(Kanpur, 2000; Purvanchal. 2003)*

Ans. Preparation. 1. It is prepared by warming a solution of potassium ferrocyanide with 50% HNO_3 till a brown solution is obtained. Potassium nitrate crystallises out on cooling. Mother liquor after neutralization with soda ash is filtered and concentrated. Sodium nitroprusside is obtained on cooling the filtrate.

$$K_4Fe(CN)_6 + 4HNO_3 \longrightarrow H_4[Fe(CN)_6] + 4KNO_3$$

$$3H_4[Fe(CN)_6] + HNO_3 \longrightarrow 3H_3[Fe(CN)_6] + NO + 2H_2O$$

$$H_3[Fe(CN)_6] + [NO \longrightarrow H_2[Fe(NO)(CN)_5] + HCN$$

$$H_2[Fe(NO)(CN)_5] + Na_2CO_3 \longrightarrow Na_2[Fe(NO)(CN)_5] + H_2O + CO_2\uparrow$$

2. It can be obtained by passing nitric oxide into an acidified solution of sodium ferricyanide.

$$Na_3Fe(CN)_6 + NO \longrightarrow Na_2[Fe(NO)(CN)_5] + NaCN$$

Properties. 1. Sodium nitroprusside exists as red rhombic shaped crystals.

2. It is soluble in water.
3. It gives a purple colour with freshly prepared solution of an alkali sulphide.

$$Na_2Fe(NO)(CN)_5 + Na_2S \longrightarrow \underset{\text{Purple colour}}{Na_4Fe(NOS)(CN)_5}$$

4. On treatment with hydrochloric acid, it gives the following reaction:

$$Na_2Fe(NO)(CN)_5 + 2HCl \longrightarrow H_2Fe(NO)(CN)_5 + 2NaCl$$

5. With silver nitrate a white precipitate is obtained due to the formation of silver nitroprusside.

$$Na_2Fe(NO)(CN)_5 + 2AgNO_3 \longrightarrow \underset{\text{(White ppt.)}}{Ag_2Fe(NO)(CN)_5} + 2NaNO_3$$

6. Action with alkalis:

$$6Na_2[Fe(CN)_5NO] + 14NaOH \longrightarrow 5Na_4[Fe(CN)_6] + Fe(OH)_2 + 6NaNO_3 + 6H_2O$$

7. Action with sulphite ions:

$$Na_2[Fe(CN)_5NO] + Na_2SO_3 \longrightarrow \underset{\text{Red colour}}{Na_4[Fe(CN)_5NO\cdot SO_3]}$$

$$\left[\begin{array}{c} CN \\ CN\quad CN \\ Fe \\ CN\quad CN \\ NO \end{array}\right]^{2-}$$

Fig. 18.14. Structure of sodium nitroprusside.

Uses. It finds use in

(*i*) detection of sulphide ion in inorganic analysis.

(*ii*) detection of ketones as it gives a violet colour with ketones.

Structure. The six groups are attached to iron atom in an octahedral manner as shown in Fig. 18.14.

Q. 21. Describe the preparation and properties and structure of sodium cobaltinitrite $Na_3[Co(NO_2)_6]$.

Ans. Preparation. 1. It can be prepared by mixing cobalt nitrate with sodium nitrite solution. A 5% solution of acetic acid is added followed by a current of air.

$$Co(NO_3)_2 + 5NaNO_2 + 2HNO_2 \longrightarrow Na_3[Co(NO_2)_6] + 2NaNO_3 + NO + H_2O$$

The solution is filtered and alcohol is added to it. Sodium cobaltinitrite is precipitated.

Properties. 1. It is a yellow powder readily soluble in water.

2. It is used as an analytical reagent for the analysis of potassium salts.

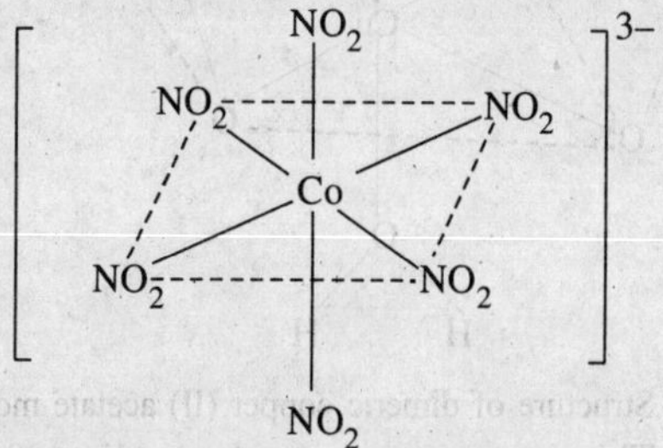

Fig. 18.15. Structure of cobaltinitrite ion

$$Na_3[Co(NO_2)_6] + K^+ \longrightarrow K[Co(NO_2)_6] + Na^+$$
Yellow ppt.

3. $2NH_4Cl + Na_3[Co(NO_2)_6] \longrightarrow (NH_4)_3[Co(NO_2)_6] + 3NaCl$

Structure. Cobaltinitrite ion (Hexanitrito cobalate (III) ion) has octahedral structure as illustated in Fig. 18.15.

Q. 22. Give IUPAC name of sodium cobaltinitrite and give any three of its important properties.

Ans. IUPAC name of sodium cobaltinitrite is sodium hexanitritocobaltate (III).

For properties, see Q. 21.

Q. 23. Give the preparation, properties and structure of copper acetate monohydrate, $Cu(CH_3COO)_2.H_2O$.

Ans. Preparation. It is prepared by dissolving copper carbonate or copper oxide in acetic acid and evaporating the solution to crystallisation.

$$2CuCO_3 + 4CH_3COOH \longrightarrow [Cu(CH_3COO)_2 . H_2O]_2 + 2CO_2$$

$$CuO + CH_3COOH \longrightarrow Cu(CH_3COO)_2 . H_2O$$

Properties. 1. It is a dark green crystalline solid soluble in alcohol and water.

2. It is antiferromagnetic having magnetic moment of 1.43 BM at 25°C.

3. It is dimeric in crytalline form and undergoes dissociation in solution.

Structure. Copper (II) monohydrate exists as dimer and its structure is shown in Fig. 18.16. The magnetic moment of the compound is only 1.43 BM which is less than the normal value of 1.75–2.2

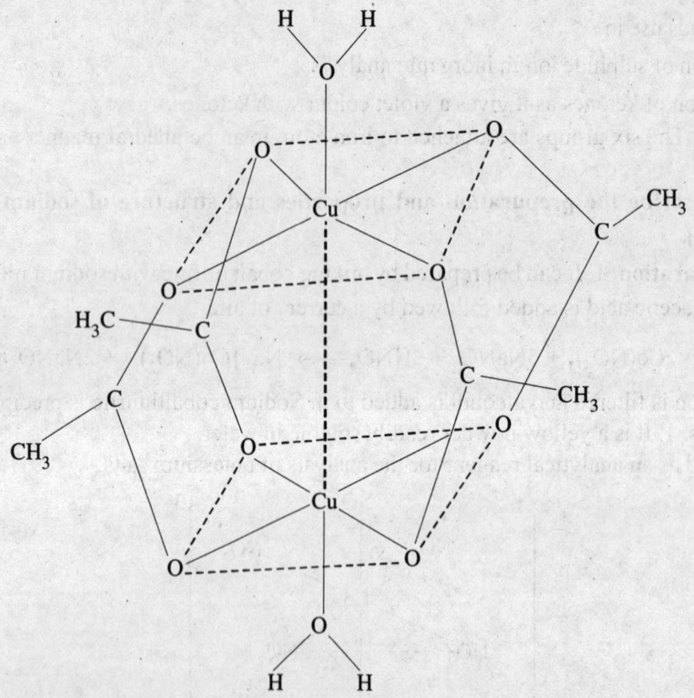

Fig. 18.16. Structure of dimeric copper (II) acetate monohydrate.

BM for one unpaired electron. This suggests some weak coupling of spins of unpaired electrons on the two copper atoms in the dimeric structure. In its structure, the acetate groups act as a bidentate bridging groups between the two copper atoms. Therefore, each copper atom is surrounded by four oxygen atoms in a square planar arrangement. The fifth position around each Cu atom is occupied by the oxygen atom from a water molecule. The other copper atom occupies the sixth of the octahedral positions. The structure is similar to that of chromous acetate.

Q. 24. Describe the preparation, properties and uses of potassium dichromate ($K_2Cr_2O_7$).

(*Meerut 2000; Kanpur, 2000*)

Ans. Preparation. 1. It is manufactured from chromite ore through the following steps:

(*i*) Preparation of sodium chromate. Finely powdered chromite is mixed with soda-ash and quick lime and then roasted in a furnace in presence of excess of air.

$$2Fe(CrO_2)_2 + O \longrightarrow Fe_2O_3 + 2Cr_2O_3$$

$$[Cr_2O_3 + 2Na_2CO_3 + 3O \longrightarrow 2NaCrO_4 + 4CO_2] \times 2$$

$$2Fe(CrO_2)_2 + 4Na_2CO_3 + 7O \longrightarrow 4Na_2CrO_4 + Fe_2O_3 + 4CO_2$$

or

$$4Fe(CrO_2)_2 + 8Na_2CO_3 + 7O_2 \longrightarrow 8Na_2CrO_4 + 2Fe_2O_3 + 8CO_2$$

Function of quick lime is to keep the mass porous and thus help oxidation by air. The yellow mass is extracted with water when ferric oxide is left behind.

(*ii*) **Conversion of chromate into dichromate.** Sodium chromatic solution obtained in (*i*) *is* treated with conc. H_2SO_4 when it is converted into sodium dichromate.

$$2Na_2CrO_4 + H_2SO_4 \longrightarrow Na_2Cr_2O_7 + Na_2SO_4 + H_2O$$

On concentration the less soluble sodium sulphate, $Na_2SO_4 \cdot 10H_2O$ crystallizes out. This is filtered hot and allowed to cool when sodium dichromate, $Na_2Cr_2O_7$, separates on standing.

***(iii)* Conversion of sodium dichromate into potassium dichromate.** Hot concentrated solution of sodium dichromate is treated with a calculated quantity of KCl when $K_2Cr_2O_7$, being less soluble, crystallizes out on cooling.

$$Na_2Cr_2O_7 + 2KCl \longrightarrow K_2Cr_2O_7 + 2NaCl$$

2. It is prepared by adding a calculated quantity of H_2SO_4 to a saturated solution of potassium chromate.

$$2K_2CrO_4 + H_2SO_4 \longrightarrow K_2Cr_2O_7 + K_2SO_4 + H_2O$$

Properties. 1. Potassium dichromate is obtained as orange red crystals with metling point 671K. It is soluble in water.

2. *It decomposes on heating as follows:*

$$4K_2Cr_2O_7 \longrightarrow 4K_2CrO_4 + 2Cr_2O_3 + 3O_2$$

3. *With concentrated sulphuric acid.*

(*i*) in the cold, CrO_3 is formed.

$$K_2Cr_2O_7 + 8H_2SO_4 \longrightarrow 2CrO_7 + 2KHSO_4 + H_2O$$

(*ii*) *On heating, oxygen is evolved.*

$$2K_2Cr_2O_7 + 8H_2SO_4 \longrightarrow 2K_2SO_4 + 2Cr_2(SO_4)_3 + 8H_2O + 3O_2$$

4. With an alkali, it gives a chromate. Chromates and dichromates are interconvertible in solution. Even weak acids cause the change from yellow chromate to orange dichromate, while this is reversed by addition of alkali or a carbonate solution.

$$2CrO_4^{2-} + 2H^+ \longrightarrow 2HCrO_4^- \rightleftharpoons H_2O + Cr_2O_7^{2-}$$

$$HCrO_4^- + OH^- \longrightarrow H_2O + CrO_4^{2-}$$

$$2HCrO_4^- + CO_3^{2-} \longrightarrow 2CrO_4^{2-} + H_2O + CO_2$$

5. Potassium dichromate solution gives insoluble chromates with soluble salts of lead and barium.

$$2Pb(NO_3)_2 + K_2Cr_2O_7 + H_2O \longrightarrow \underset{\text{Yellow ppt.}}{2PbCrO_4} + 2KNO_3 + 2HNO_3$$

6. Solid $K_2Cr_2O_7$ when heated with conc. H_2SO_4 and a chloride gives orange-red vapour of a volatile oily liquid, **chromyl chloride (CrO_2Cl_2).**

$$K_2Cr_2O_7 + 2H_2SO_4 \longrightarrow 2KHSO_4 + 2CrO_3 + H_2O$$

$$NaCl + H_2SO_4 \longrightarrow NaHSO_4 + HCl\,] \times 4$$

$$CrO_3 + 2HCl \longrightarrow CrO_2Cl_2 + H_2O\,] \times 2$$

$$K_2Cr_2O_7 + 4NaCl + 6H_2SO_4 \longrightarrow 2CrO_2Cl_2 + 2KHSO_4 + 4NaHSO_4 + 3H_2O$$

7. Potassium dichromate is a strong oxidising agent. In the neutral or acidic medium, it furnishes nascent oxygen and thus acts as an excellent oxidising agent.

$$K_2Cr_2O_7 + 4H_2O \longrightarrow 2KOH + 2Cr(OH)_3 + 3O$$

$$K_2Cr_2O_7 + 4H_2SO_4 \longrightarrow K_2SO_4 + Cr_2(SO_4)_3 + 4H_2O + 3O$$

(*i*) *It liberates iodine from KI solution.*

$$K_2Cr_2O_7 + 4H_2SO_4 \longrightarrow K_2SO_4 + Cr_2(SO_4)_3 + 4H_2O + 3O$$

$$2KI + H_2SO_4 \longrightarrow K_2SO_4 + 2HI\,] \times 3$$

$$2HI + O \longrightarrow H_2O + I_2\,] \times 3$$

$$K_2Cr_2O_7 + 7H_2SO_4 + 6KI \longrightarrow K_2SO_4 + Cr_2(SO_4)_3 + 7H_2O + 3I_2$$

or $$Cr_2O_7^{2-} + 6I^- + 14H^+ \longrightarrow 2Cr^{3+} + 7H_2O + 3I_2$$

(*ii*) *It oxidises ferrous sulphate to ferric sulphate.*

$$K_2Cr_2O_7 + 4H_2SO_7 \longrightarrow K_2SO_4 + Cr_2(SO_4)_3 + 3Fe_2(SO_4)_3 + 4H_2O + 3(O)$$

$$2FeSO_4 + H_2SO_4 + O \longrightarrow Fe_2(SO_4)_3 + H_2O\,] \times 3$$

$$K_2Cr_2O_7 + 7H_2SO_4 + 6FeSO_4 \longrightarrow K_2SO_4 + Cr_2(SO_4)_3 + 3Fe_2(SO_4)_3 + 7H_2O$$

or $$Cr_2O_7^{2-} + 14H^+ + 6Fe^{2+} \longrightarrow 2Cr^{3+} + 7H_2O + 6Fe^{3+}$$

(*iii*) *It oxidizes SO_2 to H_2SO_4.*

$$Cr_2O_7^{2-} + 2H^+ + 3SO_2 \longrightarrow 2Cr^{3+} + 3SO_4^{2-} + H_2O$$

(*iv*) *It oxidizes H_2S to S.*

$$Cr_2O_7^{2-} + 8H^+ + 3H_2S \longrightarrow 2Cr^{3+} + 3S + 7H_2O$$

(*v*) *It oxidizes sulphite ion $(SO_3)^{2-}$ to sulphate ion (SO_4^{2-})*

$$Cr_2O_7^{2-} + 3SO_3^{2-} + 8H^+ \longrightarrow 2Cr^{3+} + 3SO_4^{2-} + 4H_2O$$

(*vi*) *It oxidizes alcohol first to aldehydes and then to acids.*

$$\underset{\text{Ethyl alcohol}}{CH_3CH_2OH} \xrightarrow{[O]} \underset{\text{Acetaldehyde}}{CH_3CHO} \xrightarrow{[O]} \underset{\text{Acetic acid}}{CH_3COOH}$$

Uses. It is used:

(*i*) For preparation of other chromium compounds like chrome-alum, chrome red, zinc chromate and lead chromate.

(*ii*) As an oxidizing agent in organic chemistry.

(*iii*) In chrome tanning.

(*iv*) In dyeing and calico printing.

(*v*) As a volumetric reagent for esstimation of ferrous ions, iodine etc.

(*vi*) Mixture of $K_2Cr_2O_7$ and conc. H_2SO_4 is used as a cleaning agent for glasswares.

(*vii*) In photography.

Q. 25. Describe the structure of dichromate ion.

Ans. The dichromate ion $(Cr_2O_7)^{2-}$ consists of two CrO_4 tetrahedra linked through one common oxygen atom at a common corner. Each of the Cr atoms is at the centre of a tetrahedron (Fig. 18.17). The Cr–O–Cr bond angle is 115° and the common O–Cr bond distance is 1.9 Å. The other O–Cr bonds are shorter being 1.6 Å each as shown in Fig. 18.17.

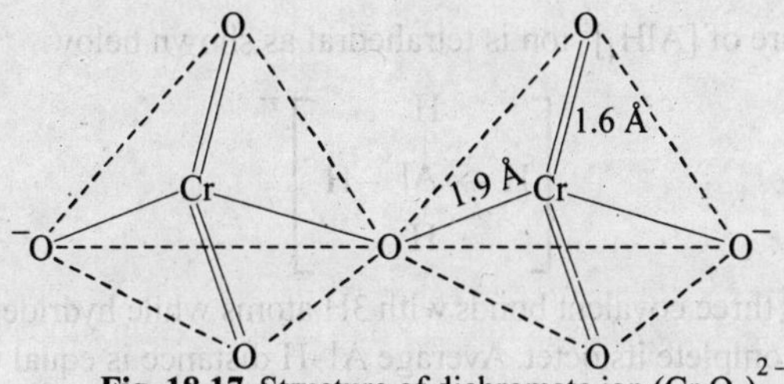

Fig. 18.17. Structure of dichromate ion $(Cr_2O_7)^{2-}$.

Q. 26. Describe the preparation, properties and structure of Lithium aluminium hydride.

(*Meerut, 2000; Kanpur, 2001; Vidyasagar, 2007*)

Ans. Preparation. (*i*) By the direct synthesis of elements.

$$Li + Al + 2H_2 \longrightarrow Li\,[Al\,H_4\,]$$

(*ii*) By the direct combination of Lithium hydride and aluminium chloride in ether solution.

$$4LiH + A1C1_3 \xrightarrow{\text{ether}} Li\,[Al\,H_4\,] + 3Li\,C1_3$$

(*iii*) By double decomposition. LiC1 and sodium aluminium hydride react to give lithium aluminium hydride.

$$LiC1 + Na\,[A1H_4\,] \longrightarrow NaC1 + Li\,[Al\,H_4]$$

Properties. (*i*) It is non-volatile, crystalline solid. It is stable below 398 K and is soluble in ether.

(*ii*) **Reducing properties.** It is a strong reducing agent and reduces certain halides of groups III, IV and V to covalent hydrides of the elements concerned.

$$4BF_3 + 3Li\,[Al\,H_4] \longrightarrow \underset{\textbf{Diborane}}{2B_2\,H_4} + 3LiF + 3A1F_3$$

$$A1C1_3 + 3Li\,[A1H_4] \longrightarrow 4A1H_3 + 3LiCl$$

$$SiC1_4 + Li\,[A1H_4] \longrightarrow \underset{\text{Silane}}{Si\,H_4} + LiC1 + AlC13$$

(*iii*) Reduction of organic compounds. It is an ideal reducing agent for aldehydes, ketones and carboxylic acids.

$$RCHO \xrightarrow{LiAlH_4} \underset{\textbf{Alcohol}}{RCH_2\,OH}$$

$$RCOR \xrightarrow{LiAlH_4} \underset{\textbf{Sec. alcohol}}{RCHOHR}$$

$$RCOOH \xrightarrow{LiAlH_4} \underset{\text{Alkane}}{RCH_3} + 2H_2O$$

(*iv*) Hydrolysis. It is hydrolysed by water evolving hydrogen.

$$Li\,[Al\,H_4\,]\,2H_2O \longrightarrow LiA1O_2\,4H_2$$

Uses. It is used as powerful reducing agent for both organic and inorganic substances.

Structure. The structure of $[AlH_4]^-$ ion is tetrahedral as shown below.

$$\left[\begin{array}{c} H \\ | \\ H \rightarrow Al - H \\ | \\ H \end{array}\right]^-$$

Here aluminium forms three covalent bonds with 3H atoms while hydride ion, $H:^-$ donates a pair of electrons to Al atom to complete its octet. Average Al–H distance is equal to 1.55 Å.

Q. 27. Describe the preparation, properties, uses and structure of aluminium chloride ($AlCl_3$).

Ans. Preparation. (*i*) Aluminium chloride is obtained by dissolving alumina Al_2O_3 or $Al(OH)_3$ in hydrochloric acid and crystallizing the solution.

$$Al(OH)_3 + 3HCl \longrightarrow AlCl_3 + 3H_2O$$

(*ii*) By passing carbonyl chloride over heated Al_2O_3.

$$Al_2O_3 + 3COCl_2 \longrightarrow 2AlCl_3 + 3CO_2$$

(*iii*) It may be obtained by heating a mixture of Al_2O_3 and carbon in a current of chlorine at about 1273 K.

$$Al_2O_3 + 3C + 3Cl_2 \longrightarrow 2AlCl_3 + 3CO$$

(*iv*) Anhydrous aluminium chloride is obtained by passing dry Cl_2 or HCl gas over heated aluminium.

$$2Al + 3Cl_2 \longrightarrow 2AlCl_3$$

$$2Al + 6HCl \longrightarrow 2AlCl_3 + 3H_2$$

Aluminium chloride volatilizes and condenses in the receiver.

Properties. (*i*) Anhydrous $AlCl_3$ is a white crystalline substance which absorbs moisture on exposure to air. It sublimes below 473 K. The vapour density of the compound near its sublimation point corresponds with the formula Al_2Cl_6. At higher temperature, it dissociates into simple molecules.

$$Al_2Cl_6 \rightleftharpoons 2AlCl_3$$

(*ii*) It dissolves in water and the solution formed is strongly acidic due to hydrolysis as follows:

$$AlCl_3 + 3H_2O \longrightarrow Al(OH)_3 + 3HCl$$

(*iii*) It absorbs ammonia forming $AlCl_3.6NH_3$.

(*iv*) It is also soluble in benzene, CS_2 etc.

Uses. (*i*) $AlCl_3$ (Anhydrous aluminium chloride) is used as a catalyst in many organic reactions, *e.g.*, Friedel and Craft's reaction.

(*ii*) In the refining of mineral oils.

(*iii*) In the manufacture of dyes, drugs and perfumes.

Structure. Aluminium chloride exists as dimeric in vapour state (up to 673°K) and in non-polar solvents. The configuration of Cl atoms about Al is roughly tetrahedral. Thus each Al atom is tetrahedrally bonded to four Cl atoms; two of these Cl atoms are each bonded to two Al atoms. In Al_2Cl_6 dimer, one Cl atom of each monomeric covalent molecule donates a pair of electrons to be shared with the Al atom of the other molecule and thus two "Cl-bridges" are formed. In this way each Al atom completes its octet.

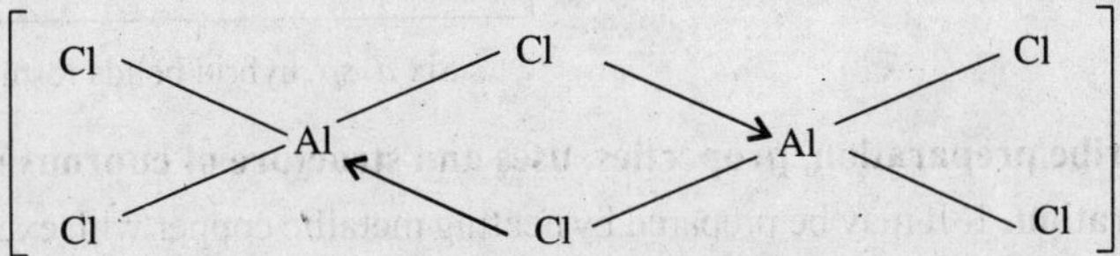

Q. 28. Describe the preparation, properties and structure of hexamine cobalt (III) chloride $[Co(NH_3)_6]Cl_3$.

Ans. Preparation. 1. It is prepared by passing a current of air through an aqueous solution of cobaltous chloride containing ammonium chloride and ammonia for several hours. The reaction is catalysed by activated charcoal.

$$4CoCl_2 + 4NH_4Cl + 20NH_3 + O_2 \longrightarrow 4\,[Co\,(NH_3)_6\,]\,Cl_3 + 2H_2O$$

2. It is obtained by adding 20 volume hydrogen peroxide to a solution containing ammonium chloride, cobaltous chloride, ammonia and activated charcoal cooled to 283 K. The mixture solution is heated to 323-333 K and filtered when crystals separate out on cooling.

$$2CoCl_2 + 2NH_4Cl + 10NH_3 + H_2O_2 \longrightarrow 2\,[Co\,(NH_3)_6\,]Cl_3 + 2H_2O$$

Properties. 1. It is a golden brown crystalline compound and is soluble in water.

2. Hexammine cobaltic chloride is very sensitive to hydroxyl ion and is decomposed to cobaltic oxide, Co_2O_3.

$$2\,[Co\,(NH_3)_6\,]\,Cl_3 + 6\,NaOH \longrightarrow Co_2O_3 + 6NH_3 + 6NaCl + 3H_2O$$

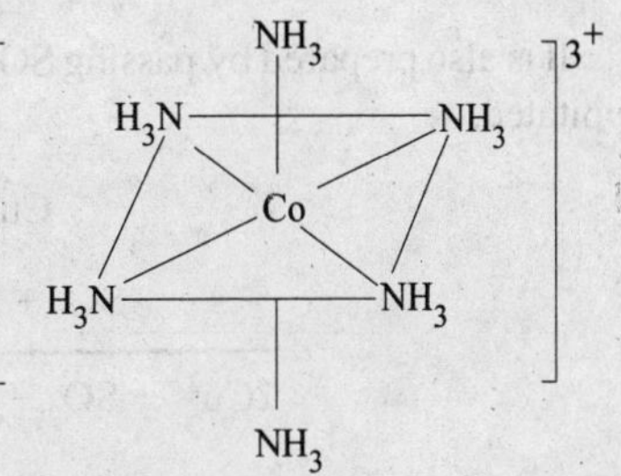

Fig. 18.18. Octahedral structure of $[Co(NH_3)_6]^{3+}$ ion.

3. It is extremely stable and is not decomposed even by conc. H_2SO_4 which converts chloride to sulphate.

$$2\,[Co\,(NH_3)_6\,]\,Cl_3 + 3H_2SO_4 \longrightarrow [Co\,(NH_3)_6]_2\,(SO_4)_3 + 6HCl$$

Structure. $[Co\,(NH_3)_6]^{3+}$ ion is an octahedral complex formed by d^2sp^3 hybridisation (Fig. 18.18). The complex has no unpaired electron and therefore, it is diamagnetic.

	3d					4s	4p		
(*i*) Co Atom (Ground State)	↑↓	↑↓	↑	↑	↑	↑↓			
(*ii*) Co^{3+} ion	↑↓	↑	↑	↑	↑				
(*iii*) Co^{3+} hybridized	↑↓	↑↓	↑↓						

Six empty d^2sp^3 hybrid orbitals

(*iv*) $[Co(NH_3)_6]^{3-}$

↑↓	↑↓	↑↓	×	×		×		×	×	×

Six d^2sp^3 hybrid bonds formed

Q. 29. Describe preparation, properties, uses and structure of cuprous chloride (CuCl).

Ans. Preparation. 1. It may be prepared by heating metallic copper with excess of conc. HC1 in presence of a little $KClO_3$ or HNO_3.

$$Cu + 2HCl + O \xrightarrow{KClO_3} CuCl_2 + H_2O$$

$$CuCl_2 + Cu \longrightarrow 2CuCl$$

The reaction mixture is poured in water when cuprous chloride is precipitated.

2. It is obtained by heating cupric oxide, cupric chloride or copper sulphate with excess of copper turnings and conc. hydrochloric acid.

$$CuO + Cu + 2HCl \longrightarrow 2CuCl + H_2O$$

$$CuCl_2 + Cu \longrightarrow 2CuCl$$

$$CuSO_4 + Cu + 2HCl \longrightarrow 2CuCl + H_2SO_4$$

3. It is also formed when $CuCl_2$ is strongly heated (710 K).

$$2CuCl_2 \longrightarrow 2CuCl + Cl_2$$

4. It is also prepared by passing SO_2 through a solution of $CuSO_4$ containing NaC1 when CuCl is precipitated.

$$Cu^{2+} + e^- \longrightarrow Cu^+] \times 2$$

$$SO_2 + 2H_2O \longrightarrow SO_4^{2-} + 4H^+ + 2e^-$$

$$2Cu^{2+} + SO_2 + 2H_2O \longrightarrow 2Cu^+ + SO_4^{2-} + 4H^+$$

$$Cu^+ + Cl^- \longrightarrow CuCl$$

Properties. 1. Action of heat. On heating white CuCl becomes deep blue at 450 K and melts to a deep green liquid.

2. It is a white solid almost insoluble in water but soluble in conc. HCl with which it gives the complex ions.

$$CuCl + 3Cl^- \rightleftharpoons [CuCl_4]^{3-}$$

3. Action with ammonia. It dissolves in aqueous ammonia forming a colourless solution due to the formation of complex $[Cu(NH_3)_4]$ Cl.

4. Action of air. In air it is slowly oxidized to green basic cupric chlroide.

$$3CuO.CuCl_2.3H_2O$$

5. Action with acetylene. It also reacts with acetylene to give a red precipitate of cuprous acetylide, Cu_2C_2.

6. Action with hydrogen sulphide. Hydrogen sulphide reacts with CuCl to give a black precipitate of cuprous sulphide.

$$2CuC1 + H_2S \longrightarrow Cu_2S + 2HCl$$

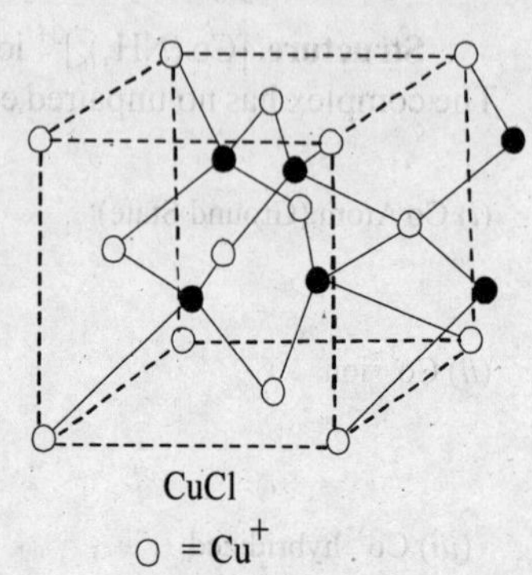

Fig. 18.19. Structure of CuCl.

7. Action with sodium hydroxide. Cuprous chloride reacts with NaOH forming a yellow precipitate of cuprous oxide which turns red on boiling.

$$2CuCl + 2NaOH \longrightarrow Cu_2O + 2NaCl + H_2O$$

8. Action of oxidizing agent. When treated with oxidizing agents, it is converted into cupric chloride

$$2CuCl + 2HCl + (O) \longrightarrow 2CuCl_2 + H_2O$$

Uses. Cuprous Chloride is used in the gas analysis for absorbing carbon monoxide and acetylene.

Structure. CuCl has a covalent, tetrahedral zinc blende [ZnS] structure with the metal atom tetrahedrally surrounded by halogen atoms. Cu^+ ions occupy half of the suitable tetrahedral sites in a face centred cubic lattice formed by Cl^- ions. Thus each Cl^- ion is also surrounded tetrahedrally by four Cu^+ ions. (Fig. 18.19).

In vapour state, CuCl is polymeric and contains a cyclic ring of alternating Cu and Cl atoms.

Q. 30. Write the reactions involved in chromyl chloride test. *(Calcutta, 2007)*

Ans. Chromyl chloride test is performed to identify chloride in a sample. The sample is heated along with solid potassium dichromate and concentrated sulphuric acid when red vapours of chromyl chloride are evolved. The reactions may be represented as:

$$K_2Cr_2O_7 + 2H_2SO_4 \longrightarrow 2KHSO_4 + 2CrO_3 + H_2O$$
$$[NaCl + H_2SO_4 \longrightarrow NaHSO_4 + HCl] \times 4$$
$$[CrO_3 + 2HCl \longrightarrow CrO_2Cl_2 + H_2O] \times 2$$

$$\underset{\text{Pot. dichromate}}{K_2Cr_2O_7} + 4NaCl + 6H_2SO_4 \longrightarrow \underset{\text{Chromyl chloride}}{2CrO_2Cl_2} + 2KHSO_4 + 4NaHSO_4 + 3H_2O$$

Vapours of chromyl chloride are passed through water. A yellow solution is obtained. To the yellow solution is added lead acetate.A yellow precipitate of lead chromate is formed.

$$CrO_2Cl_2 + H_2O \longrightarrow \underset{\substack{\text{Chromic acid}\\ \text{(yellow)}}}{H_2CrO_4} + 2HCl$$

$$H_2CrO_4 + (CH_3COO)_2Pb \longrightarrow \underset{\substack{\text{Lead chromate}\\ \text{(yellow ppt.)}}}{PbCrO_4} + 2CH_3COOH$$

Q. 31. Potassium dichromate solution oxidises KI into I_2 in presence of dilute HCl. Balance the equation for the above reaction by ion electron method. Find the equivalent weight of $K_2Cr_2O_7$. *(Calcutta, 2007)*

Ans. The balanced equation for the oxidation of KI into I_2 in presence of dil HCl is :

$$Cr_2O_7^{2-} + 6I^- + 14H^+ \longrightarrow 2Cr^{3+} + 3I_2 + 7H_2O$$

As six moles of iodide ions are converted into iodine with one mole of dichromate ion, the equivalent weight of potassium dichromate will be one sixth of the molecular weight.

$$\text{Equivalent weight of } K_2Cr_2O_7 = \frac{\text{Molecular wt.of } K_2Cr_2O_7}{6}$$

$$= \frac{(2\times 39) + (2\times 52) + (7\times 16)}{6}$$

$$= \frac{294}{6} = 49$$

THE *f*-BLOCK ELEMENTS

1	2	3	4	5	6	7	8	9	10	11	12	13	14	15	16	17	18
H																	He
Li	Be	Transition Elements										B	C	N	O	F	Ne
Na	Mg											Al	Si	P	S	Cl	Ar
K	Ca	21 Sc	22 Ti	23 V	24 Cr	25 Mn	26 Fe	27 Co	28 Ni	29 Cu	30 Zn	Ga	Ge	As	Se	Br	Kr
Rb	Sr	39 Y	40 Zr	41 Nb	42 Mo	43 Tc	44 Ru	45 Rh	46 Pd	47 Ag	48 Cd	In	Sn	Sb	Te	I	Xe
Cs	Ba	57 La*	72 Hf	73 Ta	74 W	75 Re	76 Os	77 Ir	78 Pt	79 Au	80 Hg	Tl	Pb	Bi	Po	At	Rn
Fr	Ra	89 Act	104 Ku														

f- block elements

Lanthanide series (4*f*)	58 Ce	59 Pr	60 Nd	61 Pm	62 Sm	63 Eu	64 Gd	65 Tb	66 Dy	67 Ho	68 Er	69 Tm	70 Yb	71 Lu
Actinide series (5*f*)	90 Th	91 Pa	92 U	93 Np	94 Pu	95 Am	96 Cm	97 Bk	98 Cf	99 Es	100 Fm	101 Md	102 No	103 Lr

LANTHANIDES

Q. 1. What are lanthanides? Discuss their position in the periodic table.

(*Lucknow, 2000; Bombay, 2002; Nagpur, 2008*)

Or

What are lanthanides? Why are they called inner transition elements?

(*Kurukshetra, 2002; Lucknow, 2002*)

Ans. The fifteen elements from lanthanum to lutetium (at numbers 57 to 71) are called lanthanides or rare earths (see periodic table). These elements involve the filling of 4*f* orbitals. Strictly speaking, the first member lanthanum (configuration $6s^2\ 5d^1 4f^0$) and the last member, lutetium (configuration $6s^2\ 5d^1 4f^{14}$) have no partially filled 4*f* orbital in their ground state yet they are considered to belong to lanthanide series, because their chemical properties are very closely similar to the other thirteen elements of the series.

These elements are also called the *inner transition elements* because the filling of electrons takes place in the 4*f* subshell, *i.e.*, two levels below the outer 5*p* and 6*s* subshells which are already filled in these elements. The 4*f* electrons aren't exposed to their surroundings as they don't extend out to the surface of the atoms or ions. As a result, the properties of lanthanides are very similar to one another and don't show much variation. Decrease in their radii as we move across from La to Lu is responsible for small variation in their properties.

Q. 2. Discuss the electronic configuration of lanthanides.

(*Punjab, 2000; Lucknow, 2002; Guwahati, 2002; Purvanchal, 2007*)

Ans. In lanthanum (57) the incoming electron enters the 5*d* subshell and therefore it has the configuration [Xe]: $6s^2\ 5d^1\ 4f^0$. However, as we move one step further, the 4*f* subshell becomes more stable than the 5*d* subshell and the next electron in cerium enters the 4*f* subshell. Cerium has the configuration [Xe]: $6s^2\ 4f^1 5d^1$. The filling of 4*f* subshell continues till we reach ytterbium which has the configuration [Xe]: $6s^2\ 4f^{14}\ 5d^0$. Now after 6*s* and 4*f* subshells have been filled, the 5*d* subshell is the next choice for filling electrons. At lutetium the electron enters the 5*d* subshell and it has configuration [Xe]: $6s^2\ 4f^{14}\ 5d^1$. The electronic configuraiton and oxidation states of lanthanides are listed in Table 19.1. In general, their electronic configuraiton may be written as:

$$[Xe]: 6s^2\ 4f^{0-14}\ 5d^{0-1}$$

Since the lanthanides have similar electronic configuration, they all exhibit +3 oxidation state and have similar chemical properties. Therefore they are all grouped together at one position of the periodic table.

Table 19.1. Electronic configuration of lanthanides

Atomic number	*Name*	*Symbol*	*Electronic configuration*	*Oxidation states*
57	Lanthanum	La	[Xe] $4f^0\ 5d^1\ 6s^2$	+3
58	Cerium	Ce	[Xe] $4f^1\ 5d^1\ 6s^2$	+3, +4
59	Praseodymium	Pr	[Xe] $4f^3\ 6s^2$	+3, +4
60	Neodymium	Nd	[Xe] $4f^4\ 6s^2$	+2, +3
61	Promethium	Pm	[Xe] $4f^5\ 6s^2$	+3
62	Samarium	Sm	[Xe] $4f^6\ 6s^2$	+2, +3
63	Europium	Eu	[Xe] $4f^7\ 6s^2$	+2, +3
64	Gadolinium	Gd	[Xe] $4f^7\ 5d^1\ 6s^2$	+3
65	Terbium	Tb	[Xe] $4f^9\ 6s^2$	+3, +4
66	Dysprosium	Dy	[Xe] $4f^{10}\ 6s^2$	+3, +4
67	Holmium	Ho	[Xe] $4f^{11}\ 6s^2$	+3
68	Erbium	Er	[Xe] $4f^{12}\ 6s^2$	+3
69	Thulium	Tm	[Xe] $4f^{13}\ 6s^2$	+2, +3
70	Ytterbium	Yb	[Xe] $4f^{14}\ 6s^2$	+2, +3
71	Lutetium	Lu	[Xe] $4f^{14}\ 5d^1\ 6s^2$	+3

Q. 3. Discuss oxidation states of lanthanide elements.

(*Guwahati, 2002; Nagpur, 2005; Purvanchal, 2007; Lucknow, 2010*)

Ans. The common oxidation state shown by all the lanthanides is +3. Some elements display +2 and +4 states, but these are always less stable than the group valency +3. The unusual oxidation states such as +2 or +4 exhibited by the lanthanides can be explained on the basis of extra stability associated with empty, half-filled or fully filled 4*f* subshell. Some examples are as follows:

***(i)* Eu and Yb exhibit +2 oxidation states.** This is because in +2 oxidation state europium and ytterbium acquire f^7 and f^{14} electronic configuration as:

$$Eu^{2+} : [Xe]\ 4f^7,\ Yb^{+2} : [Xe]\ 4f^{14}$$

***(ii)* Ce and Tb exhibit +4 oxidation states.** This is because in +4 oxidation state cerium and terbium acquire f^0 and f^7 electronic configuration which are supposed to be stable configurations

$$Ce^{4+} : [Xe]\ 4f^0,\ Tb^{4+} : [Xe]\ 4f^7$$

***(iii)* La, Gd and Lu exhibit +3 oxidation states.** This is because in +3 oxidation states La^{3+} has [Xe] $4f^0$, Gd^{3+} has [Xe] $4f^7$ and Lu^{3+} has [Xe] $4f^{14}$ electronic configurations which are stable due to empty, half-filled and fully filled $4f$ subshell respectively.

Exceptions. There are a few elements which show +2 and +4 oxidation states but their existence can't be easily justified. For example, Sm and Tm show +2 state having f^6 and f^{13} electrons, when they are only one electron short of the necessary half-filled or completely filled configuration. We have other examples where Pr and Nd show +4 state even though they are left with f^1 and f^2 electrons. Oxidation states of lanthanides are shown in Table 19.1.

Q. 4. Give the important oxidation states of lanthanides. How would you account for them? Why is +3 most stable?

Ans. In reactions of lanthanides, it is the two electrons from $6s$ orbital and one electron from either the $5d$ or $4f$ that are removed. That is why these elements exhibit +3 oxidation state. In some cases +2 and +4 oxidation states are also experienced (see Q.3). It is on energy consideration of orbitals that +3 oxidation state is most stable.

Q. 5. In transition series starting from La (57) the next element is Hf (72). Why is this jump observed?

Ans. In the transition series from La (57) to Hg (80) the filling of electrons takes place in *5d* orbitals, La containing one electron and Hg containing 10 electrons in *5d* orbital. After La, filling starts taking place in $4f$ orbitals on energy consideration and build up of elements from cerium to lutetium takes place. It is only after Lu (71) that the electrons are again filled up in $5d$ orbital and the next element is Hf (72).

Q. 6. Can lanthanum (Z = 57) ion exist in the +4 oxidation state? Justify your answer.

Ans. Lanthanum has electronic configuration [Xe] $5d^1\ 6s^2$ and forms La^{3+} by the removal of these electrons leaving behind stable inert gas configuration. To form La^{4+} ion, stable configuration will have to be disturbed. Hence La^{4+} ion doesn't exist.

Q. 7. M^+ and M^{2+} are less stable than M^{3+} in lanthanide series. Discuss.

Ans. See Q.3.

Q. 8. How is the observed oxidation state of lanthanide related to their electronic configuration?

Ans. See Q.3.

Q. 9. Give electronic configuration for Nd (60), Gd (64) and Lu (71).

Ans. See Table 19.1.

Q. 10. Give electronic configuration for Pr (59), Gd (64) and Tm (69).

Ans. See Table 19.1.

Q. 11. Discuss anomalous oxidation states of lanthanides.

Ans. See Q. 3.

Q. 12. Ytterium is taken as a practical member of lanthanide series though it doesn't belong to this series. Explain.

Ans. Y and La belong to group 3 of the periodic table. They have similar outer electronic configuration, that of Y being $4d^1\ 5s^2$ and that of La being $5d^1\ 6s^2$. Therefore they show identical properties. Moreover La along with other lanthanides show identical properties. Therefore *Y* is taken as a practical member of lanthanide series.

Q. 13. Discuss spectral properties of lanthanides? (*M. Dayanand, 2000; Mumbai, 2004*)

Ans. The absorption spectra of the compounds of trivalent lanthanide ions show very sharp linelike bands in the ultraviolet, visible or near infrared regions. These sharp bands obtained as a result of transitions in the partly filled 4*f* subshell electrons are known as *f–f* transitions. The bands of lanthanide ions are so sharp that they are very useful for characterising the lanthanides and for their quantitative estimations.

Since 4*f* orbitals lie deep in the atoms of these elements, therefore they are shielded from perturbation by complexing agents. Thus the complexing agents have little effect on the spectral bands.

Q. 14. Discuss magnetic properties of lanthanides?

(*M. Dayanand, 2002; Bombay, 2004; Lucknow, 2010; Ranchi 2010*)

Ans. Magnetism is a property associated with unpaired electrons. All lanthanide ions with the exception of La^{3+}, Lu^{3+}, Yb^{3+} and Ce^{4+} are paramagnetic because they contain unpaired electrons in the 4*f* orbital. Their magnetic moments don't obey the simple "spin only" formula (except for f^6, f^7 and f^{14} cases) for the magnetic moment.

$$\mu_{\text{spin only}} = \sqrt{n(n+2)}$$ where *n* is equal to the number of unpaired electrons.

In case of lanthanides, the magnetic effect arising from the orbital motion of the electron, as well as that arising from the electrons spinning on its axis, contributes to the total magnetic moment. Therefore magnetic moments of lanthanides are calculated by taking into consideration spin as well as orbital contributions and a more complex formula.

$$\mu = \sqrt{4S(S+1) + L(L+1)}$$

which involves the orbital quantum number, L and spin quantum number, *S* is used.

Magnetic moment values of lanthanides vary depending upon the atomic number. La^{3+} is diamagnetic (due to f^0) and the value of magnetic moment increases attaining maximum value at neodymium. It then suddenly drops to 1.47 for samarium. It increases again reaching maximum value for dysprosium and holmium. Thereafter, it again falls touching zero at lutetium which is diamagnetic because it contains 14 electrons in its *f* orbitals. Change in magnetic moment with atomic number is shown in Fig. 19.1.

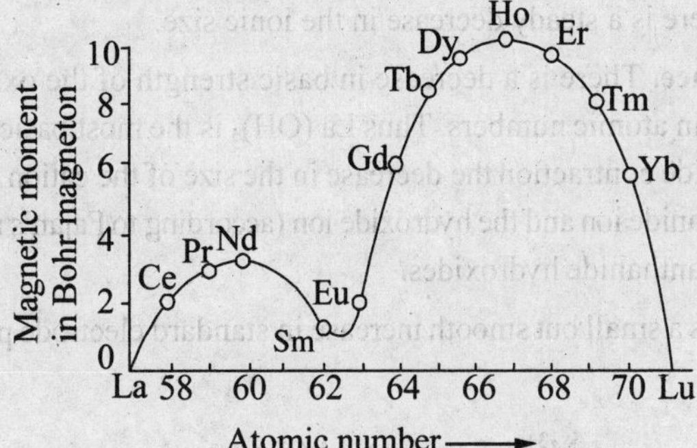

Fig. 19.1. Experimental values of the magnetic moments of the lanthanide ions.

Q. 15. Discuss lanthanide contraction giving causes and its consequences.
(*Delhi, 2003; Purvanchal, 2007; Nagpur, 2008; Lucknow 2010; Ranchi 2010*)

Ans. From Table 19.2 and Fig. 19.2 we find that as we move along the lanthanide series from lanthanum to lutetium there is a continuous decrease in the size of lanthanide ion with increase in atomic number. This decrease is known as *lanthanide contraction.* Lanthanide contraction may be defined as the steady decrease in size of lanthanide ions with increase in atomic number. Thus in the lanthanide series, lanthanum has the largest radius while lutetium has the smallest radius.

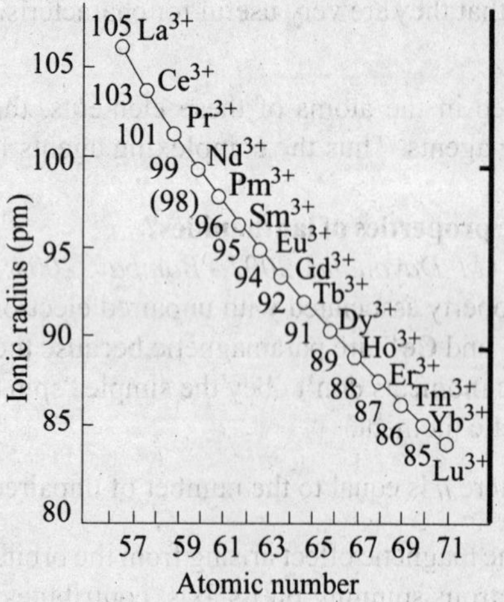

Fig. 19.2. Ionic radii of trivalent lanthanide ions.

Cause of Lanthanide contraction. The cause of lanthanide contraction is generally attributed to imperfect shielding of one 4*f* electron by another in the same shell. Thus as we move along the lanthanide series, the nuclear charge and the number of 4*f* electrons increase by one unit at each step. However due to imperfect shielding (because the shape of *f* orbitals is very much diffused) the net effective nuclear charge increases which causes contraction in the size of the electron charge cloud and thus each ion shrinks in comparison with its predecessor. For example, in going from La^{3+} to Lu^{3+} the ionic radii decrease from 106 pm to 85 pm.

Consequences of Lanthanide Contraction

(*i*) **Electronegativity.** There is a slight increase in electronegativity of trivalent ions.

(*ii*) **Ionic radius.** There is a steady decrease in the ionic size.

(*iii*) **Basicity difference.** There is a decrease in basic strength of the oxides and hydroxide of lanthanides with increase in atomic numbers. Thus La $(OH)_3$ is the most basic while Lu $(OH)_3$ is the least basic. Due to lanthanide contraction the decrease in the size of the cation increases the covalent character between the lanthanide ion and the hydroxide ion (according to Fajan's rules) thereby reducing the basic character of the lanthanide hydroxides.

(*iv*) **E° value.** There is a small but smooth increase in standard electrode potential values, E° for the process.

$$M^{3+} + 3e \longrightarrow M(g)$$

ranging from –2.52 V for lanthanum to –2.25 V for lutetium.

Table 19.2. Ionic radii of lanthanides.

Atomic number	*Name*	*Symbol*	*Electronic configuration*	M^{3+} *radii (pm)*
57	Lanthanum	La	[Xe] $4f^0\ 5d^1\ 6s^2$	106
58	Cerium	Ce	[Xe] $4f^1\ 5d^1\ 6s^2$	103
59	Praseodymium	Pr	[Xe] $4f^3\ 6s^2$	101
60	Neodymium	Nd	[Xe] $4f^4\ 6s^2$	100
61	Promethium	Pm	[Xe] $4f^5\ 6s^2$	98
62	Samarium	Sm	[Xe] $4f^6\ 6s^2$	96
63	Europium	Eu	[Xe] $4f^3\ 6s^2$	95
64	Gadolinium	Gd	[Xe] $4f^7\ 5d^1\ 6s^2$	94
65	Terbium	Tb	[Xe] $4f^9\ 6s^2$	92
66	Dysprosium	Dy	[Xe] $4f^{10}\ 6s^2$	91
67	Holmium	Ho	[Xe] $4f^{11}\ 6s^2$	89
68	Erbium	Er	[Xe] $4f^{12}\ 6s^2$	88
69	Thulium	Tm	[Xe] $4f^{13}\ 6s^2$	87
70	Ytterbium	Yb	[Xe] $4f^{14}\ 6s^2$	86
71	Lutetium	Lu	[Xe] $4f^{14}5d^1 6s^2$	85

Q. 16. What is the consequence of lanthanide contraction on the basic strength of lanthanide hydroxides?

Ans. Due to the lanthanide contraction the size of ion decreases from Ce to Lu as a general trend. According to Fajan's rules smaller positive ions can cause greater polarisation of the anion. Based on this generalisation decrease in basic strength of the oxide in going from Ce to Lu is predicted.

Q. 17. Basic strength of lanthanide oxide M_2O_3 decreases with rise in atomic number. Explain. (*Rohilkhand, 2003*)

Ans. See Q. 16.

Q. 18. Discuss the colour of lanthanide compounds.

Ans. The lanthanide metals are silvery white but the trivalent lanthanide ions show different colours. Colour of the ions depends on the number of unpaired electrons because the elements with *xf* electrons often have a similar colour to those of $(14 - x)\, f$ electrons. This is evident from Table 17.3. The colour of lanthanide ions is due to the presence of partly filled *f*-orbitals. As a result it is possible to absorb certain wavelength from the visible region of the spectrum. This results in transition from one 4*f* orbital to another 4*f* orbital known as *f*-*f* transition. The colour transmitted is the colour shown by the substance.

Table 19.3.

No. of unpaired electrons ions		*Colour*
0	La^{3+}, Lu^{3+}	Colourless
1	Ce^{3+}, Yb^{3+}	Colourless
2	Pr^{3+}, Tm^{3+}	Green
3	Nd^{3+}, Er^{3+}	Red
4	Pm^{3+}, Ho^{3+}	Pink and yellow respectively
5	Sm^{3+} Dy^{3+}	Yellow
6	Eu^{3+}, $T6^{3+}$	Pink
7	Gd^{3+}	Colourless

Q. 19. Discuss the variation in melting and boiling points of lanthanides?

Ans. Lanthanides have fairly high melting and boiling points. However, their melting and boiling points don't show definite trend as we move from La to Lu. Both Eu and Yb form +2 ions and so it results in slightly weaker metallic bonding as there are fewer than three electrons per atom in the metallic bonding. As a consequence, melting and boiling points have lower values for these two elements as compared to the values for other lanthanide elements. Also abnormally greater radii for these two elements resulting in weaker packing account for lower melting and boiling points. Melting and boiling points of lanthanides are given in Table 19.4.

Table 19.4.

Atomic number	*Name*	*Density (g/ml)*	*Melting point (K)*	*Boiling point (K)*
57	Lanthanum	6.17	1193	3742
58	Cerium	6.77	1068	3741
59	Praseodymium	6.78	908	3400
60	Neodymium	7.00	1297	3300
61	Promethium	7.20	–	–
62	Samarium	6.93	1345	2173
63	Europium	5.24	1099	1712
64	Gadolinium	7.95	1585	3273
65	Terbium	8.33	1629	3073
66	Dysprosium	8.56	1680	2873
67	Holmium	8.76	1734	2873
68	Erbium	9.16	1770	3173
69	Thulium	9.35	1818	2000
70	Ytterbium	7.01	1097	1700
71	Lutetium	9.74	1925	3600

Q. 20. Explain that

(i) Ce^{3+} is more basic than Ce^{4+}

(ii) La^{3+} is colourless but Lu^{4+} is orange red.

Ans. (i) Configuration of Ce^{3+} and Ce^{4+} are:

Ce^{3+} : [Xe] $4f^1$

Ce^{4+} : [Xe]

Ce^{3+} has a tendency to lose $4f^1$ electron to attain the inert gas structure, whereas Ce^{4+} wouldn't lose the electron because that would disrupt inert gas structure. Therefore Ce^{3+} is more basic than Ce^{4+}.

(ii) Configuration of La^{3+} and Lu^{4+} are:

La^{3+} : [Xe] $4f^0$

Lu^{4+} : [Xe] $4f^{13}$

Transition within the *f* orbitals is not possible in La^{3+} as there are no electrons in *f*-orbitals. Hence no absorption of visible light is possible; it is therefore colourless.

In the case of Lu^{4+}, there is one vacancy in 4*f* orbitals. Hence *f*- *f* transition takes place absorbing a radiation from visible light and transmitting orange-red light.

Q. 21. What are *d–d* and *f–f* transitions? How do they differ?

Ans. Transitions within the *d*-orbitals of an element are called *d–d* transition, while transitions within *f* orbitals of an element are called *f–f* transitions. These transitions can be brought about by the absorption of radiation from the visible light and transmission of the complementary colour. The compound thus looks coloured.

Q. 22. Discuss the variation of following properties of lanthanides:

(i) Metallic radii

(ii) Density

(iii) Ionisation energy

(iv) Electronegativity values.

Ans. (i) Metallic radii. The metallic radii (see Table 19.5) decrease with increase in atomic number. The metallic radii of lanthanides fall in the range 174-208 pm which are quite comparable with those of *s*-block elements, particularly calcium having metallic radius of 174 pm. This indicates fairly large size of the atoms of lanthanide elements. Eu and Yb show very surprisingly irregular sizes. The reason is that these elements suffer repulsions between greater number of *f*-electrons. Oxidation state of +2 for these elements is also responsible for anomalies.

(ii) Density. Lanthanides have low densities which range between 6.77 and 9.74 g cm^{-3}. However, densities of lanthanides don't show definite trends with rise in atomic number. Eu and Yb have low values of density than expected (see Table 19.4).

(iii) Ionisation energies. Lanthanides have fairly low ionisation energy values (IE_I is about 585 kJ/mol and IE_{II} is about 1150 kJ mol^{-1}) due to their fairly large size which are quite comparable with those of alkaline earth metals particularly calcium (IE_I = 589 kJ/mol, IE_{II} = 1145 kJ mol^{-1}).

(iv) Electronegativity values. Their electronegativity values fall in the range 1.0 –1.15 (Allred and Rochow scale) and are quite comparable with the electronegativity values of the *s*-block elements, particularly calcium. Therefore, lanthanide ions are expected to form ionic compounds.

Table 19.5. Metallic radii of lanthanides.

Name	*Symbol*	*Electronic configuration*	*Metallic radii (pm)*	*E°(V) M^{3+}/M*
Lanthanum	La	[Xe] $4f^0$ $5d^1$ $6s^2$	188	–2.52
Cerium	Ce	[Xe] $4f^1$ $5d^1$ $6s^2$	182	–2.48
Praseodymium	Pr	[Xe] $4f^3$ $6s^2$	183	–2.47
Neodymium	Nd	[Xe] $4f^4$ $6s^2$	182	–2.44
Promethium	Pm	[Xe] $4f^5$ $6s^2$	183	–2.42
Samarium	Sm	[Xe] $4f^6$ $6s^2$	180	–2.41
Europium	Eu	[Xe] $4f^7$ $6s^2$	208	–2.41
Gadolinium	Gd	[Xe] $4f^7$ $5d^1$ $6s^2$	180	–2.40
Terbium	Tb	[Xe] $4f^9$ $6s^2$	178	–2.39
Dysprosium	Dy	[Xe] $4f^{10}$ $6s^2$	177	–2.35
Holmium	Ho	[Xe] $4f^{11}$ $6s^2$	177	–2.32
Erbium	Er	[Xe] $4f^{12}$ $6s^2$	176	–2.30
Thulium	Tm	[Xe] $4f^{13}$ $6s^2$	175	–2.28
Ytterbium	Yb	[Xe] $4f^{14}$ $6s^2$	193	–2.27
Lutetium	Lu	[Xe] $4f^{14}$ $5d^1$ $6s^2$	173	–2.25

Q. 23. Chemistry of all lanthanides is so identical. Explain.

(*M. Dayanand, 2000; Calcutta, 2005*)

Ans. All the lanthanides have similar outer electronic configuration and display mainly +3 oxidation state in their compounds, therefore lanthanides have exceedingly similar chemical properties. Their similarity is much closer than that of ordinary transition elements because lanthanides differ mainly in the number of 4*f* electrons which are buried deep in the atoms of lanthanides and thus don't influence their properties. Moreover, due to lanthanide contraction there is a very small difference in the size of all the fifteen trivalent lanthanide ions. Thus, for all practical purposes, the size of these ions is almost identical which results in similar chemical properties of these elements.

Q. 24. Describe chemical properties of lanthanides.

Ans. **(*i*) Action with air.** They are easily tarnished in moist air. If ignited in air, they burn to give +3 or +4 oxides like Ln_2O_3 or CeO_2.

(*ii*) Action with nitrogen. They react with nitrogen to form metal nitrides of the general formula MN.

(*iii*) Action with halogens. They combine directly with halogens to give trihalides MX_3.

(*iv*) Action with water. Their standard electrode potential values, E°, indicate that these metals should react vigorously with water. In fact they do liberate hydrogen slowly from cold water and rapidly from hot water.

(*v*) Action with hydrogen. They react with hydrogen especially at elevated temperatures and form non-stoichiometric hydrides having composition MH_2 and MH_3.

Q. 25. Lanthanides have a poor tendency to form complexes. Explain. *(Nagpur, 2008)*

Ans. Although the lanthanide ions have a high charge (+3), yet the size of their ions is very large (yielding small charge to size ratio). As a consequence, they have poor tendency to form complexes. They form complexes mainly with chelating agents such as β-diketones, EDTA, L-L′ dipyridyl and β-hydroxyquinoline. Complex formation tendency and stability increases with increasing atomic number. This fact is utilized in the separation of lanthanides.

Q. 26. Compare the characteristics of *d* and *f* block elements.

Ans. (*i*) Size. The sizes of the atoms and ions of *d*-block elements are relatively smaller than those of the *f*-block elements.

(*ii*) Basic character. The oxides and hydroxides of *d*-block elements are less basic than those of *f*-block elements.

(*iii*) Ionisation energies. The first ionization values of *d*-block elements are much higher as compared to those of *f*-block elements.

(*iv*) Complex formation. The elements of *d*-block form complexes easily as compared to *f*-block elements. The latter mainly form metal chelates with strong chelating agents.

(*v*) Reactions with unsaturated hydrocarbons. Both *d*- and *f*-block elements react with unsaturated hydrocarbons such as olefins and cyclopentadienes. The former form π-type of complexes while the latter form salts such as Lu $(C_5H_5)_3$ known as lanthanide cyclopentadenides.

(*vi*) Electronic spectra. Electronic spectra of salts and complexes of *d*-block elements show broad bands while those of *f*-block elements exhibit sharp line-like bands.

(*vii*) Magnetic moment values. In case of *d*-block elements, especially for the first transition series, magnetic properties are simple to interpret and the magnetic moments can be calculated from spin-only formula:

$$\text{spin magnetic moment} = 2\sqrt{S(S+1)}$$

where *S* is half the number of unpaired electrons. However in the case of *f*-block elements the magnetic properties are quite difficult to interpret. Their magnetic moment values are calculated by taking into consideration spin as well as orbital contributions by using a more complex formula:

$$\mu = \sqrt{4S(S+1) + L(L+1)}$$

Q. 27. Describe the extraction of lanthanides from monazite sands.

(Lucknow, 2000; M. Dayanand, 2002)

Ans. Monazite sand consists of mixture of Th, Ce, Nd, Pr and La phosphates along with impurities. The mineral is powdered and digested with concentrated H_2SO_4 until a paste of lanthanide sulphates containing phosphoric acid and an excess of sulphuric acid is formed. The paste is centrifuged and then treated with cold water when silica being insoluble precipitates out. The solution is neutralised with a previously prepared mixture of lanthanide oxides when thorium, zirconium and titanium precipitate out as pyrophosphates. Sodium sulphate is added to the clear mother liquor so that the light lanthanides (La to Sm) may precipitate out as double salts while the heavy lanthanides (Gd to Lu) still remain in solution.

Hot sodium hydroxide is added to the precipitated double sulphates of light lanthanides to form a mixture of hydrated oxides. This is washed (till free form Na_2SO_4) and dried in air at 373 K whereby cerium is completely oxidized to CeO_2 while other lanthanides change into their oxides. At this stage, the approximate composition of residual solids is CeO_2 (50%), Nd_2O_3 (20%), La_2O_3 (17%), Pr_2O_3 (8%), Sm_2O_3 (5%). This mixture is reacted with dilute nitric acid when all the lanthanide oxides (except CeO_2) get dissolved and filtered. The crude CeO_2 is dissolved in 85% HNO_3 and cerium is precipitated from it as the red basic nitrate, $Ce(OH)(NO_3)_3 . 2H_2O$ by the addition of excess of dilute H_2SO_4. After the removal of cerium, the filtrate containing lanthanides are separated into individual components.

The method has been summed up in the following flow sheet.

Monazite Sand

containing a mixture of phosphates of lanthanides, thorium, zirconium, etc.

Digest with conc. H_2SO_4 at 200°C

A Pasty Mass

containing sulphates of lanthanides, etc., + phosphoric acid

Treat with cold water

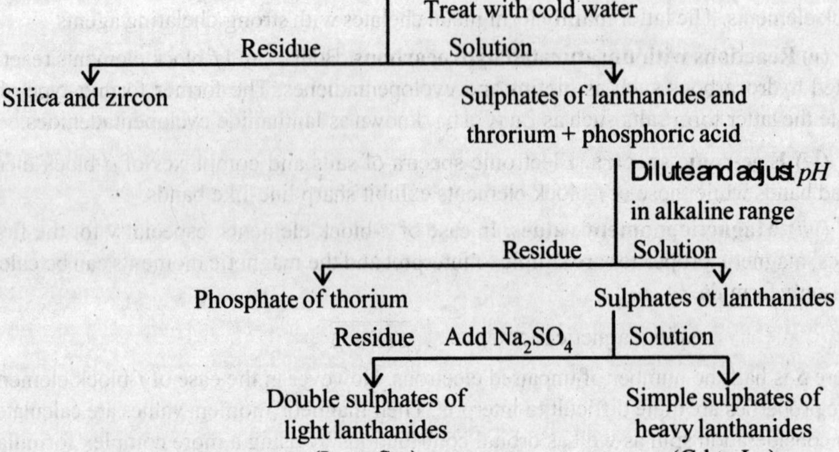

Q. 28. Describe the separation of lanthanides by ion exchange method.

(*Guwahati, 2002; Rohilkhand, 2003; Nagpur, 2005; Purvanchal, 2007; Andhra, 2012*)

Ans. This is the most rapid and effective method for the separation and purification of the lanthanides. This method is based on the following principles:

(*i*) Lanthanide ions on contact with synthetic resins containing –COOH or $-SO_3H$ group undergo proton exchange.

$$Ln^{3+} + 3H\text{–resin} \rightarrow Ln\,(resin)_3 + 3H^+$$

(*ii*) The bonding of the lanthanide ion to the resin depends on its size, *i.e.*, the smaller the size of the lanthanide ion, the more firmly it is bound to the resin and vice-versa. Since lanthanide ions are hydrated, therefore size of the hydrated ions should be considered for binding purpose. Hydration of the ions depends upon size, *i.e.*, smaller the size of the ion, greater will be the hydration. Therefore in case of lanthanide ions, the smallest lanthanide ion, namely Lu^{3+}, will be the most heavily hydrated. Thus it will have the maximum size and therefore the least firmly bound to the resin while reverse will be the case with La^{3+} which will be the most firmly bound to the resin.

A solution containing several lanthanide ions is dropped slowly down a column of synthetic ion exchange resin so that the lanthanide ions are bound less firmly to the resin in the order

La^{3+} to Lu^{3+}. They are then eluted from the column by using a solution containing citric acid and ammonium citrate. Firstly, the ammonium ions elute the metal ions from the resin as:

$$3NH^+_4 + Ln\,(resin)_3 \rightarrow 3NH_4\,resin + Ln^{3+}$$

The metal ions then form a complex with the citrate ions.

$$Ln^{3+} + \text{citrate ions} \rightarrow \text{Ln citrate complex}$$

Since Lu^{3+} is the least firmly bound to resin, therefore on elution, Lu citrate complex is obtained first from the bottom of the column while La citrate complex emerges last of all from the bottom of the column. Complexing agents such as EDTA, aminocarboxylic acids and hydroxy carboxylic acids have also been found to be convenient elutants.

Q. 29. Describe solvent extraction method for the separation of lanthanides.

(*Mumbai, 2004*)

Ans. This method is based on differential solubilities of lanthanide compounds. (*i*) Solubility of lanthanide nitrates in aqueous and organic solvents such as kerosene oil containing trialkyl phosphates is different. (*ii*) The extent of extraction of lanthanide elements by organic solvents increases with increase in atomic number because increase in nuclear charge results in decrease in ionic radius which in turn increases complexation.

Solution of lanthanide nitrates in nitric acid is made to flow counter to a solution of tri-n-butyl phosphate in kerosene and thus some of the lanthanide nitrates migrate to the organic solvent layer. Thus by using a continuous counter current method, it is possible to separate lanthanides such as Gd from other lanthanides on a kilogram scale. Pure Gd (95%) has been prepared by this method.

Q. 30. Describe change of oxidation state method to separate cerium (Ce) and europium (Eu) from a mixture of lanthanides.

Ans. Some of the lanthanides show +2 and +4 oxidation states in addition to the +3 state which is most characteristic of all the elements of the family. The properties of the M^{2+} and M^{4+} ions are different from those of the usual M^{3+} ions. To take an example, cerium can be separated from a mixture of trivalent ions of all other Ianthanides by oxidising it to Ce^{4+} state by reacting it with alkaline $KMnO_4$. Now, Ce^{4+} ion has a greater charge than Ce^{3+} ion. Therefore, it is smaller in size and, therefore, less basic and less soluble. It can be precipitated as $Ce(OH)_4$ by the addition of a small amount of an alkali leaving all the trivalent lanthanide ions in solution. It is possible to obtain 99 per cent pure cerium from a mixture containing only 40 per cent cerium by this method.

Europium can be separated from a mixture of trivalent lanthanide ions by reducing it first to the +2 state by means of zinc amalgam. Eu^{2+} can then be precipitated as $EuSO_4$ which is insoluble in water. The sulphates of all the trivalent lanthanides, on the other hand, are soluble in water.

Q. 31. How can we obtain a lanthanide metal from its salt?

Ans. (*i*) Reduction of anhydrous chlorides with sodium. Lighter lanthanides such as La, Ce and Gd can be easily prepared by the reduction of anhydrous lanthanide chlorides with sodium at 1273 K.

$$La\,Cl_3 + 3Na \rightarrow La + 3NaCl$$

(*ii*) Reduction of anhydrous fluorides with calcium or magnesium. Heavier lanthanides such as Lu are generally prepared by the reduction of anhydride fluorides with calcium or magnesium metal at 1273 K.

$$2LuF + 3Mg \rightarrow 3MgF + 2Lu$$

(*iii*) Electrolysis of fused chloride. Electrolysis of fused lanthanide chlorides is carried out. This method is similar to that of preparation of calcium by the electrolysis of calcium chloride.

Q. 32. Describe preparation and properties of (*i*) Lanthanum oxide, La_2O_3, (*ii*) Lanthanum halides, LaX_3.

Ans. Lanthanum Oxide, La_2O_3. It is obtained as a snow-white powder by heating the carbonate or oxalate. When dropped in water, it changes into hydroxide evolving a large amount of heat. This behaviour is similar to that of lime.

$$La_2(CO_3)_3 \xrightarrow{\text{Heat}} La_2O_3 + 3CO_2 \uparrow$$

It is sparingly soluble in water. It is basic in character and dissolves readily in dilute acids. It is used in the manufacture of the so-called 'Crookes' glass as it can absorb ultraviolet rays to some extent.

Lanthanum Halides, LaX_3. Lanthanum fluoride is obtained as a white gelatinous precipitate by adding hydrofluoric acid to a lanthanum salt solution. If the solution is allowed to stand for some time, the crystals of composition $LaF_3.3H_2O$ separate out. It is insoluble in dilute but soluble in concentrated mineral acids.

$$La_2(SO_4)_3 + 6HF \longrightarrow 2LaF_3 + 3H_2SO_4$$

Lanthanum chloride is obtained as anhydrous $LaCl_3$ on heating a mixture of lanthanum oxide and carbon in a current of chlorine. It is a white deliquescent powder and is readily soluble in water. It crystallises out from aqueous solution as the heptahydrate, $LaCl_3.7H_2O$.

Lanthanum bromide and ***lanthanum iodide*** are also known and are prepared by similar methods.

Q. 33. List important uses of lanthanides. (*Kanpur, 2001*)

Ans. (*i*) Cerium glass cuts off heat and ultraviolet light and so is used in glare-reducing spectacles.

(*ii*) Cerium-magnesium alloys are used in flashlight powders.

(*iii*) Neodymium oxide dissolved in selenium oxychloride is one of the most powerful liquid lasers known so far.

(*iv*) Ceria and thoria are used in gas lamp mantles.

(*v*) Cerium salts are used in analysis, dyeing, lead accumulators, medicines and as catalysts.

(*vi*) Lanthanide elements and their compounds are being used in nuclear control, shielding and fluxing devices.

(*vii*) Since lanthanides improve the workability of steel when heated, so alloys of lanthanides with iron are well known.

(*viii*) Gadolinium sulphate heptahydrate has been used to produce very low temperature.

ACTINIDES

Q. 34. What are actinides? (*Kanpur, 2001*)

Ans. The fifteen elements from actinium (Z = 89) to lawrencium (Z = 103) in which the 5*f* orbitals are being filled up are called actinide series. The name actinide is derived from actinium, the very first member of the series. These are also known as **actinons**. Only the first four elements of this series, namely, actinium, thorium, protoactinium and uranium, occur in nature, the other elements are made artificially by nuclear bombardment. All the actinides are toxic to humans.

Q. 35. Discuss the electronic configuration of actinides. (*Kanpur, 2001; Nagpur, 2008; Ranchi, 2010*)

Ans. After mercury we follow six elements from thalium to radon in which electrons enter the 6*p* subshell. Radon has the configuration [Xe] $6s^2\,4f^{14}\,5d^{10}\,6p^6$. At this stage the 7*s* and 6*d* subshells have smaller energy relative to the 5*f* subshell and the next two electrons are added to the 7*s* subshell and we get francium and radium which are *s*-block elements. We find that the 5*f* and 6*d* subshells penetrate the 7*s* orbital quite significantly and are therefore stabilized relative to the 7*p* subshell. The next few electrons enter either of these two subshells, *i.e.*, 5*f* and 6*d*. It soon becomes obvious that as we fill electrons from actinium onwards (*i.e.*, actinide series), the energies of 6*d* and 5*f* subshell don't vary significantly and the allocation of exact configuration becomes difficult. It is because of this very reason that different authors have tried to suggest different electronic configurations for actinides. However, the electronic configurations for actinides set out in Table 19.6 have been generally accepted. An important point is that around actinium both 6*d* and 5*f* subshells are of very similar energies, but as we keep filling the electrons the 5*f* subshell definitely acquires greater stability. From uranium onwards there is no doubt that the 5*f* subshell is more stable. The general electronic configuration of actinides is

$$[Rn]\ 5f^{0-14}\ 6d^{0-2}\ 7s^2$$

Table 19.6. Properties of actinides.

Element	*Symbol*	*Atomic number*	*Electronic configuration*	*Oxidation states*	*Ionic radii (pm)* M^{3+}	*Ionic radii (pm)* M^{4+}
Actinium	Ac	89	$[Rn]\ 5f^0\ 6d^1\ 7s^2$	3, –	111	–
Thorium	Th	90	$[Rn]\ 5f^0\ 6d^2\ 7s^2$	4	108	99
Protactinium	Pa	91	$[Rn]\ 5f^2\ 6d^1\ 7s^2$	3, 4, 5	105	96
Uranium	U	92	$[Rn]\ 5f^3\ 6d^1\ 7s^2$	3, 4, 5, 6	103	93
Neptunium	Np	93	$[Rn]\ 5f^5\ 6d^0\ 7s^2$	3, 4, 5, 6, 7	101	92
Plutonium	Pu	94	$[Rn]\ 5f^6\ 6d^0\ 7s^2$	3, 4, 5, 6, 7	100	90
Americium	Am	95	$[Rn]\ 5f^7\ 6d^0\ 7s^2$	3, 4, 5, 6	99	89
Curium	Cm	96	$[Rn]\ 5f^7\ 6d^1\ 7s^2$	3, 4	98	88
Berkelium	Bk	97	$[Rn]\ 5f^9\ 6d^0\ 7s^2$	3, 4	96	–
Californium	Cf	98	$[Rn]\ 5f^{10}\ 6d^0\ 7s^2$	3, 4	95	–
Einsteinium	Es	99	$[Rn]\ 5f^{11}\ 6d^0\ 7s^2$	2, 3	–	–
Fermium	Fm	100	$[Rn]\ 5f^{12}\ 6d^0\ 7s^2$	2, 3	–	–
Mendelevium	Md	101	$[Rn]\ 5f^{13}\ 6d^0\ 7s^2$	2, 3	–	–
Nobelium	No	102	$[Rn]\ 5f^{14}\ 6d^0\ 7s^2$	2,3	–	–
Lawrencium	Lr	103	$[Rn]\ 5f^{14}\ 6d^1\ 7s^2$	3	–	–

Q. 36. The electronic configuration and position of most of the actinide elements are controversial. Comment.

Ans. As there isn't much difference between 5*f* and 6*d*, it becomes difficult to know whether the electron has entered 5*f* or 6*d*. This makes predicting electronic configuration difficult.

Q. 37. Why do actinides show higher oxidation states than lanthanides?

(*M. Dayanand, 2000; Punjab, 2000; Bangalore, 2002; Purvanchal, 2007; Nagpur 2008*)

Ans. It is found that most stable oxidation states of the elements up to uranium is the one involving all the valency electrons. Neptunium forms the +7 state using all the valency electrons but this is oxidising and the most stable state is +5.

Plutonium also shows states up to +7 and americium up to +6 but the most stable state drops to Pu (+4) and Am (+3) and latter elements tend to be stable in +3 oxidation state.

Berkelium in +4 state is strongly oxidising but is more stable than curium and americium in +4 state due to f^7 configuration. Similarly, nobelium is markedly stable in +2 state due to its f^{14} configuration. Oxidation states of actinides are given in Table 19.6.

Q. 38. Actinides form oxocation but lanthanides don't. Explain.

Ans. When the oxidation state of an actinide increases to +6, the ion is no longer a simple ion. There is a high charge density on the ion which leads to the formation of oxygenated ion, *e.g.*, UO_2^{2+}. Such oxycations are stable in acid and aqueous solutions.

Q. 39. There is actinide contraction similar to that of lanthanide contraction. Explain.

(*Ranchi, 2012*)

Ans. It is seen from Table 19.6 that there is a regular decrease in ionic radii with increase in atomic number from actinium to lawrencium. This is called actinide contraction. This contraction is similar to the lanthanide contraction.

The actinide contraction is caused due to imperfect shielding of one 5*f* electron by another in the same shell. Thus as we move along the actinide series, the nuclear charge and the number of 5*f* electrons increase by one unit at each step. However, due to imperfect shielding (because the shape of *f* orbitals is very much diffused) the effective nuclear charge increases which causes contraction in the size of the electron cloud.

However in actinide contraction there are bigger jumps in contraction between the consecutive elements as compared to lanthanides. This is due to lesser shielding of 5*f* electrons which results in stronger pull by the nucleus.

Q. 40. Discuss the variation in m.p. and b.p., density, magnetic properties and reducing power, reactivity and colour of actinides. (*Rohilkhand, 2003*)

Ans. (*i*) Melting and boiling points. There is no regular trend in the melting and boiling points of actinides as seen in Table 19.7.

Table 19.7. Melting points, boiling points and densities of some actinides.

Elements	*Ac*	*Th*	*Pa*	*U*	*Np*	*Pu*	*Am*	*Cm*
m.p. (K)	1323	2033	1503	1405	910	913	1273	1613
b.p. (K)	—	4123	—	4086	—	3508	2880	—
Density (g/cm^3)	—	11.72	15.37	19.07	19.5	19.82	11.7	7.0

(*ii*) Density. Densities of actinides increase from left to right in Table 19.7

(*iii*) Magnetic properties. The magnetic moments of actinide ions are smaller than the theoretically predicted values. This is due to the fact that 5*f* electrons of actinides are less effectively shielded which results in quenching of orbital contribution.

(*iv*) Reducing power. E° values. All the actinides have relatively high E° values (about 2 volts). Therefore they are strong reducing agents.

(*v*) **Reactivity.** They are very reactive metals. The reactions of metals with oxygen, helogens and acids resemble those with lanthanides.

(*vi*) **Coloured ions.** Ions of actinides are generally coloured and the colour depends upon the number of 5*f* electrons. The colour exhibited by actinide ions is due to *f-f* transitions. Electrons change their positions from one *f* orbital to another *f* orbital. The colours shown by actinide ions in aqueous solution are given in Table 19.8.

Table 19.8. Colour of actinide ions in aqueous solutions.

Ion	Ac^{3+}	U^{3+}	Np^{3+}	Pu^{3+}	Am^{3+}	Cm^{3+}
f-electron configuration	F^o	f^3	f^4	f^5	f^6	f^7
Colour	Colourless	Red	Blue to purple	Blue to violet	Pink	Colourless
Ion	Th^{4+}	Pa^{4+}	U^{4+}	Np^{4+}	Pu^{4+}	Am^{4+}
f- electron configuration	f^0	f^1	f^2	f^3	f^4	f^5
Colour	Colourless	Colourless	Green	Yellow green	Orange	Rose

Q. 41. Actinides have a greater tendency to form complexes than lanthanides. Explain.

(*M. Dayanand, 2002; Rohilkhand, 2003; Purvanchal, 2007*)

Ans. Actinides have a greater tendency to form complexes because of higher nuclear charge and small size of their atoms. Complexes formed by actinides are described below.

1. Small and highly charged M^{4+} ions exhibit greatest tendency towards complex formation, *e.g.*, Pu^{4+} forms very strong anion complexes.

2. Most actinide halides form adducts with alkali metal halides. For example, $ThCl_4$ forms the complexes $MThCl_5$, M_2ThCl_6 and M_3ThCl_7 where M is the metal alkali.

3. With pyridine, $ThCl_4$ as well as $ThBr_4$ form monopyridine complexes.

4. They also form complexes with acetylacetone, oxine, EDTA, such as tertrakis — (acetyl acetonato) thorium, Th $(acac)_4$ and $UO_2(C_9H_6NO)_2 \cdot C_9H_7ON$.

5. UO^{2+}_2 forms rather unstable complexes with EDTA as compared to lanthanides.

6. The degree of complex formation in case of actinides decreases in the order

$$M^{4+} > MO_2^{2+} > M^{3+} > MO_2^{+}$$

where M is the element of actinide series.

Q. 42. Compare the properties of lanthanides and actinides. (*Andhra, 2010*)

Ans. Similarities. (*i*) Oxidation state of +3 dominates in both the series.

(*ii*) In both the series *f* orbitals are being progressively filled.

(*iii*) Just like lanthanide contraction we have actinide contraction.

(*iv*) The nitrates, perchlorates and sulphates of trivalent actinides as well as lanthanides are soluble while the hydroxides, fluorides and carbonates are insoluble.

(*v*) In the absorption spectra of the elements of both the series sharp line-like bands appear due to *f–f* transitions.

Differences

Lanthanides	Actinides
(*i*) Binding energy of 4*f* are higher.	Binding energies of 5*f* are lower.
(*ii*) 4*f* electrons have greater shielding effect.	5 *f* electrons have poor shielding effect.
(*iii*) Maximum oxidation state exhibited bylanthanides is +4, *e.g.*, Ce^{4+}.	Due to lower binding energies they show higher oxidation states such as +4, +5 and +6.
(*iv*) They don't form oxocations.	They form oxocations such as UO_2^{2+}, UO_2^+ NpO_2^+, PuO_2^+.
(*v*) Except prometheium they are non-radioactive.	All of them are radioactive.
(*vi*) They do not form complexes easily. Their complexes with π-bonding ligands are not known.	They form complexes even with π-bonding ligands such as alkyl phosphines, thioethers etc.
(*vii*) They are paramagnetic but magnetic properties can be easily explained.	They are also paramagnetic but it is difficult to explain magnetic properties.

Q. 43. Describe the extraction of thorium from monazite sands.

(*Kerala, 2000; Kanpur, 2001*)

Ans. Various steps involved in the extraction of thorium from monazite sands are as follows:

(*a*) Conversion into thoria. Monazite is powdered and digested with conc. H_2SO_4 until pasty mass of sulphates become soluble in water. Since thorium is more basic than other rare earths, addition of sodium hydroxide or magnesium oxide to the acidic solution of the salts results in the precipitation of thorium phosphate. The washed thorium phosphate is heated with sodium carbonate when it changes into crude thoria ThO_2. The crude thoria contains lanthanides as impurities and is purified by conversion into soluble oxalate, recrystallization as nitrate and, finally, calcination into thoria.

(*b*) Conversion to $ThCl_4$ and then into thorium. Pure thoria is heated with phosgene when thorium tetrachloride is formed which on reduction with calcium forms thorium.

$$ThO_2 + 2COCl_2 \longrightarrow ThCl_4 + 2CO_2$$

$$ThCl_4 + 2Ca \longrightarrow 2CaCl_2 + Th$$

(*c*) Refining of crude thorium by Van Arkel method. The metal thus prepared is slightly impure and contains thoria as impurity. It is purified by heating in iodine vapours when it forms volatile ThI_4. The vapours of ThI_4 are dissociated into the metal on a hot filament (1400-1500 K).

$$Th + 2I_2 \longrightarrow ThI_4$$

$$ThI_4 \xrightarrow{1400-1500K} \underset{\text{Pure}}{Th} + 2I_2$$

Q. 44. Give various steps involved in obtaining uranium from its ore pitchblende.

(*Bangalore, 2002, Calcutta, 2005; Purvanchal, 2007; Patna, 2010*)

Ans. Pitchblende contains uranium as U_3O_8 or UO_2. The various steps are as under:

(*a*) Concentration. The ore is crushed to a fine powder and then concentrated by gravity separation in order to remove lighter impurities.

(*b*) Roasting. The concentrated ore is then roasted in a reverberatory furnace in the presence of excess of air in order to remove volatile impurities such as sulphur, arsenic and antimony.

$$S + O_2 \longrightarrow SO_2 \uparrow$$

$$4As + 3O_2 \longrightarrow 2As_2O_3\uparrow$$

$$4Sb + 3O_2 \longrightarrow 2Sb_2O_3\uparrow$$

The roasted ore is then mixed with sodium carbonate and sodium nitrate. The whole mass is then fused on the hearth of a reverberatory furnace resulting in the formation of sodium uranate.

$$U_3O_8 + 3Na_2CO_3 + [O] \rightarrow 3Na_2UO_4 + 3CO_2\uparrow$$

(c) Leaching with sulphuric acid. The fused mass is leached with dilute sulphuric acid whereby sodium uranate changes into soluble uranyl sulphate while impurities such as $PbSO_4$ and $RaSO_4$ remain insoluble and are filtered off.

$$Na_2UO_4 + 2H_2SO_4 \longrightarrow \underset{\text{Uranyl sulphate}}{UO_2SO_4} + Na_2SO_4 + 2H_2O$$

(d) Treatment with sodium carbonate. The solution containing uranyl sulphate is treated with excess of sodium carbonate to precipitate iron, aluminium, nickel, cobalt and manganese as their insoluble carbonates while uranyl sulphate reacts with sodium carbonate to form soluble sodium tricarbonatouranate $Na_4[UO_2(CO_3)_3]$.

$$UO_2SO_4 + 3Na_2CO_3 \longrightarrow Na_4[UO_2(CO_3)_3] + Na_2SO_4$$

(e) Treatment with sulphuric acid. The insoluble carbonates are removed by filtration and the filtrate is treated with dilute sulphuric acid, boiled and evaporated, when heavy yellow precipitate of sodium diuranate $Na_2U_2O_7\cdot 6H_2O$ is obtained.

$$2Na_4[UO_2(CO_3)_3] + 3H_2SO_4 \rightarrow Na_2U_2O_7\cdot 6H_2O + 3CO_2 + 3Na_2SO_4$$

(f) Treatment with ammonium carbonate. The yellow precipitate of sodium diuranate is digested with ammonium carbonate resulting in the formation of soluble ammonium uranyl carbonate $(NH_4)_4[UO_2(CO_3)_3]$ while any insoluble impurity is removed by filtration.

The filtrate is boiled when lemon yellow crystals of ammonium diuranate $(NH_4)_2U_2O_7$ are obtained.

(g) Conversion into U_3O_8 and then UO_2. Ammonium diuranate on calcination yields black oxide of uranium, U_3O_8 which on heating with charcoal or hydrogen is reduced to uranium dioxide UO_2.

(h) Conversion of UO_2 into U. It can be achieved by either of the two processes:

(*i*) UO_2 is heated with chlorine to form yellow crystals of uranyl chloride UO_2Cl_2 This on heating with carbon in presence of chlorine yields volatile uranium tetrachloride UCl_4. This on heating with sodium metal gives uranium metal.

$$UO_2 + Cl_2 \longrightarrow UO_2Cl_2$$

$$UO_2O_2 + Cl_2 \xrightarrow{\text{Carbon}} UCl_4 + O_2$$

$$UCl_4 + 4Na \longrightarrow U + 4NaCl$$

(*ii*) UO_2 is heated in hydrofluoric acid vapours to form uranium tetrafluoride UF_4 which on heating with magnesium in pressure vessel yields uranium metal.

$$UO_2 + 4HF \longrightarrow UF_4 + 2H_2O$$

$$UF_4 + 2Mg \longrightarrow U + 2MgF_2\ \text{(slag)}$$

Q. 45. Describe occurrence, production, properties and uses of plutonium.

(*Purvanchal, 2007; Lucknow, 2010*)

Ans. Occurrence. It occurs in minute concentration in uranium-bearing ores like Canadian Pitchblende. The isotope involved is Pu^{239}. No other isotope has been found to occur in nature. The important ores of plutonium are *Colorado Pitchblende* and *Camolite*.

Production of Plutonium. It is produced mainly from uranium in nuclear reactors. Natural uranium contains a large proportion of U^{238} whereas U^{235} is present only to the extent of 0.7%. The interaction of slow neutrons with U^{238} present in natural uranium gives rise to plutonium in a series of following nuclear reactions:

$$_{92}U^{238} + {_0}n^1 \longrightarrow {_{92}}U^{239} + \gamma$$

$$_{92}U^{239} \longrightarrow {_{93}}Np^{239} + \beta$$

$$_{93}Np^{239} \xrightarrow{\text{2-3 days}} {_{94}}Pu^{239} + \beta$$

Plutonium, being chemically different from uranium, can be separated relatively easily. Plutonium undergoes nuclear fission changing into U^{235} thereby releasing a tremendous amount of energy. Energy released by a gram of plutonium is equal to that released by about 20 tons of T.N.T. This forms the basis of the manufacture of plutonium bomb. The fission of plutonium can also be carried out in controlled manner. The energy thus released can be utilised for several peaceful purposes.

Properties : 1. There are six allotropic forms of metallic plutonium below its melting point of 912 K.

2. It forms intermetallic compounds with Al, Be, Co, Fe, Mn, Ni and Ag.

3. In aqueous solution it has four oxidation states, *viz.*, + 3, + 4, + 5, + 6. The ions in aqueous solution possess characteristic colours, Pu^{3+} is blue lavender, Pu^{4+} is yellowish brown, PuO_2^{2+} is pink-orange. No solution of pure PuO_2^{2-} has been prepared. The formation of complexes alters the colours of the ions and the aqueous ions formed are strongly paramagnetic.

Uses : 1. It is used for the manufacture of plutonium bomb.

2. Pu^{239} is mainly used for fuelling atomic reactors. The heat of radioactive decay is used for the generation of electricity.

Q. 46. Give a brief account of nuclear fuels. (*Punjab, 2000*)

Or

Discuss briefly the uses of uranium and plutonium. (*Kolhapur, 2000, Purvanchal, 2007*)

Or

Discuss the scope of various nuclear fuels used in India.

Ans. Uranium (U-233, U-235) and plutonium (Pu-239) are fusion fuels that can be directly used in atomic reactors for fission while uranium (U-238) and thorium (Th-232) can't be used as fission fuels directly. However, by nuclear reactions they give fission fuels. U-238 gives Pu-239 while Th-232 gives U - 233.

In all naturally occurring uranium the isotopic abundance of fission fuel U-235 is only 0.714%; the balance being U-238. Enrichment of U-235 is generally carried out by gaseous diffusion using UF_6. Since India has no uranium enrichment facilities, therefore natural uranium as fuel for reactor is used with moderators such as heavy water, graphite, whereby U-235 undergoes fission as:

$$^{235}_{92}U + {^1_0}n \rightarrow {^{141}_{56}}Ba + {^{91}_{36}}Kr + 3{^1_0}n + \text{energy}$$

U-238 changes into Pu-239 as:

$$^{238}_{92}U + ^{1}_{0}n \longrightarrow ^{239}_{92}U \xrightarrow{-\beta} ^{239}_{93}Np \xrightarrow{-\beta} ^{239}_{94}Pu \text{ (fission fuel)}$$

Large-scale separation of plutonium from uranium and other fission products is carried out by solvent extraction with kerosene containing tributyl phosphate (TBP) wherein uranyl and plutonyl nitrate migrate into the organic solvent layer by forming a complex $MO_2(NO_3)_2$ 2 TBP (where M = U or Pu while the fission products remain in the aqueous phase.

The organic solvent layer containing uranium and plutonium is separated, washed with 0.1 M HNO_3 containing a reducing agent whereby Pu^{VI} is reduced to Pu^{III} and migrates to the aqueous phase from which it is recovered while U^{VI} remains in the organic solvent layer and is obtained by washing it with water.

Another fission fuel U-233 which is not available in nature but can be produced by neutron irradiation of Th-232 in breeder reactors as:

$$^{232}_{90}Th + ^{1}_{0}n \xrightarrow{-\beta} ^{233}_{91}Pa \xrightarrow{-\beta} ^{232}_{92}U$$

Thorium is abundantly available in India and is likely to be the major fuel for future atomic reactors.

Q. 47. Write a note on transuranic elements.

(Punjab, 2000; Kerala, 2000; Kolhapur, 2000)

Or

What are uranides? Discuss their uses as nuclear fuel.

Ans. Elements with atomic numbers higher than 92 (that of uranium) are called transuranic elements. In minerals only Np and Pu occur and that too in small amounts. Their extraction is tedious and expensive. Elements from neptunium to lawrencium have been synthesised by artificial means. There are two main nuclear reactions which are used in the synthesis of transuranium elements:

(*i*) *By capture of neutron by the nucleus of an element* which increases its neutron-proton ratio. This results in instability of "host" nucleus. Then a neutron is converted to a proton with emission of a β-particle and the increase of atomic number by one unit. For example,

$$^{238}_{92}U + ^{1}_{0}n \longrightarrow ^{239}_{92}U \xrightarrow{-\beta} ^{239}_{93}Np \xrightarrow{-\beta} ^{239}_{94}Pu$$

(*ii*) *By the capture of unstable nuclei of light elements* ranging from helium to neon which increases atomic number by several units in one step. Bombardment with alpha particles has been used quite often to synthesise trans-uranium elements. For example,

$$^{242}_{95}Am + ^{4}_{2}He \longrightarrow ^{243}_{97}Bk + 3^{0}_{1}n$$

To synthesise very heavy elements, bombarding nuclei with heavier than alpha particles is required. Stripped boron and carbon atoms accelerated in the cyclotron have recently been used to synthesise elements with atomic numbers 102 and 103.

$$^{246}_{96}Cm + ^{12}_{6}C \longrightarrow ^{254}_{102}No + 4^{1}_{0}n$$

$$^{250}_{98}Cf + ^{11}_{5}B \longrightarrow ^{257}_{103}Lr + 4^{1}_{0}n$$

Q. 48. What are transuranic elements? Name the elements having atomic number 95, 97, 99. Give examples.

Ans. See Q. 47.

Elements with atomic number 95, 97, 99 are given below:

At. No.	*Element*	*Symbol*
95	*Americium*	*Am*
97	*Berkelium*	*Bk*
99	*Einsteinium*	*Es*

Q. 49. Write a brief account of conversion of ammonium diuranate to uranium.

Ans. (*i*) Ammonium diuranate $(NH_4)_2U_2O_7$ is calcinated to get U_3O_8.

$$3(NH_4)_2U_2O_7 \longrightarrow 2U_3O_8 + 6NH_3 + 3H_2O + O_2$$

(*ii*) U_3O_8 is then reduced by heating with charcoal or hydrogen to uranium dioxide UO_2 as:

$$U_3O_8 + 2H_2 \longrightarrow 3UO_2 + 2H_2O$$

(*iii*) UO_2 is then heated with HF at 670-770 K to form uranium tetrafluoride UF_4 as

$$UO_2 + 4HF \longrightarrow UF_4 + 2H_2O$$

(*iv*) UF_4 is then heated with calcium or magnesium to form uranium as:

$$UF_4 + 2Ca \longrightarrow 2CaF_2 + U$$

Q. 50. What happens when

(*i*) Uranyl sulphate is treated with excess of Na_2CO_3.

(*ii*) Cerium (III) nitrate is treated with alk.$KMnO_4$.

(*iii*) Ammonium diuranate is calcined.

(*iv*) Sodium diuranate is digested with ammonium carbonate.

Ans. (*i*) Soluble sodium tricarbonatournate is formed

$$UO_2SO_4 + 3Na_2CO_3 \longrightarrow Na_4[UO_2(CO_3)_3] + Na_2SO_4$$

(*ii*) Ce (III) is oxidised to Ce (IV)

$$MnO_4^- + 3Ce^{3+} + 2H_2O \longrightarrow MnO_2 + 3Ce^{4+} + 4\,OH^-$$

(*iii*) On calcination, it gives U_3O_8

$$(NH_4)_2U_2O_7 \longrightarrow 2U_3O_8 + 6NH_3 + 3H_2O + O_2$$

(*iv*) $$Na_2U_2O_7 \xrightarrow{(NH_4)_2CO_3} (NH_4)_4[UO_2(CO_3)_3] \longrightarrow (NH_4)_2U_2O_7$$

Q. 51. Why is europium (II) more stable than cerium (II)?

Ans. Europium (II) has electronic configuration [Xe] $4f^7\,5d^0$ while cerium (II) has electronic configuration [Xe] $4f^1\,5d^1$.

Since in Eu^{2+}, $4f$ sub-shell is half-filled while $5d$ sub-shell is empty, Eu^{2+} ion is more stable than Ce^{2+} in which neither $4f$ sub-shell nor $5d$ sub-shell is half-filled or fully-filled or empty.

Q. 52. Among La (57), Sm (62), Gd (64) and Yb (70)

(*i*) which element will give coloured ion?

(*ii*) which element will give paramagnetic ion?

(*iii*) what are the possible oxidation states?

Ans. (*i*) Sm^{3+} is coloured whereas others are colourless.

(*ii*) Sm^{3+} and Gd^{3+} are paramagnetic while La^{3+} and Yb^{3+} are not. Paramagnetism arises because of unpaired electrons in 4*f* orbitals.

(*iii*) La and Gd have oxidation state of +3 only.

$La^{3+} = [Xe]\,4f^0,\ Gd^{3+} = [Xe]\,4f^7$

Yb has an additional oxidation state of +2 also besides +3

$Yb^{2+} = [Xe]\,4f^{14}$

Sm has also an oxidation state of +2 besides +3

$Sm^{2+} = [Xe]\,4f^6$

Sm^{2+} is an exception as it doesn't belong to a category of half-filled or filled orbitals.

Q. 53. (*i*) Which is more basic: Gd_2O_3 or YbO?

(*ii*) Which is thermally more stable: La $(NO_3)_3$ or $Tm(NO_3)_3$?

(*iii*) Which forms complexes more easily: Pr^{3+} or Er^{2+}? Explain your answers.

Ans. (*i*) Gd_2O_3 is more basic than YbO. Basic strength decreases with increase in atomic number. Due to lanthanide contraction the decrease in size with increase in atomic number increases the covalent character thereby reducing the basic character.

(*ii*) La $(NO_3)_3$ is more stable thermally than Tm $(NO_3)_3$. Thermal stability of nitrates decreases along the series because of the increasing polarising power of the cations with decrease in size along the series.

(*iii*) $Er^{3+} > Pr^{3+}$. The ability to form complexes increases with decrease in size along the series.

Q. 54. Explain the following:

(*i*) Which is more basic: Nd_2O_3 or Dy_2O_3 ?

(*ii*) Which is thermally more stable: $Sm(NO_3)_3$ or $Tm(NO_3)_3$?

(*iii*) La^{3+}, Lu^{3+}, Yb^{2+}, Ce^{4+} are diamagnetic while Sm^{+3} has low paramagnetism.

(*M. Dayanand, 2002*)

Ans. (*i*) Nd_2O_3 is more basic than Dy_2O_3 · Nd^{3+} has greater ionic radius than Dy^{3+}. Thus Dy_2O_3 will have greater covalent character (Dy^{3+} being smaller) and will thus be less basic than Nd_2O_3.

(*ii*) A crystal with smaller ion has always more stable lattice having higher m.p. Tm^{3+} being smaller than Sm^{3+}, Tm $(NO_3)_3$ should form a more stable lattice and thus be more thermally stable.

(*iii*) Sm^{3+} has $4f^5$ configuration having unpaired electrons whereas La^{3+}, Lu^{3+}, Yb^{2+} and Ce^{4+} have empty or complete 4*f* orbitals.

Q. 55. Write the following in order of decreasing strength: Eu^{2+}, Sm^{2+}, Yb^{2+}.

Ans. The order is

$$Sm^{2+} > Eu^{2+} > Yb^{2+}$$

Q. 56. Describe general characteristics of *f*- block elements. (*Meerut, 2000*)

Ans. See Qs. 14, 18, 19 and 22.

NOBLE GASES

Q. 1. Discuss the occurrence of noble gases.

Ans. Noble gases exist as under:

(*a*) In the earth's atmosphere. The non-radioactive gases are present in the atmosphere to the extent of 1% by volume. The composition of these gases in air is as follows:

100 litres of air contains 9.34 ml of argon, 1.82 ml of neon, 0.524 ml of helium, 0.114 ml of krypton and 0.0087 ml of xenon.

(*b*) In the stars and interstellar space. It has been established that the main constituents of the sun are hydrogen and helium. The masses of the major planets Jupiter and Saturn are estimated to be due to 23% and 37% of helium respectively. The atmosphere of Mars consists mostly of nitrogen and argon.

(*c*) In minerals and springwaters. Noble gases occur in minerals in the occluded form. Except helium, the concentration of noble gases in minerals is quite low. The concentration of helium in thorianite has been found to be 10.5 ml/g.

(*d*) In natural gas. Helium along with traces of neon are found to occur in natural gas in the wells of certain oil fields of U.S.A. In the natural gas, helium concentration varies from 2 to 7%. At present most of the helium is commercially recovered from natural gas which is a much cheaper source than atmosphere.

Q. 2. Describe the isolation of noble gases by fractionation of liquid air.

(*Kumaon, 2001; Lucknow, 2002*)

Ans. Principle. This method is based on the difference in the boiling points of noble gases. The gases are separated from one another and also from oxygen and nitrogen (which are the other major constituents of air) by fractional distillation. The boiling points of noble gases along with those of oxygen and nitrogen are given in Table 20.1.

Table 20.1. Boiling points of some gases

GAS	He	Ne	N_2	Ar	O_2	Kr	Xe
Boiling point, K	4.2	27.2	77	87.3	90.3	121	165

Some important conclusions from the above table are:

(*i*) Helium and neon have very low boiling points. In other words, during the process of liquefying, air, helium and neon remain in gaseous state while the other constituents change into liquid.

(*ii*) There is only a difference of +3 K in the boiling points of oxygen and argon. So these two gases are collected more or less as a single fraction.

(*iii*) Krypton and xenon have fairly high boiling points as compared to those of nitrogen, oxygen and argon. So krypton and xenon are always found in the less volatile portion of liquid air and are always easy to separate from other gases.

The plant used for isolation of noble gases is shown in Fig. 20.1.

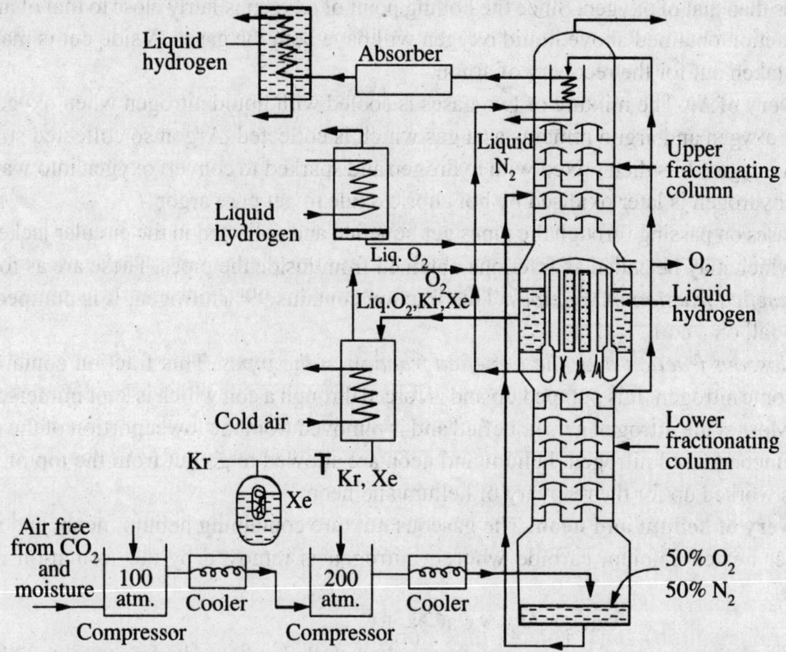

Fig. 20.1. Isolation of noble gases from air.

Procedure. Air free from carbon dioxide and moisture is compressed to 100 atmospheric pressure. Air is then cooled to remove heat produced by compression. The air is again compressed to about 200 atmospheric pressure and again cooled by a mixture of salt and ice. It is then passed into liquefier in which it escapes through a valve and expands to a pressure of about 20 atmospheres. Then air cools down. The cold compressed air enters the lower fractionating column near its bottom (Fig. 20.1). The air expands and as a result cools down further to an extent that oxygen and nitrogen change into liquid and collect at the bottom of the fractionating column. This liquid fraction consists of equal proportions of oxygen and nitrogen. It is pumped up and is dropped from the middle of the upper fractionating column.

The uncondensed gases rise up and reach the middle portion where they pass through pipes surrounded by liquid oxygen. Four different fractions are obtained, here two fractions are obtained from outside the pipes and two fractions are obtained from condensed material in the pipes. These fractions are worked up as follows:

Two fractions outside the pipes are:

(*i*) Liquefied fraction in liquid oxygen;

(*ii*) Gaseous fraction over liquid oxygen.

(*i*) Krypton and xenon get cooled and are dissolved in liquid oxygen. This liquid fraction of O_2, Kr and Xe is taken out for recovery of Kr and Xe.

(*i*) Recovery of Kr and Xe. The liquid is taken to another rectifying column and is dropped from above while cold air is being circulated through it. Consequently, liquid oxygen changes into vapours and it is collected from the top of this column while krypton and xenon still remain liquids and are collected from the bottom of this column. From the residual liquid containing krypton and xenon, the two gases are separated either by selective low temperature absorption or by fractionation in a separate rectifier.

***(ii)* Gaseous fraction over the liquid oxygen.** Argon remains in the gaseous state as its boiling point is less than that of oxygen. Since the boiling point of oxygen is fairly close to that of argon so the gaseous fraction obtained above liquid oxygen will have both the gases. A side cut is made and the fraction is taken out for the recovery of argon.

Recovery of Ar. The mixture of two gases is cooled with liquid nitrogen when oxygen changes into liquid oxygen and argon remains as a gas which is collected. Argon so collected still contains about 20% oxygen. It is then mixed with hydrogen and sparked to convert oxygen into water and the unreacted hydrogen is later oxidised by hot cupric oxide to get pure argon.

The gases on passing through the pipes get liquefied and collected in the circular jacket from two fractions which may be called as fractions obtained from inside the pipes. These are as follows:

(*i*) *Liquefied fraction in the pipes.* This fraction contains 99% nitrogen. It is pumped up and is allowed to fall on a coil.

(*ii*) *Gaseous fraction over the liquefied fraction in the pipes.* This fraction contains helium, neon and some nitrogen. It is pumped up and is forced through a coil which is kept immersed in liquid nitrogen. Most of the nitrogen gets liquefied and is removed from the lower portion of the cooled coil while the uncondensed nitrogen, helium and neon are allowed to go out from the top of the cooled coil and is worked up for the recovery of helium and neon.

Recovery of helium and neon. The gaseous mixture containing helium, neon, and nitrogen is passed over heated calcium carbide wherein nitrogen is removed by the formation of calcium cyanamide.

$$CaC_2 + N_2 \rightarrow CaCN_2 + C$$

The outgoing mixture of helium and neon is then chilled in liquid hydrogen (b.p. 20 K) wherein neon changes into a liquid while helium still remains a gas and the two are collected.

Q. 3. Describe chemical method of extraction of noble gases from air.

(Indore, 2001; Bangalore, 2005)

Ans. The methods involves the following two steps:

(*a*) Removal of nitrogen and oxygen from air. Any of the following method may be used to remove nitrogen and oxygen from pure dry air.

(*i*) Rayleigh's method. This apparatus is shown Fig. 20.2. It consists of a 50-litre round-bottomed flask and is fitted with a cork having five holes. Leads for platinum are fitted though two holes. A mixture of pure dry air and oxygen in the ratio of 9:11 is introduced through a tube from the third hole. Long glass tubes are fitted through the remaining two holes, one of these tubes is used as an inlet for sodium hydroxide solution while the other is used as an exit for sodium hydroxide solution.

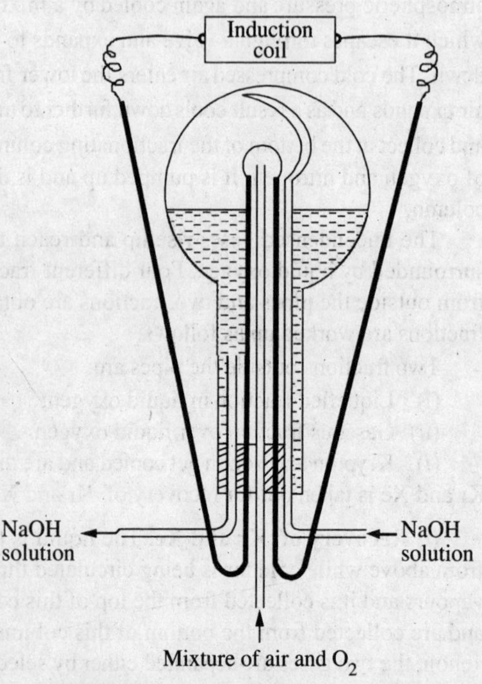

Fig. 20.2. Isolation of noble gases from air.

Mixture of air and oxygen is introduced. Sodium hydroxide solution is forced into the flask. An electric discharge from an induction coil of 6000 – 8000 volts between the platinum electrodes is passed. Due to combination of nitrogen and oxygen, two oxides of nitrogen are formed.

$$N_2 + O_2 \longrightarrow 2NO$$

$$2NO + O_2 \longrightarrow 2NO_2$$

These oxides of nitrogen get adsorbed in concentrated sodium hydroxide solution.

$$2NaOH + 2NO_2 \longrightarrow NaNO_2 + NaNO_3 + H_2O$$

$$2NaOH + NO + NO_2 \longrightarrow 2NaNO_2 + N_2O$$

As a result, most of oxygen and nitrogen are removed from the air in the flask. The residual gases contain noble gases with traces of oxygen. The mixture of residual gases is passed through tubes containing alkaline pyrogallol where oxygen gets adsorbed and the mixture of noble gases, free from oxygen, comes out.

***(ii)* Ramsay's method.** In this method, pure and dry air is passed slowly over hot finely divided copper contained in a long tube to remove oxygen.

$$2Cu + O_2 \longrightarrow 2CuO$$

Thereafter, the air (free from oxygen) is passed over heated magnesium ribbon when nitrogen is removed due to the formation of magnesium nitride.

$$3Mg + N_2 \longrightarrow Mg_3N_2$$

Though this method gives very satisfactory results but it takes a very long time to complete.

***(iii)*Fischer-Ringe's method.** In this method, pure dry air is passed over a mixture of 90% calcium carbide and 10% calcium chloride at 800°C when oxygen and nitrogen are removed according to the following reactions:

$$CaC_2 + N_2 \longrightarrow CaNCN + C$$

$$C + O_2 \longrightarrow CO_2$$

$$2C + O_2 \longrightarrow 2CO$$

The resulting gases are passed over hot cupric oxide to oxidise carbon monoxide to carbon dioxide.

$$CO + CuO \longrightarrow Cu + CO_2$$

Therefore, the residual gases are passed through potash bulbs to absorb carbon dioxide.

$$2KOH + CO_2 \longrightarrow K_2CO_3 + H_2O$$

The gases are dried by bubbling them through concentrated sulphuric acid and phosphorus pentoxide. The gaseous mixture now obtained contains only the non-radioactive noble gases.

***(b)* Separation of Noble Gases by Dewar's Charcoal Method. Principle.** This method is based on the principle that the adsorption capacity of these gases at low temperatures by coconut charcoal increases with increase in their atomic weights. Thus, helium has the lowest adsorption capacity while xenon has the maxmum adsorption capacity. Moreover, adsorption capacity depends on temperature, *i.e.*, the lower the atomic weight of the noble gas, the lower is the temperature needed to adsorb it.

Procedure. The mixture of noble gases is brought in contact with coconut charcoal packed in a bulb maintained at –100°C in a Dewar's flask for one hour (Fig. 20.3). Xenon, Krypton and argon are adsorbed by the coconut charcoal while helium and neon remain unadsorbed and are pumped out of the bulb. The mixture of helium and neon is then introduced into another bulb containing coconut charcoal and maintained at –180°C when neon gets adsorbed while helium is pumped out. Thereafter, this charcoal is warmed to recover neon.

The first charcoal in which xenon, krypton and argon had been absorbed is brought in contact with another charcoal sample which had earlier been cooled to the liquid air temperature. Argon being a lighter gas diffuses into third charcoal unit and is recovered by warming it.

Krypton and xenon remain absorbed on the first charcoal which is slowly warmed so that its temperature rises to –90°C. At this temperature krypton does not remain adsorbed and so it starts coming out of the charcoal and is collected. Xenon remains adsorbed on the charcoal which is released on further warming and is collected.

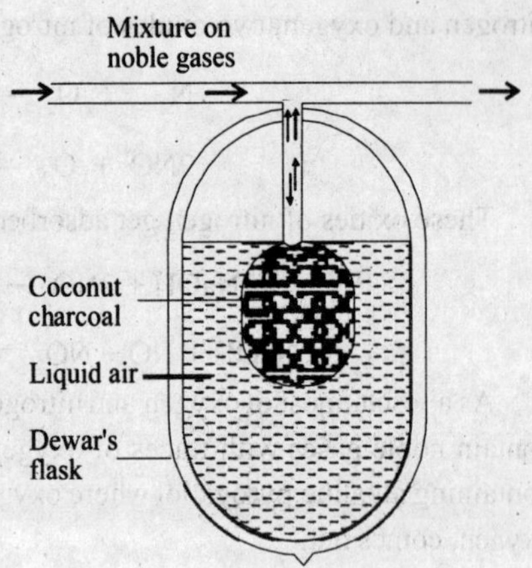

Fig. 20.3. Fractional adsorption of noble gases by activated charcoal.

Q. 4. Why are noble gases unreactive?

(Indore, 2001)

Ans. Noble gases are not very reactive. This may be ascribed to the following reasons:

(*a*) They have high ionization energies.

(*b*) Their electron affinities are almost zero.

Both these factors are a direct result of their electronic configuraion. At each noble gas the electronic sub-level is just complete ($ns^2\ np^6$) and this particular electronic configuration resists change or disturbance of any kind. Therefore, all noble gases are extremely unreactive as compared to the other elements.

Q. 5. Write the reaction which led to the first noble gas compound.

Or

Discuss the factor that led to the formation of first noble gas compound by Bartlett.

Ans. The first chemical compound of the noble gases was reported by N. Bartlett in 1962. The preparation of this compound, $XePtF_6$, was based on the chance discovery of another compound formed by the reaction between oxygen gas and platinum hexafluoride (deep red vapours).

$$O_2(g) + PtF_6(g) \longrightarrow O_2PtF_6(s)$$

Dioxygen hexafluoroplatinate(V)

The above compound is formulated as $O_2^+(PtF_6)^-$. This word means that platinum hexafluoride is a very strong oxidising material, and could even attract electron from molecular oxygen. Since the ionization energy of oxygen gas (1165.6 kJ mol^{-1}) is comparable to that of xenon gas (1170 kJmol^{-1}) and as their molecular diameters are also similar, Bartett thought that xenon would also be susceptible to oxidation with platinum hexafluoride. Therefore, he proceeded to mix xenon and platinum hexafluoride (both gases) at room temperature and obtained on his first attempt an orange yellow crystalline compound of composition, $XePtF_6$.

$$Xe(g) + PtF_6(g) \longrightarrow Xe^+(PtF_6)^-\ (s)$$

Q. 6. Xenon is known to form some compounds but not He and Ar. Explain why?

Or

Xenon compounds are more stable than those of other noble gases. Explain.

Or

Prove that the noble gases are not so inert.

Ans. We may expect only the heavier noble gases which have low ionization energy values to show some reactivity as a result of electron loss. Ionization energies of noble gases are given in Table 20.2.

Table 20.2. Comparison of ionization energy values of noble gases with other reactive elements.

Element	*Ionization energy value. kJ mol*$^{-1}$
Helium	2372.3
Neon	2080.7
Argon	1520.5
Krypton	1350.7
Xenon	1170.4
Radon	1037.0

From the above ionization energy values, it is obvious that we may expect radon to go into chemical combination with the greatest ease. But radon is radioactive and we know very little about its ability to form chemical compounds. Xenon forms the largest number of compounds as compared to any other noble gas because xenon is the stable element which has the smallest ionization energy. Krypton tends to form compounds but with difficulty as compared to xenon. The lighter noble gases, helium, neon and argon which have very high ionisation energies, are expected to show very little reactivity. So far experimental evidence also suggests that no chemical compounds of these gases have been reported.

Q. 7. Why most of noble gas compounds involve only fluorine and oxygen?

(Meerut, 2002)

Ans. Any chemical reactivity shown by these gases may be attributed to their tendency to lose electrons. For this reason, the combining atom must be highly electronegative, so as to accept these electrons. Xenon forms stable compounds only with the most electronegative elements; fluorine (X = 4.0) and oxygen (X = 3.5) or with highly electronegative groups such as $OSeF_5$ and $OTeF_5$ that contain both oxygen and fluorine.

Q. 8. Describe the preparation and properties of Xenon difluoride, XeF_2.

(Nagpur, 2003; Rohilkhand, 2003; Purvanchal, 2007, Vidyasagar, 2007; Patna, 2010; Garhwal, 2010)

Ans. Preparation. (*i*) It can be prepared quantitatively by photosynthesis on exposing to sunlight mixtures of xenon and oxygen difluoride (OF_2) or xenon and excess of oxygen monofluoride (O_2F_2) at −118°C.

(*ii*) It is best prepared in the laboratory in high yields and purity by heating xenon and fluorine in the molar ratio of 2:1 (excess of xenon is taken to avoid the formation of any xenon tetrafluoride).

The reaction is carried out in a nickel or Monel vessel at 400°C. The reaction products are quenched to room temperature and XeF_2 is isolated by vacuum sublimation.

$$Xe + F_2 \xrightarrow[1\ atm.]{400°C} XeF_2 \text{ (Xe in excess)}$$

Properties. (*i*) It is a colourless crystalline solid melting at 140°C.

(*ii*) The fluorine of XeF_2 may be substituted by highly electronegative ligands by reacting it with protonic acids.

$$XeF_2 + HX \longrightarrow FXeX + HF$$

$$FXeX + HX \longrightarrow XeX_2 + HF$$

(*iii*) The xenon-F bond in XeF_2 has almost the lowest bond energy among the known fluorides. Due to this reason XeF_2 is a strong oxidising and fluorinating agent.

(*iv*) It reacts with sulphur trioxide to give Xe.

$$XeF_2 + 2SO_3 \longrightarrow Xe\uparrow + S_2O_6F_2$$

(*v*) It is reduced by hydrogen to xenon

$$XeF_2 + H_2 \longrightarrow Xe\uparrow + 2HF$$

(*vi*) It reacts with NO to give nitrosyl fluoride.

$$2NO + XeF_2 \longrightarrow 2NOF + Xe\uparrow$$

It appears that in all of the above fluorination and oxidation reactions, the first step involves the ionization of XeF_2 to XeF^+ followed by electron transfer to give XeF. The XeF species is very weakly bound and is a potential source of fluorine atoms.

(*vii*) It dissolves in HF (hydrofluoric acid) without any reaction. The solution does not show any change in conductivity.

(*viii*) It reacts with SbF_5 or AsF_5 to give solid compounds in which it reacts as a fluoride ion donor.

$$XeF_2 + SbF_5 \longrightarrow XeF_2 \cdot SbF_5$$

$$XeF_2 + AsF_5 \longrightarrow XeF_2 \cdot AsF_5$$

It has now been established that the compounds contain ionic species, $XeF^+.AsF_6^-$ and XeF^+ SbF_6^-. The reaction can also take place as follows:

$$XeF_2 + 2SbF_5 \longrightarrow XeF_2 \cdot 2SbF_5$$

XeF_2 also reacts with AsF_3 in a different stoichiometry:

$$2XeF_2 + AsF_5 \longrightarrow 2XeF_2 . AsF_5 \longrightarrow (Xe_2F_3)^+ (AsF_6)^-$$

The ion $Xe_2F_3^+$ has the expected linear geometry around xenon and an angular geometry at the central fluorine.

F
151°
Xe: Xe: 190 pm
F F
$[Xe_2F_3]^+$

(*ix*) It oxidizes iodine in the presence of fluoride ion acceptor to give IF.

$$I_2 + XeF_2 \xrightarrow{BE_3} 2IF + Xe\uparrow$$

$$IF + XeF_2 \xrightarrow{BE_3} IF_3 + Xe\uparrow$$

Q. 9. Discuss the structure and bonding in XeF_2, xenon difluoride.

(*Kerala, 2000; Kanpur, 2000; Indore, 2001; Purvanchal, 2007; Vidyasagar, 2007*)

Ans. (*i*) Infrared and Raman spectra show that XeF_2 is a linear molecule in the vapour state.

$$F — Xe — F$$

(*ii*) Crystal structure of solid XeF_2 shows that XeF_2 consists of linear units.

Bonding. Xenon difluoride may be represented in terms of Lewis electron dot representation as:

The actual structure is a resonance hybrid of the following two structures:

$$[:\ddot{F}\times\ \ \overset{\times\times}{\underset{\times\times}{Xe}}{}^{\times}_{\times}]^{+}\ \ {}^{\times}\ddot{F}:^{-} \longleftrightarrow :\ddot{F}^{\times}\ \ [{}^{\times}_{\times}\overset{\times\times}{\underset{\times\times}{Xe}}\ \ {}^{\times}\ddot{F}:]^{+}$$

Valence bond treatment. In the valence bond approach, the valence shell of xenon is expanded by promotion of an electron into the higher energy vacant $5d$ orbitals:

	5s	5p			5d				
Ground state of Xe : [Kr] $5s^2\,4d^{10}\,5p^6$	↑↓	↑↓	↑↓	↑↓					
Valence state of Xe : [Kr] $5s^2\,4d^{10}\,5p^5\,5d^1$	↑↓	↑↓	↑↓	↑	↑				

sp^3d

The five hybridised orbitals are oriented towards the corners of an equivalent triangle and along the two vertices to give the shape of trigonal bipyramid (Fig. 20.4). The orbitals along the triangle accommodate lone pairs and those along the vertices contain single electrons. These two orbitals form bonds with fluorine. The shape of XeF_2 is given below in Fig. 20.4.

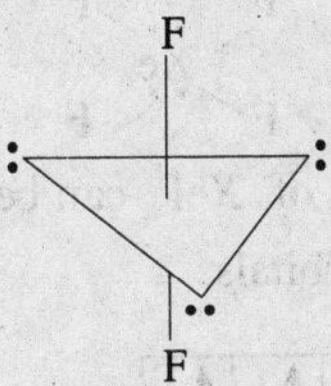

Fig. 20.4. Shape of XeF_2.

Q. 10. Describe the preparation and properties of xenon tetrafluoride, XeF_4.

(*Purvanchal, 2007; Vidyasagar, 2007; Patna, 2010; Garhwal, 2010*)

Ans. Preparation. (*i*) It is prepared by heating a mixture of xenon and fluorine in a molar ratio 1:5 at 400°C and 5-6 atmosphere in an enclosed nickel vessel.

$$Xe + 2F_2 \xrightarrow[6\ atm]{400°C} XeF_4$$

(*ii*) By passing an electric discharge in a reaction vessel maintained at –80°C containing xenon and fluorine in the molar ratio 1:2. The pressure is maintained between 2-15 mm of Hg.

Properties. (*i*) It is a colourless crystalline solid, melting at 117.1°C.

(*ii*) It is reduced by mercury.

$$XeF_4 + 2Hg \longrightarrow Xe + 2HgF_2$$

(*iii*) It is also reduced by hydrogen gas at 130°C.

$$XeF_4 + 2H_2 \longrightarrow Xe + 4HF$$

(*iv*) It forms an addition compound with antimony pentafluoride.

$$XeF_4 + SbF_5 \longrightarrow XeF_4SbF_5 \longrightarrow (XeF_3)^+ (SbF_6)^-$$

$$XeF_4 + 2SbF_5 \longrightarrow XeF_4.2SbF_5$$

$$\longrightarrow (Xe\,F_3)^+ (Sb_2F_{11})^-$$

The XeF_3^+ ions have the expected T-shaped structure as shown in Fig. 20.5.

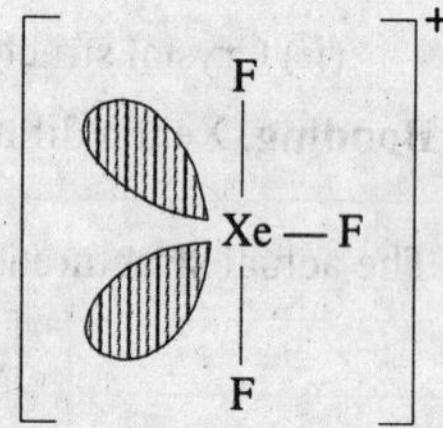

Fig. 20.5. Structure of $(XeF_3)^+$

(*v*) It reacts with boron trichloride to give boron trifluoride.

$$3XeF_4 + 4BCl_3 \longrightarrow 4BF_3 + 3Xe + 6Cl_2$$

(*vi*) It is readily hydrolysed by water. The reaction can be controlled to yield different products.

$$XeF_4 + H_2O \xrightarrow{-80°C} \underset{\text{Xenon oxyfluoride}}{XeOF_2} + 2HF$$

$$XeF_4 + H_2O \longrightarrow \underset{\text{Xenon trioxide}}{XeO_3}$$

Q. 11. Discuss the structure and bonding in xenon tetrafluoride, XeF_4.

(*Kerala, 2000; M. Dayanand, 2000; Indore, 2001; Andhra, 2002; Calcutta, 2005; Nagpur, 2005; Guwahati, 2006, Purvanchal, 2007, Vidyasagar, 2007*)

Ans. (*i*) Vibrational spectroscopy and electron diffraction data have established that XeF_4 is a planar molecule:

```
F        F
  \ ∙∙ /
    Xe
  / ∙∙ \
F        F
```

(*ii*) The valence bond representation of XeF_4 can be explained by the promotion of electron from the 5*p* sub-level to high energy 5*d* orbitals.

Ground state configuration of Xe : [↑↓] [↑↓|↑↓|↑↓]

Excited state configuration of Xe : [↑↓] [↑↓|↑|↑] [↑|↑| | |]

Six sp^3d^2 hybrid orbitals

The 5*s*, three 5*p* and two 5*d* atomic orbitals of xenon hybridize to produce six sp^3d^2 hybridized orbitals. The four singly occupied hybridized orbitals are used by fluorine for bond formation to xenon. The other two orbitals contain lone pairs. The molecule has octahedral geometry with two transpositions occupied by lone pairs (Fig. 20.6).

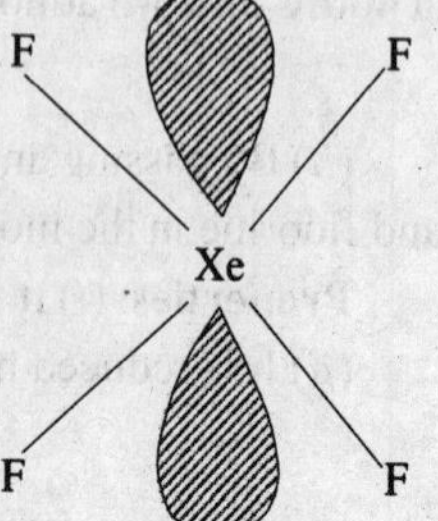

Fig. 20.6. Structure of XeF_4.

Q. 12. How is xenon oxide difluoride, $XeOF_2$, prepared? Discuss its structure.

Ans. Preparation. It is prepared by the hydrolysis of XeF_4 at – 80°C.

$$XeF_4 + H_2O \xrightarrow{-80°C} XeOF_2 + 2HF$$

Structure. (*i*) It has T-shaped structure.

F
|
O=Xe
|
F

(*ii*) The valence bond structure of $XeOF_2$ may be explained if we assume that two electrons from 5*p* sub-level of xenon are promoted to 5*d* sub-level.

Ground state configuration of Xe :

Excited state configuration of Xe :

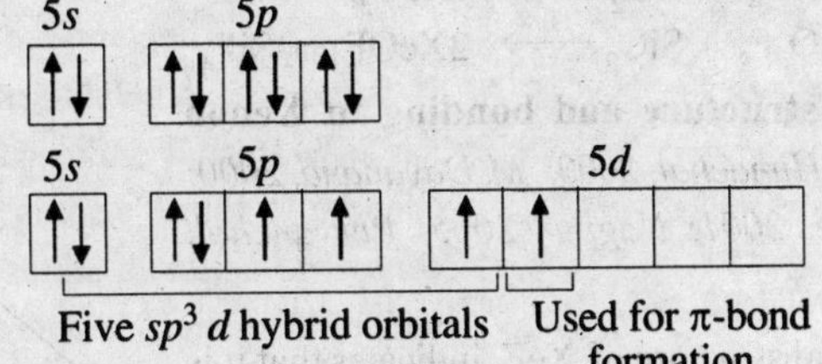

The 5*s*, three 5*p* and 5*d* orbitals of xenon hybridize to form five sp^3d hybridized orbitals. The three singly occupied orbitals are used for σ - bond formation of two fluorine atoms and one oxygen atom. The two positions of trigonal bipyramidal structure are occupied by lone pairs. One of the electrons in the 5*d* orbital which is not used in the hybridization scheme is used for π-bond formation to oxygen atom. It may be mentioned that two of orbitals lying along the corners of a triangle are occupied by lone pairs.

F

O ═ Xe

F

Fig. 20.7. Structure of $XeOF_2$

Q. 13. Describe the preparation and properties of xenon hexafluoride, XeF_6. (*Calcutta, 2000; Kolhapur, 2000; Himachal, 2000; Indore, 2001; Guwahati, 2002; Vidyasagar, 2007; Purvanchal, 2007*)

Ans. Preparation. (*i*) Xe F_6 may also be prepared by the oxidation of XeF_4 with O_2F_2 at temperatures below −80°C.

(*ii*) It may also be prepared by the electric discharge of a mixture of fluorine (3 parts) and xenon (1 part) at low temperature.

(*iii*) It is best prepared by reacting xenon and fluorine in the ratio 1:20 in a nickel vessel. The reaction is carried out at 300°C and at 60 atmospheric pressure.

$$Xe + 3F_2 \xrightarrow[60\text{ atm.}]{300°C} XeF_6$$

Properties. (*i*) It is a colourless solid melting at 49.5°C.

(*ii*) It is very rapidly hydrolysed.

(*iii*) It is a stronger oxidation and fluorinating agent than both XeF_2 and XeF_4.

(*iv*) It reacts with fluoride ion acceptors to produce $(XeF_5)^+$ species as

$$XeF_6 + PtF_5 \longrightarrow XeF_6 . PtF_5 \longrightarrow (XeF_5)^+ (PtF_6)^-$$
$$XeF_6 + AsF_5 \longrightarrow XeF_6 . AsF_5 \longrightarrow (XeF_5)^+ (AsF_6)^-$$
$$XeF_6 + SbF_5 \longrightarrow XeF_6 . SbF_5 \longrightarrow (XeF_5)^+ (SbF_6)^-$$
$$XeF_6 + BF_3 \longrightarrow XeF_6 . BF_3 \longrightarrow (XeF_5)^+ (BF_4)^-$$

(*v*) It reacts with water to give xenon trioxide which is highly explosive and makes the handling of XeF_6 extremely difficult.

$$XeF_6 + 3H_2O \longrightarrow XeO_3 + 6\,HF$$

(*vi*) It is reduced to xenon by the action of H_2, HCl and NH_3.

$$XeF_6 + 3H_2 \longrightarrow Xe + 6HF$$
$$XeF_6 + 6\,HCl \longrightarrow Xe + 3Cl_2 + 6HF$$
$$XeF_6 + 8NH_3 \longrightarrow Xe + 6NH_4F + N_2$$

(*vii*) It cannot be stored in glass or quartz vessels owing to the following reactions.

$$2XeF_6 + SiO_2 \longrightarrow 2XeOF_4 + SiF_4$$
$$2XeOF_4 + SiO_2 \longrightarrow 2XeO_2F_2 + SiF_4$$
$$2XeO_2F_2 + SiO_2 \longrightarrow 2XeOF_3 + SiF_4$$

Q. 14. Discuss structure and bonding in Xenon hexafluoride, XeF_6. (*Himachal, 2000; M. Dayanand, 2000; Kerala, 2000; Indore, 2001; Nagpur, 2005; Purvanchal, 2007*)

Ans. (*i*) The crystal structure of XeF_6 indicates that it is generally present as $(XeF_5)^+F^-$ in the solid state. The pyramidal XeF_5^+ cations are bridged by free fluoride ions. The determination of exact structure of gaseous XeF_6 molecule has proved difficult. It is now well established that it is not a regular octahedral molecule but the arrangement of the fluorines is distorted from the regular octahedral arrangement in such a way as to suggest that the non-bonding electron pair occupies a position in the centre of one face of an octahedron so that the fluorines at the corners of this face are forced apart, as shown in Fig. 20.8.

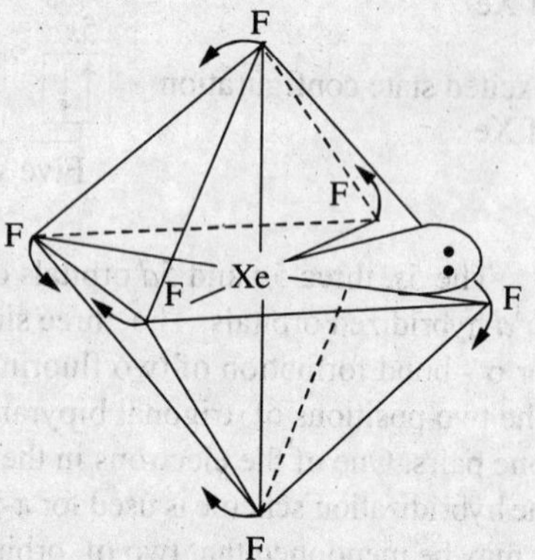

Fig. 20.8. Distorted octahedral structure of XeF_6.

(*ii*) The valence bond description predicts pentagonal bipyramid structure of XeF_6 with one transposition occupied by a lone pair. It may be assumed that three electrons from the 5*p* sublevel are promoted to 5*d* sublevel as:

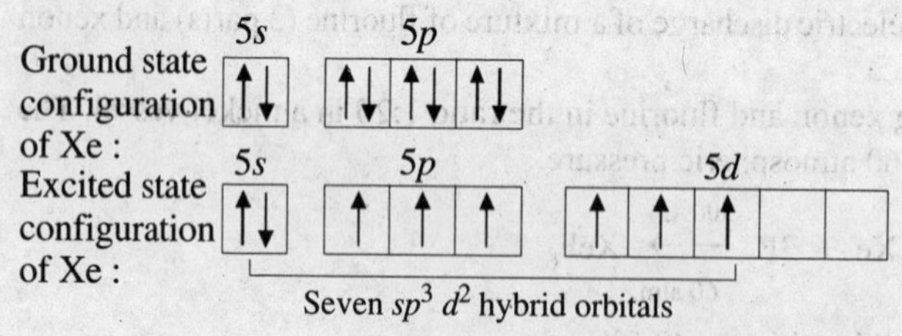

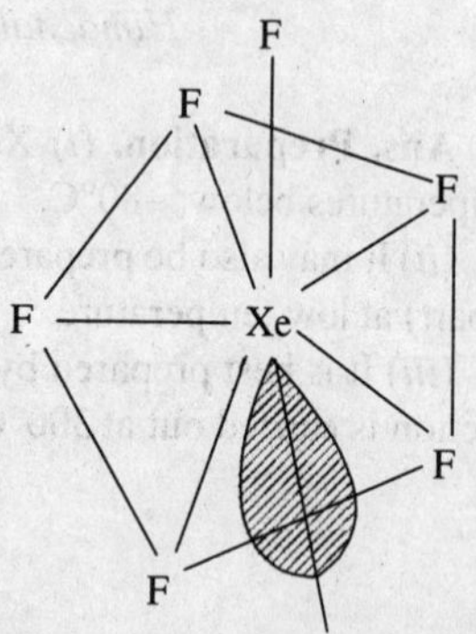

Fig. 20.9. Pentagonal bipyramidal structure of XeF_6.

The 5*s*, three 5*p* and three 5*d* orbitals hybridize to produce seven sp^3d^3 hybrid orbitals. The singly occupied orbitals are used for bond formation to six fluorine atoms. The seventh orbital contains a lone pair. (Fig. 20.9).

Q. 15. Describe the preparation and properties of Xenon oxide tetrafluoride, $XeOF_4$.

Ans. Preparation. (*i*) It can be prepared by the reaction of SiO_2 with XeF_6.

$$2XeF_6 + SiO_2 \longrightarrow 2\,XeOF_4 + SiF_4$$

(*ii*) It may be prepared by the partial hydrolysis of XeF_6.

$$XeF_6 + H_2O \longrightarrow XeOF_4 + 2HF$$

(*iii*) By heating a mixture of xenon (1 part) and fluorine (4 parts) with large excess of oxygen round 230°C.

$$Xe + 2F_2 + O \xrightarrow{230°C} XeOF_4$$

(*b*) **Properties.** (*i*) It is a colourless substance melting at –46°C.

(*ii*) It is reduced by hydrogen gas to xenon.

$$XeOF_4 + 3H_2 \longrightarrow Xe + H_2O + 4HF$$

(*iii*) It reacts with water or SiO_2 to give XeO_2F_2 and XeO_3.

$$XeOF_4 + H_2O \longrightarrow XeO_2F_2 + 2HF$$

$$XeO_2F_2 + H_2O \longrightarrow XeO_3 + 2HF$$

$$2XeOF_4 + SiO_2 \longrightarrow XeO_2F_2 + SiF_4$$

$$2XeO_2F_2 + SiO_2 \longrightarrow 2XeO_3 + SiF_4$$

Q. 16. Discuss the structure and bonding in $XeOF_4$. (*Nagpur, 2000*)

Ans. (*i*) The structure of molecule $XeOF_4$ is square pyramid as shown below:

(*ii*) The valence bond description of $XeOF_4$ may be given by assuming that three 5*p* electrons are promoted to the 5*d* sub-level.

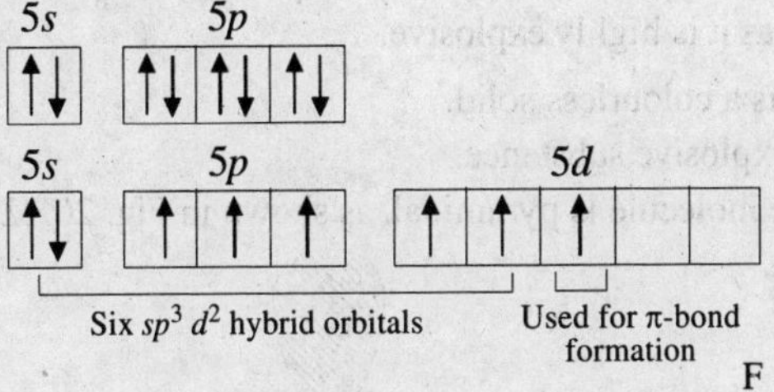

The 5*s*, three 5*p* and two 5*d* atomic orbitals hybridize to give six sp^3d^2 hybrid orbitals. The five singly occupied orbitals are used for σ-bond formation to four fluorine and one oxygen atom. The sixth hybrid orbital is occupied by the lone pair. The single electron in a *d*-orbital is used for π-bond formation to oxygen atom. This explains the required geometry of $XeOF_4$ (Fig. 20.10).

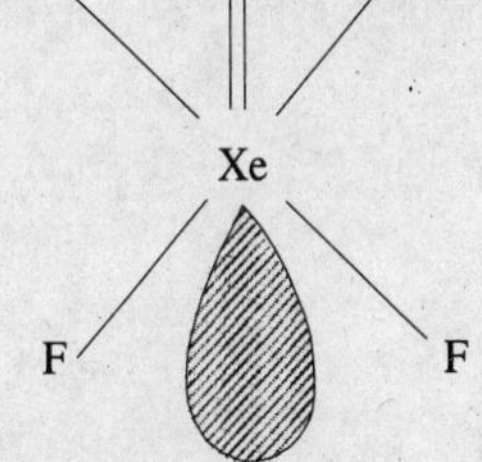

Fig. 20.10. Structure of $XeOF_4$.

Q. 17. Describe the preparation, properties and structure of Xenon dioxide difluoride, XeO_2F_2. (*Calcutta, 2005*)

Ans. Preparation. It is prepared by mixing XeO_3 and $XeOF_4$ at low temperatures (–78°C).

Properties. (*i*) It is a colourless solid which melts at 30°C.

(*ii*) It is easily hydrolysed to give XeO_3.

Structure and bonding. (*i*) It has a see-saw structure.

(*ii*) The valence bond structure suggests trigonal bipyramid geometry with one equatorial position occupied by a lone pair.

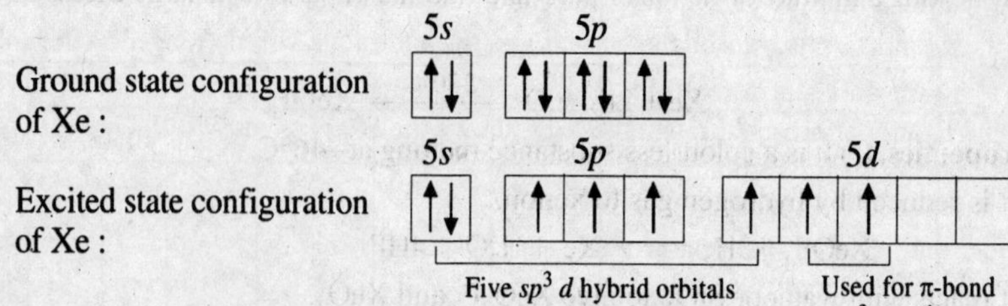

The three $5p$ electrons are promoted to higher energy $5d$ sub-level. The $5s$, three $5p$ and one $5d$ orbitals hybridize to give five sp^3d hybrid orbitals. The four singly occupied orbitals are used for bond formation to two fluorine and two oxygen atoms. The fifth hybrid orbital contains the lone pair. The other two $5d$ electrons of xenon which do not take part in the hybridization scheme are involved in π-bond formation to two oxygen atoms. The structure of XeO_2F_2 is represented in Fig. 20.11.

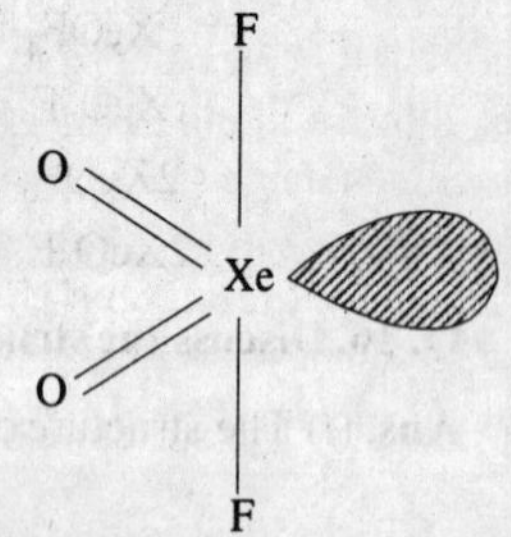

Fig. 20.11 Structure of XeO_2F_2

Q. 18. Describe the preparation, properties and structure of Xenon trioxide, XeO_3.

(*Kolhapur, 2000*)

Ans. Preparation. It is prepared by the hydrolysis of XeF_6 and XeF_4. Great care is to be taken during its preparation as it is highly explosive.

Properties. (*i*) It is a colourless solid.

(*ii*) It is a highly explosive substance.

Structure. (*i*) The molecule is pyramidal, as shown in Fig. 20.12.

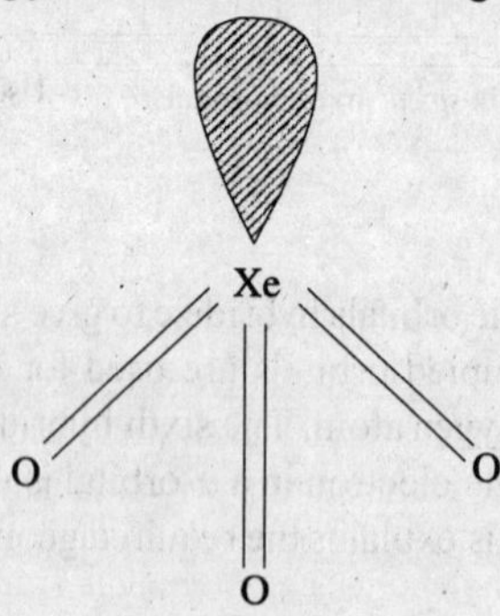

Fig. 20.12. Structure of XeO_3

(*ii*) The valence bond description suggests xenon to acquire sp^3 hybridization and one of the hybrid orbitals contains a lone pair.

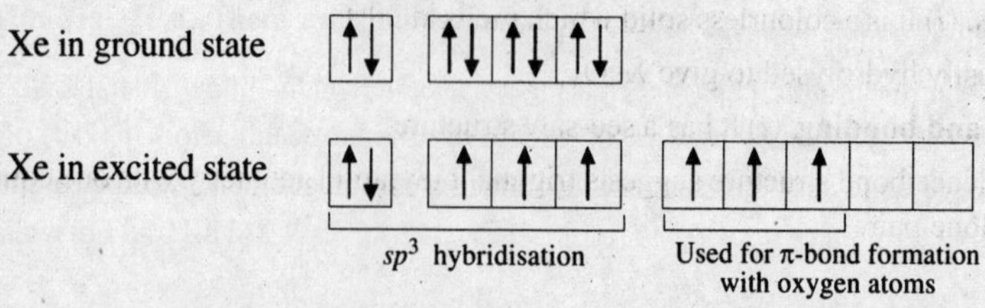

It is a case of sp^3 hybridisation giving rise to tetrahedral shapes. One of the tetrahedral positions is occupied by the lone pair ($5s^2$) of electrons. Three orbitals overlap with the orbitals of oxygen to form three Xe-O single bonds. Three $5d$ orbitals which don't participate in hybridisation form π-bonds by overlapping laterally with three singly occupied orbitals of oxygen. The shape of XeO_3 is given in Fig. 20.12.

Q. 19. Describe the preparation, properties and structure of Xenon tetraoxide, XeO_4.

(*Kolhapur, 2000*)

Ans. Preparation. It is prepared by the action of conc. H_2SO_4 on sodium or barium xenate (Na_4XeO_6 ; Ba_2XeO_6) at room temperature. It may be purified by vacuum sublimation at −78°C.

Properties. It is quite unstable and decomposes to xenon and oxygen.

Structure. The molecule is tetrahedral. In this case, the electrons from 5*s* and 5*p* orbitals are excited to 5*d* orbital, thus giving rise to eight orbitals having single electrons. Out of these, one *s* and three *p* orbitals hybridise. Hybridised orbitals form four σ-bonds with oxygens. Unhybridised orbitals (4 in no.) form four π-bonds with oxygen to give XeO_4 as shown in Fig. 20.13.

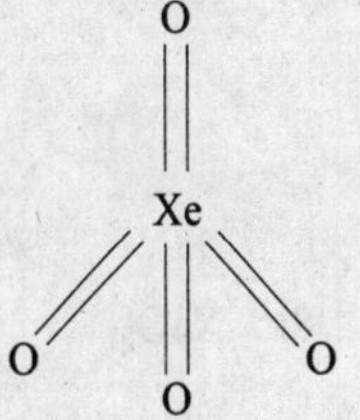

Fig. 20.13. Shape of XeO_4.

Q. 20. Give a comprehensive list of hybridisation and shape of xenon compounds.

Ans. The hybridisation and shape of xenon compounds are given in the table below:

Molecule	*Shape*	*Description*
1. XeF_2	F — Xe — F	Linear; three nonbonding electron pairs at the points of an equilateral triangle. sp^3d hybridisation.
2. XeF_4		Square planar; two nonbonding electron pairs one above and one below the plane of the molecule. sp^3d^2 hybridisation.
3. XeF_6	or	Distorted octahedron; one nonbonding electron pair either at the centre of a face or the midpoint of an edge. sp^3d^3 hybridisation.
4. $XeOF_4$		Square pyramidal; one nonbonding electron pair coming out from the base of the pyramid. sp^3d^2 hybridisation.
5. $XeOF_2$		T-shaped; two nonbonding electron pairs pointing towards the corners of an equilateral triangle. sp^3d hybridisation.

(*Contd.*)

Molecule	Shape	Description
6. XeO_2F_2	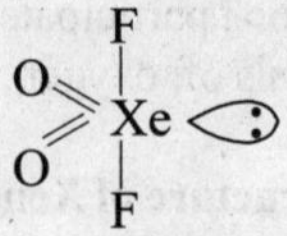	See-saw; one nonbonding electron pair pointing towards the corner of an equilateral triangle. sp^3d hybridisation.
7. XeO_3	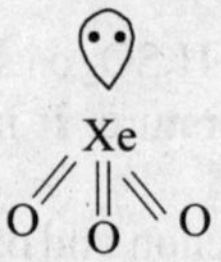	Trigonal pyramidal; one nonbonding electron pair pointing out from the apex of the pyramid. sp^3 hybridisation.
8. XeO_4	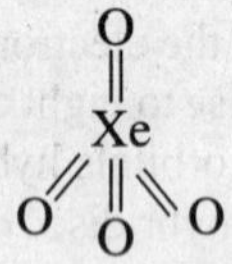	Tetrahedron sp^3 hybridisation.
9. XeF_5^-	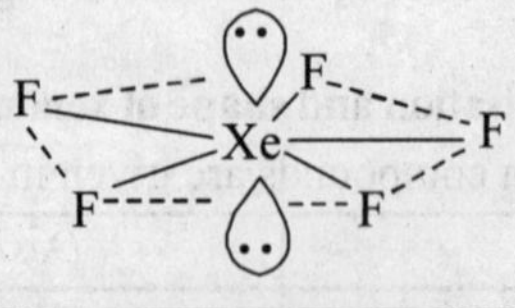	Pentagonal planar; two nonbonding electron pairs one above the plane and the other below the plane. sp^3d^3 hybridisation.

Q. 21. What are clathrates? How are they prepared? (*Lucknow, 2000; Himachal, 2000*)

Ans. According to Powell, clathrate compounds consist of atoms or molecules of one substance called the 'guest' held in the crystalline cages formed by the molecules of a second substance called the 'host'.

The only type of interaction in these "cage compounds" is the relatively weak van der Waals forces and there is no chemical bonding. It is not necessary that all the cavities in the host structure are filled. As a result, the clathrates are normally non-stoichiometric compounds.

Essential Conditions for Clathrate Formation

(*a*) Presence of cavities of appropriate size in the crystal lattice of the 'host'.

(*b*) Size of the 'guest' particle should be such as to fit in the cavities of the 'host' without bringing any atoms closer together than which correspond to the van der Waals radii of the atoms.

Stability of Clathrate Compounds

Clathrate compounds once formed are quite stable due to the following reasons:

(*a*) The 'guest' molecules tightly fit in the cavities of the 'host' molecules.

(*b*)The 'guest' molecules within the cages are at a minimum potential energy.

The 'guest' can escape from the 'host' only when the forces holding the molecular cages together are overcome. This can be achieved by the following two methods:

(*i*) By heating the crystals.

(*ii*) By dissolving them in a suitable solvent such as alcohol.

Preparation of Clathrates

Clathrate compounds of argon, krypton and xenon have been isolated by locking up these gases in the cages of water, phenol or quinol. However, clathrate compounds of helium and neon have not yet been prepared because helium atoms are thought to be too small and difficult to deform or polarize as required by van der Waals bonding to be imprisoned in the cages. The neon is borderline element and its clathrate could possibly be formed at pressures in excess of 160 atmospheres. It is predicted that in clathrate of neon, if formed, the neon atoms (being so small) will diffuse through the crystal from one cavity to another.

Uses of Clathrates

(*i*) Krypton-85 clathrate provides a safe and useful source of beta radiation which are useful for measuring thickness of gauges.

(*ii*) Xenon-133 clathrate provides a compact source for gamma radiation.

(*iii*) In the separation of noble gases. For example, neon can be separated from argon, krypton and xenon because neon is the only gas that does not form clathrates with quinol.

(*iv*)Clathrates play an important role in some physiological actions, *e.g.*, it is thought that the anaesthetic action of xenon and many other anaesthetics is due to aqueous clathrate in physiologically strategic spots. When the anaesthetic is no longer administered, the clathrate equilibrium is destroyed, the clathrate decomposes and the consciousness returns.

Q. 22. Complete the following reactions:

(*i*) $\mathbf{XeF_4 + SbF_5 \longrightarrow}$

(*ii*) $\mathbf{XeF_4 + H_2O \longrightarrow}$

(*iii*) $\mathbf{XeF_6 + H_2O \longrightarrow}$

Ans. See Qs. 10 and 13.

Q. 23. Which of the following compounds is most difficult to prepare and why?

$$\mathbf{XeF_2, XeF_4, XeF_6}$$

Ans. XeF_6 is most difficult to prepare. Maximum number of electrons are required to be excited from 5*p* to 5*d* orbitals to prepare this compound, hence large amount of energy is needed.

Q. 24. Why xenon fluorides cannot be stored in quartz vessel?

Ans. Xenon fluorides react with silicon dioxide present in quartz to form SiF_4.

$$2XeF_6 + SiO_2 \longrightarrow 2XeOF_4 + SiF_4$$

Q. 25. Complete the following equations:

(*i*) $\mathbf{XeF_2 + I_2 \longrightarrow}$

(*ii*) $\mathbf{XeF_2 + SO_3 \longrightarrow}$

(*iii*) $\mathbf{XeF_2 + NO \longrightarrow}$

(*iv*) $\mathbf{XeF_2 + HSO_3F \longrightarrow}$

(*v*) $\mathbf{XeF_2 + H_2 \longrightarrow}$

Ans. See Q. 8.

Q. 26. What happens when (give equation also)

(*a*) Xe and F_2 are heated in a sealed Ni vessel at 400°C in the ratio of 1:20.

(*b*) XeF_4 is treated with benzene.

Ans. See Qs. 10 and 13.

Q. 27. How does XeF_4 react with:

(*a*) HCI (*b*) H_2O (*c*) H_2 (d) NO?

Ans. See Q. 8.

Q. 28. What is the stability of clathrates due to?

Ans. See Q. 21.

Q. 29. Draw the structures of XeF_3^+?

Ans. See Q. 10 (Fig. 20.5).

COORDINATION COMPOUNDS

Q. 1. What is the difference between double salt and coordination compounds?

(*Calcutta, 2007; Lucknow, 2010*)

Ans. (*i*) Those compounds which lose their identity in solution are called double salts.

(*ii*) Those compounds which do not lose their identity in solution are called *coordination* or *complex* compounds. The double salts when dissolved in water give tests for each constituent ion *e.g.*, a solution of carnallite (KCl .MgCl.6 H_2O) gives tests for Mg^{++} and Cl^- ions. A solution of Mohr's salt similarly gives tests for Fe^{2+}, NH_4^+ and SO_4^{2-} ions. They only exist in crystals. The coordinate compounds or complex compounds when dissolved in water do not give tests of each constituent ion, *e.g.*, solution of potassium ferrocyanide $K_4Fe(CN)_6$ in water does not give the usual test of Fe^{2+} and CN^- ions, rather a new ferrocyanide ion is $[Fe(CN)_6]^{4-}$ formed and similarly the solution of potassium argentocyanide in water does not give test of Ag^+ and CN^- ions but instead a new argentocyanide ion is formed. Ferrocyanide and argentocyanide ions are called complex ions and compounds containing such ions are called *complex or coordination compounds.*

Q. 2. List some important applications of coordination compounds.

Ans. 1. A well known Ziegler-Natta catalyst, a complex of aluminium and titanium, is used for low pressure polymerisation of ethylene and propene to propare polyalkenes and rubber.

2. Complex formation is used extensively in analytical chemistry for detection, estimation and separation of metal ions.

3. Coordination compounds are also used in soil treatment, water softening, corrosion control, extraction of metal, electroplating etc.

4. Chlorophyll which is a vital pigment in photosynthesis is a magnesium complex and haemoglobin which carries O_2 to animal cell is a complex of iron.

Q. 3. Explain that $Fe(NH_4)_2(SO_4)_2$ (Mohr salt) is a salt but $K_4Fe(CN)_6$ is a complex.

Ans. $Fe(NH_4)_2(SO_4)_2$ is a salt because its solution gives reactions of Fe^{2+}, NH_4^+ and SO_4^{2-}. $K_4Fe(CN)_6$ is not a salt but it is a complex because the solution of this compound furnishes only K^+ and $[Fe(CN)_6]^{4-}$. It does not give tests of Fe^{2+} and CN^-.

Q. 4. How will you distinguish between the following?

(*i*) $[Cr(H_2O)_6]Cl_3$ and $[Cr(H_2O)_5Cl]Cl_2 \cdot H_2O$

I II

(*ii*) $[Co(NH_3)_5Br]SO_4$ and $[Co(NH_3)_5SO_4]Br$?

I II

Ans. (*i*) Compound I will give precipitate of AgC1 with $AgNO_3$ that will correspond to the presence of three chloride ions whereas compound II will give precipitate corresponding to only two chloride ions.

(*ii*) Compound I gives the test for sulphate ions and not for bromide ion whereas compound II gives the test for bromide ion not for sulphate ions.

Q. 5. Define the following terms:

(*i*) Ligands **(*ii*) Coordination number** **(*iii*) Coordination sphere.**

(*Lucknow, 2002; Rohilkhand, 2003; Bangalore, 2004; Bhopal, 2004; Nagpur 2008*)

Ans. (*i*) Ligands: The ion or neutral molecule attached to central atom in a coordination compound is called Ligand. Ligand may be neutral molecule such as ammonia, water, ethyle-nediamine. It may also be an ion such as Cl^-, CN^-, NO_2^-, SCN^- etc.

(*ii*) Coordination number. Total number of ligands which are directly attached to the central metal atom is known as coordination number of that metal ion. Coordination number varies from 1-12 but more common are four and six. The coordination number of silver in $[Ag(NH_3)_2]^+$ is two and that of platinum in $[Pt(Cl)_6]^{-2}$ is 6. A metal atom may exhibit more than two coordination numbers in different complexes. For example, in complexes $[FeCl_4]^-$ and $[Fe(CN)_6]^{3-}$ iron atom shows co-ordination numbers 4 and 6 respectively. Similarly copper with ethylenediamine (abbreviated as en) shows coordination number 6 in $[Cu(en)_3]^{2+}$ but with ammonia in $[Cu(NH)_4]^{2+}$ it shows coordination number 4.

(*iii*) Coordination sphere. The central metal atom and ligands that are directly attached to it enclosed in square brackets are collectively known as coordination sphere. The ligands and metal atom inside the square brackets behave as single constituent unit.

Q. 6. How is the charge on the complex ion calculated?

Ans. The charge carried by a complex ion is the algebraic sum of the charges carried by central metal ion and the ligands coordinated to the central metal ion. For example, in complex ion $[Cu(NH_3)_4]^{2+}$ Cu ion carries +2 charge and NH_3 molecule is neutral therefore the net charge on the complex ion is +2. Similarly in the complex ion $[Ag(CN)_2]^{-1}$, Ag ion carries a charge of +1 and each CN carries a charge of −1. Therefore the net charge on the complex ion $[Ag(CN)_2]$ is + 1 −2 = − 1.

Some basic rules for the determination of oxidation number of central atom in the complex compound are given below:

(*i*) In a complex compound, group I elements (Li, Na, K, etc.) have an oxidation state of +1 and group II elements (Be, Mg, Ca, etc.) have an oxidation state of + 2.

(*ii*) In a compound oxidation number of hydrogen is generally + 1 except in hydride where it is −1.

(*iii*) In a compound oxidation number of halogen is generally taken as −1.

(*iv*) The sum of the charges carried by atoms, ions or molecules in a complex compound is equated to the actual charge of the complex ion.

Q. 7. What are different types of ligands?

Ans. If a ligand is attached to the central metal atom only by one coordination bond then the ligand is called as *monodentate or unidentate ligand.* Example of unidentate ligands are Cl^-, NO_2^-, Br^-, CN^-, OH^-, H_2O, NH_3, CO, etc.

Ligands having two or more than two donor atoms are called *polydentate ligands.* Depending upon the number of donor atoms or coordinating atoms, the ligand may be classified as bidentate (two), tridentate (three donor atoms) and those with four, five and six atoms are called tetra, penta and hexa dentate ligands respectively.

For example, in tetramine copper (II) complex ion, *i.e.*, $[Cu(NH_3)_4]^{2+}$, I, NH_3 is a monodentate ligand and ethylene diamine is a bidentate ligand in ethylene diamine copper (II) complex ion II.

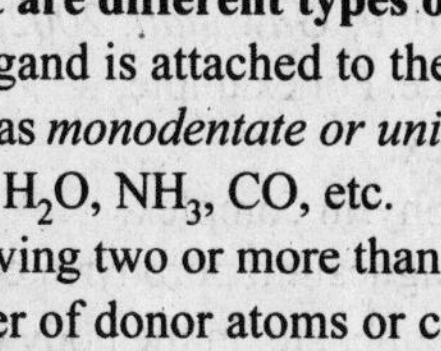

NH_3 NH_3 / Cu / NH_3 NH_3 $]^{2+}$

I

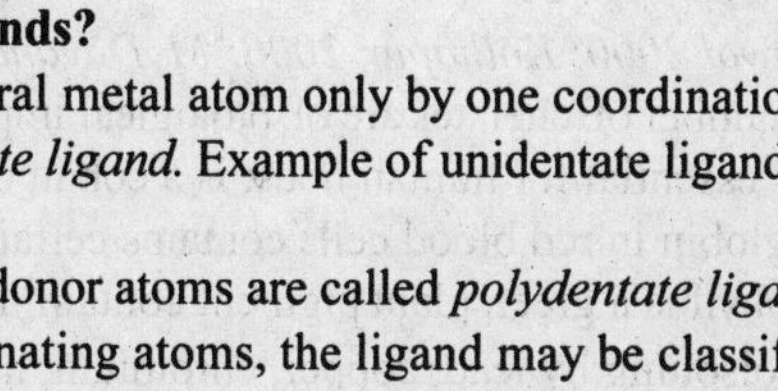

$CH_2—H_2N$ $NH_2—CH_2$ / Cu / $CH_2—H_2N$ $NH_2—CH_2$ $]^{2+}$

II

ethylenediamine ligand $\left(\begin{array}{l} CH_2 - \ddot{N}H_2 \\ | \\ CH_2 - \ddot{N}H_2 \end{array}\right)$ is a typical bidentate ligand.

Q. 8. What are chelates? (*M. Dayanand, 2002; Kolhapur, 2000; Garhwal, 2000; Guwahati, 2002; Nagpur 2008; Lucknow, 2010; Pune, 2010*)

Ans. Chelates are the special types of coordination compounds containing one or more rings in their structure. The ring structures are formed linking metal ion with polydentate ligand. The ring structure compounds are called chelates. The polydentate ligands are called chelating agents and the process of formation is called *chelation.*

$$Cu^{2+} + 2 \begin{array}{l} CH_2 - NH_2 \\ | \\ CH_2 - NH_2 \end{array} \longrightarrow \left[\begin{array}{ccccc} CH_2 - NH_2 & & & & NH_2 - CH_2 \\ | & \searrow & & \swarrow & | \\ & & Cu & & \\ | & \nearrow & & \nwarrow & | \\ CH_2 - NH_2 & & & & NH_2 - CH_2 \end{array}\right]^{2+}$$

For example, Ethylenediamine, a bidentate ligand, forms a complex with Cu ions.

This complex contains a pair of five-membered ring and this is called a chelate.

Another familiar example of chelation is nickel-DMG complex which is formed when nickel ion reacts with DMG in an ammoniacal medium

$$2 \begin{array}{l} CH_3 - C = NOH \\ \quad\quad | \\ CH_3 - C = NOH \end{array} + Ni^{2+} \longrightarrow \begin{array}{ccccc} & & H & & \\ & O & & O & \\ & \uparrow & & | & \\ CH_3 - C = N & & & & N = C - CH_3 \\ | & \searrow & & \swarrow & | \\ & & Ni & & \\ | & \nearrow & & \nwarrow & | \\ CH_3 - C = N & & & & N = C - CH_3 \\ & | & & \downarrow & \\ & O & & O & \\ & & H & & \end{array} + 2H^+$$

Dimethyl glyoxime (DMG)

Chelate

Chelates have been classified as bidentate, tridentate, quadridentate and hexadentate depending upon whether the chelating agent is bidentate, tridentate, quadridentate or hexadentate respectively.

Q. 9. Describe some applications of chelate compounds.

(*Garhwal 2000; Kolhapur, 2000; M. Dayanand, 2000; Guwahati, 2002; Nagpur 2008*)

Ans. (1) A number of chelates are of biological importance. For example,

(*i*) Vit. B_{12}, essential for human body, is a cobalt complex.

(*ii*) Haemoglobin in red blood cells contains certain porphyrin complex.

(*iii*) Chlorophyll is a green plant pigment containing a magnesium porphyrin complex.

(*iv*) Metal poisoning by lead, copper, chromium, nickel, zinc, etc. in the body can be removed by treating with EDTA. It forms metal-EDTA chelate which is easily excreted in the urine.

(2) In analytical chemistry chelates are used in both qualitative and quantitative analysis. For example, Ni^{2+}, Mg^{2+}, Cu^{2+} etc. are quantitatively precipitated by chelating agent as insoluble *compounds.*

(3) Standard EDTA, a bidendate ligand solution, is widely used in volumetric analysis to determine hardness of water.

Q. 10. Describe the postulates of Werner's theory of coordination compounds.

(*Purvanchal, 2007; Nagpur, 2008; Ranchi, 2010*)

Ans. Important postulates of this theory are:

(1). Metals possess two types of valencies, *primary or ionizable* valency and *secondary or non-ionizable* valency.

(*a*) Primary valencies of the central metal ion are satisfied by negative ions. For example, in simple salts such as $CoC1_3$, $CuSO_4$ and AgC1, the primary valencies of Co, Cu and Ag are 3, 2 and 1 respectively. In complex compounds its attachment to metal ion is shown by dotted lines. In modem terminology it corresponds to oxidation state of metal atom or ion and is *ionizable.*

(*b*) The secondary valencies are satisfied by either –ve ions or neutral molecules. Its attachment to metal is shown by thick lines and is non-ionizable. It corresponds to coordination number of metal ion or atom.

(2) Every metal ion has a tendency to satisfy both its primary and secondary valencies. In order to meet this requirement, a negative ion must satisfy both types of valencies.

(3) Every metal has fixed number of secondary valencies.

(4) Secondary valencies (or ligands satisfying the secondary valencies) are always directed towards fixed positions in space around the central metal ion.

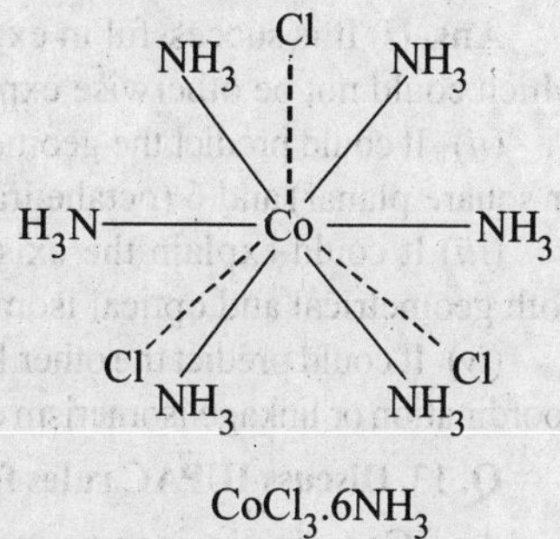

$CoCl_3.6NH_3$

The geometry of a complex is determined by the number and arrangement of such ligands in space. For example, if a metal ion has secondary valency of four, the geometry of complex can be tetrahedral or square planar. On the other hand, if secondary valency is six, the complex is expected to be *octahedral.*

The primary valencies are non-directional and play no important role in determining the geometry of a complex compound. Thus on the basis of Werner's theory the complex $CoCl_3 \cdot 6NH_3$ may be represented as shown in the figure.

Cobalt has three primary or ionizable valencies (shown by dotted lines) satisfied by chlorine. It has six secondary valencies (shown by dark lines) and are satisfied by ammonia molecules.

In modern notation the complex ion is shown in square brackets with ionable ligands outside the coordination sphere. The primary valencies are ionic bonds, whereas secondary valencies are coordinate bonds. Thus the complexes can be rewritten as shown in Table 21.1.

Table 21.1.

S. No.	*Werner formula*	*Modern formula*	*Cation*	*Anion*	*Total no. of ions*
1.	$Co\,C1_3 . 6NH_3$	$[Co(NH_3)_6]C1_3$	$[Co(NH_3)_6]^{3+}$	$3Cl^-$	4
2.	$Co\,C1_3 . 5NH_3$	$[Co(NH_3)_5C1]Cl_2$	$[Co(NH_3)_5\,Cl]^{2+}$	$2Cl^-$	3
3.	$Co\,C1_3 . 4NH_3$	$[Co(NH_3)_4C1_2]C1$	$[Co(NH_3)_4C1_2]^+$	Cl^-	2
4.	$Co\,C1_3 . 3NH_3$	$[Co(NH_3)_3C1_3]$	non-ionizable	—	—

Q. 11. Give evidence to support Werner's theory of coordination compounds.

(Nagpur, 2005)

Ans. The following evidence supports Wemer's theory:

1. Validity of Werner's Co(III)-Ammonia complexes (Structure 1 to 4) given in Table 21.1 is supported by chemical reactivity or by precipitation data, *i.e.*, by precipitating the number of chlorine as chloride ion by adding silver nitrate solution to freshly prepared solution of the complexes. The number of chloride ions precipitated are 3, 2, 1 and zero respectively providing primary or ionizable valencies of complexes as given by Werner.

2. Measurement of molar conductivities of solutions of coordination compounds also helps to estimate the number of ions in solution. For example, a value of 500 ohm^{-1} or more indicates the presence of 4 ions, between 200-250 ohm^{-1} indicates the presence of 3 ions and about 100 ohm^{-1} indicates 2 ions. It may also be noted that when the number of coordinate anions is same as the oxidation number of central cation, the coordination compound is a non-electrolyte.

The validity of formulating the Werner's compounds $CoCl_3$. $6NH_3$ and $CoCl_3$ $5NH_3$ as $[Co(NH_3)_6]$ Cl_3 and $[Co(NH_3)_5$ Cl] Cl_2 respectively is supported by conductance measurement. Experimentally the conductivities of $CoCl_3.6NH_3$ and $CoCl_3$ $5NH_3$ fall in the range expected for 4 and 3 ions respectively and hence Werner's compounds are represented as $[Co(NH_3)_6Cl_3]$ (4 ions) and $[Co(NH_3)_5Cl]Cl_2$ (3 ions).

Q. 12. Give the applications of Werner's theory of coordination compound.

Ans. (*i*) It is successful in explaining the formation and behaviour of coordination compounds which could not be otherwise explained on the basis of electronic theory of valency.

(*ii*) It could predict the geometry of different complexes with coordination number 4 (tetrahedral or square planar) and 6 (octahedral).

(*iii*) It could explain the existence of isomerism in coordination compounds. It could predict both geometrical and optical isomerism in coordination compound.

(*iv*) It could predict the other forms of isomerism in coordinational compounds, *e.g.*, ionisation or coordination or linkage isomerism etc. which are different from both geometrical and optical isomerism.

Q. 13. Discuss IUPAC rules for nomenclature of coordination compound.

Ans. Coordination compounds are named according to following rules:

1. The order of naming ions. (*i*) *The positive ions (whether in a complex ion or in a simple salt) is named first followed by negative ions.* For example.

HCl-hydrogen chloride, $FeCl_2$ -iron (II) chloride, KCl-potassium chloride, $FeCl_3$-Iron (III) chloride $[Cu(NH_3)_4 SO_4]$– tetrammine copper (II) sulphate, $K_4[Fe(CN)_6]$–Potassium hexacyanoferrate (II).

(*ii*) *Non-ionic or molecular complexes are given a one-word name*

$[Co(NH_3)_3Cl_3]$ Triamminetrichloridocobalt (III)

$[Pt(NH_3)_2Cl_2]$ Diamminedichloridoplatinum (II)

$[Pt(NH_3)_2Cl_4]$ Diammnetetrachloridoplatinum (IV)

2. Naming of ligand. (*i*) **Neutral ligands** have no special ending except that NH_3 is ammine, H_2O is aqua, CO-carbonyl, NO-nitrosyl. Rest are used as such, *e.g.*, ethylenediamine (en) and pyridine (py).

(*ii*) **Negative ligands.** *Names of negative ligands end in –o and positive ligands end in ium.*

Example of Negative ligands are: –CN–cyano, Cl^-–chlorido, Br–bromido, I^- – iodido, NO_2^- nitrito-N, NO_3^- –nitrato, O NO^-–nitrito–O, NH_2^- –amido, NH^{2-} – imido, OH^- –hydroxo, SO_4^{2-} – sulphato, O^{2-}–oxo, $C_2O_4^{2-}$– oxalato, SCN^- –thiocyanato, NCS^-–isothiocyanato, $S_2O_3^{2-}$ –thiosulphato, CO_3^{2-}–carbonato, O_2^{2-}–peroxo, ClO_4^- – perchlorato, O_2^-–superoxo, CH_3COO^- –acetato etc.

(*iii*) **Positive ligands** (which are very few) end in ium, *e.g.*, $NH_2 - NH_3^+$–hydrazinium, NO_2^+– nitronium, NO^+ –nitrosonium.

(*iv*) *Prefixes di-, tri-, tetra- etc. are used to indicate the no. of ligands of one type.*

For complex ligands which already have di-, tri-, etc. on their names, the prefixes bis, tris, tetrakis etc. are used and the group is placed within brackets. For example $[Cu(en)_2]Cl_2$ is named as bis (ethylenediamine) copper (II) chloride.

3. Naming of a complex ion. (*a*) Ligands are named first followed by central metal atom at the end. The ligands are listed in alphabetical order regardless of charge and number of each. The name of ligand is treated as a unit and the prefixes di- tri- etc. are not considered for alphabetial order. For example, "diammine" is listed under *'a'* not *'d'*.

According to old rules, negative ligands were written first followed by neutral and positive ligand.

(*b*) When a coordination sphere or complex is a cation or a neutral molecule, the name of central metal atom is written as such, followed immediately by a Roman numeral in parenthesis to indicate its oxidation state, *e.g.*, copper (II), cobalt (II) etc. When the complex is anionic, the suffix **ate** is added to the name of metal, along with the oxidation state, *e.g.*, ferrate (II), argentate (I), manganate (II), nickelate (I) etc.

The complete name may be given as shown in examples:

$[Cr(H_2O)_4Cl_2]Cl$ – Tetraaquadichloridochromium (III) chloride.

$[Co(NH_2)_6]Cl_2$ – Hexaamminecobalt (III) chloride

$K_4[Fe(CN)_6]$ – Potassium hexacyanoferrate (II)

$[Co(NH_3)_6][CuCl_5]$ – Hexaamminecobalt (III) pentachloridocuperate (I)

$K_4[Ni(CN)_4]$ – Potassium tetracynonickelate (0)

$K_4[Mo(CN)_8]$ – Potassium octacyanomolybdate (IV)

(*c*) If the complex contains two or more metal ions, it is called as polynuclear complex. The ligands which link the two metal atoms are called bridge groups and are separated from rest of the complex by hyphens and denoted by Greek letter 'μ'.

$$\left[(H_2O)_4Fe\begin{matrix}OH\\ \diagup\ \diagdown\\ \diagdown\ \diagup\\ OH\end{matrix}Fe(H_2O)_4\right](SO_4)_2$$

Octaaqua-μ-dihydroxodiiron (III) sulphate

$$\left[(NH_3)_4Co\begin{matrix}NH_2\\ \diagup\ \diagdown\\ \diagdown\ \diagup\\ NO_2\end{matrix}Co(NH_3)_4\right](NO_3)_4$$

Octaammine-μ-amido-μ-nitrodicobalt (III) nitrate

(*d*) Geometrical isomers are named by term **cis** or **trans** depending upon whether similar ligands are on the same side or on opposite side of the metal ion, *e.g.*, $[Pt(NH_3)_2Cl_2]$ exists as

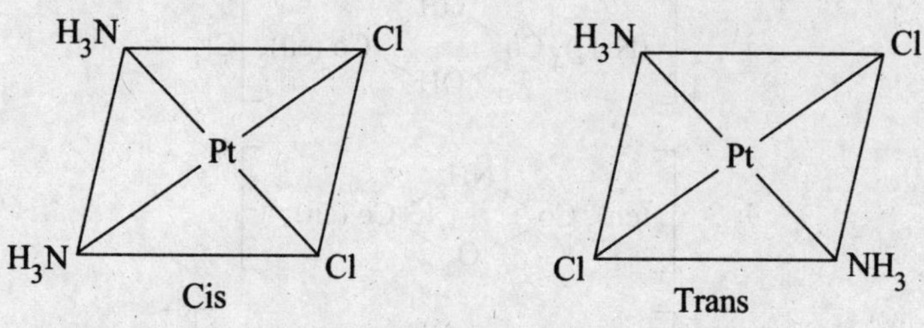

Diamminedichloridoplatinum (II)

(*e*) Dextro and laevorotatory optically active compounds are designated by (+) – or d- and (–) – or l- respectively, *e.g.*, *d*-$K_3[Cr(C_2O_4)_3]$ – potassium trioxalatochromate (III).

(*f*) Complicated ligands are gener ally denoted by abbreviations such as 'en' for ethylenediamine, 'py' for pyridine, 'phen' for 1, 10 phenanthroline etc.

Q. 14. Give the names of following coordination compounds:

(1) $[Co(ONO)(NH_3)_5]SO_4$, (2) $K(BF_4)$, (3) $[Cr(NH_3)_6]Cl$, (4) $Na[Co(CO)_4]$, (5) $[(CH_3)_4N]_2[Co(NCS)_4]$, (6) $Na_3[Ag(S_2O_3)_2]$, (7) $K_3[Al(C_2O_4)_3]$, (8) $[Co(en)_3]_2(SO_4)_3$, (9) Cis-$[Cr(H_2O)_4Cl_2]Cl$, (10) $(NH_4)_2TiCl_6$, (11) $[Pt(NH_3)_4][PtCl_4]$, (12) $H[AuCl_4]$, (13) $Na_2[Fe(CN)_5NO]$, (14) $K[Ag(CN)_2]$, (15) $[Co(NH_3)_6][CdCl_5]$, (16) $[Co(NH_3)_4.H_2O.Br](NO_3)_2$.

(Kerala, 2000; Garhwal, 2000; Lucknow, 2010)

Ans. 1. Pentamminenitrito cobalt (III) sulphate.
2. Potassium tetrafluoridoborate (III)
3. Hexamminechromium (III) chloride
4. Sodium tetracarbonylcobaltate (I)
5. Tetramethylammonium tetrathiocyanato–N–cobaltate (II)
6. Sodium dithiosulphato argentate (I)
7. Potassium trioxalatoaluminate (III)
8. Tris (ethylenediamine) cobalt (III) sulphate
9. Cis-tetraaquadichloridochrominium (III) chloride
10. Ammonium hexachloridotitanate (IV)
11. Tetrammine platinum (II) tetrachloridoplatinum (II)
12. Hydrogen tetrachloridoaurate (III)
13. Sodium pentacyanonitrosylferrate (III)
14. Potassium dicyanoargenate (I)
15. Hexaamminecobalt (III) pentachloridocadmate (II)
16. Tetraammineaquabromidocobalt (III) nitrate.

Q. 15. Write down the structures of following polynuclear complexes:

1. **Decaammine-μ-hydroxodichromium (III) bromide**
2. **Tetrabis (ethylenediammine)-μ-amido-μ-hydroxodicobalt (III) sulphate**
3. **Tetrammine-μ-dihydroxo bis (ethylenediammine) dicobalt (III) chloride**
4. **Bis (ethylenediammine) cobalt (III)-μ-amido-μ-peroxo-bis (ethylenediammine)**
5. **Triammine cobalt (III)-μ-trihydroxo-triammine cobalt (III) ion.**

Ans. 1. $[(NH_3)_5Cr-OH-Cr(NH_3)_5]Br_5$

2. $\left[(en)_2Co\begin{matrix} NH_2 \\ \\ OH \end{matrix}Co(en)_2\right](SO_4)_2$ (Co atoms bridged by NH_2 and OH)

3. $\left[(NH_3)_4Co\begin{matrix} OH \\ \\ OH \end{matrix}Co(en)_2\right]Cl_4$ (Co atoms bridged by two OH)

4. $\left[(en)_2Co\begin{matrix} NH_2 \\ \\ O_2 \end{matrix}Co(en)_2\right]^{4+}$ (Co atoms bridged by NH_2 and O_2)

5. $\left[(NH_3)_3Co\begin{matrix} OH \\ -OH- \\ OH \end{matrix}Co(NH_3)_3\right]^{3+}$ (Co atoms bridged by three OH)

Q. 16. Write down the chemical formulae of the following compounds:

(*i*) Chloridonitrito-N- bis (ethylenediammine) cobalt (III) ion.
(*ii*) Cis-chloridonitrito-N-diammineplatinum (II)

(*iii*) Sodium trioxalatoaluminate (III)

Ans. (*i*) $[CoC1 (NO_2) (en)2]^+$

(*ii*) $\left[\begin{array}{c} Cl \quad NO_2 \\ \searrow Pt \swarrow \\ \nearrow \quad \nwarrow \\ NH_3 \quad NH_3 \end{array}\right]$

(*iii*) $Na_3 [Al (C_2O_4)_3]$

Q. 17. Write down the IUPAC nomenclature of the following compounds:

(*i*) $[Co(NH_3)_6] Cl_3$, (*ii*) $K_4 Fe(CN)_6$ and (*iii*) $[Cr(NH_3)_6]Cl_3$ (*Calcutta, 2007*)

Ans. (*i*) Hexaamminecobalt (III) Chloride

(*ii*) Potassium hexacyanoferrate (II)

(*iii*) Hexaamminechromium (III) Chloride

Q. 18. Discuss the following isomerisms in coordination compounds. (*Lucknow, 2010*)

1. Ionisation Isomerism (*Nagpur, 2000*)

2. Hydrate Isomerism (*Pune, 2010*)

3. Linkage Isomerism (*Nagpur, 2005; Guwahati, 2005*)

Ans. 1. Ionisation isomerism. This type of isomerism occurs due to exchange of groups between the complex ion and ion outside it. Such compounds behave differently when dissolved in water. For example $[Co(NH_3)_5Br] SO_4$ is red-violet in solution and gives a precipitate with $BaCl_2$ confirming the presence of SO_4^{2-} ion.

On the other hand, $[Co(NH_3)_5 SO_4] Br$ is red and does not give tests for sulphate ion in solution but instead gives precipitate of AgBr with $AgNO_3$. These two compounds are ionisation isomers.

2. Hydrate isomerism. These isomers differ in number of water molecules attached to metal ion as ligands. Thus three isomers of $CrCl_3$ $6H_2O$ are known as given below. From the conductivity measurement and quantitative precipitation of ionized chlorine, they have the following formulae:

(*i*) $[Cr(H_2O)_6]C1_3$ Violet (three ionic chlorine)

(*ii*) $[Cr(H_2O)_5C1]C1_2 . H_2O$ Green (two ionic chlorine)

(*iii*) $[Cr(H_2O)_4C1_2] C1_2 .2H_2O$ Dark green (one ionic chlorine)

3. Linkage isomerism. This type of isomerism occurs when more than a single atom in a coordinated group may act as a donor. For example, in NO_3^- ion, either oxygen or nitrogen can act as donor giving two different isomers. Two isomers of cobalt having ligands ammonia and NO_2^- showing linkage isomerism are $[Co(NH_3)_5NO_2]^{2+}$ pentamminenitrito-N-cobalt (III) ion and $[Co (NH_3)_5 ONO]^{2+}$ pentammine nitrito-O-cobalt (II) ion. They are linkage isomers of each other.

Another ligand which can exhibit linkage isomerism is thiocyanate ion SCN^-, *e.g.*, It can form M–SCN and M–NCS.

Q. 19. Discuss the following with reference to coordination compounds:

(*i*) Coordination isomerism (*Calcutta, 2000; Lucknow, 2000; Nagpur, 2008*)

(*ii*) Coordination position isomerism.

(*iii*) Polymerisation isomerism. (*Nagpur, 2003*)

Ans. (*i*) Coordination isomerism. When both positive and negative ions are complex, isomerism is possible due to the interchange of ligands between the two complex ions. This type of isomerism is called coordination isomerism and is illustrated by the following examples:

$[Co(NH_3)_6][Cr(CN)_6]$ and $[Cr(NH_3)_6][Co(CN)_6]$

$[Co(NH_3)_6][Cr(NO_2)_6]$ and $[Cr(NH_3)_6][Co(NO_2)_6]$

In $[Co(NH_3)_6]$ $[Cr(CN)_6]$, $[Co(NH_3)_6]^{3+}$ is a cation and $[Cr(CN)_6]^{2-}$ is an anion. Ligands interchange their positions to form $[Cr(NH_3)_6]$ $[Co(CN)]_6$. Cation is written first.

***(ii)* Coordinate position isomerism.** In polynuclear complexes, an interchange of ligands between two metal atoms gives rise to another type of isomerism called coordination position isomerism, *e.g.*,

$$\left[(NH_3)_4Co\begin{matrix} NH_2 \\ O_2 \end{matrix}Co(NH_3)_2Cl_2\right]Cl_2$$

Tetrammine cobalt (III) μ-amido-μ superoxodiamminidichlorocobalt (III) chloride and

$$\left[Cl(NH_3)_3Co\begin{matrix} NH_2 \\ O_2 \end{matrix}Co(NH_3)_3Cl\right]Cl_2$$

Triamminechlorocobalt (III)-μ-amido-μ-superoxotriamminechlorocobalt (III) are coordination position isomers. Here O_2, superoxo, is one bridging group and the other is amido (NH_2^-).

***(iii)* Polymerization isomerism.** This is not a true isomerism since it occurs between compounds having the same empirical formula but different molecular formula. Thus $[Pt(NH_3)_2Cl_2]$, $[Pt(NH_3)_4]$ $[PtCl_4]$, are polymerization isomers and have same empirical formula but have molecular weight in the ratio of 1:2.

Q. 20. Discuss geometrical isomerism in coordination compounds.

(*Meerut, 2000; Awadh, 2000; Nagpur, 2005; Lucknow, 2000; Punjab, 2000; Garhwal, 2000*)

Ans. This type of isomerism is exhibited by complexes in which same two ligands occupy positions adjacent to each other (cis-isomer) or opposite to each other (trans-isomer). This type of isomerism is not possible in complexes having coordination number 2, 3 and tetrahedral complexes (C.N. 4). This is due to the fact that in these arrangements all the positions are adjacent to each other.

This type of isomerism occurs in four coordinated square planar complexes and six coordinated octahedral complexes.

(A) Cis-trans isomerism in square planar complexes

(*i*) *MA_2B_2 Type*. It was first shown by Werner that Pt (II) complex $[Pt(NH_3)_2Cl_2]$ existed in two geometrical forms and it must have a planar structure. The cis- and trans- forms of diammine dichloro platinum (II) complex are shown below.

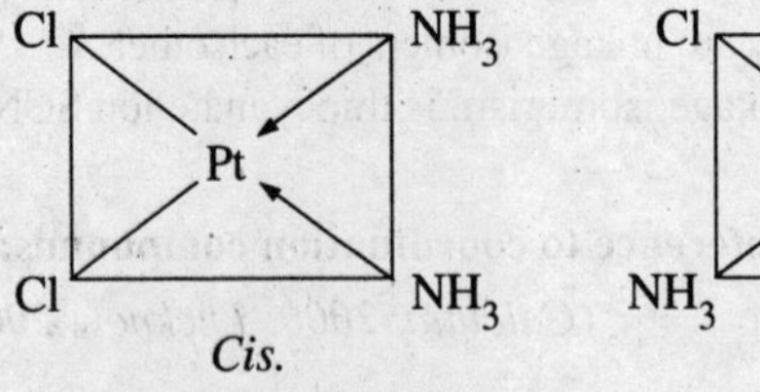

Cis. (two Cl on the same side)

Trans. (two Cl on the opposite side)

(*ii*) *MA_2BC type of complexes* also exist in cis- and trans- forms *e.g.*,

$[Pt(NH_3)_2ClNO_2]$ can exist in the following two geometrical forms.

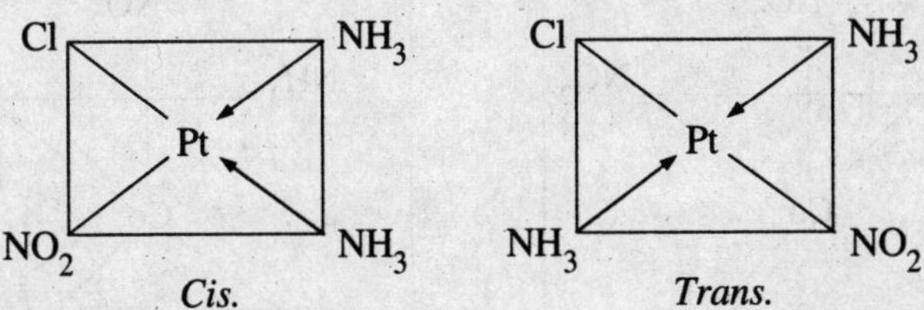

Cis. *Trans.*

In cis form the two ammine groups are on the same side of Pt whereas in trans form, the two ammine groups are on the opposite side of Pt.

(*iii*) *Complexes of the type MABCD* exhibit three isomers, *e.g.*, [Pt (Py) (NH_3) BrCl] can exist as:

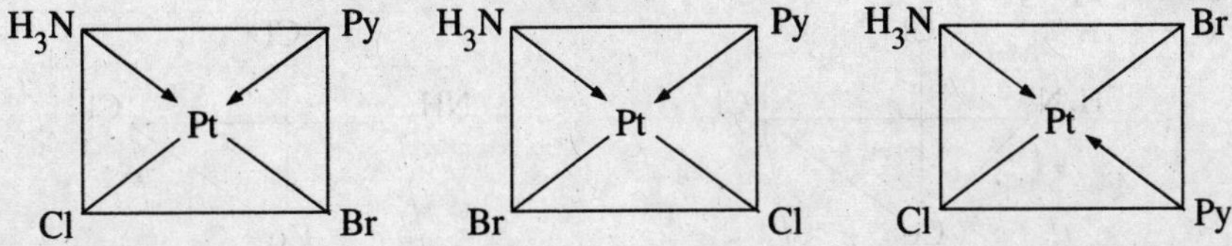

These can be written by fixing any one ligand to one position (say NH_3) and then placing other ligands trans to it. Three isomers of the complex [Pt(Py) (NH_3) (NH_2OH) NO_2] have actually been isolated and identified.

(*iv*) *Geometrical isomers are also found in square planar complexes* which have unsymmetrical bidentate ligands, *e.g.*, the Pt (II)-glycinate complex exists in cis- and trans-forms as shown below:

$CH_2 — NH_2$ $NH_2 — CH_2$ $CH_2 — NH_2$ O —— CH_2
Pt Pt
CO —— O O —— CO CO —— O $NH_2 — CO$

(*v*) *Geometrical isomers is also shown by bridged binuclear planar complexes* of the type $M_2A_2X_4$, *e.g.*, [Pt Cl_2 PEt_3]$_2$ where PEt_3 is triethylenephosphine.

Cl Cl Cl Et_3P Cl Cl
Pt Pt Pt Pt
Et_3P Cl PEt_3 Cl Cl PEt_3

Cis *Trans*

(B) Geometrical Isomerism in Octahedral Complexes

(*i*) For monodentate ligands A and B, no isomers are possible for [MA_5B] or [MAB_5] type of complexes. [MA_4B_2] type of octahedral complexes exist as cis. isomer (the two monodentate ligands in 1, 2-positions) and trans- isomers (the two ligands in 1-6-positions), *e.g.*, in tetramminedinitro cobalt (III) ion. The cis- and trans- isomers in tetramminedinitro cobalt (III) ion are given below.

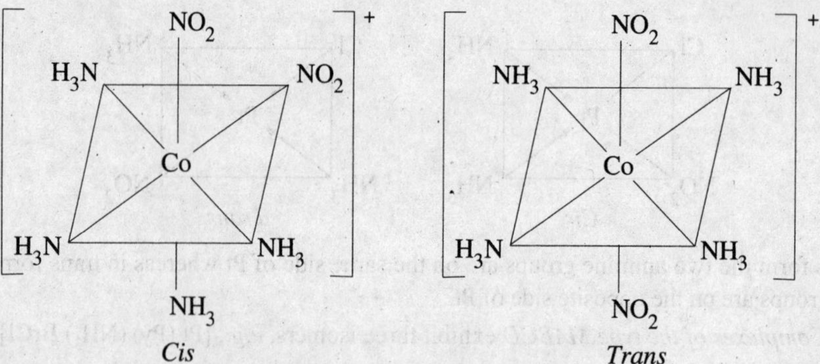

Cis *Trans*

(*ii*) $[MA_3B_3]$ type of octahedral complexes are known to give cis- and trans- isomers, *e.g.*, the complex $[Co(NH_3)_3C1_3]$ triamminetrichloridocobalt (III) exists in two forms.

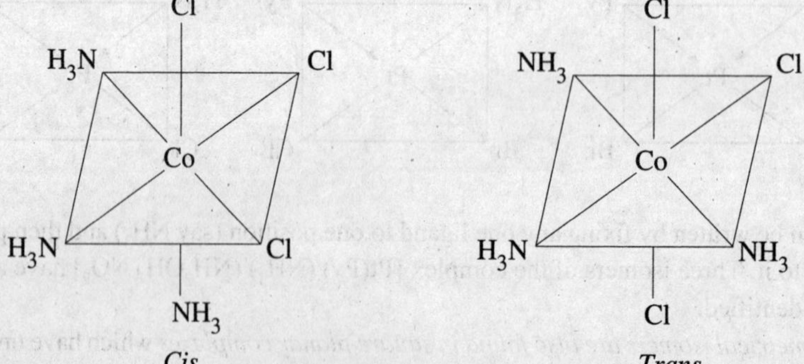

Cis *Trans*

(*iii*) Complexes of the type [M(ABCDEF)] also exhibit geometrical isomerism. Theoretically fifteen different isomers are possible but in actual practice for a complex, $[Pt (Py) (NH_3) (NO_2) Cl, Br, I]$ only three isomers have been isolated.

(*iv*) Unsymmetrical bidentate ligands in octahedral complexes also exhibit geometrical isomerism (similar to square planar complexes). For example, the chromium-triglycinato complex is found to exist in *cis*- and *trans*- forms.

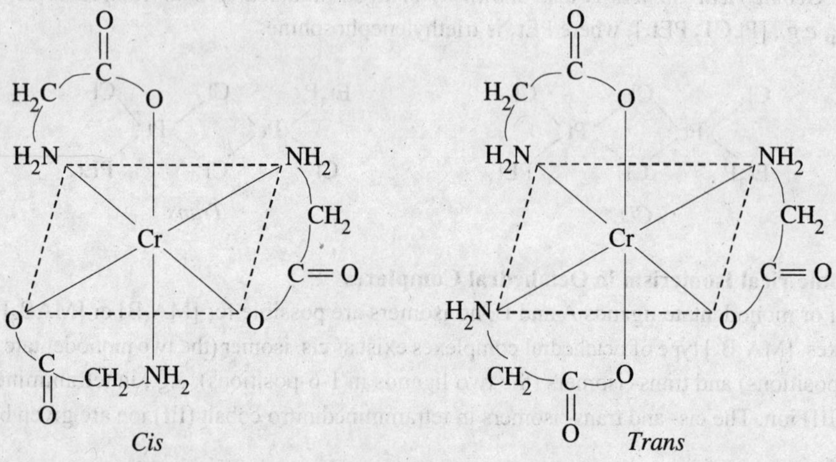

Cis *Trans*

Q. 21. Discuss optical isomerism in coordination compounds.

(*Punjab, 2000; Garhwal, 2000; Lucknow, 2000; Guwahati, 2002; Nagpur, 2008*)

Ans. Optical isomerism arises when the structure of one form is the mirror image of the other and the two are not superimposable. Such compounds are called optical isomers. For optical activity a compound must not have a plane of symmetry. An optically active compound has a property of rotating the plane of polarised light either to its right (called dextro-rotatory, *d or* (+)) or its left (called laevorotatory *l* or (–)). The two isomers *"d"* and *"l"* of a compound are called *enantiomers.*

(*i*) Square planar compounds do not exhibit optical activity as they possess a plane of symmetry. However, the property is exhibited by tetrahedral *complexes.*

In tetrahedral complexes of type (Mabcd) (analogous to tetrahedral carbon atom) it is expected that a metal complex with four different groups attached to metal ion should exhibit optical activity. However it has not been possible to isolate optically active compound of this type.

(*ii*) Tetrahedral complexes of Be, B, Cu (II), Zn (II) etc. containing unsymmetrical ligands have been resolved into optical isomers. One such example is bis (benzoylactonat), berytlium (II) complex as shown below.

Mirror

CH_3, C=O, HC, C—O, C_6H_5, Be, O=C, CH_3, CH, O—C, C_6H_5 | CH_3, C=O, CH, C—O, C_6H_5, Be, O=C, CH_3, CH, O—C, C_6H_5

(*iii*) *M*(*AA*)$_2$, X_2 *type* [*Co*(*en*)$_3$]$^{3+}$. Tris (ethylenediamine) cobalt (III) complex ion and [Cr(OX$_3$)]$^-$ trioxalato-chromate (III) ion exhibit optical isomerism as illustrated below. (AA stands for bidentate ligand).

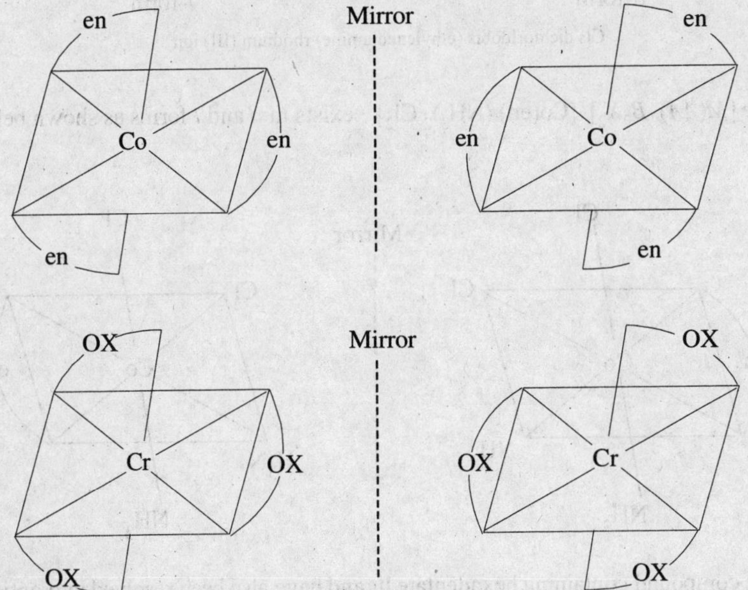

(*iv*) *Type* [*M*(*AA*)$_2$ *X*$_2$]. Such complexes show both geometrical and optical isomerism. The trans-isomers will be optically inactive due to the presence of a plane of symmetry, only cis- form is unsymmetrical and can form two optical isomers. For example, in the complex ion cis-chlorobis (ethylenediammine) cobalt (III) ion $[Co(en)_2Cl_2]^+$ three isomers are formed.

cis (object) (*cis* mirror image) ***trans***

Optically active forms Optically inactive

cis dichlorobis (ethylenediamine) cobalt (III) ion.

Cis-Dichloridobis (ethylenediamine) rhodium (III) ion also exists in two optically active forms.

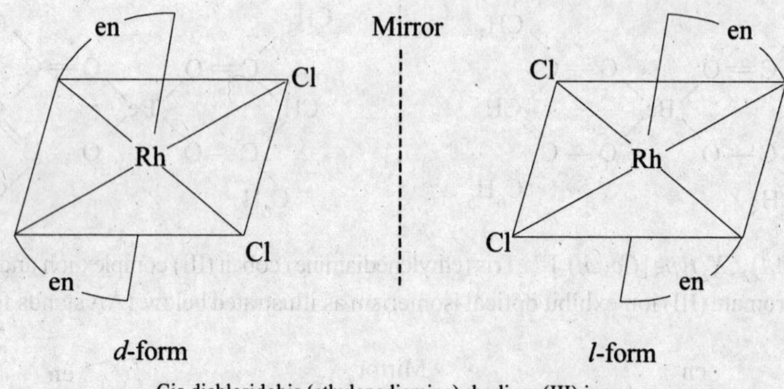

d-form *l*-form

Cis dichloridobis (ethylenediamine) rhodium (III) ion

(*v*) *Type* $[M(AA), B_2X_2]$. $[Co(en)(NH_3)_2Cl_2]^+$ exists in *d* and *l* forms as shown below

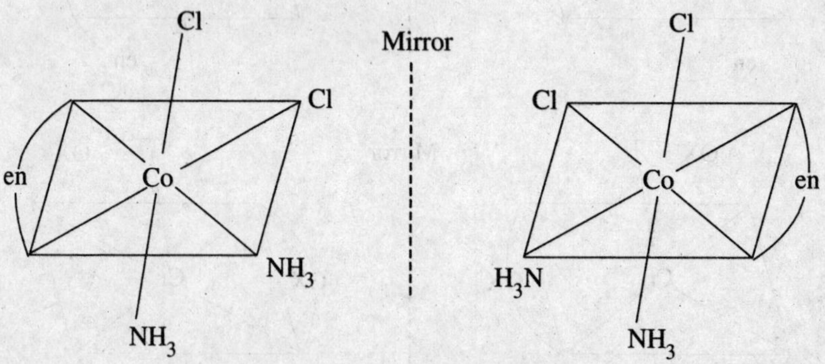

(*vi*) The compound containing hexadentate ligand have also been resolved into optical isomers. For example, the ethylenediaminetetra acetic acid (EDTA), a hexadentate ligand, forms $[Co(EDTA)]^-$, a complex amion which exists in two isomeric forms.

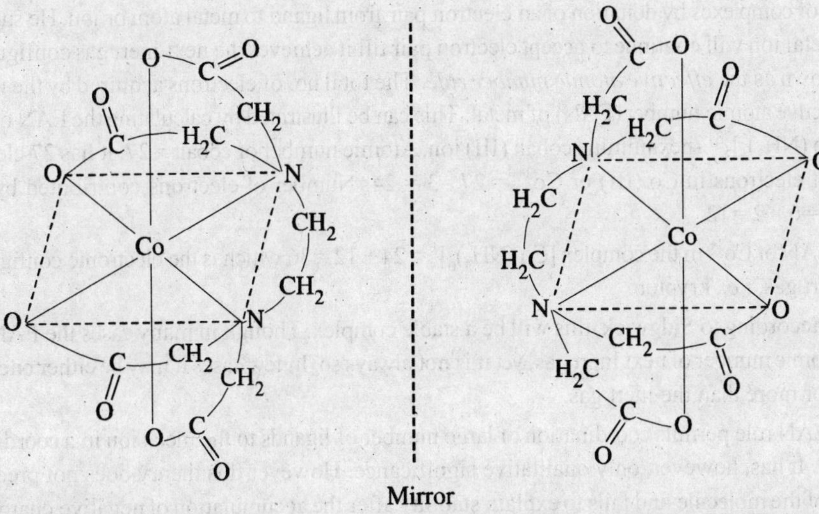

Mirror

Q. 22. (*a*) Draw the structural formula of two isomers of the complex ion $[Co(NH_3)_5 NO_2]^{2+}$. What sort of isomers are these?

(*b*) Although the no. of electrons in the outermost shell in transition metals is the same as alkali or alkaline earth metals, yet the transition elements are much less reactive. How do you account for it?

Ans. (*a*) The structural formula of the two isomers are

$[(NH_3)_5 Co-O-N=O]^{2+}$ and $[(NH_3)_5 Co-N(=O)(\rightarrow O)]^{2+}$

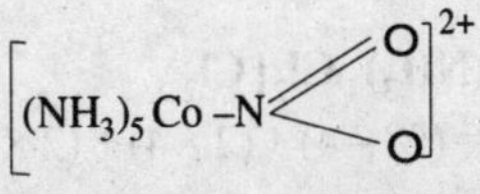

These are linkage isomers.

(*b*) Transition metals are less reactive than alkali and alkaline earth metals because of higher ionisation energies.

Q. 23. Give the names and illustrate all types of isomers that are possible in an octahedral complex compound of one cobalt (III) ion, two en molecules, two chloride ions and one nitrate ion.

Ans. The complex is $[Co(en)_2 Cl_2 NO_2]$. It can exhibit the following isomerisms:

(*i*) *Linkage isomerism.* When NO_2^- can link through oxygen or nitrogen giving nitrite and nitrito compounds respectively.

(*ii*) *Ionisation isomerism*

$[Co(en)_2 Cl_2]^+ NO_2^-$

$[Co(en)_2 Cl NO_2]^+ Cl^-$

(*iii*) *Geometrical isomerism*

$[Co (en)_2 Cl_2]^+ NO_2^-$ will exhibit *cis* and *trans* geometrical isomerism.

$[Co (en)_2 ClNO_2]^+ Cl^-$ will also exhibit cis and trans isomerism.

Both the *cis* as well as *trans* isomers will then show optical *isomerism*.

Q. 24. Describe Sidgwick theory (effective atomic no:. EAN concept) of coordination compounds. (*Bombay, 2002; Nagpur, 2005; Lucknow, 2010*)

Ans. Sidgwick in 1927 explained the bonding in coordination compounds. He considered the formation of complexes by donation of an electron pair from ligand to metal atom or ion. He suggested that the metal ion will continue to accept electron pair till it achieves the next inert gas configuration. This is known as the *effective atomic number rule.* The total no. of electrons acquired by the metal is called effective atomic number (EAN) of metal. This can be illustrated by calculating the EAN of cobalt (III) in $[Co(NH_3)_6]_2^{3+}$-hexammine cobalt (III) ion. Atomic number of cobalt = 27, it has 27 electrons. Number of electrons in Co (III) or Co^{3+} = 27 – 3 = 24. Number of electrons contributed by $6NH_3$ molecules = 6 × 2 = 12.

$\therefore$ EAN of Co^{3+} in the complex $[Co(NH_3)_6]^+$ = 24 + 12 = 36 which is the electronic configuration of next inert gas, *i.e.*, krypton.

Thus according to Sidgwick, this will be a stable complex. Though in many cases the EAN is the same as atomic number of next inert gas, yet it is not always so. In few cases it may be either one or two units less or more than the inert gas.

This EAN rule permits coordination of large number of ligands to the metal ion in a coordination compound. It has, however, only qualitative significance. However this theory does not predict the geometry of the molecule and fails to explain stability after the accumulation of negative charge from ligands.

Q. 25. Calculate the effective atomic number of the central metal ion in the following complexes and write their IUPAC names.

(*i*) $[Pt(NH_3)_4Cl_2]Cl_2$ **(*ii*) $Na_2[Fe(CN)_5NO]$**

(*iii*) $[Co(en)_2Cl_2]Cl$ **(*iv*) $K[Ag(CN)_2]$**

(*Purvanchal, 2007; Lucknow, 2010*)

Ans. (*i*) **$[Pt^{IV}(NH_3)_4Cl_2]Cl_2$**

E.A.N = (78 – 4) + (2 × 4) + (2 × 2)

of Pt^{4+} = 74 + 8 + 4 = 86

IUPAC name : Tetraamminedichloridoplatinum (IV) chloride

(*ii*) $Na_2[Fe^{III}(CN)_5NO]$

E.A.N. of Fe^{3+} = (26 – 3) + (2 × 5) + 2 = 23 + 10 + 2 = 35

There are a total of six monodentate ligands. Each ligand contributes two electrons

IUPAC name : Sodium pentacyanonitrosylferrate (III)

(*iii*) $[Co(en)_2Cl_2]Cl$

E.A.N. of Co^{3+} = (27 – 3) + (4 × 2) + (2 × 2) = 24 + 8 + 4 = 36

Each (en) contributes 4 electrons as it is a bidentate ligand.

IUPAC name : Dichloridobis ethylenediamminecobalt (III) chloride

(*iv*) $K(Ag(CN_2))$

E.A.N of Ag+ = (47 – 1) + 2 × 2 = 50

IUPAC name : Potassium dicyanoargentate (I)

Q. 26. Give the postulates of Valence-Bond theory (VBT) of coordination compounds.

(*Kumaon, 2000; Nagpur, 2003; Patna, 2010*)

Ans. The main postulates of this theory are as under:

(1) The metal atom or ion in the complex has a number of empty orbitals which can accommodate electrons donated by the ligands to form coordinate bonds, *e.g.*, the Cr^{3+} ion has 3*d*, 4*s*, and 4*p* vacant orbitals.

(2) Since the energy of the (*n* –1) *d*-orbitals are near the energies of *s* and *p* orbitals of the nth

orbit, these orbitals are together hybridised to give new equivalent orbitals called hybridized orbitals which take part in bond formation with ligand orbitals.

(3) The hybridized metal orbitals overlap with ligand orbitals to form strong bonds. Greater the extent of overlapping, greater is the stability of the complex.

(4) Each ligand donates a pair of electrons to the central metal ion.

(5) If the complex contains unpaired electrons, the complex is paramagnetic in nature whereas, if it has paired electrons, the complex is diamagnetic.

(6) The non-bonding metal electrons present in the inner orbitals do not take part in chemical bonding.

(7) Under the influence of a strong ligand, the electrons can be forced to pair up against the Hund's rule of maximum multiplicity. Either $(n-1)\,d$ inner orbitals or outer nd orbitals may be used for hybridization.

Q. 27. Discuss the formation of following complex ions on the basis of valence-bond theory:

(*a*) $[Cr(NH_3)_6]^{3+}$ *(Rohilkhand, 2003)*

(*b*) $[Co(CN)_6]^{3-}$ *(M. Dayanand, 2002)*

(*c*) $[Fe(CN)_6]^{3-}$ *(Nagpur, 2008)*

(*d*) $[Fe(CN)_6]^{4-}$

(*e*) $[Fe(H_2O)_6]^{2+}$

Ans. (*a*) Formation of $[Cr(NH_3)_6]^{3+}$ ion. The electronic configurations of only 3*d*, 4*s* and 4*p* outer orbitals ($3d^5 4s^1$) are taken into account.

(i) $_{24}Cr$ ground state configuration

3*d*					4*s*	4*p*		
↑	↑	↑	↑	↑	↑			

(ii) Cr^{3+} ion configuration.

3*d*					4*s*	4*p*		
↑	↑	↑						

(iii) $[Cr(NH_3)_6]^{3+}$ Complex ion

3*d*					4*s*	4*p*		
↑	↑	↑	X	X	X	X	X	X

d^2sp^3 Hybridization

The donated electron pairs from ligands are shown as *crosses.* (×)

The two 3*d*-orbitals, one 4*s* orbital and three 4*p* orbitals are available to form six hybrid orbitals d^2sp^3, which can accept six electron pairs (one electron pair from each nitrogen atom of NH_3) from the ligands. The complex has a octahedral geometry and should be paramagnetic due to the presence of 3-unpaired electrons and actually it is so. Other complexes of Cr with similar inner structure are $[Cr(CN)_6]^{3-}$ and $[Cr(H_2O)_6]^{3+}$. These are inner orbital complexes since inner *d*-orbitals (3*d*) are used in bond formation.

(*b*) Formation of $[Co(CN)_6]^{3-}$ ion. Cobalt (atomic number is 27) has outer electronic configuration as

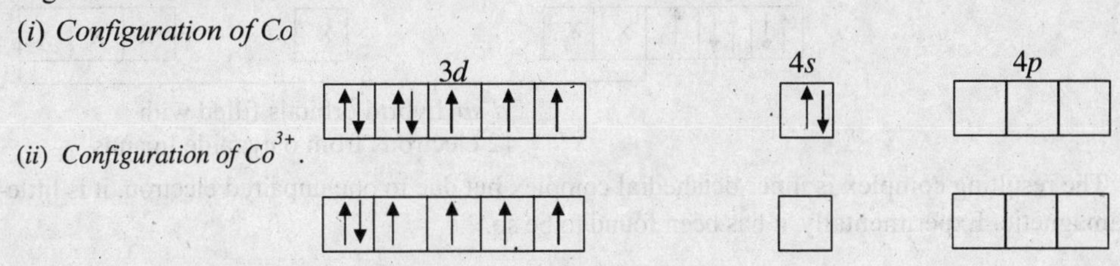

Co^{3+} does not contain any empty 3*d*-orbitals. Now according to VBT, in presence of CN^- (a strong ligand) the electrons in 3*d*-orbital are forced to pair up against the Hund's rule in order to make room for electrons donated by the ligands. Thus after pairing up against Hund's rule, the electronic configuration of Co^{3+} ion becomes as follows:

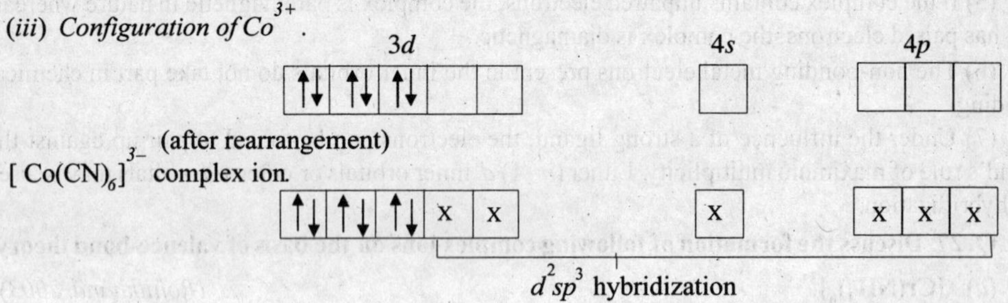

The two 3*d*-orbitals, one 4*s* orbital and three 4p orbitals are now available to form six hybrid orbitals d^2sp^3, which can accept six electron pairs (one pair from each ligand) from the ligands. d^2sp^3 hybridization gives octahedral geometry and also it has no unpaired electrons, so it is diamagnetic. The inner $(n-1)$ *d* orbitals are taking part in bond formation, hence it is inner complex ion.

Other complexes of this type are

$$[Co\,(NH_3)_6]^{3+},\ [Co\,(NO_2)_6]^{3-}$$

(c) Formation of ferricyanide ion, $[Fe(CN)_6]^{3-}$. Iron has atomic number 26, its outer electronic configuration is $3d^6$, $4s^2$.

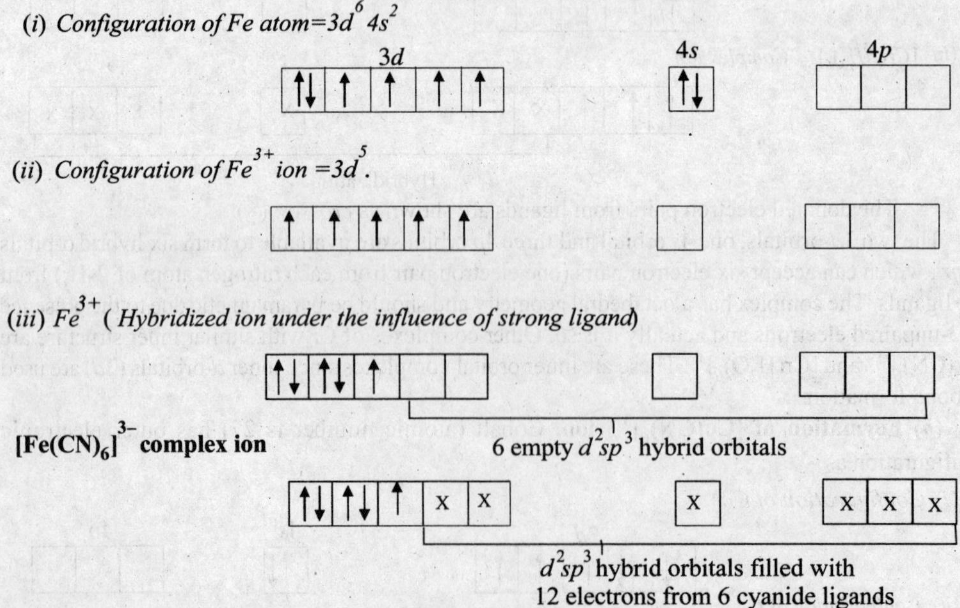

The resulting complex is inner octahedral complex but due to one unpaired electron, it is little-paramagnetic. Experimentally, it has been found to be so.

(*d*) Formation of ferrocyanide ion $[Fe(CN)_6]^{4-}$. This is explained as under:

The resulting complex is inner octahedral complex and it should be diamagnetic due to absence of unpaired electrons and actually it is so.

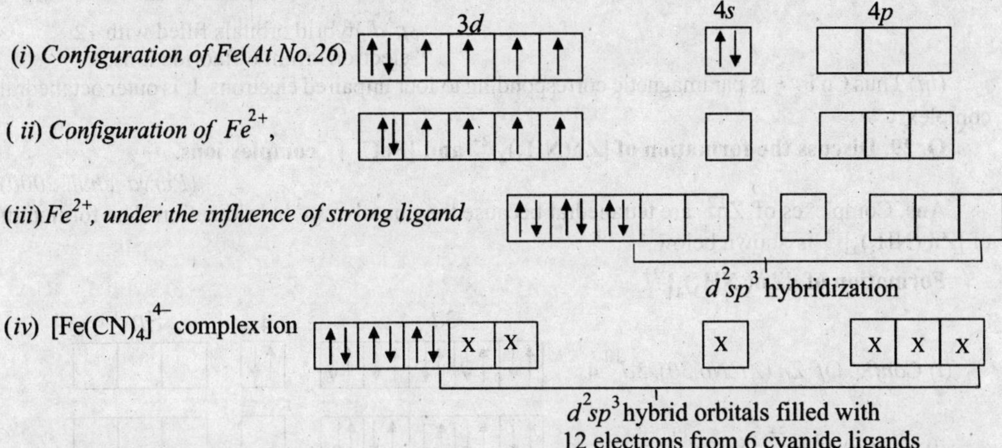

(*e*) Formation of $[Fe(H_2O)_6]^{2-}$. It has been experimentally observed that this complex is not diamagnetic. It shows paramagnetism corresponding to four unpaired electrons. This means water molecule is too weak a ligand to effect pairing of electrons in 3*d* orbitals. In such cases, outer *d* orbitals, 4*d* in this case, are used for bond formation with ligands. Thus configuration will be as under in this case:

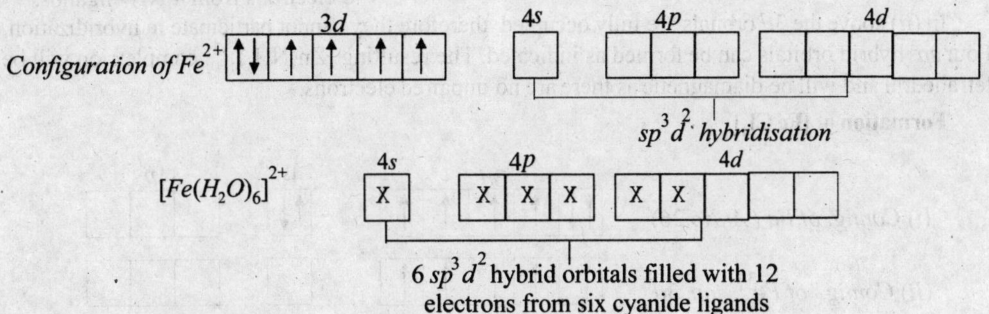

This complex is thus outer complex.

Q. 28. Discuss the formation of $[CoF_6]^{3-}$ complex ion. (*Garhwal, 2000; M. Dayanand, 2002*)

Ans. The formation of $[Co\,F_6]^{3-}$ is explained as under:

3*d* 4*s* 4*p* 4*d*

(*i*) *Configuration of Co (At. No. 27)* ↑↓ ↑↓ ↑ ↑ ↑ | ↑↓ | | |

(*ii*) *Configuration of Co^{3+} ion* ↑↓ ↑ ↑ ↑ ↑ | | | |

The complex $[(Co\,(Fe_6)]^{3-}$ has been found to be paramagnetic in nature (unlike $[Co(CN)_6]^{3-}$ or $[Co(NH_3)_6]^{3+}$ which are diamagnetic. This can be explained on the basis of weak nature of ligand. It cannot force the pairing up of electrons against Hund's rule. Hence outer 4d orbitals are used for bond formation with ligands.

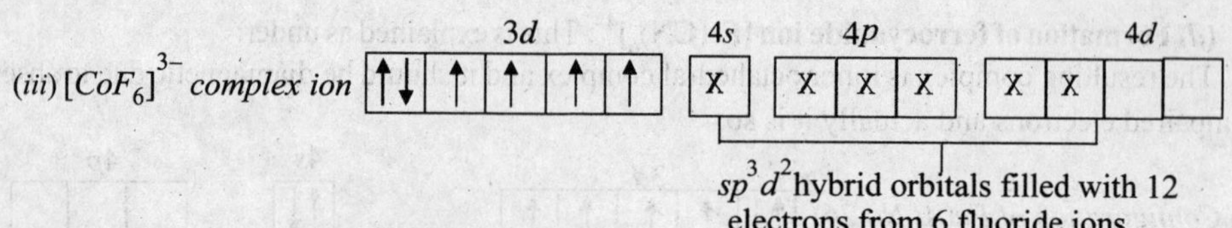

(*iii*) Thus Co F_6^{3-} is paramagnetic corresponding to four unpaired electrons. It is outer octahedral complex.

Q. 29. Discuss the formation of $[Zn(NH_3)_4]^{2+}$ and $[FeCl_4]^-$ complex ions.

(*Purvanchal, 2000*)

Ans. Complexes of Zn^{2+} are tetrahedral because they involve sp^3 hybridization. The formation of $[Zn(NH_3)_4]^{++}$ is shown below.

Formation of $[Zn(NH_3)_4]^{2+}$

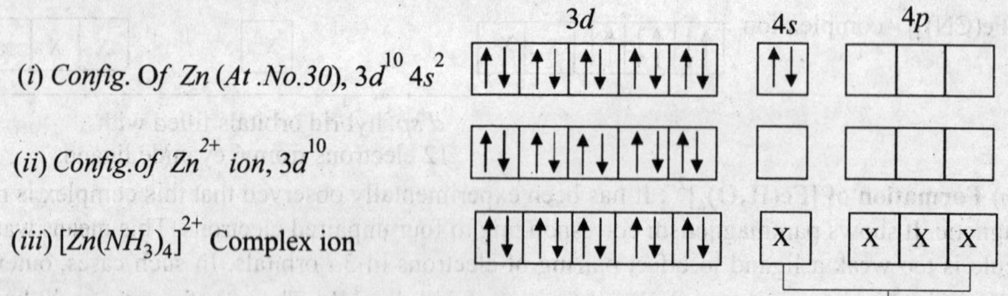

Sp^3 hybrid orbitals filled with
8 electrons from 4 NH_3 ligands.

In (*ii*) above the 3*d* orbitals are fully occupied, therefore they cannot participate in hybridization. Four sp^3 hybrid orbitals can be formed as indicated. The resulting $[Zn(NH_3)_4]^{2+}$ complex ion will be tetrahedral and will be diamagnetic as there are no unpaired electrons.

Formation of $[FeCl_4]^-$

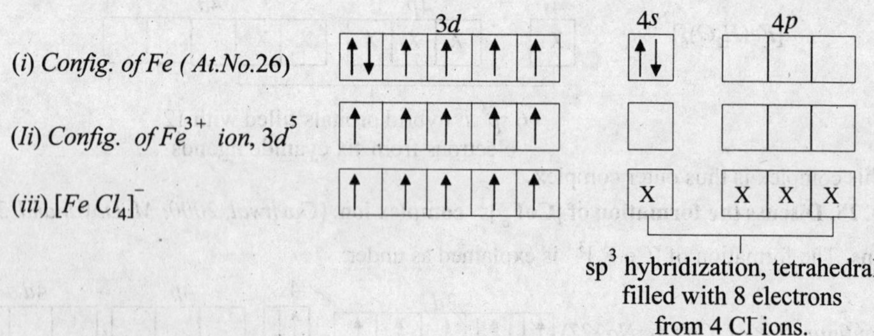

Cl^- is a weak ligand. It cannot force the pairing up of electrons against Hund's rule. Hence it is paramagnetic in nature.

Q. 30. Why do transition metal ions and their complexes show megnetic properties?

Ans. When the transition metal ions or their complexes are placed in a magnetic field, they show different behaviour. On the basis of this behaviour these substances have been classified as *paramagnetic substances, diamagnetic substances and ferromagnetic substances.*

The transition metal complexes whose central atom/ion contains one or more unpaired electrons are paramagnetic while those with no unpaired electrons behave as diamagnetic substances. The paramagnetic character increases with the increase in the number of unpaired electrons. The paramagnetism of a substance is expressed in terms of its magnetic moment (μ). Larger the number of unpaired electrons in the central atom/ion greater is its paramagnetic character and hence larger is the value of the magnetic moment. The magnetic moment is expressed in *Bohr Magnetons* (B.M.). Paramagnetic substances have some definite value of magnetic moment while diamagnetic substances have zero value. Ferromagnetism is due to large number of umpaired electrons.

Q. 31. Deduce a formula for calculating the magnetic moment of transition metal complexes?

Ans. Consider a transition metal complex of 3*d*-elements whose central metal ion contains only unpaired electrons in its 3*d* orbitals. The magnetic moment (μ) of an unpaired electron is equal to the sum of its magnetic moment due to its orbital motion, called *orbital magnetic moment* (μ_l) and that due to its spin motion, called *spin magnetic moment* (μ_s).

We may write

$$\mu = \mu_l + \mu_s$$

But $$\mu_l = \sqrt{l(l+1)}\frac{eh}{4\pi mc}$$

and $$\mu_s = 2\sqrt{s(s+1)}\frac{eh}{4\pi mc}$$

∴ $$\mu = [\sqrt{l(l+1)} + 2\sqrt{s(s+1)}]\frac{eh}{4\pi mc}$$

$$= [\sqrt{l(l+1)} + 2\sqrt{s(s+1)}]\,\text{B.M}$$

where B.M. is called Bohr Magneton which is equal to $\frac{eh}{4\pi mc}$. B.M. is the unit of magnetic moment. One B.M. is equal to 9.27×10^{-21} ergs/gauss.

Here *e* = charge on the electron,
h = Planck's quantum number,
m = mass of the electron and
c = velocity of light.

In case of transition metal ions of 3*d*-series, the ligands surrounding the metal ions restrict the movement of the electron round the nucleus in orbitals and hence the magnetic moment due to the orbital motion of the electron (μ_l) is quenched or $\sqrt{l(l+1)} = 0$.

Thus the magnetic moment of an unpaired electron residing in a 3*d*-orbital of the ions of 3*d*-series elements is given by:

$$\mu = 2\sqrt{s(s+1)}\ \text{B.M.}$$

In fact, the same equation is valid for a large no. of compounds of elements other than 3*d* elements also.

Value of *s* for one electron = $\frac{1}{2}$. Substituting for *s* in the above equation, we get,

$$\mu = 2\sqrt{\frac{1}{2}\left(\frac{1}{2}+1\right)}\ \text{B.M.}$$

With *n* unpaired electrons, $$\mu = 2\sqrt{\frac{n}{2}\left(\frac{n}{2}+1\right)}\ \text{B.M.}$$

$$= \sqrt{n(n+2)}\ \text{B.M.}$$

Since the value of μ is only due to the spin motion of *n* unpaired electrons, it is also called *spin-only magnetic moment* and is represented as μ_s. Thus:

$$\mu = \mu_s = \sqrt{n(n+2)}\ \text{B.M.} \quad .....(i)$$

Equation (*i*) shows that the magnitude of μ depends on the number of unpaired electrons (*n*) present in a given species. The value of μ for different number of unpaired electrons can be calculated with the help of equation (*i*) as shown below:

For $n = 0$ $\mu_s = \sqrt{0(0+2)} = 0.0$ B. M.

For $n = 1$ $\mu_s = \sqrt{1(1+2)} = \sqrt{3} = 1.73$ B.M.

For $n = 2$ $\mu_s = \sqrt{2(2+2)} = \sqrt{8} =$ B.M

For $n = 3$ $\mu_s = 3.87$ B.M.

For $n = 4$ $\mu_s = 4.90$ B.M.

For $n = 5$ $\mu_s = 5.92$ B.M.

Q. 32. What are high spin and low spin complexes?

(*Purvanchal, 2002; Rohilkhand; 2003; Nagpur, 2005; Patna, 2010*)

Ans. Complexes which are formed by using inner *d* orbitals such as $[Co(CN)_6]^{3-}$ and $[Fe(CN)_6]^{3-}$ have smaller number of unpaired electrons and are low spin complexes. Complexes such as $[Fe(H_2O)_6]^{2+}$ which are formed by using outer *d* orbitals have a large number of unpaired electrons and have high spin. Thus outer complexes are high spin complexes.

Q. 33. Co (III) forms paramagnetic complexes with weak field ligands whereas with strong field ligands, their complexes are diamagnetic.

Ans. See. Q. 28.

Q. 34. Discuss the formation of $[Ni(CN)_4]^{2-}$ complex ion.

(*Garhwal, 2000; M. Dayanand, 2002; Nagpur, 2005; Lucknow, 2010; Pune, 2010*)

Ans. Hybridization of one '*d*' one '*s*' and two '*p*' orbitals gives four equivalent hybrid orbitals dsp^2 which are planar and mutually at right angles pointing towards the corners of a square.

The nickelocyanide ion or tetracyanonickelate (II) ion $[Ni(CN)_4]^{2-}$ formation can be represented as follows:

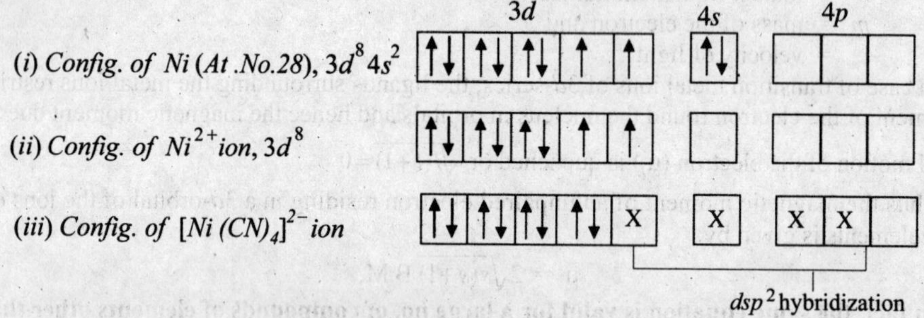

In this complex, the ligand is strong enough to force the two unpaired electrons to pair up. The result is that one 3*d* orbital becomes available for hybridization with 4*s* and two 4*p* orbitals to give four dsp^2 hybrid orbitals. These accept an electron pair from each of the four CN^- ligands. The shape of complex will be square planar and it will be diamagnetic as there is no unpaired electron in the complex.

Q. 35. What are the limitations of the valence bond theory of coordination compounds?

(*Lucknow, 2010*)

Ans. (1) The basic defect in the theory is that attention is focussed only on the orbitals of central atom and the importance of the ligands are not properly stressed.

(2) It does not explain the relative energies of different structures.

(3) It offers no explanation of the colour observed for complex ion.

(4) It cannot predict whether a 4-coordinate complex will be tetrahedral or square planar. Consider, for example, a Cu (II) complex $[Cu(NH_3)_4]^{2+}$. The VBT predicts the complex to be tetrahedral.

In order to explain the experimentally determined square planar geometrically of Cu (II) complex it is assumed that one electron from the 3*d* orbital is promoted to the 4*p* orbital to vacate a *d*–orbital for accommodating a pair of electrons from the ligand.

(5) It does not take into account the splitting of *d*-energy levels.

(6) For the same metal in some complexes electrons pair up against the Hund's rule while in others the electronic configuration is not disturbed. VBT does not offer any strong explanation for their anomaly.

(7) It does not explain the spectra of complexes.

(8) The VBT does not provide any satisfactory explanation for the existence of inner orbital and outer orbital complexes.

(9) It cannot explain why certain complexes are more labile than others. Labile complexes are those in which one ligand can be easily displaced by another ligand. On the other hand, inert complexes are those in which displacement of ligand is slow.

Q. 36. Discuss the shape of *d* orbitals.

Ans. There are five *d*-orbitals viz., d_{xy}, d_{yz}, d_{zx}, $d_{x^2-y^2}$ and d_{z^2}. Spatial orientation of *d*-orbitals are shown in Fig. 21.1. All the five *d*-orbials are not identical and there are in fact two sets t_{2g} and e_g. The t_{2g} orbitals point in between the axes *x, y* and *z* and e_g orbitals point along the axes (Fig. 21.1). Alternate names for t_{2g} and e_g are d_ε (called as *d*-epsilon, a Greek alphabet) and d_γ (called as *d*-gamma, a Greek alphabet) respectively.

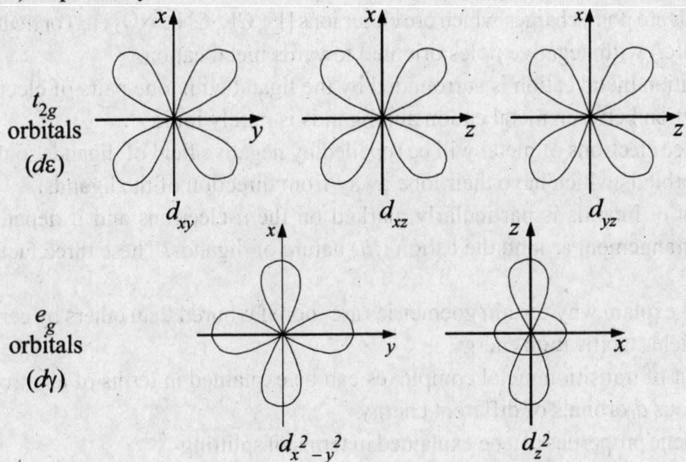

Fig. 21.1. Spatial orientation of *d*-orbitals.

Q. 37. A square planar complex of divalent metal outer electronic configuration $ns^1(n-1)\,d^{10}$ is not oxidised. Can you explain its formation of the basis on V.B. theory?

Ans. $ns^1(n-1)\,d^{10}$ refers to Cu. The divalent metal ion will have configuration d^9. X-ray studies of the complex $[Cu(NH_3)_4]^{2+}$ shows that it has square planar geometry. Thus it involves dsp^2 (or sp^2d) hybridisation. We could shift one election from 3*d* orbital to 4*p* orbital and then explain the square planar geometry by hybridisation of one 3*d*, one 4*s* and two 4*p* orbitals. One electron left in 4*p* orbital should be removed on oxidation. But that does not happen.

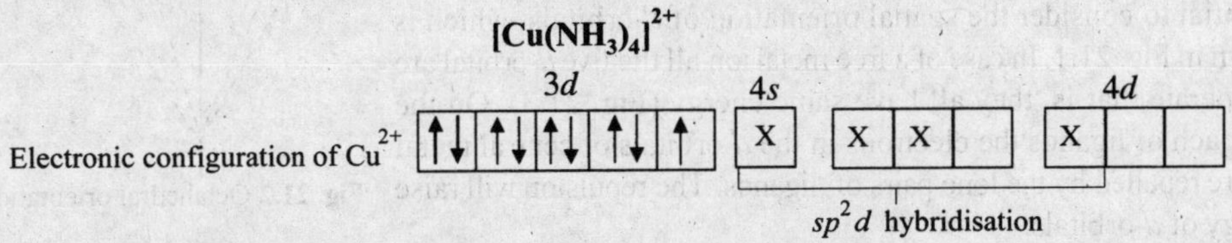

Formation of $[Cu(NH_3)_6]^{2+}$ is explained by assuming that one unpaired electron remains in 3*d*. Hybridisation of one 4*s*, two 4*p* and one 4*d* orbitals takes place (sp^2d) to give square planar shape. As 3*d* electron possesses higher energy oxidation of $[Cu(NH_3)_4]^{2+}$ or removal of 3*d* electron does not take place.

This is how the shape of the molecule is explained.

Q. 38. Explain why $Ni(CO)_4$ is diamagnetic but tetrahedral. (*Purvanchal, 2007*)

Ans. The oxidation state of Ni in $Ni(CO)_4$ is zero. The electronic configuration of Ni atom is

	3d					4s	4p		
Ni(0)	↑↓	↑↓	↑↓	↑	↑	↑↓			

However 4*s* electrons are shifted 3*d* and sp^3 hybridisation takes place as follows.

	3*d*					4*s*	4*p*		
$Ni(CO)_4$	↑↓	↑↓	↑↓	↑↓	↑↓	X	X	X	X

sp^2d hybridisation

Thus there are no unpaired electrons and the complex is diamagnetic. The shape of the complex is tetrahedral due to sp^3 hybridisation.

Q. 39. Give the main points of crystal field theory of coordination compounds.

(*Lucknow, 2002; Guwahati, 2002; Rohilkhand, 2003; Rajasthan, 2010*)

Ans. The main points of theory are given below:

1. The ligands are point charges which are either ions [F^-, Cl^-, CN^-, NO_2^- etc.) or neutral molecules. (NH_3, H_2O, en, etc.) with negative poles oriented towards metal cation.

2. The transition metal cation is surrounded by the ligand with lone pairs of electrons.

3. The attraction between metal cation and ligands is purely ionic.

4. The valence electrons of metal will be repelled by negative field of ligands so these electrons occupy those *d*-orbitals which have their lobe away from direction of the ligands.

5. The effect of ligands is particularly marked on the *d*-electrons and it depends upon their (*i*) number (*ii*) arrangement around the cation (*iii*) nature of ligands. These three factors determine the crystal field.

6. It can also explain why certain goemetries are more favoured than others by certain metals in terms of crystal field stabilization energy.

7. The colour of transition metal complexes can be explained in terms of electronic transition between the various *d*-orbitals of different energy.

8. The magnetic properties can be explained in terms of splitting of *d*-orbitals in different crystal field.

9. Different crystal fields will have different effect on the relative energies of the five *d*-orbitals.

Q. 40. Discuss splitting of *d*-orbitals in octahedral field according to crystal field theory. (*Bombay, 2002; Ranchi, 2010; Lucknow, 2010; Pune, 2010; Rajasthan, 2010*)

Ans. In an octahedral complex the metal ion is at the centre of the octahedron and the ligands are at six corners of octahedron (Fig. 21.2). Before considering the effect of ligands on metal ion it is essential to consider the spatial orientation of *d*-orbitals, which is shown in Fig. 21.1. In case of a free metal ion all the five *d*-orbital are degenerate that is, they all have same energy (Fig. 21.3). On the approach of ligands the electrons in the *d*-orbitals of central metal ion are repelled by the lone pairs of ligands. The repulsion will raise energy of *d*-orbitals.

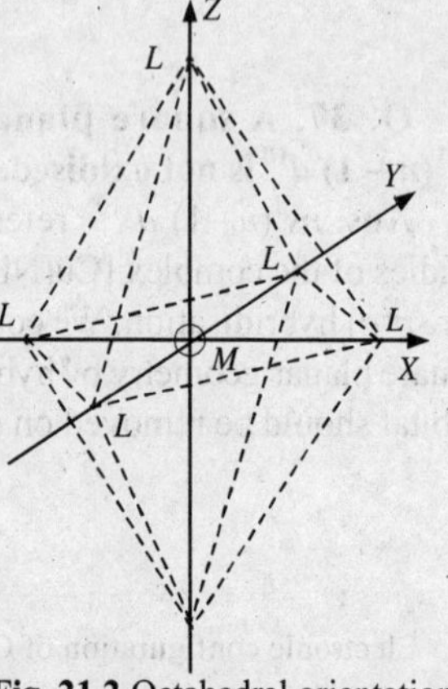

Fig. 21.2 Octahedral orientation.

If all the *d*-orbitals are spherically symmetrical, the energy of each orbital will increase by same amount. But the axes of two orbitals $d_{x^2-y^2}$ and d_{z^2} often designated as e_g orbitals are more concentrated on the axes along which the ligands are approaching. Therefore, in an octahedral complex, when six ligands are brought near the central metal ion along the axes. (*i.e.*, along the *X*, *Y*, *Z*, $-X$, $-Y$ and $-Z$ directions) the electrons in these orbitals suffer a greater repulsion then the t_{2g} orbitals (d_{xy}, d_{yz} and d_{zx}) which are situated between the axes. Thus in are octahedral field, the *d*-subshell is split into two sets of orbitals, one set e_g has higher energy and the other set (t_{2g}) has lower energy as shown in Fig. 21.3.

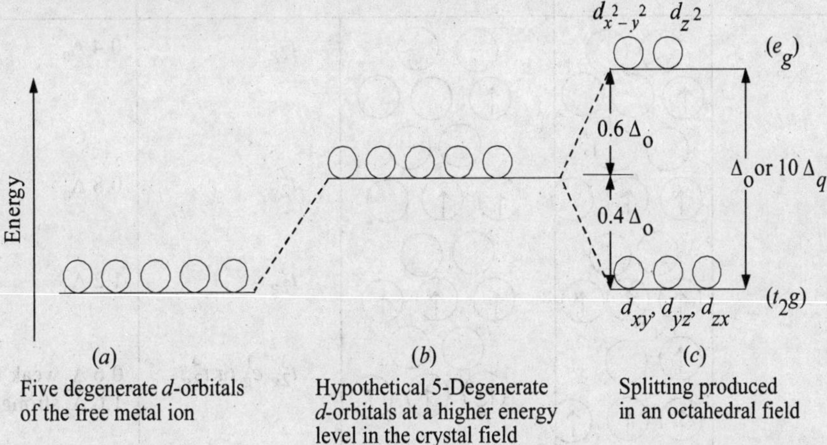

Fig. 21.3. Spatial orientation of *d*-orbitals.

The splitting of five degenerate *d*-orbitals of metal ion under the influence of approaching ligands, into two sets of orbitals having difference energies is called **crystal field splitting.**

The magnitude of difference in energy between the two sets of orbitals (e_g and t_{2g}) is usually designated as Δ_0 (where the subscript (o) stands for octahedral. Δ_0 is called the crystal field splitting energy is an octahedral field. This is also measured in terms of another parameter D_q. The magnitude of splitting is arbitrarily fixed as 10 D_q so that $\Delta_0 = 10\ D_q$.

The magnitude of Δ_0 depends upon the metal and the oxidation state of metal and the dipole moment of ligands. Some ligands produce greater crystal field splitting (Δ_0) than the others. Such ligands are called **strong ligands** while the others which produce less splitting energy are called **weak ligands.** On the basis of crystal field splitting energy, some common ligands in increasing order of their field strength are listed below. It is known as spectrochemical series.

$$I^- < Br^- < Cl^- < F^- < OH^- < C_2O_4^{2-} < H_2O < NH_3 < en < NO_2^- < CN^- < CO.$$

The geometry of an octahedral system shows that the energy of the t_{2g} orbital is 2/5 Δ_0 or 0.4Δ_0 (or 4 D_q) lower than that of the hypothetical degenerate *d*-orbitals. The energy of e_g orbital is 3/5 Δ_0 or 0.6 Δ_0 (or 6 D_q) higher that of the hypothetical degenerate of *d*-orbitals. Electrons will tend to occupy the lower energy t_{2g} orbitals in order to achieve stability. Each electron entering the t_{2g} orbitals stabilizes the complex ion by 0.4.Δ_0 and each electron entering the higher energy, e_g orbitals destabilizes by an amount of 0.6 Δ_0.

The gain in energy achieved by preferential filling up of orbitals by electrons is known as **crystal field stabilization energy (CFSE).** Greater the amount of CFSE of the complex is, greater is its stability. Crystal field stabilization energies of transition metal ions in octahedral complexes have been calculated and are given in Table 21.2.

Table 21.2. CFSE in octahedral complexes.

No. of electrons in d orbitals	*Configuration*		*No. of electrons in t_{2g} and e_g orbitals*	*CFS energy (E)*
	Weak Field	*Strong Field*		
1	( ) ( ) (↑) ( ) ( )	( ) ( ) (↑) ( ) ( )	t_{2g}^1	0.4 Δ_0
2	( ) ( ) (↑) (↑) ( )	( ) ( ) (↑) (↑) ( )	t_{2g}^2	0.8 Δ_0
3	( ) ( ) (↑) (↑) (↑)	( ) ( ) (↑) (↑) (↑)	t_{2g}^3	1.2 Δ_0
4	(↑) ( ) (↑) (↑) (↑)	( ) ( ) (↑↓) (↑) (↑)	$t_{2g}^3\, e_g^1$ or t_{2g}^4	0.6 Δ_0 weak field 1.6 Δ_0 strong field
5	(↑) (↑) (↑) (↑) (↑)	( ) ( ) (↑↓) (↑↓) (↑)	$t_{2g}^3\, e_g^2$ or t_{2g}^5	0 weak field 2.0 Δ_0 strong field
6	(↑) (↑) (↑↓) (↑) (↑)	( ) ( ) (↑↓) (↑↓) (↑↓)	$t_{2g}^5\, e_g^2$ or $t_{2g}^6 e_g^1$	0.4 Δ_0 weak field 2.4 Δ_0 strong field
7	(↑) (↑) (↑↓) (↑↓) (↑)	(↑) ( ) (↑↓) (↑↓) (↑↓)	$t_{2g}^5\, e_g^2$ or $t_{2g}^6 e_g^1$	0.8 Δ_0 weak field 1.8 Δ_0 strong field
8	(↑) (↑) (↑↓) (↑↓) (↑↓)	(↑) (↑) (↑↓) (↑↓) (↑↓)	$t_{2g}^6\, e_g^2$	1.2 Δ_0
9	(↑↓) (↑) (↑↓) (↑↓) (↑↓)	(↑↓) (↑) (↑↓) (↑↓) (↑↓)	$t_{2g}^6\, e_g^3$	0.6 Δ_0
10	(↑↓) (↑↓) (↑↓) (↑↓) (↑↓)	(↑↓) (↑↓) (↑↓) (↑↓) (↑↓)	$t_{2g}^6\, e_g^4$	0

Q. 41. Discuss the splitting of *d*-orbitals in tetrahedral field according to CF theory.

(*Rohilkhand 2003; Nagpur 2008; Garhwal, 2010*)

Ans. In tetrahedral field, the four ligands may be imagined as occupying alternate corners of a cube and the metal ion at the centre as shown in Fig. 21.4.

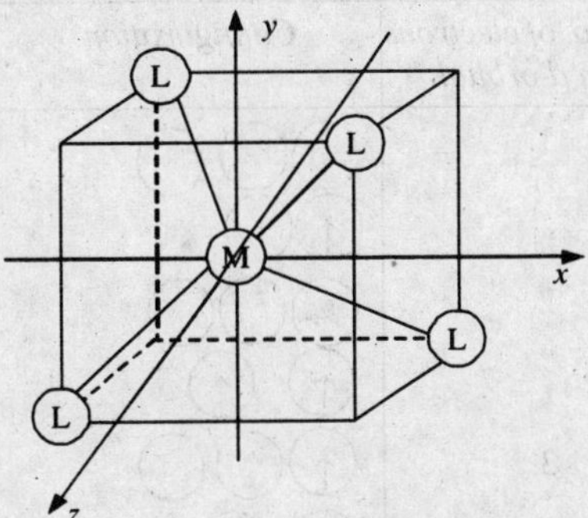

Fig. 21.4 Tetrahedral orientation.

The directions *X, Y, Z* point to the centres of faces of cube. The e_g ($d\gamma$) orbitals are oriented along *X, Y* and *Z* and the t_{2g} ($d\varepsilon$) orbitals are situated in between *X, Y, Z*. The direction of approach of the ligands does not coincide exactly with either of the e_g or the t_{2g} orbitals. But t_{2g} orbitals are nearer to the direction of ligands than the e_g orbitals. Therefore, the approach of the ligands raises the energy of the d_{xy}, d_{yz}, d_{zx} orbitals more than d_{z^2} and $d_{x^2-y^2}$ orbitals. Thus the t_{2g} orbitals are now of higher energy and e_g orbitals are of lower energy. This is in the reverse order to what has been observed in an octahedral field. The energy difference between t_{2g} and e_g orbitals for tetrahedral field is shown in Fig. 21.5.

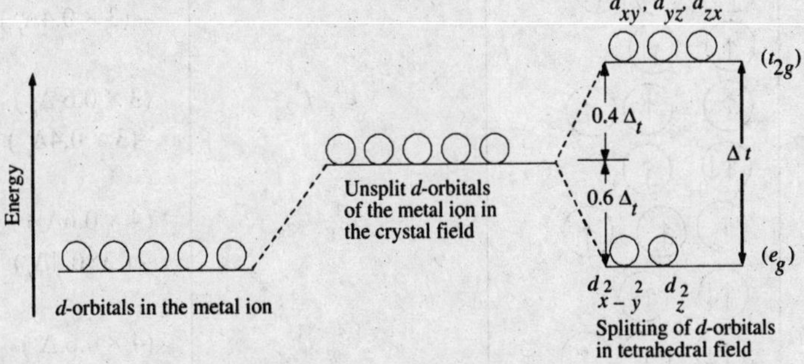

Fig. 21.5. Splitting of *d*-orbitals in tetrahedral field.

The magnitude of crystal field splitting is denoted as Δ_t, where t stands for tetrahedral. The difference Δ_t is less than that for an octahedral field. This is partly due to the number of ligands as only four ligands are there in the tetrahedral field as compared to six in the octahedral field. Moreover there is a ligand along each axis in the octahedral field while no ligand exists along any axis in the tetrahedral field. For a given metal ion and a given ligand the crystal field splitting Δ_t for a tetrahedral field is about one half of Δ_0. In fact it has been observed that

$$\Delta_t = \frac{4}{9}\Delta_0$$

In view of higher the crystal field splitting the formation of octahedral complexes is favoured. Since an electron prefers to go into a region of lower energy for reason of stability, it follows that if a tetrahedral complex contains one *d*-electron it will occupy one of the e_g orbitals.

An electron occupying e_g orbitals will be stabilized by 0.6 Δ_t and that occupying a t_{2g} orbital will be destabilized by 0.4 Δ_t. On this basis of the crystal field stabilization energies in tetrahedral complexes have been calculated as given in Table 21.3.

Table 21.3. CFSE in tetrahedral field.

No. of electrons in d-orbital	*Configuration*	*Electrons in e_g or t_{2g}*	*CFSE(E)*
1.	( ) ( ) ( ) (↑) ( )	e_g^1	$1 \times 0.6\Delta_t = 0.6\ \Delta t$
2.	( ) ( ) ( ) (↑) (↑)	e_g^2	$2 \times 0.6\Delta_t = 1.2\ \Delta_t$
3.	(↑) ( ) ( ) (↑) (↑)	$e_g^2 t_{2g}^1$	$(2 \times 0.6\Delta_t) - (1 \times 0.4\ \Delta_t) = 0.8\Delta_t$
4.	(↑) (↑) ( ) (↑) (↑)	$e_g^2\ t_{2g}^2$	$(2 \times 0.6\ \Delta_t) - (2 \times 0.4\Delta_t) = 0.4\Delta_t$
5.	(↑) (↑) (↑) (↑) (↑)	$e_g^2\ t_{2g}^3$	$(2 \times 0.6\Delta_t) - (3 \times 0.4\Delta_t\ ) = 0$
6.	(↑) (↑) (↑) (↑↓) (↑)	$e_g^3\ t_{2g}^3$	$(3 \times 0.6\ \Delta_t\ ) - (3 \times 0.4\Delta_t\ ) = 0.6\Delta_t$
7.	(↑) (↑) (↑) (↑↓) (↑↓)	$e_g^4\ t_{2g}^3$	$(4 \times 0.6\Delta_t) - (\ 3 \times 0.4\Delta_t) = 1.2\ \Delta_t$
8.	(↑↓) (↑) (↑) (↑↓) (↑↓)	$e_g^4\ t_{2g}^4$	$(4 \times 0.6\ \Delta_t) - (4 \times 0.4\Delta_t) = 0.8\ \Delta_t$
9.	(↑↓) (↑↓) (↑) (↑↓) (↑↓)	$e_g^4\ t_{2g}^5$	$(4 \times 0.6\Delta_t) - (5 \times 0.4\Delta_t) = 0.4\Delta_t$
10.	(↑↓) (↑↓) (↑↓) (↑↓) (↑↓)	$e_g^4\ t_{2g}^6$	$(4 \times 0.6\Delta_t) - (6 \times 0.4\ \Delta_t) = 0$

Since the magnitude of crystal field spitting in tetrahedral field is small and always less than the pairing energy, therefore pairing of electrons is not energetically favourable. Therefore, all the tetrahedral complexes are high spin complexes.

Q. 42. Discuss splitting of *d*-orbitals in square planar field according to CF theory.

Ans. The square planar arrangement of ligands may be considered to be one derived from the octahedral field by removing two trans-ligands located along the Z-axis. In the

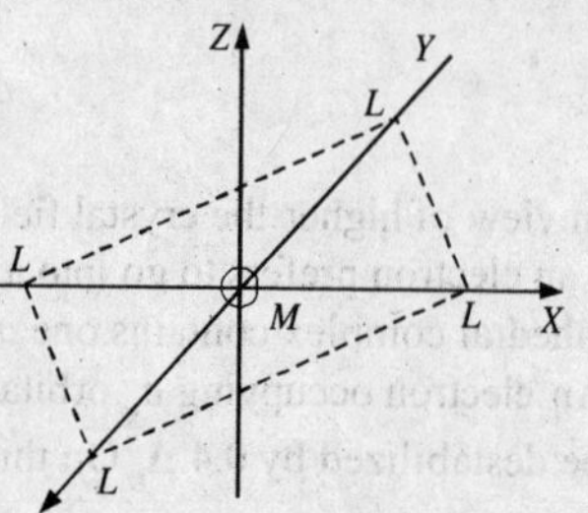

Fig. 21.6 Square planar arrangement.

process, the e_g and t_{2g} sets of orbitals is lifted *i.e.*, these orbitals will no longer be degenerate.

The four ligands in a square planar arrangement around the central metal ion are shown in Fig. 21.6. As the ligands approach through the axes, they would have greatest influence on $d_{x^2-y^2}$ orbital, so that the energy of this orbital, will be raised most. The d_{xy} orbital, lying in the same plane, but between the ligands will also have a greater energy through the effect will be less than that on the $d_{x^2-y^2}$ orbitals. On the other hand, due to absence of ligands along Z-axis, the d_{z^2} orbital becomes stable and has energy lower than that of d_{xy} orbital. Similarly d_{yz} and d_{zx} become more stable. The energy level diagram may be represented as in Fig. 21.7 along with tetrahedral and octahedral fields.

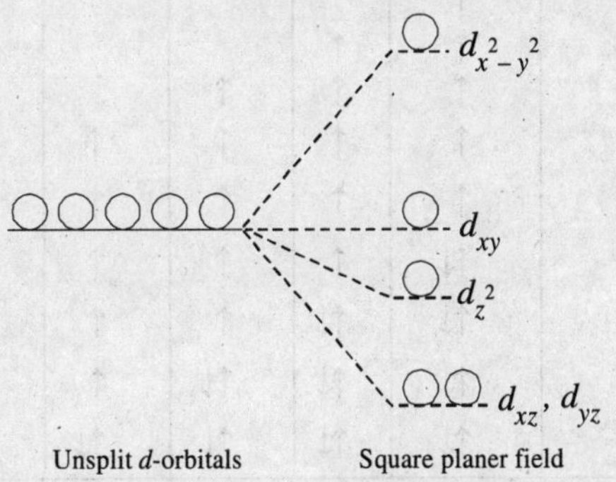

Fig. 21.7. Splitting of d- orbitals in square plannar field.

Q. 43. How does crystal field theory explainthe diamagnetic behaviour of $Ni(CN)_4]^{2-}$ and paramagnetic behaviour of $[NiCl_4]$. *(Guwahati, 2002)*

Ans. Cyanide ion is a strong field ligand. It splits the *d* orbitals into three sets of orbitals with varying energies as follows:

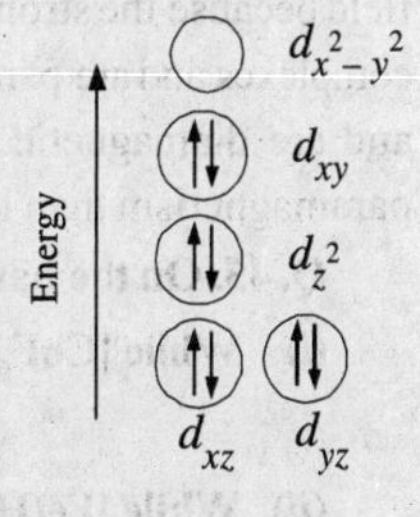

Eight electrons in Ni^{2+} ions are filled in the orbitals as shown above. Because of large value of splitting energy, electrons don't occupy higher energy $d_{x^2-y^2}$ orbitals. Thus, there are no unpaired electrons and the complex is diamagnetic. Chloride is a weak ligand causing only a small splitting. Electrons occupy different orbitals giving tetrahedral geometry. Complex is paramagnetic due to presence of unpaired electrons.

Q. 44. Discuss magnetic properties of octahedral complexes.

Ans. The filling of *d*-orbitals in presence of weak ligand and strong ligand is shown in Table 21.4. In a weak ligand field *d*-orbital will be occupied singly before any pairing of electrons takes place. This is because the energy difference between t_{2g} and e_g orbitals will be too small to affect the issue.

In the field of weak ligands (such as F^-, OH^-, Cl^-, Br^-. I^- etc.) the crystal field splitting is small so less energy is required to feed the fourth electron into a e_g orbital than to pair it with one of the t_{2g} orbitals. Hence it will be more favourable energetically to occupy the upper e_g level and have a high spin complex rather than to pair of electrons with one of t_{2g} orbitals. Here Δ_0 (octahedral) is less than pairing energy *P*, *i.e.*, $\Delta_0 < P$ or pairing energy *P* is more than splitting energy (Δ_0) *i.e.*, $P > \Delta_0$.

In a field of strong ligands such as CN^-, NO_2, en, NH_3 or CNS^-, the crystal field splitting is quite large. Thus Δ_0, the difference between the energies of e_g and t_{2g} orbitals is quite significant, so electron pairing takes place. The electrons would rather pair in lower energy t_{2g} orbitals rather than occupying higher energy e_g orbitals. Splitting energy Δ_0 is more than pairing energy *i.e.*, $\Delta_0 > P$ or paring energy is less than splitting energy *i.e.*, $P < \Delta_0$.

There is a significant difference in the arrangement of d^4, d^5, d^6 and d^7 metal ion configurations in a weak and a strong field as shown in Table. 21.4.

Table 21.4

No. of Electrons	Weak Field (High spin) t_{2g}			e_g		Strong Field (Low Spin) t_{2g}			e_g	
1	↑					↑				
2	↑	↑				↑	↑			
3	↑	↑	↑			↑	↑	↑		
4	↑	↑	↑	↑		↑↓	↑	↑		
5	↑	↑	↑	↑	↑	↑↓	↑↓	↑		
6	↑↓	↑	↑	↑	↑	↑↓	↑↓	↑↓		
7	↑↓	↑↓	↑	↑	↑	↑↓	↑↓	↑↓	↑	
8	↑↓	↑↓	↑↓	↑	↑	↑↓	↑↓	↑↓	↑	↑
9	↑↓	↑↓	↑↓	↑↓	↑	↑↓	↑↓	↑↓	↑↓	↑

In these arrangements the number of unpaired electrons is greater in a weak fie d than in a strong field because the strong field forces the electrons to pair. Complexes with weak ligands are high spin complexes and are paramagnetic in nature while those with strong ligand fields are low spin complexes and are diamagnetic in nature, High spin complexes having more unpaired electrons have high paramagnetism than low-spin complexes which have fewer unpaired electrons.

Q. 45. On the basis of crystal field theory, account for the following:

(*i*) While $[CoF_6]^{3-}$ is paramagnetic, $[Co(CN)_6]^{3-}$ is diamagnetic.

(*Awadh, 2000; M. Dayanand 2002; Rajasthan, 2010*)

(*ii*) While $[Fe(H_2O)_6]^{3+}$ is strongly paramagnetic $[Fe(CN)_6]^{3-}$ is less paramagnetic.

(*Awadh, 2000*)

Ans. (*i*) The $[CoF_6]^{3-}$ ion is an example of high spin (outer orbital) complex while $[Co(CN)_6]^{3-}$ is a low spin (inner orbital) complex. Both have octahedral geometry and the magnetic preperties are explained below on the basis of the field strength of the CN^- ligand.

In a free Co atom ($3d^7\ 4s^2$) and a Co^{3+} ion ($3d^6\ 4s^0$) the electrons are arranged as follows:

	3*d*					4*s*
Co atom	↑↓	↑↓	↑	↑	↑	↑↓
Co^{3+} ion	↑↓	↑	↑	↑	↑	

In an octahedral complex, however the five *d*-orbitals split up into t_{2g} and e_g orbitals in the $[CoF_6]^{3-}$ ion. The energy difference Δ_0 between the two sets of orbitals is not great enough (as F^- is a weak field ligand) to cause any spin-pairing ($\Delta_0 < P$) so that the arrangement of the five *d*-orbitals is as shown in Fig. 21.8. There are four unpaired electrons and the complex ion is strongly paramagnetic.

In the $[Co\ (CN)_6]^{3-}$ ions, the stronger field of CN^- ion causes a greater energy difference (Δ_0) between t_{2g} and e_g orbitals. As a result spin pairing occurs because $P < \Delta_0$. The five *d*-orbitals are arranged as show in Fig. 21.8 (*c*).

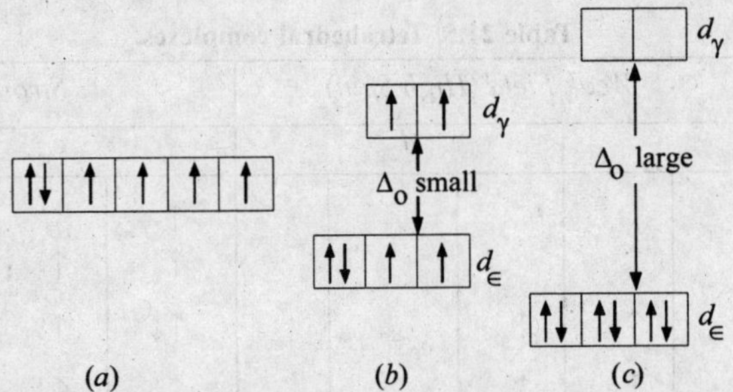

Fig. 21.8. Arrangement of *d* electrons in free Co^{3+} ion and under different ligand fields (*a*) Free ion (*b*), In a weak octahedral field, (*c*) In a strong octahedral field.

There is no unpaired electron in the resulting complex ion and so it is diamagnetic.

(*ii*) Similarly in the complex ions $[Fe\ (H_2O)_6]^{3+}$ and $[Fe\ (CN)_6]^{3-}$ it can be shown that the former ion is very strongly paramagnetic while the latter ion is less paramagnetic. This is because H_2O is a weak ligand and $\Delta_0 < P$ and therefore no spin pairing takes place and the complex obtained is strongly paramagnetic. In the case of CN^- which is a strong ligand $\Delta_0 > P$ and, therefore, electrons pair up leaving only one *d*-electron. This gives a weak paramagnetic complex. This is shown in **Fig. 21.9.**

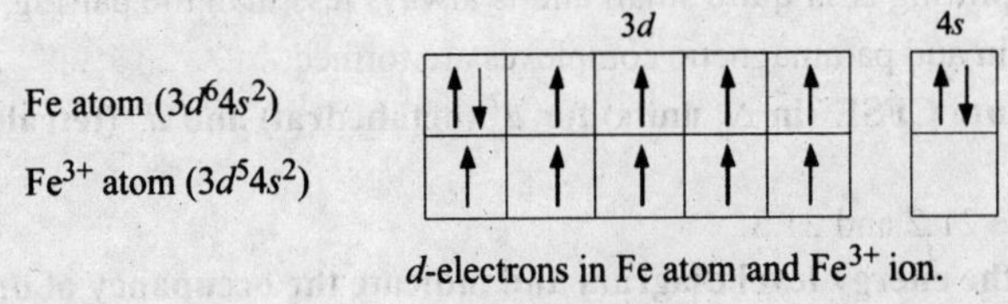

d-electrons in Fe atom and Fe^{3+} ion.

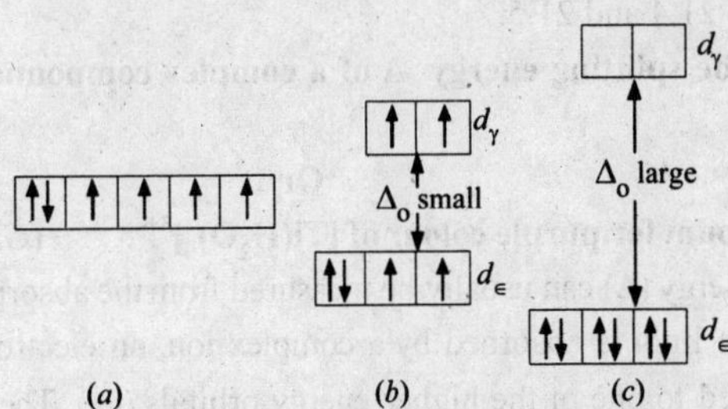

Fig. 21.9. Occupation of *d*-electrons in Fe^{3+} ion under different ligand fields (*a*) Free ion, (*b*), In weak octahedral field, *e.g.*, $[Fe(H_2O)_6]^{3+}$, (*c*) In strong cotahedral field *e.g.*, $[Fe(CN)_6]^{3-}$

Q. 46. Discuss magnetic properties in tetrahedral complexes on the basis of crystal field theory.

Ans. The filling of *d*-orbitals in certain ions surrounded by tetrahedral ligands in weak and strong fields is shown in Table 21.5.

Table 21.5. Tetrahedral complexes.

No. of Electrons	*Weak Field (High Spin)*					*Strong Field (Low Spin)*				
	e_g		t_{2g}			e_g		t_{2g}		
1	↑					↑				
2	↑	↑				↑	↑			
3	↑	↑	↑			↑↓	↑			
4	↑	↑	↑	↑		↑↓	↑↓			
5	↑	↑	↑	↑	↑	↑↓	↑↓	↑		
6	↑↓	↑	↑	↑	↑	↑↓	↑↓	↑	↑	
7	↑↓	↑↓	↑	↑	↑	↑↓	↑↓	↑	↑	↑
8	↑↓	↑↓	↑↓	↑	↑	↑↓	↑↓	↑↓	↑	↑
9	↑↓	↑↓	↑↓	↑↓	↑	↑↓	↑↓	↑↓	↑↓	↑

It is clear from the table that the configurations of d^3, d^4, d^5 and d^6 ions may form high and low spin complexes. But in actual practice only high spin complexes are formed. This is due to the fact that crystal field splitting Δ_t is quite small and is always less than the pairing energy, *i.e.*, $\Delta_t < P$ .Therefore high spin and paramagnetic complexes are formed.

Q. 47. Compute CFSE (in Δ_0 units) for d^7 (octahedral) and d^7 (tetrahedral) in a strong ligand field.

Ans. See Table 21.2 and 21.3.

Q. 48. Draw the energy level diagram and indicate the occupancy of orbitals (splitted) in d^6 (octahedral low spin) and d^6 (tetrahedral).

Ans. See Tables 21.4 and 21.5.

Q. 49. How is the splitting energy Δ of a complex compound obtained from absorption spectrum? *(Ranchi, 2010)*

Or

How do you account for purple colour of $[Ti(H_2O)_6]^{3+}$? *(Guwahati, 2002; Nagpur, 2008)*

Ans. Splitting energy (Δ) can usually be measured from the absorption spectra of complex ions. In simple cases when light is absorbed by a complex ion, an electron in one of the lower energy orbitals (t_{2g}) is excited to one of the higher energy orbitals (e_g). The energy corresponding to the frequency of absorbed light is equal to Δ_0. For many transitions metal complexes this absorption lies in the visible region of the spectrum (λ = 4000 to 7000 Å or wave number $\bar{\nu}$ = 25000 to 14300 cm^{-1} $\left(\bar{\nu} = \frac{1}{\lambda}\right)$. It gives rise to the bright colours associated with transition metal compounds. For example in case of hydrated Ti^{3+} complex ion (Ti^{3+} has only one *d*-electron) the light of wavelength of about 5000 Å (wavenumber 20300 cm^{-1}) is absorbed as shown in the absorption spectrum of $Ti(H_2O)_6]^{3+}$ in Fig. 21.10 (*a*).

The absorbed light excites an electron in the lower energy t_{2g} orbitals to the higher energy e_g orbitals as in the case of Ti^{3+} [Fig. 21.10 (*b*)]. The energy corresponding to the frequency of abosrbed light is equal to Δ In this case Δ = E= 243 kJ/mol.

The absorption is in the range of 5000 to 6000 Å *i.e.*, green and the yellow components of the white light (VIBGYOR) are absorbed and the transmitted colour will be the mixture of remaining colours (mainly violet and red) so the colour of this complex is purple.

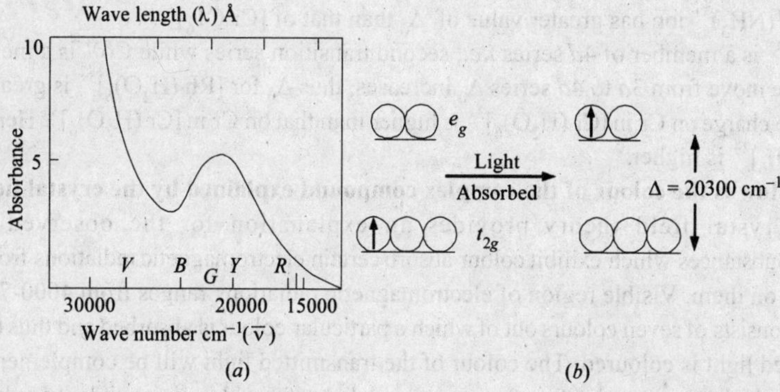

Fig. 21.10. (*a*) Absorption spectrum of $[Ti(H_2O)_6]^{3+}$, (*b*) Excitation of electron.

Q. 50. What are the factors that affect the crystal field splitting energy (Δ).

(*Nagpur, 2000*)

Ans. The magnitude of Δ depends on the (*i*) nature of the ligands, (*ii*) oxidation state of the metal

ion, (*iii*) type of *d*-electrons, (*iv*) geometry of the complex.

(*i*) Nature of the ligand. Ligands which cause only a small splitting are called weak ligands, and those which bring a large splitting are called strong ligands. It is possible to arrange ligands according to the magnitude of Δ observed with a given metal ion. The sequence for common ligands in order of increasing Δ values is as follow:

$I^- < Br^- < Cl^- < OH^- < F^- < C_2O_4^{2-} < H_2O < CNS^- < NH_3 < en < NO_2^- < CN^- < CO$

Weak ligands Strong ligands

This series is known as the spectrochemical series.

(*ii*) Oxidation state of the metal ion. The magnitude of Δ increases as the oxidation state of the metal ion increases. The splitting produced by +3 oxidation state is roughly 1.5 times or double than that of +2 oxidation states.

(*iii*) Type of *d*-electrons. The magnitude of Δ varies little through the first transition series (3*d*) for a particular oxidation state of the metal ions but become considerably large i.e., about 30 to 50 per cent in the second (4*d*) and the third transition series (3*d*) for a particular ligand. This may be explained on the basis that 4*d* and 5*d* orbitals are bigger in size as compared to 3*d* orbitals and extend further out from the metal ion and thus interact more strongly with the negative ligands or the negative end of the dipolar molecule.

(*iv*) Geometry of the complex. The magnitude of Δ varies with the geometry of the complex. For example, Δ values for tetrahedral complexes is about 40–50% less than that of octahedral complexes for the same ligand and it has been found that Δ for square planar complexes is more than that for octahedral complexes.

Q. 51. Which complex of the following pairs has a larger value of Δ_0?

(*i*) $[Cr(CN)_6]^{3-}$ **and** $[Cr(H_2O)_6]^{3+}$

(*ii*) **$[Cr(NH_3)_6]^{3+}$** **and** **$[CrCl_6]^{3-}$**
(*iii*) **$[Co(H_2O)_6]^{3+}$** **and** **$[Rh(H_2O)_6]^{3+}$**
(*iv*) **$[Cr(H_2O)_6]^{3+}$** **and** **$[(Cr(H_2O)_6]^{2+}$**

Ans. (*i*) Since CN^- is a stronger ligand than H_2O (*i.e.*, CN^- ion has greater field strength than H_2O). Therefore, Δ_0 for $[Cr(CN)_6]^{3-}$ is greater than Δ_0 for $[Cr(H_2O)_6]^{3+}$.

(*ii*) $[Cr(NH_3)]^{3+}$ ion has greater value of Δ_0 than that of $[Cr\,C1_6]^{3-}$.

(*iii*) Rh^{3+} is a member of 4*d* series *i.e.*, second transition series while Co^{3+} is a member of 3*d* series. As we move from 3*d* to 4*d* series Δ_0 increases, thus Δ_0 for $[Rh(H_2O)_6]^{3+}$ is greater.

(*iv*) The charge on Cr in $[Cr(H_2O)_6]^{3+}$ is higher than that on Cr in $[Cr(H_2O)_6]^{3+}$ Hence Δ_0 value for $[Cr(H_2O)_6]^{3+}$ is higher.

Q. 52. How is the colour of the complex compound explained by the crystal field theory?

Ans. Crystal field theory provides an explanation for the observed colour in complexes.Substances which exhibit colour absorb certain electromagnetic radiations from the white light falling on them. Visible region of electromagnetic radiations ranges from 4000-7000 Å. The white light consists of seven colours out of which a particular colour is absorbed and thus the reflected or transmitted light is coloured. The colour of the transmitted light will be complementary to that which is absorbed. For example, if a substance absorb lights from blue region, it looks red; conversely, if light corresponding to red wavelength is absorbed then the substance will appear blue. Most of the complexes are coloured because the energy of light in the visible region is just the right value to promote *d*-electrons from lower t_{2g} level to higher e_g level in an octahedral complex and from lower e_g level to higher t_{2g} level is tetrahedral complex. These complexes have metal ions with incomplete *d* sub-shell. The transition metal ions having completely filled or completely empty sub-shell of *d*-electrons (no unpaired electrons) will give colourless solutions, *e.g.*, Cu^+, Ti^{4+}, Zn^{2+}, Ag^+ etc. The energy (Δ) required for such a promotion will vary according to the ligand so that the colour of a complex is dependent on the ligand it contains. For example $(Ti(H_2O)_6]^{3+}$ ion which absorbs green and yellow components of the white light in the wavelength region of 5000 to 6000 Å is purple (as green and yellow radiations are absorbed and the complementary colour *i.e.*, purple is transmitted). The hydrated cobalt (II) salts containing the octahedral cation $[Co(H_2O)_6]^{2+}$ absorb blue part of visible light for its electronic transitions and thus looks pink.

For a given central ion replacement of the ligand by others to the right in the spectrochemical series will give larger values of Δ. This will lead to a lowering of the wavelength of the absorbed light and as a result shift from red towards blue in the colour of the absorbed light. The colour or transmitted light will be complementary to that which is absorbed as shown below:

Colour observed (Complementary colour)	*Green*	*Blue*	*Violet*	*Red*	*Orange*	*Yellow*
Colour absorbed	Red	Orange	Yellow	Green	Blue	Violet
Wavelength absorbed	7000	6000	5000		4000	

Q. 53. List some limitations of crystal field theory?

Ans. (1) In considers only the metal ion *d*-orbitals and gives no consideration at all to other metal orbitals (such as *s*, p_x, p_y and p_z orbitals).

(2) It is unable to account satisfactorily for the relative strengths of ligands for example it gives no explanation as to why H_2O is a stronger ligand than OH^- in the spectrochemical sereis.

(3) According to this theory, the bond between the metal and ligands are purely ionic. It gives no account on the partly covalent nature of the metal ligand bonds.

(4) The CFT cannot account for the π bonding in complexes.

Q. 54. Draw a comparison between valence-bond and crystal field theories of coordination compounds.

Sl No.	*V.B.T.*	*Crystal field theory*
1.	The bonding between the metal to ligands is considered to be covalent as well as ionic.	The bonding is considered to be purely electrostatic between the positive metal ion and the negative ions or negative ends of the ligands.
2.	It does not take into account the splitting of *d*-orbitals of the metal.	It takes into account the splitting of *d*-orbitals of the metals.
3.	It involves the concept of hybridization of orbitals of the central metal ion before complex formation.	It does not involve this concept.
4.	The colour of complexes are not clearly explained.	The colour of complexes are explained in terms of electronic transitions between the various d-orbitals of different energies.
5.	The magnetic properties are not clearly explained.	The magnetic properties of complexes depend upon the splitting of *d*-orbitals of the central cation in different crystal fields.

Q. 55. Calculate the CFSE of the following complexes

$[Co\ Fe_6]^{3-}$, $[Fe\ (CN)_6]^{4-}$, $[Cu\ (NH_3)_4]^{2+}$

Ans. $[Co\ F_6]^2$ is weak field high spin d^6 case. The configuration is d^2_{xy}, d^1_{xz} d^1_{zy} $d^1_{x^2-y^2}$ $d^1_{x^2}$.
CFSE $=0.4\ \Delta_0$.

$[Fe\ (CN)_6]^{4-}$ is a strong field low spin d^6 case. The configuration is $d^2_{xy} d^2_{xz} d^2_{yz}$ CFSE $= 2.4\ \Delta_0$.

$[Cu\ (NH_3)_4]^{2+}$ is a d^9 case with configuration $t^6_{2g}\ e^3_g$, CFSE $= 0.6\ \Delta$.

Q. 56. Calculate spin magnetic moment of the following complexes according to V.B. theory as well as CF theory.

$[(Cr\ (H_2O)_6]^{3+}$, $[Fe\ (NH_3)_6]^{2+}$, $[Mn\ (CN)_6]^{4-}$

Ans. (*i*) $[Cr\ (H_2O)_6]^{3+}$

3*d*					4*s*	4*p*		
↑	↑	↑	X	X	X	X	X	

This compound exhibits $d^2\ sp^3$ hybridization. It has three unpaired electrons.

In terms of C.F. theory the configuration is d^1_{xy}, d_{yz}, d^1_{xz} in octahedral field with 3 unpaired electrons. Here n = number of unpaired electrons.

$$\mu_{spin} = \sqrt{n\times(n+2)} = \sqrt{3\times5} = 3.87$$

(*ii*) $[Fe\ (NH_3)_6]^{3+}$

3*d*					4*s*	4*p*			4*d*				
↑	↑	↑	↑	↑	X	X	X	X					

According to V.B.T. there are 5 unpared electrons. The complex exhibits $d^2\ sp^3$ hybridization. According to C.F. theory it is a high spin complex with configuration $t^3_{2g} e^2_g$ having 5 unpaired electrons.

$$\mu = \sqrt{n\times(n+2)} = \sqrt{5\times7} = 5.9$$

(*iii*) $[Mn(CN)_6]^{4-}$

3*d*					4*s*	4*p*				
↑↓	↑↓	↑	X	X	X	X	X	X		

According to the V B. theory the complex exhibits d^2sp^3 hybridisation with one unpaired electron.

According to C.F. theory, it is a low spin complex with configuration $d^2_{xy}, d^1_{yz}\ d^1_{xz}$ due to strong field. It has one unpaired electron. $\mu = \sqrt{1 \times 3} = 1.73$.

Q. 57. Determine the number of unpaired electrons and CFSE for

(*i*) $[Fe\ (H_2O)_6]^{3+}$

(*ii*) $[Cr\ (NH_3)_6]^{3+}$

(*iii*) $[Co\ Cl_4]^{2-}$

Ans. (*i*) $[Fe\ (H_2O)_6]^{3+}$ The compound has octahedral geometry. Five *d* orbitals are split into two sets, one consisting of $d_{x^2-y^2}$ and d_{z^2} orbitals of high energy and other consisting of the three orbitals d_{xy}, d_{xz} and d_{yz} of lower energy. Water is a weak ligand. We therefore predict this complex to be high spin. The electrons occupy the orbitals as shown below.

↑ ↑ e_g

Δ_o ↑ ↑ ↑ t_{2g} $\Delta_o < P$

Thus $[Fe\ (H_2O)_6]^{2+}$ has five unpaired electrons in it.

$CFSE = 3 \times 0.6\ \Delta_0 - 2 \times 0.4\Delta_0$

$= \Delta_0$

(*ii*) $[Cr\ (NH_3)_6]^{3+}$. The compound has octahedral geometry, five *d*-orbitals are split into two sets e_g and t_{2g}.

○ ○ e_g

↑ ↑ ↑ t_{2g}

The complex $[Cr\ (NH_3)_6]^{3+}$ has 3 unpaired electrons.

$CFSE = 3 \times\ 0.6\ \Delta_0 = 1.8\Delta_0$

(*iii*) $[CoCl_4]^{2-}$. Here Co is in +2 oxidation state. The compound has tetrahedral geometry. It is a weak field high spin d^7 case

↑ ↑ ↑ t_{2g}

↑↓ ↑↓ e_g

$CFSE = 1.2\ \Delta_t$

Q. 58. Explain with example, the effect of nature of ligand on the magnitude of CFSE (Δ).

Ans. See Q. 50.

Q. 59. Why does NH_3 readily form complex but NH_4^+ does not?

Ans. NH_3 contains a lone pair of electrons which coordinate with the metal ion to form the complex compound. This lone pair of electrons is bound to H^+ in NH_4^+ and thus is not available for linking to the metal ion. Therefore NH_4^+ does not form a complex readily.

Q. 60. What are CFSE for the following systems: d^1-octahedral, d^1-tetrahedral, d^5-low spin octahedral and thigh spin octahedral.

Ans. See Tables 21.2 & 21.3.

Q. 61. How does crystal field theory explain:

(*i*) $[Ti(H_2O)_6]^{3+}$ is coloured and

(*ii*) Zn^{2+} is colourless

Ans. See Q. 52.

Q. 62. In what respects does an inner orbital complex differ from an outer orbital complex?

Ans. A complex obtained by using inner $(n-1)\,d$ orbitals is called inner orbital complex. On the other hand if outer *nd* orbitals are used in bond formation an outer complex is formed.

An inner complex is obtained usually by pairing of electrons in order to create some $(n-1)d$ orbitals. Such complexes are either diamagnetic or weakly paramagnetic.

On the other hand, no such pairing of electrons is required in outer complexes. Thus such complexes are generally strongly paramagnetic.

Q. 63. Draw all possible isomers of $[Cr(en)_2Cl_2]Cl$.

Ans. See Q. 21 (IV) take Cr in place of Co.

Q. 64. Why square planar complexes do not show optical isomerism and tetrahedral complexes do not show geometrical isomerism.

Ans. Square planar complexes do not show optical isomerism because they possess a plane of symmetry. For a compound to exhibit optical isomerism, it is necessary that it should not possess a plane of symmetry. Tetrahedral complexes do not show geometrical isomerism because all positions are adjacent to each other.

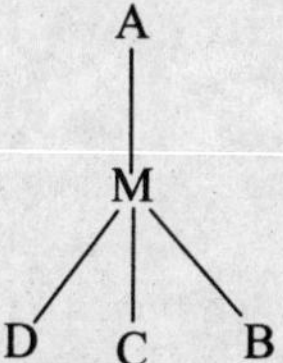

Q. 65. The magnetic moment of $[Fe(H_2O)_6]^{3+}$ is 5.92 B.M. and that of $[Fe(CN)_6]^{3-}$ is 1.73 B.M. Explain this on the basis of crystal field theory.

Ans. $[Fe(H_2O)_6]^{3+}$ is a weak field high spin octahedral complex. Five *d* electrons are distributed as

(↑) (↑) e_g

(↑) (↑) (↑) t_{2g}

Thus the complex has 5 unpaired electrons

$$\mu_s = \sqrt{n\times(n+2)} = \textbf{B.M}$$

$$= \sqrt{5\times7} = 5.92 \ \textbf{B.M.}$$

$[Fe(CN)_6]^{-3}$ is a strong field low spin octahedral complex. Five electrons are distributed as

( ) ( ) e_g

(↑↓) (↑↓) (↑) t_{2g}

It is thus left with one unpaired electron.

$$\mu_s = \sqrt{n\times(n+2)} = \sqrt{1\times3} = 1.73 \ \textbf{B.M.}$$

Q. 66. Tetrahedral complexes are generally high spin. Why?

Ans. Tetrahedral complexes have generally high spin even when strong field ligands are involved. This is because the ligands hardly approach towards the orientation of *d*-orbitals (See. Q. 41). Hence the extent of splitting energy is small. The electrons prefer to obey Hund's rule and occupy separate orbitals as far as possible. Therefore, such complexes are high spin.

Q. 67. Explain with the help of valence bond theory that nickel forms a tetrahedral complex with CO group and a square planar complex with a cyano group.

Ans. See Qs. 34 and 38.

Q. 68. What is the significance of 10 D_q?

Ans. This is an arbitrary measure of the splitting energy. For example in the case of octahedral complexes,

$$10D_q = \Delta_0$$

where Δ_0 = octahedral splitting energy.

Q. 69. What is the oxidation state, type of hybridisation and geometry of the following molecules?

$$[Ni(CN)_4]^{2-},\ [MnCl_4]^{2-}$$

Ans. $[Ni(CN)_4]^{2-}$

Oxidation state: $x - 4 = -2$ or $x = 2$

Hybridisation: sp^2d

Geometry: Square planar.

$[MnCl_4]^{-2}$

Oxidation state $x - 4 = 2$ or $x = 2$

Hybridisation: sp^3

Geometry: Tetrahedral.

Q. 70. Give the number of unpaired electron in strong and weak octahedral field for Cr^{3+} and Fe^{2+}.

(*G. Nanakdev, 1996*)

Ans. Cr^{3+} is d^3 system

Fe^{2+} is d^6 system

See Table 21.4.

Q. 71. Explain the given magnetic moment of the following compound on the basis of valence bond and crystal field theories.

Compound	***Magnetic moment (B M.)***
$K_3(FeF_6)$	5.9
$[Co(NH_3)_6]Cl_3$	0.0
$[Ni(H_2O)_6]Cl_2$	2.9
$K_4[Mn(CN)_6]$	1.9
$K_4[MnF_6]$	5.9

Ans. $K_3[FeF_6]$. Fluoride ion is a weak field high spin ligand. Iron has oxidation state of +3. Fe^{3+} has five unpaired electrons in 3*d* orbitals.

$\therefore$ Magnetic moment $= \sqrt{n \times (n+2)}$ or $\sqrt{35} = 5.9$

$[Co(NH_3)_6]Cl_3$. It is a strong field, low spin complex. Cobalt has oxidation state +3. Co^{3+} has electrons in *d* orbitals which get paired due to strong ligand leaving no unpaired electron. Therefore, magnetic moment $= \sqrt{0 \times 0} = 0$

$[Ni(H_2O)_6]Cl_2$. Nickel is in +2 oxidation state. Ni^{2+} has 8 electrons in 3*d* orbitals distributed as follows:

↑↓	↑↓	↑↓	↑	↑

Water being a weak ligand cannot force pairing of electrons. Hence there are two unpaired electrons.

Magnetic moment = $\sqrt{2 \times 4}$ = 2.9 **B.M.**

$K_4[Mn(CN)_6]$. Mn is in +2 oxidation state but cyanide is a strong ligand. Mn^{2+} has 5 electrons in 3*d* orbitals which get paired leaving only one unpaired electron.

Hence, magnetic moment = $\sqrt{1 \times 3}$ = 1.9 **B.M.**

$K_4[MnF_6]$. Mn^{2+} has 5 electrons in 3*d* orbital, but fluoride being weak ligand, pairing of electrons does not take place.

Therefore, magnetic moment = $\sqrt{5 \times 7}$ = 5.9 **B.M.**

Q. 72. Give an account of modified crystal field theory or ligand field theory.

Ans. Crystal field theory predicts the splitting of *d* orbitals due to the electrostatic potential of the ligands to which the *d* electrons of the metal ions are exposed. The crystal field theory can explain the spectra of the metal ions and of complexes on the assumption that these arise from the transition of electrons from lower energy *d* orbitals to higher energy *d* orbitals. However, the positions and intensities of spectral bands calculated on the basis of crystal field theory do not always coincide with those determined experimentally. There are frequent deviations. Apart from this, a pure electrostatic interaction between the metal ion and the ligands fails to explain the relative positions of ligands in the spectrochemical series. In fact, the bond between central atom and the ligand is not purely ionic. There is partial covalent nature also associated here.

There is clear evidence that *covalent bonding too makes a significant contribution towards the metal-ligand bonding.*

This is supported by the following evidence.

1. Nuclear Magnetic Resonance (NMR). The NMR spectral studies clearly indicate that the spin of an unpaired electron of a metal ion interacts with the nuclear spin of ^{19}F in the fluoro complex of a paramagnetic metal ion. This electron spin-nuclear spin interaction is possible only if the unpaired electron spends more than negligible time on the ^{19}F-nucleus. This is possible through the overlapping of orbitals of the ligand and the metal ion, *i.e.,* through covalent bonding.

2. Electron Spin Resonance (ESR). The ESR spectral studies conducted on the complexes of paramagnetic metal ions reveal the presence of unpaired electrons in the molecular orbitals formed due to the overlapping of orbitals of the ligand and the metal ion. This is in support of covalence in metal-ligand bonding.

3. Interelectronic Repulsion. Electrons in the partly filled *d* orbitals of a metal ion repel one another and give rise to a number of energy levels depending upon the arrangement of these electrons in the *d* orbitals. The energy of each of these levels can be expressed in terms of some interelectronic repulsion parameters. The energy gap between two such energy levels which have the same or different spin multiplicity can be expressed in terms of repulsion parameters. It is observed experimentally that the magnitude of these interelectronic repulsion parameters always decreases on the complexation of the metal ion.

4. Lande's Splitting Factor, *g*. The Lande's splitting factor *g* which is employed to predict the magnetic behaviour of transition metal complexes can also determine the extent of delocalisation (or penetration) of *d* electrons of the metal into the orbitals of the ligand. The value of *g* determined experimentally shows that the metal *d* electrons are almost always delocalised to some extent into the ligand orbitals. This can happen only due to the overlapping of the metal *d* orbitals with the ligand orbitals through covalent bonding.

All the above observations point to the role played by covalent bonds in metal-ligand linkage.

Q. 73. What is meant by stability constant of a complex ion?

Ans. Complex ions dissociate, in aqueous solutions to establish an equilibrium with the central metal ion and the ligands from which they are derived. Thus, in solution containing silver-ammonia complex ion, $[Ag(NH_3)_2]^+$, the following equilibrium exists :

$$[Ag(NH_3)_2]^+ \rightleftharpoons Ag^+ + 2NH_3$$

The equilibrium constant, *K,* given by the following expression,

$$K = \frac{[Ag^+][NH_3]^2}{[Ag(NH_3)_2]^+}$$

is 6×10^{-8}. This shows that the degree of dissociation of the complex ion is extremely small. In other words, the complex ion is highly stable.

The above equilibrium constant is called the *dissociation constant* of the complex ion. Smaller the value of the dissociation constant, greater will be the stability of the complex. The *reciprocal of dissociation constant* is **stability constant** of the complex ion. Greater the value of the stability constant, greater will be the stability of the complex ion.

The stability constant of the complex ion $[Ag(NH_3)_2]^+$ is the equilibrium constant of the reverse reaction, $Ag^+ + 2NH_3 \rightleftharpoons [Ag(NH_3)_2]^+$.

Thus,

$$K' = \frac{1}{K} = \frac{[Ag][NH_3)_2]^+}{[Ag^+][NH_3]^2}$$

The stability constant *K'* of the complex ion $[Ag(NH_3)_2]^+$ is $1/(6 \times 10^{-8}) = 1.67 \times 10^7$.

Q. 74. Discuss the factors which effect the stability of a complex ion.

(Ranchi, 2010; Garhwal, 2010; Rajasthan, 2010)

Ans. Main factors that affect the stability of a complex are discussed as under:

1. The nature of the central ion. In general, *greater the charge density on the central ion, greater is the stability of its complexes.* In other words, *greater the charge and the smaller the size of an ion, i.e.,* the larger the charge/radius ratio of an ion, *greater is the stability of its complexes.* For instance, Fe^{3+} ion carries a higher charge and has a smaller size than Fe^{2+} ion. Hence, charge density is higher on Fe^{3+} than on Fe^{2+} ion. The complexes of Fe^{3+} ion thus show greater stability. This is illustrated by the fact that while stability constant of $[Fe(CN)_6]^{3-}$ is 10^{31}, that of $[Fe(CN)_6]^{4-}$ is only about 10^6.

2. Electronegativity. Electronegativity of the central ion also influences the stability of its complexes. *Greater the electronegativity of the central ion, greater is the stability of its complexes.* This is because the bonding between a central ion and a ligand takes place due to the donation of electron pairs by the ligands. Hence, *a strong electron-attracting central ion will give stable complexes.*

3. The nature of the ligand. In general, *more basic a ligand, greater is the ease with which it can donate its lone pairs of electrons and, therefore, the greater is the stability of the bonds formed by it.* Thus, CN^- and F^- ions and NH_3 molecules, which are strong bases, are good ligands to form stable complexes.

For anionic ligands, *higher the charge and smaller the size, more stable is the complex formed.* Thus, F^- ion gives more stable complexes than Cl^- ion.

4. The chelate effect. The complexes containing chelate rings are found to be *more stable* than those containing similar but unidentate ligands. This effect is known as **chelate effect.**

Consider the formation of complexes from the hydrated cadmium ion $[Cd(H_2O)_4]^{2+}$ with methylamine and ethylenediamine (en) as the ligands, in aqueous medium. The reactions can be represented as under:

$$[Cd(H_2O)_4]^{2+} + 4CH_3NH_2 \rightleftharpoons [Cd(CH_3NH_2)_4]^{2+} + 4H_2O \quad .....(1)$$

$$[Cd(H_2O)_4]^{2+} + 2(en) \rightleftharpoons [Cd(en)_2]^{2+} + 4H_2O \quad .....(2)$$

The complex formed by ethylenediamine is a chelate, as shown below :

$H_2C - \overset{H_2}{N} \rightarrow Cd \leftarrow \overset{H_2}{N} - CH_2$; $H_2C - \underset{H_2}{N} \rightarrow Cd \leftarrow \underset{H_2}{N} - CH_2$ (each H_2C pair joined, forming two chelate rings around Cd)

The stability constants of the complexes with CH_3NH_2 and ethylene diamine are known to be 3.10×10^6 and 3.2×10^{10}, respectively. Evidently, the chelate is 1000 times more stable than the other complex. This is explained by increase in entropy in the case of complex formation with ethylenediamine.

The value of $\Delta H°$ for both the above reactions are known to be practically the same, *viz.,* about – 57 kJ. Thermodynamically, the stability of a specie depends upon the value of $\Delta G°$ which should be large and negative. $\Delta G°$, in turn, is equal to $\Delta H° - T\Delta S°$. The greater stability of $[Cd(en)_2]^{2+}$ complex than that of $[Cd(CH_3NH_2)_4]^{2+}$ complex should be due to its greater positive value of $\Delta S°$ since $\Delta H°$ is the same for both the reactions.

In the first reaction, 4 molecules of CH_3NH_2 displace 4 molecules of H_2O coordinated to Cd^{2+} ion to form the complex ion $[Cd(CH_3NH_2)_4]^{2+}$ so that the number of molecules on the products side is the *same* as the number of molecules on the reactants side. As a consequence, *there is very little change in the extent of randomness in going from reactants to products* ; that is, *$\Delta S°$ for the reaction is nearly zero.* Therefore, for this reaction, the contribution of $\Delta S°$ to $\Delta G°$ is nearly zero so that $\Delta G° \simeq \Delta H° \simeq -57$ kJ. The driving force for reaction (*i*) thus comes mainly from the decrease in the enthalpy of the reaction.

In the second reaction, two molecules of ehtylenediamine displace four molecules of H_2O. There is thus release of *two extra molecules resulting in an increase in the number of molecules on the products side and hence an increase in the disorder or entropy of the system.* $\Delta S°$ is thus positive for this reaction. Consequently, $\Delta G°$ is equal to -57 kJ $- T\Delta S°$. Evidently, $\Delta G°$ *has a larger negative value. This accounts for the greater stability of the* $[Cd(en)_2]^{2+}$ *complex.*

> **Note :** A summary of shape and magnetic nature of the complexes of Cr, Mn, Fe, Co, Ni and Cu has been provided on the following pages. This will be found as a quick and useful ready reckoner on complexes, by the reader. Electronic configuration of the metal ions, oxidation state of the metal, type of hybridisation and number of unpaired electrons have also been given alongside.

Summary of shape and magnetic nature of complexes of Cr, Mn, Fe, Co, Ni and Cu.

Atom/ion/complex	Electronic configuration	Oxidation state of metal involved	Type of hypridisation	Shape of the molecule	No.of unpaired, electrons	Magnetic nature
$Cr^{3+}(d^3)$	3d: [↑][↑][↑][ ][ ]; 4s: [ ]; 4p: [ ][ ][ ]	+3			3,	Paramagnetic
$[Cr(NH_3)_6]^{3+}$	3d: [↑][↑][↑][:][:]; 4s: [:]; 4p: [:][:][:] — d^2sp^3	+3	d^2sp^3(Inner)	Octahedral	3	Paramagnetic
$[Cr(H_2O)_6]^{3+}$	3d: [↑][↑][↑][ ][ ]; 4s: [:]; 4p: [:][:][:]; 4d: [:][:][ ][ ][ ] — sp^3d^2	+3	$sp3d^2$ (Outer)	Octahedral	3,	Paramagnetic
$Mn^{2+}(d^5)$	3d: [↑][↑][↑][↑][↑]; 4s: [ ]; 4p: [ ][ ][ ]	+2			5,	Paramagnetic
$[Mn(CN)_6]^{4-}$	3d: [↑↓][↑↓][↑↓][:][:]; 4s: [:]; 4p: [:][:][:] — d^2sp^3	+2	d^2sp^3 (Inner)	Octahedral	1,	Paramagnetic
$[MnCl_4]^{2-}$	Rearrangement 3d: [↑][↑][↑][↑][↑]; 4s: [:]; 4p: [:][:][:] — sp^3	+2	sp^3	Tetrahedral	5,	Paramagnetic
$Fe^{2+}(d_6)$	3d: [↑↓][↑][↑][↑][↑]; 4s: [ ]; 4p: [ ][ ][ ]	+2			4,	Paramagnetic

$[Fe(CN)_6]^{4-}$	[↑↓][↑↓][↑↓][:][:] [:] [:][:][:] Rearrangement; d^2sp^3	+2	d^2sp^3(Inner)	Octahedral	0	Diamagnetic
$[Fe(H_2O)_6]^{2+}$	[↑↓][↑][↑][↑][↑] [:] [:][:][:] [:][:][ ][ ][ ]; sp^3d^2	+2	sp^3d^2(Outer)	Octahedral	4	Paramagnetic
$[Fe(NH_3)_6]^{2+}$	same as above	+2	sp^3d^2 (Inner)	Octahedral	4	Paramagnetic
$Fe^{3+}(d^5)$	[↑][↑][↑][↑][↑] [ ] [ ][ ][ ]	+3			5	Paramagnetic
$[Fe(CN)_6]^{3-}$	[↑↓][↑↓][↑][:][:] [:] [:][:][:]; d^2sp^3	+3	d^2sp^3 (Inner)	Octahedral	1	Paramagnetic
Fe	[↑↓][↑][↑][↑][↑] [↑↓] [ ][ ][ ]	0			4	Paramagnetic
$Fe(CO)_5$	[↑][↑][↑][↑][:] [:] [:][:][:]; dsp^3	0	dsp^3(Inner)	Trigonal bipyramidal	0	Diamagnetic
$Co^{3+}(d^6)$	[↑↓][↑][↑][↑][↑] [ ] [ ][ ][ ]		+3		4	Paramagnetic

	3d	3s	3p	3d (outer)					
$[CoF_6]^{3-}$	↑↓ ↑ ↑ ↑ ↑	:	: : :	: : (sp³d²)	+3	sp^3d^2(Outer)	Octahedral	4	Paramagnetic
$[Co(NH_3)_6]^{3+}$	↑↓ ↑↓ ↑↓ : : (Rearrangement)	:	: : : (d^2sp^3)		+3	d^2sp^3 (Inner)	Octahedral	0	Diamagnetic
$Co^{2+}(d^7)$	↑↓ ↑↓ ↑ ↑ ↑				+2			3	Paramagnetic
$[Co(H_2O)_6]^{2+}$	↑↓ ↑↓ ↑ ↑ ↑	:	: : :	: : (sp^3d^2)	+2	sp^3d^2 (Outer)	Octahedral	3	Paramagnetic
$Ni^{2+}(d^8)$	↑↓ ↑↓ ↑↓ ↑ ↑				+2			2	Paramagnetic
$[NiCl_4]^{2-}$	↑↓ ↑↓ ↑↓ ↑ ↑	:	: : : (sp^3)		+2	sp^3	Tetrahedral	2	Paramagnetic
$[Ni(CN)_4]^{2+}$	↑ ↑ ↑ ↑ : (Rearrangement)	:	: : (dsp^2)		+2	dsp^2	Square Planer	0	Diamagnetic
Ni	↑↓ ↑↓ ↑↓ ↑ ↑	↑↓			0			2	Paramagnetic

$Ni(CO)_4$	3d: ↑↓ ↑↓ ↑↓ ↑↓ ↑↓ (Rearrangement); 4s: : ; 4p: : : : (sp^3)	0	sp^3	Tetrahedral	0	Diamagnetic
$[Ni(NH_3)_6]^{2+}$	3d: ↑↓ ↑↓ ↑↓ ↑ ↑; 4s: : ; 4p: : : : ; 4d: : : (empty, empty, empty) (sp^3d^2)	+2	sp^3d^2 (Outer)	Octahedral	2	Paramagnetic
$Cu^{2+}(d^9)$	↑↓ ↑↓ ↑↓ ↑↓ ↑; (empty); (empty, empty, empty)	+2			1	Paramagnetic
$[CuCl_4]^{2-}$	↑↓ ↑↓ ↑↓ ↑↓ ↑; : ; : : : (sp^3)	+2	sp^3	Tetrahedral	1	Paramagnetic
$[Cu(NH_3)_4]^{2+}$	↑↓ ↑↓ ↑↓ ↑ : ; : ; : : ↑ (dsp^2) one electron is shifted from 3d- to 4p- orbtial.	+2	dsp^2	Square planer	1	Paramagnetic

In the diagrams, the orbital boxes are labelled 3d, 3s, 3p and 4d on the $[Ni(NH_3)_6]^{2+}$ row.

22 SPECTRAL PROPERTIES OF COMPOUNDS

Q. 1. How can you obtain crystal field splitting energy from the absorption spectrum of a compound? Illustrate taking the example of $[Ti(H_2O)_6]^{3+}$. *(Rajasthan, 2010; Garhwal, 2010)*

Ans. Absorption spectrum of $[Ti(H_2O)_6]^{3+}$ is shown as under in Fig. 22.1.

Metal ion possesses d^1 configuration in this complex ion. The electron occupies one of the t_{2g} orbitals in the ground state. Absorption of energy takes place at 5000 Å as seen from Fig. 22.1. Calculation of energy corresponding to this wavelength can be done as under:

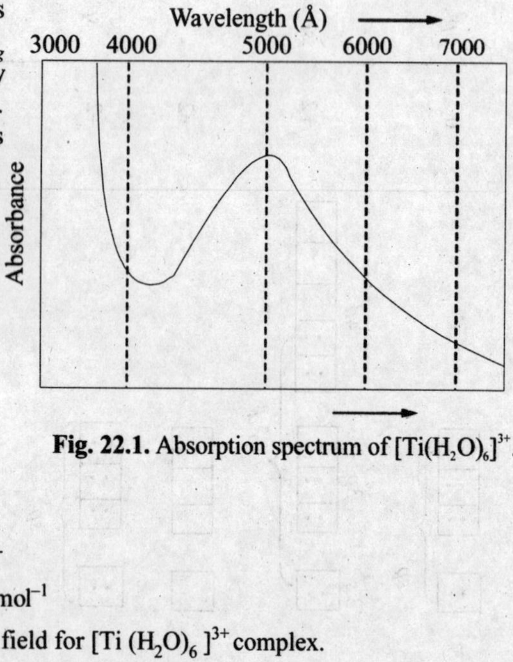

$$E = N_0 \times h \times \nu \ \text{or} \ \frac{N_0 \times h \times c}{\lambda}$$

The notations are explained as under:

N_0 = Avogadro's number

h = Planck's constant

c = Velocity of light

Substituting the value of N_0, h and c in SI units we have

Fig. 22.1. Absorption spectrum of $[Ti(H_2O)_6]^{3+}$.

$$E = \frac{6.023 \times 10^{23} \times 6.626 \times 10^{-34} \times 3 \times 10^{8}}{5000 \times 10^{-10}}$$

$$= 239 \text{ kJ mol}^{-1}$$

This is the value of C.F.S.E. in octahedral field for $[Ti(H_2O)_6]^{3+}$ complex.

And this corresponds to the promotion of single electron from a lower t_{2g} orbital to higher energy e_g orbital. Green and yellow lights corresponding to the above energy are absorbed out of white light leaving blue and red radiations which are emitted. Thus the complex looks purple.

Q. 2. Explain the following:

(*a*) Copper sulphate hydrated ($CuSO_4.5H_2O$) is blue in colour.

(*b*) Cuprous compounds are colourless.

(*c*) Anhydrous copper sulphate is white.

Ans. (*a*) The Cu^{2+} ion is a d^9 system. The spectrum of d^9 shows the same behaviour in a crystal field as does d^1 ion. But in this case the system is viewed by means of the *hole concept.* In this we may say that Cu (II) has spherically symmetrical d^{10} system with a hole or a missing electron. The hole behaves exactly in the same way as electron does. When radiations of visible light fall on it, the $CuSO_4, 5H_2O$ solution absorbs in the red region corresponding to 5750-5900 Å. Therefore, it transmits blue colour.

(*b*) The cuprous compounds (Cu^+) have the electronic configuration $3d^{10}$ and therefore no *d-d* transitions are possible and these compounds are colourless.

(*c*) Anhydrous copper sulphate is white. It is a $3d^9$ system but in anhydrous form, there is no crystal field of ligands. Therefore, all the five 3*d* orbitals are of equal energy and no transition is possible.

Q. 3. What is meant by term symbol? *(Ranchi, 2010; Garhwal 2010)*

Ans. An electron has a spin angular momentum(*s*) and orbital angular momentum (*l*), which may interact or couple to give a *resultant angular momentum* for the entire multi-electron atom. The resultant angular momentum and the energy of the system is expressed in a *term symbol.* Thus, *term symbols are used to indicate the electronic configuration and the resultant angular momentum of an atom.*

The term symbol for a particular atomic state is expressed as:

$$^{2S+1}L_j$$

where S denotes the total spin quantum number and 2S + 1 gives the *multiplicity of the state,* L is the total orbital angular momentum and J gives the total angular quantum number. Like the symbols *s, p, d, f* ... for *l* = 0, 1, 2, 3...., the different values of L are designated as L = 0 (S), L = 1 (P), L = 2 (D), L = 3(F) .... etc.

Q. 4. Describe LS or Russell Saunder scheme of coupling of spin and angular momenta of electrons. *(Kanpur, 2000)*

Ans. This is applicable to systems in which spin orbital interactions are relatively small. In this scheme, the individual orbital angular momenta of the electrons couple to give a resultant angular momentum represented by the quantum number, L for the state. The individual electron spin momenta also couple to give a resultant spin momentum described by the quantum number, S. The L and S values determine the total angular momentum, J, which can take quantized positive values ranging from $|L - S|$ to $|L + S|$. In this method, the consecutive values are separated by one and | | indicates the absolute value of $|L - S|$ and no regard is paid to the sign, so that J is always ≥ 0.

Rules for determining the term symbol according to L – S coupling scheme.

(*a*) Determination of total spin angular momentum (ss coupling). For a single electron, spin quantum number, *s* = 1/2. When two or more electrons are present in a subshell, then their magnetic fields couple to produce resultant spin quantum number, S. The resultant spin quantum number is given as:

$$S = (s_1 + s_2), (s_1 + s_2 - 1) \; ..... \; |s_1 - s_2|$$

where the modules sign | | indicates the positive value. For example, for a subshell having two electrons in *p* or *d* orbitals.

$$S = 1/2 + 1/2\;,\;\; 1/2 + 1/2 - 1$$

or

$$S = 1 \qquad , \; 0$$

The spin multiplicity is given by the relation 2S + 1 and is written in the upper left hand corner of the symbol for the state. Depending upon the spin multiplicity, we call *singlet state* (spin multiplicity = 1), *triplet state* (spin multiplicity = 3) and so on.

(*b*) Determination of resultant orbital angular momentum (*l* – *l* coupling). When two electrons with angular quantum number l_1 and l_2 interact, the resultant angular quantum number, L are given by the relationship:

$$L = (l_1 + l_2), (l_2 + l_2 - 1), (l_1 + l_2 - 2) \; .... \; | l_1 - l_2 |$$

where the modulus sign | | indicates the positive value.

For example, for a *p* subshell having two electrons

$l_1 = 1$ and $l_2 = 1$ so that

$$L = (1 + 1)\ ,\quad (1 + 1 - 1)\ ,\ \ldots\ldots\quad (1 - 1)$$
$$= 2\ ,\quad 1\ ,\quad 0$$
$$= D\ ,\quad P\ ,\quad S$$

Therefore, p^2 electrons will be represented by three energy states, S, P and D corresponding to L = 0, 1 and 2 respectively.

Similarly, for d^2 configuration,

$l_1 = 2$ and $l_2 = 2$

$$L = (2 + 2),\quad (2 + 2 - 1),\quad 2 + 2 - 2,\quad 2 + 2 - 3,\quad 2 - 2$$
$$= 4\quad 3\quad 2\quad 1\quad 0$$
$$= G\quad F\quad D\quad P\quad S$$

Thus, d^2 configuration is represented by five energy states S, P, D, F and G corresponding to L = 0, 1, 2, 3 and 4 respectively.

(c) Spin-orbital or L - S Coupling. In an atom, the magnetic effects of the resultant angular momentum, L, and the resultant spin moment S couple together to give new quantum number J called the *total angular quantum number.* The J values can be obtained by the vectorial combination of L and S and the coupling scheme is called *Russel Saunders or LS coupling.* The J can acquire all quantized positive values ranging from |L – S| to | L + S| as :

$$J = (L + S),\ (L + S - 1),\ (L + S - 2)\ \ldots\ldots\ |L - S|$$

The J values are separated by 1. The modules sign | | indicates the absolute value of | L – S | so that $J \geq 0$.

The J value is written as a subscript in the term symbol.

Q. 5. Derive term symbols for p^2 configuration.

Ans. For both *p*-electrons, $l_1 = 1$ and $l_2 = 1$

$$L = (1 + 1)\ ,\quad (1 + 1 - 1)\ ,\quad (1 + 1 - 2)$$
$$= 2\ ,\quad 1\ ,\quad 0$$
$$= D\ ,\quad P\ ,\quad S$$

The energy states are S, P and D.

For two electrons,

$$S = (+ 1/2 + 1/2)\ ,\quad (+ 1/2 + 1/2 - 1)$$
$$= 1\ ,\quad 0$$

The spin multiplicity (2S + 1) is:

For S = 1, spin multiplicity = 3 (Triplet)

For S = 0, spin multiplicity = 1 (Singlet)

Therefore, we have six states; ^3S, ^3P, ^3D, ^1S, ^1P and ^1D, three corresponding to singlet and three corresponding to triplet.

Let us calculate J values;

(*i*) When L = 2, S = 1, *i.e.,* ^3D *state*

$$J = (2 + 1),\ (2 + 1 - 1),\ (2 + 1 - 2)$$
$$= 3,\quad 2,\quad 1$$

∴ Complete spectroscopic term symbols are:

$$^3D_3,\ ^3D_2, \text{ and } ^3D_1$$

(*ii*) When L = 1 and S = 1, *i.e.*, ^3P *state*

$$J = (1 + 1),\ (1 + 1 - 1),\ (1 + 1 - 2)$$
$$= \quad 2, \qquad 1, \qquad 0$$

∴ Complete spectroscopic terms are:

$$^3D_2,\ ^3D_1, \text{ and } ^3P_0$$

(*iii*) When L = 0, S = 1, *i.e.*, ^3S *state*

$$J = 1 + 0$$
$$= 1$$

Complete spectroscopic term is:

$$^3S_1$$

(*iv*) When L = 2, S = 0, *i.e.*, ^1D *state*

$$J = 2 + 0$$
$$= 2$$

Complete spectroscopic term:

$$^1D_2$$

(*v*) When L = 1 and S = 0, *i.e.*, ^1P *state*

$$J = 1 + 0$$
$$= 1$$

Complete spectroscopic term is:

$$^1P_1$$

(*vi*) When L = 0 and S = 0, *i.e.*, ^1S *state*

$$J = 0$$

Complete spectroscopic term is:

$$^1S_0$$

Thus, the total term symbols for p^2 configuration are:

$^3D_3$, $^3D_2$, $^3D_1$, $^3P_2$, $^3P_1$, $^3P_0$, $^3S_1$, $^1D_2$, $^1P_1$ and $^1S_0$

Each of these values corresponds to an electronic arrangement. For example, $^3D_3$ symbol indicates a D state (L = 2) having spin multiplicity three so S = 1 (two unpaired electrons) and total quantum number, J = 3.

Similarly, $^1S_0$ indicates a S state (L = 0) having spin multiplicity one, *i.e.*, S = 0 (no unpaired electron) and the resultant quantum number, J = 0.

All the spectroscopic terms derived for a p^2 configuration would occur for an excited state of carbon ($1s^2\,2s^2\,2p^1\,3p^1$) in which the two *p* electrons belong to different subshells. However, in the ground state of carbon atom ($1s^2\,2s^2\,2p^2$) the number of states is restricted by the *Pauli exclusion principle.* According to Pauli exclusion principle, no two electrons in the same atom can have the same value for all the four quantum numbers. In the ground state configuration of carbon, the two *p* electrons both have the same value of *n* and *l* (*n* = 2 and *l* = 1), so they must differ in at least one of the remaining quantum numbers out of m and s. This restriction reduces the number of term from ^1S, ^1P, ^1D, ^3S, ^3P, ^3D to ^1S, ^3P and ^1D only. This can be easily understood by writing down only those electronic arrangements of *m* and *s* which do not violate the Pauli exclusion principle. Table 27.1 gives all the fifteen allowed values of M_L and M_S for the p^2 configuration. It can be shown that these 15 arrangements can be represented by only three terms ^1S, ^3P and ^1D by taking suitable values of M_L (2L + 1) and M_S (2S + 1) as explained hereunder.

For *p* electrons, the angular quantum number $l = 1$ and magnetic quantum number *m* may have values from $+l$ .... 0 .... $-l$, giving in this case values of $m = +1, 0, -1$. The spin quantum number can have values of +1/2 and −1/2 for each value of *m*. The total spin and total orbital quantum numbers can be obtained by adding the appropriate values of *m* and *l* as explained earlier:

M_L has values from + L ....0 ....– L (a total of 2L + 1 values).

M_S has values from + S ....0 ....– S (a total of 2S + 1 values).

The *l* and S quantum numbers associated with each electronic configuration or spectroscopic term symbol can be worked out from M_L and M_S quantum numbers as shown in Table 22.1.

Table 22.1. Allowed values of M_L and M_s for p^2 configuration.

S.No.	*m* = +1	0	−1	M_s, Σm_s	M_L, Σm_l	Term symbol
1.	↑↑			0	2	^1D
2.			↑↑	0	−2	^1D
3.		↑↑		0	0	
4.	↑		↓	0	0	^3P, ^1D, ^1S
5.	↓		↑	0	0	
6.	↑	↓		0	1	^3P, ^1D
7.	↓	↑		0	1	
8.		↑	↓	0	−1	^3P, ^1D
9.		↓	↑	0	−1	
10.	↑	↑		1	1	^3P
11.	↑		↑	1	0	^3P
12.		↑	↑	1	−1	^3P
13.	↓	↓		−1	1	^3P
14.	↓		↓	−1	0	^3P
15.		↓	↓	−1	−1	^3P

Steps to Assign Spectroscopic Term Symbol

(*i*) Write the different values of M_S and M_L for the given combination of valence shell electron.

(*ii*) Select the maximum M_S value and select the maximum M_L value associated with it. In the present case, it is $M_S = 1$ and $M_L = 1$ (number 10 in Table 22.1). This corresponds to a group of terms where L = 1 and S = 1. Since L = 1, it must be a P state and since S = 1, the multiplicity (2S + 1) = 3 so that it is a ^3P (triplet P) state.

Now if L = 1, M_L may have values + 1, 0 and –1 and if S = 1, M_S may have the values + 1, 0 and –1.

Thus, there are nine combinations of M_L and M_S values:

$M_L = +1$ $M_S = +1, 0, -1$

$M_L = 0$ $M_S = +1, 0, -1$

$M_L = -1$ $M_s = +1, 0, -1$

Inspection of Table 22.1 shows that nine of the allowed values could be assigned ^3P term symbol.

From the remaining unassigned M_S and M_L values, we pick out the next maximum M_S and M_L value. In this case, $M_S = 0$ and $M_L = 2$. Since L = 2, it must be a D state and S = 0 gives a multiplicity of 2S + 1 = 1, so it is a ^1D (singlet D) state.

Now, if L = 2, M_L may have values of + 2, + 1, 0, –1, –2 and if S = 0, M_S may have 0 values.

This gives five combinations of M_L and M_S as:

$M_L = +2$ $M_S = 0$

$M_L = +1$ $M_S = 0$

$M_L = +0$ $M_S = 0$

$M_L = -1$ $M_S = 0$

$M_L = -2$ $M_S = 0$

Inspection of Table 22.1 shows that five of the allowed values could be assigned a ^1D term symbol.

The ^3P and ^1D make 9 + 5 = 14 combinations. The remaining configuration corresponds to $M_S = 0$ and $M_L = 0$. This gives a singlet S state ^1S.

Thus, all the 15 electronic arrangements are represented by three states:

^1S, ^3P and ^1D

Q. 6. Derive term symbols for d^2 configuration.

Ans. For *d* electrons, the subsidiary quantum number *l* is 2. For two *d*-electrons having $l_1 = 2$ and $l_2 = 2$, we have

L = (2 + 2),	(2 + 2 – 1),	(2 + 2 – 2),	(2 + 2 – 3),	(2 + 2 – 4)
4	3	2	1	0

These values of L represent S, P, D, F and G states corresponding to L = 0, 1, 2, 3, and 4 respectively.

S = (+ 1/2 + 1/2),	(1/2 + 1/2 – 1)
1	0

Therefore, spin multiplicity (2S + 1) is 3 and 1. So we have ^1S, ^1P, ^1D, ^1F, ^1G and ^3S, ^3P, ^3D, ^3F and ^3G. There are 45 ways in which two *d*-electrons may be arranged which do not violate the Pauli exclusion principle. These arrangements are given in Table 22.2.

Table 22.2. Allowed values of M_S and M_L for the d^2 configuration

S.No.	m = +2	+1	0	−1	−2	M_s, Σm_s	M_L Σm_l	Term symbol
1.	↑↑					0	4	^1G
2.	↑	↑				1	3	^3F
3.	↑	↓				0	3	
4.	↓	↑				0	3	$^1G, ^3P$
5.	↓	↓				−1	3	^3F
6.	↑		↑			1	2	^3F
7.	↑		↓			0	2	
8.	↓		↑			0	2	$^1G, ^3F, ^1D$
9.		↑↓				0	2	
10.	↓		↓			−1	2	^3F
11.	↑			↑		1	1	
12.		↑	↑			1	1	$^3F, ^3P$
13.	↓			↓		−1	1	
14.		↓	↓			−1	1	$^3F, ^3P$
15.	↑			↓		0	1	
16.	↓			↑		0	1	
17.		↑	↓			0	1	$^1G, ^3F, ^1D, ^3P$
18.		↓	↑			0	1	
19.	↑				↑	1	0	
20.		↑		↑		1	0	$^3F, ^3P$
21	↑				↓	0	0	
22.	↓				↑	0	0	$^1G, ^3F, ^1D, ^3P, ^1S$

(*Contd*)

S.No.	*m* = +2	+1	0	−1	−2	M_s, Σm_s	M_L Σml	*Term symbol*
23		↑		↓		0	0	^1G, ^3F, ^1D, ^3P, ^1S
24.		↓		↑		0	0	
25.			↓↑			0	0	
26.	↓				↓	−1	0	^3F, ^3P
27.		↓		↓		−1	0	
28.		↑			↓	0	−1	^1G, ^3F, ^1D, ^3P
29.		↓			↑	0	−1	
30.			↑	↓		0	−1	
31.			↓	↑		0	−1	
32.		↑			↑	1	−1	^3F, ^3P
33.			↑	↑		1	−1	
34.		↓			↓	− 1	−1	^3F, ^3P
35.			↓	↓		−1	−1	
36.			↑		↑	1	−2	^3F
37.			↑		↓	0	−2	^1G, ^3F, ^1D
38.			↓		↑	0	−2	
39.				↑↓		0	−2	
40.			↓		↓	−1	−2	^3F
41.				↑	↑	1	−3	^3F
42.				↑	↓	0	−3	^1G, ^3F
43.				↓	↑	0	−3	
44.				↓	↓	−1	−3	^3F
45.					↑↓	0	−4	^1G

However, it may be shown that these arrangements can be represented by only five states; ^1S, ^3P, ^1D, ^3F and ^1G. This can be done in a similar way as for p^2 electrons, in the following steps.

Steps to Assign Spectroscopic Term Symbols

(*i*) Maximum value of $M_S = 1$ and the maximum value of M_L associated with $M_S = 1$ is 3 (corresponds to No. 2 in Table 22.2). Since this corresponds to L = 3, it represents F state. This occurs with $M_S = +1, 0, -1$ suggesting a triplet F state, *i.e.*, ^3F.

Now if L = 3, M_L has values

$$M_L = +3, +2, +1, 0, -1, -2, -3$$

Thus, there are 21 combinations of M_L and M_S values associated with ^3F term:

$M_L = +3$ $M_S = +1, 0, -1$
$M_L = +2$ $M_S = +1, 0, -1$
$M_L = +1$ $M_S = +1, 0, -1$
$M_L = 0$ $M_S = +1, 0, -1$
$M_L = -1$ $M_S = +1, 0, -1$
$M_L = -2$ $M_S = +1, 0, -1$
$M_L = -3$ $M_S = +1, 0, -1$

(*ii*) The highest unassigned M_L value is 4 and it arises if L = 4 corresponding to a G state. Since $M_S = 0$, S must be 0 and it is a singlet G term, ^1G

Now, if L = 4, M_L can have values +4, +3, +2, +1, 0, −1, −2, −3, −4. If S = 0, M_S can have only one value (*i.e.*, 0 value).

Thus, there are nine configurations associated with this term.

(*iii*) For M_L values of 2, maximum $M_S = 0$. This $M_L = 2$ corresponds to D state. Since $M_S = 0$ for this state, it corresponds to spin multiplicity of 1 suggesting a single D state, *i.e.*, ^1D.

Now, if L = 2, M_L may have values + 2, + 1, 0, −1, −2
and if S = 0, M_S may have 0 value (only one value)

This explains five combinations corresponding to ^1D state.

Now 45 − (21 + 9 + 5) = 10 unassigned combinations are left. For these we have maximum M_S = 1. So S = 1 and the maximum M_L value associated with $M_S = 1$ is 1. Therefore, it corresponds to a P state. Since S = 1, spin multiplicity is 2 × 1 + 1 = 3. So it corresponds to a triplet P state, ^3P.

Now, if L = 1, M_L may have values + 1, 0, −1 if S = 1, M_S may have the values + 1, 0, −1.

Thus there are nine combinations of M_L and M_S values:

$M_L = +1$ $M_S = +1, 0, -1$
$M_L = 0$ $M_S = +1, 0, -1$
$M_L = -1$ $M_S = +1, 0, -1$

(*iv*) So far we have considered 21 + 9 + 5 + 9 = 44 combinations and only one combination is left. For this $M_S = 0$ and maximum M_L value associated with it is 0. Therefore, L = 0 and it corresponds to singlet. Thus, the term is singlet S, ^1S.

Since L = 0 and S = 0, it corresponds to only one term.

Thus, all the allowed values of M_L and M_S for d^2 configuration may be represented by ^1S, ^3P, ^1D, ^3F and ^1G states.

Q. 7. Give the rules for determining the order of energy for different terms.

Ans. The rules are as under:

1. The terms are arranged depending upon their spin multiplicities, *e.g.*, their S-values. The most stable state has the highest S-value and stability decreases as the value of S decreases. This means that the most stable state (ground state) has maximum unpaired electrons because this gives the minimum electrostatic repulsion. This is in accordance with Hund's rule.

2. For a given value of S, the state with the highest value of L is the most stable state. This means that if two or more terms have the same value of S (same spin multiplicity), the state with highest value of L will have the lowest energy.

3. For a given value of S and L (*i.e.*, having same S and L values), the terms with smallest J value is most stable if the subshell is less than half-filled and the term with maximum J value is most stable if the subshell is more than half-filled.

Q. 8. Describe selection rules for electronic spectra. *(Ranchi, 2010; Rajasthan, 2010)*

Ans. The main selection rules for electronic spectra are:

1. *Transitions in which number of unpaired electrons changes in going from a lower to a higher energy state are referred* to as **spin** or **multiplicity forbidden.** This means that transitions to only those excited states are to considered which have the same spin multiplicity as the ground state, *i.e.*, transitions in which $\Delta S = 0$. This is also known as $\Delta S = 0$ rule. For example, for a d^2 configuration, the only transitions which are seen are from ^3F (ground state) to ^3P (excited state) because both ^3F and ^3P terms have same spin multiplicity 3. The transitions in which $\Delta S \neq 0$ (*e.g.*, $\Delta S = \pm 1$) are forbidden or very weak.

2. *Transitions which do not involve a change in the subsidiary quantum number ($\Delta l = 0$) are forbidden if molecule has centre of symmetry. These forbidden transitions are called* **Laporte forbidden transitions.** The transitions which involve a change in the subsidiary quantum number, *i.e.*, $\Delta l = \pm 1$ are **Laporte allowed transitions** and have a high absorbance. This rule suggests that transitions from one *d*-level to another *d*-level (*i.e.*, *d* - *d* transitions) in transition metal complexes are not allowed (Laporte forbidden transitions) because the change in $l = 0$. For example, in case of Ca ([Ar] $4s^2$), the transition occurs $s^2 \longrightarrow s^1p^1$ in which $\Delta l = 1 - 0 = +1$ and the molar absorbance coefficient $\in$ is 5000 – 10000 litres per mol per centimeter. In contrast to this, *d* - *d* transitions are Laporte forbidden but spectra of much lower absorbance are observed ($\in$ = 5 to 10 l mol^{-1} cm^{-1}) because of slight relaxation in Laporte rule.

Q. 9. Discuss the spectra of d^1 and d^9 ions. *(Rajasthan, 2010)*

Ans. (A) Octahedral complexes of metal ion with d^2 and d^9 configurations: The simplest example of a d^1 complex is Ti (III) in octahedral complexes such as $[Ti(H_2O)_6]^{3+}$ or $[TiCl_6]^{3-}$. The splitting of *d*-orbitals is shown in Fig. 22.2 (*a*). In the ground state the single electron occupies the lower t_{2g} level and only one transition from $t_{2g} \longrightarrow e_g$ is possible. Consequently, the absorption spectrum of $[Ti(H_2O)_6]^{3+}$ shows only one band with a peak at 20300 cm^{-1} as shown in Fig. 22.2 (*b*). The energy corresponding to this wavelength corresponds to green and yellow light which are absorbed from the white light while the blue and red portions of the light are emitted. Therefore, solution of complex $[Ti(H_2O)_6]^{3+}$ looks purple.

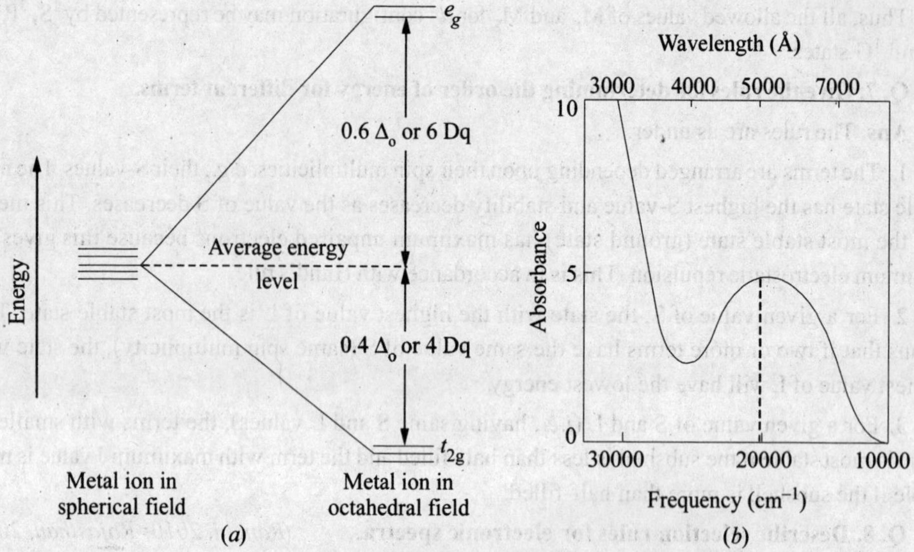

Fig. 22.2 (*a*) Splitting of energy levels for d^1 configuration in an octahedral field and (*b*) visible spectrum of $[Ti(H_2O)_6]^{3+}$ complex ion.

Since the magnitude of the splitting (Δ_0) depends upon the nature of the ligands and affects the energy of the transition and therefore, the frequency of maximum absorption vary in the spectra of these complexes. For example, $[TiCl_6]^{3-}$ absorbs at 13,000 cm^{-1}, $[TiF_6]^{3-}$ absorbs at 18900 cm^{-1} and $[Ti(H_2O)_6]^{3+}$ absorbs at 20300 cm^{-1}, while $[Ti(CN)_6]^{3-}$ absorbs at 22300 cm^{-1}. Therefore, these complexes exhibit different colours. The amount of splitting caused by various ligands is related to their position in spectrochemical series.

Octahedral complexes of ions with a d^9 configuration such as $[Cu(H_2O)_6]^{2+}$ can be explained in a similar way to the Ti^{3+} octahedral complexes with a d^1 arrangement. In the d^1 complex there is a single electron in the lower T_{2g} level while in the d^9 complex, there is a single hole in the upper E_g level. Thus, the transition in the d^1 case corresponds to the promotion of an electron from the T_{2g} to E_g level, whereas the transition in the d^9 case corresponds to the promotion of an electron as the transfer of a hole from E_g to T_{2g}. *The energy level diagram or d^9 is, therefore, the inverse of that for a d^1 configuration* (Fig. 22.3).

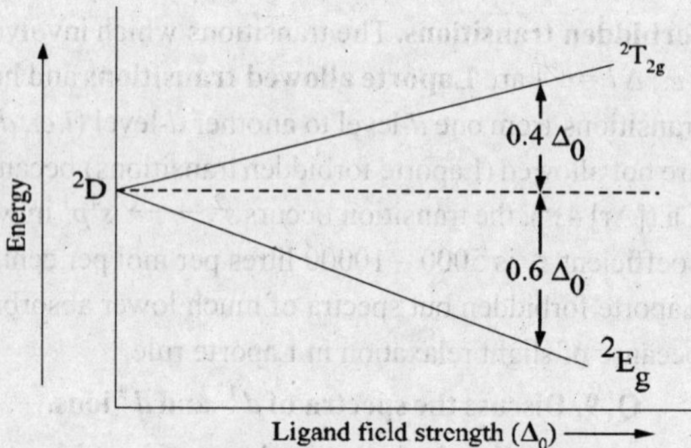

Fig. 22.3. Splitting of energy levels for d^9 complex in octahedral field.

(B) Tetrahedral complexes of metal ions with d^1 and d^9 configurations. In the case of tetrahedral ligand field, the energy level diagram is inverse of that of octahedral field. The *d*-orbitals split into two e_g orbitals of lower energy and three t_{2g} orbitals of higher energy.

Obviously, this energy level diagram for d^1 configuration in tetrahedral field is similar to the d^9 octahedral field (Fig. 22.3). The main difference between tetrahedral field and octahedral field is that the splitting is only 4/9 as that of octahedral field ($\Delta_t = 4/9\Delta_0$).

These splitting can be described in terms of a diagram known as **Orgel diagram.** It is apparent that d^1, d^4, d^6 and d^9 ions should also give only one *d-d* absorption band. The Orgel diagram for d^1, d^4, d^6 and d^9 ions is shown in Fig. 22.4. In this diagram, the splitting of the states as a function of Δ_0 for octahedral complexes with d^1 and d^6 electron configurations and tetrahedral complexes with d^4 and d^9 electron configurations are described on the right hand side. The spectra of these complexes contain only one band arising from *d - d* transitions and is assigned as $T_{2g} \longrightarrow E_g$.

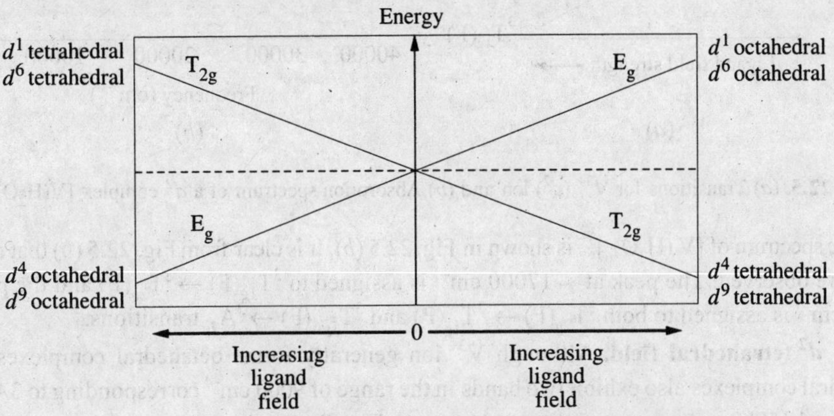

Fig. 22.4. Orgel diagram of d^1, d^6 and d^9 complexes.

The left hand side of the Orgel diagram applies to octahedral complexes with d^4 and d^9 electron configurations and tetrahedral complexes with d^1 and d^6 electron configurations. The spectra of these complexes contain only one band arising from single *d-d* transition and is assigned as $E_g \longrightarrow T_{2g}$. It may be noted that for tetrahedral complexes, subscripts are removed because a tetrahedron does not have a centre of symmetry.

Q. 10. Draw combined Orgel diagram for d^1, d^4, d^6 and d^9 complexes.

Ans. See Q. 9.

Q. 11. Discuss the spectra of d^2 and d^8 ions.

Ans. (*a*) d^2 octahedral field. In the ground state for a d^2 configuration, the two electrons are present in different orbitals. In an octahedral field, the *d*-orbitals are split into three t_{2g} orbitals of lower energy and two e_g orbitals of higher energy. The two *d*-electrons occupy t_{2g} orbitals because of their lower energy. The spectroscopic terms for d^2 electron configuration are ^1S, ^3P, ^1D, ^3F and ^1G. If we apply Hund's rule, we observe that ^3F is the ground state with lowest energy and ^3P, ^1G, ^1D and ^1S are excited states. The transitions from the level ^3F are allowed to another triplet state, *i.e.*, ^3P. The transitions from ^3F to ^1S, ^1D and ^1G states are spin forbidden and will be very weak and are not observed. As already discussed, the *p*-orbitals are not split by an octahedral field and so P states are not split but are transformed into a $^3T_{1g}$ state. On the other hand, the *f*-orbitals are split into three levels and so ^3F state splits into $^3A_{2g} + {}^3T_{1g} + {}^3T_{2g}$. The energy level diagram for these is shown in Fig. 22.5 (*a*).

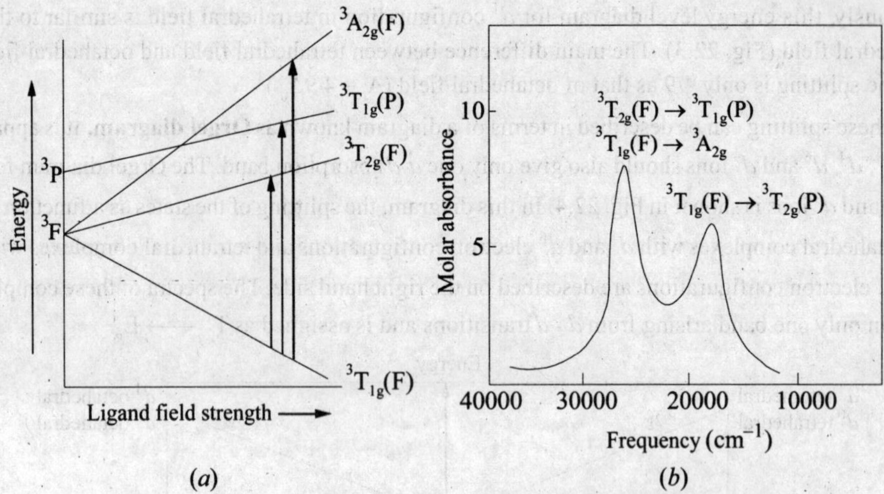

Fig. 22.5. (*a*) Transitions for V^{3+} (d^2) ion and (*b*) Absorption spectrum of a d^2 complex $[V(H_2O)_6]^{3+}$.

The spectrum of $[V(H_2O)_6]^{3+}$ is shown in Fig. 22.5 (*b*). It is clear from Fig. 22.5 (*b*) that only two peaks are observed. The peak at ~ 17000 cm^{-1} is assigned to $^3T_{1g}(F) \rightarrow {}^3T_{2g}(P)$ and the peak at– 25000 cm^{-1} is assigned to both $^3T_{2g}(F) \rightarrow {}^3T_{1g}(P)$ and $^3T_{1g}$ $(F) \rightarrow {}^3A_{2g}$ transitions.

(*b*) d^2 tetrahedral field. Although V^{3+} ion generally forms octahedral complexes, a few tetrahedral complexes also exhibit two bands in the range of 9000 cm^{-1} corresponding to $3A_{2g} \longrightarrow {}^{3T}{}_{1g}$(F) and 15000 cm^{-1} corresponding to $^3A_{2g} \longrightarrow {}^3T_{1g}$ (P) transitions.

(c) d^8 octahedral. The complexes of metals with d^8 configuration such as $[Ni(H_2O)_6]^{2+}$ can be treated similar to d^2 octahedral complexes. In such cases, there are two holes in the e_g level and therefore, promoting one electron is equivalent to transferring a hole from e_g to t_{2g} level. This is inverse of d^2 case and is shown in Fig. 22.6. It may be noted that ^3P state is not split and is not inverted but the ^3F state is split into three states and is inverted. Hence the ground state term for Ni^{2+} is A_{2g}. It may be noted that in both d^2 and d^8 configurations, the ^3F state is the lowest energy state. Three spin allowed transitions are observed in spectra of $[Ni(H_2O)_6]^{2+}$, $[Ni(NH_3)_6]^{2+}$, $[Ni(NH_2CH_2CH_2NH_2)_3]^{2+}$ etc. complex ions. These transitions correspond to

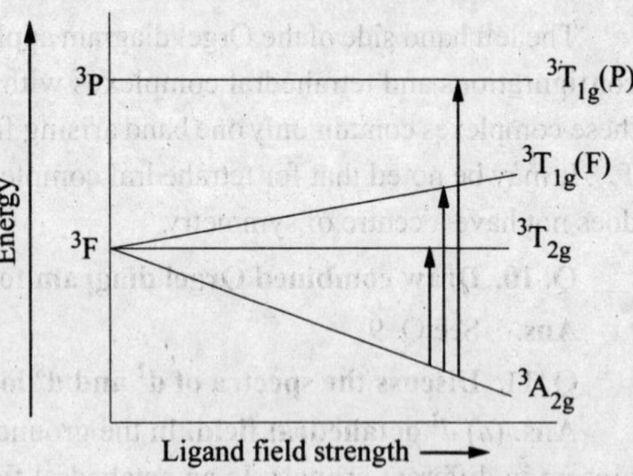

Fig. 22.6. Energy level diagram for d^8 ion.

$$3A_{2g} \longrightarrow {}^3T_{2g}\ (F)$$
$$3A_{2g} \longrightarrow {}^3T_{1g}\ (F)$$
$$3A_{2g} \longrightarrow {}^3T_{1g}\ (P)$$

Q. 12. Discuss the electronic spectra of Cr (III) octahedral complexes.

Ans. Chromium (III) ion has a d^3 electron configuration. In the octahedral complex, such as $[Cr(H_2O)_6]^{3+}$ it has electron configuration: t_{2g}^3. The d^3 arrangement gives rise to two state ^4F and ^4P.

In an octahedral field ^4F state splits into $^4A_{2g}$ (F), $^4T_{2g}$ (F) and $^4T_{1g}$ (F) states while ^4P state does not split but transforms into a $^4T_{1g}$ (P) (Fig. 22.7). The spectra of Cr (III) complexes would be expected to show three absorption peaks corresponding to the following transitions:

$$^4A_{2g} \longrightarrow {}^4T_{2g}\ (F)$$

$$^4A_{2g} \longrightarrow {}^4T_{1g}\ (F)$$

$$^4A_{2g} \longrightarrow {}^4T_{1g}\ (P)$$

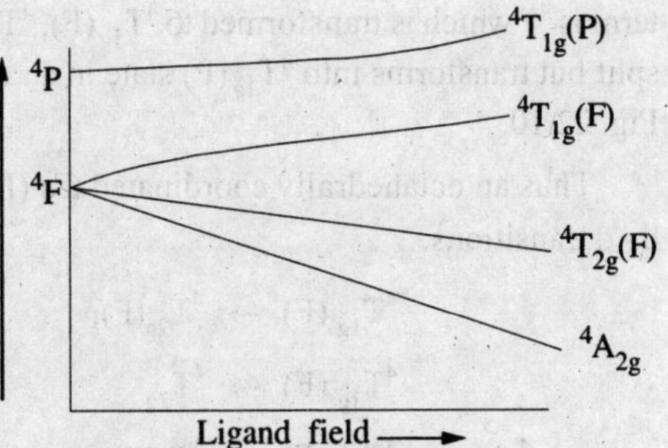

Fig. 22.7. Energy level diagram for a d^3 ion.

These three bands in $[Cr(H_2O)_6]^{3+}$ ion are observed at 17,400, 24,700 and 37,000 cm^{-1}.

Q. 13. Discuss the spectra of manganese (II) octahedral complex ions.

Ans. Manganese (II) has a d^5 configuration. In high spin octahedral complexes formed with weak ligands such as $[MnF_6]^{4-}$, $[Mn(H_2O)_6]^{2+}$, it has t_{2g}^3, e_g^2 electronic configuration. The ground state term is ^6S and there are five unpaired electrons with parallel spins. Any electronic transition within *d* level must involve a reversal of spins and therefore these *d* - *d* transitions are spin *forbidden transitions* and are very weak. This explains the pale pink colour of Mn (II) compounds.

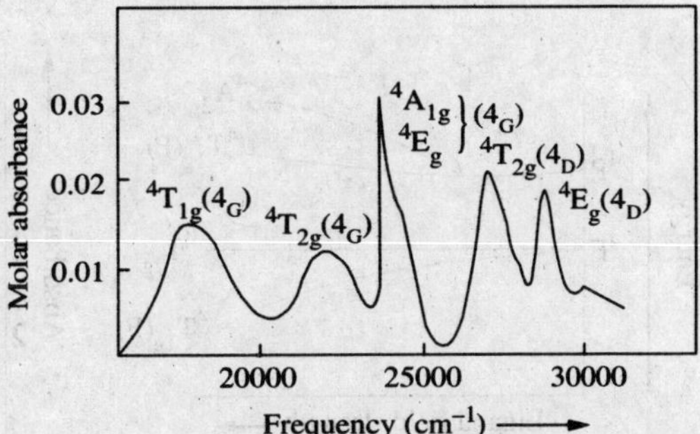

Fig. 22.8. Electronic spectrum of $[Mn(H_2O)_6]^{2+}$.

The spectrum of $[Mn(H_2O)_6]^{2+}$ is shown in Fig. 22.8. The main features of the spectrum are:

(*i*) The bands are very weak. The molar absorption coefficient, ∈, is about 0.02 – 0.03 *l* mol^{-1} cm^{-1} as compared with 5–10 *l* mol^{-1} cm^{-1} for spin allowed transitions.

(*ii*) There are large number of bands.

(*iii*) Some of the bands are sharp and others are broad.

For the ground state term ^6S, there are 11 excited states (^4G, ^4F, ^4D, ^4P, ^2I, ^2H, ^2G, ^2F, ^2D, ^2P, ^2S; refer table 22.2). Out of these 11 excited states only four quartets, ^4G, ^4F, ^4D and ^4P, involve the reversal of only one spin. The other seven states are doublets and are *doubly spin forbidden* and are unlikely to be observed. In octahedral field, these four states split into ten states and therefore, up to ten weak absorption bands may be observed (Fig. 22.9).

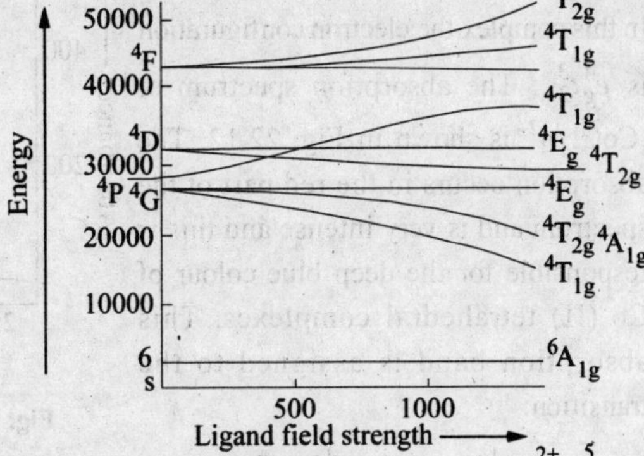

Fig. 22.9. Orgel energy level diagram for Mn^{2+} (d^5) octahedral

Q. 14. Discuss the spectra of cobalt (II) octahedral complex ions.

Ans. Cobalt (II) has d^7 electronic configuration. In octahedral field of weak ligands, it has the electronic configuration as : t_{2g}^5, e_g^2 and therefore, has three unpaired electrons. The ground state term is ^4F which is transformed to $^4T_{1g}$(F), $^4T_{2g}$ and $^4A_{2g}$. The other quartet state is ^4F which does not split but transforms into $^4T_{1g}$(P) state in an octahedral field. The energy level diagram is shown in Fig. 22.10.

Thus an octahedrally coordinated Co (II) ion should exhibit the following three spin allowed d - d transitions:

$$^4T_{1g}(F) \rightarrow {}^4T_{2g}(F)$$

$$^4T_{1g}(F) \rightarrow {}^4T_{2g}$$

$$^4T_{1g}(F) \rightarrow {}^4T_{2g}(P)$$

The absorption spectrum is shown in Fig. 22.11. The absorption spectrum is weak and occurs in the blue part of the visible region and therefore accounts for the pale pink colour of the $[Co(H_2O)_6]^{+2}$ ion.

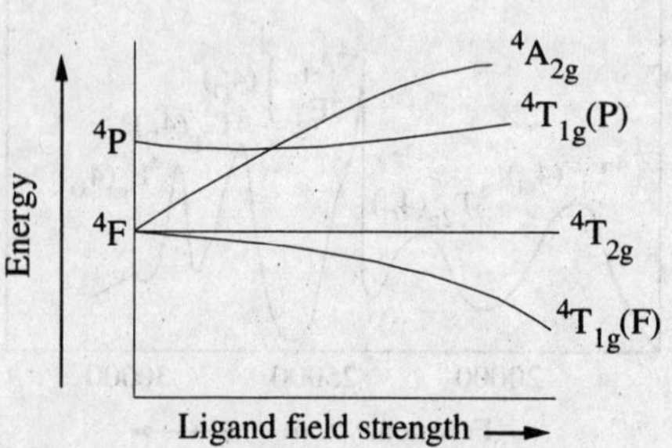

Fig. 22.10. Energy level diagram for Co (II) octahedral complexes

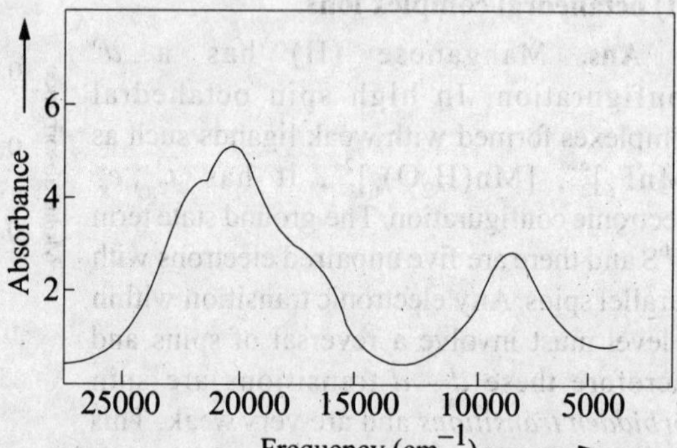

Fig. 22.11 The absorption spectrum of octahedral complex of Co (II)

Q. 15. Discuss the spectrum of Co (II) tetrahedral complex ions.

Ans. Co^{2+} in the tetrahedral complex $[CoCl_4]^2$ has d^7 configuration. In this complex the electron configuration is $e_g^4 t_{2g}^3$. The absorption spectrum of $[CoCl_4]^{2-}$ is shown in Fig. 22.12. The absorption occurs in the red part of the spectrum and is very intense and this is responsible for the deep blue colour of Co (II) tetrahedral complexes. This absorption band is assigned to the transition:

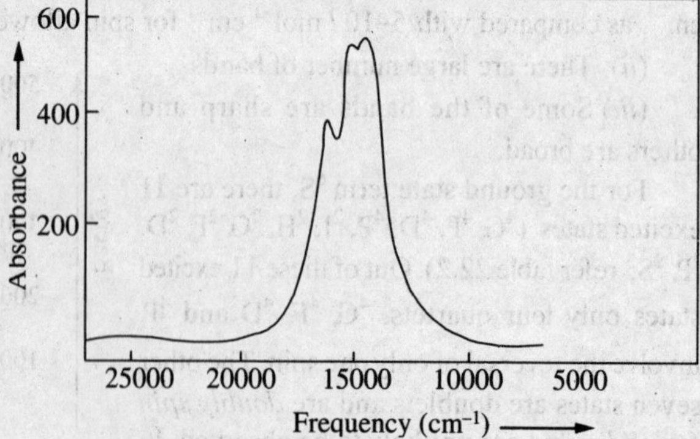

Fig. 22.12. Absorption spectrum of $[CoCl_4]^{2-}$ ion.

$$^4A_{2g} \longrightarrow {}^4T_{1g}(P)$$

The fine structure (number of shoulder bands) are caused by spin orbit coupling. The other transitions are:

$$^4A_{2g} \longrightarrow {}^4T_{2g} \text{ and}$$

$$^4A_{2g} \longrightarrow {}^4T_{1g}(F)$$

These fall outside the visible region in the infrared red region.

Q. 16. Discuss the spectrum of nickel (II) octahedral complex ions.

Ans. Ni (II) in octahedral complexes such as $[Ni(H_2O)_6]^{2+}$ has d^8 configuration. Their spectra can be easily understood by considering the energy level diagram (Fig. 22.13). Three spin allowed transitions are expected and these are assigned to the following transitions:

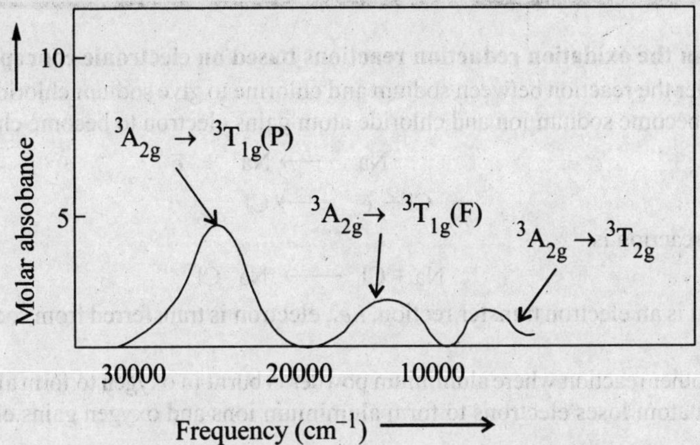

Fig. 22.13. Absorption spectrum of $[Ni(H_2O_6]^{2+}$ ion.

$^3A_{2g} \rightarrow {}^3T_{2g}$ 9000 cm^{-1}

$^3A_{2g} \rightarrow {}^3T_{1g}(F)$ 14,000 cm^{-1} for $[Ni(H_2O)_6]^{2+}$

$^3A_{2g} \rightarrow {}^3T_{1g}(P)$ 25,000 cm^{-1}

The bands are quite weak as expected. The absorption spectrum is shown in Fig. 22.13.

OXIDATION AND REDUCTION (INCLUDING LATIMER AND FROST DIAGRAMS)

Q. 1. Explain the oxidation reduction reactions based on electronic concept.

Ans. Consider the reaction between sodium and chlorine to give sodium chloride. Sodium atom loses electron to become sodium ion and chloride atom gains electron to become chloride ion.

$$Na \longrightarrow Na^+ + \bar{e}$$

$$Cl + \bar{e} \longrightarrow Cl^-$$

The overall reaction is:

$$Na + Cl \longrightarrow Na^+ Cl^-$$

This reaction is an electron transfer rection, *i.e.*, electron is transferred from sodium to chlorine atom.

Consider another reaction where aluminium powder is burnt in oxygen to form aluminium oxide. Here, aluminium atom loses electrons to form aluminium ions and oxygen gains electrons to form oxide as follows:

$$[Al \longrightarrow Al^{3+} + 3e] \times 4$$

$$O + 2e \longrightarrow O^{2-}] \times 6$$

The electron produced in the first reaction are used up in the second and no electrons are observed in the net reaction:

$$4Al + 3O_2 \longrightarrow 2Al_2O_3$$

Such electrons transfer reactions are commonly called oxidation reduction reactions where oxidation and reduction reactions occur side by side in terms of change of electrons. This is also commonly stated as: *"oxidation and reduction reactions go hand in hand in every electron transfer reaction."* There can be no oxidation without reduction and vice-versa.

Q. 2. Explain oxidation process in terms of electronic concept.

Ans. A substance is said to have been oxidised if it loses electrons. Oxidation is a process which involves the loss of electrons. According to this definition of oxidation, substances like oxygen, halogens and other non-metals act as oxidizing agents. Any substance that accepts electron is called an *oxidising agent.* If the chlorine gas is passed through an iodide solution, the solution becomes brown due to the formation of iodine.

$$Cl_2 + 2I^- \longrightarrow 2Cl^- + I_2$$

This means, chlorine oxidizes iodide ion to iodine. The loss of electrons during oxidation may result in:

(*i*) Increase of positive charge on atom or ion undergoing oxidation.

For example,

$$Na - e \longrightarrow Na^+$$

$$Al - 3e \longrightarrow Al^{+3}$$

$$Fe^{2+} - e \longrightarrow Fe^{3+}$$

$$Sn^{2+} - 2e \longrightarrow Sn^{4+}$$

***(ii)* Decrease of negative charge on atom or ion undergoing oxidation.**

For example,

$$2I^- - 2e \longrightarrow I_2$$

$$S^{2-} - 2e \longrightarrow S$$

$$MnO_4^{2-} - e \longrightarrow MnO_4^-$$

$$[Fe(CN_6)]^{4-} - e \longrightarrow [Fe(CN)_6]^{3-}$$

Q. 3. Explain reduction process in terms of electronic concept.

Ans. Reduction is a process which involves the gain of electrons. Any substance from which electrons can be readily removed acts as a reducing agent. A reducing agent is a substance that can give electrons. The gain of electrons during reduction may lead to the following situation:

***(i)* Decrease of positive charge on atom or ion undergoing reduction.**

For example,

$$Na^+ + \bar{e} \longrightarrow Na$$

$$Fe^{3+} + e \longrightarrow Fe^{2+}$$

$$Sn^{4+} + 2e \longrightarrow Sn^{2+}$$

$$O_2 + 2e \longrightarrow 2O^{2-}$$

***(ii)* Increase of negative charge on atom or ion undergoing reduction.**

For example,

$$MnO_4^- + e \longrightarrow MnO_4^{2-}$$

$$[Fe(CN_6]^{3-} + e \longrightarrow [Fe(CN)_6]^{4-}$$

Definition of oxidation and reduction and of oxidising agents can be summarised by an equilibrium:

$$\text{Reducing agent} \underset{\text{Reduction}}{\overset{\text{Oxidation}}{\rightleftharpoons}} \text{Oxidising agent} + \text{Electrons}$$

Reduction oxidation reactions are also abbreviated as Redox reactions.

Q. 4. What is oxidation number? How is it determined? Explain.

(Nagpur, 2000; Kumaon, 2000; M. Dayanand, 2002)

Ans. The oxidation number is a measure of the extent of oxidation of an element in its compounds. It is determined by using a set of simple rules:

(i) An element in any of its allotropic forms is assigned an oxidation number of zero.

(ii) The oxidation number of a monoatomic ion is equal to its charge.

The Na^+ is assigned an oxidation number +1 and magnesium ion Mg^{+2} is assigned a number equal to +2. Singly charged monoatomic, negative ions such as Cl^- are assigned an oxidation number of –1 and double charged negative ions such as O^{2-} have an oxidation number of –2.

(iii) In any neutral molecule the sum of the oxidation numbers is zero; in an ion sum of the oxidation numbers is equal to the charge on the ion.

(iv) Fluorine, the most electronegative element, has an oxidation number of –1 in all its compounds.

(v) Oxygen has an oxidation number –2 in its compounds. However, there are three exceptions to this generalization:

(a) When oxygen is combined with fluorine as in the compounds F_2O, its oxidation number is +2.

(b) In compounds containing oxygen-oxygen bonds, such as H_2O_2, the oxidation number of oxygen is –1.

(c) In super oxides like KO_2, the oxidation number of oxygen atom is –1/2.

(*vi*) The halogens other than fluorine (*i.e.*, chlorine, bromine, iodine) have an oxidation number of –1 except when they are combined with a more electronegative element, that is, with oxygen or with a more electronegative halogen.

(*vii*) Hydrogen in its compounds exhibits an oxidation number of +1. The only exceptions occur in metal hydrides in which hydrogen has an oxidation number of –1, for example, in sodium hydride, $Na^{+1} H^{-1}$.

(*viii*) When an atom is bonded to another atom of same electronegativity (for example, a molecule consisting of two identical atoms (H_2 or O_2), then the bond electrons are divided equally between the two atoms. Each atom may thus have an oxidation number equal to zero. Similarly, phosphorus in P_4 or sulphur in S_8 have oxidation number zero.

The concept of oxidation number gives us another definition of oxidation and reduction: *Oxidation may be defined as any change in which the oxidation number of an atom increases. Reduction may be defined as any change in which the oxidation number of an atom decreases.*

(*ix*) For complex ions the sum of oxidation numbers of all the atoms is equal to the charge on the ion. For example, in NO_3^- ion, the sum of the oxidation numbers of nitrogen atom and 3 oxygen atoms must be equal to –1.

Calculation of oxidation number. Keeping in view the above rules, we can calculate the oxidation number of any atom in a molecule/ion.

Step 1. Write down the formula of the given molecule/ion leaving some space between the atoms.

Step 2. Write oxidation number of each atom on its top. Write *x* for the oxidation number of the atom whose oxidation number has to be found out.

Step 3. Determine the total oxidation number of each element by multiplying oxidation no. of each atom with the number of atoms of that kind in the molecule/ion.

Step 4. Add up the oxidation number of each element and equate it equal to charge on the molecule/ion.

Step 5. Solve for the value of *x*.

Q. 5. Determine oxidation number of

(*i*) S in H_2SO_3

(*ii*) P in PO_4^{3-}

(*iii*) Cr in $Cr_2O_7^{2-}$ and Cr O_4^{2-} (*Garhwal, 2000*)

(*iv*) Fe in Fe_2O_3

Ans. (*i*) H_2SO_3. Since the oxidation number of hydrogen is +1 and that of oxygen is –2, we have

$$\overset{+1}{H_2}\quad\overset{x}{S}\quad\overset{-2}{O_3}$$

∴ Total oxidation numbers of all the atoms in H_2SO_3.

$$= (+1 \times 2) + (x \times 1) + (-2 \times 3)$$
$$= 2 + x - 6$$
$$= x - 4$$

Since the sum of oxidation numbers of all the atoms of a neutral molecule like H_2SO_3 is zero, therefore,

$$x - 4 = 0$$
$$x = +4$$

Hence, the oxidation number of sulphur is H_2SO_3 is +4.

(ii) ***PO_4^{3-} ion.*** Let the oxidation number of P be x and that of oxygen is –2. We have

$$\overset{x}{\text{P}}\ \overset{-2}{\text{O}}_4$$

∴ Total oxidation numbers of all the atoms in anion like PO_4^{3-} is equal to –3; therefore,

$$x - 8 = -3$$
$$x = -3 + 8$$
$$x = +5$$

Hence the oxidation number of phosphorus is PO_4^{3-} ion is +5.

(iii) ***$Cr_2O_7^{2-}$ ion.*** Let the oxidation number of Cr be x. Oxygen has oxidation number –2. Then we have

$$\overset{x}{\text{Cr}_2}\ \overset{-2}{\text{O}_7}$$

∴ Total oxidation number of all the atoms in $Cr_2O_7^{2-}$ ion

$$= (x \times 2) + (-2 \times 7)$$
$$= 2x - 14$$

Since the sum of the oxidation numbers of all the atoms in an ion $Cr_2O_7^{2-}$ is equal to (–2), therefore,

$$2x - 14 = -2$$
$$2x = -2 + 14$$
$$2x = +12$$
$$x = \frac{12}{2} = 6$$
$$x = +6$$

Hence, oxidation number of chromium in $Cr_2O_7^{2-}$ is + 6. Similarly, oxidation number of Cr in CrO_4^{2-} is also + 6.

(iv) ***Fe_2O_3.*** Let the oxidation number of Fe be x, that of oxygen is –2. We have

$$\overset{x}{\text{Fe}}_2\ \overset{-2}{\text{O}}_3$$

∴ Total oxidation numbers of all the atoms in Fe_2O_3

$$= (x \times 2) + (-2 \times 3)$$
$$= 2x - 6$$

Since the sum of the oxidation number of all the atoms in a neutral molecule like Fe_2O_3 is zero,

Therefore $2x - 6 = 0$

or $x = +3$

Q. 6. Assign oxidation number (O.N.) to each of the elements in the following compounds or ionic species.

(*a*) BaF_2 (*b*) CO_2 (*c*) SCl_2 (*d*) S_2Cl_2

Ans. (*a*) BaF_2 is an ionic compound: $Ba^{+2}(F^-)_2$. Therefore O.N. of Ba is +2 and O.N. of F is –1.

(*b*) Each oxygen has O.N. = –2 and therefore, C has O.N. = +4.

(*c*) Each chlorine has O.N. = –1 and therefore S has an O.N. +2.

(*d*) Chlorine has an oxidation number = –1, so

$$2(-1) + 2(x) = 0$$

or $$x = 1$$

Hence oxidation number of S = 1.

Q. 7. Find out the oxidation number of carbon in each of its compounds.

(*i*) CH_4 (*ii*) CH_3Cl (*iii*) CH_2Cl_2 (*iv*) $CHCl_3$ (*v*) CCl_4

Ans. (*i*) $\overset{+1}{C}H_4$: $x + 4 = 0$ or $x = -4$

(*ii*) $C\overset{+1}{H_3}\overset{-1}{Cl}$: $x + 3 - 1 = 0$ or $x = -2$

(*iii*) $C\overset{+1}{H_2}\overset{-1}{Cl_2}$: 2H = + 2, 2Cl = –2 ∴ $x + 2 - 2 = 0$ or $x = 0$

(*iv*) $C\overset{+1}{H}\overset{-1}{Cl_3}$: $x + 1 - 3 = 0$ or $x = 2$

(*v*) CCl_4^{-1} : $x - 4 = 0$ or $x = 4$

Q. 8. Give the oxidation number of sulphur in each of the following:

(*i*) $H_2S_2O_2$ (*ii*) $H_2S_2O_4$ (*iii*) $H_2S_2O_6$ (*iv*) $H_2S_2O_8$ (*v*) $S_4O_6^{2-}$ (*vi*) $S_2O_7^{2-}$

Ans. (*i*) $\overset{+1}{H_2}S_2\overset{-2}{O_2}$: $2 + 2x - 4 = 0$ or $x = 1$

(*ii*) $\overset{+1}{H_2}S_2\overset{-2}{O_4}$: $2 + 2x - 8 = 0$ or $x = 3$

(*iii*) $\overset{+1}{H_2}S_2\overset{-2}{O_6}$: $2 + 2x - 12 = 0$ or $x = 5$

(*iv*) $\overset{+1}{H_2}S_2\overset{-2}{O_8}$: $2 + 2x - 16 = 0$ or $x = 7$

However this is not true because this compound contains peroxo O–O group which contributes –2(–1 per oxygen atom). Consequently, 6 oxygens contribute –12 and 1 peroxo group contributes –2 charges to make a total contribution of –14 from 8 oxygen atoms. Hence oxidation number of S in $H_2S_2O_8$ is +6.

(*v*) $S_4O_6^{2-}$: $4x - 12 = -2$ or $x = 2.5$

(*vi*) $S_2O_7^{2-}$: $2x - 14 = -2$ or $x = 6$

Q. 9. Find out the oxidation of the central atom in each of the following fluro derivatives:

(*i*) ClF_3 (*ii*) OF_2 (*iii*) $XeOF_2$ (*iv*) HSO_3F (*v*) IF_7

Ans. (*i*) $Cl\overset{-1}{F_3}$: $x - 3 = 0$ or $x = 3$

(*ii*) $O\overset{-1}{F_2}$: $x - 2 = 0$ or $x = 2$

(*iii*) $Xe\overset{-2}{O}\overset{-1}{F_2}$: $x - 2 - 2 = 0$ or $x = 4$

(*iv*) $\overset{+1}{H}S\overset{-2}{O_3}\overset{-1}{F}$: $1 + x - 6 - 1 = 0$ or $x = 6$

(*v*) $\overset{-1}{I}F_7$; $x - 7 = 0$ or $x = 7$

Q. 10. What is the oxidation number of

(*i*) N in HNO_3, NO, NO_2

(*ii*) P in PH_3, P_2O_5, H_3PO_4

(*iii*) S in $Na_2S_2O_3$, H_2SO_4 (*G. Nanakdev, 2000*)

Ans. (*i*) +5, +2 and +4

(*ii*) –3, +5 and +5

(*iii*) +2, +6

Q. 11. Calculate the oxidation number in the following:

(*i*) Si in $Be_3Al_2Si_6O_{18}$

(*ii*) P in $HP_2O_7^-$

(*iii*) U in UO_2SO_4

(*iv*) Cr in CrO_2Cl_2 and K_2CrO_4.

Ans. (*i*) +4 (charge on Si_6O_{18} is –12)

(*ii*) +5 (charge on P_2O_7 is – 4)

(*iii*) +6

(*iv*) +6, +6

Q. 12. In the given reactions, indicate (*i*) oxidation, (*ii*) reduction, (*iii*) oxidising agent, and (*iv*) reducing agent:

(*a*) $2\,FeCl_3 + SnCl_2 \longrightarrow 2\,FeCl_2 + SnCl_4$

(*b*) $3\,Fe + 4H_2O \longrightarrow Fe_3O_4 + 4H_2$

Can we classify these as electron-transfer reactions?

Ans. (*a*) (*i*) Oxidation of $SnCl_2$ occurs to form $SnCl_4$.

(*ii*) Reduction of $FeCl_3$ occurs to form $FeCl_2$.

(*iii*) $FeCl_3$ acts as an oxidizing agent.

(*iv*) $SnCl_2$ acts as a reducing agent.

(*b*) (*i*) Oxidation of Fe occurs to form Fe_3O_4.

(*ii*) Reduction of H_2O occurs to form H_2.

(*iii*) H_2O acts as an oxidising agent.

(*iv*) Fe acts as a reducing agent. Yes we can classify these as electron-transfer reactions.

Q. 13. Give the points of distinction between valency and oxidation number.

Ans. (*M. Dayanand, 2000; Nagpur, 2000*)

Valency	*Oxidation number*
(1) It refers to combining capacity of an element. It refers to number of hydrogen atoms which combine with one atom of an element.	It refers to the residual charge which an atom appears to have when all other atoms are removed from it as ions.
(2) It is always a whole number.	It can be even in fractions, *e.g.*, oxidation number of iron in Fe_3O_4 is 2.60 and S in $S_4O_6^{2-}$ is 2.5.

(*Cont.*)

Valency	*Oxidation number*
(3) In general, elements like C, N and S have fixed valency. For example, the valency of nitrogen in all its compounds is always three.	Even elements like C, N, S have variable oxidation states depending upon the compounds formed by them. For example, oxidation no. of nitrogen in various compounds are : *Compound* — *Oxidation no.* N_2O — +1 NO — +2 N_2O_3 — +3 NO_2 — +4 N_2O_5 — +5 N_2 — zero NH_3 — −3

Q. 14. Define oxidation and reduction in terms of oxidation number? (*G. Nanakdev, 2000*)

Ans. In terms of oxidation number, oxidation and reduction may be defined as follows:

Oxidation. It is a process in which the oxidation number of an atom shows an increase. For example, in the oxidation of ferrous Fe^{+2} to Fe^{+3} ferric the oxidation number of iron increases from +2 to +3. Similarly, conversion of iodide I^- to I_2 iodine is oxidation because the oxidation number of I increases from −1 to 0.

Reduction. It is reverse of oxidation and refers to a process in which the oxidation number shows decrease. For example, in the reduction of stannic Sn^{4+} to stannous Sn^{2+}, the oxidation no. of tin decreases from +4 to +2. Similarly, conversion of chlorine Cl_2 to chloride Cl^- is reduction because the oxidation number of Cl decreases from zero to −1.

Since oxidation and reduction go hand in hand, therefore, in a chemical reaction whenever there is an increase in oxidation number of any species, there is a corresponding decrease in oxidation number of some other species. This is illustrated below in the chemical reaction between chlorine and potassium iodide.

Here, the oxidation number of Cl decreases from zero in Cl_2 to −1 in Cl^-. This refers to reduction of Cl_2 to Cl^-. There is a corresponding increase in oxidation number of I from −1 in I^- to zero in I_2. This indicates oxidation of iodide to iodine.

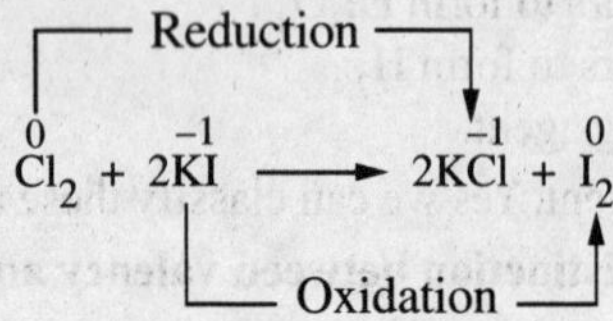

An oxidizing agent is sometimes abbreviated as **oxidant** and a reducing agent as **reductant.** Changes in oxidation number in oxidation, reduction, oxidising agent or reducing agent are summarised as under:

Oxidation: Increase of oxidation number or loss of electrons.

Reduction: Decrease of oxidation number or gain of electrons.

Oxidising agent: Shows decrease in oxidation no. or is an electron acceptor.

Reducing agent: Shows increase in oxidation no. or is an electron donor.

Q. 15. Identify atoms undergoing oxidation-reduction and the oxidants and reductants in the following equations:

(*a*) $4NH_3 + 5O_2 \longrightarrow 4NO + 6H_2O$

(*b*) $4Sn + 10HNO_3 \longrightarrow 4Sn(NO_3)_2 + 3H_2O + NH_4NO_3$

(*c*) $2FeSO_4 + H_2SO_4 + H_2O_2 \longrightarrow Fe_2(SO_4)_3 + 2H_2O$

Ans (*a*)

Reduction

$$\underset{\text{(Red.)}}{4NH_3} + \underset{\text{(Oxid.)}}{5\overset{0}{O_2}} \longrightarrow 4N\overset{+2-2}{O} + 6H_2\overset{-2}{O}$$

Oxidation

Reductant and oxidant are abbreviated as Red. and Oxid. respectively.

(*b*)

Oxidation

$$\underset{\text{(Red.)}}{4\overset{0}{Sn}} + \underset{\text{(Oxid.)}}{10H\overset{+5}{N}O_3} \longrightarrow 4\overset{+2}{Sn}(NO_3)_2 + \overset{-3}{N}H_4NO_3 + 3H_2O$$

Reduction

In this reaction oxygen is reduced to oxide (O^{-2}) present both in NO as well as H_2O.

Nitric acid HNO_3 (O.N of N = + 5) behaves as an oxidant. All N atoms on the right hand side remain the same. (O.N. = +5) except one N of ammonium ion in NH_4NO_3 which shows an oxidation state = –3.

(*c*)

Oxidation

$$\underset{\text{(Red.)}}{2\overset{+2}{Fe}SO_4} + H_2SO_4 + \underset{\text{(Oxid.)}}{H_2\overset{-1}{O_2}} \longrightarrow \overset{+3}{Fe_2}(SO_4)_3 + 2H_2\overset{-2}{O}$$

Reduction

Q. 16. Give various steps involved in balancing redox equations by oxidation number method.

Ans. Various steps for balancing a redox equation by this method are as follows:

Step I. Assign oxidation number of atoms undergoing oxidation reduction and write separate equations for such atoms only.

Step II. Add electrons on the right side for oxidation and on the left side for reduction. The number of electrons added in each case should be equal to the change in oxidation number.

Step III. Balance the number of electrons in the two equations and add the two equations to cancel out electrons.

Step IV. Then write the complete equation in its original form taking the coefficients of the oxidants and reductants in step II. Balance other atoms.

Step V. In acid medium, if oxygen remains unbalanced add water on the side deficient in O. If hydrogen still remains unbalanced then add H^+ on the required side. Be sure that all atoms and charges are balanced in the fmal equation.

Step VI. In basic medium, if oxygen remains unbalanced then add same number of H_2O as the unbalanced atoms on the side with excess and add double the number of OH^- on the other side. This will balance oxygen.

In basic medium, if H remains unbalanced, then add same number of OH^- as the difference on the side with the excess H and add same number of H_2O on the other side.

Be sure that all the participants and charges are balanced in the final equation.

Q. 17. Complete and balance the following equation:

$$\mathbf{Zn + HNO_3 \longrightarrow N_2O + Zn(NO_3)_2 + \ldots\ldots}$$

Ans. Step I. $\overset{0}{Zn} + H\overset{+5}{N}O_3 \longrightarrow \overset{+1}{N_2}O + \overset{+2}{Zn}(NO_3)_2$

$$\overset{0}{Zn} \longrightarrow \overset{2+}{Zn} \text{ (oxidation)}$$

$$2\overset{+5}{N} \longrightarrow 2\overset{+1}{N} \text{ (reduction)}$$

Step II. $\overset{0}{Zn} \longrightarrow \overset{+2}{Zn} + 2e$ ...(*i*)

$$2\overset{+5}{N} + 8e \longrightarrow 2\overset{+1}{N} \quad \ldots(ii)$$

Step III. Multiply (*i*) by 4 and add

$$4\overset{0}{Zn} \longrightarrow 4\overset{2+}{Zn} + 8\bar{e}$$

$$2\overset{+5}{N} + 8\bar{e} \longrightarrow 2\overset{+1}{N}$$

$$4\overset{0}{Zn} + 2\overset{+5}{N} \longrightarrow 4\overset{+2}{Zn} + 2\overset{+1}{N}$$

Step IV. $4Zn + 2HNO_3 \longrightarrow 4Zn(NO_3)_2 + N_2O + \ldots..$

H_2O would be formed in the reaction and balancing H on the two sides,

$$4Zn + 1OHNO_3 \longrightarrow 4Zn(NO_3)_2 + N_2O + 5H_2O$$

Q. 18. Balance the following reaction occurring in basic medium:

$$\mathbf{CrO_4^{2} + SO_3^{2-} \longrightarrow Cr(OH)_4^- + SO_4^{2-}}$$

Ans. Step I. $\overset{+6}{Cr}O_4^{2-} + \overset{+4}{S}O_3^{2-} \longrightarrow \overset{+3}{Cr}(OH)_4^- + \overset{+6}{S}O_4^{2-}$

$$\overset{+6}{Cr} \longrightarrow \overset{+3}{Cr} \text{ (reduction)}$$

$$\overset{+4}{S} \longrightarrow \overset{+6}{S} \text{ (oxidation)}$$

Step II. $\overset{+6}{Cr} + 3e \longrightarrow \overset{+3}{Cr}$ ... (*i*)

$$\overset{+4}{S} \longrightarrow \overset{+6}{S} + 2\bar{e} \quad \ldots(ii)$$

Step III. Multiply (*i*) by 2 and (*ii*) by 3 to cancel electrons and add

$$2\overset{+2}{Cr} + 6e \longrightarrow 2\overset{+3}{Cr}$$

$$3\overset{+4}{S} \longrightarrow 3\overset{+6}{S} + 6e$$

$$2\overset{+6}{Cr} + 3\overset{+4}{S} \longrightarrow 2\overset{+3}{Cr} + 3\overset{+6}{S}$$

Step IV. $2CrO_4^{2-} + 3SO_3^{2-} \longrightarrow 2Cr(OH)_4^- + 3SO_4^{2-}$

Step V. In basic medium as per rule VI add $3H_2O$ on R.H.S. and $6O\bar{H}$ on L.H.S. to balance oxygen atoms.

$$2CrO_4^{2-} + 3SO_3^{2-} + 6OH^- \longrightarrow 2Cr(OH)_4^- + 3SO_4^{2-} + 3H_2O$$

In order to balance H as per rule VI, there is an excess of 8H on R.H.S. and $8OH^-$ has to be added on R.H.S. Add $8H_2O$ on L.H.S. to balance total hydrogen atoms.

$$2CrO_4^{2-} + 3SO_3^{2-} + 6OH^- + 8H_2O \longrightarrow 2Cr(OH)_4^- + 3SO_4^{2-} + 3H_2O + 8OH^-$$

Rearranging we get

$$2CrO_4^{2-} + 3SO_3^{2-} + 5H_2O \longrightarrow 2Cr(OH)_4^- + 3SO_4^{2-} + 2OH^-$$

Q. 19. Balance the following equation taking place in acidic medium:

$$\mathbf{Cr_2O_7^{2-} + H_2SO_3 \longrightarrow Cr^{+3} + HSO_4^-}$$ *(G. Nanakdev, 2000)*

Ans. Step I. $Cr_2O_7^{2-} + H_2\overset{+4}{S}O_3 \longrightarrow \overset{+3}{Cr} + H\overset{+6}{S}O_4^-$

$$2\overset{+6}{Cr} \longrightarrow 2\overset{+3}{Cr} \text{ (reduction)}$$

$$\overset{+4}{S} \longrightarrow \overset{+6}{S} \text{ (oxidation)}$$

Step II. $$2Cr^{+6} + 6e \longrightarrow 2Cr^{+3} \quad ....(i)$$

$$S^{4+} \longrightarrow S^{+6} + 2e \quad ...(ii)$$

Step III. To cancel electrons multiply *(ii)* by 3

$$2Cr^{+6} + 6e \longrightarrow 2Cr^{+3}$$

$$3S^{+4} \longrightarrow 3S^{+6} + 6e$$

$$2Cr^{+6} + 3S^{+4} \longrightarrow 2Cr^{3+} + 3S^{+6}$$

Step IV. $$Cr_2O_7^{2-} + 3H_2SO_3 \longrightarrow 2Cr^{3+} + 3HSO_4^-$$

Step V. Except oxygen and hydrogen atoms, all other atoms are balanced. To balance oxygen, add water on the side deficient in oxygen [*i.e.*, R.H.S.]. We get

$$Cr_2O_7^{2-} + 3H_2SO_3 \longrightarrow 2Cr^{3+} + 3HSO_4^- + 4H_2O$$

Since this reaction occurs in acidic medium, therefore, add H^+ on L.H.S. to balance hydrogen atom.

$$Cr_2O_7^{2-} + 3H_2SO_3 + 5H^+ \longrightarrow 2Cr^{3+} + 3HSO_4^- + 4H_2O$$

Q. 20. Balance the equation:

$$\mathbf{HCl + KMnO_4 \longrightarrow Cl_2 + MnCl_2 + KCl + H_2O}$$ *(M. Dayanand, 2000)*

Ans. Step I. $H\overset{-1}{Cl} + K\overset{+7}{Mn}O_4 \longrightarrow \overset{0}{Cl_2} + \overset{+2}{Mn}Cl_2 + KCl + H_2O$

$\overset{-1}{Cl}$ is oxidised to $\overset{0}{Cl}$ and $\overset{+7}{Mn}$ is reduced to $\overset{+2}{Mn}$

$$\overset{-1}{Cl} \longrightarrow \overset{0}{Cl_2} \text{ (oxidation)}$$

$$\overset{+7}{Mn} \longrightarrow \overset{2+}{Mn} \text{ (reduction)}$$

Step II. $$2Cl^- \longrightarrow Cl_2 + 2e \quad ...(i)$$

$$\overset{+7}{Mn} + 5e \longrightarrow \overset{2+}{Mn} \quad ...(ii)$$

Step III. To cancel electrons, multiply *(i)* by 5 and *(ii)* by 2 and then add

$$10\,Cl^- \longrightarrow 5Cl_2 + 10e$$

$$2\overset{+7}{Mn} + 10e \longrightarrow 2\overset{+2}{Mn}$$

$$2\overset{+7}{Mn} + 10Cl^- \longrightarrow 2\overset{+2}{Mn} + 5Cl_2$$

Step IV. $2KMnO_4 + 10HCl \longrightarrow 2MnCl_2 + 5Cl_2 + 2KCl + H_2O$

Balance K by multiplying KCl by 2.

$$2KMnO_4 + 10HCl \longrightarrow 2MnCl_2 + 5Cl_2 + 2KCl + H_2O$$

Then balance 16Cl on R.H.S. by making 16 HCl on the L.H.S. and we get

$$2KMnO_4 + 16HCl \longrightarrow 2MnCl_2 + 5Cl_2 + 2KCl + H_2O$$

Both oxygen and hydrogen get balanced by multiplying H_2O with 8. We get the final balanced equation

$$2KMnO_4 + 16\,HCl \longrightarrow 2MnCl_2 + 2KCl + 5Cl_2 + 8H_2O$$

Q. 21. Balance the following equation: $KMnO_4 + NH_3 \longrightarrow KNO_3 + MnO_2 + KOH + H_2O$

Ans. Step I. $K\overset{+7}{Mn}O_4 + \overset{-3}{N}H_3 \longrightarrow K\overset{+5}{N}O_3 + \overset{+4}{Mn}O_2 + OH^- + H_2O$

$$\overset{+7}{Mn} \longrightarrow \overset{+4}{Mn}\text{(reduction)}$$

$$\overset{-3}{N} \longrightarrow \overset{+5}{N}\text{ (oxidation)}$$

Step II. $\overset{+7}{Mn} + 3e \longrightarrow \overset{+4}{Mn}$ ... (*i*)

$\overset{-3}{N} \longrightarrow \overset{+5}{N} + 8\overline{e}$ ...(*ii*)

Step III. Multiply (*i*) by 8 and (*ii*) by 3 and add

$$8\overset{+7}{Mn} + 3\overset{-3}{N} \longrightarrow 8\overset{+4}{Mn} + 3\overset{+5}{N}$$

Step IV. $8KMnO_4 + 3NH_3 \longrightarrow 8MnO_2 + 3KNO_3 + KOH + H_2O$

To balance K we have 5KOH on R.H.S.

$$8KMnO_4 + 3NH_3 \longrightarrow 8MnO_2 + 3KNO_3 + 5KOH + H_2O$$

In order to balance hydrogen and oxygen atoms, we add one H_2O on R.H.S.

$$8KMnO_4 + 3NH_3 \longrightarrow 8MnO_2 + 3KNO_3 + 5KOH + 2H_2O$$

Q. 22. Balance the following equation:

$$\mathbf{KClO_3 + H_2SO_4 \longrightarrow KHSO_4 + HClO_4 + ClO_2 + H_2O}$$

Ans. Step I. $K\overset{+5}{Cl}O_3 + H_2SO_4 \longrightarrow KHSO_4 + H\overset{+7}{Cl}O_4 + \overset{+4}{Cl}O_2 + H_2O$

$$\overset{+5}{Cl} \longrightarrow \overset{+7}{Cl}\text{ (oxidation)}$$

$$\overset{+5}{Cl} \longrightarrow \overset{+4}{Cl}\text{ (reduction)}$$

Step II. $\overset{+5}{Cl} \longrightarrow \overset{+7}{Cl} + 2e$ ... (*i*)

$\overset{+5}{Cl} + e \longrightarrow \overset{+4}{Cl}$ ...(*ii*)

Step III. Multiply (*ii*) by 2 and add, we get

$$3\overset{+5}{Cl} \longrightarrow \overset{+7}{Cl} + 2\overset{+4}{Cl}$$

Step IV. One $\overset{+7}{Cl}$ comes from 1 $HClO_4$ and $2\overset{+4}{Cl}$ from $2ClO_2$.

$$\therefore \qquad 3KClO_3 + H_2SO_4 \longrightarrow KHSO_4 + HClO_4 + 2ClO_2 + H_2O$$

This is balanced.

Q. 23. Balance and complete the following equations:

(*i*) $\mathbf{Mn^{+2} + PbO_2 \longrightarrow MnO_4^- + H_2O + Pb^{+2}}$ **(acidic medium)**

(*ii*) $\mathbf{ClO_2 + SbO_2^- \longrightarrow ClO_2 + Sb(OH)_6^-}$ **(basic medium)**

Ans. (*i*) Step I. $\overset{+2}{Mn} + \overset{+4}{Pb}O_2 \longrightarrow \overset{+7}{Mn}O_4^- + H_2O + \overset{+2}{Pb}$

$$\overset{+2}{Mn} \longrightarrow \overset{+7}{Mn} \text{(oxidation)}$$

$$\overset{+4}{Pb} \longrightarrow \overset{+2}{Pb} \text{ (reduction)}$$

Step II.

$$\overset{+2}{Mn} \longrightarrow \overset{+7}{Mn} + 5e \qquad ...(i)$$

$$\overset{+4}{Pb} + 2e \longrightarrow \overset{+2}{Pb} \qquad ...(ii)$$

Step III. Multiply (*i*) by 2 and (*ii*) by 5, then add:

$$2\overset{+2}{Mn} + 5\overset{+4}{Pb} \longrightarrow 2\overset{+7}{Mn} + 5\overset{+2}{Pb}$$

Step IV. $2Mn^{+2} + 5PbO_2 \longrightarrow 2MnO_4^- + 5Pb^{2+} + H_2O$

Step V. Add H_2O on the side deficient in O atoms. To balance H we add $4H^+$ on L.H.S. and the final equation becomes:

$$2Mn^{2+} + 5PbO_2 + 4H^+ \longrightarrow 2MnO_4^- + 5Pb^{2+} + 2H_2O$$

This balances all the charges and all the atoms.

(*ii*) Step I. $\overset{+4}{Cl}O_2 + \overset{+3}{Sb}O_2^- \longrightarrow \overset{+3}{Cl}O_2^- + \overset{+5}{Sb}(OH)_6^-$

$$\overset{+4}{Cl} \longrightarrow \overset{+3}{Cl} \text{ (reduction)}$$

$$\overset{+3}{Sb} \longrightarrow \overset{+5}{Sb} \text{(oxidation)}$$

Step II.

$$\overset{+4}{Cl} + e \longrightarrow \overset{+3}{Cl} \qquad ...(i)$$

$$\overset{+3}{Sb} \longrightarrow \overset{+5}{Sb} + 2e \qquad ...(ii)$$

Step III. Multiply (*i*) by 2 and add

$$2\overset{+4}{Cl} + \overset{+3}{Sb} \longrightarrow 2\overset{+3}{Cl} + \overset{+5}{Sb}$$

Step IV. $2ClO_2 + SbO_2^- \longrightarrow 2ClO_2^- + Sb(OH)_6^-$

Step V. As per rule V we add $4H_2O$ on R.H.S. and $8OH^-$ on L.H.S.

$$2ClO_2 + SbO_2^- + 8OH^- \longrightarrow 2ClO_2^- + Sb(OH)_6^- + 4H_2O$$

In order to balance H atoms we add $6OH^-$ on R.H.S. As it has six excess H, to balance this, we add $4H_2O$ on L.H.S.

$$2ClO_2 + SbO_2^- + 8OH^- + 6H_2O \longrightarrow 2ClO_2^- + Sb(OH)_6^- + 4H_2O + 6OH^-$$

Rearranging we get

$$2ClO_2^- + SbO_2^- + 2OH^- + 2H_2O \longrightarrow 2ClO_2^- + Sb(OH)_6^-$$

Q. 24. Balance the following equation:

$$\mathbf{KMnO_4 + MnSO_4 + H_2O \longrightarrow MnO_2 + K_2SO_4 + H_2SO_4}$$

Ans. Step I.

$$K\overset{+7}{Mn}O_4 + \overset{+2}{Mn}SO_4 + H_2O \longrightarrow \overset{+4}{Mn}O_2 + K_2SO_4 + H_2SO_4$$

$$\overset{+7}{Mn} \longrightarrow \overset{+4}{Mn}\text{ (reduction)}$$

$$\overset{+2}{Mn} \longrightarrow \overset{+4}{Mn}\text{ (oxidation)}$$

Step II.

$$\overset{+7}{Mn} + 3e \longrightarrow \overset{+4}{Mn} \qquad ...(i)$$

$$\overset{+2}{Mn} \longrightarrow \overset{+4}{Mn} + 2e \qquad ...(ii)$$

Step III. Multipling (*i*) by 2 and (*ii*) by 3, we get

$$2\overset{+7}{Mn} + 3\overset{+2}{Mn} \longrightarrow 5\overset{+4}{Mn}$$

Step IV. $2KMnO_4 + 3MnSO_4 + H_2O \longrightarrow 5MnO_2 + K_2SO_4 + H_2SO_4$

To balance SO_4 on the two sides we have $2H_2SO_4$ on R.H.S. That is,

$$2KMnO_4 + 3MnSO_4 + H_2O \longrightarrow 5MnO_2 + K_2SO_4 + 2H_2SO_4$$

Both H and O get balanced if we add one H_2O on L.H.S.

$$2KMnO_4 + 3MnSO_4 + 2H_2O \longrightarrow 5MnO_2 + K_2SO_4 + 2H_2SO_4$$

Q. 25. Balance the following equations:

(*i*) $Mn^{2+} + S_2O_8^{2-} \longrightarrow MnO_4^- + HSO_4^-$ (in acidic medium)

(*ii*) $AsO_3^{3-} + MnO_4^- \longrightarrow AsO_4^{3-} + MnO_2$ (in alkaline medium)

Ans. (*i*) Step I.

$$\overset{+2}{Mn} + \overset{+7}{S}_2O_8^{2-} \longrightarrow \overset{+7}{Mn}O_4^- + H\overset{+6}{S}O_4^-$$

$$\overset{+2}{Mn} \longrightarrow \overset{+7}{Mn}\text{ (oxidation)}$$

$$2\overset{+7}{S} \longrightarrow 2\overset{+6}{S}\text{ (reduction)}$$

Step II.

$$\overset{+2}{Mn} \longrightarrow \overset{+7}{Mn} + 5e \qquad ...(i)$$

$$2\overset{+7}{S} + 2e \longrightarrow 2\overset{+6}{S} \qquad ...(ii)$$

Step III. Multiply (*i*) by 2 and (*ii*) by 5 and add

$$2\overset{+2}{Mn} + 10\overset{+7}{S} \longrightarrow 2\overset{+7}{Mn} + 10\overset{+6}{S}$$

Step IV. $2Mn^{+2} + 5S_2O_8^{2-} \longrightarrow 2MnO_4^- + 10HSO_4^-$

Step V. Add $8H_2O$ on L.H.S. to balance deficient O

$$2Mn^{2+} + 5S_2O_8^{2-} + 8H_2O \longrightarrow 2MnO_4^- + 10HSO_4^-$$

Balance H by adding $6H^+$ on R.H.S.

$$2Mn^{+2} + 5S_2O_8^{2-} + 8H_2O \longrightarrow 2MnO_4^- + 10HSO_4^- + 6H^+$$

***(ii)* Step I.** $\overset{+3}{As}O_3^{3-} + \overset{+7}{Mn}O_4^{3-} \longrightarrow \overset{+5}{As}O_4^{3-} + \overset{+4}{Mn}O_2$

$$\overset{+3}{As} \longrightarrow \overset{+5}{As} \text{ (oxidation)}$$

$$\overset{+7}{Mn} \longrightarrow \overset{+4}{Mn} \text{ (reduction)}$$

Step II. $\overset{+3}{As} \longrightarrow \overset{+5}{As} + 2e$ ...(*i*)

$\overset{+7}{Mn} + 3e \longrightarrow \overset{+4}{Mn}$ ... (*ii*)

Step III. Multiply (*i*) by 3 and (*ii*) by 2 and add

$$3\overset{+3}{As} + 2\overset{+7}{Mn} \longrightarrow 3\overset{+5}{As} + 3\overset{+4}{Mn}$$

Step IV. $3AsO_3^{3-} + 2MnO_4^- \longrightarrow 3AsO_4^{3-} + 2MnO_2$

Step V. One oxygen atom on L.H.S. is in excess, therefore one H_2O is added on L.H.S. and $2OH^-$ on R.H.S.

$$3AsO_3^{3-} + 2MnO_4^- + H_2O \longrightarrow 3AsO_4^{3-} + 2MnO_2 + 2OH^-$$

Q. 26. Balance the following equations:

(*i*) $\mathbf{MnO_4^- + H^+ + H_2O \longrightarrow Mn^{+2} + O_2 + H_2O}$

(*ii*) $\mathbf{Zn + NaNO_3 + NaOH \longrightarrow Na_2ZnO_2 + NH_3 + H_2O}$

Ans. (*i*) Step I. $\overset{+7}{Mn}O_4^- + H^+ + H_2\overset{-1}{O_2} \longrightarrow \overset{+2}{Mn} + \overset{0}{O_2} + H_2O$

$$Mn^{+7} \longrightarrow Mn^{+2} \text{ (reduction)}$$

$$2\overset{-1}{O} \longrightarrow \overset{0}{O_2} \quad \text{(oxidation)}$$

Step II. $Mn^{+7} + 5e \longrightarrow \overset{+2}{Mn}$ ...(*i*)

$2\overset{-1}{O} \longrightarrow O_2 + 2\bar{e}$ ...(*ii*)

Step III. Multiply (*i*) by 2 and (*ii*) by 5 and add

$$2\overset{+7}{Mn} + 10\overset{-}{O} \longrightarrow 2\overset{+2}{Mn} + 5O_2$$

Step IV. $2MnO_4^- + 5H_2O_2 + H^+ \longrightarrow 2\overset{+2}{Mn} + 5O_2 + H_2O$

Step V. Add water on the side deficient in oxygen. Consequently, we add $7H_2O$ on R.H.S. to get a total of $8H_2O$.

$$2MnO_4^- + 5H_2O_2 + H^+ \longrightarrow 2Mn^{2+} + 5O_2 + 8H_2O$$

Add H^+ on L.H.S. to balance hydrogens.

$$2MnO_4^- + 5H_2O_2 + 6H^+ \longrightarrow 2Mn^{2+} + 5O_2 + 8H_2O$$

***(ii)* Step I.** $\overset{0}{Zn} + Na\overset{+5}{N}O_3 + NaOH \longrightarrow Na_2\overset{+2}{Zn}O_2 + \overset{-3}{N}H_3 + H_2O$

$\overset{0}{Zn} \longrightarrow \overset{+2}{Zn}$ (oxidation) ...(*i*)

$$N^{+5} \longrightarrow N^{-3} \text{ (reduction)} \quad ...(ii)$$

Step II. $\overset{0}{Zn} \longrightarrow \overset{+2}{Zn} + 2e$ ...(i)

$N^{+5} + 8e \longrightarrow N^{-3}$ ...(ii)

Step III. Multiply (*i*) by 4 and add (*ii*) to get

$$4\overset{0}{Zn} + \overset{+5}{N} \longrightarrow 4\overset{+2}{Zn} + \overset{-3}{N}$$

Step IV. $4Zn + NaNO_3 + NaOH \longrightarrow 4Na_2ZnO_2 + NH_3 + H_2O$

Balancing sodium atoms,

$$4Zn + NaNO_3 + 7NaOH \longrightarrow 4Na_2ZnO_2 + NH_3 + H_2O$$

Add one H_2O on R.H.S. to balance all the atoms.

$$4Zn + NaNO_3 + 7NaOH \longrightarrow 4Na_2ZnO_2 + NH_3 + 2H_2O$$

Q. 27. Complete and balance the following equations:

(*i*) $K_2Cr_2O_7 + KI + HCl \longrightarrow KCl + H_2O + ... + ...$

(*ii*) $MnO_4^- + Fe^{2+} \longrightarrow Mn^{2+} + H_2O +$ ....(in acidic medium)

(*iii*) $(NH_4)_2SO_4 + NO + NO_2 \longrightarrow H_2SO_4 + H_2O + ...$

(*iv*) $Zn + HNO_3 \longrightarrow ......+ H_2O + N_2O$

(*v*) $CrO_4^{2-} + CN^- \longrightarrow Cr(OH)_3 +$ ...... (in alkaline medium)

Ans. (*i*) $K_2\overset{+6}{Cr_2}O_7 + K\overset{-1}{I} + HCl \longrightarrow KCl + H_2O + \overset{0}{I_2} + \overset{+3}{Cr}Cl_3$

Balancing oxidation number, we get

$$K_2Cr_2O_7 + 6KI + HCl \longrightarrow KCl + H_2O + 3I_2 + 2CrCl_3$$

Balancing K and CI, we get

$$K_2Cr_2O_7 + 6KI + 14HCl \longrightarrow 8KCl + H_2O + 3I_2 + 2CrCl_3$$

Hydrogen and oxygen are balanced by adding H_2O on R.H.S.

$$K_2Cr_2O_7 + 6KI + 14HCl \longrightarrow 8KCl + 7H_2O + 3I_2 + 2CrCl_3$$

(*ii*) $MnO_4^- + 5Fe^{2+} \longrightarrow Mn^{2+} + H_2O + Fe^{3+}$

Balancing oxidation numbers, we get

$$MnO_4^- + 5Fe^{2+} \longrightarrow Mn^{2+} + 5Fe^{3+} + H_2O$$

Balancing oxygen by adding H_2O on R.H.S. and the H by adding H^+ on L.H.S. (acidic medium)

$$MnO_4^- + 5Fe^{2+} + 8H^+ \longrightarrow Mn^{2+} + 5Fe^{3+} + 4H_2O$$

(*iii*) $(\overset{-3}{N}H_4)_2SO_4 + \overset{+2}{N}O + \overset{+4}{N}O_2 \longrightarrow H_2SO_4 + H_2O + ...$

The oxidation numbers on nitrogen add up to give total $(-3 \times 2 + 2 + 4 = 0)$ of zero. Therefore, nitrogen on R.H.S. is in the form of N_2.

$$\therefore \quad (NH_4)_2SO_4 + NO + NO_2 \longrightarrow H_2SO_4 + 3H_2O + 2N_2$$

(*iv*) $Zn + HNO_3 \longrightarrow ..... + H_2O + N_2O$

Product of zinc and nitric acid would be zinc nitrate

$$\overset{0}{Zn} + H\overset{+5}{N}O_3 \longrightarrow \overset{+2}{Zn}(\overset{+5}{N}O_3)_2 + H_2O + \overset{+1}{N_2}O$$

Balancing oxidation numbers, we get

$$4Zn + 2HNO_3 \longrightarrow 4Zn(NO_3)_2 + N_2O + H_2O$$

Balancing nitrogen atoms, we get

$$4Zn + 10HNO_3 \longrightarrow 4Zn(NO_3)_2 + N_2O + H_2O$$

Balancing (O) by adding H_2O, we get

$$4Zn + 10HNO_3 \longrightarrow 4Zn(NO_3)_2 + N_2O + H_2O$$

(*v*) $$Cr\,O_4^{2-} + 3CN^- \longrightarrow 2Cr(OH)_3 + CNO^-$$

In this equation Cr^{+6} is reduced to Cr^{+3} and CN^- is oxidised to CNO^-.

Balancing these, we get

$$2CrO_4^{2-} + 3CN^- \longrightarrow 2Cr(OH)_3 + 3CNO^-$$

Balancing oxygen in alkaline medium, we get

$$2CrO_4^{2-} + 3CN^- + 2OH^- \longrightarrow 2Cr(OH)_3 + 3CNO^- + H_2O$$

Balancing H by adding $6H_2O$ on L.H.S. and $6OH^-$ on R.H.S., we get

$$2CrO_4^{2-} + 3CN^- + 2OH^- + 6H_2O \longrightarrow 2Cr(OH)_3 + 3CNO^- + H_2O + 6OH^-$$

$$\text{Or}_2CrO_4^{2-} + 3CN^- + 5H_2O \longrightarrow 2Cr(OH)_3 + 3CNO^- + 4OH^-$$

Q. 28. Describe ion electron method of balancing chemical equations.

(*Lucknow, 2002*)

Ans. Various steps used in the balancing of equation by this method are given below.

Step I. Divide equation into oxidation half and reduction half.

Step II. Balance atoms in each half equation. Then balance oxygen by adding H_2O and hydrogen by adding H^+. In case of alkaline medium use OH^- and H_2O as in oxidation number method.

Step III. Balance charges in each equation by adding electrons.

Step IV. Use appropriate integers to make loss of electrons equal to gain of electrons and then add, so that electrons get cancelled.

Step V. Check balancing of all the atoms and charges.

Q. 29. Balance the following equations for reactions taking place in alkaline medium:

(*i*) $$\mathbf{Cl_2 + KIO_3 \longrightarrow Cl^- + KIO_4}$$

(*ii*) $$\mathbf{Zn + NO_3^- \longrightarrow ZnO_2^{2-} + NH_3}$$

Ans. (i) Step I.

$$Cl_2 \longrightarrow 2Cl^-$$

$$IO_3^- \longrightarrow IO_4^-$$

Step II.

$$Cl_2 \longrightarrow 2Cl^-$$

$$IO_3^- + 2OH \longrightarrow IO_4^- + H_2O \text{ (being alkaline)}$$

Step III.

$$Cl_2 + 2e \longrightarrow 2Cl^-$$

$$IO_3^- + 2OH^- \longrightarrow IO_4^- + H_2O + 2e$$

Step IV. Adding, we get

$$Cl_2 + IO_3^- + 2OH^- \longrightarrow 2Cl^- + IO_4^- + H_2O$$

$$Cl_2 + KIO_3 + 2OH^- \longrightarrow 2Cl^- + KIO_4 + H_2O$$

(*ii*) **Step I.**

$$Zn \longrightarrow ZnO_2^{2-}$$

$$NO_3^- \longrightarrow NH_3$$

Step II.

$$Zn + 4OH^- \longrightarrow ZnO_2^{2-} + 2H_2O$$

$$NO_3^- + 3H_2O \longrightarrow NH_3 + 6OH^-$$

To balance H atom, we add hydroxide on R.H.S. and water on L.H.S.

$$NO_3^- + 3H_2O + 3H_2O \longrightarrow NH_3 + 6OH^- + 3OH^-$$

Step III. $$Zn + 4OH^- \longrightarrow ZnO_2^{2-} + 2H_2O + 2e \quad ...(i)$$

$$NO_3^- + 6H_2O + 8e \longrightarrow NH_3 + 9OH^- \quad ...(ii)$$

Step IV. Multiplying (*i*) by 4 and adding to (*ii*), we get

$$4Zn + 16OH^- + NO_3^- + 6H_2O \longrightarrow 4ZnO_2^{2-} + 8H_2O + NH_3 + 9OH^-$$

or $$4Zn + 7OH^- + NO_3^- \longrightarrow ZnO_2^{2-} + NH_3 + 2H_2O$$

Q. 30. Balance the following equations:

(*i*) $K_2Cr_2O_7 + H_2SO_4 + (COOH)_2 \longrightarrow K_2SO_4 + H_2O + Cr_2(SO_4)_3 + CO_2$

(*ii*) $Zn + H_2SO_4 + H_2MoO_4 \longrightarrow ZnSO_4 + H_2O + Mo_2(SO_4)_3$

Ans. (*i*) Step I. $$Cr_2O_7^{2-} \longrightarrow 2Cr^{3+}$$

$$C_2O_4^{2-} \longrightarrow 2CO_2$$

Step II. $$Cr_2O_7^{2-} + 14H^+ \longrightarrow 2Cr^{3+} + 7H^2O$$

$$C_2O_4^{2-} \longrightarrow 2CO_2$$

Step III. $$Cr_2O_7^{2-} + 14H^+ + 6e \longrightarrow 2Cr^{+3} + 7H_2O$$

$$C_2O_4^{2-} \longrightarrow 2CO_2 + 2e$$

Step IV. $$Cr_2O_7^{2-} + 14H^+ + 6e \longrightarrow 2Cr^{3+} + 7H_2O$$

$$3C_2O_4^{2-} \longrightarrow 6CO_2 + 6e$$

$$Cr_2O_7^{2-} + 14H^+ + 3C_2O_4^{2-} \longrightarrow 2Cr^{3+} + 7H_2O + 6CO_2$$

Step V. While writing in the form given in the question, we take care not to write H_2SO_4.

$$K_2Cr_2O_7 + 14H^+ + 3(COOH)_2 \longrightarrow Cr_2(SO_4)_3 + 7H_2O + 6CO_2$$

Accounting for $6H^+$ added for oxalic acid we are left with $8H^+$ which mean $4H_2SO_4$.

$$K_2Cr_2O_7 + 4H_2SO_4 + 3(COOH)_2 \longrightarrow Cr_2(SO_4)_3 + 7H_2O + 6CO_2$$

(*ii*) $$H_2MoO_4 + Zn + H_2SO_4 \longrightarrow Mo_2(SO_4)_3 + ZnSO_4 + H_2O$$

Step I. $$Zn \longrightarrow Zn^{2+}$$

$$MoO_4^{2-} \longrightarrow Mo^{3+}$$

Step II. $$Zn \longrightarrow Zn^{2+}$$

$$MoO_4^{2-} + 8H^+ \longrightarrow Mo^{3+} + 4H_2O$$

Step III. $$Zn \longrightarrow Zn^{2+} + 2e$$

$$MoO_4^{2-} + 8H^+ + 3e \longrightarrow Mo^{3+} + 4H_2O$$

Step IV. $$3\,Zn \longrightarrow 3Zn^{2+} + 6e$$

$$2MoO_4^{2-} + 16H^+ + 6e \longrightarrow 2Mo^{3+} + 8H_2O$$

$$3Zn + 2MoO_4^{2-} + 16H^+ \longrightarrow 3Zn^{2+} + 2Mo^{3+} + 8H_2O$$

Step V. $3Zn + 2H_2MoO_4 + 16\,H^+ \longrightarrow 3ZnSO_4 + Mo_2(SO_4)_3 + 8H_2O$

Balancing H and SO_4, we get $6\,H_2SO_4$ on L.H.S. because $4H^+$ are taken care of by $2H_2MoO_4$

$$3Zn + 2H_2\,MoO_4 + 6H_2SO_4 \longrightarrow 3ZnSO_4 + Mo_2(SO_4)_3 + 8H_2O$$

Q. 31. Balance the following equations by oxidation number method:

(*i*) $CH_4 + O_2 \longrightarrow CO_2 + H_2O$
(*ii*) $FeCl_3 + SnCl_2 \longrightarrow SnCl_4 + FeC_2$
(*iii*) $P_4 + HNO_3 \longrightarrow H_3PO_4 + NO_2 + H_2O$ (*G. Nanakdev, 2000*)
(*iv*) $Zn + HNO_3 \longrightarrow Zn(NO_3)_2 + NO + H_2O$
(*v*) $NH_3 + O_2 \longrightarrow NO + H_2O$
(*vi*) $Mn^{2+} + BiO_3^- + H^+ \longrightarrow HMnO_4 + Bi^{3+} + H_2O$
(*vii*) $Fe^{2+} + H_2O_2 + H^+ \longrightarrow Fe^{3+} + H_2O$

Ans. (*i*) $CH_4 + 2O_2 \longrightarrow CO_2 + 2H_2O$
(*ii*) $2FeC1_3 + SnC1_2 \longrightarrow SnC1_4 + 2FeC1_2$
(*iii*) $P_4 + 20HNO_3 \longrightarrow 4H_3PO_4 + 20NO_2 + 4H_2O$
(*iv*) $3Zn + 8HNO_3 \longrightarrow 3Zn(NO_3)_2 + 2NO + 4H_2O$
(*v*) $4NH_3 + 5O_2 \longrightarrow 4NO + 6H_2O$
(*vi*) $2Mn^{2+} + 5BiO_3^- + 16H^+ \longrightarrow 2HMnO_4 + 5Bi^{3+} + 7H_2O$
(*vii*) $2Fe^{2+} + H_2O_2 + 2H^+ \longrightarrow 2Fe^{3+} + 2H_2O$

Q. 32. Balance the following equations by oxidation number method:

(*i*) $H_2S + MnO_4^- + H^+ \longrightarrow S + Mn^{2+} + H_2O$
(*ii*) $Cr_2O_7^2 + H^+ + Fe^{2+} \longrightarrow 2Cr^{3+} + Fe^{3+} + H_2O$

(*Nagpur, 2000; M. Dayanand, 2000*)

(*iii*) $AsO_3^{3-} + IO_3^- \longrightarrow AsO_4^{3-} + I^-$
(*iv*) $HNO_3 + I_2 \longrightarrow HIO_3 + NO_2 + H_2O$
(*v*) $MnO_4^- + Cl^- + H^+ \longrightarrow Mn^{2+} + Cl_2 + H_2O$

Ans. (*i*) $5H_2S + 2MnO_4^- + 6H^+ \longrightarrow 5S + 2Mn^{2+} + 8H_2O$
(*ii*) $Cr_2O_7^{2-} + 6Fe^{2+} + 14H^+ \longrightarrow 2Cr^{3+} + 6Fe^{3+} + 7H_2O$
(*iii*) $3AsO_3^{2-} + IO_3^- \longrightarrow 3AsO_4^{3-} + I^-$
(*iv*) $I_2 + 10HNO_3 \longrightarrow 2HIO_3 + 10NO_2 + 2H_2O$
(*v*) $2MnO_4^- + 10Cl^- + I6H^+ \longrightarrow 5C1_2 + 2Mn^{2+} + 8H_2O$

Q. 33. Balance the following by oxid. No. and Ion electron methods:

(*i*) $IO_3^- + Cl^- \longrightarrow Cl_2 + I_2$
(*ii*) $Zn + NO_3^- \longrightarrow Zn^{2+} + NH_4^+$
(*iii*) $MnO_2 + BiO_3^- \longrightarrow MnO_4^- + Bi^{3+}$ (acidic medium)
(*iv*) $Ni(CN)_4^{2-} + Br_2 \longrightarrow Ni_2O_3 + Br^- + CN^-$
(*v*) $Cl_2 + OH^- \longrightarrow Cl^- + C1O_3^-$
(*vi*) $C_6H_{12}O_6 + H_2SO_4 \longrightarrow SO_2 + CO_2$
(*vii*) $NO_2^- \longrightarrow NO + HNO_3$
(*viii*) $H_3AsO_4 + KI + HCl \longrightarrow H_3AsO_3 + I_2 + KCl + H_2O$
(*ix*) $K_2MnO_4 + H_2O \longrightarrow MnO_2 + KMnO_4 + KOH$

Ans. (*i*) $2IO_3^- + 10Cl^- + 12H^+ \longrightarrow I_2 + 5C1_2 + 6H_2O$
(*ii*) $4Zn + 10H^+ + NO_3^- \longrightarrow NH_4^+ + 4Zn^{2+} + 3H_2O$
(*iii*) $2MnO_2 + 3BiO_3 + 12H+ \longrightarrow 2HMnO_4 + 3Bi^{3+} + 5H_2O$

(*iv*) $2Ni(CN)_4^{2-} + 6OH^- + Br_2 \longrightarrow Ni_2O_3 + 3H_2O + 2Br^- + 8CN^-$

(*v*) $3Cl_2 + 6OH^- \longrightarrow 5Cl^- + ClO_3^- + 3H_2O$

(*vi*) $C_6H_{12}O_6 + 12H_2SO_4 \longrightarrow 12SO_2 + 6CO_2 + 18H_2O$

(*vii*) $3NO_2^- + 3H^+ \longrightarrow NO + HNO_3 + H_2O$

(*viii*) $H_3AsO_4 + 2HCl + 2KI \longrightarrow H_3AsO_3 + I_2 + H_2O + 2KCl$

(*ix*) $3K_2MnO_4 + 2H_2O \longrightarrow 2KMnO_4 + MnO_2 + 4KOH$

Q. 34. Complete and balance the following equations:

(*i*) $PCl_5 + NO_2^- \longrightarrow PCl_3 + Cl^- + ...$

(*ii*) $P + HNO_3 \longrightarrow H_3PO_4 + H_2O + ...$

(*iii*) $S + OH^- \longrightarrow S_2O_3^{2-} + 2S^{2-}$

(*iv*) $MnO_4^{2-} \longrightarrow MnO_4^- + MnO_2$ **(acidic medium)**

Ans. (*i*) Nitrogen dioxide is formed in this redox reaction.

$$PCl_5 + 2NO_2^- \longrightarrow PCl_3 + 2Cl^- + NO_2$$

(*ii*) NO_2 is evolved in this reaction

$$P + HNO_3 \longrightarrow H_3PO_4 + H_2O + NO_2$$

Without going into each step, we have

Oxidation $P \longrightarrow PO_4^{3-}$

$$P + 4H_2O \longrightarrow PO_4^{3-} + 8H^+ + 5e \quad ...(i)$$

Reduction $NO_3^- \longrightarrow NO_2$

$$NO_3^- + 2H^+ \longrightarrow NO_2 + H_2O + e^- \quad ...(ii)$$

Multiply (*ii*) by 5 and add to (*i*), and rearrange:

$$P + 5HNO_3 \longrightarrow H_3PO_4 + 5NO_2 + H_2O$$

(*iii*) $S + OH^- \longrightarrow S^{2-} + S_2O_3^{2-}$

This reaction takes place in basic medium.

Oxidation $2S^0 \longrightarrow S_2O_3^{2-}$ (O.N. of S in $S_2O_3^{2-}$ is + 2)

$$2S + 6OH^- \longrightarrow S_2O_3^{2-} + 3H_2O + 4e \quad ...(i)$$

Reduction $S \longrightarrow S^{2-}$

$$S + 2e \longrightarrow S^{2-} \quad ...(ii)$$

Multiplying (*ii*) by 2 and adding to (*i*), we get

$$4S + 6OH^- \longrightarrow 2S^{2-} + S_2O_3^{2-} + 3H_2O$$

(*iv*) $MnO_4^{2-} \longrightarrow MnO_4^- + MnO_2$

Oxidation $MnO_4^{2-} \longrightarrow MnO_4^- + e \quad ...(i)$

Reduction $MnO_4^{2-} + 4H^+ + 2e \longrightarrow MnO_2 + 2H_2O \quad ...(ii)$

Myltiplying (*i*) by 2 and adding to (*ii*)

$$3MnO_4^{2-} + 4H^+ \longrightarrow 2MnO_4^- + MnO_2 + 2H_2O$$

Q. 35. Write a note on auto-oxidation. *(G. Nanakdev, 2000)*

Ans. It is a process in which a substance undergoes oxidation and reduction by itself. It may be compared with *disproportionation* reaction but in case of auto-oxidation the process occurs slowly without any action of heat, light or electricity. Some simple auto-oxidation reactions are given below:

(i) $$3HNO_2 \longrightarrow HNO_3 + 2NO + 2H_2O$$

Here nitrous acid undergoes simultaneous oxidation and reduction into HNO_3 and NO respectively.

$$\overset{+3}{N}O_2^- \longrightarrow \overset{+5}{N}O_3^- \text{(oxidation)}$$

$$\overset{+3}{N}O_2^- \longrightarrow \overset{+2}{N}O \text{ (reduction)}$$

(ii) $$H_2\overset{+6}{Mn}O_3 \longrightarrow 2H\overset{+7}{Mn}O_4 + \overset{+4}{Mn}O_2 + 2H_2O$$

Manganic acid undergoes self-oxidation and reduction into peramanganic acid and manganese dioxide.

$$\overset{+6}{Mn}O_4^{2-} \longrightarrow \overset{+7}{Mn}O_4^- \text{ (oxidation)}$$

$$\overset{+6}{Mn}O_4^{2-} \longrightarrow \overset{+4}{Mn}O_2 \text{ (reduction)}$$

(iii) Hypochlorous acid undergoes auto-oxidation as follows:

$$3H\overset{+1}{Cl}O \longrightarrow H\overset{+5}{Cl}O_3 + 2H\overset{-1}{Cl}$$

Here $$Cl^{+1} \longrightarrow Cl^{+5} \text{ (oxidation)}$$

$$Cl^{+1} \longrightarrow Cl^{-1} \text{ (reduction)}$$

Q. 36. Explain working of galvanic cell based on redox reactions. *(Kerala, 2000)*

Ans. Consider the redox reaction

$$Zn + CuSO_4 \longrightarrow ZnSO_4 + Cu$$

Taking place in a galvanic cell. The above reaction in ionic form may be written as

$$Zn + Cu^{+2} \longrightarrow Zn^{+2} + Cu$$

which can be split up into two half reactions as given below:

$$Zn \longrightarrow Zn^{+2} + 2e^- \quad \text{(oxidation half reaction)}$$

$$Cu^{+2} + 2e^- \longrightarrow Cu \quad \text{(reduction half reaction)}$$

On adding the two half reactions, we get the overall reaction again. In order to generate electricity, we allow this reaction to take place in an indirect manner in a cell. We may proceed as under:

Take two beakers A and B containing dilute solution of zinc sulphate and copper sulphate respectively in them (Fig. 23.1). Put a zinc metal rod in beaker A and a copper metal rod in beaker B.

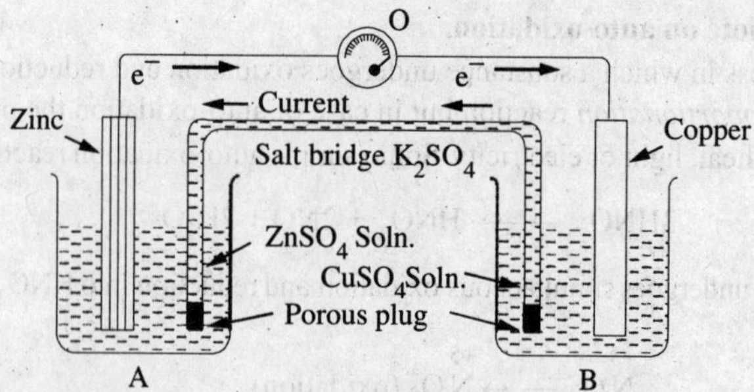

Fig. 23.1. Zn-$CuSO_4$ reaction in a Galvanic cell.

The two rods are connected by a wire through an ammeter and a key externally. The two solutions are connected by a salt bridge (a U-shaped tube containing concenterated solution of an electrolyte like KCl or K_2SO_4 or partially solidified solution of these electrolytes in agar-agar and gelatin). As soon as the connections are made by inserting the key, the current starts flowing which is indicated by the deflection produced in ammeter. Some important features of the device called cell may be summarised as below:

(*i*) The direction of the flow of electrons is from zinc to copper rod. However, the direction of conventional current (positive electricity) is from copper rod to zinc rod as indicated by the direction of deflection in the ammeter.

(*ii*) The zinc rod at which oxidation takes place is called **anode** and the copper rod at which reduction takes place is called **cathode.** Each of the beakers is called a **half cell or electrode.**

(*iii*) *Since* electrons are produced at the zinc electrode ($Zn \longrightarrow Zn^{+2} + 2e^-$), it is negative pole whereas the electrons are consumed at copper electrode ($Cu^{+2} + 2e^- \longrightarrow Cu$), it acts as the +ve pole of cell.

(*iv*)The chemical reaction $Zn + Cu^{+2} \longrightarrow Zn^{+2} + Cu$ starts taking place as soon as the connections are made. No heat is produced whereas large amount of heat (≈ 12.5 kJ per mole) is produced when this reaction is done directly in a single vessel. The obvious inference is that an approximately equivalent amount of electrical energy is produced when the reaction is done in a galvanic cell.

(*v*) The two-half reaction taking place separately at the zinc rod and copper rod can be represented as:

$$Zn \longrightarrow Zn^{+2} + 2e^- \qquad \text{(oxidation half reaction)}$$
$$Cu^{+2} + 2e^- \longrightarrow Cu \qquad \text{(reduction half reaction)}$$

(*vi*) The oxidation of Zn into Zn^{+2} ions produces excess of Zn^{2+} ions and the reduction of Cu^{+2} ions into Cu atoms leave the excess of SO_4^{2-} ions in the two half cells A and B respectively. The salt bridge maintains electrical neutrality in the two half cells by allowing the diffusion of the excess of ions to the other beakers respectively.

(*vii*) The concentration of copper sulphate solution decreases with the operation of the cell due to reduction of Cu^{+2} ions into copper atoms and migration of SO_4^{-2} ions to the other side. Consequently the current strength falls with the passage of time. The cell based on the above reaction is commonly called *Daniel Cell.* The salt bridge is replaced by a porous pot through which SO_4^{2-} ions migrate from copper sulphate solution (taken in porous pot) to zinc sulphate solution. In the similar manner many other galvanic cells based upon different redox reactions can be set up. For example,

$$Cu + 2AgNO_3 \longrightarrow Cu(NO_3)_2 + 2Ag$$
$$Zn + 2HCl \longrightarrow ZnCl_2 + H_2$$

Q. 37. Explain the representation of a galvanic cell. *(Kerala, 2000)*

Ans. A galvanic cell can be represented by a shorthand notation called cell diagram. The convention adopted for it are given below:

(*i*) The interfaces across which a potential difference exists are shown by a semicolon ; or a single verticle line. For example, a zinc half cell can be denoted as:

$$Zn; Zn^{+2} \text{ or } Zn/Zn^{+2}$$

(*ii*) The contact between two solutions with the help of a salt bridge can be denoted by a double vertical line (||) between the two solutions. In the above cell Zn^{+2} ions and Cu^{+2} ions solution are connected by a salt bridge; it can be shown as:

$$Zn^{+2} \parallel Cu^{+2}$$

(*iii*) The half cell in which oxidation takes place is written on left hand side and the half cell in which reduction takes place is written on R.H.S. with the metal electrodes always on the outside extreme.

(*iv*) The concentration of the two solutions, partial pressure of gases involved and their physical states are shown by suitable signs within brackets.

(*v*) The polarity of the two electrodes are indicated by (–) and (+) signs on the metal electrodes.

(*vi*) The e.m.f. of the cell is stated in volts along with the temperature at which it is applicable. Thus the cell diagram for the Daniel cell is.

$$(+)Zn \mid Zn^{+2} \parallel Cu^{+2} \mid Cu(-) \qquad E_{298} = 1.1 \text{ volts.}$$

Q. 38. What is the function of salt bridge in a galvanic cell?

Ans. The salt bridge serves three important functions in a galvanic cell:

(*i*) It prevents any net charge accumulation in other beaker by permitting the excess of ions to diffuse from one half cell to other through the salt bridge. Thus, it maintains electrical neutrality in the two half cells.

(*ii*) It physically separates the two electrodes, but completes the electrical circuit with electrons flowing from one electrode to another externally and the ions flowing between the two compartments through salt bridge.

(*iii*) It minimizes the liquid function potential [voltage produced where two dissimilar solutions are in contact] between the two solutions connected through salt bridge.

Q. 39. Explain the term electrode potential.

Ans. When a metal rod is dipped in water or the aqueous solution of its ions (Fig. 23.2), the chemical potential of the metal ions in the metal rod is higher than in aqueous phase. This leads to a flow of metal ions from the metal rod to the aqueous phase. The electrons are not solubilized in aqueous phase and, therefore, are left on the metal surface. This creates an electric field between the metal rod and the aqueous solution, the metal being negatively charged and the solution is positively charged.

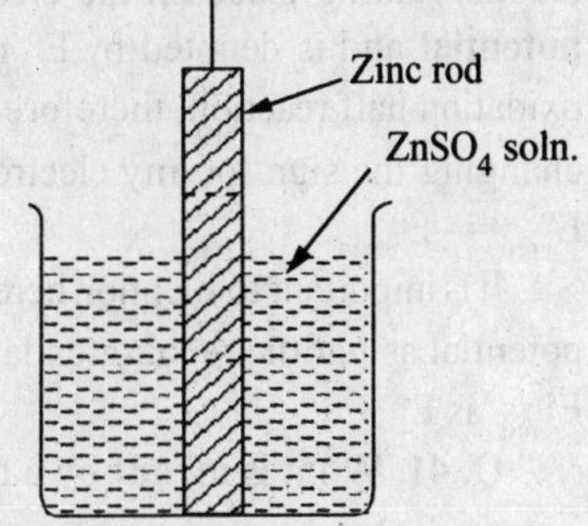

Fig. 23.2. Zn metal electrode.

$$M \longrightarrow M^{+n} + ne^-$$

An equilibrium state is established after a short while due to the reverse reaction initiated by the collisions of metal ions (M^{+n}) with the surface of metal rod and combining with electrons available on it.

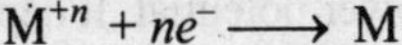

$$M^{+n} + ne^- \longrightarrow M$$

At this stage a definite potential difference is developed between the metal rod and the solution. The equilibrium potential difference between the metal electrode and its surrounding solution is called **electrode potential.** Thus it is the tendency of an electrode to lose electrons when kept in contact with the solution of its own ions.

In a galvanic cell two electrodes are combined together so as to oppose their tendencies of losing electrons. Oxidation takes place at one electrode and reduction at the other. In other words, one of the electrodes has greater tendency of losing the electrons than the other. The other electrode can be considered as to have greater tendency of gaining the electrons than the former. *Thus, the tendency of an electrode to lose or gain electrons in a cell is called electrode potential.*

The tendency of losing the electrons is called the oxidation electrode potential and the tendency of gaining the electrons is called the reduction potential of the electrode. They can be denoted by $E_{oxid.}$ and $E_{red.}$ respectively. The magnitude of electrode potential of an electrode depends upon several factors such as:

(*i*) Nature of the electrode.

(*ii*) Concentration of metal cations.

(*iii*) Temperature.

Under similar condition the two tendencies of an electrode are equal in magnitude but opposite in direction. This is because the equilibrium state remains unaltered from whatever direction it is approached. Therefore,

$$E_{red} = -E_{oxid}$$

Q. 40. What is meant by standard electrode potential?

(*Kanpur, 2000; Garhwal, 2000*)

Ans. If in an electrode set up, all active constituents or reactants are at unit activity, it is said to be **normal or standard electrode.** This can be achieved, when the concenteration of the electrolyte taken is one molar and the electrode is a pure solid or gas at a pressure of one atmosphere. The electrode is set up at a temperature of 298K. The electrode potential now developed is called **standard electrode potential**. It is denoted by E°, such an electrode is called standard or normal electrode. For example, a standard zinc electrode is set up by dipping a zinc metal rod into one molar zinc sulphate solution kept at 298K and one atmosphere pressure. It can be represented as $Zn(s) \mid Zn^{+2}$ (1M).

The electrode potential developed in a standard electrode due to oxidation reaction taking place on the electrode in a galvanic cell is called **standard oxidation electrode potential** and is denoted by E° oxidation. Similarly, the electrode potential developed in a standard electrode due to reduction reaction taking place on the electrode in a galvanic cell is called **standard reduction electrode potential** and is denoted by E° reduction. Since the reduction half reaction is just the reverse of oxidation half reaction, therefore reduction potential is obtained from oxidation potential by simply changing the sign for any electrode as these are equal in magnitude but opposite in direction. So $E°_{red} = -E°_{oxid}$.

It is important to mention here that it is a convention to express all electrode potentials as reduction potential as per the recommendations of IUPAC. Therefore, E_{red} may be expressed as E only and $E°_{red}$ as E°.

Q. 41. What is meant by e.m.f. of a cell? (*Garhwal, 2000*)

Ans. The electromotive force of a galvanic cell is the potential difference, which causes the flow of current from the electrode at higher potential to the electrode at lower potential. It depends upon the chemical nature of the reactants and products. In other words, it represents the driving force of the chemical reaction taking place in the cell. This is because when the cell is opposed by a greater e.m.f. the spontaneous reaction is reversed and electrolysis takes place. The e.m.f. of a cell is denoted by E_{cell}.

Thus the e.m.f. of galvanic cell is the sum of the tendencies of the two electrodes of transporting the electrons in the same direction in the galvanic cell. The tendency of transporting the electrons by anode can be measured by its reduction electrode potential. Therefore,

$$\text{E.M.F. of cell} = E_{oxid} \text{ of anode } + E_{red} \text{ of cathode}$$
$$E_{cell} = E_{oxid.} \text{(Anode)} + E_{red} \text{ (cathode)}$$
$$\text{or } E_{cell} = E_{red} \text{ (cathode)} - E_{red} \text{ (anode)}$$
$$E_{cell} = E_{cathode} - E_{anode} \quad [\because E_{oxid} = -E_{red}]$$

In the cell diagram of a galvanic cell the cathode is written are right hand side and the anode is written on left hand side. Therefore, the e.m.f. of a cell can be expressed as:

$$E_{cell} = E_{right} - E_{left} = E_R - E_L$$

If both the electrodes are standard electrodes, the e.m.f. of such a cell is called standard e.m.f. and is denoted by E° cell. Hence

$$E°_{cell} = E°_{oxid.} \text{ (anode)} + E°_{(Red)} \text{ (cathode)}$$
$$E°_{cell} = E°_{cathode} - E°_{Anode}$$
$$= E°_{right} - E°_{left} = E°_R - E°_L$$

Q. 42. How is the e.m.f. of a cell measured?

Ans. The potential difference between the two electrodes of a galvanic cell can be measured by connecting a voltmeter between them. The value of e.m.f. so obtained is not very correct due to the following reasons:

(*i*) The voltmeter draws some current from the cell, thereby it changes the e.m.f. of the cell.

(*ii*) With appreciable flow of current a part of e.m.f. will be utilized in overcoming the internal resistance of the cell.

Therefore, it is evident that e.m.f. of a cell can be measured by a method which involves little or no flow of current. The method is based upon Poggendorff s compensation principle and is called potentiometric method.

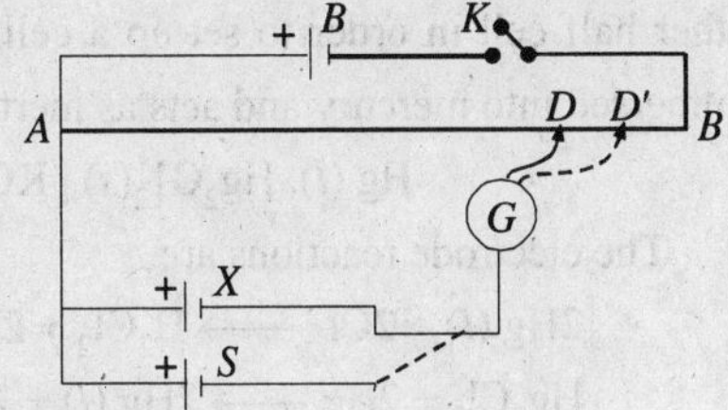

Fig. 23.3 Potentiometric method of measurement of e.m.f.

Procedure. In this method the unknown e.m.f. is opposed by another known e.m.f. until the two become equal and there is no net flow of current in the circuit as shown in Fig. 23.3.

(*i*) It consists of a uniform wire *AB* of high resistance and connected with a storage battery *B* of e.m.f. larger than the e.m.f. of the cell to be measured.

(*ii*) The cell *X* whose e.m.f. is to be measured is connected to the end A and the sliding contact *D,* through a galvanometer in such a manner that it sends e.m.f. in a direction opposite to that of storage battery *B*.

(*iii*) The sliding contact D is moved along the wire AB till at a particular point D no current flows through the galvanometer. The e.m.f. of the cell say E_x is proportional to the length AD, *i.e.,*

$$E_x \propto AD$$

(*iv*) The cell X is now replaced by a standard cell the e.m.f. of which say E_s is known. The null point is obtained now at D′.

Then $E_s \propto AD'$.

Comparing the two equations

$$E_x = \frac{AD}{AD'} \times E_s$$

Knowing the value of E_S, AD and AD′ the value of E_x can be calculated, which is the e.m.f. of the cell. The standard cell must give a constant and reproducible e.m.f. without any variation with temperature. A Weston cadmium standard cell can be used for this purpose.

Q. 43. Write note on

(*i*) Hydrogen electrode, (*ii*) Calomel electrode.

Ans. Hydrogen electrode. Hydrogen gas at one atmospheric pressure is bubbled into 1 molar hydrochloric acid solution in which a platinized platinum foil remains dipped (Fig. 23.4). The electrode can be represented as:

$$Pt, H_2 (P = 1 \text{ atm}) \mid H^+ (1 \text{ M})$$

and electrode reactions taking place on the electrode as:

$$H_2 \rightleftharpoons 2H^+ (1M) + 2e^-$$

Oxidation takes place when it acts as anode and reduction when acts as cathode.

When it acts as cathode, following reaction takes place:

$$2H^+ + 2e^- \rightleftharpoons H_2 (g)$$

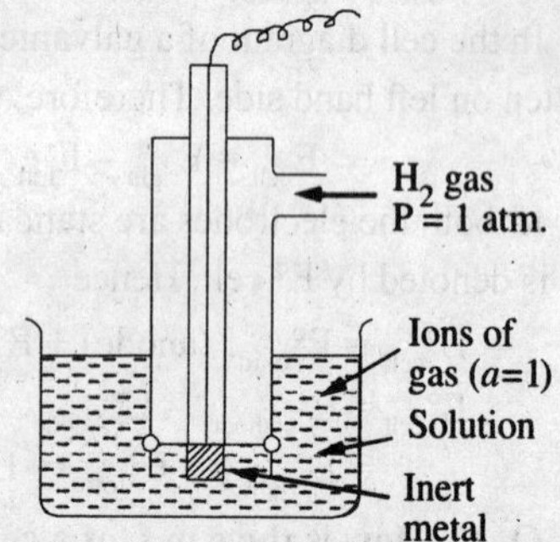

Fig. 23.4. Normal hydrogen gas electrode.

Calomel electrode. Hg (*l*), Hg_2Cl_2 (*s*)/KCl (*g*) combination is known as calomel electrode. The electrode is set up by taking pure mercury along with mercurous chloride paste at the bottom of the tube (Fig. 23.5) which is then filled with KCl solution of desired concenteration (0.1 N, 1 N or saturated) from the side tube (*A*). Another side tube *B* ending in a fine jet is used for connecting to the other half cell in order to set up a cell. A platinum wire is dipped in the tube so that it remains immersed into mercury and acts as inert metal electrode. The electrode can be represented as:

$$Hg (l), Hg_2Cl_2(s) \mid KCl (.1 \text{ N})$$

The electrode reactions are

$$2Hg (l) + 2Cl^- \longrightarrow H_2Cl_2 + 2e^- \text{ (as anode)}$$

$$Hg_2Cl_2 + 2e^- \longrightarrow 2Hg (l) + 2Cl^- \text{ (as cathode)}$$

Q. 44. How is the standard potential of an electrode measured?

Ans. The standard electrode potential is an important characteristic of a metal or non-metal. It predicts the relative ease of oxidation or reduction of the various metals and non-metals.

It is not possible to measure the electrode potential of a single electrode, because as soon as metallic conductor is placed in the solution of electrolyte to make the electrical circuit, a second electrode comes into operation. Hence in order to measure the standard electrode potential of an electrode, a standard electrode is set up and is

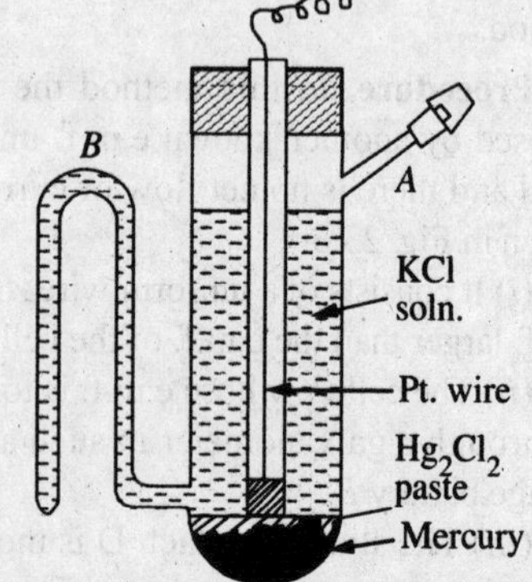

Fig. 23.5. Calomel electrode (metal metal insoluble salt anion electrode).

coupled with a normal hydrogen electrode (NHE) to form a galvanic cell. The NHE is a reference electrode and its electrode potential is arbitrarily taken as zero. The e.m.f. of the constituted cell, which is the difference of potential between two electrodes can be experimentally measured. Since the electrode potential of one of them is zero, that of the other is equal to the observed e.m.f.

$$E^\circ = E^\circ_{Cathode} - E^\circ_{Anode}$$
$$= E^\circ_{right} - E^\circ left$$

[where E° is the cell voltage of the cell. $E^\circ_{cathode}$ and E°_{anode} are standard electrode potentials of the two electrodes expressed as reduction potential.]

Hence the standard electrode potential of the electrode is numerically equal to the e.m.f. of the cell, constituted above, the sign of standard electrode potential of the electrode is determined by the polarity of the electrode or nature of the reaction occurring at the electrode. The polarity of the electrode can be determined by the direction of the flow of current, which can be noted from the deflection in ammeter connected in the circuit.

For example, the standard electrode potential of zinc electrode can be found by coupling it with NHE. The cell voltage is found to be 0.763 volt and electrons flow from zinc metal electrode to the hydrogen electrode. Therefore zinc electrode acts as anode and hydrogen gas electrode as cathode.

e →

$$Zn \mid Zn^{+2} (a=1) \parallel H^+ (a=1) \mid H_2 (g) (1 \text{ atm}), Pt$$

$$E^\circ = E^\circ_{cathode} - E^\circ_{anode}$$
$$0.763 = E^\circ_{hydrogen} - E^\circ_{zinc}$$
$$= 0 - E^\circ_{Zn}$$
$$\therefore \quad E^\circ_{Zn} = -0.763 \text{ volt}$$

If the electrode under experiment acts as negative pole when coupled with NHE, it will have –ve electrode potential and if it acts as a +ve pole then it will have +ve electrode potential. The standard electrode potential of copper electrode can be found in a similar manner. The galvanic cell formed when it is coupled with NHE is:

e →

$$Pt, H_2 (g) (1 \text{ atm}) \mid H^+ (a = 1) \parallel Cu^{++} (a= 1) \mid Cu$$

The cell voltage is measured as 0.34 volt and the electrons flow from hydrogen electrode to copper electrode,

$$E^\circ = E^\circ_{Cathode} - E^\circ_{Anode}$$
$$= E^\circ_{Copper} - E^\circ_{Hydrogen}$$
$$0.34 = E^\circ_{Copper} - 0$$
$$E^\circ_{Copper} = 0.34 \text{ volt}$$

Since the copper electrode acts as positive pole in the constituted cell, it will have positive potential.

Q. 45. What is electrochemical series?

Ans. The table that gives the relative reactivities of metals or relative tendencies of reactions is called electrochemical series. In fact this is the table of standard electrode reduction potential. (See Table 23.1.) Reduction potential of an electrode denotes the tendency of the relevant reaction to take place in the direction of reduction (*i.e.*, to get reduced). It may be remembered that taking up of

Table 23.1. Some standard reduction potentials.

Half -Reaction				$E°, V$
$3/2N_2$	$+H^+ + e$	⇌	HN_3 (aq)	– 3.09
Li^+	$+ e$	⇌	Li	– 3.045
K^+	$+e$	⇌	K	–2.925
Cs^+	$+e$	⇌	Cs	–2.923
Ba^{2+}	$+ 2e$	⇌	Ba	–2.906
Sr^{2+}	$+2e$	⇌	Sr	–2.888
Ca^{2+}	$+2e$	⇌	Ca	–2.866
Na^+	$+e$	⇌	Na	–2.714
Mg^{2+}	$+2e$	⇌	Mg	–2.363
$1/2H_2$	$+e$	⇌	H	–2.25
Be^{2+}	$+2e$	⇌	Be	–1.79
Al^{3+}	$+3e$	⇌	Al	–1.662
V^{2+}	$+2e$	⇌	V	–1.186
Mn^{2+}	$+ 2e$	⇌	Mn	– 1.180
TiO^{2+}	$+ 2H^+ + 4e$	⇌	$Ti + H_2O$	–0.88
Zn^{2+}	$+2e$	⇌	Zn	–0.763
Cr^{3+}	$+3e$	⇌	Cr	–0.744
As	$+ 3H^+ + 3e$	⇌	AsH_3	–0.607
Fe^{2+}	$+ 2e$	⇌	Fe	– 0.440
Cr^{3+}	$+ e$	⇌	Cr^{2+}	– 0.41
Cd^{2+}	$+2e$	⇌	Cd	–0.403
Co^{2+}	$+2e$	⇌	Co	–0.277
H_3PO_4	$+ 2H^+ + 2e$	⇌	$H_3PO_3 + H_2O$	–0.276
V^{3+}	$+e$	⇌	V^{2+}	–0.256
$V(OH)_4{}^+$	$+ 4H^+ + 5e$	⇌	$V + 4H_2O$	–0.254
Ni^{2+}	$+2e$	⇌	Ni	–0.25
Sn^{2+}	$+ 2e$	⇌	Sn	– 0.136
O_2	$+H^+ + e$	⇌	HO_2	–0.13
Pb^{2+}	$+2e$	⇌	Pb	–0.126
$2H^+$	$+ 2e$	⇌	H_2	0.0000
S(*r*)	$+ 2H^+ + 2e$	⇌	H_2S	+0.142
Sn^{4+}	$+ 2e$	⇌	Sn^{2+}	+ 0.154
Cu^{2+}	$+ e$	⇌	Cu^+	+ 0.153
$SO_4{}^{2-}$	$+ 2H^+ + 2e$	⇌	$SO_3{}^{2-} + H_2O$	+ 0.172
Hg_2C1_2	$+ 2e$	⇌	2Hg + 2C1	+ 0.2676

← Oxidising Power decreases (upward)

Reducing Power decreases → (downward)

(*Contd.*)

Half-Reaction				$E°,V$
Cu^{2+}	$+2e$	$\rightleftharpoons$	Cu	+0.337
VO^{2+}	$+2H^+ + e$	$\rightleftharpoons$	$V^{3+} + H_2O$	+0.36
$[Fe(CN)_6]^{3+}$	$+e$	$\rightleftharpoons$	$[Fe(CN)_6]^{4-}$	+0.36
Cu^+	$+e$	$\rightleftharpoons$	Cu	+0.52
I_2	$+2e$	$\rightleftharpoons$	$2I^-$	+0.5355
MnO_4^-	$+e$	$\rightleftharpoons$	MnO_4^{2-}	+0.564
$2Hg^{2+}$	$+2e$	$\rightleftharpoons$	Hg^{2+}	+0.64
O_2	$+2H^+ + 2e$	$\rightleftharpoons$	H_2O_2	+0.6824
Fe^{3+}	$+e$	$\rightleftharpoons$	Fe^{2+}	+0.771
$Hg_2{}^{2+}$	$+2e$	$\rightleftharpoons$	Hg	+0.788
Ag^+	$+e$	$\rightleftharpoons$	Ag	+0.7991
$2NO_3^-$	$+4H^+ + 2e$	$\rightleftharpoons$	$N_2O_4 + 2H_2O$	+0.803
NO_3^-	$+4H^+ + 3e$	$\rightleftharpoons$	$NO + 2H_2O$	+0.96
Pd^{2+}	$+2e$	$\rightleftharpoons$	Pd	+0.987
N_2O_4	$+4H^+ + 4e$	$\rightleftharpoons$	$2NO + 2H_2O$	+1.03
Br_2 (l)	$+2e$	$\rightleftharpoons$	$2Br^-$	+1.0652
Br_2 (aq)	$+2e$	$\rightleftharpoons$	$2Br^-$	+1.087
O_2	$+4H^+ + 4e$	$\rightleftharpoons$	$2H_2O$	+1.229
MnO_2	$+4H^+ + 2e$	$\rightleftharpoons$	$Mn^{2+} + 2H_2O$	+1.23
ClO_4^-	$+2H^+ + 2e$	$\rightleftharpoons$	$ClO_3^- + H_2O$	+1.230
$Cr_2O_7^{2-}$	$+14H^+ + 6e$	$\rightleftharpoons$	$2Cr^{3+} + 7H_2O$	+1.33
Cl_2	$+2e$	$\rightleftharpoons$	$2Cl^-$	+1.3595
PbO_2	$+4H^+ + 2e$	$\rightleftharpoons$	$Pb^{2+} + 2H_2O$	+1.455
Au^{3+}	$+3e$	$\rightleftharpoons$	Au	+1.50
MnO_4^-	$+8H^+ + 5e$	$\rightleftharpoons$	$Mn^{2+} + 4H_2O$	+1.51
$2BrO_3^-$	$+12H^+ + 10e$	$\rightleftharpoons$	$Br_2 + 6H_2O$	+1.51
Ce^{4+}	$+e$	$\rightleftharpoons$	Ce^{3+}	+1.61
Au^+	$+e$	$\rightleftharpoons$	Au	+1.69
MnO_4^-	$+4H^+ + 3e$	$\rightleftharpoons$	$MnO2 + 2H_2O$	+1.695
H_2O_2	$+2H^+ + 2e$	$\rightleftharpoons$	$2H_2O$	+1.776
Co^{3+}	$+e$	$\rightleftharpoons$	Co^{2+}	+1.81
Ag^{2+}	$+e$	$\rightleftharpoons$	Ag^+	+1.98
$S_2O_8^{2-}$	$+2e$	$\rightleftharpoons$	$2SO_4^{2-}$	+2.01
$O_3 + 2H^+$	$+2e$	$\rightleftharpoons$	$O_2 + H_2O$	+2.07
F_2 (g)	$+2e$	$\rightleftharpoons$	$2F^-$	+2.87

← Oxidising Power decreases (upward)

Reducing Power decreases → (downward)

electrons means reduction. A negative value of reduction potential for an electrode denotes that there is little tendency for the electrode reduction. Rather such electrodes show oxidation tendencies, *i.e.*, to lose the electrons. For example, potassium electrode (see Table 23.1) has a potential of –2.925 volts. Thus it has little tendency for reduction. But fluorine electrode has a potential of +2.87 volts. Thus it has strong tendency for reduction and fluorine takes up electrons to change into fluoride ions.

Q. 46. Give some applications of electrochemical series (electrode potentials).

(*Kanpur, 2000; M. Dayanand, 2002*)

Ans. Applications of electrochemical series are described below.

***(i)* Calculation of standard e.m.f. of the cell.** It is possible to determine the e.m.f. of a cell using the electrochemical series. The electrode having the higher reduction potential is taken as cathode and the one having lower value is taken as anode. E.m.f. of the cell E° is then obtained by using the relation.

$$E° = E°_{Cathode} - E°_{Anode}$$

where $E°_{cathode}$ and $E°_{anode}$ are the potentials of the cathode and anode respectively.

***(ii)* Comparison of the reactivity of metals.** The metals which occupy higher position in electrochemical series have lower reduction potentials. This indicates that ions of such metals will not be reduced to metals easily. On the contrary, such metals would be easily oxidised to their ions by losing electrons. Therefore, from the position of a metal in the electrochemical series, it is possible to predict the relative reactivities of metals. A metal having smaller reduction potential (*i.e.*, greater negative reduction potential) can displace metals having greater reduction potential (*i.e.*, smaller negative reduction potential) from the solution of their salts. In other words, a metal occupying higher position in the series can displace the metals lying below it from the solution of their salts.

We conclude that the metals occupying higher positions in the electrochemical series are more reactive in displacing the other metals from the solutions of their salts. For example, zinc lies above copper in the series and hence, it can displace the copper from copper sulphate solution. Copper cannot displace zinc from zinc sulphate solution because it lies below zinc in the series and hence is less reactive.

***(iii)* Predicting the feasibility of a redox reaction.** With the help of electromotive series we can predict whether a given redox reaction will take place or not. From the given equation the substances undergoing oxidation and reduction are identified. The substance undergoing oxidation will act as anode and the substance undergoing reduction will act as cathode. The e.m.f. of this hypothetical cell is calculated as under:

$$\text{E.M.F.} = E°_{cathode} - E°_{anode}$$

If e.m.f. comes out to be +ve the given redox reaction will take place and if e.m.f. comes out to be –ve the given redox reaction will not take place.

Q. 47. Predict whether the following redox reaction is feasible or not under standard conditions:

$$\mathbf{Sn^{+2} + Cu \longrightarrow Sn + Cu^{+2}}$$

Given that $E°_{Sn^{+2}/Sn}$ = – 0.136 volts and $E°_{Cu^{+2}/Cu}$ = 0.34

Ans. The above reaction can be split into two half reactions as given below:

$Cu \longrightarrow Cu^{+2} + 2e^-$ (oxidation) Assuming that the above

$Sn^{+2} + 2e^- \longrightarrow Sn$ (reduction) reaction takes place

Since at copper electrode oxidation takes place, so it acts as anode and at tin electrode reduction takes place so it acts as cathode.

$$\text{EMF} = E^\circ_{cathode} - E^\circ_{anode}$$
$$= E^\circ_{Sn^{+2}/Sn} - E^\circ_{Cu^{+2}/Cu}$$
$$= -0.136 - 0.34$$
$$= -0.476 \text{ volts}$$

Since the e.m.f. comes out to be –ve, the reaction as given above will not take place. However the reverse reaction,

$$\mathbf{Sn + Cu^{+2} \longrightarrow Sn^{2+} + Cu}, \text{ does take place.}$$

Q. 48. Given:

(*i*) $Cr_2O_7^{2-} + 14H^+ + 6e \rightleftharpoons 2Cr^{3+} + 7H_2O$ $E^\circ = 1.33$ V

$S_2O_8^{2-} + 2e \rightleftharpoons 2SO_4^{2-}$ $E^\circ = 2.01$ V

(*ii*) $Cl_2 + 2e \rightleftharpoons 2Cl^-$ $E^\circ = 1.3595$ V

$2I^- \rightleftharpoons I_2 + 2e$ $E^\circ = -0.5355$ V

(*iii*) $Fe^{3+} + e \rightleftharpoons Fe^{2+}$ $E^\circ + 0.771$ V

$Sn^{2+} \rightleftharpoons Sn^{4+} + 2e$ $E^\circ + -0.154$ V

Write balanced equations for the three pairs and also give the appropriate cell potentials.

Ans. (*i*) Since $S_2O_8^{2-}/SO_4^{2-}$ has higher reduction potential, therefore, $S_2O_8^{2-}$ acts as a stronger oxidising agent and the cell equation would be

$$[S_2O_8^{2-} + 2e \rightleftharpoons 2SO_4^{2-}] \times 3 \qquad E^\circ = 2.01 \text{ V}$$
$$2Cr^{3+} + 7H_2O \rightleftharpoons Cr_2O_7^{2-} + 14H^+ + 6e \qquad E^\circ = -1.33 \text{ V}$$

On addition we get,

$$2Cr^{3+} + 3S_2O_8^{2-} + 7H_2O \rightleftharpoons Cr_2O_7^{2-} + 6SO_4^{2-} + 14H^+$$
$$E^\circ_{cell} = 2.01 - 1.33 = 0.68 \text{ V}$$

Note: While multiplying electrochemical equations, E° values are not multiplied and taken as such.

(*ii*) In this case the second equation does not represent standard reduction potential but is written to indicate standard oxidation potential. On addition of the two, we get

$$Cl_2 + 2I^- \rightleftharpoons I_2 + 2Cl^-$$
$$E^\circ_{cell} = 1.3595 - 0.5355 = 0.8240 \text{ V}$$

(*iii*) Multiply eq. (1) by 2 and add to eq. (2)

$$2Fe^{3+} + Sn^{2+} \rightleftharpoons 2Fe^{2+} + Sn^{4+}$$
$$E^\circ_{cell} = 0.771 \text{V} - 0.154 \text{ V} = 0.617 \text{ V}$$

Q. 49. Given: $Ce^{4+} + e \longrightarrow Ce^{3+}$ $E^\circ = 1.61$ V ...(*i*)

$MnO_4^- + 8H^+ + 5e \rightleftharpoons Mn^{2+} + 4H_2O$ $E^\circ = 1.51$V ...(*ii*)

Comment on the feasibility of the reaction.

Ans. E° values indicate that Ce^{4+} is a stronger oxidant than acidic MnO_4^-. Therefore acidic MnO_4^- cannot oxidise Ce^{3+} to Ce^{4+} but theoretically Ce^{4+} can oxidise Mn^{2+} to MnO_4^-. That is, theoretically the following reaction is possible:

$$Ce^{4+} + e \rightleftharpoons Ce^{3+}] \times 5$$
$$Mn^{2+} + 4H_2O \rightleftharpoons MnO_4^- + 8H^+ + 5e$$
$$5Ce^{4+} + Mn^{2+} + 4H_2O \rightleftharpoons 5Ce^{3+} + MnO_4^- + 8H^+$$

The cell potential for such a reaction is:

$E°_{call} = 1.61 - 1.51 = 0.10$ V

Since $E°_{cell}$ is +ve therefore this reaction is theoretically possible. However $E°_{cell} = 0.10$ V has a very small cell potential. Therefore such a reaction would be slow or may not occur under normal conditions.

Q. 50. Given: $Co^{3+} + e \rightleftharpoons Co^{2+}$ **E° = 1.81 V**

$Fe^{3+} + e \rightleftharpoons Fe^{2+}$ **E° = 0.77 V**

Write the appropriate redox reaction and the correct cell potential.

Ans. Based on E° values the correct redox reaction would be:

$Co^{3+} + e \rightleftharpoons Co^{2+}$ E° = 0.181 V

$Fe^{2+} \rightleftharpoons Fe^{3+} + e$ E° = – 0.77 V

Adding to cancel out electrons, we get

$$Co^{3+} + Fe^{2+} \rightleftharpoons Co^{2+} + Fe^{3+}$$

E° = 1.81 – 0.77 = 1.04 V

A positive value E° = 1.04 V indicates that the cell is properly set up.

Q.51. How can we determine the equilibrium constant of a reaction from redox potential?

Ans. The oxidation reduction reactions are reversible in nature and therefore equilibrium is established between oxidising agent and reducing agent. The value of equilibrium constant gives the extent to which the reaction proceeds in the forward direction. Consider the oxidation reduction reaction of Zn and Cu. The cell reaction is:

$$Zn\ (s) + Cu^{2+}\ (aq) \rightleftharpoons Zn^{2+}(aq) + Cu\ (s)$$

Applying law of chemical equilibrium, we have

$$K = \frac{[Cu(s)][Zn^{2+}]}{[Zn(s)][Cu^{2+}]}$$

Concentrations of solids Cu (s) and Zn (s) remain constant and may be assumed equal to 1.

$$K = \frac{[Zn^{2+}]}{[Cu^{2+}]} \quad \text{... } (i)$$

The equilibrium constant for any oxidation reduction reaction is related to $E°_{cell}$ as :

$$E°_{cell} = \frac{2.303\ RT}{nF} \log K_c$$

or

$$E°_{cell} = \frac{0.059}{n} \log K_c \text{ at } 25°C$$

Here $E°_{cell} = E°_{cathode} - E°_{anode}$ and n is the number of electrons involved in the reaction. For the reaction:

$$Zn\ (s) + Cu^{2+}\ (aq) \rightleftharpoons Zn^{2+}\ (aq) + Cu(s)$$

The two-half cell reactions the electrode potentials are:

$$Zn^{2+} + 2e^- \rightleftharpoons Zn\ (s) \qquad E° = -0.763\ V$$

$$Cu^{2+} + 2e^- \rightleftharpoons Cu\ (s) \qquad E° = +0.337\ V$$

$$E°_{cell} = E°_{cathode} - E°_{anode}$$

$$= 0.337 - (-0.763) = 1.10V$$

Now $$E°_{cell} = \frac{0.059}{n} \log K \qquad ...(ii)$$

Substituting for $E°_{cell}$ and n, we have

$$1.10 = \frac{0.059}{n} \log K$$

or $$\log K = \frac{2 \times 1.10}{0.059} = 37.2$$

or $$K = 10^{37..2}$$

Since the value of K is very large, it means that the concentration of Zn^{2+} [Eq (i)] is very large and that of Cu^{2+} is very small. In other words, the reaction virtually goes to completion.

Similarly let us study the oxidation of Fe^{2+} by MnO_4^- ions:

$$MnO_4^- + 8H^+ + 5Fe^{2+} \rightleftharpoons Mn^{2+} + 5Fe^{3+} + 4H_2O$$

$$E°_{cell} = 1.510 - 0.771V = 0.739\ V$$

$$\log K = \frac{5 \times 0.739}{0.059} = 62.6$$

$$K = 10^{62.6}$$

As the value of K is again very large, Fe^{2+} ions are oxidised to a large extent in the presence of acidified MnO_4^- ions.

Q. 52. How are the free energy changes in a reaction related to redox potential?

Ans. The standard electrode potential E°, of the cell is related to the standard Gibbs free energy change, $\Delta G°$ of the reaction by the equation:

$$\Delta G° = -n FE° \qquad ...(i)$$

The free energy change $\Delta G°$ is related to enthalpy change $\Delta H°$ and entropy change of the reaction, $\Delta S°$ at temperature T as :

$$\Delta G° = \Delta H° - T\Delta S° \qquad ...(ii)$$

Since the overall radox reaction is the difference of two half reactions, the standard Gibbs free energy change of the overall reaction is the difference of the standard Gibbs free energies of the two half reactions. The overall reaction is spontaneous or favourable (*i.e.* K > 1) in the direction which gives to the negative value of the resulting overall $\Delta G°$. We choose one-half reaction which has $\Delta G° = 0$ and express all other values relative to it. As in case of electrode potential, the half reaction of reduction of hydrogen ions is chosen for which $\Delta G° = 0$ at all temperatures:

$$2H+ (aq) + 2e^- \longrightarrow H_2(g) \qquad \Delta G°=0$$

This can be used for calculating $\Delta G°$ for redox reaction. For example for the redox reaction,

$$Zn^{2+}\ (aq) + H_2(g) \longrightarrow Zn\ (s) + 2H^+\ (aq) \qquad \Delta G° = +\ 147\ kJ\ mol^{-1}$$

The H^+ reduction half reaction has zero value of Gibbs free energy (according to the convention), therefore,

$$Zn^{2+}(aq) + 2e^- \longrightarrow Zn(s) \qquad \Delta G^\circ = +147 \text{ kJ mol}^{-1}$$

Standard Gibbs free energies may be measured by setting up an electrochemical cell having one hydrogen electrode. The potential difference between the electrodes is measured. From this, Gibbs free energy is calculated using the relation:

$$\Delta G^\circ = -nFE^a \qquad ...(iii)$$

The potential E° is called standard potential.

Since the standard Gibbs energy is the difference of ΔG° for the half reactions, E° for the overall reaction is also the difference of the two standard potentials of the reduction half reactions contributing to the overall reaction. Thus, from the half reactions:

$$2H^+(aq) + 2e^- \longrightarrow H_2(g) \qquad E^\circ = 0.0 \text{ V}$$

$$Zn^{2+}(aq) + 2e^- \longrightarrow Zn(s) \qquad E^\circ = -0.76\text{V}$$

The difference of the two-half cells is

$$2H^+(aq) + Zn(s) \longrightarrow H_2(g) + Zn^{2+}(aq) \qquad E^\circ = 0-(-0.76) = 0.76 \text{ V}$$

Since $\Delta G^\circ = -nFE^\circ$, the reaction is favourable if $E^\circ > 0$. For the above reaction, $E^\circ > 0$ therefore, zinc thermodynamically reduces H^+ ions under standard condition (acidic solution, $pH = 0$ and unit activity of Zn^{2+}). All the couples having negative electrode potentials can reduce H^+ ions.

Q. 53. Trace the different steps involved in a redox cycle.

Ans. The electrode potential of a metal depends upon a number of factors. The conversion of M to M^+ in aqueous solution may be considered to occur in a series of steps:

(1) Sublimation. This converts solid metal to the gaseous state. The energy required for this process is enthalpy of sublimation represented by ΔH°_s.

(2) Ionization. The energy equal to ionization energy is required to remove electron from the gaseous atom. It is represented by I.

(3) Hydration. When gaseous ion is formed, it gets hydrated in aqueous solution. The amount of energy released is enthalpy of hydration represented by ΔH_{aq}.

For the formation of a cation in aqueous solution, we may consider a cycle as shown below:

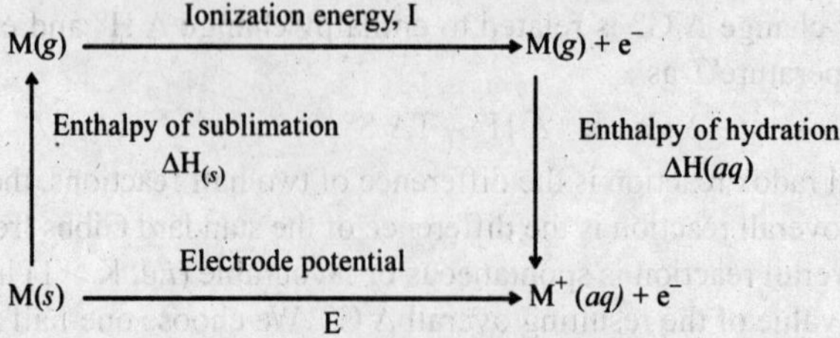

The overall, energy change E which gives electrode potential may be written as:

$$E = \Delta H_s + I + \Delta H(aq)$$

The enthalpy of sublimation and the ionization energy are positive because energy required to sublime and ionise while enthalpy of hydration is negative because energy is evolved.

If we are considering formation of an anion in solution the ionization energy may be replaced by electron affinity, EA when the gaseous atoms goes to gaseous anion. Thus, for the anion, the following relation will hold:

$$E = \Delta H_s + EA + \Delta H\,(aq)$$

It may be realised that the electrode potential values as obtained above are from enthalpy changes only. We need entropy values also for calculating Gibbs free energy changes.

When similar systems are being compared (e.g. $Fe^{2+} | Fe$ and $Zn^{2+} | Zn$) the entropy changes will be very similar in each so that the difference in entropy contribution to the systems are negligible.

Therefore, for approximate treatments, the entropy changes may be neglected especially for comparison purposes. Thus, we may use the approximate relation:

$$-nFE° = \Delta H°$$

Variation in electrode potential in a group

Alkali metals have low melting points and hence low enthalpies of sublimation and also the atoms are large and therefore, have low ionization energies. Thus, the electrode potentials are high and the metals are very reactive. The decrease in electrode potential takes place from sodium to cesium due to decrease in ionization energy and sublimation energy with increasing size down the group. Lithium shows anomalous behaviour because of very large hydration energy released for the small cation.

In case of halogens, the sum of the atomization energy (half the bond energy of X_2 molecules) and the electron affinity is approximately constant from F to I. The decrease in electrode potential from fluorine to iodine is due to decrease in hydration energy from F to I as the size of the ion increases.

Most transition metals have high melting points and therefore, their sublimation energies are high. Because of their small size, their ionization energies are also high. Therefore, they have low values of electrode potential and such metals have less tendency to lose electrons.

Q. 54. Discuss redox stability in a solvent.

The ion or molecule may undergo oxidation or reduction depending upon the type of the solvent. For example, strong oxidizing agents will oxidize water to O_2 while strong reducing agents will reduce water to H_2. Thus, strong oxidizing and reducing agents may not exist in aqueous solutions. As a result, water may be oxidized or reduced depending upon the nature of the ions or molecules present.

Water may be reduced to hydrogen and therefore, act as an oxidizing agent according to the reaction :

$$2H_2O + 2e^- \longrightarrow H_2 + 2OH^- \quad E° = -1.23V - (0.059V)\,pH$$

The electrode potential depends upon the *p*H of the solution. In different solutions, the following half reactions occur:

Reduction of Water

Neutral solution: $H_2O + e^- \longrightarrow OH^- + \frac{1}{2}H_2$ $E° = -0.414V$

1.0 M acid solution: $H_3O^+ + e^- \longrightarrow H_2O + \frac{1}{2}H_2$ $E° = 0.00V$

1.0 M basic solution: $H_2O + e^- \longrightarrow OH^- + \frac{1}{2}H_2$ $E° = -0.828V$

Oxidation of water

Neutral solution: $\frac{1}{2}O_2 + 2H^+ + 2e^- \longrightarrow H_2O$ $E° = +0.185$ V

1.0 M acid solution: $\frac{1}{2}O_2 + 2H^+ + 2e^- \longrightarrow H_2O$ $E° = +1.229$V

1.0 M basic solution : $\frac{1}{2}O_2 + H_2O + 2e^- \longrightarrow 2OH^-$ $E° = +0.401$V

From the above data minimum reduction potentials required to oxidize water to dioxygen should be more than + 0185 V in neutral solution, more than +1.229V in 1.0 M acid solution and more than +0.401V in 1.0 M basic solution.

Similarly, the half reactions with reduction potentials less than 0.00V (*i.e.* negative) in acid solution, less than –0.414 V (E° < – 0.414V) in neutral solution and less than –0.828 V (E° < – 0.828 V) in basic solution, should reduce water to H_2.

Reactions of metals with water or aqueous acids are usually the oxidation of metal by water or hydrogen ions. The overall reaction may be represented as:

$$M\,(s) + H_2O\,(l) \longrightarrow M^+(aq) + \frac{1}{2}H_2\,(g) + OH^-(aq)$$

or

$$M\,(s) + H^+\,(aq\ ) \longrightarrow M^+(aq) + \frac{1}{2}H_2\,(g)$$

In view of the reduction potentials given above, these reactions are favourable for metals of s-block and first transition series from Group 4 to Group 7.

As for as oxidation is concerned, only a few oxidizing agents can oxidize water to give appreciable of O_2 evolution. The standard reduction potentials greater than +1.23 V occur for redox couples such as acidified dichromate couple, $Cr_2O_7^{2-} \mid Cr^{3+}$ (1.38V) and acidified permanganate couple, $MnO_4^- \mid Mn^{2+}$ (+1.51 V).

Representation of potential data

Electrode potentials data help us to understand the relative oxidising or reducing abilities of couples arranged in same diagram. They also provide information about the oxidation states of an element in a convenient way. These are commonly known as diagrams. Three types of diagrams are described below.

Q. 55. What are Latimer diagrams? Explain taking the case of iron.

These are the simplest diagrams which give the reduction potentials for the half reactions corresponding to different oxidation states that an element exhibits in a single reduction potential diagram. When an element can exist in several oxidation states, it is convenient to display the reduction potentials corresponding to various half reactions in a diagrammatic form. In this diagram,

(i) highest oxidation state is written at the left and lowest oxidation state is written at the right and species such as electrons, H^+ and H_2O are omitted.

(ii) The value of the standard potential (in volts) is written over a horizontal line connecting species with the element in different oxidation states.

Latimer diagram of Iron

Iron exists in different oxidation states like Fe (VI) in FeO_4^{2-} , Fe (III) in Fe^{3+}, Fe (II) in Fe^{2+} and Fe (0) in Fe. The half cell reactions involving these changes are:

$$Fe^{2+} + 2e^- \longrightarrow Fe(s) \qquad E° = -0.47V$$

$$Fe^{3+} + 3e^- \longrightarrow Fe(s) \qquad E° = -0.057V$$

$$Fe^{3+} + e^- \longrightarrow Fe^{2+} \qquad E° = 0.77V$$

$$FeO_4^{2-} + 3e^- + 8H^+ \longrightarrow Fe^{3+} + 4H_2O \qquad E° = +2.20V$$

The Latimer diagram for iron giving all the above information can be represented as :

$$\begin{array}{ccccccc} VI & & III & & II & & 0 \\ FeO_4^{2-} & \xrightarrow{+2.20} & Fe^{3+} & \xrightarrow{0.77} & Fe^{2+} & \xrightarrow{-0.47} & Fe \end{array}$$

$$Fe^{3+} \xrightarrow{-0.057} Fe$$

The species such as H_2O, H^+, OH^- and e^- are omitted in the diagram. The oxidation numbers are written under or over the species. Potential for $Fe^{3+} \longrightarrow Fe$, according to the above diagram should be (0.77 –0.44) *i.e* 0.30 V. Actually, it is shown as – 0.057V. It may be noted that the potentials may not be added for half reactions because the electrons may not balance. In such cases potential are converted into free energies using the relation $\Delta G° = -nFE°$ (where n is the number of electrons involved in the reaction and F is the Faraday). Since Gibbs free energy is a thermodyanmic function, the free energies may be added and the resulting Gibbs free energy is converted back to E° value as explained below:

$$Fe^{3+} + e^- \longrightarrow Fe^{2+} \qquad E° = 0.77V$$

$$\Delta G° = -1\,(1+0.77)\,F = -0.77\,F$$

$$Fe^{2+} + 2e^- \longrightarrow Fe \qquad E° = -0.47V$$

$$\Delta G° = -2\,(-0.47)\,F = -0.94\,F$$

Adding

$$Fe^{3+} + 3e^- \longrightarrow Fe \qquad \Delta G° = (-0.77+0.94)\,F \text{ or } +0.17F$$

Now, E° for the above reaction , $Fe^{3+} + 3e^- \longrightarrow Fe$ may be calculated by converting free energy change into potential as follows:

$$E° = \frac{\Delta G°}{-nF} = \frac{+0.17F}{-3F} = -0.057$$

Q. 56. How do we calculate standard potential for non-adjacent species?

Ans. The basis of calculation is that the overall ΔG° for two successive steps is the sum of the individual values that is

$$\Delta G° \text{ (overall)} = \Delta G°_1 + \Delta G°_2$$

Where $\Delta G°_1$ and $\Delta G°_2$ are individual free energy changes for the two steps. We convert individual E° values to ΔG° values by multiplying with relevant factor $-n$F and add them. Then the sum is converted back to E° for the non-adjacent couple by dividing by $-n$F for the overall electron transfer. Thus,

$$-nFE° = -n_1FE_1° - n_2FE_2°$$

where $$n = n_1 + n_2$$

$$-nFE° = -(n_1E°_1 + n_2E°_2)F \quad \text{or} \quad E° = \frac{n_1E°_1 + n_2E°_2}{n}$$

$$\text{or } E° = \frac{n_1E°_1 + n_2E°_i}{n_1 + n_2}$$

or $$E° = \frac{\sum n_i E_2^o}{\sum n_i}$$

This general procedure is followed to calculate the electrode potential for any non-adjacent species.

The electrode potential for the reduction of FeO_4^{2-} to Fe^{3+} is + 2.20V.

$$FeO_4^{2-} \xrightarrow{2.20} Fe^{3+}$$

As $\Delta G° = -nFE°$, $\Delta G°$ for this change will be large and negative. Thus, the reaction is thermodynamically possible and **FeO_4^{2-} can be easily reduced.** In order words, **FeO_4^{2-} acts as a strong** oxidizing agent.

The electrode potentials are measured on a scale with

$$H^+ (aq) + e^- \longrightarrow \frac{1}{2}H_2 \qquad E° = 0.0V$$

We know that reactions with negative value for E° are more strongly reducing than hydrogen and we call them as strongly reducing. The E° values for the changes $Fe^{3+} \longrightarrow Fe$ and $Fe^{2+} \longrightarrow Fe$ are both negative so that $\Delta G°$ is positive. Thus, neither Fe^{3+} nor Fe^{2+} has any tendency to get reduced to Fe.

Latimer diagram for chlorine in acidic solution

VII	V	III	I	0	–I

$$ClO_4^- \xrightarrow{+1.20} ClO_3^- \xrightarrow{+1.18} HClO_2 \xrightarrow{+1.65} HClO \xrightarrow{+1.67} Cl_2 \xrightarrow{+1.36} Cl^-$$

Latimer diagram of redox potentials of Cl_2 (volts) at pH = 0

Information conveyed by Latimer diagram of chlorine

(i) $ClO^-_4 + 2H^+ 2e^- \longrightarrow ClO_3^- + H_2O$ $\qquad E° = + 1.20$ V

Similarly

(ii) $ClO_3^- \xrightarrow{1.18} HClO_2$ represents the following half reaction:

$2\,ClO_3^- + 6H^+ + 4e^- \rightarrow 2HClO_2 + 2H_2O$ $\qquad E° = +1.18V$

(iii) $HClO_2 \xrightarrow{1.65} HClO$ represent the following half reaction:

$HClO_2 + 2H^+ + 2e^- \longrightarrow HClO + H_2O$ $\qquad E° = +1.65V$

(iv) $HClO \xrightarrow{1.67} Cl_2$+ represents the following half reaction:

$2HClO + 2H^+ + 2e^- \longrightarrow Cl_2 + 2H_2O$ $\qquad E° = 1.67$ V

(v) $Cl_2 \xrightarrow{1.36} Cl^-$ represents the following half reaction:

$Cl_2 + 2e^- \longrightarrow 2Cl^-$ $\qquad E° = + 1.36V$

The half reactions have been balanced by H^+ or H_2O as explained earlier. Latimer diagram for chlorine in basic medium (pH = 14)

VII	V	III	I	0	–I

$$ClO_4^- \xrightarrow{+0.36} ClO_3^- \xrightarrow{+0.33} ClO_2^- \xrightarrow{+0.66} OCl^- \xrightarrow{+0.42} \frac{1}{2}Cl_2 \xrightarrow{+1.36} Cl^-$$

The equations in basic medium may be balanced by the appropriate number of H_2O and OH^- as explained in the previous section.

(i) $ClO^-_4 \xrightarrow{+0.36} ClO_3^-$ represents the following half reaction:

$$ClO^-_4 + H_2O + 2e^- \longrightarrow ClO_3^- + 2OH^- \qquad E° = +0.36\ V$$

(ii) $$ClO^-_3 + H_2O + 2e^- \longrightarrow ClO_3^- + 2OH^- \qquad E° = +0.33\ V$$

and so on.

Q. 57. What is meant by Disproportionation of an ionic species? Explain by taking some examples.

Ans. *Disproportionation is a process in which one oxidation state decomposes forming some ions in a higher oxidation state and some in a lower oxidation state.* One of the most important information which can be obtained from Latimer diagram is whether any of the oxidation states are unstable with respect to disproportionation. This can be understood in terms of electrode potentials.

Consider the disproportionation reaction:

$$2M^+ (aq) \longrightarrow M (s) + M^{2+} (aq) \qquad ...(i)$$

The reaction will be spontaneous only if E° > 0. Let us express the overall reaction as the difference of two-half reactions.

$$M^+ (aq) + e^- \longrightarrow M (s) \qquad E° (R) \qquad ...(ii)$$

$$M^{2+} (aq) + e^- \longrightarrow M^+ (aq) \qquad E° (L) \qquad ...(iii)$$

Eq (*i*) can be obtained by subtracting eq. (*iii*) from eq (*ii*)

The deignation R and L refer to relative positions right and left of the couple in the Latimer diagram. More highly oxidized species is always placed on the left. The standard potential for the overall reaction is:

$$E° = E°(R) - E° (L)$$

It will be positive only if E° (R) > E° (L).

Thus, a *species has a thermodynamic tendency to disproportionate into its neighbours if the potential on the right of the species in the Latimer diagram is higher than the potential on the left.*

Latimer diagram of iron shows that reduction potential values become progressively less positive (or more negative) on moving left to right and therefore, Fe^{3+} and Fe^{2+} are stable with respect to disproportionation, that is, they do not disproportionate.

Consider reduction potential diagram for copper in acid solution. The Latimer diagram is:

$$\begin{array}{ccccc} II & & I & & 0 \\ Cu^{2+} & \xrightarrow{+0.15} & Cu^{+} & \xrightarrow{+0.50} & Cu \end{array}$$

$$Cu^{2+} \xrightarrow{+0.35V} Cu$$

The potential and hence the free energy released when Cu^{2+} is reduced to Cu^+ are both very small and therefore Cu^{2+} is not an oxidizing agent but is stable. On moving from left to right the potentials become more positive. Hence the species in the middle is unstable to disproportionation. In other words, it behaves both as a self oxidizing and self reducing agent because it is energetically favourable. This can also be shown as:

$$Cu^+ + e^- \longrightarrow Cu \qquad E° = +0.50\ V$$

$$Cu^{2+} + e^- \longrightarrow Cu^+ \qquad E° = +0.15\ V$$

Overall $$2Cu^+ \longrightarrow Cu + Cu^{2+} \qquad E° = 0.35V$$

Since E° > 0, the reaction takes place i.e, Cu^+ disproportionates to Cu^{2+} and Cu in solution and hence Cu^+ is only found in the solid state.

Consider Latimer diagram for O_2.

$$\overset{0}{O_2} \xrightarrow{+0.682} \overset{-I}{H_2O_2} \xrightarrow{+1.776} \overset{-II}{H_2O}$$
$$O_2 \xrightarrow{+1.229} H_2O$$

In this case, on moving from left to right, the reduction potential increases and therefore H_2O_2 is unstable with respect to disproportionation. This can be verified by considering the two-half reactions:

$$H_2O_2(aq) + 2H^+ + 2e^- \longrightarrow 2H_2O(l) \qquad E° = 1.776 \text{ V} \qquad ...(i)$$

$$O_2(g) + 2H^+ + 2e^- \longrightarrow H_2O_2\ (aq) \qquad E° = +\,0.682 \text{ V} \qquad ...(ii)$$

Subtracting (*ii*) from (*i*), we get

$$2H_2O_2(aq) \longrightarrow 2H_2O(l) + O_2\ (g)\ E° = 1.094\text{V}$$

Since E° is more than 0, *the disproportion takes place.*

Consider the Latimer diagram for Cl_2 in basic medium.

$$\overset{VII}{ClO_4^-} \xrightarrow{+0.36} \overset{V}{ClO_3^-} \xrightarrow{+0.33} \overset{III}{ClO_2} \xrightarrow{+0.66} \overset{I}{OCl^-} \xrightarrow{+0.42} \overset{0}{\tfrac{1}{2}Cl_2} \xrightarrow{+1.36} \overset{-I}{Cl^-}$$

From the above diagram we can infer that both ClO_2 and Cl_2 should disproportionate. ClO_2 should disproportionates into ClO_3^- and OCl^- and Cl_2 disproportionates into OCl^- and Cl^-.

The diagram indicates that OCl^- should be stable with regards to disproportionation. But it is not true. Now, if we calculate electrode potentials for the changes ClO_3^- to OCl^- and OCl^- to Cl^-, by the procedure explained earlier for non-adjacent species, we get the Latimer diagram as:

$$\overset{VII}{ClO_4^-} \xrightarrow{+0.36} \overset{V}{ClO_3^-} \xrightarrow{+0.33} \overset{III}{ClO_2} \xrightarrow{+0.66} \overset{+I}{OCl^-} \xrightarrow{+0.42} \overset{0}{\tfrac{1}{2}Cl_2} \xrightarrow{+1.36} \overset{-I}{Cl_2}$$
$$ClO_3^- \xrightarrow{+0.50} OCl^- \xrightarrow{+0.88} Cl_2$$

From this diagram it is clear that the potentials around OCl^- and ClO_3^- do not decrease from left to right as:

$$ClO_3^- \xrightarrow{+0.50} OCl^- \xrightarrow{+0.88} Cl^-$$

Therefore, OCl^- is unstable to disproportionation and it disproportionates into ClO_3^- and Cl^-

Similarly, for the sequence,

$$ClO_4^- \xrightarrow{+0.36} ClO_3^- \xrightarrow{+0.50} OCl^-$$

We find that ClO_3^- *is unstable with respect to disproportionation into* ClO_4^- *and* OCl^-.

Another utility of Latimar diagrams is in predicting the products of the reactions in which the element have several oxidation states. For example, consider the reaction between acidified solution of $KMnO_4$ and KI. The reduction potential diagrams are:

VII VI IV III II 0

$$MnO_4^- \xrightarrow{+0.56} MnO_4^{2-} \xrightarrow{+2.26} MnO_2 \xrightarrow{+0.95} MnO^{3+} \xrightarrow{+1.51} Mn^{2+} \xrightarrow{-1.19} Mn$$

$MnO_4^- \xrightarrow{+1.69} MnO_2$; $MnO_2 \xrightarrow{+1.23} Mn^{2+}$; $MnO_4^- \xrightarrow{+1.51} Mn^{2+}$

VII V III 0 −1

$$IO_4^{-1} \xrightarrow{+165} IO_3^- \xrightarrow{+1.34} HOI^* \xrightarrow{+1.44} \frac{1}{2} I_2(s) \xrightarrow{+0.54} I^-$$

$H_5IO_6 \xrightarrow{+160} IO_3^-$; $IO_3^- \xrightarrow{+1.19} \frac{1}{2} I_2(s)$; $HOI \xrightarrow{+0.99} I^-$

We can rule out certain species that will not be obtained as products because they have a tendency for disproportionation. We observe that there are three species, HOI, Mn^{3+} and MnO_4^{2-} in this diagram which are unstable towards disproportionation. Therefore, these are not formed. The half reaction, $Mn^{2+} \longrightarrow Mn$ has a large negative value and therefore, $\Delta G°$ will have large positive value for this. Hence this step can be ignored. Thus, the reduction potential diagram may be simplified as:

VII V IV 0 II –I

$$MnO_4^- \xrightarrow{+1.70} MnO_2 \xrightarrow{+1.23} Mn^{2+}$$

$$IO_4^- \xrightarrow{+1.65} IO_3^- \xrightarrow{+1.19} I_2(s) \xrightarrow{+0.54} I^-$$

$H_5IO_6 \xrightarrow{+1.60} IO_3^-$

The reduction potentials of some common couples are given in Table 3.2

Table Redox potential data for some common couples

In acid solution

1. $Cr_2O_7^{2-} \xrightarrow{1.33} Cr^{3+} \xrightarrow{-0.41} Cr^{2+} \xrightarrow{-0.91} Cr$

$Cr^{3+} \xrightarrow{-0.74} Cr$; $CrCl_2^+ \xrightarrow{-0.76} Cr$; $Cr_2O_7^{2-} \xrightarrow{0.295} Cr$

2. $MnO_4^- \xrightarrow{0.56} MnO_4^{2-} \xrightarrow{2.26} MnO_2 \xrightarrow{0.95} Mn^{3+} \xrightarrow{1.51} Mn^{2+} \xrightarrow{-1.19} Mn$

$MnO_4^{2-} \xrightarrow{(2.0)} MnO_3^- \xrightarrow{(2.5)} MnO_2$; $MnO_2 \xrightarrow{1.23} Mn^{2+}$

3. $TiO^{2+} \xrightarrow{(0.1)} Ti^{3+} \xrightarrow{-0.37} Ti^{2+} \xrightarrow{(-1.63)} Ti$

$TiO^{2+} \xrightarrow{(-0.86)} Ti$

Contd.

4. $TcO_4^- \xrightarrow{(0.7)} TcO_2 \xrightarrow{(0.6)} Tc^{2+} \xrightarrow{(0.4)} Tc$; $TcO_4^- \xrightarrow{(0.6)} Tc$; $TcO_2 \xrightarrow{(0.5)} Tc$

5. $V(OH)_4^+ \xrightarrow{1.0} VO^{2+} \xrightarrow{0.36} VO^{3+} \xrightarrow{-0.25} V^{2+} \xrightarrow{-1.2} V$; $V(OH)_4^+ \xrightarrow{-0.253} V$

6. $H_2MoO_4 \xrightarrow{(+0.4)} MoO_2^+ \xrightarrow{(+0.0)} Mo^{3+} \xrightarrow{-0.2} Mo$; $H_2MoO_4 \xrightarrow{(0.0)} Mo$

7. $RuO_4 \xrightarrow{1.00} RuO_4^- \xrightarrow{0.59} RuO_4^{2-} \xrightarrow{0.3} Ru^{2+} \xrightarrow{0.45} Ru$

8. $Ti^{3+} \xrightarrow{1.25} Ti^+ \xrightarrow{0.3363} Ti$

9. $BrO_4^- \xrightarrow{1.82} BrO_3 \xrightarrow{1.49} HOBr \xrightarrow{1.59} Br_2 \xrightarrow{1.07} Br^-$; $BrO_3 \xrightarrow{1.52} Br_2$; $Br_2(aq) \xrightarrow{1.087} Br^-$

10. $ReO_4^- \xrightarrow{(0.4)} ReO_3 \xrightarrow{(0.6)} ReO_2 \xrightarrow{(0.2)} Re^{3+} \xrightarrow{(0.3)} Re \xrightarrow{(-0.4)} Re^-$; $ReO_4^- \xrightarrow{0.51} ReO_2$; $ReO_2 \xrightarrow{0.252} Re$

11. $H_5IO_6 \xrightarrow{1.7} IO_3^- \xrightarrow{1.14} HOI \xrightarrow{1.45} I_2 \xrightarrow{0.535} I^-$; $IO_3^- \xrightarrow{1.20} I_2$

12. $FeO_4^{2-} \xrightarrow{1.9} Fe^{3+} \xrightarrow{0.771} Fe^{2+} \xrightarrow{-0.440} Fe^\circ$

13. $Hg^{2+} \xrightarrow{0.920} Hg_2^{2+} \xrightarrow{0.789} Hg$; $Hg^{2+} \xrightarrow{0.854} Hg$

14. $ClO_4^- \xrightarrow{1.19} ClO_3^- \xrightarrow{1.21} HClO_2 \xrightarrow{1.645} HOCl \xrightarrow{1.63} Cl_2 \xrightarrow{1.3595} Cl^-$; $ClO_3^- \xrightarrow{1.15} ClO_2 \xrightarrow{1.27} HClO_2$; $ClO_3^- \xrightarrow{1.47} Cl_2$

15. $Cd^{2+} \xrightarrow{< -0.6} Cd_2^{2+} \xrightarrow{> -0.2} Cd$; $Cd^{2+} \xrightarrow{-0.403} Cd$

Contd.

16. $In^{3+} \xrightarrow{-0.45} In_2^{4+} \xrightarrow{-0.35} In^{+} \xrightarrow{-0.25} In$; $In^{3+} \xrightarrow{-0.342} In$

17. $Ga^{3+} \xrightarrow{-0.65} In_2^{4+} \xrightarrow{-0.45} Ga$; $Ga^{3+} \xrightarrow{-0.53} Ga$

In Basic Solution

1. $BrO_3^- \xrightarrow{0.54} BrO^- \xrightarrow{0.45} Br_2^\circ \xrightarrow{1.07} Br^-$; $BrO_3^- \xrightarrow{0.61} Br^-$; $BrO^- \xrightarrow{0.76} Br^-$

2. $FeO_4^{2-} \xrightarrow{0.9} Fe(OH)_3 \xrightarrow{-0.56} Fe(OH)_2 \xrightarrow{-0.887} Fe$

3. $Cu(OH)_2 \xrightarrow{-0.08} Cu_2O \xrightarrow{0.358} Cu$

4. $ClO_4^- \xrightarrow{0.36} ClO_3^- \xrightarrow{0.33} ClO_2^- \xrightarrow{0.66} ClO^- \xrightarrow{0.42} Cl_2 \xrightarrow{1.3595} Cl^-$; $ClO_3^- \xrightarrow{-0.50} ClO_3^- \xrightarrow{1.16} ClO_2^-$; $ClO_3^- \xrightarrow{0.50} ClO^-$; $ClO^- \xrightarrow{0.88} Cl^-$

5. $H_3IO_6^{2-} \xrightarrow{(0.7)} IO_3^- \xrightarrow{0.14} IO^- \xrightarrow{0.45} I_2(s) \xrightarrow{0.535} I^-$; $IO^- \xrightarrow{-0.49} I^-$

6. $MnO_4^- \xrightarrow{0.564} MnO_4^{2-} \xrightarrow{0.60} MnO_2 \xrightarrow{0.2} Mn(OH)_3 \xrightarrow{0.1} Mn(OH)_2 \xrightarrow{-1.55} Mn$; $MnO_4^{2-} \xrightarrow{0.34} MnO_3^- \xrightarrow{0.84} MnO_2$; $MnO_4^- \xrightarrow{0.588} MnO_2$; $MnO_2 \xrightarrow{0.05} Mn(OH)_2$

7. $CrO_4^{2-} \xrightarrow{-0.13} Cr(OH)_3 \xrightarrow{-1.1} Cr(OH)_2 \xrightarrow{-1.4} Cr$; $Cr(OH)_3 \xrightarrow{-1.3} Cr$; $CrO_2^- \xrightarrow{-1.2} Cr$

8. $CoO_2 \xrightarrow{0.7} Co(OH)_3 \xrightarrow{0.14} Co(OH)_2 \xrightarrow{-0.72} Co$

Example 58. The Latimer diagram is

$$ClO^- \xrightarrow{+0.42} Cl_2 \xrightarrow{+1.36} Cl^-$$

Calculate the value for the reduction of ClO^- to Cl^- in aqueous basic medium.

Solution. The half reactions may be written as:

$$ClO^- + H_2O + e^- \longrightarrow \frac{1}{2}Cl_2 + 2OH^- \qquad E° = +0.42$$

$$\frac{1}{2}Cl_2 + e^- \longrightarrow Cl^- \qquad E° = +1.36$$

Here $n_1 = 1$ and $n_2 = 1$

∴ E° for ClO^- to Cl^- is calculated by the formula

$$E° = \frac{n_1 E_1^\circ + n_2 E_2^\circ}{n_1 + n_2}$$

Subustituting the values, we have

$$E° = \frac{(1 \times 0.42) + (1 \times 1.36)}{2} = +0.89 \text{ V}$$

Q. 59. Latimer Diagram for Iron is:

$$Fe^{3+} \xrightarrow{0.771 \text{ V}} Fe^{2+} \xrightarrow{0.44 \text{ V}} Fe \qquad (Fe^{3+} \to Fe: -0.04 \text{ V})$$

How do we derive the potential for $Fe^{3+} + 3e^- \longrightarrow Fe$

Ans. Convert E° values into ΔG° values = $-nFE°$ or $E° = -\Delta G°/nF$)

$$\Delta G°_{net} = -(1)(F)(+0.771 \text{ V}) + [-2(F)(-0.44 \text{ V})]$$
$$= +0.109 (F) \text{ V}$$
$$E°_{net} = [+0.109 (F)/3F$$
$$= -0.0363\text{V}$$

Q. 60. The Latimer diagram for manganese system is:

$$\mathbf{MnO_4^- \xrightarrow{0.56} MnO_4^{2-} \xrightarrow{2.26} MnO_2 \xrightarrow{0.95} Mn^{3+} \xrightarrow{0.51} Mn^{2+}}$$

Calculate the E° value for $MnO_4^{2-} \mid Mn^{2+}$ couple.

Ans. The process ocurs as:

$$MnO_4^{2-} \xrightarrow[n=2]{2.26} MnO_2 \xrightarrow[n=1]{0.95} Mn^{3+} \xrightarrow[n=1]{0.51} Mn^{2+} \qquad (MnO_4^{2-} \to Mn^{2+}: E° = ?,\ n = 4)$$

As we know

$$\Delta G° \text{ (Overall)} = \Delta G_1° + \Delta G_2° + \Delta G_3°$$

$$E° = \frac{n_1 E°_1 + n_2 E°_2 + n_3 E°_3}{n}$$

$$= \frac{2 \times 2.26 + 1 \times 0.95 + 1 \times 0.51}{4}$$

$$= 1.75 \text{ V}$$

Q. 61. Show that disproportionation of H_2O_2 into O_2 and H_2O is spontaneous under acidic conditions. Given that

$$O_2 \xrightarrow{+0.70} H_2O_2 \xrightarrow{+1.76} H_2O$$

Solution. The disproportionation of H_2O_2 can be written as:

$$2H_2O_2 \longrightarrow 2H_2O + O_2$$

The reaction may be split into two half reactions as:

$$H_2O_2 + 2H^+ + 2e^- \longrightarrow 2H_2O \qquad E° = +1.76\ V$$

$$O_2 + 2H^+ + 2e^- \longrightarrow H_2O_2 \qquad E° = 0.70\ V$$

Subtracting the two equations, we have

$$2H_2O_2 \longrightarrow 2H_2O + O_2 \qquad E° = 1.06\ V$$

Since E° is positive, the reaction is spontaneous.

Thus, H_2O_2 disproportionates into H_2O and O_2

Q. 62. What are Frost diagrams? What is their utility?

A **Frost diagram** of an element is a **plot of product of oxidation state (N) and reduction potential (E°) against oxidation number of the element (N).** For example, for an element X, it is plot of NE° for the couple X(N) | X (0) against the oxidation number, N.

$$X(N) + Ne^- \longrightarrow X(0) \quad E°$$

We know that NE° proportional to the standard reaction Gibbs energy for the conversion of the species X(N) to the element (NE° = – Δ G°/F). A Frost diagram can also be regarded as *a plot of standard reaction Gibbs energy against oxidation number* (Fig. 23.6).

In general, the following observations may be made from Frost diagram:

1. Frost diagram is regarded as a plot of Gibbs free energy against oxidation number. Therefore, **the most stable oxidation state of an element corresponds to the species that lies lowest in the Frost diagram.**

For example, the most stable oxidation state in a Frost diagram shown below corresponds to point B.

2. The slope of the line joining any two points in the Frost diagram is equal to the standard potential of the couple formed by the two species represented by the points. We can easily predict the spontaneity of the reaction between any two couples by comparing the slopes of the corresponding lines.

- More positive the slope, the greater is the oxidizing power of the couple.
- More negative the slope, the greater is the reducing power of the couple.

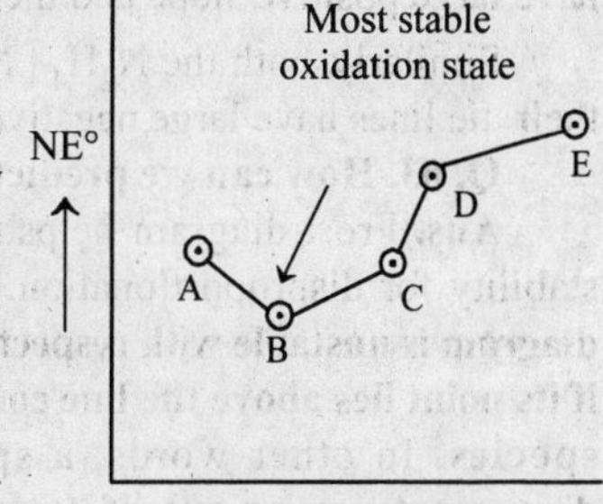

Fig. 23.6. Frost diagram

These diagram are plotted with reference to the $H^+ | H_2$ or $OH^- | H_2$ couple having slope of zero. The Frost diagram for nitrogen under basic conditions is shown in Fig. 23.7.

(*i*) The reducing agent in the couple with less positive slope (i.e. less positive E°) for example NO_2^- is likely to undergo oxidation.

(*ii*) The oxidising agent in the couple with more positive slope (i.e. more positive E°) for example NO_3^- is likely to undergo **reduction**.

(*iii*) Steeper the line joining two points in a Frost diagram, higher is the potential of the corresponding couple.

(*iv*) Predicts disproportionation and comproportionation.

For example, a steep slope connecting HNO_3 (NO_3^-) to lower oxidation state in Fig indicates that HNO_3 is good oxidizing agent.

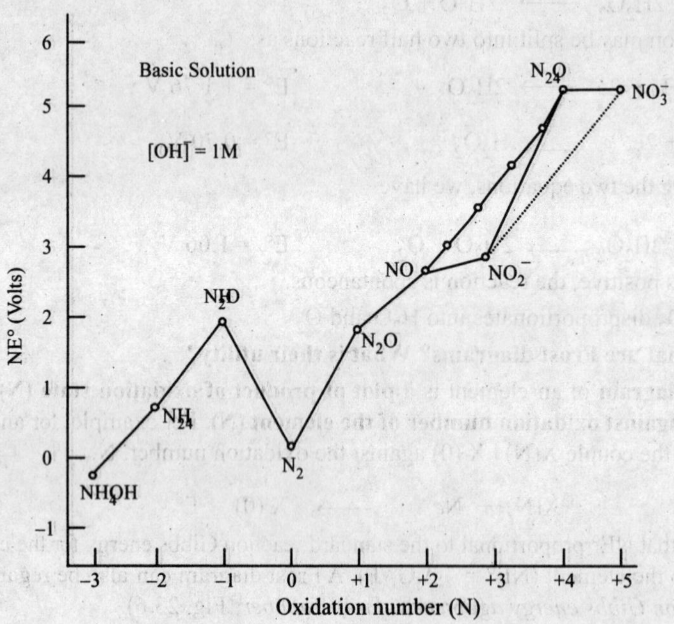

Fig. 23.7. Frost diagram for nitrogen

The oxidizing and reducing ability of any couple can be easily predicted from the Frost diagram on the basis of sign and magnitude of the slope. The couples $N_2O_4 \mid NO_2^-$, $NO \mid N_2O$ and $NO_3^- \mid NO_2^-$ have large positive slope and therefore are **strong oxidising agents**.

Similarly, both the $N_2H_4 \mid N_2$ and $NH_2OH \mid N_2$ couples are strong reducing couples because their tie lines have large negative slopes.

Q. 63. How can we predict disproportionation with the help of frost diagram?

Ans. Frost diagram helps to predict the relative stability for disproportionation. **A species in a Frost diagram is unstable with respect to disproportionation if its point lies above the line connecting two adjacent species**. In other words, **a species will undergo disproportionation only if its point lies above the line connecting two adjacent species.**

For example, as shown in Fig. 23.8, N_2O_4 lies above the joining NO_3^- and NO_2^-. Therefore, it is unstable with respect to disproportionation. Disproportionation takes place as under :

$$N_2O_4 + 2OH \longrightarrow NO_2^2 + 2OH$$

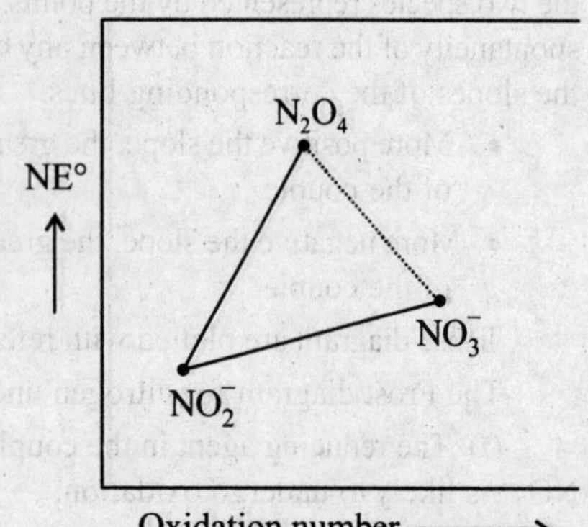

Fig. 23.8. Relative stability for disproportionation

Reverse of disproportionation is called comproportionation. It may also be predicted from the Frost diagram. Two species will tend to comproportionate into an intermediate species that lies below the straight line joining the terminal species. For example, the reaction between nitrogen (II) oxide and dinitrogen tetraoxide to form nitrite ion (NO_2^-) is predicted to be thermodynamically favourable because of the location of these species in the diagram.

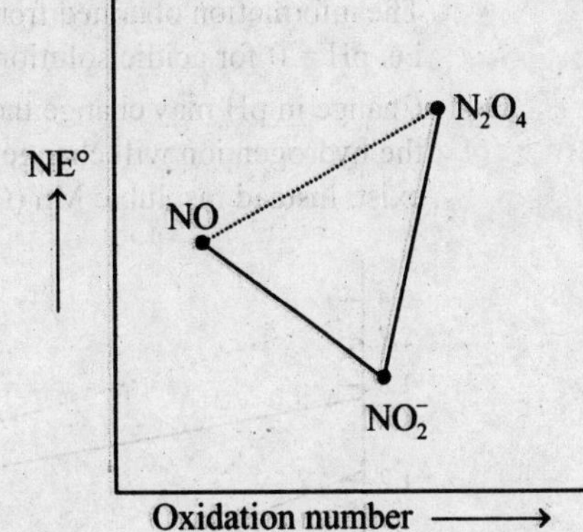

Fig. 23.9. Predicting comproportionation

$$NO + \frac{1}{2} N_2O_4 + 2OH^- \longrightarrow 2NO_2^- + H_2O$$

Frost diagram of Mn is given in Fig. 23.10.

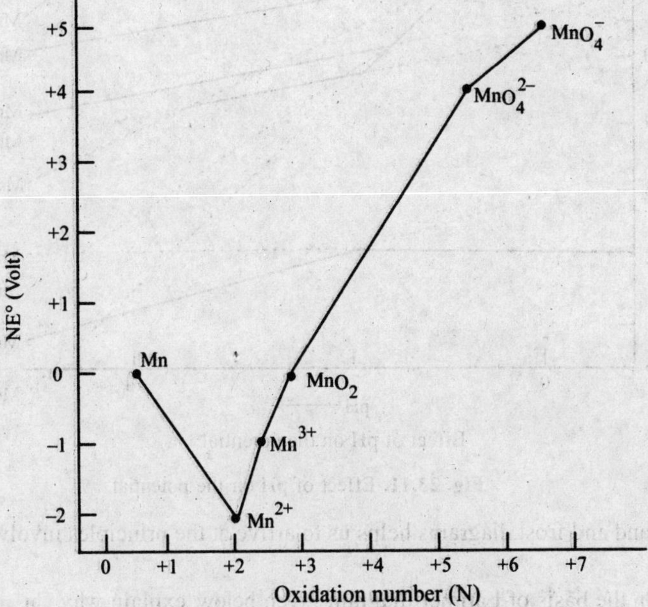

Fig. 23.10. Frost diagram for manganese.

The following inferences can be drawn from the Frost diagram of Mn

- The lower a species is present on the diagram, the more thermodynamically stable it is. Thus Mn (II) is the most stable species.
- A species located on the convex curve can undergo disproportionation. Thus, MnO_4^{2-} and Mn (III) have a tendency to disproportionate.
- The species on a concave curve shows no tendency towards disproportionation. Thus, MnO_2 does not proportionate.

  Any species located on the upper right side of the diagram will be a strong oxidising agent. Thus, MnO_4^- is a strong oxidising agent.
- A species located on the upper left side of the diagram will work as a reducing agent. Thus, Mn metal is a moderate reducing agent.

  It may be mentioned that while a given species may be thermodynamically unstable towards reduction, the kinetics of the reaction may be slow.

- The information obtained from a Frost diagram is for species under standard conditions i.e. pH = 0 for acidic solution and pH = 14 for basic solution.
- Change in pH may change the relative stability. The potential of any process involving the hydrogen ion will change with pH. Under basic conditions aqueous Mn^{2+} does not exist. Instead insoluble Mn $(OH)_2$ is formed. (Fig. 23.10).

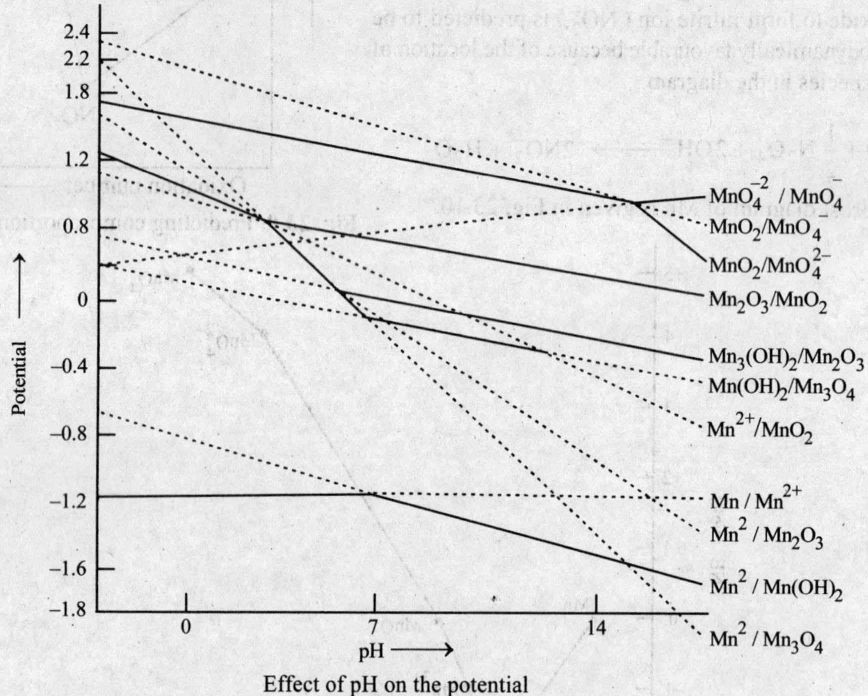

Fig. 23.11. Effect of pH on the potential.

Latimer and and frost diagrams helps us to arrive at the principles involved in the extraction of elements.

Q. 64. On the basis of Latimer diagram given below, explain why Cu^+ disproportionates in aqueous solution.

$$Cu^{2+} \xrightarrow{+0.15} Cu^+ \xrightarrow{+0.50} Cu$$

Ans. We know that a species has a thermodynamic tendency to disproportionate into its two neighbours if the potential on the right of species is higher than potential on the left. In this case, the potential on the right is higher than on the left of Cu^+, therefore Cu^+ has a tendency to disproportionate.

This can also be shown as:

	$Cu^+ + e^-$	$\longrightarrow$ Cu	$E° = + 0.50$ V
	$Cu^{2+} + e^-$	$\longrightarrow$ Cu^+	$E° = + 0.15$ V
Subtracting	$2\ Cu^+$	$\longrightarrow$ $Cu + Cu^{2+}$	$E° = + 0.35$ V

Since $E° > 0$ in solution, Cu^+ disproportionates into Cu^{2+} and Cu.

Example 65. Construct a frost diagram from the Latimer diagram given below:

$$O_2 \xrightarrow{+0.70} H_2O_2 \xrightarrow{1.76} H_2O \quad (O_2 \rightarrow H_2O: +1.23)$$

According to the diagram, which species is thermodynamically least stable and hence will undergo disproportionation.

Solution. The Frost diagram can be obtained by plotting NE° against oxidation number of the element, N.

Oxidation numbers of oxygen in O_2, H_2O_2 and H_2O are 0, –1 and – 2 respectively.

(i) For change of oxidation number from 0 to –1 (O_2 to H_2O_2), E° = + 0.70 and N = –1 so that

NE° = –1 × 0.70 = – 0.70 V. The coordinates of this point are (–1, –0.7)

(ii) For change of oxidation number from 0 to – 2 (O_2 to H_2O), E° = +1.23 V and N = –2 so that

NE° = –2 × 1.23 = – 2.46V. The coordinates of this point are (–2, –2.46)

On plotting, we get the following diagram

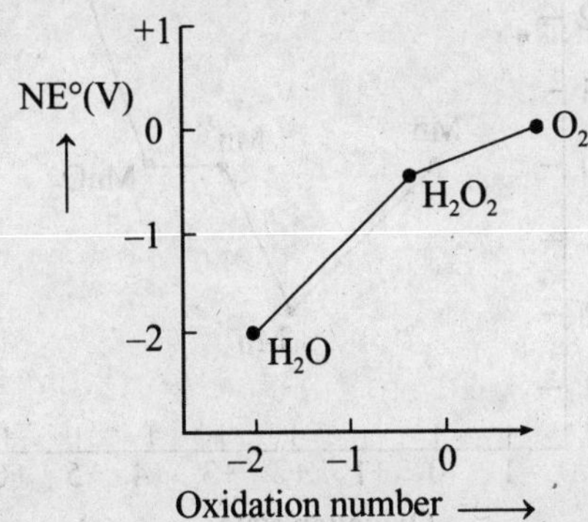

From the above figure, we find that the point for H_2O_2 lies above the line connecting two adjacent species H_2O and O_2. Therefore, it will undergo disproportionation as:

$$2H_2O_2 \longrightarrow 2H_2O + O_2$$

Q. 66. Consider the Latimer diagram for thallium:

$$Tl^{3+} \xrightarrow{+1.26\ V} Tl^{+} \xrightarrow{-0.34V} Tl \quad (Tl^{3+} \rightarrow Tl: +0.73)$$

Construct a Frost diagram.

***(i)* Predict the stability or unstability of Tl^+**

***(ii)* Which of these species is strong oxidant?**

Ans. The oxidation number of thallium in Tl^{3+}, Tl^+ and Tl is + 3, +1 and 0 respectively.

For change of Tl (0) to Tl^+ (I), E ° = – 0.34V and N = 1 so that NE° = 1 × (0.34) =–0.34V

For change of Tl (0) to Tl^{3+} (III), E° = + 0.73 and N = 3 so that

NE° = 3 × 0.73 = 2.19 V.

Coordinate of points corresponding to Tl, Tl^+ and Tl^{3+} are (0, 0), (1, – 0.34) and (3, 2.19) respectively.

Plotting the graph, we get the Frost diagram:

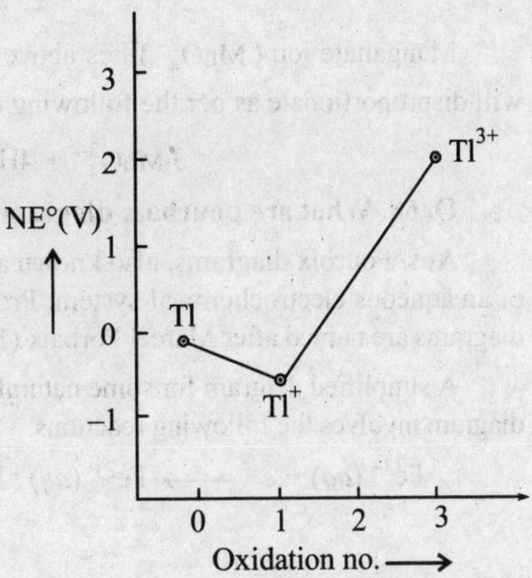

(*i*) Tl^+ is stable with respect to disproportionation because it does not lie above the line joining two adjacent species (Tl and Tl^{3+}). Since it lies lowest in the diagram, it is most stable oxidation state.

(*ii*) Tl^{3+} is strong oxidant because the slope of the line connecting it with either lower oxidation state is large and positive.

Q. 67. Consider the Frost diagram for manganese in acidic medium.

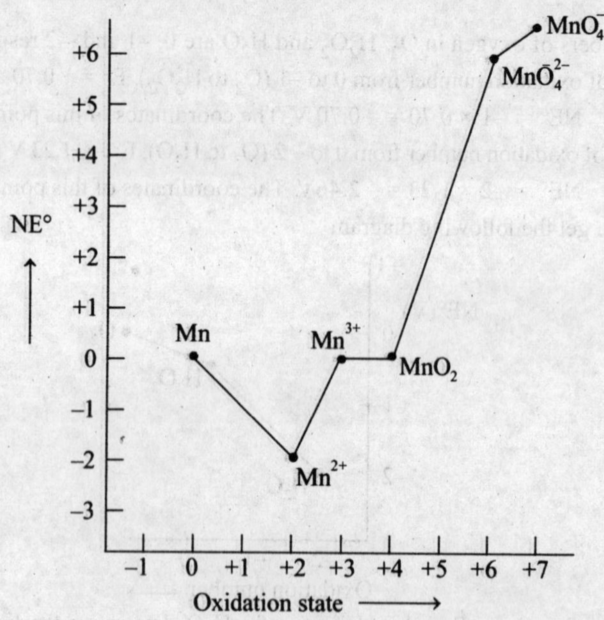

Comment on the stability of manganese (III) and manganate ion with respect to disproportionation. Write the reactions.

Ans. Mn^{3+} lies above the line joining Mn^{2+} and MnO_2. Therefore, it will disproportionate as per the following equation:

$$2Mn^{3+} + 2H_2O \longrightarrow Mn^{2+} + MnO_2 + 4H^+$$

Manganate ion (MnO_4^{2-}) lies above the line joining MnO_2 and Mn (MnO_4^-). Therefore, it will disproportionate as per the following reaction:

$$3\,MnO_4^{2-} + 4H^+ \longrightarrow MnO_2 + 2\,MnO_4^- + 2H_2O$$

Q. 68. What are pourbaix diagrams? What is their significance?

Ans. Pourbix diagrams, also known as potential/pH diagrams map out possible stable phases of an aqueous electrochemical system. Predominant ion boundaries are represented by lines. The diagrams are named after Marcel Porbaix (1904 - 98), the Russian born chemist, who invented them.

A simplified diagram for some naturally occurring compounds of iron is given in Fig. 3.9. This diagram involves the following reactions:

$$Fe^{3+}\,(aq) + e^- \longrightarrow Fe^{2+}\,(aq) \quad E° = +\,0.77V$$

The reduction half reaction does not involve H^+ ions and therefore, its potential is independent of *p*H. This gives horizontal line in the diagram. If the solution contains a couple having potential above this line (*i.e.* more positive and more oxidizing couple) then oxidized species, Fe^{3+} will be the major product. Hence the horizontal line at the top left of the diagram represents a boundary that seperates the regions of Fe^{3+} and Fe^{2+}.

Consider the reaction

$$Fe^{3+}\ (aq)\ 3H_2O(l) \longrightarrow Fe(OH)_3\ (s) + 3H^+\ (aq)$$

This does not represent a redox reaction because there is no change in the oxidation number of any element.

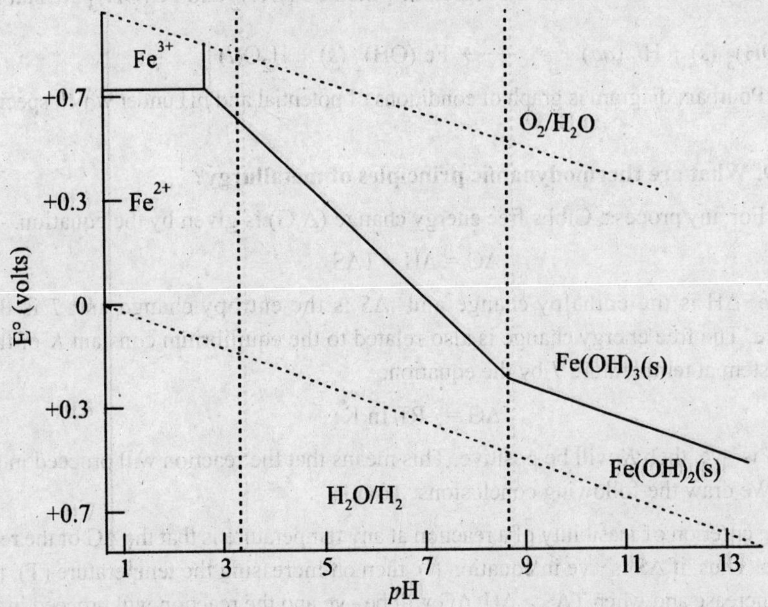

Fig. 23.12. Pourbix diagram for some important naturally occurring compounds of iron.

The boundary between the region where Fe^{3+} is soluble and the region where Fe $(OH)_3$ precipitates is independent of any redox couples. However, this boundary depends upon *p*H. Fe^{3+} (*aq*) is favoured by low *p*H while the Fe $(OH)_3$ is favoured by high *p*H. The equilibrium concentration varies with *p*H and the vertical boundary represents the *p*H at which Fe^{3+} becomes dominant according to this definition.

Now consider the reaction.

$$Fe\ (OH)_3\ (s) + 3H^+\ (aq) + e^- \longrightarrow Fe^{2+}\ (aq) + 3H_2O\ (l)$$

This reaction depends upon the *p*H because H^+ ion concentration appears in the half reaction as shown by Nernst equation:

$$E = E° - 0.059\ V\ \log \frac{(Fe^{2+})}{(H^+)^3}$$

$$= E° - 0.059\ V\ \log\ (Fe^{2+}) - (0.177\ V\ pH)$$

Fig. 23.12. shows that the potential falls lineraly as the pH increases. The regions of potential and pH that lie above this line correspond to conditions in which the oxidized species Fe $(OH)_3$ is stable. The regions below this line correspond to conditions under which the reduced species Fe^{2+} is stable. We observe that a strong oxidizing agent is required to oxidize and precipitate Fe^{2+} ions in acidic media than in basic media.

The vertical line at $pH = 9$ divides the regions in which either the reactants or products are stable in the reactions:

$$Fe^{2+} (aq) + 2H_2O (l) \longrightarrow Fe (OH)_2 (s) + 2H^+ (aq)$$

Similarly, the line separates the region in which Fe $(OH)_2$ and Fe $(OH)_3$ are stable.

$$Fe (OH)_3 (s) + H^+ (aq) + e^- \longrightarrow Fe (OH)_2 (s) + H_2O(l)$$

Thus Pourbaix diagram is graph of conditions of potential and pH under which species are stable in water.

Q. 69. What are thermodynamic principles of metallurgy?

Ans. For any process, Gibbs free energy change (Δ G) is given by the equation,

$$\Delta G = \Delta H - T\Delta S \qquad ...(i)$$

where ΔH is the enthalpy change and ΔS is the entropy change and T is the absolute temperature. The free energy change is also related to the equilibrium constant K of the reactant-product system at temperature T by the equation.

$$\Delta G = -RT \text{ In } K \qquad ...(ii)$$

If ΔG is –ve, then K will be positive. This means that the reaction will proceed in the forward direction. We draw the following conclusions:

1. The criterion of feasibility of a reaction at any temperature is that the ΔG of the reaction must be negative. Thus, if ΔS is +ve in equation (i), then on increasing the temperature (T), the value of T ΔS will increase and when $T\Delta S > \Delta H$, ΔG will be –ve and the reaction will proceed in the forward direction.

2. A reaction with ΔG positive can be still made to occur by coupling it with another reaction having large negative ΔG so that the net ΔG of the two reactions in negative.

Such coupling reactins can be explained in terms of Ellingham diagram, i.e., $\Delta G°$ vs T plots for formation of oxides.

Q. 70. What is Ellingham diagram? What is its utility?

Ans. Ellingham diagram consists of posts of $\Delta_f G°$ vs T for formation of oxides of element (Fig 3.10). Similar diagrams can also be constructed for sulphides and halides. Such diagrams help us in predicting the feasibility of thermal reduction of an ore.

Let us consider the formation of a metal oxide from the metal

$$2xM (s) + O_2 (g) \longrightarrow 2M_xO (s) \qquad ...(i)$$

In this reaction, dioxygen – a gas has been used up and metal oxide – a solid is formed. Gases have higher entropy than liquids and solids. Therefore, during this reaction, ΔS becomes negative.

Thus, if temperature is increased, TΔS becomes more and more negative. Therefore, ΔG becomes less and less negative according to equation ΔG = ΔH – T ΔS

Thus, ΔG increases or becomes less – ve with increase in temperature (normally, ΔG decreases with increase in temperature). In other words, $\Delta_f G°$ vs T lines have +ve slopes for most of the reactions involving the formation of metal oxides, $M_xO(s)$ as shown in Ellingham diagram. Each plot is a straight line except when some change in phase (solid → liquid or liquid → gas) takes places. The temperature at which such a change occurs, is indicated by an increase in the slope on the +ve side. For example in the Zn – ZnO plot, at the boiling point of zinc (1180 K), there is an abrupt increase in the +ve slope of the curve. Similarly, the +ve slope of the Mg – MgO curve increases

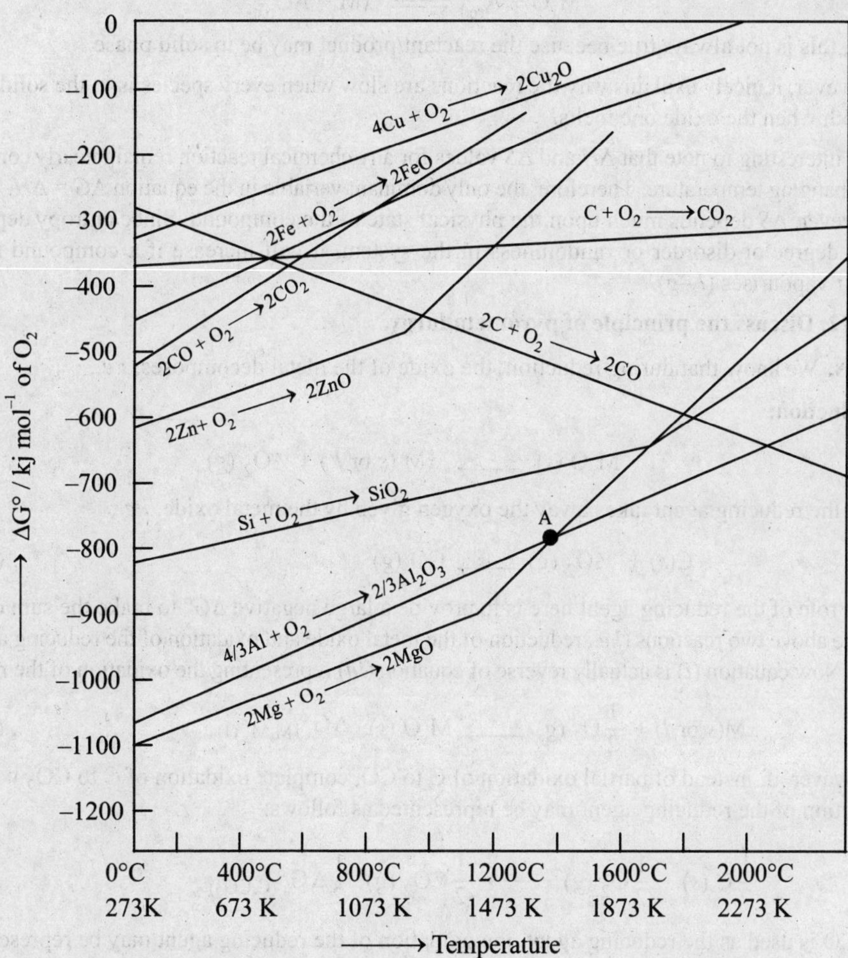

Fig. 23.13 Gibbs energy (ΔG°) vs T plots (schematic) for formation of some oxides (Ellingham diagram)

abruptly at the melting point of Mg at 1363 K and that of Si–SiO_2 curve increases abruptly at the melting point of Si at 1966 K. The curves in such cases move upwards suddenly.

Q.71. State the limitations of Ellingham diagram.

Ellingham diagram although quite useful have some limitations as well:

(i) Kinetics of reduction: Ellingham diagrams are based on thermodynamic concepts, therefore, these diagrams simply suggest whether the reduction of a given metal oxide with a particular reducing agent is possible or not. It, however, does not tell anything about the rate of the reduction process.

(ii) Reactant/product equilibrium: The interpretation of $\Delta G°$ is based upon K (i.e., $\Delta G° = -RT \ln K$). Thus, it is assumed that reactants and products are in equilibrium.

$$M_xO + A_{red} \rightleftharpoons xM + AO_{ox}$$

But, this is not always true because the reactant/product may be in solid phase.

However, it nicely explains why the reactions are slow when every species is in the solid state and smooth when the oxide one melts.

It is interesting to note that ΔH and ΔS values for any chemical reaction remain nearly constant even on changing temperature. Therefore, the only dominant variable in the equation $\Delta G = \Delta H - T\Delta S$ is T. However, ΔS depends much upon the physical state of the compound. Since entropy depends upon the degree of disorder or randomness in the system, it will increase if a compound melts ($s \rightarrow l$) or vapourises ($l - g$).

Q. 72. Discuss the principle of pyrometallurgy.

Ans. We know that during reduction, the oxide of the metal decomposes, *i.e.*,

Reduction:

$$M_xO\,(s) \longrightarrow xM\,(s \text{ or } l) + \tfrac{1}{2}O_2\,(g) \qquad ...(i)$$

and the reducing agent takes away the oxygen given by the metal oxide, *i.e.*,

$$C(s) + \tfrac{1}{2}O_2\,(g) \longrightarrow CO\,(g) \qquad ...(ii)$$

The role of the reducing agent here is to provide a large negative $\Delta G°$ to make the sum of the $\Delta G°$ of the above two reactions (*i.e.*, reduction of the metal oxide and oxidation of the reducing agent) negative. Now equation (*i*) is actually reverse of equation (*iii*) representing the oxidation of the metal.

$$xM(s \text{ or } l) + \frac{1}{2}O_2\,(g) \longrightarrow M_xO\,(s);\ \Delta G°_{(M,M_xO)} \qquad ...(iii)$$

However, if, instead of partial oxidation of C to CO, complete oxidation of C to CO_2 occurs, the oxidation of the reducing agent may be represented as follows:

$$\frac{1}{2}C\,(s) + \frac{1}{2}O_2\,(g) \longrightarrow \frac{1}{2}CO_2\,(g);\ \frac{1}{2}\Delta G°_{(C,CO_2)} \qquad ...(iv)$$

If CO is used as the reducing agent, the oxidation of the reducing agent may be represented as follows:

$$CO\,(g) + \frac{1}{2}O_2\,(g) \longrightarrow CO_2\,(g);\ \Delta G°_{(CO,CO_2)} \qquad ...(v)$$

On subtracting equation (*iii*) from equations (*ii*), (*iv*) and (*v*) respectively, we have

$$M_xO\,(s) + C\,(s) \longrightarrow xM\,(s \text{ or } l) + CO\,(g) \qquad ...(vi)$$

$$M_xO\ (s) + \frac{1}{2}C\ (s) \longrightarrow xM(s \text{ or } l) + \frac{1}{2}CO_2\ (g) \qquad ...(vii)$$

$$M_xO\ (s) + CO\ (g) \longrightarrow xM(s \text{ or } l) + CO_2\ (g) \qquad ...(viii)$$

These three equations (*vi*), (*vii*) and (*viii*) describe the actual reduction of the metal oxide, M_xO to the metal, M. The $\Delta G°$ values for these reactions, in general, can be obtained by subtraction of the corresponding $\Delta G°$ values obtained from the Ellingham diagram. If these values are –ve, the reduction is feasible otherwise not. Without actual calculations for $\Delta_r G°$, we can predict the feasibility of a reduction process simply by looking at the Ellingham diagram. Metals for which free energy of formation ($\Delta_f G°$) of their oxides is more negative can reduce those metals for which the free energy of formation ($\Delta_f G°$) of their respective oxides is less negative.

For example, at any given temperature, any metal will reduce the oxide of other metals which lie above it in the Ellingham diagram because the free energy change ($\Delta_f G°$) for the overall redox reaction comes out to be negative.

Greater the difference (Net ΔG) easier is the reduction. For example, Al reduces ZnO, FeO and CuO more readily than Zn reduces FeO and Cu_2O.

Q.73. Discuss the effect of temperature on the free energy change ($\Delta_r G°$) of the overall reduction process.

Ans. For any process,

$$\Delta G = \Delta H - T\Delta G$$

On increasing the temperature, the value of ΔH and ΔS nearly remain constant, therefore, the value of $\Delta_r G°$ becomes more negative. This means that if a particular reduction process does not occur at a lower temperature, it may occur at a higher temperature. We need to select the temperature in such a way that the $\Delta_r G°$ of the overall redox reaction becomes –ve.

In the Ellingham diagram, this temperature is indicated by the intersection of the two curves *i.e.*, curve for formation of M_xO and the curve for the oxidation of the reducing agent. At the point of intersection of the two curves, $\Delta G° = 0$. Below this temperature, the $\Delta G°$ is –ve compared so that of the reducing agent and hence the oxide is stable. Above this temperature, $\Delta G°$ is +ve compared to that of the reducing agent and hence the oxide is unstable, *i.e.*, it would give metal and oxygen. For example, the temperature at the intersection point 'A' of the Al $\longrightarrow$ Al_2O_3 and Mg $\longrightarrow$ MgO curves is approx. 1623 K. Therefore, below 1623 K, MgO is more stable and hence Mg can reduce Al_2O_3 to Al but above 1623 K, Al_2O_3 is more stable and hence Al can reduce MgO to Mg.

Q.74. How do we extract non-metals by oxidation?

Ans. Non-metals occur in the combined state in the reduced form. Therefore, they are generally isolated by oxidation of their compounds.

An example is discussed below:

Isolation of chlorine: A very common example of extraction based on oxidation is the isolation of chlorine from brine.

$$2Cl^-(aq) + 2H_2O(l) \longrightarrow 2OH^-(aq) + H_2(g) + Cl_2\ (g)$$

Since oxidation cannot be carried out by ordinary chemical methods, it is accomplished by electrolysis. The $\Delta G°$ for this reaction is + 422 kJ. The minimum potential difference required for oxidation is calculated as under:

$$\Delta G° = -nFE°$$

$$\text{or } 422 \times 1000\text{J} = -2 \times 96500 \times E°$$

$$\text{or } E° = -2.18 \text{ V} - \simeq -2.2\text{V}.$$

However, due to overvoltage, actual electrolysis requires an external emf which is greater than 2.2 V.

During electrolysis, Cl_2 is liberated at the cathode and H_2 at the anode while NaOH is obtained in the solution. In other words, H_2 and NaOH are the by products of this electrolysis.

Electrolysis of moltent NaCl can also be carried out but in this case Na metal is liberated at the cathode.

Q. 75. Given $\quad Sn^{4+} + 2e^- \rightleftharpoons Sn^{2+} \quad E° = 0.154$ V

$$Fe^{3+} + e^- \rightleftharpoons Fe^{2+} \quad E° = 0.771 \text{ V}$$

Determine the cell voltage.

Ans. Since E° of Fe^{3+} is greater it will act as an oxidising agent and will oxidise Sn^{2+} to Sn^{4+}. The cell potential can be determined as:

$$E°_{cell} = E°_{cathode} - E°_{anode}$$

$$= 0.771 - 0.154 = 0.617 \text{ V}$$

Q. 76. Given : $\quad Ce^{4+} + e^- \rightleftharpoons Ce^{3+} \quad E°=1.61$ V $\quad E° = 1.61$ V

$$MnO_4^- + 8H^+ + 5e^- \rightleftharpoons Mn^{2+} + 4H_2O \quad E° = 1.51 \text{ V}$$

Comment on the feasibility of the reaction.

Ans. E° values indicate that Ce^{4+} is a stronger agent than acidic MnO_4^-. Therefore, acidic MnO_4^- cannot oxidise Ce^{3+} to Ce^{4+} but Ce^{4+} can oxidise Mn^{2+} to MnO_4^-, theoretically.

For this reaction,

$$[Ce^{4+} + e^- \rightleftharpoons Ce^{3+}] \times 5$$

$$Mn^{2+} + 4H_2O \rightleftharpoons MnO_4^- + 8H^+ + 5e^-$$

$$5Ce^{4+} Mn^{2+} + 4H_2O \rightleftharpoons 5Ce^{3+} + MnO_4^- + 8H^+$$

The cell potential is:

$$E°_{cell} = 1.61 - 1.51 = 0.10 \text{ V}$$

Since $E°_{cell}$ is positive, the reaction is theoretically feasible. However, as the value of $E°_{cell}$ is very small, such reaction would be very slow.

Q. 77. Will it be possible to oxidise Cl^- and Co^{2+} with acidic $Cr_2O_7^{2-}$?

$E° (Cl_2|2Cl^-) = 1.3595$ V, $E° (Co^{3+}|Co^{2+}) = 1.81$ V and $E° (Cr_2O_7^{-2}|2Cr^{3+}) = 1.33$ V?

Ans. As reduction potentials of both Cl^- and Co^{2+} are greater than that of $Cr_2O_7^{-2}|Cr^{3+}$ acidified dichromate cannot oxidise Cl^- to Cl_2 or Co^{2+} to Co^{3+}.

Q. 78. Will permangate ion oxidise H_2O to evolve O_2 in acidic solution given that:

$$MnO_4^- + e^- \rightleftharpoons MnO_4^{2-} \qquad E° = +0.564 \text{ V}$$

$$\frac{1}{2}O_2 + 2H^+ + 2e^- \rightleftharpoons H_2O(l) \qquad E° = 1.23 \text{ V}$$

Ans. We want to check the possibility of the following reaction:

$$2MnO_4^- + H_2O \longrightarrow 2MnO_4^{2-} + \frac{1}{2}O_2 + 2H^+$$

Given that

$$2MnO_4^{-2} + 2e^- \rightleftharpoons 2MnO_4^{2-} \quad E° = 0.564 \text{ V} \qquad ...(i)$$

$$\frac{1}{2}O_2 + 2H^+ + 2e^- \rightleftharpoons H_2O \qquad E° = 1.23 \text{ V} \qquad ...(ii)$$

Subtracting eq. (*ii*) from (*i*), we have

$$2MnO_4^- + H_2O \rightleftharpoons 2MnO_4^- + \frac{1}{2}O_2 + 2H^+ \qquad E° = -0.666 \text{ V}$$

Since E° is negative, the above reaction will not occur.

Q. 79. The Latimer diagram for amercium is:

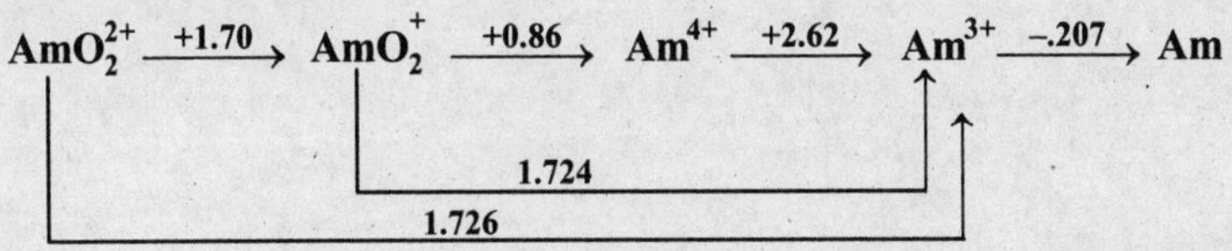

$$AmO_2^{2+} \xrightarrow{+1.70} AmO_2^+ \xrightarrow{+0.86} Am^{4+} \xrightarrow{+2.62} Am^{3+} \xrightarrow{-.207} Am$$

($AmO_2^+ \to Am^{3+}$: 1.724; $AmO_2^{2+} \to Am^{3+}$: 1.726)

Predict the species which can disproportionate and into which species.

Ans. AmO_2^+ will disproportionate to AmO_2^{2+} and Am^{3+} because the potential the potential for $AmO_2^{2+} \longrightarrow AmO_2^+$ is less and for $AmO_2^+ \longrightarrow Am^{3+}$ is more.

Similarly, Am^{4+} will disproportionate into AmO_2^+ and Am^{3+}

Q. 80. The reduction potentials are:

$$Cu^{2+} + 2e \longrightarrow Cu^+ \qquad E° = 0.15 \text{ V}$$

$$Cu^+ + e \longrightarrow Cu \qquad E° = +0.50 \text{ V}$$

Calculate the value for $Cu^{2+} \longrightarrow Cu$ and draw Latimer diagram.

Ans. Given

$$Cu^+ + e \longrightarrow Cu \qquad E° = 0.50$$

$$Cu^{2+} + e \longrightarrow Cu^+ \qquad E° = 0.15\ V$$

Subtracting (*ii*) from (*i*)

$$2Cu^{2+} \longrightarrow Cu^{2+} + Cu \qquad E° = 0.35\ V$$

Latimer diagram is

$$Cu^{2+} \xrightarrow{+0.15} Cu^+ \xrightarrow{+0.50} Cu$$

$$Cu^{2+} \xrightarrow{+0.35} Cu$$

24 CHEMICAL INDUSTRIES

CEMENT INDUSTRY

Q. 1. What is Portland cement? Name the raw materials required for the manufacture of Portland cement. (*M. Dayanand, 2000*)

Ans. Cement is a finely ground mixture of calcium aluminates and silicates which sets to a hard mass when treated with water. It is also called **Portand cement** because of its resemblance to a famous variety of building stone mined near Portland, England. Portland cement is made by fusing together following types of materials:

(*a*) **Calcareous** such as limestone, chalk, alkali waste, shells of marine animals etc.

(*b*) **Argillaceous** (or rich in alumina and silica) such as clay, shale, slate, blast furnace slag etc.

Q. 2. Give the composition of Portland cement. (*M. Dayanand, 2002; Calcutta, 2005*)

Ans. The relative percentage of the raw materials naturally depends upon type of cement desired. The following table gives the composition of normal Portland cement.

Constituents	*Per cent*
Lime (CaO)	62
Silica (SiO_2)	22
Alumina (Al_2O_3)	7.5
Iron Oxide (Fe_2O_3)	2.5
Magnesia (MgO)	2.5
Sulphur trioxide (SO_3)	1.25
Alkalis (Na_2O) from clay	Traces

Q. 3. Discuss the manufacture of cement. (*Calcutta, 2005*)

Ans. Manufacture of cement involves the following steps:

(*i*) Treatment of raw materials. The raw materials are worked up either by dry or wet process before being fed into kiln.

(*a*) **Dry process.** The calcareous and argillaceous materials are mixed in the ratio 3:1, crushed, dried, sieved and then ground to a fine powder. The mixture is now ready for the kiln.

(*b*) **Wet process.** The rocky matter is crushed and water is sprinkled over it. The resulting material is ground in the wet condition to a slurry. After grinding the slurry which contains 35–40% of water, it is stored in large tanks equipped with mechanical stirrers. It is then fed into the kiln.

(*ii*) Calcination in rotary kiln. Dry mixture from the dry process or the slurry from the wet process is fed into the rotary kiln (Fig. 24.1) for calcination. Rotary kiln consists of a steel cylinder lined with firebricks. It is 45–120 m long and has about 3 m internal diameter. The kiln is inclined at an angle of about 15 degrees to the horizontal and it rotates on its axis at the rate of 1/2 to 1 revolution per minute. The charge which is introduced at upper end travels down as kiln rotates.

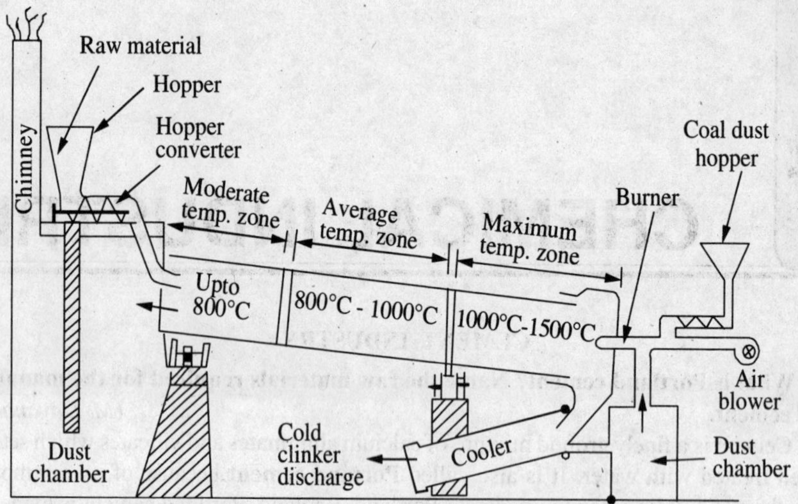

Fig. 24.1. Manufacture of portland cement.

The charge is heated by burning powdered coal, oil, or natural gas which enters at lower end with the help of a blower. As the charge moves forward, it meets higher temperatures till it reaches the lower end where the temperature is highest (around 1400°C). The charge takes 2 to 3 hours to cover the journey in the kiln. Reactions take place in the rotary kiln in three major zones as:

(*a*) **Moderate Temperature Zone (up to 800°C).** At 100°C, the removal of free moisture takes place. At 500°C, kaolin ($Al_2O_3 \cdot 2SiO_2 \cdot 2H_2O$) present in clay breaks up into amorphous silica and alumina.

(*b*) **Average Temperature Zone (800°C-1000°C).** Limestone decomposes to form lime and carbon dioxide.

$$CaCO_3 \longrightarrow CaO + CO_2$$

(*c*) **Maximum Temperature Zone (1000°C-1500°C).** The main reactions between lime, alumina and silica takes place near the lower end of kiln resulting in formation of calcium silicates and aluminates as shown below.

$$2CaO + SiO_2 \longrightarrow 2CaO\ .\ SiO_2$$

Dicalcium silicate

$$3CaO + SiO_2 \longrightarrow 3CaO\ .\ SiO_2$$

Tricalcium silicate

$$3CaO + Al_2O_3 \longrightarrow 3CaO\ .\ Al_2O_3$$

Tricalcium aluminate

$$4CaO + Al_2O_3 + Fe_2O_3 \longrightarrow 4CaO.\ Al_2O_3\ .\ Fe_2O_3$$

Tetracalcium alumina ferrite

Moreover, due to high temperature in this zone, about 20 – 30% of mass melts and combines with solid mass to form pebbles (which vary in size) and are called *cement clinkers*.

The burnt gases (containing a good amount of dust) are made to pass through a dust chamber, where most of dust is retained and gases are allowed to escape through chimney.

***(iii)* Cooling of clinkers.** The hot clinkers from the rotary kiln go to coolers which consist of series of tubes parallel to kiln in which air is forced through. As a result of this, the clinkers cool down while the air gets heated. The hot air is then used for the combustion of the fuel as an economy measure.

(*iv*) **Grinding and packing.** The cooled clinkers are mixed with 3–5% of gypsum $CaSO_4 . 2H_2O$, along with certain dispersing and waterproofing agents and then ground to a fine powder. Since pure finely powdered clinkers set very rapidly, therefore, gypsum has to be added to retard the setting of cement. The powdered material is passed through fine sieves and finally it is sent to the automatic packing machines where it is packed in jute or polythene bags.

Q. 4. Write a note on setting of cement. (*Himachal, 2000; M. Dayanand, 2002*)

Ans. When a paste of cement in water is allowed to stand for some time, it changes to a solid hard mass. This phenomenon is called *setting of cement.*

Setting of cement is a complicated process. However, it has been concluded that setting of cement is a physicochemical process. The important physical changes during setting of cement are gel formation and crystallisation while the chemical changes involved are hydration and hydrolysis as discussed below:

(***a***) **During first 24 hours.** After the cement is mixed with water, the following reactions start taking place:

(*i*) Tricalcium aluminate absorbs water to form a gel of composition $3CaO . Al_2O_3 . 6H_2O$

$$3CaO . Al_2O_3 + 6H_2O \longrightarrow \underset{\text{Hydrated colloidal gel}}{3CaO . Al_2O_3 . 6H_2O}$$

The gel so formed starts crystallising slowly.

(*ii*) Tricalcium aluminates reacts with gypsum (which is added to retard the setting of cement) to form sulphoaluminate crystals.

$$3CaO . Al_2O_3 + CaSO_4 + x\, H_2O \longrightarrow \underset{\text{Sulphoaluminate crystals}}{3CaO. Al_2\, O_3 .CaSO_4.xH_2O}$$

(***b***) **Between 1 and 7 days.** Within 24 hours, the hydrolysis of tricalcium silicate begins and is generally complete within 7 days according to equation

$$3CaO . SiO_2 + H_2O \longrightarrow Ca(OH)_2 + \underset{\text{Dicalcium silicate (amorphous)}}{2CaO . SiO_2}$$

$Ca(OH)_2$ formed in the above reaction changes into needle-shaped crystals.

Moreover, during this period some of the tricalcium aluminate also gets hydrolysed.

$$3CaO . Al_2O_3 + 6H_2O \longrightarrow 3Ca(OH)_2 + 2A1(OH)_3$$

Maximum hydroxide plugs the interstices resulting in hardening of mass.

(***c***) **Between 7 and 28 days.** Amorphous dicalcium silicate begins to hydrate (a very slow reaction) resulting in formation of hydrated colloidal gel.

$$2CaO\ SiO_2 + x\, H_2O \longrightarrow \underset{\text{Hydrated colloidal gel}}{2CaO\ SiO_2\, x\, H_2O}$$

Needles of $Ca(OH)_2$ get studded in hydrated colloidal gel of dicalcium silicate and thus impart strength to it.

Moreover tetracalcium alumino ferrite gets hydrated and forms a hydrated colloidal gel.

$$4CaO . Al_2O_3 . Fe_2O_3 + 6H_2O \longrightarrow \underset{\text{Hydrated colloidal gel.}}{3CaO . Al_2O_3. H_2O + CaO . Fe_2O_3}$$

The gel slowly starts losing water partly by evaporation and partly by forming hydrates with unhydrated constituents as a result of which the cement sets to a hard mass.

Q. 5. Discuss the role of water and gypsum during setting of cement.

Ans. Role of water during setting of cement. During the setting of cement, water plays an important role because hydration as well as hydrolysis involve reaction with water. Water is essential for the formation of gels. Since hydration is an exothermic reaction, therefore, cement structures have to be cooled during setting either by spraying water or circulating cold water through pipes embedded in the heavy structures such as big dams.

Role of gypsum during setting of cement. The addition of gypsum to powdered clinkers helps in the removal of the fast setting tricalcium aluminate forming calcium sulpho-aluminate.

$$3CaO \,.\, Al_2O_3 + 3CaSO_4 + 2H_2O \longrightarrow 3CaO \,.\, Al_2O_3 \,.\, 3CaSO_4 \,.\, 2H_2O$$

Consequently, the process of setting of cement gets retarded. In fact, slow setting of cement is preferred because it provides us better results as far as strength of constructed mass is concerned.

Q.6. What is white cement? How is it made? Discuss some of its uses.

Ans. White cement is similar to ordinary grey Portland cement in all respects except for its high degree of whiteness. To obtain this colour, we require a substantial modification in the method of manufacture. Because of that it is fairly expensive compared to the grey cement.

Manufacture

The Characteristic greenish-grey colour of ordinary cement is due to the presence of chromium, manganese and iron. The content of these metals is smaller in white cement. Cr_2O_3 is kept below 0.003%, Mn_2O_3 is kept below 0.03% and Fe_2O_3 is kept below 0.35% in the clinker. Ordinary cement is generally made from low quality raw materials containing substantial amounts of these metals. We use special quality of raw materials containing smaller amounts of these metals to obtain white cement.

In general, the rotary kiln used to combine the raw materials are operated at higher temperatures (1450 – 1500°C) for white cement compared to 1400 – 1450°C for grey cement. This requires higher fuel combustion and results in lower kiln output.

Uses white cement is used in combination with white aggregates to produce white concrete for prestige construction projects and decorative work. White cement is also used in combination with inorganic pigments to produce brightly coloured concretes and mortars.

FERTILIZER INDUSTRY

Q. 7. Discuss the role of nitrogen, phosphorus and potassium in the healthy growth of plants.

Ans. Nitrogen. (*i*) It produces rapid growth of plants.

(*ii*) It increases protein content of the plants.

(*iii*) It imparts dark green colour to the leaves.

(*iv*) It increases yield of the plants.

(*v*) It improves quality of the plants.

Phosphorus. (*i*) It stimulates early root formation and rapid growth.

(*ii*) It helps in blooming and seed formation.

(*iii*) It hastens maturity.

(*iv*) It increases resistance to frost and insects by aiding the production of sound plant tissues.

(*v*) It increases yield and quality of the products.

Potassium. (*i*) It plays the same role in the plants as salt plays in the blood of animals. Thus it retains water in the plant cells by osmotic pressure enabling the plants to resist withering and frost.

(*ii*) It helps in the development of healthy root system.

(*iii*) It makes stalks stronger.

(*iv*) It increases vigour and disease resistance in the plants.

(*v*) It helps in the formation of sugar, starch and cellulose.

(*vi*) It aids the ripening of cereals and fruits.

Q. 8. What are the essential qualities of a good fertilizer? *(M. Dayanand, 2000)*

Ans. (*i*) The fertilizer should be soluble in water so that it can penetrate the soil and thus it may be easily assimilated by the plants.

(*ii*) The elements N, P and K present in it must be readily available to the plants. Therefore, the mere presence of N, P and K in a chemical compound does not make it a good fertilizer, if these elements are present in such a state of combination that the plants may not be able to assimilate them readily.

(*iii*) It should have a high percentage of nitrogen, phosphorus and potassium so that minimum cost is involved in its handling, freight and package.

(*iv*) The fertilizer should be stable so that the elements present in it may be source of steady supply of food to the plants for a long time.

(*v*) It should be helpful in correcting acidity of the soil.

(*vi*) Its storage properties must be good, with little or no tendency to setting or deliquescence.

Q. 9. Describe the manufacture of ammonium sulphate as a nitrogenous fertilizer.

Ans. It is manufactured by two processes:

(*i*) From ammoniacal liquor. During the dry distillation of coal, 'ammoniacal liquor' is obtained as an important by-product. The 'ammoniacal liquor' contains various ammonium salts, *e.g.*, $(NH_4)_2CO_3$, $(NH_4)_2S$, NH_4CN as the volatile salts and $(NH_4)_2SO_4$, $(NH_4)_2SO_3$, NH_4Cl as the stable salts. The total ammonia in combination is about 17 grams/litre. The 'ammoniacal liquor' is heated in stills first with steam and then with milk of lime, *i.e.*, $Ca(OH)_2$, in order to decompose volatile as well as stable salts.

$$(NH_4)_2CO_3 \longrightarrow 2NH_3 + CO_2 + H_2O$$

$$(NH_4)_2S \longrightarrow 2NH_3 + H_2S$$

$$(NH_4)_2SO_4 + Ca(OH)_2 \longrightarrow 2NH_3 + CaSO_4 + 2H_2O$$

$$2NH_4Cl + Ca(OH)_2 \longrightarrow 2NH_3 + CaCl_2 + 2H_2O$$

The mixture of ammonia and steam coming out of the still is bubbled through 60% H_2SO_4 contained in a lead lined tank wherein crystals of ammonium sulphate separate out.

$$2NH_3 + H_2SO_4 \longrightarrow (NH_4)_2SO_4$$

This method is being followed at all the major steel plants in India as "ammoniacal liquor" is obtained as a by-product in the manufacture of coke (from coal) needed for steel production.

(*ii*) Sindhri process. Air and steam are passed over red hot coke when a mixture of nitrogen, hydrogen and carbon monoxide is formed.

$$C + \frac{1}{2}O_2 + N_2 \longrightarrow CO + N_2$$

$$C + H_2O \longrightarrow CO + H_2$$

The mixture of the three gases is then passed over hot Fe_3O_4 and Cr_2O_3 when carbon monoxide gets oxidised to carbon dioxide. The mixture of CO_2, H_2 and N_2 is then compressed to about 20 atmospheric pressure in presence of water wherein carbon dioxide gets dissolved and is thus removed from the mixture. The composition of the remaining gases, namely, nitrogen and hydrogen, is adjusted in the ratio of 1:3 and then converted into ammonia by Haber' s Process (450 – 550°C, 150 –200 atmospheric pressure using Fe_2O_3 as a catalyst along with Al_2O_3 as parameter). Carbon dioxide gas is passed through a suspension of calcium sulphate in aqueous ammonia resulting in the formation of ammonium sulphate as

$$2NH_3 + CO_2 + H_2O + CaSO_4 \longrightarrow (NH_4)_2SO_4 + CaCO_3\downarrow$$

The resulting slurry is filtered on rotary drum vacuum filters to remove $CaCO_3$. The filtrate is concentrated by evaporation under vacuum and cooled when crystals of ammonium sulphate appear. The crystals are dried by centrifuging at a high speed. They are packed in jute bags and sold in market under the name *Sindhri Fertilizer.*

Q. 10. Describe characteristics of ammonium sulphate as a fertilizer.

Ans. It contains 24–25% NH_3 which is converted by nitrifying bacteria in the soil into nitrates in which form it is made available to the plants. It is less hygroscopic than urea or ammonium nitrate and as a result it has better storage properties.

It is a very valuable nitrogenous fertilizer for all types of crops and soils except the acidic soils where it should be used in combination with ground limestone or chalk — wherein lime neutralises the acidity. It should always be used in conjunction with bulky organic manures like farmyard manure, compost etc. Addition to sodium chloride is found to increase its fertilizing power.

It can be applied prior to sowing or at the sowing time or during the growing period of the crop. After it is applied to the soil, it is absorbed and retained by the soil particles and hence there is no danger of its being lost from the soil in drainage.

Q. 11. Describe the manufacture of calcium ammonium nitrate (CAN) or nitrolim.

(*M. Dayanand, 2000*)

Ans. Following steps are involved in the manufacture of calcium ammonium nitrate, also known as *nitrolim:*

***(i)* Production of ammonia.** Nitrogen (obtained from liquefaction of the air) and hydrogen (obtained by the electrolysis of water) are mixed in the ratio of 1:3 and are converted into ammonia by Haber's process. Mixture of the gases is heated to 450–550°C, compressed to 150 – 200 atmospheric pressure and passed over hot Fe_2O_3, Al_2O_3 mixture.

$$N_2 + 3H_2 \rightleftharpoons 2NH_3$$

***(ii)* Production of nitric acid.** A part of ammonia produced above is used in the manufacture of nitric acid by the Ostwald process. Ammonia and air are mixed in the ratio of 1:10 by volume. The mixture is passed into a converter having a cage of fme platinum gauge which is kept electrically heated between 820 –920°C wherein nitric oxide is formed.

$$4NH_3 + 5O_2 \rightleftharpoons 4NO + 6H_2O$$

The outgoing gases are cooled and mixed with air when NO_2 is formed. Nitrogen dioxide is dissolved in water to provide nitric acid.

$$3NO_2 + H_2O \longrightarrow 2HNO_3 + NO$$

Unreacted nitric oxide is reoxidized and changed into nitrogen dioxide which is converted into nitric acid. By this method about 54% concentrated nitric acid can be obtained.

***(iii)* Neutralization.** Nitric acid obtained in the above step is heated to about 75°C and is allowed to fall in the form of a fme spray from the top of a tower called neutralizer while ammonia gas preheated to about 70°C is introduced from the bottom of the neutralizer which results in the formation of ammonium nitrate.

$$NH_3 + HNO_3 \longrightarrow NH_4NO_3$$

The liquor obtained at the bottom of the neutralizer contains 84% ammonium nitrate. It is taken to vacuum concentrator where its concentration reaches about 92%.

***(iv)* Formation of final fertilizer.** The 92% ammonium nitrate solution is stirred with fmely powdered limestone in a granulator when rounded pellets of CAN are obtained. These are screened through a sieve and cooled limestone is added as a filter. Its function is to keep ammonium nitrate safe to handle as it is an explosive material.

***(v)* Coating.** Since CAN is very hygroscopic material, therefore, it needs protection from atmospheric moisture during its transportation and storage. The rounded pellets obtained above and stirred with concentrated solution of fmely powdered soapstone (sodium silicate). Thus a thin layer of soapstone is coated on the pellets which are dried and packed in polythene lined bags.

Q. 12. List characteristics of calcium ammonium nitrate as a fertilizer. (*Himachal, 2000*)

Ans. (*i*) It contains about 20% nitrogen.

(*ii*) It is directly assimilated by plants without having to undergo any change in the soil.

(*iii*) It is highly soluble in water and therefore assimilated by the soil very easily.

Q. 13. Briefly describe the manufacture of urea.

Ans. It is manufactured by reacting liquefied ammonia with liquefied carbon dioxide at 170–200°C and 100 – 300 atmospheric pressure. Urea formation takes place in two stages: Ammonium carbamate is formed first which decomposes into urea and water.

$$2NH_3 + CO_2 \rightleftharpoons \underset{\text{Ammonium Carbamate}}{H_2NCOONH_4} \rightleftharpoons \underset{\text{Urea}}{H_2NCONH_2} + H_2O$$

The above reactions are reversible and the equilibrium depends on the pressure of various compounds. The urea solution is concentrated in vacuum evaporators. The molten urea is then rapidly cooled and solidified by spraying down a tower. The granules are treated with diatomaceous earth and amines or dyes to give the product good handling and storage properties.

Q. 14. Describe merits and demerits of using urea as fertilizer.

Ans. Merits. (*i*) It is a concentrated source of nitrogen having 46% nitrogen while ammonium sulphate and CAN have about 21% nitrogen only. Concentration results in economy of freight and packing.

(*ii*) Its cost of production is relatively low as carbon dioxide needed for its manufacture is obtained from the ammonia plant.

(*iii*) It has no fire or explosion hazards.

(*iv*) It is suitable for application to all crops and soils.

Demerits. (*i*) It is very soluble in water and is hygroscopic. Therefore, it requires a better quality package.

(*ii*) It is not as stable as other solid nitrogenous fertilizers. Its decomposition occurs at lower temperatures releasing ammonia which results in serious loss.

(*iii*) In case sufficient precautions with regard to temperature control are not taken during its manufacture, the overheating leads to the formation of biuret and other impurities. The urea to be used as fertilizer should not contain more than 1% of these impurities as these are toxic to certain crops, particularly citrus crops.

Q. 15. Describe the preparation of calcium cyanamide ($CaCN_2$).

Ans. Calcium cyanamide is a very good fertilizer. It is prepared by heating a mixture of crushed calcium carbide with a little NaF at 1000 – 1100°C in an atmosphere of nitrogen. The reaction occurs as:

$$CaC_2 + N_2 \longrightarrow CaCN_2 + C$$

The product is grey because of carbon contamination. The complete process involves the fol lowing steps:

(*i*) **Preparation of calcium carbide.** Lime and coke are heated in an electric furnace at about 2000°C. The molten mass of CaC_2 is tapped from the furnace and cooled to a solid mass which is then broken into smaller lumps and then powdered.

$$CaO + 3C \longrightarrow CaC_2 + CO\uparrow$$

(*ii*) **Nitrification.** The powdered CaC_2 is packed in the electric oven (Fig. 24.2) with

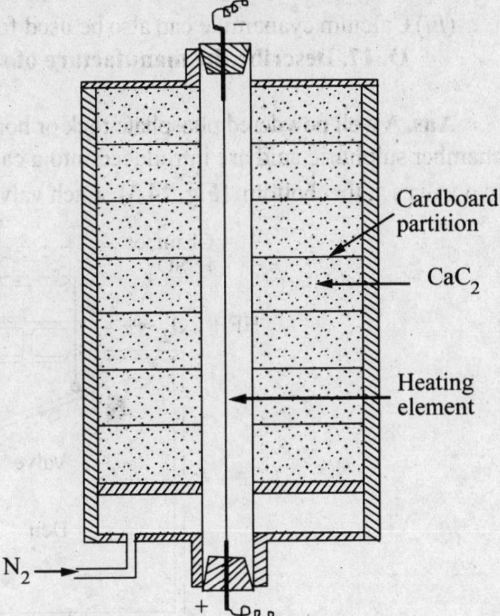

Fig. 24.2. Preparation of calcium cyanamide.

partitions of cardboard. A thin carbon rod fixed at the centre of the oven is used as a heating element. The number of heating elements depends upon the size of the oven. As many as eight such elements may be fixed in an oven. The current is switched on to heat the oven and a stream of pure nitrogen gas is passed through the oven. As the charge is heated, the cardboard gets charred and becomes porous. This prevents the charge from fusion into one mass. The temperature inside the oven is maintained around 1000°C. The charge is maintained at this temperature for 30 - 35 hours. The material, thus formed, is cooled and the sintered $CaCN_2$ is taken out of the oven.

(*iii*) **Removal of CaC_2.** $CaCN_2$, thus obtained, contains an impurity of CaC_2. The solid mixture is broken into smaller lumps which are then powdered in a ball mill. For use as a fertilizer, the carbide contents should not exceed 0.1%. To remove excess CaC_2 from $CaCN_2$, the product is washed with water which decomposes CaC_2 immediately. The commercial product is sold as Nitrolim in the market. The commercial product contains about 22% fixed nitrogen and a little lime.

Q. 16. Describe characteristics of calcium cyanamide as a fertilizer.

Ans. (*i*) Calcium cyanamide is extensively used as a fertilizer in the soil. It reacts with water slowly to undergo a series of changes giving successively cyanamide, urea, ammonia, and finally nitrates which are easily assimilated by the plants.

$$CaCN_2 + 2H_2O \xrightarrow{\text{Cold}} Ca(OH)_2 + \underset{\text{Cyanamide}}{H_2CN_2}$$

$$H_2CN_2 + H_2O \xrightarrow{\text{Warm}} \underset{\text{Urea}}{NH_2-CO-NH_2}$$

$$CaCN_2 + 3H_2O \longrightarrow CaCO_3 + 2NH_3$$

Nitrolim is a slow acting fertilizer. Therefore, it is preferred to soluble compounds like $NaNO_3$ or $(NH_4)_2SO_4$ because it provides fertility of a more permanent nature. It is a long lasting fertilizer. It must be applied as a manure below the surface of the soil before sowing and should not be applied to plants after they have grown.

(*ii*) Before the discovery of Haber process for manufacture of ammonia, cyanamide was used for the preparation of NH_3 on large scale.

$$CaCN_2 + 3H_2O \xrightarrow{\text{Steam}} CaCO_3 + 2NH_3$$

(*iii*) Calcium cyanamide can also be used for preparing urea, another nitrogenous fertilizer.

Q. 17. Describe the manufacture of superphosphate of lime, a phosphate fertilizer.

(*M. Dayanand, 2002*)

Ans. A well powdered phosphate rock or bone ash along with calculated quantity of concentrated chamber sulphuric acid are introduced into a cast iron chamber fitted with a mechanical stirrer and two valves at the bottom (Fig. 24.3). Each valve opens in a big chamber called den. The mixture is

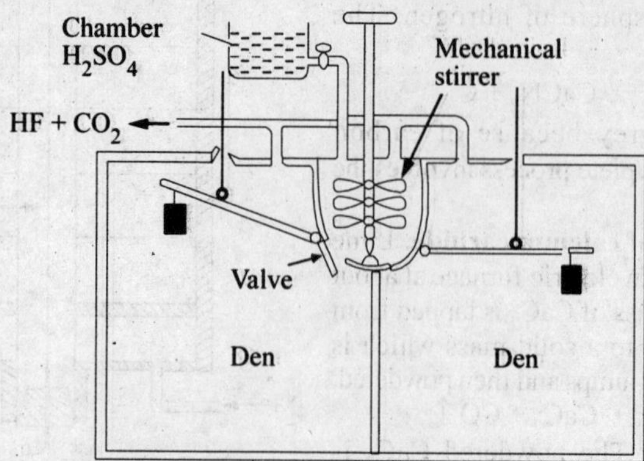

Fig. 24.3. Manufacture of superphosphate of lime.

stirred for 2 to 5 minutes and then dumped mechanically into one of the dens through the valve. The mass is kept in the den for about 24 hours.

The following reactions take place in the den:

(*i*) Calcium phosphate changes into monocalcium phosphate

$$Ca_3(PO_4)_2 + 2H_2SO_4 + 4H_2O \longrightarrow Ca(H_2PO_4)_2 + 2CaSO_4 . 2H_2O + \text{heat}$$

(*ii*) The hydrated calcium sulphate produces the setting of the reaction mixture.

(*iii*) The temperature of the mass gradually rises up to 100°C because of the exothermic nature of the reaction.

(*iv*) Since rock phosphate invariably contains calcium carbonate and calcium fluoride as impurities, therefore, they react with sulphuric acid evolving CO_2 and HF.

$$CaCO_3 + H_2SO_4 \longrightarrow CaSO_4 + CO_2 + H_2O$$
$$CaF_2 + H_2SO_4 \longrightarrow CaSO_4 + 2HF$$

The evolution of CO_2 and HF from the mass makes the charge porous and capable of being easily dried and pulverised. However, large quantities of fluorides are harmful because they produce dangerous gases and consume a good amount of sulphuric acid. As a result of the above reactions, the mixture gradually stiffens and fmally solidifies to a porous dry mass. The charge is then removed from the den and is allowed to stand in the open for 4–6 weeks. During this period, the excess of sulphuric acid is completely consumed and the mass dries up. The solid mass is crushed to a fine powder. While the reactions are going on in one den, the second den can be put to use for next lot.

Q. 18. What are the merits and demerits of using superphosphate of lime as a fertilizer?

Ans. Merits. It contains about 18% P_2O_5 and its active ingredient is monocalcium phosphate which is soluble in water. Thus it provides its phosphate in a water soluble form which is readily assimilated by the plants for their growth. In this respect it has an advantage over normal calcium phosphate which is insoluble in water and cannot be assimilated by the plants.

Demerits. (*i*) It is relatively dilute fertilizer containing a small percentage of P_2O_5.

(*ii*) Being a dilute product it requires relatively heavy freight and handling charges. Thus it is not cost effective.

(*iii*) It has a setting tendency and forms hard lumps.

(*iv*) The acidity of the product causes the package to rot and contents to spill.

Q. 19. Briefly describe manufacture of phosphate slag or Thomas Meal and discuss its use as a fertilizer.

Ans. Manufacture. It is formed during the manufacture of steel. Iron ore (containing a fairly good amount of phosphorus) on being blasted with limestone in presence of air forms calcium phosphate and calcium phosphosilicates which float on the top of the molten metal as slag. It is formed according to the following reactions:

$$4P + 5O_2 \longrightarrow 2P_2O_5$$
$$3CaO + P_2O_5 \longrightarrow Ca_3(PO_4)_2$$
$$CaO + SiO_2 \longrightarrow CaSiO_3$$

The slag is separated and cooled when it forms a solid mass. It is ground to a fine powder, packed and supplied as a fertilizer.

Its use as a fertilizer. (*i*) It contains 14–22% of P_2O_5. On being added to the soil it decomposes in due course of time giving a steady supply of phosphoric acid for plant growth.

(*ii*) It contains about 40% of lime which is readily available for the neutralization of acidity in soils.

(*iii*) It is thought to be a double salt of tricalcium phosphate and calcium silicate, $Ca_3(PO_4)_2$. $CaSiO_3$.

Q. 20. Write short notes on

(*i*) Bone meal

(*ii*) Nitrophosphate

Ans. (*i*) Bone meal. The bones are degreased and the gelatine removed. The bones are ground and the bone meal obtained thus is ready for use as phosphatic fertilizer.

The bone meal contains phosphorus as dicalcium phosphate which is not soluble in water, therefore, it is not very widely used. It has special uses in horticulture.

(*ii*) Nitrophosphate. It is manufactured by a process similar to that used for superphosphate but nitric acid is used instead of sulphuric acid. It supplies both nitrogen and phosphorus to the plants.

Q. 21. Write a note on potash fertilizers.

Ans. The important potassium fertilizers are muriate (KCl, the highest grade with 60% K_2O), potassium sulphate with 50% K_2O and potassium nitrate (contains about 50% K_2O). In India potassium nitrate occurs as an efflorescence on the soil. It is also known as Indian saltpetre. It is obtained from such "nitrous earth" by extraction with water and the solution is then evaporated by spreading in the fields. Heat of the sun helps in evaporation and after a few days crystals of potassium nitrate are collected from the fields. This provides a cheap method of obtaining potassium nitrate. It is an excellent fertilizer because it contains potassium and nitrogen both.

Q. 22. What are mixed or NPK fertilizers? (*Calcutta, 2005*)

Ans. Fertilizers containing more than one of the elements, *i.e.*, nitrogen, phosphorus and potassium, are referred to as *Mixed Fertilizers* or *Fertilizer mixtures.* In the fertilizer industry, fertilizer mixtures are obtained by mixing the nitrogen (N), phosphorus (P) and potassium (K) fertilizers together to give balanced fertilizers. The chief sources of nitrogen are ammonia, ammonium sulphate, ammonium nitrate, CAN and urea. The sources of phosphorus are superphosphate of lime, triple superphosphate and phosphate rock and those of potassium are muriate of potash, the sulphate or Indian saltpetre. Small amounts of other elements such as boron, magnesium, cobalt, copper and zinc are added to meet special requirements of the soil. Such fertilizers are commonly known as *NPK fertilizers.*

Q. 23. (*a*) Name any three essential nutrients for plant growth.

(*b*) Differentiate between manure and fertilizer.

Ans. (*a*) Besides air and water, the three essential nutrients for plant growth are compounds that can supply nitrogen, phosphorus and potassium to the plants.

(*b*) Manures come from natural sources like plants and animals. It is usually decayed plants, animal waste, etc.

Fertilizers come from artificial sources which are chemical compounds containing N, P or K. For example, urea, superphosphate of lime and potassium nitrate.

Q. 24. (*a*) Among CAN, carbamide, nitrolim and ammonium sulphate, which one has

(*i*) the highest N content.

(*ii*) the least hygroscopic nature.

(*iii*) the slowest fertilizing action?

(*b*) Name any three K and P fertilizers.

Ans. (*a*) (*i*) Carbamide $CO(NH_2)_2$

(*ii*) $(NH_4)_2SO_4$

(*iii*) Nitrolim $CaCN_2$

(*b*) Potash fertilizers are muriate (KCl) with 60% K_2O.

Potassium sulphate K_2SO_4 with 50% K_2O

Potassium nitrate KNO_3 with 50% K_2O

Common phosphatic fertilizers are:

Superphosphate of lime with 18% P_2O_5

Phosphate slag with 14–20% P_2O_5

Triple superphosphate with 42– 46% P_2O_5.

Q. 25. (*a*) What is the tentative composition of ammoniacal liquor?

(*b*) What happens when a mixture of air and steam is passed over red hot cake?

Ans. (*a*) Ammoniacal liquor comprises two types of ammonium salts:

(*i*) $(NH_4)_2CO_3$, $(NH_4)_2S$ and NH_4Cl

(*ii*) $(NH_4)_2SO_4$, $(NH_4)_3SO_3$ and NH_4Cl

(*b*) A mixture of CO, N_2 and H_2 gases is obtained

$$C + \frac{1}{2}O_2 + N_2 \longrightarrow CO + N_2$$

$$C + H_2O \longrightarrow CO + H_2$$

Q. 26. What are biofertilizers?

Ans. A biofertilizer is a substance which contains living microorganisms which when applied to seed, plant surface or soil, promotes growth by increasing the supply or availability of nutrients to the host plant. Biofertilizers add nutrients through the natural process of nitrogen fixation. Biofertilizers can be expected to reduce the use of chemical fertilizers and pesticides. The microorganisms in biofertilizers restore the soil's nutrient cycle and build soil organic matter.

Rhizobium, Azotobacter, Azospirillum and blue green algae (BGA) are some examples of biofertilizers which have been in use for a long time. Rhizobiuminoculant is used for leguminous crops. Azotobacter can be used for crops like wheat, maize, mustard, cotton etc. Azospirillum inoculations are recommended mainly for millets, maize, sugarcane and wheat.

Microorganism function is in long duration, causing improvement of soil fertility. It maintains the natural habitat of the soil. It increases crop yield by 20 – 30%, replaces chemical nitrogen and phosphorus by 25% and stimulates plant growth.

GLASS INDUSTRY

Q. 27. Which raw materials are used for the manufacture of glass? (*Garhwal, 2000*)

Ans. Major raw materials used for glass manufacture are as follows:

(*i*) Soda ash (Na_2CO_3). It supplies Na_2O for the glass. The cheaper substitutes of soda ash are sodium bicarbonate, sodium nitrate and a mixture of salt cake (Na_2SO_4) and carbon. Sodium nitrate is specially useful for oxidising iron and in accelerating the melting.

(*ii*) Sand (SiO_2). It forms silicates on fusion with the other ingredients. The sand used in glass manufacture should be pure and free from iron as it imparts colour to the glass. The sand is washed before use in order to make it free from impurities.

(*iii*) Limestone ($CaCO_3$). It provides CaO. Sometimes burnt dolomite containing CaO and MgO is employed as a substitute of lime.

(*iv*) Potassium carbonate (K_2CO_3). It is used for making hard and superior glasses.

(*v*) Borax ($Na_2B_4O_7 . 10H_2O$). It is a minor ingredient. Besides its high fluxing power, borax lowers the expansion coefficient of glass and adds to its chemical durability.

(*vi*) **Feldspar**. It is a naturally occurring mineral with a formula $Na_2O . Al_2O_3 . 6SiO_2$ (soda feldspar) or $K_2O . Al_2O_3 . 6SiO_2$ (Potash feldspar). It is cheap, pure and easily fusible. Feldspar is used as a flux, *i.e.*, to lower the melting point, and to retard devitrification of glass.

(*vii*) **Arsenious oxide (As_2O_3)**. It is generally used for removing air bubbles from glass.

(*viii*) **Cullet.** It is the crushed glass from imperfect or defective articles or their trimmings. It makes the melting easy and utilizes the waste.

(*ix*) **Lead oxide.** Litharge (PbO) or red lead (Pb_3O_4) is used for the production of heavy and high quality flint glass, which is used in optical instruments, tableware and decorative articles.

(*x*) **Colouring materials.** These are the materials employed for imparting colour to glass. Following substances are used for imparting different colours:

Cobalt oxide	—Blue	Manganese dioxide	— Violet
Cuprous oxide	— Red	Cadmium sulphate	— Yellow
Chromium oxide	— Green	Ferric oxide	— Black

Q. 28. Give the composition of common glass. (*Garhwal, 2000*)

Ans. The ordinary soda and potash glass can be represented by approximate formula $Na_2O . CaO . 5SiO_2$ and $K_2O . CaO . 5SiO_2$ respectively. The ratio of the individual constituents is therefore fixed according to these formulae. For an ordinary *Soda glass* the proportion of the raw materials is as follows:

Sand	—	100 parts
Cullet	—	50 parts
Limestone	—	27 parts
Soda ash	—	35 parts
Salt cake	—	10 parts
Powdered charcoal	—	0.5 parts

Colouring materials and minor substances — variable small amounts.

Q. 29. Discuss manufacture of glass elaborating different steps.

Ans. The manufacturing process of glass consists of four major operations: (1) Melting, (2) Shaping, (3) Annealing and (4) Finishing. Each operation is discussed separately.

1. Melting. The ingredients called batch materials are mixed in the right proportion and heated to fusion in a furnace. Many designs of glass furnace are in use. The two most commonly used furnaces are (*i*) Tank furnace and (*ii*) Pot furnace.

(*i*) **Tank furnace.** It consists of a large rectangular tank built of fireclay blocks, measuring 125 by 30 by 5 feet and delivers about 1400 tons of glass per day. The batch materials are fed into the tank and producer gas is used as a fuel in the furnace (Fig. 24.4)

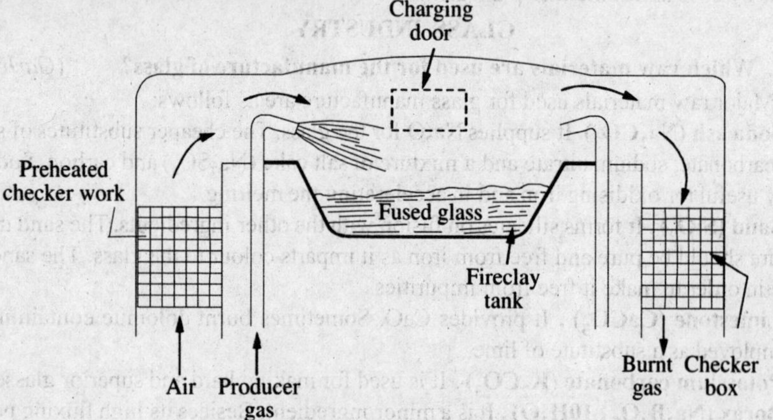

Fig. 24.4. Tank furnace for glass.

In order to economise on heat a regeneration system is provided, which consists of a chamber with a fireclay checker work on either side of the furnace. The system works in two alternate cycles of 20–30 minutes each. In one cycle the hot burnt gases from the furnace are swept through one of the chambers to heat up the checker work. In the second cycle the secondary air used for burning the producer gas is passed through the hot checker.

The charge is heated at 1400 –1500° for 10–12 hours and the following reactions take place:

$$Na_2CO_3 + CaCO_3 + 2SiO_2 \longrightarrow Na_2SiO_3 + CaSiO_3 + 2CO_2\uparrow$$

$$2Na_2SO_4 + 2SiO_2 + C \longrightarrow 2Na_2SiO_3 + CO_2\uparrow + 2SO_2\uparrow$$

During the melting lot of frothing is caused owing to the evolution of the gases like CO_2, SO_2, O_2 etc. When the frothing subsides, the temperature is raised and the molten glass is allowed to stand for some time. This is called refining and its object is to form a homogeneous mass free from gas bubbles and bits of undissolved material or batch stones. Lighter impurities called glass gall float on the surface of the melt and are skimmed off by means of silica laddles known as floaters. The lower layer of fused glass is then taken out for the next step.

Tank furnace is a continuous process and usually employed for the production of large quantities of only one variety of glass at a time.

***(ii)* Pot furnace.** It employs small fireclay pots or crucibles. The pots may be open or closed (Fig. 24.5), the latter is used when the glass is to be protected from the products of combustion. Pots of different sizes are used according to the type and quantity of the glass. Smaller pots hold 3 to 12 kg charge, while the bigger ones have a capacity of 2 tons or so. Like the tank furnace the pot furnace is also fired with producer gas and is provided with a heat regenerative system.

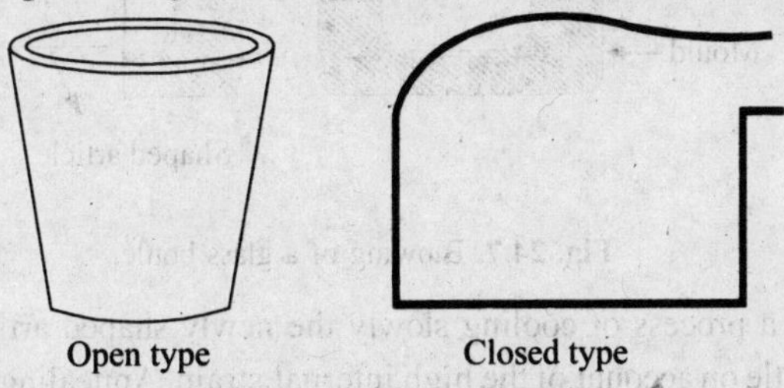

Fig. 24.5. Pots for making glass.

The batch materials are put in the pots. They are placed in a circle inside a furnace and heated by burning producer gas around them (Fig. 24.6). The reactions take place as in tank furnace. When the fusion is complete the pots are removed from the furnace and the fused plastic mass is taken out for

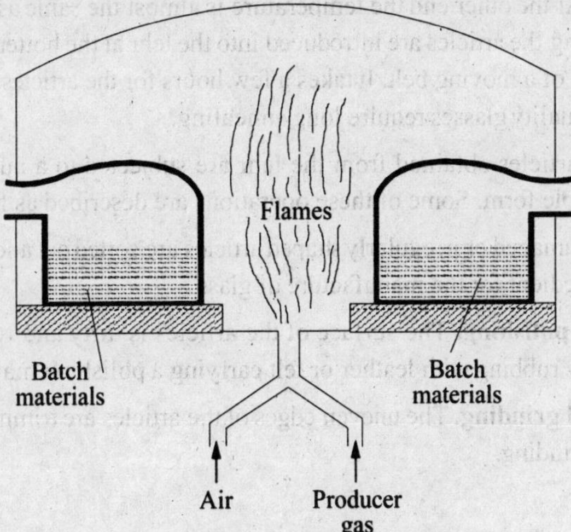

Fig. 24.6. Pot furnace for glass.

shaping. Pot furnace is employed for the production of high quality glass, since the charge is protected from the products of combustion.

2. Shaping. The plastic glass formed in the furnace is next shaped or formed into the desired articles. It is accomplished by *blowing* from mouth or by means of a machine. Glass blowing is a skilful art. The blowing of a glass bottle is done as illustrated in Fig. 24.7. A lump of the plastic glass is taken on a long iron pipe. It is elongated under its weight when hung downwards. The elongated lump is introduced into a mould and is inflated by blowing air into it from the mouth. On cooling, the bottle is taken out by removing the two halves of the mould.

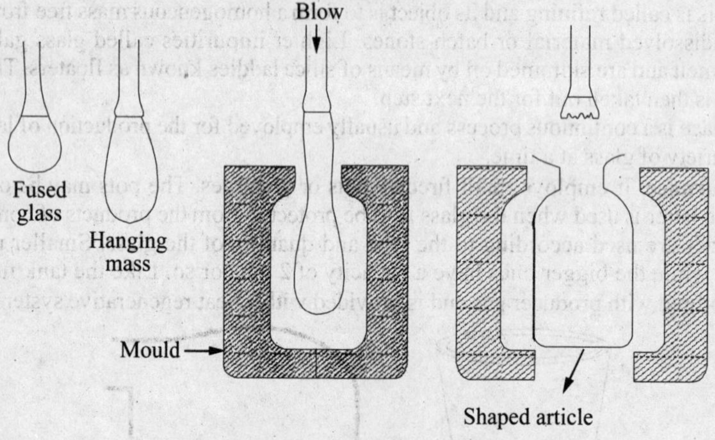

Fig. 24.7. Blowing of a glass bottle.

3. Annealing. It is a process of cooling slowly the newly shaped articles. If they are cooled quickly they become brittle on account of the high internal strain. Annealing allows the molecules to arrange themselves such that there is no internal strain when the mass is cooled.

Annealing is done in a tunnel-like oven called **lehr** which is 50 to 60 feet long. At one end the temperature is a little below the softening point of glass, *i.e.*, 500–600°, and it gradually falls along the length of the oven. At the other end the temperature is almost the same as the room temperature. Immediately after shaping the articles are introduced into the lehr at the hotter end and travel towards the cooler end by means of a moving belt. It takes a few hours for the articles to move along through the tunnel. Some high quality glasses require long annealing.

4. Finishing. The articles obtained from the lehr are subjected to a number of operations for bringing them to a useable form. Some of these operations are described as follows:

***(i)* Sorting.** The damaged or irregularly shaped articles are sorted out and rejected. This material called Cullets is an ingredient for the manufacture of glass.

***(ii)* Cleaning and polishing.** The surface of the articles is dirty and rough. It is cleaned and smoothed or polished by rubbing with leather or felt carrying a polishing material.

***(iii)* Trimming and grinding.** The uneven edges of the articles are trimmed to a proper size and then made smooth by grinding.

(*iv*) Rounding edges. Tiny flames are played on the sharp edges to make them blunt or rounded.

Q. 30. Describe different varieties of glass. (*Garhwal, 2000; Bangalore, 2002*)

Ans. The chief varieties of glass are: (*i*) *Lime Glass,* (*ii*) *Borosilicate Glass,* (*iii*) *Flint or Lead Glass and* (*iv*) *Coloured glass.*

1. Lime glass. This glass involves the use of lime or limestone along with the alkali carbonates (Na_2CO_3 and CO_3). This is further classified into: (*i*) Soft Glass and (*ii*) Hard Glass.

(*i*) Soft glass. It is made by fusing soda ash, sand and limestone together. It is believed to have an approximate composition $Na_2O . CaO . 5SiO_2$ (or $Na_2O . CaO . 6SiO_2$). It fuses at a low temperature and cannot stand mechanical injuries. It is used for making bottles, tumblers, window panes, cheap glassware etc.

(*ii*) Hard glass. It is obtained by fusing potassium carbonate and limestone together. It is understood to have a formula $K_2O . CaO . 5SiO_2$ (or $K_2O . CaO . 6SiO_2$). It is harder and stronger than the soda glass. It is used for making apparatus capable of standing high temperature and pressure.

2. Borosilicate glass. In this glass a part of the silica is replaced by boric oxide (B_2O_3). This is super cooled solution of various borates and silicates and hence its name borosilicate glass. This is made by fusing a mixture of sand, lime, borax and alkali carbonates. Sometimes a little alumina (Al_2O_3) and zinc oxide are also added to them. The former makes the glass more durable and the latter adds to its thermal and chemical resistance. Pyrex and Jena are two well-known trade marks of borosilicate glasses, which are widely used for making laboratory apparatus, thermometers, kitchenware, tableware, insulators etc.

3. Flint or lead glass. This is made from potassium carbonate lead oxide or red lead and sand. This is heavy, shining and extraordinarily durable. This is used for making optical instruments like lenses, prisms, etc. and cut glass articles, like objects of decoration, imitation diamonds or jewels.

4. Coloured glass. Glass is coloured in various shades by mixing some oxides or salts of certain heavy metals such as copper, chromium, cobalt, iron and so on. These metals form their silicates with almost the same colour shades as produced in a borax bead. Some substances disperse in the glass in the colloidal state imparting characteristic colours to the glass.

Q.31. What is clay? Give some characteristics of clay.

Ans. Clay is a general term for combinations of one or more clay minerals with traces of metal oxides. Geological clay deposits are composed of phyllosilicate minerals containing varying amounts of water.

Clay minerals are typically formed over long periods of time by gradual chemical weathering of rocks by low concentrations of carbonic acid and other solvents. Some clay minerals are also formed by hydrothermal activity. Clays are distinguished from other fine grained soils by difference in size and mineralogy.

Clays exhibit plasticity when mixed with water in certain proportion. When dry, clay becomes firm and when fired in a kiln, permanent physical and chemical changes occur. These reactions cause the clay to be converted into a ceramic material. Because of these properties, clay is used for making pottery items, both utility and decorative. Different types of clay when used with different minerals and firing conditions are used to produce earthenware, stoneware and porcelain.

Q.32. What is feldspar? Give some of its applications

Ans. Feldspars are a group of rock forming tectosilicate minerals which make up as much as 60% of earth's crust. They are aluminosilicates like K Al Si_3 O_8, NaAl Si_3 O_8 and Ca Al_2 Si_2 O_8. About 20 million tones if feldspar was produced in 2010 mostly by three countries: Italy (4.7 Mt), Turkey (4.5. Mt) and China (2Mt).

Feldspar is a common raw material used in glass making, ceramics and to some extent as filler and extender in paints, plastics and rubbers. In glass making, alumina from feldspar improves product hardness, durability and resistance to corrosion. In ceramics, the alkalis in felspar CaO, Na_2O and K_2O act as a flux lowering the melting temperature of the mixture. Fluxes melt at an early stage in the firing process, forming a glassy matrix that bonds the other components of the system together. In earth sciences and archaeology, felspars are used for K-Ar dating and optical dating.

Q.33. What is porcelain? How is it made?

Ans. Porcelain is a ceramic material made by heating raw materials generally including clay in the form of kaolin, in a kiln to temperatures between 1200°C to 1400°C. The toughness, strength and translucence of porcelain arise mainly from the formation of glass and the mineral mullite within the fired body at these high temperatures. Porcelain can informally be referred to as 'China' as China was the birthplace of porcelain making. Properties associated with porcelain include low permeability and elasticity, considerable strength, hardness, toughness, whiteness, translucency and resonance. It offers a high resistance to chemical attack and thermal shock. Porcelain is used to make household wares, decorative items and objects of fine art.

Kaolin clay is the primary material from which porcelain is made, other raw materials can include feldspar, ball clay, glass, bone ash, etc.

Normally, porcelain does not require glazing to render it impermeable to liquids. However for making decorative articles, porcelain is subjected to glazing to make it resistant to dirt and stains. Porcelainwares may be decorated under the glaze using pigments that include cobalt or copper.

Q.34. What is glazing? How is it carried out?

Ans. Glaze is a layer or coating of a vitreous substance which has been fused to a ceramic object through firing. The process to achieve this is called glazing. Glaze can serve to colour, decorate, strengthen or waterproof an object. Glazing is functionally important for earthenwares which would otherwise be unsuitable for holding liquids due to porosily Glazes also enhance a design or texture which may be either the natural texture of the clay or painted design.

Ceramic glazes generally contain silica to form glass in combination with a mixture of metal oxides such as Na_2O, K_2O and CaO which act as flux and allow the glaze to melt at a particular temperature. Alumina stiffens the glaze and prevents it from running off the piece. Iron oxide, copper carbonate and cobalt carbonate are added as colourants Glazing may be achieved by applying the glaze by dry dusting a mixture over the surface of the clay body or by inserting salt or soda into the kiln at high temperature to create a sea of sodium vapours that interact with aluminium and silica oxides in the body to form and deposit glass. Liquid glazes – suspension of various powdered minerals and metal oxides can be applied by dipping pieces directly into the glaze, pouring the glaze over the piece, spraying it onto the piece with an airbrush or any other tool that will give the desired result.

Q.35. Discuss vitrification in the light of glass transition temperature.

Ans. Vitrification is the transformation of a substance into glass. Usually, it is achieved by rapid cooling of a liquid through the glass transition. Certain chemical reactions also result into the formation of glasses. Vitrification is characteristic for amorphous substances or disordered systems and occurs when bonding between elementary particles (atoms, molecules and forming blocks) becomes higher than a certain threshold value. Thermal fluctuations break the bonds, therefore, the lower the temperature, the higher the degree of connectivity. Because of that amorphous materials have a characteristic threshold temperature termed glass transition temperature (T_g). Below T_g amorphous materials are glassy whereas above T_g they are molten. In a wider sense, the embedding of the material in a glassy matrix is also called vitrification.

Ordinary soda-lime glass used in windows and drinking containers is created by the addition of sodium carbonate and lime (CaO) to silicon dioxide. Without these additives, silicon dioxide would require very high temperature to obtain a melt and subsequently a glass.

25 ENVIRONMENTAL CHEMISTRY

Q. 1. Define environmental chemistry? What are the different components of environment?

Ans. Environmental chemistry is defined as a branch of chemistry that deals with the study of sources, reactions and effects of chemical species on air, water and soil. It is a multidisciplinary science involving chemistry, physics, life sciences, medicine, agriculture and public health. The accessible components of the environment comprise the following:

(*i*) Atmosphere, (*ii*) Hydrosphere, (*iii*) Lithosphere, (*iv*) Biosphere

Each is separately described as under.

(*i*) Atmosphere. The atmosphere refers to gaseous layers around earth which support life on the earth and saves us from the environments of outer space. The atmosphere absorbs most of the harmful radiations from the outer space and keeps heat balance of the earth to sustain life on the earth.

(*ii*) Hydrosphere. It comprises all kinds of sources of water, such as oceans, rivers, lakes, glaciers, polar ice caps, groundwater, etc. Most of the earth's water (96%) is contained in oceans and this water is unfit for human consumption due to high salt content.

(*iii*) Lithosphere. It refers to earth's crust containing the outer mineral cover up to a depth of about 16 km. It comprises soil, minerals, air, water and some organic matter.

(*iv*) Biosphere. It refers to the domain of living organisms in contact with the atmosphere, hydrosphere and lithosphere. Activities of living organisms influence environments and vice versa. For example, oxygen and carbon dioxide content depends a great deal on animal and plant kingdom through photosynthesis, decay and respiration.

Q. 2. How do you divide the atmosphere into various components?

Ans. Atmosphere can be divided into the following components:

(*i*) *Troposphere,* (*ii*) *Stratosphere,* (*iii*) *Ionosphere,* (*iv*) *Mesosphere,* (*v*) *Thermosphere.*

(*i*) Troposphere. The portion of the atmosphere closest to the earth is called the troposphere. The troposphere extends upwards to a distance of about 11 kilometres. The troposphere contains about 70% of the mass of the atmosphere. The important chemical constituents of the troposphere are N_2, O_2, CO_2, noble gases and water vapours. Its density decreases with increase in altitude. Normally, in the lower air layer of troposphere, the temperature decreases as one rises above the earth.

(*ii*) Stratosphere. Sometimes temperature (instead of decreasing) may increase to some extent with altitude. Such a situation is known as *temperature inversion.* In temperature inversion a layer of cold air is trapped by a warmer upper air mass. The pollutants in the lower air cannot rise because they are in a colder air mass that is being held down by the warmer upper air mass. Although such temperature inversions do not occur very often, yet whenever they occur, the accumulated pollutants can lead to discomfort, disease and death.

Region beyond 11 km up to an altitude of about 40 km is known as stratosphere. Here the temperature increases with increase in altitude. Ozone is the important chemical compound of this region which absorbs ultraviolet radiation and raises the temperature in the zone. This act of ozone protects us from the harmful effects of sun's ultraviolet radiation.

***(iii)* Ionosphere.** The region beyond stratosphere is dominated by +ve ions such as O_2^+, O^+ and NO_2^+ and free electrons. It is known as ionosphere. Ions can persist independently so long as ultraviolet radiations from the sun are available. In the absence of ultraviolet radiations at night, +ve ions recombine with electrons to form neutral species in regions of higher concentration of ions.

***(iv)* Mesosphere.** Mesosphere extends from about 40 km to an altitude of about 80 km. Here, once again, the temperature decreases with increase of altitude due to low level of ozone.

***(v)* Thermosphere.** It extends from about 80 km. Here atmospheric gases split into atoms and undergo ionisation after absorption of solar radiations. Here temperature increases with increase in altitude and shows a maximum of 1200°C.

Q. 3. What is the difference between pollutant and contaminant?

Ans. Pollutant. If the concentration of a substance already present in nature is increased to undesirable proportion by human action, so that it spoils the environment and reduces quality of life or affects health, then it is called as a pollutant. Carbon monoxide, sulphur dioxide, oxides of nitrogen, pollen, dust etc. are pollutants.

Contaminant. A substance that does not occur in nature but is introduced in significant amounts into the environment by human action is known as a contaminant. A contaminant also reduces quality of life and affects our health.

Methylisocyanate (MIC) CH_3NCO does not occur in nature. In December 1984, three tanks of MIC of Union Carbide exploded in Bhopal killing over 10,000 persons. This man-made substance, because of its dangerous effects, is a contaminant.

Q. 4. Define the terms:

***(i)* Receptor *(ii)* Sink**

Ans. *(i)* Receptor. It is an element or organism affected by a pollutant. For example, human being is the receptor of smog as it causes irritation in eyes and breathing problems. It is also a receptor of contaminated water because cholera and gastroenteritis are caused by it.

***(ii)* Sink.** The element or surroundings that interact and consume a long lived pollutant is known as a sink. A marble deposit acts as a sink for atmospheric acid and in the process gets damaged.

$$CaCO_3 + H_2SO_4 \longrightarrow CaSO_4 + H_2O + CO_2$$

Ocean are sinks for atmospheric carbon dioxide. Groundwater and subsoil water act as sinks for pesticides used in agriculture.

Q. 5. Explain the following terms:

(*a*) Threshold limiting value (TLV), (*b*) Dissolved oxygen (DO), (*c*) Biological oxygen demand (BOD), (*d*) Chemical oxygen demand (COD) *(Calcutta, 2005)*

Ans. (*a*) Threshold limiting value (TLV). This indicates the permissible level of a toxic pollutant in atmosphere to which a healthy worker is exposed during eight hours a day or 40 hours a week for lifetime without any adverse effect. TLV of a pollutant is determined by experimentation on animals, medical knowledge and experience and environmental studies. TLV values of some common pollutants are given as under:

Pollutant	Arsenic	Benzene	CO	CCl_4	Cl_2	$CHCl_3$
TLV (ppm)	0.2–0.5	10	50	5	1	10
Pollutant	DDT	H_2S	O_3	NO	Pb	Hg
TLV (ppm)	1	10	0.1	25	0–15	.01–.05

(*b*) Dissolved oxygen (DO). It is an important parameter to determine water quality because it is vital for aquatic life. The optimum value for good water quality is 4-6 mg/lit of DO. Lower DO values indicate polluted water unfit for human or aquatic consumption.

(*c*) Biological oxygen demand (BOD). The capacity of the organic matter in the sample of natural water to consume oxygen is called its BOD. It is evaluated experimentally by determining dissolved O_2 at the beginning and at the end of a 5-day period in a sealed sample. The BOD equals the amount of oxygen consumed in the period as a result of the oxidation of dissolved organic matter in the sample.

(*d*) Chemical oxygen demand (COD). This is an important water quality parameter. It is an index of the organic content of water because the most common substance oxidised by dissolved oxygen in H_2O is organic matter having a biological origin, such as, dead plant and animal waste. Dissolved oxygen in H_2O is also consumed to some slight extent for oxidation of inorganic matter such as dissolved ammonia, and ammonium salts into nitrates. COD is determined with the help of acidified dichromate solution. In practice, excess dichromate solution is added to the sample and the resulting solution is back titrated with Fe^{2+} ion.

Q. 6. What is greenhouse effect? (*Bangalore, 2002; Rohilkhand, 2003; Madras, 2004*)

Or

What can be the effect of increasing concentration of CO_2 on the atmosphere?

Ans. Some gases in air such as CO_2 temporarily absorb thermal infrared light ($\lambda = 4000–5000$ nm) and, therefore, not all the IR emitted from the Earth's surface and atmosphere escape directly into space. Shortly after its absorption by CO_2 molecules, this thermal IR light is re-emitted in all directions. Thus some thermal IR is redirected back towards the Earth's surface and consequently further heats both the surface and the air. *This phenomenon of redirection of thermal IR towards the Earth is called the greenhouse effect.* The fact that our planet is not covered by a thick sheet of ice is due to the natural operation of the greenhouse effect. The surface is as much warmed by this mechanism as it is by the solar energy it receives directly. The atmosphere operates in the same way as a blanket, retaining within the immediate region some of the heat released by a body and thereby increasing the local temperature. A matter of concern, however, is that increasing the concentration of the trace gases such as CO_2, CH_4, N_2O, CFC's in air that absorb thermal IR, would result in the redirection of even more of the outgoing thermal IR energy and would thereby increase the average surface temperature beyond +15°C. This phenomenon is referred to as the *enhanced greenhouse effect* to distinguish its effect from the one that has been operating naturally for millions of years.

The principal gases of the atmosphere N_2, O_2, and Ar do not absorb infrared light. The atmospheric gases that have predominantly produced greenhouse warming till recently are water (2/3 of effect) and CO_2 (responsible for 1/4 of effect). The greenhouse gases whose concentration has been small in atmosphere but have ability to warm the air substantially are CH_4, N_2O, $CFC1_3$, CF_2Cl_2 and O_3. Their concentration in atmosphere is increasing with industrialisation and excessive use of these materials.

Chlorofluoro carbons (CFC's) have the greatest potential to cause global warming due to their higher efficiency of absorbing thermal IR. Each molecule of CFC has the potential to cause the same amount of global warming as do tens of thousands of CO_2 molecules. CFC's are used excessively in freezers, refrigerators and airconditioners.

The net effect of increased level of all these gases is increased greenhouse effect, *i.e.*, increased global warming which may lead to increased average global temperature.

Q. 7. What are the implications of increased (enhanced) Greenhouse effect?

Ans. (*i*) Melting of Antarctic Greenland ice sheets and breaking of icebergs into oceans would cause a major increase in sea level. This will pose a threat to submerge many coastal countries.

(*ii*) The total amount of global rainfall will increase but some regions will receive less rainfall.

(*iii*) Tropical storms, hurricanes, etc. will be stronger and more frequent and hence more devastation.

(*iv*) The number of days having intense showers and high temperatures both will increase.

(*v*) Ocean temperature may change which may affect marine life adversely.

(*vi*) Human health will be affected by global warming. Increased number of hot days and extreme weathers may cause increased chronic respiratory diseases, heart disease, high blood pressure, etc.

(*vii*) Some areas may become more humid or wetter but some other areas will become drier.

(*viii*) Increase in greenhouse gases is expected to cause a cooling of the stratosphere. This may happen firstly because most thermal IR radiations will be absorbed at low altitudes and little is left to warm the stratosphere. Secondly, at stratospheric temperatures, CO_2 emits more thermal IR to space than it absorbs. Cooling of stratosphere would mean an enhanced greenhouse effect, that is, greater global warming.

AIR POLLUTION

Q. 8. Write a note on air quality standards.

Ans. Advanced countries of the world have set standards for air quality for the sake of healthy living. Air quality standards for some primary pollutants are given in Table 25.1. According to air quality standards, it is not appropriate to evaluate pollutant gases in terms of their tonnage produced because it does not take into account the possibility of one pollutant being more harmful than another. Thus, particulates are the most toxic pollutants in the list (Table 25.1) although tonnagewise CO is the most produced among primary pollutants. The relative toxicity value, though qualitative, indicates seriousness of particulate pollution.

Table 25.1. Air quality standards for primary pollutants.

Pollutant	*Tolerance Limit*	*Levels* $\mu.gm^3$	*Relative Toxicity*
CO	9.0* ppm	10,000	1.0
Hydrocarbons	—	19,300	2.1
SO_x	0.50	1,430	28.0
NO_x	0.25	514	78.0
Particulates	—	375	107.0

*Exposure not to exceed more than once/year for 8-hour period.

Q. 9. Trace the sources and sinks of CO in the atmosphere. How does CO act as pollutant if inhaled in excess? (*Himachal, 2000*)

Or

What are the different sources of CO pollution? What are the effects of continuous exposure of CO on human beings? (*Lucknow, 2010*)

Or

Discuss air pollution caused by CO.

Ans. Sources. Automobile engines and furnaces are responsible for an increased release of CO into the atmosphere.

(*i*) Reaction of CO_2 with coke at higher temperature in industrial furnaces such as blast furnace.

$$CO_2 + C \longrightarrow 2CO$$

(*ii*) Incomplete combustion of fuel, carbon compounds and coal and burning of gasoline in vehicles are the main sources of CO.

$$2C + O_2 \longrightarrow 2CO\ (g)$$

(*iii*) Some natural processes like volcanic activity, natural gas emission, seed germination, electrical discharge during storm etc. release some amount of CO into the atmosphere.

Sinks. Some micro-organisms in the soil remove CO from the air under ambient conditions and thereby act as major CO sink. Of these, fungi appear to be quite active in the removal of CO.

Pollutant. Its danger as air pollutant comes from the fact that it binds itself readily with iron atom of the haemoglobin (Hb). Consequently, when inhaled along with air, it combines with haemoglobin (Hb) in the capillaries that contact the air sacs of the lungs. Stable compound carboxyhaemoglobin (Hb–CO) is obtained as a result of this combination.

$$Hb + CO \longrightarrow Hb - CO$$

Haemoglobin is needed to bind O_2 molecules and transport them from lungs to muscles where they are delivered to myoglobin molecules which store O_2 until it is required for metabolic action. Since CO combines with Hb more readily than O_2, the amount of Hb available for O_2 transport decreases. It has been found that people exposed to low concentration of CO have difficulty in distinguishing varying degrees of brightness and in estimating time intervals accurately. This results in lack of O_2 supply for metabolic action. Higher concentration of CO causes dizziness and, if exposure to CO continues, it leads to death. CO poisoning can be cured by exposing the affected person to fresh supply of O_2, whereby the reverse reaction occurs:

$$COHb + O_2 \rightleftharpoons O_2\,Hb + CO$$

Q. 10. Discuss measures for the control of CO pollution.

(*Punjab, 2000; Himachal, 2000; Calcutta, 2005*)

Ans. Since vehicular exhaust is the main source of CO content in the atmosphere, efforts to check CO pollution should be directed at improving the functioning of gasoline internal combustion engine so that the exhaust is as harmless as possible. The exhaust of internal combustion engine consists of mixture of CO, oxides of nitrogen NO_x, hydrocarbons HC, particulate matter.

The following measures may be adopted for checking CO pollution:

***(i)* Use of anti-pollution exhaust catalysts.** Automobile exhaust is treated with an anti-pollution catalyst before discharging into the atmosphere. The process consists of two steps:

(*a*) In the first step, oxides of nitrogen NO_x are reduced mainly to N_2 and traces of NH_3. This is done by reacting the exhaust gases over finely divided Pt catalyst. The formation of NH_3 pollutant is kept to a minimum.

(*b*) In the second step, air is introduced to oxidise CO and hydrocarbons into CO_2 and H_2O in the presence of finely divided Pt catalyst.

The anti-pollution catalyst is not very practical because its surface gets coated with Pb particles and consequently the catalyst becomes ineffective. Lead is present in gasoline as $Pb\,(C_2H_5)_4$ which is added in small amounts as an anti-knocking agent. However, car engines have been redesigned to accept unleaded petrol. This will go a long way to check the menace of lead on the exhaust catalyst.

***(ii)* Modification of internal combustion engines.** As an alternative to catalytic purification of vehicular exhaust, starified charge engines are coming up. It consists of an extra combustion chamber where fuel-rich mixture is ignited at low temperature to minimize the formation of NO_x. This burning mixture is then fed into a bigger combustion chamber, where a fuel-deficient mixture (containing excess air) is ignited. The presence of excess air ensures complete combustion of CO and hydrocarbon into CO_2 and water.

***(iii)* Substitute fuels.** Natural gas in compressed form (CNG), liquefied petroleum gas (LPG) and liquefied hydrogen as (LHG) have been tested. These are relatively pollution-free fuels but less economical and difficult to store. Alcohol-gasoline mixtures are used as fuels but burning of alcohols leads to aldehydes as combustion product. These fumes irritate eyes.

***(iv)* Alternative power sources.** Sources such as solar cells, steam and electric power are possible but none of these has been put to practice because of economic non-viability.

Q. 11. A gas is colourless, tasteless and odourless but highly toxic. What are its harmful effects?

Ans. The gas is carbon monoxide. For harmful effects, see Q. 10.

Q. 12. Discuss sources and sinks of carbon dioxide in the atmosphere.

Or

Carbon dioxide is inert and harmless gas, yet it is considered to be a serious pollutant. Explain.

Or

Discuss the role of CO_2 as a pollutant. (*M. Dayanand, 2000*)

Ans. Sources. The main sources of CO_2 are:

(*i*) Complete combustion of fuels and carbon compounds and carbonates.

$$C + O_2 \longrightarrow CO_2$$

$$CaCO_3 \xrightarrow{\Delta} CO_2 + CaO$$

(*ii*) Auto-exhaust also makes significant contribution in increasing levels of CO_2 in atmosphere.

(*iii*) Biological decay and release from ocean are the two important natural processes for the formation of CO_2.

Sinks. (*i*) The most important sink of CO_2 is the ocean. Oceans contain bulk of dissolved CO_2 in the form of carbonates and bicarbonates and organic compounds.

(*ii*) Another sink of CO_2 are the living green plants in which the photosynthesis rate is accelerated with increase in CO_2 level.

Pollutant. Carbon dioxide is a colourless gas with faint smell and weakly acidic taste in solution in H_2O. It is one and a half times as heavy as air. It is moderately soluble in water and more so under pressure. The increasing concentration of CO_2 (presently 325 ppm) can lead to increase in earth's temperature (refer to Q. 6 on greenhouse effect). A slight increase in earth's temperature can cause havoc in times of change in world climate, melting of glaciers, flooding of the coastal plains and change in biological activity of the ocean.

Plants help in maintaining a balance in the atmospheric CO_2 level by using it in photosynthesis.

$$6CO_2 + 6H_2O \longrightarrow C_6H_{12}O_6 + 6O_2$$

The atmospheric level of 290 ppm of CO_2 in 1870 has risen to 325 ppm due to enhanced combustion of fossil fuel and large-scale deforestation. Consumption of oxygen in combustion has almost no effect on the large oxygen level in the atmosphere but it contributes a lot towards an increase in the CO_2 content of the atmosphere.

Control of CO_2 pollution. CO_2 level can be maintained by reducing its release from industrial processes and controlling vehicular exhaust. An increased rate of photosynthesis from increasing forest cover is the most practical way of maintaining CO_2 level and ensuring moderate climatic conditions.

Q. 13. Name the oxides of nitrogen present in the atmosphere. What are the sources and sinks of NO_x? What is their polluting action on the plants and aquatic life?

Or

What are the principal environmental effects of NO_2?

Or

Discuss air pollution caused by oxides of nitrogen.

Ans. Oxides of nitrogen present in the atmosphere are nitrous oxide N_2O, nitric oxide NO and nitrogen dioxide NO_2.

Nitrous oxide arises from microbiological process in soil and occurs in trace amounts in unpolluted air and does not influence chemical reactions. At higher altitudes it can undergo photo-initiated reactions and deplete the ozone layer.

$$N_2O \xrightarrow{h\nu} N_2 + O$$
$$N_2O + O \longrightarrow 2\,NO$$
$$NO + O_2 \longrightarrow NO_2 + O$$

Nitric oxide and nitrogen dioxide are important constituents of polluted air and are collectively designated as NO_x.

Sources of NO_x. (*i*) An aerobic breakdown of nitrogenous compounds in soil by bacteria, forest fires and lightning constitute the natural sources of NO_x. Distribution of NO_x from soil is more or less uniform but that from man-made sources varies from area to area.

$$N + xO \longrightarrow NO_x$$

(*ii*) The major man-made sources of NO_x are combustion of coal, oil, natural gas and gasoline. The basic reactions are:

$$N_2 + O_2 \longrightarrow 2NO$$
$$2NO + O_2 \longrightarrow 2NO_2$$

Sinks of NO_x. Natural processes and photochemical reactions take care of NO_x in the atmosphere and the end product is nitric acid, HNO_3. The reactions for the formation of HNO_3 are:

(*i*) $$2NO + O_2 \longrightarrow 2NO_2$$
$$3NO_2 + H_2O \longrightarrow 2HNO_3 + NO$$

$$4NO_2 + 2H_2O + O_2 \longrightarrow 4HNO_3$$

(*ii*)
$$O_3 + NO_2 \longrightarrow NO_3 + O_2$$

$$NO_3 + NO_2 \longrightarrow N_2O_5$$

$$N_2O_5 + H_2O \longrightarrow 2HNO_3$$

Nitric acid acts as a temporary sink and comes down either in the form of acid rain or precipitates as nitrate salts after coming in contact with bases like ammonia, lime etc.

Pollutant (Environmental effects). Nitrogen oxides are pungent chemicals of the polluted air. They bring about bifacial necrosis leading to collapse of leaves, enhancement of green colour followed by chlorosis and extensive leaf drops as well as enhanced fruit drop. In man it produces general pulmonary odema and haemorrhage.

Exposure of NO_2 in milder doses can lead to inflammation of lung tissues, whereas over-exposure can cause death.

Acid rain (HNO_3) can cause the pH of water to drop to 4 or 5. This can affect vegetation, building materials and various species of freshwater fish.

Like CO, nitric oxide can bond with haemoglobin and reduce oxygen transport efficiency of haemoglobin.

Q. 14. Describe various steps for the control of NO_x pollution.

(*Punjab, 2000; Bombay, 2002*)

Ans. (*i*) Power plants and industrial processes emit large amounts of NO_x. Such an emission can be reduced by almost 90% with the help of 2-step combustion process as given below:

(*a*) Fuel is fired at relatively higher temperatures using less than stoichiometric amounts of air to reduce the yield of NO.

(*b*) Fuel burnout is completed at relatively lower temperature in excess air to hinder formation of NO.

(*ii*) NO_x in vehicular exhaust is converted into N_2 and trace amounts of NH_3 with the help of platinum catalyst in stage I of the anti-pollution exhaust catalyst.

(*iii*) Acidic scrubbing process for the removal of NO_2 can also be worked out as under.

(*a*) Fuel gas containing NO, NO_2 and SO_2 is introduced with an oxidiser.

$$NO_2 + SO_2 + H_2O \longrightarrow H_2SO_4 + NO$$

(*b*) NO and NO_2 react to form N_2O_3 which is scrubbed by H_2SO_4 as follows:

$$NO_2 + NO \longrightarrow N_2O_3$$

$$N_2O_3 + 2H_2SO_4 \longrightarrow 2NOHSO_4 + H_2O$$

The cleaned fuel gas is released into air.

(*c*) $NOHSO_4$ is then decomposed in separate chamber and the resulting H_2SO_4 is reused.

$$2NOHSO_4 + 1/2\, O_2 + H_2O \longrightarrow 2H_2SO_4 + 2NO_2$$

(*d*) NO_2 is then converted into HNO_3.

$$3NO_2 + H_2O \longrightarrow 2HNO_3 + NO$$

and excess NO and NO_2 are recycled into the oxidiser in step (*a*).

(*iv*) NO_x in fuel gases can be removed by alkaline scrubbing solutions containing $Ca(OH)_2$ and $Mg(OH)_2$.

Q. 15. Describe sources, sinks, and polluting effects of SO_2.

Or

What are the principal environmental effects of SO_2?

Or

How does SO_2 cause pollution? *(Gujarat, 2003)*

Or

List two health problems caused by SO_2. *(M. Dayanand, 2002)*

Ans. Sources. (*i*) Most of SO_2 is formed from the combustion of sulphur containing coal and oil.

$$S + O_2 \longrightarrow SO_2$$

Because of the relatively high sulphur content of most coal deposits and the difficulty of removing it before combustion, most of SO_2 entering the air originates from coal. Lesser amounts come from the burning of oil.

(*ii*) Significant amounts of SO_2 are formed from the roasting of metallic sulphide ores.

$$2FeS_2 + 5O_2 \longrightarrow 2FeO + 4SO_2$$
$$2ZnS + 3O_2 \longrightarrow 2ZnO + 2SO_4$$
$$Cu_2S + O_2 \longrightarrow 2Cu + SO_2$$
$$2PbS + 3O_2 \longrightarrow 2PbO + 2SO_2$$

Sulphur dioxide in air is always contaminated with sulphur trioxide, SO_3 due to photolytic and catalytic oxidation reactions.

$$2SO_2 + O_2 \longrightarrow 2SO_3$$

Sinks. Sulphur dioxide in air can undergo several reactions. It can combine with water droplets to form sulphurous acid.

$$SO_2 + H_2O \longrightarrow H_2SO_3$$

Sulphur dioxide can also undergo oxidation with O_3 in air by photolytic and catalytic processes. Sulphur trioxide combines with water vapour to form droplets of H_2SO_4.

$$\left.\begin{array}{l} SO_2 + O_3 \xrightarrow{-O_2} \\ SO_2 + 1/2O_2 \end{array}\right\rangle \longrightarrow SO_3 + H_2O \longrightarrow H_2SO_4$$

Sulphuric acid comes down in the form of acid rain, which, in turn, reacts with ammonia emitted from biological process and the limestone or marble.

$$2NH_3 + H_2SO_4 \longrightarrow (NH_4)_2SO_4$$
$$CaCO_3 + H_2SO_4 \longrightarrow CaSO_4 + H_2O + CO_2$$

Thus marble statues or structures, limestone mineral and sources of ammonia act as sinks for atmospheric oxide of sulphur.

Pollutant. (*i*) Sulphuric acid droplets and various sulphate compounds formed by SO_2 affect visibility and lead to a reduction in brightness contrast between objects.

(*ii*) Droplets of H_2SO_4 and SO_2 can cause nose, eye and lung irritations. If SO_2 and very small H_2SO_4 droplets or sulphate aerosols penetrate the innermost passage of the lungs, they can cause broncho-constriction and decrease in size of the air postage in the lungs.

(*iii*) In addition to affecting humans, SO_2 and its related compounds attack vegetation also. The green pigment chlorophyll is affected by SO_2 and this causes many leaves to lose their colour. The extent of plant damage depends upon the age, plant type and part of the exposed plant to the air pollutant. Vegetable fibres like cotton, linen, and hemp as well as rayon and nylon are attacked by SO_2. Materials composed of limestone or marble are attacked by H_2SO_4. Surface of marble faced buildings are eaten away by acidic pollutants. Damage to Taj Mahal by such pollutants has been highlighted by scientists.

Q. 16. Discuss measures for the control of SO_2 pollution. (*Punjab, 2000*)

Ans. SO_x pollution can be prevented by the following methods.

(*i*) **Removal of S from fuel before combustion.** If we remove pyrites (Fe_2S_2) from coal before combustion, one of the biggest pollution problems would be solved. Pyrites in coal can be eliminated by metallurgical techniques, such as grinding followed by washing. Organic sulphur in coal is removed by processes like carbonization, liquefaction or gasification although this is not so easy.

(*ii*) **Removal of SO_x from smokestacks.** When coal containing undesirable amount of S is burnt, SO_2 is produced. A variety of devices are now available to remove SO_2 to a great extent. The process of removal of SO_2 is known as flue gas desulphurization (FGD). The most widely used SO_2 removal method is lime- stone method, wherein $CaCO_3$ is added direct to the boiler where quick lime is formed.

$$CaCO_3 \longrightarrow CaO + CO_2$$

Quick lime CaO reacts with SO_x in the flue gas to form calcium sulphite and sulphate.

$$CaO + SO_2 \longrightarrow CaSO_3$$

$$CaO + SO_3 \longrightarrow CaSO_4$$

Calcium sulphite and calcium sulphate are removed and disposed of.

Q. 17. Describe pollution caused by hydrogen sulphide (H_2S).

Ans. Sources. (*i*) Volcanic eruptions and biological decay of protein materials in stagnant water and swamps are the two primary natural sources of H_2S.

(*ii*) It is formed as a by-product from paper mills, oil refineries, natural gas plants and chemical plants manufacturing sulphur compounds.

Sink. It is easily oxidised in air to SO_2. Consequently there is hardly any sink of H_2S itself except that it may be trapped as insoluble sulphides.

Pollutant. Hydrogen sulphide like organo-sulphur, such as, CH_3SH, dimethyl sulphide $(CH_3)_2S$, dimethyl disulphide $(CH_3)_2S_2$ etc. has rotten and objectionable smell even at very low concentrations. It has a high relative toxicity and wild exposure of H_2S leads to leaf lesions, mottling, defoliation and reduced growth.

Q. 18. Give a brief account of particulate pollution. How can this be checked?

Ans. Small-sized solid particles and liquid droplets are collectively termed as particulates. Of the various particles emitted into the atmosphere, carbon (soot) particles are common ones. Soot particles can deposit on clothing, automobiles, buildings and vegetation. It can act as a carrier for gaseous pollutants. Other particulate pollutants in the atmosphere are metals, metal salts, fly ash, dust hydrocarbons, asbestos powder, mineral particles, etc.

Sources of particles. Soot particles are produced when fossil fuels (coal, fuel oil, natural gas, wood, etc.) burn in an insufficient supply of oxygen.

Particles are also emitted from volcanic eruptions, blowing of dust and soil by the wind and spraying of salts.

Fly ash particles are released into the atmosphere from power plants, smelters and mining operations.

When hydrocarbon molecules present in coal and oil are oxidised, various organic molecules enter the atmosphere. Structures of these molecules are complex. Some of these are polynuclear aromatic hydrocarbons (PAH) containing benzene rings fused together.

Some common particulates are mentioned here.

(1) *Soot (carbon) particulates.* The most common particles released into the atmosphere are carbon (soot) particles. Gases like SO_2, PAH particles, toxic metals (Be, Cd, Cr, Pb, etc.) are also adsorbed on the surface of soot. It carries adsorbed toxic metals and toxic organic PAH molecules on its surface.

(2) *Dust particulates.* Volcanic eruptions, blowing of dust and solid by wind, spraying of salts and other solid particles by oceans, crushers, mining operations, etc. throw a lot of dust into the atmosphere. The proportion of elements in the dust show a wide variety depending upon the source. Some specific dust particles are given below.

(*i*) *Fly ash.* Smaller ash particles, called fly ash, emerge from the chimney when high ash fossil fuels are burnt. Fly ash emissions from a variety of coal-combustion units differ in composition.

(*ii*) *Inorganic particulate matter.* Among the inorganic particulate matter, lead pollution in the atmosphere invites the greatest attention. It originates from the combustion of leaded gasoline. Lead tetraacetate $Pb(C_2H_5)_4$ in gasoline reacts with O_2 and halogenated alkanes to yield lead halides which emerge through the exhaust systems.

$$Pb(C_2H_5)_4 + O_2 + C_2H_4Cl_2 + C_2H_4Br_2 \longrightarrow$$
$$CO_2 + H_2O + PbCl_2 + PbClBr + PbBr_2$$

These emissions have been controlled considerably these days with the use of unleaded gasoline.

(*iii*) *Metal oxides form* another important class of inorganic particles released into the atmosphere where fossil fuels containing these metals are burnt. As, for example, combustion of pyrite contaminated coal gives Fe_3O_4.

$$3FeS_2 + 8O_2 \longrightarrow Fe_3O_4 + 6SO_2$$

Effects of particulates. A larger concentration of dust particulates in the atmosphere can reduce visibility and disturb heat balance, influence weather by acting as nuclei upon which water condensation can occur. Soot, fumes and mist carry a variety of chemicals which may be corrosive and highly toxic. Besides, they can cause damage to buildings, sculptures, monuments, paints, etc.

A higher concentration of particulates affects human health and activity. Finer particles have ability to penetrate deep into air passage and damage lungs. Coarser particles are held in the nose and throat and get eliminated. Particles in the lungs cause breathing problems. Coal miners get black-lung disease.

Prolonged exposure to air-borne asbestos fibre dust can be very dangerous because of increasing chances of *asbestosis.*

Control of Particulates

Electrostatic precipitator. Particles acquire electrical charges when subjected to an electric field. An electrostatic precipitator operates within a smokestack. Two electrodes are fitted against the inside walls of the stack (Fig. 25.1). When a high voltage is applied between the electrodes, an electric

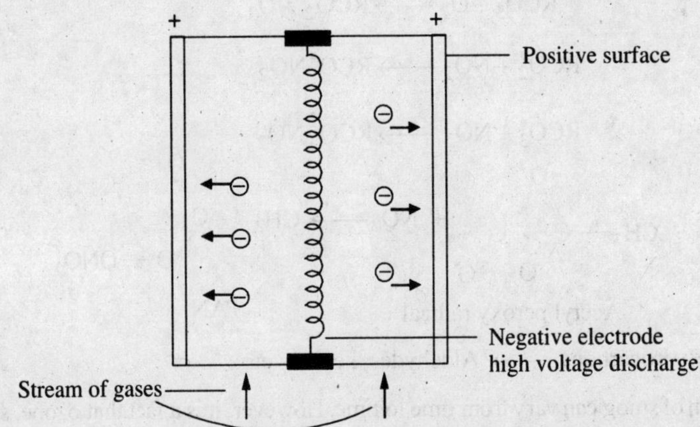

Fig. 25.1. Electrostatic precipitator.

discharge takes place across the stack and the air in the stack is ionized. The ionized air consists of the ions and free electrons. The free electrons become attached to gaseous particles moving up the stack and give them a negative charge. The negatively charged particles are attached to the positive electrode located on an inner wall of the stack. At this electrode they are collected. Later, water or the vibration of the electrodes will dislodge the particles and they will fall into a reservoir and be collected. A properly operating electrostatic precipitator can remove up to 99% of all the particles moving up the smokestack.

Q. 19. What is smog? How does it cause pollution?

(*Indore, 2001; Guwahati, 2002; Madras, 2004*)

Ans. Combination of smoke and fog is called *smog.* Early in the morning before sunrise, the traffic begins to increase with vehicles emitting CO, hydrocarbons and oxides of nitrogen. Power plants and other industries also release these chemicals. As the sun rises, ultraviolet and visible radiations fall on the earth. Molecules of NO_2 are very good absorbers of ultraviolet radiation and this starts a series of reactions.

$$NO_2 + UV - \text{radiations} \longrightarrow NO + O$$

Oxygen atoms are highly reactive and combine with O_2 to form ozone.

$$O + O_2 \longrightarrow O_3$$

Ozone then reacts with NO to form NO_2 and O_2

$$O_3 + NO \longrightarrow NO_2 + O_2$$

The NO_2 again absorbs ultraviolet radiation and the entire cycle starts again. Reactive oxygen atoms also combine with hydrocarbons coming out of the exhaust of vehicles. Oxygen atoms easily attack the double bonded molecules like C_2H_4 and 1, 3-butadiene to form highly reactive molecules. The reactive molecules undergo further reactions to produce ozone and assorted irritating molecules as, for example, peroxyacetyl nitrate (PAN) which even at low concentrations attacks plants.

Some aldehydes and ketones absorb light energy and become free radicals. They react with O_2 as

$$RCO^{\bullet} + O_2 \longrightarrow RCO_3^{\bullet}$$

These free radicals react further with O_2, NO and hydrocarbons as:

$$RCO_3^{\bullet} + O_2 \longrightarrow RCO_2^{\bullet} + O_3$$

$$RCO_2^{\bullet} + NO_2 \longrightarrow RCO^{\bullet}NO_2$$

$$RCO_3^{\bullet} + NO_2 \longrightarrow RCO_3^{\bullet}NO_2$$

$$CH_3-C(=O)-O-O^{\bullet} + NO_2 \longrightarrow CH_3-C(=O)-O-ONO_2$$

Acetyl peroxy radical PAN

$$RCO_3^{\bullet} + \text{Hydrocarbons} \longrightarrow \text{Aldehydes, ketones etc.}$$

Composition of smog can vary from time to time. However, it is a fact that ozone, some aldehydes and certain hydrocarbons are always present in smog. Smog reduces visibility, irritates eyes, nose and throat and damages plant wealth.

Q. 20. What is ozone layer ? How does it get destroyed by chlorofluorocarbons ?

OR

Describe the effect of freons on environment. (*Calcutta, 2005*)

Ans. Chlorofluorocarbons such as $CFCl_3$, CF_2Cl_2 and CF_3Cl are commonly known as freons. These are used as refrigerants, aerosol propellants, solvents for cleaning of electronic components, fire- fighting agents and packing materials.

Sources. These are generally made by the following reactions in the presence of a catalyst:

$$CCl_4 + HF \xrightarrow{SbFCl_4} CFCl_3 + HCl$$

$$CFCl_3 + HF \xrightarrow{SbFCl_4} CF_2Cl_2 + HCl$$

Pollutant. Because of spillage from industry in the process of manufacture, extensive use of spray cans, losses during leakage and filling of freon refrigerants, their extensive use as solvents for cleaning of electronic components and as fire-fighting reagents, they find their way into the upper atmosphere. Stratosphere between 15-40 km above earth has **ozone** as the important chemical compound. This **ozone** layer around the globe absorbs ultraviolet light in the range 220-330 nm and protects life on earth from severe damage.

It has been observed during the last few years that the ozone layer diminishes over the south pole year after year. Consequently holes or gaps develop in ozone layer over Antarctica and excessive short wave solar radiations reach the ground. Scientists believe that the reason for this phenomenon is that in the Antarctic spring the sun is very low over the south pole. Its radiation is insufficient to produce ozone but it is enough to release chlorine by means of various chemical processes from the trace substances like Freons that have collected there and chlorine breaks down ozone. Since the breakdown of ozone is then not counteracted by any equivalent buildup of ozone, therefore, a hole develops.

Thus freons are responsible for breaking the ozone layer around the earth. Depletion of ozone poses severe threat to mankind. Medical scientists believe that every 1% decrease in ozone leads to

2% increase in skin cancer due to exposure to sun's ultraviolet rays. Ultraviolet rays also damage immune system which may lead to increased viral infection. Increased exposure to ultraviolet radiations also damages marine plants and threatens marine animals and fish which form an important part of human food. Excessive ultraviolet radiations also damage land's plants and crops.

Q. 21. What is the importance of ozone layer? What are the different sources which are responsible for its depletion?

(*M. Dayanand, 2000; Rohilkhand, 2003; Guwahati, 2002; Calcutta, 2005*)

Ans. See Q. 20.

WATER POLLUTION

Q. 22. What are the principal sources of water pollution? (*Madras, 2004*)

Ans. The principal sources of water pollution may be broadly classified under the following categories:

(*i*) *Organic matter*
(*ii*) *Organic chemicals*
(*iii*) *Oils*
(*iv*) *Plant nutrients*
(*v*) *Sediments*
(*vi*) *Heat*
(*vii*) *Inorganic chemicals and minerals*
(*viii*) *Pesticides*
(*ix*) *Detergents*
(*x*) *Radioactive wastes.*

Q. 23. Describe source, action and breakdown of organic matter as water pollutant.

(*M. Dayanand, 2002*)

Ans. Oraganic matter comprises decomposable materials like domestic sewage, animal manure, decaying plants, food processing wastes, slaughter house wastes, paper discards and anything that can be broken down by various micro-organisms. Since micro-organisms can attack and break down the organic waste in water, therefore, *the decomposable organic matter is referred to as biodegradable.* In order to attack the degradable material, the sewage contains both aerobic and anaerobic bacteria which act under different conditions.

Aerobic bacteria need free or dissolved molecular oxygen to bring about oxidation of degradable organic matter. The oxidation products are inoffensive smelling, non-putrifying nitrites, nitrates, sulphates, phosphates etc. This kind of oxidation of sewage is called **aerobic oxidation**.

If O_2 molecules are not present in the waste, aerobic bacteria cannot utilize food and they die. On the other hand, when dissolved or free oxygen supply is below a certain value, the sewage is called stable and anaerobic bacteria bring about chemical change producing methane, H_2S, $(NH_4)_2S$ and PH_3 which give offensive smell. This kind of oxidation of sewage is called **anaerobic oxidation**. In anaerobic oxidation, the bacteria break apart combined oxygen present in organic matter nitrates, nitrites and sulphates of sewage. When anaerobic breakdown is taking place, the sewage is known as septic sewage.

The amount of oxygen needed by bacteria in order to oxidise their food is known as biochemical oxygen demand (B.O.D.). Quantitatively, *the B.O.D. is the amount of oxygen in milligrams per litre or parts per million used by micro-organism to consume degradable material.* A high B.O.D. indicates that water is polluted.

If the untreated sewage is disposed into rivers, lakes and allowed to seep into subsoil water, it has several undesirable effects. This can lead to water-borne diseases like typhoid, dysentery, cholera, hookworm infection, interviral infection, malaria, polio infections and hepatitis.

Q. 24. What are the sources of water pollution? What is the self-cleaning mechanism of water bodies?

Ans. See Qs. 22 and 23.

Q. 25. Describe water pollution caused by (*i*) oils, (*ii*) synthetic organic chemicals, and, (*iii*) plant nutrients.

Ans. (*i*) Oils. Oil pollution of seas and rivers has increased during recent years due to increased use of oil-based technology, excessive offshore mining and massive transportation of oil through waterways. The sources of oil pollution are oil spills from tankers into waterways, losses during offshore exploration and accidents, leakage from pipelines over and under waterways.

(*ii*) Synthetic organic chemicals. Chemical industry produces thousands of organic chemicals which include fuels, solvents, plastics, plasticizers, fibres, elastomers, detergents, paints, insecticides, food additives, pharmaceuticals, lubricating oils. etc. Their presence in water imparts offensive tastes, odours and colours even if they are present in low concentrations. Every effort should be made to prevent their entry into water because most of these are difficult to remove from water. Their presence in water leads to offensive taste and odour to aquatic plants, fish and sea food which enter the human food chain.

(*iii*) Plant nutrients. This includes man-made fertilizers, runoffs from farmlands, sewage treatment discharge into lakes and many chemicals that help aquatic organism like algae and weeds to grow. The presence of large quantities of nutrients can lead to surface aquatic organism rapidly reproduce and use up a great deal of dissolved oxygen. At a reduced oxygen concentration, some species of fish cannot survive. In some lakes, the rapid loss of dissolved oxygen can lead to water becoming cloudy and having an odour.

Q. 26. Describe water pollution due to lead. How can lead poisoning be cured?

(*G. Nanakdev, 2001*)

Ans. The major source of air-borne lead is lead gasoline which releases lead in the form of $Pb(CH_3)_4$, $Pb(C_2H_5)_4$, $PbCl_2$, $PbBr_2$, PbBrCl, $PbSO_4$, Pb_3O_4, PbO, PbS and Pb particles. Lead is removed from the atmosphere by wet and dry deposition processes. Rain water washes lots of lead compounds into rivers, lakes, ponds and oceans. Another source of Pb in drinking water is the lead pipes.

Lead in water is present in the form of soluble compounds, colloids and particulates. Lead in sea water is largely colloidal in nature. In fresh water of pH 6, lead is mainly present as $Pb(OH)_2$, $PbCO_3$ particles. Alkyl lead in lake and pond waters is quite stable as $Pb(CH_3)_4$, $Pb(CH_3)_3^+$ and $Pb(CH_3)_2^{2+}$.

Lead can cause disruption of the synthesis of haemoglobin leading to lack of utilization of O_2 and glucose for life sustaining processes. A higher level of Pb in blood leads to loss of appetite, anaemia, kidney disfunction, nervous disorder and brain damage. Lead is a neurotoxin.

Lead poisoning can be cured by feeding the patient with an aqueous solution of Ca complex of EDTA. Lead ions displace calcium in the EDTA complex to form chelated lead and Ca^{2+}. The soluble lead chelate is excreted in the urine.

$$\underset{\text{Soluble}}{Ca-EDTA} + Pb^{2+} \longrightarrow Ca^{2+} + \underset{\text{Soluble}}{Pb-EDTA}$$

Q. 27. Discuss how excessive use of chlorine can bring about water pollution.

Ans. Raw waters need effective chlorination before release for public water supply. Sometimes waste water is also chlorinated before discharge into rivers and other waterways. Chlorine acts a disinfectant and germicide.

Bleaching powder is added to water (1 kg per 1000 kilolitres of H_2O) and the mixture is allowed to stand undisturbed for several hours. The chemical reaction produces HClO which is a powerful germicide and disinfecting action occurs due to Cl_2 formed by the hydrolysis of bleaching powder, $CaOCl_2$.

$$CaOCl_2 + H_2O \longrightarrow Ca(OH)_2 + Cl_2$$

$$Cl_2 + H_2O \longrightarrow HCl + HClO$$

Use of bleaching powder suffers from certain drawbacks:

(*i*) It introduces Ca^{2+} ions in water, thereby making it hard.

(*ii*) Excess of bleaching powder in water can give bad taste and pungent smell to treated water.

Chlorination with chlorine has its advantages:

(*i*) It is effective and economical.

(*ii*) It can be used at low or high temperatures.

(*iii*) It produces no harmful salts in water.

In recent years, disinfection of drinking water with chlorine has received criticism. The basis for criticism was the discovery of minute quantities of organo-chlorine compounds in chlorine treated water. Some of the organo-chlorine molecules are potential carcinogenic (cancer causing).

Chlorine pollution has acquired serious proportion in some waterways due to industrial discharges from DDT and chloralhydrate CCl_3CHOH_2O plants. Both DDT and chloralhydrate in waste water are highly toxic and inhibit fish growth.

Q. 28. What are (*i*) Persistent pesticides, (*ii*) Biodegradable pesticides? (*Punjab, 2000*)

Ans. Persistent pesticides. Pesticides are applied as aerial sprays, direct to foliage or to soil. Finally they reach the soil which serves as a reservoir. However, from the land they move to air and water or are degraded. The ultimate depository is water. In case of DDT, about 25% of total DDT is likely to have been transferred to the oceans. There is significant accumulation of DDT in food chains. The average level of DDT, the most widely spread pesticide, is estimated at 5-10 ppm in human tissues.

Biodegradable pesticides. The biodegradation of pesticides in aquatic and terrestrial environments is important for better environments. In soil, a pesticide is transported into various sectors of the environment, such as, (*i*) absorbed by soil, (*ii*) leached by rain water, (*iii*) picked up by plants and animals, (*iv*) evaporates or is carried away by wind into the atmosphere. This means if the pesticide does not degrade, the sum total of pesticide in environment remains the same. It is important to degrade the pesticide to improve the quality of environment. The process of degradation is mediated by micro-organism, animals, plants and sunlight.

The major groups of soil micro-organisms are actinomycetes, fungi and bacteria. They can degrade pesticides through oxidation, ether cleavage, ester and amide hydrolysis, epoxidation, reduction etc.

Q. 29. Discuss water pollution caused by detergents. (*Guwahati, 2002*)

Ans. Detergents are cleaning agents and have been in extensive use because of their better cleansing properties compared to traditional soaps especially in hard waters. A detergent comprises the following constituents:

(*i*) Surfactant (10 to 30%), the cleanser.

(*ii*) Builder (10 to 12%) to improve action of surfactant.

(*iii*) Other ingredients such as anticorrosive sodium silicate, amide foams stabilizer, soil suspending corboxymethylcellulose, diluent sodium sulphate and water.

Surfactants are sodium salts of organic sulphonic acid R– SO_3 H. Synthetic detergents are much more soluble and, therefore, more effective than soaps in water.

$$\mathrm{HO-\overset{\overset{\Large R}{|}}{\underset{\underset{\Large O}{||}}{S}}=O}$$

(Organosulphonic acid)

They do not form insoluble precipitates with calcium or magnesium ions in hard water.

Nonbranched chain of carbon atoms is also used to bond to the benzene ring attached to SO_2 unit. Such surfactants are biodegradable easily. For example, an anionic linear alkyl surfactant as shown below

Cationic surfactants are also known and comprise quaternary ammonium salt were substituents

So^-_3

(Linear alkyl sulphonates)

R_1, R_2, R_3 on N are short chain hydrocarbons and R_4 an aromatic and the anion X is a halogen atom or an acidic group.

$$\left[\mathrm{R_4-\overset{\overset{R_1}{|}}{\underset{\underset{R_3}{|}}{N}}-R_2}\right]^+ X^-$$

The surfactant lowers the surface tension of water and gives a stable emulsion with soil particles which need removal.

The *builders* added to detergents, complex with Ca^{2+} or Mg^{2+} ions and react with H_2O to form alkaline solution which provides optimum medium for the function of surfactant. Sodium tripolyphosphate $Na_5P_3O_{10}$ is the most popular builder.

$$\left[\mathrm{O-\overset{\overset{O}{||}}{\underset{\underset{O}{|}}{P}}-O-\overset{\overset{O}{||}}{\underset{\underset{O}{|}}{P}}-O-\overset{\overset{O}{||}}{\underset{\underset{O}{|}}{P}}-O}\right]^{-5}$$

Polyphosphate builders undergo fast biodegradation by hydrolysis:

$$P_3O_{10}^{-5} + 2H_2O \longrightarrow 2HPO_4^{2-} + H_2PO_4^-$$

These hydrolysis products do not pose any threat to aquatic life but act as nutrients for plants leading to their excessive growth in ponds, lakes and rivers. Thus they act as pollutants in water.

Q. 30. Discuss radioactive pollution. How can it be prevented?

Ans. Nuclear power plants generate some pollutants as given below:

(*a*) *Low level liquid radioactive waste.*

(*b*) *Liquid and gaseous waste from fuel elements.*

(*c*) *Fission products.*

Similarly nuclear weapon testing in air as well as leakage from underground nuclear explosions give rise to radioactive fallout. This has far reaching consequences on the environment and mankind. Sr^{90} is a component of nuclear fallout and is similar to Ca chemically. It accompanies Ca into soil, plants, animals and finally into humans in bones and teeth. The presence of Sr^{90} in bones leads to disorder in blood-cell formation and causes anaemia or more serious disorders.

Radioactive nuclei in water originate from reactors, explosions and leaching of minerals. They pose a serious threat to environment.

Radioisotopes are important tool in medicine, agriculture and industry. Their use for the welfare of mankind cannot be avoided. Unnecessary exposure, their safe disposal are again as important as their use for welfare.

Q. 31. What is acid rain? (*Himachal, 2000*; *Rohilkhand, 2003*)

Ans. When fossil fuels are burned, a great deal of CO_2 enters the atmosphere. The gaseous CO_2 dissolves in water droplets to produce carbonic acid.

$$CO_2 + H_2O \longrightarrow H_2CO_3 \rightleftharpoons H^+ + HCO_3^-$$

Rain water normally is not acidic but the weak carbonic acid can give water a pH value of about 5.6. If no other acid enters the waterway the pH of water should remain between 5.6 and 7.0. However, the acidity of rain water in many parts of the world has increased significantly. This is attributed to large quantities of SO_2 and oxides of nitrogen in the air. These gases react to produce H_2SO_4 and HNO_3.

$$NO + O_3 \longrightarrow NO_2 + O_2$$

$$2NO_2 + O_2 \longrightarrow 2NO_3$$

$$NO_2 + NO_3 \longrightarrow N_2O_5$$

$$N_2O_5 + H_2O \longrightarrow 2HNO_3$$

Alternatively:

$$2NO + O_2 \longrightarrow 2NO_2$$

$$3NO_2 + H_2O \longrightarrow 2HNO_3 + NO$$

$$4NO_2 + 2H_2O + O_2 \longrightarrow 4HNO_3$$

Sulphuric acid is formed as

$$SO_2 + 1/2O_2 + H_2O \xrightarrow[\text{partciles}]{\text{Soot}} H_2SO_4$$

Suphuric acid and nitric acid are brought down from the atmosphere during rain and snow and can cause the pH of water to drop to 4 or even less than that.

Acid rain can affect vegetation, building material and various species of fish in fresh water. The Taj Mahal is threatened by acid vapour from industry.

Q. 32. Describe quality parameters and standards for drinking water.

Ans. Drinking water fit for human consumption should have the following essential parameters for water quality:

(*i*) It should be colourless, odourless, pleasant in taste and clear.

(*ii*) Its pH should be between 6.0 and 9.0.

(*iii*) It should be free from objectionable chemicals.

(*iv*) It should be free from disease producing micro-organisms. Dissolved oxygen content should be 4-6 ppm.

(*v*) Its turbidity should be less than 10 ppm.

(*vi*) Its total dissolved solids should not be more than 500 ppm.

(*vii*) It should be reasonably soft.

NON-AQUEOUS SOLVENTS

Q. 1. What are ionizing and non-ionizing solvents?

Ans. (*i*) Ionizing solvents. These are polar in nature and dissolve ionic compounds. In these solvents, ionic reactions occur. They also show self-ionization. Ammonia, sulphur dioxide and hydrofluoric acid are examples of ionizing solvents.

$$NH_3 + NH_3 \rightleftharpoons NH_4^+ + NH_2^-$$

$$SO_2 + SO_2 \rightleftharpoons SO^{2+} + SO_3^{2-}$$

$$3HF \rightleftharpoons H_2F^+ + HF_2^-$$

(*ii*) Non-ionizing solvents. These solvents are non-polar and dissolve only non-polar or neutral compounds. No ionic reactions occur in these solvents. They do not undergo self-ionization. Benzene and carbon tetrachloride are examples of non-ionizing solvents.

Q. 2. What are protic and aprotic solvents?

(*Mumbai, 2002; Rohilkhand, 2003; Nagpur, 2005; Purvanchal, 2007; Patna, 2010*)

Ans. Solvents such as NH_3 or HF which have hydrogen as their constituent are called *protic solvents.* Solvents such as SO_2 which have no hydrogen in their molecule are *aprotic solvents.*

Q. 3. Discuss auto-ionization (solvent system) theory of solvent. (*Patna, 2010*)

Ans. This theory is based on the mode of self-ionization of the solvent. Each solvent is considered to consist of an acidic part and a basic part and on auto-ionization is split into these two parts, *i.e.*, acidic part and basic part. A substance that produces or increases the concentration of acidic part of the solvent behaves as an acid. Similarly a substance that increases the concentration of basic part of the solvent behaves as a base. The behaviour of any solute is thus determined in terms of the release of a characteristic cation or anion of that solvent. For example, ammonium chloride (NH_4Cl) is defined as an acid and sodium amide ($NaNH_2$) as a base in liquid ammonia.

$$2NH_3 \rightleftharpoons NH_4^+ + NH_2^- \text{ (Self–ionization)}$$

$$NH_4Cl \rightleftharpoons NH_4^+ + Cl^- \text{ (Acid)}$$

$$NaNH_2 \rightleftharpoons Na^+ + NH_2^- \text{ (Base)}$$

Thus ammonium chloride acts as acid and sodomide as base.

Similarly, the behaviour of acidic and basic solutes in phosphorus oxy-chloride ($POCl_3$) solvent may be studied as:

$$2POCl_3 \rightleftharpoons POCl_2^+ + POCl_4^- \text{ (Self -ionization)}$$

$$FeCl_3 + POCl_3 \rightleftharpoons POCl_2^+ + FeCl_4^- \text{ (Acid)}$$

Thus $FeCl_3$ acts as an acid in $POCl_3$ solvent

$$(CH_3)_4 N^+ Cl^- + POCl_3 \rightleftharpoons (CH_3)_4 N^+ + POCl_4^- \text{ (base)}$$

Thus tetramethyl ammonium chloride acts as a base in $POCl_3$ solvent.

The reaction between tetramethyl ammonium chloride and ferric chloride may be investigated as a titration in $POCl_3$.

$$(CH_3)_4 N^+ Cl^- + FeCl_3 \xrightarrow{POCl_3} (CH_3)_4 N^+ + FeCl_4^-$$

Behaviour of acidic and basic solutes is determined by the self-ionization species of the solvents. Therefore, self-ionization of the solvent to give ionic species is very essential in solvent system concept.

Q. 4. Discuss coordination model of solvent.

Ans. Doubtful existence of ionic species in some solvents led to the development of coordination concept of solvent. Coordination of solutes by solvent molecules is quite important. When a solute dissolves in a solvent, dissolution is always accompanied by solvent coordination. However, behaviour of solutes in solvents has generally ignored the importance of coordination by the solvent and more emphasis has been placed on the self-ionization of the solvent. Lately a number of chemists have criticized the overemphasis on ionic reactions in solutions as laid down by the proponents of solvent system concept. Meek and Drago found that the reaction between tetramethyl ammonium chloride and ferric chloride could be as well carried out in triethyl phosphate $PO(OC_2H_5)_3$ as in phosphorus oxychloride $POCl_3$. This is very surprising since triethyl phosphate did not contain any chloride ion and yet $FeCl_4^-$ ion is formed.

$$FeCl_3 + POCl_3 \xrightarrow{POCl_3} POCl_2^+ + FeCl_4^-$$

Meek and Drag suggested the 'Coordination Model' as an alternative way to 'solvent system' to account for these observations. They suggested that in triethyl phosphate the chloride ion transfer takes place from one $FeCl_3$ molecule to another with the formation of a cationic iron (III) species.

$$2FeCl_3 \xrightarrow{PO(OC_2H_5)_3} [FeCl_2[OP(OC_2H_5)_3]_4]^+ + FeCl_4^-$$

Similarly, ionisation of $FeCl_3$ in $POCl_3$ solvent may be represented as:

$$2FeCl_3 \xrightarrow{POCl_3} [FeCl_2(OPCl_3)_n]^+ + FeCl_4^-$$

Thus, they could easily account for the presence of $FeCl_4^-$ species both in $POCl_3$ and $PO(OC_2H_5)_3$ solutions simply by suggesting that the solvent molecules coordinate with interacting solutes. These authors emphasized that similarities in physical properties of the two solvents, mainly their coordinating ability and dielectric constants, are more important than the presence or absence of self-ionization species.

Q. 5. Compare the physical properties of ammonia and water.

Ans. Some physical properties of water and ammonia as solvent are given in Table 26.1.

Table 26.1. Physical properties of ammonia and water.

Property	*Ammonia*	*Water*
1. Freezing point or melting point K	195.3	273
2. Boiling point K	239.6	373
3. Dielectric constant	22 at 239.6 K	81 at 291 K
4. Ionic product at 298 K	1.9×10^{-3} at 223 K	1×10^{-14}
5. Specific conductance	9.0×10^{-11} at 195 K	6×10^{-6} at 298 K
6. Dipole moment	1.148 at 233 K	1.84 at 291 K
7. Latent heat of fusion J/kg	0.34	0.33

LIQUID NH_3 AS A SOLVENT

Q. 6. Give important characteristics of ammonia as a solvent.

Ans. Some important points about liquid ammonia are:

(*i*) Like water, ammonia molecules are associated through hydrogen bonding in liquid ammonia as it contains a strongly electronegative element-nitrogen, which is responsible for hydrogen bonding in it.

(*ii*) The lower value of its dielectric constant as compared to that of water shows that liquid ammonia is a poorly electrolytic solvent than water.

(*iii*) Lower viscosity of liquid ammonia tends to promote greater ionic mobilities and thereby overcome, to some extent, the dielectric constant effect. Most salts actually conduct electric current better when dissolved in liquid ammonia than when dissolved in water. This is not due to high ionizing power of liquid ammonia but due to low viscosity which increases the mobility of ions.

(*iv*) The melting point and boiling point of ammonia are abnormal with respect to PH_3, AsH_3 and SbH_3 due to intermolecular association occurring in it as a result of H-bonding. The freezing and boiling points of liquid ammonia are lower than those of water. Therefore, more precautions are to be taken while handling liquid ammonia.

(*v*) Among the soluble compounds in liquid ammonia are nitrates, nitrites, thiocyanates and most cyanides. The majority of fluorides are insoluble, as are oxides, hydroxides, sulphates, carbonates, phosphates, sulphites and most sulphides. Among the halides, other than the fluorides, solubility decreases from the iodides to the chlorides. Most iodides (even that of silver) are soluble, the bromides are less soluble and among the chlorides only the ammonium, beryllium and sodium compounds are soluble.

(*vi*) Because of its ability to donate lone pair of electrons to form a covalent coordinate link as well as by ion-dipole attractions, ammonia forms many solvates which are analogues of salt-hydrates.

(*vii*) In general, solubilities of covalent organic compounds in liquid ammonia are considerably higher than in water probably due to greater dispersion energy interactions.

(*viii*) The specific heat of ammonia is higher than that of water.

(*ix*) Liquid ammonia possesses unusual high thermal coefficient of expansion, *i.e.*, its density decreases with rise in temperature.

(*x*) It has high critical temperature and pressure.

(*xi*) Liquid ammonia is a very poor conductor of electricity.

(*xii*) Liquid ammonia is an ionizing solvent, *i.e.*, undergoes ionization though to a lesser extent as compared to water.

Q. 7. Discuss auto-ionization of liquid ammonia. Show that NH_4Cl acts as strong acid in liquid NH_3. (*Purvanchal, 2007*)

Auto-ionization of liquid ammonia. The ionization of ammonia is supposed to take place just like that of water,

$$2NH_3 \rightleftharpoons \underset{\text{Ammonium ion}}{NH_4^+} + \underset{\text{Amide ion}}{NH_2^-}$$

$$2H_2O \rightleftharpoons \underset{\text{Hydronium ion}}{H_3O^+} + \underset{\text{Hydroxide ion}}{OH^-}$$

But the ionization of ammonia is very small as is evident from its low conductance and also from its ionic product, *i.e.*,

$$K_{H_2O} = [H_3O^+] \times [OH^-] = 1 \times 10^{-14} \text{ at } 291 \text{ K}$$

$$K_{NH_3} = [NH_4^+] \times [NH_2^-] = 1.9 \times 10^{-33} \text{ at } 223 \text{ K}$$

Ammonium ion in liquid ammonia is analogous to hydronium ion in aqueous solution and amide ion to hydroxyl ion. Hence any substance which gives ammonium ions in liquid ammonia behaves as acid and is known as ammonia acid, *e.g.*, NH_4Cl. Any substance which yields amide ion acts as base and is known as ammonia base, *e.g.*, sodamide $NaNH_2$.

On the basis of Bronsted-Lowery concept, ammonium ions would behave as acids, and amides as bases in liquid ammonia.

$$NH_3 + NH_3 \rightleftharpoons NH_4^+ + NH_2^-$$

$$\underset{\text{Acid I}}{H_2O} + \underset{\text{Base II}}{H_2O} \rightleftharpoons \underset{\text{Acid II}}{H_3O^+} + \underset{\text{Base I}}{OH^-}$$

Q. 8. Describe various types of reactions in liquid ammonia as a solvent.

(*M. Dayanand, 2002; Rohilkhand, 2003; Nagpur, 2008*)

Ans. Various reactions are described as under:

(1) Acid-base reactions. As substances producing H_3O^+ ions in water are acids, in the same way ammonium salts giving NH_4^+ ions in liquid ammonia are acids. Similarly, substances producing OH^- ions in water and NH_2^- (amide) ions in liquid ammonia are bases in that solvent. Thus acid base neutralization can occur in both solvents.

$$\underset{\text{Acid}}{HCl} + \underset{\text{Base}}{NaOH} \rightleftharpoons NaCl + H_2O \text{ (in water)}$$

$$\underset{\text{Acid}}{NH_4Cl} + \underset{\text{Base}}{NaNH_2} \longrightarrow \underset{\text{Salt}}{NaCl} + 2NH_3 \text{ (in ammonia)}$$

The course of neutralization reactions in liquid ammonia may be followed by means of conductometric titrations, and in certain cases through the indicators. For example, phenolphthalein is colourless in liquid ammonia but develops an intense red colour in the presence of alkali amides.

(2) Redox Reactions. Redox reactions take place in liquid ammonia just as they do in water as illustrated hereunder:

(*i*) Solutions of alkali and alkaline earth metals in liquid ammonia may act as electron donors by yielding their conduction electrons to electron acceptors. Hence the addition of metallic sodium in liquid ammonia to NH_4Br is a redox reaction.

$$2Na + 2NH_4Br \xrightarrow{\text{Liq. } NH_3} 2NaBr + 2NH_3 + H_2$$

(*ii*) Nitrous oxide oxidizes metallic potassium in liquid ammonia to potassium ion and is itself reduced to nitrogen in the process.

$$N_2O + 2K + NH_3\ (l) \rightleftharpoons KNH_2 + KOH + N_2$$

(3) Precipitation Reactions or Metathesis Reactions. Because of difference in solubilities of various substances in liquid ammonia and water, a large number of reactions, which do not occur in water, have been reported to be carried out in ammonia. The courses of these reactions are determined by the solubilities of products. For example,

(*i*) Lithium chloride may be precipitated by the reaction of ammonium chloride and lithium nitrate in liquid ammonia.

$$NH_4\,Cl + LiNO_3 \xrightleftharpoons{NH_3(l)} NH_4\,NO_3 + LiCl \downarrow$$

(*ii*) When solutions of potassium iodide and ammonium chloride in liquid ammonia are brought together, a white ppt. of KCl is produced.

$$KI + NH_4\,CI \xrightleftharpoons{\text{Liq. } NH_3} KCl \downarrow + NH_4I$$

(*iii*) The precipitation of barium sulphide occurs from ammonium solution of barum nitrate and ammonium sulphide.

$$Ba\,(NO_3)_2 + (NH_4)_2\,S \xrightarrow{\text{Liq.}NH_3} BaS \downarrow + 2NH_4\,NO_3$$

(*iv*) When solutions of metal nitrates and ammonium halides in liquid ammonia are brought together, iodides and bromides get precipitated.

$$2NH_4\,I + Zn\,(NO_3)_2 \xrightleftharpoons{NH_3(l)} Zn\,I_2 \downarrow + 2NH_4\,NO_3$$

$$2NH_4\,Br + Sr\,(NO_3)_2 \xrightleftharpoons{NH_3(l)} Sr\,Br_2 \downarrow + 2NH_4\,NO_3$$

(4) Complex Formations. Some reactions which have been studied are quite similar to complex formation occurring in the water system, *e.g.*, the zinc ion is precipitated as insoluble zinc amide which dissolves in excess of amide to form the ammonozincate.

$$Zn\,(NO_3)_2 + 2KNH_2 \xrightleftharpoons{NH_3(l)} \underset{\text{Insoluble}}{Zn(NH_2)_2} + 2KNO_3$$

$$Zn(NH_3)_2 + \underset{\text{Excess}}{2KNH_2} \xrightleftharpoons{NH_3(l)} \underset{\text{Soluble}}{K_2[Zn(NH_2)_4]}$$

(5) Solvolysis Reactions. These reactions are usually referred to as *ammonolysis or ammonolytic* reactions. Reactions of this type are analogous to the familiar hydrolysis reactions of water. In other words, an ammonolytic reaction is one of double decomposition in which ammonia is a reactant.

(*i*) Alkali metal and alkaline earth metal hydrides are ammonolysed to the corresponding metal amide and hydrogen.

$$M^+H^- + NH_3\ (l) \longrightarrow M^+\,NH_2^- + H_2$$

(*ii*) Many covalent halides undergo ammonolysis in liquid ammonia. In some cases elevated temperature and pressure are maintained to carry out the reactions. Some examples are given below:

$$HgCl \xrightarrow{\text{Liq.}NH_3} Hg\,NH_2\,Cl$$

An ammonolytic reaction, familiar to most chemists, is the action of ammonia on mercurous chloride.

$$Hg_2\,Cl_2 + 2NH_3\,(l) \longrightarrow Hg + HgNH_2\,Cl + N\,H_4^+ + Cl^-$$

$$SiCl_4 \xrightarrow{\text{Liq.}NH_3} \underset{\text{Tetramide silane}}{Si(NH_2)_4}$$

$$Si\,H_3\,Cl \xrightarrow{\text{Liq.}NH_3} (Si\,H_3)_3\,N$$

$$CCl_4 \xrightarrow[\text{elevated temp and pressure}]{\text{Liq.}NH_3} C\,(NH)\,(NH_2)_2$$

$$PCl_3 \xrightarrow{\text{Liq.}NH_3} P\,(NH_2)_3 \xrightarrow{273\ K} P_2(NH)_2$$

(*iii*) When organic halides are ammonolysed at the boiling point of liquid ammonia, primary, secondary and tertiary amines are formed.

$$RX + 2NH_3\,(l) \longrightarrow RNH_2 + NH_4X$$

$$RX + RNH_2 + NH_3\,(l) \longrightarrow R_2\,NH + NH_4X$$

$$RX + R_2\,NH + NH_3\,(l) \longrightarrow R_3\,N + NH_4\,X$$

(6) Reaction with Metals. An important property of liquid ammonia is its ability to dissolve the alkali and alkaline earth metals (except Be and Mg) to give homogeneous solution. Dilute metal ammonia solutions are blue but when more metal is added, a bronze phase (copper-coloured) separates and floats on the blue solution. The metal can be recovered unchanged by evaporating the solvent.

$$2M + 2NH_3\,(l) \longrightarrow 2MNH_2 + H_2 \quad \text{...(1)}$$

(M = an alkali metal)

$$M' + 2NH_3(l) \longrightarrow M'\,(NH_2)_2 + H_2 \quad \text{...(2)}$$

(M′ = Alkaline earth metal)

Studies have led to the theory that they contain solvated metal cations and solvated electrons.

$$M \rightleftharpoons M^+ + e^- \text{ or}$$

$$M + x\,NH_3 \longrightarrow M\left(NH_3\right)_x^+ + e^-$$

$$e^- + y\,NH_3 \longrightarrow e\left(NH_3\right)_y^-$$

The blue colour in these metal solutions can be attributed to the presence of free electrons. The fact that the metals may be reclaimed by evaporation of the solvent shows that these electrons are returned to their parent ions. The blue solutions of metals in liquid ammonia are unique as they conduct an electric current and this bears evidence to the presence of free electrons. It is also true that the same free electrons render their solutions as most powerful reducing agents.

(7) Formation of Ammoniates. Ammonia forms ammoniates with many inorganic salts, such as $MgI_2.6NH_3$, $NiI_2 \cdot 6NH_3$, $NaI \cdot 4NH_3$ in a similar way as water forms hydrates. Ammonia molecules are coordinated to the cation in these ammoniates.

(8) Extensive Dissociation of Weak Acids. Because liquid ammonia accepts protons readily, it enhances the ionization of so called weak acids such as acetic acid.

$$NH_3\,(l) + CH_3\,COOH \rightleftharpoons CH_3\,COO^- + NH_4^+$$

Removal of H^+ by NH_3 causes the reaction to proceed in the forward direction. Thus acetic acid has a pH value of 5 in water but is almost completely ionized in liquid ammonia. Ammonia thus reduces the difference between strength of acids and is called a *levelling solvent.*

Q. 9. What happens when alkali metal is dissolved in liquid NH_3?

(*Purvanchal, 2007; Lucknow, 2010*)

Ans. See Q. 8 (6).

Q. 10. Discuss applications of liquid ammonia as a solvent.

Ans. (*i*) *Metal ammonia solutions are powerful reducing agents* because they add electrons or produce nascent hydrogen. Thus it is possible to synthesize unusual compounds which are otherwise difficult to prepare. Metal ammonia solutions react with substances that are acidic in liquid ammonia to form metal salts and hydrogen gas. Many covalent hydrides and their derivatives react with metal ammonia solutions to form metal salts that are stable in liquid ammonia.

$$As\,H_3 + Na \xrightarrow{NH_3(l)} Na\,As\,H_2 + \frac{1}{2}H_2$$

$$PH_3 + Na \xrightarrow{NH_3(l)} Na\,PH_2 + \frac{1}{2}H_2$$

$$PH_3 + 2Na \xrightarrow{NH_3(l)} Na_2\,PH + H_2$$

(*ii*) The most useful property of liquid ammonia as a reaction medium in organic chemistry comes from its ability to form stable solutions of alkali metals that are highly reducing and from its unusual ability to dissolve a variety of organic compounds. These properties allow reduction reactions to be carried out in a homogeneous phase than by catalytic hydrogenation. In organic synthesis, metal ammonia solutions react with conventional transition metal complexes to yield compounds in which the metal atom has a formal valency of zero, assuming K = + 1 and CN = – 1.

$$K_3\,[Cr\,(CN)_6] + 3K \xrightarrow{NH_3(l)} K_6[Cr\,(CN)_6]$$

$$K_2[Ni\,(CN)_4] + 2K \xrightarrow{NH_3(l)} K_4[Ni\,(CN)_4]$$

$$K_2[Pt\,(CN)_4] + 2K \xrightarrow{NH_3(l)} K_4\,[Pt\,(CN)_4]$$

Oxidation state of Cr, Ni and Pt in these products comes out to be zero.

Q. 11. Discuss advantages and disadvantages of liquid ammonia as a solvent.

Ans. Advantages of using liquid Ammonia as solvent. (*i*) One of the greatest advantages of using liquid ammonia as solvent is that liquid ammonia dissolves alkali metals. The dissolved alkali metals can be recovered simply by evaporating the alkali metal-ammonia solution.

(*ii*) If a reducing agent stronger than hydrogen is to be used in an aqueous medium, it will react with H_2O to liberate H_2 and hence cannot be used in aqueous solution. In such cases, many alkali-ammonia solutions which contain mobile ammoniated electrons e^- (NH_3) can be used for reducing the materials soluble in liquid ammonia.

Disadvantages of using Liquid Ammonia. (*i*) The use of liquid ammonia as a solvent and a reaction medium requires elaborate equipment and special techniques are to be used.

(*ii*) Due to the hygroscopic nature of liquid ammonia, reactions are done in sealed tubes so as not to allow ammonia to come in contact with moisture.

(*iii*) The boiling point of liquid ammonia is 239.5 K and hence while working with this solvent, a lower temperature is to be maintained, if high pressures are to be avoided.

LIQUID SO_2 AS A SOLVENT

Q. 12. Discuss characteristics of liquid sulphur dioxide as a solvent.

(*M. Dayanand, 2002; Nagpur, 2008*)

Ans. Some of the properties by liquid SO_2 as a solvent are given below:

(*i*) It boils at 263 K and freezes at 196.5 K. It is clear that its liquid range is quite high, therefore, it can work as a good solvent.

(*ii*) The bond angle of 119.5° in O – S – O indicates that SO_2 molecule is in triangular valence state, *i.e.*, sp^2 hybridization with two hybrid orbitals to form bonds with oxygen and the third orbital occupied by a lone pair of electrons. The S – O bond length of 1.46 Å indicates some double bond character in the molecule. The SO_2 molecule may be shown to exist in the following two resonating structures.

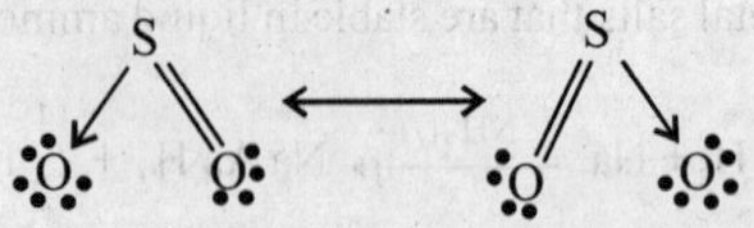

As S-atom contains a vacant bonding orbital, it can act as an electron acceptor and form molecular complexes.

(*iii*) The auto-ionization of liquid SO_2 takes place as follows:

$$SO_2 + SO_2 \rightleftharpoons \underset{\text{Thionyl ion}}{SO^{2+}} + \underset{\text{Sulphite ion}}{SO_2^{2-}}$$

(*iv*) Solubility of inorganic substances in liquid SO_2. Iodides are found to be most soluble in liquid SO_2. In case of alkali metal halides the order of solubility is found to be MI > MBr > MC1 > MF. Alkali metal sulphites, sulphates, acetates, cyanides and thiocyanates are also found to be soluble. Metal sulphates, sulphites, oxides and hydroxides are sparingly soluble. Ammonium, mercury and thalium salts are also soluble.

Many substances like halogens, interhalogens (ICI, IBr etc.), phosphorus halides ($PC1_3$, PBr_3 etc.), oxyhalides of phosphorus and sulphur ($POCl_3$, $SOCl_2$, etc.),their halides like $CC1_4$, $SiC1_4$, $GeCl_4$, $SnCl_4$, $AsCl_3$ and CS_3 are found to be soluble in liquid SO_2.

(*v*) Conductivity of salt solutions in SO_2. Because of its high dipole moment value, it dissolves many inorganic compounds in it. Therefore, it behaves as a moderately good conductor for many alkali metals, ammonium and tetra alkyl ammonium chlorides, bromides, iodides and thiocyanates while di-, tri-alkyl ammonium salts are very poor conductors.

It is found that the conductivity of electrolyte solutions in liquid SO_2 increases with the size of the cation. Thus the following order of conductivity is found:

$$Na^+ < NH_4^+ < K^+ < (CH_3)_3\,S^+ < (CH_3)_4\,N^+ \text{ etc.}$$

The conductivity of the anions is found to increase in the order:

$$SCN^- < ClO_4^- < Cl^- < I^- \text{ etc.}$$

(*vi*) Solubility of organic compounds in liquid SO_2. SO_2 is found to be a better medium for the dissolving of covalent compounds. Aromatic hydrocarbons and alkenes are more soluble than aliphatic hydrocarbons. SO_2 has also been used as a solvent for carrying out many organic reactions such as Friedel-Craft's reactions, sulphonation, bromination etc.

Q. 13. Discuss various chemical reactions in liquid SO_2.

(Rohilkhand, 2003; Mumbai, 2004; Nagpur, 2005; Lucknow, 2010; Ranchi, 2010; Patna, 2010)

Ans. These reactions are discussed under the following headings.

1. Acid-Base Reactions. Liquid SO_2 undergoes auto-ionization as follows:

$$\underset{\text{Acid}_1}{SO_2} + \underset{\text{Base}_2}{SO_2} \rightleftharpoons \underset{\text{Acid}_2}{SO^{2+}} + \underset{\text{Base}_1}{SO_3^{2-}}$$

Thus according to solvent system concept, any substance which makes available SO^{2+} ions would behave as an acid, whereas the one making available SO_3^{2-} would behave as base. The compounds such as $SOCl_2$ and $SOBr_2$ behave as acids in liquid SO_2 and Na_2SO_3 would behave as a base. The titration of potassium sulphite (K_2SO_3) against thionyl chloride or acetate can be followed conduc- tometrically.

$$\underset{\text{Acid}}{SOCl_2} + \underset{\text{Base}}{K_2SO_3} \longrightarrow \underset{\text{Salt}}{2KCl} + \underset{\text{Solvent}}{2SO_2}$$

Some other acid-base neutralization reactions are:

$$\underset{\text{Acid}}{SOCl_2} + \underset{\text{Base}}{Cs_2SO_3} \longrightarrow \underset{\text{Salts}}{2CsCl} + \underset{\text{Solvent}}{2SO_2}$$

$$SOBr_2 + [N(CH_3)_4]_2SO_3 \longrightarrow 2[N(CH_3)_4]Br + 2SO_2$$

$$SO(SCN)_2 + K_2SO_3 \longrightarrow 2K(SCN) + 2SO_2$$

Amphoteric behaviour is shown by the zinc (II) or aluminium (III) salts in liquid SO_2.

$$ZnCl_2 + K_2SO_3 \xrightarrow{SO_2(l)} ZnSO_3 + 2KCl \text{ (Acidic character)}$$

$$ZnSO_3 + K_2SO_3 \xrightarrow{SO_2(l)} K_2[Zn(SO_3)_2] \text{ (Basic character)}$$

$$Al_2(SO_3)_2 + 3K_2SO_3 \xrightarrow{SO_2(l)} 2K_3[Al(SO_3)_3] \text{ (Basic character)}$$

$$2AlCl_3 + 3K_2SO_3 \xrightarrow{SO_2(l)} Al_2(SO_3)_3 + 6KCl \text{ (Acidic character)}$$

2. Redox Reactions. Many reactions of gaseous sulphur dioxide are known where it behaves both as oxidizing and as reducing agent. But sulphur dioxide, in liquid state, does not have any marked reducing or oxidising tendency and acts as a medium for many redox reactions. Iodine is liberated from KI in the presence of $SbCl_5$.

$$6KI + 3SbCl_5 \xrightleftharpoons{SO_2(l)} 3I_2 + SbCl_3 + 2K_3[SbCl_6]$$

3. Solvolysis Reactions. Certain covalent halides are found to undergo solvolysis reactions in liquid SO_2. For example,

$$PCl_5 + SO_2(l) \longrightarrow POCl_3 + SOCl_2$$

$$AsCl_5 + SO_2(l) \longrightarrow AsOCl_3 + SOCl_2$$

4. Metathetical reactions. The precipitation of a substance depends upon its solubility in the solvent. There are many chemical reactions which have been carried out in this solvent.

$$BaI_2 + Zn(CNS)_2 \xrightarrow{SO_2(l)} Ba(CNS)_2\downarrow + ZnI_2$$

$$SbCl_3 + 3LiI \xrightarrow{SO_2(l)} SbI_3\downarrow + 3LiCl$$

$$PbF_2 + Li_2SO_4 \xrightarrow{SO_2(l)} PbSO_4\downarrow + 2LiF$$

Thionyl chloride in liquid SO_2 has helped to prepare many compounds, *e.g.*,

$$2CH_3COOAg + SOCl_2 \xrightarrow{SO_2(l)} 2AgCl\downarrow + SO(OCOCH_3)_2 \text{ Thionyl acetate}$$

$$2KBr + SOCl_2 \xrightarrow{SO_2(l)} 2KCl\downarrow + SOBr_2$$

$$2NH_4(SCN) + SOCl_2 \xrightarrow{SO_2(l)} 2NH_4Cl\downarrow + SO(SCN)_2$$

5. Sulphur dioxide of crystallization. These reactions are also called solvation reactions. Salts having sulphur dioxide of crystallization have been obtained.

$$NaI + 4SO_2(l) \longrightarrow NaI . 4SO_2$$

$$KI + 4SO_2(l) \longrightarrow KI . 4SO_2$$

6. Complex formation. Complex formation in the presence of liquid SO_2 has been reported by Walden and coworkers. Sulphur dioxide itself does not take part in complexation, but provides medium for the purpose.

$$SbCl_3 + 3KCl \xrightarrow{SO_2(l)} K_3^+[SbCl_6]^{3-}$$

$$SbCl_5 + KCl \xrightarrow{SO_2(l)} K^+[SbCl_6]^-$$

$$SbCl_5 + [N(CH_3)_4]^+Cl^- \xrightarrow{SO_2(l)} [N(CH_3)_4^-]^+[SbCl_6]^-$$

$$SbCl_5 + CH_3COCl \xrightarrow{SO_2(l)} [CH_3CO]^+[SbCl_6]^-$$

7. Organic reactions in liquid sulphur dioxide. Organic reactions where relatively inert and non-inflammable solvents are required, liquid SO_2 serves the best medium. Some of the important reactions are:

(*a*) *Sulphonation.* Good yields have been reported for the sulphonation of aromatic compounds in liquid SO_2.

$$C_6H_6 + \underset{\text{Chloro sulphonic acid}}{ClSO_2H} \xrightarrow{SO_2(l)} C_6H_5{-}SO_3H + HCl$$

(*b*) *Bromination.* Both for addition and substitution reactions with bromine, liquid SO_2 has been used.

$$\underset{\text{Styrene}}{C_6H_5{-}CH{=}CH_2} + Br_2 \xrightarrow{SO_2\,(l)} C_6H_5{-}CH(Br){-}CH_2Br + HCl$$

$$\underset{\text{Phenol}}{C_6H_5OH} + Br_2 \xrightarrow{SO_2\,(l)} \underset{p\text{-Bromophenol}}{p\text{-}BrC_6H_4OH} + HBr$$

(*c*) *Friedel-Crafts reaction.* $AlCl_3$ is highly soluble in liquid SO_2 and is, therefore, a preferred solvent over ether which is highly inflammable.

$$C_6H_6 + \underset{\text{Benzoyl chloride}}{C_6H_5COCl} \xrightarrow[SO_2(l)]{AlCl_3} \underset{\text{Benzophenone}}{C_6H_5{-}CO\,C_6H_5} + HCl$$

Q. 14. Give equation of auto-ionization of liquid NH_3 and SO_2. What are the species characteristic of acids and bases in liquid NH_3 and liquid SO_2 ? (*Purvanchal, 2007*)

Ans. See Q. 7 & 13.

Q. 15. Discuss the reaction of $FeCl_3$ with $(CH_3)_4$ NCl in $POCl_3$ as a solvent as well as in triethyl orthophosphate $PO(OC_2H_5)_3$.

Ans. According to solvent system concept, we have

$$2POCl_3 \rightleftharpoons POCl_2^+ + POCl_4^-$$

$$FeCl_3 + POCl_3 \rightleftharpoons POCl_2^+ + FeCl_3^- \quad \text{(acid)}$$

$$(CH_3)_4\,N^+\,Cl^- + POCl_3 \rightleftharpoons (CH_3)_4\,N^+ + POCl_4^- \quad \text{(base)}$$

Therefore, neutralisation reaction in $POCl_3$ occurs as:

$$(CH_3)_4\,N^+\,Cl^- + FeCl_3 \xrightarrow{POCl_3} [(CH_3)_4\,N^+] + FeCl_4^-$$

The same reaction could also be carried out in $PO(OC_2H_5)_3$ and although $PO(OC_2H_5)_3$ does not contain any Cl^- ions, even then $FeCl_4^-$ is formed. This was explained by Meek and Drago on the basis of coordination Model. They suggested that Cl^- transfer takes place from one $FeCl_3$ to another with the formation of cationic iron (III) as:

$$2FeCl_3 \xrightarrow{PO(OC_2H_5)_3} [FeCl_2[OP(OC_2H_5)_3]_4]^+ + FeCl_4^-$$

Q. 16. Under what circumstances the solvent system model fails?

Ans. The main difficulty in the application of solvent system arises when there is no proof for the existence of self -ionization species. This concept has been applied even in almost non-conducting solvents like SO_2 where the existence of species SO^{2+} is doubtful. Reaction of $SOCl_2$ with Cs_2SO_3 in terms of solvent system is explained as:

$$2SO_2 \rightleftharpoons SO^{2+} + SO_3^{2-}$$

$$Cs_2SO_3 \rightleftharpoons 2Cs^+ + SO_3^{2-} \quad \text{(base)}$$

$$SOCl_2 \rightleftharpoons SO^{2+} + 2Cl \quad \text{(acid)}$$

Neutralisation occurs as:

$$2Cs^+ + SO_3^{2-} + SO^{2+} + 2Cl^- \longrightarrow 2CsCl + 2SO_2$$

However, there is no experimental support for the existence of SO^{2+}. Probably self-ionization is not required and the reaction can be explained in terms of coordination model.

$$2Cs^+ + SO_3^{2-} + SOCl_2 \rightleftharpoons 2Cs^+ + (SO_3\,SOCl_2)^{2-}$$

$$2SO_2 + 2CsCl \longleftarrow 2Cs^+ + SO_3 \cdot SO + 2Cl^- \longleftarrow (SO_3 \cdot SOCl)^- + Cl^- + 2Cs^+$$

Q. 17. (*a*) Explain that liquid NH_3 is an associated liquid.

(*b*) Name some important properties that determine the utility of liquid NH_3 as a solvent.

(c) Mention some important advantages which liquid SO_2 possesses as a solvent.

Ans. (*a*) Ammonia is an associated solvent (like water) due to presence of small and strongly electronegative nitrogen which is responsible for strong hydrogen bonds in it.

(*b*) Important properties determining utility of ammonia as a solvent better than water are:

(*i*) Specific heat of NH_3 is greater than that of water.

(*ii*) Most salts conduct electric current better than in H_2O.

(*iii*) The solvent itself is a poor conductor of electricity.

(*iv*) It has high critical temperature and pressure.

(*c*) Special advantage of SO_2 as a solvent.

(*i*) Low boiling point allows its easier evaporation.

(*ii*) Low melting point allows reaction at low temperatures.

(*iii*) Inert solvent.

(*iv*) Liquid SO_2 has unusual solvating power for both covalent and ionic compounds.

(*v*) Used as a diluent for super acids (HSO_3 F–SbF_3) which help in reducing its viscosity without affecting its acidity.

Q. 18. (*a*) Give two examples to show the utility of metathetical reactions in non-aqueous solvents.

(*b*) Alkali metals and alkaline earth metals dissolve in liquid NH_3. How do these processes differ?

(*c*) Classify the following as acids and bases in suitable non-aqueous solvents:

$(NH_4)_2SO_4$, $NaNH_2$, SOF_2, $Cr_2(SO_3)_3$, $[(CH_3)_3N]_2SO_3$.

Give one reaction in each case to justify your answer.

Ans. (*a*) Water soluble $BaCl_2$ can be precipitated from the solution of silver chloride and $Ba(NO_3)_2$ in liquid ammonia.

$$Ba(NO_3)_2 + 2AgCl \xrightarrow{NH_3} BaCl_2\downarrow + 2AgNO_3$$

Similarly, LiCl can be precipitated out of the reaction of NH_4Cl and $LiNO_3$ in liquid NH_3.

$$NH_4Cl + LiNO_3 \xrightarrow{NH_3} NH_4NO_3 + LiCl\downarrow$$

Metathetical reactions in SO_2 are:

$$SOCl_2 + 2CH_3COOAg \xrightarrow{SO_2} 2AgCl\downarrow + SO(OOCCH_3)_2$$

$$2NH_4SCN + SOCl_2 \xrightarrow{SO_2} 2NH_4Cl\downarrow + SO(SCN)_2$$

(*b*) Alkaline earth metals are soluble in ammonia giving blue solutions but on evaporation of the solvent they are found to be present as ammoniates $Ca(NH_3)_6$, $Sr(NH_3)_6$, $Ba(NH_3)_6$. On the other hand, alkali metals are obtained as such on evaporation of ammonia.

(*c*) Ammonium salts $(NH_4)_2SO_4$ act as acids and amides as bases in liquid ammonia $(2NH_3 \rightleftharpoons NH_4^+ + NH_2^-)$.

$$(NH_4)_2SO_4 + 2NaNH_2 \xrightarrow{NH_3} \underset{\text{Salt}}{Na_2SO_4} + 4NH_3$$

In SO_2 which ionises as $2SO_2 \rightleftharpoons SO^{2+} + SO_3^{2-}$, SOF_2 acts as acid and $Cr_2(SO_3)_2$ or $[(CH_3)_4N]_2SO_3$ as a base.

$$3SOF_2 + Cr_2(SO_3)_3 \xrightarrow{SO_2} 2CrF_3 + 6SO_2$$

$$SOF_2 + [(CH_3)_4N]_2SO_3 \xrightarrow{SO_2} 2(CH_3)_4NF + 2SO_2$$

Q. 19. Complete the following reactions:

(*i*) $SO_2 + SO_3 \xrightarrow{liq.SO_2}$

(*ii*) $SOCl_2 + Cs_2SO_3 \xrightarrow{liq.SO_2}$

(*iii*) $AgNO_3 + KNH_2 \xrightarrow{liq.NH_3}$

(iv) $Na_2O + NH_4Cl \xrightarrow{liq.NH_3}$

(v) $NbCl_5 + SO_2 \xrightarrow{liq.SO_2}$

(vi) $CH_3COOAg + SOCl_2 \xrightarrow{liq.SO_2}$

(vii) $AlCl_3 + [(CH_3)_4N]_2 SO_3 \xrightarrow{liq.SO_2}$

(viii) $SnI_4 + KNH_2 \xrightarrow{liq.SO_2}$

(ix) $C_5H_5N + SOCl_2 \xrightarrow{liq.SO_2}$

Ans. (*i*) SO_3 dissolves in SO_2.

(*ii*) $SOCl_2 + Cs_2SO_2 \xrightarrow{SO_2} 2CsCl + 2SO_2$

(*iii*) $AgNO_3 + KNH_2 \xrightarrow{NH_3} KNO_3 + AgNH_2$

(*iv*) $Na_2O + 2NH_4Cl \xrightarrow{NH_3} 2NaCl + 2NH_3 + H_2O$

(*v*) $NbCl_5 + SO_2 \xrightarrow{SO_2} NbOCl_3 + SOCl_2$

(*vi*) $2CH_3COOAg + SOCl_2 \xrightarrow{SO_2} 2AgCl + SO(OOCCH_3)_2$

(*vii*) $2AlCl_3 + 3[(CH_3)_4N]_2SO_3 \xrightarrow{SO_2} 6(CH_3)_4NCl + Al_2(SO_3)_3$

Q. 20. (*a*) Account for the following:

(*i*) K_2SO_3 is soluble in SO_2.

(*ii*) SO_2 acts both as a Lewis base and acid.

(*b*) Complete the following reactions:

(*i*) $PbI_2 + KNH_2 \xrightarrow{liq.NH_3}$

(*ii*) $SO_2Cl_2 + NH_3 \xrightarrow{liq.NH_3}$

(*iii*) $C_6H_6 + SO_3 \xrightarrow{liq.SO_2}$

(*iv*) $NH_4Cl + NaNH_2 \xrightarrow{liq.NH_3}$

(*v*) $Na + NH_4Cl \xrightarrow{liq.NH_3}$

Ans. (*a*) (*i*) It dissolves because of solvate formation.

(*ii*) SO_2 can behave both as Lewis acid and Lewis base. It acts as a donor (Lewis base) in the following solvates: $CaI_2 . 4SO_3, LiI . 4SO_2$: where it donates electrons through an oxygen atom in SO_2. It acts as an acceptor (Lewis acid) in the following: $C_5H_5N \cdot SO_2$, $C_6H_7N \cdot SO_2$ where it accepts electrons through sulphur atom in the molecule SO_2.

(*b*) (*i*) $PbI_2 + KNH_2 \xrightarrow{NH_3} PbNH$

(*ii*) $SO_2Cl_2 + NH_3 \xrightarrow{NH_3} SO_2(NH_2)_2$

(*iii*) $C_6H_6 + SO_3 \xrightarrow{SO_2} C_6H_5SO_3H$

(*iv*) $NH_4Cl + NaNH_2 \xrightarrow{NH_3} NaCl + 2NH_3$

(*v*) $NH_4Cl + Na \xrightarrow{NH_3} NaCl + NH_3 + 1/2\ H_2$

Q. 21. (*a*) The electrical conductivity of liquid ammonia is increased when ammonium chloride is dissolved in it.

(*b*) A liquid ammonia solution of phenolphthalein is colourless but becomes red coloured on addition of KNH_2.

(*c*) A number of metals dissolve in liquid ammonia to give strongly reducing solutions but react with water.

Ans. (*a*) Liquid ammonia ionises as follows:

$$2NH_3 \rightleftharpoons NH_4^+ + NH_2^-$$

Substances yielding ammonium ions in liquid ammonia will act as ammono acids. Addition of NH_4Cl increases NH_4^+ concentration in liquid ammonia with the result that the solution shows increased conductance.

(*b*) Addition of KNH_2 to ammonia increases amide ion NH_2^- concentration and behaves as ammono base. As soon as concentration of NH_2^- ions increases in solution, the solution behaves as a base and imparts red colour to phenolphthalein just as an aqueous basic solution imparts red (pink) colour.

(*c*) Metals with low ionization energy and higher solvation energy, that is, with higher electrode potential, dissolve in liquid ammonia (without chemical reaction) to give blue solutions which show strong reducing properties and high electrical conductance. Ammoniated electron is responsible for strong reducing action and high conductance.

$$Na \rightleftharpoons Na^+ + e^-$$

$$Na^+ + (NH_3)_x \rightleftharpoons [Na(NH_3)_x]^+$$

$$e^- + x(NH_3) \rightleftharpoons [e(NH_3)_x]^-$$

Such metals react with water because water is much stronger oxidising agent than ammonia.

$$Na + H_2O \longrightarrow NaOH + 1/2\ H_2$$

Q. 22. (*a*) Suggest reasons for reverse order of solubility for alkali halides in liquid SO_2 and liquid NH_3.

(*b*) With the help of examples, point out the difference between solvation and solvolysis reactions.

(*c*) Liquid NH_3 is a poor electrical conductor, yet salt solutions in this solvent conduct electricity better than their aqueous solution.

Ans. (*a*) Alkali halides are soluble in liquid ammonia due to complexation. For example, NaI dissolves in ammonia due to the formation of complex $NaI.4NH_3$. However, alkali halides are not soluble in liquid SO_2 due to the inability of SO_2 to form complexes with alkali metal halides.

(*b*) Solvation involves coordination of the solvent molecules to the metal ion. Solvation in water is usually stated as hydration. Take the case of $MgI_2.6NH_3$. Here ammonia molecules are bound to MgI_2. This complex is called a solvate or an ammoniate. Similarly $MgC1_2 . 6H_2O$ is called a solvate or a hydrate. Therefore, solvation refers to simple coordination on complexation. On the other hand, solvalytic reaction in liquid ammonia means attachment of ammonium NH_4^+ or amide NH_2^- ions to the reactant. This is analogous to hydrolysis in water. The following are solvolytic reactions in NH_3.

$$SiCl_4 \xrightarrow{+NH_3} Si\,(NH_2)_4$$

$$SO_2\,Cl_2 \xrightarrow{+NH_3} SO_2\,(NH_2)_2$$

(*c*) Most salts conduct electric current better when dissolved in liquid NH_3 than when dissolved in H_2O. This is not due to higher ionizing power of liquid NH_3. (It is actually 1/3 of H_2O) but due to its low viscosity which greatly increases the mobility of ions.

Q. 23. Give two examples of solvolytic reactions in liquid ammonia.

Ans. See Q. 8.

Q. 24. Write equations to explain the behaviour of the following:

(*i*) KNH_2 in liquid NH_3

(*ii*) PCl_5 in liquid SO_2

Ans. See Qs. 8 and 13.

Q. 25. Solution of sodium metal in liquid NH_3 is blue and good electrical conductor. Explain.

Ans. See Q. 8.

Q. 26. Liquid ammomic exhibits association whereas liquid SO_2 does not. Explain.

Ans. Liquid NH_3 associates itself like water through hydrogen bonding. Nitrogen is sufficiently electronegative to cause hydrogen bonding. This passbility does not exist with sulphur. Hence SO_2 does not associate.

Q. 27. Complete the following in liquid SO_2

(*i*) $HgI_2 + KI \longrightarrow$, (*ii*) $SbCl_5 + NOCl \longrightarrow$, (*iii*) $Cr_2SO_3 + SOCl_2 \longrightarrow$

Ans. See. Q. 13.

Q. 28. In what respects NH_3 is a better solvent than water?

Ans. See Q. 11.

Q. 29. Write a note on failure of theory of auto-ionisation.

Ans. See Q. 4.

Q. 30. Discuss the characteristics of liquid hydrogen fluoride as a solvent.

Ans. Important characteristics of liquid hydrogen fluoride are given as under.

1. Hydrogen fluoride is a water-like solvent since it is a liquid at tempratures below 19.4°C up to – 89.4°C.

2. Its boiling point (+ 19.4°C) is very high as compared to the boiling points of other non-aqueous solvents.

3. Its dipole moment (1.90 D) and dielectric constant (83.6 at 0°C) are also quite close to those of water and as such, it serves as an excellent ionising solvent.

4. Its use as a non-aqueous solvent is quite limited because it is poisonous. It can dissolve only a few substances without reacting with them because it possesses low specific conductance.

Q. 31. Describe chemical reactions in liquid hydrogen fluoride.

Ans. Some of the important chemical reactions in liquid hydrogen fluoride are discussed below.

1. Acid-Base Reactions. The high specific conductance of liquid hydrogen fluoride suggests a high degree of its auto-ionisation.

$$3HF \rightleftharpoons \underset{\text{Acid}}{H_2F^+} + \underset{\text{(Base)}}{HF_2^-}$$

Any substance which can yield H_2F^+ ion would behave as an acid and any substance which can yield HF_2^- or F^- ion would behave as a base.

Accordingly, even strong acids such as HNO_3 and H_2SO_4 behave as bases in hydrogen fluoride since they release F^- ions as shown by the following reactions:

$$\underset{\text{Base}}{HNO_3} + \underset{\text{Acid}}{HF} \rightarrow H_2NO_3^+ + F^-$$

$$H_2SO_4 + HF \rightarrow H_3SO_4^+ + F^-$$

Perchloric acid, which behaves as a very strong acid in aqueous solution, exhibits only an amphoteric character in liquid hydrogen fluoride. Thus,

$$\underset{\text{Base}}{HClO_4} + HF \rightarrow H_2ClO_4^+ + F^-$$

$$HClO_4 + HF \rightarrow ClO_4^- + H_2F^+$$

However substances such as BH_3, AsF_5 and SnF_4, which behave as Lewis acids in aqueous solution, function as acids in liquid hydrogen fluoride. Thus,

$$\underset{\text{Acid}}{SbF_5} + \underset{\text{Base}}{2HF} \rightarrow H_2F^+ + SbF_6^-$$

$$\underset{\text{Acid}}{BF_3} + \underset{\text{Base}}{2HF} \rightarrow H_2F^+ + BF_4^-$$

2. Protonation of organic compounds. Organic compounds such as benzene and ethanol get protonated when dissolved in BF. Thus,

$$C_6H_6 + HF \rightarrow C_6H_7^+ + F^-$$

$$C_2H_5OH + HF \rightarrow C_2H_5OH_2^+ + F^-$$

Due to the formation of ions, organic compounds mentioned as above give conducting solutions in HF. The aromatic hydrocarbons give coloured solutions in HF and their solubilities can be increased further by adding fluoride acceptors as BF_3 and SbF_5.

3. Formation of Addition Compounds

Metallic fluorides form addition compounds with HF. Thus potassium fluoride and ammonium fluoride form the addition compounds. KF.2HF, KF.3HF, NH_4F . HF.

4. Precipitation Reactions. Fluorides of non-alkali metals dissolved in liquid hydrogen fluoride, on treatment with solutions of sulphates, perchlorates and periodates of alkali metals produce precipitate of sulphates, perchlorates and periodates of such non-alkali metals.

$$Na_2SO_4 + NiF_2 \rightarrow \underset{\text{Ppt.}}{NiSO_4} + 2NaF$$

$$NaClO_4 + TlF \rightarrow NaF + TlClO_4$$

$$KIO_4 + AgF \rightarrow AgIO_4 + KF$$

Q. 32. Describe chemical reactions in liquid dinitrogen tetroxide.

Ans. The auto-ionisation equilibrium of dinitrogen tetroxide can be represented as under :

$$N_2O_4 \rightleftharpoons NO^+ + NO_3^-$$

Although there is no physical evidence for the above type of auto-ionization occurring in N_2O_4, it has been observed that the electrical conductance of N_2O_4 increases on adding a donor or a polar solvent to it. This is perhaps due to the increased stabilisation of cations and anions (NO^+ and NO_3^-) of N_2O_4 as a result of the ion-dipole interaction of these ions with the molecules of the added solvent.

1. Acid-Base reactions. Thus assuming that N_2O_4 undergoes auto-ionisation as mentioned above, it can be said that any chemical species which gives the cation NO^+ in this solvent will behave as an acid and any chemical species which gives the anion NO_3^- in this solvent will behave as a base. Thus, NOCl behaves as an acid in N_2O_4, as it releases NO^+ in solution.

$$2NOCl \xrightleftharpoons{N_2O_4} 2NO^+ + 2Cl^-$$

$$2NO^+ + Sn \xrightleftharpoons{N_2O_4} 2NO + Sn^{2+}$$

$$Sn^{2+} + Cl^- \xrightleftharpoons{N_2O_4} SnCl_2$$

$$2NOCl + Sn \rightleftharpoons 2NO + SnCl_2$$

There is a similarity in the acid behaviour of NOCl in N_2O_4, when it reacts with Sn, and the behaviour of HCl in water when it reacts with Zn as illustrated by the following reaction.

$$HCl + 2H_2O \xrightleftharpoons{H_2O} 2H_3O^+ + 2Cl^-$$

$$2H_3O^+ + Zn \rightleftharpoons Zn^{2+} + 2H_2O + H_2$$

$$Zn^{2+} + 2Cl^- \rightleftharpoons ZnCl_2$$

$$Zn + 2HCl \rightleftharpoons ZnCl_2 + H_2$$

An example of a compound acting as a base in liquid N_2O_2 is therefore, acts as a base in N_2O_4. A typical reaction of $[EtNH_3]^+ [NO_3^-]$ with zinc metal in N_2O_4 is similar to the reaction of zinc with $NaNH_2$ in liquid ammonia, as illustrated by the reactions :

$$2N_2O_4 \rightleftharpoons 2NO^+ + 2NO_3^- \text{ (auto-ionisation)}$$

$$Zn + 2NO^+ \rightleftharpoons Zn^{2+} + 2NO$$

$$2\,[EtNH_3]^+\,[NO_3]^- + Zn^{2+} + 2NO_3^- \rightleftharpoons [EtNH_3]_2^+\,[Zn\,(NO_3)_2]^{2-}$$

$$Zn + 2N_2O_4 + 2\,[EtNH_3]^+\,[NO_3]^- \rightleftharpoons [EtNH_3]_2^-\,[Zn\,(NO_3)_4]^{2-} + 2NO$$

$$4NH_3 \rightleftharpoons 2NH_4^+ + 2NH_2^- \text{ (auto-ionisation)}$$

$$Zn + 2NH_4^+ \rightleftharpoons 2NH_3 + H_2 + Zn^{2+}$$

$$Zn^{2+} + 2NaNH_2 + 2NH_2^- \rightleftharpoons Na_2\,[Zn\,(NH_2)_4]$$

$$Zn + 2NH_3 + 2NaNH_2 \rightleftharpoons Na_2\,[Zn\,(NH_2)_4] + H_2$$

2. Solvolysis. The solvolysis of metal oxides in N_2O_4 takes place similar to their solvolysis in liquid ammonia, as illustrated by the following reactions.

$$2N_2O_4 \rightleftharpoons 2NO_+ + NO_3^- \text{ (auto-ionisation)}$$

$$CaO + 2NO^+ \rightleftharpoons Ca^{2+} + N_2O_3$$

$$Ca^{2+} + 2NO_3^- \rightleftharpoons Ca\,(NO_3)_2$$

$$CaO + 2N_2O_4 \rightleftharpoons Ca\,(NO_3)_2 + N_2O_3$$

$$4NH_3 \rightleftharpoons 2NH_4^+ + H_2O + 2Na^+$$

$$2Na^+ + 2NH_2^- \rightleftharpoons 2NaNH_2$$

$$Na_2O + 4NH_3 \rightleftharpoons 2NaNH_2 + 2NH_3 + H_2O$$

In fact, solvolysis of the halides of metals in N_2O_4 provides an excellent method for preparing anhydrous metal nitrates which are otherwise extremely difficult to prepare. Thus,

$$2N_2O_4 \rightleftharpoons 2NO^+ + 2NO_3^-$$

$$ZnX_2 + 2NO^+ \rightleftharpoons 2NOX + Zn^{2+}$$

$$Zn^{2+} + 2NO_3^- \rightleftharpoons Zn\,(NO_3)_2$$

The NOX produced in the reaction decomposes to NO and X_2 and, therefore, it does not participate in the reverse equilibrium reaction. The overall reaction is, thus,

$$2N_2O_4 + ZnH_2 \rightleftharpoons 2NO + X_2 + \underset{\text{(anhydrous)}}{Zn(NO_3)_2}$$

Several metal carbonyls are also solvolysed in a similar manner. Thus,

$$N_2O_4 \rightleftharpoons NO^+ + NO_3^-$$

$$3N_2O_4 \rightleftharpoons 3NO^+ + 3NO_3^-$$

$$Fe\,(CO)_5 + 3NO_3^- \rightleftharpoons Fe\,(NO_3)_3 + 5CO + 3e^-$$

$$3NO^+ + 3e^- \rightleftharpoons 3NO$$

$$Fe\,(CO)_5 + 3N_2O_4 \rightleftharpoons Fe\,(NO_3)_3 + 5CO + 3NO$$

N_2O_4 is also known to form solvated species of the type shown below:

$$2N_2O_4 \rightleftharpoons 2NO^+ + 2NO_3^-$$

$$Ca + 2NO^+ + 2NO_3^- + N_2O_4 \rightleftharpoons [Ca\,(NO_3)_2.N_2O_4] + 2NO$$

$$Ca + 3N_2O_4 \rightleftharpoons [Ca\,(NO_3)_2.N_2O_4] + 2NO$$

27 QUALITATIVE INORGANIC ANALYSIS

Q. 1. How is the dry-heating test for analysing an inorganic compound performed? Describe, with the help of chemical equations, the gases, sublimates and residues that are obtained from various salts.

Ans. A little of the salt is heated first gently and then strongly in a dry test tube, by keeping it slightly slanting; when one or more of the following changes occur, they should be carefully noted:

1 Evolution of a gas. Certain salts decompose to give gases on heating. The decomposition is also sometimes brought about by the atmospheric oxygen or water molecules present in the hydrated salts:

(*a*) *Colourless and odourless gas.* Alkali metal nitrates on heating give oxygen gas which rekindles a glowing splinter.

$$2NaNO_3 \longrightarrow 2NaNO_2 + O_2$$

Carbonates and bicarbonates of non-alkali metals on dry heating give CO_2.

$$ZnCO_3 \longrightarrow ZnO + CO_2$$

$$CaCO_3 \longrightarrow CaO + CO_2$$

$$2NaHCO_3 \longrightarrow Na_2CO_3 + CO_2 + H_2O$$

[**Note.** Some oxalates evolve a mixture of carbon monoxide and carbon dioxide.]

$$\underset{\text{Cal-oxalate}}{(COO)_2\,Ca} \longrightarrow CaO + CO + CO_2$$

(*b*) *Colourless gas with odour.* (*i*) Ammonium salts on heating evolve NH_3 gas.

$$(NH_4)_2\,SO_4 \longrightarrow NH_4\,HSO_4 + NH_3$$

(*ii*) Sulphides give H_2S gas due to hydrolysis.

$$Na_2S + 2H_2O \longrightarrow 2NaOH + H_2S$$

Thiosulphates and sulphites on heating liberate sulphur dioxide gas.

$$CaSO_3 \longrightarrow CaO + SO_2$$

(*iii*) Certain hydrated chlorides give HC1 gas on heating due to partial hydrolysis.

$$MgC1_2 + H_2O \longrightarrow MgO + 2HC1$$

$$CaC1_2 + 2H_2O \longrightarrow Ca\,(OH)_2 + 2HC1$$

(*iv*) Acetates give vineger smelling acetic acid vapours.

$$CH_3COONa + H_2O \longrightarrow CH_3COOH + NaOH$$

(*c*) *Coloured gas.* (*i*) Chloride on heating yield greenish yellow chlorine.

$$CuCl_2 + H_2O \longrightarrow CuO + 2HCl$$

$$CuO + 2HCl \longrightarrow Cu + H_2O + Cl_2$$

Bromides on heating yield reddish brown bromine.

Iodides on heating yield violet iodine.

$$2CdI_2 + O_2 \text{ (from air)} \longrightarrow 2CdO + I_2$$

$$CdBr_2 + O_2 \text{ (from air)} \longrightarrow 2CdO + Br_2$$

Nitrates other than those of alkali metals yield brown coloured nitrogen dioxide and oxygen.

$$2Pb(NO_3)_2 \longrightarrow 2PbO + 4NO_2 + O_2$$

$$2Cu(NO_3)_2 \longrightarrow 2CuO + 4NO_2 + O_2$$

2. Fusion or melting. Salts of alkali metals (Na, K, etc.) generally melt without decomposition, when heated. Certain salts containing water of crystallisation appear to have melted because on heating they give up their water and dissolve in this water forming a liquid.

3. Decrepitation. Crystals of certain substances do not contain any water of crystallisation but still they contain minute quantities of mother liquor enclosed in the crystals. Examples are $Pb(NO_3)_2$, NaCl, KI etc. On heating, the enclosed liquid vapourises and causes the crystals to burst out with a sharp cracking sound. This phenomenon is called *decrepitation.*

4. Swelling. Alums, borates and phosphates on heating swell up, due to loss of water of crystallisation.

$$\underset{\text{Boax}}{Na_2B_4O_7 \cdot 10H_2O} \longrightarrow Na_2B_4O_7 + 10H_2O$$

5. A sublimate is formed. Some salts vaporise themselves on heating and others give a volatile product, when decomposed. The vapours condense on the cooler parts of the test tube to form a *sublimate.*

(*a*) Halides of Hg, Sb, and As and NH_4 when volatilised produce white sublimate.

(*b*) HgI_2 forms a yellow sublimate which when rubbed with glass rod turns red.

(*c*) Salts of mercury when heated give a grey sublimate with globules of metallic mercury.

$$2Hg(NO_3)_2 \longrightarrow 2HgO + 4NO_2 + O_2$$

$$2HgO \longrightarrow 2Hg + O_2$$

(*d*) Iodides, when heated, decompose liberating violet iodine vapours which are condensed to form black sublimate.

6. Colour of Residue. (*a*) Certain salts on heating decompose yielding oxides, which have got different colours than the original salt, *e.g.*,

$$\underset{\text{White}}{2Zn(NO_3)_2} \longrightarrow \underset{\text{Yellow when hot, White when cold}}{ZnO} + 4NO_2 + O_2$$

$$\underset{\text{White}}{CaCO_3} \longrightarrow \underset{\text{Reddish brown}}{CdO} + CO_2$$

$$\underset{\text{Bluish green}}{CuCO_3} \longrightarrow \underset{\text{Black}}{CuO} + CO_2$$

(*b*) Change in colour in some cases is due to dehydration, for example,

$$\underset{\text{Blue}}{CuSO_4{\cdot}5H_2O} \longrightarrow \underset{\text{White}}{CuSO_4} + 5H_2O$$

$$\underset{\text{Pink}}{CoCl_2{\cdot}6H_2O} \longrightarrow \underset{\text{Blue}}{CoCl_2} + 6H_2O$$

$$\underset{\text{Green}}{NiCl_2{\cdot}6H_2O} \longrightarrow \underset{\text{Yellow}}{NiCl_2} + 6H_2O$$

Q. 2. How is flame test for analysing cations performed? Give the colours imparted by various cations (*a*) with naked eyes and (*b*) through blue glass.

Ans. This test is based upon the fact that salts of certain metals, particularly the *chlorides,* when brought in contact with flame, impart characteristic colouration.

Make a loop at the end of a platinum wire. Clean it by dipping it in conc. HCl and heating in non-luminous flame. Repeat the process till the wire imparts no colour to the flame. Now prepare a paste of salt with conc. HCl on a watch glass. Take a small quantity of paste on clean loop and introduce it into the flame. If a colour is imparted to the flame, observe first with naked eyes and then through cobalt blue glass.

Flame Test

S.	*Colour of the flame*		
No.	*With naked eyes*	*Through blue glass*	*Cation*
1.	Persistent brilliant golden yellow	Invisible	Sodium (Na)
2.	Pale violet	Pink	Potassium (K)
3.	Brick red	Light yellow	Calcium (Ca)
4.	Crimson red, flashes up quickly	Crimson	Strontium (Sr)
5.	Persistent grassy green	Visible	Barium (Ba)
6.	Bluish green or blue	Visible	Copper (Cu)

Q. 3. What type of salts are subjected to borax-bead test? Give the colours obtained in oxidising and reducing flames with various cations, supported by chemical equations.

(*Garhwal, 2000; Lucknow, 2002; Calcutta, 2005*)

Ans. This test is applied for the identification of metals which form *coloured salts*. It is based upon the fact that when salts of certain metals are heated with borax bead, a characteristic coloured bead is obtained.

The test is performed as follows:

A loop is made at the clean end of platinum wire and heated in flame. The red hot loop is dipped in powdered borax, when some of it adheres to the loop. On heating, borax swells up and gives a transparent bead. Now touch it while hot with the salt to be examined and heat it in flame again. Note the colour of the bead both in oxidising and reducing flames.

Borax Bead Test

Oxidising flame	*Reducing flame*	*Inference*
Deep blue	Deep blue	Cobalt
Deep green	Deep green	Chromium
Pinkish	Colourless or opaque	Manganese
Green in hot, blue in cold	Colourless or red opaque	Copper
Reddish brown (when hot), pale yellow (when cold)	Bottle green	Iron
Brown (in hot), pale (in cold)	Grey or black opaque	Nickel

On heating, borax is converted into sodium metaborate, boric anhydride and water. Boric anhydride combines with coloured salts to form corresponding metaborates having characteristic colours.

$$\underset{\text{Borax}}{Na_2B_4O_7 \cdot 10H_2O} \longrightarrow Na_2B_4O_7 + 10H_2O$$

$$Na_2B_4O_7 \xrightarrow{\Delta} \underset{\text{Sodium metaborate}}{2NaBO_2} + \underset{\text{Boric anhydride}}{B_2O_3}$$

$$Cr_2(SO_4)_3 + 3B_2O_3 \longrightarrow \underset{\text{Chromium metaborate (green)}}{2Cr(BO_2)_3} + 3SO_3$$

$$CoSO_4 + B_2O_3 \longrightarrow \underset{\text{Blue}}{Co(BO_2)_3} + SO_3$$

$$Fe_2O_3 + 3B_2O_3 \longrightarrow \underset{\text{Yellow}}{2Fe(BO_2)_3}$$

In reducing flame, ferric metaborate is reduced to ferrous metaborate.

$$2Fe(BO_2)_3 + C \longrightarrow \underset{\text{(Bottle green)}}{2Fe(BO_2)_2} + B_2O_2 + CO$$

$$CuSO_4 + B_2O_3 \longrightarrow Cu(BO_2)_2 + SO_3 \qquad \text{(Oxidising flame)}$$

$$2Cu(BO_2)_2 + C \longrightarrow \underset{\text{Colourless}}{CuBO_2} + B_2O_3 + CO \qquad \text{(Reducing flame)}$$

Q. 4. Describe charcoal cavity test for analysing inorganic compounds, giving chemical equations.

Ans. In this test the powdered salt is mixed with about twice its bulk of solid sodium carbonate and placed in a cavity made in charcoal piece. It is moistened with water and strongly heated in a reducing flame with the help of a blowpipe. From the bead formed in the cavity or from the colour of the deposit round the cavity called *incrustation,* the basic radical present in the salt can be identified.

Charcoal Cavity Test

Incrustation or Residue	*Bead*	*Inference*
Residue—Yellow (hot) white (when cold)	—	Zinc
Residue—Reddish brown, (cold and hot)	—	Cadmium
Incrustation—Dirty yellow (hot) white (cold)	White malleable bead, does not mark the paper	Tin
Residue—Reddish brown (hot) yellow (cold)	White malleable bead (marks paper black)	Lead
Residue—Orange (hot) yellow (cold)	—	Bismuth
Residue—Black	Red Scales	Copper

It is based upon the fact that when certain salts are heated with $Na_2 CO_3$, their metal carbonates are formed, which decompose on heating to give the corresponding metal oxides. The oxide in certain cases are reduced to give metal, in the form of a bead, with a deposit of oxide around the cavity,

$$CuSO_4 + Na_2CO_3 \longrightarrow CuCO_3 + Na_2SO_4$$

$$Cu\,CO_3 \longrightarrow CuO + CO_2$$

$$CuO + C \longrightarrow Cu + CO$$

$$CuO + CO \longrightarrow \underset{\text{Reddish metallic bead}}{Cu} + CO_2$$

Q. 5. Give the procedure and chemistry of cobalt nitrate test.

Ans. It is a good test for the detection of certain acid and basic radicals. It is performed only in those cases in which a white infusible residue is left in the charcoal cavity. For this test, the white residue, obtained on heating mixture of salt with Na_2CO_3 in charcoal cavity test, is moistened with a few drops of cobalt nitrate solution and again strongly heated with the help of a blowpipe in an oxidising flame. The colour of residue indicates the radical.

Cobalt Nitrate Test

Colour of residue	*Inference*
Green	Zinc
Blue	Aluminium, borate or phosphate
Light pink	Magnesium
Dirty green	Tin

It is based upon the fact that on strongly heating the salt with Na_2CO_3, metal carbonates are formed, which decompose to give oxides.

$$ZnSO_4 + Na_2CO_3 \longrightarrow ZnCO_3 + Na_2SO_4$$

$$ZnCO_3 \longrightarrow ZnO + CO_2$$

These oxides give characteristic coloured compound with cobalt oxide.

$$2Co(NO_3)_2 \longrightarrow 2CoO + 4NO_2 + O_2$$

$$CoO + ZnO \longrightarrow \underset{\text{Green}}{CoO \cdot ZnO}$$

$$CoO + Al_2O_3 \longrightarrow \underset{\text{Blue}}{CoO} \cdot Al_2O_3$$

Q. 6. Which radicals respond to dil. H_2SO_4 test? Explain giving equations.

Ans. Carbonate, bicarbonate, nitrite, sulphide, sulphite and thiosulphate are identified by heating the salt with dil. H_2SO_4 or dil. HCl. Dil. H_2SO_4 or HCl decomposes the salts of very weak and much more voltatile acids. In each case a gas is evolved which is characteristic of the particular acid radical.

1. Carbonate $\left(CO_3^{2-}\right)$ or bicarbonate $\left(HCO_3^-\right)$. They evolve colourless, odourless CO_2 gas, with a brisk effervescence. The gas is tested by passing it through lime water, which turns milky.

$$Na_2CO_3 + 2HCl \longrightarrow 2NaCl + CO_2 + H_2O$$

$$Na_2CO_3 + H_2SO_4 \longrightarrow Na_2SO_4 + CO_2 + H_2O$$

$$NaHCO_3 + HCl \longrightarrow NaCl + CO_2 + H_2O$$

$$NaHCO_3 + H_2SO_4 \longrightarrow NaHSO_4 + CO_2 + H_2O$$

$$\underset{\text{Lime water}}{Ca(OH)_2 + CO_2} \longrightarrow \underset{\text{Milky}}{CaCO_3} + H_2O$$

2. Nitrite (NO_2^-). Nitric oxide, a colourless gas, is evolved which immediately turns brown (NO_2) due to oxidation by air. NO gas is identified by passing it through $FeSO_4$ solution, which turns black.

$$2KNO_2 + H_2SO_4 \longrightarrow Na_2SO_4 + 2HNO_2$$

$$2HNO_2 \longrightarrow NO + H_2O + NO_2$$

$$2NO + O_2 \longrightarrow 2NO_2$$

$$FeSO_4 + NO \longrightarrow FeSO_4.NO \text{ (black)}$$

3. Sulphide (S^{2-}). Colourless gas (H_2S) with rotten eggs' smell is evolved. It is tested with lead acetate paper, which turns black.

$$Na_2S + 2HCl \longrightarrow 2NaCl + H_2S$$

$$Na_2S + H_2SO_4 \longrightarrow Na_2SO_4 + H_2S$$

$$(CH_3COO)_2Pb + H_2S \longrightarrow \underset{\text{Black}}{PbS} + 2CH_3COOH$$

4. (Sulphite SO_3^{2-}). Colourless gas with burning sulphur smell due to evolution of SO_2 is produced. It turns potassium dichromate paper green.

$$Na_2SO_3 + H_2SO_4 \longrightarrow Na_2SO_4 + SO_2 + H_2O$$

$$K_2Cr_2O_7 + H_2SO_4 + 3SO_2 \longrightarrow \underset{\text{Green}}{Cr_2(SO_4)_3} + K_2SO_4 + H_2O$$

Thiosulphate $\left(S_2O_3^{2-}\right)$. In addition to SO_2 (as identified in case of SO_3^{2-}), there occurs formation of colloidal sulphur, due to which the whole of the solution turns yellow.

$$Na_2S_2O_3 + 2HCl \longrightarrow 2NaCl + SO_2 + S + H_2O$$
$$Na_2S_2O_3 + H_2SO_4 \longrightarrow Na_2SO_4 + SO_2 + S + H_2O$$

Q. 7. Which acid radicals respond to conc. H_2SO_4 test? Explain giving equations.

Ans. This test is used for the identification of fluorides chlorides, bromides, iodides, nitrates, acetates and oxalates. On heating the salts with conc. H_2SO_4, certain acids which are either weak or more volatile are obtained as primary product, which react further with H_2SO_4, to give certain gases Reaction with fluorides is typical.

1. Fluoride (F^-). Action of conc. H_2SO_4 on a fluoride liberates hydrogen fluoride (HF). This reacts with the silica of glass to give silicon of glass to give silicon tetrafluoride. Hydrogen fluoride gives fumes in moist air. Its corrosive action on the glass gives a greasy appearance on the inner walls of the test tube.

$$NaF + H_2SO_4 \longrightarrow NaHSO_4 + \underset{\text{(Fumes in air)}}{HF}$$

$$SiO_2 + 4HF \longrightarrow \underset{\text{(Greasy appearance)}}{SiF_4} + 2H_2O$$

2. Chloride (Cl^-). Colourless gas, with pungent smell (HC1) is evolved, which gives white dense fumes with NH_4OH and white ppt. when passed through silver nitrate solution which is soluble in ammonia.

(*a*) $$NaCl + H_2SO_4 \longrightarrow NaHSO_4 + HCl$$

$$NH_3 + HCl \longrightarrow \underset{\text{Fumes}}{NH_4Cl}$$

(*b*) $$AgNO_3 + HCl \longrightarrow AgC1 + HNO_3$$

$$AgC1 + 2NH_4OH \longrightarrow \underset{\text{Soluble complex}}{[Ag(NH_3)_2]Cl} + 2H_2O$$

3. Bromide (Br^-). Reddish brown vapours of Br_2 are evolved, which turn starch paper yellow and have no effect on $FeSO_4$ solution.

$$KBr + H_2SO_4 \longrightarrow KHSO_4 + HBr$$
$$2HBr + H_2SO_4 \longrightarrow Br_2 + SO_2 + 2H_2O$$
$$Br_2 + \text{starch} \longrightarrow \text{yellow colour}$$

4. Iodide (I^-). Violet vapours of iodine are evolved, which turn starch paper blue.

$$KI + H_2SO_4 \longrightarrow KHSO_4 + HI$$
$$2HI + H_2SO_4 \longrightarrow SO_2 + I_2 + 2H_2O$$
$$I_2 + \text{starch} \longrightarrow \text{Blue}$$

5. Nitrate (NO_3^-). Brown gas (NO_2) with copious fumes are evolved on adding cupper chips or a paper piece. The gas is tested by passing through $FeSO_4$ soln. which turns black.

(*a*) $$NaNO_3 + H_2SO_4 \longrightarrow NaHSO_4 + HNO_3$$

$$4HNO_3 \longrightarrow 4NO_2 + 2H_2O + O_2$$

(*b*) $$4HNO_3 + Cu \longrightarrow Cu(NO_3)_2 + 2NO_2 + 2H_2O$$

(*c*) $$2HNO_3 \longrightarrow 2NO + H_2O + 3O$$

$$FeSO_4 + NO \longrightarrow \underset{\text{(Black)}}{FeSO_4.NO}$$

6. Acetate (CH_3COO^-). Colourless vapours of acetic acid with vinegar-like smell are evolved.

$$2CH_3COONa + H_2SO_4 \longrightarrow Na_2SO_4 + \underset{\text{Vinegar smell}}{2CH_3COOH}$$

7. Oxalate ($C_2O_4^{2-}$). On heating oxalate with conc. H_2SO_4, carbon dioxide and carbon monoxide gases are produced. CO_2 is identified by passing it through lime water, which turns milky and carbon-monoxide burns with a blue flame.

$$\underset{\text{Sod.oxalate}}{\begin{array}{l}COONa\\ |\\ COONa\end{array}} + H_2SO_4 \longrightarrow \begin{array}{l}COOH\\ |\\ COOH\end{array} + Na_2SO_4$$

$$\begin{array}{l}COOH\\ |\\ COOH\end{array} \xrightarrow{-H_2O} CO + CO_2$$

$$CO_2 + \underset{\text{Lime water}}{Ca(OH)_2} \longrightarrow \underset{\text{Milky}}{CaCO_3} + H_2O$$

Q. 8. Give wet (confirmatory) tests for carbonate, bicarbonate, sulphide, sulphite, thiosulphate and nitrite radicals. *(Andhra 2010)*

Ans. Carbonate and bicarbonate. Shake little of the salt with distilled water. Add dil. HCl to the water extract. Brisk effervescence produced indicates soluble *carbonate* or *a bicarbonate.*

$$Na_2CO_3 + 2HCl \longrightarrow 2NaCl + CO_2 + H_2O$$

$$NaHCO_3 + 2HCl \longrightarrow 2NaCl + CO_2 + H_2O$$

Tests for soluble carbonate and bicarbonate. 1. *$MgSO_4$ test.* To the salt solution in water, add $MgSO_4$ soln. *White ppt* in *cold* confirms *carbonate* and *white ppt* on *heating* confirms *bicarbonate.*

$$Na_2CO_3 + MgSO_4 \longrightarrow \underset{\text{ppt.}}{MgCO_3} + Na_2SO_4$$

$$2NaHCO_3 + MgSO_4 \longrightarrow Na_2SO_4 + Mg(HCO_3)_2$$

$$Mg(HCO_3)_2 \xrightarrow{\text{Heat}} \underset{\text{ppt.}}{MgCO_3} + CO_2 + H_2O$$

2. *Phenolphthalein test.* To the salt soln., add a few drops of phenolphthalein –*Pink colour* shows *soluble carbonate* and *no colour* indicates presence of *bicarbonate.*

Sulphide (S^2)$^-$. 1. *Sodium nitroprusside test.* To the salt solution in water add NaOH and sod. nitroprusside solution. – *violet colouration.*

$$Na_2S + Na_2[Fe(CN)_5NO] \longrightarrow \underset{\substack{\text{Sod-sulphonitroprusside}\\ \text{(violet)}}}{Na_4[Fe(CN)_4 NOS]}$$

2. *Lead acetate test.* To the salt solution in water add lead acetate solution – *black ppt.* insoluble in acetic acid.

$$Na_2S + (CH_3COO)_2Pb \longrightarrow \underset{\text{Black ppt.}}{PbS} + 2CH_3COONa$$

Sulphite $\left(SO_3^{2-}\right)$

1. *Barium chloride test.* To the salt solution in water add $BaCl_2$ solution; *white ppt* soluble in dil. HCl.

$$Na_2SO_3 + BaCl_2 \longrightarrow \underset{\text{White ppt.}}{BaSO_3} + 2NaCl$$

2. *$K_2Cr_2O_7$ test.* To the salt solution add $K_2Cr_2O_7$ solution and dil. HS_2O_4. Solution turns *green.*

3. *$FeCl_3$ test.* To the salt solution in water, add $FeCl_3$ solution–*deep red colouration.*

$$2Na_2SO_3 + 2FeCl_3 \longrightarrow 6NaCl + Fe_2(SO_3)_3$$

Thiosulphate $\left(S_2O_3^{2-}\right)$. 1. *Silver nitrate test.* To the salt solution in water add silver nitrate solution. A white ppt. is formed, which quickly changes to orange, brown and finally black.

$$Na_2\,S_2O_3 + 2AgNO_3 \longrightarrow \underset{\text{White ppt.}}{Ag_2S_2O_3} + 2NaNO_3$$

$$Ag_2S_2O_3 + H_2O \longrightarrow \underset{\text{Black ppt.}}{Ag_2S} + H_2SO_4$$

2. *Iodine test.* Add iodine solution to a portion of water-extract. *Brown* colour is discharged.

$$2Na_2S_2O_3 + I_2 \longrightarrow Na_2S_4O_6 + 2NaI$$

Nitrite $\left(NO_2^-\right)$. 1. *Diphenylamine test.* To the salt solution in water, add diphenylamine solution–*deep blue* colouration is obtained.

2. *$FeSO_4$ test.* To the salt solution in water, add $FeSO_4$ solution *–Dark brown colouration.*

$$FeSO_4 + NO \longrightarrow \underset{\text{Nitrosoferrous sulphate}}{FeSO_4.NO}$$

3. *$KMnO_4$ test.* To the salt solution add dil. H_2SO_4 and $KMnO_4$. *Pink colour* is discharged.

$$2KMnO_4 + 3H_2SO_4 + 5KNO_2 \longrightarrow K_2\,SO_4 + 2MnSO_4 + 3H_2O + 5KNO_3$$

Q. 9. Describe wet (confirmatory) tests for fluoride, chloride, bromide, iodide and nitrate acetate and oxalate. *(Indore, 2001; Andhra, 2010)*

Ans. Fluoride (F^-). Mix a little sand with mixture in a dry test tube, add to it 1–2 ml conc. H_2SO_4. Warm the test tube gently. Bring a rod moistured with water inside the test tube without touching the test tube and the solution. A white deposition on the rod occurs. Also, there is a greasy appearance in the test tube.

$$NaF + H_2SO_4 \longrightarrow NaHSO_4 + HF$$

$$SiO_2 + 4HF \longrightarrow \underset{\text{Greasy appearance}}{SiF_4} + 2H_2O$$

$$3SiF_4 + 4H_2O \longrightarrow 2H_2\,[SiF_6] + \underset{\text{white depositon on the glass rod.}}{H_4SiO_4 \text{ (silicic acid)}}$$

Chloride (Cl^-). 1. *Silver nitrate test.* To the salt solution in distilled water, add $AgNO_3$ solution. *White ppt.* soluble in NH_4OH, is obtained.

$$NaCl + AgNO_3 \longrightarrow AgCl + NaNO_3$$

$$AgCl + 2NH_4OH \longrightarrow \underset{\text{Soluble complex}}{Ag(NH_3)_2Cl_2} + 2H_2O$$

2. *Chromyl chloride test.* Warm the chloride salt with some solid $K_2Cr_2O_7$ and a few ml of conc. H_2SO_4. Pass red vapours through NaOH solution. Add acetic acid and lead acetate solution to the yellow solution – *yellow ppt. of lead chromate* soluble in NaOH is obtained.

$$K_2Cr_2O_7 + 6H_2SO_4 + 4KCl \longrightarrow 6KHSO_4 + 3H_2O + 2CrO_2Cl_2$$

$$CrO_2Cl_2 + 2NaOH \longrightarrow \underset{\text{Yellow soln.}}{Na_2CrO_4} + 2NaCl + 2H_2$$

$$Na_2CrO_4 + (CH_3COO)_2Pb \longrightarrow \underset{\text{Yellow ppt.}}{PbCrO_4} + 2CH_3COONa$$

Bromide (Br^-). 1. *$AgNO_3$ test.* To the salt solution in distilled water, add $AgNO_3$ solution. Light yellow ppt. partly soluble in NH_4OH, is obtained.

$$NaBr + AgNO_3 \longrightarrow \underset{\text{Pale yellow ppt.}}{AgBr} + NaNO_3$$

2. *CS_2 test.* To the salt solution in water, add fresh chlorine water and few drops of carbon disulphide. Shake well and allow to stand. An orange colour in CS_2 layer.

$$2NaBr + Cl_2 \longrightarrow 2NaCl + \underset{\text{Imparts orange colour to } CS_2 \text{ layer}}{Br_2}$$

Iodide (I^-). 1. *$AgNO_3$ test.* To the salt solution in distilled water, add $AgNO_3$ solution. *Yellow ppt.* insoluble in NH_4OH is obtained.

$$NaI + AgNO_3 \longrightarrow AgI + NaNO_3$$

2. *CS_2 test.* To the salt solution in water, add chlorine water and then add few dops of CS_2 and shake. Allow to stand–*violet coloured* CS_2 layer is obtained.

$$2KI + Cl_2 \longrightarrow 2KCl + I_2$$

$$I_2 + CS_2 \longrightarrow \text{violet coloured layer}$$

Nitrate $\left(NO_3^-\right)$. 1. *Copper chips test.* Heat the salt with a few chips of copper and conc. H_2SO_4. Dense *brown fumes* are produced.

$$KNO_3 + H_2SO_4 \longrightarrow KHSO_4 + HNO_3$$

$$Cu + 3HNO_3 \longrightarrow Cu(NO_3)_2 + \underset{\text{Brown gas}}{2NO_2} + 2H_2O$$

2. *Ring test.* To the water extract, add freshly prepared $FeSO_4$ solution. Then conc. H_2SO_4 along the sides of test tube. A dark-brown ring is formed at the junction of the two layers.

$$KNO_3 + H_2SO_4 \longrightarrow KHSO_4 + HNO_3$$

$$6FeSO_4 + 3H_2SO_4 + 2HNO_3 \longrightarrow 3Fe_2(SO_4)_3 + 4H_2O + 2NO$$

$$FeSO_4 + NO \longrightarrow \underset{\text{(Nitrosoferrous sulphate)}}{\underset{\text{Brown ring}}{FeSO_4.NO}}$$

Acetate (CH_3COO^-)

1. Ferric chloride test. To 3 ml of water extract (all acetates are soluble in water), add a few drops of freshly prepared neutral $FeCl_3$. Red colour indicates acetate, Boil. Reddish brown precipitate confirms acetate.

To prepare neutral $FeCl_3$. Due to hydrolysis, $FeCl_3$ is slightly acidic as it contains free HCl. Take about 5ml of bench $FeCl_3$ solution, add dilute NH_4OH solution dropwise till a slight precipitate persists. Then boil till a clear brown solution of neutral $FeCl_3$ is obtained.

$$3CH_3COONa + FeCl_3 \longrightarrow \underset{\substack{\text{Ferric acetate} \\ \text{(Red colour)}}}{(CH_3COO)_3Fe} + 3NaCl$$

2. Oxalic acid test. Mix a small amount of mixture with equal amount of solid oxalic acid on a watch glass. Moisten the mixture with 2–3 drops of water. Rub with the finger and smell. A characteristic vinegar smell is given out.

$$2CH_3COOK + \begin{matrix} COOH \\ | \\ COOH \end{matrix} \longrightarrow \begin{matrix} COOK \\ | \\ COOK \end{matrix} + \underset{\substack{\text{(Acetic acid)} \\ \text{(Vinegar smell)}}}{2CH_3COOH}$$

3. Ester test. Warm a litle of the mixture with conc. H_2SO_4 and a few drops of alcohol in a test tube. A pleasant fruity smell is obtained.

$$2CH_3COONa + H_2SO_4 \longrightarrow 2CH_3COOH + Na_2SO_4$$

$$CH_3COOH + C_2H_5OH \longrightarrow \underset{\substack{\text{Ethyl acetate} \\ \text{(Fruity smell)}}}{CH_3COOC_2H_5} + H_2O$$

Oxalate $\left(\begin{matrix} COO^{2-} \\ | \\ COO \end{matrix}\right)$

Calcium chloride test. Neutralise a little of sodium extract with acetic acid. Add $CaCl_2$ solution. White precipitate indicates the presence of oxalate. Filter, warm the white precipitate with dil. H_2SO_4 and add a drop of $KMnO_4$ solution. Pink colour is discharged.

$$\underset{\text{Sod. oxalate}}{\begin{matrix} COONa \\ | \\ COONa \end{matrix}} + CaCl_2 \longrightarrow \underset{\text{Cal. acetate}}{\begin{matrix} COO \\ \\ COO \end{matrix}\!\!>Ca} + 2NaCl$$

$$\begin{matrix} COO \\ \\ COO \end{matrix}\!\!>Ca + H_2SO_4 \longrightarrow \underset{\text{Acetic acid}}{\begin{matrix} COOH \\ | \\ COOH \end{matrix}} + CaSO_4$$

$$\begin{matrix} COOH \\ | \\ COOH \end{matrix} + O \longrightarrow \underset{\text{(Pink colour discharged)}}{2CO_2 + H_2O}$$

Q. 10. Describe, with chemical equations, wet tests for sulphate, phosphate and borate.

Ans. Sulphate (SO_4^{2-}). 1. *Barium chloride test.* To the salt solution in water, add $BaCl_2$ solution. *White ppt* insoluble in conc. HC1 is obtained.

$$Na_2SO_4 + BaC1_2 \longrightarrow \underset{\text{White ppt.}}{BaSO_4} + 2NaCl$$

2. *Lead acetate test.* To the salt solution in water, add lead acetate solution, *white ppt* soluble in hot ammonium acetate solution.

$$Na_2SO_4 + (CH_3COO)_2Pb \longrightarrow \underset{ppt.}{PbSO_4} + 2CH_3COONa$$

$$PbSO_4 + 2CH_3COONH_4 \longrightarrow (CH_3COO)_2Pb + (NH_4)_2SO_4$$

Phosphate (PO_4^{3-}). 1. *Ammonium molybdate test.* Heat the salt with few drops of conc. HNO_3, till brown vapours are given off. Add excess of ammonium molybdate solution and again heat. *Yellow ppt.* or *yellow coloured* solution is formed.

$$2Na_3PO_4 + 3(NH_4)_2MoO_4 \longrightarrow 2(NH_4)_3PO_4 + 3NaMoO_4$$

$$Na_2MoO_4 + 2HNO_3 \longrightarrow \underset{\text{Molybdic acid}}{H_2MoO_4} + 2NaNO_3$$

$$H_2MoO_4 \longrightarrow MoO_3 + H_2O$$

$$(NH_4)_3PO_4 + 12MoO_3 + 6H_2O \longrightarrow (NH_4)_3PO_4\,12MoO_3 6H_2O$$

Yellow ppt.
Amm. phosphomolybdate

2. *Magnesia mixture test.* To the salt solution in water, add magnesia mixture ($MgSO_4 + NH_4Cl + NH_4OH$) and shake. *White ppt* is formed.

$$Na_3PO_4 + MgSO_4 + 3NH_4Cl \longrightarrow \underset{ppt.}{Mg(NH_4)PO_4} + 3NaCl + (NH_4)_2SO_4$$

Borate (BO_3^{3-}). Take solid salt in a China dish. Add conc. H_2SO_4 and alcohol. Ignite the vapours. A *green edged* flame is obtained.

$$Na_2B_4O_7 + H_2SO_4 + 4H_2O \longrightarrow 4H_3BO_3 + Na_2SO_4$$

$$H_3BO_3 + 3C_2H_5OH \longrightarrow (C_2H_5)_3BO_3 + 3H_2O$$

Ethyl borate
burns with green
flame

Q. 11. How are the cations of first group detected? Give their confirmatory tests with equations.

Ans. This group includes lead, silver and mercury, the chlorides of which are insoluble in acidified cold water.

$$Pb(NO_3)_2 + 2HCl \longrightarrow PbCl_2 + 2HNO_3$$

ppt. soluble
in hot H_2O

$$2AgNO_3 + 2HCl \longrightarrow 2AgCl + 2HNO_3$$

ppt. soluble
in NH_4OH

$$Hg_2(NO_3)_2 + 2HCl \longrightarrow Hg_2Cl_2 + 2HNO_3$$

Turns black
in NH_4OH

Procedure for analysis. To the original aqueous solution of salt add dil. HCl. Insoluble chlorides of these cations are precipitated. Filter the ppt. and boil with water. $PbCl_2$ dissolves if present. The separation of these chlorides depends upon the following features:

(*a*) *$PbCl_2$ is soluble in hot water while AgCl, Hg_2Cl_2 are insoluble.*
(*b*) *AgCl is soluble in NH_4OH, while Hg_2Cl_2 is not soluble.*
(*c*) *Hg_2Cl_2 reacts with NH_4OH forming black ppt.*

Confirmation of Pb^{2+} (ous). (1) On cooling the portion of above hot solution under tap water needle-shaped crystals of $PbCl_2$ separate out.

(2) *K_2CrO_4 test.* To the salt solution, add potassium chromate solution. *Yellow ppt.* soluble in NaOH is obtained.

$$PbCl_2 + K_2CrO_4 \longrightarrow \underset{\text{Yellow ppt.}}{PbCrO_4} + 2KCl$$

(3) *KI test.* To the salt solution, add KI. *Yellow ppt.* is obtained.

$$PbCl_2 + 2KI \longrightarrow \underset{\text{Yellow ppt.}}{PbI_2} + 2KCl$$

If ppt. are insoluble in hot water, Pb^{2+} absent Then treat the ppt. with NH_4OH, shake and warm. AgCl dissolves, whereas Hg_2Cl_2 turns black.

$$AgCl + NH_4OH \longrightarrow \underset{\text{Soluble}}{[Ag(NH_3)_2]\,Cl} + 2H_2O$$

$$Hg_2\,Cl_2 + 2NH_4OH \longrightarrow Hg(NH_2)Cl + NH_4Cl + \underset{\text{Black}}{Hg} + 2H_2O$$

Confirmation of Ag^+. (1) *K_2CrO_4 test*. To the salt solution add K_2CrO_4 solution, brick red ppt. is obtained.

$$2AgNO_3 + K_2CrO_4 \longrightarrow \underset{\text{Red ppt.}}{Ag_2CrO_4} + 2KNO_3$$

(2) *KI test.* To the salt solution, add KI solution–yellow ppt.

$$AgNO_3 + KI \longrightarrow \underset{\text{ppt.}}{AgI} + KNO_3$$

Confirmation of Hg_2^{2+}. (1) *$SnCl_2$ test.* To the salt solution, add $SnCl_2$, white ppt. turning grey.

$$Hg_2\,Cl_2 + SnCl_2 \longrightarrow \underset{\text{Black}}{2Hg} + SnCl_4$$

(2) *Caustic soda test.* To the salt solution, add NaOH in excess, black ppt. is obtained.

$$Hg_2\,(NO_3)_2 + 2NaOH \longrightarrow \underset{\text{Grey}}{Hg_2O} + 2NaNO_3 + H_2O$$

(3) *Copper turning test*. To the original solution, add copper turning, white deposit of mercury on copper is formed.

$$Hg_2(NO_3)_2 + Cu \longrightarrow Cu(NO_3)_2 + 2Hg$$

(4) *KI test.* To the original solution, add KI, scarlet red ppt. which dissolves in excess of reagent.

$$Hg_2(NO_3)_2 + 2KI \longrightarrow \underset{\text{ppt.}}{Hg_2I_2} + 2KNO_3$$

$$Hg_2I_2 + 2KI \longrightarrow 2KHgI_4$$

Q. 12. How are the cations of IIA and IIB groups analysed? Give the confirmatory (wet) tests of cations of these groups along with the reactions. *(G. Nanakdev, 2000)*

Ans. This group comprises mercury, lead, bismuth, copper, cadmium, arsenic, antimony and tin. The sulphides of their metals are insoluble in dil. HCl. This group is further divided into two sub-groups IIA and IIB.

Sub. group IIA – (Hg^{2+}, Pb^{2+}, Bi^{3+}, Cu^{2+} and Cd^{2+})

Sub-group IIB – (As^{3+}, Sb^{3+}, Sn^{2+} and Sn^{4+})

The separation of IIA and IIB sub-groups is based upon the features that members of IIA sub-group are insoluble in yellow ammonium sulphide while those of IIB sub-group are soluble in yellow ammonium sulphide.

The separation of the sulphides of IIA group is based upon the following features:

(*a*) *HgS is insoluble in dil. HNO_3, while sulphides of Pb, Bi, Cu, Cd are soluble forming nitrates:*

(*b*) *$PbSO_4$ is insoluble in water, while sulphates of other metals are soluble.*

(*c*) *$Bi(OH)_3$ is insoluble in NH_4OH, while Cu, Cd hydroxides form soluble complexes.*

(*d*) *Cu and Cd separation is based upon the fact that in presence of KCN, only Cd is precipitated as sulphide on passing H_2S.*

Procedure for analysis of IIA group radicals. The basic radicals of II group are precipitated as their sulphides by passing H_2S gas through their salt solution prepared in dil. HCl.

$$HgCl_2 + H_2S \longrightarrow \underset{\text{Black ppt.}}{HgS} + 2HCl$$

$$PbCl_2 + H_2S \longrightarrow \underset{\text{Black ppt.}}{PbS} + 2HCl$$

$$2BiCl_3 + 3H_2S \longrightarrow \underset{\text{Black ppt.}}{Bi_2S_3} + 6HCl$$

$$CuCl_2 + H_2S \longrightarrow \underset{\text{Black ppt.}}{CuS} + 2HCl$$

$$CdCl_2 + H_2S \longrightarrow \underset{\text{Yellow ppt.}}{CdS} + 2HCl$$

$$2AsCl_3 + 3H_2S \longrightarrow \underset{\text{Yellow ppt.}}{As_2S_3} + 6HCl$$

$$2SbCl_3 + 3H_2S \longrightarrow \underset{\text{Orange ppt.}}{Sb_2S_3} + 6HCl$$

$$SnCl_2 + H_2S \longrightarrow \underset{\text{Brown ppt.}}{SnS} + 2HCl$$

Boil the portion of the ppt. with yellow ammonium soluphide. If ppt. remains insoluble, then analyse the ppt. for IIA group radicals and if the ppt. dissolves, then proceed for the analysis of IIB group radicals.

Boil the ppt. of IIA group with 50% HNO_3. Sulphides of all cations except Hg dissolve as nitrates.

$$3PbS + 8HNO_3 \longrightarrow 3Pb(NO_3)_2 + 2NO + 4H_2O + 3S$$

$$Bi_2S_3 + 8HNO_3 \longrightarrow 2Bi(NO_3)_3 + 2NO + 4H_2O + 3S$$

$$3CuS + 8HNO_3 \longrightarrow 3Cu(NO_3)_2 + 2NO + 4H_2O + 3S$$

$$3CdS + 8HNO_3 \longrightarrow 3Cd(NO_3)_2 + 2NO + 4H_2O + 3S$$

Confirmation of Hg^{2+}. A black ppt. insoluble in hot 50% HNO_3 indicates Hg^{2+}. Dissolve the ppt. in aqua-regia by boiling.

$$3HgS + 2HNO_3 + 6HCl \longrightarrow 3HgCl_2 + 2NO + 4H_2O + 3S$$

Confirm Hg^{2+} with above solution as in group I.

Confirmation of Pb^{2+}. The solution of metal nitrate is tested for Pb^{2+}, Bi^{+2} and Cu^{2+} as under. To a portion of the solution add dil. H_2SO_4. Pb^{2+} if present is precipitated as $PbSO_4$.

$$Pb(NO_3)_2 + 2H_2SO_4 \longrightarrow \underset{\text{ppt.}}{PbSO_4} + 2HNO_3$$

Dissolve the ppt. of $PbSO_4$ in ammonium acetate.

(1) *Potassium chromate test.* To a part of above solution, add K_2CrO_4 solution. *Yellow ppt.* is obtained.

$$PbSO_4 + 2(CH_3COONH_4) \longrightarrow (NH_4)_2SO_4 + \underset{\text{Soluble}}{(CH_3COO)_2Pb}$$

$$(CH_3COO)_2Pb + K_2CrO_4 \longrightarrow \underset{\text{Yellow ppt.}}{PbCrO_4} + 2CH_3COOK$$

(2) *Potassium iodide test.* Add KI solution to another part of solution, *yellow ppt.* is formed

$$(CH_3COO)_2Pb + 2KI \longrightarrow \underset{\text{Yellow ppt.}}{PbI_2} + 2CH_3COOK$$

Confirmation of Bi^{3+}. A *white ppt* on adding excess of NH_4OH to solution of metal nitrate (group ppt. in dil. HNO_3) indicates Bi.

$$Bi(NO_3)_3 + 3NH_4OH \longrightarrow \underset{\text{White ppt.}}{Bi(OH)_3} + 3NH_4NO_3$$

Dissolve the white ppt. in minimum quantity of dil. HCl and divide the solution in two parts.

$$Bi(OH)_3 + 3HCl \longrightarrow BiCl_3 + 3H_2O$$

(1) *Dilution test.* To one part of the solution, add excess of water, *white ppt.* or *milkiness* is obtained.

$$BiCl_3 + H_2O \longrightarrow \underset{\text{White ppt.}}{BiOCl} + 2HCl$$

(2) *Sodium stannite test.* To the second part of solution add sodium stannite solution ($SnCl_2$ + NaOH) — a *black ppt.* is obtained.

$$BiCl_3 + 3NaOH \longrightarrow Bi(OH)_3 + 3NaCl$$

$$2Bi\,(OH)_3 + 3Na_2SnO_2 \longrightarrow \underset{\text{(Black)}}{2Bi} + 3Na_2SnO_3 + 3H_2O$$

Confirmation of Cu^{2+}. *Deep blue* coloured solution, obtained on adding, NH_4OH to the solution of the metal nitrate, indicates copper.

$$Cu(NO_3)_2 + 4NH_4OH \longrightarrow \underset{\substack{\text{Soluble complex} \\ \text{(deep blue colour)}}}{[Cu(NH_3)_4]\,(NO_3)_2} + 4H_2O$$

Acidify the deep blue coloured solution, with acetic acid and divide in two parts.

(1) *Potassium ferrocyanide test.* Add potassium ferrocyanide to one part of above solution. Chocolate coloured ppt. is obtained.

$$2Cu(NO_3)_2 + K_4[Fe(CN)_6] \longrightarrow \underset{\text{Chocolate ppt.}}{Cu_2[Fe(CN)_6]} + 4KNO_4$$

(2) *NaOH solution test.* To the second part of the solution add NaOH, *blue ppt* is obtained.

$$Cu(NO_3)_2 + 2NaOH \longrightarrow \underset{\text{Blue ppt.}}{Cu(OH)_2} + 2NaNO_3$$

(3) *Perform borax bead test. Bluish green bead.*

Confirmation of Cd^{+2}. *Yellow ppt* insoluble in ammonium sulphide, obtained on passing H_2S gas through the salt solution indicates Cd^{2+}.

(1) *Potassium ferrocyanide test.* Add potassium ferrocyanide solution to the salt solution. A *white* or *bluish white ppt* of cadmium ferrocyanide is obtained.

$$2Cd(NO_3)_2 + K_4[Fe(CN)_6] \longrightarrow \underset{\substack{\text{Bluish white}\\ \text{ppt.}}}{Cd_2[Fe(CN)_6]} + 4KNO_3$$

(2) *Ammonium hydroxide test.* On adding NH_4OH, to very dil. solution of the salt–*white ppt.* of cadium hydroxide soluble in excess of reagent is obtained.

$$Cd(NO_3)_2 + 2NH_4OH \longrightarrow \underset{\text{White ppt.}}{Cd(OH_2)} + 2NH_4NO_3$$

$$Cd(OH_2) + 4NH_4OH \longrightarrow \underset{\text{Soluble complex}}{[Cd(NH_3)_4](OH_2)} + 4H_2O$$

Identification of Basic Radical of IIB Group

The sulphides of arsenic, antimony and tin are soluble in yellow ammonium sulphide, forming thiosalts.

The separation of the sulphides of IIB group is based upon the following features:

(*a*) *Arsenic sulphide is insoluble in conc. HCl while Sb and Sn sulphides are soluble.*

(*b*) *Antimony sulphide is precipitated in the presence of oxalic acid. Tin under such conditions forms a soluble complex.*

Confirmation of As^{3+}. Yellow ppt. insoluble in conc. HCl, indicates As_2S_2. Dissolve the yellow ppt. in hot conc. HNO_3 and proceed as follows:

(1) *Ammonium molybdate test.* Add ammonium molybdate solution to portion of above solution. *Yellow ppt* due to ammonium arseno-molybdate is obtained.

$$AsS_3 + 10HNO_3 \longrightarrow 2H_3AsO_4 + 3S + 2H_2O + 10NO_2$$

$$H_3AsO_4 + 12(NH_4)_2MoO_4 + 21HNO_3 \longrightarrow \underset{\text{Yellow ppt.}}{(NH_4)_3AsO_3 \cdot 12MoO_3} + 21NH_4NO_3 + 12H_2O$$

(2) *Magnesia mixture test.* Add magnesia mixture ($MgSO_4 + NH_4Cl + NH_4OH$) to another portion of above solution, *white ppt.* is obtained.

$$H_3AsO_4 + MgSO_4 + 3NH_4OH \longrightarrow \underset{\text{White ppt.}}{MgNH_4AsO_4} + (NH_4)_2SO_4 + 3H_2O$$

Confirmation of Sb^{3+}. *Orange ppt* indicates antimony sulphide. Dissolve the ppt. in conc. HCl and boil off H_2S gas. Use the solution to confirm Sb^{3+}.

(1) *Dilution test.* Dilute a part of the solution, *white ppt.* is obtained.

$$SbCl_3 + H_2O \longrightarrow \underset{\text{White ppt.}}{SbOCl} + 2HCl$$

(2) *Tin metal test.* To the other part of the solution, put some tin pieces. A *black deposit* on tin metal pieces is formed.

$$2SbCl_3 + 3Sn \longrightarrow 2SnCl_2 + \underset{\text{Black}}{2Sb}$$

(3) *Oxalic acid test.* To the salt solution add NH_4OH, just to make it alkaline and add 1-2 gram of solid oxalic acid. Pass H_2S gas through the solution–*Orange ppt.* is obtained.

Confirmation of tin. Original ppt. *yellow* or *brown* indicates Tin. Treat the ppt. with dil. HCl and zinc pieces. Allow it to stand for some time to reduce stannic chloride to stannous chloride and filter.

$$SnS_2 + 4HCl \longrightarrow SnCl_4 + 2H_2S$$

$$SnS + 2HCl \longrightarrow SnCl_2 + H_2S$$

$$Zn + 2HCl \longrightarrow ZnCl_2 + 2H$$

$$SnCl_4 + 2H \longrightarrow SnCl_2 + 2HCl$$

(1) *Mercuric chloride test.* To one portion of the above filtrate, add mercuric chloride–*white ppt* turning grey.

$$SnCl_2 + 2HgCl_2 \longrightarrow \underset{\text{White ppt.}}{Hg_2Cl_2} + SnCl_4$$

$$Hg_2Cl_2 + SnCl_2 \longrightarrow \underset{\text{Grey}}{2Hg} + SnCl_2$$

(2) *Ammonium molybdate test.* To the second portion of filtrate, add ammonium molybdate solution–*blue ppt.* is obtained.

$$SnCl_2 + 3(NH_4)_2MoO_4 + 8HCl \longrightarrow 6NH_4Cl + \underset{\text{Blue ppt.}}{Mo_3O_8} + SnCl_4 + 4H_2O$$

Q. 13. Give the chemistry of analysis of cations of third group.

Ans. This group comprises iron, aluminium and chromium, the hydroxides of which are insoluble in ammonia. The separation of these hydroxides depends upon the following features:

(*a*) $Al(OH)_3$ is soluble in excess of NaOH, forming sodium aluminate, while $Fe(OH_3)$ and $Cr(OH)_3$ are insoluble.

(*b*) $Cr(OH)_3$ is soluble in NaOH and bromine water, while $Fe(OH)_3$ is insoluble.

Boil the salt solution with a drop of conc. HNO_3. Add solidNH_4Cl followed by excess of NH_2OH. The cations are precipitated as their hydroxides.

$$FeCl_3 + 3NH_4OH \longrightarrow \underset{\text{Brown ppt.}}{Fe(OH)_3} + 3NH_4Cl$$

$$AlCl_3 + 3NH_4OH \longrightarrow \underset{\text{White gelatinous}}{Al(OH)_3} + 2NH_4Cl$$

$$CrCl_3 + 3NH_4OH \longrightarrow \underset{\text{Green ppt.}}{Cr(OH)_3} + 3NH_4Cl$$

From the colour of the ppt. the radical can be identified.

Reddish brown ppt. $\longrightarrow Fe^{3+}$
White gelatinous ppt. $\longrightarrow Al^{3+}$
Dirty green ppt. $\longrightarrow Cr^{3+}$

Confirmation of iron. A *brown ppt* indicates $Fe(OH)_3$. Dissolve the ppt. in dil. HCl and divide the solution into two parts:

(1) To one part of the solution add *Potassium ferrocyanide–prussian blue colour* due to ferric ferrocyanide is produced.

$$Fe(OH)_3 + 3HCl \longrightarrow FeCl_3 + 3H_2O$$

$$4FeCl_3 + 3K_4[Fe(CN)_6] \longrightarrow \underset{\text{Blue}}{Fe_4[Fe(CN)_6]_3} + 12KCl$$

(2) *Sulphocyanide test.* To the second portion of the solution and amm-sulphocyanide or pot. sulphocyanide. *Blood red colouration* is produced.

$$FeCl_3 + 3NH_4CNS \longrightarrow Fe(CNS)_3 + 3NH_4Cl$$

or

$$FeCl_3 + 3KCNS \longrightarrow \underset{\text{Blood red.}}{Fe(CNS)_3} + 3KCl$$

Confirmation of Cr^{3+}. A *green ppt.* indicates $Cr(OH)_3$. Suspend the ppt in water and add sod.-peroxide carefully. A *yellow solution* is obtained.

$$2Cr(OH)_3 + 3Na_2O_2 \longrightarrow \underset{\text{(Yellow)}}{2Na_2CrO_4} + 2NaOH + 2H_2O$$

Divide the solution into two parts and perform the following tests.

(1) *Lead acetate test.* To one portion of the yellow solution, add acetic acid and lead acetate. A *yellow ppt* is obtained.

$$Na_2CrO_4 + (CH_3COO)_2Pb \longrightarrow \underset{\text{Yellow ppt.}}{PbCrO_4} + 2CH_3COONa$$

(2) *Hydrogen peroxide test.* Acidify second portion of above yellow solution with dil. H_2SO_4. Then add small amount of ether and hydrogen peroxide. *Blue* coloured solution obtained.

$$2Na_2CrO_4 + 4H_2O_2 + 2H_2SO_4 \longrightarrow \underset{\substack{\text{Chromic}\\ \text{peroxide}\\ \text{(blue)}}}{2CrO_5} + 2Na_2SO_4 + 6H_2O$$

Confirmation of Al^{3+}. A *white gelatinous* ppt. indicates $Al(OH)_3$. Dissolve the ppt. in dil. HCl and divide the solution in two parts.

$$Al(OH)_3 + 3HCl \longrightarrow AlCl_3 + 3H_2O$$

(1) *NaOH test.* To one part of solution add NaOH, *white ppt* soluble in excess is obtained.

$$AlCl_3 + 3NaOH \longrightarrow Al(OH)_3 + 3NaCl$$

$$\underset{\text{(White ppt.)}}{Al(OH)_3} + NaOH \longrightarrow \underset{\text{(Soluble)}}{NaAlO_2} + 2H_2O$$

(2) *Lake test.* To the second portion of solution, add a drop of litmus solution, followed by the addition of ammonium hydroxide dropwise. *Blue ppt* floating in colourless solution is obtained.

Gelatinous white ppt. of aluminium hydroxide has got the property of absorbing the colour. Therefore, it takes up the blue colour of the litmus solution, and leaves the solution colourless.

(3) *Perform cobalt nitrate charcoal cavity test.* Blue mass.

Q. 14. Give the chemistry of analysis of cations of fourth group. (*G. Nanakdev, 2000*)

Ans. This group comprises zinc, manganese, cobalt and nickel. These radicals are precipitated as their sulphides. The separation of these sulphides is based upon the following features:

(*a*) ZnS and MnS are soluble in cold dil. HCl, while CoS and NiS are insoluble.

(*b*) $Zn(OH)_2$ is soluble in excess of NaOH, while $Mn(OH)_2$ is insoluble.

(*c*) Cyanides of Ni and Co behave differentially in presence of excess of KCN when treated with NaOH and bromide water.

(*d*) Nickel salt in presence of excess of NH_4OH forms a red ppt. with dimethyl glyoxime while cobalt salt gives a blue colour in the alcoholic layer when treated with NH_4CNS and amyl alcohol.

Procedure for analysis. The fourth group radicals are precipitated as their sulphides by passing H_2S gas through their salt solution after adding ammonium chloride and ammonium hydroxide solutions.

$$ZnCl_2 + H_2S \longrightarrow \underset{\substack{\text{White}\\\text{ppt.}}}{ZnS} + 2HCl$$

$$MnCl_2 + H_2S \longrightarrow \underset{\text{Buff.}}{MnS} + 2HCl$$

$$CoCl_2 + H_2S \longrightarrow \underset{\text{Black}}{CoS} + 2HCl$$

$$NiCl_2 + H_2S \longrightarrow \underset{\text{Black}}{NiS} + 2HCl$$

Note the colour of ppt. formed and proceed as follows for the confirmation of these radicals.

Confirmation of Zn^{2+}. *White ppt* indicates ZnS. Dissolve the ppt. in dil. HCl and divide the solution in two parts.

(1) *NaOH test.* To one part of the above solution add NaOH, *white ppt.* soluble in excess of NaOH is obtained.

$$ZnCl_2 + 2NaOH \longrightarrow Zn(OH)_2 + 2NaCl$$

$$Zn(OH)_2 + \underset{\text{Excess}}{2NaOH} \longrightarrow \underset{\text{(Soluble)}}{Na_2ZnO_2} + 2H_2O$$

(2) *Potassium ferrocyanide test.* To the second part of the solution, add potassium ferrocyanide solution *Bluish white* ppt. is obtained.

$$2ZnCl_2 + K_4[Fe(CN)_6] \longrightarrow \underset{\substack{\text{Zinc ferrocyanide}\\\text{bluish white ppt.}}}{Zn_2[Fe(CN)_6]} + 4KCl$$

(3) Perform charcoal cavity *cobalt nitrate test–a* green mass is obtained.

Confirmation of Mn^{2+}. *Flesh* or *buff* coloured ppt. indicates MnS.

(1) *NaOH test.* To the original salt solution, add enough NaOH solution and bromine water and shake–*brown ppt.* is obtained.

$$MnCl_2 + 2NaOH \longrightarrow Mn(OH)_2 + 2NaCl$$

$$Br_2 + H_2O \longrightarrow 2HBr + O$$

$$Mn(OH)_2 + O \longrightarrow \underset{\text{Manganic hydroxide brown ppt.}}{MnO(OH)_2}$$

(2) *PbO_2 test.* Boil the buff ppt. with conc. NHO_3 and add lead peroxide, PbO_2 On dilution, a *purple coloured* solution is obtained.

$$2Mn(NO_3)_2 + 5PbO_2 + 6HNO_3 \longrightarrow \underset{\text{Purple permanganic acid}}{2HMnO_4} + 5Pb(NO_3)_2 + H_2O$$

Confirmation of Ni^{+2}. *Black ppt.* obtained on passing H_2S gas, indicates *CoS* and *NiS.* Dissolve the ppt. in conc. HCl, and heat to dryness. Note the colour of the residue left. A *yellow* mass indicates nickel. Dissolve it in H_2O. Perform the following confirmatory tests.

(1) *Dimethyl glyoxime test.* To a part of the above solution add NH_4OH and then dimethyl glyoxime solution. A *bright red ppt* is obtained.

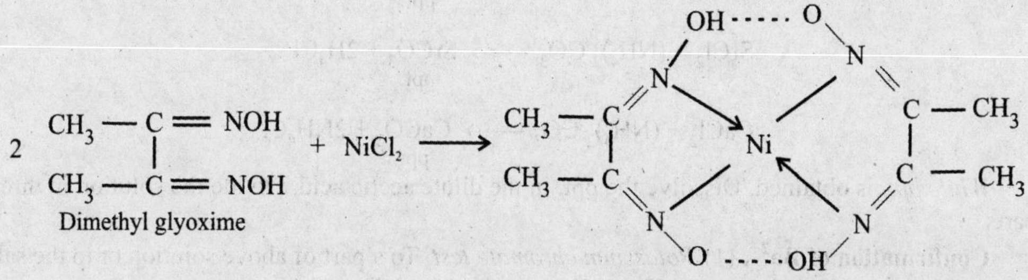

$$2\begin{array}{l} CH_3-C=NOH \\ \quad\quad\ | \\ CH_3-C=NOH \end{array} + NiCl_2 \longrightarrow \text{Nickel Dimethyl glyoxime}$$

Dimethyl glyoxime

Nickel Dimethyl glyoxime
red ppt.

(2) *NaOH and Br_2 water test.* To another part of the solution, add NaOH and bromine water and boil. A *black* ppt. is obtained.

$$NiCl_2 + 2NaOH \longrightarrow Ni(OH)_2 + 2NaCl$$

$$Br_2 + H_2O \longrightarrow 2HBr + O$$

$$2Ni(OH)_2 + H_2O + O \longrightarrow \underset{\text{Nickel hydroxide black}}{2Ni(OH)_3}$$

Confirmation of Co^{+2}. If the residue left is of *blue colour,* it indicates presence of Co^{+2}. Dissolve it in H_2O, and confirm Co^{+2} as

(1) *Cobaltinitrite test.* To a portion of the above solution add KNO_2 (s) and CH_3COOH–*yellow ppt.* is obtained.

$$CoC1_2 + 6KNO_2 + CH_3COOH \longrightarrow \underset{\text{Pot. cobaltinitrite (yellow)}}{K_3[Co(NO_2)_6]} + 2KC1 + CH_3COOK$$

(2) *Sulphocyanide test.* To another part of the solution, add solid ammonium sulphocyanide and amyl alcohol. Shake well. Blue layer is formed.

$$CoCl_2 + 2NH_4CNS \longrightarrow Co(CNS)_2 + 2NH_4Cl$$

$$Co(CNS)_2 + 2NH_4CNS \longrightarrow \underset{\text{Amm. cobaltic thiocyanide blue colour}}{(NH_4)_2[Co(CNS)_4]}$$

Q. 15. Give the chemistry of analysis of cations of fifth group. *(Garhwal, 2000)*

Ans. This group comprises *barium, strontium* and *calcium,* the carbonates of which are insoluble in ammoniacal solution. The separation of these three radicals utilizes the following features:

(*a*) *Barium chromate is almost insoluble in dil. acetic acid while $SrCO_3$ and $CaCO_3$ are soluble.*

(*b*) *Strontium sulphate is insoluble in a saturated solution of ammonium sulphate in which $CaSO_4$ is soluble forming a complex.*

Procedure for analysis. If fourth group is absent, concentrate the solution by boiling. Now make it alkaline by adding ammonium hydroxide and then add ammonium carbonate solution in slight excess.

$$BaCl_2 + (NH_4)_2CO_3 \longrightarrow \underset{\text{ppt.}}{BaCO_3} + 2NH_4Cl$$

$$SrCl_2 + (NH_4)_2CO_3 \longrightarrow \underset{\text{ppt.}}{SrCO_3} + 2H_4Cl$$

$$CaCl_2 + (NH_4)_2CO_3 \longrightarrow \underset{\text{ppt.}}{CaCO_3} + 2NH_4Cl$$

White ppt. is obtained. Dissolve the ppt. in the dilute acetic acid. Divide the solution in three parts.

Confirmation of Ba^{2+}. (1) *Potassium chromate test.* To a part of above solution or to the salt solution add K_2CrO_4. *Yellow ppt.* of $BaCrO_4$ is obtained.

$$(CH_3COO)_2Ba + K_2CrO_4 \longrightarrow \underset{\text{Yellow ppt.}}{BaCrO_2} + 2CH_3COOK$$

(2) *Flame test.* Perform flame test with group ppt. or the original salt–*grassy green flame.*

Confirmation of Sr^{2+}. If Ba^{+2} is absent, then

(1) To the second part of the solution add $(NH_4)_2SO_4$–*White ppt* is obtained.

$$(CH_3COO)_2Sr + (NH_4)_2SO_4 \longrightarrow \underset{\text{White ppt.}}{SrSO_4} + 2CH_3COONH_4$$

(2) *Flame test.* Perform flame test with group ppt. or original salt–*crimson red* flame.

Confirmation of Ca^{2+}. If Ba^{2+} and Sr^{2+} are absent, then

(1) To the third part of solution obtained above, add *ammonium oxalate* solution. White ppt. of *cal-oxalate* is obtained.

$$\underset{\text{Calcium acetate}}{(CH_3COO)_2Ca} + (COONH_4)_2 \longrightarrow \underset{\text{White ppt.}}{(COO)_2Ca} + 2CH_3COONH_4$$

(2) *Flame test.* Perform flame test with group ppt. or salt–*Brick red flame.*

Q. 16. Give the chemistry of analysis of cations of sixth group. *(Calcutta, 2005)*

Ans. This group comprises magnesium, sodium, potassium and ammonium. There is no common reagent for this group and so they are tested individually.

Confirmation of Mg^{2+}. *Ammonium phosphate test.* To the original salt solution add NH_4Cl, NH_4OH and *ammonium phosphate* or *disodium hydrogen* phosphate. Heat and scratch the sides of the test tube, *white ppt* is obtained.

$$MgCl_2 + NH_4OH + Na_2HPO_4 \longrightarrow \underset{\text{ppt.}}{Mg(NH_4)PO_4} + 2NaCl + H_2O$$

(2) *Cobalt nitrate test.* Perform charcoal cavity-cobalt nitrate test with the given salt–*Pink mass.*

Confirmation of Na^+. (1) *Potassium pyroantimonate test.* To the salt solution in water, add potassium pyroantimonate–A *white ppt.* is obtained.

$$2NaCl + K_2H_2Sb_2O_7 \longrightarrow \underset{\substack{\text{Sod. pyroantimonate} \\ \text{white ppt.}}}{Na_2H_2Sb_2O_7} + 2KCl$$

(2) *Flame test.* Perform flame test–persistent *golden yellow flame,* invisible through blue glass.

Confirmation of K^+. (1) *Sodium cobaltinitrite test.* To the salt solution in water, add acetic acid and solid cobaltinitrite solution. A *yellow ppt.* is obtained.

$$3KCl + Na_3Co(NO_2)_6 \longrightarrow 3NaCl + \underset{\text{Yellow ppt.}}{K_3[Co(NO_2)_6]}$$

(2) *Tartaric acid test.* To the salt solution in water, add tartaric acid, shake *well–white ppt.* is obtained.

$$KCl + \underset{\text{Tartaric acid}}{\begin{array}{c} CHOHCOOH \\ | \\ CHOHCOOH \end{array}} \longrightarrow \underset{\text{White ppt.}}{\begin{array}{c} CHOHCOOK \\ | \\ CHOHCOOK \end{array}} + 2HCl$$

(3) *Picric acid test.* To the salt solution in water, add picric acid–*yellow ppt.*

$$KCl + \underset{\text{Picric acid}}{C_6H_2(NO_2)_3OH} \longrightarrow \underset{\substack{\text{Pot. picrate} \\ \text{yellow ppt.}}}{C_6H_2(NO_2)_2OK} + HCl$$

(4) *Flame test.* Perform flame test–*pale violet flame.*

Confirmation of NH_4^+. (1) *NaOH test.* Heat the salt with NaOH solution. NH_3 gas is evolved, which gives *dense white fumes* with a rod dipped in conc. HCl placed over the mouth of test tube.

$$NH_4Cl + NaOH \longrightarrow NaCl + NH_3 + H_2O$$

$$NH_3 + HCl \longrightarrow \underset{\text{Dense fumes}}{NH_4Cl}$$

(2) *Nessler's reagent test.* Pass the ammonia gas through Nessler's reagent, a brown ppt. is obtained.

$$\underset{\text{Nessler's soln.}}{2K_2HgI_4} + NH_3 + 3NaOH \longrightarrow \underset{\text{Brown ppt.}}{NH_2HgO \cdot HgI} + 3NaI + 4KI + 2H_2O$$

Q. 17. How will you detect carbonate ion in presence of sulphite ion?

Ans. In the presence of sulphite the test of carbonate is influenced because sulphite and carbonate when treated with dil. sulphuric acid liberate sulphur dioxide and carbon dioxide respectively and these gases, when passed through lime water, turn it milky due to the formation of calcium sulphite and calcium carbonate respectively. So lime water test for carbonate is not a reliable one in presence of sulphite and hence a special procedure is adopted.

Reactions:

$$K_2CO_3 + H_2SO_4 \longrightarrow K_2SO_4 + H_2O + CO_2$$

$$K_2SO_3 + H_2SO_4 \longrightarrow K_2SO_4 + H_2O + SO_2$$

$$SO_2 + Ca(OH)_2 \longrightarrow CaSO_3 + H_2O$$

$$CO_2 + Ca(OH)_2 \longrightarrow CaCO_3 + H_2O$$

Procedure. To the solid mixture add about 1 g of powdered pot. dichromate and dil. sulphuric acid. Heat the contents and pass the evolved gas through lime water. If it turns *milky* it indicates the presence of *carbonate.* The solution in the first test tube becomes *green* due to the formation of *chromic sulphate* and so confirms sulphite.

Explanation. Acidified pot. dichromate oxidises only sulphur dioxide while carbon dioxide is not affected.

$$K_2Cr_2O_7 + H_2SO_4 + 3SO_2 \longrightarrow Cr_2(SO_4)_3 + K_2SO_4 + H_2O$$

Hence $K_2Cr_2O_7$ does not allow SO_2 to go along with CO_2.

Q. 18. How will you detect carbonate in the presence of nitrite?

Ans. In this combination, nitrite is tested as usual by acidified pot. iodide-starch paper. But lime water test for carbonate is not a reliable test in presence of nitrite because when a mixture of both is treated with dil. H_2SO_4, carbon dioxide and nitrogen peroxide gases are evolved respectively and on passing these gases through lime water the turbidity produced is dissolved by HNO_3 and HNO_2 (formed by the reaction of NO_2 with water). Hence in presence of nitrite no turbidity will be produced even if carbonate is present. Therefore, a special method is adopted.

Reactions:

$$K_2CO_3 + H_2SO_4 \longrightarrow K_2SO_4 + H_2O + CO_2$$

$$KNO_2 + H_2SO_4 \longrightarrow KHSO_4 + HNO_2$$

$$3HNO_2 \longrightarrow HNO_3 + H_2O + 2NO$$

$$2NO + O_2 \text{ (form air)} \longrightarrow 2NO_2$$

$$Ca(OH)_2 + CO_2 \longrightarrow \underset{\text{White ppt.}}{CaCO_3} + H_2O$$

$$H_2O + 2NO_2 \longrightarrow HNO_3 + HNO_2$$

$$CaCO_3 + 2HNO_3 \longrightarrow Ca(NO_3)_2 + H_2O + CO_2\uparrow$$

$$CaCO_3 + 2HNO_2 \longrightarrow Ca(NO_2)_2 + H_2O + CO_2\uparrow$$

Procedure. To the solid mixture in a dry test tube add an equal amount of powdered solid $K_2Cr_2O_7$ and 1-2 ml dil. H_2SO_4. Heat and pass the evolved gas through lime water. If it turns milky it confirms the presence of carbonate.

Explanation. Acidified $K_2Cr_2O_7$ oxidizes nitrite to nitrate while CO_3^{2-} remains unaffected.

$$K_2Cr_2O_7 + 4H_2SO_4 \longrightarrow K_2SO_4 + Cr_2(SO_4)_3 + 4H_2O + 3O$$

$$KNO_2 + O \longrightarrow KNO_3$$

Q. 19. How will you analyse nitrite in the presence of iodide?

Ans. When a mixture of both nitrite and iodide is treated with dil. H_2SO_4, violet fumes of iodine are evolved.

Reactions:

$$2KNO_2 + H_2SO_4 \longrightarrow K_2SO_4 + 2HNO_2$$

$$2KI + H_2SO_4 \longrightarrow K_2SO_4 + 2HI$$

$$2HNO_2 + 2HI \longrightarrow 2H_2O + 2NO + I_2$$

So evolution of violet fumes with dil. H_2SO_4 confirms both nitrite and iodide because in the absence of nitrite, iodide is decomposed only by conc. H_2SO_4. Both can also be confirmed by the following test:

Procedure. Acidify S.E. with dil. H_2SO_4 and add a few drops of starch solution. A blue colour confirms the presence of both nitrite and iodide.

Explanation. In this test the blue colour is due to the reaction between starch and free iodine evolved from the mixture of nitrite and iodide by the action of dil. H_2SO_4.

$$I_2 + \text{Starch} \longrightarrow \text{Blue colour}$$

Q. 20. How will you detect nitrate in the presence of nitrite? (*Garhwal, 2000*)

Ans. Nitrite is tested as usual by acidified pot. iodide–starch paper. But nitrate cannot be detected in presence of nitrite because nitrous acid set free from nitrite by sulphuric acid is partly oxidized to nitric acid which is detected even if no nitrate was originally present.

Reactions:

$$KNO_2 + H_2SO_4 \longrightarrow KHSO_4 + HNO_2$$

$$3HNO_2 \longrightarrow HNO_3 + H_2O + 2NO$$

$$2NO + O_2 \longrightarrow 2NO_2$$

Similarly nitrate cannot be tested by ring test because the solution turns black in the presence of nitrite when ferrous sulphate and sulphuric acid are added.

Reactions:

$$KNO_2 + H_2SO_4 \longrightarrow KHSO_4 + HNO_2$$

$$3HNO_2 \longrightarrow HNO_3 + H_2O + 2NO$$

$$FeSO_4 + NO \longrightarrow \underset{\text{Black}}{[FeSO_4 \cdot NO]}$$

Therefore, nitrite must be completely decomposed. Nitrite is removed by the following methods:

(*a*) **NH_4Cl method.** By boiling the mixture solution with amm. chloride until there is no more effervescence.

$$KNO_2 + NH_4Cl \longrightarrow NH_4NO_2 + KCl$$

$$NH_4NO_2 \longrightarrow 2H_2O + N_2\uparrow$$

(*b*) **Urea method.** By warming the mixture solution with urea and dilute sulphuric acid until the evolution of gases stops.

$$2KNO_2 + H_2SO_4 \longrightarrow K_2SO_4 + 2HNO_2$$

$$NH_2CONH_2 + 2HNO_2 \longrightarrow 3H_2O + CO_2 + 2N_2$$

Procedure. Boil the W.E. with NH_4Cl solution until there is no more evolution of any gas (Test for the complete removal of nitrite by acidified KI–starch paper) KI starch paper is placed at the mouth of the test. When it gives no blue colour, nitrite is absent. Now filter the solution and divide the filtrate into two parts.

(*a*) To one part add zinc dust and dil. sulphuric acid. Heat the contents and place a paper moistened with pot. iodide starch solution over the mouth of the test tube. If the paper turns blue, nitrate is confirmed.

(*b*) Perform ring test. If positive, nitrate is confirmed.

Explanations with Reactions. In the first test the blue colour is due to the reaction between starch and free iodine.

$$Zn + H_2SO_4 \longrightarrow ZnSO_4 + 2H$$

$$NaNO_3 + 2H \longrightarrow NaNO_2 + H_2O$$

$$2KI + 2NaNO_2 + 2H_2SO_4 \longrightarrow 2HI + 2HNO_2 + K_2SO_4 + Na_2SO_4$$

$$2HI + 2HNO_2 \longrightarrow 2H_2O + 2NO + I_2$$

$$I_2 + \text{Starch} \longrightarrow \text{Blue colour}$$

Q. 21. How will you detect nitrate in presence of bromide/iodide? (*Kanpur, 2001*)

Ans. In this combination nitrate does not interfere in the usual tests of bromide. But bromide and iodide interfere in test of nitrate. This is due to the following reasons:

Firstly, on treatment with conc. sulphuric acid, the deep brown vapours from a bromide and violet vapours from an iodide mask the light brown vapours from a nitrate. So it is difficult to distinguish between the colours of vapours of both.

Reactions:

$$2KNO_3 + H_2SO_4 \longrightarrow K_2SO_4 + 2HNO_3$$

$$4HNO_3 \longrightarrow 2H_2O + O_2 + \underset{\text{(Light brown fumes)}}{4NO_2}$$

$$2KBr + H_2SO_4 \longrightarrow K_2SO_4 + 2HBr$$

$$2HBr + H_2SO_4 \longrightarrow 2H_2O + SO_2 + \underset{\text{Deep brown fumes}}{Br_2}$$

$$2KI + H_2SO_4 \longrightarrow K_2SO_4 + 2HI$$

$$2HI + H_2SO_4 \longrightarrow 2H_2O + SO_2 + \underset{\text{(Violet)}}{I_2}$$

Secondly, nitrate cannot be tested by ring test because the liberation of free bromine from a bromide with conc. sulphuric acid destroys the ring.

$$2KNO_3 + 2KBr + 6FeSO_4 + 4H_2SO_4 \rightarrow$$
$$2K_2SO_4 + 2Fe_2(SO_4)_3 + 4H_2O + 2[Fe\ SO_4 \cdot NO] + Br_2$$

Hence a special procedure is adopted.

Procedure. (*i*) *NaOH and zinc dust* (*or Al powder*) *method.*

Add sod. hydroxide solution to the solid mixture. Heat the contents until ammonium salt, if present, is completely decomposed. [It is judged by the evolution of ammonia. When the evolution of ammonia ceases ammonium salt is completely decomposed.] Now add zinc dust and warm the solution gently. *Evolution of ammonia gas again confirms nitrate.*

(ii) Dil. H_2SO_4 and zinc dust test. To W.E. add sulphuric acid and a pinch of zinc dust. Heat and place pot. iodide-starch paper over the mouth of the test tube. If the paper turns blue, nitrate is confirmed. *If iodide is present along with the nitrate, then in this test free iodine will also be liberated which will mask brown fumes.*

(iii) **Silver sulphate method.** Water extract is heated with excess of silver sulphate solution. Bromide and iodide are removed as silver halides. Filtrate is used to test for nitrate by ring test.

Explanation. *(i)* In the first test the evolution of ammonia is due to the reduction of nitrate to ammonia by zinc and sodium hydroxide.

$$Zn + 2NaOH \longrightarrow Na_2ZnO_2 + 2H$$

$$NaNO_3 + 8H \longrightarrow NaOH + 2H_2O + NH_3 \uparrow$$

$$3NaNO_3 + 5NaOH + 8Al + 2H_2O \longrightarrow 8NaAlO_2 + 3NH_3 \uparrow$$

(ii) In the second test the blue colour is due to the reaction between starch and free iodine.

$$Zn + H_2SO_4 \longrightarrow ZnSO_4 + 2H$$

$$KNO_3 + 2H \longrightarrow KNO_2 + H_2O$$

$$2KI + 2KNO_2 + 2H_2SO_4 \longrightarrow 2K_2SO_4 + 2HNO_2 + 2HI$$

$$2HI + 2HNO_2 \longrightarrow 2H_2O + 2NO + I_2$$

$$I_2 + \text{Starch} \longrightarrow \text{Blue colour}$$

(iii)

$$Ag_2SO_4 + 2NaBr \longrightarrow 2AgBr + Na_2SO_4$$

$$Ag_2SO_4 + 2NaI \longrightarrow 2AgI + Na_2SO_4$$

Q. 22. How will you analyse oxalate in the presence of carbonate? *(Garhwal, 2000)*

Ans. In this combination oxalate will not interfere in the test of carbonate because when the mixture is treated with dil. sulphuric acid, carbonate decomposes liberating CO_2 which turns lime water milky. But if the mixture is treated with conc. sulphuric acid, both the oxalate and carbonate decompose liberating CO_2 and hence we cannot decide whether it is from oxalate or carbonate. Moreover, oxalate after decomposition also produces CO which should burn with a blue flame. But the amounts of CO_2 produced is so much that burning of CO at the mouth of the test tube becomes almost impossible. Hence oxalate is tested as follows:

Procedure. *(i)* To the mixture in a dry test tube add excess of dil. sulphuric acid and heat the contents until there is no more effervescence. Now add a pinch of manganese dioxide and heat. If effervescence starts again, oxalate is present.

(ii) Neutralize S.E. with acetic acid and add calcium choride solution. A white ppt. is formed. Filter and disolve the ppt. in the hot dil. sulphuric acid and a few drops of ppt. permanganate solution. Pink colour is discharged.

Explanation with reactions. In the first test, effervescence on heating the mixture with dil. sulphuric acid is due to the decomposition of carbonate.

$$K_2CO_3 + H_2SO_4 \longrightarrow K_2SO_4 + H_2O + CO_2 \uparrow$$

Effervescence on adding manganese dioxide is due to the decomposition of oxalate.

$$K_2C_2O_4 + MnO_2 + 2H_2SO_4 \longrightarrow K_2SO_4 + MnSO_4 + 2H_2O + 2CO_2 \uparrow$$

In the second test, the white ppt. on adding $CaCl_2$ to S.E. acidified with acetic acid is obtained due to the formation of cal. oxalate.

$$K_2C_2O_4 + CaCl_2 \longrightarrow CaC_2O_4 + 2KCl$$

This ppt. when dissolved in dil. sulphuric acid and treated with pot. permanganate solution decolourises the latter.

$$CaC_2O_4 + H_2SO_4 \longrightarrow H_2C_2O_4 + CaSO_4$$

$$2KMnO_4 + 3H_2SO_4 + 5H_2C_2O_4 \longrightarrow 2MnSO_4 + K_2SO_4 + 8H_2O + 10CO_2$$

Q. 23. How will you test chloride in presence of nitrate?

Ans. In conc. H_2SO_4 test both the radicals react to form HCl and HNO_3 vopours respectively with pungent smell.

Dense white fumes are obtained due to both with ammonium hydroxide. On adding copper chips or paper pallet colourless HNO_3 changes to reddish brown fumes showing the presence of NO_3 Nitrate does not interfere with chromyl chloride test, also Cl^- does not interfere in Ring test.

Reactions:

$$\left.\begin{array}{l} KCl + \underset{\text{Conc.}}{H_2SO_4} \longrightarrow KHSO_4 + HCl \\ KNO_3 + \underset{\text{Conc.}}{H_2SO_4} \longrightarrow KHSO_4 + HNO_3 \end{array}\right\} \text{Colourless pungent smell}$$

$$\left.\begin{array}{l} HCl + NH_4OH \longrightarrow H_2O + NH_4Cl \\ HNO_3 + NH_4OH \longrightarrow H_2O + NH_4NO_3 \end{array}\right\} \text{Dense white fumes}$$

$$Cu + 4HNO_3 \longrightarrow Cu(NO_3)_2 + 2NO_2\uparrow + 2H_2O$$

$$\underset{\text{(Paper pallet)}}{C} + 4HNO_3 \longrightarrow CO_2 + 4NO_2 + 2H_2O$$

Q. 24. How will you analyse sulphide, sulphite, sulphate and thiosulphate in presence of one another? *(Meerut, 1995)*

Ans. Systematic analysis is required for detection of S^{2-}, $S_2O_3^{2-}$, SO_3^{2-} and SO_4^{2-} ions when present together, because some of these interfere in tests of other. For example, if the mixture contains S^{2-} and SO_3^{2-}, addition of acid liberates H_2S and SO_2 simultaneously and the two gases immediately interact.

$$SO_2 + 2H_2 \longrightarrow 3S\downarrow + 2H_2O$$

As a result of this reaction only the gas present in excess will be detected. If this gas is SO_2, addition of acid would be accompanied by the same effects (formation of SO_2 and S) as would be observed if the solution contained $S_2O_3^{2-}$. Thiosulphate would therefore be indicated wrongly. It is obvious that a mixture of SO_3^{2-} and $S_2O_3^{2-}$ would behave on acidification in the same way as $S_2O_3^{2-}$ by itself.

$$Na_2SO_3 + 2HCl \longrightarrow 2NaCl + SO_2\uparrow + H_2O$$

$$Na_2S_2O_3 + 2HCl \longrightarrow 2NaCl + SO_2\uparrow + S\downarrow + H_2O$$

Finally in presence of S^{2-}, the test for SO_3^{2-} with acidified $K_2Cr_2O_7$ cannot be carried out (H_2S also reduces acidified $K_2Cr_2O_7$ to green chromium salt). Similarly, detection of SO_3^{2-} is difficult in presence of $S_2O_3^{2-}$ or mixture of S^{2-} and SO_3^{2-}. In both cases a white ppt. of sulphur is formed when the solution is acidified with acid, because it is insoluble in acids, it may be mistaken for $BaSO_4$.

The following procedure is adopted in dealing with such mixtures.

Procedure of Analysis

The systematic analysis of mixtures of S^{2-}, $S_2O_3^{2-}$, SO_3^{2-} and SO_4^{2-}, described in detail below, is based on (*a*) precipitation of S^{2-} by the action of $CdCO_3$, when SO_3^{2-}, SO_4^{2-} and $S_2O_3^{2-}$ remain in solution; (*b*) precipitation of SO_3^{2-} and SO_4^{2-} by $BaCl_2$ which does not precipitate $S_2O_3^{2-}$, (*c*) the fact that $BaSO_4$ is almost insoluble while $BaSO_3$ is soluble in dilute acids.

Take the sodium carbonate extract and add excess of freshly prepared $CdCO_3$, shake well and filter.

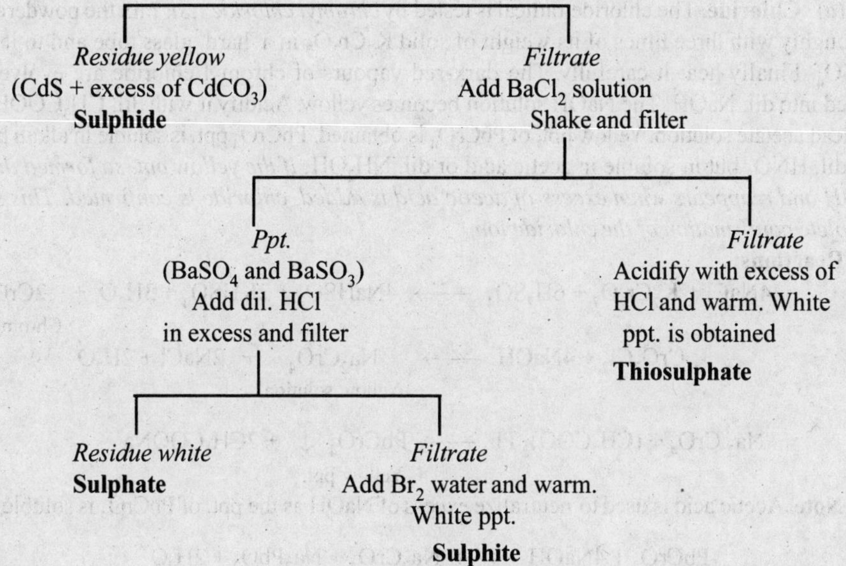

Reactions:

$$Na_2S + CdCO_3 \longrightarrow \underset{\text{Yellow ppt.}}{CdS \downarrow} + Na_2CO_3$$

On addition of $BaCl_2$, SO_3^{2-} and SO_4^{2-} both are precipitated but HC1 dissolves $BaSO_3$.

1. $Na_2SO_3 + BaCl_2 \longrightarrow BaSO_3 \downarrow + 2NaCl$

2. $Na_2SO_4 + BaCl_2 \longrightarrow BaSO_4 \downarrow + 2NaCl$

Br_2 water converts SO_3^{2-} to SO_4^{2-} on boiling

$$Br_2 + H_2O \longrightarrow 2HBr + [O]$$

$$BaSO_3 + [O] \longrightarrow BaSO_4 \downarrow$$

Though the above systematic scheme gives promising results, some selective methods are also available to test these ions individually.

(*i*) **Thiosulphate.** Boil the mixture with water and filter. Cool the filtrate, add to it a few drops of ferric chloride solution. Dark violet colour shows the presence of thiosulphate. The colour disappears after a short interval of time and leaving an almost colourless solution.

$$2Na_2S_2O_3 + FeCl_3 \longrightarrow \underset{\text{Dark violet}}{Na[Fe(S_2O_3)_2]} + 3NaCl$$

$$Fe^{3+} + [Fe(S_2O_3)_2]^- \longrightarrow S_4O_6^{2-} + 2Fe^{2+}$$

Note: The dark violet colour also disappears on warming.

(*ii*) **Sulphide.** This is tested by addition of $PbCO_3$ to soda extract solution which gives black ppt. of PbS.

Sulphide can also be tested by sodium nitroprusside, $Na_2[Fe(CN)_5NO]$. Add 1 ml of this reagent to 1 ml soda extract (alkalised by NaOH). A red-violet colour, due to formation of complex $Na_4[Fe(CN)_5(NOS)]$, prove the presence of S^{2-}. In this test SO_3^{2-}, SO_4^{2-} and $S_2O_3^{2-}$ do not interfere.

Q. 25. How will you analyse a mixture of chloride bromide and iodide present together.

(*M. Dayanand, 2002*)

(*a*) **Chloride.** The chloride radical is tested by *chromyl chloride test;* mix the powdered mixture thoroughly with three times of its weight of solid $K_2Cr_2O_7$ in a hard glass tube and to it add conc. H_2SO_4. Finally heat it carefully. The dark-red vapours of chromyl chloride are evolved and are passed into dil. NaOH. The NaOH solution becomes yellow. Acidify it with dil. CH_3COOH and then add lead acetate solution. Yellow ppt. of $PbCrO_4$ is obtained. $PbCrO_4$ ppt. is soluble in alkali hydroxides and dil. HNO_3 but in soluble in acetic acid or dil. NH_2OH. *If the yellow ppt. so formed dissolves in NaOH and reappears when excess of acetic acid is added, chloride is confirmed. This step is the complete confirmation of the chloride ion.*

Reactions:

$$4NaCl + K_2Cr_2O_7 + 6H_2SO_4 \longrightarrow 4NaHSO_4 + 2KHSO_4 + 3H_2O + \underset{\text{Chromyl chloride}}{2CrO_2Cl_2}$$

$$CrO_2Cl_2 + 4NaOH \longrightarrow \underset{\text{Yellow solution}}{Na_2CrO_4} + 2NaCl + 2H_2O$$

$$Na_2CrO_4 + (CH_3COO)_2Pb \longrightarrow \underset{\text{Yellow ppt.}}{PbCrO_4 \downarrow} + 2CH_3COONa$$

Note. Acetic acid is used to neutralize excess of NaOH as the ppt. of $PbCrO_4$ is soluble in NaOH:

$$\underset{\text{Yellow ppt.}}{PbCrO_4} + 4NaOH \longrightarrow \underset{\text{Soluble}}{Na_2CrO_4} + \underset{\text{Soluble}}{Na_2PbO_2} + 2H_2O$$

(*b*) **Iodide and bromide.** Acidify 1 ml soda extract with dil. HCl or dil. H_2SO_4 and a few drops of $CHCl_3$(CCl_4 or CS_2). Add freshly prepared chlorine water drop by drop while shaking. If $CHCl_3$ layer becomes violet – Iodide confirmed. Add excess of chlorine water with vigorous shaking. If orange-brown colour in $CHCl_3$ layer is obtained – Bromide confirmed.

Chlorine liberates Br_2 and I_2 from bromide and iodide.

Reactions. Since out of three halogens iodine is least electronegative, it is liberated first by adding chlorine water. Liberated iodine being soluble in $CHCl_3$, CCl_4 and CS_2, produces violet colour in organic layer.

$$2NaI + Cl_2 \longrightarrow 2NaCl + I_2$$

$$I_2 \text{ in chloroform} \longrightarrow \text{Violet colour}$$

Chlroine also displaces bromine from bromide salts. The iodine which imparts violet colour to organic layer obscures the orange brown colour of bromine (liberated from the bromide salt). The excess of chlorine water oxidises the free iodine to iodic acid which is colourless and then the bromine colour can be noted.

$$I_2 + 5Cl_2 + 6H_2O \longrightarrow \underset{\text{(Iodic acid)}}{2HIO_3} + 10HCl$$

On further addition of chlorine water, more bromine is liberated from bromide and dissolves into organic layer and a clear orange-brown colour is developed.

$$2NaBr + Cl_2 \longrightarrow 2NaCl + Br_2$$

$$Br_2 + CHCl_3 \longrightarrow \text{Orange-brown colour}$$

If only bromide and iodide are present, the procedure (*b*) above may be followed.

Q. 26. How will you test oxalate and fluoride in the presence of each other?

Ans. *First method.* Add CH_3COOH to soda extract till it becomes distinctly acidic. Add $CaCl_2$ in excess when oxalate precipitates as CaC_2O_4 and fluoride as CaF_2, both are obtained as white ppt.

Reactions:

$$2NaF + CaCl_2 \longrightarrow \underset{\text{ppt.}}{CaF_2} + 2NaCl$$

$$\underset{\text{Sod. oxalate}}{(COONa)_2} + CaCl_2 \longrightarrow \underset{\text{ppt.}}{Ca(COO)_2} + 2NaCl$$

Add dil. HCl.

(*a*) If the precipitate is soluble, only oxalate is present.

(*b*) If the ppt. is not at all soluble, it should be *fluoride.*

This is the method to distinguish between the two radicals. However, if both are present, we resort to another method.

Second method. Neutralise soda-extract with dil. HNO_3, add $AgNO_3$. Filter the precipitate and proceed as follows:

(*i*) *Residue treatment.* Add to it dil. H_2SO_4 and a few drops of $KMnO_4$ and warm. The colour is discharged – **oxalate present.**

(*ii*) *Filtrate treatment.* Add barium nitrate to this filtrate. This yields white ppt. of barium fluoride –**fluoride confirmed.**

Q. 27. How will you test sulphate in the presence of fluoride?

Ans. Fluoride is tested in usual maner, *i.e.*, Mixture + conc. H_2SO_4 $\longrightarrow$ white deposit on the rod moistened with water.

If fluoride is present, $BaCl_2$ test for sulphate cannot be performed. By addition of $BaCl_2$ solution to S.E., acidified with dil. HCl, a white ppt. of barium fluoride is obtained which is insoluble in HCl.

Hence to test SO_4^{2-} in the presence of F^-, we proceed as below:

Acidify S.E. with excess of acetic acid, add lead acetate solution and warm. A white ppt. is produced which confirms the presence of sulphate.

Reactions:

$$Na_2SO_4 + (CH_3COO)_2Pb \longrightarrow \underset{\text{(Insoluble)}}{PbSO_4} + 2CH_3COONa$$

$$2NaF + (CH_3COO)_2Pb \longrightarrow \underset{\text{(Soluble)}}{PbF_2} + 2CH_3COONa$$

Q. 28. How will you test phosphate, arsenate and arsenite in the presence of each other?

Ans. All these radicals give yellow ppt. with conc. HNO_3 and ammonium molybdate.

Procedure. Acidify the S.E. with dil. H_2SO_4 and boil off CO_2. Filter, if necessary. Then add a small quantity of $MgCl_2$. Finally make it alkaline by adding NH_4OH. Stir and allow to stand for some time. The formation of white ppt. indicates the presence of PO_4^{3-} and arsenate in the mixture. Filter the solution.

Reactions:

$$Na_2HPO_4 + MgCl_2 \longrightarrow MgHPO_4 + 2NaCl$$

$$MgHPO_4 + NH_4OH \longrightarrow \underset{\text{White ppt.}}{MgNH_4PO_4\downarrow} + H_2O$$

$$Na_3AsO_4 + H_2SO_4 \longrightarrow Na_2HAsO_4 + NaHSO_4$$

$$Na_2HAsO_4 + MgCl_2 \longrightarrow MgHAsO_4 + 2NaCl$$

$$MgHAsO_4 + NH_4OH \longrightarrow \underset{\text{White ppt.}}{MgNH_4AsO_4\downarrow} + H_2O$$

Filtrate – may contain arsenite

Precipitate – may contain PO_4^{3-} and/or arsenate

Test for arsenite (AsO_3^{3-}). Acidify the filtrate with dil. HCl and pass H_2S in cold. Formation of yellow ppt. indicates the presence of arsenite.

Test for arsenate AsO_4^{3-} . Dissolve the white ppt. in conc. HCl. Heat the solution and pass H_2S gas. A yellow ppt indicates the presence of arsenate. Filter this solution and perform usual tests for PO_4^{3-} with the filtrate.

Q. 29. Which acid radicals interfere in the analysis of basic radicals? Discuss the theory of interference. (*Kumaon, 2000; Awadh, 2000*)

Ans. Fluoride, oxalate, borate tartarate and phosphate radicals interfere in the analysis of basic radicals.

Theory of interference. The fluorides, oxalates, borates and tartarates and phosphates of cations of III, IV, V groups and Mg are soluble in acidic medium but insoluble in alkaline medium. Thus they remain in solution as long as the medium is acidic. But when the medium is made alkaline by addition of excess of NH_4OH, in group III of basic radicals these will get precipitated along with hydroxides of cations of 3rd group. Hence these cations (of IV, V groups and Mg) would not only be completely missed in their respective groups but would also create complication in the analysis of cations of 3rd group. This can be illustrated by the following example. Suppose the mixture contains a phosphate together with Mg^{2+}.

Although the salt is insoluble in water, it dissolves in HCl and phosphoric acid is set free.

$$Mg_3(PO_4)_2 + 6HCl \rightleftharpoons 3MgCl_2 + \underset{\text{Weakly ionised}}{2H_3PO_4}$$

Phosphoric acid, which is formed above, is non-volatile and weak and remains as such in solution. Because of the common ion effect, its ionisation is suppressed in the presence of strong acids like HCl.

$$HCl \rightleftharpoons H^+ + Cl^- \quad ...(1)$$

$$H_3PO_4 \rightleftharpoons 3H^+ + PO_4^{3-} \quad ...(2)$$

Consequently the concentration of PO_4^{3-} ions in solution is quite low so that the ionic product of $[Mg^{2+}]$ and $[PO_4^{3-}]$ does not exceed the solubility product of $Mg_3(PO_4)_2$ and thus magnesium phosphate remains in solution. This is the position up to II group.

During analysis of 3rd group, when the solution is made alkaline by adding excess of NH_4OH, the OH^- ions from NH_4OH combine with H^+ ions to form almost unionised water molecules.

$$NH_4OH \rightleftharpoons NH_4^+ + OH^-$$

$$OH^- + H^+ \rightleftharpoons H_2O$$

As a result, the equilibrium (2) is shifted towards the right hand side and the concentration of PO_4^{3-} ions increases. Consequently the ionic product of $[Mg^{2+}]$ and $[PO_4^{3-}]$ now exceeds the solubility product of $Mg_3(PO_4)_2$ which is, therefore, precipitated in 3rd group.

This is true for other interfering radicals and also for the other cations.

Elimination (Removal) of the interfering acid radicals. From the above discussion it follows that in order to carry out the analysis of basic radicals in the usual manner, it is necessary to first remove the interfering acid radicals.

Q. 30. Describe the method of removal of oxalate and tartarate in order to analyse basic radicals.

Ans. Removal of oxalates and tartarate. To the filtrate from the 2nd group, add 2-3 ml of conc. HNO_3 and evaporate it to almost dryness in a china dish. Add more of conc. HNO_3 to the residue and evaporate the contents again to dryness. If oxalate or tartarate is present, it gets decomposed to carbon dioxide and is thus removed.

Reactions:

$$BaC_2O_4 + 2HNO_3 \longrightarrow Ba(NO_3)_2 + H_2C_2O_4$$

$$H_2C_2O_4 \longrightarrow H_2O + 2CO_2\uparrow$$

$$\underset{\text{Tartaric acid}}{H_2C_4H_4O_6} + 8HNO_3 \longrightarrow 8NO_2\uparrow + 3CO_2\uparrow + CO\uparrow + 7H_2O$$

Extract the residue in a china dish with dil. HCl and use this solution for analysis of cations as usual.

Q. 31. Describe the method of removal of borate and fluoride in order to analyse basic radicals.

Ans. Removal of borate and fluoride. Evaporate the filtrate from group II to dryness. Add conc. HCl to the residue and evaporate again to dryness. Repeat the above process at least two times. Extract the residue with dil. HCl.

Reactions:

$$CaF_2 + 2HCl \longrightarrow CaCl_2 + 2HF\uparrow$$
(Volatile with HCl)

$$Ca_3(BO_3)_2 + HCl \longrightarrow 3CaCl_2 + 2H_3BO_3\uparrow$$
(Volatile with steam)

If borate is present alone add to the filtrate from 2nd group, 2-3 ml of ethyl alcohol and 4-5 mol of conc. HCl and evaporate the mixture of dryness on a water bath. Finally extract with dil. HCl.

Reactions:

$$Ca_3(BO_3)_2 + 6HCl \longrightarrow 3CaCl_2 + 2H_3BO_3$$
$$H_3BO_3 + 3C_2H_5OH \longrightarrow (C_2H_5)_3BO_3 + 3H_2O$$
(Ethyl borate volatile)

Q. 32. Discuss chemistry of interference of phosphate radical in inorganic analysis. How is phosphate radical removed? (*Meerut, 2002; Kumaon, 2000; Lucknow, 2000; Calcutta, 2005*)

Ans. It may be removed by either of the two methods discussed below:

1. Acetate buffer ferric chloride method. Take about 1 ml of the filtrate from IInd group and boil off H_2S from it. Add to it a few drops of conc. HNO_3, boil for a few minutes. Add about 1 g of solid NH_4Cl, boil and cool. Add excess of NH_4OH to make the solution alkaline. If no precipitate is formed, cation of IIIrd group are absent and then there is no need of removal of phosphate. If ppt. is formed and phosphate has been detected in the mixture, then proceed as follows:

Filter the solution, preserve the filtrate for Na^+ and K^+ (It may also contain sometimes Co^{+2}, Ni^{2+} and Zn^{2+}).

Dissolve the above ppt in the minimum quantity of dil. HCl. Neutralise the excess acid by adding NH_4OH solution dropwise with shaking. Go on adding NH_4OH till a slight turbidity persists on shaking.

Add to it 2-3 ml of acetic acid and 1-2 g of solid ammonium acetate.

In case a precipitate is produced, it shows the presence of group III, reject it. Note the colour of the solution. If it is not reddish, add neutral $FeCl_3$ solution dropwise with constant shaking. Continue the addition of $FeCl_3$ till the solution becomes reddish brown in colour. Dilute the solution with some water, boil and filter.

Residue	*Filtrate*
It may contain phosphates of cations of 3rd group. If 3rd group present, proceed as usual for its analysis. Test it for Cr^{3+} and Al^{3+}, if it is not tested earlier.(In general, no residue is obtained.)	Concentrate the solution by evaporation so as to reduce the volume to about 10 ml. Add about 1g of solid NH_4Cl and boil. Then add an excess of NH_4OH solution. Filter, if needed. Analyse it for IV, V group and Mg^{2+}.

Explanations. (*i*) Phosphates of iron, aluminium and chromium do not dissolve in acetic acid in the presence of ammonium acetate while phosphates of metals belong to IV and V group and Mg are soluble.

(*ii*) Phosphates of other metals, if present, can be removed as ferric phosphate by the addition of neutral $FeCl_3$.

Neutral solution of ferric chloride is employed since it produces free HCl on hydrolysis which, if present, would dissolve some of the phosphates of Fe, Al and Cr.

$$FeCl_3 + 3H_2O \rightleftharpoons Fe(OH)_3 + 3HCl$$

(*ii*) Excess of iron present as soluble ferric acetate can be removed easily as basic ferric acetate on dilution and boiling.

$$FeCl_3 + 3CH_3COONa \longrightarrow (CH_3COO)_3Fe + 3NaCl$$

$$(CH_3COO)_3Fe + 2H_2O \longrightarrow \underset{\text{(Basic ferric acetate brown ppt.)}}{Fe(CH_3COO)(OH)_2\downarrow} + 2CH_3COOH$$

Excess of Fe^{3+} ions present in the solution can be removed by precipitating it as $Fe(OH)_3$ on the addition of NH_4Cl and NH_4OH.

2. Zirconyl nitrate method. Boil off H_2S gas from the filtrate of II group. Then add a little of dil. HCl and concentrate the solution by evaporaton. Add about 1 g of solid NH_4Cl and shake it to dissolve. Then add zirconyl nitrate, stir, with constant shaking till the precipitation is complete. Heat the reaction mixture to boiling, cool and filter.

Residue	*Filtrate*
Contains zirconyl phosphate, reject it.	It may consist of cations of III and subsequent groups. Test it for complete removal of phosphate by adding a drop of zirconyl nitrate. To get the clear solution add 1g of solid NH_4Cl. Heat it to boil. Then add NH_4OH solution dropwise to make it ammoniated. Heat and filter.
May contain hydroxides of cations of 3rd group. Analyse them as usual.	Analyse for cations of IV, V and VI group.

Explanation. Phosphate ions are completely precipitated as zirconyl phosphate.

$$ZrO(NO_3)_2 + 2Na_2HPO_4 + 2HCl \longrightarrow \underset{\text{(White ppt.)}}{ZrO(H_2PO_4)_2} + 2NaCl + 2NaNO_3$$

Q. 33. How are the following reagents used in the detection of cations in qualitative analysis?

(*i*) Sodium bismuthate (*ii*) Nessler's reagent (*iii*) Yellow ammonium sulphide.

(*Indore, 2001*)

Ans. (*i*) Sodium bismuthate $NaBiO_3$ is used as an oxidant in the confirmation of Mn^{2+}. It oxidises Mn^{2+} into MnO_4^-.

(*ii*) Nessler's reagent is used in the detection of NH_3. It forms a brown ppt. of $3HgO.Hg(NH_3)_2I_2$ and $H_2NHg_2I_3$.

(*iii*) Yellow ammonium sulphide is used to separate group IIB from IIA.

Q. 34. Give the theory of

(*a*) Lake test for Al

(*b*) DMG test for Ni

Ans. See Questions 13 and 14.

Q. 35. Give the theory of the following tests:

(*a*) Green-edged flame for borates (*Calcutta, 2007*)

(*b*) Chromyl chloride test for chlorides

(*M. Dayanand, 2000; Indore, 2000; Punjab, 2000; Calcutta, 2007*)

(*c*) Ring test for nitrates

Ans. See Qs. 9 and 10.

Q. 36. Point out the common errors in the detection of Bi^{3+} and Sb^{3+}.

Ans. A solution of these ions in conc. HCl on dilution with water gives precipitates of BiOCl and SbOCl respectively. These precipitates may be mistaken for the chlorides of first group radicals. In such cases, add a little conc. HCl to dissolve BiOCl and SbOCl and then proceed further.

Q. 37. An anion reacts with dil. H_2SO_4 to give pungent smell and a yellow precipitate. How does it react with $AgNO_3$?

Ans. Thiosulphate reacts with dil. H_2SO_4 to give pungent smelling SO_2 and a yellow precipitate of sulphur. It reacts with $AgNO_3$ to give a yellow ppt. which changes to orange, brown and finally black.

$$Na_2S_2O_3 + H_2SO_4 \longrightarrow Na_2SO_4 + SO_2 + S + H_2O$$

$$Na_2S_2O_3 + 2AgNO_3 \longrightarrow Ag_2S_2O_3 + 2NaNO_3$$

$$Ag_2S_2O_3 \longrightarrow Ag_2S + SO_2 + O$$

Q. 38. ECl_3 gives a milky solution on dilution with water. Give the tests for E^{3+} ions.

Ans. Bi^{3+} and Sb^{3+} solutions on ditution with water give a white ppt. of BiOCl and SbOCl respectively.

For tests of Bi^{3+} and Sb^{3+}, see question 12.

Q. 39. H_2SO_4 is not used for preparing O.S. for analysis of cation. Explain.

Ans. H_2SO_4 is avoided as a solvent for preparing original solution as it would produce a precipitate if Pb^{2+}, Ba^{2+}, Sr^{2+} and Ca^{2+} are present in the mixture.

Q. 40. An alkaline earth metal forms soluble chloride, sulphide, hydroxide and carbonate in qualitative analysis. How will you detect it?

Ans. Out of alkaline earth metals Be, Mg, Ca, Sr and Ba, only magnesium forms soluble chloride, suphide, hydroxide and carbonate. It is precipitated as MgH_2PO_4 in the 6th group.

Q. 41. Two anions X^{2-} and Y^{2-} react with dil. H_2SO_4 to give gases that turn lime water milky. How can you detect these in the presence of each other?

Ans. CO_3^{2-} and SO_3^{2-} react with dil. H_2SO_4 to give CO_2 and SO_2 both of which turn lime water milky. Solid $K_2Cr_2O_7$ is added to the mixture and it is heated with dil. HCl. Sulphite is retained in the test tube and vapours of carbon dioxide escape which turn lime water milky.

Q. 42. Explain.

(*i*) Why conc. HNO_3 is added in group III analysis.

(*ii*) Why Zn^{2+} does not precipitate with Cd^{2+} in II group.

Ans. (*i*) Conc. NHO_3 is added in order to oxidise Fe^{2+} into Fe^{3+} and then Fe^{3+} is precipitated as $Fe(OH)_3$.

(*ii*) Zn does not precipitate in II group because the conc. of S^{2-} is small and the ionic product $[Zn^{2+}] \times [S^{2-}]$ remains less than the solubility product of ZnS.

Q. 43. Discuss detection of Cr^{3+} Al^{3+} and Fe^{3+} in a mixture taking care that there is no interfering radical.

Ans. See Q 13.

Q. 44. Explain

(*i*) NH_4OH and not NaOH is used as a precipitant in the qualitative analysis of 3rd group.

(*ii*) Cations of 4th group are not precipitated as their sulphides in acidic solution.

Ans. (*i*) Sodium hydroxide dissolves the hydroxides of Al, Fe and Cr.

(*ii*) In the acidic medium, the ionisation of H_2S is suppressed. Therefore, there is only a small concentration of S^{2-} ions. Ionic product of sulphides of 4th group cations is not able to cross the solubility products. Hence these are not precipitated.

Q. 45. (*i*) Discuss the theory of ring test as performed for the confirmation of NO_3^- in qualitative analysis.

(*ii*) Discuss the chemistry of chromyl chloride test as used for the confirmation of chloride in qualitative analysis.

(*iii*) How is soluble carbonate confirmed? Give theory of the test.

(*iv*) How is nickel detected in presence of cobalt? Describe the theory of the test.

Ans.

(*i*) See Q. 8.

(*ii*) See Q. 9.

(*iii*) See Q. 8.

(*iv*) See Q. 14.

Q. 46. Give the chemistry involved in the analysis of cobalt and nickel in a mineral.

Ans. See. Q. 14.

Q. 47. State, with equation, what happens when

(*a*) CCl_4 is added to an aqueous solution of KI and then excess of chlorine water is added dropwise with shaking? (*Calcutta, 2007*)

(*b*) Excess stannous chloride solution is added dropwise to a mercuric chloride solution?

Ans. (*a*) Violet colour is imparted to $CC1_4$ layer

$$2KI + Cl_2 \longrightarrow 2KCl + I_2$$

Iodine evolved imparts violet colour to CCl_4 layer

(*b*)
$$2HgCl_2 + SnCl_2 \longrightarrow \underset{\text{White ppt.}}{Hg_2Cl_2} + SnCl_4$$

$$Hg_2Cl_2 + SnCl_2 \longrightarrow \underset{\text{Grey}}{2Hg} + SnCl_4$$

Thus on adding $SnCl_2$ to mercuric chloride solution, first a white ppt. of Hg_2Cl is formed which turns grey due to further reduction of Hg_2Cl_2 to metallic mercury.

Q. 48. What is Nessler's reagent? Give its use. (*Vidyasagar, 2007*)

Ans. Nessler's reagent has the formula K_2HgI_4. It is used in the detection of NH_3 and ammonium salts as per the following reaction.

$$2K_2HgI_4 + NH_3 + 3NaOH \longrightarrow \underset{\text{Brown ppt.}}{NH_2HgO.HgI} + 3NaI + 4KI + 2H_2O$$

Q. 49. What happens when a mixture of borax, conc. sulphuric acid and methyl alcohol is heated and the outcoming vapour is ignited. (*Calcutta, 2007*)

Ans. See Q. 10 Test for borate.

Q. 50. What happens when a mixture of calcium fluoride, silica and concentrated sulphuric acid is warmed in a lead crucible and a drop of water is held in the issued vapour ? (*Calcutta, 2007*)

Ans.

$$CaF_2 + H_2SO_4 \longrightarrow CaSO_4 + 2HF$$

$$\underset{\text{silica}}{SiO_2} + 2HF \longrightarrow \underset{\text{silicon}}{SiF_4} + 2H_2O$$

$$3SiF_4 + 4H_2O \longrightarrow \underset{\text{Fluosilicic acid}}{2H_2[SiF_6]} + \underset{\text{silic acid (white deposit glass rod)}}{H_4SiO_4}$$

Thus, a white deposit of silicic acid on glass rod occurs.

28 QUANTITATIVE ANALYSIS

Q. 1. What is relative supersaturation?

Ans. According to Von Weimarn, relative supersaturation is given by the relation:

$$\text{Relative supersaturation} = \frac{Q-S}{S}$$

where Q = Concentration of the solute in solution at any instant

S = Equilibrium concentration of the solute in a saturated solution

A thorough investigation of the effect of relative supersaturation on the particle size of various substances was made by Von Weimarn. He inferred that increasing relative supersaturation caused a marked decrease in the size of particles.

A gelatinous precipitate of barium sulphate of very small size was obtained when saturated solutions of barium thiocyanate and manganous sulphate were mixed rapidly at a relative supersatu-ration of 1,75,000. On the other hand, crystals of barium sulphate of size 0.15 mm were obtained when the dilute solutions of the two substances were mixed slowly at relative supersaturation of 3.

Q. 2. Discuss conditions for precipitation from a solution. *(M. Dayanand, 2000)*

Ans. The following conditions apply for precipitation:

1. Dilute solutions may be used for precipitation purposes.

2. Precipitation should be done from hot solutions as this increases the value of S. Precipitate obtained in the hot solution is cooled to make precipitation quantitative.

3. Precipitating agent should be added slowly and with constant stirring. This helps to keep the value of Q low.

4. Precipitation should be carried out at a low value of pH as this yields quantitative precipitation. If the precipitate dissolves in acidic medium, then an optimum pH may be arrived at.

5. Slower rate of precipitation decreases chances of impurities getting trapped. The larger crystal size means smaller surface area (for a given mass of crystals) and hence lesser chance of adsorbing impurities.

6. When the precipitation is done, a slight excess of precipitating agent is added to decrease the solubility of the precipitate by common ion effect and to ensure complete precipitation. A large excess of precipitating agent is avoided to keep its adsorption on the precipitate under control.

Gravimetric precipitate is always contaminated with other ions present in solution. However, we have to reduce the level of such contamination to a very low level so as to get accurate results. The contamination of the precipitate can occur by any one of the following processes:

Q. 3. Write notes on

***(i)* Simultaneous precipitation**

***(ii)* Co-precipitation** *(Andhra, 2002; Mumbai, 2005)*

***(iii)* Post-precipitation** *(Garhwal, 2000; Kanpur, 2001; Andhra, 2010)*

Ans. (*a*) Simultaneous precipitation. If the solubility product of some compound other than the desired also gets crossed during the addition of precipitating reagent, the other substance simultaneously forms a precipitate along with the desired compound. Thus the desired precipitate will be contaminated with the second precipitate. For example, a mixture solution containing mainly sodium chloride along with impurities of sodium bromide on adding silver nitrate will produce a precipitate of silver chloride contaminated with silver bromide precipitate. This is an example of simultaneous precipitation.

***(b)* Co-precipitaton.** Soluble substances cling to the insoluble precipitates and thus contaminate the precipitate. *This phenomenon by which normally soluble substances are carried out of solution by an insoluble precipitate is called co-precipitation.* Co-precipitation has been found to be of following types:

(*i*) *Surface adsorption.* This is the most common form of contamination. This type of co-precipitation is caused by the interaction of the counter ions of opposite charge to the adsorbed primary layer of ions present in excess in solution. For example, sulphate ion is precipitated as $BaSO_4$ by the addition of a slight excess of a solution of $Ba(NO_3)_2$ to K_2SO_4 solution. The excess Ba^{2+} in solution can adsorb on to the neutral surface of $BaSO_4$ to create a primary layer of Ba^{2+} ion which can further attract the counter ions, that is, nitrate anions. This results in an adsorbed layer of barium nitrate $Ba(NO_3)_2$ onto the surface of $BaSO_4$.

(*ii*) *Isomorphic inclusion* (or *mixed crystal formation*). Here the particles of co-precipitated compound has dimensions and charge which can fit into the crystal structure of the main precipitate, *e.g.*, K^+ in NH_4MgPO_4 precipitation.

(*iii*) *Occlusion.* Here, the impurities get entrapped as imperfections within the crystals of the precipitate. For example, water may get trapped in pockets when $AgNO_3$ crystals are formed.

(c) **Post-precipitation.** On the addition of a reagent to a mixture of two or more ions, some ions get precipitated readily while other ions are precipitated slowly. Therefore, it is possible to separate the two ions which form precipitates with the same reagent by removing the precipitate of the first before second ion had time to precipitate. For example, a mixture solution containing calcium and magnesium ions on adding oxalate ions, at once from a precipitate of calcium oxalate while the precipitate of magnesium oxalate is formed after an hour or so. If calcium oxalate is not filtered immediately after its formation, then magnesium oxalate will contaminate it. Such *contamination by subsequent formation of a second insoluble precipitate* is called **post-precipitation.**

Q. 4. What is meant by digestion of precipitate? (*Kanpur, 2001*)

Ans. Very small crystals with a large specific surface area have a higher surface energy and a higher apparent solubility than larger crystals. However, when a precipitate is allowed to stand in contact with its mother liquor, larger crystals grow at the expense of the smaller ones. This process is called **digestion.** The smaller particles tend to dissolve and reprecipitate on the surfaces of the larger crystals. In addition to that, individual crystals can also come together to form bigger crystals. This is called **agglomeration.**

Digestion of precipitates is done usually at higher temperatures to speed up the formation of larger crystals. Larger crystals are easier to filter and are mostly purer.

Q. 5. What are the desirable properties of precipitates?

Ans. There are numerous reactions which are accompanied by the formation of precipitate, yet all of them are not used in analysis. Precipitate should have the following desirable properties:

(*i*) *Filterability.* Particle size of the precipitate should be quite large so that its filtration is easy and rapid.

(*ii*) *Solubility.* Its solubility should be extremely small.

(*iii*) *Purity.* The physical properties of the precipitate should be such that can be readily freed from normally soluble contaminants by a simple treatment.

(*iv*) *Selective precipitation.* Precipitating agent should be able to precipitate only one component from a mixture solution of many components.

Q.6. Describe the factors governing the solubility of a precipitate.

Ans. (*i*) *pH value of the reaction medium.* The solubility of the precipitates of many substances depends upon the pH value of the reaction medium. This is particularly so for the formation of metallic hydroxides.

(*ii*) *Temperature.* The solubility of a precipitate generally increases with rise in temperature.

(*iii*) *Composition of the solvent.* The solubility of most inorganic compounds decreases markedly in aqueous solution of organic solvents than in pure water.

(*iv*) *Electrolyte concentration.* The solubility of most compounds is greater in solutions of an electrolyte than in pure water. This is due to an increased attraction for the ions of the solute by the dissolved electrolyte in the solvent.

Q.7. What is common ion effect? Discuss its importance.

(*M. Dayanand, 2000; Calcutta, 2000*)

Ans. On dissolving a weak electrolyte such a weak acid (*e.g.*, CH_3COOH), or a weak base (*e.g.*, NH_4OH) in water, an equilibrium gets established between the ions and the undissociated molecule as:

$$CH_3COOH \rightleftharpoons CH_3COO^- + H^+$$

$$NH_4OH \rightleftharpoons NH^+_4 + OH^-$$

If to the solution of a weak electrolyte a strong electrolyte containing a common ion is added, the concentration of the common ion increases. For example, if NH_4Cl (a strong electrolyte) is added to a solution of NH_4OH (a weak electrolyte), NH_4Cl being strong electrolyte dissociates and provides a large number of NH_4+ ions (which are common ions).

$$NH_4OH \rightleftharpoons NH_4^+ + OH^- \quad \text{[Weak dissociation]}$$

$$NH_4Cl \rightleftharpoons NH_4^+ + Cl^- \quad \text{[Strong dissociation]}$$

According to Le Chatelier's Principle, with increase in concentration of NH_4^+ ions the equilibrium of dissociation of NH_4OH gets shifted towards the left and thus the dissociation of ammonium hydroxide is suppressed. In other words, the degree of dissociation of a weak electrolyte is suppressed by the addition of the solution of a strong electrolyte having a common ion. This is known as **common ion effect.**

Importance of Common Ion Effect

(*a*) In qualitative analysis of cations, by the addition of an appropriate amount of the strong electrolyte having common ion, the dissociation of a weak electrolyte is suppressed to a desired extent. Consequently, conditions are created to precipitate certain cations while other cations remain in solution.

(*b*) Common ion effect is quite important in gravimetric analysis. A gravimetric precipitate has slight solubility in the precipitating medium. If the solubility is not very low then this method cannot be used for gravimetric estimation. The solubility of a precipitate can be decreased by the use of common ion effect. This is achieved mostly by adding a small excess of the precipitating agent having a common ion. For example, in the determination of Ag^+ as AgCl, a chloride ion is added to the solution of silver salt. If exactly equivalent amount of Cl^- ion is added, then the resulting saturated solution of AgC1 will contain 0.0015 g/L of AgC1 in aqueous medium (that is, 0.4% solubility). However, if a slight excess of Cl^- ion (0.01 M) is added during precipitation, then the solubility of AgC1 is reduced to 0.004%. Therefore, a very accurate estimation of Ag^+ is accomplished if a slight excess of Cl^- ion is added during precipitation.

Q. 8. What is solubility product of a substance?

(*M. Dayanand, 2000; Calcutta, 2000; Garhwal, 2000*)

Ans. There are many ionic solids which are sparingly soluble in water, *e.g.*, AgC1, Ag_2S, $PbSO_4$ and $BaSO_4$. A sparingly soluble salt on being placed in water dissolves slightly forming a saturated solution. Consequently, an *equilibrium* is established between the undissolved ionic solid and the ions of the dissolved salt. For example, in case of AgC1, we have

$$\underset{\text{(solid)}}{AgCl} \rightleftharpoons \underset{\text{(in soln.)}}{AgCl} \rightleftharpoons Ag^+ + Cl^-$$

Applying the law of chemical equilibrium, we get

$$K = \frac{[Ag^+][Cl^-]}{[AgCl]}$$

or $$[Ag^+][Cl^-] \rightleftharpoons K \times [AgCl]$$

The concentration of the dissolved silver chloride is constant at a specified temperature, *i.e.*, [AgCl] is constant. Therefore,

$$[Ag^+][Cl^-] \rightleftharpoons K \times \text{constant} = K_{sp}$$

where K_{sp} is called **solubility product** and the expression $[Ag^+][Cl^-]$ is known as the *ionic product.* Since the solubility of a salt varies with temperature, the numerical value of K_{sp} changes with temperature.

In general, if the sparingly soluble salt is A_xB_y in equilibrium with saturated solution of its ions, then

$$\underset{\text{(solid)}}{A_x B_y} \rightleftharpoons \underset{\text{(in soln.)}}{A_x B_y} \rightleftharpoons xA^+ + yB^-$$

Then the solubility product of the salt, $A_x B_y$, is

$$K_{sp} = [A^+]^x [B^-]^y$$

Therefore, *solubility product of a salt is equal to the product of the concentration of its ions, each raised to the appropriate power, in a saturated solution at that temperature.* Solubility products of some sparingly soluble salts are expressed as under:

$$BaCO_3 \rightleftharpoons Ba^{2+} + CO_3^{2-} \quad K_{sp} = [Ba^{2+}][CO_3^{2-}]$$

$$Bi_2S_3 \rightleftharpoons 2Bi^{3+} + 3S^{2-} \quad K_{sp} = [Bi^{3+}]^2[S^{2-}]^3$$

$$Ag_2CrO_4 \rightleftharpoons 2Ag^+ + CrO_4^{2-} \quad K_{sp} = [Ag^+]^2[CrO_4^{2-}]$$

Applications of solubility product. The concept of solubility product is very useful in explaining a number of phenomena such as dissolution and precipitation in analytical chemistry as discussed below:

Q. 9. How can we calculate solubility of a sparingly soluble salt using solubility product?

Ans. Knowing the solubility product of a sparingly soluble salt like AgC1, CaF_2, Ag_2CrO_4 at a particular temperature, the solubility of that salt can be calculated as follows:

(*i*) For 1: 1 electrolytes such as AgC1, AgBr, AgI, $PbSO_4$, $BaSO_4$, $SrSO_4$ and $AlPO_4$.

$$\text{Solubility of the salt} = \sqrt{K_{sp}} \text{ moles per litre}$$

$$= \sqrt{K_{sp}} \times \text{molecular mass gram per litre}$$

(*ii*) For 1: 2 electrolytes such as CaF_2, $Mg(OH)_2$, $PbC1_2$, PbI_2, Ag_2CrO_4, and HgI_2.

$$\text{Solubility of the salt} = \sqrt[3]{\frac{K_{sp}}{4}} \text{ mole per litre}$$

$$= \sqrt[3]{\frac{K_{sp}}{4}} \times \text{molecular mass gram per litre}$$

(*iii*) For 1: 3 electrolytes such as Fe $(OH)_3$, $Cr(OH)_3$ and $Al(OH)_3$.

$$\text{Solubility of the salt} = \sqrt[4]{\frac{K_{sp}}{27}} \text{ mole per litre}$$

$$= \sqrt[4]{\frac{K_{sp}}{27}} \times \text{molecular mass gram per litre}$$

Q.10. Calculate the solubility of silver chloride

(*i*) in moles per litre

(*ii*) in gram per litre at a particular temperature when the solubility product of AgCI is 1.6×10^{-10}.

Ans. The solubility equilibrium equation is

$$AgC1 \rightleftharpoons Ag^+ + Cl^-$$

Let the solubility of silver chloride in water be *c* moles per litre, then

	AgC1	$\rightleftharpoons$	Ag^+	+	Cl^-
	1 mole		1 mole		1 mole
or	*c* moles		*c* moles		*c* moles

Ionic product of AgC1

$$= [Ag^+] [Cl^-]$$

$$= c \times c = c^2$$

Since AgC1 is a sparingly soluble salt, its ionic product is equal to its solubility product. Thus

$$c^2 = K_{sp}$$

or $$c = \sqrt{K_{sp}}$$

$$= \sqrt{1.6 \times 10^{-10}}$$

$$= 1.26 \times 10^{-5} \textbf{ moles/ litre}$$

$$= 1.26 \times 10^{-5} \times 143.5 = 1.8 \times 10^{-3} \textbf{ gram / litre.}$$

Q. 11. Calculate the solubility of silver chromate, Ag_2CrO_4 in moles per litre if the solubility product of Ag_2CrO_4 is 1.9×10^{-12}.

Ans. The solubility equilibrium equation is

$$Ag_2CrO_4 \rightleftharpoons 2Ag^+ + CrO_4^{2-}$$

Let the solubility of silver chromate in water be c moles per litre, then

	Ag_2CrO_4 $\rightleftharpoons$	$2Ag^+$ +	CrO_4^{2-}
	1 mole	2 moles	1 mole
or	*c* moles	2c moles	c moles

Ionic product of $Ag_2CrO_4 = [Ag^+]^2\,[CrO_4^{2-}]$

$$= (2c)^2\,(c) = 4c^3$$

Since Ag_2CrO_4 is a sparingly soluble salt, its ionic product is equal to its solubility product.

Thus

$$4c^3 = K_{sp}$$

or

$$c = \sqrt[3]{\frac{K_{sp}}{4}} = \sqrt{\frac{1.9\times10^{-12}}{4}}$$

$$= 7.8 \times 10^{-5} \textbf{ moles / litre.}$$

Q.12. Making use of solubility product, how can we predict a precipitation reaction?

Ans. Knowing the solubility product of a salt, it is possible to predict whether on mixing the solutions of its ions, a precipitate will be formed or not. For example, it is possible to predict whether a precipitate of lead iodide will be formed or not if a solution containing lead ions is mixed with a solution containing iodide ions.

Based on the concept of solubility product, it is evident that an *electrolyte forms a precipitate if its ionic product is greater than its solubility product.* Therefore, to predict the precipitation reaction, the ionic product is calculated to find out whether it is greater than K_{sp} or not. If ionic product remains less than solubility product, then precipitation does not take place.

Q.13. What happens when 10 ml of 0.2 M solution of sodium sulphate is mixed with an equal volume of 0.1 M barium chloride solution? Will a precipitate of barium sulphate be formed or will it not be formed? Given that K_{sp} for barium sulphate is 1×10^{-10}.

Ans. $$BaSO_4 \rightleftharpoons Ba^{2+} + SO_4^{2-}$$

On mixing equal volumes of the two solutions, the total volume becomes double so that the concentration of each gets reduced by half. Therefore, the new concentration of the salts are as follows:

$\therefore$ Concentration of $Na_2SO_4 = \frac{0.2M}{2} = 1\times10^{-1}$ M

Concentration of $BaCl_2 = \frac{0.1M}{2} = 5 \times 10^{-2}$ M

Since Na_2SO_4 ionizes as

$$Na_2SO_4 \rightleftharpoons 2Na^+ + SO_4^{2-}$$

1 mole $\qquad$ 1 mole

$1 \times 10^{-1}M$ $\qquad$ 1×10^{-1} M

$\therefore \qquad [SO_4^{2-}] = 1 \times 10^{-1}$ M

Since $BaCl_2$ ionizes as

$$BaCl_2 \rightleftharpoons Ba^{2+} + 2Cl^-$$

1mole $\qquad$ 1 mole

5×10^{-2} M $\qquad$ 5×10^{-2} M

$\therefore \qquad [Ba^{2+}] = 5 \times 10^{-2}$ M

Now, ionic product of $BaSO_4 = [Ba^{2+}1\ [SO_4^{2-}]$

$= (5 \times 10^{-2})\ (1\times 10^{-1}) = 5 \times 10^{-3}$

But the solubility product of $BaSO_4$ is 1.0×10^{-10}.

This means that ionic product of $BaSO_4$ is greater than its solubility product. Therefore, on mixing the two solutions a precipitate of $BaSO_4$ will be formed.

Q. 14. Discuss the role of solubility product in the precipitation of various cations in different groups in qualitative analysis. *(Calcutta, 2000; Garhwal, 2000)*

GROUP I

Ans. The cations of group I are Pb^{2+}, Ag^+ and Hg. These cations are easily precipitated as their chlorides ($PbCl_2$, AgCl and Hg_2Cl_2) by adding dilute HCl to the solution of the mixture. This is due to the fact the chlorides of the cations of group I have very low solubility products so that their ionic products easily cross their solubility products, resulting in their precipitation. On the other hand, chlorides of the cations of succeeding groups are not precipitated in the group I as they have very high solubility products which are not crossed by their ionic products.

GROUP II

The cations of group II are Hg^{2+}, Pb^{2+}, Bi^{3+}, Cu^{2+}, Cd^{2+}, As^{3+}, Sb^{3+} and Sn^{2+}. These are precipitated as their sulphides by passing hydrogen sulphide through their solution containing hydrochloric acid.

$$Cu^{2+} + H_2S \longrightarrow \underset{\text{Black ppt.}}{CuS} + 2H^+$$

Function of HCl. *It suppresses the degree of ionisation of H_2S (a weak electrolyte) by common ion effect as explained below:*

HC1 is a strong acid and furnishes a large number of hydrogen ions as:

$$HCl \longrightarrow H^+ + Cl^-$$

On the other hand, H_2S is a weak acid and furnishes a small number of hydrogen ions as

$$H_2S \longrightarrow 2H^+ + S^{2-}$$

On account of the presence of common H^+ ions in both the equilibria, **common ion effect** operates. This results in suppressing the degree of dissociation of H_2S. Thus, *the concentration of sulphide ions is reduced very much in the solution.* However, even this reduced concentration of sulphide ions is sufficient to precipitate the cations of group II. This is because the solubility products of the sulphides of cation of group II are small and ionic products easily cross the solubility products.

However, the sulphides of the cations of next groups are not precipitated in group II because they have very high solubility products which are not exceeded by their ionic products due to very small concentration of sulphide ions in the solution.

GROUP III

The cations of group III are Fe^{3+}, Cr^{3+} and Al^{3+}. These are precipitated as their hydroxides by adding NH_4OH in presence of NH_4Cl.

Function of NH_4Cl. *It suppresses the degree of ionisation of NH_4OH* (a weak electrolyte) *by common ion effect* as explained below:

NH_4Cl is a strong electrolyte and furnishes a large number of NH_4^+ ions as:

$$NH_4Cl \longrightarrow NH_4^+ + Cl^-$$

On the other hand, NH_4OH is a weak electrolyte and furnishes a small number of NH_4^+ and OH^- ions as:

$$NH_4OH \rightleftharpoons NH_4^+ + OH^-$$

On account of the presence of common NH_4^+ ions in both the equilibria, common ion effect operates. This results in suppressing the degree of dissociation of NH_4OH. In other words, the *concentration of hydroxide ions is reduced very much in the solution.* However, even this reduced concentration of hydroxide ions is sufficient to precipitate the cations of group III. This is because the solubility products of the hydroxides of the cations of group III are very low and ionic products easily exceed the solubility products.

However, the hydroxides of the cations of succeeding groups are not precipitated in group III because they have very high solubility products which are not exceeded by their ionic products due to very small concentration of hydroxide ions in the solution.

GROUP IV

The cations of group IV are Ni^{2+}, Co^{2+}, Mn^{2+} and Zn^{2+}. These are precipitated as their sulphides by passing H_2S through their solution containing ammonium hydroxide.

Function of NH_4OH. *It helps to increase the ionisation of H_2S and thereby increases the concentration of sulphide ions* as explained below:

NH_4OH ionises as

$$NH_4OH \rightleftharpoons NH_4^+ + OH^- \quad ...(i)$$

On the other hand, H_2S ionises as

$$H_2S \rightleftharpoons 2H^+ + S^{2-} \quad ...(ii)$$

The hydroxide ions from equilibrium reaction (*i*) combine with H^+ ions from equilibrium reaction (*ii*) *to* form undissociated molecules as:

$$H^+ + OH^- \longrightarrow H_2O$$

Consequently, more H_2S molecules dissociate and in the process concentration of sulphide ions increases. As the concentration of sulphide ions increases, correspondingly the ionic products of the sulphides of the cations of group IV increases. Ultimately, the ionic products of the sulphides of the cations of group IV get crossed by their solubility products and these sulphides are obtained as precipitates.

GROUP V

The cations of group V are Ba^{2+}, Sa^{2+} and Ca^{2+}. These are precipitated as their carbonates by adding $(NH_4)_2\ CO_3$ to their solution containing NH_4Cl and NH_4OH.

Function of NH_4Cl. *It suppresses the degree of ionisation of* $(NH_4)_2\ CO_3$ *by common ion effect as* explained below:

NH_4Cl which is a strong electrolyte furnishes a large number of NH_4^+ ions and ionises as:

$$NH_4Cl \longrightarrow NH_4^+ + Cl^-$$

$(NH_4)_2CO_3$ (being a weak electrolyte) ionises as:

$$(NH_4)_2\ CO_3 \rightleftharpoons 2\ NH_4^+ + CO_3^{2-}$$

On account of the presence of common NH_4^+ ions in both the equilibria, common ion effect operates. This results in suppressing the degree of dissociation of $(NH_4)_2CO_3$. In other words, the concentration of carbonate ions is reduced in the solution. However, even this reduced concentration of carbonate ions is sufficient to precipitate the cations of group V as their solubility products are low and ionic products cross their solubility products.

However, magnesium carbonate is not precipitated in group V as it has high solubility products which is not exceeded by its ionic product due to small concentration of carbonate ions in solution.

Function of NH_4OH. *It converts ammonium bicarbonate present in ammonium carbonate solution to ammonium carbonate* as:

$$NH_4\ HCO_3 + NH_4\ OH \longrightarrow (NH_4)_2\ CO_3 + H_2O$$

In case bicarbonate ions are present in solution, then bicarbonates of Ba^{2+}, Sr^{2+} and Ca^{2+} are formed which are soluble in water. Consequently, the quantity of the precipitate formed in group V decreases very much.

GROUP VI

The cations of group VI are Mg^{2+}, Na^+, K^+ and NH_4^+. These cations are tested independently as there is no group reagent for them.

Q.15. Discuss the role of solubility product in quantitative analysis.

Ans. The solubility product concept is fairly accurate in its application to quantitative analysis if the concentrations are small. The solubility product is the ultimate value which is attained by the ionic product when equilibrium has been established between the solid phase and the solution. If the experimental conditions are such that the ionic product is different from the solubility product, then the system will attempt to adjust itself in such a way that ionic product and solubility product are equal. Thus if ionic product in solution is made to exceed the solubility product, then the adjustment of the system to equilibrium results in precipitation of the solid salt provided supersaturation is excluded. If the ionic concentration is made less than the solubility product that the solute will pass into solution until solubility product is attained.

Q.16. Discuss the role of pH in precipitation reactions.

Ans. pH value plays an important role in chemical reactions which are carried out in the laboratory. For example, in the gravimetric estimation of aluminium, methyl red is used as indicator to control the pH of the medium so that excess of NH_4OH is not added to precipitate aluminium as aluminium hydroxide. As soon as the colour of the solution containing methyl red changes from red to yellow, pH attains a value of 6.5 and the addition of NH_2OH is stopped.

Similarly, pH is important in the quantitative precipitation of calcium oxalate which is insoluble in basic medium. However, if the solution is acidic then the oxalate ions tend to combine with hydrogen ions to form a weak acid, oxalic acid. In such cases pH is maintained above 7.

Control of pH can also help in selective precipitation from a mixture. For example, in a mixture of magnesium and aluminium ions, the latter can be precipitated quantitatively with the help of 8-hydroxyquinoline (oxine) at a pH = 4. At this pH = 4, the concentration of anion form of oxine is so low that it does not precipitate out magnesium ions. At higher pH, the concentration of anion of oxine increases and magnesium is also precipitated quantitatively.

N, OH ⇌ N, O^- + H^+

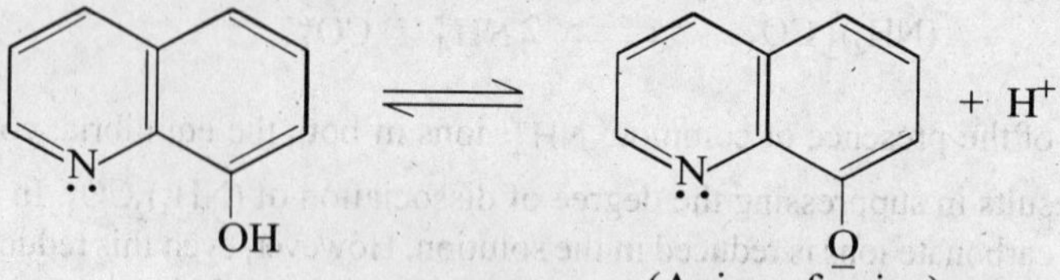

(Anion of oxine)

Q. 17. Write notes on (*i*) masking, (*ii*) demasking as applied to selective precipitation.

Ans. (*i*) Masking is a process in which a substance, without physical separation of it or its reaction products, is so transformed that it does not enter into a particular reaction. For example, tartarate acts as a masking agent in the separation of titanium from iron. In this case titanium is prevented from precipitation as its sulphide TiS_2 by the addition of tartarate to the titanium solution before passing H_2S. This principle is used for the separation of titanium from iron. Iron precipitates as its sulphide (FeS), whereas titanium remains in solution.

Similarly copper is not precipitated as $Cu(OH)_2$ when NaOH solution is added to a solution of Cu^{++} ions containing tartarate. This provides an excellent method for preventing precipitation of copper from a basic solution. Tartarate is a masking agent in such cases.

Phenomenon of masking works on complexation reaction. Tartarate forms a complex with iron or copper and thus precipitation with a reagent is prevented.

(*ii*) Demasking. It is a process in which the masked substance regains its ability to enter into a particular reaction. For example, the cyanide complexes of Zn and Cd can be demasked with HCHO-acetic acid solution or better with chloral-hydrate.

$$[Zn(CN)_4]^{2-} + 4H^+ + 4HCHO \longrightarrow Zn^{2+} + 4\ HO—CH_2—CN$$

The use of masking and selective demasking agents helps in performing the successive titration of many metals. Thus a solution of Mg, Zn and Cd can be titrated as follows:

(*a*) Add excess standard EDTA solution and back titrate the excess EDTA against standard Mg^{2+} solution using Solochrome Black as indicator. This gives combined concentration of all metals.

(*b*) Treat a portion with excess KCN and titrate as before. This gives Mg only because Zn and Cd are masked.

(*c*) Add excess of chloral-hydrate or HCHO—CH_3COOH solution to the titrated solution in order to demask Zn from its cyanide complex. Titrate until indicator turns blue. This gives Zn. Copper is found by difference.

Q.18. What is solvent extraction? Describe the principle on which it works.

(*Kurukshetra, 2001; Calcutta, 2005*)

Ans. Solvent extraction *is defined as the process by which a substance may be extracted from dilute solutions* (usually aqueous or inorganic solutions) *into an immiscible solvent* (usually organic solvent) *with or without the use of a complexing agent.* The organic solvents used for extraction may be polar like chloroform, nitrobenzene, tetrahydrofuran, butyl acetate, ethers, methyl isobutyl ketone, etc. or non-polar like benzene, carbon tetrachloride, petroleum ether, kerosene, xylene, *n*-hexane, etc.

Nerst distribution law governs the partition of a solute or substance say X between two immiscible solvents expressed by the equilibrium:

$$X_{aq} \rightleftharpoons X_{org}$$

The ratio of concentration of X in organic (*org*) or aqueous (*aq*) phases is constant and independent of the total concentration of X at a given temperature.

$$K = \frac{[X_{org}]}{[X_{aq}]}$$

K = Partition coefficient which is independent of the total solute concentration in either of the phases.

$[X_{org}]$ = Concentration of X in organic phase.

$[X_{aq}]$ = Concentration of X in aqueous phase.

At constant temperature, the value of K is constant. Quite often, K is approximately equal to the ratio of the solubility of X in each solvent.

Generally, ionic substances are insoluble in non-polar solvents but soluble in polar solvents with high dielectric constant due to the solvation of ions. On the other hand, covalent compounds are mostly soluble in non-polar solvents. Therefore, the extraction of an ion from aqueous solution into organic phase can take place only if the ion forms a species showing preference for organic phase either by

(*i*) Chelate formation, (*ii*) Solvate formation, (*iii*) Ion-pair formation.

(*i*) **Chelate formation.** Organic bases form an important class of chelating agents which can interact with a metal atom/ion from more than one position in the liquid molecule in such a way that the metal atom/ion is bound in a stable 4 or 5 or 6 membered ring. Such complexes are called *chelates* and organic bases forming these complexes are called *Chelating agents*. A widely used

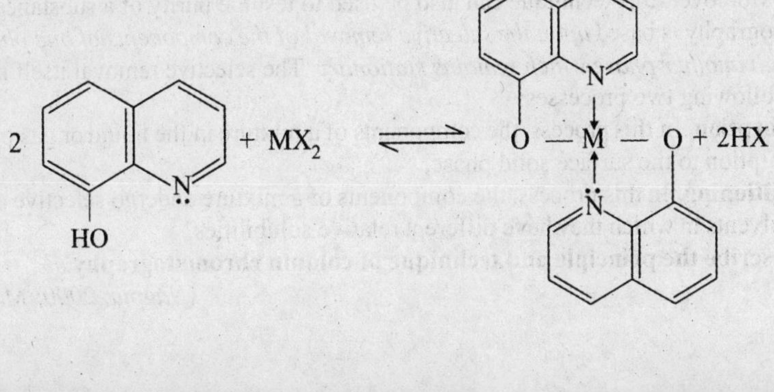

2 (HO, N) $+ MX_2 \rightleftharpoons$ O — M — O $+ 2HX$ (N → M ← N)

chelating agent is 8-hydroxyquinoline (C_9H_7NO). Most of its chelates are soluble in organic solvents. It combines with say divalent metal halides MX_2 in aqueous medium to form the complex involving two 5-membered chelate rings as shown below:

The complex is soluble in organic solvents and the equilibrium is pH dependent. As a consequence, metals having different formation constants with the ligand can be separated through pH control of the aqueous phase. This way even trace amounts of metal can be separated. Chloroform is used as the solvent in such cases.

(*ii*) **Extraction by solvation.** It involves solvation of the extracted species into the organic phase. There are several kinds of solvating solvents, for example, ether, ethylacetate, methylisobutyl ketone (MIBK), tributylphosphate $(C_4H_9O)_3$ PO(TBP), tributylphosphine oxide $(C_4H_9)_3PO$ etc. This method is frequently used for the extraction of transition metals and inner transition metals. For example, uranium can be conveniently separated from lead and thorium if the nitrate in water is extracted with ethereal phase as a solvate $[UO_2(NO_3)_2.2H_2O.2Et_2O]$. Lead and thorium are left in aqueous phase. Uranium can be extracted on industrial scale using TBP as an extractant (for solvate formation) in kerosene phase. This method is based on the principle of solvation of the metallic species into the organic phase.

(*iii*) **Ion-pair formation.** Another category of extractants involves extraction by ion-pair formation. The extractants are mostly high molecular weight amines. As for example, tertiary amines like trioctylamine TOA $(C_8C_{17})_3N$, tri-isooctylamine TIOA $(i\text{-}C_8H_{17})_3N$, trioctylamine oxide $(C_8H_{17})_3NO$ are best extractants for anionic complexes with mineral acids. A typical example showing mechanism of extraction of a solution of $FeC1_3$ in concentrated hydrochloric acid $HFeCl_a$ is as follows:

$$Fe^{3+}_{aq} + 4Cl^-_{aq} \rightleftharpoons FeCl_4^-{}_{aq}$$

$$R_3NO_{org} + H^+_{aq}Cl^-_{aq} \rightleftharpoons R_3NH_{org} + Cl_{org}$$

$$R_3NH_{org} + Cl^-_{org} + FeCl_4^-{}_{aq} \rightleftharpoons \left[R_3NH^+FeCl_4^-\right]_{org} + Cl_{aq}$$

Ion-pairs involving big metallic cations or anions with ions of the opposite charge, such as, tetraphenylarsonium perrhenate $[(C_6H_5)_4As]^+ ReO_4^-$ or tetraphenylarsonium permanganate $[(C_6H_5)_4^- As]^+ MnO_4^-$ undergo easy extraction with organic phase of chloroform.

Q.19. What is chromatography? Describe the principle on which it works.

(*Guwahati, 2002; Nagpur, 2003; Calcutta, 2005*)

Ans. Tswett described a new technique known as chromatography to separate a mixture of coloured substances into individual components. The original technique has been modified and extended so that it has been used to separate a mixture of coloured as well as colourless substances into individual components. Moreover, this technique can also be used to test the purity of a substance.

Chromatography is based *upon the selective removal of the components of one phase from it as it flows through another phase which remains stationary.* The selective removal itself may be due to either of the following two processes:

(*i*) **Adsorption.** In this process, the components of a mixture in the liquid or gas phase undergo selective adsorption to the surface solid phase.

(*ii*) **Partitioning.** In this process, the components of a mixture undergo selective dissolution in immiscible solvents in which they have different relative solubilities.

Q.20. Describe the principle and technique of column chromatography.

(*Nagpur, 2000; Mumbai, 2005*)

Ans. Principle. The various constituents of the mixture get adsorbed to different extents on the fixed solid phase such as alumina or silica gel which is packed as a long column. The constituents get separated into distinct **bands** which are removed with the help of a solvent. The extracted component along with the solvent is known as **elute.** From the elute, the solvent is removed to get the individual component.

It is believed that there is always a competition between solute and solvent molecules for the adsorption sites. Consequently, a dynamic equilibrium gets established at the interface where the solute and solvent molecules get attracted and settled at the solid surface for a short while and then leave the solid surface to re-enter the mobile liquid phase. However the desorbed molecules which have greater affinity for the solid surface flow into the mobile layer slowly while those desorbed molecules which have lesser affinity for the solid surface flow into the mobile layer quickly.

Commonly used adsorbents are *activated alumina, activated charcoal, magnesium oxide, silica gel, starch and Fuller's earth.*

As a result of experimental investigations, it has been found out that greater the number of double bonds and hydroxyl groups present in a molecule, greater is the adsorption. Thus we have adsorption series as:

Acids, bases > alcohols > ketones, aldehydes > unsatured hydrocarbons > saturated hydrocarbons.

Adsorption chromatography is mainly used to separate a mixture of organic compounds such as a mixture of ortho and para nitroanilines; a mixture of blue and red dyes etc.

Q.21. Give the principle and advantages of paper chromatography.

(*Awadh, 2000; M. Dayanand, 2002; Mumbai, 2005; Nagpur, 2008*)

Ans. Principle. This technique is based on the principle of adsorption and partition both. A mixture of the solute is placed on a strip of chromatographic paper such as Whatmann No. 1 or 3 and a solvent is allowed to move along the paper strip. The solvent extracts the solute because of distribution of the solute between the two solvents.

Advantages of Paper Chromatography. Its main advantages are:

(*i*) The procedure is very simple and precise.

(*ii*) It is reasonably rapid process.

(*iii*) Only small quantities of the material are required.

(*iv*) No costly apparatus as well as reagents are required.

Q.22. What is thin layer chromatography? What are its advantages?

(*M. Dayanand, 2000*)

Ans. In this technique, a solid adsorbent such as alumina or silica gel is mixed with a little binder such as hydrated calcium sulphate or starch. It is then spread as a thin layer on a glass plate or plastic sheet. The plate thus formed is known as **chromatoplate.** By means of a fine capillary, solution of the sample to be analysed is applied on chromatoplate and then dried. Then it is allowed to stand in a jar containing solvent and covered with lid. After some time, spots are located either through naked eye or by applying suitable reagent.

Principle of thin layer chromatography is the same as that of paper chromatography.

Advantages of Thin Layer Chromatography

(*i*) It is a very rapid process as compared to paper chromatography.

(*ii*) It is useful for large number of sample separations.

(*iii*) In this method sharp spots are obtained while in paper chromatography diffused spots are formed.

(*iv*) In this method, even a solution containing acid or alkali can be separated whereas paper chromatography cannot be used for such solutions.

Q.23. What are (*i*) cation exchangers and (*ii*) anion exchangers? (*Nagpur, 2000*)

Ans. (*i*) **Cation exchangers.** They are acidic resins as they conain sulphonic acid groups ($RSO_3^- H^+$) or carboxylic acid groups (RCOOH). The resins containing sulphonic acid groups are known as *strong acid resins* and those containing carboxylic acid groups are known as *weak acid resins.*

This type of resins are made by first copolymerizing styrene and divinyl benzene, the polymer

$— CH — CH_2 — CH — CH_2 — CH — CH_2 — CH — CH_2 —$

$SO_3^- H^+$ $SO_3^- H^+$ $SO_3^- H^+$

$CH — CH_2 —$

$SO_3^- H^+$

is then sulphonated to get the requisite type of cation exchange resin whose typical formula is given below :

(*ii*) **Anion exchangers.** They are basic resins as they are either quaternary or tertiary amines or contain hydroxyl groups. The resins which are quaternary or tertiary amines are known as *strongly basic anion* exchangers. On the other hand, resins which contain —OH groups are known as *weakly basic anion exchangers.* Structurally —SO_3H group of the cation resin is replaced with —CH_3—NR_3^+ or —OH^- group in anion exchangers.

Q. 24. Explain the mechanism of ion exchange. Give some applications of ion exchange.
(*M. Dayanand, 2001; Kanpur, 2001; Mumbai, 2005; Nagpur, 2008*)

Ans. Different views have been put forth to explain the mechanism of ion exchange. These are explained as under:

(*i*) **Crystal lattice exchange theory.** According to this view, the ion exchange in resins is quite similar to exchange of crystal lattice ions. The exchange of ions on the resins occurs throughout the entire gel structure of the resin and not on the exposed surface only.

(*ii*) **Double layer theory.** According to this theory, as in colloids, a double layer exists with inner fixed layer and outer mobile layer of charges. The charged layers are due to absorbed ions and the concentrations of the ions on the outer layer depends upon the concentration and pH of the external solution. As soon as the foreign ions present in the external solution come in contact with the outer layer, new ions enter and replace the old ions in the outer layer and consequently, new equilibrium is established.

(*iii*) **Donnan membrane theory.** According to this theory, there is unequal distribution of ions on two sides of a membrane. In the same way, in ion exchange equilibria, the interface between solid and liquid phases is considered as membrane. There is greater activity of ions on one side of the membrane as compared to the other. The exchange of ions occurs so that the activity ratio becomes equal on both sides.

Some Applications of Ion Exchange Method

In the recent years, ion exchange methods have become quite popular and have numerous applications in analytical chemistry as well as in industry. Some of their applications are listed below:

(*i*) This method was first of all used for softening or deionising water and is still used quite extensively.

(*ii*) Using EDTA solutions, it has been possible to separate quantitatively a mixture of alkali and alkaline earth metals.

(*iii*) Using ammonium citrate-citric acid, buffered at pH ~ 5.5 cation exchange resins are used to separate lanthanide ions.

Q. 25. A student mixes 10 ml of M/100 AgNO, solution with 20 ml of M/1000 $CaCl_2$ solution. Will he get a precipitate of AgCl ?

$$(K_{sp} \text{ of AgCl} = 1.6 \times 10^{-10})$$

Ans. 10 ml of M/100 $AgNO_3$ is mixed with 20 ml of M/1000 solution of $CaCl_2$.

Therefore the total volume will be = 30 ml

$$\therefore \quad \text{Conc. of } AgNO_3 = \frac{10}{30} \cdot \frac{M}{100} = \frac{10^{-2}}{3} M$$

$$\text{and} \quad \text{Conc. of } CaCl_2 = \frac{20}{30} \cdot \frac{M}{1000} = \frac{2\times 10^{-3}}{3} M$$

$$AgNO_3 \rightleftharpoons Ag^+ + NO_3^-$$

$$\therefore \ [Ag^+] = 0.33 \times 10^{-2} \text{ M}$$

$$\underset{\text{c moles}}{CaCl_2} \rightleftharpoons \underset{\text{c moles}}{Ca^{2+}} + \underset{\text{2c moles}}{2Cl^-}$$

$$\therefore \quad [Cl^-] = 2c = 2\times\frac{2}{3}\times 10^{-3}\text{ M} = \frac{4}{3}\times 10^{-3}\text{M}$$

Ionic product of AgCl = $[Ag^+]\,[Cl^-]$

$$= .33 \times 10^{-2} \times \frac{4}{3} \times 10^{-3}$$

$$= .44 \times 10^{-5}$$

$$K_{sp} = 1.6 \times 10^{-10} \text{ (given)}$$

Ionic product is higher than solubility product. Therefore AgCl will precipitate out.

Q. 26. What is the role of NH_4Cl in the analysis of III group of basic radicals?

(*Lucknow, 2000*)

Ans. See Qs. 7 and 14.

INORGANIC POLYMERS

Q. 1. Give the general properties of inorganic polymers ?

Ans. Some properties of inorganic polymers are given below :

1. With a few exceptions, inorganic polymers do not burn. They soften or melt at high temperatures.

2. Most of the inorganic polymers are built up of highly polar substances. Most of these polymers, however, react with the solvents. There are thus only a few inorganic polymers which actually dissolve in solvents properly.

3. Inorganic polymers having cross-linked structures with a high density of covalent bonds are generally stiffer and harder than the organic polymers.

4. The chain segments between cross links in polymers having cross-linked structures are usually short. Consequently, these structures are not flexible enough to permit interaction of solvent molecules.

5. Inorganic polymers are generally much less ductile than the organic polymers. Thus, while organic polymers such as polyethylene can extend by about 20 per cent or more before breaking, inorganic polymers break even when stretched by about 10 per cent.

6. Inorganic polymers, in general, are stronger, harder and more brittle than the organic polymers.

7. Inorganic polymers can usually be obtained in pure crystalline as well as in pure amorphous forms. Organic polymers, on the other hand, have partial crystalline and partial amorphous structure.

Q. 2. Give a detailed account of glass transition temperature. T_g.

Ans. The glass transition temperature is defined as the temperature at which the internal energy of polymer molecules increases to such an extent that the chain segments of the polymer molecules leave their lattice positions. At this temperature, the chain segments start moving apart from one another even on applying a very small strain to the polymer.

The glass transition temperature depends upon the chain length and the degree of cross linking. It also depends upon the barrier restricting internal rotation round the chain links. Below the glass transition temperature, the chain segments of the polymer are fixed on the lattice sites. The polymer in this condition is hard, brittle and hence breakable like glass. Above the glass transition temperature, the segments of the polymer begin to diffuse like the molecules of a liquid. At temperatures sufficiently higher than T_g, the polymer is present in the molten state. Consider the case of a polyphosphate polymer.

The above exchange reaction shows breaking of P – O bonds at some places and the formation of new P – O bonds at other places giving sufficient mobility to the chain segments of the polymer in the molten state.

The cross-linked inorganic polymers show a wide range in their glass transition temperatures. In this temperature range, they change from rigid solids to leathery solids, from leathery solids to rubbery solids and finally to highly viscous solids.

The glass transition temperature of a linear polymer is fairly sharp. This is because the movement of chain segments from one site to another does not involve the exchange of bonds as happens in the case of cross-linked inorganic polymers.

Q. 3. What are polyphosphonitrilic chlorides (polyhalophosphazenes) ? Give their methods of preparation and properties. *(Ranchi, 2010; Rajasthan, 2010; Garhwal, 2010)*

Ans. These polymers constitute inorganic rubbers. They exhibit high elasticity and can be stretched manifold with almost complete recovery. However, they undergo degradation under normal atmospheric conditions more rapidly. Hence, these polymers do not fmd much commercial use.

Preparation. Polyphosphonitrilic chlorides can be prepared as follows :

$$n\text{PCl}_5 + n\text{NH}_4\text{Cl} \xrightarrow{C_2H_2Cl_4} \text{Cl}\left(\begin{matrix}\text{Cl}\\ | \\ \text{P}\\ | \\ \text{Cl}\end{matrix} = \text{N}\right)_{4n-4}\text{Cl} + \underset{\text{Cyclic trimer}}{(\text{PCl}_2\text{N})_3} + \underset{\text{Cyclic tetramer}}{(\text{PCl}_2\text{N})_4} + (4n-1)\text{HCl}$$

The above reaction yields a mixture of cyclic trimers and tetramers along with a series of linear polymers. The linear polymerisation is terminated by H and Cl atoms.

The cyclic trimer or tetramer on heating in vacuum at 250°C changes into a chain polymer as shown below :

$$n(\text{PCl}_2\text{N})_4 \longrightarrow \underset{\text{Chain polymer}}{\text{Cl}\left(\begin{matrix}\text{Cl}\\ | \\ \text{P}\\ | \\ \text{Cl}\end{matrix} = \text{N}\right)_{4n-4}\text{Cl}} + \text{P}_4 + 2\text{N}_2 + 3\text{Cl}_2$$

The resulting chain polymer is somewhat cross-linked and hence has limited solubility in inorganic solvents. Polyphosphonitrilic chloride is a good elastomer like vulcanised rubbler.

Properties. The freshly prepared polyphosphonitrilic chlorides are soluble in chloroform but insoluble in petroleum ether. When the solutions are allowed to stand, the polymers get cross-linked and slowly gel formation takes place.

When samples of highly elastic inorganic rubber are stored in absence of air, there is no change in their elastic properties. However, storage in air makes the material brittle. This is due to chain so that there is reduction in average length of chain segments at lattice sites, as depicted below.

$$\begin{matrix}\text{Cl} & & \text{Cl} & & \text{Cl} & & \\ | & & | & & | & & \\ \sim\text{P} & =\text{N}- & \text{P} & =\text{N}- & \text{P} & =\text{N}-\text{H} \\ | & & | & & | & & \\ \text{Cl} & & \text{Cl} & & \text{Cl} & & \\ \text{Cl} & & \text{Cl} & & \text{Cl} & & \\ | & & | & & | & & \\ \sim\text{P} & =\text{N}- & \text{P} & =\text{N}- & \text{P} & =\text{N}-\text{H} \\ | & & | & & | & & \\ \text{Cl} & & \text{Cl} & & \text{Cl} & & \end{matrix} \xrightarrow{H_2O} \begin{matrix}\text{Cl} & & \text{Cl} & & \text{Cl} & & \\ | & & | & & | & & \\ \sim\text{P} & =\text{N}- & \text{P} & =\text{N}- & \text{P} & =\text{N}-\text{H} \\ | & & | & & | & & \\ \text{Cl} & & | & & \text{Cl} & & \\ \text{Cl} & & \text{O} & & \text{Cl} & & \\ | & & | & & | & & \\ \sim\text{P} & =\text{N}- & \text{P} & =\text{N}- & \text{P} & =\text{N}-\text{H} \\ | & & | & & | & & \\ \text{Cl} & & \text{Cl} & & \text{Cl} & & \end{matrix} + 2\,\text{HCl}$$

The Cl atoms of polymerising units of chain polymers $\left(\begin{array}{c} Cl \\ | \\ -P=N- \\ | \\ Cl \end{array}\right)_n$ can be replaced by other substituents such as OCH_3, OC_2H_5 etc. These substituted polyphosphazines are generally chain polymers. But, when $\left(\begin{array}{c} Cl \\ | \\ -P=N- \\ | \\ Cl \end{array}\right)_n$ reacts with ammonia it produces cross-linked polymers as depicted below.

$$\begin{array}{c} \begin{array}{ccccc} Cl & & Cl & & Cl \\ | & & | & & | \\ \sim P{=}N & - & P{=}N & - & P{=}N \sim \\ | & & | & & | \\ Cl & & Cl & & Cl \end{array} \\ \begin{array}{ccccc} Cl & & Cl & & Cl \\ | & & | & & | \\ \sim P{=}N & - & P{=}N & - & P{=}N \sim \\ | & & | & & | \\ Cl & & Cl & & Cl \end{array} \end{array} + 3NH_3 \longrightarrow \begin{array}{ccccc} Cl & & Cl & & Cl \\ | & & | & & | \\ \sim P{=}N & - & P{=}N & - & P{=}N \sim \\ | & & | & & | \\ Cl & & | & & Cl \\ & & NH & & \\ Cl & & | & & Cl \\ | & & | & & | \\ \sim P{=}N & - & P{=}N & - & P{=}N \sim \\ | & & | & & | \\ Cl & & Cl & & Cl \end{array} + 2NH_4Cl$$

Q. 4. Write a note on Polydimethoxy and Polydiethoxyphosphazines.

$$\left(\begin{array}{c} \mathbf{OCH_3} \\ | \\ \mathbf{-P=N-} \\ | \\ \mathbf{OCH_3} \end{array}\right)_n \qquad \left(\begin{array}{c} \mathbf{OC_2H_5} \\ | \\ \mathbf{-P=N-} \\ | \\ \mathbf{OC_2H_5} \end{array}\right)_n$$

Ans. Polydimethoxy and Polydiethoxyphosphazine can be prepared by reacting phosphos-nitrilic chloride with sodium methoxide and sodium ethoxide, respectively, when the chloride gets replaced by the alkoxy groups.

$$\left(\begin{array}{c} Cl \\ | \\ -P=N- \\ | \\ CH_3 \end{array}\right)_n + 2nCH_3ON \longrightarrow \left(\begin{array}{c} OCH_3 \\ | \\ -P=N- \\ | \\ OCH_3 \end{array}\right)_n + 2nNaCl$$

Polydimethoxy and Polydiethoxyphosphazines are colourless, transparent film-forming thermoplasts. They, however, degrade when heated above 100ºC, and get converted into cyclic $[P(OR)_2N]_3$ or $[P(OR)_2N]_4$ polymers where R is CH_3 or C_2H_5.

Q. 5. Write notes on *(a)* Polyphosphoryl chlorides, *(b)* Vitreous polyphosphates, *(c)* Maddrell's salt, *(d)* Kuroll's salts and *(e)* Polyphosphoryl acids.

Ans. *(a)* **Polyphosphoryl Chlorides.** These polymers have the general formula $P_nO_{2n-1}Cl_{n+2}$ and the general structure is represented below :

$$\begin{array}{ccccccc} Cl & & Cl & & Cl & & Cl \\ | & & | & & | & & | \\ \sim P & -O- & P & -O- & P & -O- & P-Cl \\ \| & & \| & & \| & & \| \\ O & & O & & O & & O \end{array}$$

Preparation. Polyphosphoryl chlorides are prepared by reducing phosphorus oxychloride with phosphorus pentaoxide.

$$(n+2)\ POCl_3 + (n-1)\ P_2O_5 \longrightarrow 3\ P_nO_{2n-1}Cl_{n+2}$$

Properties. The nature of the product depends upon the Cl/P ratio. The polymer obtained at Cl/P ratio of 1:1 is a clear amorphous mass exhibiting considerable easticity. It can be stretched to a large extent without breaking.

Polyphosphoryl chlorides are strongly acidic being similar in properties to acid chlorides. In air, they give fumes at HCl due to their interaction with moisture. They act as good chlorinating agents in organic reactions.

(*b*) **Vitreous Polyphosphates (Phosphate Glasses).** In vitreous polyphosphates, the charges on phosphate anions are neutralised by the charges on cations like Na^+, K^+, etc.

```
~~O     O       O       O~~
   \   / \     / \     /
    P       P       P
   / ||    / ||    / ||
 O⁻  O   O⁻  O   O⁻  O
 Na⁺     Na⁺     Na⁺
```

Vitreous sodium polyphosphate can be readily prepared by heating $Na_4P_2O_7$ or Na_3PO_4 with P_2O_5.

The polyphosphate anions in these polymers have long straight chains. But, on an average, one P atom out of every thousand P atoms is a part of a branching unit, thereby forming a partial cross-link between the chains. These cross-links, however, disappear within 12 hours when the polymer is dissolved in water or in a neutral aqueous salt solution.

(*c*) **Maddrell's Salts.** Sodium polymetaphosphates $NaPO_3$–II and $NaP0_3$-III are known as Maddrell's salts. These salts are slightly soluble in water and in saline solutions at room temperature. But, at elevated temperatures and in the presence of salts of alkali metal ions (other than Na^+ ions) and ammonium ions in solution, these metaphosphates dissolve at a higher rate. Also $NaPO_3$-II dissolves more readily than $NaPO_3$-III under the conditions mentioned above. The potassium Maddrell's salts KPO_3-II and KPO_3-III behave in a similar manner.

The X-ray studies on Maddrell's salt, $NaPO_3$-II, reveal that the polymer consists of long – P– O – P – chains made up of interconnected PO_4 groups. However, the pattern repeats itself after every three units and the chains do not spiral.

(*d*) **Kuroll's Salts.** The polymetaphosphates Na (or K) PO_3-IV are known as Kuroll's salts. These are more soluble than the corresponding Maddrell's salts in pure water. If a Kuroll's salt is kept in ice cold distilled water for a few days or in hot water for a few hours, its crystals swell and become gummy. Eventually they dissolve completely to give a highly viscous solution. The dissolution of sodium Kuroll's salt is heeped up by the presence of salts having cations other than Na^+ ions. The potassium Kuroll's salt behaves in a similar manner.

X-ray studies on rubidium Kuroll's salt reveal that its structure consists of – P – O – P –chains running in the direction of fibrous cleavage and consisting of interconnected PO_4 groups. Each unit cell contains a portion of two different chains which spiral in opposite directions around the screw axis of the crystal. The pattern repeats itself after every two P atoms in the chains. Thus, there are four units of $RbPO_3$ per unit cell. The P – O distances within the chains are 1.62 ± 0.03Å and the angles O – P – O and P – O – P in the chains are 99.5° and 129°, respectively.

(*e*) **Polyphosphoric Acids.** The polyphosphoric acids consist of long straight chain of $-p-o-p-$ units as shown.

$$\sim P(=O)(OH)-O-P(=O)(OH)-O-P(=O)(OH)-O\sim$$

The H_2O/P_2O_5 molar ratio of these amorphous solid is 1. These polyacids when put in sealed containers start flowing slowly after a few days and assume the shape of the container. The amorphous polyacids are tough and not rubber-like because of hydrogen bonding in the chains. They are highly hygroscopic and hydrolyse rapidly.

Q. 6. Describe phosphorous-based network polymers.

Ans. Various phosphorous-based network polymers are described as under :

1. Polymer Phosphorous Pentoxide. Polymer phosphorous pentoxide exists in three crystalline forms. Two of them are sheet polymers and the third is glassy material having a three-dimensional network structure.

All the three forms contain P – O – P linkages. One sheet form consists of large rings containing ten P and ten O atoms each while the other sheet form consists of rings containing six P and six O atoms each.

In all the three forms of polyphorous pentaoxide, each P atom is tetrahedrally surrounded by four oxygen atoms, three oxygen being shared with adjacent PO_4 tetrahedral.

2. Polyorthophosphates of Boron, Aluminium, Silver and Iron. The crystal structure of boron polyphosphate is similar to that of silica. The P – O distance is 1.55 Å and B – O distance is 1.44 Å. A -two dimensional picture of the structure is given below.

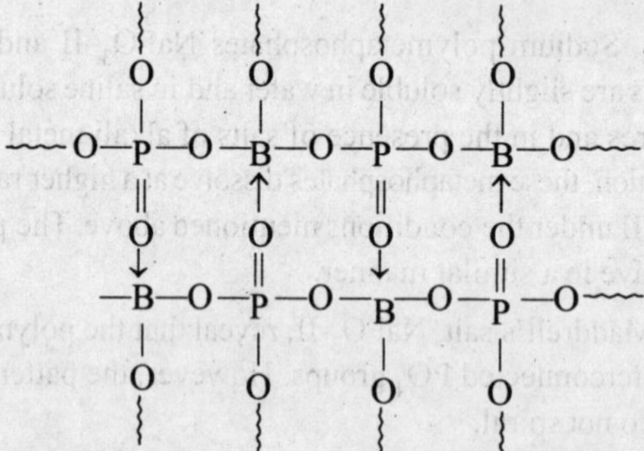

Polymeric aluminium orthophosphate $(AlPO_4)_n$ is similar in structure to polymeric boron orthophosphate. Like SiO_2, the polymeric aluminium phosphate also exists in several crystalline modifications.

Polymeric silver phosphate $(Ag_3PO_4)_n$ is a yellow crystalline substance in which each Ag atom is surrounded by four O atoms (Ag – O distance is 2.34 A). In addition to these O atoms, each Ag atom has two other Ag atoms as its neighbours.

3. Sulphur-Phosphorous Polymers. The liquid sulphur at 160 – 250°C, when rapidly cooled, gets transformed into highly elastomeric form Sμ. This form contains long chains of S atoms.

The elastic properties of Sμ can be changed by the addition of a small percentage of P. P vulcanises elastic Sμ in about the same way as S vulcanises natural rubber, *i.e.*, through cross-linking. Ten or higher percentage of P increases the cross-linking of Sμ to such an extent that the resulting material loses elastomeric properties almost completely.

4. Ultraphosphate Glasses. These are amorphous polyphosphates in which the molar ratio of phosphate anion/metal cation is greater than unity. They can be considered to be derived from P_2O_5 polymer by substitution of a portion of P – O – P bonds by metal cations and hydroxy groups or from linear metaphosphates $(MPO_3)_x$ by pairs of M^+ ions by an oxygen cross-link.

The ultraphosphate glasses resemble the glassy organic plastic in processability.

Amongst the alkali metal ultraphosphates, sodium ultraphosphate is prepared by heating a mixture of NaH_2PO_4 and H_3PO_4 or a mixture of NaH_2PO_4 and $NH_4H_2PO_4$. Thus, when NaH_2PO_4 and $NH_4H_2PO_4$ are heated together, polycondensation occurs to form linear polyphosphate.

$$n\mathrm{NaH_2PO_4} + 2n\mathrm{NH_4H_2PO_4} \xrightarrow{400\text{–}500^\circ\mathrm{C}} \left(\mathrm{O{-}P(=O)(O^-Na^+){-}O{-}P(=O)(OH){-}O{-}P(=O)(HO)} \right)_n + 2n\,\mathrm{NH_3} + 3n\mathrm{H_2O}$$

Further heating of $(NaH_2P_3O_9)_n$ at still higher temperatures produces cross-links by removing H_2O molecules from hydroxy groups of different polyphosphate chains to yield sodium ultraphosphate, as illustrated below :

$$\begin{array}{c} \mathrm{{\sim}P(=O)(O^-Na^+){-}O{-}P(=O)(OH){-}O{-}P(=O)(OH){\sim}} \\ \mathrm{{\sim}P(=O)(O^-Na^+){-}O{-}P(=O)(OH){-}O{-}P(=O)(OH){\sim}} \end{array} \xrightarrow{-2\mathrm{H_2O}} \text{cross-linked chains (P–O–P links between chains)}$$

5. Borophosphate Glasses. The borophosphate glasses are produced by melting together H_3PO_4, boric oxide and appropriate quantities of alkali metal carbonates or oxides. The optimum temperature for polycondensation is around 700°C.

The structure of borophosphate glasses with boron atoms occupying three coordination sites and four coordination sites are shown below :

Borophosphate glass with three coordinated boron atoms.

Borophosphate glass with four coordinated boron atoms.

The addition of 5 mole percent B_2O_3 to an ultraphosphate polymer with composition : Li_2O 10 mole %, Na_2O 10 mole % and BaO 5 mole %, increases the durability of the resulting polymer towards running water. A number of multicomponent borophosphate glasses have been prepared which are as durable as ordinary silica glass at similar temperatures. For instance, the rate of dissolution in running water of borophosphate glass with composition : P_2O_5 55 mole %, B_2O_3 5 mole %, MgO 9 mole %, CaO 9 mole %, Li_2O 7 mole % and Na_2O 15 mole %, is as low as 5×10^{-6} mole per year at 20°C. This rate is the same as that of a reasonably durable silica glass. However, the rate of attack by water on borophosphate glasses increases with increase in temperature much more rapidly than on silica glass.

The water vapours on borophosphate glasses condense in the form of a thin layer and not in the form of minute water droplets. The visibility through borophosphate glasses is thus not impaired. This is the reason why these glasses are ideal for the manufacture of spectacles and other lenses.

Applications of Phosphorous-Based Polymers

1. The borophosphate glasses are used for manufacturing optical lenses.

2. The potassium Kuroll's salt is employed in the manufacture of sausages to prevent water bars.

3. The long chain vitreous sodium phosphate polymers are employed in food processing industry. Their use in this industry is based on the interaction of phosphate with properties.

4. The short chain vitreous phosphate polymers are used for deflocculation of proteins.

Q.7. Describe the following inorganic polymers :

(*a*) Polymeric sulphur, (*b*) Polymeric sulphur nitride.

Ans. (*a*) Polymeric Sulphur. It is prepared by heating rhombic sulphur (S_8) to 165° – 180°C and then suddenly cooling the melt in ice bath. The cyclic S_8 structure of rhombic sulphur opens up to produce $\cdot S - S_6 - S\cdot$ radical which recombines with S_8 unit to produce $\cdot S - S_{14} - S\cdot$ radical which further combines with S_8 unit and so on. This is depicted as under :

$$\cdot S - S_6 - S\cdot + S_8 \xrightleftharpoons{165 - 180°C} \cdot S - S_{14} - S\cdot$$

$$\cdot S - S_{14} - S\cdot + S_8 \rightleftharpoons \cdot S - S_{22} - S\cdot \text{ and so on.}$$

The final product is washed with carbon disulphide to remove the traces of S_8 rhombic sulphur.

The pure polymeric sulphur has a glass transition temperature of 75°C. Above this temperature, the movement of polymeric sulphur chains involves the dissociation of S – S bonds, as shown below

$$\sim S - S\cdot + \begin{matrix} S - S \sim \\ \vdots \\ S - S \sim \end{matrix} \longrightarrow \begin{matrix} \sim S - S - S - S \sim \\ + \cdot S - S \sim \end{matrix}$$

At temperatures below the melting point of sulphur, *viz*., 114°C, the polymeric sulphur is unstable and reverts back to rhombic sulphur, S_8.

(*b*) Polymeric Sulphur Nitride $(SN)_n$. Pure $(SN)_n$ can be preapred by passing S_4N_4 vapour over silver wool at 220°C. The gaseous S_2N_2 formed is condensed to solid state by passing through liquid nitrogen. It is then sublimed into an ice-cooled trap from which it is slowly allowed to warm up to the room temperature. The S_2N_2 during this process of slow warming gets polymerised to $(SN)_n$.

Polymeric sulphur nitride consists of layers of fibres stacked parallel to one another. The polymer shows metallic conductivity and metallic ductility. It is soft and malleable. The polymer is slowly attacked by moisture when exposed for several months. It decomposes in vacuum at 140°C to give sulphur and nitrogen.

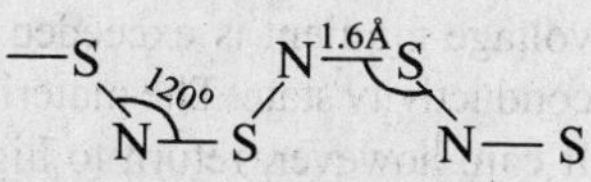

The structure of the polymer comprises nearly planar chains of S – N bonds of approximately the same length (1.6 Å).

The electrical conductivity of the polymer $(SN)_n$ is in the range of 1200 –3700 S cm^{-1} and increases 50-200 times when the temperature is lowered to 4.2 K. It has been found recently that the polymer becomes a superconductor at 0.26 K. The polymer is the first example of a non-metallic superconductor.

Q. 8. Describe chalcogenide glasses. How does switching take place in these polymers ?

Ans. The chalcogenide glasses are a family of amorphous cross-linked polymers formed by components of chalcogens (S, Se, Te) with one of the elements such as antimony, arsenic, bismuth, germanium, indium, phosphorus, tin and thalium. Best known example of two component chalcogenide glass is $(As_2S_3)_n$.

Chalcogenide polymers are generally prepared by the fusion of the elements under conditions selected for minimising oxidation and loss of volatilisation of the components.

The X-ray studies reveal that chalcogenide glasses consist of short chains of chalcogen atoms linked together by polyvalent atoms which serve as branching and cross-linking sites. A typical As-S polymer network is as shown.

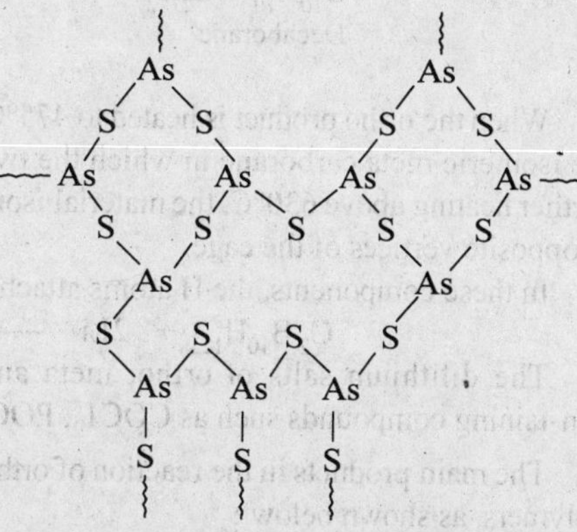

These polymers resemble organic polymers. Compared to inorganic polymers such as ultra-phosphate and borophosphate glasses, these polymers have lower softening points, higher refractive indices and lower tensile strengths.

Chalcogenide glasses are fairly stable to acids but are attacked by concentrated alkalies. Most of the polymers begin to oxidise in air at about 3000°C. But some of these polymers can be distilled in vacuum without showing any sign of decomposition – a property most unlikely of polymers. This happens because of the dissolution of polymeric network into small fragments in vapour phase in vacuum which recombine spontaneously when the vapours condense. In the range of 5-15 mole % As and 3-8 mole % Ge, the chalcogenide glasses show distinct rubbery properties although over a small range of temperature. In this respect the chalcogenide polymers differ markedly from the oxide glasses (*e.g.*, borate, phosphate and silica glasses) which do not exhibit rubber-like properties in any temperature range.

Chalcogenide glasses are deeply coloured. However, most of them are good infrared transmitting materials and have been used for the fabrication of IR windows for army and civil optical devices.

Switching Phenomenon in Chalcogenide Glasses. The chalcogenide glasses undergo reversible change from low conductivity to high conductivity state under applied voltage. This phenomenon is known as **'switching'**. It can occur in two forms as discussed below.

In **threshold switching,** the chalcogenide glass maintains a high resistance until a certain threshold voltage gradient (*i.e.,* 10^4 volts/cm) is reached. At this voltage gradient the chalcogenide glass changes suddenly into a low resistance state which has about a thousand-fold greater conductivity than the state prior to the application of threshold voltage. The chalcogenide glass continues to remain in this high conductivity state as long as a certain minimum current is maintained. If the current falls below the holding current value, the chalcogenide glass returns to the high resistance state once again.

In the **memory switching,** the chalcogenide glass exhibits a high resistance until a threshold voltage gradient is exceeded. At this stage, the chalcogenide glass suddenly changes into a high conductivity state. The material then remains in this condition even after the current is switched off. It can, however, return to high resistance state by passing a short current pulse that exceeds the original threshold voltage gradient.

The switching phenomenon is probably due to the fact that the current of high voltage causes decomposition of bonds in the chalcogenide glass so that when the voltage gradient becomes high enough, a conducting path within the glass is produced.

The chalcogenide glasses have a great potential for their use in ultrasonic delay lines, high energy particle detector multipliers, memory devices for computers.

Q.9. Discuss the characteristics of polycarboranes.

Ans. When acetylene is added to decaborane, $B_{10}H_{10}$, it forms a closed icosahedral molecule $C_2B_{10}H_{12}$ with a diameter of about 40 Å in which the two carbon atoms are adjacent to each other. This product is named orthodicarboclosododecacarborane.

$$\underset{\text{Decaborane}}{B_{10}H_{10}} + C_2H_2 \longrightarrow \underset{\text{Orthodicarbacloso dodecacarborane}}{C_2B_{10}H_{12}}$$

When the ortho product is heated to 475°C, it undergoes an intramolecular rearrangement to give the isomeric meta carborane in which the two carbon atoms are separated by one boron atom. On further heating above 630°C, the material isomerises to para carborane in which the carbon atoms lie at opposite vertices of the cage.

In these components, the H atoms attached to C atoms can be easily replaced by Li atoms :

$$C_2B_{10}H_{12} + 2Li \longrightarrow C_2B_{10}H_{10}Li_2 + H_2$$

The dilithium salts of ortho, meta and para carboranes then easily react with halogen-con-taining compounds such as $COCl_2$, $POCl_3$ etc., to give cyclic or linear polycarboranes.

The main products in the reaction of ortho $C_2B_{10}H_{10}Li_2$ with $COCl_2$ and $POCl_3$ are called cyclic polymers, as shown below :

The lithium derivatives of meta and para carboranes react with $(CH_3)_2SnCl_2$ to give crystalline polymers of reasonable chain lengths containing 20-30 repeat units.

$$2B_{10}H_{10}C_2Li_2 \xrightarrow{2COCl_2} OC\begin{matrix} \diagup CB_{10}H_{10}C \diagdown \\ \diagdown CB_{10}H_{10}C \diagup \end{matrix}CO + 4LiCl$$

$$2B_{10}H_{10}C_2Li_2 \xrightarrow{2POCl_3} ClPO\begin{matrix} \diagup CB_{10}H_{10}C \diagdown \\ \diagdown CB_{10}H_{10}C \diagup \end{matrix}OPCl + 4LiCl$$

$$nLi_2C_2B_{10}H_{10} + n\,(CH_3)_2SnCl_2 \longrightarrow (-CB_{10}H_{10}CSn(CH_3)_2)_n- + 2nLiCl$$

The most interesting polycarboranes are those in which the carboranes are bound by polysiloxane bridges. If meta $Li_2C_2B_{10}H_{10}$ is made to react with $(CH_3)_2SiC_2$, it produces $ClSi(CH_3)_2C(B_{10}H_{10}CSi(CH_3)_2Cl$, as shown below :

$$\underset{\text{III (meta)}}{LiCB_{10}H_{10}CLi} + 2\,(CH_3)_2SiCl_2 \longrightarrow \underset{\text{IV (meta)}}{ClSi(CH_3)_2C(B_{10}H_{10})CSi(CH_3)_2Cl} + 2LiCl$$

The product IV is then reacted with methanol to give the methoxy derivative :

$$\text{IV (meta)} + CH_3OH \longrightarrow \underset{\text{V (meta)}}{CH_3OSi(CH_3)_2C(B_{10}H_{10})CSi(CH_3)_2OCH_3}$$

Compounds IV and V then condense at 140°C in the presence of ferric chloride (catalyst) to yield the long chain polycarborane :

$$\underset{\text{IV}}{n\text{Cl}-\underset{\text{CH}_3}{\overset{\text{CH}_3}{\text{Si}}}-\text{C}(\text{B}_{10}\text{H}_{10})\text{C}-\underset{\text{CH}_3}{\overset{\text{CH}_3}{\text{Si}}}-\text{Cl}} + \underset{\text{V}}{n\text{CH}_3\text{O}-\underset{\text{CH}_3}{\overset{\text{CH}_3}{\text{Si}}}-\text{C}(\text{B}_{10}\text{H}_{10})\text{C}-\underset{\text{CH}_3}{\overset{\text{CH}_3}{\text{Si}}}-\text{OCH}_3}$$

$$\Big\downarrow \text{Condensation} \; -n\text{CH}_3\text{Cl}$$

$$\sim\underset{\text{CH}_3}{\overset{\text{CH}_3}{\text{Si}}}-\text{O}-\underset{\text{CH}_3}{\overset{\text{CH}_3}{\text{Si}}}-\text{C}(\text{B}_{10}\text{H}_{10})\text{C}-\underset{\text{CH}_3}{\overset{\text{CH}_3}{\text{Si}}}-\text{O}-\underset{\text{CH}_3}{\overset{\text{CH}_3}{\text{Si}}}-\text{C}(\text{B}_{10}\text{H}_{10})\text{C}-\underset{\text{CH}_3}{\overset{\text{CH}_3}{\text{Si}}}-\text{O}-\underset{\text{CH}_3}{\overset{\text{CH}_3}{\text{Si}}}-\text{O}-\underset{\text{CH}_3}{\overset{\text{CH}_3}{\text{Si}}}\sim$$

VI

long chain polymer

All polycarborane containing Si – O – Si linkages are partially crystalline elastomers whose glass transition temperatures decrease with increasing number of Si atoms in the linking groups. To make these elastomers useful as rubbers, a small quantity of 1-vinyl 2-methyldichloro 1,2 dicarbaclosododecacarborane is added to the condensation reaction. The resulting elastomer is cross-linked with peroxide initiators. The properties of the cross-linked material as rubber can be further improved by compounding it with finely divided silica fillers.

Q.10. Discuss the characteristics of polymeric boron nitride.

Ans. The polymeric boron nitride, $(BN)_n$, exists in two forms, *viz.*, as a layer polymer and as a three-dimensional network polymer.

A. The Layer Polymer. It can be obtained by a variety of methods. A common method involves heating of boric acid with urea in an atmosphere of nitrogen at 600°C followed by the treatment of the product with ammonia at 950°C. The fmal product is an amorphous powder.

The structure of boron nitride resembles that of graphite. It consists of infinite sheets of six-membered rings of alternating boron and nitrogen atoms stacked in layers. There is, however, an important difference between the structure of polymeric boron nitride and graphite. In boron nitride,

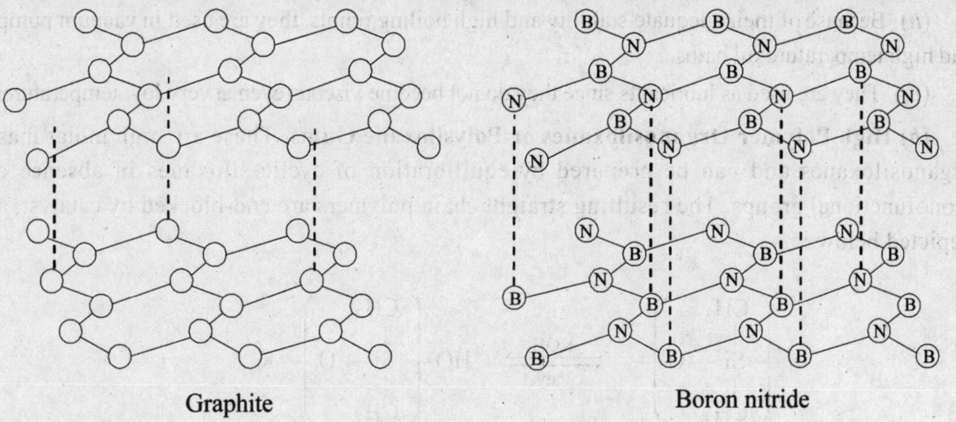

Fig. 29.1 Layer structures of graphite and boron nitride.

the sheets are stacked over one another in such a manner that the atoms in each successive layer are directly superimposed, whereas in graphite, the atoms in one layer lie opposite to the centres of the six-membered rings in the next layer, as shown in the figure.29.1.

Unlike graphite, the polymeric boron nitride is an electrical insulator. This is because the nitrogen atom in one layer of $(BN)_n$ is always adjacent to a boron atom so that the lone pair of electrons on each N atom in a layer is bound to the electron-deficient B atom of the adjacent layer. In other words, the electrons in boron nitride are highly localised making the polymer non-conducting.

Boron nitride is soft and can be easily machined. The articles fabricated from this polymer, either by sintering or by machining, can be safely exposed to atmosphere up to a temperature of about 800°C.

B. The Three-Dimensional Network Polymer. Layer form of polymeric $(BN)_n$ at 1500-2000°C under high pressure in the presence of catalyst gets converted into a three-dimensional network polymeric form. This form is isomorphic with diamond and is comparable with the latter in hardness. It is, therefore, used as a substitute for diamond in jewellery as also for the fabrication of cutting tools.

Q.11. Describe the following :

(*a*) Liquid polysiloxanes, (*b*) Polysiloxane gums.

Ans. (*a*) Liquid Polysiloxanes. These are low molar mass straight chain polymers which are obtained by the equilibration of cyclopolysiloxanes with disiloxane, as shown by the following reaction :

$$3\,[(CH_3)_2SiO]_4 \;(\text{cyclic tetramer}) + H_3C-\overset{CH_3}{\underset{CH_3}{Si}}-O-\overset{CH_3}{\underset{CH_3}{Si}}-CH_3 \xrightleftharpoons{H^+} H_3C-\overset{CH_3}{\underset{CH_3}{Si}}-O\left(\overset{CH_3}{\underset{CH_3}{Si}}-O\right)_{12}\overset{CH_3}{\underset{CH_3}{Si}}-CH_3$$

These polysiloxanes can also be prepared by the hydrolysis of dichlorosilanes.

Uses. (*i*) Liquid polysiloxanes are used as insulators in electrical appliances.

(*ii*) Because of their adequate stability and high boiling points, they are used in vacuum pumps and high temperature oil baths.

(*iii*) They are used as lubricants since they do not become viscous even at very low temperatures.

(*b*) High Polymer Organosiloxanes or Polysiloxane Gums. These are high molar mass organosiloxanes and can be prepared by equilibration of cyclic siloxanes in absence of monofunctional groups. The resulting straight chain polymers are end-blocked by catalyst, as depicted below :

$$n\left(\overset{CH_3}{\underset{CH_3}{Si}}-O\right)_4 \underset{\text{catalyst}}{\overset{KOH}{\rightleftharpoons}} HO\left(\overset{CH_3}{\underset{CH_3}{Si}}-O\right)_{4n}K$$

Cyclictetramer   Straight chain Polymer

The reaction take place through the following steps :

$$(i)\quad \left[\begin{array}{c} CH_3 \\ | \\ -Si-O- \\ | \\ CH_3 \end{array}\right]_4 \xrightleftharpoons{KOH} HO\left[\begin{array}{c} CH_3 \\ | \\ Si-O \\ | \\ CH_3 \end{array}\right]_3 \begin{array}{c} CH_3 \\ | \\ Si-O^-K^+ \\ | \\ CH_3 \end{array}$$

$$(ii)\quad HO\left[\begin{array}{c} CH_3 \\ | \\ Si-O \\ | \\ CH_3 \end{array}\right]_3 \begin{array}{c} CH_3 \\ | \\ Si-O^-K^+ \\ | \\ CH_3 \end{array} + \left[\begin{array}{c} CH_3 \\ | \\ -Si-O- \\ | \\ CH_3 \end{array}\right]_4 \rightleftharpoons HO\left[\begin{array}{c} CH_3 \\ | \\ Si-O \\ | \\ CH_3 \end{array}\right]_7 \begin{array}{c} CH_3 \\ | \\ Si-O^-K^+ \\ | \\ CH_3 \end{array}$$

$$\Big\Updownarrow \left[\begin{array}{c} CH_3 \\ | \\ -Si-O- \\ | \\ CH_3 \end{array}\right]_4$$

$$HO\left[\begin{array}{c} CH_3 \\ | \\ Si-O \\ | \\ CH_3 \end{array}\right]_{15} \begin{array}{c} CH_3 \\ | \\ Si-O^-K^+ \\ | \\ CH_3 \end{array} \overset{\left[\begin{array}{c} CH_3 \\ | \\ -Si-O- \\ | \\ CH_3 \end{array}\right]_4}{\rightleftharpoons} HO\left[\begin{array}{c} CH_3 \\ | \\ Si-O \\ | \\ CH_3 \end{array}\right]_{11} \begin{array}{c} CH_3 \\ | \\ Si-O^-K^+ \\ | \\ CH_3 \end{array}$$

Kinetic studies prove that the rate of the polymerisation reaction is related to the square root of the concentration of the cyclic tetramer. It is also related to the ring strain in the cyclic polymer taken for equilibration. Thus, cyclic hexamer is more easily equilibrated than cyclic tetramer.

Uses. Polysiloxane gums are used as lubricants and as antifoams. They are also used in textile finishing.

Q.12. Describe the preparation and uses of (*a*) Silicon Rubber, (*b*) Silicon Resins.

(Lucknow, 2010, Garhwal, 2010)

Ans. (*a*) Silicon Rubber or Elastomers. Preparation of silicon rubbers involves two steps, *viz.*, compounding and cross-linking.

Compounding. It consists of milling of polysiloxane gum, a filler and a cross-linking agent together with certain additives. A typical formulation comprises 100 parts of benzene-soluble polysiloxane gum, 20–50 parts of SiO_2 as filler, 6 parts of benzoyl peroxide as vulcanising agent and suitable amounts of additives.

Cross-Linking. This step involves the cross-linking which connects polymer molecules with one another so that it becomes an elastomer of desired properties.

The polysiloxane gum required for making silicon rubber should be a linear polymer of high molar mass since the tensile strength of the elastomer increases with increase in the size of the gum polymer. Fillers are added to reinforce the polysiloxane gum which, in itself, is soft and weak. Silica, which is used as a filler, is in a very fine state.

Cross-linking is achieved by the addition of reagent like benzoyl peroxide. The vulcanising temperature ranges between 120° and 130° C.

The mechanism of cross-linking step is explained as under :

$$\left(\underset{\text{Benzol perodide}}{Ph-\overset{O}{\overset{\|}{C}}-O}\right)_2 \longrightarrow 2\,Ph-\overset{O}{\overset{\|}{C}}-O\bullet$$

$$Ph-\overset{O}{\overset{\|}{C}}-O\bullet + \left[\begin{array}{c} CH_3 \\ | \\ Si \\ | \\ CH_3 \end{array} -O\right]_3 \longrightarrow -\begin{array}{c} CH_3 \\ | \\ Si \\ | \\ \bullet CH_2 \end{array}-O-\left[\begin{array}{c} CH_3 \\ | \\ Si \\ | \\ CH_3 \end{array} -O\right]_{n+1} + PhCOOH$$

$$-\begin{array}{c} CH_3 \\ | \\ Si \\ | \\ \bullet CH_3 \end{array}-O-\left[\begin{array}{c} CH_3 \\ | \\ Si \\ | \\ CH_3 \end{array} -O\right]_{n-1}^{\bullet} + \left[\begin{array}{c} CH_3 \\ | \\ Si \\ | \\ CH_3 \end{array} -O\right]_m \xrightarrow[-PhCOOH]{Ph-\overset{O}{\overset{\|}{C}}-O\bullet} \begin{array}{l} \begin{array}{c} CH_3 \\ | \\ Si \\ | \\ CH_2 \end{array}-O-\left[\begin{array}{c} CH_3 \\ | \\ Si \\ | \\ CH_3 \end{array} -O\right]_{n-1}^{\bullet}- \\ \begin{array}{c} CH_2 \\ | \\ Si \\ | \\ CH_3 \end{array}-O-\left[\begin{array}{c} CH_3 \\ | \\ Si \\ | \\ CH_3 \end{array} -O\right]_{m-1}^{\bullet} \end{array}$$

Uses. Silicon rubbers are extremely useful materials on account of their high thermal stability, low temperature performance, chemical stability, water repelling power and flexibility.

(*b*) **Silicon Resins.** Silicon resins possess ring structures and have a much higher cross-link density than silicon elastomers. In a stretched elastomer, the molecular chains uncoil through rotation about the - Si— O— bonds. In a resin, such rotation is highly restricted and the response to stress occurs largely through bending and stretching of bonds. The properties of resins are dependent upon R/Si ratio where R is the alkyl group. As the R/Si ratio is lowered from 2.0 to 1.0, the polymers become progressively less fluid, less soluble and less fusible. Most of the resins have R/Si ratio between 1.00 and 1.6 and their properties are intermediate between those of fluid silicones and those expected of a highly cross-linked structure with R/Si ratio equal to 1.00.

Preparation. For the preparation of silicon resins, a mixture of chlorosilones such as CH_3SiCl_3, $(CH_3)_2SiCl_2$, $C_6H_5SiCl_3$ $(C_6H_5)_2SiCl_2$ are first hydrolysed. During hydrolysis about 90 per cent of silanol groups, Si – OH, condense to form siloxane linkages. In this process of random condensation, some ring closure invariably occurs because silanol groups condense more slowly and with increasing difficulty.

After hydrolysis the product is shaken with an organic solvent and the organic layer is separated. This organic layer is then washed thoroughly to remove the Cl^- ions in order to check the reversibility of the hydrolysis.

The silanol groups (*i.e.*, Si – OH) left uncondensed in the finished resin impart typical properties to the resin. These groups contribute towards viscosity of the product which is an important property that determines the usefulness of resin. For instance, the viscosity of resin used for laminating purposes should be relatively low so that it may impregnate the glass fibres better. The silanol groups also determine the shelf life of the resin.

The nature of the substituents of Si alters the characteristics of the resin. For example, while CH_3 groups give flexibility, C_2H_5, C_3H_7, etc. groups make the resin softer, increase its solubility in organic solvents and increase its water-repellence. The phenyl groups increase resistance towards heat and so on.

Copolymeric resins of desired properties can be obtained by condensing two polymers, as shown below :

$$\sim\!\!\sim Si-OH \;+\; \left(C_6H_4 \begin{array}{l} -\overset{O}{\overset{\|}{C}}- \\ -\underset{O}{\underset{\|}{C}}-OCH_2\underset{OH}{\underset{|}{C}H}CH_2-O- \end{array} \right)_n \xrightarrow[-H_2O]{\text{Condensation}} \left(C_6H_4 \begin{array}{l} -\overset{O}{\overset{\|}{C}}- \\ -\underset{O}{\underset{\|}{C}}-CH_2\underset{\underset{\sim\sim Si \sim\sim}{|}}{\underset{O}{\underset{|}{C}H}}CH_2-O- \end{array} \right)_n$$

Silane polymer of Resin

The silicon resins can be copolymerised with phenolic polyester, epoxy cellulosic or other functional resins. The properties of these copolymeric resins depend upon the nature of the condensing polymers and the degree of condensation.

Uses. Silicon resins are used as electrical insulators. They are also used for lamination purposes and for coating cooking pans, etc.

ROLE OF ORGANIC AND INORGANIC REAGENTS IN INORGANIC ANALYSIS

Q. 1. What are the common features of organic reagents?

Ans. Organic reagents have the following general features:

1. Organic reagents form coloured precipitate with a number of cations.

2. These reagents import colour to the solution at the end point in a titration.

3. Extraction in organic solvent can be made at controlled pH.

4. They get either oxidised or reduced to other valency state (oxidation state) in a chemical test.

5. Organic reagents also act as masking reagent.

6. They exhibit different volatility.

Q. 2. What are the advantages of using organic reagents?

Ans. Advantages of using organic reagents are listed below:

1. These reagents are more *sensitive* and are capable of detecting ions in micrograms.

2. These are more *specific* than the conventional reagents. Thus, these can be used to detect a single type of ion in a complex mixture.

3. These reagents are suitable for quantitative analysis as well. Precipitates given by these reagents are stable and insoluble in water at a particular pH and can be quantitatively separated.

4. Those organic reagents which give coloured soluble products show a gradation in colour intensity with concentration of the ions under test. Thus, these reagents can be used in colorimetric experiments.

Q. 3. Discuss advantages and disadvantages of organic reagents over inorganic reagents.

Ans. Advantages

1. By adjusting the pH of the solution, it is possible to affect precipitation of one constituent alone. By changing the pH condition, the other constituent can be precipitated. This does not happen generally with inorganic reagents.

2. Organic reagents are very specific in nature. For example, mixtures of cobalt and nickel, aluminium and iron and cadmium and copper can be separated easily using organic reagents.

3. It is generally easier to dry precipitates at lower temperature because of lower affinity for water. This helps prevent decomposition of the precipitate.

4. Co-precipitation is checked. For example, co-precipitation of Na and K along with Mg takes place when we use the inorganic phosphate or oxalate reagent. But this is checked when the organic reagent 8-hydroxyquinoline is used.

Disadvantages

1. It is very difficult to obtain an organic reagent of high degree of purity. If there is a contamination of another substance, the specificity of the reagent is ruled out. For example, dimethyl glyoxime is a specific reagent for nickel. But, it is the α-form of dimethy1glyoxime which is specific for the analysis of nickel. During preparation of dimethyl glyoxime, β-and γ - forms are also obtained which need to be removed from the mixture, so that it can work satisfactorily.

2. Some of the metallic complexes which are precipitated in analytical chemistry are volatile above 130 – 150°C and decompose above this temperature. The drying of the precipitate has to be done below 130°C to prevent decomposition. This is a limitation of using organic reagents.

3. Another disadvantage of using organic reagents is that they have a low solubility in water.

Q. 4. Give the analytical applications of oxine, C_9H_7ON.

Ans. Oxine has the structure as given below.

It was introduced by Halm and Berg as analytical reagent in the year 1927.

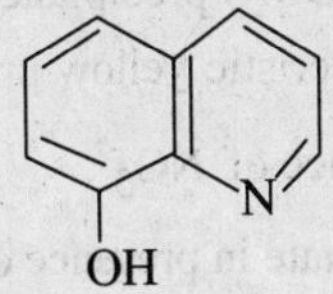

[Structure of oxine: quinoline ring with N and OH]

Although oxine can work as a reagent for quantitative analysis of Mg, Zn, Cu, Cd, Pb, Al, Fe, B, Ga, Th and Zr, it is best suited for Mg and Al. The divalent metal like Zn forms a complex with oxine as shown below.

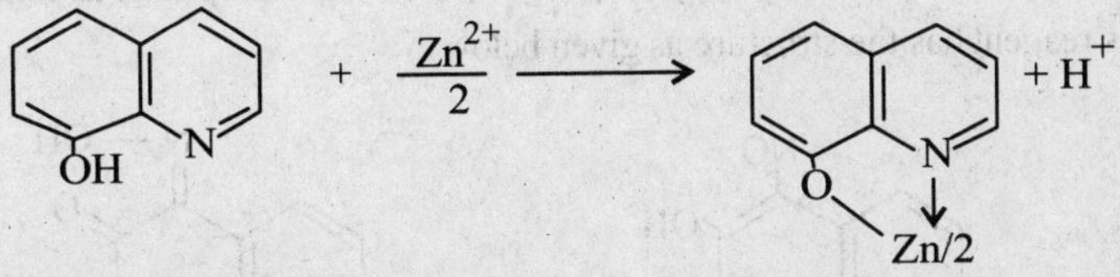

$$\text{Oxine (OH, N)} + \frac{Zn^{2+}}{2} \longrightarrow \text{(O–Zn/2 ← N complex)} + H^+$$

1. The **gravimetric** estimation of the ion is carried out by treating the solution of the ion with oxine. The complex formed is filtered through sintered crucible and dried at 100°C to remove water. Thus, the mass of the metal-oxine complex obtained from a known quantity of the sample is obtained. Concentration of the metal in the sample is then calculated.

2. In **volumetric** estimation, the precipitate of the complex is dissolved in hydrochloric acid and the resulting solution is titrated against standard potassium bromate-bromide solution.

3. In spectroscopic method, metal ion is estimated by measuring optical density of the complex. A 2% solution of oxine in 2N HCl is used for the estimation.

Q. 5. Give the analytical application of dimethyl glyoxime.

Ans. Dimethyl glyoxime has the structure as given below.

$$\begin{array}{l} \qquad\quad\ \ OH \\ \qquad\quad\ \ | \\ CH_3-C=N \\ \qquad\quad\ \ | \\ CH_3-C=N \\ \qquad\quad\ \ | \\ \qquad\quad\ \ OH \end{array}$$

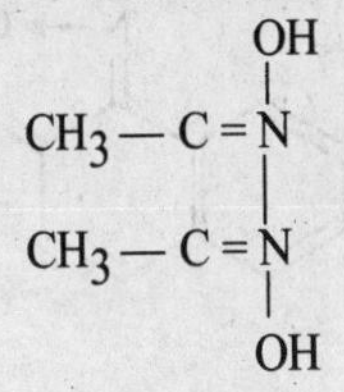

This reagent was discovered by L. Tschngaff in the year 1905. It was used by O. Brunck in 1907 for the determination of nickel in steel. We obtain a red chelate compound when the reagent reacts with Ni^{2+} in ammoniacal solution.

$$2\ \begin{matrix} OH \\ | \\ CH_3-C=N \\ | \\ CH_3-C=N \\ | \\ OH \end{matrix} + Ni^{2+} \longrightarrow \begin{matrix} O\cdots H\cdots O \\ CH_3-C=N \quad N=C-CH_3 \\ Ni \\ CH_3-C=N \quad N=C-CH_3 \\ O^{-}\cdots H\cdots O^{-} \end{matrix} + 2H^{+}$$

Ni – DMG
Chelate

In gravimetric method, the precipitate of Ni–DMG is filtered through a sintered crucible and dried at 100°C. From the weight of Ni–DMG precipitate, the amount of Ni can be calculated.

Solution of Pd salts give a characteristic yellow precipitate in dil HCl or H_2SO_4. Nitric acid cannot be used because of the interference of NO_3^-.

Bismuth salts give a yellow precipitate in presence of EDTA at pH 11.0 to 11.5.

While using spectroscopic method, the precipitate is extracted into chloroform which gives yellow colour. Optical density of the solution is measured to arrive at the concentration of the substance.

For analytical work, 1% solution of dimethyl glyoxime in alcohol is used.

Q. 6. Discuss the use of α-nitroso-β-naphthol (1-nitroso-2-naphtol) as analytical reagent.

Ans. This reagent has the structure as given below

NO, OH (on naphthalene ring) OR N—OH, O (on naphthalene ring)

It is an orange-brown powder having melting point 109°C.

It is soluble is organic solvents, but insoluble in water.

The reagent is used for the precipitation of ferric iron, copper cobalt, palladium and uranyl salts in acid solution.

Nickel and magnesium are not precipitated with this reagent. Therefore the reagent in acid solution can be used for the separation of nickel from cobalt. The structure of the chelate complex formed with Co^{3+} is given below.

$[\,\bar{N}\rightarrow O,\ O$ (on naphthalene ring)$\,]_3$ Co

The reagent is employed as saturated solution in 50% acetic acid.

The reagent finds use in spot test of cobalt and spectrophotometric determination of cobalt.

Q. 7. Discuss cupferron as analytical reagent.

Ans. A small portion of cupferron exists in nitrosohydroxy form (a) but a larger portion exists in tautomeric amine oxide form (b)

$$C_6H_5 - \underset{\underset{N=O}{|}}{N} - OH \qquad\qquad C_6H_5 - \underset{\underset{N-OH}{||}}{\overset{+}{N}} - O^-$$

(*a*) (*b*)

In strongly acidic solution, it precipitates Fe, Ce, Nb, Ta, W, Ga, Sn, Cu, etc.

Precipitation is carried out is cold, since the reagent is decomposed into nitrobenzene on heating.

Also, the precipitate is filtered immediately after formation as cupferron is not very stable in acid solution. The precipitate cannot be weighed after drying (because of decomposition to nitrobenzene) and, hence, it is ignited to obtain the oxide. From the weight of the oxide obtained, the concentration of the metal can be calculated.

The chelate complex with Cu^{2+} can be written as under.

$$2\ C_6H_5 - \underset{\underset{N-OH}{||}}{N^+} - O^- + Cu^{2+} \longrightarrow \text{Chelated complex} + 2H^+$$

(Chelated complex: two $C_6H_5-N^+(-O^-)=N-O$ units coordinated to Cu through O atoms)

Chelated complex

Q. 8. What notes on

(a) Aluminon

(b) Rubeanic acid (dithioxamide)

Ans. (a) Aluminon

It has the structure as given below

(Structure: H_4NOOC, HO-substituted phenyl and $COONH_4$, OH-substituted phenyl attached to C, which is double-bonded to a quinonoid ring bearing H_4NOOC and =O)

It is the ammonium salt of aurine tricarboxylic acid. It is a brownish red powder, freely soluble in water. It forms brilliant coloured lakes with certain metals, particularly aluminium with which it gives a bright red lake.

The reagent is also used for the estimation of Al by spectrophotometric method. By keeping the pH value below 7.8, it is possible to estimate 0.005 mg of Al even in the presence of Ca.

The reagent is prepared by dissolving 0.1 g of the substance in 100 ml water.

(b) Rubeanic acid (Also called dithioxamide)

The reagent exists as a mixture of two tautomeric forms given below.

$$\begin{matrix} S{=}C{-}NH_2 \\ | \\ S{=}C{-}NH_2 \end{matrix} \rightleftharpoons 2H^+ + \left[\begin{matrix} S{-}C{=}NH \\ | \\ S{-}C{=}NH \end{matrix}\right]^{2-}$$

(I) (II)

It is the diamide of di-thiooxalic acid

The *acid*-form (II) combines with Ni, Co and Cu to give coloured water insoluble inner complexes which can be represented as under.

$$M\begin{matrix} \diagup S - C{=}NH \\ \qquad | \\ \diagdown S - C{=}NH \end{matrix}$$

Complex formed between M (M = Ni, Co, Cu) and (II)

These metals can be estimated from strongly ammoniacal solution. In acidic medium, the reagent can be used for the estimation of platinum salts. Rubeanic acid forms coloured compounds with Ag, Hg(I), Bi and Fe (III).

It is also used in spot test analysis and spectrophotometric and gravimetric analysis of metals.

Q. 9. Write notes on

(a) Rhodamine -B

(b) Benzidine

Ans. Both the substances are analytical reagents.

(a) Rhodamine -B

It is tetraethyl rhodamine and has the structure given below.

O, $(C_2H_5)_2N$, $N(C_2H_5)_2$, C, COOH

Sb (V) salts in strong HCl react with rhodamine-B. Aqueous solution of the reagent changes from red to violet (or blue) and finely divided precipitate separates out. Traces of Sb can be detected in the presence of large quantities of Sn.

Hg, Au, and Pd are precipitated while molybdates and tungstates in acid solution give colour changes as with antimony salt.

A freshly prepared benzene solution of the reagent is used in preference to aqueous solution.

(b) Benzidine

The structure of benzidine is given below.

$$H_2N-C_6H_4-C_6H_4-NH_2$$

It is a basic substance (because of $-NH_2$ groups). Its acidic solution as dihydrochloride is used for the estimation of sulphate ions. The precipitate has the formula $[C_{12}H_{12}N_2H_2]^{2+}\ SO_4^{2+}$.

It also precipitates anions like phosphates and tungstates.

It gives crystalline precipitates with a number of metal ions such as Ag, Cu, Cd, Au and Zn.

With Co(III) and Ti(III), benzidine gives a blue coloration in alkaline medium. It is an important spot test for these metals.

Q. 10. Describe analytical use of the following:

(a) Cupron or α-benzoin

(b) Magneson.

Ans. (a) Cupron

Cupron has the structure as given below.

$$\begin{array}{l} C_6H_5-CH-OH \\ \qquad\quad | \\ C_6H_5-C=N-OH \end{array}$$

It is a white crystalline solid, sparingly soluble in water but fairly soluble in alcohol.

It is used as a reagent for quantitative determination of Cu, Mo and W.

Alcoholic solution of cupron gives a green precipitate with ammoniacal solution of copper salt.

Precipitation of other metals in ammoniacal solution is prevented in the presence of tartaric acid.

The green precipitate is soluble in dilute mineral acids.

The green precipitate is a complex compound formed by the following reaction.

$$\begin{array}{l} C_6H_5-CH-OH \\ \qquad\quad | \\ C_6H_5-C=N-OH \end{array} + Cu^{2+} \longrightarrow \begin{array}{l} C_6H_5-CH-O \\ \qquad\quad | \qquad\ \ \diagdown Cu \\ C_6H_5-CH=N \diagup \\ \qquad\qquad\quad\ \downarrow \\ \qquad\qquad\quad\ O \end{array} + 2H^+$$

Green precipitate

Copper can be estimated both by the gravimetric and volumetric methods.

Mo and V can be quantitatively precipitated in strongly acidic solution using cupron. Vanadium gives a yellow precipitate while molybdenum gives a white precipitate. W and Pd are also quantitatively precipitated.

(b) Magneson

It is a versatile reagent for the detection of magnesium. It is also called p-nitrobenzene azoesorcinol and has the structure as given below.

$O_2N-C_6H_4-N=N-C_6H_3(OH)_2$

In alkaline solution, it forms a bright blue lake with magnesium ions. 0.5% solution of the reagent in 0.25 N NaOH is used for analysis.

Presence of ammonium salts must be avoided as this interferes with the action of the reagent.

Q. 11. Discuss the analytical capatibity of the following reagents:

(a) Alizarin – S

(b) Nitron

(c) Salicylaldoxime

Ans. (a) Alizarin –S

It is alizarin sulphonic acid having the following structure:

(Anthraquinone structure with OH, OH and SO_3H substituents; O, O at carbonyl positions)

It is a dye whose sodium salt is soluble in water and employed as analytical reagent.

Alizarin –S gives a red precipitate with aluminium salt in ammoniacal solution. The complex formed between the two is called **adsorption** complex. An adsorption complex obtained by the reaction between metal hydroxide and a dye stuff (like alizarin –S) is called **lake.**

Zirconium (IV) forms a red-violet lake with the reagent in alkaline medium.

Boric acid also forms brilliant red colour in UV light in the presence of H_2SO_4.

Because of the colour formation with metal ions, alizarin– S finds use as a spectrophotometric reagent also.

(b) Nitron

It is a yellow solid, insoluble in water but fairly soluble in acetic acid. It has the structure as given below.

(Structure: C_6H_5 N——N, C_6H_5, N, HC, C, N, C_6H_5)

It is an excellent reagent for the quantitative estimation of nitrate. It forms a sparingly soluble crystalline precipitate with nitrate ions in the presence of acetic acid or sulphuric acid.

It also acts as a reagent for the qualitative and quantitative determination of perchlorate, perrhenate, tetrafluborate and tungstate. Anions like bromide, iodide, chlorate, thiocyanate, nitrite and chromate which interfere in the analysis are required to be removed by normal methods before proceeding with the reagent.

A 10% solution of the reagent in 5% acetic acid is generally employed.

(c) Salicylaldoxime

It is a white crystalline solid having melting point 57°C. It is sparingly soluble in water.

The reagent is unstable in solution form and should be used within 72 hours. It has the structure as given below.

$$C_6H_4(-CH{=}N{-}OH)(-OH)$$

Salicyladoxime gives a yellow precipitate of copper salicylaldoxime with copper ions in the presence of acetic acid.

$$C_6H_4(CH{=}N{-}OH)(OH) + Cu^{2+}/2 \longrightarrow C_6H_4(CH{=}N{-}OH\downarrow)(O{-}Cu/2) + H^+$$

Yellow greenish precipitate

5% solution of salicylaldoxime in 50% ethyl alcohol is used for analytical work.

INORGANIC REAGENTS

Q. 11. Discuss common inorganic reagents for the precipitation of metal ions.

Ans. Common inorganic reagents are discussed as under:

Aqueous solution of NH_3. The metal ions that can be precipitated with this reagent are Al^{3+}, Sc^{3+}, Fe^{3+}, Cr^{3+}, etc. Hydroxide of the metal is obtained which is ignited in a silica crucible to obtain metal oxide.

$$MCl_3 + 3NH_4OH \longrightarrow M(OH)_3 + 3\,NH_4Cl$$
$$(M = Al, Fe, Cr, etc)$$

$$2M(OH)_3 \xrightarrow{\Delta} M_2O_3 + 3H_2O$$

Hydrogen sulphide (H_2S). Hydrogen chloride is passed through acidic solution in case of metals like Cu and in ammoniacal solution in case of metals like Zn. The sulphide obtained is ignited to obtain the oxide. From the mass of the oxide obtained the concentration of metal can be worked out.

$$CuCl_2 + H_2S \longrightarrow CuS + 2HCl$$

$$CuS \xrightarrow{[O]} CuO + SO_2\uparrow$$

$$Zn\,Cl_2 + H_2 \longrightarrow \underset{\text{Dirty white ppt.}}{ZnS} + 2HCl$$

$$ZnS \xrightarrow{[O]} ZnO + SO_2$$

Hydrochloric acid (HCl). This reagent is used for quantitative analysis of metal ions like Ag^+, Hg_2^{2+}, etc.

$$AgNO_3 + HCl \longrightarrow \underset{\text{white ppt}}{AgCl}\downarrow + HNO_3$$

$$Hg_2(NO_3)_2 + 2HCl \longrightarrow \underset{\text{White ppt}}{Hg_2Cl_2}\downarrow + 2HNO_3$$

The precipitate of AgCl or Hg_2Cl_2 is filtered and ignited in a silica crucible to obtain the chloride in dry state from the mass of the dry chloride, concentration of the metal ion can be worked out.

Sulphuric acid (H_2SO_4). This reagent is used for the quantitative analysis of metal ions like Pb^{2+}, Ba^{2+}, Sr^{2+}, Cd^{2+}, etc.

$$PbCl_2 + H_2SO_4 \longrightarrow \underset{\text{White ppt}}{PbSO_4} + 2HCl$$

$$Ba(NO_3)_2 + H_2SO_4 \longrightarrow \underset{\text{White ppt}}{BaSO_4} + 2HNO_3$$

Silver Nitrate (Ag NO_3). Halogens can be estimated quantitatively using silver nitrate

$$NaCl + AgNO_3 \longrightarrow \underset{\substack{\text{Silver chloride}\\ \text{(White ppt)}}}{AgCl} + NaNO_3$$

Barium chloride ($BaCl_2$). Sulphate radical can be quantitatively estimated using $BaCl_2$.

$$K_2SO_4 + BaCl_2 \longrightarrow \underset{\substack{\text{Barium sulphate}\\ \text{(White ppt)}}}{BaSO_4} + 2KCl$$

Ammonium carbonate, $(NH_4)_2CO_3$. Bismuth ions react with a solution of $(NH_4)_2CO_3$ to give a precipitate of bismuth carbonate which is ignited in a crucible to obtain bismuth oxide (Bi_2O_3).

$$2BiCl_3 + 3(NH_4)_2CO_3 \longrightarrow Bi_2(CO_3)_3 + 6NH_4Cl$$

$$Bi_2(CO_3)_3 \xrightarrow{\Delta} Bi_2O_3 + 3CO_2$$

Oxalic acid ($C_2H_2O_4$). It is a suitable reagent for the precipitation of ions like Ca^{2+} and Sr^{2+}. The oxalate obtained first is ignited to carbonate and on further ignition, the oxide is formed.

$$CaCl_2 + C_2H_2O_4 \longrightarrow \underset{\text{Cal. oxalate}}{CaC_2O_4} + 2HCl$$

$$\underset{\text{Cal. oxaltae}}{CaC_2O_4} \xrightarrow{450°C} CaCO_3 \xrightarrow{850°C} CaO$$

Ammonium hydrogen phosphate, $(NH_4)_2HPO_4$. It is a suitable reagent for the precipitation of Mg^{2+}, Zn^{2+} and Al^{3+}. Reactions with Mg^{2+} are given below.

$$MgSO_4 + (NH_4)_2HPO_4 + 6H_2O \longrightarrow \underset{\text{Mag.ammo. phosphate}}{Mg(NH_4)PO_4.6H_2O} + NH_4HSO_4$$

$$2Mg(NH_4)PO_4.6H_2O \xrightarrow{\Delta} \underset{\substack{\text{Magnesium}\\ \text{pyrophosphate}}}{Mg_2P_2O_7} + 2NH_3 + 7H_2O$$

Magnesia mixture ($MgCl_2 + NH_4Cl$). This mixture is used for the precipitation of PO_4^{3-}

$$Na_3PO_4 + MgCl_2 + NH_4Cl \longrightarrow \underset{\text{Mag.amm. phosphate}}{Mg(NH_4)PO_4} + 3NaCl$$

$$2Mg(NH_4)PO_4 \xrightarrow{\text{Ignite}} \underset{\text{Mag. pyrophosphate}}{Mg_2P_2O_7} + 2NH_3 + H_2O$$

31 BIOINORGANIC CHEMISTRY

Q. 1. What are essential and trace elements? Give their classification.

Ans. An essential element is that element which is required for the maintenance of life. Its absence results in death or a severe malfunction of the organism. Of the 112 known elements, 30 elements are believed to be essential for life processes in plants and animals. Nineteen of thirty elements are trace elements, out of which twelve are transition elements. Biological activity of about one half of the essential elements including Fe, Zn, Cu, Mn, Mo, Co, Ni and Se has been properly explained by bioinorganic chemistry while the action of V, Cr, Cd, Pb, Sn, Li, F, Si, As, and B is still to be understood completely.

30 essential elements have been classified as under (required in milligrams or micrograms)

1. *Bulk elements* H, C, N, O, P, S
2. *Macrominerals* Na, K, Mg, Ca, Cl,
3. *Trace elements* Fe, Zn, Cu
4. *Ultratrace elements*
   - Non-metals F, I, Se, Si, As, B
   - Metals Mn, Mo, Co, Cr, V, Ni, Cd, Sn, Pb, Li

An element is considered essential when a deficient intake produces an impairment of function and when restoration of level of that elements cures the impaired function.

Q. 2. What are the criteria for an essential element?

Ans. Criteria for an essential element are listed below.

1. An element is considered essential when a deficient intake produces an impairment of function and when restoration of level of that elements cures the impaired function.

2. A specific biochemical function is associated with a particular element. It is quite difficult to define their concentration limits in the system. This varies from system to system. The same element may prove to the inadequate at a particular concentration in one biological system while it may be over dose in another system.

All the essential and trace elements are not required by every plant or animal. Sodium is vital for humans and higher animals but it is not essential for bacteria. Some higher plants require Al, B and V, which is not so much required for humans. Definitely, some essential elements prove to be toxic, even fatal, when taken in higher concentration.

Q. 3. Discuss the role of Na^+ and K^+ ions in biological systems. *(Lucknow, 2010; Garhwal, 2010)*

Ans. Sodium

Sodium is the major component of extracellular fluid. It mainly exists as chloride and bicarbonate.

1. Its role in biological system is to regulate acid-base equilibrium.

2. It also helps in maintaining osmotic pressure of the body fluid, thereby protecting the body against fluid loss.

3. It has a role to play in preservation of normal irritability of muscle and permeability of the cell.

4. Many enzyme reactions are controlled by Na^+.

5. The injectable medicines are dissolved in sodium chloride before they are injected into human body.

Potassium

Potassium is present as cation in intracellular fluid as well as extracellular fluid.

1. It influences acid base equilibrium like sodium in extracellular fluid.

2. It controls osmotic pressure and water retention.

3. It is important for metabolic functions like protein biosynthesis by ribosomes.

4. A number of enzymes such as glycolytic enzyme pyruvate kinase require K^+ for maximum activity.

Na^+ and K^+ are present in the red blood cells. The ratio of these ions in mammals such as human beings, rabbits, rats and horses is 1:7. In cats and dogs, this ratio is 15:1.

To establish this ratio also called *concentration gradient* in the cell, biologists have suggested different mechanisms involving sodium pump and potassium pump.

1. The concentration gradient controls the development and functioning of the nerve cells.

2. In restful state, potential of the nerve cell is linked to K^+ concentration across the membranes.

3. During activation of the nerve cells, a chemical called *acetylcholine* is discharged. This discharge is transmitted through the length of the nerve cell by electric impulse.

Q. 4. Discuss the role of Mg^{2+} and Ca^{2+} in biological systems.

(*Nagpur, 2008; Garhwal, 2010; Rajasthan, 2010; Lucknow, 2010*)

Ans. 1. Mg^{2+} ions are present in chlorophyll which is the green colouring matter present in plants. Chlorophyll absorbs light from the sun and carries out the process of photosynthesis in plants.

2. Magnesium ions are present in the enzyme *phosphatase* which act upon organic phosphates to hydrolyse them into free phosphates.

3. Magnesium ions are present in the enzyme *aminopeptidase* which hydrolyses polypeptides at the free amino acid end of the chain to form lower peptides and sometimes free amino acids.

4. Calcium ions are present as phosphates in the bones of human beings and animals. These ions play an important role in muscle contraction. The malnutrition in children is due to the deficiency of Ca^{2+}.

5. About 99% of the body calcium is in the skeleton where it is maintained as deposits of calcium phosphates in soft and fibrous matrix. A very small quantity of calcium not present in skeletal structure, is in the body fluids where it is partly ionized. This small amount of ionized calcium in the body fluid is of great importance in blood coagulation, in maintaining the normal excitability of the heart, muscles and nerves and in the differential aspects of membrane permeability.

6. The major inorganic constituent of bone is comprised of a crystalline from of calcium phosphate resembling the mineral *hydroxyapatite*. Additionally, the bone contains a substantial amount of non-crystalline, amorphous calcium phosphate. It appears that the amorphous material is predominant is early life but later on the crystalline form takes over in adult life.

Some noteworthy points regarding Ca^{2+} and Mg^{2+} ions are :

1. Calcium ions are needed to bring about **blood clotting** and contract of muscles such as **heart beat**.

2. Deficiency of Ca^{2+} ions causes **tetany**, while excess of it causes **Calcification**.
3. Both Mg^{2+} and Ca^{2+} ions are excreted via the large intestine instead of kidney. Calcium deficiency might occur due to its precipitation as calcium oxalate, after reaction with soluble oxalates present in the body.

Magnesium, which is essential to all organisms is relatively present in greater concentration in red cells than in plasma. It is a constituent of chlorophyll which plays an important role in photosynthesis.

It has been observed that transphosphorylation reactions involving ATP (adenosine triphosphate) proceed smoothly in the presence of Mg^{2+} ions. ATP is an energy-rich molecule which drives most of the cell reactions. Mg plays a vital role in these reactions.

Free energy of conversion of ATP to ADP (adenosine diphosphate) through hydrolysis is -31.0 kJ mol^{-1}

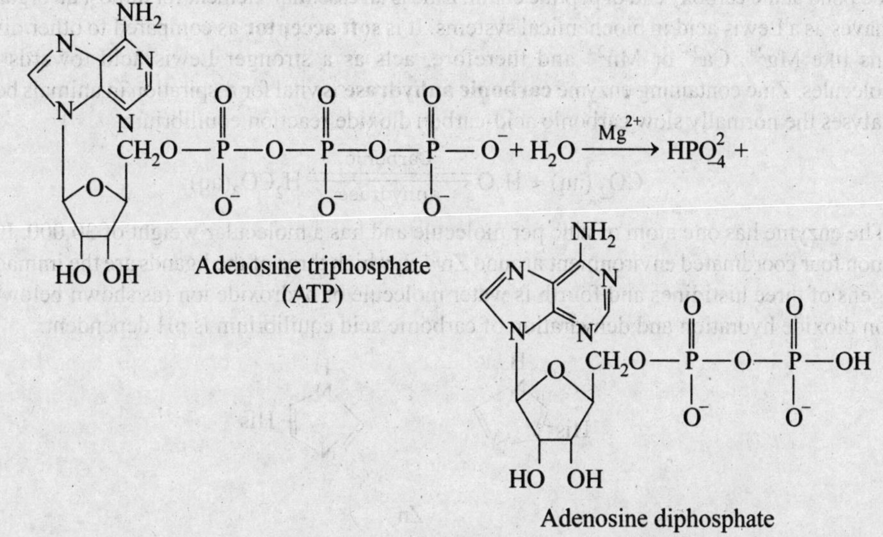

$$\text{ATP} + H_2O \xrightarrow{Mg^{2+}} HPO_4^{2-} + \text{ADP}$$

Adenosine triphosphate (ATP)

Adenosine diphosphate (ADP)

Inorganic phosphorous compounds are involved in the incorporation and release of phosphate groups from organic compounds and skeletal materials (bones). The important phosphate compounds are nucleotides, polynucleotides and phospholipids. Phosphate plays its role in the metabolism of carbohydrates, fats, proteins and vitamins. The presence of phosphorous in bones and its function in regulation of pH of the blood cannot be overlooked. Relative amounts of dietary phosphorus, and calcium in addition to vitamin D, control the formation of normal bones.

Q. 5. Discuss the role of copper in biological systems?

Ans. Copper is essential to all organisms and is a constituent of redox enzymes and hemocyanin. Copper in hemocyanin is **oxygen carrier** and supplies oxygen to certain aquatic creatures. Copper containing *enzymes play an important role in the pigmentation of skin and functioning of brain*. The deficiency of copper in human system develops an anemia characterized not only by marked decrease in the total iron and heme content in blood and tissues but also by an increased amount of free porphyrin in the erythrocytes.

The excess of copper in human body causes a disease known as ***Wilson disease***. It upsets the copper metabolism of human system in such a way that it absorbs excessive copper from intestinal track and gets deposited in excess in the liver, kidney or brain. Patients suffering from this disease are treated with controlled doses of EDTA which forms a water soluble complex with Cu^{2+} which get excreted through urine. It is a constituent of the enzyme phosphatase which converts organic phosphatase into free phosphates.

About 2 mg of copper per day is the recommended dietary intake for an adult.

Q. 6. Discuss the role of cobalt and zinc in biological systems?

Ans. Cobalt is also essential trace element which is essential for many organisms including mammals. It **activates a number of enzymes**. It is constituent of vitamin B_{12} which is required for the formation of haemoglobin in *vivo*. However, it is highly toxic to plants and moderately toxic to mammals when injected intravenously. Deficiency of cobalt in soil adversely affects the health of grazing animals. Adding cobalt salts to soil, improves their health.

Zinc

Zinc is a constituent of enzyme carboxypeptidase which catalyses the hydrolysis of terminal peptide bond at the carboxy end of peptide chain. Zinc is an essential element for almost all organisms. It behaves as a Lewis acid in biochemical systems. It is **soft acceptor** as compared to other divalent cations like Mg^{2+}, Ca^{2+} or Mn^{2+} and therefore, acts as a stronger Lewis acid towards many biomolecules. Zinc containing enzyme **carbonic anhydrase** is vital for respiration in animals because it catalyses the normally slow carbonic acid-carbon dioxide reaction equilibrium.

$$CO_2\,(aq) + H_2O \xrightleftharpoons[\text{anhydrase}]{\text{Carbonic}} H_2CO_3(aq)$$

The enzyme has one atom of zinc per molecule and has a molecular weight of 30,000. It has a common four coordinated environment around Zn^{2+} in which three of the ligands are the immadazole nitrogens of three histidines and fourth is water molecule or hydroxide ion (as shown below). The carbon dioxide hydration and dehydration of carbonic acid equilibrium is pH dependent.

H H
His N N His
Zn
H_2O N NH His

Carboxypeptidase is also zinc containing enzyme having some 300 amino acid residues including three methionyl residues. Their functions are the catalytic hydrolysis of peptide bonds. Its activity is directed specially towards carboxyl terminal peptide bonds.

$$-\overset{\overset{\large R}{|}}{C}H-\overset{\overset{\large O}{\|}}{C}-NH-\overset{\overset{\large R'}{|}}{C}H-\overset{\overset{\large O}{\|}}{C}-NH-\overset{\overset{\large R''}{|}}{C}H-COO^- \xrightarrow[\text{Carboxy peptidase}]{H_2O}$$

$$-\overset{\overset{\large R}{|}}{C}H-\overset{\overset{\large O}{\|}}{C}-NH-\overset{\overset{\large R'}{|}}{C}H-COO^- + H_2N-\overset{\overset{\large R''}{|}}{C}H + COO^-$$

This selective enzyme hydrolyses those polypeptides in which the terminal amino acid segment has an aromatic or a branch chain aliphatic substituent, R″.

The deficiency of this enzyme may cause loss of appetite. It is toxic only in very large amounts. Zinc is essential for normal growth, reproduction and life expectancy of animals and has a major role to play in tissue repair and wound healing.

Q. 7. Discuss the role of selenium, molybdenum, Chromium and manganese in biological systems

Ans. Selenium is essential trace element for mammals and some higher plants. It is a component of glutathione peroxidase. It is an essential constituent of some enzymes and protects biological systems against free radical oxidants and stresses. Livestocks grown on selenium deficient pastures suffer from white muscle disease. However, if grazing in a soil having higher selenium concentration, they suffer from **central nervous system toxinosis**.

The minimum dietary intake of selenium is 70 mg. Sea foods, meats and onion family are the sources of selenium. Selenium deficiency in humans results in degenerative condition of the heart tissue, known as **Keshan disease**.

Molybdenum

It is an essential trace element for all organisms except green algae. It is a constituent of several enzymes used by nature for nitrogen fixation and nitrate reduction.

It is moderately toxic. Molybdenum excesses in biological systems may cause gout like syndrome.

Chromium

This element is present in adult human body in ultra trace amounts. It is involved in glucose metabolism and diabetes. Cr (III) and insulin both maintain the correct level of glucose in the blood.

In +6 oxidation state chromium is carcinogenic.

Manganese

It is essential to all organisms. It **activates** numerous enzymes. The Mn (II) enzyme produced in the liver converts nitrogenous waste products into urea which is carried by the blood to the kidneys which excrete it out into the urine.

The deficiency of manganese in soil inhibits plant growth and leads to infertility in mammals which consume such plants.

Q. 8. Discuss the role of phosphorus, chloride an dcodine in biological systems

Ans. Phosphorus : It is an essential element for life. It is an important part of energy rich molecules, ATP. It also is an important component of the bones as apatite $Ca_5(PO_4)_3$ in the skeleton of vertebrates. It is also strength ensuring material of the teeth. Partial formation of fluorapatite, $Ca_5(PO_4)_3F$ strengthens the structure and makes it less soluble in the acid formed from fermenting organic material in teeth, and thus protects tooth decay.

Chloride

Chloride ion has a vital role in ion balance. It is essential for higher plants and mammals as NaCl electrolyte. In HCl, it is essential in digestic juices. The deficiency of chloride in infants can cause impaired growth.

Iodine

It is an essential constituent of the thyroid hormone *e.g.* thyroxine which is important in metabolism and growth regulation.

I I
HO–⟨O⟩–O–⟨O⟩–CH_2–CH–COOH
I I |
NH_2

Thyroxine

Nearly 75% of iodine present in our body is located in thyroid gland. Iodine deficiency causes abnormal functioning of thyroid gland and causes goitre.

Q. 9. Write a note on harmful effect of excess intake of metals on human body.

Ans. Small quantities of the essential elements are necessary and useful for normal functioning of the biological system. However, it has been observed that excessive amounts of these metals might prove harmful or even fatal. Thus they have the toxic effect also.

1. The recommended dose of copper intake is 2 mg per day for an adult. If the intake rises to about 100 times of this, it can cause *necrotic hepatitis* and *hemolytic anaemia.* Also excessive intake of copper metal causes *Wilson disease.* The patient experiences nervous disorders, tremors, difficulty in breathing and stiff joints.

2. Controlled intake of iron helps in the formation of haemoglobin and thus cures anaemia. However, if taken in excess, this metal can cause *siderosis*.

3. Too much of manganese (in excess of 2.5 –5 mg/day for an adult) causes *psychic* and *neurological* disorders.

4. Both insufficient and excess of iodine causes thyroid problems.

Q. 10. What is the role of metal chelates in living system?

Ans. Most of the essential elements in the human body do not exist in free state. They form complex compounds (called chelates) with other groups. These chelates perform the following functions.

1. They help in the storage and transfer of the metal ions and serve as agents for transmission of energy in the animal and plant metabolism.

2. They catalyse metabolic processes involving (*a*) acid-base reactions in which the metal is the Lewis acid that combines with a substrate to accelerate a reaction (*b*) reduction -oxidation reaction with the change in the valency of the metal in the chelate.

Q. 11. What are metalloporphyrins? What is their role in biological systems?

Ans. Porphyrins are one of the most important groups of bioinorganic compounds in which a metal ion is surrounded by the four nitrogens of porphin ring. Porphines are made of four pyrrole rings linked together through methene bridges. Therefore, porphines have macrocylic pyrrole system with conjugated double bonds as shown in Fig. 1 These porphines act as tetradentate ligands with four nitrogen donor sites. Two of these are tertiary nitrogen donor positions which can form coordinate bonds by donating a pair of electrons each to the metal ion. The other two are secondary nitrogen donor positions, each of which lose a proton in forming a coordinate bond with a metal ion. Thus, a porphin ring acts as a tetradentate dinegative ligand (or dianion). Dipositive cations such as Mg^{2+}, Fe^{2+}, or Ni^{2+} are capable of forming neutral complexes with porphine as shown Fig. 2.

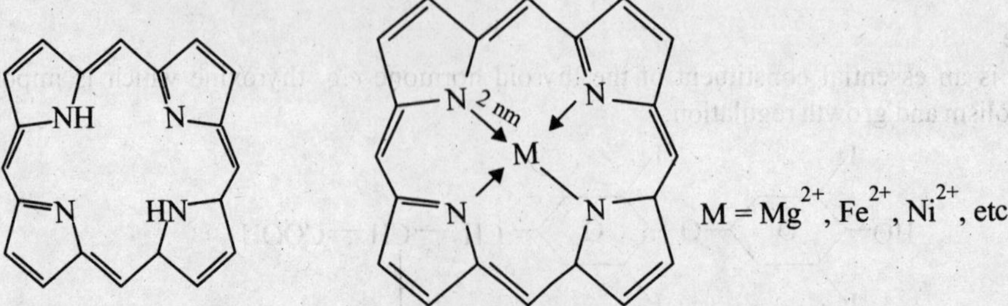

Fig. 1 Porphin ligand **Fig. 2** Metal complex with porphin ligand

Four pyrrole rings of porphin carrying substituents other than hydrogen are called porphyrins. The complexes in which a metal ion is held in the porphyrin ring system are called **metalloporphyrins**.

Such complexes play a vital role in biological systems.

With delocolisation of electrons in the pyrrole rings we obtain stable porphyrin system. The best value of the size of the central hole has been estimated to be of radius 0.2 nm. The size of the central **hole** in the centre of the porphyrin ring is ideal for accommodating metals of first transition series or lighter alkaline earth metals. The metal ion of appropriate size are surrounded by four nitrogens of porphine ring in a square planar geometry and the axial sites are available for other ligands. However, if the size of the metals ion is smaller than required (or ideal) size, the ring becomes ruffled to allow closer approach of the nitrogen atoms to the metal. On the other hand, if the size of the metal ion is lager than the size of the hole, the metal ion cannot fit into the hole and stays out of the plane of four nitrogen atoms. In such a case the metal sites above the ring which becomes domed and acquires square pyramidal configuration.

Q. 12. Discuss the structure of chlorophyll and explain the mechanism of photosynthesis.

Ans. Chlorophyll is a green coloured pigment present in green plants and algae. The basic unit in the molecule of chlorophyll is porphyrin ring. Chlorophyll is a square planar complex with Mg^{2+} at the centre. It is a naturally occurring chelate complex of Mg.

It has been noticed that Mg-atom in chlorophyll lies slightly above the plane of ring. Two water molecules can be added axially to the molecule of chlorophyll. These coordinated water molecules provide an opportunity to associate with other chlorophyll molecules through hydrogen bonding.

Two types of chlorophyll are found in higher plants, chlorophyll *a* and chlorophyll *b*, which differ in a substituent on ring II, while algae contain four chlorophyll (*a*, *b*, *c*, *d*). Chlorophyll *a* and *b* function most effectively in the blue region and the red region of the visible light at wavelengths of 400 – 500 and 600 – 700 nm but weakly between 500 – 600 nm. Structure of the most abundant chlorophyll *a* is shown in the figure.

Four nitrogen atoms are situated at the four corners of a square.

All types of chlorophyll contain porphyrin ring in which a double bond in one of the pyrrole rings is reduced and a fused cyclopentanone ring is also present.

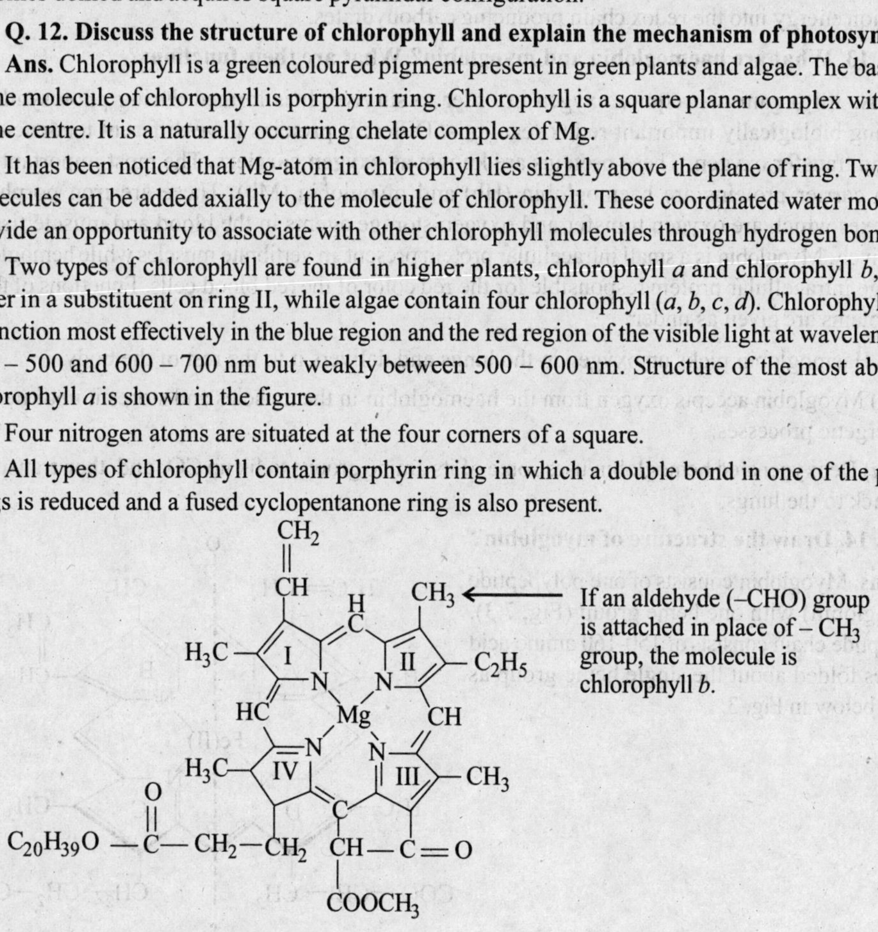

If an aldehyde (–CHO) group is attached in place of – CH_3 group, the molecule is chlorophyll *b*.

Structure of chlorophyll *a*.

Mechanism of Photosynthesis

The process in which carbon dioxide and moisture in the atmosphere combine in the presence of chlorophyll and sunlight to produce carbohydrates is called *photosynthesis*. The reaction for the process can be represented as under:

$$6CO_2 + 6H_2O \xrightarrow[\text{Sunlight}]{\text{Chlorophyll}} C_6H_{12}O_6 + 6O_2 \quad .....(i)$$

Tracer studies have proved that all the oxygen evolved in the above reaction comes from water and not from carbon dioxide. Hence the following mechanism for photosynthesis is proposed:

First Step. Chlorophyll absorbs solar energy which in used in forming energetic ATP molecules and H_2O molecules are oxidised to O_2.

$$12H_2O \longrightarrow 12H_2 + 6O_2 \quad .....(ii)$$

Second step. Energetic ATP molecules formed above reduce CO_2 into glucose

$$6\,CO_2 + 12H_2 \longrightarrow C_6\,H_{12}\,O_6 + 6H_2O \quad .....(iii)$$

On adding equations (*ii*) and (*iii*), we get equation (*i*). Magnesium makes the entire molecule rigid such that energy is not lost thermally and it also increases the rate at which short lived singlet excited state is transformed into triplet state. Triplet state has a longer life and thus can transfer its excitation energy into the redox chain producing carbohydrates.

Q. 13. What are haemoglobin and myoglobin? What are their functions?

Ans. All organisms require oxygen in order to survive. It is formed during photosynthesis involving biologically important redox reactions. Different proteins have different tendencies to bind and transfer oxygen. These proteins are known as **oxygen carriers.** The most important two oxygen carrier proteins are haemoglobin (Hb) and myoglobin (Mb). These are iron porphyrin complexes which are oxygen transfer and oxygen storage agents in the blood and muscle tissues respectively. Myoglobin is a small intracellular protein present in vertibrate muscles while hemoglobin is a large intracellular protein responsible for the red color of the red blood cells. Functions of these two proteins are given as under:

(*i*) Haemoglobin picks up oxygen in the lungs and delivers it to the rest of the body.

(*ii*) Myoglobin accepts oxygen from the haemoglobin in the muscles and stores it until needed for energetic processes.

(*iii*) Deoxygenated hemoglobin uses some of its amino groups to fix up CO_2 and then transports CO_2 back to the lungs.

Q. 14. Draw the structure of myoglobin?

Ans. Myoglobin consists of one polypeptide chain (globin) with one heme group (Fig. 7.3). The peptide chain consists of 150-160 amino acid residues folded about the single heme group as shown below in Fig. 3.

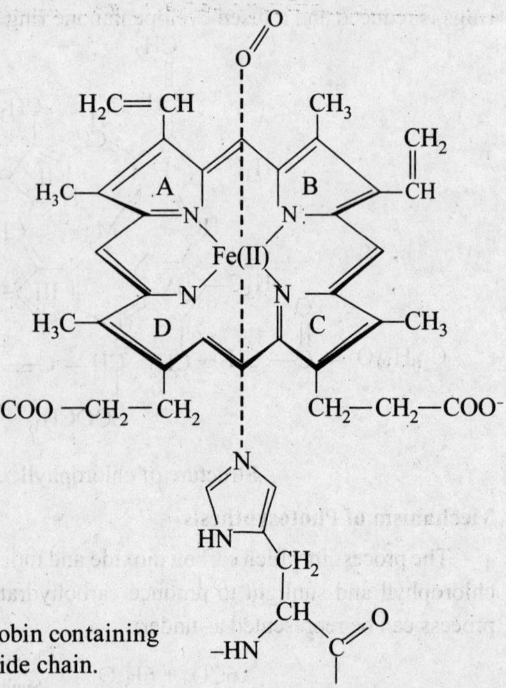

Fig. 3. The structure of myoglobin containing heme group and polypeptide chain.

The heterocyclic ring system of heme is porphyrin derivative containing four pyrrole groups joined by methene bridges. The Fe (II) atom present at the centre of the heme is bonded by four porphyrin nitrogen atoms and one nitrogen atom from imidazole side chain of histidine residue which is a part of long protein chain of amino acid residues. This polypeptide chain plays an important role in biological fixation of O_2.

Q. 15. Discuss the structure of haemoglobin.

Ans. About 65% of iron in the human body (roughly 4g) is present as haemoglobin. Haemoglobin is regarded as red prigment in the red blood cells (RBC's). Haemoglobin binds O_2 at lungs and then arteries carry blood to different parts of the body such as muscles where O_2 is required.

100 ml of the blood of a normal male contains approximately 15 g of haemoglobin. Haemoglobin contains about 0.35% of iron. Thus 100 ml of the blood contains 50 mg Fe. Total iron content in blood volume of 5000 ml in the human body is thus 2.5 g approximately.

Haemoglobin is a conjugated protein having a molecular weight of 64500. It contains four identical units arranged roughly tetrahedrally. Each unit contains one heme group whose structure is shown below.

Structure of heme group·

Each haemoglobin molecule has four *heme* group which are bound to globin (a protein) on its surface. Schematic representation of structure of haemoglobin is shown below.

Schematic representation of the structure of hemoglobin.

Fe (II) is in high spin state ($3d^6$ or $t_{2g}^4 e_g^2$). Thus, iron has four unpaired electrons in haemoglobin molecule.

Thus, haemoglobin is an octahedral complex of iron (II). Iron occupies the central position and The four corners of the square base are occupied by four N-atom of histidine while the other axial position is occupied by H_2O molecule.

Q. 16. Discuss the role of haemoglobin in living systems. *(Ranchi, 2010)*

Ans. When we breathe in air, oxygen of the air combines with haemoglobin (Hb) present in blood to form oxyhaemoglobin (HbO_2)

$$\text{Haemoglobin} + O_2 \rightleftharpoons \text{Oxyhaemoglobin} + H_2O$$

$$Hb + O_2 \rightleftharpoons HbO_2 + H_2O$$

This process is known as oxygenation of haemoglobin. This takes place in the lungs. Iron is in +3 oxidation state. H_2O molecule present at one of the axial positions in haemoglobin is reversibly replaced by O_2 in this process.

The blood runs through the arteries to the tissues in the body. In the course of this process, oxygen pressure decreases and oxygen linked to HbO_2 is set free. Free oxygen diffuses into body cells where it combines with glucose and oxidises it to carbon dioxide and water.

$$C_6H_{12}O_6 + 6O_2 \longrightarrow 6CO_2 + 6H_2O + 38 \text{ ATP (Energy)}$$

Thus, oxidation of glucose (food) is an exothermic process in which energy in the form of ATP is produced. Living organisms utilize this energy to perform various metabolic process and to maintain body temperature. Water produced in the reaction is retained in the body while carbon dioxide combines with amino group of haemoglobin to form carbamino haemoglobin which later decomposes to give haemoglobin. Carbon dioxide is set free and exhaled out. Haemoglobin goes back to lungs for reuse. Thus haemoglobin supplies oxygen to various parts of the body or its acts as **oxygen carrier.**

It may be mentioned that in the oxygenation process, a proton (H^+) is also produced. This proton reacts with bicarbonate dissolved in blood to liberate CO_2.

$$HCO_3^- + H^+ \rightleftharpoons H_2CO_3 \rightleftharpoons H_2O + CO_2 \uparrow$$

Q. 17. Discuss carbon dioxide transport and Bohr effect?

Ans. A graph between percentage saturation and partial pressure of O_2 is shown in Fig. 7.7.

The graph shows that Hb and Mb (myoglobin) have almost similar binding affinity for O_2 at high O_2 pressure but Hb is much poor O_2 binder at lower pressures of O_2 in muscles. As a result, Hb passes its O_2 onto Mb as required. Moreover, the need for O_2 is greatest in tissues which have already consumed oxygen and simultaneously produced CO_2. The CO_2 lowers the pH ($2H_2O + CO_2 \rightleftharpoons HCO_3^- + H_3O^+$) and the increased acidity favours the release of O_2 from oxyhaemoglobin to Mb. The oxygen affinity of haemoglobin varies with the pH of the medium. This pH **sensitivity effect** is called **Bohr effect**. In other words the variation of oxygen affinity with the pH of the medium is called Bohr effect. It can be explained in terms of the effect of pH on the interaction between the heme group and the ionizable groups of the protein.

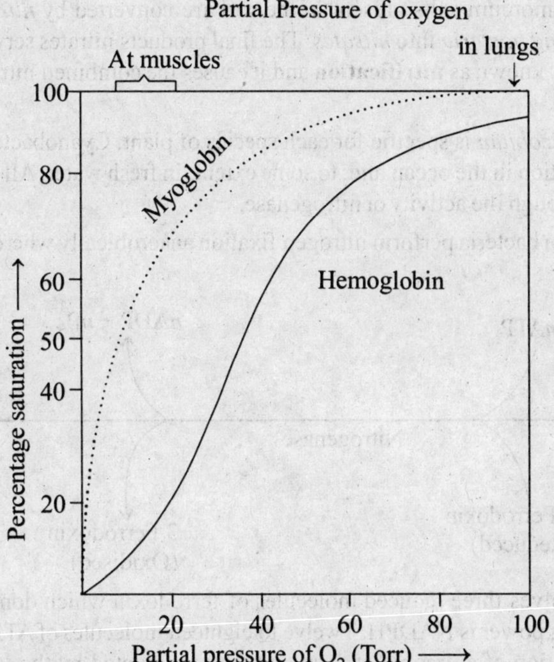

Fig. 4 O_2 binding curves for Mb and Hb with partial pressure of O_2.

Q. 18. Discuss nitrogen fixation and nitrogen cycle.

Ans. Nitrogen cycle in Nature. Nitrogen is a necessary constituent of all animal and vegetable growth. Neither animals nor the plants can live without nitrogen, but nitrogen cannot be directly assimilated by them from air. Nitrogen passes from the atmosphere to the soils in the form of nitrates and ammonium salts by special processes. From the soils, it passes on to the plants which are eaten by the animals and from the animals it passes back to the atmosphere and thus the percentage of nitrogen in the atmosphere remains same.

Nitrification or conversion of atmospheric nitrogen to plant food in the soil. This conversion is being brought about in a number of ways.

1. By symbiotic bacteria. Certain leguminous plants like peas and beans have some symbiotic bacteria (*rhizobium*) living on their roots. On their roots are nodule which contain these nitrogen fixing bacteria as shown in the figure.

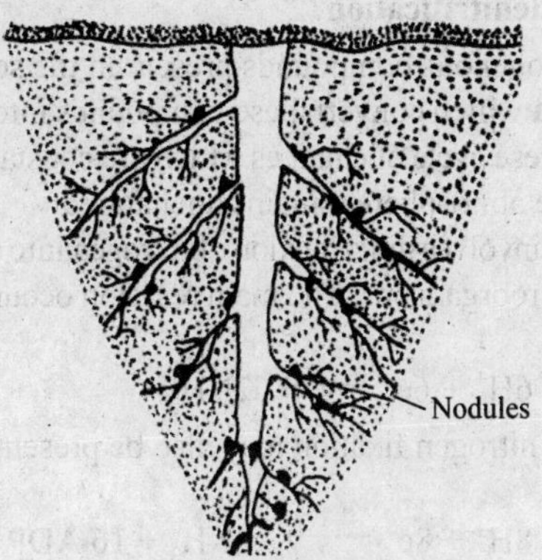

Fig. Leguminous plants.

Nitrogen and ammonium salts present in the soil are converted by *nitrosifying bacteria* into *nitrites* and by *nitrifying bacteria* into *nitrates.* The final products nitrates serve as a plant food. This process of oxidation is known as **nitrification** and it causes the combined nitrogen to be available to the plants.

The species of *rhizobium* is specific for each species of plant. Cyanobacteria (blue-green algae) perform nitrogen fixation in the ocean and, to some extent, in fresh water. All these microorganisms produce ammonia through the activity of nitrogenase.

Most free nitrogen bacteria perform nitrogen fixation anaerobically whereas cyanobacteria do it aerobically.

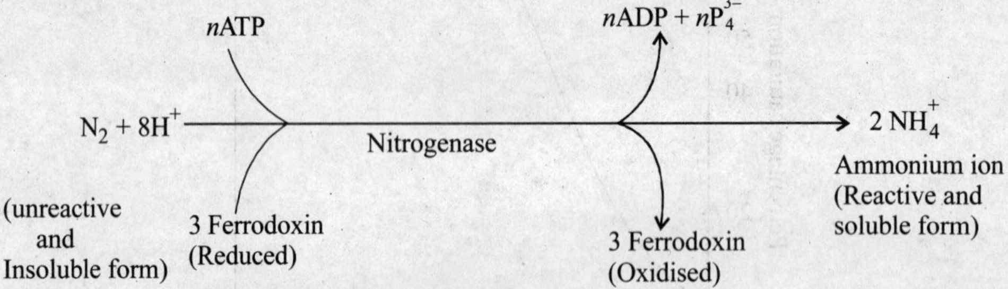

The reaction involves three reduced molecules of ferrodoxin which donate six electrons. The source of this reducing power is NADPH. Twelve to eighteen molecules of ATP supply the necessary energy for the production of ammonia at a rate which is regulated by the inhibitory effect of the product on the association of the two components of the nitrogenase.

2. By lightning Discharges. Nitrogen and oxygen, present in air, combine to from nitric oxide under the influence of lightning discharges. Nitric oxide reacts with excess of oxygen in the atmosphere to give nitrogen pentoxide which dissolves in water to produce nitric acid. Rain washes down nitric acid to the soil where there is a reaction with limestone to produce nitrates. Nitrates are stored in the soil as plant food. Various reaction may be represented as below:

$$N_2 + O_2 \xrightarrow{\text{Lightning}} 2NO$$

$$4NO + 3O_2 \longrightarrow 2N_2O_5$$

$$N_2O_5 + H_2O \longrightarrow 2HNO_3$$

$$2HNO_3 + CaCO_3 \longrightarrow Ca(NO_3)_2 + CO_2 + H_2O$$

About 250000 tons of the acid are said to be formed in this way in 25 hours.

Q. 19. What is meant by denitrification?

Ans. Denitrification. Nitrogenous compounds present in the soil are acted upon by a bacteria known as **denitrifying bacteria** which convert these compounds into free nitrogen which goes back to the atmosphere. Because of these cycle of changes which are constantly going on in nature the total average amount of the gas in the atmosphere remains unchanged.

Biological nitrogen fixation involves the reduction of nitrogen into ammonia by a biological process with the help of enzymes or microorganisms. The main reaction occurring in the process is

$$N_2 + 6H^+ + 6e^- \longrightarrow 2NH_3$$

The reaction for biological nitrogen fixation may also be presented as:

$$N_2 + 16\ ATP + 8H^+ + 8e^- \longrightarrow 2NH_3 + 16\ ADP + 16\ PO_4^{3-} + H_2$$

The ammonia may be further converted into nitrate or nitrite or directly used in the synthesis of amino acids or other essential compounds. This reaction occurs at atmospheric conditions of temperature and pressure in Rhizobia, bacteria in nodules on the roots of leguminous plants such as peas and beans. In contrast, industrial synthesis of ammonia require high temperatures and pressures and catalysts.

Q. 20. Give a labelled diagram of nitrogen cycle.

Ans. The cycles of nitrification and denitrification together constitute nitrogen cycle which is given below:

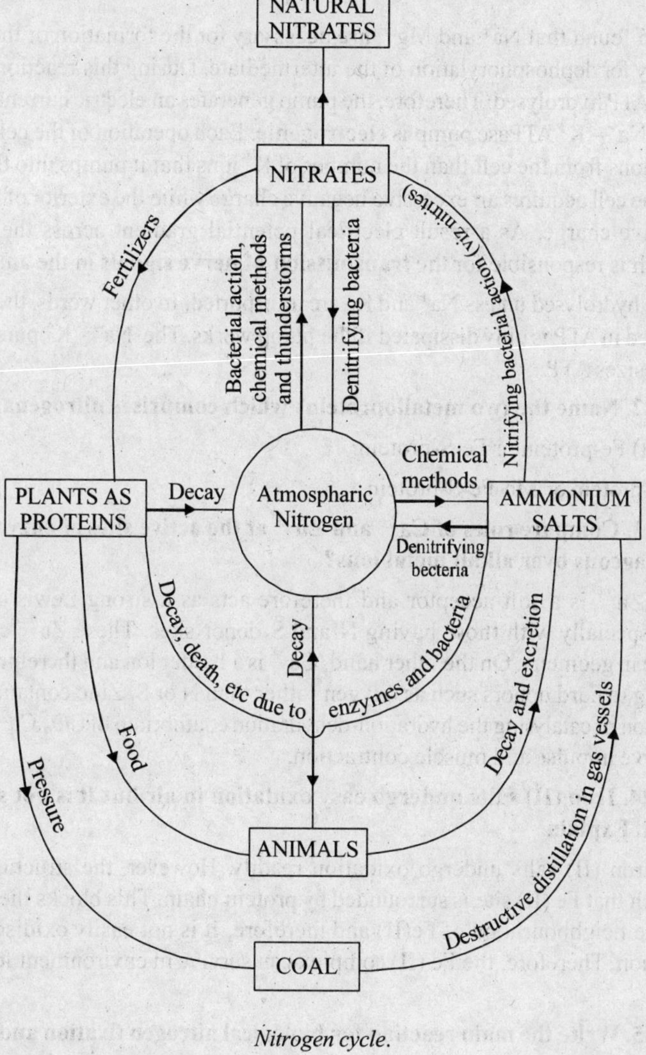

Nitrogen cycle.

MISCELLANEOUS SOLVED CONCEPTUAL PROBLEMS

Example 21. Explain the mechanism of $Na^+ - K^+$ pump.

Solution. The basis of the mechanism is that the $Na^+ - K^+$ AT Pase is phosphorylated by ATP in the presence of Na^+ and Mg^{2+} ions. The site for phosphorylation is the side chain of a specific **asparate** residue, E.

The phosphorylated intermediate (E–P) is then hydrolysed. It gets dephosphorylated in the presence of K^+ and gives the original phosphoprotein.

$$E + ATP \xrightleftharpoons{Na^+,\ Mg^{2+}} \underset{\text{Intermediate}}{E - P} + ADP \qquad \text{Phosphorylation}$$

$$E - P + H_2O \xrightleftharpoons{K^+} E + HPO_4^{2-} \qquad \text{Dephosphorylation}$$

It has been found that Na^+ and Mg^{2+} are necessary for the formation of the intermediate and K^+ are necessary for dephosphorylation of the intermediate. During this reaction $3Na^+$ and $2K^+$ are transported per ATP hydrolysed. Therefore, the pump generates an electric current across the plasma. In other words, $Na^+ - K^+$ ATPase pump is **electrogenic**. Each operation of the cell, pumps out larger number of Na^+ ions from the cell than the number of K^+ ions that it pumps into the cell. As a result, the interior of the cell acquires an excessive negative charge while the exterior of the cell acquires an excessive positive charge. As a result electrical potential gradient across the cell membrane is developed which is responsible for the **transmission of nerve signals** in the animals.

ATP is not hydrolysed unless Na^+ and K^+ are transported. In other words, the system is coupled. The energy stored in ATP is only dissipated if the pump works. The $Na^+ - K^+$ pump is also reversible so that it synthesizes ATP.

Example 22. Name the two metalloproteins which comprises nitrogenase.

Solution. (i) Fe-protein or Fe-S-protein.

(ii) Mo-Fe-protein or Mo-Fe-S-protein.

Example 23. Compare roles of Ca^{2+} and Zn^{2+} at the active sites of enzymes. In what ways is Ca^{2+} advantageous over alkali metal ions?

Solution. Zn^{2+} is a soft acceptor and therefore acts as a strong Lewis acid towards many biomolecules especially with those having N and S donor sites. These Zn^{2+} complexes are very labile and facile in geometry. On the other hand, Ca^{2+} is a harder ion and therefore, is more selective towards bonding to hard donors such as oxygen rather than N or S. Zinc containing enzymes serve important function of catalysing the hydration-dehydration equilibrium of CO_2.Ca^{2+} is an intermediary between the nerve impulse and muscle contraction.

Example 24. Iron (II) salts undergo easy oxidation in air but it is not so in haemoglobin and myoglobin. Explain.

Solution. Iron (II) salts undergo oxidation readily. However, the structure of deoxy-Mb or deoxy-Hb is such that Fe (II) site is surrounded by protein chain. This blocks the approach of larger molecules to the neighbourhood of Fe(II) and therefore, it is not easily oxidised. This chain also prevents solvation. Therefore, the Fe (II) complex can survive in environment long enough to bind and release O_2.

Example 25. Write the main reaction for biological nitrogen fixation and name the enzyme involved.

Solution. $N_2 + 16\ ATP + 8e^- + 8H^+ \longrightarrow 2NH_3 + 16\ ADP + 16\ PO_4^{3-} + H_2$.

The enzyme nitrogenase is involved in the reaction.

Example 26. In what ways in Ca^{2+} advantageous over alkali metal ions?

Solution. Ca^{2+} has higher charge to radius ratio and therefore has a higher affinity for most hard ligands than alkali metal ions which are also hard acceptors. Therefore, Ca^{2+} should be advantageous when especially high reaction rates are needed alongwith high affinity for oxygen ligands.

Example 27. The reduction of N_2 to NH_3 is thermodynamically favourable ($\Delta G = -ve$). In biological systems, still ATP molecule is required for energy requirement. Comment.

Solution. N≡N bond has very high bond dissociation energy (945 kJ mol^{-1}). This N≡N bond must be broken during nitrogen fixation. Therefore, a very large energy input is required to overcome the large activation energy barrier even though $\Delta G°$ is negative. In biological systems, this energy barrier is overcome by the energy provided by ATP.

Example 28. Give names of three essential trace elements and three essential ultra trace elements.

Solution. Trace elements : Iron, copper, zinc

Ultra trace elements : fluorine, iodine, selenium

Example 29. Oxygen molecule behaves as a π-acceptor ligand during interaction with Fe(II) in haemoglobin or myoglobin. Can other well known π-acceptors like CO, NO or CN^- etc. bind themselves with Mb or Hb?

Solution. Yes any π-acceptor can bind to haemoglobin but to a limited extent. For example CO is most toxic because of its greater π-acceptor character. However, the toxic effect of CO is limited by the steric effects of protein chain which prevents appropriate geometry to the bound CO.

Example 30. What is the oxidation state of iron in haemoglobin and myoglobin?

Solution. The iron in haemoglobin and myoglobin is + 2 oxidation state.

Example 31. Why nitrogenase enzyme perform better than the Haber process in the fixing of nitrogen?

Solution. Nitrogenase enzyme fixes nitrogen to produce NH_4^+ salts from N_2 and water in the presence of O_2 at the temperature of soil and atmospheric pressure. On the other hand, in the Haber process, the fixing of N_2 to NH_3 is carried out in the presence of catalyst at a very high temperature and high pressure. Thus, nitrogenase enzymes perform better than the Haber process.

Example 32. Sodium pump is electrogenic in nature. Comment.

Solution. Sodium pump acts as a transport system to maintain $Na^+ - K^+$ concentration gradient. For each ATP molecule hydrolysed, $3Na^+$ are transported out of the cell and $2K^+$ ions are transported into the cell according to the reaction:

$ATP^{4-} + H_2O + 3Na^+$ (inside) $+ 2K^+$ (outside) $\longrightarrow 2K^+$ (inside) $+ 3Na^+$ (outside) $+ ADP^{3-} + H_2PO_4^-$. The net result is the movement of one positive charge outward per cycle and hence sodium pump is electrogenic in nature.

32 EVALUATION OF ANALYTICAL DATA

Q. 1. Explain what is meant by significant figures ? Discuss the rules for rounding off numerical expressions involving addition, subtraction, multiplication and division.

Ans. *The significant figures in a numerical expression are defined as all those digits whose values are known with certainty with one additional digit whose value is uncertain.*

For example, if the length of the object measured by using scale capable of measuring accurately to the nearest millimetre is reported to be 1.56786 m, then only first four digits are meaningful. The last digit 6 is superfluous.Last but one digit 8 only indicates that length of the object is more than 1.567 m, but less than 1.568. Thus, the expression 1.56786 has only five significant figures, four of which are certain and fifth is uncertain.

The position of decimal has nothing to do with determining the number of significant figures. Thus, the numerical expressions 0.2391, 2.391, 23.91, 239.1 and 2391 all have four significant figures. A 'zero' is not a significant figure when used for locating a decimal but becomes significant when used at the end of the expression. Thus, the expression .0804 has three significant figures, whereas the expression .08040 has four significant figures.

The number of significant figures can be expressed most clearly, if we write the number using exponential notation. Thus, a number written as 8.04×10^3 has three significant figures, whereas a number written as 8.040×10^3 has four significant figures, whereas the number .008040 written as 8.040×10^{-3} has four significant figures.

Rules for Rounding off the Numerical Expression

1. The digits to be rounded off are all removed together and not one at a time.
2. If the figure following the last figure to be retained is less than five, the last digit retained remains unchanged.
3. If the figure following the last figure to be retained is more than five, the last digit retained is increased by one.
4. If the figure following the last figure to be retained is five and at least one of the other digits to be removed is non-zero, the last digit retained is increased by one.
5. If the figure following the last figure to be retained is five and all other digits to be removed are zero, the last digit retained remains *unchanged,* if it is even but is *increased* by one, if it is odd.

The following two *additional rules* are applied when addition, subtraction, multiplication and division are involved:

1. For multiplication and division, the number of significant figures in the result is the same as in the measurement containing least significant figures.
2. The result of addition or subtraction cannot contain more digit to right of decimal than the figure containing least number of digits to the right of decimal.

Q. 2. Round off the following numbers to three significant figures:

(*a*) 65.8437 (*b*) 69.873 (*c*) 68.752 (*d*) 67.250 (*e*) 63.35

Ans. (*a*) Applying rules 1 and 2, 65.8437→65.8
(*b*) Applying rules 1 and 3, 69.873 → 69.9
(*c*) Applying rules 1 and 4, 68.752 → 68.8
(*d*) Applying rules 1 and 5, 67.250 → 67.2
(*e*) Applying rules 1 and 5, 63.35 → 63.4

Q. 3. Evaluate the following expressions rounding off the answer to the appropriate number of significant figures:

(*a*) 42.71 g + 9.643 g + 18.0 g
(*b*) 0.16 m^3 × 10.487 kg m^{-3}

Ans. (*a*) 42.71 g + 9.643 g + 18.0 g = 70.4 g
This is because 18.0 has only one figure to the right of the decimal point.
(*b*) .16 m^3 × 10.487 kg m^{-3} = 1.67792 kg
This is because 0.16 has only two significant figures.

Q. 4. What is meant by the term 'error' ? What are its different types ?
(*Himachal, 2000; Nagpur, 2000; Calcutta, 2005; Mumbai, 2005*)

Ans. Error is defined as the numerical difference between a measured value and the absolute or true value of an analytical determination. The absolute or true value of a quantity is, however, never known. All that we can use is only an *accepted Value.* The value for any quantity is 'accepted' when the only uncertainty in this value is less than the uncertainty in some other quantity with which the other quantity is to be compared.

Error may be represented in the following two ways :

1. **Absolute Error**. The absolute error E in a measurement is expressed as :

$$E = x_i - x_t$$

where x_i = Measured value.
x_t = True value (accepted) for given measurement.

2. **Relative Error.** The relative error E_r in a measurement, is expressed as :

$$E_r = \frac{x_i - x_t}{x_t}$$

where x_i x_i have the same significance as mentioned above.

Relative error may also be expressed as per cent or as parts per thousand (ppt) Thus,

$$E_r = \frac{x_i - x_t}{x_t} \times 100\%$$

$$E_r = \frac{x_i - x_t}{x_t} \times 1000 \text{ ppt}$$

Q. 5. In an experiment, the concentration of zinc in a given sample was found to be 20.17 ppm. Taking the accepted value as 20.00 ppm, calculate the absolute error as well as the relative error as per cent and as parts per thousand.

Ans. Measured value = x_i = 20.17 ppm.
True or accepted value = x_t = 20.00 ppm.
∴ Absolute error, $E = x_i - x_t$
$= 20.17 \text{ ppm} - 20.00 \text{ ppm} = 0.17 \text{ ppm}$

Relative error, $$E_r = \frac{x_i - x_t}{x_t} \times 100 = \frac{0.17 \text{ ppm} \times 100}{20.00 \text{ ppm}} = 0.85\%$$

$$\frac{x_i - x_t}{x_t} \times 1000 = \frac{0.17 \text{ ppm} \times 1000}{20.00 \text{ ppm}} = 8.5 \text{ ppm.}$$

Q. 6. Discuss various types of error. (*Kanpur, 2000; Mumbai, 2005 Andhra 2010*)

Ans. The various types of error that usually occur in measurements can be classified into two broad categories, *i.e.*,

(1) Determinate or Systematic Errors

(2) Indeterminate or Random Errors.

1. **Determinate Errors.** Determinate errors have a finite source which can be identified. A given determinate error is generally unidirectional with respect to the true value and thus makes measured value either lower or higher than the true value. These errors would evidently render the results of all the replicate measurements either low or high.

Determinate errors can be further divided into following types :

(*i*) **Instrument Errors.** These errors arise from the imperfections in measuring device. For instance, measuring devices such as pipettes, burettes, measuring flasks, etc., deliver volumes that are slightly different from those indicated by their graduations. There can be several reasons for these differences like the use of the glassware at a temperature which is significantly different from the temperature at which glassware was calibrated, errors in the original calibration etc.

Instruments powered by electricity are susceptible to determinate errors because of the fall in voltage of battery operated instruments or increased resistance in circuit due to unclean electrical contacts. But these errors can be easily detected and corrected.

(*ii*) **Method Errors.** These errors generally arise from the non-ideal behaviour of reagents and reactions involved. The non-ideality originates from slowness of reactions, incompleteness of reactions, instability of reactions or non-specificity of reagents etc. These errors cannot be easily detected and corrected.

(*iii*) **Personal Errors.** These errors arise from incorrect personal judgement. Many experimental measurements, such as the estimation of the position of pointer between two scale divisions. Judgement of the colour of solution at the end point in a titration, judgement of level of a liquid with respect to graduation on a burette or a pipette are sources of personal errors.

Determinate errors may also be classified into *constant errors* and *proportional errors.*

(*i*). **Constant Errors.** The magnitude of constant error is independent of the size of the sample or size of the quantity that is being measured. It is also independent of the concentration of the substance being analysed. For example, in volumetric analysis, the excess of the titrant (say, 1 drop) that has to be added to bring about a change of colour at the end point remains the same whether we titrate 5 ml or 25 ml of the solution.

Relative error would increase five times if we use a 5 ml pipette in place of a 25 ml pipette, *i.e.*, if we take 5 ml of the solution for titration in place of 25 ml of the solution.

It follows from the above that the effect of constant error can be reduced to minimum by increasing the size of the sample to maximum within the permissible limits.

(*ii*) **Proportional Errors.** Proportional errors take place due to the presence of interfering impurities in the sample. The magnitude of such an error depends upon the fraction of the impurity and is independent of the size of the sample.

Correction of Determinate Errors. The determinate instrument errors are detected and corrected by periodic calibration of the instruments. We observe that response of most of the instruments changes with time because of wear, corrosion or mistreatment. A periodic calibration in therefore a must.

Determinate personal errors can be reduced to a minimum by care and self-discipline. It would be certainly desirable to develop a habit of checking instrument's readings. Most important requisite of avoiding personal error is to keep away from pre-conceptions.

2. **Indeterminate Errors.** These are random or accidental errors whose sources cannot be positively identified. As a result of these errors, the data from replicate measurements fluctuate randomly around the mean of the set. Most frequently, the deviation from the mean is very small. Also, there is almost equal probability of the occurrence of positive and negative errors with the result that overall magnitude of indeterminate errors becomes almost insignificant.

Q. 7. What is meant by accuracy of a measurement ? *(Calcutta, 2005)*

Ans. Accuracy of a measurement denotes closeness of an *experimental* value or the mean value of a set of measurements to the *true* value.

In other words, *it is the difference between the experimental value or the mean value of a set of measurements and the true value.*

Accuracy = Mean value – True value

We can say, smaller the difference between the mean value and the true value, *larger* is the accuracy.

Suppose a sample contains 20.50 per cent of sulphur. The chemcial analysis of the sample by one method gives the mean value of 20.40 and by another method 20.65 per cent of sulphur. Then,

Accuracy of first method = 20.40 – 20.50 = 0.10% (Neglecting the negative sign)

Accuracy of second method = 20.65 – 20.50 = 0.15%

Thus, the first method gives more accurate determination of the percentage of sulphur than the second method.

However, it may be realised that in certain cases, the true value of a quantity is not known. In such cases, it becomes difficult to calculate the accuracy of the measurement. We measure *precision* of the measurement in such situations.

Q. 8. What is meant by the term precision ? *(Himachal,2000; Nagpur,2000; Kerala, 2000; Calcutta, 2005; Mumbai, 2005)*

Ans. *The degree of agreement between two or more replicate measurements made on a sample in an identical manner is known as the precision of the measurement.* Precision reflects the closeness among replicate measurements of the same quantity. For illustration, let us suppose that the percentages of an element in an alloy determined by one analyst are 2.63, 2.61 and 2.62 while those determined by another analyst are 2.70, 2.75, 2.80. Evidently, the precision of the first set of results is better than that of the second.

Frequency of occurrence of a given error

Values of a single quantity measured

Fig. 32.1. The error distribution curve.

If we make a large number of observations of single quantity and then plot the number of times a given value occurs against the value of the quantity itself, we obtain a curve known as *Error Distribution Curve* as shown in Fig. 32.1.

These curves have two useful qualitative features, *viz.*, the *height* of the peak of the distribution curve and the *spread* of the distribution curve (also called *dispersion*).

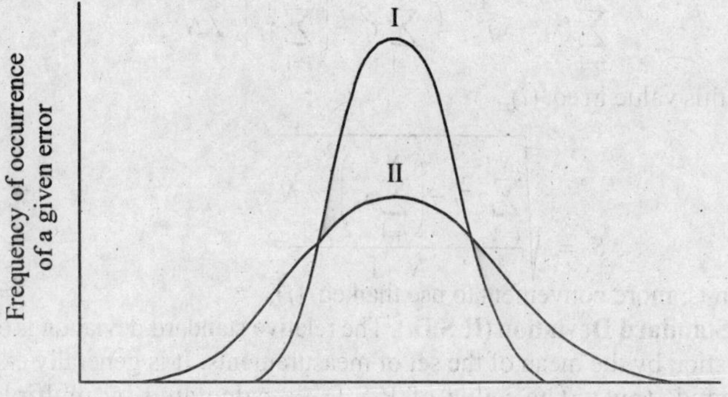

Fig. 30.2. Characteristics of less precise and more precise result

The precision of two sets of measurements of the same quantity can thus be easily compared with the help of their error distribution curves. Thus, in Fig. 32.2, the curve I with lesser spread and higher peak is characteristic of more precise measurements as compared to the curve II which represents less precise measurement.

Q. 9. Write notes on :

(1) Average deviation from the mean.

(2) Standard deviation and *(Calcutta, 2005)*

(3) Relative standard deviation. *(Nagpur, 2003)*

Ans. (1) Average Deviation from the Mean. The precision of data can be expressed conveniently in terms of average deviation from the mean. This quantity for a set of results is obtained by summation of the individual deviations from the mean regardless of sign (because the signs are retained, the mathematical summation would be equal to zero) and dividing the summation by the number of measurements made in the set. Thus, the average deviation from the mean $<d>$ may be expressed as

$$<d> = \sum_{i=1}^{i=N} (x_i - \bar{x}) / N$$

where x_i is the individual measurement, $\bar{x}$ is the mean of the measurements and N is the total number of measurements.

(2) Standard Deviation. The term standard deviation is commonly called in statistics as a measure of precision. This quantity is obtained by the summation of the squares of the individual deviations from the mean, dividing the sum by $N-1$ where N is the number of measurements and then taking the square root. The standard deviation S is thus expressed mathematically as

$$S = \sqrt{\frac{\sum_{i=1}^{N}(x_i - \bar{x})^2}{N-1}} \quad ...(i)$$

The divisor $(N-1)$, in fact, represents the degree of freedom for the set of N measurements. It indicates that there can be $(N-1)$ independent deviations from the mean.

By definition, the mean $\bar{x}$ of the set N measurements is given by

$$\bar{x} = \sum_{i=1}^{N} x_i / N. \quad ...(ii)$$

Mathematically, the numerator under the square root in eq. (*ii*) may be expressed as follows :

$$\sum_{i=1}^{N}(x_i - \bar{x})^2 = \sum_{i=1}^{N} x_i^2 - \left(\sum_{i=1}^{N} x_i\right)^2 / N \quad ...(iii)$$

Substituting this value in eq. (*i*),

$$S = \sqrt{\frac{\sum_{i=1}^{N} x^2{}_i - \left(\sum_{i=1}^{N} x_i\right)^2 / N}{N-1}} \quad ...(iv)$$

This equation is, more convenient to use than eq. (*i*).

(3) Relative Standard Deviation (R.S.D.). The relative standard deviation is obtained by dividing the standard deviation by the mean of the set of measurements. It is generally expressed in terms of 'parts per thousand' (ppt). The value of R.S.D. is calculated by multiplying the quotient $S / \bar{x}$ by 1000. Thus,

$$\text{R.S.D.} = (S / \bar{x}) \times 1000 \text{ ppt}$$

R.S.D. is also expressed in terms of per cent by multiplying $S / \bar{x}$ by 100. The R.S.D. thus obtained is called coefficient of variation (CV).

Q. 10. For titrating 10 ml of a solution with the help of microburette, the volumes of the titrant used are 9.98, 9.99, 9.98, 9.95, 10.00 and 10.20 ml. Calculate the standard deviation.

Ans.

Trial	x_i	x_i^2
1	9.98	99.6004
2	9.99	99.8001
3	9.98	99.6004
4	9.95	99.0025
5	10.00	100.0000

$$\sum x_i = 49.90, \sum x_i^2 = 498.0034$$

$$\frac{\left(\sum x_i\right)^2}{N} = \frac{(49.90)^2}{5} = 498.0020$$

$$S = \sqrt{\frac{498.0034 - 498.0020}{5-1}}$$

$$= \sqrt{\frac{0.0014}{4}} = \sqrt{0.00035} = 0.0187$$

Q. 11. In a set of measurements, the following concentrations of Zn (ppm) were reported : 20.2, 20.4, 20.3, 20.1, 19.9, 20.0.

Calculate : 1. Average Deviation from Mean, 2. Standard Deviation, 3. Relative Standard Deviation and 4. Coefficient of Variation.

Ans. The data may be tabulated as follows :

	Conc. of Zn (ppm) x_i	*Deviation from mean* $x_i - \bar{x}$	$(x_i - \bar{x})^2$
x_1	20.2	.03	.0009
x_2	20.4	.17	.0289
x_3	20.3	.07	.0049
x_4	20.1	.13	.0169
x_5	19.9	.33	.1089
x_6	20.0	.27	.0729
	$\sum x_i = 121.4$	$\sum (x_i - \bar{x}) = 1.00$	$\sum (x_i - \bar{x})^2 = 0.2334$

$$\bar{x} = \frac{\sum x_i}{N} = \frac{121.4}{6} = 20.23$$

1. Average deviation from mean $= \sum_{i=1}^{N} (x_i - \bar{x})/N = \frac{1.00}{6} = 0.167$ ppm Zn.

2. Standard deviation, $S = \sqrt{\frac{0.2334}{5}} = 0.216$

3. Relative standard deviation, R.S.D. $= (S / \bar{x}) \times 1000$ ppm.

$$\text{R.S.D} = \frac{.216}{20.23} \times 1000 = 10.68$$

4. Coefficient of variation, (CV) = $(S / \overline{x}) \times 100\%$

$$= \frac{.216}{20.23} \times 100 = 1.068\%$$

Q. 12. In the measurement of the weight of a sample with an analytical balance which has an uncertainty of .1 mg, the following data are recorded :

Weight of weighing bottle + sample = 11.8733 g

Weight of weighing bottle = 11.4221 g

Calculate the weight of the sample indicating the uncertainty in the final result.

Ans. Since the balance has an uncertainty of 0.1 mg = .0001 g

∴ Weight of weighing bottle + Sampl = 11.8733 ± .0001 g

Weight of weighing bottle = 11.4221 ± .0001 g,

Since the calculation of the weight of the sample from two actual measurements involves the mathematical operation of subtraction, the total uncertainty in the final result will be given by the sum of the absolute uncertainties in individual measurements.

∴ Weight of sample = (11.8733 ± .0001) – (11.4221 ± .0001)

= 0.4512 ± 0.0002

Q. 13. Calculate the absolute error in the following :

(15.02 ± .02) ml × (.2000 ± .0002) mmol ml^{-1}

Ans. When calculations involve operations of multiplication and division, the relative determinate errors are transmitted directly into final result.

$$\textbf{Relative Error.} (\%) = \pm \frac{.02}{15.02} \times 100 = \pm 0.13$$

$$\text{Relative error in second term } (\%) = \pm \frac{.0002}{.2000} \times 100 = \pm 0.10$$

∴ Total relative error (%) = ± (0.13 + 0.10) = 0.23

Magnitude of total calculated quantity = 15.02 × 0.2000= 3.004

$$\therefore \quad \text{Absolute error} = \pm \frac{.23 \times 3.004}{100} = \pm 0.007$$

Q. 14. Calculate the maximum uncertainty in each of the following expressions :

(*a*) (17.3 ± 0.2) ml – (9.7 ± 0.2) ml + (11.6 ± 0.1) ml

(*b*) (16.6 ± 0.1) g –2 (16.7 ± 0.1) g + 3 (7.3 ± 0.1) g

Ans.

(*a*)	(*b*)
17.3 ± 0.2	16.6± 0.1
– 9.7 ± 0.2	–33.4± 0.2
11.6 ± 0.1	21.9 ± 0.3
19.2±.5ml	5.1 ± 0.6 g

Q. 15. Calculate the result of following expression :

(23.4 ± 0.2) (17.7 ± 0.05)

Ans. (*i*) From the given data, calculate percentage uncertainties.

(*ii*) Add up these percentage uncertainties.

(*iii*) Convert the percentage uncertainties into uncertainty.

Step 1. $(23.4 \pm 0.1)(17.7 \pm 0.05) = 414.18 \pm ?$

Step 2. Percentage uncertainties of two multipliers are :

$$\pm\frac{0.2}{23.4}\times 100 \quad \text{and} \quad \pm\frac{0.5}{17.7.}\times 100$$

Step 3. Summation of percentage uncertainties = ± 1.13

The result may, therefore, be written as 414.18 ± 1.13%

Step 4. Uncertainties of the result $= \pm\frac{1.13}{100.} \times 414.18 = 4.68$

The final result is 414.18 ± 4.68

Q. 16. What is meant by sample population, sample mean and population mean ?

Ans. The limited set of data of finite number of value is called the 'Sample' which is regarded as a subset of a universe of data that exist only in imagination and is called the'Population'.

The Sample Mean and the Population Mean. The sample mean is the mean of the 'Sample' of measurements,*i.e.*,a finite number of measurements drawn from a 'Population'of data, *i.e.*,from an infinite number of measurements. It is denoted by the symbol $\bar{x}$ and is mathematically represented as :

$$\bar{x} = \sum_{i=1}^{N} x_i / N$$

where N is usually a relatively small number. The population mean is the mean of the entire population of data. It is denoted by the sample and is mathematically represented as :

$$\mu = \sum_{i=1}^{N} x_i / N$$

where N approaches to infinity.

Q. 17. What is Q-Test (quotient test) ? (*Nagpur, 2003*)

Ans. It is a test which has been widely used by analysts to decide whether to retain or reject the suspect result. In this test :

1. The results of various observations in a given set are arranged in the decreasing order of their numerical values.
2. The difference between the suspected result x_s and its nearest neighbour x_n (ignoring the sign ±) is divided by spread w of the entire set (*i.e.*, the difference between the minimum and maximum value in the entire set) to give a quantity $Q_{exp.}$. Thus
$$Q_{exp.} = (x_s - x_n)/w$$
3. The ratio $Q_{exp.}$ is then compared with critical values of the rejection ratios obtained statistically at appropriate confidence levels and tabulated as a function of the number of observations. These ratios are termed as $Q_{critical}$.

If $Q_{exp.} > Q_{critical.}$ The suspect result is rejected.

Q. 18. In the analysis of an ore, the percentages of an element were found to be 65.55, 65.90, 67.85, 66.85, 69.90 and 65.00. The value 69.90 appears to be suspect. Ascertain whether this should be retained or rejected. The $Q_{critilical}$ for 6 observations at 90% confidence level is 0.56.

Ans. 1. Arranging the various experimental values in the decreasing order of magnitude, we have :

69.90, 67.85, 66.85, 65.90, 65.55, 65.00

2.The difference between the suspect result (x_s)and the nearest neighbour x_n.

i.e, $(x_s - x_n) = 69.90\% - 67.85\% = 2.05\%$

3. Spread of the entire set, $w = 69.90\% - 65.00\% = 4.90\%$

4. $Q_{exp.} = 2.5/4.90 = 0.418$

5. $Q_{critical} = 0.56$ (given)

Since $Q_{exp} < Q_{critical}$ hence, the suspect result cannot be rejected.

Q. 19. Describe *F*-test to determine the difference in precision of the new method and the standard method.

Ans. This test is based on measurements of standard deviation, *S*. The various steps involved in this test are as follows :

1.Determination of standard deviation by employing the standard deviation *S* method. The standard deviation is determined with the following equation :

$$S = \sqrt{\frac{\sum (x_i - \bar{x})^2}{N-1}} \quad ...(i)$$

where $N-1$ gives the number of degree of freedom which is less than the number *N* of the results. The value of *N* is sufficiently large. The deviation obtained by standard method is denoted by $S_{std.}$

2. Determination of standard deviation by employing the newly developed method. The degree of freedom (or number of results) used in this determination is small, generally less than 20.Let the standard deviation obtained by the new method be denoted by S_n.

3. Determination of *F*-value with the help of the expression

$$F = S_{std}^2 / S_n^2 \quad ... (ii)$$

The square of standard deviation is also known as *variance.*

4. Tabulation of critical values of *F* determined statistically at appropriate levels of the confidence by extending the degrees of freedom (*N* value) to infinity both for the determination of S_{std} and S_n.

5. Comparison of the value of *F* as obtained from eq. (*ii*) with the critical value of *F* corresponding to *N* as infinity.

If the value of $F <$ critical value, there is no significant difference in the precision of the new method and the standard method.

Q. 20. Two sets of results, in cm/sec., one obtained by the standard method and the other by the new method, are given below:

Standard Method	**31**	**27**	**26**	**35**	**23**	**31**	**33**	
New Method	**25**	**23**	**24**	**29**	**24**	**28**	**30**	**25**

Determine whether the precision of the new method differs significantly from that of the standard method. The critical value of *F* for 7 degree of freedom is 2.23.

Ans.

New Method			Standard Method		
x_i	$(x_i - \bar{x})$	$(x_i - \bar{x})^2$	x_i	$(x_i - \bar{x})$	$(x_i - \bar{x})^2$
25	1	1	30	1	1
23	3	9	27	2	4
24	2	4	25	4	16
29	3	9	35	6	36
24	2	4	23	6	36
28	2	4	31	2	4
30	4	16	32	3	9
25	1	1			
$\Sigma x = 208$		$\Sigma(x_i - \bar{x})^2 = 48$	$\Sigma x_i = 203$		$\Sigma(x_i - \bar{x})^2 = 106$

$$\bar{x} = \frac{208}{8} = 26 \qquad \bar{x} = \frac{203}{7} = 29.0$$

$$S_n = \sqrt{\frac{\sum(x_i - \bar{x})^2}{N-1}} = \sqrt{\frac{48}{7}} \qquad S_{std} = \sqrt{\frac{106}{6}}$$

$$F = \frac{S_{std}^2}{S_n^2} = 2.57$$

Critical value of $F = 2.23$ (given)

Since the value of $F <$ critical value, there is no significant difference in the precision of the new method and the standard method.

Q. 21. What is *t*-test ? *(Calcutta, 2005)*

Ans. This test is employed to determine the difference in precision between a new method and the standard method.

In this test, different samples of material under test, having different concentrations are analysed by the new and the standard methods from the results, the t value is obtained with the help of the relation

$$t = \frac{\bar{d}}{S_d}\sqrt{N} \qquad \text{... (i)}$$

S_d = Standard deviation defined by the equation

$$S_d = \sqrt{\frac{\sum(d_i - \bar{d})^2}{N-1}} \qquad \text{...(ii)}$$

In eq. (*ii*), N is the number of measurements, d_i is the individual difference in results yielded by two methods and $\bar{d}$ is the mean of the individual differences.

The critical values of t are also obtained statistically and tabulated as function of the number (N) of the measurements made. From the table, the critical value of t corresponding to any degrees of freedom ($N-1$) can be readily obtained.

The experimental value of t as obtained from eq. (*i*) is then compared with the critical value of t for the same degree of freedom. If the experimental value is less than the critical value of t, there is no significant difference between the two methods.

LEAST SQUARE METHOD

Q.22. What is meant by the method of least squares for curve fitting.

Ans. The best fitting curve with minimal deviation from all data points can be obtained by the method of least squares.

The method of least squares

The method of least squares assumes that the best fitting curve is the curve that has the minimal sum of deviations squared (least square error) from a given set of data.

Suppose the data points are $(x_1, y_1), (x_2, y_2), ... (x_n, y_n)$ where x is the independent variable and y is the dependent variable. The fitting curve $f(x)$ has the deviation or error d from each data point *i.e.* $d_1 = y_1 - f(x_1), d_2 = y_2 - f(x_2) ... d_n = y_n - f(d_n)$. According to the method of least square, the best fitting curve has the property that

$$\pi = d_1^2 + d_2^2 + ... + d_n^2 = \sum_{i=1}^{n} d_i^2 = \sum_{i=1}^{n} [y_1 - f(x_i)]^2 = a \text{ minimum}$$

The least squares line. The least square line method uses a straight line $y = a + bx$ to approximate the given set of data

$$(x_1, y_1), (x_2, y_2), \dots (x_n, y_n) \text{ where } n \geq 2$$

The least squares parabola. The least squares parabola uses a second degree curve $y = a + bx + cx^2$ to approximate the given set of data $(x_1, y_1), (x_2, y_2), \dots, (x_n, y_n)$ where $n \geq 3$.

The least squares mth degree polynomials. The least square *m*th degree polynomial method uses *m*th degree polynomials $y = a_0 + a_1x + ax_2^2 + ... + a_mx^m$ to approximate the given set of data (x_1, y_1), $(x_2, y_2) \dots (x_n, y_n)$ where $n \geq m + 1$

Q. 23. Derive the equation for least squares line

Ans. The least squares line uses a straight line

$$y = a + bx$$

to approximate the given set of data $(x_1, y_1), (x_2, y_2), ... (x_n, y_n)$ where $x \geq 2$

The best fitting curve $f(x)$ has the least square error. Thus

$$\pi = \sum_{i=1}^{n} [y_i - f(x_i)]^2 = \sum_{i=1}^{n} [y_i - (a + bx_i)]^2 = \text{Minimum}$$

All x_i and y_i values are given, while a and b are unknown coefficients. In order to obtain the least square error, the coefficients a and b must give first derivatives as zero. Thus

$$\frac{\partial \pi}{\partial a} = 2\sum_{i=1}^{n} [y_1 - (a + bx_i)] = 0$$

$$\frac{\partial \pi}{\partial b} = 2\sum_{i=1}^{n} x_i[y_i - (a + bx_i) = 0$$

On expanding the above equations, we get

$$\sum_{i=1}^{n} y_i = a\sum_{i=1}^{n} 1 + b\sum_{i=1}^{n} x_i$$

$$\sum_{i=1}^{n} x_i y_i = a\sum_{i=1}^{n} x_i + b\sum_{i=1}^{n} x_i^2$$

From the above equations, values of a and b are obtained as under.

$$a = \frac{(\Sigma y)(\Sigma x^2) - (\Sigma x)(\Sigma xy)}{n\Sigma x^2 - (\Sigma x)^2}$$

$$b = \frac{n\Sigma xy - (\Sigma x)(\Sigma xy)}{n\Sigma x^2 - (\Sigma x)^2}$$

where Σ.... stands for $\sum_{i=1}^{n}$......

Q. 24. Derive the equation for least squares parabola.

Ans. We use the second degree curve for least squares parabola as

$$y = a + bx + cx^2$$

for the approximation of the data $(x_1, y_1), (x_2, y_2), \ldots (x_n, y_n)$ where $n \geq 3$.

The best fitting curve $f(x)$ has the least square error given by

$$\pi = \sum_{i=1}^{n} [y_i - f(x_i)]^2 = \sum_{i=1}^{n} [y_i - (a + bx_i + cx_i^2)]^2 = \text{Minimum}$$

All x_i and y_i values are given and a, b and c are unknown coefficients. The coefficients a, b and c must give first derivatives as zero, in order to obtain least square error.

$$\frac{\partial \pi}{\partial a} = 2\sum_{i=1}^{n} [y_i - (a + bx_i + cx_i^2)] = 0$$

$$\frac{\partial \pi}{\partial b} = 2\sum_{i=1}^{n} x_i[y_i - (a + bx_i + cx_i^2)] = 0$$

$$\frac{\partial \pi}{\partial c} = 2\sum_{i=1}^{n} x_i^2[y_i - (a + bx_i + cx_i^2)] = 0$$

Expanding the above equations,

$$\sum_{i=1}^{n} y_i = a\sum_{i=1}^{n} 1 + b\sum_{i=1}^{n} x_i + c\sum_{i=1}^{n} x_i^2$$

$$\sum_{i=1}^{n} x_i y_i = a\sum_{i=1}^{n} x_i + b\sum_{i=1}^{n} x_i^2 + c\sum_{i=1}^{n} x_i^3$$

$$\sum_{i=1}^{n} x_i^2 y_i = a\sum_{i=1}^{n} x_i^2 + b\sum_{i=1}^{n} x_i^3 + c\sum_{i=1}^{n} x_i^4$$

The quantities a, b, c can be obtained by solving the above equations

VOLUMETRIC ANALYSIS

Q. 1. What is Volumetric analysis? What are the conditions associated with volumetric analysis?

Ans. It is a technique of quantitative analysis recording the volumes of various solutions involved in a chemical process.

In volumetric analysis, a known volume of a solution (*called titrant*) is taken in a conical flask. To this is added drop wise a second solution (*called titre*) whose strength is to be estimated. When the reaction is complete, the volume of second solution used for a known volume of the first solutions is noted. Knowing the volumes of both the solutions and strength of one of them, the strength of other can be estimated. The process is called *titration* and the flask used in the titration is known as *titration flask.* The point at which the reaction is completle is called *end point* or *equivalence point.* The end point is indicated by some marked change such as appearance or disappearance of colour. The reagent which is used to detect this change is called *indicator.*

It may be noted that in actual practice it is difficult to attain the extract point but it will be very near to exact one.

Conditions for Volumetric Analysis

The following requirements must be fulfilled in order that a reaction can be used as the basis of a volumetric determination.

(*a*) The reaction involved must be definitely known and should be representable by a chemical equation.

(*b*) The reaction must be very fast and instantaneous so that the titration can be performed with a reasonable period of time.

(*c*) The reaction should proceed to completion.

(*d*) The end point should be well defined either by reacting solutions or by use of another substance called indicator

Q. 2. Explain the terms (*i*) concentration (*ii*) standard solution

Ans.

1. **Concentration.** The concentration of solution is the amount of solute dissolved in a specified amount of the solvent of the solution. It is usually expressed as gram of the solute dissolved per litre of the solution.
2. **Standard Solution.** It is defined as a solution containing a known weight of the solute dissolved in a known volume of it. Strength is often expressed in g/litre.
   For example, if 6 g of a substance are dissolved in 1000 ml, its strength is 6g per litre.
   Similarly, if 2.5 g of a substance are dissolved in 250 ml of a solution, its strength is 10 g per litre.

In general the concentration of solution refers to the weight of a solute dissolved in a unit volume of the solution. There are many modes of expressing concentration of which the following are very common.

Q. 3. What are the different modes of expressing concentration of a solution?

Ans. Different modes of expressing concentration are discussed as under:

(*i*) **Percentage Composition by Weight.** It is the number of grams of the substance dissolved in 100 g of its solution. This term is generally used in dilute solutions of electrolytes in water *e.g.,* A 5% solution of a given salt is prepared by dissolving 5 g of the salt in 95 g of water.

(*ii*) **Percentage Composition by Volume.** In this method, the volume ratios of solute and solvent are considered. So a 25% of a solution of alcohol is prepared by mixing 25 ml of alcohol with 75 ml of water.

(*iii*) **Molarity and Molar solution.** It is the number of moles of the substance dissolved per litre of its solution. If one gm mole of a substance dissolved in a solvent and the solution is made upto the one litre mark of the flask, such solution is a Molar solution. If W g of a substance is dissolved in V ml of the solution and mol. wt. of the substance is M then

$$\text{Molarity} = \frac{\text{Strengh of solute in g/litre}}{\text{Mol. wt. of the solute}}$$

$$= \frac{W \times 1000}{M \times V}$$

(*iv*) **Molarity and Molal solution.** It is the number of moles of the solute dissolved per 1000 grams of the solvent and is symbolised as 'm'. A solution is known as 1 molal if one mole of the substance is dissolved in 1 kg of the solvent

or
$$\text{Molarity} = \frac{\text{Strengh of solute/1000 g of solvent}}{\text{Mol. wt. of the solute}}$$

(*v*) **Normality.** It is the number of moles of the solute dissolved per litre of the solution. If one gm equivalent of an acid, base or salt is dissolved in water and the solution is made up one litre, such solution is called 'Normal solution'

If W gm of a substance is dissolved in V ml of the solution and equivalent wt. of the substance is E.

$$\text{Normality} = \frac{\text{Strength of substance in g/litre}}{\text{Equivalent weight}}$$

$$= \frac{W \times 1000}{M \times V}$$

(*vi*) **Formality.** It is the number of gram formula weight of the substance dissolved per litre of solution.

$$\text{Formality} = \frac{\text{Strength of substance in g/litre}}{\text{Equivalent weight}}$$

Q. 4. How do we calculate the equivalent weight of different kinds of substances?

Ans. The chemical equivalent with reference to some standard is known as equivalent weight. Equivalent weight of hydrogen is taken as unity. Equivalent weight of an element is defined as the number 0, parts by weight of it that combine or displace one part by weight of hydrogen or 8 parts by weight of oxygen or 35.5 parts by weight of chlorine.

$$\text{Eq. wt. of an acid} = \frac{\text{Mol. wt.}}{\text{basicity}}$$

$$\text{Eq. wt. of the base} = \frac{\text{Mol. wt.}}{\text{acidity}}$$

$$\text{Eq. wt. of salt} = \frac{\text{Mol. wt.}}{\text{Hydrogen atoms equivalent to total number of cations or anions}}$$

Eq. wt. of an electrolyte = Eq. wt. of cation + Eq. wt. of anion

Equivalent weight is not a constant number. It may change from reaction to reaction. Such variations are often observed in the reaction involving oxidation-reduction.

Gram Equivalent Weight. It is the equivalent weight of a substance expressed in grams. For example, equivalent weight of hydrogen chloride is 36.5. Therefore its gram equivalent weight is 36.5 gram.

Q. 5. What is normality equation? Derive this equation?

Ans. The expression which co-relates the normalities and volumes of the two reacting solution in a titration is known as 'Normality Equation'. It is written as

$$\boxed{N_1 \times V_1 = N_2 \times V_2}$$

where N_1 and V_1 represent normality and volume used for one solution, N_2 and V_2 stand for normality and volume used for the second solution.

Derivation of Normality Equation. Consider a neutralisation reaction in which V_1 ml of a solution (I) of normality N_1 require V_2 ml of a second solution (II) of normality N_2. On the completion of the chemical reaction, they should have reacted in the ratio of their gram equivalents. The number of gm. equivalents of each of them can be calculated as below:

For Solution I

$$\text{Normality} = \frac{\text{No. of g. eq.}}{\text{Volume in litre}}$$

or

$$\text{No. of g. eq.} = \text{Normality} \times \text{Volume in litres}$$

$$= N_1 \times \frac{V_1}{1000}$$

For Solution II

$$\text{No. of g. eq.} = N_2 \times \frac{V_2}{1000}$$

By the law of chemical equivalents

$$N_1 \times \frac{V_1}{1000} = N_2 \times \frac{V_2}{1000}$$

$$N_1 V_1 = N_2 V_2$$

In the above equation if three quantities out of N_1, N_2, V_1 and V_2 are known, the fourth can be calculated. This is the big use of this equation.

Q. 6. Describe the commonly used apparatus in volumetric analysis.

Ans. The commonly used apparatus in volumetric analysis are given below.

1. Burette

The most important item in volumetric analysis is burette. It is a long graduated cylindrical tube having a bore of uniform cross section. At the lower end it is fitted with a glass stopper or a short rubber tubing carrying a jet and pinch cock which controls the flow of liquid, (See Fig. 33.1). The commonly used burette has a capacity of 50 ml. Each ml mark is further divided into 10 equal parts. Each part signifying 1/10th of ml. By carefully controlling the stop cock or pinch cock, the liquid can be allowed to flow and hence known volume of it can be delivered out.

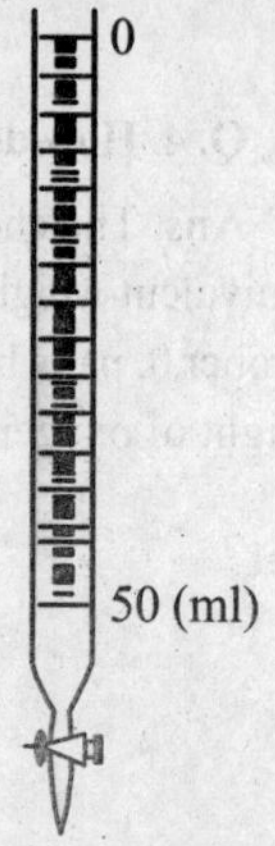

Fig. 33.1. Burette

How to use a burette ?

Before filling a burette it is washed with water to check that on draining water comes out, and no drops stick to the inside wall. Immediately after washing, the burette is rinsed twice with the solution to be filled in. After draining out the rinsing liquid pour in the liquid into the burette (through clean funnel) till the level is fairly high above the zero mark. Remove the funnel and open the pinch cock very slowly and allow the liquid just to drop out after filling the jet. See that no air bubble is left anywhere inside the jet.

Open the stock cock and let the meniscus of the liquid come within the range of graduations. In case of colourless liquids the lower meniscus should be noted but in case of coloured liquids upper meniscus should be noted. To make the meniscus distinct to eye a small *anti parallax* card may be used. It is simply a piece of paper the lower portion of which is blackened.

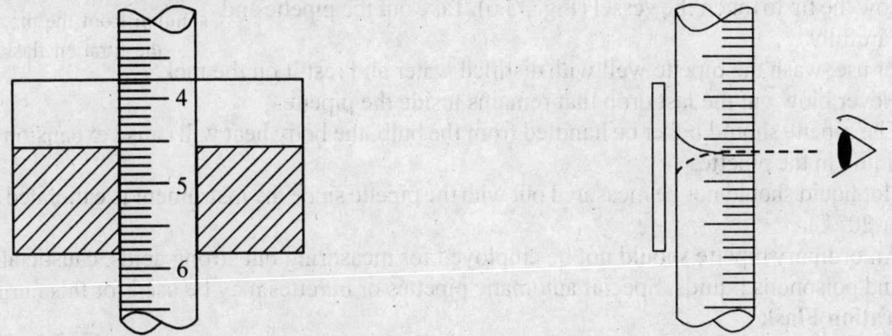

Fig. 33.2. Use of antiparallax card **Fig. 33.3.** Correct method of noting burette reading

While reading the burette, see that you do not put your eye above or below the meniscus of the liquid. The eye should be maintained in line with the meniscus. Read whereever the meniscus of the liquid stands (Fig. 33.3).

Now run the liquid in the titration flask and find the end point. Again, read the burette, the difference between the two reading gives the volume of liquid used up.

Lower meniscus should be read in case of colourless liquids and upper meniscus in case of coloured liquds.

2. Pipette

It is a long narrow tube with the bulb in the centre and drawn into a jet at the lower end. On the upper part of the narrow tube, there is a circular etched mark. The capacity of a pipette is marked on the bulb (Fig. 33.4)

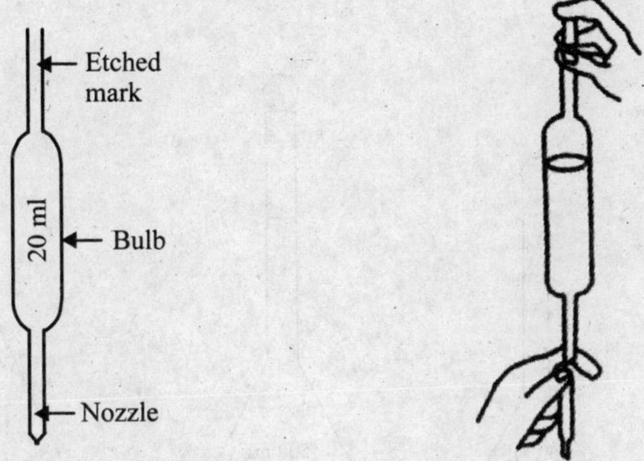

Fig. 33.4. Pipette **Fig. 33.5.** How to hold pipette?

Use of the Pipette

Before a pipette is used it is made scrupulously clean and free from grease and dirt using tap water followed by distilled water. Rinse the pipette with the solution to be pipetted. Now the pipette is ready for use. Dry the upper part of the pipette with your finger tips. Dip the tip of pipette well inside the solution and suck the liquid to a point well above the etched mark. Close the upper end with your forefinger quickly (Fig. 33.5). The liquid is now allowed to fall down slowly by releasing the forefinger pressure carefully and press again to hold the liquid so that its lower meniscus is in the level of the index mark. Hold the pressure and do not allow the liquid to fall. Take out the pipette and place it inside the conical flask. When the whole of the liquid has been transferred, lift the pipette and again allow the tip to touch the vessel (Fig. 33.6). Take out the pipette and keep it carefully.

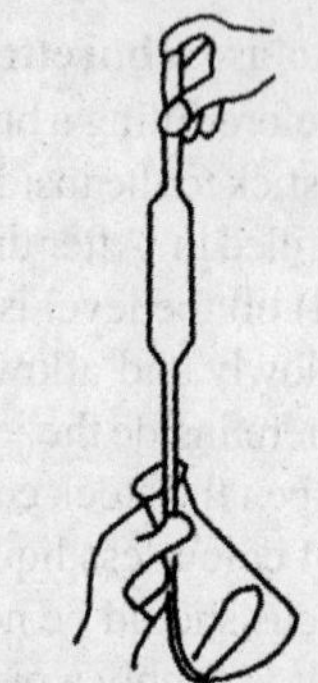

Fig. 33.6. How to deliver solution from the pipette to the titration flask.

After use, wash the pipette well with distilled water and rest it on the rack.

1. Never blow out the last drop that remains inside the pipette.
2. The pipette should never be handled from the bulb, the body heat will cause expansion of the liquid in the pipette.
3. Hot liquid should not be measured out with the pipette since the instrument is calibrated at 15° or 20° C.
4. An ordinary pipette should not be employed for measuring out strong acids, caustic alkalies and poisonous liquids. Special automatic pipettes or burettes may be used for this purpose.

3. Titration Flask

It is a special type of flask of convenient size used for carrying out titrations. It is usually a conical flask with a short neck and sloping sides. Its bottom is flat (Fig. 33.7). It not only permits a thorough shaking of the contents without any fear of their splashing out but also it can be heated when required. Before using the titration flask, thoroughly wash it with water.

It must never be rinsed with the liquid to be titrated.

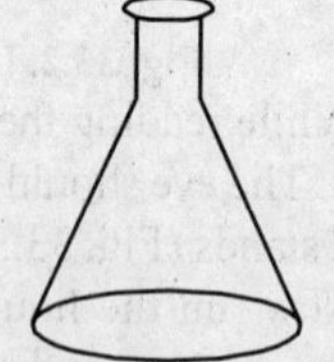

Fig. 33.7. Titration flask.

4. Measuring Flask

It is a flat bottomed flask with a narrow neck. Its mouth is fitted with an air tight glass stopper. The stem of the flask has an engraved mark on it. The volume of liquid or solution when filled upto this mark is equal to the volume printed on it. The measuring flasks usually available in chemical laboratory are of 100 ml, 250 ml, 500 ml and 1000 ml. (Fig. 33.8)

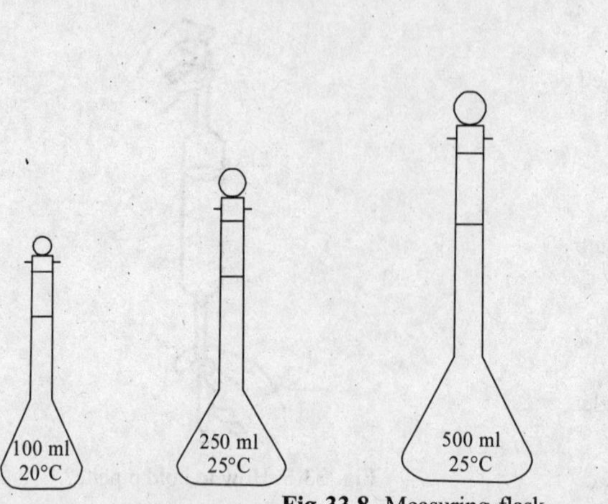

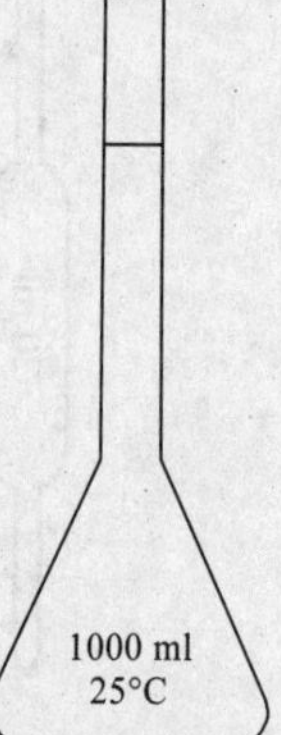

Fig 33.8. Measuring flask

Q. 7. (*a*) Explain the terms (*i*) Primary standard (*ii*) Secondary standard

(*b*) How is the standard solution of a substance prepared?

(*a*) (*i*) Primary Standard

The substance whose standard solution is prepared by dissolving directly its known amount in a definite volume of solvent is known as primary standard.

A few examples of primary standard substances are oxalic acid, Na_2CO_3, Mohr's salt, $K_2Cr_2O_7$, NaCl.

Conditions for a Substance to act as a Primary Standard

1. It should be easily available in pure and dry form and should be stable under the titration conditions.
2. It should not be hygroscopic, deliquescent efflorescent.
3. For a hydrated substance (Crystals), water content should remain the same.
4. It should not be oxidized by air.
5. It should be readily soluble in water or the solvent employed.
6. It should have high equivalent weight so that weighing errors are negligible.
7. Its standard solution should react in stoichiometric ratio with the volumetric reagent (titrant).

(*ii*) Secondary Standard

A substance whose standard solution cannot be prepared directly by weighing its definite amount and dissolving it in a definite volume of water is called as secondary standard. This is because such substances are not available in pure form.

The solution of such type of substance is prepared with approximate strength, which is then standardized with a standard solution of a primary standard. Potassium permanganate and sodium hydroxide are examples of secondary standard.

(*b*) Following apparatus is needed for preparing a standard solution of a substance. Measuring flask, funnel, beaker, wash-bottle, analytical balance, weight box, fractional weight box, weighing bottle or watch glass.

The substance is generally weighed on a watch glass or a weighing bottle. Weight of the empty bottle is taken first. The desired amount of the substance is then taken in the weighing bottle and the bottle is weighed again, difference in the two volumes of weight gives the weight of the substance. The substance is then transferred to the measuring flask, particles sticking to the bottle are removed with a spray from wash bottle.

Q. 8. Give a brief account of different types of titrations.

Ans.

1. Acid-alkali titrations

These involves neutralisation of an acid with an alkali. Thus

$$2NaOH + \begin{array}{c} COOH \\ | \\ COOH \end{array} \longrightarrow \begin{array}{c} COONa \\ | \\ COONa \end{array} + 2H_2O$$

$$Na_2CO_3 + 2HCl \longrightarrow 2NaCl + H_2O + CO_2$$

The operation in which an acid is estimated by titrating it with a standard alkali is termed as acidimetry, while the estimation of an alkali by titrating against a standard acid is called alkalimetry.

2. Redox titrations

These involve reduction and oxidation reactions. Since both these reactions take place simultaneously; the titrations are termed as redox titrations.

These can be further classified into three types:

(*a*) **Permanganate titrations.** In these titrations a standard solution of potassium permanganate is added to an acidified solution of a (Fe^{2+}) salt or a soluble oxalate. Chemical reactions involved are as given below:

$$2KNnO_4 + 3H_2SO_4 \longrightarrow K_2SO_4 + 2MnSO_4 + 3H_2O + 5[O]$$

$$2FeSO_4 + H_2SO_4 + O \longrightarrow Fe_2(SO_4)_3 + H_2O$$

$$\begin{matrix} COONa \\ | \\ COONa \end{matrix} + H_2SO_4 \longrightarrow \begin{matrix} COOH \\ | \\ COOH \end{matrix} + Na_2SO_4$$

$$\begin{matrix} COOH \\ | \\ COOH \end{matrix} + O \longrightarrow 2CO_2 + H_2O$$

(*b*) **Dichromatic titrations.** In these titrations a standard solution of potassium dichromate $K_2Cr_2O_7$ is added to an acidified solution of ferrous (Fe^{2+}) salt.

$$K_2Cr_2O_7 + 4H_2SO_4 \longrightarrow K_2SO_4 + Cr_2(SO_4)_3 + 4H_2O + 3[O]$$

$$2FeSO_4 + H_2SO_4 + O \longrightarrow Fe_2(SO_4)_3 + H_2O$$

(*c*) **Iodine titrations.** These titrations involve oxidation reactions by iodine. These can be divided into the following types:

(*i*) *Iodimetry.* It covers titrations of thiosulphate, sulphite, nitrite, arsenite, tartaremetic, etc., with a standard solution of Iodine.

$$\underset{\text{Sod. thiosulphate}}{2Na_2S_2O_3} + I_2 \longrightarrow \underset{\text{Sod. tetrathionate}}{Na_2S_2O_6} + 2\,NaI$$

$$\underset{\text{Sod. sulphite}}{Na_2SO_3} + 2\,I_2 + 2H_2O \longrightarrow Na_2SO_4 + 2\,HI$$

$$\underset{\text{Arsenious oxide}}{As_2O_3} + 2\,I_2 + 2H_2O \longrightarrow As_2O_5 + 4HI$$

$$2\underset{\text{Tartaremetic}}{\begin{matrix} CH(OH)COOK \\ | \\ CH(OH)COO(SbO) \end{matrix}} + I_2 + 3H_2O \longrightarrow 2\begin{matrix} CH(OH)COOK \\ | \\ CH(OH)COOH \end{matrix} + Sb_2O_5 + 2HI$$

(*ii*) *Iodometry.* It deals with titration of iodine liberated from potassium iodide by the action of an oxidizing agent like $KMnO_4$, $K_2Cr_2O_7$, chlorine water, hydrogen peroxide, copper sulphate, etc. with a standard solution of sodium thiosulphate.

$$2KI + H_2O + O \longrightarrow 2\,KOH + I_2$$

$$2KI + H_2O_2 \longrightarrow 2\,KOH + I_2$$

$$2KI + Cl_2 \longrightarrow 2\,KCl + I_2$$

$$2CuSO_4 + 4KI \longrightarrow Cu_2I_2 + 2K_2SO_4 + I_2$$

Then

$$I_2 + 2Na_2S_2O_3 \longrightarrow Na_2S_4O_6 + 2\,NaI$$

3. Complexometirc Titrations

These are the titrations which involve the reaction of a metal ion with a ligand to form a soluble complex. The reactants in this titration should be such that the rate of reaction is fast enough so that the titration could be carried out in a reasonable time and no precipitation should occur during titration. In such titrations we use EDTA as a chelating ligand for the estimation of Zn^{2+}, Mg^{2+}, Ca^{2+}, etc.

$$\begin{matrix} HOOC-CH_2 \\ HOOC-CH_2 \end{matrix}\!\!>N-CH_2-CH_2-N\!\!<\!\!\begin{matrix} CH_2COOH \\ CH_2COOH \end{matrix}$$

EDTA

Q. 9. What are the indicators used in acid base titrations?

Ans. Phenolphthalein and methyl orange are the two indicators used in acid-base titrations. Phenolphthalein gives a pink colour in alkaline medium and is colourless in acidic medium. Methyl orange gives a pink colour in acidic medium and is yellow in alkaline medium. However, which indicator to use in a particular acid-base combination depends upon the type of acid and base. The following table will prove useful.

Strong acid (like HCl, HNO_3, H_2SO_4)	Strong base (like NaOH, KOH)	Methyl orange or phenolphthalein
Strong acid	Weak base (like NH_4OH, Na_2CO_3)	Methyl orange
Weak acid (like acetic acid, citric acid)	Strong base	Phenolphthalein

Q. 10. How will you estimate a mixture of Na_2CO_3 and $NaHCO_3$ by acid-base titration?

Ans. Theory

The mixture of Na_2CO_3 and $NaHCO_3$ is dissolved in a known volume of water. The solution is first titrated with a standard solution of HCl using methyl orange indicator. This gives us the total normality due to Na_2CO_3 and $NaHCO_3$. Then the mixture solution is titrated with standard solution using phenolphthalein indicator. The neutralisation of Na_2CO_3 takes place in two steps.

$$Na_2CO_3 + HCl \longrightarrow NaHCO_3 + NaCl$$

$$NaHCO_3 + HCl \longrightarrow NaCl + CO_2 + H_2O$$

As soon as CO_2 is evolved, it reacts with phenolphthalein indicator. Phenolphthalein stops working as indicator. However, methyl orange is not affected by the evolution of CO_2.

Thus the normality that we shall obtain with methyl orange indicator will give the total normality due to Na_2CO_3 and $NaHCO_3$. But the normality that we shall obtain with phenolphthalein indicator will be half the normality due to Na_2CO_3 only. By calculation we can determine the amount of Na_2CO_3 and $NaHCO_3$ in the mixture.

Calculation

Let the weight of the mixture be $= x$ g

Let us say this mass was dissolved in 250 ml of water

First Titration (using Methyl orange as indicator). Let the volume of mixture solution taken each time = 20 ml.

Volume of N/10 HCl required for neutralisation $= v_1$ ml

Apply normality equation

$$\underset{\text{Mix}}{N_1 V_1} = \underset{\text{HCl}}{N_2 V_2}$$

Substituting the values

$$N_1 \times 20 = \frac{N}{10} \times V_2$$

or $$N_1 = \frac{N}{10} \times \frac{v_1}{20} \qquad (\because V_2 = v_1)$$

where N_1 is the total normality due to Na_2CO_3 and $NaHCO_3$

Second Titration (using phenolphthalein as indicator)

Let the volume of mixture solution taken each time = 20 ml

Volume of N/10 HCl required for neutralisation $= v_2$ ml

Apply normality equation

$$\underset{\text{Mix}}{N_1'V_1'} = \underset{\text{HCl}}{N_2'V_2'}$$

Substituting the values

$$N_1' \times 20 = \frac{N}{10} \times v_2 \qquad (\because V_2' = v_2)$$

or

$$N_1' = \frac{N}{10} \times \frac{v_2}{20}$$

where N_1' is equal to half the normality due to Na_2CO_3

$\therefore$ Normality due to $Na_2CO_3 = 2N_1'$

Normality due to $NaHCO_3$ = Total normality – normality due to Na_2CO_3

$$= N_1 - 2N_1'$$

Strength /litre of Na_2CO_3 = Normality × Eq. weight

$$= 2N_1' \times 53 \text{ g}$$

Strength in 250 ml of Na_2CO_3 Solution $= \frac{2N_1' \times 53}{4} = \frac{N_1' \times 53}{2}$

Strength/ litre of $NaHCO_3$ = Normality × Eq. weight

$$= (N_1 - 2N_1') \times 84 \text{ g}$$

Strength in 250 ml of $NaHCO_3$

$$\text{Solution} = (N_1 - 2N_1') \times \frac{84}{4}$$

$$= 21\,(N_1 - 2N_1')\text{ g}$$

After obtaining the weights of Na_2CO_3 and $NaHCO_3$, we can determine the percentage of each in the mixture.

POTASSIUM PERMANGANATE TITRATIONS

Q. 11. How do we prepare a standard solution of $KMnO_4$?

Ans. Potassium permanganate is not a primary standard substance. It is always impure and contaminated with manganese dioxide. Ordinary distilled water is associated with organic matter which may reduce potassium permanganate to manganese dioxide. The presence of manganese dioxide (MnO_2) is unwanted because it catalyses the auto decomposition of permanganate solutions on standing. Because of these reasons a standard solution of permanganate cannot be prepared by exact weighing directly. Its standard solution is obtained by an indirect method.

The solution of $KMnO_4$ is first prepared, kept for some time, filtered through purified glass wool and is standardized against a primary standard. The substances which are commonly used standardize $KMnO_4$ are hydrated oxalic acid, $H_2C_2O_4.2H_2O$, sod. oxalate, $(COONa)_2$ and ferrous ammonium sulphate. $FeSO_4(NH_4)_2SO_4.6H_2O$. They are employed as primary standard as they are available in pure form and are stable. Sodium oxalate is the best of all.

Q. 12. How do we perform $KMnO_4$ titrations?

Ans. The following points may be kept in mind while performing $KMnO_4$ titrations.

1. $KMnO_4$ solution is always taken in a burette.
2. Avoid the use of burette with a rubber pinch cock as potassium permanganate attacks rubber.
3. If you forget to add dilute sulphuric acid to the solution in the titration flask, a brown precipitate will appear on running $KMnO_4$ solution into it due to formation of MnO_2 which cannot be destroyed by adding the acid now. In such a case throw away the solution and start afresh.
4. No indicator is added since the pot. permanganate acts as self indicator.

5. In case oxalic acid or an oxalate is the reducing agent to be titrated, add dil. Sulphuric acid as directed above and heat the flask at 60-70°C on a tripod stand with a wire gauge before adding permanganate. To ascertain that the solution in the flask has been heated to the desired temperature, heat it for 3–4 minutes, hold with the left hand, touch its bottom with your right hand as shown in Fig. 33.9. If it is unbearable to touch, it is in order, otherwise heat more. May be that you take long to complete the titration and the solution get cold. Heat the titration flask again to the same temperature. Make sure that the reaction mixture is duly hot at the end point.

Fig. 33.9. Judging the temperature of the flask.

6. Never run in large amount $KMnO_4$ solution at a time. This may form a brown precipitate of hydrated manganese dioxide in a titration flask and spoil the titration.
7. Unlike oxalic acid and oxalates, the oxidation of ferrous ion proceeds in cold, hence the solutions of ferrous salts are titrated against $KMnO_4$ solution in cold.
8. As $KMnO_4$ solution is run into the hot solution of oxalic acid in the titration flask, the temperature of the latter solution will come down. Therefore, it is advisable to see that the pink colour obtained at the end of the titration does not disappear on re-heating the titration flask.

Q. 13. Give in a tabular form the equivalent weights of different reducing agents in $KMnO_4$ titrations.

Ans. Equivalent weight of substance is given by

$$\text{Eq. weight} = \frac{\text{Molecular weight}}{\text{No. of electrons lost or gained}}$$

The values for $KMnO_4$ and some reducing agents are given in the table below. The ion electron equations and the calculation of eq. weight is also explained.

Substance	Ion electron equation	Eq. wt. = Mol. wt./No. of electrons lost or gained.
1. Potassium permanganate ($KMnO_4$)	$MnO_4^- + 8H^+ + 5e^- \longrightarrow Mn^{2+} + 4H_2O$	158/5 = 31.6
2. Oxalic acid ($C_2H_2O_4\ 2H_2O$)	$C_2O_4^{2-} \longrightarrow 2CO_2 + 2e^-$	126/2 = 63
3. Sodium oxalate ($Na_2C_2O_4$)	--------- DO --------	134/2 = 67
4. Ferrous sulphate ($FeSO_4 \cdot 7H_2O$)	$Fe^{2+} \longrightarrow Fe^{3+} + e^-$	278/1 = 278
5. Mohr's salt Ferrous ammonium sulphate [$FeSO_4\ (NH_4)_2\ SO_4 \cdot 6H_2O$]	$Fe^{2+} \longrightarrow Fe^{3+} + e^-$	392/1 = 392
6. Ferrous oxalate [$FeC_2O_4 . 2H_2O$)	$Fe^{2+} + C_2O_4^{2-} \longrightarrow Fe^{3+} + 2CO_2 + 3e^-$	180/3 = 60
7. Manganese dioxide [MnO_2]	$MnO_2 + 4H^+ + 2e^- \longrightarrow Mn^{2+} + 2H_2O$	87/2 = 43.5
8. Hydrogen peroxide [H_2O_2]	$H_2O_2 + 2H^+ + 2e^- \longrightarrow 2H_2O$	34/2 = 17

Note: After performing the titration and using normality equation, we can solve problems on percentage purity, percentage composition, number of water molecules of crystallisation, atomic weight of unknown elements, percentage of dehyration, percentage of oxidation of a sample etc.

POTASSIUM DICHROMATE TITRATIONS

Q. 13. What are potassium dichromate titrations?

Ans. Potassium dichromate is a strong oxidising agent in acidic medium and it is used for estimation of many reducing salts, especially ferrous salts.

$$K_2Cr_2O_7 + 4H_2SO_4 \longrightarrow K_2SO_4 + Cr_2(SO_4)_3 + 4H_2O + 3(O)$$

$$Cr_2O_7^{2-} + 14\,H^+ + 6e^- \longrightarrow 2Cr^{3+} + 7H_2O$$

Unlike potassium paramaganate solution, potassium dichromate is available in absolutely pure form and solutions of potassium dichromate are very stable and can be stored indefinitely without any change. Moreover, $KMnO_4$ cannot be used with chlorides because it may be used up in oxidising HCl. $K_2Cr_2O_7$ is, however, free from this disadvantage and may be used in the presence of HCl or its salts.

Q. 15. Give a detailed account of the indicators used in $K_2Cr_2O_7$ titrations.

Ans. $K_2Cr_2O_3$ it does not act as self-indicator, there being no sharp changes in colour during the process of titrations, moreover the final product possesses a green colour due to Cr^{3+} ions.

Two kinds of indicators are used in potassium dichromate titrations:

1. External indicator
2. Internal indicator

1. External Indicator

About 1% solution of pur potassium ferricyanide is used as an external indicator. This may be prepared as : Take a crystal of pot. ferricyanide in a clean test tube and wash the crystal twice with few ml of distilled water to remove any pot. ferrocyanide adhering on the surface. Reject washings. Then, dissolve the crystal in about 2 ml of distilled water.

Use of External Indicator

Place a series of small drops of freshly prepared indicator solution on a dry white tile by means of a thin glass rod. To 20 ml of the solution to be titrated, add half a test tube of dil. H_2SO_4. Run in dichromate solution from the burette in 1 ml lots and after every addition shake the reaction mixture and take out a drop of the reaction mixture from the conical flask with the help of a thin clean glass rod, mix it with one of the drops of ferricyanide on the test-plate (see Fig. 33.10). If a blue colour is produced it shows the presence of ferrous ions in the solution.

$$3Fe^{2+} + 2\,[Fe\,(CN)_6]^3 \longrightarrow Fe_3\,[Fe\,(CN)_6]_2 \downarrow$$

Blue Complex

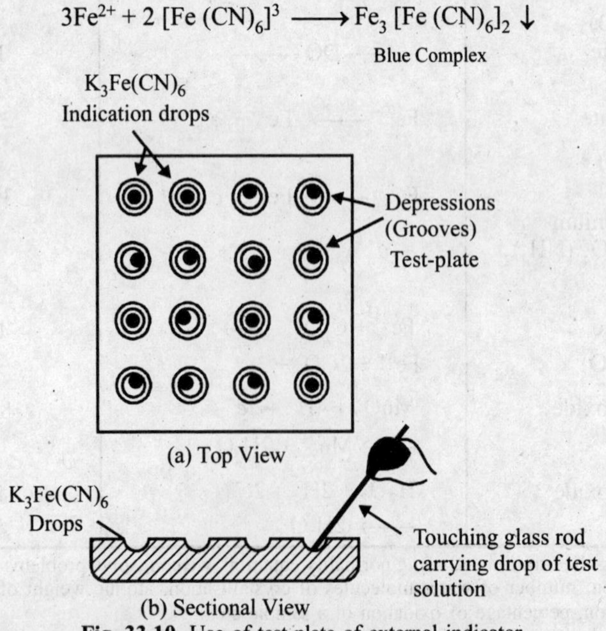

Fig. 33.10. Use of test-plate of external indicator

Continue the addition of the dichromate until a droplet of the reaction mixture from the titration flask ceases to give any blue colour with the indicator drop.

The first titration which is obviously a rough one, gives the rough range in which the true reading lies. In the subsequent titrations, test is carried out after each addition of 0.1 ml.

Notes:

(1) The indicator solution of potassium ferricyanide must be free from ferrocyanide radial. Its absence is tested with ferric chloride solution when no blue colour is produced.

(2) Drops of the indicator are arranged on a white procelain tile. They are yellowish coloured. A glass rod wetted with the reaction solution is touched with a drop. After each test, the glass rod is thoroughly washed before performing the next test.

(3) Freshly prepared solution of potassium ferricyanide should always be used.

(4) Near the end point, the bluish green colour in the indicator develops slowly and is advisable to allow the test-drop on the tile to stand for 3–4 minutes before going for the completion of titration.

(5) Potassium ferricyanide indicator is not added into the titration flask containing ferrous salt solution. In that case a deep bluish colour is produced at once.

(6) Frequent testing will lead to inaccurate results.

2. Internal Indicator

There are various internal indicators which are in use. The three commonly employed are:

(*i*) *1% solution (in conc. H_2SO_4) of diphenyl amine.*

This indicator has no influence on the colour of the reaction mixture, but when a slight excess of potassium dichromate is present, the latter oxidises the indicator producing an intensely blue coloured compound, which indicates the end point. However ferric ions (formed in the course of titration of ferrous salts) are likely to cause the oxidation of the diphenyl amine prematurely *i.e.* $K_2Cr_2O_7$ is in excess. In order to prevent this some syrupy phosphoric acid is added to solution of ferrous salt before titration is carried out. The ferric ions produced by the oxidation are taken up by phosphoric acid giving a ferric phosphate complex which is almost undissociated and thus the ferric ions are prevented from oxidising diphenyl amine.

For titrating 20 ml of solution of ferrous ions, add about 20 ml dil. H_2SO_4, 5 – 6 drops of indicator and then 5 ml syrupy phosphoric acid, now $K_2Cr_2O_7$ sol. is added from the burette until the last drop causes a change from green to blue-violet which persists on shaking. This is the end point.

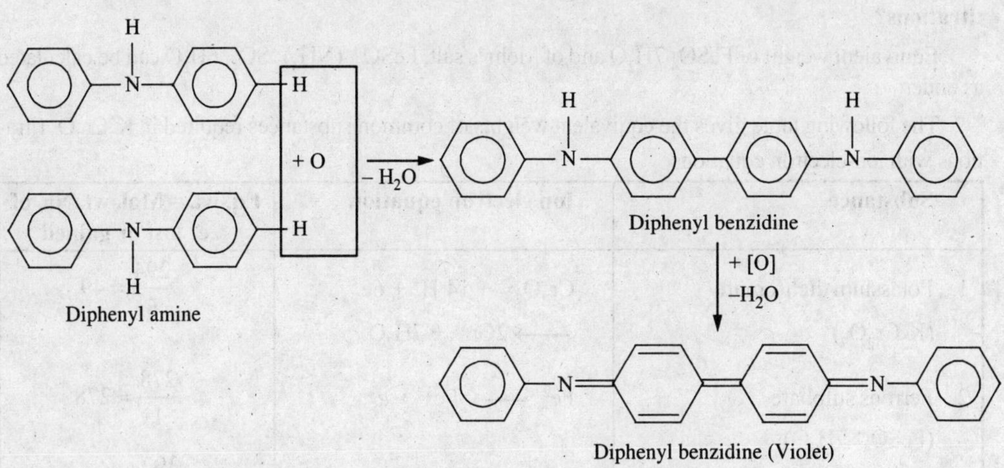

+ O
$-H_2O$
Diphenyl amine
Diphenyl benzidine
+ [O]
$-H_2O$
Diphenyl benzidine (Violet)

(*ii*) 1% solution (in conc. H_2SO_4) of diphenyl benzidine.

Diphenyl benzidine gets oxidised to violet diphenyl benzidine and hence it can be conveniently used as an indicator in redox titrations involving pot. dichromate as the oxidant (For reactions refer above). 1% solution of diphenyl benzidine in conc. H_2SO_4 is employed as an indicator. The titration carried out in the presence of phosphoric acid which helps in lowering the concentration of the ferric ions thus giving a sharp end point (blue violet or purple colouration) 2 or 3 drops of the indicator are used. Rest of the details are the same as given above in (*i*)

(*iii*) *N-phenyl anthranilic acid*

HN — COOH

N-Phenyl anthranilic acid

This is found to be a better indicator than the others.

How to prepare this indicator?

(*a*) 0.25 g of the solid N-phenyl anthranilic acid is dissolved in 12 ml of 0.1 N NaOH solution and the volume is raised to 250 ml by diluting with water and this solution is ready to be used as indicator.

or

(*b*) 0.25 g of the solid N-phenyl anthranilic acid is dissolved in 10 ml of 5% Na_2CO_3 solution and the volume is raised to 250 ml by diluting with water.

Use of N-phenyl anthranilic acid as an indicator

When using N-phenyl anthranilic acid as internal indicator there is no need to add phosphoric acid. In such a case, to 20 ml of the Fe^{++} salt solution in the titration flask are added about 100 ml of 2N H_2SO_4 (5 test tube of bench dil H_2SO_4 diluted two and a half time) and 8 – 10 drops of the indicator. This indicator is better than the first two indicators as it is:

(*i*) less expensive

(*ii*) addition of phosphoric acid is not necessary, and

(*iii*) the end point is quite sharp within one drop

Q. 16. How do we calculate the eq. weights of reducing agents in potassium dichromate titrations?

Equivalent weight of $FeSO_4$ $7H_2O$ and of Mohr's salt, $FeSO_4$, $(NH_4)_2$ SO_4. $6H_2O$ can be calculated as under:

The following table gives the equivalent weights of common substances required in $K_2Cr_2O_7$ titrations with ion-electron equations

Substance	Ion electron equation	Eq. wt. = Mol. wt./No. of e^- lost or gained
1. Potassium dichromate ($K_2Cr_2O_7$)	$Cr_2O_7^{2-} + 14H^+ + 6e^- \longrightarrow 2Cr^{3+} + 7H_2O$	$\frac{294}{6} = 49$
2. Ferrous sulphate ($FeSO_4 \cdot 7H_2O$)	$Fe^{2+} \longrightarrow Fe^{3+} + e^-$	$\frac{278}{1} = 278$
3. Ferrous ammonium sulphate [$FeSO_4$ $(NH_4)_2SO_4 \cdot 6H_2O$]	$Fe^{2+} \longrightarrow Fe^{3+} + e^-$	$\frac{392}{1} = 392$

Q. 17. Which are the primary standards in $K_2Cr_2O_7$ titrations?

Ans.

1. **Potassium dichromate:** It is available in a pure form and is quite stable, so its standard solution can be prepared by dissolving a known wt. of it in a given volume of its solution.
2. **Mohr's salt :** [$FeSO_4$. $(NH_4)_2$ SO_4 $6H_2O$]. It is a double salt containing stable ferrous sulphate. Its standard solution is prepared by dissolving a known wt. of it (Eq. wt. of Mohr's salt is 392) in distilled water to a known volume of the solution.

   A few drops of H_2SO_4 are added to make the solution clear and to check hydrolysis before diluting it to a known volume.

Q. 18. Describe the general procedure in potassium dichromate titrations.

Ans. The following points should be kept in mind while performing $K_2Cr_2O_7$ titrations.

1. Potassium dichromate solution is always taken in the burette and its upper meniscus noted as it is coloured yellowish orange.
2. To a known volume of ferrous salt in the titration flask an equal volume of dil H_2SO_4 is added and titration is carried out in cold.
3. In case, diphenyl benzidine is used as internal indicator, then to 20 ml of the Fe^{++} salt solution in the titration flask, an equal volume of dil. H_2SO_4 (5N) (One full test tube of bench dilute acid) about 5 ml of syrupy phosphoric acid and 2 to 3 drops of indicator are added.

   Towards the end of titration (when ratio of Fe^{3+} to Fe^{2+} ions is high) the ferric ion exerts a slight oxidising action on the indicator. The addition of phosphoric acid reduces the above effect due to the formation of a complex ion.
4. While using N-phenyl anthranilic acid as internal indicator, there is no need to add phosphoric acid. In such a case, to 20 ml of the Fe^{2+} salt solution in the titration flask are added about 100 ml of 2N H_2 SO_4 (5 test tubes of bench dil H_2SO_4 diluted two and a half times) and 10 drops of the indicator

Q. 19. The given solution contains 20.5 g/litre of an impure sample of Mohr's salt. Determine volumetrically the percentage purity of the sample. Provided solid K_2 Cr_2 O_7.

Theory

A standard solution of $K_2Cr_2O_7$ is prepared by weighing. Mohr's salt solution is then titrated with standard $K_2Cr_2O_7$ solution and its normality and strength calculated. Percentage purity of the sample is then found out as discussed in calculations.

Reactions:

$$K_2Cr_2O_7 + 4H_2SO_4 \longrightarrow K_2SO_4 + Cr_2(SO_4)_3 + 4H_2O + 3\,[O]$$

$$2Fe\,SO_4\,(NH_4)_2SO_4 + H_2SO_4 + (O) \longrightarrow Fe_2\,(SO_4)_3 + 2(NH_4)_2\,SO_4 + H_2O$$

$$2Fe\,SO_4 + H_2SO_4 + (O) \longrightarrow Fe_2\,(SO_4)_3 + H_2O$$

Indicator

N-phenyl anthranilic acid.

End Point

Green to violet -red.

Procedure

(*i*) *Preparation of standard potassium dichromate solution*

Weigh 0.6125 g powdered potassium dichromate on a watch glass. Transfer the solid to a clean beaker. Transfer any crystals sticking to the watch glass also to the beaker with the help of a wash bottle. Dissolve the solid in the beaker in about 50 ml of distilled water. Transfer the solution to a 250 ml measuring flask with the help of a funnel. Give washings to the funnel and the beaker and transfer the washings also to the measuring flask. Complete the volume in the flask upto the mark with distilled water. Stopper and shake the measuring flask. Thus N/20 potassium dichromate solution is obtained.

(*i*) *Titration of Mohr's salt solution against potassium dichromate solution.* Rinse and fill the burette with potassium dichromate solution. Pipette out 20 ml of Mohr's salt solution into a titration flask. Add about 100 ml of 2N H_2SO_4. Add 10 drops of indicator and titrate against pot. dichromate solution from burette. Addition of $K_2Cr_2O_7$ sol. is continued till the last drop changes the colour of the solution from green to violet-red. Repeat the procedure to get concordant set of readings.

IODINE TITRATION

Q. 20. What are iodine titrations? How are they further classified? Explain with the help of chemical equations.

Ans. Titrations which involve the use of iodine directly or indirectly are called iodine titrations. These titration can be classified as under:

1. Iodimetric titrations

Iodimetric titration are those titrations in which a standard iodine solution is used as an oxidant and is directly titrated with a reducing agent. Iodimetric procedures are used for determination of reducing agents like sulphites, arsenites, stannous chloride, etc by titrating them against standard sol. of iodine run from a burette. Some oxidation reduction reactions are given as under:

(*i*) $$2Na_2S_2O_3 + I_2 \longrightarrow Na_2S_4O_6 + 2NaI$$

Ionic equation

$$2S_2O_3^{2-} + I_2 \longrightarrow S_4O_6^{2-} + 2e^-$$

(*ii*) $$SnCl_2 + I_2 + 2HCl \longrightarrow SnCl_4 + 2HI$$

Ionic equation

$$Sn^{2+} + 2H^+ + I_2 \longrightarrow Sn^{4+} + 2HI$$

(*iii*) $$Na_3AsO_3 + I_2 + H_2O \longrightarrow Na_3AsO_4 + 2HI$$

Ionic equation

$$AsO_3^{3-} + I_2 + H_2O \longrightarrow AsO_4^{3-} + 2H^+ + 2I^-$$

(*iv*) $$Sb_2O_3 + 2I_2 + 2H_2O \rightleftharpoons Sb_2O_5 + 4HI$$

Ionic equation

$$Sb^{3+} \rightleftharpoons Sb^{5+} + 2e^-$$

and $$2I_2 + 4e^- \longrightarrow 4I^-$$

2. Iodometric titrations

Iodometric titrations are those titrations in which some oxidising agent liberates iodine from an iodide and the liberated iodine is titrated against standard solution of reducing agent added from a burette. In such titrations a netural or an acidic solution of oxidising agent is employed. The amount of liberated from an iodide (*i.e.* KI) is proportional to the quantity of the oxidising agent present. Iodometric titration are carried out for determination of $CuSO_4$, $K_2Cr_2O_7$, $KMnO_4$, ferric ions, antimonate ions, H_2O_2, MnO_2, bromine, chlorine, etc. The equations for some of the reactions are as follows:

(*i*) $$2KMnO_4 + 3H_2SO_4 \longrightarrow K_2SO_4 + 2MnSO_4 + 3H_2O + 5[O]$$

$$2KI + H_2SO_4 \longrightarrow K_2SO_4 + 2HI\ ] \times 5$$

$$2HI + [O] \longrightarrow H_2O + I_2] \times 5$$

$$2KMnO_4 + 8\,H_2SO_4 + 10KI \longrightarrow 6K_2SO_4 + 2MnSO_4 + 8H_2O + 5I_2$$

Ionic equation

$$2MnO_4^- + 10\,I + 16H^+ \longrightarrow 2Mn^{2+} + 8H_2O + 5I_2$$

(*ii*) $$K_2Cr_2O_7 + 14H^+ + 6I \longrightarrow K_2SO_4 + Cr_2(SO_4)_3 + 7H_2O + 3I_2$$

Ionic equation

$$Cr_2O_7^{\ 2-} + 14H^+ + 6I^- \longrightarrow 2Cr^3 + 7H_2O + 3I_2$$

(*iii*) $$CuSO_4 + 4KI \longrightarrow K_2SO_4 + CuI_2] \times 2$$

$$2\,CuI_2 \longrightarrow Cu_2I_2 + I_2$$

or $$2CuSO_4 + 4KI \longrightarrow 2K_2SO_4 + CuI_2 + I_2$$

white ppt

Ionic equation

$$2Cu^{2+} + 2I^- \longrightarrow Cu_2^{\ 2+} + I_2$$

(*iv*) $$I_2 + 2\,Na_2\,S_2O_3 \longrightarrow Na_2S_2O_3 + 2NaI$$

Ionic equation

$$I_2 + 2S_2O_3^{\ 2-} \longrightarrow S_2O_6^{\ 2-} + 2I^-$$

Q. 21. Discuss the indicator and end point in iodine titrations.

Ans. In titrations, involving iodine (free or liberated) 1% *starch solution* is used as an indicator. Iodine forms a deep *blue complex* with starch.

(*i*) *In iodimetry*, when iodine is taken in the burette, starch solution is added to the reducing agent taken in the titration flask in the beginning and end point is the *appearance of blue colour*.

(*ii*) *In iodometry*, when the solution of a reducing agent in run from the burette into titration flask containing liberated iodine, starch solution is added just before the approach of end point *i.e.*, when the colour of iodine fades from brown to pale yellow. The end point in this case is disappearance of blue colour.

Q. 22. Discuss standard substances in iodine titrations.

Ans. *Iodine*. In the preparation of a standard solution of iodine following difficulties arise:

(*a*) Iodine is almost insoluble in water.

(*b*) Iodine is very volatile and its equeous solution has an appreciable vapour pressure. Both these difficulties are overcome by dissolving iodine in an aqueous solution of potassium iodide where dissolution occurs due to the formation of potassium tri-iodide.

$$KI + I_2 \rightleftharpoons KI_3$$

Greater the concentration of iodide solution, more is the solubility of iodine in it. This tri-iodide liberates iodine so readily during titration that the solution behaves as if the dissolved iodine were all free iodine.

From the equation

$$I + e^- \rightleftharpoons I^-$$

it follows that the Eq. wt. of iodine as an oxidising agent is the same as its atomic weight *i.e.* 127

(*ii*) **Sodium thiosulphate.** $Na_2S_2O_3 \cdot 5H_2O$ (Hypo). It is readily available in pure form and its standard solution may be prepared by direct weighing. However, due to efflorescent nature, there is no clarity about its exact water content. It is therefore, not suitable as a primary standard.

(*iii*) **Potassium dichromate.** It is a primary standard and is employed for standardizing sodium thiosulphate (Hypo). It is used in iodometry.

(*iv*) **Arsenious oxide.** It is available in a state of purity (99.9%) and an excellent primary standard for the standardization of iodine solution.

$$H_3AsO_3 + I_2 + H_2O \rightleftharpoons H_3AsO_4 + 2HI$$

There exists reversible relationship between arsenious oxide and iodine.

In case hydrogen iodide is not removed from the solution, as it is formed, the reaction proceeds in the backward direction. To check this, $NaHCO_3$ is added. Sodium hydroxide and sodium carbonate cannot be used for this purpose since they react with iodine.

Q. 23. How do we calculate the equivalent weights of iodine and the reducing agents used in I_2 titrations?

Ans. 1. **Iodine I_2,** It acts as a weak oxidizing agent in aqueous solution

$$I_2 + H_2O \longrightarrow 2\,HI + [O]$$

$$\underset{2\times127}{2I} = \underset{16}{[O]}$$

$$\text{Eq. wt.} = \frac{2\times127}{16}\times8 = 127$$

Ionic equation

$$I_2 + 2e^- \rightleftharpoons 2I^-$$

$$\text{Eq. wt.} = \frac{\text{Mol. wt}}{2} = \frac{254}{2} = 127$$

Hence, equivalent weight is the same as atomic weight viz, 127.

2. Potassium dichromate $K_2Cr_2O_7$

Ionic equation

$$Cr_2O_7^{2} + 14H^+ + 6e \longrightarrow 2Cr^{3+} + 7H_2$$

$$\text{Eq. wt} = \frac{\text{Mol. wt.}}{6} = \frac{294}{6} = 49$$

3. Potassium permanganate, $KMnO_4$. The equation involved is

$$2KMnO_4 + 8H_2SO_4 + 10KI \longrightarrow 6K_2SO_4 + 2MnSO_4 + 8H_2O + 5I_2$$

or $$\underset{2\times158}{2KMnO_4} \equiv \underset{10\times127}{10\,I}$$

$$\therefore \quad \text{Eq. wt.} = \frac{2\times158}{10\times127}\times127 = 31.6.$$

Hence, the equivalent weight is one-fifth of its molecular weight.

Ionic equation

$$MnO_4^- + 8H^+ + 5e^- \longrightarrow Mn^{2+} + 4H_2O$$

$$\therefore \quad \text{Eq. wt.} = \frac{\text{Mol. wt.}}{5} = \frac{158}{5} = 31.6.$$

4. **Copper sulphate, $CuSO_4.5H_2O$.** The equation involved is

$$2CuSO_4 + 4KI \longrightarrow Cu_2I_2 + I_2 + 2K_2SO_4$$

$$\underset{2 \times 249.5}{2CuSO_4.5H_2O} \equiv \underset{2 \times 127}{2I}$$

$$\therefore \quad \text{Eq. wt.} = \frac{2 \times 249.5}{2 \times 127} \times 127 = 249.5$$

Ionic equation

$$Cu^{2+} + e^- \longrightarrow Cu^+$$

$$\therefore \quad \text{Eq. wt.} = \frac{\text{Mol. wt.}}{1} = \frac{249.5}{1} = 249.5$$

Hence, the equivalent weight is the same as the molecular weight.

The equivalent weights of some common compounds which find use in iodine titration are grouped in Table below:

Table

Compound	Mol. wt.	Eq. wt = Mol. wt./ No. of electrons lost or gained
1. Iodine, I_2	254	127
2. Sod. thiosulphate, $Na_2S_2O_3.5H_2O$	248	$\frac{\text{Mol. wt.}}{1} = 248$
3. Sodium thiosulphate $Na_2S_2O_3$ (anhy).	158	$\frac{\text{Mol. wt.}}{1} = 158$
4. Sodium arsenite, Na_3AsO_3	192	$\frac{\text{Mol. wt.}}{2} = 96$
5. Arsenious oxide, As_2O_3	198	$\frac{\text{Mol. wt.}}{4} = 49.5$
6. Arsenic, As	75	$\frac{\text{At. wt.}}{2} = 37.5$
7. Tartaremetic CHOHCOOK | CHOHCOO (OSb). $\frac{1}{2}$ H_2O	334	$\frac{\text{Mol. wt.}}{2} = 167$
8. Antimonous oxide, Sb_2O_3	292	$\frac{\text{Mol. wt.}}{4} = 73$
9. Antimony, Sb	122	$\frac{\text{At. wt.}}{2} = 61$
10. Copper sulphate, $CuSO_4.5H_2O$	249.5	$\frac{\text{Mol. wt.}}{1} = 249.5$
11. –do– (Anhydrous)	159.5	$\frac{\text{Mol. wt.}}{1} = 159.5$
12. Potassium dichromate, $K_2Cr_2O_7$	294	$\frac{\text{Mol. wt.}}{6} = 49$
13. Potassium permanganate, $KMnO_4$	158	$\frac{\text{Mol. wt.}}{5} = 31.6$
14. Stannous chloride, $(SnCl_2.H_2O)$	292	$\frac{\text{Mol. wt.}}{4} = 73$

In iodometry, iodine is involved from an iodine by an oxidising agent in acid or neutral medium. Following should be followed in iodometry.

Q. 24. Give the general procedure for carrying out iodometric titrations.

Ans.

1. Always take hypo solution in the burette.
2. Take a known volume (say 20 ml) of the oxidising agent in the titration flask.
3. Add a full test tube of dil. H_2SO_4 if the solution taken in the titration flask is $KMnO_4$ or $K_2Cr_2O_7$.
4. Add half a test tube of 10% KI to the oxidising agent in the flask and shake it well to liberate iodine.
5. Evolution of iodine by interaction between an iodide and the oxidising agent is slow. So, a period of 1 - 2 minutes should be allowed for the completion of the reaction.
6. The titration flask should be covered with a watch glass to prevent the loss of iodine due to its volatile nature.
7. Run down hypo solution from the burette dropwise while shaking the flask, till the reddish-brown colour fades to pale yellow (straw colour).
8. HI (from excess of KI and acid) is readily oxidised by air especially in the presence of Cr^{+3} salts. To overcome this difficulty about 1g solid $NaHCO_3$ should be added to the solution. The CO_2 evolved keeps the atmospheric oxygen off. Keep the flask covered as far as possible

$$2KI + H_2SO_4 \longrightarrow K_2SO_4 + 2HI$$

$$2HI + (O) \text{ from air} \longrightarrow H_2O + I_2 \uparrow$$

9. The solution should be diluted before titration to check volatility of iodine during the titration.
10. On adding 1 - 2 ml of *freshly prepared* starch solution a blue colour is obtained.
11. Continue the addition of hypo till the blue colour is discharged which shows the end point.
12. The indicator (starch) should be added near the end point otherwise a *stable starch iodide complex* is formed and the blue colour does not disappear on adding more of hypo solution.
13. In titration between acidified $K_2Cr_2O_7$ containing KI and hypo a green colour persists at the end point due to Cr^{3+} ions.

Q. 25. Standardise the given solution of sodium thiosulphate. Provided solid potassium dichromate.

Ans.

Theory

A standard solution (say N/20) of $K_2Cr_2O_7$ is first of all prepared by direct weighing. The given solution of sodium thiosulphate is then titrated against the standard solution of potassium dichromate. From this the normality and hence strength of sodium thiosulphate solution can be calculated.

The exercise involves the following steps:

(*a*) *Preparation of N/20 $K_2Cr_2O_7$ solution.*

Weigh exactly 0.6125 g of $K_2Cr_2O_7$ to prepare exactly 250 ml solution.

Observations:

Wt. of watch glass = w_1g

Wt. of watch glass + $K_2Cr_2O_7$ = w_2g

∴ Wt. of $K_2Cr_2O_7 = (w_2 - w_1) = 0.6125$ g

Volume of solution prepared = 250 ml.

∴ Normality of $K_2Cr_2O_7$ solution = N/20

(*b*) *Titration of sodium thiosulphate solution against N/20 $K_2Cr_2O_7$ solution.*

Equations:

$$K_2Cr_2O_7 + 4H_2SO_4 \longrightarrow K_2SO_4 + Cr_2(SO_4)_3 + 4H_2O + 3[O]$$

$$2KI + H_2SO_4 \longrightarrow K_2SO_4 + 2HI$$

$$2HI + O \longrightarrow I_2 + H_2O$$

$$2Na_2S_2O_3 + I_2 \longrightarrow Na_2S_4O_6 + 2NaI$$

Indicator Starch solution.

End point Blue to light green solution (sodium thiosulphate solution in the burette).

Procedure

Pipette out 20 ml of the dichromate solution into a titration flask. Add to it one test tube of dil. H_2SO_4 and half a test tube of 10% KI solution. Add 1 - 2 g of solid $NaHCO_3$. Cover with a watch glass and allow to stand for 1 - 2 minutes. Then dilute the solution with about 50 ml of distilled water. Titrate the iodine thus liberated against hypo. Go on adding hypo little by little with shaking till the solution becomes pale yellow. Add 1 ml of freshly prepared starch solution. The solution becomes blue. Continue the addition of hypo solution dropwise till the colour of the solution changes to light greenish. Repeat to get three concordant readings.

Observations

Titration of hypo ($Na_2S_2O_3$) *against* $K_2Cr_2O_7$.

Volume of $K_2Cr_2O_7$ solution taken each time = 20 ml

Let the volume of $Na_2S_2O_3$ solution used in titration = V ml

General Calculations

(*i*) *To calculate the normality of* $K_2Cr_2O_7$ *solution.*

250 ml of the solution contains = 0.6225 g

1000 ml of the solution contains = 4 × 0.6125 g

$$\text{Hence normality of solution} = \frac{4 \times 0.6125}{49} = \frac{N}{20}$$

(*ii*) *To calculate normally of sodium thiosulphate solution.*

Let Vml of sodium thiosulphate solution ≡ 20 ml of N/20 $K_2Cr_2O_2$

Applying normality equation, we get

$$N_1 \times V = 20 \times 1/20$$

$$\therefore N_1 \text{ (Normality of sodium thiosulphate solution)} = \frac{1}{V}$$

(*iii*) *To calculate the strength of sodium thiosulphate in g/litre.*

Strength in g/litre = Normality × Eq. wt.

$$= \frac{1}{V} \times 248 \text{ (Eq. wt. of } Na_2S_2O_3.5H_2O = 248)$$

= *a* g per litre (say)

COMPLEXOMETRIC TITRATIONS

Q. 26. What is EDTA? What are its characteristics?

Ans. Ethylene diamine tetra acetic acid abbreviated as EDTA is one of the most suitable and widely used chelating ligand.

$$\begin{matrix} HOOC-CH_2 \\ HOOC-CH_2 \end{matrix} \Big> N-CH_2-CH_2-N \Big< \begin{matrix} CH_2COOH \\ CH_2COOH \end{matrix}$$

It is a hexadentate ligand which coordinates with a metal ion through two nitrogen atoms and four oxygen atoms as shown below:

$$\begin{matrix} {}^{-}O-\overset{O}{\overset{\|}{C}}\,CH_2 \\ {}^{-}O-\underset{O}{\underset{\|}{C}}\,CH_2 \end{matrix} \Big> N-CH_2-CH_2-N \Big< \begin{matrix} CH_2-\overset{O}{\overset{\|}{C}}-O^{-} \\ CH_2-\underset{O}{\underset{\|}{C}}-O^{-} \end{matrix} \quad (\text{coordinated to } M^{2+})$$

EDTA is a versatile complexing agent. It can form complexes with a large no. of metals at proper pH conditions which can be achieved with the help of buffer solutions.

Buffer solutions of EDTA can be used to carry out titrations to estimate the amount of Ca^{2+} and Mg^{2+} present in water in order to determine its hardness. It forms a complex with Mg^{2+} which has the following structure:

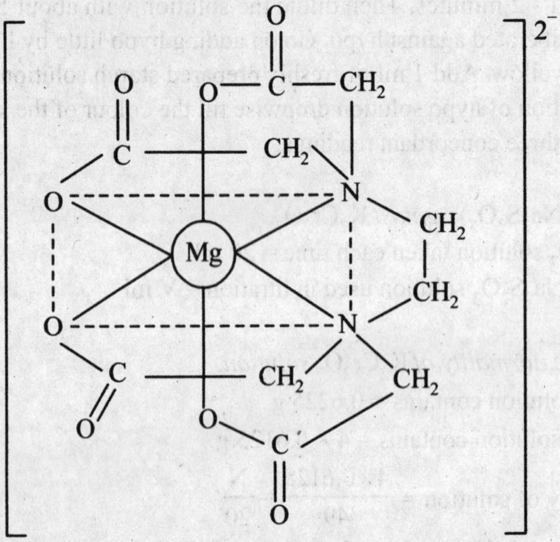

EDTA has the following notable properties:

(*i*) It is a white crystalline powder with a limited solubility in water.

(*ii*) It can form complexes instantaneously and stoichiometrically. As such it can be directly employed as a titrant to perform the complexometric titrations for the estimation of calcium, magnesium, zinc and other metals.

(*iii*) The stability of EDTA metal complexes is markedly influenced by pH. It has been found that the complexes of bivalent metals are stable in ammoniacal solution but tend to decompose in acidic medium. However the complexes of trivalent transition metals are stable even in the presence of strong mineral acids.

(*iv*) Almost all such complexes are soluble in water.

It may, however, be pointed out here that these complexometric titrations are highly versatile having their merits as well as demerits. The merit lies in the fact that the same reagent can be used for the estimation of a large number of ions. The demerit is that on account of the high reactivity of the reagent, there is a greater probability of interference by other ions. Consequently this leads to low selectivity. However, the selectivity can be markedly improved by use of appropriate experimental conditions such as *controlling the pH and by using buffer solutions*.

Q. 27. How do we prepare standard EDTA solution?

Ans. EDTA is usually employed as its sodium salt viz, disodium dihydrogen ethylene diamine tetra acetic acid $[Na_2H_2(C_{10}H_{12}O_8N_2).2H_2O]$ commonly known as disodium EDTA or more commonly as EDTA only. The commercially available AR reagent may contain some traces of moisture. It is dried at 80°C and the salt can now be directly employed as a primary standard. As such its standard solution can be prepared by weighing the compound directly and dissolving in double distilled water. Generally EDTA solutions having concentrations 0.01 M and 0.05 M are employed. These can be prepared by dissolving respectively 3.72 g and 18.6 g of EDTA (mol. wt. = 372) per litre of the solution 0.01 N and 0.05 N solutions are prepared by dissolving 1.86 and 9.3 g of EDTA respectively in one litre of the solution.

Q. 28. What are the different indicators used in complexometric titrations?

Ans. In general, *metallochromic indicators* are employed to determine the end point in EDTA titrations. These indicators are actually organic dye-stuffs which can form intensely coloured complexes with metal ions. Since the indicator colour is markedly influenced by the pH of the solution, therefore, in all EDTA titrations, the pH of the solution must be maintained within some optimum range by the use of buffers. The metallochromic indicator selected, should be one which forms complex with the metal ion to be titrated which is quite stable at the pH employed but sufficiently unstable so as to get readily decomposed by the titrant. The reaction involved at the end point is —

$$\text{Metal-indicator complex} + \text{titrant} \longrightarrow \text{Metal-titrant complex} + \text{Indicator}$$

Thus, the colour observed will actually be the colour of the free indicator.

It may be mentioned here that if the indicator complex is not sufficiently unstable, its reaction with the titrant will be slow. Hence, in such a case the titration should be carried out slowly near the end point, or otherwise, the solution may be warmed to accelerate the reaction.

Some commonly employed metallochromic indicators alongwith their pH range are given below in Table 33.1.

Table 33.1. Some commonly used complexometric indicators.

Indicator	**Useful pH range**	**Colour at the end point**	**Metal ion for which employed**
Eriochrome black T	8 – 10.5	Sky blue	Mg, Cd, Zn
Murexide	6 – 13	Deep purple	Ca, Cu, Co, Ni
Xylenol orange	1 – 5	Lemon yellow	Pb, Bi, Co, Th

These indicators are briefly described as under:

Eriochrome Black T. It is also known as *Solochrome* Black T. Chemically it is sodium 1-(1-hydroxy-2-naphthylazo)-6-nitro-2-naphthol-4-sulphonate:

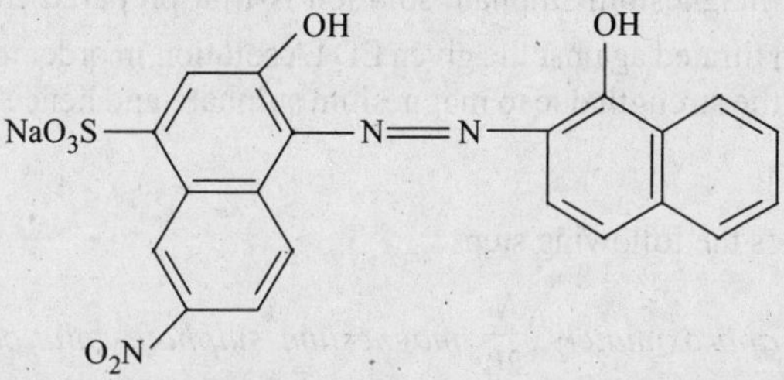

OH OH

NaO_3S—N═N—

O_2N

This indicator has been used successfully in the estimation of Mg, Cd, Hg, Pb, Cu, Zn, Mn, Fe, Al, Co, Ni and Pt metals. It works perfectly in the pH range of 8-10.5. Buffer solutions are employed to maintain the pH. The indicator solution is red below pH 5.5, blue in the pH ragne 7-11 and yellowish orange above pH 11.5. At the end in EDTA titrations, the indicator is set free and the colour will change from red to sky-blue.

To prepare the indicator solution, dissolve 0.1 g of Eriochrome Black in 7-8 ml of triethylamine and 2-3 ml of absolute alcohol.

Alternatively 0.4 per cent solution of Eriochrome Black T in methyl alcohol can also be used.

Murexide. It was probably the first metal indicator to be used in EDTA titrations. It is the ammonium salt of purpuric acid and has the following structure.

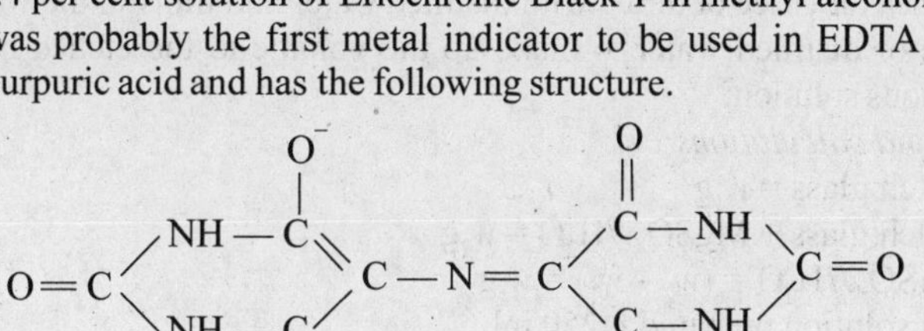

```
                O⁻                        O
                |                         ||
          NH — C                     C — NH
         /        \\                 /        \
   O = C            C — N = C                 C = O
         \        /                 \        /
          NH — C                     C — NH
                ||                        ||
                O                         O
```

The optimum pH range of this indicator is 6-13. The indicator solution is red-violet below pH 9; violet in pH range of 9-11 and blue if the pH is above 11. It forms complexes with Ca, Cu, Co, Ni etc. Freshly prepared saturated (about 1%) aqueous solution of murexide is used as indicator.

Xylenol Orange. Chemically it is 3, 3′ bis[N-N-di (carboxymethyl), aminomethyl *o*-cresol sulphonaphthalein.]

OH
H_3C — CH_2 — N (CH_2COOH)(CH_2COOH)
CH_3
C
OH
CH_2 — N (CH_2COOH)(CH_2COOH)
O_2S — O

It is employed in the determination of Pb, Cd, Bi, Zn, Co and Th metals. pH range in which it works smoothly is 1-5. In acid solution, it is lemon yellow in colour and forms red coloured complexes with the above mentioned metals. At the end point in EDTA titrations, the indicator is set free and the colour changes from red to yellow.

To prepare the indicator solution, dissolve 0.5g of the indicator in 100 ml of distilled water.

Q. 29. Determine the percentage purity of the given sample of $MgSO_4.7H_2O$. Provided $\frac{N}{2}$ EDTA solution.

Ans.

Theory

Approximately $\frac{N}{20}$ magnesium sulphate solution is first prepared from the given sample by direct weighing. It is then titrated against the given EDTA solution in order to determine its normality. Knowing the normality, the strength due to magnesium sulphate and hence its percentage purity can be calculated.

Procedure

The exercise involves the following steps:

(*i*) *Preparation of approximately $\frac{N}{20}$ magnesium sulphate solution.* Molecular weight of $MgSO_4.7H_2O$ is 246 and Equivalent weight is 123. Hence, mass of magnesium sulphate required to prepare 250 ml of $\frac{N}{20}$ solution is : $123 \times \frac{1}{20} \times \frac{1}{4} = 1.5375$ g

1. Weigh about 1.5 g of the given sample of magnesium sulphate and transfer it into a beaker containing 40-50 ml of distilled water.
2. Dissolved by stirring and transfer the solution to a 250 ml measuring flask.
3. Wash the beaker twice or thrice with distilled water and transfer the washings to the flask.
4. Add more of distilled water to make up the volume to the etched mark, shake to get a homogeneous solution.

Observations and calculations:

Wt. of watch glass = w_1g
Wt. of watch glass + $MgSO_4.7H_2O = w_2$g
Wt. of $MgSO_4.7H_2O = (w_2 - w_1) = w$ g
Volume of solution prepared = 250 ml
∴ Strength/litre of $MgSO_4.7H_2O$ solution = 4 *w* g.

(ii) Preparation of buffer solution of pH 10. Add 57 ml of concentrated ammonia solution (sp. gravity = 0.88) to 7g of A.R. of ammonium chloride and then dilute to 100 ml by adding redistilled water.

(iii) Titration of magnesium sulphate solution against EDTA solution.

Reactions

$$Mg^{2+} + Na_2(H_2\ EDTA) \longrightarrow 2Na^+ + (Mg\ EDTA)^{2-} + 2H^+$$

Indicator Eriochrome Black T.

End Point Red to sky blue coloured solution.

Procedure

1. Pipette out 20 ml of the magnesium sulphate solution, prepared as above, into a titration flask.
2. Add to it an equal volume of distilled water followed by 2 ml of buffer solution (pH = 10) and 2-3 drops of indicator solution.
3. Run down the EDTA solution from the burette with constant shaking of the flask till the end point is reached.

Repeat to get a set of three concordant readings.

General Calculations:

Let the volume of EDTA required = V ml for titration

(i) To calculate the normality of magnesium sulphate solution.

Let V ml of EDTA solution $\equiv$ 20 ml of $MgSO_4$ solution

Applying normality equation, we have

$$\frac{1}{20} \times V = N_1 \times 20$$

$$\therefore\ N_1 \text{ (normality of } MgSO_4.7H_2O \text{ solution)} = \frac{V}{400}$$

(ii) To calculate the percentage purity of the given sample of $MgSO_4.7H_2O$

Strength per litre of $MgSO_4.7H_2O$ solution = Normality × Eq. wt.

$$= \frac{V}{400} \times 123 = a \text{ g (say)}$$

$$\therefore\quad \text{Percentage purity of the sample} = \frac{a}{4w} \times 100$$

Q. 30. One gram of dolomite ($CaCO_3.MgCO_3$) was dissolved in minimum quantity of dil. HCl and the solution was diluted to one litre. Determine volumetrically the amount of calcium and magnesium ions in the sample. Provided 0.01 N EDTA solution.

Ans.

Theory

The total amount of calcium and magnesium in the sample is determined by titration against EDTA using Eriochrome Black T as indicator at pH 10. Again a known volume of mixture solution is titrated against EDTA for calcium using murexide as indicator after precipitating magnesium as magnesium hydroxide by adding 2N sodium hydroxide. The difference in the volume of EDTA used corresponds to the amount of magnesium in the sample.

Chemical Equations

$$Mg^{2+} + Na_2(H_2EDTA) \longrightarrow 2Na^+ + (Mg\ EDTA)^{2-} + 2H^+$$

$$Ca^{2+} + Na_2(H_2\ EDTA) \longrightarrow 2Na^+ + (Ca\ EDTA)^{2-} + 2H^+$$

Indicator

Titration I (EDTA vs Ca^{2+} and Mg^{2+} ions) - Eriochrome Black T

Titration II (EDTA vs Ca^{2+} ions) - Murexide.

End point

Titration I-Wine red to blue.

Titration II-Red to purple.

Procedure

Titration I : EDTA vs Ca^{2+} and Mg^{2+} ions

1. Rinse and fill the burette with 0.01 M EDTA solution.
2. Rinse the pipette and pipette out 20.0 ml of the given solution into the titration flask and dilute it with 50.0 ml water.
3. Add 5.0 ml of NH_4Cl/NH_4OH buffer (pH = 10) and 4-5 drops of Eriochrome Black T indicator.
4. Note the initial reading of the burette and run EDTA solution into the titration flask slowly with stirring till the colour changes from wine red to blue.
5. Repeat the titration to get three concordant readings.

Titration II: *EDTA vs Ca^{2+} ions*

1. Pipette out 20 ml of the solution into the titration flask. Dilute it with 20.0 ml.
2. Add 5 ml of 50% NaOH solution to precipitate magnesium as magnesium hydroxide.
3. Add 3-4 drops of murexide indicator.
4. Run 0.01 M EDTA solution from burette into the titration flask till the colour changes from red to purple.
5. Note the final reading of the burette.
6. Repeat the titration to get three concordant readings.

Calculations

1st titration (Total normality due to $CaCO_3$ and $MgCO_3$)

Apply the normality equation

$$\underset{\text{(Dolomite sol)}}{N_1V_1} = \underset{\text{EDTA}}{N_2V_2}$$

Substituting the values, we have

$$N_1 \times 20 = \frac{N}{100} \times V_2 \quad (V_2 \text{ is the volume of EDTA solution required for titration})$$

or
$$N_1 = \frac{N}{100} \times \frac{V_2}{20}$$

Thus we obtain total normality due to $MgCO_3$ and $CaCO_3$

2nd titration (Normality due to $CaCO_3$ only)

Again use the normality equation

$$N_1'V_1' \equiv N_2'V_2'$$

Substitute the values

$$N_1' \times 20 = \frac{N}{100} \times V_2' \quad (V_2' \text{ is the volume of EDTA solution required for titration})$$

or
$$N_1' = \frac{N}{100} \times \frac{V_2'}{20}$$

N_1' is the normality due to $CaCO_3$ only

Normality due to $MgCO_3 = N_1 - N_1'$

To obtain the strength/litre of $MgCO_3$ and $CaCO_3$, multiplying the normalities by the respective equivalent weights.

Note. A similar method is adopted to analyse Ca^{2+} and Mg^{2+} ions in hard water.

Summary of Volumetric Analysis

Type of volumetric analysis	Indicator	End Point	Reactions	General Procedure
$KMnO_4$ Titrations				
Estimation of ferrous sulphate/Mohr salt	$KMnO_4$ itself	Appearance of pink colour	$2KMnO_4 + 3H_2SO_4 \rightarrow K_2SO_4 + 2MnSO_4 + 3H_2O + 5[O]$ $2FeSO_4 + H_2SO_4 + O \rightarrow Fe_2(SO_4)_3 + H_2O$ In Mohr salt $FeSO_4\,(NH_4)_2\,SO_4.6H_2O$, only $FeSO_4$ react as shown above.	Take 20 ml of $FeSO_4$/Mohr salt solution in a titration flask. Add an equal volume (20ml) of dil. H_2SO_4. Perform the titration with $KMnO_4$ taken in the burette. Just appearance of pink colour is the end point
Estimation of oxalic acid/oxalates	As above	As above	$2KMnO_4 + H_2SO_4 \rightarrow K_2SO_4 + 2MnSO_4 + 3H_2O + 5[O]$ $\begin{matrix} COOH \\ \vert \\ COOH \end{matrix} + O \rightarrow 2CO_2 + H_2O$ $\begin{matrix} COOK \\ \vert \\ COOK \end{matrix} + H_2SO_4 \rightarrow \begin{matrix} COOH \\ \vert \\ COOH \end{matrix} + K_2SO_4$	Take 20 ml of oxalic acid/oxalate solution in a titration flask. Add an equal volume of dil. H_2SO_4. Heat the flask to 60-70° and perform the titration with $KMnO_4$ taken in the burette till a permanent pink colour is obtained.
Estimation of ferric salts	As above	As above	$Fe_2(SO_4)_3 + 2[H] \xrightarrow[H_2SO_4]{Zn} 2FeSO_4 + H_2SO_4$ $2FeSO_4 + H_2SO_4 + O \rightarrow Fe_2(SO_4)_3 + H_2O$	Take 100 ml of ferric salt solution into a conical flask. Add 50 ml dil. H_2SO_4 and 5 g zinc dust or granules. Fit a Bunsen valve and heat it to flame till free from ferrous ions, test with KCNS or NH_4CNS. Filter and pipette out 20 ml of this solution. Proceed as in case of ferrous sulphate.
$K_2Cr_2O_7$ Titrations				
Estimation of $FeSO_4$/Mohr Salt	*N*-Phenyl Anthranilic acid (Interna indicator)	Green to violet red	$K_2Cr_2O_7 + 4H_2SO_4 \rightarrow K_2SO_4 + Cr_2\,(SO_4)_3 + 4H_2O + 3[O]$ $2FeSO_4 + H_2SO_4 + [O\} \rightarrow Fe_2\,(SO_4)_3 + H_2O$ In case of Mohr salt, only $FeSO_4$ reacts according to the above equation.	Pipette out 20 ml of $FeSO_4$/Mohr salt solution into a titration flask. Add 100 ml of 2N H_2SO_4 and 10 drops of the indicator. Titrate with $K_2Cr_2O_7$ taken in the burette till the colour changes from green to violet red.

Estimation of $FeSO_4$/Mohr salt	$K_3[Fe(CN)_6]$ (External Indicator)	No blue colour with $K_3[Fe(CN)_6]$	Above Equations Plus $3Fe^{2+} + 2[Fe(CN)]_6^{3-} \rightarrow Fe_3\,[Fe(CN)_6]_2$ Blue complex	Pipette out 20 ml of $FeSO_4$/Mohr salt solution into a titration flask. Add an equal volume of dil H_2SO_4. Run down $K_2Cr_2O_7$ solution from the burette into the titration flask till a drop from the mixture in the titration flask does not give a blue colour on touching with $K_3[Fe(CN)_6]$ on a tile.
Estimation of ferric sulphate	N-Phenyl Anthranilic Acid	Green to violet red	$Fe_2(SO_4)_3 + 2HCl + SnCl_2 \rightarrow 2FeSO_4 + SnCl_4 + H_2SO_4$ $2HgCl_2 + SnCl_2 \rightarrow SnCl_4 + Hg_2Cl_2$ $K_2Cr_2O_7 + 4H_2SO_4 \rightarrow K_2SO_4 + Cr_2(SO_4)_3 + 4H_2O + 3[O]$ $2FeSO_4 + H_2SO_4 + [O\} \rightarrow Fe_2(SO_4)_3 + H_2O$	Pipette out 20 ml of ferric sulphate solution in a titration flask. Add 2-3 ml of con. HCl and heat to boil. Add $SnCl_2$ solution dropwise with shaking till yellow colour disappears. Cool, add 10 ml of saturated solution of mercuric chloride to remove excess of $SnCl_2$. After reduction, add 100 ml of dil. H_2SO_4 and 10 drops of N-Phenyl anthranilic acid indicator. Titrate with $K_2Cr_2O_7$.
Iodine Titrations Estimation of oxidising agents like $KMnO_4$ or $K_2Cr_2O_7$	Starch solution	Disappearance of blue colour and appearance of light green colour (with $K_2Cr_2O_7$)	$K_2Cr_2O_7 + 4H_2SO_4 \rightarrow K_2SO_4 + Cr_2(SO_4)_3 + 4H_2O + 3[O]$ $2KMnO_4 + 3H_2SO_4 \rightarrow K_2SO_4 + 2MnSO_4 + 3H_2O + 5[O]$ $2KI + H_2SO_4 \rightarrow K_2SO_4 + 2HI$ $2HI + [O] \rightarrow H_2O + I_2$ $2Na_2S_2O_3 + I_2 \rightarrow Na_2S_4O_6 + 2NaI$	Pipette out 20 ml of $KMnO_4$ or $K_2Cr_2O_7$ solution into a titration flask. Add 1 test tube of dil. H_2SO_4 and half test of 10% KI solution. Add 1-2 g solid $NaHCO_3$. Wait for 1-2 minutes. Dilute with 50 ml of water. Titrate with hypo taken in the burette. When the solution in the titration flask becomes yellow, add 1-2 ml starch and continue titration till the solution becomes colourless (in case of $KMnO_4$) and light green (in case of $K_2Cr_2O_7$).
Estimation of $CuSO_4$	Starch solution	Disappearance of blue colour and appearance of white ppt.	$CuSO_4 + 2KI \rightarrow CuI_2 + K_2SO_4$ $2CuI_2 \rightarrow Cu_2I_2 + I_2$ $2Na_2S_2O_3 + I_2 \rightarrow Na_2S_4O_6 + 2NaI$	Pipette out 20 ml of copper sulphate solution into a titration flask. Add a pinch of Na_2CO_3 till a turbidity appears. Add dil. acetic acid dropwise till the turbidity disappears. Add half test tube of 10% KI solution. Wait for 1-2 minutes. Dilute the solution with 50 ml of water. Perform the titration with hypo till the colour becomes pale yellow. Add 1-2 ml starch solution and continue the titration till a white precipitate is obtained.

Complexometric Titrations				
Estimation of Mg^{2+}	Eriochrome Black T	Red to sky blue coloured solution	$Mg^{2+} + Na_2\,(H_2\,EDTA) \rightarrow 2Na^+ + (Mg\ EDTA)^{2-} + 2H^+$	Pipette out 20 ml of magnesium sulphate solution in a titration flask. Add to it equal volume of distilled water followed by 2 ml of buffer solution. (pH = 10) and 2-3 drops of indicator solution. Run down EDTA solution from the burette with constant shaking of the flask till the end point is reached.
Estimation of Zn^{2-}	Eriochrome Black T	Red to sky blue Colour	$Zn^{2+} + Na_2\,(H_2\,EDTA) \rightarrow 2Na^+ + (Zn\ EDTA)^{2-} + 2H^+$	As above TakeZn^{2+} solution in place of Mg^{2+} solution.
Estimation of Ca^{2+}	Murexide	Red to purple	$Ca^{2+} + Na_2\,(H_2\,EDTA) \rightarrow 2Na^+ + (Zn\ EDTA)^{2-} + 2H^+$	Pipette out 20 ml of the solution. Add to it equal volume of distilled water. Add 5 ml of NaOH solution to remove magnesium if present. Add 3-4 drops of murexide indicator. Perform the titration with standard EDTA solution.

34 GRAVIMETRIC ANALYSIS

Q. 1. What is meant by gravimetric analysis?

Ans. Gravimetric analysis is a mode of quantitative analysis of substances, mainly metallic salts, by the method of weighing. The substance to be estimated is taken in a small but known mass, it is brought into the solution state and then by the addition of a suitable reagent converted into an insoluble known substance. The precipitate of the newly formed substance is filtered and washed and then dried (or *ignited)* in a crucible. From the mass of the precipitate, quantitative analysis of the initial substance can be done.

Gravimetric estimation requires a great deal of skill and practice in order to obtain accurate results. Although it involves greater efforts and consumes more time compared to volumetric analysis, it is capable of providing results of precision if carried out metriculously.

An important aspect involved in gravimetric analysis is the availability of good balances and weights (including fractional weights), their handling and the time required to accomplish weighing. This hurdle of balances and weighing is, to some extent, being solved with the advent of electronic balances. These electronic balances, though quite expensive, are being provided in increasing numbers in the laboratories to enable the experimenter to carry out gravimetric analysis with the maximum accuracy and least hassle. In order to determine the weight, one has simply to place the object on it and read the weight displayed in digital form upto three or four decimal places depending upon the precision of the balance.

Q.2. Describe the common apparatus used in gravimetric analysis.

Ans. Common apparatus used in gravimetric analysis is described as under:

1. Weighing Bottles. Solids or liquid can be weighed in special bottles provided with stoppers as shown in Fig. 34.1. These can also be weighed on watch glass for this purpose if the substance is a solid. Weighing bottle or watch glass should be cleaned and dried before use.

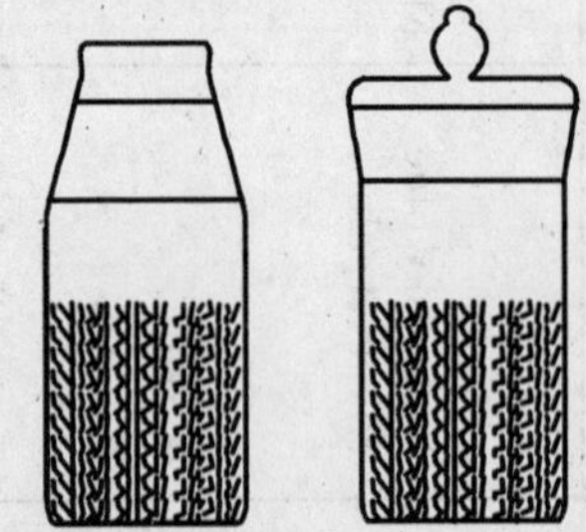

Fig. 34.1. Weighing Bottles

2. Beaker. Beaker is a key apparatus in gravimetry. It should be heat and scratch resistant. Generally 400 or 500 ml. pyrex or corning glass beakers are used for carrying out gravimetric work. The beaker to be used should be thoroughly cleaned and washed freely with water and then rinsed with distilled water. Water drops sticking on the exterior should be wiped off. While heating, the beaker should be placed on the wire gauge and not directly on the flame lest it should break.

3. Glass-rod (stirrer). It is required to dissolve a solid in water (or in any other solvent) or to mix a precipitating agent to the original solution or to transfer the precipitate to the filter paper in a funnel in gravimetric operations. For this purpose we use a glass rod whose edges have been smoothened by rotating in a flame.

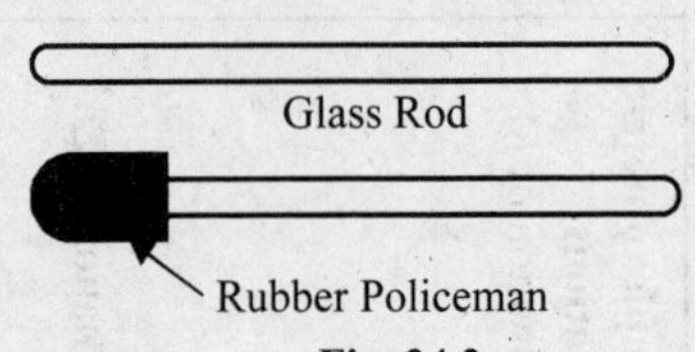

Fig. 34.2

Precipitate sticking to the beaker can be removed and transferred by capping one end of the rod with a piece of rubber tube of appropriate diameter which fits onto the glass rod (Fig. 34.2). This rubber piece is called *policeman.*

This rubber piece rubs off any precipitate sticking to the beaker and with the used of a wash bottle can be swayed to the desired place. While boiling the solution, the glass rod should never be placed in the beaker as there is a chance of the beaker breaking due to spurting of the rod. Also the rubber end (*policeman*) should not remain dipped in the solution as it could be attacked by the chemicals. The glass-rod should never be placed on the table as it could catch dirt. It should rest horizontally on the beaker or on the watch glass kept on the beaker.

4. Wash-bottle. In order to use a forceful steam of water to transfer the solid completely from the watch glass to the beaker or to transfer the precipitate adhering to the interior of the beaker to the filter paper, a wash bottle is used. Two types of bottles are in use. (Fig. 34.3)

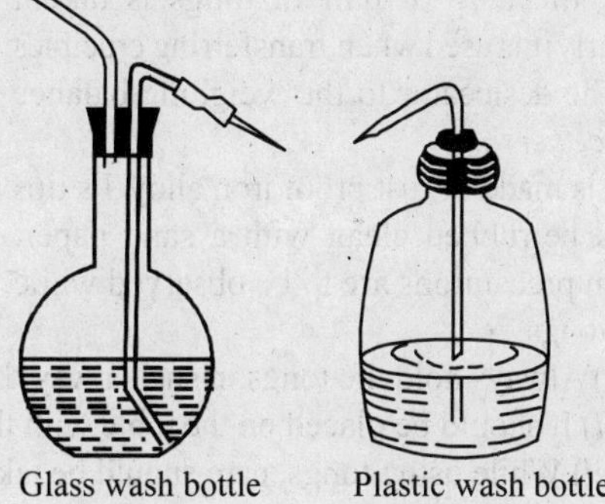

Fig. 34.3. Types of wash bottles.

Glass wash bottle can be assembled in the laboratory by using the components generally available in a chemistry laboratory whereas the plastic bottles are available readymade in different capacities like 200 ml, 400 ml, 500 ml, or 1 litre.

While using the glass bottle, air has to be blown with the mouth and the stream of water coming out is to be focused on the desired spot. To use the plastic bottle, it has simply to be pressed enough to obtain a stream of water. If hot water is needed, then it is advisable to use glass wash bottle.

5. Crucible

(*i*) Silica crucible. Certain gravimetric exercises require the used of silica crucible (Fig. 34.4). It is also provided with a lid. This material is heat resistant and can stand temperatures upto 1200°C. The precipitate is transferred (sometimes alongwith the filter paper) to the silica crucible and the crucible is ignited to obtain the dry substance.

(*ii*) Sintered glass crucible. For certain other gravimetric exercises, a sintered glass crucible (Fig. 34.4) is used. It made of pyrex glass whose bottom is provided with a sintered disc. The porous bottom is of varying porosity and is numbered as 1, 2, 3 and 4. Pore size decreases as the number increases.

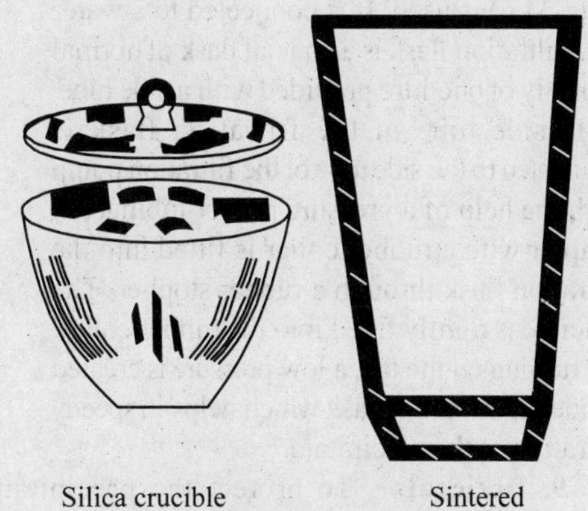

Fig. 34.4. Crucibles

Thus sintered crucible No. 4 will have very fine pores suitable for filtering fine precipitate such as barium sulphate. These crucibles are fitted onto vacuum filtration flask. The solvent is removed quickly due to reduced pressure and the precipitate is drained on the porous bottom of the sintered crucible.

6. Filter paper. The precipitate obtained after adding the precipitating agent is to be filtered. Some of these precipitates are coarse while others are fine. Therefore in order to filter them, filter papers of suitable porosity are needed. For gravimetric work, Whatmann filter paper no. 40, 41, 42 are used depending upon the particle size of the precipitate. To filter very fine particle, we use Whatmann filter paper No. 42, while to filter gelatinous precipitate, filter paper No. 40 can be used conveniently.

The material of the filter paper is such that on ignition it produces negligible mass of the ash. These are generally available in circular discs.

7. Tongs. A pair of tongs (Fig. 34.5) is used for handling silica crucible and sintered glass crucible. While the crucible is being ignited, its position is changed on the flame in order to ensure complete ignition of the contents. For changing the position on the flame, the help of pair of tongs is taken. Similarly it is used when transferring crucibles from the desiccator to the weighing balance and *vice versa.*

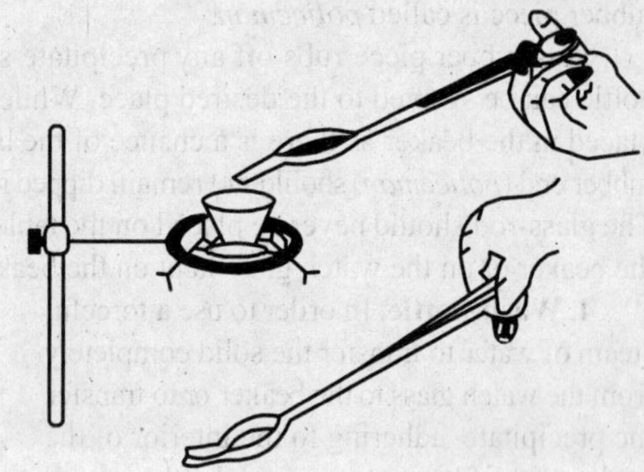

Fig. 34.5. How to hold the tongs

It is made of rust proof iron alloy. Its tips should be rubbed clean with a sand paper. Certain precautions are to be observed while using tongs.

(*i*) Always hold the tongs in such a way that the tips point to the back of your palm.

(*ii*) It should be placed on the table with the tips pointing upwards.

(*iii*) While using tongs, care should be taken to see that the tips do not touch the contents of the crucible.

8. Filtration Flask and Filtration Pump. Filtration of the precipitate is usually a slow process and takes a lot of time.

This time of filtration can be shortened if the filtration is carried out under reduced pressure. For the purpose, a filtration pump. (Fig. 34.6) is used. It is connected to a water tap. Filtration flask is a concial flask of normal capacity of one litre provided with a side tube. This side tube of the filtration flask is connected to the side tube of the filtration pump with the help of a pressure rubber tubing. An adapter with a rubber collar is fitted into the filtration flask through a rubber stopper. The crucible is tightly fixed into the rubber collar. On turning on the tap, a low pressure is created inside the filtration flask which helps in speedy filtration of the precipitate.

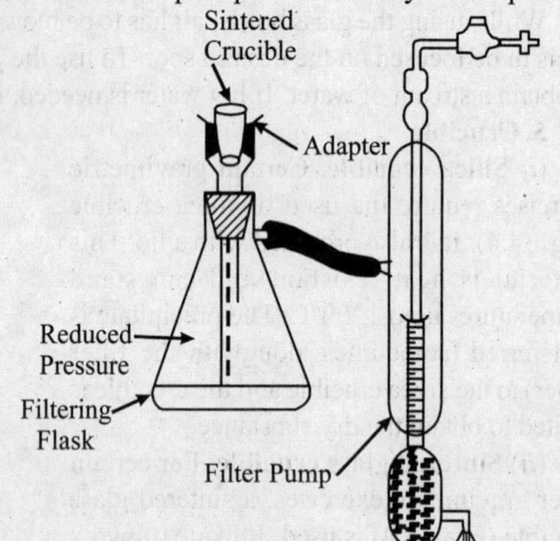

Fig. 34.6. Filtration flask and filtration pump

9. Desiccator. To protect the precipitate from moisture and atmospheric gases the crucible containing the precipitate is placed in a desiccator (Fig. 34.7). It is a glass apparatus having two compartments. The lower compartment is filled with anhydrous calcium chloride or phosphorus pentoxide. In the upper compartment, a number of crucibles can be placed on a china disc having big holes. It is provided with a heavy lid which fixes tightly on the rim of the desiccator when grease is applied on it. Some tips for using a desiccator are given below:

(*i*) Don't place very hot crucible in the desiccator. Allow it to cool a bit outside.

(*ii*) Remove and place the lid on the desiccator by sliding action. This will be convenient and will ensure complete closure.

Fig. 34.7. Desiccator

(*iii*) Replace calcium chloride periodically.

(*iv*) Hold the desiccator in both hands while carrying from one place to another.

10. Water Bath. Sometimes the precipitate requires digestion to get coagulated. This is done by placing the beaker containing the precipitate on a water bath (Fig. 34.8). A coagulated precipitate is easier to filter and the risk of fine particles of the precipitate crossing the filter paper is also eliminated. A water bath is a vessel made of copper provided with a number of concentric rings. A few rings can be removed from the bath, depending upon the size of the beaker or flask, for proper heating. It should be half filled with water.

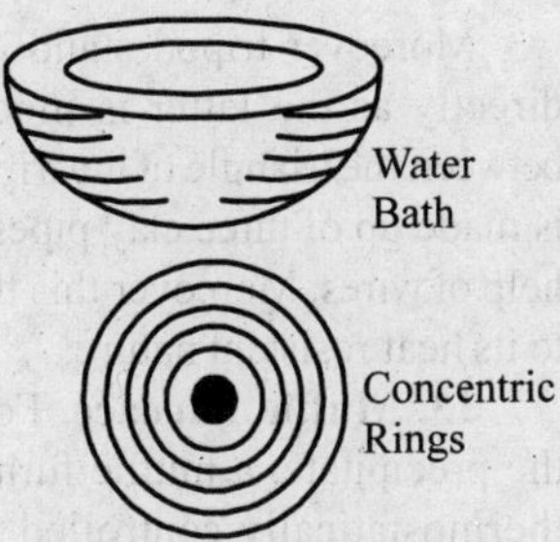

Fig. 34.8. Water bath

11. Electric Oven. The sintered crucibles containing the precipitate are dried in an electric oven. (Fig. 34.9). It is thermostatically controlled to achieved the desired temperatures. It can produce a temperature upto 250°C. For removing moisture and water from the precipitate, a temperature of 120°C for one hour is normally enough for good result.

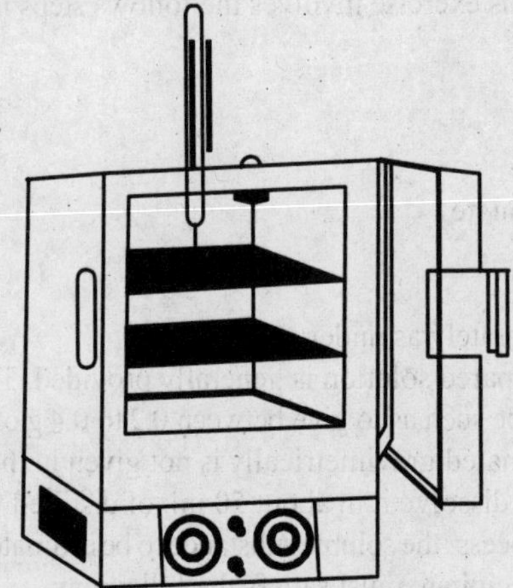

Fig. 34.9. Electric oven

12. Clay Pipe Triangle. When the ignition of the crucible is done on a burner, a clay-pipe triangle (Fig. 34.10) is placed on the tripod stand and then the crucible is placed on the clay-pipe triangle.

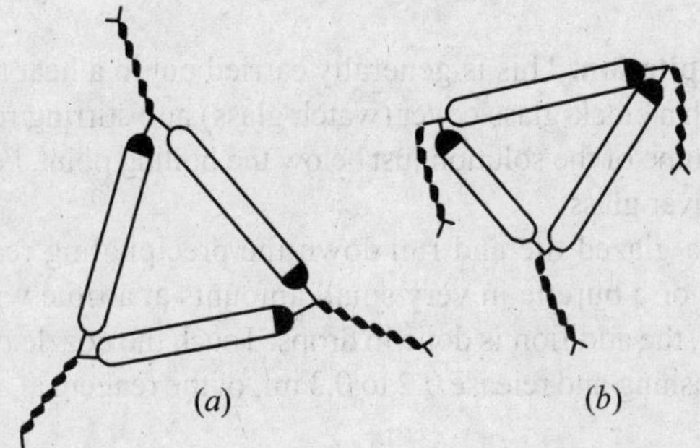

Fig. 34.10. (*a*) Clay pipe triangle, (*b*) Clay pipe triangle stand.

Since tripod stand is mostly rusted, the rust could pass on to the crucible. Hence the crucible is not heated directly on the tripod stand.

Moreover tripod stand cannot hold the crucible directly as the latter is much smaller than the gap between the triangle of the tripod. The clay pipe triangle is made up of three clay pipes joined together with the help of wires. Moreover this triangle remains clean due to its heat resistant nature.

13. Muffle Furnace. For convenient ignition of the precipitate, a muffle furnace (Fig. 3.11) which is thermostatically controlled is used. It can produce temperatures upto 1200-1300°C. Depending upon the requirement of the experiment, a required temperature is maintained. To cope with the high temperature, a long pair of tongs are used to place or take out the silica crucible.

Fig. 34.11. A muffle furnace

Q.3. Write the different steps that are performed in a gravimetric analysis experiment.

Ans. A gravimetric analysis exercise involves the follows steps in the order given.

(*i*) Solution making
(*ii*) Precipitation
(*iii*) Digestion
(*iv*) Filtration
(*v*) Washing of the precipitate
(*vi*) Drying and ignition
(*vii*) Weighing

Each step is described separately as under:

1. Solution making. A prepared solution is generally provided. The amount dissolved in 20 ml or 25 ml of the solution should be such as to give between 0.2 to 0.4 g of ignited residue for weighing. In case the substance to be estimated gravimetrically is not given in the form of a solution weigh out exactly a suitable quantity and dissolve it in about 50 ml. of distilled water in a clean beaker.

2. Precipitation. In this process, the soluble substance to be estimated is converted into a insoluble substance (precipitate). The precipitate must satisfy the following:

(*a*) It must have a low solubility product otherwise complete precipitation of the given ion or element is not possible.

(*b*) It must consists of large particles to allow the rapid filtration and not choke the pores of filter paper.

(*c*) It must be stable, having a definite composition or easily convertible to such a substance by ignition.

Procedure for precipitation. This is generally carried out in a heat resistant glass beaker 400 ml. capacity provided with a clock-glass cover (watch glass) and stirring rod of suitable length.

Heat the required volume of the solution just below the boiling point, keeping the beaker covered with the watch glass or cover glass.

Place the beaker on a glazed tile and run down the precipitating reagent by the spout of the breaker or from a pipette or a burette in very small amounts at a time with regular stirring of the solution. Towards the end, the addition is done in drops. Touch the nozzle of the pipette with the side of the beaker to avoid splashing and release 0.2 to 0.3 ml. of the reagent at a time until it is slightly in excess.

When the precipitation is believed to be completely effected, the precipitate should be allowed to settle down and then a few drops of the precipitant are added to the clear supernatant liquid to see whether the precipitation is complete. If no more precipitate is formed the precipitation is taken to be completed.

3. Digestion of the precipitate. The purpose of digestion is to obtain complete precipitation in a form which can be easily filtered. Most precipitates require digestion before filtration. During digestion, the precipitate is kept in contact with the liquid from which it was formed. Digestion is carried out at a temperature just below boiling point over a water bath at low flame (Fig. 34.12).

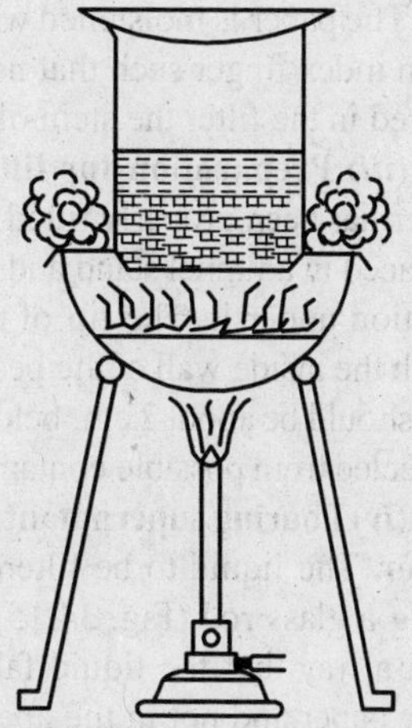

Fig. 34.12. Digestion of the precipitate on a water bath

In some cases, the beaker is kept at the room temperature for 12-24 hours. Digestion reduces the amount of absorbed impurities and makes the precipitate more granular easy for filtration. During digestion of the precipitate, the following changes occur:

(*a*) The very small particles pass back into the solution and ultimately re-deposit upon the larger particles.

(*b*) The rapidly formed crystals which are of irregular shape and possess a comparatively large surface become regular in shape and dense, leading to the decrease in area of the surface and finally reduction of adsorption.

The net outcome of digestion is to reduce co-precipitation and to increase the size of the particles, making filtration easier.

Digestion is not always necessary. Some precipitates which are colloidal such as $Fe(OH)_3$, $Al(OH)_3$ do not require digestion. With such precipitates a few minutes of boiling the contents will bring about the coagulation.

4. Filtration and washing of the precipitate. The object of filtration is to separate a precipitate from the mother liquor. In this process, the precipitate is retained and collected on the filter paper (or the filtering medium in case of filtering crucibles) while mother liquor passes through it. Washing is done to separate the last traces of mother liquor from the precipitate.

There are four main filtering devices: (*a*) Filter paper, (*b*) Gooch crucible, (*c*) Sintered glass crucible, (*d*) Porous porcelain filtering crucible. The selection of the devices depends upon the nature of the precipitate.

A. Technique of filtration with paper

The size and porosity of filter paper must be considered at before starting the process of filtration. It should be remembered that the size is determined by the amount of precipitate rather than volume of liquid to be filtered. The precipitate should normally fill not more than a third of the filter at the end of filtration. Circles of 11 cm. diameter are most frequently used. The filtration of the precipitates is carried out as follows:

(*i*) Preparation of the filter paper. To begin with, the filter paper is folded exactly in half and then once again so that the corners of the two halves do not coincide but are 1-2 mm apart (Fig. 34.13).

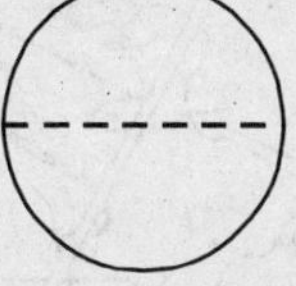

Filter Paper Circle

Fold in Half

Folds in Quarters not Coinciding

Fig. 34.13. How to fold filter paper.

When this folded paper is opened, to give a cone with three layers on one side and one on the other and placed in a 60° funnel it will fit into it exactly.

(*ii*) Fixing the filter paper in the funnel. The filter paper is now placed in a funnel of proper size; the top of the filter paper when set in the funnel should be about 1 cm. below the rim of funnel (Fig. 34.14)

Fig. 34.14. Correctly fixed filter paper.

The paper is moistened with water and is pressed down tightly to the sides of the funnel with the clean index finger such that no air space between the paper and the sides is left. When water is now poured in the filter the stem of the funnel should remain filled without any air bubbles.

***(iii)* Preparation for filtration.** When the filter paper has been properly fitted into the funnel, the later is placed in a funnel stand and a clean beaker is put into position under it. The tip of the funnel steam should touch the inside wall of the beaker to prevent splashing and should be about 2 cm. below the rim. The filtrate is protected from possible contamination by a watch glass.

***(iv)* Pouring supernatant liquid through the filter paper.** The liquid to be filtered is then poured down along a glass rod (Fig. 34.15) into the filter paper in such a way that the liquid falls along the side of the filter paper and not in the apex. The lower end of the glass rod should be near but not touching the filter paper. The filter paper is filled completely with the liquid and the level of the liquid should be 1 cm. below the top of the filter paper. In transferring liquid to the filter, cares should be taken not to disturb the precipitate. Even if the supernatant liquid appears to be entirely clear, it must always be decanted through the filter. When most of the supernatant liquid has been poured off, particles of precipitate will come over and the speed of the filtration will gradually decrease.

Fig. 34.15. How to pour the liquid through the filter paper.

(*v*) Washing the precipitate by decantation. When possibly the whole of the supernatant liquid has been poured through and a relatively small volume of liquid with precipitate is left, washing by decantation is begun. The wash liquid about 20-50 ml. from the wash bottle is directed against the sides of the beaker to wash down adhering particles, the precipitate is stirred up and then allowed to settle. When the supernatant liquid is clear, maximum amount of the clear liquid is decanted through the filter paper. The operation is repeated three or four times.

(*vi*) Transferring of precipitate to filter paper. After washing the precipitate by decantation, the main bulk of the precipitate is first transferred to the filter paper by mixing with wash solution, and pouring off the suspension, the process being repeated until most of the solid is removed from the beaker. The precipitate adhering to the sides and bottom of the precipitation beaker is removed in the following manner:

The beaker is held in the left hand and the stirring rod is put across its top and held in place with the index finger (Fig. 34.16). The rod should project 2 or 3 cm., beyond the tip of the beaker. The beaker is then inclined over the funnel and the wash bottle is held in the right hand, a steam of wash solution or water is directed against the inside of the beaker so that the particles of precipitate are washed out into the filter paper.

Even after this treatment, there will be small amounts of precipitate adhering to the walls. These are removed by rubbing gently with a policeman.

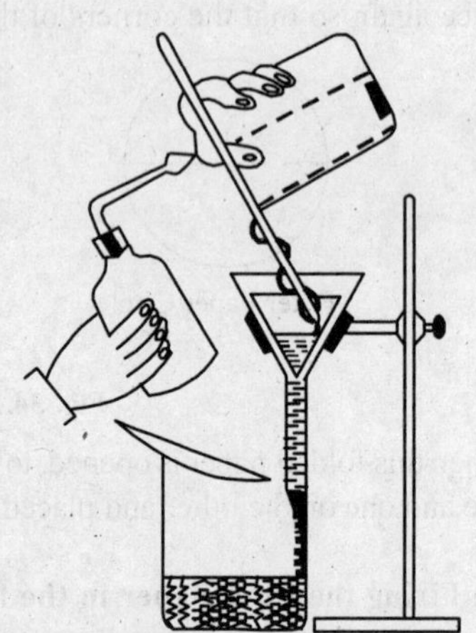

Fig. 34.16. How to transfer the precipitate to the filter paper.

When all particles have been dissolved, the policeman is rinsed and removed from the beaker. The small amount of precipitate thus loosened is washed into the filter paper in the same manner as before.

(*vii*) Final washing of the precipitate in the filter paper. After all the precipitate has been transferred to the filter paper, the last stage of the washing is begun. In this case, the wash solution is put along the rim and then gradually towards the apex of the cone where the precipitate has collected. In all cases, the completion of the washing is checked. A small portion of the filtrate direct from the funnel is taken and an appropriate qualitative test is applied. A precipitate should not be allowed to stand on the filter for any length of time unwashed. The mass will crack and cake together, and it will be impossible to wash it free of soluble salt. The wash liquid should have the following characteristics:

(*i*) The precipitate to be washed by a wash liquid should be insoluble in it, but the impurities should dissolve easily.

(*ii*) The wash liquid should not lead to peptisation.

(*iii*) The wash liquid should not contain an electrolyte which may interfere with final determination in the filtrate.

(*iv*) It should be easily volatile at the temperature of drying.

(*v*) It should not form any volatile or insoluble product with the precipitate.

Precautions during washing

(*i*) Minimum volume of the washing liquid be used to remove impurities since all precipitate dissolve to some extent in water.

(*ii*) Washing should be done with a number of small portions of the wash solution.

(*iii*) Always allow the whole of wash to drain out before adding a fresh lot.

(*iv*) The filtration of gelatinous precipitates like those of $Al(OH)_3$ and $Fe(OH)_3$ is difficult, for these tend to choke the pores of the filter paper. This difficulty is avoided by adding some filter paper pulp to the precipitate before filtering. Paper pulp or filter pulp is prepared as given below:

Paper pulp. The filter paper pulp is prepared by moistening quantitative filter paper with concentrated hydrochloric acid in a beaker, allowing the acid to remain in contact with the paper for two to three minutes (not longer) and then washing the paper under tap water and then a spray of distilled water. The washed paper is torn into pieces and shaken with distilled water in a conical flask when it distintegrates into pulp. This suspension can be stored for ready use.

Alternatively, it may be prepared by vigorously shaking an ordinary filter paper, torn into pieces, with hot water in a stoppered conical flask till it gives a pulp.

B. Technique of filtration with sintered glass crucible

The crucible is cleaned and heated at a temperature at which the precipitate is to be dried. A new sintered glass crucible is washed with concentrated hydrochloric acid and then with distilled water. After use, ammonia solution is used to remove AgCl; acetic acid to remove $PbSO_4$ and $PbCrO_4$; and conc. H_2SO_4 to remove $BaSO_4$, while aqua regia is used to remove metal sulphides.

Precipitates of nickel dimethyl glyoxime and copper thiocyanate can be removed by adding concentrated hydrochloric acid to the crucible. If required, the crucibles may be placed in hot conc. HCl taken in a beaker for some time followed by washing with water.

The filter crucibles (Fig. 34.17) have the following advantages over the use of filter paper during filtration:

(*i*) Filtration and washing take less time under reduced pressure created with a filtration pump.

(*ii*) Washing is much easier and more perfect.

(*iii*) There is no danger of change in the composition of the precipitate by the action of reducing gases and carbon as in the case of using filter paper.

(*iv*) Drying or heating the precipitate requires no special attention and is rapid.

The apparatus used for filtering the precipitate using sintered crucible is shown below (Fig. 34.18).

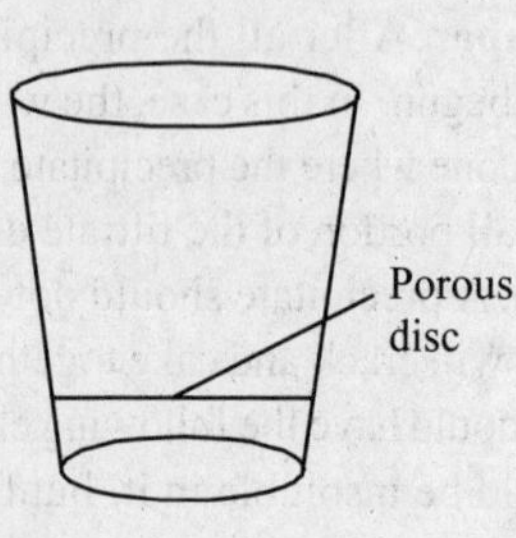

Fig. 34.17. Sintered crucible

Fig. 34.18. Filtration flask

5. Drying and ignition of the precipitate. When the precipitate has been washed and freed from soluble impurities, it must be obtained in a dried state for weighing. In some cases, the precipitate originally obtained is not suitable for direct weighing, but gives another more suitable compound when ignited. Precipitates such as $Fe(OH)_3$ are ignited to get Fe_2O_3 which is then weighed. Some precipitates like $BaSO_4$ or $PbSO_4$ need to be dried before weighing.

Thus the further technique depends upon the nature of the precipitate and the filtering medium. This treatment involves drying or igniting the precipitate. No general rule can be given as to the temperature necessary for the drying or ignition of a precipitate. Drying is significant when the temperature is below 250°C and ignition is significant when temperature is above 250°C and upto about 1200°C.

***(i)* To dry the precipitate obtained on the filter paper.** First remove the water in the stem of the funnel with a clear piece of filter paper. Place the funnel on a drying cone which is kept on a hot sand bath placed on a tripod stand (Fig. 34.19). The stem of the funnel should not touch the hot sand bath. If the stem touches the sand bath, raise it by placing another drying cone over the first. Drying is complete when the filter paper begins to separate from the funnel. Avoid overheating otherwise the filter paper may be charred.

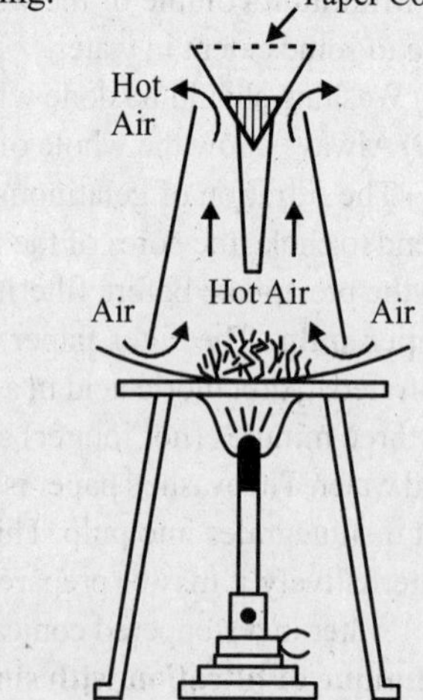

Fig. 34.19. Heating the funnel on drying cone.

Alternatively, the filter paper and its contents may be dried by placing in an oven (Fig. 34.20) at a temperature between 100°C and 110°C. The oven is also suitable for general use.

***(ii)* Drying and ignition in a filtering crucible.** The crucible containing the precipitate is removed from its adapter and is placed in an electric oven at a temperature of 120–150°C for a specific period. For temperature upto about 250°C, a thermostatically controlled electric oven is used.

A Incineration (complete burning) of the filter paper along with the precipitate. This procedure is adopted when the precipitate is not even partly reduced by carbon of the paper during heating and/or the reduced substance is again converted into the original substance by oxidation.

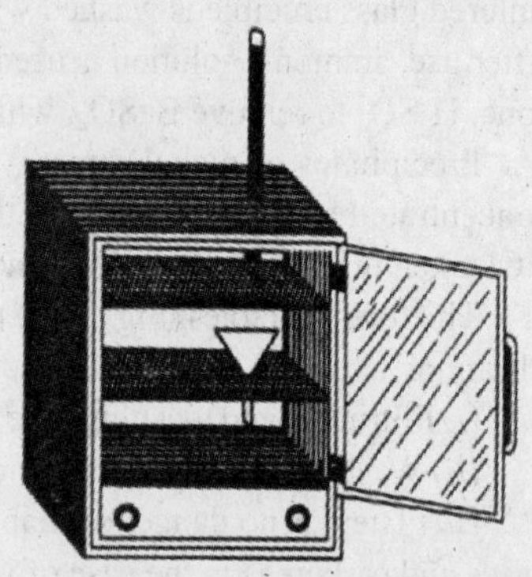

Fig. 34.20. Electric oven

When the incineration of filter paper along with the precipitate is to be carried out, it is unnecessary tó dry the funnel containing filter paper. The filter paper may be incinerated in the presence of the precipitate in the case of $Al(OH)_3$, $Fe(OH)_3$ and $BaSO_4$ as no reduction of the precipitate take place here.

The incineration of the filter paper with precipitate involves the following steps:

(*i*) Heating the empty crucible to a constant weight (taring). The silica crucible is treated with concentrated HCl or aqua regia to remove any previously adhering precipitate (from previous experiment). The crucible is then washed with distilled water and is placed on a clay triangle of proper size. The triangle should be placed on a ring fitted to a stand as its is easier to adjust its height. The crucible is then heated first with a low flame and then strongly for about half an hour at the same temperature at which the precipitate is intimately heated. It is then cooled in desiccator to the room temperature and is weighed. The heating, cooling and weighing is repeated till the crucible has a constant weight.

(*ii*) Transferring the precipitate into the crucible. The dried filter cone and the precipitate or the well-drained filter cone and the precipitate are carefully detached from the funnel and lifted out.

The finger should never touch the inside of the paper. The filter paper is then folded over the top so as to completely enclose the precipitate.

The packet is the placed in weighed crucible; if necessary it is pressed down gently into the bottom of the crucible.

(*iii*) Dring the paper and crucible. The crucible containing the packet is set vertically in a clay triangle and heated over a small Bunsen flame. The crucible should be partially covered, so that steam can escape; the drying in this way proceeds quickly and without risk.

(*iv*) Charring of paper. When the contents of the crucible are free from moisture, the flame is increased slightly so that the paper starts charring slowly. When charring commences, the paper should not be allowed to inflame, as this may cause fine particles of the precipitate to fly out owing to the rapid escape of the products of combustion. If by chance, it does catch fire, the flame should be extinguished by placing the cover on the mouth of the crucible with a pair of crucible tongs. Charring takes about 15-20 minutes.

(*v*) Oxidation of carbon. When the paper is completely charred, as revealed by its black colour and no more vapours escape, the crucible is tilted on the triangle and the cover is placed against the mouth. (Fig. 34.21)

The flame is moved to the bottom of the crucible and the carbon slowly burnt off while the flame is gradually increased.

The crucible is heated to a redness at the bottom. If the carbon on the lid is oxidized only slowly, the cover may be heated separately in a flame. During the oxidation of carbon, the crucible should be turned from one side to the other so as to expose a fresh portion to the heat till the whole of black ash is oxidized to white ash. At this point, the precipitate is also ignited. **Thermostatically controlled furnace called muffled furnace is used for convenient and quick work.**

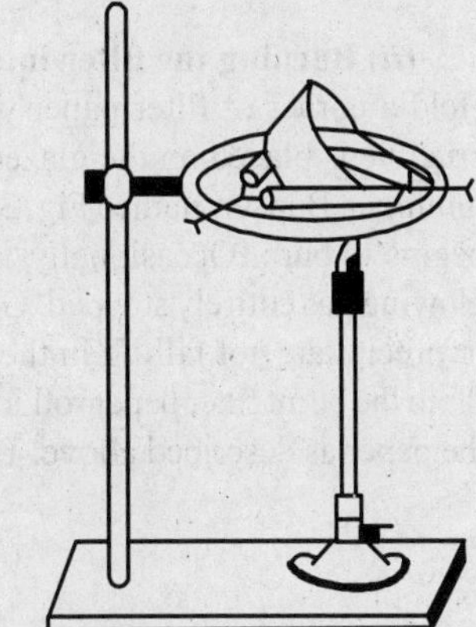

Fig. 34.21. How to heat the crucible

(*iv*) Placing the crucible in a desiccator and weighing. When the ignition is over, the flame is removed and after 1-2 minutes, the crucible and the lid are placed in a desiccator containing a suitable desiccant and then allowed to cool for about 20-25 minutes. The crucible and lid are then weighed. The crucible and contents are then ignited at the same temperature for 10-25 minutes, allowed to cool in a desiccator as before and weighed again. The ignition and weighing is repeated until a constant weight is attained. Crucibles should be handled with clean tongs.

B. Incineration of the filter paper apart from the precipitate.

The filter paper may be incinerated separately if:

(*a*) the precipitate is fusible or volatile at the temperature of incineration, *e.g.*, silver chloride

OR

(*b*) the precipitate suffers reduction to the metallic state during the charring of the filter paper, e.g., silver chloride, lead, sulphate, zinc carbonate.

In this case, the filter paper and the precipitate must be completely dried, as already described, before incinerating the filter. The following procedure is carried out:

(*i*) Transferring the dried precipitate to the watch glass. A sheet of glazed paper (25 cm. × 25 cm.) of a suitable colour, coloured for white and white for coloured precipitates is placed on the table. The paper should have no creases and should be smooth. A clear dry watch glass is placed on it. The dried filter paper is removed from the funnel and the precipitate is detached by rubbing the side of the filter paper together. The precipitate is then dropped on the watch glass.

The small particles of the precipitate still clinging to the filter paper are brushed with a camel-hair brush into the watch glass. If any particles of the precipitate happen to fall on the glazed paper, transfer the same also to the watch glass by means of camel-hair brush. Cover the watch glass resting on a glazed paper with an inverted funnel (Fig. 34.22).

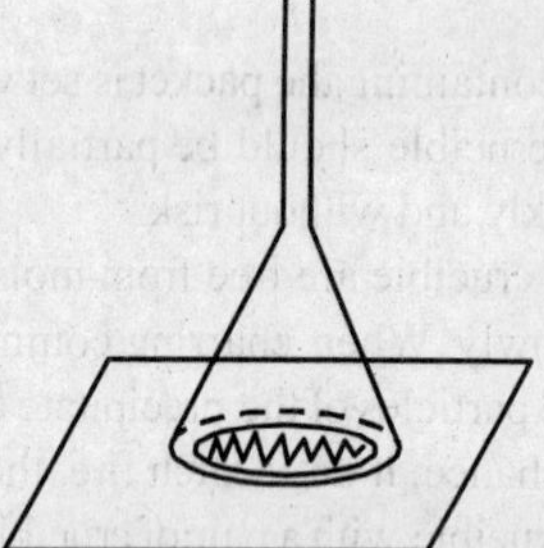

Fig. 34.22. Covering the precipitate with a glass funnel

Fig. 34.23. Use of a tong

(*ii*) Burning the filter into the crucible. The filter paper is now given one or two more folds. Hold a corner of filter paper with clean nickel or stainless steel tongs (Fig. 34.23). The weighed crucible is placed on the glazed paper and the rolled paper is held over it in the tip of a clear non-luminous Bunsen flame (Fig. 34.24). The dry paper catches fire. Remove the flame when the paper begins to burn. Occasionally, introduce the flame in the mass until it is completely burnt and the glowing has entirely stopped. Gently tap the ash against the side of the crucible. Small amounts of ash or precipitate not falling in the crucible are also transferred to the crucible with camel-hair brush. Place the burnt filter paper roll into the crucible, and keeping the crucible partially uncovered, incinerate the paper as described above. The crucible is cooled and again heated until the ash turns white.

Fig. 34.24. Incineration of filter paper.

(*iii*) Transferring the precipitate from the watch glass of the crucible containing filter ash. The crucible is now placed upon the glazed paper and the main bulk of the precipitate on the watch glass is carefully transferred to the crucible with a dry camel-hair brush (Fig. 34.25). The particles fallen, if any on the glazed paper are also swept into the crucible with a camel-hair brush.

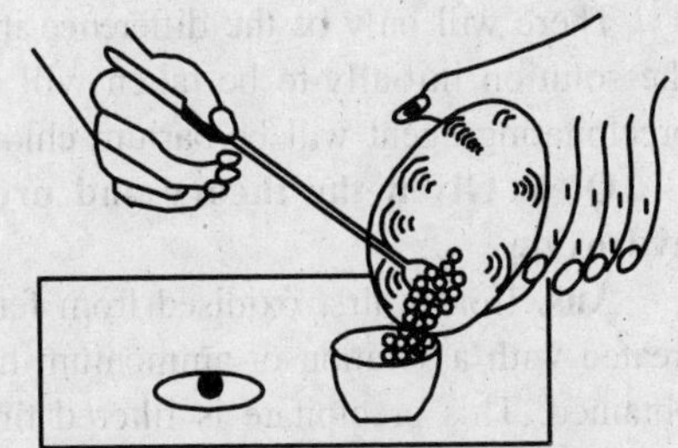

Fig. 34.25. Transferring the precipitate to the crucible.

(*iv*) Ignition. The precipitate, together with the filter paper ash is then ignited to a constant weight as described earlier, over a burner or in a muffle furnace.

6. Weighing. The crucible should be cooled in a desiccator before weighing. Pick up the crucible from the desiccator with the help of a pair of tongs and place it on the left hand pan of the balance. Find out the weight of the crucible containing the precipitate as explained earlier. Subtract from it, the weight of empty crucible to obtain the weight of the precipitate.

Note. This estimation can also be carried out using sintered glass crucible.

Q. 4. Give the theory and procedure for the estimation of barium chloride as barium sulphate.

Ans. ***Theory*** Barium is conveniently precipitated as barium sulphate $BaSO_4$. The precipitate of barium sulphate is filtered, ignited and weighed in a silica crucible or filtered, dried and weighed in a sintered glass crucible.

$$BaCl_2 + H_2SO_4 \longrightarrow \underset{ppt}{BaSO_4} + 2HCl$$

Procedure

1. Pipette out 20 ml of the given solution into a clean 400 ml beaker.
2. Dilute it with about 100 ml distilled water and add a few drops of conc. HCl.
3. Place the beaker in a water bath. Stir the solution in the beaker with a glass rod. Add dilute H_2SO_4 to it till no more precipitation takes place.
4. Allow the precipitate to digest and settle down. Remove it from the water bath and place it on the table.
5. Add a few drops of dilute H_2SO_4 from the sides of the beaker. Observe whether a turbidity (precipitation) occurs. If no more precipitation takes place, wait for some time.
6. If more of precipitation takes place, add 1 – 2 ml of dil H_2SO_4 and again place it in the water bath. Digest the precipitate for some more time.
7. Filter the precipitate of $BaSO_4$ through whatman filter paper No. 42. Give washings to the precipitate with distilled water till the washings are free from chloride ions (test with $AgNO_3$ solution) and sulphate ions (test with barium chloride solution)
8. Ignite the precipitate alongwith the paper in a weighed silica crucible.
9. Determine the mass of the crucible after ignition and cooling.
10. The difference will give the mass of $BaSO_4$ formed.
11. Do the calculations as required
    Calculations can be done for percentage purity, percentage composition, number of water of crystallisation, atomic weight of the metal, etc.

Q. 5. Give the theory and procedure for the estimation of sulphate using barium chloride.

Ans. The principle and procedure for the above estimation is the same as for the estimation of barium chloride using dil. H_2SO_4. The equation for this estimation may be written as

$$K_2SO_4 + BaCl_2 \longrightarrow 2KCl + \underset{ppt}{BaSO_4}$$

or

$$H_2SO_4 + BaCl_2 \longrightarrow 2HCl + \underset{ppt}{BaSO_4}$$

There will only be the difference in calculations. Procedure will remain the same except that the solution initially to be taken will be that of potassium sulphate or sulphuric acid and the precipitating agent will be barium chloride.

Q. 6. Given the theory and procedure for the estimation of iron using ammonium hydroxide.

Ans. Iron is first oxidised from ferrous to ferric state. The solution containing ferric salt is treated with a solution of ammonium hydroxide when a brown precipitate of ferric hydroxide is obtained. This precipitate is filtered through Whatmann filter paper No. 40. The precipitate is washed to remove nitric acid and ammonia. The precipitate is ignited in a silica crucible to ferric oxide.

$$Fe^{2+} \xrightarrow[HNO_3]{[O]} Fe^{3+}$$

$$Fe^{3+} + 3OH^- \longrightarrow Fe(OH)_3$$

$$2Fe(OH)_3 \xrightarrow[-H_2O]{\Delta} Fe_2O_3$$

Procedure

1. Pipette out 20 ml of the iron solution into a 400 ml beaker.
2. Add 100 ml of distilled water and 2-3 ml of conc. HNO_3 and heat slowly on the flame for about 15 minutes. Ferrous ions will be converted into ferric ion.
3. Remove the beaker from the flame and let it cool.
4. Add ammonia solution in small lots with constant sterring till there is a distinct smell of ammonia in the beaker. Add 1-2 ml more of ammonia solution.
5. Place the beaker in hot water bath and let the precipitate digest so that it is easier to filter. We can add half a test tube of filter paper pulp to the beaker. This helps in the digestion of the precipitate.
6. Filter the solution and collect the precipitate on Whatmann filter paper No. 40. Wash the precipitate with distilled water to remove traces of nitric acid and ammonium hydroxide sticking to the precipitate.
7. Place the filter paper alongwith the precipitate into a weighed silica crucible and ignite. Iron hydroxide will be converted to iron oxide.
8. Find out the mass of Fe_2O_3 by difference and make the necessary calculations like percentage purity, percentage composition, number of water molecules of crystallisation, atomic weight of the metal etc.

   Note 1. We cannot use a sintered glass crucible for this experiment because gelatinous precipitates like $Fe(OH)_3$, $Al(OH)_3$ choke the pores of filter disk.

   Note 2. We can estimate aluminium and chromium by precipitating them as hydroxide and igniting them into Al_2O_3 and Cr_2O_3 respectively in silica crucible.

Q. 7. What is gravimetric factor? How does it help in gravimetric calculations?

Ans. Gravimetric factor (GF) is the appropriate ratio of the formula mass of the substance analysed to that of the substance weighed

$$\text{Gravimetric factor} = \text{GF} = \frac{\text{Formula mass of the substance analysed (taken)}}{\text{Formula mass of the substance weighed}}$$

Gravimetric factors as applied to certain experiments are given in the table on the next page.

Substance analyzed	Precipitate formed	Precipitate weighed	Gravimetric factor
Fe	$Fe(OH)_3$	Fe_2O_3	$\frac{2Fe \text{ (at. mass)}}{Fe_2O_3 \text{(formula mass)}} = \frac{2\times 56}{160} = 0.6994$
Al	$Al(OH)_3$	Al_2O_3	$\frac{2Al \text{ (at. mass)}}{Al_2O_3 \text{ (formula mass)}} = \frac{2\times 27}{102} = 0.5291$
Mg	$Mg(C_9H_6ON)_2.2H_2O$	$Mg(C_9H_6ON)_2$	$\frac{Mg \text{ (at. mass)}}{Mg(C_9H_6ON)_2 \text{ (formula mass)}} = \frac{24}{312} = 0.0769$
Ni	$Ni(dmg)_2$	$Ni(dmg)_2$	$\frac{Ni \text{ (at. mass)}}{Ni(dmg)_2 \text{(formula mass)}} = \frac{58.7}{288.7} = 0.0203$
Cu	CuCNS	CuCNS	$\frac{Cu \text{ (at. mass)}}{CuCNS \text{ (formula mass)}} = \frac{63.5}{121.5} = 0.5226$
SO_4^{2-}	$BaSO_4$	$BaSO_4$	$\frac{SO_4^{2-} \text{ (formula mass)}}{BaSO_4 \text{ (formula mass)}} = \frac{137}{233} = 0.4115$

Knowing the weight of the residue formed, amount of the substance taken may be calculated using relations from chemical equations. Some useful relations or calculations are summarised below:

1. Estimation of iron and iron salts as Fe_2O_3

$$\underset{160 \text{ parts}}{Fe_2O_3} \equiv \underset{2\times 56 = 112 \text{ parts}}{2Fe} \equiv \underset{2\times 72 = 144 \text{ parts}}{2FeO}$$

2. Estimation of aluminium and its salts as Al_2O_3

$$\underset{102 \text{ parts}}{Al_2O_2} \equiv \underset{2\times 27 = 54 \text{ parts}}{2Al} \equiv \underset{2\times 133.5 = 267 \text{ parts}}{2AlCl_3}$$

3. Estimation of magnesium as magnesium oxinate $Mg(C_9H_6ON)_2$

$$\underset{348 \text{ parts}}{Mg(C_9H_6ON)_2.2H_2O} \equiv \underset{312 \text{ parts}}{Mg(C_9H_6ON)_2} \equiv \underset{24 \text{ parts}}{Mg}$$

4. Estimation of nickel as nickel dimethyl glyoxime $Ni(CH_3—C=N=OH)_2$ or $Ni(C_4H_7O_2N_2)_2$

$$\underset{288.70 \text{ parts}}{(C_4H_7O_2N_2)Ni} \equiv \underset{58.70 \text{ parts}}{Ni^{2+}} \equiv \underset{58.70 \text{ parts}}{Ni^{2+}}$$

$$\equiv \underset{394.70 \text{ parts by wt.}}{NiSO_4(NH_4)_2SO_4.6H_2O}$$

Q. 8. Describe the theory and procedure for the estimation of copper as cuprous thiocyanate.

Ans.

Theory

Copper is conveniently precipitated and estimated as cuprous thiocyanate CuCNS or $Cu_2(CNS)_2$. First the Cu(ic) salt is reduced to cuprous salt with sulphurous acid or ammonium bisulphite solution.

Ammonium thiocyanate is an excellent precipitant for Cu(ous) ion as it forms insoluble $Cu_2(CNS)_2$. The precipitate is washed, dried and weighed in a sintered crucible.

$$2CuSO_4 + 2H_2O + SO_2 \longrightarrow Cu_2SO_4 + 2H_2SO_4$$

$$2CuSO_4 + H_2O + NH_4HSO_3 \longrightarrow Cu_2SO_4 + NH_4HSO_4 + H_2SO_4$$

$$Cu_2SO_4 + 2NH_4CNS \longrightarrow \underset{\text{White ppt.}}{2CuCNS} + (NH_4)_2SO_4$$

Procedure

1. Pipette out 20 ml of the given solution into a clean 400 ml beaker.
2. Add few drops of HCl followed by 2 or 3 ml of freshly prepared saturated solution of sulphurous acid or 25 ml of 10% ammonium bisulphite solution.
3. Dilute the contents with 150 ml of distilled water. Heat the contents to boiling. Remove the beaker from the burner and place it on an asbestos pad.
4. Add freshly prepared 10% ammonium thiocyanate solution dropwise with constant stirring. After adding 5-7 ml of the precipitant stir the liquid briskly and allow to stand for some time.
5. When there is no blue colour in the supernatant liquid, then assume that the precipitation is complete. Ensure the complete precipitation by adding a few drops of the precipitant in the supernatant liquid.
6. Cover the beaker with watch glass. Keep it undisturbed for 2-3 hours preferably overnight.
7. Filter the precipitate through sintered glass crucible of porosity No. 4.
8. Wash the precipitates with 1% cold dil. solution of ammonium bisulphite in water, 10-12 times until the filtrate becomes free from CNS^- ion (No red colour with $FeCl_3$ solution).
9. Finally wash the precipitate 2-3 times with 20% alcohol. Heat the crucible in the electric oven at 120-130°C to constant weight.

Precautions to be Observed

1. Solution to be precipitated must be free from oxidising agents like nitrates, chromates.
2. Non-alkali metals usually interfere; therefore the sample should be free from such metals.
3. Excess of thiocyanate and high acidity should also be avoided. They effect the precipitation of Cu(ous) thiocyanate.
4. Solubility of the precipitate is greater at low pH. Thus the precipitation should be carried out in weakly acidic medium.
5. To prevent the oxidation of Cu(ous) salt, washing must be done by dil. NH_4HSO_3.
6. Excess of the precipitant should be avoided as it increases the solubility of the CuCNS due to the formation of a complex thiocyanate ion.

Preparation of Sulphurous Acid

It can be prepared freshly by adding dil. HCl or dil. H_2SO_4 to solid Na_2SO_3 or $Na_2S_2O_3$ in a flask and passing SO_2 formed into distilled water by means of a delivery tube.

Preparation of NH_4HSO_3 Solution

This is prepared by diluting to 10 times by volume commercial conc. solution which has a specific gravity of 1.33 containing 54% of SO_2.

Instructions for Laboratory Assistant

1. Dissolve 20 g/litre of $CuSO_4.5H_2O$. Ask the students to pipette 20 ml. of given solution.
2. Freshly prepared saturated solution of H_2SO_3 or NH_4HSO_3.
3. Sintered glass crucible of porosity No. 4.
4. 10% ammonium thiocyanate solution.
5. Washing solution (100 ml. water + 1 ml. sulphurous acid.)

Q.9. You are provided with $CuSO_4.5H_2O$ solution, 20 g of which are dissolved/litre. Find out gravimetrically the percentage of Cu in given solution.

Ans. Procedure. Copper is precipitated as cuprous thiocyanate after reducing Cu salt with H_2SO_3 into Cu(ous) salt and adding ammonium thiocyanate solution. For complete procedure read Q. 8.

Observations and Calculations:

Volume of solution taken for precipitation = 20 ml

Strength of copper sulphate = 20 g/litre

Weight of $Cu_2(CNS)_2$ formed = w g

$$\text{Weight of } Cu_2(CNS)_2 \text{ obtained from one litre} = \frac{w \times 1000}{20} = 50 \text{ w g.}$$

$$\underset{(243)}{Cu_2(CNS)_2} = \underset{(127)}{2 \times Cu}$$

243 g Cu(ous) thiocyanate is obtained from = 127 g of copper

$$\text{50 w g of Cu(ous) thiocyanate is obtained from} = \frac{127 \times 50w}{243} = a \text{ g/of copper (say)}$$

20 g of copper sulphate contains = a g of copper

$$\text{100 g of copper sulphate contains} = \frac{a \times 100}{20} \text{ g Cu}$$

$$\therefore \quad \text{Percentage of copper in sample} = \frac{a \times 100}{20} = 5a$$

Note. Similarly we can compute percentage purity, percentage composition, number of water molecules of crystallisation, atomic weight of metals, etc.

Q. 10. Describe the theory and procedure for the estimation of nickel as nickel dimethylglyoxime (Ni DMG).

Ans.

Theory

Nickel is estimated as nickel dimethylglyoxime whose brilliant red precipitates are formed by adding alcoholic solution of dimethyl glyoxime to a hot and slightly acidic solution of the nickel-salt and then adding a slight excess of dilute aqueous ammonia solution. Nickel dimethyl glyoxime is a coordinate chelate compound which is highly coloured with a low degree of dissolution. The complex gets precipitated within the pH range of 5-9. Higher the pH more complete the precipitation. To maintain the pH nearly to 9 the precipitation should be done in presence of NH_4Cl and NH_4OH. Precipitates are then filtered out through a sintered glass crucible, washed with water, dried at 110–120°C and weighed.

Preparation of Dimethyl Glyoxime Solution

Dimethyl glyoxime is insoluble in water, 1% solution of it is obtained by dissolving 1 g of it in 100 ml of absolute alcohol. One ml of this solution is enough to precipitate 0.0025 g of nickel.

Procedure

1. Pipette out 20 ml of the given nickel salt solution in a clean 400 ml beaker provided with a stirring glass rod fitted with "*policeman*".
2. Add 5 ml of dilute (1 : 1) hydrocholric acid and dilute it by adding 200 ml of distilled water.
3. Heat the contents to 70-80°C.
4. Remove the beaker from the flame and keep it on an asbestos pad.
5. Add about 30 ml of 1% alcoholic solution of dimethyl glyoxime with stirring.
6. Now immediately add dilute solution of ammonia in a fine stream with stirring of the solution. Stop adding ammonia solution when the contents of the beaker give smell of ammonia and red ppt. appear. Cover the beaker with a watch glass.
7. Digest the precipitates over a water bath for half an hour, allow to cool undisturbed for some time.
8. When the precipitates settle down add a little of the dimethyl glyoxime solution to the supernatant liquid to check for the complete precipitation. Allow the precipitates to stand for one hour.
9. Filter the precipitates through a weighed sintered crucible of porosity No. 3 fitted on a suction filtration of flask.
10. Pour the supernatant liquid in the crucible and then transfer the precipitates to the crucible with the help of *policeman.* Wash the ppt. with cold water several times until the filtrate is free from chloride ions *i.e.* stops giving white ppt. with dil. HNO_3 and $AgNO_3$. Dry the precipitates in an electric oven for 40–50 minutes at 110–120°C. Allow to cool in a desiccator and weigh. Repeat drying, heating and cooling until a constant weight is obtained.

Precautions to Observe

1. Avoid adding excess of dimethyl glyoxime because it may itself get precipitated due to its low solubility.
2. Excess of alcohol should be avoided because some of precipitates may dissolve when the concentration of the alcohol is more.
3. NH_4OH solution should be added in a thin stream with constant stirring so as to give larger crystals.
4. Strong heating must be avoided. High temperature causes decomposition of the precipitates.
5. Pure salt of nickel should be provided. Nickel salts commonly have iron impurity. It interferes in the process.

Q. 11. Determine the percentage purity of $NiSO_4.7H_2O$ gravimetrically, 20 g of which are dissolved/litre.

Ans.

Procedure

Pipette out 20 ml of the given solution into a beaker. Precipitate nickel as nickel dimethyl glyoxime by adding alcoholic solution of dimethyl glyoxime to the aqeuous solution of nickel salt under specified conditions of pH range 9 as described in Q. 10.

Observations and Calculations

Vol. of the solution taken for precipitation = 20 ml

Strength of $NiSO_4.7H_2O$ = 20 g/litre

Weight of Ni-dimethyl glyoxime = w g

Weight of Ni-dimethyl glyoxime obtained from 1 litre

$$\text{of the solution} = \frac{1000 \times w}{20} = 50\,w$$

$$\underset{(288.7)}{Ni(C_4H_7O_2N_2)_2} \equiv \underset{(280.7)}{NiSO_4.7H_2O}$$

$$288.7 \text{ g of Ni-dimethyl glyoxime is obtained from } = 280.7 \text{ g of } NiSO_4.7H_2O$$

$$50 \text{ w g of Ni-dimethyl glyoxime is obtained from } = \frac{280.7 \times 50w}{288.7} \text{ g } NiSO_4.7H_2O$$

$$= a \text{ g}/NiSO_4.7H_2O \text{ (say)}$$

$$20 \text{ g of impure } NiSO_4.7H_2O \text{ contain} = a \text{ g of pure } NiSO_4.7H_2O$$

$$100 \text{ g of impure } NiSO_4.7H_2O \text{ contain} = \frac{a \times 100}{20}$$

$$\text{Percentage purity of the sample} = \frac{a \times 100}{20}$$

APPENDIX
Chemical Galaxy
Or
Stewart Periodic Table

When the elements are arranged in order of the increasing nuclear charge or the increasing number of orbital electrons around it, they form a sequence in which the properties of the elements are repeated periodically. Elements with similar properties are located in the same group. The Periodic Table is shown by chopping the sequence of elements into sections and arranging them as rectangular blocks.

There is another possibility for the arrangement of elements. We can wind the sequence into a spiral. Because the properties are repeated at longer and longer intervals, increasing number of elements have to be fitted on the coils. P.J. Stewart, an ecologist at the university of Oxford in England recently (2005) devised a Periodic table which resembles a galaxy. In this version, the size of the successive turns is made to increase at a constant rate.

The *spokes* in the periodic table are curved in order to keep the inner elements reasonably close together while making room for the extra elements in the outer turns. The gap begins to open even in the second turn of the spiral between beryllium and boron, but this is balanced by a corresponding though smaller gap between oxygen and fluorine which closes up as you move outwards. The eight main groups lie on four curved that run through the centre joining them in four pairs.

Some characteristics of Stewart Galactic Periodic Table are:

Neutronium (neutron) has been placed at the centre of the **galaxy**. It may be regarded as the elements of zero atomic number. Its atoms are neutrons, the electrically neutral particles which combine with protons to form atomic nuclei of different elements heavier than ordinary hydrogen. When the density of the very early universe was too heavy for protons and electrons to exist separately, this must have been the original form of matter. It still exists inside neutron stars where nuclear fusion can no longer counteract gravity. Neutronium has caught the imagination of science fiction fans in novels such as *Dragon's Egg and Starquake* by Robert Forward and the *Neutronium Alchemist* by Peter F Hamilton.

Hydrogen finds its place half way between the beginning and end of the first turn of the spiral in Stewart's Periodic Table. Scientists hold different opinion regarding the position of hydrogen in the conventional periodic table. Some would place in the alkali group while the others would like to place it in the halogen group. According to Stewart, it sits comfortable above carbon which it resembles in many, whereas it has hardly any resemblance to alkali metals or halogen except in the valency. Both hydrogen and carbon have half-filled outer shell of electrons. Both of these elements share the electrons instead of losing or gaining them. Carbon and hydrogen have a great affinity for each other and are found in a vast number of compounds.

Lutetium and Lawrencium pose a problem in the conventional periodic table which has separate space for *f*-block elements. The main idea behind chemical galaxy is to amalgamate all the elements into a single entity. Should Lu and Lw be regarded as the last of the lanthanides or the first of the next block of transition elements? The problem is solved with the spiral arrangement. The two elements can be visualised as both.

CHEMICAL GALAXY
(STEWART'S PERIODIC TABLE)

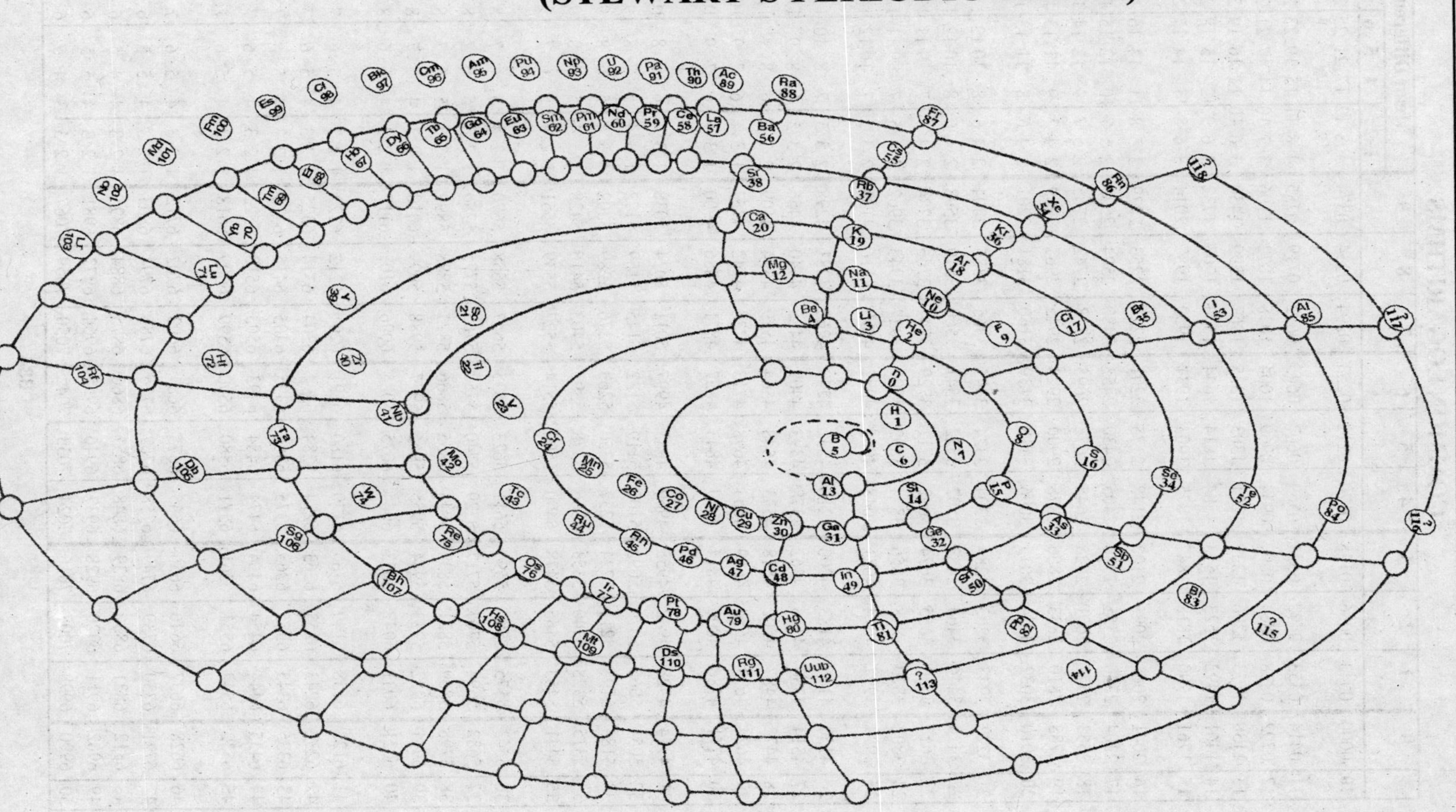

LOGARITHMS TABLES

COMMON LOGARITHMS

	0	1	2	3	4	5	6	7	8	9	Mean Differences								
											1	2	3	4	5	6	7	8	9
10	0000	0043	0086	0128	0170	0212	0253	0294	0334	0374	4	8	12	17	21	25	29	33	37
11	0414	0453	0492	0531	0569	0607	0645	0682	0719	0755	4	8	11	15	19	23	26	30	34
12	0792	0828	0864	0899	0934	0969	1004	1038	1072	1106	3	7	10	14	17	21	24	28	31
13	1139	1173	1206	1239	1271	1303	1335	1367	1399	1430	3	6	10	13	16	19	23	26	29
14	1461	1492	1523	1553	1584	1614	1644	1673	1703	1732	3	6	9	12	15	18	21	24	27
15	1761	1790	1818	1847	1875	1903	1931	1959	1987	2014	3	6	8	11	14	17	20	22	25
16	2041	2068	2095	2122	2148	2175	2201	2227	2253	2279	3	5	8	11	13	16	18	21	24
17	2304	2330	2355	2380	2405	2430	2455	2480	2504	2529	2	5	7	10	12	15	17	20	22
18	2553	2577	2601	2625	2648	2672	2695	2718	2742	2765	2	5	7	9	12	14	16	19	21
19	2788	2810	2833	2856	2878	2900	2923	2945	2967	2989	2	4	7	9	11	13	16	18	20
20	3010	3032	3054	3075	3096	3118	3139	3160	3181	3201	2	4	6	8	11	13	15	17	19
21	3222	3243	3263	3284	3304	3324	3345	3365	3385	3404	2	4	6	8	10	12	14	16	18
22	3424	3444	3464	3483	3502	3522	3541	3560	3579	3598	2	4	6	8	10	12	14	15	17
23	3617	3636	3655	3674	3692	3711	3729	3747	3766	3784	2	4	6	7	9	11	13	15	17
24	3802	3820	3838	3856	3874	3892	3909	3927	3945	3962	2	4	5	7	9	11	12	14	16
25	3979	3997	4014	4031	4048	4065	4082	4099	4116	4133	2	3	5	7	10	12	14	16	15
26	4150	4166	4183	4200	4216	4232	4249	4265	4281	4298	2	3	5	7	8	10	11	13	15
27	4314	4330	4346	4362	4378	4393	4409	4425	4440	4456	2	3	5	6	8	9	11	13	14
28	4472	4487	4502	4518	4533	4548	4564	4579	4594	4609	2	3	5	6	8	9	11	12	14
29	4624	4639	4654	4669	4683	4698	4713	4728	4742	4757	1	3	4	6	7	9	10	12	14
30	4771	4786	4800	4814	4829	4843	4857	4871	4886	4900	1	3	4	6	7	9	10	11	13
31	4914	4928	4942	4955	4969	4983	4997	5011	5024	5038	1	3	4	6	7	8	10	11	12
32	5051	5065	5079	5092	5105	5119	5132	5145	5159	5172	1	3	4	5	7	8	9	11	12
33	5185	5198	5211	5224	5237	5250	5263	5276	5289	5302	1	3	4	5	6	8	9	10	12
34	5315	53285	5340	5353	5366	5378	5391	5403	5416	5428	1	3	4	5	6	8	9	10	11
35	5441	5453	5465	5478	5490	5502	5514	5527	5539	5551	1	2	4	5	6	7	9	10	11
36	5563	5575	5587	5599	5611	5623	5635	5647	5658	5670	1	2	4	5	6	7	8	10	11
37	5682	5694	5705	5717	5729	5740	5752	5763	5775	5786	1	2	3	5	6	7	8	9	10
38	5798	5809	5821	5832	5843	5855	5866	5877	5888	5899	1	2	3	5	6	7	8	9	10
39	5911	5922	5933	5944	5955	5966	5977	5988	5999	6010	1	2	3	4	5	7	8	9	10
40	6021	6031	6042	6053	6064	6075	6085	6096	6107	6117	1	2	3	4	5	6	8	9	10
41	6128	6138	6149	6160	6170	6180	6191	6201	6212	6222	1	2	3	4	5	6	7	8	9
42	6232	6243	6253	6263	6274	6284	6294	6304	6314	6325	1	2	3	4	5	6	7	8	9
43	6335	6345	6355	6365	6375	6385	6395	6405	6415	6425	1	2	3	4	5	6	7	8	9
44	6435	6444	6454	6464	6474	6484	6493	6503	6513	6522	1	2	3	4	5	6	7	8	9
45	6532	6542	6551	6561	6571	6580	6590	6599	6609	6618	1	2	3	4	5	6	7	8	9
46	6628	6637	6646	6656	6665	6675	6684	6693	6702	6712	1	2	3	4	5	6	7	7	8
47	6721	6730	6739	6749	6758	6767	6776	6785	6794	6803	1	2	3	4	5	5	6	7	8
48	6812	6821	6830	6839	6848	6857	6866	6875	6884	6893	1	2	3	4	4	5	6	7	8
49	6902	6911	6920	6928	6937	6946	6955	6964	6972	6981	1	2	3	4	5	5	6	7	8
50	6990	6998	7007	7016	7024	7033	7042	7050	7059	7067	1	2	3	3	4	5	6	7	8

LOGARITHMS (Contd.)

	0	1	2	3	4	5	6	7	8	9	Mean Differences								
											1	2	3	4	5	6	7	8	9
51	7076	7084	7093	7101	7110	7118	7126	7135	7143	7152	1	2	3	3	4	5	6	7	8
52	7160	7168	7177	7185	7193	7202	7210	7218	7226	7235	1	2	3	3	4	5	6	7	7
53	7243	7251	7259	7267	7275	7284	7292	7300	7308	7316	1	2	3	3	4	5	6	6	7
54	7324	7332	7340	7348	7356	7364	7372	7380	7388	7396	1	2	3	3	4	5	6	6	7
55	7404	7412	7419	7427	7435	7443	7451	7459	7466	7474	1	2	3	3	4	5	5	6	7
56	7482	7490	7497	7505	7513	7520	7528	7536	7543	7551	1	2	2	3	4	5	5	6	7
57	7559	7566	7574	7582	7589	7597	7604	7612	7619	7627	1	2	2	3	4	5	5	6	7
58	7634	7642	7649	7657	7664	7672	7679	7686	7694	7701	1	1	2	3	4	5	5	6	7
59	7709	7716	7723	7731	7738	7745	7752	7760	7767	7774	1	1	2	3	4	5	5	6	7
60	7782	7789	7796	7803	7810	7818	7825	7832	7839	7846	1	1	2	3	4	5	5	6	6
61	7853	7860	7868	7875	7882	7889	7896	7903	7910	7917	1	1	2	3	4	4	5	6	6
62	7924	7931	7938	7945	7952	7959	7966	7973	7980	7987	1	1	2	3	4	4	5	6	6
63	7993	8000	8007	8014	8021	8028	8035	8041	8048	8055	1	1	2	3	4	4	5	5	6
64	8062	8069	8075	8082	8089	8096	8102	8109	8116	8122	1	1	2	3	4	4	5	5	6
65	8129	8136	8142	8149	8156	8162	8169	8176	8182	8189	1	1	2	3	4	4	5	5	6
66	8195	8202	8209	8215	8222	8228	8235	8241	8248	8254	1	1	2	3	3	4	5	5	6
67	8261	8267	8274	8280	8287	8293	8299	8306	8312	8319	1	1	2	3	3	4	5	5	6
68	8325	8331	8338	8344	8351	8357	8363	8370	8376	8382	1	1	2	3	3	4	4	5	6
69	8388	8395	8401	8407	8414	8420	8426	8432	8439	8445	1	1	2	3	3	4	4	5	6
70	8451	8457	8463	8470	8476	8482	8488	8494	8500	8506	1	1	2	3	3	4	4	5	6
71	8513	8519	8525	8531	8537	8543	8549	8555	8561	8567	1	1	2	2	3	4	4	5	5
72	8573	8579	8585	8591	8597	8603	8609	8615	8621	8627	1	1	2	2	3	4	4	5	5
73	8633	8639	8645	8651	8657	8663	8669	8675	8681	8686	1	1	2	2	3	4	4	5	5
74	8692	8698	8704	8710	8716	8722	8727	8733	8739	8745	1	1	2	2	3	4	4	5	5
75	8751	8756	8762	8768	8774	8779	8785	8791	8797	8802	1	1	2	2	3	3	4	5	5
76	8808	8814	8820	8825	8831	8837	8842	8848	8854	8859	1	1	2	2	3	3	4	5	5
77	8865	8871	8876	8882	8887	8893	8899	8904	8910	8915	1	1	2	2	3	3	4	4	5
78	8921	8927	8932	8938	8943	8949	8954	8960	8965	8971	1	1	2	2	3	3	4	4	5
79	8976	8982	8987	8993	8998	9004	9009	9015	9020	9025	1	1	2	2	3	3	4	4	5
80	9031	9036	9042	9047	9053	9058	9063	9069	9074	9079	1	1	2	2	3	3	4	4	5
81	9085	9090	9096	9101	9106	9112	9117	9122	9128	9133	1	1	2	2	3	3	4	4	5
82	9138	9143	9149	9154	9159	9165	9170	9175	9180	9186	1	1	2	2	3	3	4	4	5
83	9191	9196	9201	9206	9217	9217	9222	9227	9232	9238	1	1	2	2	3	3	4	4	5
84	9243	9248	9253	9258	9263	9269	9274	9279	9284	9289	1	1	2	2	3	3	4	4	5
85	9294	9299	9304	9360	9315	9320	9325	9330	9335	9340	1	1	2	2	3	3	4	4	5
86	9345	9350	9355	9360	9365	9070	9375	9380	9385	9390	1	1	2	2	3	3	4	4	5
87	9395	9400	9405	9410	9415	9420	9425	9430	9435	9440	0	1	1	2	2	3	3	4	4
88	9445	9450	9455	9460	9465	9469	9474	9479	9484	9489	0	1	1	2	2	3	3	4	4
89	9494	9499	9504	9509	9513	9518	9523	9528	9533	9538	0	1	1	2	2	3	3	4	4
90	9542	9547	9552	9557	9562	9566	9571	9576	9581	9586	0	1	1	2	2	3	3	4	4
91	9590	9595	9600	9605	9609	9614	9619	9624	9628	6712	0	1	1	2	2	3	3	4	4
92	9638	9643	9647	9652	9657	9661	9666	9671	9675	6803	0	1	1	2	2	3	3	4	4
93	9685	9689	9694	9699	9703	9708	9713	9717	9722	6893	0	1	1	2	2	3	3	4	4
94	9731	9736	9741	9745	9750	9754	9759	9763	9768	6981	0	1	1	2	2	3	3	4	4
95	9777	9782	9786	9791	9795	9800	9805	9809	9814	9818	0	1	1	2	2	3	3	4	4
96	9823	9827	9832	9836	9841	9845	9850	9854	9859	9863	0	1	1	2	2	3	3	4	4
97	9868	9872	9877	9881	9886	9890	9894	9899	9903	9908	0	1	1	2	2	3	3	4	4
98	9912	9916	9921	9926	9930	9934	9939	9943	9948	9952	0	1	1	2	2	3	3	4	4
99	9956	9961	9965	9969	9974	9978	9983	9987	9991	9996	0	1	1	2	2	3	3	3	4

ANTILOGARITHMS

	0	1	2	3	4	5	6	7	8	9	Mean Differences								
											1	2	3	4	5	6	7	8	9
.00	1000	1002	1005	1007	1009	1012	1014	1016	1019	1021	0	0	1	1	1	1	2	2	2
.01	1023	1026	1028	1030	1033	1035	1038	1040	1042	1045	0	0	1	1	1	1	2	2	2
.02	1047	1050	1052	1054	1057	1059	1062	1064	1067	1069	0	0	1	1	1	1	2	2	2
.03	1072	1074	1076	1079	1081	1084	1086	1089	1091	1094	0	0	1	1	1	1	2	2	2
.04	1096	1099	1102	1104	1107	1109	1112	1114	1117	1119	0	1	1	1	1	2	2	2	2
0.5	1122	1125	1127	1130	1132	1135	1138	1140	1143	1146	0	1	1	1	1	2	2	2	2
.06	1148	1151	1153	1156	1159	1161	1164	1167	1169	1172	0	1	1	1	1	2	2	2	2
.07	1175	1178	1180	1183	1186	1189	1191	1194	1197	1199	0	1	1	1	1	2	2	2	2
.08	1202	1205	1208	1211	1213	1216	1219	1222	1225	1227	0	1	1	1	1	2	2	2	3
.09	1230	1233	1236	1239	1242	1245	1247	1250	1253	1256	0	1	1	1	1	2	2	2	3
.10	1259	1262	1265	1268	1271	1274	1276	1279	1282	1285	0	1	1	1	1	2	2	2	3
.11	1288	1291	1294	1297	1300	1303	1306	1309	1312	1315	0	1	1	1	2	2	2	2	3
.12	1318	1321	1324	1327	1330	1334	1337	1340	1343	1346	0	1	1	1	2	2	2	2	3
.13	1349	1352	1355	1358	1361	1365	1368	1371	1374	1377	0	1	1	1	2	2	2	3	3
.14	1380	1384	1387	1390	1393	1396	1400	1403	1406	1409	0	1	1	1	2	2	2	3	3
.15	1413	1416	1419	1422	1426	1429	1432	1435	1439	1442	0	1	1	1	2	2	2	3	3
.16	1445	1449	1452	1455	1459	1462	1466	1469	1472	1476	0	1	1	1	2	2	2	3	3
.17	1479	1483	1486	1489	1493	1496	1500	1503	1507	1510	0	1	1	1	2	2	2	3	3
.18	1514	1517	1521	1524	1528	1531	1535	1538	1542	1545	0	1	1	1	2	2	2	3	3
.19	1549	1552	1556	1560	1563	1567	1570	1574	1578	1581	0	1	1	1	2	2	3	3	3
.20	1585	1589	1592	1596	1600	1603	1607	1611	1614	1618	0	1	1	1	2	2	3	3	3
.21	1622	1626	1629	1633	1637	1641	1644	1648	1652	1656	0	1	1	2	2	2	3	3	3
.22	1660	1663	1667	1671	1675	1679	1683	1687	1690	1694	0	1	1	2	2	2	3	3	3
.23	1698	1702	1706	1710	1714	1718	1722	1726	1730	1734	0	1	1	2	2	2	3	3	4
.24	1738	1742	1746	1750	1754	1758	1762	1766	1770	1774	0	1	1	2	2	2	3	3	4
.25	1778	1782	1786	1791	1795	1798	1803	1807	1811	1816	0	1	1	2	2	2	3	3	4
.26	1820	1824	1828	1832	1837	1841	1845	1849	1854	1858	0	1	1	2	2	3	3	3	4
.27	1862	1866	1871	1875	1879	1884	1888	1892	1897	1901	0	1	1	2	2	3	3	3	4
.28	1905	1910	1914	1919	1923	1928	1932	1936	1941	1945	0	1	1	2	2	3	3	3	4
.29	1950	1954	1959	1963	1968	1972	1977	1982	1986	1991	0	1	1	2	2	3	3	3	4
.30	1995	2000	2004	2009	2014	2018	2023	2028	2032	2037	0	1	1	2	2	3	3	3	4
.31	2042	2046	2051	2056	2061	2065	2070	2075	2080	2084	0	1	1	2	2	3	3	4	4
.32	2089	2094	2099	2104	2109	2113	2118	2123	2128	2133	0	1	1	2	2	3	3	4	4
.33	2138	2143	2148	2153	2158	2163	2168	2173	2178	2183	0	1	1	2	2	3	3	4	4
.34	2188	2193	2198	2203	2208	2213	2218	2223	2228	2234	1	1	2	2	3	3	3	4	5
.35	2239	2244	2249	2254	2259	2265	2270	2275	2280	2286	1	1	2	2	3	3	3	4	5
.36	2291	2296	2301	2307	2312	2317	2323	2328	2333	2339	1	1	2	2	3	3	4	4	5
.37	2344	2350	2355	2360	2366	2371	2377	2382	2388	2393	1	1	2	2	3	3	4	4	5
.38	2399	2404	2410	2415	2421	2427	2432	2438	2443	2449	0	1	2	2	3	3	4	4	5
.39	2455	2460	2466	2472	2477	2483	2489	2495	2500	2506	0	1	2	2	3	3	4	5	5
.40	2512	2518	2523	2529	2535	2541	2547	2553	2559	2564	0	1	2	2	3	4	4	5	5
.41	2570	2576	2582	2588	2594	2600	2606	2612	2618	2624	1	1	2	2	3	4	4	5	5
.42	2630	2636	2642	2649	2655	2661	2667	2673	2679	2685	1	1	2	2	3	4	4	5	6
.43	2692	2698	2704	2710	2716	2723	2729	2735	2742	2748	1	1	2	3	3	4	4	5	6
.44	2754	2761	2767	2773	2780	2786	2793	2799	2805	2812	1	1	2	3	3	4	4	5	6
.45	2818	2825	2831	2838	2844	2851	2858	2864	2871	2877	1	1	2	3	3	4	5	5	6
.46	2884	2891	2897	2904	2911	2917	2924	2931	2938	2944	1	1	2	3	3	4	5	5	6
.47	2951	2958	2965	2972	2979	2985	2992	2999	3006	3013	1	1	2	3	3	4	5	5	6
.48	3020	3027	3034	3041	3048	3055	3062	3069	3076	3083	1	1	2	3	4	4	5	6	6
.49	3090	3097	3105	3112	3119	3126	3133	3141	3148	3155	1	1	2	3	4	4	5	6	6
.50	3162	3170	3177	3184	3192	3199	3206	3214	3221	3228	1	1	2	3	4	4	5	6	7

ANTILOGARITHMS (Contd.)

	0	1	2	3	4	5	6	7	8	9	Mean Differences								
											1	2	3	4	5	6	7	8	9
.51	3236	3243	3251	3258	3266	3273	3281	3289	3296	3304	1	2	2	3	4	5	5	6	7
.52	3311	3319	3327	3334	3342	3350	3357	3365	3373	3381	1	2	2	3	4	5	5	6	7
.53	3388	3396	3404	3412	3420	3428	3436	3443	3451	3459	1	2	2	3	4	5	6	6	7
.54	3467	3475	3483	3491	3499	3508	3516	3524	3532	3540	1	2	2	3	4	5	6	6	7
.55	3548	3556	3565	3573	3581	3589	3597	3606	3614	3622	1	2	2	3	4	5	6	7	7
.56	3631	3639	3648	3656	3664	3673	3681	3690	3698	3707	1	2	3	3	4	5	6	7	8
.57	3715	3724	3733	3741	3750	3758	3767	3776	3784	3793	1	2	3	3	4	5	6	7	8
.58	3802	3811	3819	3828	3837	3846	3855	3864	3873	3882	1	2	3	4	5	5	6	7	8
.59	3890	3899	3908	3917	3926	3936	3945	3954	3963	3972	1	2	3	4	5	5	6	7	8
.60	3981	3990	3999	4009	4018	4027	4036	4046	4055	4064	1	2	3	4	5	6	6	7	8
.61	4074	4083	4093	4102	4111	4121	4130	4140	4150	4159	1	2	3	4	5	6	7	8	9
.62	4169	4178	4188	4198	4207	4217	4227	4236	4246	4256	1	2	3	4	5	6	7	8	9
.63	4266	4276	4285	4295	4305	4325	4315	4335	4345	4355	1	2	3	4	5	6	7	8	9
.64	4365	4375	4385	4395	4406	4416	4426	4436	4446	4457	1	2	3	4	5	6	7	8	9
.65	4467	4477	4487	4498	4508	4519	4529	4539	4550	4560	1	2	3	4	5	6	7	8	9
.66	4571	4581	4592	4603	4613	4624	4634	4645	4656	4667	1	2	3	4	5	6	7	9	10
.67	4677	4688	4699	4710	4721	4732	4742	4753	4764	4775	1	2	3	4	5	6	7	9	10
.68	4786	4797	4808	4819	4831	4842	4853	4864	4875	4887	1	2	3	4	5	6	7	9	10
.69	4898	4909	4920	4932	4943	4955	4966	4977	4989	5000	1	2	3	4	5	6	7	9	10
.70	5012	5023	5035	5047	5058	5070	5082	5093	5105	5117	1	2	3	4	5	6	7	9	11
.71	5129	5140	5152	5164	5176	5188	5200	5212	5224	5236	1	2	3	4	5	6	7	9	11
.72	5248	5260	5272	5284	5297	5309	5321	5333	5346	5358	1	2	3	4	5	6	7	9	11
.73	5370	5383	5395	5408	5420	5433	5445	5458	5470	5483	1	2	3	4	5	6	8	9	11
.74	5495	5508	5521	5534	5546	5559	5572	5585	5598	5610	1	2	3	4	5	6	8	9	12
.75	5623	5636	5649	5662	5675	5689	5702	5715	5728	5741	1	2	3	4	5	7	8	9	12
.76	5754	5768	5781	5794	5808	5821	5834	5848	5861	5875	1	3	4	5	7	8	9	11	12
.77	5888	5902	5916	5929	5943	5957	5970	5984	5998	6012	1	3	4	5	7	8	10	11	12
.78	6026	6039	6053	6067	6081	6095	6109	6124	6138	6152	1	3	4	6	7	8	10	11	13
.79	6166	6180	6194	6209	6223	6237	6252	6266	6281	6295	1	3	4	6	7	9	10	11	13
.80	6310	6324	6339	6353	6368	6383	6397	6412	6427	6442	1	3	4	6	7	9	10	12	13
.81	6457	6471	6486	6501	6516	6531	6546	6561	6577	6592	2	3	5	6	8	9	11	12	14
.82	6607	6622	6637	6653	6668	6683	6699	6714	6730	6745	2	3	5	6	8	9	11	12	14
.83	6761	6776	6792	6808	6823	6839	6855	6871	6887	6902	2	3	5	6	8	10	11	13	14
.84	6918	6934	6950	6966	6982	6998	7015	7031	7047	7063	2	3	5	6	8	10	11	13	15
.85	7079	7096	7112	7129	7145	7161	7178	7194	7211	7228	2	3	5	7	8	9	12	13	15
.86	7244	7261	7278	7295	7311	7328	7345	7362	7379	7396	2	3	5	7	8	10	12	13	15
.87	7413	7430	7447	7464	7482	7499	7516	7534	7551	7568	2	3	5	7	9	10	12	14	16
.88	7586	7603	7621	7638	7656	7674	7691	7709	7727	7745	2	4	5	7	9	11	12	14	16
.89	7762	7780	7798	7816	7834	7852	7870	7889	7907	7925	2	4	5	7	9	11	13	14	16
.90	7943	7962	7980	7998	8017	8035	8054	8072	8091	8110	2	4	6	7	9	11	13	15	17
.91	8128	8147	8166	8185	8204	8222	8241	8260	8279	8299	2	4	6	9	9	11	13	15	17
.92	8318	8337	8356	8375	8395	8414	8433	8453	8472	8492	2	4	6	8	10	12	14	15	17
.93	8511	8531	8551	8570	8590	8616	8630	8650	8670	8690	2	4	6	8	10	12	14	16	18
.94	8710	8730	8750	8770	8790	8810	8831	8851	8872	8892	2	4	6	8	10	12	14	16	18
.95	8913	8933	8954	8974	8995	9016	9036	9057	9078	9099	2	4	6	8	10	12	15	17	19
.96	9120	9141	9162	9183	9204	9226	9247	9268	9290	9311	2	4	6	8	11	13	15	17	19
.97	9333	9354	9376	9397	9419	9441	9462	9484	9506	9528	2	4	7	9	11	13	15	17	20
.98	9550	9572	9594	9616	9638	9661	9683	9705	9727	9750	2	4	7	9	11	13	16	18	20
.99	9772	9793	9817	9840	9863	9886	9908	9931	9954	9977	2	5	7	9	11	14	16	18	20